D1269267

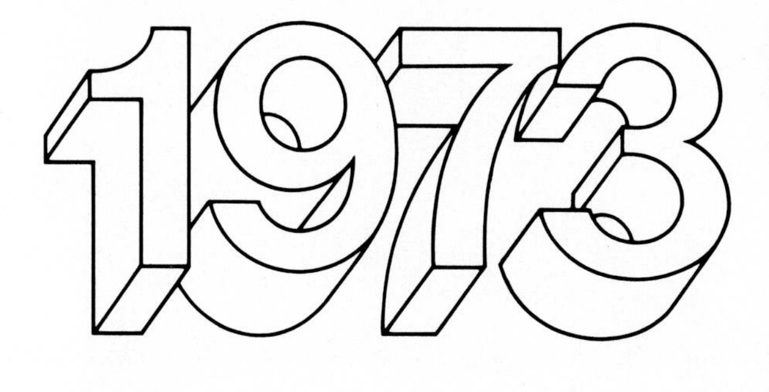


# Britannica Yearbook of Science and the Future

Encyclopædia Britannica, Inc.
William Benton, Publisher

Chicago Toronto London
Geneva Sydney Tokyo Manila
Johannesburg

# 1973 Britannica Yearbook of Science and the Future

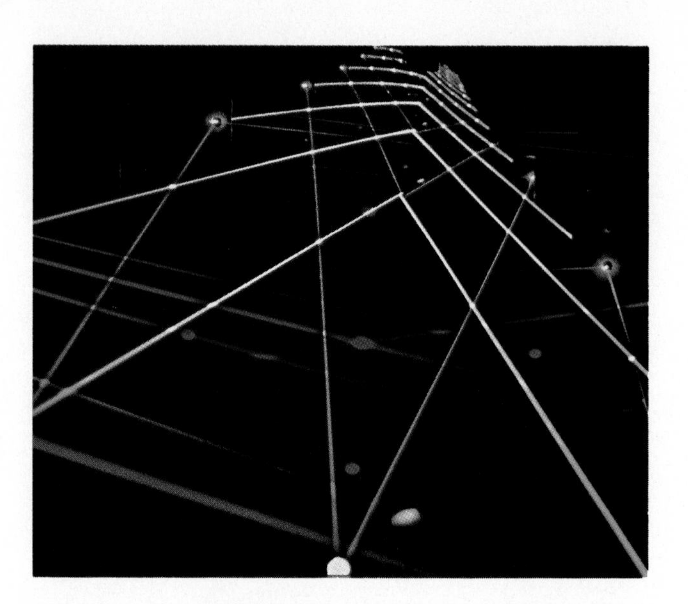






Library of Congress Catalogue Number: 69-12349 Standard Book Number: 0-85229-276-7

THE UNIVERSITY OF CHICAGO
The Britannica Yearbook of Science and the Future is published with the editorial advice of the faculties of the University of Chicago

# Editorial Advisory Board

# Shaping Our Lives Through Science and Technology

© Karsh, Ottawa

*The world is cool, peaceful, green. Birds sing in the trees. They seem to answer man's call. For there, he is a friend. Man and his environment are in harmony. Nature fulfills his material needs and he does not abuse it. If man is unhappy, he does not know it or show it.*

Those words might be said to describe the life of a society of just 25 people whose existence became known to us only in 1971. And if ignorance is bliss, is it folly to be wise? For perhaps centuries, their discoverers tell us, the Tasaday tribe has been isolated from the rest of the world in a dense rain forest in the Philippines. There, the Tasaday are still living in caves like our primitive ancestor, Stone Age man. Even the discovery in 1902 of a group of similar people in the Celebes does not equal the importance of the Tasaday in adding to man's knowledge of his distant past. (The Celebes group, called Toalas, had begun giving up caves 30 years before they became known to the outside world.)

As the description suggests, the Tasaday live a serene life. As far as we know, they have no natural enemies. They are food gatherers, rather than food growers, and apparently have to spend as little as three hours a day providing themselves with food that is in abundance and close by. The rest of the day the Tasaday is usually free to do as he pleases. He and his mate play with their children and teach them about the jungle. Or the Tasaday man may make music. Or swing on vines or climb trees. Or just sit and look contentedly into the valleys of his tropical rain forest.

The discovery of a unique people is of scientific significance, an event of exciting dimensions to anthropologists. The layman, too, must celebrate this find with the scientist. We here at Encyclopædia Britannica share in the concern felt all over the world for the ultimate fate of a Stone Age people suddenly exposed to our technological civilization.

I congratulate Philippine President Ferdinand Marcos, Imelda Marcos, his wife, herself a beloved figure among her people, and the PANAMIN Foundation, headed by Manuel Elizalde, Jr., for taking quick and decisive steps to assure that the Tasaday culture will not be destroyed. The Foundation, which assists all minority groups in the Philippines, is a good, indeed an excellent, example of government and the private sector working together for the greater good of the people. It has a simple guiding principle. As Elizalde told Britannica's editors, "We are for choice, the people's choice, not ours." Assistance to the minority peoples must be coupled with acceptance of their right to be different, says Elizalde. "The minorities themselves must retain the option of preserving their traditional way of life, or, if they so desire, changing it at the pace and in the direction they choose."

And this happily applies to the Tasaday. Already the Philippine government has set aside 46,299 acres of government land as a preserve for the Tasaday and their neighbors, the Manobo Blit. This will prevent encroachments by loggers, miners, and other industrial segments of society.

We are grateful to the Philippine government and to the PANAMIN Foundation for making the story about the Tasaday possible in this edition of the *Britannica Yearbook of Science and the Future.* The fascinating report that begins on page 50 was assembled by members of the PANAMIN staff, all of them anthropologists. As publisher of the Britannica, concerned about the future, I am grateful for the foresight shown by the government of the people of the Philippines in dealing so sensibly with such a challenging matter.

I suppose like most men of this rampaging society, I rather envy the Tasaday the choice they have to make. I wonder, if we all had it to do our-

John Nance from Magnum

*This rare photograph shows members of the Tasaday tribe outside their caves.*

selves, to make the choice consciously, whether we would elect the tranquil existence of a Tasaday rain forest or exposure to the rest of the world that seems full of so much stress and tension.

Can't we on our part seek to use the power through our own science and technology to make our world into our own version of a tranquil rain forest, where the world is cool, peaceful, and green, where birds sing in the trees, answering man's call?

Every year in the Science Yearbook we see proof in developing science and technology that man has the skills to shape his life as he wants it. One can see that from the variety of things reported in this year's edition. We know, for example, that we are nearing the time when babies can be conceived in test tubes (page 398); that crimes can be solved scientifically (page 374). We are just beginning to learn about the scientific developments in the People's Republic of China and the progress of its 800,000,000 citizens, whose cultural, scientific, and technological advancements have largely been a matter of guesswork up to now (page 30). We are learning more and more about nuclear fusion as a possible new and endless source of power that will not pollute the environment (page 110). We know now that the presence of mercury, lead, and cadmium in our systems and our environment (page 360) is as much a danger to us as it is to the animals of the earth, including those especially that are in imminent danger of extinction (page 10).

It's all there, really, the ability, the knowledge, the technology that can make our present and our future a comparative paradise.

With nature's surrender to man, with our ability to destroy everything on earth with the push of a button, we now face the necessity of conscious action. It is a time that T George Harris of *Psychology Today* has termed the "Age of Conscious Action." Those things that we have since the beginning of time left to nature, accident, tradition, now must be faced as conscious acts. Our society has finally reached the point that our own technology is forcing us to learn how to face up to the consequences of technology. Where are the teachers? We are prepared to be the taught. What are the next steps? Where do we go from here? This *1973 Britannica Yearbook of Science and the Future* contains some of the guideposts.

Wm Benton

**PUBLISHER**

# Contents

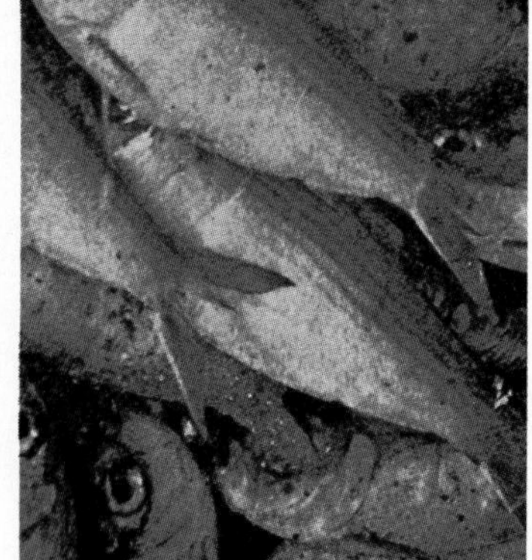

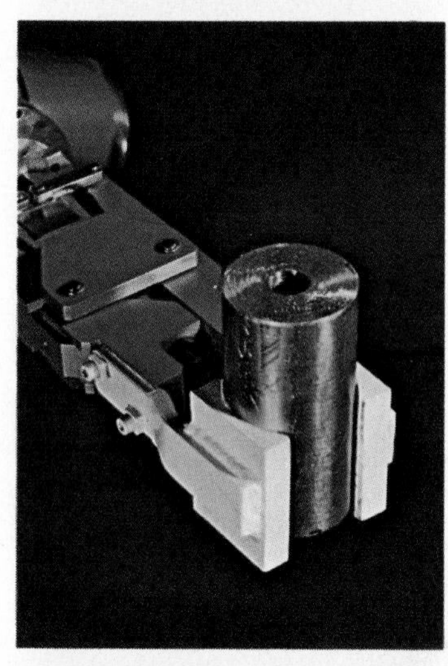

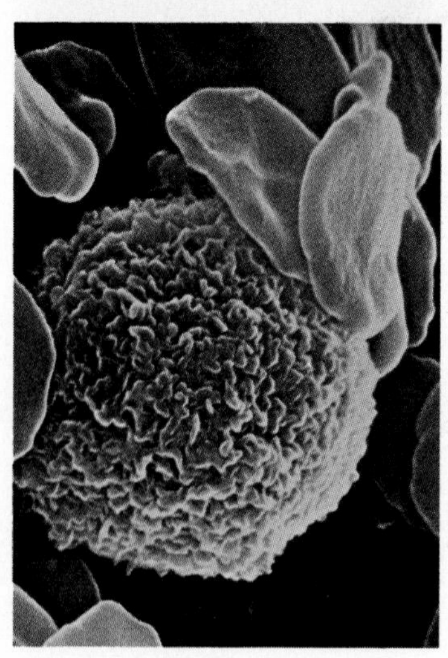

# Contributors to the Science Year in Review

**Joseph Ashbrook** *Astronomy.* Editor, *Sky and Telescope,* Cambridge, Mass.
**William J. Bailey** *Chemistry: Structural Chemistry.* Research Professor of Chemistry, University of Maryland, College Park.
**Jeremy E. Baptist** *Molecular Biology: Biophysics.* Assistant Professor of Radiation Biophysics, University of Kansas, Lawrence.
**Louis J. Battan** *Atmospheric Sciences.* Professor of Atmospheric Sciences and Associate Director of the Institute of Atmospheric Physics, University of Arizona, Tucson.
**Harold Borko** *Information Science and Technology.* Professor in the School of Library Service, University of California, Los Angeles.
**George M. Briggs** *Foods and Nutrition.* Professor of Nutrition, University of California, Berkeley.
**H. Keith H. Brodie** *Medicine: Psychiatry.* Assistant Professor of Psychiatry, Stanford University, Stanford, Calif.
**D. Allan Bromley** *Physics: Nuclear Physics.* Professor and Chairman, Department of Physics, and Director, A. W. Wright Nuclear Structure Laboratory, Yale University, New Haven, Conn.
**Morris E. Chafetz** *Medicine: Alcoholism.* Director, National Institute on Alcohol Abuse and Alcoholism, Bethesda, Md.
**Walter Clark** *Photography.* Formerly head of the Applied Photography Division, Research Laboratory, Eastman Kodak Co., Rochester, N.Y.
**Robert A. Davidson** *Chemistry: Chemical Synthesis.* Research Fellow, Department of Chemistry, Pennsylvania State University, University Park.
**John M. Dennison** *Earth Sciences: Geology and Geochemistry.* Professor and Chairman, Department of Geology, University of North Carolina, Chapel Hill.
**F. C. Durant III** *Astronautics and Space Exploration: Earth Satellites.* Assistant Director (Astronautics), National Air and Space Museum, Smithsonian Institution, Washington, D.C.
**Robert G. Eagon** *Microbiology,* Professor of Microbiology, University of Georgia, Athens.
**Edward H. Estes, Jr.** *Medicine: Paramedical Personnel.* Professor and Chairman, Department of Community Health Sciences, Duke University, Durham, N. C.
**Joseph Gies** *Architecture and Building Engineering.* Senior Editor/Technology, *Encyclopædia Britannica,* Chicago, Ill.
**Nathaniel Grossman** *Mathematics.* Associate Professor of Mathematics, University of California, Los Angeles.
**John Healy** *Earth Sciences: Geophysics.* Geophysicist, National Center for Earthquake Research, U.S. Geological Survey, Menlo Park, Calif.
**L. A. Heindl** *Earth Sciences: Hydrology.* Executive Secretary, U.S. National Committee for the International Hydrological Decade, National Academy of Sciences—National Research Council, Washington, D.C.
**Robert L. Hill** *Molecular Biology: Biochemistry.* Professor of Biochemistry, Duke University, Durham, N.C.
**Thomas H. Hunter** *Medicine: Death and Dying.* Professor of Medicine and Owen R. Cheatham Professor of Science, University of Virginia, Charlottesville.
**Daniel F. Johnson** *Behavioral Sciences: Psychology.* Associate Professor of Psychology, Virginia Polytechnic Institute and State University, Blacksburg, Va.
**Richard S. Johnston** *Astronautics and Space Exploration: Manned Space Exploration.* Deputy

Director, Medical Research and Operations, NASA Manned Spacecraft Center, Houston, Tex.
**Lou Joseph** *Dentistry.* Assistant Director, Bureau of Public Information, American Dental Association, Chicago, Ill.
**Walter S. Koski** *Chemistry: Chemical Dynamics.* Professor of Chemistry, Johns Hopkins University, Baltimore, Md.
**Jerry Lama** *Medicine: Venereal Disease.* Venereal Disease Consultant, Institute for Sex Education, Chicago, Ill.
**John H. Lannan** *The Science Year in Review: A Perspective.* Special Assistant, Office of Science and Technology, Executive Office of the President, Washington, D.C.
**Joshua Lederberg** *Molecular Biology: Genetics.* Professor of Genetics, Stanford University, Stanford, Calif.
**John G. Lepp** *Zoology.* President, Marion County Technical Institute, Marion, Ohio.
**Howard J. Lewis** *Science, General.* Director, Office of Information, National Academy of Sciences—National Academy of Engineering—National Research Council, Washington, D.C.
**Norman Metzger** *Chemistry: Applied Chemistry.* Manager, News Service, American Chemical Society, Washington, D.C.
**Rice Odell** *Environmental Sciences.* Editor, *CF Letter,* the Conservation Foundation, Washington, D.C.
**Alan J. Perlis** *Computers.* Eugene Higgins Professor of Computer Science, Yale University, New Haven, Conn.
**Willard J. Pierson, Jr.** *Marine Sciences.* Professor of Oceanography, New York University, New York, N.Y.
**Froelich Rainey** *Archaeology.* Professor of Anthropology and Director of the University Museum, University of Pennsylvania, Philadelphia.
**Robert B. Rathbone** *Agriculture.* Director, Information Division, Agricultural Research Service, U.S. Department of Agriculture, Washington, D.C.
**Peter H. Raven** *Botany.* Director, Missouri Botanical Garden, and Professor of Botany, Washington University, St. Louis, Mo.
**Marc Ross** *Physics: High-Energy Physics.* Professor of Physics, University of Michigan, Ann Arbor.
**Byron T. Scott** *Medicine: General Medicine.* Executive Editor, *Medical Opinion,* New York, N.Y.
**Mitchell R. Sharpe** *Astronautics and Space Exploration: Space Probes.* Science writer, author of *Satellites and Probes* and *Dividends from Space,* Huntsville, Ala.
**Philip Skell** *Chemistry: Chemical Synthesis.* Professor of Chemistry, Pennsylvania State University, University Park.
**Frank A. Smith** *Transportation.* Senior Vice-President-Research, Transportation Association of America, Washington, D.C.
**J. F. Smithcors,** *Veterinary Medicine.* Editor, American Veterinary Publications, Santa Barbara, Calif.
**Charles Süsskind** *Electronics.* Professor of Engineering Science, University of California, Berkeley.
**Sol Tax** *Behavioral Sciences: Anthropology.* Professor of Anthropology, University of Chicago.
**Irwin Welber** *Communications.* Associate Executive Director, Toll Transmission, Bell Laboratories, North Andover, Mass.
**James A. West** *Fuel and Power.* Chief, Branch of Interfuels and Special Studies, Division of Fossil Fuels, U.S. Bureau of Mines, Washington, D.C.

A Pictorial Essay

# Species in Peril

## by William G. Conway

**Man and his civilization have already caused many varieties of animals to become extinct. Unless preventive measures can be taken, other species also seem certain to disappear.**

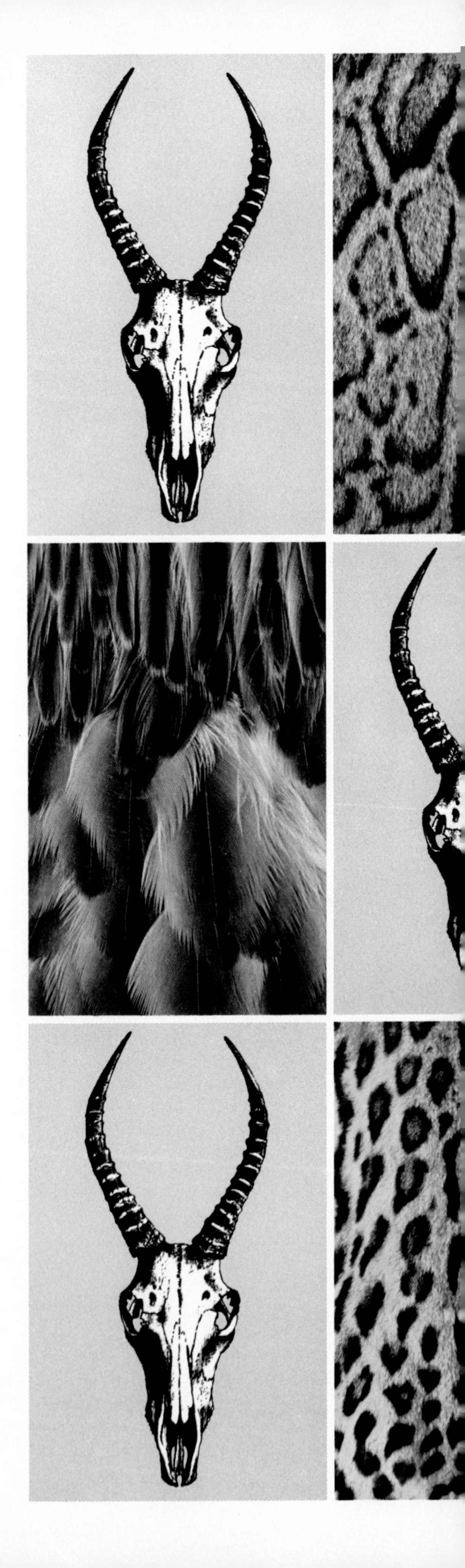

Jo Conner

*The green sea turtle of Central America (above) is gradually vanishing. It is being killed for its meat and skin, and its eggs are dug up for the market. The monk seal (opposite), long hunted for its oil, is nearly extinct.*

Within the past 400 years more than 220 mammals, birds, and reptiles have become extinct, the majority through the activities of man. Many other animal species are now declining. Among the endangered are such common animals as the grizzly bear, the polar bear, the giant panda, the wolf, several of the big cats (the cheetah, the jaguar, the ocelot), the rhinoceros, the great apes, many of the large birds of prey, the stork, the crane, the crocodile family, the large sea turtle, and the larger whales—as well as a host of the more exotic, lesser known species such as the thylacine, the gaur, the tuntong, the Philippine monkey-eating eagle, and the Komodo Island monitor, the world's largest lizard.

Until the coming of civilized man, the extinction of wild animals and the evolution of new ones proceeded surely but almost imperceptibly on a continuum of loss and renewal stretching back more than 3,000,000,000 years. A species' span on earth was not measured in centuries but in millennia—a particular animal's species might be traced back at least through several hundred thousand years.

Today this continuum of loss and renewal in forms of animal life is gravely threatened. In recent years man has sharply accelerated the pace of wildlife extinction at the same time that he has reduced the opportunities for the adaptation and evolution of new forms, especially of larger and higher animals. He has done this directly, by catching or killing animals—mainly for food, sport, pets, commercial use, or because the animals compete with him in some way—and indirectly, by displacement or by the despoliation of animal habitats.

***WILLIAM G. CONWAY*** *is general director of the New York Zoological Society and the Center for Field Biology and Conservation.*

Erwin A. Bauer

## Collecting wild animals alive

Perhaps the least important drain on wildlife caused by man is by live collecting for pets, zoos, and biomedical research. However, this drain alone is enormous. In 1970, according to the U.S. Department of the Interior, the United States imported alive 83,867,029 fish, 572,670 amphibians, 2,109,571 reptiles, 687,901 birds, and 101,302 mammals. The great majority of fishes were destined for pet shops. No one has any real idea of the effect of this vast trade upon fish stocks in nature. Most imported amphibians were frogs intended for laboratory studies and human pregnancy tests. The reptiles, which varied from crocodiles to pythons, were sold primarily as pets. Surprisingly, the huge bird importation figure did not even include canaries or such psittacine birds

Dan Budnik from Woodfin Camp

Joe Bilbao from Photo Researchers

*Destruction of their natural environments, through pesticide pollution and encroaching human habitation, threatens (left to right) the American bald eagle, brown pelican, saddle-billed stork, and Galápagos flightless cormorant.*

as parrots or parakeets, but this trade too went almost entirely for pets. Of the 101,302 wild mammals that were imported in 1970, some 85,000 were monkeys captured mostly for biomedical research. Although consumption of wild monkeys in 1970 may seem high, it was modest compared to the importations during the height of the production of poliomyelitis vaccine in 1958–60, which tallied 634,000.

The demand by zoos for wildlife is relatively modest. All of the world's zoos together contain fewer than 600,000 mammals, birds, reptiles, and amphibians, many of which were bred in captivity. Nevertheless, zoo importations occasionally affect species of exceptional rarity.

Unfortunately, U.S. wildlife import figures give only an approximate idea of the drain on wild animal populations created by the live trade. Most nations do not keep such records. Thus the real number of animals taken from the wild is much higher. And for every animal imported alive, several may have died during the stress of collecting and shipping processes. Moreover, large numbers of mammals, birds, and reptiles are consumed in their countries of origin by local demands for pets.

N. Myers from Bruce Coleman Inc.

Peterson from Photo Researchers

## Hunting for sport

For many wildlife populations, sport hunting is a far more important drain than live collecting. Nevertheless, such hunting in developed countries like the United States, which assiduously assesses and manages its "game crop," usually does not occasion extinction and may even provide funds and support for wildlife preserves. Sport hunting can be a more serious threat in developing nations that cannot afford the costs of highly trained game biologists and the constant research their work entails to determine the numbers of animals that may be annually killed without endangering the breeding base of the species.

In the United States, annual sport kills of waterfowl have ranged from 9 million to 15 million in recent years, deer kills about 2 million, and black bear more than 20,000. It is estimated that U.S. hunters shot 41.9 million mourning doves in 1965–66! The lesson in this is that some wildlife populations are enormously productive and produce enough individuals so that many may be killed without serious threat of extinction *if* they are provided with the proper habitat and protected from overhunting.

Jerry L. Hout from Photo Researchers

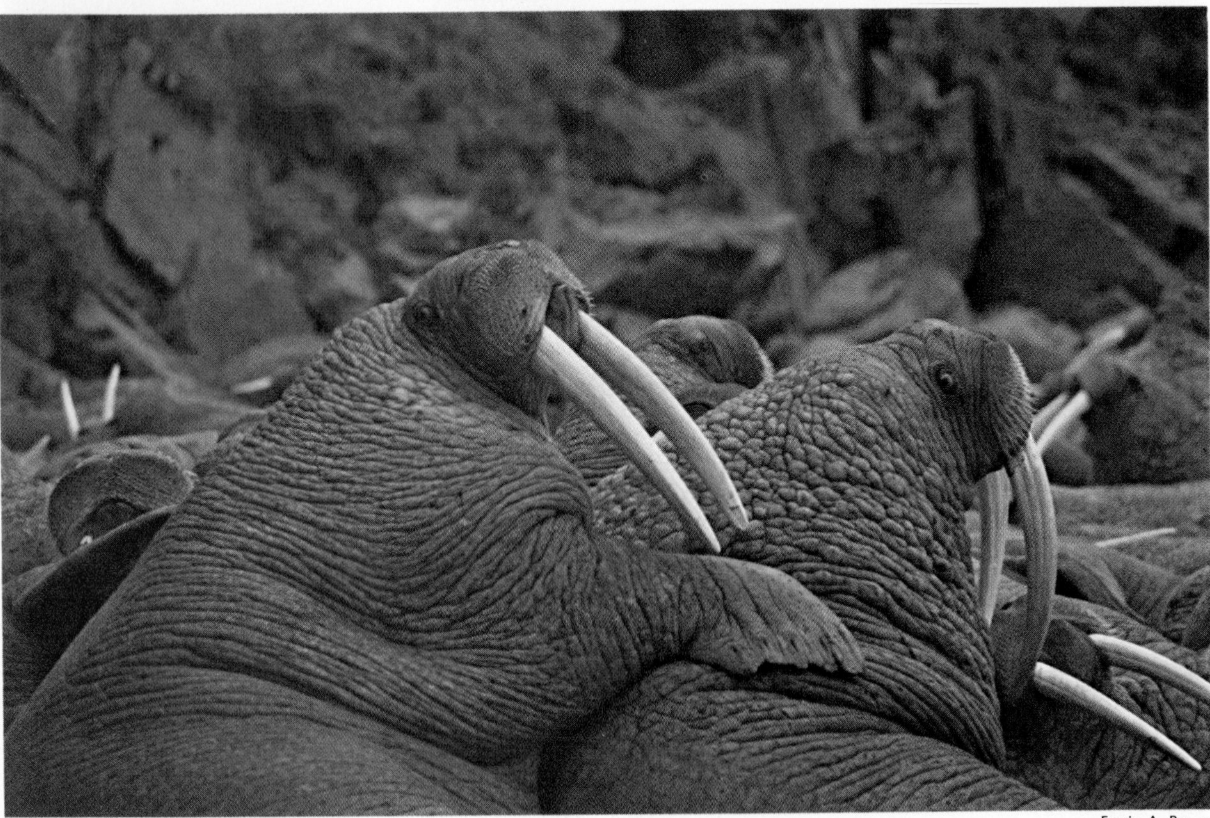

Erwin A. Bauer

### Hunting for food

The hunting of wild animals for daily food is an integral part of the livelihood of millions of people in the developing nations. The United Nations Food and Agriculture Organization recently reported, for example, that more than 80% of all fresh meat consumed in Ghana comes from game, and more than 50% in Zaire (formerly Congo [Kinshasa]). There is no doubt that the drain on wildlife populations from food hunting far outweighs that from live collecting and sport hunting. Many species are endangered by it. It was commercial hunting for food that destroyed the Eskimo curlew and passenger pigeon populations in the United States, not sport hunting.

In the fishing industry, overfishing has long been a problem affecting certain species. One indication of the seriousness of this drain on fish stocks is the fact that the 1969 world fish catch (published in 1971) showed an overall drop of about 2% from the previous year.

### The hide and fur, meat, and animal oil trades

The number of animal lives sacrificed to the hide and skin trades dwarfs all other drains on those species it affects. The big cats, the otters, the fur seals, and the crocodilians are the primary victims. Before New York's recent conservation legislation (now being followed by other states), thousands of these animals were killed for the U.S. trade alone. For example, in 1968 and 1969, U.S. furriers imported 3,168 raw cheetah skins, 17,490 leopard skins, 23,347 jaguar skins, 262,030 ocelot skins, and 99,002 otter skins. These figures do not include the animal deaths represented by the finished products imported from furriers in other countries. In 1969 alone, the United States was shipped 275,000 skins from the capybara (the world's largest rodent) and 436,000 deer skins from Brazil.

Commercial exploitation of wild animals for meat and oil affects such disparate creatures as sea turtles, kangaroos, and whales. The decline of whale stocks in the face of almost unregulated slaughter and modern technology is a well-known international disgrace. Today the trade is primarily in the hands of Soviet and Japanese whalers who appear bent upon converting the last whales into dog food, lipstick, soap, paint, transmission oil, and, to a certain extent, food for human consumption. Unhappily, few species in international waters have ever been satisfactorily protected by reasonable hunting regulations.

It is depressing to note that not one essential commercial product is made from the entire traffic in wild animal hides and furs or from whale oil for which there are not adequate substitutes.

### The destruction of predators

Animals whose homes and food supply have been usurped by man and who must therefore compete with him for the crops he plants or the domesticated animals he rears are regarded by man as predators. In the United States, studies have shown that the majority of wild predators make only limited inroads on livestock but do offer real values in

*Animals of the polar regions—the emperor penguin (below), musk-ox (opposite, top), and walrus (opposite, bottom)—are endangered by man's activities. Pesticide pollution has been detected in Antarctica, home of the penguin, while the walrus and musk-ox of the Arctic suffer from overhunting.*

Tom Myers

Mondadoripress from Pictorial Parade

the control of other competing wildlife, particularly rodents. Broad nonspecific predator control programs can rarely be justified on economic grounds. Nevertheless, shooting, trapping, and poisoning campaigns have long been conducted against wolves, foxes, coyotes, and against the large birds of prey.

However, it is unrealistic to ignore the problems of those who must live next door to tigers and crocodiles. Predators such as these must have large enough areas within parks and refuges to permit them to avoid man if they are to survive at all. In fast-growing over-crowded countries, there appears small chance that the people will decide to set aside such lands until it is too late for the great predators.

## Environmental destruction

The most serious menace to current wild animal populations comes from man indirectly, not through guns or traps. The continuing increase of human population—with its supporting technology and seemingly inevitable habitat despoliation, environmental poisoning, and pollution —dwarfs all other threats to the survival of wild creatures in the importance and finality of its effects.

Man's population increased by more than 70 million annually in the early 1970s and was expected to rise by a greater amount in succeeding years. As a consequence, more land is continually being settled and made unsuitable for wild creatures. When a marsh is drained, an estuary filled, a forest cut, a savanna plowed, a stream dammed, or a river channelized, animal habitats are eliminated. There is usually no wish to destroy wild animals in this process but it is the inevitable result. When animals are hunted directly, usually not all are killed; their homelands remain where remnants of the population can find food and shelter in which to rebuild their numbers. But when land or water is

Tom Myers

Les D. Line

*Reptiles under pressure include the giant monitor of Indonesia (opposite), the banded gecko (above, left), and the alligator (above). Collectors have depleted the supply of monitors, while the gecko suffers from its use as a pet. Alligators, long hunted for their skins, are now under protection and are making a comeback.*

made uninhabitable to wildlife, those populations it supported must then be considered extinct, whether the species dwells elsewhere or not. Canvasback ducks cannot feed and nest in wheat fields, orangutans cannot rear their families in lumber mills, and bison cannot graze on parking lots.

Pesticides and pollution are the most subtle threats to wildlife. In North America the chlorinated hydrocarbons, the best known of which is DDT, have been shown to affect the physiology of many animals, particularly birds. Pesticides applied to control insects on crops eventually wash into watercourses and the sea, where they are ingested by microorganisms which are themselves eaten by fish and other animals. These latter are, in turn, eaten by the birds. At each step in this food chain, the pesticide is further concentrated. One result of such concentrations is that affected birds produce eggs with shells so deficient in calcium that they are crushed in incubation. Because of this process, brown pelicans, cormorants, peregrines, bald eagles, and many other birds of prey are already extinct in various localities.

The threat to other animals posed by DDT is increasing. DDT or its breakdown products is now being found even in remote Antarctic populations of seals and penguins.

And there are other pervasive pollutants. Recently, PCBs (polychlorobiphenyls), the organic compounds used frequently in making plastics, have been implicated in birth defects of seabirds off the coasts of Connecticut and New York. There has also been an increase of heavy metals, particularly mercury, in lakes and rivers as a result of sewage from plants using this element in manufacturing processes. As a result, U.S. government agencies have been forced to declare certain food fish unfit to eat. (*See* Feature Article: THE NOTORIOUS TRIO: MERCURY, LEAD, AND CADMIUM.)

Fred Bruemmer

*Baby seals like the one shown above with its mother are hunted extensively for their white fur. The orangutan (opposite) has suffered greatly from the destruction of its natural habitat and from collecting for laboratories, museums, and zoos.*

Pollution of the air and waters is emerging as an increasingly serious threat to wildlife as well as to man. Secondary habitat despoliation is the most readily observable result. Various types of conifers near the sprawling urban complex of Los Angeles are dying, and the epiphytic Spanish moss, pride of many cities in the southeastern United States, also appears to be on the decline—the result of air pollution. These losses should be viewed as the forerunners of future losses in the wildlife community that will inevitably follow the changes in habitat.

Water pollution from sewage and industrial wastes has also had grave consequences for wildlife. The life of Lake Erie has been seriously depleted, and fishes and other aquatic life have disappeared from many of the streams and rivers in the United States and Europe. The effects of pollution can grow at an increasing rate as the demands imposed by bacteria breaking down sewage cause a decline of oxygen in surface waters and pathogenic bacteria reach concentrations sufficient to spread disease to man as well as wildlife. Wildlife is also threatened by rising industrial activity with the resulting increases in industrial accidents, such as oil spills. A serious oil spill off the coast of an island where a restricted colony of mammals or birds exists might eliminate a species. Finally, almost all polluting materials dumped on the surface of the earth or sprayed within its atmosphere ultimately come to rest within the oceans. Although fish are harder to count than some land creatures, there is no reason to believe that they are any less threatened.

## Animals added and subtracted

The introduction of exotic wild or domestic animals is still another way in which man has endangered species. Countless cases have been recorded where the introduction of birds, pigs, goats, cats, mongooses, dogs, rabbits, and rats have pressed animals with small populations or those confined upon islands to the verge of extinction. At the same time, the introduced have often become serious pests. Introduced animals usually compete with native forms for food and even destroy food supplies. Some simply eat the native animals.

The removal of a native animal may also have unexpected results for other species in the same area. The American alligator, for example, now recovering under protection, has been shown to be necessary to the survival of other wildlife in the Everglades during a drought. Its deeply dug den, or "well," may be the only remaining water when the glades dry up. Seed stocks of fish and amphibians survive in the well to repopulate the glades when the waters return.

The removal of a predator may result in such increases of its prey that their demands for food wreak havoc upon the vegetation of the area and thus the habitat's ability to support them. Thus populations of deer in some rural areas of the United States have expanded to such an extent, with the elimination of the wolf and puma, that they have destroyed forest understory vegetation and even barked trees in some years. Ecological interrelationships like these call for careful study and research before wildlife is killed or habitats altered.

G. B. Schaller from Bruce Coleman Inc.

*Large African mammals are among the endangered species, even though Africa has sought vigorously to preserve its wildlife. The gorilla (above) is menaced by the destruction of the rain forests in which it lives, while the pygmy hippopotamus (opposite, top) is hunted for food. The rhinoceros (opposite, bottom) is being depleted by both habitat destruction and hunting for its horns.*

**The effort to save vanishing species**

The status and probable rates of decline of wild animals vary in different parts of the world. Although our knowledge is incomplete, what is clear is that the scope of the problem is immense. Unhappily, those who would protect wild lands and animals, at best, can fight only a holding battle. They can preserve a particular forest or a rare creature only until ignorance or self-interest impels yet another challenge to its existence through further exploitation.

The hard facts faced by conservationists are that man's utilization of wild lands and parks must remain modest if the plant and animal life they support is to survive; that much of the world's most fascinating wildlife is in the developing countries, which have high rates of population growth and tremendous economic needs that might well override preservation; that preservation of wildlife may not offer immediate local economic incentives; that the profit motive in commercial exploitation of wildlife is hard to combat; and that wild creatures whose life cycles regularly take them over national boundaries or into international waters are particularly difficult to protect.

Nevertheless, strong park and conservation departments have developed in many nations, and hundreds of nonprofit organizations have taken up the conservation challenge. These organizations mainly work through public education by producing publications, holding conferences, giving lectures, and generating funds. They muster support for conservation legislation and for the establishment of parks and reserves, and they conduct research on wildlife problems. Clearly, the establishment and maintenance of parks and reserves that include appropriate habitats is the most important step that can be taken to insure the survival of most threatened species. In a few cases, wild animals have been saved or aided by captive propagation.

One thing is clear. There is no reason whatsoever for optimism about the future of wild animals. It remains an open question whether man the destroyer can become man the preserver in time to prevent many more species from becoming extinct.

Russ Kinne from Photo Researchers

Eliot Elisofon, LIFE Magazine © Time Inc.

Shelly Grossman from Woodfin Camp

Richard Keane

Ylla from Rapho-Guillumette

*White pelicans have decreased considerably because of the effect of pesticides on their calcium metabolism, causing their eggshells to break before the young are hatched. Leopards are hunted for their skins, while the pressure of increasing human population on their habitats has caused the tiger to decline in numbers from 30,000 to 2,500 in 35 years.*

S. Gillsater from Bruce Coleman Inc.

Bruce Coleman from Bruce Coleman Inc.

N. Myres from Bruce Coleman Inc.

George Holton from Photo Researchers

*Endangered large mammals include the polar bear (opposite, top), which suffers from excessive hunting, and the giant panda (opposite, bottom), a species of small numbers that lives only in western China. African hoofed mammals that are threatened include the oryx (above), hunted extensively for food, and the already rare okapi (left).*

*Monkeys are threatened both by human encroachment on their habitats and by hunting. In 1970 approximately 85,000 monkeys were imported into the United States, mainly for biomedical research. On the opposite page two endangered animals, the cheetah and open-billed stork, enjoy the protection of a game refuge in Kenya.*

Bill N. Kleeman from Tom Stack & Associates

# Science in China

## by C. H. Geoffrey Oldham

**According to Mao, science should serve the people. After almost 25 years of Communist rule, what methods have been evolved to reach this goal and how close is it to realization?**

Almost 25 years have passed since Mao Tse-tung established the People's Republic of China and committed that country along a path toward his concept of a Communist utopia. Throughout these 25 years the cornerstone of China's science policy has been that "science must serve the people." Most of the Chinese investments in science and technology have been governed by this dictum and, hence, any assessment of the current status of science and technology in China must be measured against Mao's own objectives.

Had this article been written 20 years ago it would have been much more difficult to develop this viewpoint. Then a nation's scientific prestige was measured almost entirely by its ability to produce new knowledge and, to a lesser extent, to use knowledge in producing sophisticated weapons and generating wealth. In the 1970s policymakers in almost all countries are much more concerned with the utilization of science and technology to achieve a whole gamut of objectives, frequently summed up in the phrase "improvement in the quality of life." That "science should serve the people" has now become the cornerstone of science policy in the United States and Europe and, indeed, in most countries of the world.

Although Chinese thinking about science has been dominated by a utilitarian viewpoint, the precise ways in which science should serve the people have been the subject of heated debate. The specific policies have changed frequently and sometimes dramatically. Indeed, it is these changes and the Chinese willingness to experiment with novel approaches that make a study of Chinese science and technology so interesting and yet so tantalizing. It is interesting not only because of the inherent interest of China's social experiment but because of the possible relevance of the Chinese experience to other developing countries. It is tantalizing because, although it is possible to form a reasonably good idea of Chinese policies from the Chinese press, it is extraordinarily difficult to obtain a clear picture of their success or failure in achieving the desired goals.

This article will attempt to reflect both the interest and the frustrations of a study of Chinese science and technology. It will attempt to summarize the changing Chinese policies on science and technology and, in a concluding section, it will speculate on the reasons behind Mao's strategy.

Throughout the article reference will be made to the major issues that are the subject of debate among those concerned with how less developed countries can best use science and technology as tools to aid their development. It is essential to recognize that China is the world's largest developing country in both area and population, and that it exists in a world where 98% of expenditure on research and experimental development takes place in the industrialized countries. This maldistribution of world research endeavor has profound implications for the less developed countries. Not only is much of the world's research irrelevant to the needs of poor countries, especially in the tropics, but some of it, such as research on the development of syn-

***C. H. GEOFFREY OLDHAM*** *is a senior research fellow of the Science Policy Research Unit at the University of Sussex in England. He has traveled extensively in China.*

Courtesy, Smithsonian Institution, Freer Gallery of Art, Washington D.C.

*China's knowledge of the rudder more than 1,000 years before it appeared in Europe, suggested by the painting (above) copied in the 11th or 12th century A.D. from a 4th-century design, was confirmed in 1958 by archaeological evidence. Civilization in China has been traced back many thousands of years. The excavations in Sian (below) date from the Stone Age.*

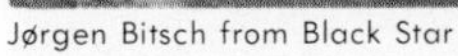

Jørgen Bitsch from Black Star

thetics, is positively harmful to their prospects for trade in raw materials. It is against this background that the Chinese priorities and experience must be viewed.

## Historical background

Although China's commitment to modern science is relatively new, it has a long history of scientific discovery and technological invention. As Joseph Needham of Cambridge University has documented so well in his *Science and Civilisation in China*, China can lay claim to more technological firsts than almost any other civilization. It was the Chinese who invented the seismograph in the 2nd century A.D. The magnetic compass, gunpowder, and the printing press with movable type are further examples of technologies first developed in China and rediscovered in the Western world many years later. However, for some as yet unknown reason, the scientific revolution took place in Western Europe rather than in China. And following the scientific revolution came the industrial revolution and the growth in material wealth of Europe and then the United States. Meanwhile, China and the rest of what we now call the developing world changed only slowly.

Apart from a few early contacts with missionaries such as Matteo Ricci, China remained isolated from the changes that were sweeping through Europe until its defeat at the hands of the British during the Opium Wars in the 19th century. From that point the slow, painful process of modernization took on a new dimension as some Chinese reformers urged that China begin to adopt what they thought of as

*Chinese industrial policy appears to give equal emphasis to small-scale, local factories and to heavy industry. The steel mill (below) is in Anshan in Manchuria. Automobile plants include those in Shanghai (opposite page, left) and Changchun.*

Western science and Western technology. Their voices had but little effect, and it was not until after the Manchu dynasty was overthrown in 1911 that the efforts to transplant modern science onto Chinese soil produced any significant results.

Even then it was not until the 1920s that the major intellectual break with Confucianism was made and the politicians committed themselves to the building of a new China based on modern science and technology. By the mid-1930s the Nationalists had begun to institutionalize science and had established several research institutes and universities. Although this scientific effort was severely disrupted during the Japanese and civil wars, there still existed in China in 1949 a nucleus of competent scientists and engineers, and a start had been made toward building an indigenous scientific system.

## The organization of science

The importance of the fact that Mao Tse-tung "inherited" a nucleus of scientists and an organizational framework, albeit in a fragmented form, should not be underestimated. For many years "experts" concerned with development put a low priority on the need for less developed countries to have their own indigenous scientific capability. They argued that the industrialized countries were a veritable supermarket of technologies and all the less developed countries had to do was to shop around and choose what they wanted. Mao has always completely rejected this argument, and now most other less developed countries are following his line.

(Left) Jørgen Bitsch from Black Star; (above and opposite page) Emil Schulthess from Black Star

Emil Schulthess from Black Star

To begin with, most of the industrial technologies developed in the advanced countries are designed for conditions of large-scale production and a situation where capital is relatively abundant and labor is scarce. This is hardly the situation in China or other developing countries. Even more inappropriate for less developed countries is most of the agricultural and medical research done in the developed world. But probably the single most important reason why China and, now, the rest of the developing countries want their own scientific establishments is that otherwise they will be dependent on foreigners to make all the important technological decisions affecting them. This, they find, is economically unpalatable and—what is even more important—politically unacceptable.

The network of research institutions and facilities for training scientists and engineers that grew up in China in the 1950s was based largely on the Soviet system. The Academies of Sciences, Agricultural Sciences, and Medical Sciences each operated its own research institutes and together they carried out most of the advanced research being done in the country. Overall strategy was dictated by the State Scientific and Technological Commission, with applied research sponsored and carried out by the relevant ministry. Very few of the universities carried out advanced research, and most of the new research workers received their training on-the-job at the various institutes of the academies.

After an initial period in the early 1950s when the academies determined their own research priorities, the work of the various institutes became subject to an overall planning authority. In 1956 a 12-year plan for science was drawn up with a great deal of Soviet help. It was claimed that in the 10 priority areas identified in the plan China would aim to catch up with advanced-world levels by the end of the plan period. However, during the campaign of mass mobilization called the Great Leap Forward (1958–60), the earlier concept of central planning of research priorities seems to have been abandoned in favor of encouraging all innovations that originated with peasants and workers in whatever field they occurred. During the 1960s, although the State Scientific and Technological Commission continued to exist and continued to be the main policy-making body for science and technology, there was little reference to it in the Chinese press. It either operates in a very secret way or is largely ineffective.

The Great Leap Forward coincided with another event that was to have a major effect on Chinese science—as well as on foreign policy. This was the withdrawal of all Soviet aid. Between 1958 and 1961 all Soviet nationals were called home, and the massive Soviet technical assistance program came to a halt. The Soviets even took the blueprints of unfinished plants with them. This event still rankles very deep with the Chinese. Even in 1972 visitors to China were being told with evident bitterness about the fickle Soviets, and there can be little doubt that the Chinese attitude to the Soviet Union is heavily influenced by this event.

*Industrial production suffered less than scientific research and education during the convulsions of the Cultural Revolution. The 12,000-ton hydraulic press (opposite page) was photographed in Shanghai.*

The effect on Chinese science policy was immediate and long lasting. It made the Chinese distrustful of foreign aid, and "self-reliance" became the key slogan of the early 1960s. More and more resources were injected into the Chinese science system, and by 1965 it began to seem that the Chinese aim of catching up with advanced-world levels in certain priority areas was well on the way to reality. In 1964 and 1965 contacts with Western scientists were considerable, and formal scientific exchanges were arranged with many Western European countries.

The reports of the foreign scientists who visited China indicated that China was creating a solid foundation in most scientific disciplines. The Chinese scientists were apparently well informed of work going on in their respective disciplines in other parts of the world. In a few areas they were already making important new contributions themselves, although by and large there were few surprises—the most commonly used adjective of the foreign visitors was "sensible." The quality of the scientific papers published in the Chinese journals had become sufficiently high to warrant cover-to-cover translation of several of them by U.S. commercial enterprises. By the mid-1960s, judged by Western scientific eyes, it appeared that the Chinese science system was extremely healthy and vigorous and doing good work. But apparently Mao felt otherwise.

When the Cultural Revolution first began in 1966, the scientists were specifically excluded. This exclusion did not last long, however, and by early 1967 the activities of the scientists came under close scrutiny. For several months the Chinese press was full of accounts of upheavals within the institutes of the Academy of Sciences. The test was whether the institutes really had "served the people." Many scientists were accused of being too greatly influenced by the foreign scientific literature and too little influenced by Chinese problems; they were too eager for scientific awards and titles and sought personal fame rather than devoting themselves to relatively unglamorous work. In short, scientific research was not sufficiently integrated with the needs of society.

By the early 1970s these same charges were being made against the scientific community in most of the world's developing countries and, increasingly, in the more industrialized countries. The Chinese solution has been to put the direction of the research institutes under the control of "revolutionary committees," made up of representatives of the People's Liberation Army, the political cadres, and "revolutionary" workers. They have also experimented with different ways of integrating research with education and production, but too little is known about the details of these experiments to permit evaluation of their results.

Contacts with foreign scientists were broken off during the Cultural Revolution and, since the publication of Chinese scientific journals also stopped, it became very difficult to assess the effect of the Cultural Revolution on Chinese scientific research. In the early 1970s, however, China made a dramatic reentry into international affairs, and scientists were once again able to visit the research institutes of the Academy of

Rene Burri from Magnum

Sciences. They found that the institutes were working on more applied problems than had been the case in the mid-1960s and that, despite the criticism of scientists during the Cultural Revolution and the fact that many had spent time working in communes, most were now back doing research. They also found that, although foreign scientific journals were being received, there seemed to be less familiarity, especially on the part of the younger workers, with the latest research in the rest of the world than had been the case before the Cultural Revolution.

It is still too early to judge the long-term effects of China's isolation from the mainstream of world scientific research. It is possible that in some research areas Chinese scientists may have pioneered entirely novel approaches. But the rate of growth of scientific knowledge in the world is such that, by cutting itself off from much of this knowledge, China can only delay the time when it reaches its stated goal of catching up with the advanced countries of the world.

*Like many developing countries, China faces the problem of whether a sophisticated, capital-intensive industrial sector can coexist with labor-intensive technologies that provide employment and involve workers in the system.*

Brian Brake from Rapho Guillumette

### Scientific and technical education

Although the Communist government inherited a nucleus of scientific institutions when it came to power in 1949, the actual number of fully qualified scientists and engineers was small. A major effort was launched to increase the complement of technical personnel, and certainly the number of persons with some technical training has risen dramatically. It is very difficult, however, to make comparisons either between China and other countries or between the situations that have existed at different times within China, largely because there seem to be no consistent manpower definitions. The best manpower estimates, based on admittedly conflicting Chinese evidence, indicate that the stock of scientists and engineers rose from 600,000 in 1955 to 2.4 million in 1967. Approximately 670,000 of the 2.4 million had been trained in full-time institutes of higher education in China after 1949. The rest either were promoted to the rank of scientist or engineer without receiving full-time training or were already trained in 1949.

It has been estimated that the number of college-trained scientists and engineers had reached 1.3 million when the Cultural Revolution erupted and all the schools and universities were closed. The universities remained closed for four years and only began teaching scientific subjects again in 1971. This represents a major loss of technical manpower. But even though the schools and colleges were shut down, the research institutes of the Academy of Sciences remained open, and a number of them tripled their staffs during the Cultural Revolution. Since most of the new recruits were secondary-school graduates, the Academy of Sciences must have played an important role in continuing to provide a scientific education, at least for some boys and girls.

When the colleges and universities were reopened in 1971, there had been many reforms. The new entrants were drawn mainly from families of working-class origin, and it was far more difficult for the children of the former bourgeois classes to gain entry. The period of instruction was to be cut from four and five years to two or three years, and higher education was to be linked more closely with production. This meant not only changes in the curricula but also that students would spend part of their time working in factories or communes. Students had engaged in periods of manual work before the Cultural Revolution, but now such work was to be integrated more systematically into the students' schedules. These policies mean that the new scientists and engineers will enter the work force more as scientific and technical generalists than as specialists.

As important as the education of the most highly trained members of the society may be, perhaps equally important in a developing country is the scientific education given to the population as a whole. The Chinese have accorded this a high priority. Not only do full-time secondary schools stress scientific and technical subjects but a wide variety of spare-time and part-time educational programs have been provided especially for workers in factories. In fact, one of the most important educational accomplishments has been the training of para-technical personnel. This has been particularly successful in the medical sphere, where large numbers of "barefoot doctors" have brought rudimentary health care to millions of peasants. Also, great emphasis has been placed on giving all members of society some elementary scientific knowledge. This is seen by the Chinese as particularly important, since they look to the workers to provide many of the technical innovations taken up by industry. It is claimed that workers' innovations are more relevant to production than those generated by the "experts." This is a controversial issue in China, but certainly during the Cultural Revolution workers' innovations received the greatest publicity.

### Science, technology, and society: industry

So far this article has focused attention on the ways in which China has built up its indigenous science system. In particular, it has concentrated on the organization of scientific research and on the training of scientific and technical manpower. However, research carried out in a country's own laboratories is only one way in which the production sectors of that country can acquire new technologies. The other way is by importation from abroad.

One of the most important science and technology policy decisions for any country involves how much and what to import and how much and what to generate at home. In the past, most developing countries have relied heavily on foreign imports. Now, however, it is recognized that such reliance can lead to economic and political dependence on foreign countries, and many developing countries are trying to find ways to diversify their sources of technology.

China has made a big issue of "self-reliance." Just how self-reliant has it been and to what extent has China, too, depended on foreign technology? Also, in their choice of techniques, to what extent have Chinese policy-makers been concerned with goals other than economic growth? The remaining sections of the article will attempt to answer some of these questions as they apply to different sectors of the economy. Examples will be given that are illustrative of more general policy decisions and, wherever possible, comparison will be made with the policies of other less developed countries.

It is in the industrial sector that the policy conflicts over choice of techniques and issues of importing foreign technology can be seen most clearly. But here, as in almost all phases of Chinese development strategy, there has been no single approach that has dominated Chinese thinking over the entire period of Communist rule. Rather, there have been three or four periods, during each of which different policies had predominance.

In the first period, immediately after 1949, Chinese industrial policy clearly aimed at copying the Soviet model. Emphasis was put on heavy industry, and technologies—mostly imported from the Soviet Union—were for the most part capital intensive. Rapid economic growth was China's principal goal. This period lasted until 1958, when problems of employment and income distribution assumed greater importance and a major switch was made to small-scale industries, especially in rural areas. Greatest publicity has been given to the backyard blast furnaces introduced at this time. They proved to be an economic failure, since the quality of the iron was too low to be of much practical use, but other local industries initiated in the communes—for making farm implements, for example—thrived. During this period, which lasted about three years, tremendous publicity was given to workers' innovations, and the role of the expert declined.

The next period began in 1961 and continued until the Cultural Revolution in 1966. It started with the severing of Soviet aid, the withdrawal of the Soviet technicians, and the canceling by the Chinese of orders for new Soviet plants. The enormous flow of Soviet technology into China, which had characterized the 1950s, became only a trickle, and "self-reliance" was the order of the day. This call for self-reliance led to a boost in indigenous research and restored the prestige of the research worker and expert, but it did not mean that China became autarchic. Although cut off from Soviet technology, China began to buy modern technology from Japan and Western Europe. Orders were placed not only for equipment and machinery but also for entire plants.

The Cultural Revolution once again led to new policies. Egalitarianism became the principal goal. The scientist and engineer were criticized for becoming a new "elite" and were accused of being the forerunners of a return to capitalism. The earlier policies were ascribed to the chief of state of the republic, Liu Shao-ch'i, and were deemed to be wrong. Once again emphasis was given to innovations that originated with workers. The importation of foreign technology declined from its previous levels but, despite the emphasis on self-reliance, it never stopped.

Finally, in 1971, the policy began to change again. The importation of sophisticated technologies increased in volume, and the policy of "walking on two legs," with roughly equal emphasis being given to

*Agricultural laborers (right) still cultivate rice paddies as their ancestors have done for centuries. While much Chinese farming is carried on by traditional methods, the government has given high priority to the use of fertilizer and the construction of dams such as the San Men Dam near Peking (below).*

Frank Fischbeck—LIFE Magazine © Time Inc.

Brian Brake from Rapho Guillumette

Emil Schulthess from Black Star

China's greatest industrial concentration is in Manchuria. This steam turbine is in a factory in Harbin, in Heilungkiang Province.

modern, large-scale industrialization and local, small-scale industries, appeared to be the main policy guideline.

What lessons can be learned from these changing Chinese policies that may have some relevance for other developing countries?

First, undue reliance on one source of foreign technology may have disastrous short-term economic effects should that source dry up.

Second, the short-term economic disruption caused by cessation of a single-source foreign technology may have longer-term beneficial effects, since it stimulates self-reliance and encourages the process of shopping around for alternative sources of foreign technology that, in turn, may lead to imports more appropriate to local conditions. Great self-reliance and inventiveness were needed in China to cope with a situation in which plants were left unfinished, blueprints were taken away, and there was even an embargo on the importation of spare parts. In a different context, Cuba was placed in a similar situation vis-à-vis the United States when the latter country put an embargo on the export of goods to Cuba in 1960. Some Cubans argue that, in the long run, the benefits of the "technical learning effect" that this generated among Cuban engineers will far outweigh the short-run economic losses that the embargo caused.

Third, the goals of job creation and egalitarianism will probably call for different sorts of technologies than those that maximize short-term economic growth. The Chinese approach has been to utilize both sorts of technologies, but it is doubtful whether relatively inefficient labor-intensive technologies can exist side by side with modern sophisticated technologies in a market economy. Thus, although there is great interest in job-creating technologies in many less developed countries today, it is by no means certain that the Chinese experience is particularly relevant.

Fourth, the Chinese experience certainly indicates that considerable economic and social advantages can result from involving the workers in "innovation." On the other hand, the vacillating policy toward experts indicates that, although China appears to have as much trouble as most other countries in getting its scientists and engineers to work on "relevant" problems, there is little evidence to suggest that the Chinese solution of political thought reform has been any more successful than the less dramatic solutions posed in other countries.

### Agriculture

In the early years of Communist rule little attention was given to developing new agricultural technologies. In the late 1950s, when the decision was made to give agriculture first priority, two technical issues predominated. The first was how to increase yields, and the second was the more ideological issue of agricultural mechanization.

Technology plays an important part in increased production through its contribution to irrigation, fertilizers, pesticides, and new seeds. During the Great Leap Forward, when the communes were first introduced, the greatest emphasis was put on irrigation. Labor was mobilized on a massive scale to dig ditches and build dams. In the 1960s much more stress was placed on improving production by producing and applying more chemical fertilizer. The need for fertilizer has led to the invention of new techniques that enable small-scale plants to manufacture ammonia and urea efficiently. By the end of 1971 nearly half of all chemical fertilizer was being produced by local industry. It has been estimated that most of the increased grain output in the 1960s was due to the application of chemical fertilizer.

There is as yet no evidence that Chinese scientists have been able to breed new strains of rice and wheat as high yielding as those developed by the International Rice Research Institute in the Philippines and by the Wheat Research Institute in Mexico, which have contributed so much to the Green Revolution in much of the rest of Asia. These new rice and wheat seeds are not particularly appropriate to Chinese conditions. It would be logical to assume that the Chinese will put high priority on research to develop suitable high-yielding varieties since these, used together with water and fertilizer, will be needed if China is to meet its grain requirements over the next decades.

The mechanization issue is more clouded by ideological considerations. Mao has advocated that, in the long run, China should aim for the maximum possible mechanization of agriculture. The tactics to achieve this long-term objective were subjected to heated debate during the Cultural Revolution. The supporters of Liu Shao-ch'i argued that the agricultural machinery to be allocated by the state should be concentrated in those areas most suitable for mechanized operations—*i.e.*, large areas of arable land where draft animals are scarce. They also favored the concentration of research resources in a few centralized research institutes. Mao's supporters, on the other hand, were in favor of decentralization and of giving the initiative to the communes. If a particular commune wanted to mechanize and acquire its own tractors, it should be encouraged to do so.

Similarly, Mao's supporters favored decentralization of research. Instead of just one or two large research stations, Mao favored the establishment of dozens of semimechanization research stations, so that if a peasant had a worthwhile innovation he would be able to try it out in one of the research stations near his own commune. This not only would encourage a flow of new innovations but would give all the peasants a sense of involvement in the development of the country. Mao did not completely rule out advanced research—some resources, he argued, should be used to support it—but the bulk should be used to support the semimechanization institutes. The whole question became a controversial issue during the mid-1960s, and the extent of the debate was revealed in the Chinese press during the Cultural Revolution. As in almost all such debates, the reader is left in no doubt as to whether Mao's ideas were "correct" and are now being implemented.

There are obvious lessons here for other developing countries. Not only is Mao's policy more appropriate for generating labor-intensive technologies, it also has the further social benefit of providing more opportunities for individual peasants to become involved with development.

### Medicine

It is probably in the field of public health and medicine that the Chinese experience has greatest relevance for other countries. There are several particularly notable aspects to the Chinese approach. First, the Chinese have stressed preventive rather than curative medicine. This is the reverse of what has been done in almost all other developing countries. Second, they have given a high priority to training paramedical personnel. Peasants are given from a few months to a couple of years of medical training in rural areas, and these workers provide the backbone of the medical service. Third, in certain branches of curative medicine the Chinese have surpassed even the more industrially advanced nations. Particularly notable have been the successes in rejoining severed limbs and in the treatment of burns. Fourth, the Chinese have achieved quite remarkable successes by combining traditional with modern medicine. Herbal drugs have been shown to have considerable therapeutic value, and the new successes in using acupuncture to replace anesthetics in several types of major surgery have gained worldwide attention.

The most significant and important features, however, are the concentration on preventive medicine and the introduction of a curative

Dr. Victor W. Sidel, Montefiore Hospital, Bronx, New York

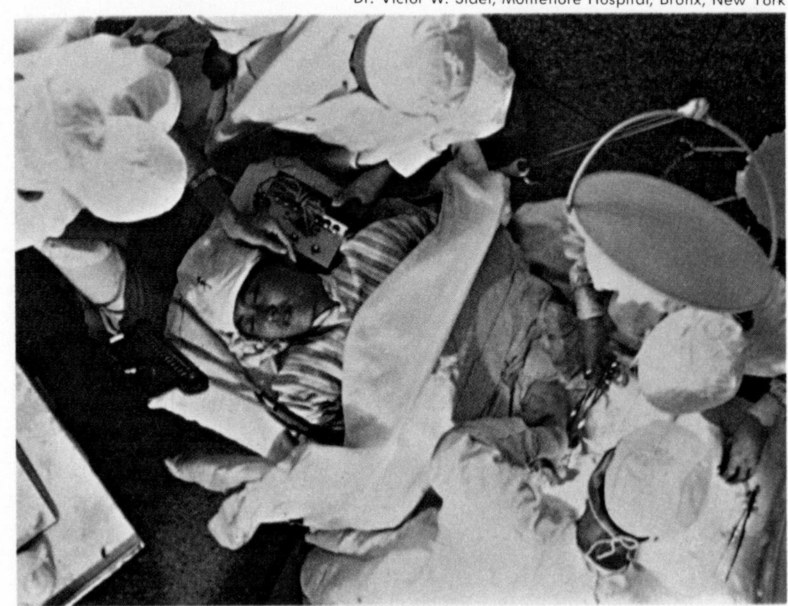

*Chinese medical practitioners have achieved considerable success by combining modern practice with such traditional methods as acupuncture, in which needles are inserted at specified points in the body. An ovarian cystectomy performed on the woman (right) under acupuncture anesthesia was witnessed by a group of American physicians at the Third Teaching Hospital in Peking. The patient was conscious and able to talk throughout the operation. Acupuncture charts (opposite page, left) are used by doctors in placing the needles for anesthesia. The "plastic man" (opposite page, right) is on display at the Industrial Exhibit Hall in Shanghai.*

service in rural areas. Just how this has been done is graphically described by Joshua Horn, who recently returned to England after working in China as a doctor for 15 years. In his book *Away with All Pests,* he describes what happened in his hospital in Peking after the start of the Cultural Revolution, when the decision was made to speed up the rate at which rural areas were being provided with health care. Every urban hospital was ordered to make one-third of its staff available for duty in rural health centers. Horn's hospital sent its people to a center in Hopei. At the end of a year some chose to stay permanently in the country, but the rest returned to Peking to be replaced by another third of the hospital staff.

Horn describes in some detail the work of the hospital staff after they had arrived in the countryside. The first function of the rural teams was to accent the role of preventive medicine by stressing public health measures and sanitation. To this end, they helped organize campaigns to eradicate such pests as flies and mosquitoes. Next, the teams set up training courses for the barefoot doctors. During the first winter the new recruits from the communes received five months of medical training. At the end of that time they were able to give inoculations and first aid and to know when to call the hospital doctors. They returned for a second five-month spell of training the next year and were scheduled to undergo three such sessions. After training they returned to their communes to provide grass-roots medical service. The rural teams also provided instruction in family planning and worked closely with existing medical units in the area to upgrade their capabilities in both traditional and modern medicine.

This widespread effort to bring health services to rural areas must be reckoned as one of the really outstanding of all Chinese accomplishments. Nowhere else has the movement of doctors from poorer to richer areas and from poorer to richer countries been reversed.

Brian Brake from Rapho Guillumette

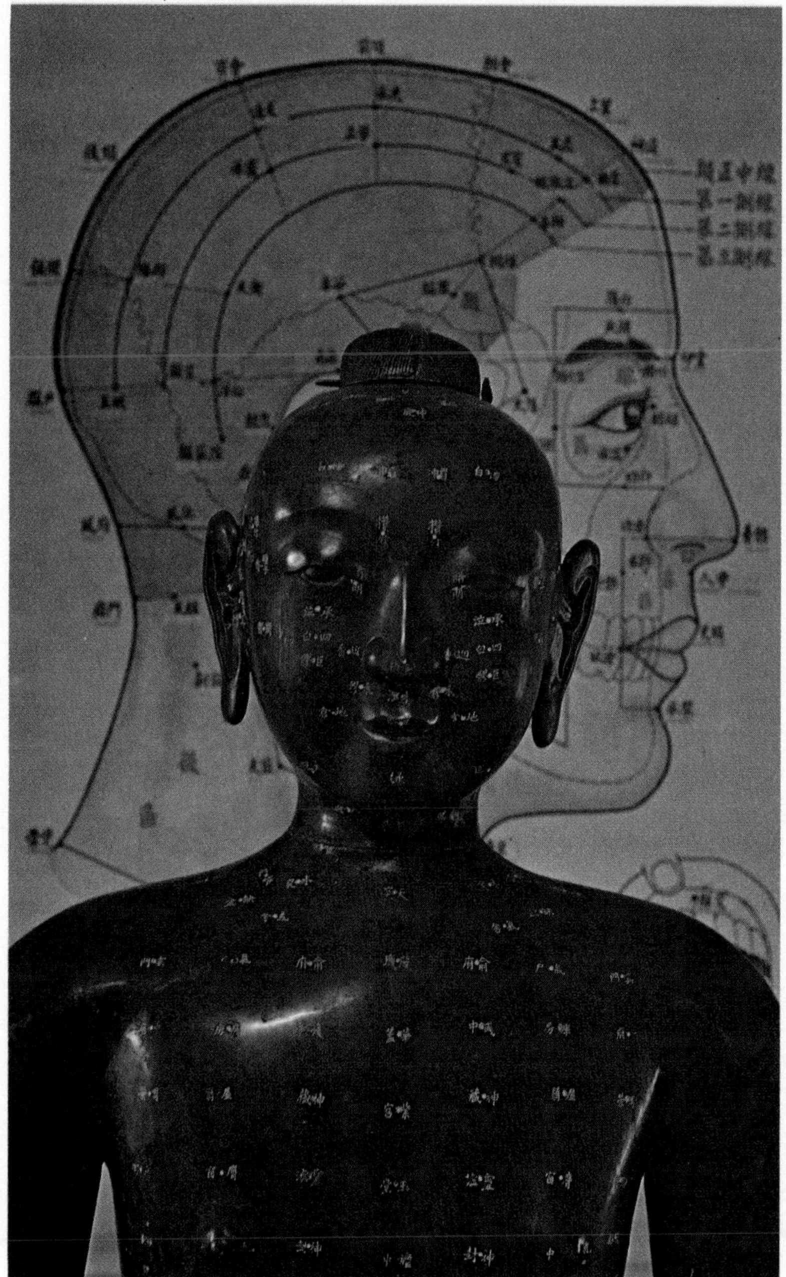

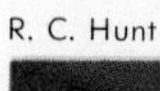

R. C. Hunt

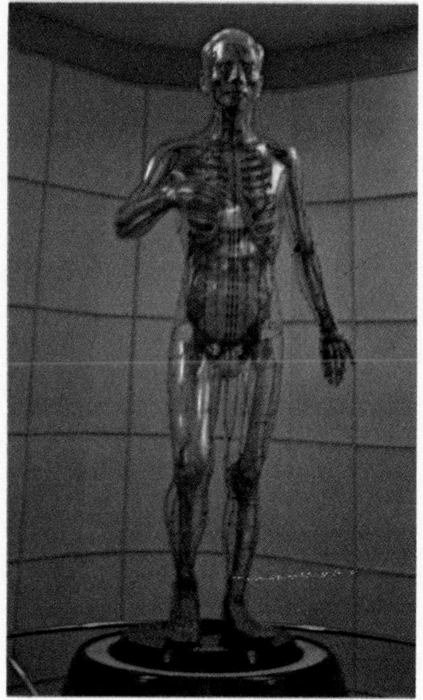

## Defense

Although the application of science and technology in the economic and health fields has greatest relevance for other less developed countries, it is the Chinese technical achievements in the development of military technologies that have gained the most notice. The timing of the events is well known, although the technical details are still vague. The first Chinese hydrogen bomb was successfully exploded only eight years after the inauguration of the first research reactor and less than three years after the first nuclear explosion. Chinese rockets have launched a guided missile and put two space satellites into orbit.

The scientific and technical work behind these successes was led by Ch'ien Hsueh-sen and Ch'ien San-chiang, both of whom had worked in the United States but who decided to return to China in the mid-1950s. Obviously, their work has been given massive research and engineering support, and although defense science and technology did not entirely escape political criticism during the Cultural Revolution, they do not seem to have undergone the same sort of disruption that civilian science experienced. The existence of such a high-technology enclave in China has ramifications far beyond its military value. Especially important is the fact that it must demand high quality, technically sophisticated products from Chinese industry. It is interesting to speculate on the extent to which lessons learned in the process of linking industrial production with research and with defense requirements have been transferred to the civilian sector.

## Conclusion

Sufficient examples of Chinese science and technology policies have now been given to enable a synthesis to be made. What appear to be contradictory and at times even irrational policies make sense if Mao's long-term objective is kept in mind. This is, and has been ever since the People's Republic was founded, to set China firmly on the road to a truly utopian Communist society, a society in which there is no exploitation, no dichotomy between manual and mental labor or between urban and rural areas, and where everyone is both "red and expert" (*i.e.*, both ideologically committed and technically competent).

That Mao was disappointed with the progress being made toward this goal was clear as early as 1964–65. He apparently decided that a major midcourse correction was needed and in mid-1966 launched the Cultural Revolution. In the years preceding the Cultural Revolution, Mao had taken the "experts," many of whom had been trained abroad, and tried to make them into "reds," but with only partial success. It seems that as China developed economically, an elitist technocracy began to emerge—one that was pragmatic rather than idealistic. Mao believed that the actions of the pragmatists would lead China back into the capitalist camp.

Clearly, from Mao's point of view, drastic action was needed if the younger generation was not to be misled permanently. If he could not make reds out of the experts, then he must try to make a new kind of expert out of the reds. This was to be accompanied by an even greater effort to reeducate the scientists and other intellectuals in what seems to be an almost desperate attempt to attune them mentally to Mao's utopia.

The growing dichotomy between manual and mental labor appears to have been one of the key reasons for starting the Cultural Revolution and explains much of what has happened in the research institutes and universities. Once the revolution was under way, however, the other dichotomy—between city and village—took on increasing importance. Mao's efforts to decentralize education and health services and to encourage small-scale industries and intermediate technologies are consistent with his aim of narrowing the gaps in development between different parts of China itself.

That the revolution has caused a major interruption in education and scientific research there can be no doubt. There appears to have been less of an interruption in industrial and agricultural production, and in some sectors that are of military significance there have been major technological feats. There have also been major gains in public health. The interruptions in education and research appear to have been temporary. The objective of self-reliance means that Mao's determination to use science and technology to build a modern China has not faltered. But whether a really modern country can be built without, at the same time, creating a technocratic elite remains to be seen.

FOR ADDITIONAL READING:

Horn, Joshua, *Away with All Pests* (Hamlyn, 1969).

Needham, J., *Science and Civilisation in China*, vol. i–iv (Cambridge University Press, 1954–65).

Signer, E., and Galston, A. W., "Education and Science in China," *Science* (Jan. 7, 1972, pp. 15–23).

Wu Yuan-li *et al.*, *The Organization and Support of Scientific Research and Development in Mainland China* (Praeger, 1970).

# The Tasaday
# Stone Age Tribe of Mindanao

**In the mountain jungles of the Philippines a group of scientists in 1971 first encountered the Tasaday, people who lived in a Stone Age culture with almost no knowledge of other human beings.**

The recent discovery of the Tasaday Manobo—a tribe of food gatherers living in caves and still using stone tools—is providing dramatic new data on human ecology and on the ability of man to adapt to a highly specialized environment. For the Tasaday inhabit a vast, high-altitude rain forest located in the mountains of South Cotabato Province, in the southern part of the island of Mindanao in the Philippines. Here they have survived as a tribe for centuries even though they have been virtually without trade or other relationships with the rest of mankind.

Official contact with these unique people was made by Manuel Elizalde, Jr., president of PANAMIN, the Private Association for National Minorities, Inc., a nonprofit foundation undertaking critically needed medical and socioeconomic development projects among cultural minorities in the Philippines. Research among the Tasaday during periods of contact beginning in June 1971 and extending through May 1972 provide the basis for this statement about their discovery, their physical traits and language, and their society and culture.

When first contacted in 1971, the Tasaday had no knowledge of any cultivated plant such as rice, taro, the sweet potato, corn, cassava, or millet. The Tasaday language has no words for these widespread cultivated plants, nor for any tools or other notions associated with agricultural activities. The Tasaday apparently never had tasted salt or sugar, nor smoked tobacco, since they also lacked words for these common items. Perhaps the Tasaday are the only people now living in an agriculturally appropriate environment who neither know nor grow tobacco, sweet potatoes, corn, or cassava (plants of American Indian origin that swept the world in sailing vessels of the 16th and 17th centuries).

*The information in this article is based on research and reports by the following scientists and officials associated with the PANAMIN foundation:* ***Frank Lynch, Robert B. Fox, Teodoro Llamzon,*** *and* **Carlos Fernandez.**

*The Tasaday Manobo were discovered in a dense, high-altitude rain forest in the southern portion of the island of Mindanao. (Opposite page) Group gathers around fire, which is used both for cooking and for warmth.*

Stone tools made by the Tasaday (and used to fashion such implements of wood and bamboo as the digging stick, fire drill, and a crude bamboo knife) are among the most primitive recorded in a tribe of modern times. They include hand scrapers made from flakes of quartz or from river pebbles ground to a sharp edge on one side and attached to wooden handles. Unmodified stones are also hafted as hammers; indeed, any handy cobble-sized stone may be picked up and used for pounding.

The archaeological record (supported by carbon-14 dates) indicates that such stone flakes and edge-ground scrapers still used by the Tasaday were effectively replaced throughout the Philippines by metal implements about 2,000 years ago. Study of the stoneworking techniques of the Tasaday may lead scholars back to an understanding of an art once thought to be lost.

The exploration of their forest habitat in March 1972 further disclosed that they inhabit or utilize three caves found in a limestone conglomerate, located 400–500 feet up a steep slope above a creek that provides drinking water and their major source of protein food. The manner in which the Tasaday live in these caves (said by them also to have been occupied by their forefathers) will provide singular new data on the period, tens of thousands of years ago, when ancient man characteristically inhabited caves and rock shelters.

The Tasaday at first seemed unaware that there were other societies of people, although they told of having occasionally glimpsed alien hunters or trappers or heard their voices in the forest. They were ignorant even of the names of such neighboring tribes as the Manobo Blit, Ubo, and T'boli. Later, however, they told of having recent intermittent contacts with neighboring forest groups, the Sanduka and the Tasafang. Indeed, it appeared that Tasaday men traditionally married Tasafang women.

Tasaday territory is in one of the most isolated, mountainous regions of the Philippines. All of this broken and rugged terrain is above 3,000 feet elevation, and is blanketed by hundreds of square kilometers of ancient (climax) rain forest. Those who daily fly airplanes (described by the Tasaday as "large birds") over the area found it difficult to believe that this vast sea of undisturbed forest was inhabited by man. Crowns of towering wing-seeded trees (dipterocarps) form a ceiling over the whole environment; even the small streams and creeks along which the Tasaday live and move are not visible from the air. Their world is the relatively warm floor of the forest, dark and damp under clouds that accumulate around the mountain tops; and their home caves. The sun, which Tasaday call the "eye of the day," filters through the forest only briefly at its zenith. Thus, features of earth and sky known to most people (such as lakes, oceans, open fields, and constellations of stars) are unfamiliar to the Tasaday and have no expression in their language. By contrast, most unique features of their forest habitat have a place name, and their knowledge of the local plants is most detailed.

Dolf Herras from Nancy Palmer Photo Agency

*The only albino in the local group of Tasaday, a retarded child, suffers from rashlike sores over much of his body and is carried constantly by his widower father.*

The Tasaday have remained apart from other men by successfully adapting to a forest environment that has resisted intrusion from outside. So far, the forest has provided them with a stable and favorable ecology that yields life's basic needs. Now, however, along with other great climax forests of Southeast Asia and the Philippines, Tasaday lands are being threatened. Loggers and miners are intruding, and lowland and neighboring mountain agriculturists faced by exploding populations are pushing into the mountains, cutting and burning clearings in the forests for their crops of rice and corn. It is estimated by United Nations sources that the primary forests in the Philippines will be wiped out in perhaps 35 years. And without the forests, the Tasaday (perhaps the last people on earth who live by collecting wild foods in a high-altitude forest) can only disappear as a unique tribal culture. Indeed, their survival even as a distinct biological subgroup of people seems imperiled.

### Discovery of a lost tribe

Persistent rumors of a primitive people living in the forest had been reaching the PANAMIN staff at Kemato, the southern Mindanao location of the T'boli Development Project, from a few Ubo and Manobo Blit visitors. The Ubo and Manobo Blit are other isolated and conservative mountain-dwelling minority tribes who rarely have contacts with the populous lowland peoples. None of these mountain dwellers, however, offered more than hearsay until 1971. In that year Kemato was often visited by a Manobo Blit trapper named Dafal—popularly known as "The Bird," with a reputation of "walking through the forest like the wind." The trapper revealed to Manuel Elizalde that he had met primitive people in the forest and had stayed with them overnight.

Dafal told how he had first seen the Tasaday in 1967 (or 1966) while placing traps for wild pigs in the forest. At that time, he surprised a small group of naked Tasaday preparing wild yams (edible tubers) along a creek. They fled at his approach but finally stopped when he continued to run after them shouting in Blit: "I am good! I am good! I can do you no harm!" According to the Tasaday, this was their first contact with anyone from a human group different from theirs.

Dafal stated (as was later checked with the Tasaday themselves) that when he first met them they had no metal tools, cloth, beads, or any of the usual trade goods that have reached all other hunter-gatherer societies known in Southeast Asia. Indeed, he observed that they were exclusively a food-gathering society with no hunting or trapping devices. Dafal added that he visited the Tasaday in the forest about a dozen times while trapping wild pigs or gathering the pith of palms. He gave them a few metal items, the most important being two small bolos (knives for hacking vegetation), only one of which, broken and badly nicked, remained in 1971, and wire for brass earrings. He also introduced the group to the bow and arrow, a few types of baskets, rags of cloth, and even contributed their only musical instrument—a mouth harp—which one Tasaday man learned to play.

Dolf Herras from Nancy Palmer Photo Agency

*Family group gathers around fire by the side of a stream. The Tasaday get much of their food, notably crabs and tadpoles, from such creeks.*

Plans then were made by Elizalde to seek the Tasaday by helicopter during a socioeconomic survey of the Manobo Blit settlements near the forest where Dafal said the Tasaday lived. Dafal was sent ahead with a party of T'boli workers who would clear a landing place on the edge of the forest while Dafal himself searched for the Tasaday. Then, on June 7, 1971, shortly after the helicopter landed, Dafal emerged into the clearing from the forest with four frightened Tasaday men. They seemed terrified by the whirling blades of the "large bird" and by the "god" (Elizalde) who Dafal told them rode inside. Dafal stated that the Tasaday believed in a kind of messiah who, according to their ancestral

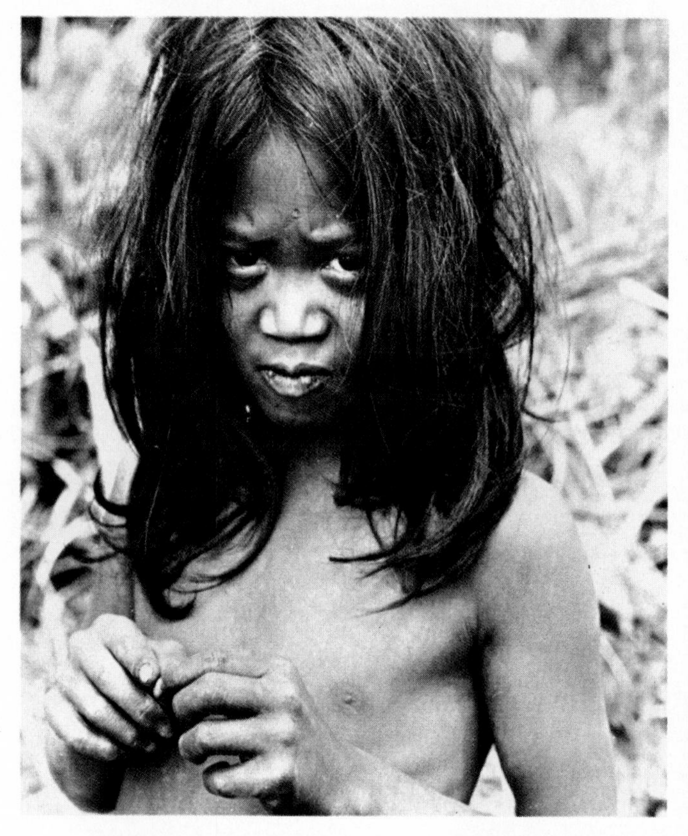

*Children in the local group of the Tasaday number 11 boys and 2 girls; all those pictured above and on the opposite page are boys. Hands of the teen-aged boy above are wrinkled and aged from constant food gathering.*

tradition, would one day appear to help them find more food. Afterward, the entire local group of Tasaday—25 men, women, and children—came out of the forest to join their companions at the landing place.

The Tasaday later described this dramatic event as a "great noise" —sudden, thunderous, terrifying, and inexplicable. Nevertheless, even that day and during all periods of subsequent contact, they have been friendly, tolerant, and understanding toward the stream of visiting anthropologists, linguists, news reporters, and photographers; albeit often shy and sometimes bewildered. Their warm behavior with strangers suggests that they never have known unfriendly outsiders; aggression to them seems inconceivable.

Even at first sight the Tasaday were clearly identifiable as Mongoloid in physical type. Related racially to their neighbors and to the dominant Filipino population, the Tasaday are relatively short with light to medium dark skin, straight or wavy hair, and prominent cheekbones. The other societies in the Philippines that still are dependent upon hunting and gathering are the pygmoid Negritos (Aeta), found primarily on the island of Luzon and racially quite distinct from the Mongoloid majority.

The following data were obtained by Elizalde, anthropologist Robert Fox, and linguist Teodoro Llamzon during the first period of contact at the helicopter clearing on the edge of the forest, and later by anthro-

pologists Frank Lynch and Carlos Fernandez during a second exploration in March and April 1972 to the forest interior and to the caves in which the Tasaday live permanently. On both occasions, the researchers communicated best through a Manobo Blit woman named Igna who also spoke T'boli. Her information then was translated by T'boli members of the survey team who were familiar with the Philippine national tongue (Tagalog) or English. It was early established that the Tasaday, as do all peoples in the Philippines, spoke an Austronesian language, one specifically related to the many Manobo tongues on Mindanao. It appears that the Tasaday language has its closest ties to Manobo Blit; nevertheless, Igna was able to understand less than half of their first answers to endless questions, although she later gained fluency.

### Habitation and social organization

The Tasaday permanently occupy or utilize a complex of three limestone caves, located about a day's walk for them from the helicopter clearing where they were first contacted. Access to the caves is difficult, as they are located 400–500 feet up a steep slope from a small creek. According to the Tasaday, there are no other habitable caves in this area of the forest. They further state that these caves were also used by their forefathers, and evidences of a refuse heap in front of the uppermost cave (the stone floor is swept daily with leaves) suggest

a lengthy period of past occupation. Other peoples in the Philippines still live in caves seasonally, even subgroups of horticulturists such as the Pala-wan living in the Ransang area of southwestern Palawan Island, but there is no local evidence of the use of caves for permanent habitation since Paleolithic times, thousands of years ago. Tabon Cave, also in Palawan, was occupied by Paleolithic hunter-gatherers for more than 40,000 years. But it is truly startling to find a contemporary tribe of food gatherers who live permanently in caves.

The only local group of Tasaday contacted was composed of six elementary families (the elementary family is the social unit of father, mother, and unmarried children). Of the total of 25 individuals, there were 12 adults; 11 of the 13 infants or children were male. Four elementary families were close blood relatives; indeed, the fathers of three families were either children or nephews of the oldest couple in the group.

Orphans and unmarried adults attached themselves to one of the families. All of the families lived and slept together in the uppermost cave, a warm, light, and airy chamber about 25 feet deep, 30 feet wide across the mouth, and 10 feet high. Balayem, an unmarried young man until April 1972, usually slept in the next lower cave, sometimes accompanied by younger children. As there appeared to be an optimum balance between the present size of the local group of Tasaday and the available food resources of the nearby forest, the group rarely foraged more than a kilometer or two away from the caves and most of the group was back in the main cave with sufficient food by midday.

The largest social unit of Tasaday society is the *nasagbung* or "local group" formed by a number of closely related families. This local group is not a band, as defined by anthropologists, since there is no formal leader who is solely responsible for jural (lawfully binding) decisions. Instead, decisions that involve group activities are shared by adults and family heads, the most active and experienced among them assuming temporary leadership (in the style of a committee or board of directors). The family is the basic social and economic unit, although its functions are considerably modified by the small size of the local group, shared food-gathering activities, and common residence in a cave.

Married couples sleep independently in the cave with their infants and younger children. Father and mother share the responsibility of child care. Older children were observed to be almost inseparable, playing, gathering fruit, and also sleeping in the same area. And they might eat with any of the adults. When not busy, they were seen sitting with arms around each other; penis play between individuals was noted. Although the family also tends to be the basic economic unit, the Tasaday emphasized that food gathered each day was shared communally among families when necessary. Their intention is that all members of the local group, particularly the children, can have a meal. Food for all families is prepared at one or the other of two hearths found in the cave (a third fire being used at night for warmth). Mealtime is irregular, although leftovers or newly cooked food are usually eaten in the morning before the daily foraging begins. Food is also prepared around noon when the foragers have returned from the forest and/or before dark. Wild fruits are eaten at any time during the day.

The Tasaday may eat food raw, roasted, or cooked in the tough leaves of a river-bank plant (*Curculigo*) twisted together and placed on the coals at the edge of the fire. Their present practice of cooking palm starch and chunks of meat in bamboo tubes, widespread among other gatherer-hunters in Southeast Asia who possess metal tools, was unquestionably introduced by Dafal. One set of flint and steel was also given to them by Dafal, but they continue to make fire when necessary by rotating a wooden drill between the palms of their hands against another piece of soft wood and small bits of tinder. This type of fire-making drill has not been observed by the writer among any other Filipino people. The Tasaday usually carry a live ember when planning to stay away from the cave overnight or when moving from one place to another. Permanent fires are maintained in the cave. At night members of the families cluster for additional warmth around the hearths.

There is little evidence of formal division of labor between the sexes. The data of Carlos Fernandez seem to indicate that the women possibly bring more food back to the cave than the men, although the men appear to be responsible for such relatively arduous tasks as climbing trees for fruit and digging for deep wild yams.

The Tasaday rarely stay away from the caves overnight but, when necessary, shelter is found in the spaces formed between buttress roots of giant trees. Specific trees are selected near small streams and creeks, which form the principal trails through the rugged terrain.

As among all peoples in the Philippines, the Tasaday have a bilateral kinship system; that is, relationship is reckoned equally with both father's and mother's kin. The kinship terms used in reference and address are also similar to those found throughout the islands. Grandparents and grandchildren call each other by the same reciprocal term. Fathers and mothers are distinguished terminologically from uncles and aunts, but brothers and sisters are identified by the same term, which is also extended to include cousins.

Social and kinship behavior undoubtedly reflect both the small size of the group and the high mortality rate among infants and children. Unusually affectionate ties were noted between parents and children; the latter are carried constantly, even when they are three or four years of age. A characteristic and touching scene is that of a Tasaday mother walking down a stream bed with one child at her breast and another riding on her back. Parents seem reluctant to stop nuzzling and smelling their children; they do not kiss. Close and affectionate ties between husband and wife were evident during a sudden downpour, he rushing to prepare a temporary shelter of palm fronds for her and their younger children. There are signs of a deep expression of intimacy among all individuals; such close affiliation is not found generally within larger and more sophisticated groups of people.

Dolf Herras

*Stone scrapers made from flakes of quartz and ground to a sharp edge on one side are used to fashion crude bamboo knives (opposite, top and bottom). A woman (above) uses her hands to comb the hair of another; both wear the traditional skirts made from leaves of a tufted herb.*

*To make fire the Tasaday rotate a wooden drill between the palms of their hands against another piece of soft wood and small bits of tinder. Anthropologists had never before seen this type of fire-making drill among Filipino people.*

There is apparently no such formal social or occupational role as that of midwife among the Tasaday. Each mother was said to deliver her baby alone, cutting the umbilical cord herself with the sharp edge of a strip of bamboo; the father is responsible for burying the cord. Marriage is said to be arranged by parents, as among most traditional societies in the Philippines, but the advice of all adult members of the local group is considered. Wedding gifts were also mentioned. Since the group is small, there is little choice of spouse, and there may be an imbalance between the number of marriageable boys and girls. (The Tasaday group contacted had two unmarried young men; there were 11 male children, yet only 2 female youngsters.) Nevertheless, the Tasaday insisted that they practice neither polygamy nor polyandry; nor do they exchange or lend spouses, although these practices are known among other people in the Philippines who still hunt and gather in small local groups, such as the Negritos of Luzon. There seems to be no ancient moral precedent for the Tasaday to ban polygamy; indeed, their first male ancestor is said to have had two wives.

People who die of old age or injury (one man was recently killed by a falling tree) are said to be simply covered with leaves near the place where they died. People who die in the caves are carried into the nearby forest for "burial." There are no formal cemeteries.

Statements of the Tasaday indicate that the Sanduka and Tasafang still live in the surrounding forest, though the Tasaday have not contacted them in recent years and are not certain that these groups still survive. Three men were born among a local group living at Lambong, which possibly has died out as a social unit. The husband of one woman and his father traveled three nights to Sanduka territory in order to obtain the young bride. This is the longest trip away from their caves that any living Tasaday has made and was also the last contact with a neighboring group. The Tasaday have been asked by Manuel Elizalde to confirm the existence of the other groups, but they are extremely reluctant to leave the area surrounding their cave homes. The Tasaday believe that as long as they remain at the caves they will be safe.

The population of the Tasaday seems never to have been large, and to have been separated throughout its history into small local groups. The size of the local group is restricted, among other factors, by the level of technology achieved by the people, by their knowledge of how to exploit the resources of the forest, and by their ability to forage sufficient territory to support themselves. The local group of 25 Tasaday contacted seems a near minimal number for human group survival. It is consistent with norms of 20 to 30 people in local groups reported for other hunter-gatherer forest societies found in other regions of Southeast Asia.

### The quest for food

When Dafal found them five years ago, the Tasaday were exclusively food gatherers with a Stone Age technology. Without trapping or hunting devices, they were unable to feed on such larger animals in the forest as wild pigs, deer, civets, and monkeys. Dafal introduced a number of traps and the bow and arrow; these are now used by some of the men with limited success. Their major source of protein, however, is still river life (tadpoles and crabs) collected by hand. Conservationists would be pleased with the remark of one strong, active Tasaday man who does not use the new traps or the bow and arrow: "The largest animal I have ever killed is a frog." Dafal also taught them how to obtain the edible starch of wild palms and "palm cabbage" with the metal knives (bolos) he gave them. This exploitation of palms has provided the Tasaday with an important, readily available new staple. It is probable that palm starch will become their principal food, further testimony to the impact that Dafal had upon Tasaday life-ways.

The Tasaday's principal sources of food in the past, and still of great importance, are the roots (tubers) of yam vines (*Dioscorea*). These wild yams are similar in properties and food value to the cultivated potato. Used by food gatherers throughout Southeast Asia, at least nine different species of wild yams (Tasaday call them *biking*) are found in the Philippine forests and their tubers are available throughout the year. The tubers can be kept without spoiling for two weeks or more, and seem to have deep meaning for the Tasaday. Thus, when one of them was asked to name the most beautiful thing to him, he immediately answered: "For me, it is to find a large yam after digging deep in the ground." The yams are excavated with a digging stick traditionally pointed with a stone scraper (or now with a metal knife).

When gathering yams, like other peoples in the Philippines, the Tasaday may leave the head of the tuber attached to the vine in the ground—a sort of incipient agriculture. Knowing the location of the vines in their territory they can return the next year to reharvest the yams. The wild fruit of such larger plants as the climbing palmlike rattans, the wild banana, and the ground-level fruit of a wild ginger are also a significant source of seasonal food. The fruit, leaves, and flowers of many trees and bushes are also eaten.

Large and fleshy tadpoles of the giant mountain frog (*Rana*), small crabs, and a rare species of the halfbeak fish still form their major source of protein. These are abundant in the small creeks and are caught with the hands by simply feeling in crevices among rocks. The Tasaday have no weirs, nets, or other fishing devices. The giant frogs are killed for food with a striking stick, and meat (particularly of the wild pig, monkey, and rat) now occasionally is obtained with traps introduced by Dafal. Grubs from decaying fallen trees are a sought-after food.

Tasaday do not gather wild honey (although it is prized by other hunter-gatherers in the islands) since, as they say, "We do not know how to drive away the bees." They rarely exploit birds for food and have no term in their language for the egg.

In summary, the economy of the Tasaday is still based upon gathering wild vegetable and animal life, even though supplemented in a limited way by recently introduced trapping and hunting devices.

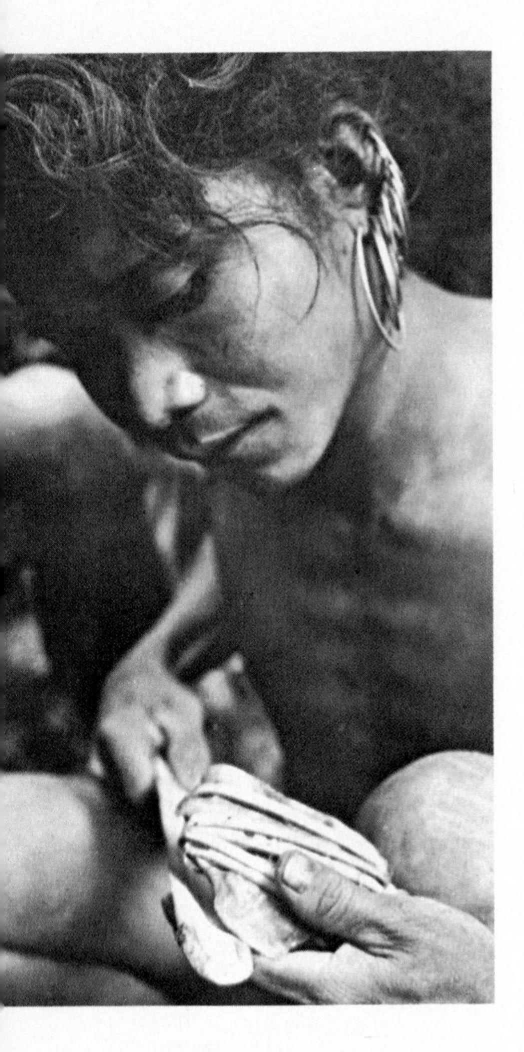

*Chief scraping tool of the Tasaday is a stone (opposite) that has been ground to a sharp edge on one side and then attached by thongs to a wooden handle; thin river pebbles are used for the stones. Man above uses an edge-ground scraper to fashion a wooden implement.*

## Health, dress, and adornment

The Tasaday appear to be in fair physical condition despite problems of food supply and the damp, high-altitude forest in which they live without permanent shelter and protective clothing. The active children and younger individuals display muscular bodies—vaulting across streams with poles, climbing trees for fruit, and moving swiftly with great agility along the slippery bottoms of creeks. The hands of younger individuals strikingly resemble those of the aged, probably the result of their continual use in the absence of tools or other technology for exploiting the environment. Older Tasaday appear wan and sleep much of the time throughout the day; life span is short, with only one couple estimated to be as old as 50 or so. Goiter is present and one young man had a severe fungous disorder of the skin. Another boy had a pair of congenitally webbed toes. A retarded albino child with multiple rashlike patches and ulcers of the skin (his mother is dead and his father must carry him throughout the day) seemed to have little chance of surviving long. Genealogical data indicate that an average of only two children per family survive to the age of two or three years; only then are they given personal names.

Dental examinations revealed no cavities (caries), but diseases of the gums (pyorrhea) were seen. Most adults chew betel nuts (from a wild Areca palm) mixed with slaked lime (from the shells of land snails) and betel leaf (from a type of pepper plant). They file blackened front teeth with abrasive stones from the streams. The Tasaday use their teeth in lieu of tools, often using them in rounding a stick for use as a fire drill, in cutting a vine to be used as a belt, or in tearing away the skin or bark from plants.

Brief skirts and small coverings for the genitals (G-strings) are fashioned from the long, tough leaves of a tufted herb (*Curculigo*). The leaves resemble those of a ground orchid (*Spathoglottis*) which may also be worn. The men wear one or two *Curculigo* leaves running between their legs and under and over a vine waistband. Women wear a bunch of similar leaves (or a broader, still unidentified leaf) held up with a vine belt around the waist, an attractive tassel effect being achieved by the pointed ends of the leaves as they fall. They now prize the cloth that Dafal gave them for G-strings and skirts.

The hair of men and women usually reaches to the waist; sometimes it is worn by women in a bun. The men with long hair may tie it at the nape with a ribbon made of a plant leaf. The men and women spend considerable time combing their long hair with their fingers, but use no implement as a comb.

Ornaments seemed limited to brass earrings worn by both men and women as a series of rings attached from the outer edge (helix) of each ear to the lobe. According to the Tasaday, these rings were formerly made of plants, and were replaced by brass wire obtained from Dafal. Not one glass, stone, or shell bead was observed among the Tasaday, further evidence of their isolation for hundreds of years from other peoples. Glass and semiprecious stone beads have been a major item of trade in the Philippines and in island Southeast Asia since about 500 B.C. Tasaday do not even use the black or white Job's-tears (seeds of the grass *Coix lachryma-jobi*) that are widely worn as necklaces by minority peoples throughout the Philippines.

Tasaday adults of both sexes tattoo themselves with thorns and dark plant sap on the inner side of the arms and legs, although the geometric designs fade rapidly. The early Spanish writers described natives of the southern Philippines as *Pintados* ("painted people") because of their extensive body tattoos. Tattooing is also characteristic of most minority groups in the islands, including the mountain neighbors of the Tasaday in Mindanao. Such evidence is to be considered along with their language in establishing the Tasaday's common cultural origins with other Filipino peoples.

## Religion and philosophy

Tasaday attribute all of their knowledge and material possessions (before Dafal) to their ancestors—the *fangul.* Their first ancestor is called Bengbang; he is said to have had two wives—Dakwe and Fiyuwe. These women bore two sons (Sanumabat and Lantoy) and a daughter (Mangi) from whom all Tasaday are said to have descended. Tasaday relationships with the supernatural primarily involve the more recently dead (*sugoy* or "soul relatives") whom they say they see in their dreams. The dead are said to dwell in the tops of trees and continually to interact with the living by helping them find food. Tasaday also speak of Salungal (the "owner of the mountains") as aiding them; he too is held to live in the crown of the forest. One informant described a type of divination in which measurements are made on the arm with the outstretched fingers of the hand. If the remeasurements of the spans of the hand on the arm are the same or longer, then the answer to a question posed is yes; if shorter, no. Questions addressed to the dead, such as where food can be found, are answered in this manner. Similar divination practices are recorded among surrounding peoples.

Rituals were not observed among the Tasaday. Their religious beliefs and practices seem poorly developed as compared with those of neighboring agricultural people, and focus primarily on relationships with the dead. The origin myth of the Tasaday is similar to although less detailed than that of the Manobo Blit and T'boli. These and other social and cultural traits the Tasaday share with other Manobo groups in southeastern Mindanao indicate a widespread, older culture basic to this region.

Living in isolation for hundreds of years with an extremely primitive technology and dependent upon food gathering, the Tasaday have a philosophy or world view that expresses deep and intimate relationships with the surrounding environment. Their universe is the canopied rain forest. Not only are the dead believed to dwell in the crowns of the towering trees, the trees also form a ceiling over their known world. In contrast, neighboring peoples with whom the Tasaday share features of an ancient culture speak of their dead and deities as living in the sky.

Food gatherers have been described as "natural man," and this the Tasaday are. They have depended entirely upon spontaneous products of nature for their wants and needs. Moreover, they live in balance with the rain forest—without disturbing ecological processes. Their life contrasts sharply to that of modern, complex technology that re-shapes the environment drastically.

**Past and future**

It seems clear that the Tasaday have survived with a Stone Age technology in a high mountain forest of southern Mindanao since prehistoric times. Evidence is that they are not a recently "banished people" who splintered from a neighboring group because of conflicts or who fled into the mountains to escape smallpox or other disease. The Tasaday retain distinctive elements of an ancient culture, indicating that they long have been isolated from the influences and developments that have changed other peoples with whom they have genetic ties.

Linguistic and cultural evidences demonstrate that the Tasaday have their closest relationships to the Manobo Blit, their mountain neighbors. The Tasaday and Blit form, in turn, a subgroup of a larger cultural-linguistic group in Mindanao described as the Manobo. The scattered distribution of the many Manobo groups, plus linguistic and cultural

*Father and child demonstrate the affectionate ties that exist between Tasaday parents and offspring. Parents frequently nuzzle their children and carry them constantly, even to the age of three or four.*

data, indicate that they once formed a closely related people who spread throughout Mindanao during the late Neolithic Period (probably before 1000 B.C.).

Jesuit priest Teodoro Llamzon, linguist at the Ateneo de Manila University, compared 150 words from 13 Mindanao languages, including Tasaday. His method measures the rate of retention of basic vocabulary, from which he has calculated the approximate date when Tasaday seem to have separated from the other Manobo speakers. Father Llamzon reports that the probable time of separation of Tasaday from Blit—the closest linguistic group—was about 700 years ago (13th century A.D.).

It is most unlikely, however, that the Tasaday separated from the Blit people as they are today. The Blit are agriculturist, and it seems improbable that in 700 years the Tasaday would have lost all knowledge of agriculture, metal tools, and hunting and trapping devices that would have added security for them in the forest. It would also mean that the Tasaday, after losing knowledge of metal tools and weapons, then would independently develop stone implements that show close relationships with similar artifacts recovered in prehistoric excavations throughout the Philippines. A more reasonable reconstruction would seem to be that the Blit were former gatherers who separated from the Tasaday to become agriculturists.

But what does the future hold for the Tasaday? News of their existence emotionally touched many who live in a troubled complex world. Some people reacted wishfully, as if the Garden of Eden had been found. It was widely insisted that the Tasaday remain free to live without external pressure. The outside world's concern for the Tasaday recalls that past contacts leading to conquests of primitive and peasant peoples have produced some of the blackest chapters in human history.

Tasaday contact with the outside world seems to have been inevitable. Their isolation was doomed to yield to the loggers and neighboring agriculturists, who are invading their forest sanctuary. The items of material culture they have already received from Dafal have created new needs and wants among the Tasaday; time has caught up with them.

In April 1972, President Ferdinand E. Marcos declared the forests the Tasaday occupy as a Philippine national reserve. The conservation of these forests is equally important to the preservation of the watersheds that support agricultural development for an expanding population in the adjacent lowlands. A program of research has been prepared by PANAMIN designed to provide fuller data on the Tasaday, one of the world's unique peoples. Hopefully the work will help the Tasaday to select a program of development that will prepare them for new experiences, and to cope with the social problems they can expect to face. Clearly, when a people are faced with change, as are the Tasaday, it should be with pride and dignity, at their own pace, and by their own choice.

XV

THE DEVIL

# The Psychic Boom

## by Samuel Moffat

**Psychic phenomena—ranging from astrology to ESP to biofeedback—have stimulated considerable interest and controversy in recent years. Do they mean anything? Are they true?**

On the opposite page is the Tarot card called the Devil—Key 15 from the Major Arcana. According to Eden Gray, author of *The Tarot Revealed,* when this card appears in a reading right side up (as shown), it means "domination of matter over spirit. Sensation divorced from understanding. . . . Bondage to the material. . . . Black magic." Turned around the other way, upside down, it indicates "the beginning of spiritual understanding."

Key 15 in its two configurations neatly illustrates the duality of our response to psychic phenomena. A real "psychic" boom has developed in the Western world during the last few years, encompassing interests from astrology to Zen meditation. Many individuals simply view such interests as modern manifestations of black magic—superstitions growing out of perennial human gullibility. Many others, however, believe psychic happenings are evidence of a cosmic force as yet poorly understood—and sometimes deliberately rejected—by modern science. Magic and superstition, or spiritual understanding—which do they represent?

(Below) Bettmann Archive; (opposite page) courtesy, U.S. Games Systems, Inc. © 1970 S. R. Kaplan

## Varieties of psychic phenomena

In the popular understanding, "psychic" covers a motley accumulation of beliefs and happenings that defy a single definition. It may include occult, apparently supernatural, and (to some) irrational topics such as fortune telling (by astrology, Tarot, the *I Ching*, palmistry, etc.); character analysis by graphology or phrenology; dowsing (water witching); faith healing (perhaps including acupuncture, which involves treatment of disease by inserting needles at specific points in the body); witchcraft; poltergeists; out-of-the-body experiences; unidentified flying objects (UFOs); and reincarnation and communication with the dead. It may include phenomena within the realm of parapsychology, which many responsible scientific investigators reject as unproven but which others accept as verified: extrasensory perception or ESP (telepathy, clairvoyance, and precognition) and psychokinesis or PK (moving objects by mental effort). And finally it generally encompasses what have recently come to be called "altered states of consciousness"—hypnosis, dreaming, drug-induced states, and established Eastern forms of meditation such as Yoga, Zen, and transcendental meditation—as well as studies involving deliberate control of brain states.

***SAMUEL MOFFAT,*** *a science writer, is a contributing editor for the* Britannica Yearbook of Science and the Future.

All of these topics have at one time or another been categorized under psychic phenomena. In fact, however, they should probably not be lumped together. From the point of view of most scientists, some have no status at all and probably no basis in fact; most (but not all)

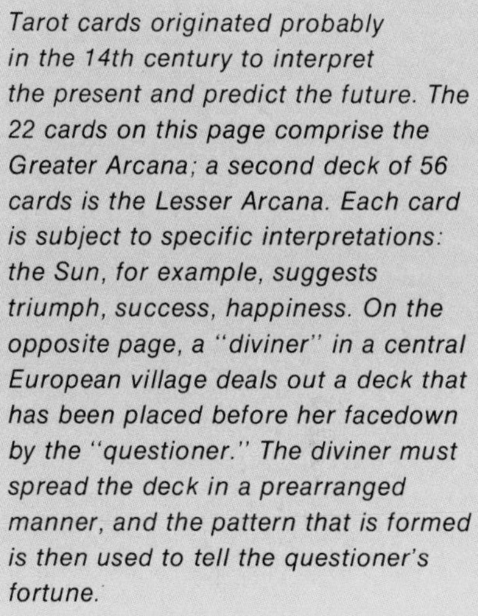

*Tarot cards originated probably in the 14th century to interpret the present and predict the future. The 22 cards on this page comprise the Greater Arcana; a second deck of 56 cards is the Lesser Arcana. Each card is subject to specific interpretations: the Sun, for example, suggests triumph, success, happiness. On the opposite page, a "diviner" in a central European village deals out a deck that has been placed before her facedown by the "questioner." The diviner must spread the deck in a prearranged manner, and the pattern that is formed is then used to tell the questioner's fortune.*

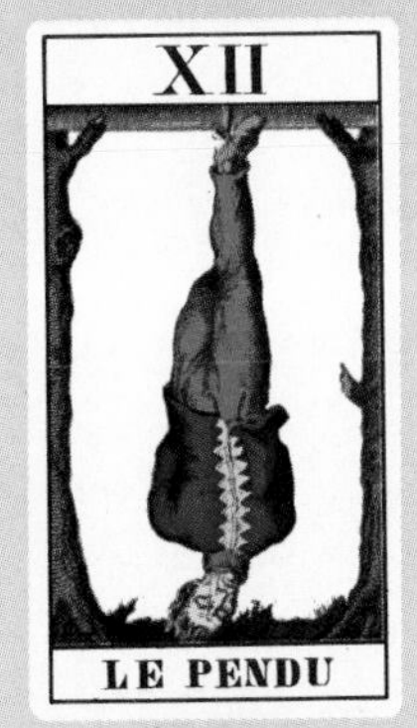


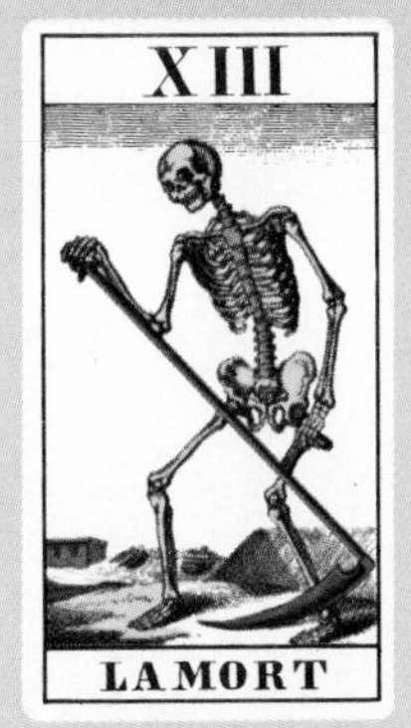

Bettmann Archive

*Each constellation making up the 12 signs of the zodiac (above) in many cultures was believed to dominate a specific organ of the human body. The first known depiction of witches on broomsticks appeared in a French book of the 15th century. The human face (opposite) was thought to be divided into sections that had specific attributes (top) and that were dominated by particular stars and constellations (bottom).*

of the occult topics fall in this category. Altered states of consciousness and deliberate control of brain states are generally accepted as fields of serious research. Parapsychology (ESP and PK) represents a controversial area, with both ardent defenders and skeptical disbelievers. The public thinks of these three broad subject areas as related even though they may not be. Any attempt to describe the psychic boom must acknowledge this confusion as part of the situation.

### A booming phenomenon

Prominent people from almost every walk of life have expressed serious interest in psychic phenomena. Actress Marlene Dietrich tried to persuade the secretary-general of the United Nations to reschedule a General Assembly meeting because the astrological portents were bad for a certain time. James Reston, one of the world's leading journalists, was treated by acupuncture for pain relief following an appendectomy in China; his impressions appeared on page one of the *New York Times* and stimulated wide interest in this ancient practice. The Beatles traveled from Britain to India for an intensive course in transcendental meditation, which took on new popularity in Europe and the United States. A prominent U.S. clergyman, Bishop James A. Pike, revealed to the world his belief that he had communicated through mediums with his dead son from "the other side."

Courtesy, Bibliotheque Nationale

The exact amount of popular interest, stimulated by events such as these, is difficult to measure. To a certain extent, however, the communications media mirror it. Almost 70% of the daily newspapers in the United States carry horoscope columns, according to one estimate. An astrologer appeared on the cover of *Time,* and a witch was featured on the cover of *Look.* A novel about the exorcism of a demon headed the best-seller list for 17 weeks. A publisher specializing in occult books reported that between 1960 and 1971 his business increased 25-fold. The practice of witchcraft was notably increased following the appearance of the movie *Rosemary's Baby,* in which a witches' coven practiced black magic. These and many other indicators make it clear that millions of Americans take some if not all psychic phenomena seriously.

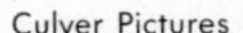
Culver Pictures

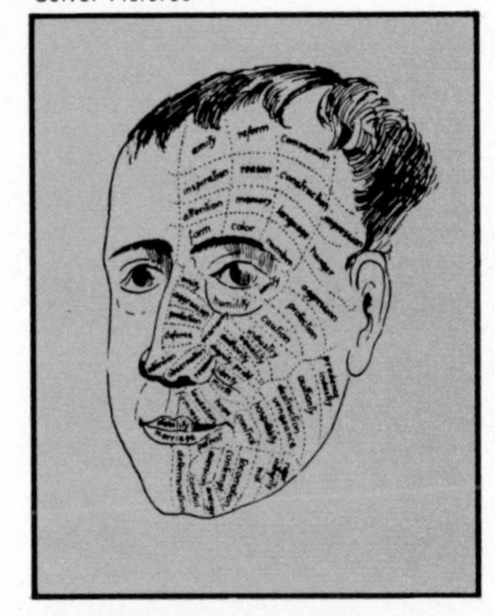

Bettmann Archive

## A product of our times

Commentators have offered many reasons for this upsurge of interest. Most observers seem to arrive at a similar conclusion: The material needs of many Americans and Europeans are being met to a greater degree than ever before. Accordingly, these people have the time and inclination to look more closely at other aspects of life—its meaning, its value, its spiritual and philosophical aspects. They often become convinced that neither the traditional religions nor modern science has solved mankind's massive social problems—the threat of nuclear war, the inequities of poverty and racial discrimination, and many others.

Searching for new ways to improve mankind, many persons have actively challenged established values and approaches. They question the findings of "rational" science and may seek new avenues to fulfillment and involvement through an exotic experience such as meditation or exposure to the occult. They may argue that if rational man cannot control his own future, why not accept astrology and let the stars do it?

The occult and psychic boom unquestionably has its faddish elements. There are charlatans and fast-buck artists capitalizing on it just as they do on any other human interest. Even people who are very open to supernatural concepts, who sincerely believe that psychic events and energies influence the lives of everyone, acknowledge that a large proportion of the activity they observe may be fakery. There is sleight-of-hand trickery or carnival showmanship passing for true ESP, they say, and there are promoters pretending to be psychic who are solely interested in commercial or individual exploitation. Nevertheless, the movement appears to be too widespread to write off as a passing fancy. It seems obvious to many observers that the interest in the occult is a response to current elemental human needs.

## The reaction of science

Given the tremendous public interest in understanding psychic events better, many laymen are puzzled that science has not taken psychic

Don Snyder

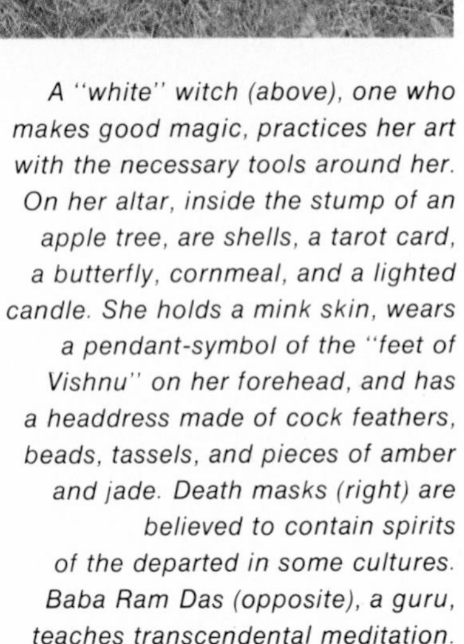

*A "white" witch (above), one who makes good magic, practices her art with the necessary tools around her. On her altar, inside the stump of an apple tree, are shells, a tarot card, a butterfly, cornmeal, and a lighted candle. She holds a mink skin, wears a pendant-symbol of the "feet of Vishnu" on her forehead, and has a headdress made of cock feathers, beads, tassels, and pieces of amber and jade. Death masks (right) are believed to contain spirits of the departed in some cultures. Baba Ram Das (opposite), a guru, teaches transcendental meditation.*

phenomena more seriously. Some scientists have clearly rejected the subject out of hand. One of the great physicists of the 19th century, Hermann Ludwig Ferdinand von Helmholtz, stated categorically, "I cannot believe . . . in the transmission of thought from one person to another independently of the recognized channels of sensation. It is clearly impossible."

Some present-day psychologists are almost as outspoken. Their point of view is well summarized in the comment on ESP contained in a college textbook published in 1970, *Psychology Today: An Introduction*, prepared by 38 contemporary psychologists. It states that "scientists today, for the most part, remain skeptical of ESP. It is axiomatic in science that a phenomenon be repeatable and predictable. To demonstrate conclusively that ESP exists, a scientist would have to be able to reproduce it whenever he liked. To date, this has proved impossible. . . . The best that can be said of ESP, then, is that it remains an unproven but fascinating phenomenon."

This is an area of heated controversy. In 1956 George R. Price of the University of Minnesota suggested in *Science* that results of some of the most widely accepted ESP experiments could have been produced fraudulently. Thirteen years later Edward U. Condon of the University of Colorado wrote, "Flying saucers and astrology are not the only pseudo-sciences which have a considerable following among us. There used to be spiritualism, there continues to be extrasensory perception, psychokinesis, and a host of others. . . . In my view, publishers who publish or teachers who teach any of the pseudo-sciences as established truth should, on being found guilty, be publicly horse-whipped, and forever banned from further activity in these usually honorable professions."

Nevertheless, some researchers who are regarded with high esteem have been convinced of the reality of certain psychic phenomena. Carl Jung, the Swiss psychoanalyst who was a major figure in modern studies of the mind, believed in the existence of spirits and himself had prophetic dreams. William James, the first great American psychologist, served as president of the Society for Psychical Research (London, England) and investigated several phenomena himself, including the accomplishments of a medium. Of this particular woman he said, "[I] believe she has supernormal powers." But he also exposed another medium, about whom he wrote, "Everyone agrees she cheats in the most barefaced manner whenever she gets an opportunity."

### An experimental milestone

This article concentrates to a certain extent on laboratory experiments involving ESP because that is where the most definitive scientific investigations of disputed psychic occurrences have been carried out. While reports of spontaneous psychic phenomena are plentiful, scientists prefer to set up experimental situations where various sources of error, such as observer bias, can be ruled out. This has been most possible to achieve with ESP. If modern science rejects ESP, it is unlikely that many of the occult topics listed earlier will become accepted areas for serious investigation for some time.

The leading investigator of ESP for several decades has been Joseph Banks Rhine, first at Duke University and then at the Foundation for Research on the Nature of Man, at Durham, N.C. According to Martin Gardner, author of a book debunking pseudoscientists, Rhine "has done more than any one man in history to give scientific respectability to the investigation of psychic forces." In countless experiments, Rhine, his wife Louisa, and their associates attempted to produce acceptable evidence of ESP, and for many years their techniques set the pattern for research in the field.

The principal technique utilized by Rhine involved a pack of cards with only five symbols. Five cards with each symbol make up a pack of 25 cards. If an individual were to guess the identity of each card in order, without seeing it, just on the basis of chance he would get one out of five correct. An individual with ESP, however, should score considerably higher. The odds against chance occurrence of any particular score can be calculated statistically. Rhine conducted what he considered a milestone experiment in 1933.

The subject, Hubert E. Pearce, Jr., had performed well on preliminary trials (once scoring 25 "hits" out of 25) and was to be given a particularly careful test with "conditions thoroughly adequate to exclude all the factors that could produce extrachance scores except ESP." Pearce was in the Duke University library, 100 yards away from a building where Rhine's assistant, J. G. Pratt, was to turn over the cards. They conducted 12 runs or 300 trials. Pure chance would have produced 60

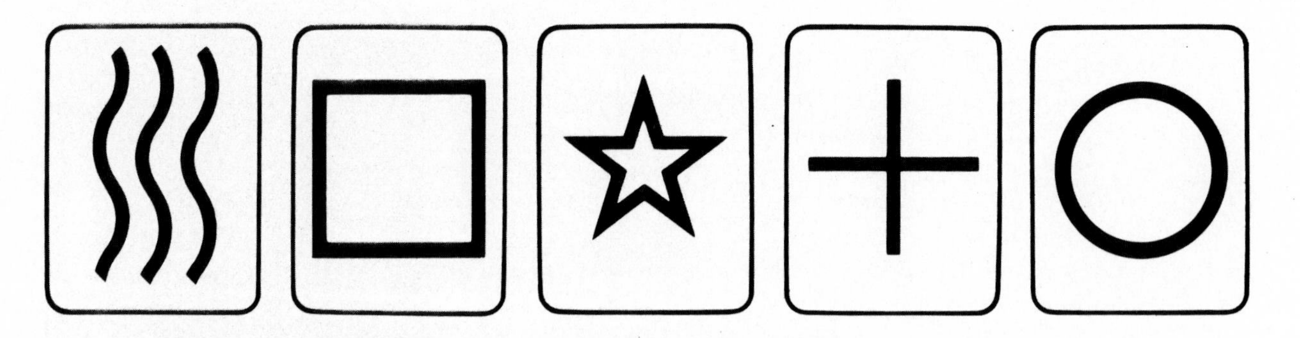

*Five of each of the above ESP cards constituted the deck used by extrasensory perception researcher J. B. Rhine. Without seeing the cards, a subject tries to guess which symbol has been turned up.*

hits, but Pearce scored 119. "A score as large as this," Rhine wrote, ". . . would be expected to occur by chance only once in approximately a quadrillion (1,000,000,000,000,000) of such experiments." Therefore, chance could be ruled out. "There are no known sensory processes that could be supposed to operate under these conditions. . . . We . . . were forced to decide that, whatever clairvoyance or the extrasensory perception of objects *is,* this was a case of it."

Experiments such as these have been criticized seriously by other scientists, however. They point out that the individuals who perform ESP poorly have been screened out so that good performers like Pearce are likely to exist simply on the basis of chance. This argument gains some strength because those who do ESP well, either with card tests or similar procedures, almost always lose their ability after a period of time—as if the law of averages were catching up with them.

In defense of selected subjects, those who believe the ESP tests are valid say that not everyone has highly refined psychic ability, just as few people can play the piano professionally, so that selection is perfectly legitimate. Regarding the second criticism, Charles T. Tart, an associate professor of psychology at the University of California, Davis, pointed out that the typical ESP card guessing procedure could be expected to destroy rather than enhance the ability to perform well.

Tart assumes that the ability to use ESP is latent in subjects and must be learned to some extent. But many experiments have shown that learning occurs when someone receives an immediate reward for desired activity (or punishment for incorrect activity). In card guessing, however, the subject does not know the result of each guess; he only finds out how well he has done at the end of one or several runs. "Any psychologist," Tart notes, "if asked to have *any* organism learn under [such conditions], will throw up his hands in disgust. . . . We have, unknowingly yet systematically, been extinguishing the operation of ESP" in those tested.

Another criticism of typical ESP results is that they are often subjected to repeated statistical analyses until the results bear fruit. And, aside from the statistics, ESP is greeted with skepticism by some because it seems to violate known principles of physics. It apparently works as well over great distances as at short range, for example, while other forms of energy transfer follow the inverse square law: at twice the distance the received energy will be one-fourth as great.

## ESP investigation in the 1970s

Some of the past criticisms may become outdated as evidence from the present generation of experimenters accumulates. They are using new techniques, including the application of automation and sophisticated electronic circuitry, to overcome past objections. There are several centers in the United States carrying out extensive research in ESP and PK, including laboratories at the University of Virginia; the American Society for Psychical Research, New York City; Maimonides Medical Center, Brooklyn, N.Y.; City University of New York; the Foundation for Research on the Nature of Man and the Psychical Research Foundation, both in Durham, N.C.; and the University of California at Los Angeles.

A great deal of research also is being conducted in other nations. Significant activity has taken place in the Soviet Union and Eastern Europe, but accurate scientific reports are difficult to obtain in the West. The publication in 1970 of the book *Psychic Discoveries Behind the Iron Curtain* by Sheila Ostrander and Lynn Schroeder stimulated considerable interest in the United States and Europe. The book contains little quantitative detail, though, and many ESP investigators in the United States are making concerted efforts to obtain data and relevant technical information from the researchers cited by these authors so that experiments can be repeated, and either confirmed or disproved.

An example of the modern approach in parapsychology was described in the journal *Science* late in 1970: "Simple equipment is now available for counting hits in predicting which of four lights will be lit at random (averaging $^1/_{10}$ of a second later), when the randomness is based on the radioactivity of strontium-90. In one recent study with such equipment 3 subjects out of 100 showed very outstanding results which, taken together, were very improbable [probability less than one in one billion]. Special precautions were taken to prevent recording errors, to check the randomness of the generated numbers, and to eliminate the possibility of fraud by the test subjects." Somewhat similar equipment has been used to teach both clairvoyance and precognition to a single subject; the machine immediately notified her she was successful when she chose the right target (in this case 35-mm color transparencies).

Researchers also are reporting increasingly that altered states of consciousness, such as occur during hypnosis or dreaming, may increase the frequency of ESP. Staff members at the William C. Menninger Dream Laboratory of Maimonides Medical Center have investigated this phenomenon for several years. Subjects are allowed to go to sleep in the laboratory while connected to electroencephalographic equipment that measures their brain waves and detects eye movements that occur during dreaming. In a separate room a "sender" attempts to influence a subject's dreams by looking at a picture.

As reported in *Science,* "Independent judges, acting blindly, have ranked the correspondences between the picture and the dream transcript significantly higher than control pictures. This has also been done with hypnotic dreams." This approach is viewed as an advance in technique because it provides a way of stimulating ESP under conditions more closely approximating those associated with spontaneous ESP reported outside the laboratory, but still provides for laboratory controls. Investigation of such "psychically favorable states" is still at an early stage of development, however.

### Is acceptance coming?

Rhine has charged that a majority of psychologists and a large number of other scientists refuse to accept ESP and PK because they cannot be explained by the present laws of physical science. Accepting a supernatural explanation "would cost [psychology] its hard-earned status as a natural science . . . and throw it back into the classification with the occult."

Is the "science establishment" being unreasonable in rejecting the evidence for the existence of ESP? Gardner Murphy, a George Washington University psychologist who was president of the American Society for Psychical Research from 1962 to 1971, views the matter more philosophically. In an interview published in *Psychic* magazine in 1970, Murphy said, "It takes an enormous amount of *powerful evidence, with a new theory,* before the orderly system can be shaken. It's not enough to have a lot of new facts, you've got to have a new theory as well. So in order to demolish the old theory, you've got to have a new one which is compatible both with the old data and the new data. . . .

"So when people question, 'Why don't scientists accept ESP?' I can only answer, 'Well, what *do* they accept that doesn't fit into an orderly scientific framework?'"

There are other reasons why scientists remain skeptical about psychic phenomena despite evidence that a minority accepts. For one thing, the whole field of psychic occurrences has been the domain not only of reputable scientists but also of quacks and frauds. More than one medium has been caught faking contact with spirits of the dead.

Another factor in scientists' skepticism is that until quite recently potentially reproducible experiments have not been possible. ESP ability faded in the laboratory, for example. But new approaches with psychically favorable states may overcome this difficulty. And other methodological improvements should enhance the overall quality of the research even further.

A final factor is the nature of psychology itself. It is not only a young science but also an exceedingly complex one. Observations in physics and chemistry can be relatively simple compared to those involving human behavior. Many observers believe that attempts to make psychology a natural science have led to premature theorizing and gross oversimplifications. This also applies to parapsychology. Much more information is needed, together with small-scale theories that can be tested, before it will be possible to draw up definitive master hypotheses. Psychologists can be considered to be at the level of Galileo when he was measuring the acceleration of falling objects; they are not ready to devise a scheme as grand as Newton's laws of gravitation and motion.

With suitable changes, parapsychology and perhaps certain other psychic phenomena may win greater acceptance among 20th-century scientists. For many years parapsychology had to remain outside the house of science in the United States, but recently it was allowed inside the front door. In December 1969 the Parapsychological Association, which includes almost all of the recognized parapsychologists outside the Communist world, became an affiliate of the American Association for the Advancement of Science, the world's largest organization of practicing scientists.

## The new frontier

Where will psychic research go next? One of the major domains undoubtedly will be altered states of consciousness (ASCs). For example, modern physiological measuring techniques are being applied extensively to the study of sleep and dreams. The scientific study of hypnosis has advanced to the point where it is largely accepted as a legitimate subject for psychologists. Several recent trends are pushing both psychology and psychic research toward greater interest in the other altered states.

One is the deliberate effort by many people to enter an altered state, either by taking such drugs as marijuana or LSD or by learning self-hypnosis or some meditative process. (Despite this activity, few scientific data have been collected about these ASCs.) A second trend is the growing use of techniques for teaching individuals to control their brain waves and therefore their state of consciousness.

It may not be sufficient just to observe ASCs from the outside. Tart argues that much current research is "relatively trivial" because it does not focus on the real experiences involved. "Experiences of ecstasy, mystical union, other 'dimensions,' rapture, beauty, space/time transcendence, transpersonal knowledge, and the like, common in ASCs, are simply not treated adequately in most conventional scientific approaches."

The solution, according to Tart, is to develop "state-specific sciences," utilizing skilled, trained researchers who are able to achieve certain states of consciousness. Within each state they can approach specific problems using the scientific method—making observations that can be confirmed by others in that state, theorizing about their findings, and then testing their theories by predicting the consequences and seeing whether the theories hold true. Observing an altered state only from man's normal, alert condition does not take into account that the pattern of communication may have changed, just the way a computer's program may be changed. Without a new approach to ASCs, Tart believes that modern psychology never will understand them adequately.

*Nudity is an important part of many mystic rituals. The man below has painted his body silver to celebrate the rites of Pan.*

Don Snyder

"Biofeedback," or voluntary control of internal states, already has been used in many laboratories to enable individuals to alter brain waves, blood pressure, blood flow, heart rate, and muscle tension. It is sometimes called autogenic training. The technique involves detecting the variable to be controlled, displaying information about it on a dial or with a light or sound, and then letting the subject practice controlling it. The procedure has been used experimentally in treating high blood pressure and migraine headaches.

While biofeedback may eventually have wide application in medicine, psychic researchers are concerned principally with its usefulness in investigating states of consciousness. Much work has been done with alpha waves, which are brain impulses that occur with a frequency of between 8 and 12 cycles per second. Most subjects are able to control their alpha waves fairly easily. About 90% of us can produce them just by closing our eyes. But with suitable training, subjects have learned to suppress them with their eyes closed, to produce them with their eyes open, or to alter the alpha frequency. The waves are detected with an electroencephalograph machine and then displayed to a subject as a sound or a light.

Alpha waves generally are associated with an alert yet passive mental state, lacking visual images. Their appearance is related to several ASCs. According to some studies, individuals who are highly susceptible to hypnosis produce alpha rhythms more often than those who are not so susceptible. Electroencephalographic studies of Zen monks have revealed that during meditation their brains produce alpha waves principally. Yogis have been shown to produce more of these waves while meditating, and individuals practicing transcendental meditation (the procedure taught the Beatles by Maharishi Mahesh Yogi) have more regular alpha waves, with increased amplitude.

Experienced Eastern mystics who truly have mastered meditation techniques can do much more than manipulate their alpha waves, however. In 1970 researchers at the Menninger Foundation in Topeka, Kan., spent several weeks examining an Indian yogi, Swami Rama of Rishikesh and the Himalayas. The swami could voluntarily maintain his production of theta and delta brain waves. Theta waves (four to seven cycles per second) often appear when a person becomes drowsy and eases toward sleep; delta waves (about one per second, with very large amplitude) are usually associated only with deep sleep. During one five-minute test, Swami Rama demonstrated theta waves 75% of the time. The next day he deliberately produced delta waves for 25 minutes; he appeared asleep and even snored gently, but afterward he could repeat almost perfectly things said in the room at that time.

These are not just exercises but have the potential for teaching science some of the as yet unexplained workings of the human brain. When in the theta state (with alpha or beta also present 50% of the time), Swami Rama became aware of many unconscious thoughts he normally suppressed. The swami maintained that when in a state of deep reverie he could heal himself of certain ailments by giving the

LIFE Magazine, courtesy, Maimonides Medical Center

*Voluntary production and suppression of the brain impulses called alpha waves are monitored at the Maimonides Medical Center in New York. Alpha waves usually are associated with an alert but passive mental state.*

body suitable instructions in the form of images. Elmer E. Green, who directed the research at Menninger, also reported that Swami Rama "could diagnose physical ailments very much in the manner of Edgar Cayce [an American psychic who made thousands of medical diagnoses while in hypnotic trances], except that he appeared to be totally conscious."

Green says: "The entrance, or key, to all these inner processes, we are beginning to believe, is a particular state of consciousness [or 'reverie' that] can be approached by means of theta brain wave training in which the gap between conscious and unconscious processes is voluntarily narrowed, and temporarily eliminated when useful. When that self-regulated reverie is established, the body can apparently be programmed at will and the instructions given will be carried out."

Another meditative procedure that is attracting scientific attention is transcendental meditation. It does not require the years of training that Yoga or Zen necessitate, nor does the practitioner need to master intricate physical positions or exercises. Because it is easy to learn and is reproducible in physiological terms, scientists believe it may provide a useful approach for further studies.

### "Radiesthesia" and "radionics"

An area of psychic research that is much more controversial involves studying the sensitivity of living organisms to radiation of any sort from any source. This sensitivity is known as radiesthesia, and involves the five commonly recognized senses. Certain investigators believe that another manifestation of radiesthesia may be involved in such psychic phenomena as reading print or identifying colors with the hands (eyeless sight), a medium's communication with someone by other than the five senses while utilizing some object associated with that person (object reading, or psychometry), and dowsing. The existence of a "sixth sense" might explain ESP and PK as well.

*Relationship between alpha waves and active attention is tested by showing a young man in an alpha state slides of flowers and a nude. When the flowers appear, the alpha state continues, but with the nude it disappears.*

Lew Merrim from Monkmeyer Press Photo Service

Modern physical scientists generally have repudiated the hypothesis that the body "transmits" and "receives" electromagnetic radiation, with the brain as a source and the body serving as an "antenna," because brain waves are far too weak even to be detected a very short distance from the head. But a limited number of scientists believe that there are unrecognized forms of energy originating from living things, including plants. Could there be "psychic forces" or "psychic fields," or some other power generator than the brain? Numerous writers have devised schemes to account for such forces and fields, but their "theories" are more in the tradition of the occult than that of science. Many are based on Eastern philosophies. Most must be accepted on faith at this point because they cannot be tested by the scientific method.

Nevertheless, some Western scientists have begun investigating radiesthesia and radionics (using instruments to detect the energy involved). At least initially they will probably follow the trail marked out by the researchers in Eastern Europe and the Soviet Union. Specifically, they are experimenting with devices capable of detecting or measuring various "psychoenergetic fields," which the Soviets have reported finding associated with the human body and with plants.

William A. Tiller of Stanford (Calif.) University was one of the first U.S. physical scientists to visit the Soviet Union with the objective of gathering technical information regarding its current psychic research. He returned from Moscow and Leningrad in 1971 with specifications for building several devices so that he could repeat some of the Soviet work. One instrument had been invented by Victor Adamenko, a Soviet physicist, who used it to identify points of low electrical resistance in the human body, in other animals, and in plants. In humans almost all of the points conform with the places where needles are placed in acupuncture therapy. After injection of radioactive phosphorus at one of the low-resistance points, radioactivity can be detected at other acupuncture points. Adamenko suggests this device has identified a new circulatory system of low-resistance circuits in the body that coincide with the meridians or channels believed to connect acupuncture points.

A second Soviet tool of interest to American psychic researchers is a photographic device invented by Semyon D. Kirlian and his wife, Valentina. It utilizes a high-frequency electrical pulse generator, somewhat like a radar power source, to produce photographs. Developed films show "flares" escaping from tissues such as the human finger or the leaf of a plant. According to Tiller the discharge involves cold electron emissions. Their characteristics change with the emotional state or health of an individual, and also with a healer when "in the healing mode" or when not healing. When up to 10% of a leaf has been cut off, the radiation pattern on film still shows the entire outline of the leaf.

The Soviets also conducted sophisticated tests on certain women they believed had demonstrated the ability to move objects by will

Courtesy, Dr. Thelma Moss, UCLA Center for the Health Sciences

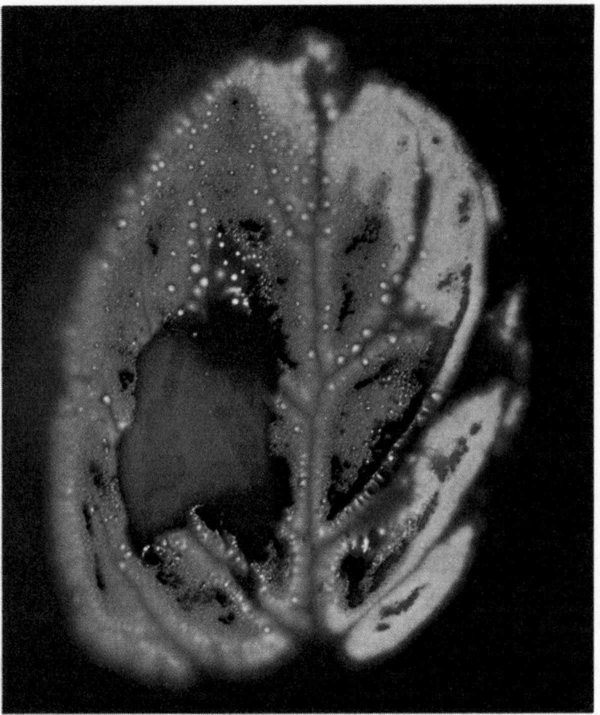

*The effects of stress and mutilation on living tissues can be seen in the photographs above and on the opposite page. They were taken with a device similar to one invented by Semyon Kirlian, which uses a high-frequency electrical pulse generator to produce its images. The leaf above was photographed in this way when freshly cut (left) and after being gashed with a needle (right). The tip of a right index finger on the opposite page was photographed when the subject was relaxed (top) and while the same subject was undergoing emotional stress (bottom).*

power (psychokinesis). Tiller believes that this work, combined with the other Soviet projects, opens exciting new possibilities for future research. He began, therefore, to duplicate Soviet equipment in order to measure the variables involved and to try to develop equations for characterizing the events.

### The "new physics"—or not?

The Oxford University zoologist Sir Alister Hardy has written: "The discovery that individuals are somehow in communication with one another by extrasensory means is, if true, undoubtedly one of the most revolutionary discoveries ever made. It is a biological phenomenon, if true, almost as fundamental as that of gravity between material bodies." Assuming that some of the phenomena investigated by the Soviet scientists and others relate to ESP, this statement takes on additional meaning.

But the "if true" qualification must be recognized. There are still other possible explanations besides the existence of a sixth sense involving psychoenergetics. Ernest Hilgard of Stanford University, co-author of a standard college textbook on psychology, has spent many years studying hypnosis and its potential medical applications and has become aware of how suggestible the human mind can be. His experiences have caused him to believe that all sorts of unusual "psychic" experiences, including the beneficial effects of faith healing or acupuncture, might be a result of suggestion.

Told how a visiting U.S. scientist had been able to do PK with the help of a Soviet volunteer, Hilgard said, "I'll bet he's hypnotizable. You don't have to go through any formal procedure if someone is highly hypnotizable." And if suggestibility is not involved, sleight-of-hand or mechanical assistance (magnets, etc.) can be, Hilgard stated.

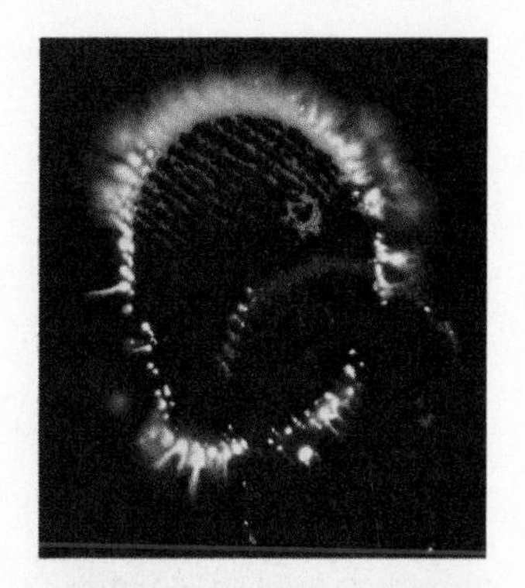

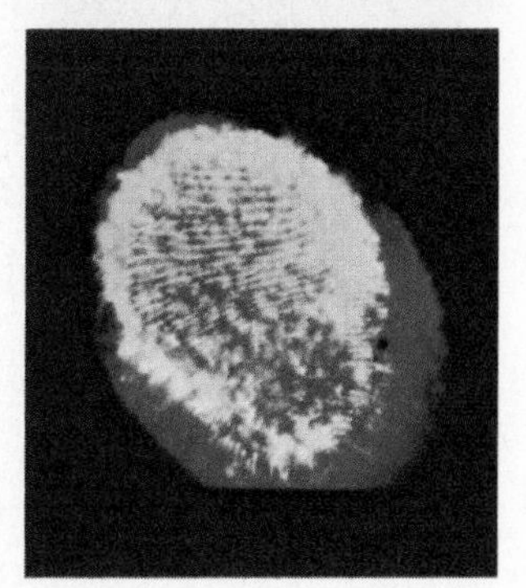

Are honest people deluding themselves, or being deluded? The experts disagree. But there seems to be a move toward increased acceptance of some psychic phenomena. Robert A. McConnell, of the University of Pittsburgh, who was the first president of the Parapsychological Association, says that there are now so many experimental reports on extrasensory perception they have "erode[d] disbelief in ESP until most psychologists are no longer willing publicly to deny the evidence for the phenomenon." Still, there is no "widely encompassing theory. . . . [ESP] is a fact without an overall theory. One suspects that it will someday cause a revolution in both psychology and physics."

Until that "someday" arrives, some words of Bertrand Russell's may serve as our guide. In his *Skeptical Essays* he wrote:

> The opinion of experts, when it is unanimous, must be accepted by non-experts as more likely to be right than the opposite opinion. . . . [However] when they are not agreed, no opinion can be regarded as certain by a non-expert.

One of the biggest benefits of the psychic boom may be that it will stimulate enough serious scientific interest to get some dependable answers about which phenomena are real, and which are not, and thus allow more of the experts to begin to agree.

FOR ADDITIONAL READING:

Gardner, Martin, *Fads and Fallacies in the Name of Science*, ch. 1, 9, and 25–26, pp. 3–15, 101–115, and 299–324 (Dover Publications, 1957).

Jahoda, Gustav, *The Psychology of Superstition* (Penguin Books, 1970).

Luce, Gay, and Peper, Erik, "Mind over Body, Mind over Mind," *The New York Times Magazine* (Sept. 12, 1971, pp. 34–35, 132–139).

McConnell, R. A., *ESP Curriculum Guide* (Simon and Schuster, 1971).

Murphy, Gardner, *Challenge of Psychical Research: A Primer of Parapsychology* (Harper and Brothers, 1961; Harper Colophon Books, 1970).

*Psychic,* bimonthly magazine, any issue (San Francisco).

Rhine, J. B., *New World of the Mind* (William Sloane Associates, 1953; Apollo Editions, 1962).

Tart, Charles T., *Altered States of Consciousness: A Book of Readings* (John Wiley and Sons, 1969; Doubleday-Anchor, 1972).

Wilson, Colin, *The Occult: A History* (Random House, 1971).

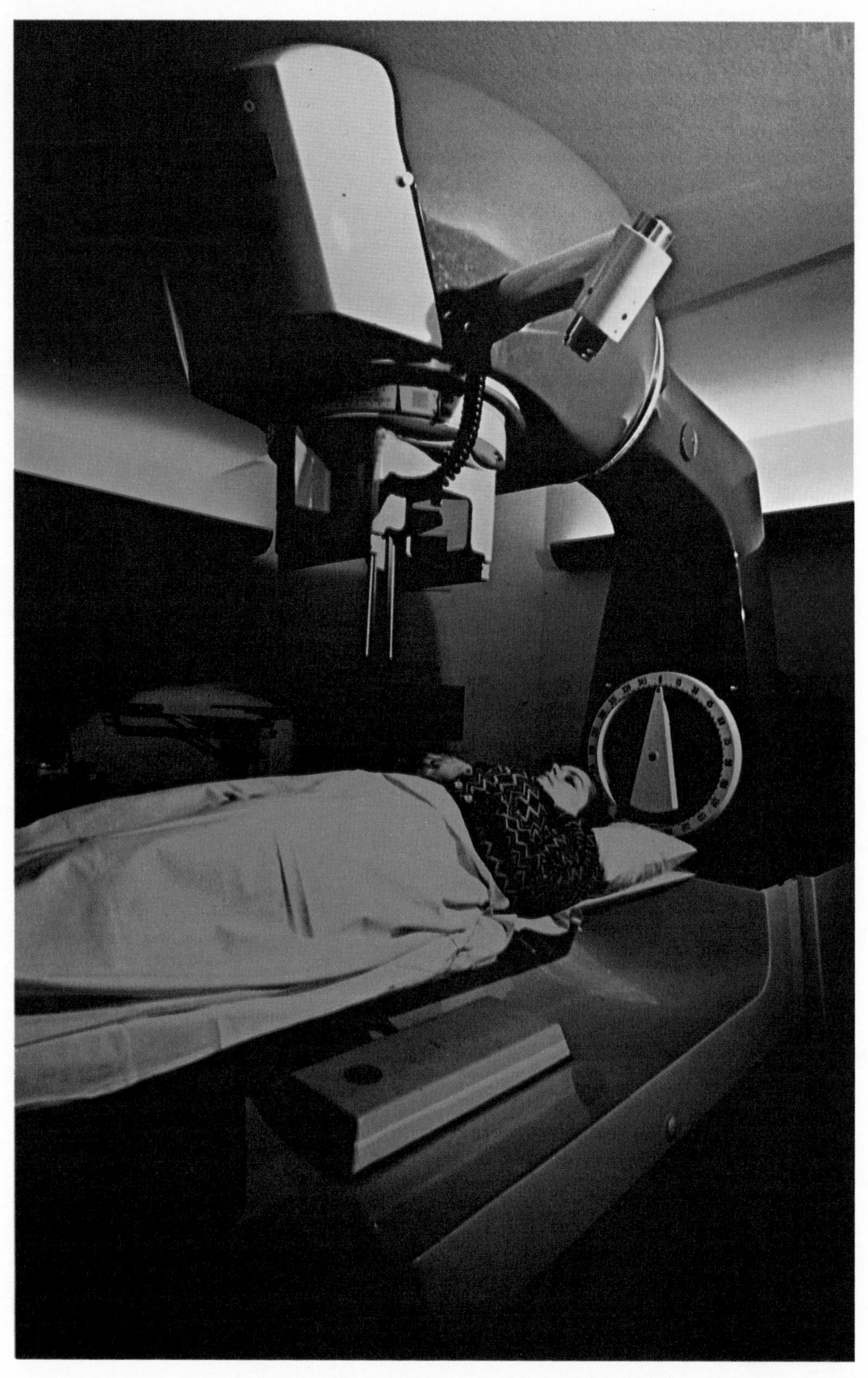

*Theratron-80, a computerized radiation therapy treatment machine. Photo, Dan Morrill, courtesy, Presbyterian-St. Luke's Hospital.*

# Cancer Under Attack

## by Ronald Kotulak

**Will ours be the last generation to suffer the ravages of cancer? As science embarks on extensive efforts that show promise of finding the cause of this dread disease, methods for detecting and treating it become increasingly refined.**

Cancer is one of man's most dreaded diseases. It will kill one out of every six Americans who die in 1972, some only after prolonged suffering. Although people became able to talk more freely about cancer in the last decade, their feeling of fear and futility in the face of its diagnosis remained. The springs of medical research, which had promised to cure cancer in the 1950s, had seemed to dry up. A frustrating stalemate settled over the laboratories. That was yesterday. Today, the gloom and disappointment that overshadowed cancer researchers is rapidly giving way to a new era of hope. Scientists, at work for decades on the cancer problem, are finally coming close to understanding the transformation of normal to malignant cells that can attack any tissue of the body. There is a genuine optimism in scientific circles that cancer is at last yielding to man's efforts to conquer it.

Almost weekly researchers report new findings, adding pieces to the cancer puzzle. Among the most important discoveries have been viruses capable of altering the hereditary behavior of mammalian cells, evidence that viruses are linked to certain human cancers, and the possibility that cancer cells may be reconvertible into normal cells. Confirmation of the link between viruses and cancer is vital: if such a link exists, then a vaccine might be developed to prevent cancers. Malignancies associated with viruses include cancers of the breast, cervix, muscles, and blood (leukemia).

As scientists unravel the mystery of cancer, they are finding that it is intertwined with the secret of life itself. Persons may be born with the seeds of their eventual cancer within them, locked up in their genes. Whether these seeds produce a full-blown malignancy may depend on time, environment, and chance. The role of the virus, which is composed of nucleic acid and protein, also the basic constituents of genetic molecules, seems to fit naturally into this scheme.

Although no one can predict just when cancer cures may be achieved, experts are confident that if enough money is available for the needed research the cures will be forthcoming. "Given the resources, the funds, and the organizational capability, I am confident that we can achieve control of this disease," asserts H. Marvin Pollard, head of gastroenterology at the University of Michigan and a spokesman for the American Cancer Society. "A really massive effort to bring this disease under control during this decade demands that not one single possibility be neglected."

Encouraged by the new scientific findings, the United States is preparing to mount an all-out effort to win the cancer battle. The president and Congress have expressed their eagerness to spend whatever is required. On December 23, 1971, Pres. Richard M. Nixon signed into law the National Cancer Act, which was to provide more than $500 million annually for the next three years to support cancer research. The investment in research could even reach $1 billion a year by 1976, rivaling the budget of the U.S. space program during the early days of the effort to land men on the moon. Justifying this kind of large-scale spending, a national panel of 26 eminent scientists and laymen, authorized by the U.S. Senate to review the cancer field, reported that advances in the fundamental understanding of cancer in the past decade have opened up far more promising areas for major advances in cancer prevention and treatment than have ever before existed. The panel concluded that the conquest of cancer is a realistic goal.

### The changing picture of cancer

Although cancer ranks second, after heart disease, in the number of victims it kills, persons fear it more than they do any other affliction, and for several reasons. It appears to rise stealthily from one's own body, frequently hidden until it becomes so widespread that treatment is futile. It causes an ugly death, robbing the patient of dignity. During the course of the illness, a family's financial reserves may be consumed to pay for what is often treatment that does little good. Not only friends and relatives, but sometimes even doctors and nurses, will avoid cancer patients out of some dark fear of the disease. And although many physicians advise a family not to tell a dying relative that he has cancer, most know without being told.

Among men, the incidence of cancer is increasing. The National Cancer Institute (NCI) has reported that among males the rate of cancer rose from 280 to 304 per 100,000 between 1947 and 1969, and that more cases of cancer of the lung, prostate, and colon were being diagnosed. The cancer rate among females during this period dropped from 294 to 256 per 100,000, with decreases in the frequency of cancers of the cervix, stomach, and rectum. The significant drop in cervical cancer, especially, was assisted by mass programs for detection of this type of cancer in its early, curable stage.

Evidence indicates that man is probably responsible for many cancers. He releases into the environment or chooses to expose himself to

***RONALD KOTULAK** is science editor of the* Chicago Tribune *and author of a prizewinning series of articles on cancer.*

Courtesy, American Cancer Society

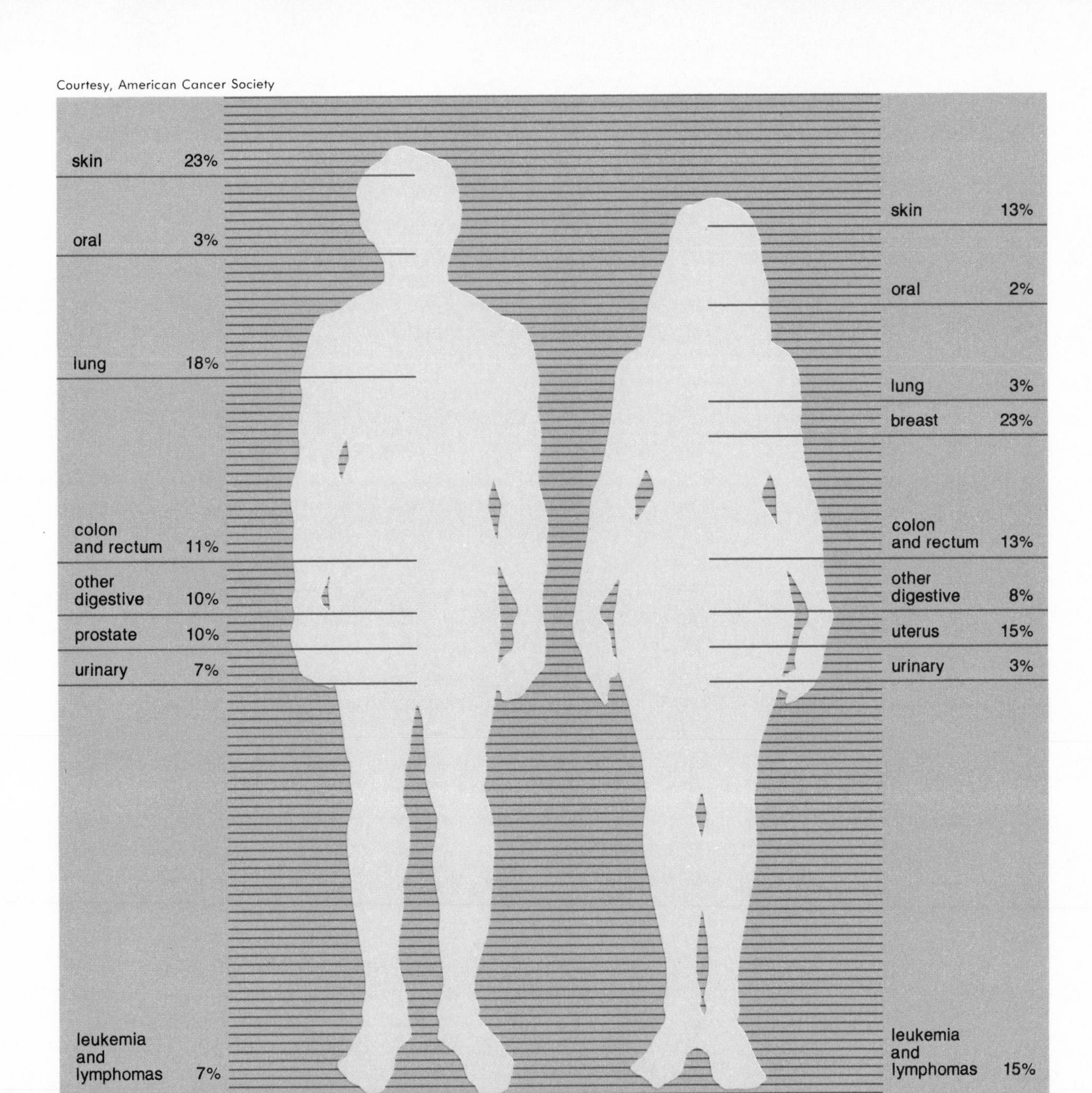


*Certain cancers occur more frequently among one sex than they do among the other. In men, cancers most often involve the skin or the lungs. Women are far less prone to these two forms but are highly susceptible to cancers of the breast and uterus.*

many known cancer-causing agents (carcinogens), including chemicals and radiation. Cigarette smoking, for example, is blamed for 60,000 lung cancer deaths annually. A definite carcinogen—N-nitrosodimethylamine—was recently identified in tobacco smoke. A chemical culprit recently identified as a carcinogen by the U.S. Food and Drug Administration is diethylstilbestrol, which is used to correct estrogen deficiency disorders in women and is widely used to fatten cattle and sheep. The FDA warned that pregnant women should not be given the compound because of the risk that their female babies may someday develop vaginal cancer. Another source of carcinogens is polluted air. Some authorities believe that lung cancer could be reduced by 20% if present levels of pollution in city atmospheres were cut in half. One of the more potent carcinogens is benzo(a)pyrene, a component of automotive and industrial soot.

But if man can cause cancers, he should also be able to prevent them by identifying and eliminating their causative agents. Exactly how carcinogens react in the body to produce cancers is not, as yet, known.

Almost everyone has been affected in some way by cancer. A relative or friend may have died of it. And one of every four living Americans will be stricken personally by cancer during his lifetime. The toll in 1971 was 339,000 cancer deaths, more than the number of U.S. troops killed in World War II. But there is some good news. The cure rate is steadily climbing. In 1930 one out of five cancer victims was cured; by 1972 the rate had risen to one in three. Many cancer experts maintain that the rate of cures could easily be boosted to one in every two cancer victims if every patient received the benefit of available early diagnosis and treatment with the best therapies now available.

But there still are needless deaths from cancer. James F. Holland, a past president of the American Association for Cancer Research, claims that doctors not specially trained to treat cancer or not interested in the disease will not ensure the best available care for their patients. He urges that any patient who has or suspects that he has cancer consult two or more physicians, including a cancer expert.

Another reason for needless deaths is the erroneous notion that all cancer is incurable. As a result, some patients delay seeing a doctor or refuse surgery or other treatment that might otherwise have saved their lives. This pessimism is being dispelled, however, by the startling advances against cancers that just a few years ago were considered incurable.

Hodgkins disease, a cancer of the infection-fighting lymph nodes, is one example. New radiation and surgical methods, along with newly developed chemotherapies, which use combinations of anticancer drugs, have boosted the five-year survival rate to 90% of the patients with early or localized advanced stages of the disease. Investigators at Stanford University are hopeful that most patients with even advanced Hodgkin's disease may soon look forward to an extension of life. The most vicious cancer of children—acute lymphocytic leukemia —has also begun to yield its deadly hold. With proper treatment (so-called total therapy), 60% of the children with acute leukemia admitted to one medical center have survived at least three years free of signs of the disease. Total therapy consists of administering combinations of anticancer drugs to destroy most of the diseased cells quickly and to prevent the patient from developing resistance to any one drug, and using radiation therapy to remove the remaining cells the drugs do not reach. Another type of cancer yielding to treatment is that of the larynx (voice box). A 35-year study at the University of Chicago showed that 95% of the localized, small cases of this cancer are curable through radiation, surgery, or both.

The three main therapies being used against cancer are surgery,

Dan McCoy from Black Star

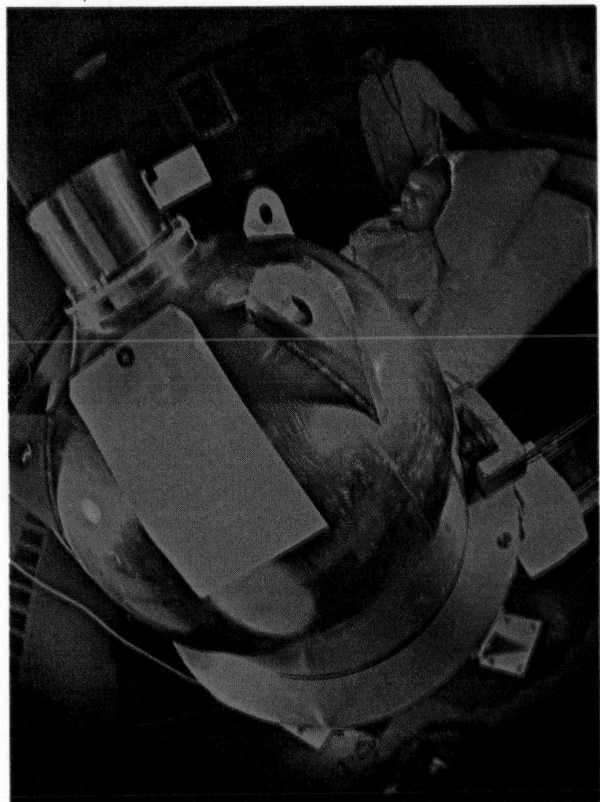

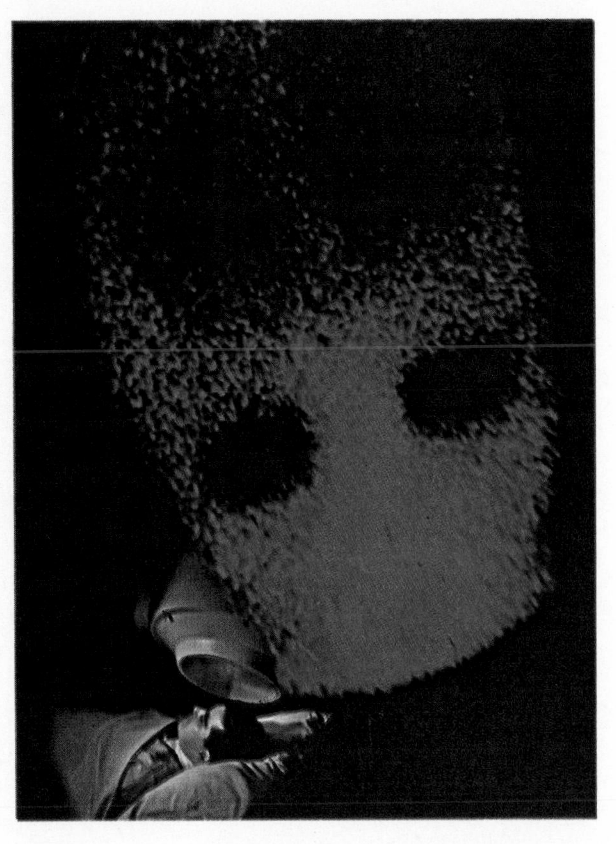

*New devices improve the physician's ability to detect and treat cancers. The steel-domed machine (left) contains a radio-cobalt source that irradiates the cancerous white cells in a leukemia patient's blood outside his body. His normal tissue is thus protected from radiation damage. The "gamma camera" (right) searches for a brain tumor in three minutes. Faint flashes of light emitted by an isotope administered to the patient are amplified, relayed by a computer, and recorded as a series of dots that indicate a tumor against a dark background of normal tissue.*

radiation, and chemotherapy (treatment with growth-inhibiting chemicals). Although therapists look forward to better anticancer drugs and improved immunotherapeutic techniques to reduce the need for surgery, surgery remains the treatment of choice in all but a few types of cancer and has scored impressive victories. Surgery is most effective when it is accomplished early to remove tumors before they have had a chance to spread cancerous cells to other parts of the body and establish new malignancies. Irradiation is used to destroy malignant cells. New drugs assist radiation therapy by preventing cells from repairing the radiation damage. For many cancers, these therapeutic techniques have produced excellent results. The NCI reports that of patients with operable tumors and no evidence of spread, 85% with uterine cancer will escape death from the disease. This holds true also for 75% of patients with ovarian cancer, and 60% of patients with cancers of the colon, rectum, breast, or bladder. With continually improving diagnostic and treatment methods, the outlook is even brighter.

Anticancer drugs can accomplish what neither surgery nor radiation can do effectively—fight spreading cancers. Today there are more anticancer compounds in various stages of development and testing than ever before. Some researchers believe that in only a few years drugs will be developed that will do to some cancers what penicillin does to bacterial infections.

One of the newest drugs is an antibiotic, bleomycin, which has had early success in treating some heretofore resistant forms of cancer. It

Dr. Keith L. Ewing, Thomas W. Davison, and James Fergason, Kent State University

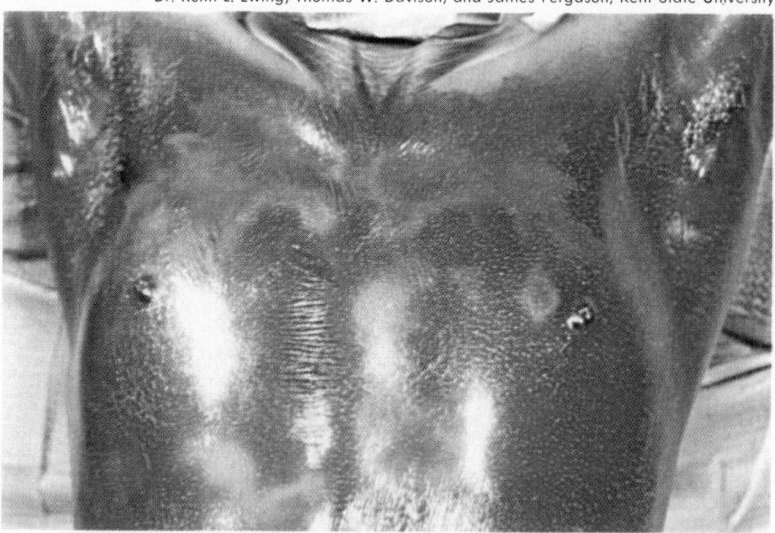

*A technique that may provide a simple means of detecting breast cancer is liquid crystal thermography. Cholesterol ester crystals applied to a patient's breasts produce patterns of iridescent colors that reflect temperature variations in the skin—reddish and green colors for cool areas, blue for warm. The blue area in what should have been a cool region of the right breast indicates a malignancy in the woman above.*

has produced improvement in patients in the terminal stage of Hodgkin's disease, and has been called the best drug so far for some highly resistant solid tumors, such as squamous cell carcinoma. Initial tests with patients suffering from advanced lymphosarcoma, for which there has been little effective treatment, reveal that a new three-drug regimen—cyclophosphamide, vincristine, and prednisone—has produced a survival rate for at least two years of 73%. Animal tests have uncovered a highly promising class of drugs that contain platinum. These drugs may have action against more types of cancer than any other yet developed—if experience with animals proves out in medical trials.

**Unraveling the malignant process**

As therapists continue to make significant strides in the treatment of cancers, researchers are coming closer to an understanding of the malignant process, on which further cures can be based. A picture of what cancer is and how it may be caused is emerging.

Cancer appears to start in a single cell, which begins to function much like embryonic cells do after conception. Each organism begins life as one fertilized cell. The cell divides in two; the two divide into four; and so on. All of these cells carry all the information necessary to reproduce the complete organism, and all of this information is being used. The cells divide very rapidly, just as cancer cells do. As the embryo begins to take shape, however, some form of cellular communication takes place and chemical substances known as repressors are called into action. The repressors attach to different sections of the genes in the chromosomes of the forming cells so that only the information the cell needs to perform the particular role it is to have in the body, whether as tissue, muscle, nerve, or bone, is allowed to send out instructions. Once the complete organism is formed, the rate of cell division slows appreciably until it is reduced to a minimum level required for replacing dead or damaged cells.

A number of agents, however, can interfere with this orderly process and produce cancer. Most of these carcinogens fall into three categories: certain chemicals and irradiation, both of which are known definitely to cause human cancers, and viruses, which most scientists suspect also cause human malignancies, although at present they are known to do so only in lower animals. While ethical considerations preclude the testing of potential cancer viruses in persons, experiments with laboratory animals have uncovered at least 100 viruses that produce cancer, and transformation of cells of human origin in tissue culture can be used in tests of suspect viruses.

Not too long ago it was thought that radiation, and probably chemicals also, caused cancer by producing a mutation, or sudden, inheritable change in a cell's genes. Hypotheses now propose that these carcinogens somehow free gene repressors so that hidden genetic information may instruct the cell to begin dividing rapidly. The cancer cells reproduce in an uncontrolled way, growing and invading without reference to the health of the body, and finally destroying their host. This change in concept was important because it suggested that cancer consisted merely of a fault in gene expression, rather than being the product of gene damage, and that it might be reversible if suitable controls could be restored. Theodore Puck, professor of biophysics at the University of Colorado Medical Center, established that the repressor concept is well founded. He showed that one line of cancerous animal cells could be returned to their normal healthy condition—within an hour—by the use of two naturally occurring body chemicals, testosterone, a male sex hormone, and cyclic AMP (adenosine monophosphate), a chemical that permits hormones to effect changes in cellular activity.

"For the first time the reversal of cancer cells is possible," said Charles Huggins of the University of Chicago, who was awarded a Nobel Prize in 1966 for his discoveries linking hormones to cancer and its control. "If they couldn't be reversed, the entire outlook for cancer would be much more pessimistic." Huggins believes that hormonal imbalances play a major role in the development of a cancer. Studies at the Imperial Cancer Research Fund in London support his claim: records from more than 5,000 women patients indicate that the onset of breast cancer appears to be preceded by abnormally low levels of two hormones, androsterone and etiocholanolone.

When a normal cell is transformed into a cancer cell, it manufactures new antigens, proteins that the body recognizes as foreign. It is now believed that each type of cancer may have its own antigen; if this is true, then methods of detecting these chemical markers could be used to make early diagnoses of cancer. Several cancer antigens have been found to resemble antigens present during fetal development, when embryonic tissue is growing rapidly. These antigens are not generally present in adult cells. Their reappearance in cancer cells may be seen as further evidence that cancer may be a manifestation of a life process that has gone awry.

The body has a natural animosity to antigens, and the same immuno-

Courtesy, Memorial Sloan-Kettering Cancer Center

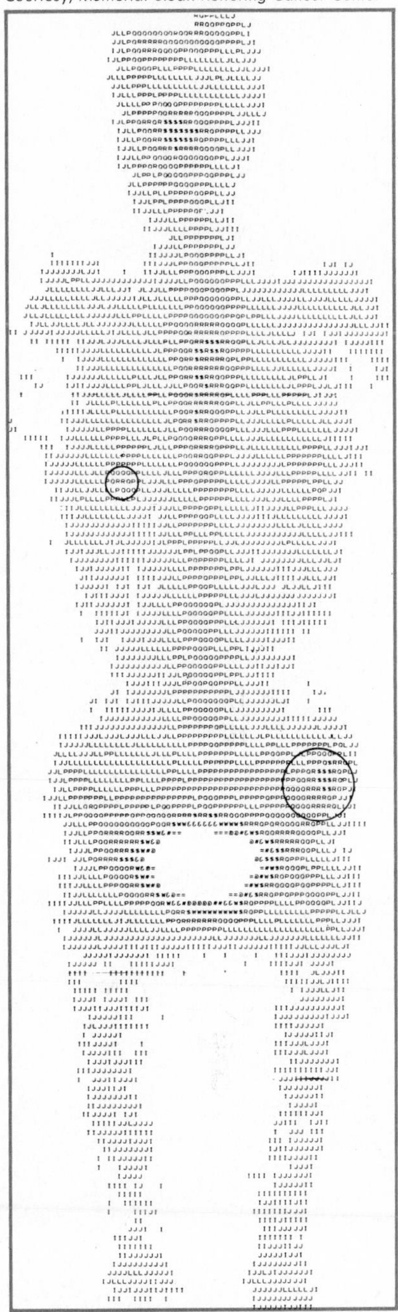

*A computer that equates each letter of the alphabet with a radiation level recorded uptake of the isotope fluorine-18 by the skeleton of this patient. Doctors reading the scan found two areas where late letters indicated higher than normal uptake, evidence of a spreading bone cancer.*

logical defense system that helps combat colds and other infections becomes aware of and attacks cancer cells that possess these abnormal proteins. Primarily, the cancer antigens trigger the production of antibodies, the specialized chemicals designed to neutralize antigens. It is possible that the transformation of normal cells into malignant ones is a common event and that the cancers that develop, called silent cancers, are destroyed by antibodies as soon as they appear. In three out of four persons, the immune response is so good that cancer never has a chance to develop. But the fourth person is in danger. His defense system may be defective because of disease, age, loss of the ability to recognize the antigens, or some other mishap that prevents it from searching out and killing the small pockets of cancer cells.

Failure of the immune system allows the cancer to grow rapidly to about the size of a pinhead, when it stops temporarily. This is the maximum size a cancer can attain by diverting nutrients from neighboring cells; to grow further it must elicit cooperation of the neighboring small blood vessels. The tiny tumor secretes a compound, called tumor angiogenesis factor (TAF), that stimulates a fantastic growth of blood vessels directly into the tumor, which previously had none. With this new source of food, the cancer begins growing again.

### Viruses, enzymes, and cancer

New areas of research into the cause and progress of cancer are taking shape. One of the most important developments stimulating these efforts was reported in 1970 by a University of Wisconsin virologist, Howard M. Temin. He claimed to have discovered that some viruses have the ability to make genetic material that can be incorporated into the chromosomes of a host cell, bringing it new hereditable information. Some investigators think that this may be the missing link in the cancer puzzle, the means by which the cell acquires the deadly instruction that orders it to reproduce in a cancerous fashion.

Currently three major theories try to explain the cancer-inducing ability of viruses. One holds that viruses from a source outside the victim infect a cell and construct a gene that eventually produces cancer. This would support the straightforward notion that perhaps most cancers are caused by viruses. The second theory, one that is gaining in popularity, suggests that everyone is born with the genetic message of a cancer virus already incorporated within his chromosomes. Robert J. Huebner of NCI proposes that such viral genetic material was probably incorporated as a result of a viral infection in man's early ancestors and has been passed down through succeeding generations. In certain carriers an environmental shock, such as exposure to chemicals, radiation, another virus, or hormonal imbalance, turns on the virus information and a cancer begins to develop. The third theory, postulated by Temin, suggests that normal human chromosomes contain specific sections that can make virus-like particles capable of infecting other cells. Once inside a fresh cell they can alter the genes so that cancer results.

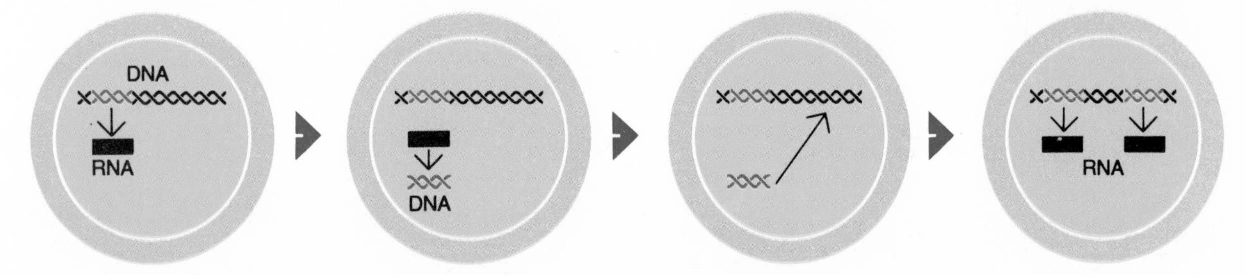


*The recent discovery of the reverse transcription process suggested that the cause of cancer may lie in one's own genes. The protovirus hypothesis proposed by Howard M. Temin suggests that sections of DNA in normal cells may be models for synthesizing RNA that has the ability to synthesize DNA, which becomes integrated into cellular DNA. Repetition of this process may produce regions of DNA that are activated to encourage cell differentiation during embryonic development, or to stimulate cancerous cell growth later in life.*

Genes are composed primarily of deoxyribonucleic acid (DNA), which constitutes the master code for the chemical language of life. The DNA carries the information for making proteins, which are the materials that build cells and other structural components and that, in the form of enzymes, speed up the cell's chemical reactions. When a protein is needed, the section of the DNA governing that protein produces a working copy of itself in the form of ribonucleic acid (RNA). The RNA then carries the instructions for making the protein to subcellular particles called ribosomes, which assemble amino acids into the protein according to the RNA blueprint.

In 1970 Temin established the existence of an enzyme (now called either RNA-directed DNA polymerase or reverse transcriptase) that can cause an RNA blueprint to make DNA—rather than protein. Geneticists were amazed! They had maintained that DNA could make RNA, but not vice versa. Temin's discovery is important to cancer research because most viruses known to cause cancer contain only RNA. It explains how RNA tumor viruses might cause a heritable, cancerous change in the cells they infect.

### Cancer detection, prevention, and treatment

Mounting evidence suggests that reverse transcriptase may be the key to controlling some cancers. Researchers have associated it with cancers, including leukemias, lymphomas, and breast cancer. The enzyme may serve as a diagnostic tool to indicate the presence of a cancer. Meanwhile, workers continue to investigate drugs to find ones that will block the action of the enzyme and perhaps control cancer. Two promising classes of compounds consist of derivatives of the antibiotics rifampicin and streptovaricin.

Reverse transcriptase has also been detected in embryonic tissue. This may mean that the same enzyme that induces cancer is also vital to the growth of fetal cells. This possibility fits nicely into both the Temin and Huebner theories: an enzyme that appears necessary for the rapid growth of fetal cells is also found to be active in rapidly growing cancer cells. The speculation is that a cancer cell, one with its genetic controls removed, essentially reverts to an embryonic state.

A remarkable piece of detective work has provided the most convincing evidence yet that a virus may cause human breast cancer. Reverse transcriptase, when mixed in a test tube with viruses known to cause animal cancers, produces DNA molecules. Since this artificially

## GLOSSARY OF CANCER TERMS

**TUMOR (or NEOPLASM):** Any new, excessive growth arising without obvious cause and serving no physiological function. Such growths can be either benign or malignant.

**BENIGN TUMOR:** Any noncancerous tumor; that is, one that does not spread rapidly from its site of origin.

**MALIGNANCY:** The tendency to cause death. Tissues are deemed malignant when, under microscopic examination, they are found to display any of the abnormalities characteristic of cancer.

**CANCER:** The common name for any of the more than 100 distinct diseases characterized by tissue cells that grow more rapidly than normal, assume abnormal shape or size, and cease functioning in a normal manner (neoplasia), and by a tendency to spread from one part of the body to another (metastasis). Cancers are generally classified either as carcinomas or sarcomas according to their site of origin; prefixes added to these designations indicate the precise tissue of origin.

**CARCINOMA:** Any cancer originating in the epithelial tissue, which lines most hollow organs. A uterine adenocarcinoma, for example, is a cancer involving ducts or glands of tissue of the uterus.

**SARCOMA:** Any cancer originating in the structural frame of the body (nonepithelial tissue), including muscle, fat, connective tissue, bone, cartilage, and tendons.

**METASTASIS:** The transmission of cancerous cells that have broken off from the original lesion to another site through the blood, lymph, or membranous surfaces.

**LESION:** Any injury, abnormal change, wound, or diseased area in an organ or tissue. In cancer, the primary lesion is the one in which the cancer originated; a secondary lesion is one established through metastasis.

**CARCINOGEN:** Any substance capable of causing a malignant tumor. In general, there are three types of carcinogens: chemical, radiation, and viruses.

**C-TYPE VIRUS:** A virus with a central core of ribonucleic acid (RNA) and a double limiting membrane secreted through a budding process from the surface of infested cells and capable of altering the genetic character of the host cell; the type of virus most commonly suspected of being a human carcinogen.

**REMISSION:** A period of return to better health in any chronic disease. In cancer this must be associated with a persistent reduction in size of measurable metastatic tumors and cessation of tumor growth.

Dr. Nurul H. Sarkar, Institute for Medical Research

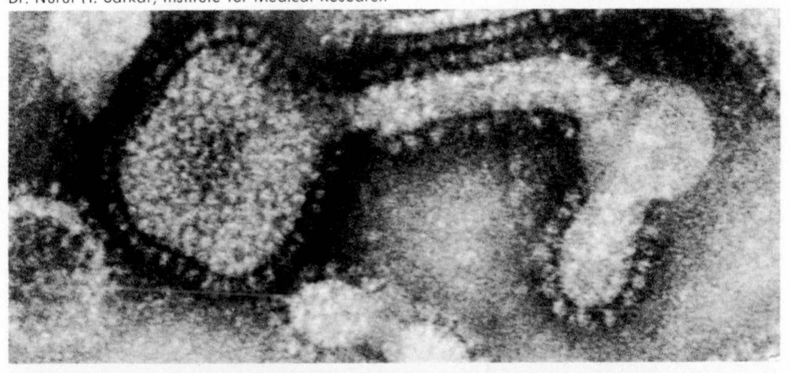

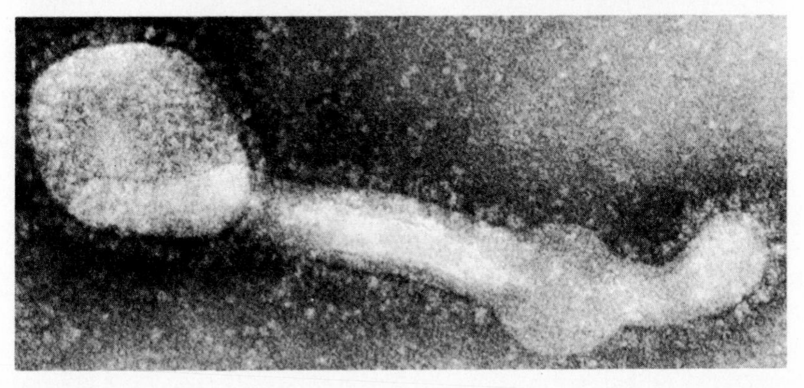

*Electron micrographs detect the close resemblance of the virus particle (top) proved to cause breast cancer in mice to another (bottom) often found in the milk of women with known high genetic risk of breast cancer. Both particles have rounded heads of similar diameter, long irregular tails, and surface projections that differ from any noted on other viruses. The human particle passed several other tests indicating it may be a viral cause of human cancer.*

created genetic material was made from known cancer-causing viruses it should hook up with complementary DNA in virus-produced cancer cells, but not with other DNA free of the virus. In 70% of the cases tested it did, revealing that the human breast cancer cells contained genetic information identical to the genetic instructions of a cancer-causing virus.

It is of considerable importance that the virus that causes breast cancer in mice is similar to a virus discovered in the milk of women with familial histories of breast cancer. Likewise, reverse transcriptase is similar in both human and other animal breast cancers. This association seems more than coincidental. Prevention of breast cancer by immunization, or its control by antiviral drugs, is no longer a mere hope. Indeed, the mouse cancer virus has been used to produce a vaccine that protects mice from breast cancer when they are subsequently exposed to the virus.

The search for a human cancer virus becomes more promising daily. Within a short period of time in late 1971 and early 1972 researchers from three separate laboratories reported finding viruses in cancer patients that might be human cancer viruses. These findings—by Robert M. McAllister of the Children's Hospital of Los Angeles; by Sarah Stewart at Georgetown University, Washington, D.C.; and by Elizabeth S. Priori and Leon Dmochowski at the M. D. Anderson Hospital and Tumor Institute, Houston, Tex.—await confirmation by other workers. All three of the suspected viruses are what scientists call C-type viruses, the kind that Huebner theorizes could be the product of cancer genes inherited from man's ancestors.

Dr. Murray B. Gardner, University of Southern California

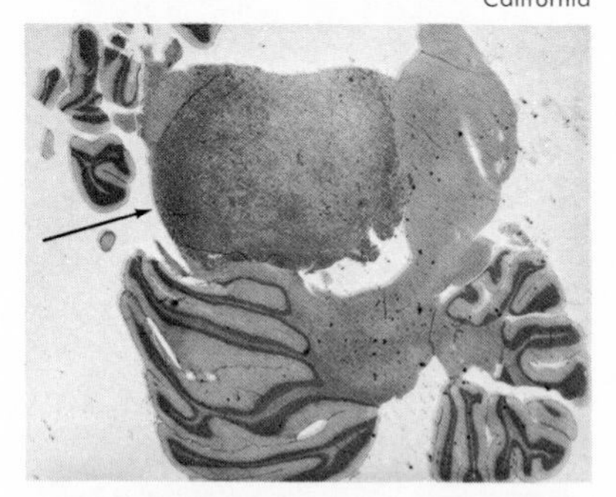

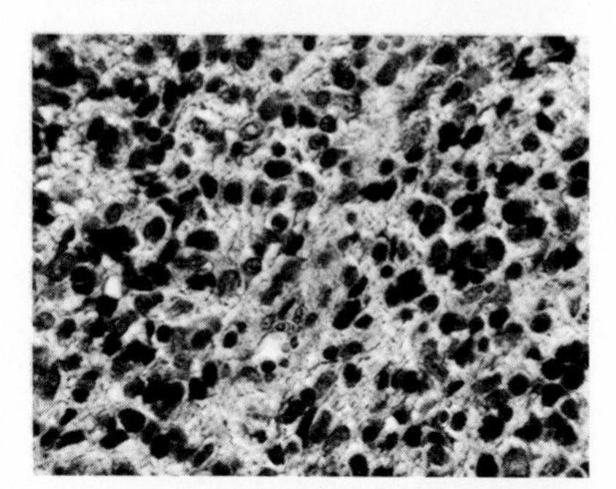

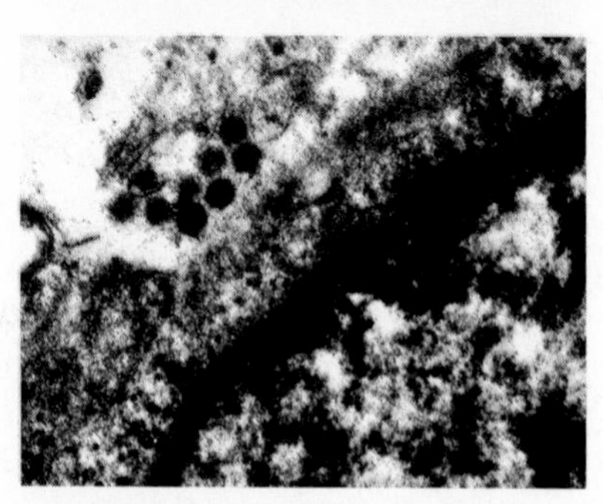

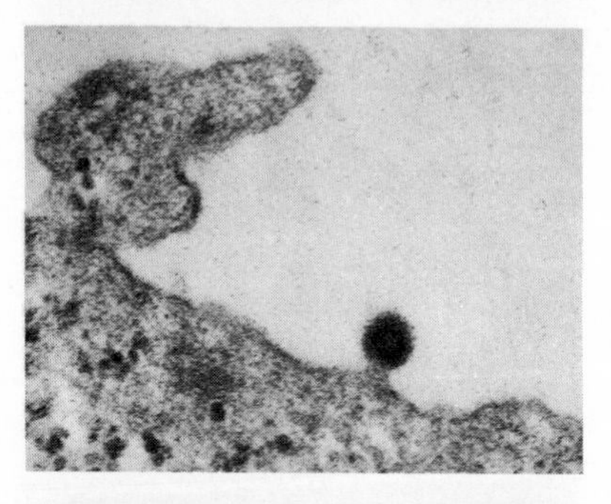

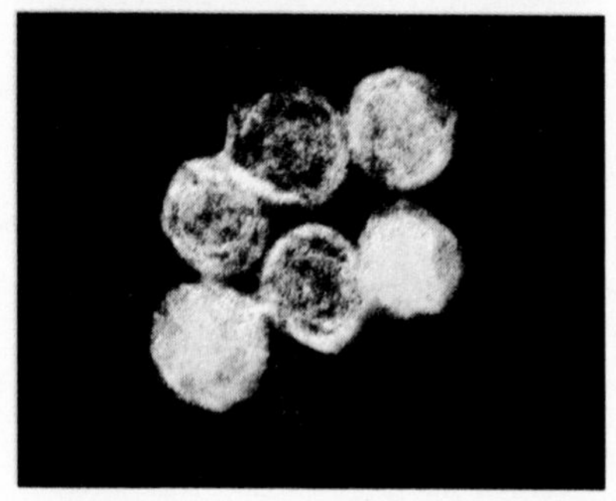

Another important bit of evidence to strengthen the virus-enzyme-cancer concept was the discovery by three other groups that two chemicals, called BUDR and 5-IUDR for short, can induce C-type viruses to appear in what are otherwise normal cells that previously had no viruses in them. It is believed that these chemicals, like an environmental shock, may jolt the cancer genes into action, again supporting the Huebner theory. The NCI is so confident that the viral cause of cancer will be substantiated that it has awarded a $1.8 million contract to Merck Sharp & Dohme Research Laboratories to develop vaccines against the viruses (including the C-type) known to cause tumors in animals.

The feeling of excitement that continues to mount in the fields of virology and molecular biology, is also being felt in immunology. Evidence that cancer flares up as a result of a breakdown in a person's immunological defense system has accumulated. Persons with immune-deficient diseases, for example, exhibit higher than normal rates of cancer, and malignancies increase in older persons, as the efficiency of the immune system diminishes. Perhaps the most incriminating information comes from the field of organ transplantation. In order to suppress the body's rejection mechanism—which would recognize a transplant as foreign matter—graft recipients are given drugs that depress their immune system. But an alarming trend has occurred: the incidence of cancers among these patients is unusually high. The drugs that successfully prevented rejection of the transplant apparently also impaired the patient's immune response against silent cancers.

A promising approach to helping the body fight cancers involves the use of a vaccine effective against tuberculosis, BCG, which seems to have the added benefit of generally strengthening the body's immune response. A special vaccine composed of BCG and cells from cancer patients has kept six leukemia patients alive for more than five years. These patients had a deadly form of leukemia that usually kills its victims within three years. These remarkable results with BCG are believed to be the first demonstration in man that an agent that stimulates cellular immunity also protects against cancer. BCG was also administered to 22 patients given up as hopeless because of malignant melanoma, a usually fatal skin cancer. Four patients underwent complete regression and eight others showed some stabilization of the disease. Promising support for the cancer-controlling properties of BCG came from the Canadian province of Quebec, where, in one town, when all the children were immunized against tuberculosis the expected incidence of acute leukemia fell off by 60%.

A second immunity-strengthening approach being employed against cancer involves the use of a compound called Poly I:C to stimulate the production of interferon, a natural cellular defense system against viruses. Mice given Poly I:C and then subjected to a known cancer-causing virus do not develop cancer. Animals not protected with the compound, however, do contract the disease. Preliminary studies indicate the compound may also be beneficial against human cancers.

One of the main goals of cancer research is the development of ways to detect cancers early, when they are most likely to be curable. A test for cancer of the colon, and possibly of the stomach and pancreas, is now undergoing clinical trials. It attempts to detect antibody response to an antigen known to be carried in the blood of colon cancer patients. The antigen is also known to be present in the gastrointestinal tissue of every unborn baby between the third and sixth months of gestation. It disappears before birth, not to be found again unless a cancer is present. It is hoped that the test for the antigen will do for colon cancer what routine use of the Pap smear has done in dramatically reducing cervical cancer in women.

Why a person's natural immune system may be powerless against cancer may have been uncovered by a husband-and-wife team, Karl and Ingegerd Hellstrom of the University of Washington. They found that patients with rampaging cancers had in their blood substances called blocking antibodies, which concealed the antigens on the cancer cells so that the defense system could not recognize and attack them. In effect, the immune system is blindfolded to the existence of the cancer. Their finding opens up a new approach for cancer therapy, for if a way could be found to remove the blocking antibodies, the patient's own immune response—already present but unaware of the cancer—may be able to destroy the tumor cells. The Hellstroms have had successful results in the test tube, and David F. Hickok of the University of Minnesota recently reported encouraging results with a similar technique in human subjects.

Because evidence of the natural history of cancers is accumulating rapidly, present treatments for malignancies may become outmoded in a few years. Also, if cancers can be spotted early enough and methods developed to block the growth of blood vessels to the tumors—a prospect currently being investigated—many cancers may be arrested and never grow bigger than pinheads. Although much remains to be done in cancer research, the massive strategy formulated in the United States may very well produce information that will not only speed man's eventual victory over cancer but also teach him more about the secrets of life.

*See* in ENCYCLOPÆDIA BRITANNICA (1972): CANCER; CANCER RESEARCH; IMMUNITY AND IMMUNIZATION.

FOR ADDITIONAL READING:

"BCG Vaccine," *American Family Physician* (February 1972).

"Chemical Control of Cancer," *Modern Medicine* (Feb. 7, 1972).

Crick, F. H. C., "The Language of Life," *1969 Britannica Yearbook of Science and the Future* (Encyclopædia Britannica, 1968).

"Virus Theory of Breast Cancer Receives Some New Support," *Journal of the American Medical Association* (May 3, 1971, pp. 751–753).

Wilson, David, *Body and Antibody: A Report on the New Immunology* (Knopf, 1972).

*Although no virus was a proved cause of human cancer, one study in which human sarcoma cells were injected into cat fetuses produced a C-type virus unlike any known in animals. Photomicrographs (opposite page, top to bottom) show human cancer cells forming a tumor (arrow) in a kitten's brain tissue, a kitten brain tumor made up of human cells, C-type virus particles (dark spots) near a human sarcoma cell in a kitten brain tumor, the virus budding off a sarcoma cell in a culture, and the isolated virus grown for 24 hours in a culture.*

# Mechanical Servants for Mankind

by John McCarthy

**Robots—mechanical beings that can take over the dull and burdensome chores of life—have long been dreamed of by man. How soon, if ever, will this dream become reality?**

A robot is an artificial intelligent being usually considered to have human form. The idea of making a robot by magic is ancient, and the idea of making one by science originated in Mary Shelley's *Frankenstein, or the Modern Prometheus* in 1818. The word "robot," from the Czech word for work, was introduced in a play *R.U.R.* by Karel Capek in 1920.

Mechanisms that imitate some of the functions of humans or animals have been constructed for several hundred years, but a real robot must have two properties that were not really feasible until the stored program digital computer was developed in the late 1940s. These are sensory input from the outside world so that the machine's actions will take the world into account, and an internal model of the outside world, the machine's goals, and the laws that determine the effects of the actions the machine may undertake. The pre-computer mechanisms, while sometimes very elaborate, essentially went through fixed, built-in sequences of actions.

### Formulating the problem

The first modern discussion of the possibilities of intelligent machines was by the British mathematician Alan Turing in 1950. Turing was well fitted for this because in 1936 he had introduced the concept of a universal computational machine and had also taken part in the design of early computers. (A universal computational machine can do any computation that can be done by any machine, because it can use a description of the other machine in order to imitate it. All present computers are universal.)

Turing's article accomplished three things: (1) It set forth as a sufficient criterion for intelligence the ability to imitate a human over a teletype line. This clarified arguments with people who believed that a machine could not be intelligent because it does what it is told. (2) It established that the difficult problem is not the construction of the robot or the computer but the programming of it. It is possible to use any stored program digital computer, but the problem is to make up the rules that determine how it will act in any situation. These rules have to be understood very precisely in order to express them as a computer program. (3) It proposed the problems of making programs for playing chess and proving mathematical theorems as major goals for research in artificial intelligence, giving reasons why these are worthwhile problems and some ideas for getting started.

Since Turing's time, artificial intelligence has slowly developed into an important scientific field with numerous branches and with numerous ideas of the fundamental problems. At present, there are still no really intelligent robots. In fact, as described below, there are fundamental scientific problems to be solved and maybe even formulated.

### Heuristics

Artificial intelligence may be divided into three sub-topics: heuristics, representation theory, and robotics. Heuristics is concerned with the search for a solution to a problem among a collection of alternatives. Representation theory is concerned with the structure of the information a computer program or a person may have about the world, the facts that allow it to determine the effects of the actions it may take, and the methods available for getting more information. Robotics is concerned with perception and action in the physical world.

The most developed of these subjects is heuristics. Most of the insight gained in this area has come from comparing the performance of computer programs in playing games and proving mathematical theorems with human performance. The most studied game has been chess. The best result in comparison with human play was achieved in the 1960s by a program written by Richard Greenblatt of Massachusetts Institute of Technology (MIT). His program played class C chess at that time. In fact, it won its class C rating by winning the class D trophy in a regular chess tournament in Boston.

This performance is quite modest considering that these programs are capable of examining several hundred positions per second on present-day computers. However, even these results were achieved with a great deal of effort, and the earlier programs were much worse. Part of the progress involved the incorporation of specific knowledge about chess, but the major defects in early programs can better be described as their lack of aspects of general human intelligence. Moreover, the weaknesses of present programs seem to come from lack of general intelligence rather than lack of chess knowledge. Programs written under the direction of chess grandmasters have not done as well as those written by people interested in heuristic programming.

***JOHN McCARTHY** is a professor of computer science at Stanford University and director of the Stanford Artificial Intelligence Laboratory.*

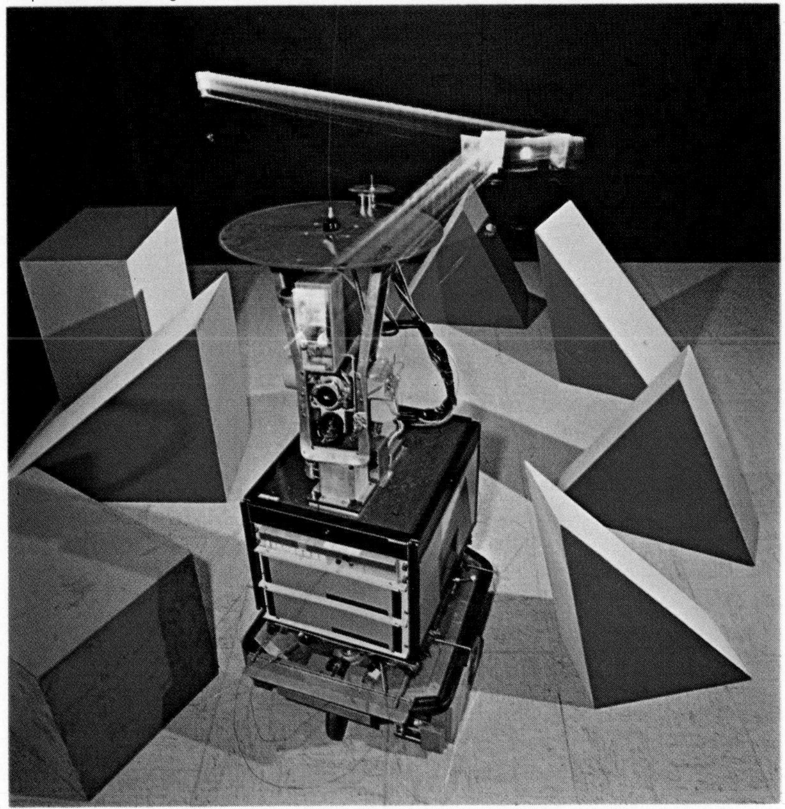

*"Shakey," named for its unsteady movements and developed by the Stanford Research Institute, has motors, wheels, feelers, a television "eye," and a range finder. These are all connected by radio to a large time-shared computer. Shakey can explore a room, remember some of its contents, and from that information achieve some simple tasks. For example, it can push a box off a platform by finding a ramp somewhere in the room, shoving it against the platform, rolling itself up the ramp, and pushing off the box.*

In order to see what a few of these intellectual mechanisms are, consider how chess programs work. They all work by exploring the so-called move tree of the game. In an initial position the machine has a number of moves which can be regarded as the branches of a tree. Each move leads to a position in which the opponent has a number of moves that give another level of branches of the tree, and so on. Ultimately the branching stops at positions in which the game is over.

It would be easy to write a computer program that would play perfect chess by examining all the positions on the move tree. The program would be smaller and use less storage than present chess programs. The only problem is that it would take too long to move, because the full move tree is quite large. In fact, if every electron in the universe visible to astronomers were a computer that could examine a position in the instant of time that it takes light to cross an electron, then the time remaining until the stars dissipate their energy would not suffice to determine the best first move in chess.

Therefore, programs look only at a small part of the move tree, and the program's level of play depends critically on the rules that determine what positions are examined. Most programs do not consider all possible moves, even at the initial level. An effort is made to consider only moves that may contribute to some goal, and for each possible goal programs are written that generate moves that may contribute to it. Goals included in typical programs are attacking the opponent's king, defending one's own, getting a good pawn structure, disrupting

Courtesy, General Electric Research and Development Center

*A four-legged "walking machine" mimics and amplifies the movements of its human operator. The right front leg of the machine is controlled by the operator's right arm, the left front leg by his left arm, the right rear leg by his right leg, and the left rear leg by his left leg. Designed by the General Electric Co. as part of a program to develop "man-amplification" machines for the military, this research prototype has walked across level ground, climbed obstacles, lifted a small vehicle out of a mudhole, and hoisted a 500-pound load onto a truck with one foot.*

the opponent's pawn structure, etc. In order to prevent the program from going deeper and deeper, there are stop rules. In general, one tries to stop the analysis at quiescent positions that can be evaluated without further looking ahead. Besides this, it pays to set an initial level of aspiration that changes as the analysis proceeds and to stop the analysis of positions that clearly cannot lead to reaching this level. It also is necessary to keep a level of aspiration for the opponent based on the alternatives that have already been examined for him.

When a computer chess program makes a bad move in a position, it can be required to print out the tree of moves that it examined in reaching its decision. It usually turns out that the trouble comes from the program not having examined an important move, either its own or its opponent's. This is much more common than having examined the right moves but incorrectly evaluating a quiescent position. On the other hand, the computer will usually have spent much time examining moves that strike the human player as not worth looking at.

When the program has failed to look at a line of play that a human would look at, this is usually readily fixed by finding a new principle for looking at moves. Of course, this principle may make the computer program look at moves a human would ignore. This is all right provided it does not waste too much time. Unfortunately, this often happens, and it is much more difficult to devise principles that will allow a program to ignore moves safely.

There are several aspects of human game playing that no one has been able to make computers accomplish satisfactorily. First, a human will often break a position down into several sub-positions, analyze them separately, and then combine the analyses. There is still not any good way of programming a computer to identify such sub-positions. Second, a move often has properties that apply to a whole class of positions; thus, an opponent's move may capture one's queen in all the positions resulting from moves that do not move the threatened queen. Third, there is not yet a system for recognizing that an aspect of a position is one for which there is a general rule; thus, no present chess program can be instructed about smother mates in any general way, let alone develop the concept from its experience.

In a few board games, computer programs play better than the best humans; these are games in which the intellectual mechanisms we know how to program, such as look-ahead and evaluation, are adequate for successful play. Such a game is Kalah, an American commercial version of a game played in Africa and Asia.

The other area to which heuristic programming has been extensively applied is proving mathematical theorems. The results are similar to the game-playing results but not as good. Certain intellectual mechanisms have been identified and programmed. In domains of mathematics where these mechanisms are adequate, the programs do rather well. Unfortunately, the scope of these domains is still very narrow compared to mathematics as a whole.

*continued on page 104*

# Robots in science fiction

Since most people have gotten their first ideas of robots from science fiction, it is worthwhile to consider how robots have been described in these works. First, while an author sometimes simply wants to tell what he thinks the future will be like, usually he has some other theme. Sometimes, he is simply writing an adventure story. Then the robots often help provide the conflict by trying to conquer the world on their own behalf or on behalf of their makers. There may be both good robots and bad robots in the story. Another common theme is the robot as the oppressed class or race. They may revolt or suffer helplessly and pathetically or be liberated because of a belief analogous to equal races living in harmony. In any case, the robots are given as many human motivational, emotional, social, or political attributes as will suit the author's purpose. Another theme is based on the story of King Midas, in which an apparent good thing is fated to turn out badly.

One of the popular science fiction concepts is Isaac Asimov's three laws of robotics—a robot must protect man, protect itself, and carry out its orders in that order of priority. Some of Asimov's plots hinge on complexities in the interpretation of these laws in particular cases, leading to neurotic behavior of the robot.

Most science fiction fails in one way or another to describe important aspects or consequences of robots. For example, in his stories Asimov ignores the possibility that a robot may do harm on account of its conception of the world being inadequate to determine the consequences of a proposed action. Similar unrealistic treatment appears in works of other writers. For example, they generally depict worlds with only one or two forms of artificial intelligence, whereas there are likely to be many. Writers have also generally found it difficult to depict an intelligence greater than man's and instead have concentrated on robots that are equally as smart.

But these shortcomings do not mean that the science fiction ideas of robots are worthless. We may get from them some ideas of opportunities to be exploited and problems to be avoided. The real development of artificial intelligence seems likely to differ from the science fictional in the following ways:

1. It is unlikely that there will be a prolonged period during which it will be possible to build machines as intelligent as humans but impossible to build them much smarter. If one can put a machine capable of human behavior in a metal skull, one can put a machine capable of acting like 10,000 coordinated people in a building.

2. The present stock of ideas is inadequate to make programs as intelligent as people even if these ideas are developed diligently over a long time. They may even be inadequate to make a program as intelligent as a dog. On the other hand, there is nothing to prevent the required new ideas from coming very soon. It may take 5 years, and it may take 500.

3. Present ideas are probably good enough to extend man's ability to have large amounts of information at one's fingertips, to check long trains of reasoning, and even to generate simple hypotheses.

*continued from page 102*

The efforts to make game-playing and theorem-proving programs have led to the identification of important intellectual mechanisms of general significance, but the programs themselves are quite narrow in their capability. Therefore, considerable effort has also gone into trying to make programs of general intellectual capability.

## Representation

One approach, which was tried in the late 1950s and early 1960s, involved simulating evolution by having a machine or computer program that gave answers to problems. The machine or program was divided into parts, and each part had a certain probability of being changed at each time. When a problem was solved correctly, the parts that operated in the correct run were "rewarded" by having their probability of change reduced, and when an incorrect answer was given the active parts were "punished" analogously. Other schemes kept the same parts but altered their probabilities of activation according to their success. When applied to problems of pattern classification, such as identifying the letter of the alphabet shown in a picture, these schemes had some success, but none of them were capable of learning any generally intelligent behavior. The problem was that in the schemes used for representing behavior, meaningful changes of behavior were not represented by small changes in the representation.

Another approach to general intelligence was the General Problem Solver (GPS) of Allen Newell and Herbert Simon. This involved representing an arbitrary problem as one of transforming one expression into another via intermediate expressions using certain allowed rules of transformation. A number of problems were put into this form and solved by the GPS, and the results were compared with the records of sessions in which humans solved the problems. In the end, however, the GPS was not general enough. In particular, the GPS could not achieve the goal of representing the problem of improving itself.

A third approach involves the use of mathematical logic. In the middle of the 19th century, the English mathematician George Boole began the development of a formal language for expressing mathematical ideas. This development was continued by such mathematicians as Gottlob Frege and Bertrand Russell. When we say a language is formal, we have three properties in mind: Effective rules are given that determine which strings of letters and digits and special characters are supposed to be sentences in the language; effective rules are given that determine when a sentence is to be regarded as an immediate consequence of several others; and, if the language is to refer to something outside itself, some rules have to be given in English that tell what the objects of the language refer to. The concept of "effective rule" used here by the mathematical logicians can be considered today to be the same as "checkable by computer."

The use of logic in artificial intelligence is in principle quite straightforward. The idea is to express what human beings know about the

*The Unimate (opposite, top) performs a great variety of heavy and delicate industrial jobs, ranging from loading machine tools to packing glass tubes. Its hydraulic-powered reach can vary from 3½ to 7½ feet, and it can carry loads up to 75 pounds while moving through an arc of 220°. The machine is "taught" by being led through its desired movements once, and its magnetic memory system has a capacity of 180 sequential commands. In the bottom photograph, Unimates are loading automotive presses.*

Courtesy, Unimation Inc.

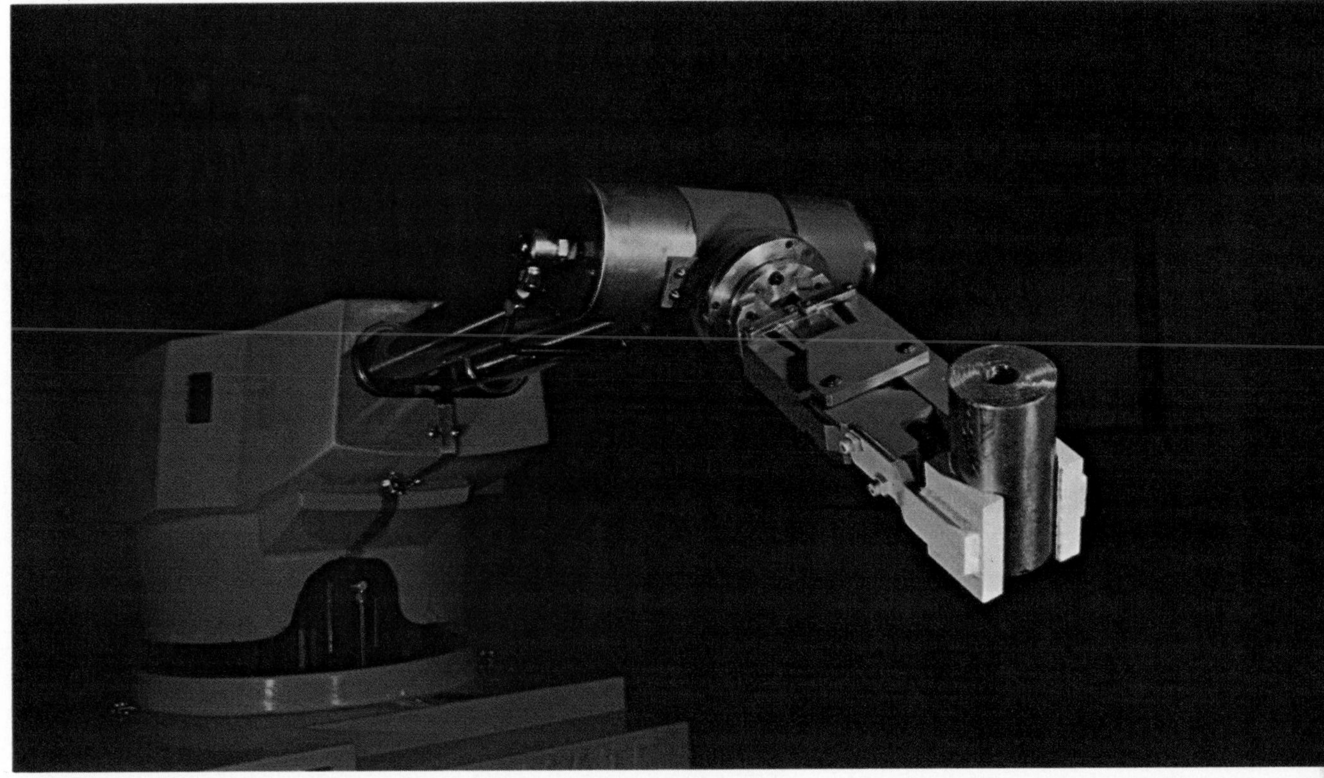

world in general and about particular situations in logical form. It is especially important to be able to express what is known about the effects of actions people may take and what is known about the future consequences of a present state of affairs. If this has been done, then the fact that a certain course of action will achieve a prescribed goal should follow from the aforementioned set of facts. Suppose we have a logical system and a particular set of facts such that it follows from these facts that a certain goal can be achieved by a certain course of action. We still need a program that, given the facts, can find the course of action and deduce that it will be successful. Different programs will have different capabilities, and no program will be able to achieve all goals whose achievability follows from the facts.

Therefore, the use of logical formalism splits the artificial intelligence problem into two parts: the problem of representing the facts in a logical system, and, given the facts, the problem of finding out what to do. The latter is a problem in heuristics as described in the previous section. The representation problem has much in common with epistemology, which is concerned with the nature and limits of knowledge. It is also connected with philosophical logic, which is concerned with the formal description of concepts like ability, causality, and knowledge.

## Robotics

Putting all this together, what progress has been made toward making a mechanical man that will clank about and either try to conquer the world or suffer oppression from the prejudiced?

Consider first the hardware situation. Nowhere in 1972 was there any mechanical man with two legs, two arms, two eyes, and a brain. It was not certain that present-day mechanical engineering could package the necessary motors, joints, etc., in a frame light enough to be moved by the motors. At any rate, no one had tried. There were, however, many separate parts in use. Specifically, artificial arms, television cameras, microphones, and radio and TV connections to vehicles that can roll around on the floor had been attached to computers. The computers were large ones, weighing many tons, but suitable small ones existed even though they were expensive and not as convenient to use as a good laboratory time-shared computer.

However, the main bottleneck in making a robot was not the hardware but the programming, and within programming the most immediate bottleneck was programming vision. In 1972 computer vision involved a television camera attached to the computer, allowing the computer to "read" a frame from the camera into its memory. The picture is represented by about $500 \times 500 = 25{,}000$ numbers, each giving the brightness of a point in the picture. If color is wanted, three frames can be taken, through red, green, and blue filters. That the picture is in the computer can be verified by printing it out again or displaying it on a television monitor connected to the computer. In fact, pictures can be improved by computer processing more flexibly than in a darkroom. The processing is used to eliminate noise, improve contrast, and transform pictures taken from different directions to a common direction and scale so that the pictures can be compared to look for changes.

None of these operations require the computer program to identify and locate the objects in the picture, but suppose we want a computer to physically build an object out of parts. In the simplest case, not requiring vision, the parts come mounted and oriented precisely on a belt. An artificial arm goes through a fixed sequence of predetermined motions in order to pick up each part and place it. Fasteners such as bolts and welding are also used, but also with precise motions. This procedure was introduced in the late 1960s by two U.S. firms, and there are now several hundred such arms in use. In particular, the new General Motors plant for producing Vega cars at Lordstown, O., uses many such manipulators, but even there they do only a small part of the work of putting together a car.

Such "robots" ordinarily are not directly computer-controlled but instead are operated by a tape or drum on which has been recorded the desired motions by a human being who had put the robot through the desired operation. The exact same motions (to the hundredth of an inch if the machinery is that accurate) are repeated each time. The only variability is in timing, since the machine can be instructed to wait after completing one operation until it receives a signal to perform the next one. These signals may be generated by switches, which detect that a part is in place before allowing the operation to proceed.

The applications of this kind of robot are somewhat limited. They

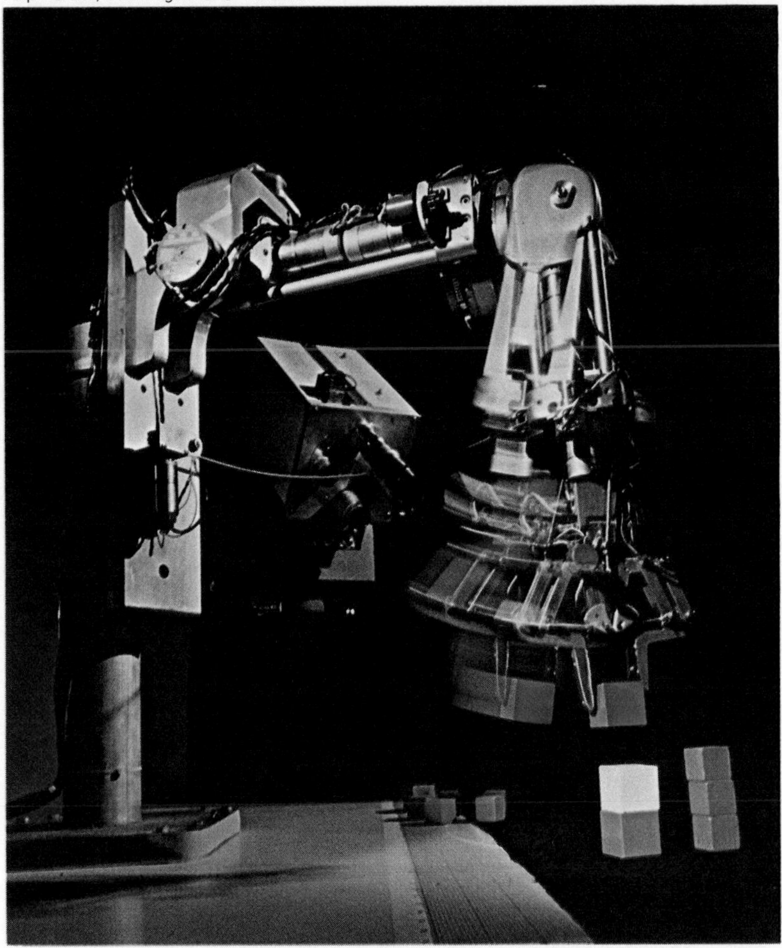

*Obeying spoken commands is a function of the machine above, which uses a manipulator arm, a television "eye," and a large computer to analyze what lies before it. When the words "Stack the small blocks at the upper right" are spoken, the computer determines what steps are required. It then transmits this information to the arm, which turns itself as necessary to grasp each block by the sides and lower it on top of a pile. The machine, however, cannot cope with irregularly shaped objects.*

are so expensive that they are used mainly for operations that are dangerous, such as putting a part in a press or a furnace. Also, it is often expensive or impossible to arrange for the parts to be so precisely made and so precisely placed that the required operations can be repeated exactly. Nonetheless, there are many applications for which this kind of first-level robot can be used, and many thousands will probably be installed in the 1970s.

The second-stage robot involves a computer-controlled arm with a television camera for an eye. The objects manipulated are simple and have flat faces. The program is designed to build a simple structure out of these objects. This work is still in the laboratory. Three examples of what recently has been accomplished there are as follows: At the MIT Artificial Intelligence Laboratory, a computer-controlled arm copied a structure made of blocks. At the Stanford University Artificial Intelligence Laboratory, a program did a puzzle requiring that a tower of four blocks with specified colors on the faces be built. At the Hitachi Central Research Laboratory, a program built a simple structure out of blocks working from a drawing giving an orthographic (involving right angles and perpendicular lines) projection of the desired structure.

All these tasks were accomplished in approximately the same way.

Courtesy, Hitachi, Ltd.

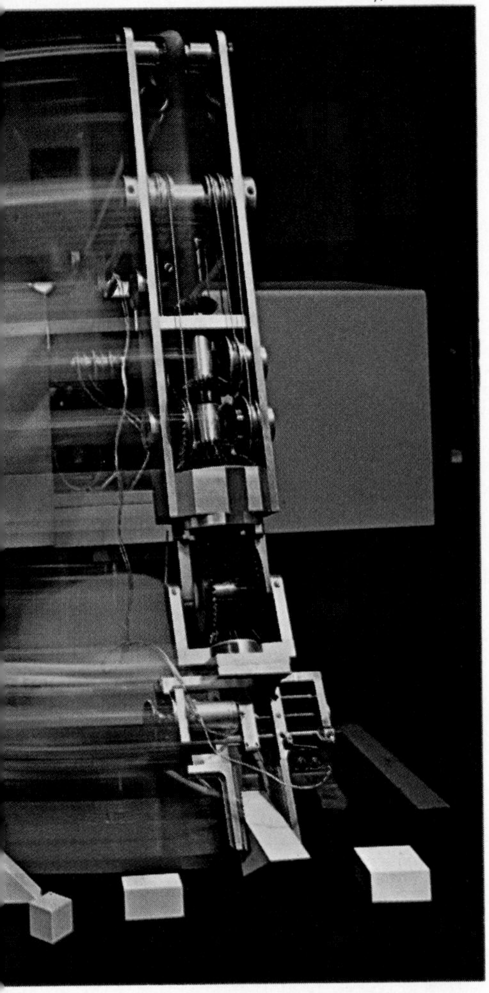

*Japan is producing robotlike machines to use in light and heavy industry. One Japanese firm, Hitachi, Ltd., has developed the "visual robot," a model with two television eyes to provide binocular vision.*

First the program found the edges of the objects in the scene by scanning the array of brightnesses of the picture that it had read from the television camera into the memory of the computer. When the program found a sudden change in brightness, it scanned in a circle about the point where the change occurred, looking for the dark-to-light transition. If it found this transition, the program took this point as the center of a new circle and thus traced along the entire dark-to-light boundary. In this way it found all the edges in the picture. (Actually edge finding is more complicated than this, because not all edges are between regions of different brightnesses; in some cases the edges themselves reflect the light. Also there are shadows to be ignored if possible.)

After the edges were found, the program determined the three-dimensional collection of objects that would give rise to this pattern of edges. Some of the objects might be partly hidden behind others, and the program had to make plausible assumptions about the hidden parts. However, present programs cannot deal with curved objects in any general way; either the objects must have all flat faces or each curved object to be found must be programmed for explicitly. Once the objects in the scene were found, the computer determined the trajectories of arm motion that would move them to the desired places. Generally the arm motion was done without visual feedback, largely because the arms were more complicated objects than the existing vision programs could handle.

Moving to the next higher level, one finds programs being developed to use more visual information than just brightness discontinuities. This work includes color-contrast edges and edges that occur when an object is seen against a background. While these techniques will make edge detection and region finding more versatile, they will not solve the basic problem of representation of visual information. An object with flat faces can be described by listing its faces, describing each face by listing its edges, and describing each edge by giving its endpoints. But how is one to describe a human head? The description must cover the subtleties of shape by which people can tell one human from another, but clearly should not be so detailed as to describe each hair. There are various ideas for dealing with this, mostly still too vague to be described here, but they are all too limited in their potential capability.

Although programming vision is perhaps the most immediate problem in robotics, even when this has been achieved we will still be far from having an intelligent machine. Suppose we want a robot capable of being a domestic servant. Then we need to program at least the following capabilities:

1. It must know the properties of physical objects: for example, to put eggs in a paper bag it must know that eggs must be placed gently and that the paper bag can change its shape as it is opened and things are put in it. It must also know about doors and drawers and stairs.

2. It must know about people and what they want and how they change their minds.

Courtesy, Hitachi, Ltd.

3. It must have some understanding of speech.
4. It must know about how to divide tasks into subtasks.
5. It must be able to recognize any harmful side effects that might result from its planned actions.
6. It must know when to give up and whether to ask for help.
7. It must have at least a primitive ability to learn from experience, at least enough so that if it is shown how to do a task it will be able to do it again in similar circumstances.

The critical reader will recognize that this list of what a robot must know is itself incomplete and may include superficial items and leave out more basic ones. This reflects the present state of our understanding of artificial intelligence.

Courtesy, Los Alamos Scientific Laboratory

# Nuclear Fusion Power Source of the Future?

## by Harold P. Furth

**Power that is clean, inexpensive, and abundant—this long-sought goal may be achieved if the great energy released by nuclear fusion can be harnessed.**

Since the explosion of the first hydrogen bomb in 1951—uncontrolled nuclear fusion—scientists have sought ways to harness and control this vast source of energy. Despite many difficulties, progress is being made. Controlled nuclear fusion reactions of the heavy isotopes of hydrogen promise to provide an economical and essentially inexhaustible new source of electrical power. Large-scale experimental work on controlled fusion is under way in the United States, the Soviet Union, Europe, and Japan. The principal scientific problem is the long-time confinement of the reacting fuel, which must be maintained at temperatures of typically 100,000,000°–1,000,000,000° C. Estimates based on recent experimental progress indicate that a demonstration of scientific feasibility of controlled fusion may be achieved toward the end of the 1970s; the first commercial application of fusion power is generally predicted for the late 1990s or for the following decade.

Fusion power has inherent economic and environmental advantages that make it the preferred solution to the world's long-range energy requirements. The basic fuel, deuterium, is inexpensive and abundant. Unlike nuclear fission power, fusion does not produce radioactive wastes except for activated structural materials of the reactor, and there is no possibility of a runaway fusion reaction. The thermal efficiency of the first fusion reactor designs is comparable to that of conventional power plants; but there is a prospect of future reactor designs with very low production of waste heat. The cost of fusion power and the required capital investment are estimated to be competitive with the corresponding figures for present-day conventional power.

***HAROLD P. FURTH*** *is a professor of astrophysical sciences and co-head of the Experimental Division of the Plasma Physics Laboratory at Princeton University.*

## Requirements for fusion power production

In order to understand the fusion problem, it is helpful to recall some basic features of the atom. The most common form of the hydrogen atom consists of a single electron revolving around a much heavier nucleus, the proton. Two rarer forms, deuterium and tritium, have nuclei that consist of a proton plus a neutron and a proton plus two neutrons, respectively. All three of these isotopes of hydrogen have nuclei that carry a single positive charge equal in magnitude to the negative charge of the electron. The positive charge of the nucleus attracts the opposite, negative charge of the electron, and thus binds the electron to the atom. On the other hand, *similar* electrical charges tend to repel one another. Thus, the fusion of two nuclei is opposed because the two positive charges repel each other.

Deuterium and tritium fuse more readily than other combinations of nuclei, and fusion reactor designs are therefore generally based on this process. When a deuterium and a tritium nucleus meet, however, they are most likely to glance off each other without fusing. Only if their energies of approach are sufficiently high is there any appreciable chance that they will come close enough to experience the non-electrical forces that give rise to nuclear fusion. The minimum energy of motion required for the deuterium-tritium reaction to take place is about 10,000 electron volts (eV). This is the energy acquired by an electron or proton in a 10,000-volt (V) accelerator, or equivalently, is the average energy of motion of particles at a temperature of about 100,000,000° C. (The term "thermonuclear reactions" refers to fusion reactions that take place as a result of this thermal motion.) Deuterium will also react with itself, but the reaction probability becomes appreciable only at temperatures above 1,000,000,000°.

*Nuclei of hydrogen isotopes (deuterium and tritium) consist of a proton plus as many as two neutrons. The positively charged nuclei repel each other (opposite, top and center), but may fuse on sufficiently close approach (bottom).*

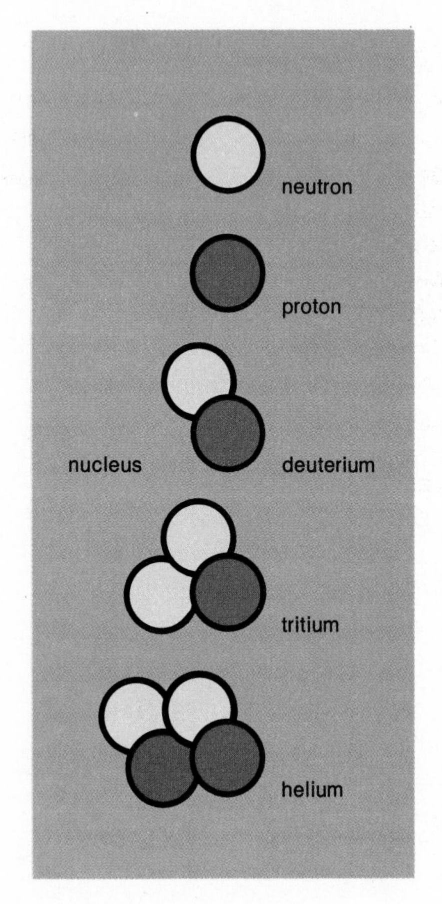


The deuterium-tritium reaction yields a helium nucleus, a neutron, and 17.6 million electron volts (MeV) of energy, most of which is carried off by the neutron. Thus potentially there is roughly a thousandfold increase over the initial investment of 10,000 eV in the energy of motion of the deuterium and tritium nuclei. The actual energy multiplication factor obtained depends on the fraction of the hot fuel that reacts before escaping or cooling.

The fundamental criterion for a successful fusion reactor is that it should confine the hot fuel long enough so that a sufficiently large fraction will react and thereby release appreciably more energy than was invested in fuel heating. This is known as Lawson's criterion. For the deuterium-tritium reaction, the fuel confinement time in seconds times the density of nuclei per cubic centimeter should exceed $10^{14}$.

In the period 1960–70 it had become relatively easy to achieve high temperatures and make fusion reactions in a laboratory experiment. To reach the Lawson criterion, however, remained extremely difficult; the best numbers that had been achieved were in the range of $10^{11}$–$10^{12}$. In order to appreciate the difficulties involved and the nature of the experimental breakthroughs that are being made, it is necessary to review some basic properties of high-temperature matter.

## Thermonuclear plasma

When a liquid is heated sufficiently, it becomes a gas: the atoms break free of each other. When a gas is heated further, it becomes a plasma: the constituent particles of the atoms, the electrons and nuclei, break free of each other. Free nuclei are also called ions, and the break-up process is called ionization. In hydrogen, ionization occurs at a temperature of about 10,000°—relatively low compared to the 100,000,000° temperatures required for fusion. On the earth natural plasmas are encountered quite rarely; for example in lightning strokes and in the northern lights. In the universe as a whole, however, the plasma state is by far the most common condition of matter. The sun and the other stars are giant fusion reactors, and the spaces surrounding them are filled with dilute plasmas.

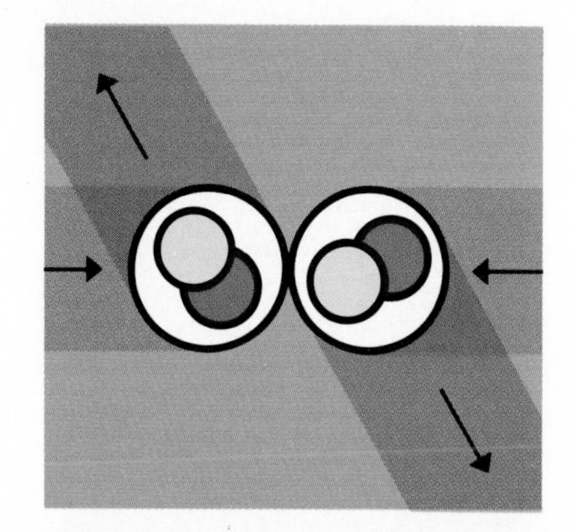

Plasma produced in a fusion reactor on the earth cannot be confined by walls made of ordinary solid matter because the contact with the walls would cool the plasma. Fortunately, there is a natural way to confine the electrically charged electrons and ions of a plasma: by means of a magnetic field. In a uniform magnetic field, a charged particle must gyrate locally around a magnetic field line, though it can travel freely along the line. This constraint on the motion of plasma particles across magnetic field lines is the basis for the concept of the "magnetic bottle."

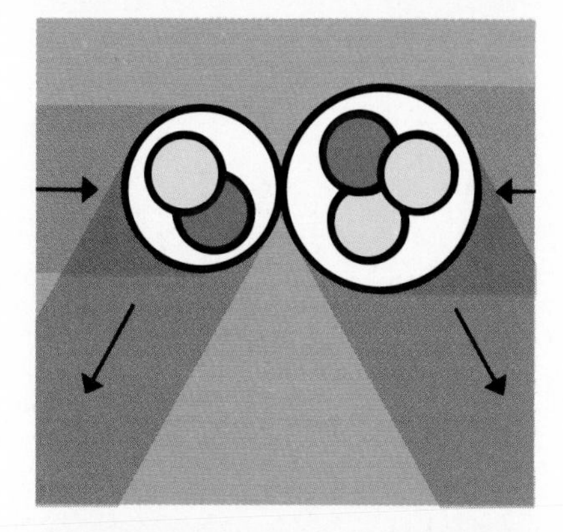

If the magnetic bottle is to confine a hot plasma and keep it away from cold walls, the magnetic field must be strong enough to bear the pressure of the plasma seeking to expand. The strongest magnetic fields technically attainable in large volumes in 1972 were about 100,000 gauss—100 times higher than the typical field strength of a small horseshoe magnet. A 100,000-gauss field can exert a maximum pressure of about 400 atmospheres (one atmosphere = approximately 14.7 pounds per square inch). To keep within this limit, a 100,000,000° plasma can be allowed to have a density of, at most, $10^{16}$ ions per cubic centimeter. Accordingly, the Lawson criterion requires that the fuel must be confined for at least a hundredth of a second. In most magnetic bottles the plasma pressure must be kept considerably smaller than the magnetic field pressure, resulting in plasma densities that range down to $10^{14}$ nuclei per cubic centimeter. At such densities the confinement times necessary to achieve effective fusion range up correspondingly to about a second. The great scientific challenge in controlled fusion research consists in making magnetic bottles that can provide such long confinement times for plasmas of 100,000,000°.

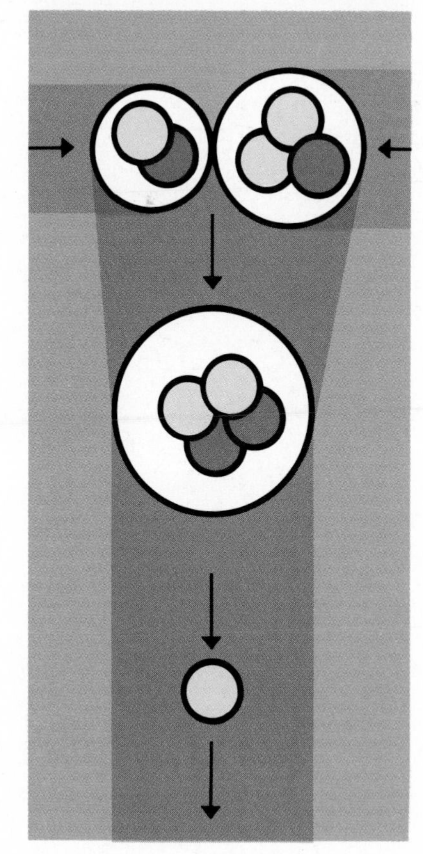

Fusion power can also be produced by freely exploding plasmas—not confined by a magnetic bottle—provided the initial plasma is sufficiently dense. For example, at the normal density of solid deuterium-tritium ($4 \times 10^{22}$ per cubic centimeter), the Lawson criterion requires a reaction time of only several times $10^{-9}$ seconds at 100,000,000°. During this time the plasma expands by only a few millimeters. The difficult problem is how to heat even a cubic millimeter of such a plasma to 100,000,000° in less than $10^{-9}$ seconds, since this requires

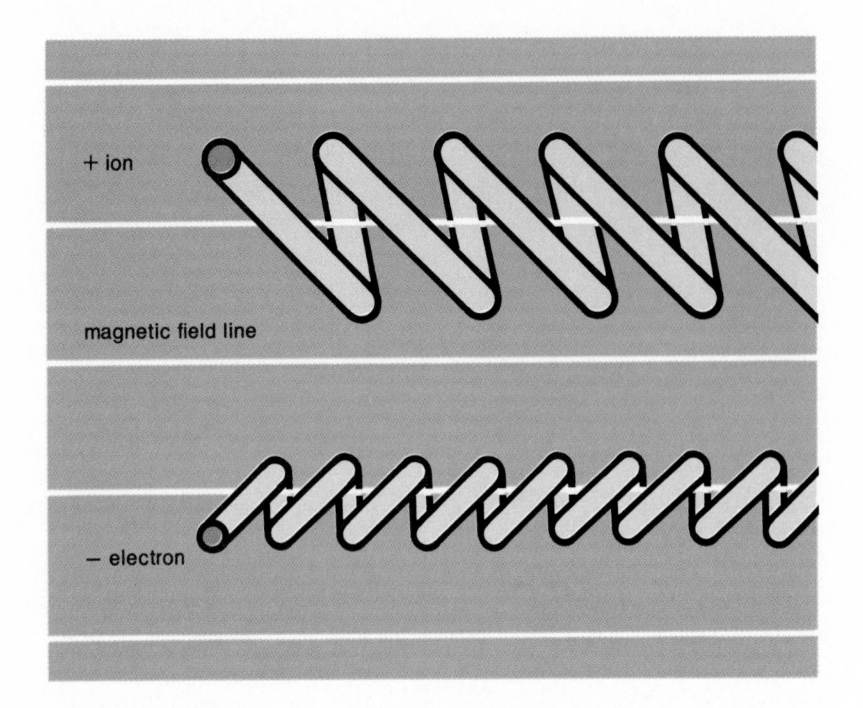


a highly focused energy input of approximately $10^5$ joules at a heating power of $10^{14}$ watts. The recent development of high-powered lasers is beginning to bring this achievement within the realm of technological feasibility.

## Magnetic bottles

How can plasma particles be kept from escaping in the direction *along* magnetic field lines? One obvious solution is to bend the magnetic field lines in a circle, making a closed toroidal (doughnut-shaped) magnetic bottle. Another solution is to leave the magnetic bottle open-ended but make the magnetic field increasingly strong toward each end of the bottle, causing the magnetic field lines to converge toward each end. This is called a "mirror machine." How the mirror machine confines a plasma particle can be understood by an analogy with a compass needle between the poles of a C-magnet. A gyrating charged particle is a small magnetic dipole, as is a compass needle: furthermore, the particle dipole is always diamagnetic (*i.e.*, like a compass needle that is oriented so as to repel the two poles of the C-magnet). Consequently, the gyrating particle stays away from the two ends of the magnetic bottle. This confinement effect does not work for particles that are simply flying straight out along a field line instead of gyrating around it. In this sense, the mirror machine is a leakier magnetic bottle than the torus, which will confine a particle no matter in what direction it is moving.

As described above, the thermonuclear reaction process consists of energetic nuclei glancing off each other many times by electrical repulsion, until they happen to make a sufficiently close collision so that they fuse. The magnetic bottle is meant to be a kind of meeting place for nuclei that are fusible. The bottle allows these nuclei to go through their bouncing and fusing interactions without being lost or cooled.

Unfortunately, the process of bouncing helps particles to escape rather rapidly from a mirror machine. Two gyrating nuclei may collide in just such a way that they both escape straight out the ends. In a toroidal bottle, the collision process gives rise to a more gradual diffusion of particles *across* the magnetic field lines. In either case scientists have calculated that plasma confinement times will be long enough to meet the Lawson criterion and make reactor operation possible. This is known as the "classical" confinement prediction for fusion reactors.

The main obstacle to progress with fusion power has been that actual plasmas escape far more rapidly from magnetic bottles than one would have expected from the classical prediction. The history of experimental advance in controlled fusion research thus far has been dominated by the process of encountering, analyzing, and partially overcoming these unexpectedly large losses. It is only quite recently that scientists have begun to approach the classically predicted plasma confinement times.

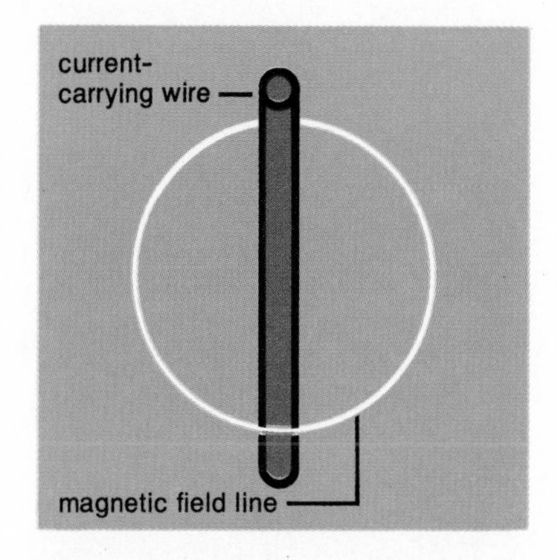


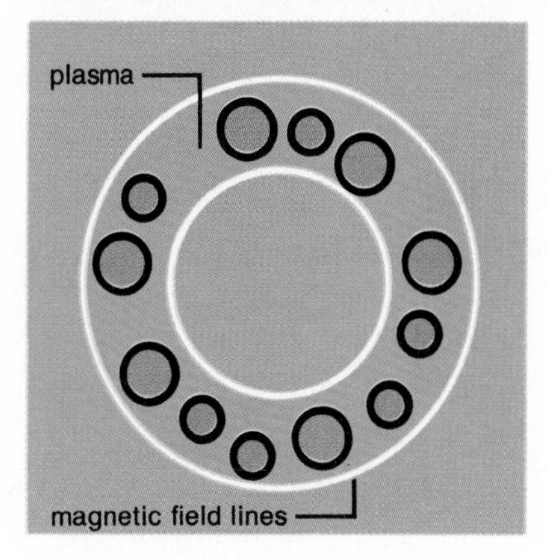


*Magnetic field lines surrounding a current-carrying wire (top) form closed circles. Toroidal magnetic bottles have a magnetic field pattern of this type as their main ingredient (above). Charged particles must gyrate locally around a magnetic field line (opposite) or travel freely along the line.*

### Mirror confinement experiments

All hot-plasma experiments are conducted inside high-vacuum systems. Mirror-type confinement must be designed to minimize particle collisions, which are most frequent at low plasma temperature and high density; accordingly, the build-up of a hot, dense plasma in the vacuum must proceed mainly by raising the density of an initially hot, dilute plasma rather than by raising the temperature of an initially dense, cold plasma. In order to reduce collisions to the lowest possible level, open-ended reactors are ultimately expected to operate at temperatures above 1,000,000,000°—rather than the mere 100,000,000° required in a closed reactor.

The favored technical approach to plasma generation in open-ended fusion reactors is the injection of beams consisting of energetic ions or neutral (uncharged) atoms produced by externally situated high-current ion accelerators. A 100-kilovolt (kV) accelerator voltage corresponds to a 1,000,000,000° plasma. Initially, only a very small fraction of the injected beam is trapped in the plasma; as the plasma density rises, however, a stage is reached where the beam interacts mainly with previously trapped beam particles and the density build-up then grows exponentially.

The plasma densities and confinement times achieved experimentally in open-ended systems have been limited primarily by unexpectedly large plasma losses due to instabilities. There are two principal classes of plasma instability: magnetohydrodynamic (MHD) modes, in which the plasma becomes grossly deformed or displaced; and microinstabilities, in which high-frequency electrical disturbances alter the velocity distribution of the plasma particles.

The most important MHD instability of open-ended systems is the so-called flute mode, in which a tube of plasma (or the entire plasma) is displaced into a region of reduced magnetic field. The equilibrium of the plasma on top of its "magnetic hill" is like that of a mechanical roller on top of a real hill: an infinitesimal displacement of the roller in either direction will increase exponentially. Such an equilibrium is called unstable. It can be shown that plasma elements lying along the same magnetic field line tend to move in unison, thereby accounting for the flutelike structure of the perturbations seen in the illustration on the following page.

*A magnetic bottle of the mirror type has nonclosed magnetic field lines that converge toward each end of the bottle (bottom), corresponding to an increase in field strength. A gyrating charged particle, which resembles a diamagnetic compass needle (below), is repelled by the ends of the bottle as the needle is by the ends of a C-magnet.*

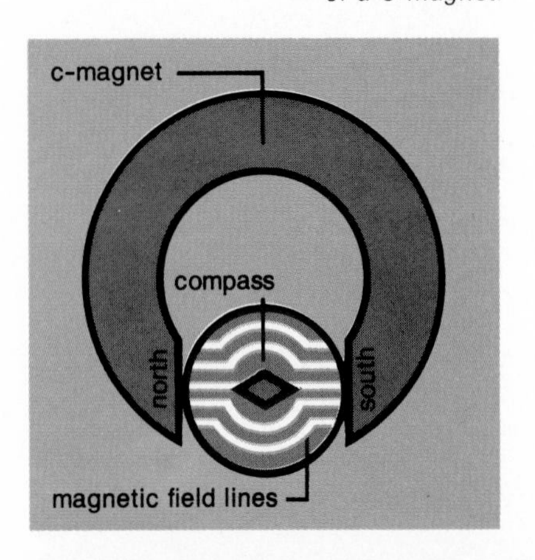


Early mirror machine experiments all encountered various forms of the flute instability. A number of flute-stabilizing techniques were found, but the most important breakthrough came in 1961, with the announcement of the first "magnetic well" experiment, carried out at the Kurchatov Institute in Moscow. As had been noted for many years by theorists in the United States and the Soviet Union, there exist special kinds of mirror configurations in which the magnetic field strength increases in *all* directions away from the plasma; a plasma confined in such a magnetic well would thus be expected to be absolutely stable against MHD modes. The Kurchatov experiment was able to operate alternately in the simple mirror machine configuration and in a magnetic well configuration. A dramatic flute-stabilization effect and increase of plasma lifetime was observed in the latter case. The typical plasma values of the PR-5, a more advanced embodiment of the Kurchatov experiment, were: ion temperature 50,000,000°, electron temperature 200,000°, density $10^{10}$ particles per cubic centimeter. The plasma lifetime was 3 milliseconds in the stable configuration and only 0.1 milliseconds in the unstable configuration.

Following the PR-5 results, the build-up of hot-ion plasmas in magnetic wells was pursued vigorously, mainly by injection and trapping of energetic neutral atom beams. The Baseball I experiment at the Lawrence Livermore Laboratory, Livermore, Calif., produced 100,000,000°

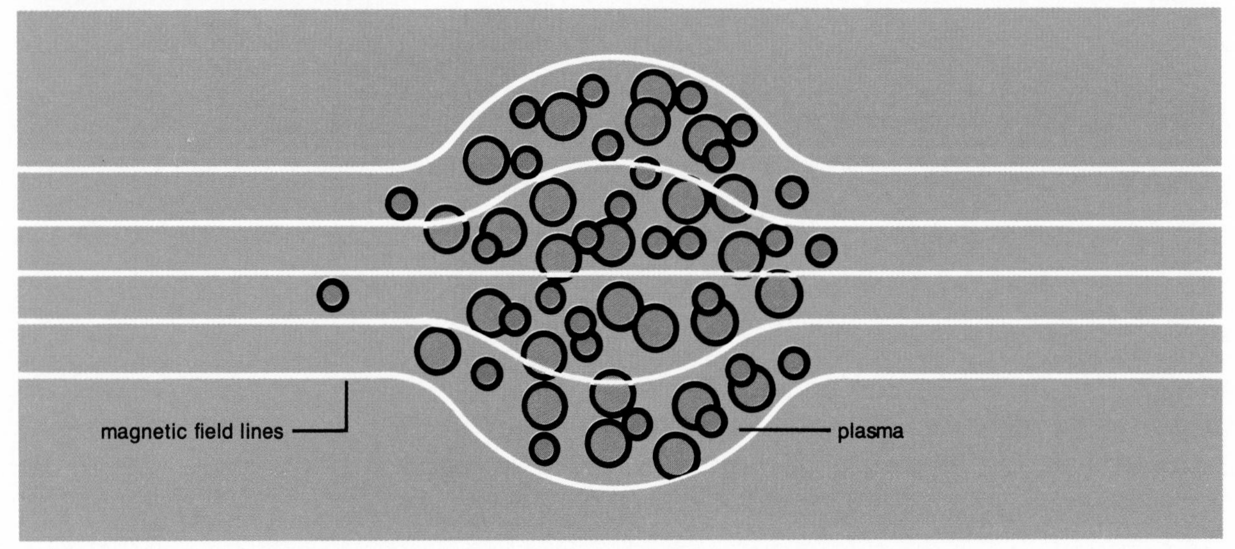

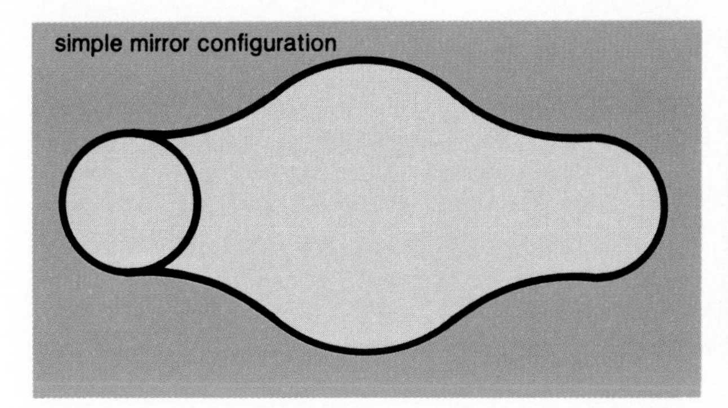


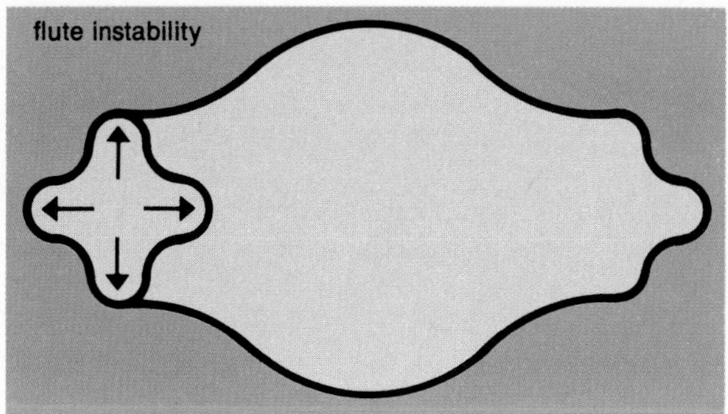


*In a simple, axially symmetric mirror machine, the magnetic field strength increases toward the two ends of the magnetic bottle but decreases radially outward. Plasma, being diamagnetic, therefore seeks to move radially outward. It can do so either as a whole (above, left) or by means of flutelike deformations (above, right).*

plasmas with densities up to $10^9$ particles per cubic centimeter and confinement times of about a second. The somewhat similar Phoenix II experiment at the Culham Laboratory in England reached densities of nearly $10^{10}$ particles per cubic centimeter, and was the first to achieve incipient exponential build-up effects. These and other magnetic well experiments confirmed the complete removal of MHD instabilities as a limiting factor in mirror confinement. Experimental mirror machine plasmas, however, were not yet able to realize the ideal process of exponential build-up toward very high, collision-limited densities. Instead, the plasma lifetimes and attainable densities were found to be limited by the onset of microinstabilities.

Microinstabilities in a mirror machine take the form of high-frequency electrical waves. These waves are generated in somewhat the same manner in which light waves are generated in a laser. To set the stage for laser action, one needs to create an abnormal distribution of excited atomic states: these can then relax in unison and drive a giant light pulse. Because of the leaky ends of the open-ended magnetic bottle, its plasma always has an "abnormal" distribution of particle velocities. Large-amplitude plasma waves can arise spontaneously and help to knock plasma particles out the ends of the bottle.

The severity of the end-losses depends on how "abnormal" the particle velocity distribution is. The worst distributions are produced by injecting a single beam of energetic particles, all with the same velocity and moving in the same direction. In the best distributions the particles are moving almost randomly in all directions, with a wide range of velocities.

In early mirror machine injection experiments, which used single beams with particles all at the same energy, severe microinstabilities and plasma losses were indeed observed, both in simple mirror geometry and in magnetic wells. In the Baseball I and Phoenix II experiments, new techniques were introduced to make the particle velocity distributions more nearly random; these techniques were shown to produce favorable effects on plasma stability. In the present generation of mirror injection devices, the favored plasma build-up plan is to maintain a nearly random particle distribution from the beginning. Calculations indicate that in this way it may be possible to realize stable conditions at plasma densities much higher than previously reached by the injection technique. Confidence in these theoretical calculations is supported by the results of the 2X magnetic compression experiment at the Lawrence Livermore Laboratory. A nearly random plasma of 80,000,000° ion temperature and $3 \times 10^{13}$ per cubic centimeter density was produced by compression of a lower-density, lower-temperature plasma in a fast-rising magnetic field of the magnetic well type. Evidence of microinstability was observed, but the typical plasma loss time of 0.3 milliseconds was found to be only about three times faster than the rate predicted for classical scattering loss.

## Toroidal confinement experiments

In a closed magnetic bottle, where plasmas are not lost rapidly by classical scattering, the favored method of build-up toward a hot, dense plasma is by raising the temperature of an initially cold, dense plasma. The plasma, with its free electrons, is an electrical conductor, and so it can be heated by inducing a current to pass through it, usually the long way around the torus. Plasma resistance results from the collisions of the current-carrying electrons with stationary ions; the power dissipation associated with the collisions causes heating of the electrons. This is the same process of "resistive" or "Ohmic" heating that is employed in an electric toaster. The heating method is effective up to electron temperatures of about 10,000,000°, at which point a hydrogen plasma has about as low a resistivity as standard copper. At still higher temperatures, the plasma resistivity becomes so low that resistive heating is too slow to be effective. Auxiliary heating methods,

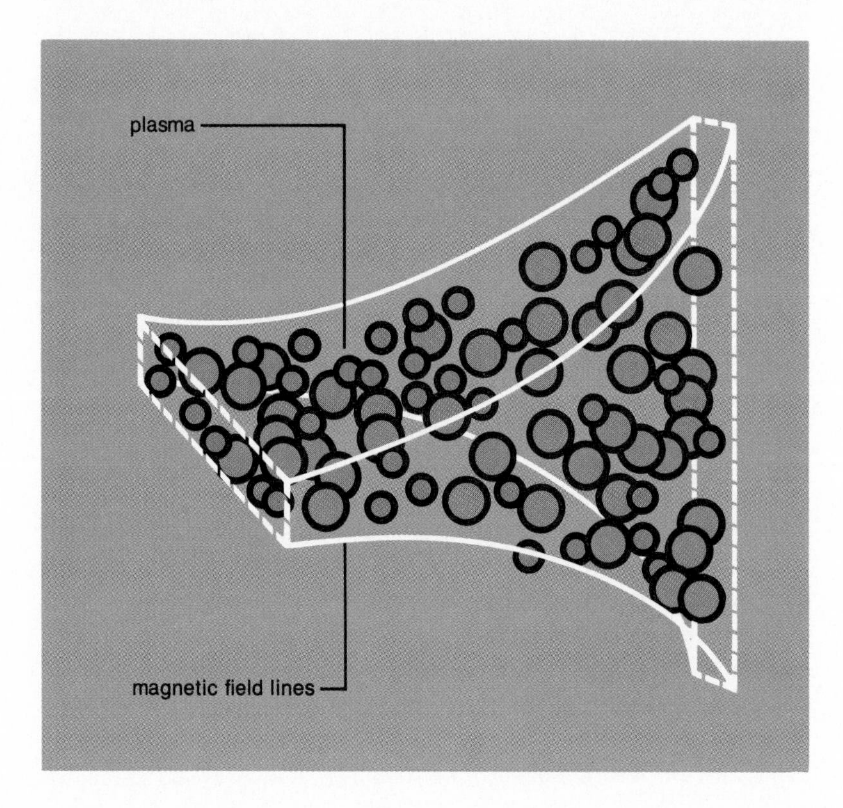


*Advanced mirror machine of the magnetic well type has a magnetic field that increases in strength in all directions away from plasma. The plasma is stable against MHD instabilities.*

therefore, must be used. Plasma compression and the injection of energetic neutral beams are considered the two most promising methods for advancing from the resistive-heating stage to the ignition of a thermonuclear reaction.

The magnetic field of a toroidal magnetic bottle must have a rather special structure in order to hold a plasma ring in equilibrium. A simple toroidal magnetic field, like that surrounding a straight wire, is not good enough: the plasma ring would expand continuously along its major radius (seeking to move in the direction of decreasing magnetic field strength). In order to provide a proper equilibrium, the toroidal magnetic field lines must be twisted into the shape of a helix. In the tokamak device, invented in the Soviet Union, this twist is provided by inducing a toroidally directed current around the plasma ring. This current also serves to heat the plasma by the electric toaster method. An alternative approach, invented at the Princeton University Plasma Physics Laboratory, is the stellarator. In this device the helical twist of the magnetic field is caused by currents in helical windings outside the plasma ring. Still other toroidal magnetic bottles use currents flowing in solid rings that are inside a hollow plasma ring, or employ circulating beams of high-energy electrons. The tokamak and stellarator, however, are generally considered the most practical approaches to a first-generation toroidal fusion reactor.

How do toruses fare in respect to MHD stability of the plasma? As in mirror machines there is a tendency toward flute instability. Even worse is the MHD "kink" mode, in which the whole plasma column buckles

Courtesy, Plasma Physics Laboratory, Princeton University

*The ST tokamak at the Princeton University Plasma Physics Laboratory has been one of the most successful toroidal confinement machines. Electron temperature is measured by the scattering of ruby laser light from the plasma.*

Sovfoto

*The T-3 tokamak at the Kurchatov Institute in Moscow provided the first breakthrough to high temperatures and long confinement times in toroidal machines. Its magnetic field strength is 35,000 gauss, and currents up to 200,000 amperes are induced in the plasma.*

in the form of a helix. The kink mode occurs when the plasma current of a torus is made too large. Stability against the kink mode is one of the advantages of the tokamak relative to the "pinch," a similar toroidal magnetic bottle with a much larger plasma current.

Experimental research directed toward a toroidal reactor began mainly with devices of the pinch type, during the early 1950s. The "pinch effect" is the name given to the magnetic self-constriction of conductors through which large currents are passed. (Evidence of pinching can sometimes be seen even on solid metal lightning rods that have been hit by large strokes.) High-current pulses of short duration, typically 100,000 amperes for several microseconds, were passed through straight plasma columns between two electrodes; or through toroidal plasmas by transformer induction the long way around the torus. In this way thin plasma columns at about 1,000,000° with densities of $10^{16}$ per cubic centimeter or more were readily produced; but these plasmas were highly unstable against kink and flute modes—just as predicted by MHD theory—and were lost typically in less than a microsecond. Following the prescriptions of the stability theory, a longitudinal (toroidal) magnetic field component was then added to stabilize the grossest MHD modes. This approach proved successful, notably in the large toroidal Zeta device at the Atomic Energy Research Establishment, Harwell, Eng., where plasma confinement times of about a millisecond were achieved at temperatures of 1,000,000°–

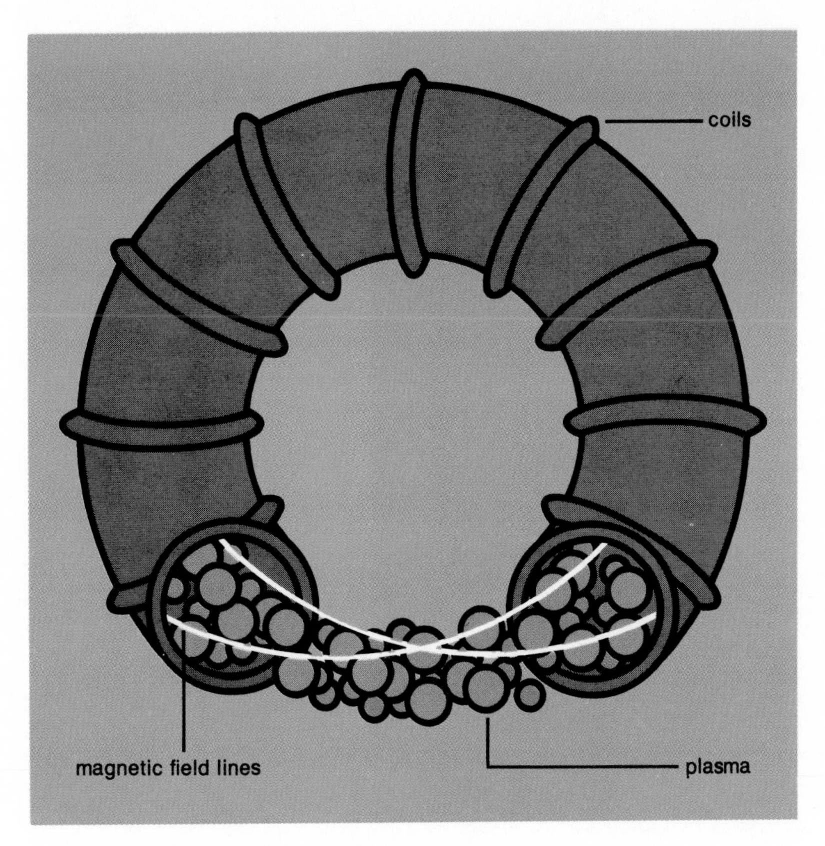


*A tokamak uses a toroidal magnetic field generated by external coils. The field lines are twisted somewhat by a poloidal magnetic field component generated by a plasma current that circulates the long way around the torus.*

2,000,000° and densities of $10^{14}$ per cubic centimeter. Small-scale MHD modes and other loss mechanisms, however, dominated the confinement time and presented an effective barrier to the achievement of higher temperatures.

During the 1950s grossly MHD-stable plasma confinement in tokamaks and stellarators was also demonstrated at the Kurchatov Institute and at the Princeton Plasma Physics Laboratory, respectively. Plasmas with temperatures of 100,000°–1,000,000° and densities of $10^{13}$–$10^{14}$ per cubic centimeter were produced by means of resistive heating currents. As in Zeta, unstable plasma fluctuations were found to persist in grossly MHD-stable regimes, and plasma confinement times were unexpectedly short. After several years of studying confinement times at the large Model C stellarator at Princeton, researchers found that all the data could be fitted, not by the predictions of classical theory but by a far more pessimistic empirical formula predicted in the late 1940s by David Bohm, on the basis of low-temperature plasma studies. It then became apparent that, if the Bohm formula continued to hold, there was no possibility of building a toroidal fusion reactor.

Later investigations revealed that, as in the case of mirror machines, the shortened confinement times resulted from the loss of plasma due to microinstabilities. In toroidal plasmas these electrical disturbances serve to transport plasma across the magnetic field rather than knocking it out the ends, as they do in a mirror machine. An important distinctive feature turned out to be that microinstabilities in toruses are most severe when the plasma particles collide very often—that is to say, in very cold, dense plasmas. On this basis, it was theoretically predicted about 1965 that higher-temperature plasmas should show much longer confinement than that given by the Bohm formula—if only one could ever break through to higher temperatures.

The greatly encouraging experimental development in the period 1965–70 was that this prediction was, in fact, correct. The most exciting results were obtained on the T-3 tokamak at the Kurchatov Institute, a device with a plasma sufficiently large to permit electron temperatures above 10,000,000° to be reached by Ohmic heating when the plasma was at a density of $10^{13}$–$10^{14}$ per cubic centimeter. Confinement times of about 20 milliseconds obtained at this record temperature were approximately 30 times better than the Bohm value, and were still rising with temperature in approximate agreement with classical diffusion theory. These remarkable results were repeated and confirmed on the ST tokamak at Princeton, a modification of Model C.

At present, larger tokamaks and stellarators are being constructed in many countries, including the Soviet Union, the United States, and Great Britain, to take advantage of the improved behavior of high-temperature plasmas. The tokamak is a relatively simple and inexpensive device, and is therefore favored for the initial approach toward developing toroidal reactors. The stellarator, on the other hand, has possible long-range advantages, such as for the confinement of high-pressure plasmas.

An important remaining technical problem is to demonstrate the heating of the plasma ions from their present typical temperature of about 5,000,000° to the required temperature of 100,000,000°. Plasma confinement times also still need to be raised to about a second, but scientists expect that this approximately hundredfold increase would be contributed automatically by the roughly tenfold increase in plasma radius when going from present experiments to the toroidal reactor regime. This is because plasma leakage from a torus takes place by diffusion, and so the loss times increase characteristically with the square of the plasma size.

### High-density plasma experiments

High plasma density is advantageous in allowing appreciable fusion reactions to be produced in relatively small and simple experiments. In the "low-density" mirror and toroidal experiments described in the preceding sections, the thermonuclear neutron yields obtained from deuterium plasmas seldom exceeded $10^6$ neutrons during any one plasma confinement time. From experiments with very dense plasmas, much larger neutron yields were obtained, even though the plasma confinement times were in the microsecond range, or below.

Initial high-density experiments were conducted during the 1950s with plasma pinches (as mentioned above). A pulsed current was passed axially between two electrodes, and the resultant azimuthal

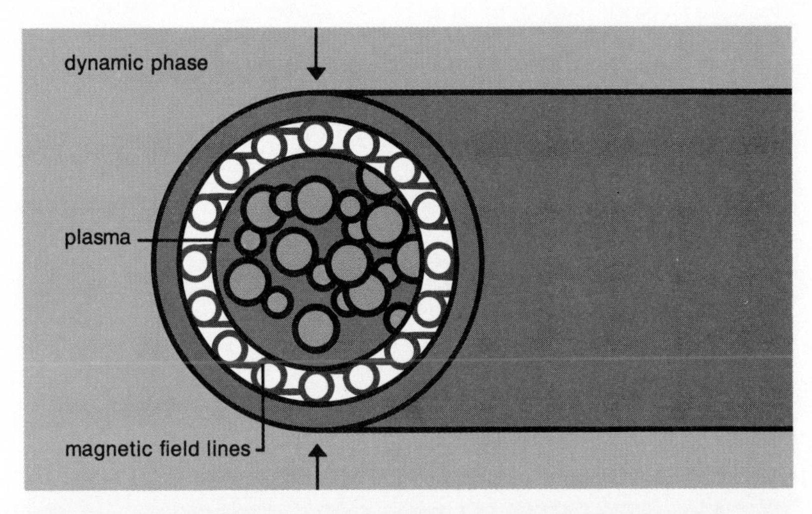


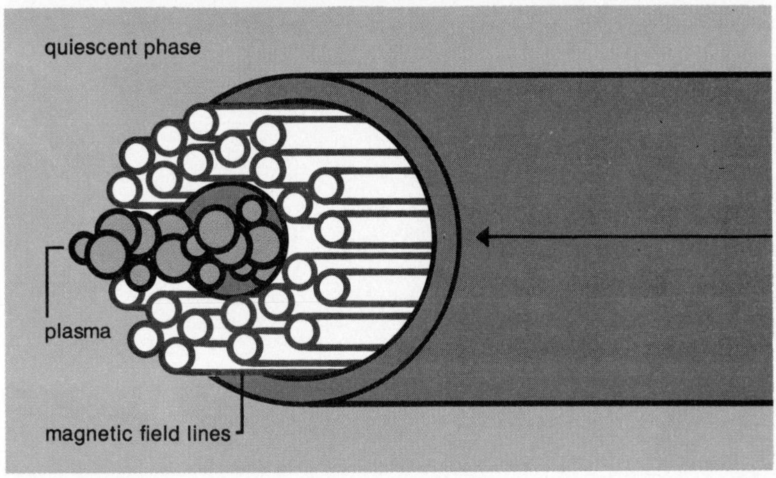


*In the linear theta pinch a strong axial magnetic field is pulsed on, compressing the plasma dynamically. The shock-heated and compressed plasma then escapes out the ends.*

magnetic field caused dynamic self-constriction—"pinching"—of the current-carrying plasma. Temperatures of a few million degrees were generated by the shock-wave and compression associated with the pinching. During the 1960s the axial, or z-pinch, technique was refined in the form of the "plasma focus" by groups at the Kurchatov Institute, the Los Alamos Scientific Laboratory, and elsewhere. Their work made it possible to produce a constriction a few millimeters in size in a z-pinch current channel in which plasmas of $10^{19}$ per cubic centimeter densities are heated to some tens of millions of degrees temperature during times of a tenth of a microsecond. The neutron yields achieved by this method in deuterium plasmas exceeded $10^{11}$ per discharge. The plasma focus constitutes a neutron source of significant practical interest, but it could not be scaled up straightforwardly into an economical fusion reactor.

A second line of development starting from the original z-pinch is the theta pinch, in which a strong azimuthal plasma current is induced on the surface of a plasma cylinder by means of a pulsed axial magnetic field. In the Scylla I-IV experimental series at Los Alamos, N.M., plasma cylinders several meters long and about a centimeter in radius

Courtesy, Lawrence Livermore Laboratory

*Possibility of direct conversion of plasma kinetic energy to electric energy is demonstrated with 20,000 eV ion beam. In experiments at Lawrence Livermore Laboratory, a conversion efficiency of 96% was achieved. This technique is especially useful in recapturing the energy of end-loss particles in mirror machines.*

were heated by shock waves and magnetic compression to temperatures as high as 50,000,000° at densities of $10^{16}$–$10^{17}$ per cubic centimeter. The plasma pressure in these straight magnetic bottles is almost as large as the pressure of the 100,000-gauss magnetic fields. Neutron yields of $10^9$ per discharge were typically obtained from deuterium plasmas in these experiments. Similar tests were carried out in West Germany and the United Kingdom. The plasma confinement times were limited typically to a few microseconds by the free expansion of the plasma in the direction along the magnetic field. (As a result of the high plasma density, the scattering rate is too high to permit long-time mirror-type confinement.) During the available confinement time there has been no sign of microinstability or anomalous diffusion. The theta-pinch approach could be applied to a reactor in several ways. A straight plasma column some kilometers in length would have a sufficiently long plasma end-loss time to allow the production of power. A more practical approach, pursued under the name Scyllac at Los Alamos, was to bend the linear theta-pinch into a stellarator-like torus of large major radius, thus eliminating the plasma end-loss altogether. The main drawback to this method is that the curved magnetic field of a torus tends to induce MHD instability in high-pressure plasmas.

The highest densities of hot plasma were being obtained not with pulsed magnetic fields but by heating solid or liquid matter with highly focused external energy sources. High-powered lasers were particularly useful for this purpose. A single neodymium glass laser system could in less than $10^{-9}$ seconds produce an infrared light pulse of about 1,000 joules focused on a millimeter target. When a solid pellet (of about $10^{23}$ per cubic centimeter density) is used as the target, the resultant dense plasma absorbs this incident light pulse with high efficiency within a submillimeter surface layer. The resultant plasma has far too high a pressure to be confined by technologically feasible magnetic bottles (at field strengths below a megagauss) and thus expands at thermal velocity. Fusion-oriented research on laser heating is being carried on vigorously in the United States, the Soviet Union, and France. Neutron yields exceeding $10^5$ were reported from deuterium targets. If the target can be precompressed by shock waves to densities higher than those of normal solids, the laser-heating task is evidently facilitated. The technical problems of adapting the laser system to the large-scale production of economical fusion power, however, are considered to be inherently severe.

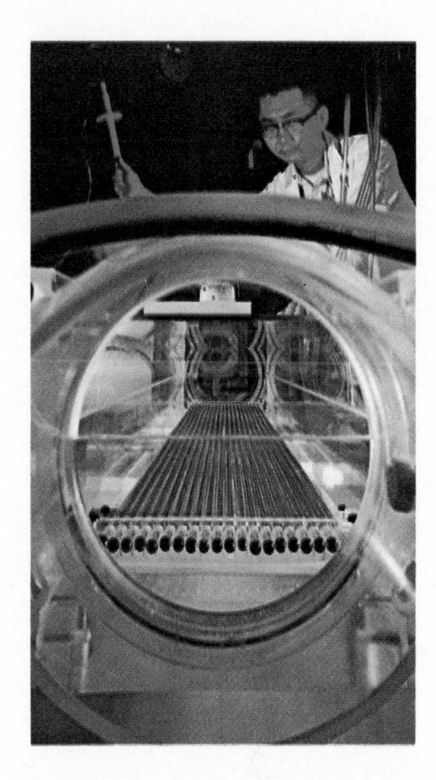

*Carbon dioxide laser is used in fusion research at the Lawrence Livermore Laboratory. The laser has a maximum energy of 100 joules.*

## Engineering aspects

Since substantial experimental progress toward the attainment of the Lawson criterion is being achieved along a number of technologically quite different routes, the fusion reactor designs now being envisaged and the practical functions toward which these designs are aimed are correspondingly diverse. The moderate-pressure toroidal deuterium-tritium reactor will serve here as a representative example of the engineering problems involved.

The deuterium-tritium reaction produces a neutron of 14.1 MeV energy and an alpha particle (helium 4 ion) of 3.5 MeV. The tritium consumed in a fusion reactor must be regenerated, since tritium, unlike deuterium, occurs naturally only in minute amounts. (It decays with a half-life of 12.4 years.) The 14.1 MeV neutron is stopped in a blanket surrounding the plasma volume; the blanket is made up largely of lithium, in which the incident neutrons breed tritium by nuclear reactions. A blanket of liquid lithium or molten lithium salts is suitable for the breeding purpose as well as for the removal of the heat generated by the energetic neutrons and their secondary reactions. Breeding ratios of 1.3 tritium atoms per incident neutron are readily obtained. The generated heat can be removed from the blanket at a temperature in the range 500°–1,000° and used to drive a conventional thermal power plant. The lithium blanket should be about two meters in thickness to prevent any appreciable neutron flux from reaching the magnet coils, which surround the blanket.

The first fusion reactor plans, proposed during the early 1950s, had to divert a substantial fraction of the electrical power output to energize

Courtesy, Gulf General Atomic

*The Doublet II device at Gulf General Atomic is a tokamak with a vertically elongated cross section. This is designed to permit stable confinement of very high-pressure tokamak plasmas.*

the magnet coil system of the magnetic bottle. However, the subsequent development of superconducting magnets with high field strengths conveniently removed this requirement. Magnetic field strengths of 50–150 kilogauss produced by superconducting coils are generally envisaged in present reactor designs. A large fraction of the fusion reactor capital cost is connected with the magnetic coil system and the blanket. In reactors designed to operate most economically the minor radius of the plasma should be at least comparable with the blanket thickness; that is, about two meters. The corresponding major radius is 5–10 meters. Such a reactor, given the plasma size, the maximum available magnetic field strength, and the maximum possible ratio of plasma pressure to magnetic field pressure (typically 0.05–0.10, for MHD stability), would provide a total electrical power output in the range of 1–3 kilomegawatts, enough to sustain a population of about two million. This is a practical size for modern power stations. The neutron flux through the vacuum chamber surrounding the plasma must not exceed about 10 megawatts per square meter of wall in order to avoid excessive rates of radiation damage.

In order to start up a fusion reactor, a plasma heating energy investment of about a hundred megajoules is required. Thereafter, the plasma "ignites": its temperature is maintained or increased further by the 3.5 MeV alpha particles of the deuterium-tritium reaction, which are trapped in the magnetic bottle. The principal process of plasma energy loss is expected to be the diffusion of hot particles. The earliest reactors will probably operate in a slow pulsed mode, burning the fuel as it leaks out and then beginning a new plasma heating cycle. The problem of true steady-state operation with continuous fuel injection appears more difficult, but can be solved in principle.

The basic fuel, deuterium, can be extracted from water at a price less than a thousandth of the present-day cost of electrical power. The dominant cost of fusion power will be due to the capital investment in the reactor, and is estimated to be competitive with the cost of both conventional and nuclear fission power. The earth's supply of deuterium would be sufficient for more than $10^{13}$ years at the present world rate of energy consumption. The known and probable land reserves of lithium would suffice for only about $10^{4}$ years at this rate, but 1,000 times more lithium could be extracted from sea water.

The environmental aspects of fusion power are favorable. The amount of fuel present in the plasma is only a few grams, and there is no way in which a nuclear explosion could occur. The only radioactive element or product of the fuel cycle is tritium, which is not a waste product but a valuable fuel to be recycled. Activated structural materials, particularly the vacuum wall of the plasma chamber, are expected to constitute the only radioactive waste. The overall biological hazard potential is estimated to be sufficiently low so that in the long run it may be possible to locate fusion reactors close to population centers and utilize the waste heat from the thermal plant.

One of the most attractive features of fusion power is that even its

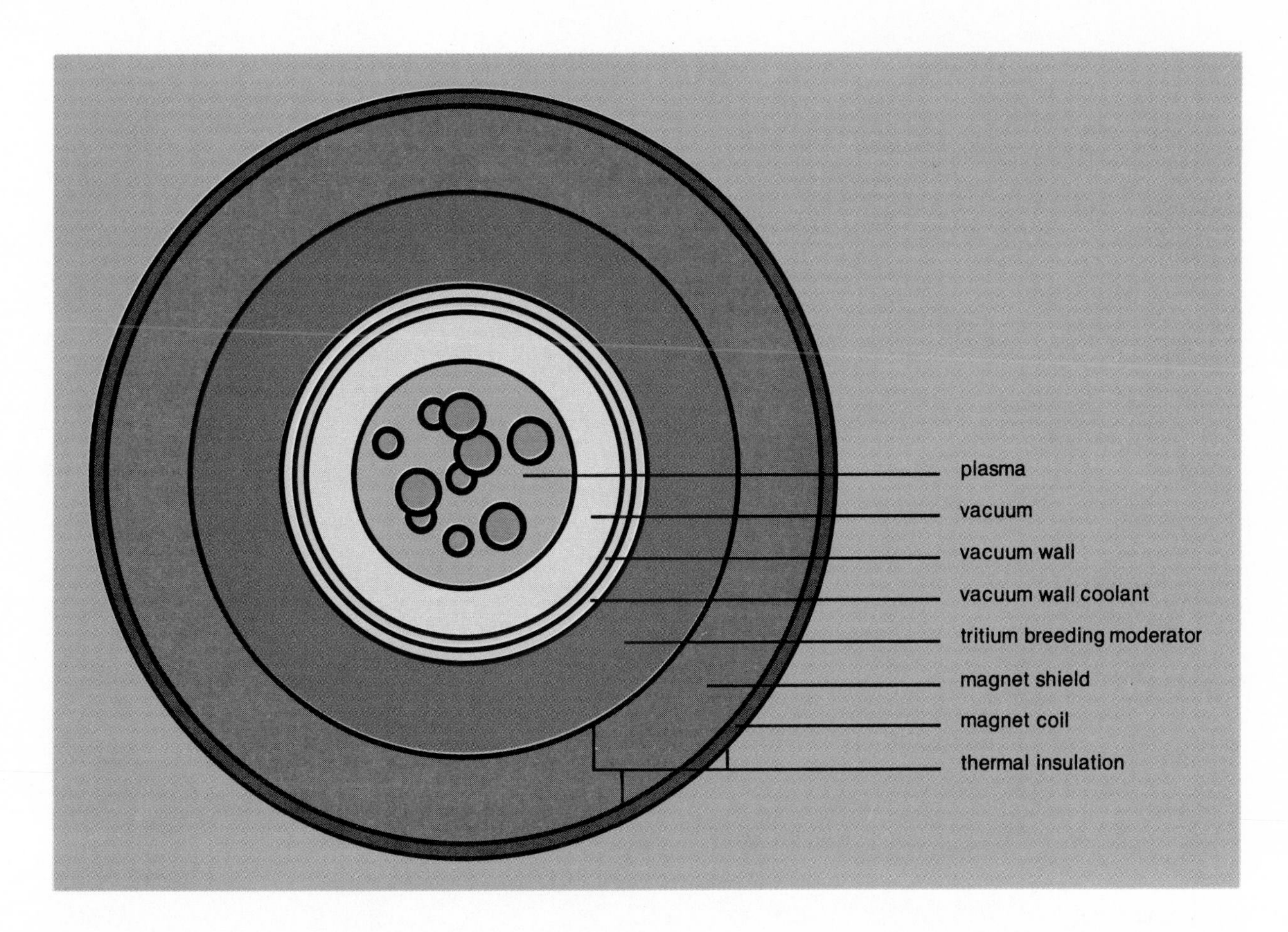


*Schematic representation of the minor cross section of a toroidal fusion reactor. The typical minor diameter of the magnet coil is 5–10 meters.*

remaining environmental defects are all of a kind that can be improved upon by future technology. With advances in fusion reactor design and size, purely deuterium-burning reactors should come into the range of economic usefulness. The problems of lithium consumption, tritium handling, and radioactive waste disposal would then be further reduced. Even the problem of waste heat becomes tractable, since a large fraction of the reactor power can be obtained in the form of charged reaction products and converted directly into electrical power.

# Nutrition and the American Diet

## by David Perlman

**A "national disaster . . . something I would not feed to my cat or dog " was the way a leading nutritionist described the diet of the average American in 1972. A possible sign of change was the growing enthusiasm for organic health foods.**

No subject in science generates more widespread popular concern than human nutrition. This is true not only in the poor, less developed countries of Asia and Africa but also in the United States and Western Europe. Among Americans, pockets of poverty produce genuine hunger and illness, infant mortality, and progressively disabling gaps in school performance. The more affluent struggle against obesity and worry about the insatiable appetites of youngsters who consume "empty calorie" snacks by the ton while ignoring all rational advice about vitamins, minerals, and adequate protein. The counter-culture fixes on the most bizarre of diets: fruit, fasting, or the brown rice of macrobiotics. At the same time, the businessman flirts with the "drinking man's diet" or battles his cholesterol level with safflower margarine, while the housewife curbs her appetite with amphetamines and "nourishes" herself with vitamin pills.

All of us put food into our bodies, and to many it becomes a mystique, a symbol, a cure, a preoccupation. Only rarely is it a guilt-free pleasure.

Science has made great strides in understanding the role that nutrients can play in keeping man healthy. The laboratory has produced valuable enrichments and fortifications for major items in the daily diet, as well as processed snack foods and advertised edibles that offer little nutrition at excessive cost. The laboratory can produce an imitation whipped cream with no food value at all, and textured soy proteins so valuable they may work a revolution in feeding the hungry. Agricultural science breeds steers with meat so well-marbled it is unhealthily rich in heart-endangering fats, but the same science breeds new high-lysine corn strains that will add critically important amino acids to meager peasant diets abroad.

A growing consumer movement in the United States demands stricter regulation of food additives, more explicit labeling of ingredients, and improved nutrition values in the marketplace. Yet millions of Americans eat fewer and fewer home-prepared "balanced" meals, while sales of pizzas, hamburgers, fried chicken, and fish-and-chips soar in the more than 30,000 franchised "fast food" outlets that crowd U.S. highways and shopping centers.

***DAVID PERLMAN,*** *science editor for the* San Francisco Chronicle, *has been president of the National Association of Science Writers and is a member of the Council for the Advancement of Science Writing.*

## Malnutrition in the United States

America's food consumption is more abundant and more varied than that of any other nation, yet malnutrition is surprisingly widespread. Hunger and serious dietary deficiencies among the poor have made headlines in recent years; less well-known are the nutritional inadequacies of the nonpoor.

The U.S. Department of Health, Education, and Welfare, for example, completed in 1971 a detailed nutrition survey in 30,000 American homes of all income levels in 10 states. While nearly 8% of the families with incomes below the poverty level showed undesirably low intakes of two or more important nutrients, 2.5% of the nonpoor families had similar inadequate diets. Deficiencies in iron, vitamin A, vitamin C, proteins, and calories were common. Approximately 25% of the poor and 12% of the nonpoor families were not receiving enough iron; vitamin A was low in 8.5% of the poor and 7.5% of the nonpoor; and vitamin C was inadequate in 7.2% of the poor and 4.3% of the nonpoor.

In a report to the U.S. Congress, the National Center for Disease Control reported even more startling figures from individual states. About 21% of the low-income families it surveyed in Texas—whether below or above the recognized poverty level—had diets that provided less than half the officially recommended calories. In New York 13% of the poverty families and 9% of the above-poverty families were equally undernourished. In Michigan 17% of the poverty group were eating less than half the recommended calorie allowance, and 10% of the above-poverty group had the same inadequate diet. Protein deficiencies were also striking: 10% of all the Texas families were consuming less than half the protein they needed. In New York severe protein deficiency was found in the diet of 11% of the poverty families, although less than 3% of the somewhat more affluent families showed the same protein deficiency. In Michigan the difference was much smaller: 5.9% of the poverty households were eating less than half the protein they needed, and 5.3% of the above-poverty families were similarly malnourished.

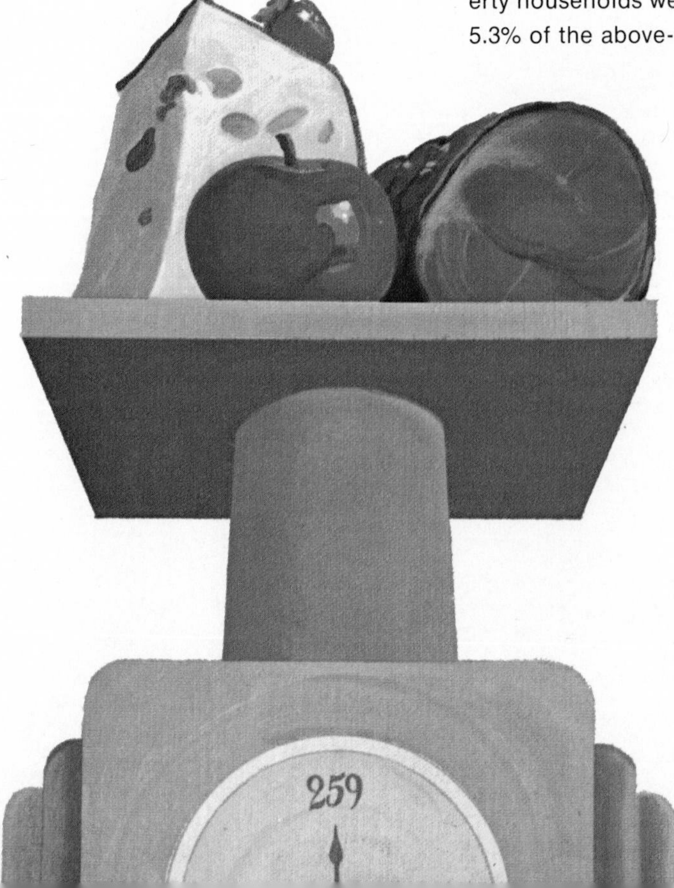

George M. Briggs, a professor at the University of California's department of nutritional sciences at Berkeley, has called the American diet a "national disaster," and insists that its deficiencies cost the public at least $30 billion a year in medical bills. "The costs of malnutrition are far greater than the cost of crime in our country," he says.

Briggs recently computed the quantities of foods that average Americans eat each year. Calculating them on a "dry weight" basis—to eliminate the essential but nutritionally valueless water content of various foods—he found this evidence for distorted nutrition:

Americans eat an average of 259 pounds per person per year of fruits, vegetables, whole cereals, meats, dairy products, and legumes, while at the same time, he said, the average diet also contains 276 pounds of sugar, white flour, fats, and polished rice. "This is a terrible diet," Briggs said. "I wouldn't feed it to my cat or dog, let alone a farm animal."

### What is malnutrition?

The Council on Foods and Nutrition of the American Medical Association (AMA) defines malnutrition as "a state of impaired functional ability or deficient structural integrity or development brought about by a discrepancy between the supply to the body tissues of essential nutrients and calories and the specific biologic demand for them."

This definition includes both bodily damage and improper intake of nutrients. It is difficult, however, to determine when an individual is suffering from the effects of poor nourishment. Many deficiencies do not demonstrate their consequences for years; in some the symptoms are obscure and manifest themselves primarily in poor resistance to disease or infection, or in delayed wound-healing.

"Malnutrition, because it can have so many manifestations, is insidious," says the AMA. "Depending on how it is defined, it can be found under many situations. Classical malnutrition exemplified by vitamin deficiency diseases is but a small segment of the spectrum. The less dramatic manifestations of malnutrition—growth retardation,

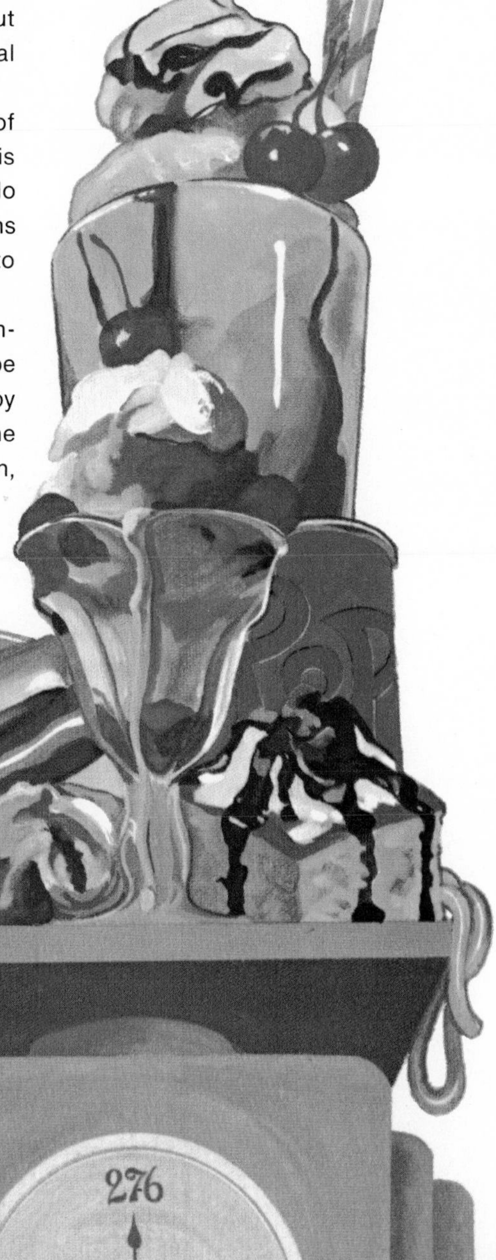

weight loss, increased burden of chronic diseases, depression, weakness, retarded convalescence from disease and trauma, poor performance in pregnancy—are widespread and of great significance. Infections in the undernourished are more devastating, persist longer, and result in a much higher death rate in malnourished children."

Since the White House Conference on Food, Nutrition, and Health in 1969, the U.S. government has moved more aggressively into the dietary picture. The Department of Agriculture distributed 1.3 billion pounds of surplus food to 3.6 million American poor people in 1971, and by early 1972, 11.2 million people were receiving government food stamps to increase their purchasing power. The food distributed by the government was fortified and balanced to include all the recommended levels of protein, minerals, and vitamins. Also by early 1972 the federal school lunch program covered more than 25 million children, of whom 8.1 million received their lunches free or at reduced prices.

The U.S. Food and Drug Administration (FDA) proposed new labeling rules requiring clearer identification of the nutrient content of processed foods. The FDA also proposed that even frozen TV dinners be required to contain at least one-third of the recommended daily nutrient allowances if they were advertised as complete meals. Scientists in the laboratories of the FDA, the Department of Agriculture, and industry all experimented with fortifying more foods, creating new foods richer in essential nutrients, and developing new products from protein sources as disparate as soybeans, fishmeal, algae, and bacteria.

### Determining proper nutrition

What is proper nutrition? Almost any reasonably educated American thinks he knows the answer to this question, and many probably do. Yet the lures of suggestive advertising, the diet fads and phobias, and the confusing display of 7,800 different items for sale in the average supermarket all combine to distort eating habits among even the most affluent, and to warp wholesome nutrition patterns for everyone.

Fredrick J. Stare, the chairman of Harvard University's department of nutrition, says, "The one and only 'magic formula' for healthful eating lies in the variety of foods eaten and in not eating too much." Stare and his colleagues listed four basic food groups:

1. Dairy products—milk, cheese, ice cream, yogurt, and other foods made from milk.
2. Meat and other high-protein sources such as fish, poultry, eggs, cheese, dried beans and peas, and nuts.
3. Cereal products—enriched and whole-grain cereals of varying types.
4. Vegetables and fruits.

These four groups of foods, according to Stare, can provide all the nutrients known to be essential to man, plus a host of trace elements whose dietary role in tiny quantities is still being explored.

In the United States a group of scientists make up the Food and Nutrition Board, which is sponsored by the National Academy of Sciences-National Research Council and is responsible for setting basic standards for the public's daily nutrient needs. These standards, called the Recommended Dietary Allowances, or RDA, are published every five years. The newest RDA list, compiled under the chairmanship of Harvard's D. Mark Hegsted, was scheduled to be published in 1973. Like the current tables, it will include recommendations for men, women, infants, and children and for pregnant and nursing mothers.

The basic document lists appropriate allowances of calories and protein, and of the known essential vitamins and minerals. The vitamins include A, D, E, and C, and the B-complex group of folacin, niacin, riboflavin, thiamine, $B_6$, and $B_{12}$. The minerals in the RDAs include calcium, phosphorus, iodine, iron, and magnesium. Other minerals that must be present in trace amounts in the diet—millionths to thousandths of a gram per day—include chromium, selenium, fluoride, molybdenum, copper, zinc, and manganese. Evidence is growing that tin, nickel, and vanadium are also essential trace minerals.

Setting allowances for these nutrients is a difficult task, and scientists by no means agree on appropriate levels. For in nutrition research, animal experiments are almost the only way of developing precise information. A rat, or a chick, or even a monkey can be deprived of a given nutrient, and the effects of the deprivation may then be closely measured as a decrease in stature, a swift onset of deficiency disease, or an alteration in blood levels of the nutrient's metabolites. But the applicability of these results to man is uncertain, and it is obvious that most human experiments to test the damaging effects of severe deficiencies are too dangerous to carry out fully even in volunteers. Studies of children and adults in malnourished families—among the American poor or the peoples of countries where starvation is endemic—can yield valuable information, but it is rarely precise.

There is ample evidence, for example, that thousands of children in the less developed countries go blind because their diets lack vitamin A. But how much vitamin A can safely be added to an average American's diet before toxic effects occur? Vitamin D can certainly prevent and cure rickets, although it is not needed at all where people receive adequate sunlight. But at what level can overenthusiastic fortification of foods with vitamin D lead to hypercalcemia and kidney problems?

Vitamin C deficiency is known to cause scurvy, and Capt. James Cook fed lime juice to his ship's crew in the 18th century to prevent the disease. Today a daily glass of orange juice, rich in vitamin C, is part of the American way of life for many. Because scientists have learned that excess amounts of vitamin C are normally discarded by the body, there is ordinarily little danger of overdosing.

A recent issue involving vitamin C is its relationship to the common cold. Can vitamin C in doses as large as several thousand milligrams prevent or cure colds, as Nobel laureate Linus Pauling claims? University of Maryland experimenters gave large doses (three grams a day) of vitamin C to a group of prisoner-volunteers and then inoculated them with cold virus (rhinovirus 44). The experimenters found no difference at all between the protective effect of the vitamin C and the effect of placebos. Many other experiments also failed to prove Pauling's claims. "The . . . efficacy of vitamin C in the treatment of the common cold and other viral infections is still an open question," says Sheldon Margen, of the University of California's department of nutritional sciences at Berkeley. "However, I also believe that there is still inadequate evidence of its efficacy and safety and that—until further well-controlled, definitive studies are undertaken—advocating large doses of ascorbic acid [vitamin C] at this time is totally unwarranted."

## Organic foods

Of all the recent trends in American nutrition, the most striking is the upsurge of interest in bizarre diets and so-called organic foods. To the true believer, organic foods are those grown without chemical fertilizers and without the use of chemicals to kill insect pests and weeds. If the organic products are processed, their natural nutrients must not be depleted—white flour, for example, is shunned—nor can the products contain additives.

In this concern for purity and naturalness lies a deep suspicion that many of the achievements of contemporary science have added unknown hazards to the environment, and that more are being concealed. The suspicion is heightened by recurrent revelations: that chlorinated hydrocarbon pesticides can damage the eggshells of birds; that cyclamate sweeteners cause bladder tumors in rats; that monosodium glutamate, the widely used flavor-enhancer, causes brain damage in infant mice.

Some processed foods deepen this suspicion: William O. Caster, a researcher at the University of Georgia, tested the nutritional value of 38 commercial dry cereal products and found two brands that promoted excellent tissue growth when rats were fed exclusively on them. But he found 20 brands—including some of the most widely sold and promoted—that could not keep his rats alive at all, not even when he supplemented the packaged cereals with a complete mixture of vitamins and minerals.

When experiments like this become widely known, suspicion increases. A recent opinion survey commissioned by the food industry found widespread public distrust: 41% of the public expressed fears that the needs of food consumers were not being protected by government or industry in the marketplace.

So the preoccupation with organic and natural foods grows. And it is reinforced by deep psychological drives. Judd Marmor, clinical professor of psychiatry at the University of California at Los Angeles, says: "For all people, what goes into their mouths or into any part of their bodies is strongly associated with good and evil. 'Good' food is 'pure, clean, wholesome'; 'bad' food is 'impure, dirty, poisonous.' To eat well is to be secure, healthy, and happy. The converse means insecurity, starvation, and death."

The public's turn toward organic foods has been marked by a tremendous increase in the number of "health food" stores, and many supermarkets now carry lines of expensive organic products. Often the label "organic" means nothing at all: chemistry defines any matter containing carbon as organic, and this includes everything of plant and animal origin. Thus, it is possible to visit an apple cider plant, for example, and watch the two-pronged process: One batch of apple juice is filtered to a clear liquid and labeled as juice; another is unfiltered and labeled "organic." Both come from the same apple crop, from the same trees—fertilized, sprayed, yet wholesome nonetheless. But one costs far more than the other.

## Diets

Far more dangerous than the interest in organic foods is the surge of extraordinary diets that has swept the country. Most are designed to attack obesity, but some aim at more profound changes: inner peace, spiritual growth, the cure of disease. Most range from useless to dangerous.

In the pursuit of slimness, Americans spend $100 million a year on reducing pills, diet books, special get-thin foods, and exercise machines. Jean Mayer, one of the world's foremost authorities on nutrition and overweight says:

"The place for an obese person to obtain a reducing diet is the office of a physician . . . or a weight-reduction clinic which is applying sound nutritional principles under medical direction. What a good many people fail to recognize, however, is that no one particular diet is appropriate for everybody. Good but personally irrelevant diets and all quack diets would be forever ignored if each of us recognized the value of a varied diet, consisting of sufficient but not too much total content, tailored to individual needs."

Two of the most widely advertised reducing diets are the "Doctor's Quick Weight-Loss Diet" and the "Doctor's Quick Inches-Off Diet," both promoted by Irwin Stillman. The first emphasizes high protein and low carbohydrate intake, plus eight glasses of water a day. According to Philip L. White, director of the AMA's department of food and nutrition, the diet proposes a dangerously low level of carbohydrate consumption that can cause excessive protein breakdown and impair the maintenance of brain and nerve cells.

The "inches-off" diet involves cutting down drastically on protein, with the promise that the dieter can lose weight in specific parts of his body. "That's a complete fallacy," says White. "Even if it were true, this kind of diet would cause loss of muscle tissue as well as fat." And he adds: "In my judgment this diet can be extremely hazardous, because it is very low in protein. So low that it can in fact cause the dieter to lose protein from his own body, to use up muscle and organ tissue and generally weaken the body. Anyone who is already unhealthy because of organ damage—liver problems, kidney damage, heart disease, anemia—can be put in a very hazardous situation by further tissue depletion."

Nutritionists raise similar objections to most other diets that offer promises of dramatic weight reduction—banana diets, grapefruit diets, the "Air Force diet" that has nothing to do with the Air Force, and the "Mayo diet" that has nothing to do with the Mayo Clinic.

In a far different arena are the diets that promise to cure disease or enhance the spirit. By far the most dramatic of these is "Zen macrobiotics," a movement that calls itself "A return to natural living." The movement was founded by Georges Ohsawa, a Japanese philosopher and writer who died in 1966. Zen macrobiotics has little to do with Zen Buddhism—in fact, practicing Buddhists scoff at the diet and particularly at its claims of curing disease.

According to Ohsawa, modern medicine cannot cure cancer because doctors do not understand "the structure of the infinite universe." To Ohsawa "no illness is more simple to cure than cancer." His treatment: meditation plus a diet consisting of nothing but whole-grain cereals —unpolished brown rice, buckwheat, wheat, corn, barley, or millet. For those of lesser faith, Ohsawa suggests the application of compresses made from ginger and yucca or taro root. Similarly, Zen macrobiotics prescribes whole-grain cereals and sesame salt for both high blood pressure and low blood pressure, and whole-grain cereals plus soy sauce and raw eggs for heart disease.

Basing its philosophy on the ancient Oriental concept of balance between opposing forces of yin (the female, passive principle) and yang (male, active), the cult of macrobiotics teaches that foods should be selected as nearly as possible to the mid-point between yin and yang. Cereals are closest to the ideal. Eggplant is extremely yin, while Japanese potatoes are extremely yang. Beets and white cabbage are balanced. Snails and pheasant are at opposite extremes; yogurt and goat's milk, pineapples and apples are also extreme. Meats and all drugs are to be avoided.

In a typical macrobiotic breakfast, half the meal would be sauteed rolled oats, while the rest might include steamed carrots and soybean paste broth garnished with green onions. Lunch would be buckwheat noodles in broth, sauteed broccoli, and seaweed with soy sauce. Dinner might be whole brown rice, with beans, baked apple, and white soybean cheese. All this might seem absurd to most people—yet increasing thousands of young people are turning to variations of the macrobiotic diet, never recognizing its gross deficiencies.

Darla Erhard, a public health specialist in California, has worked with organic vegetarian communes and macrobiotic groups in an effort to make them understand some of the basic principles of nutrition. Fortunately, she found that in discussions with young people she could stress moderation and variety, and relate the essential amino acids of diet to the principles of yin and yang. She found that macrobiotic cooking, by emphasizing speed and sparing use of water, preserved most nutrients; and that at least some nutritional balance could be achieved by encouraging nutritious combinations of yin and yang in the selection of foods. Most young people experimenting with macrobiotics and other extreme diets, however, have little knowledge of true nutritional balance and reject such advice.

"Regardless of the underlying reasons or motives, however noble and sincere, the concepts proposed in Zen macrobiotics constitute a major public health problem and are dangerous to its adherents," says the AMA's Council on Foods and Nutrition. "Cases of scurvy, anemia, hypoproteinemia (wasting of body protein due to inadequate intake), hypocalcemia (decrease in body calcium), emaciation due to starvation, and other forms of malnutrition, in addition to loss of kidney function due to restricted fluid intake, have been reported, some of which have resulted in death."

According to Fredrick Stare: "There is nothing nutritionally wrong with vegetarian diets; they provide a variety of fruits and vegetables plus frequent use of legumes, particularly soybean products and nuts. Most vegetarians consume generous quantities of milk and eggs. Good vegetarian diets are very healthful." But to Stare the extremes of macrobiotics are "no more than an excursion into make-believe Oriental cultism . . . the most dangerous food fad around."

### Nutrition and disease

Diet and disease are not linked in food fads alone; there is ample scientific evidence—and much scientific controversy—bearing on the relationship between what man eats and the illnesses he falls heir to.

Milk, the most nearly ideal food of all, can cause abdominal pain and diarrhea in many black children whose genes sometimes ordain a defect in the milk-digesting enzyme called lactase. In other children a genetic defect that prevents the breakdown of the amino acid phenylalanine is known as phenylketonuria, or PKU. Milk contains phenylalanine, and in PKU babies the penalty for milk-drinking is mental retardation. Milk-substitutes in their diet can prevent most of the brain damage if the disease is discovered early enough. Forty-three states now require PKU testing for all new-borns, and the tests discover 200 defective babies each year.

Far less clear are the relationships between diet and mental health, although evidence is strong that sound maternal nutrition before birth and adequate protein in early childhood are essential for full brain development. Some psychiatrists, such as Abram Hoffer of Saskatchewan, insist that schizophrenia is a vitamin deficiency disease, and that daily doses of the B vitamin niacin (nicotinamide or nicotinic acid) can prevent it. Linus Pauling, who coined the term "orthomolecular psychiatry," believes this too, but the vast majority of psychiatric professionals believe the evidence is tenuous if not delusional.

A new field for research is the possible relationship between diet and cancer. Denis P. Burkitt of the British Medical Research Council proposed recently that diets which are low in cellulose roughage—in other words, the typical refined-foods diets of highly industrialized Western nations—may be a major cause of cancer of the colon and rectum. This disease kills 46,000 Americans a year, and Burkitt contended that the cause may lie in the fact that refined foods move through the large intestine much more slowly than those containing roughage. Any cancer-causing elements, such as bacteria or obscure chemicals, would therefore have more time to initiate tumors in the colon before being excreted, Burkitt reasoned.

There is epidemiological evidence supporting this view. Corn-eating Bantu tribesmen in Africa have an exceptionally low incidence of colon cancer, and the incidence is eight times higher in Connecticut than in roughage-eating Cali, Colombia. Colon cancer is rare in China, but Chinese immigrants in Hawaii show rates approaching those of U.S. whites.

At present, this evidence only hints at an association, and H. Marvin Pollard, president of the American Cancer Society, warns against any changes in diet as a result of Burkitt's suggestion. "We don't really have much information on the relationship of high- and low-roughage diets and carcinoma of the colon," Pollard said as he urged a carefully controlled international study of the problem.

David G. Jose of the University of Minnesota was struck some years ago by the fact that although serious malnutrition is a major cause of infection and childhood death in Australian aborigines, these same malnourished people show an abnormally low incidence of cancer. Jose, a pediatric immunologist, wondered if protein-calorie deficiencies in the diet might selectively alter the body's immune mechanism. The deficiencies may weaken immunity against bacterial infection, Jose reasoned, while maintaining and perhaps enhancing cellular resistance to such invasive agents as viruses and malignant tumor cells.

Jose tested his reasoning by cutting calories and protein in the diets of rats and mice bred for their high susceptibility to various tumors. The animals did not grow as quickly, but their resistance to cancer was greatly increased. In the malnourished mice, in fact, tumors stopped spreading and the animals lived twice as long as normally nourished mice. "It's really premature to consider at this time any application of this type of therapy to human situations," Jose said, "but it is possible that with certain tumors and in certain situations this might be feasible."

In heart disease, despite widespread recent acceptance of the low-cholesterol, low-fats hypothesis, the role of diet in preventing heart attacks is still highly controversial. There are so many risk factors in coronary disease that no single cause-and-effect concept appears to be valid. Heredity plays a part, as do exercise (it should be moderate and regular); stress (the hard-driving and aggressive fare far worse); cigarette smoking (this shoots the heart attack rate way up); obesity (sudden cardiac death is three times as frequent among the fat); and high blood pressure (a major factor in heart disease and stroke). All told, approximately 165,000 persons under 65 die of coronary heart disease each year in the U.S.

The American Heart Association began urging diets low in the proportion of saturated fats in 1968. By 1971 most doctors were heeding the recommendations of the Inter-Society Commission for Heart Disease Resources, a group of 150 specialists from 29 leading health agencies appointed by the Department of Health, Education, and Welfare to evaluate the coronary controversies and to recommend action. The commission acknowledged that there is no conclusive proof yet that lowering fats and cholesterol in the diet will curb heart attacks. It did note that in every nation where custom or necessity dictates diets low in these elements the heart disease rate is likewise low.

In monkeys, the commission noted, the case against cholesterol has been proved experimentally; in man, Jeremiah Stamler of Chicago, one of the nation's most famed heart specialists, believes he has shown the same thing. The commission, therefore, recommended that overweight Americans reduce their caloric intake to slim down; that cholesterol in the average diet be cut in half from its present level of about 600 milligrams a day; and that dietary fats be limited to less than 35% of total calories, with only 10% coming from foods such as fatty meats, egg yolks, whole milk, butter and cheese, solid shortenings, pies, and doughnuts—all of which contain large amounts of saturated fats or cholesterol.

Significantly, the commission also made major recommendations to the government and the food industry for clearer labeling of the fat content of foods, and for the development of low-cholesterol dairy products, leaner livestock, and revision of the dietary content of school lunches and food allowances for the poor.

By 1972 some of these recommendations were being implemented: school lunch menus were changing, and the FDA was proposing that labels of processed foods be required to contain precise information as to the source of fat content—such as beef fat, cottonseed oil, and corn oil. Manufacturers would also be permitted to label foods with precise figures as to percentages of fat, the caloric content of average servings, and the proportion of unsaturated fats to total fats.

America's most intensive long-term study of heart disease has been conducted in Framingham, Mass., by the National Heart and Lung Institute. More than 5,000 of the community's residents have been checked and tested and followed closely since 1949. William B. Kannel, medical director of the Framingham project, says:

"The typical diet in North America gives every indication of being a contributor to our high atherosclerosis mortality. There is much indirect evidence that a diet too rich in saturated fat, cholesterol, sucrose, and calories is a prominent feature of our current life style contributing to the high death rate."

Like the Inter-Society Commission, Kannel is careful to qualify his conviction: "This is not to say that we have *direct* evidence that what a free-living person eats is, in fact, a major determinant of their blood lipid characteristics and whether they develop clinical atherosclerotic disease." Kannel acknowledges that most people would far rather take bitter pills or do almost anything than change their diets, or get more exercise, or give up smoking. But he has a simple and practical suggestion, which is seconded by the American Heart Association:

"Until some modification of the national diet is finally agreed upon and implemented," Kannel says, "those interested in lowering their own blood cholesterol value can attempt this by low-fat cookery which emphasizes fish, poultry, and lean meats, and by avoiding organ meats [liver, kidney, heart, etc.], butter, whole milk, and fatty cholesterol-rich cheeses, with fruit substituting for artificially sweet desserts at the end of the meal. For persons with high serum cholesterol values, particularly those with associated hypertension, impaired glucose tolerance, obesity, and cigarette habit, this becomes a matter of some urgency."

The American diet today is certainly a rich one; far too many over-eat and eat wrongly. In poverty areas deficiencies of many kinds abound, yet malnutrition is also widespread among the affluent. Scientific knowledge can now point to a more rational diet for everyone, and as the American Medical Association's Council on Foods and Nutrition says:

"The cost of a massive attack upon hunger and malnutrition will be great in money; the cost of doing nothing will be immeasurable in terms of lost human potential and social unrest."

FOR ADDITIONAL READING:

Bogert, L. J., Briggs, G. M., and Calloway, D. H., *Nutrition and Physical Fitness* (W. B. Saunders Co., 1972).

Deutsch, Ronald M., *The Family Guide to Better Food and Better Health* (Creative Home Library, 1971).

Food and Nutrition Board, *Recommended Dietary Allowances* (National Academy of Sciences-National Research Council, 1968. New edition due in 1973).

Mayer, Jean, *Overweight* (Prentice-Hall, Inc., 1969).

Stare, Fredrick J., *Eating for Good Health* (Doubleday & Co., 1964).

White, Philip L., *Let's Talk About Food* (American Medical Association, 1970).

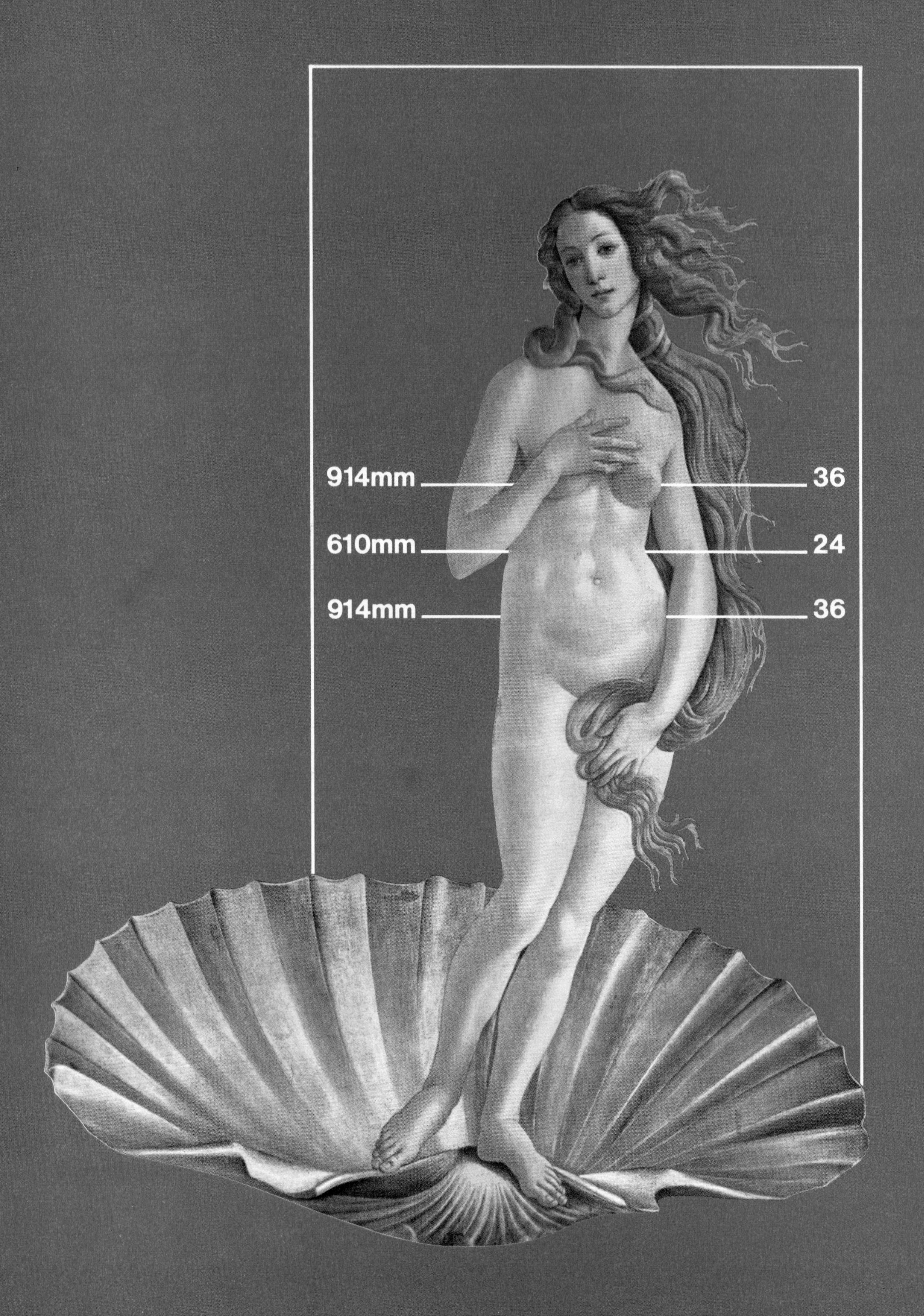
914mm
36
610mm
24
914mm
36

# Will the U.S. Go Metric?

## by Lewis M. Branscomb

**The United States remains one of the few holdouts in an increasingly metric world. Will the nation convert, and if so, how will we be affected?**

In July 1971, U.S. Secretary of Commerce Maurice H. Stans sent to Congress a report entitled "A Metric America; A Decision Whose Time Has Come." He added his own recommendation "that the United States change to the international metric system deliberately and carefully." Thus opened what may be the final chapter in a 200-year-old debate over the measurement language to be used in America. Should the United States replace its familiar pints and pounds, inches, feet, and miles with liters, kilograms, meters, and kilometers? The decision is up to Congress, but before the decision is made, many voices will be raised in public debate over an issue that has defied clear-cut resolution for the nation's entire history.

Thomas Jefferson first proposed a decimalized system of measurement in 1790. Thus the young American nation faced the difficult question of whether to establish formally the British system of weights and measures, or to adopt a new system such as Jefferson's or the metric system adopted by France in 1795.

The Constitution (Art. I, Sec. 8) authorized Congress to "fix the standard of weights and measures." President George Washington asked Congress in 1790 to address its responsibilities in this area, saying "Uniformity in the currency, weights, and measures of the United States, is an object of great importance, and will, I am persuaded, be duly attended to."

The issue was raised again in 1821, when Secretary of State John Quincy Adams submitted to Congress his "Report Upon Weights and Measures." Adams' report did not settle the issue, for although he made a very strong case for the advantages of a simple, rational, and internationally compatible system of measurement based on decimal mathematics, he noted that the principal trading partners of the United States were not agreed on a single system that would insure international compatibility. Indeed, Napoleon already had rescinded the requirement for metric usage in France. Thus Adams concluded that metric conversion was premature.

Keystone

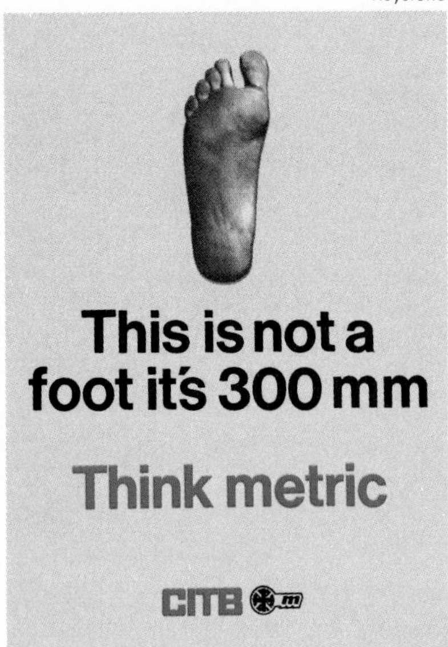


*Eye-catching posters displayed throughout Great Britain by the Construction Industry Training Board were designed to help the nation convert to the metric system. Two other examples are shown on succeeding pages. Common equivalents between the metric and present U.S. systems are shown on the opposite page.*

***LEWIS M. BRANSCOMB*** *until May 1972 was director of the U.S. National Bureau of Standards and now is chief scientist and vice-president of IBM Corp. His many publications include several articles on conversion to the metric system.*

While one European nation after another adopted the metric system, the United States continued to use inch-pound measures, and built the greatest engineering and industrial capability the world has ever seen.

Today, the task of simplifying and harmonizing the U.S. system of weights and measures remains unfinished. Congress, in the Miller-Pell-Griffin Act of 1968, asked the executive branch to conduct a three-year study of the advantages and disadvantages of U.S. adoption of the metric system. The report, prepared by the National Bureau of Standards, found many of the circumstances faced by John Quincy Adams profoundly changed.

There no longer is any doubt whether the metric system will be accepted internationally as the predominant measurement language. Since World War II, Japan, India, Greece, Egypt, Saudi Arabia, Pakistan, North and South Korea, Indonesia, Vietnam (North and South), and many other nations have changed from a variety of measurement systems to the metric system. More important to the United States, the British Commonwealth nations have also committed themselves to the change. The United Kingdom is more than half-way through a 10-year conversion process. South Africa and Australia are also in the process of conversion. Even Canada, which has close commercial and technical ties with the United States is committed to change, with only the timetable remaining to be set. The United States is one of only 13 nations, including Tonga, Trinidad, Sierra Leone, and Yemen (Aden), that have not yet adopted the metric system. All the industrialized trading partners of the United States now are committed to an internationally harmonized metric system.

### Arguments for the metric system

Why have so many nations voluntarily assumed the task and expense of changing their system of measurement? Adams noted how deeply the language of weights and measures is rooted in every phase of life. "The knowledge of them, as in established use," he wrote, "is among the first elements of education, and is often learned by those who learn nothing else, not even to read and write. This knowledge is riveted in the memory by the habitual application of it to the employments of them throughout life." Proponents of the metric system argue that the system is simpler, easier to learn, and much more amenable to arithmetic operations because the units are compatible with decimal arithmetic. The metric system, properly called the "International System of Units (SI)," is the first complete, streamlined, internally compatible system of measurement the world has known, designed to best serve the needs of scientists and engineers in all disciplines. But neither of these arguments is necessarily persuasive. Many who infrequently need to make exact calculations find convenient the great variety of customary units for the same quantity (inch, foot, yard, mile).

The strongest argument for general adoption of SI is found in the predictable and persistent inconvenience and expense of having to use both systems simultaneously. Because the United States is the approximate common equivalents

| | | | | |
|---|---|---|---|---|
| 1 inch | = 25 millimeters | | 1 millimeter | = 0.04 inch |
| 1 foot | = 0.3 meter | | 1 meter | = 3.3 feet |
| 1 yard | = 0.9 meter | | 1 meter | = 1.1 yards |
| 1 mile | = 1.6 kilometers | | 1 kilometer | = 0.6 mile |
| 1 square inch | = 6.5 square centimeters | | 1 square centimeter | = 0.16 square inch |
| 1 square foot | = 0.09 square meter | | 1 square meter | = 11 square feet |
| 1 square yard | = 0.8 square meter | | 1 square meter | = 1.2 square yards |
| 1 acre | = 0.4 hectare | | 1 hectare | = 2.5 acres |
| 1 cubic inch | = 16 cubic centimeters | | 1 cubic centimeter | = 0.06 cubic inch |
| 1 cubic foot | = 0.03 cubic meter | | 1 cubic meter | = 35 cubic feet |
| 1 cubic yard | = 0.8 cubic meter | | 1 cubic meter | = 1.3 cubic yards |
| 1 quart | = 1 liter | | 1 liter | = 1 quart |
| 1 gallon | = 0.004 cubic meter | | 1 cubic meter | = 250 gallons |
| 1 ounce | = 28 grams | | 1 gram | = 0.035 ounces |
| 1 pound | = 0.45 kilogram | | 1 kilogram | = 2.2 pounds |
| 1 horsepower | = 0.75 kilowatt | | 1 kilowatt | = 1.3 horsepower |

nation with the largest volume of foreign trade and the most pervasive economic and cultural impact on the rest of the world, the invasion of metric language and designs into our everyday lives is assured. Indeed, the use of metric measures is surprisingly widespread in the United States today. Now the dominant language for scientific and medical research, metric usage is increasing steadily in engineering. The pharmaceutical, frictionless bearing, and photographic industries are substantially metric today. Not only imported cars, but some domestic autos contain metric parts and need metric tools for their maintenance. The Olympic Games have made sports fans accustomed to metric events, and U.S. swimming pools are often made to metric sizes for this reason. A growing number of packaged goods in retail trade are labeled in metric as well as customary units, and most American school children now encounter metric units in some of their books. The testimony of thousands of respondents to the questionnaires of the U.S. Metric Study reveals that a substantial majority of U.S. businessmen believe increasing use of the metric system in the United States is in the best interest of the nation.

Therefore, with metric language becoming more commonplace, Americans will have to contend with difficult arithmetic conversions not only within the customary system, but back and forth between

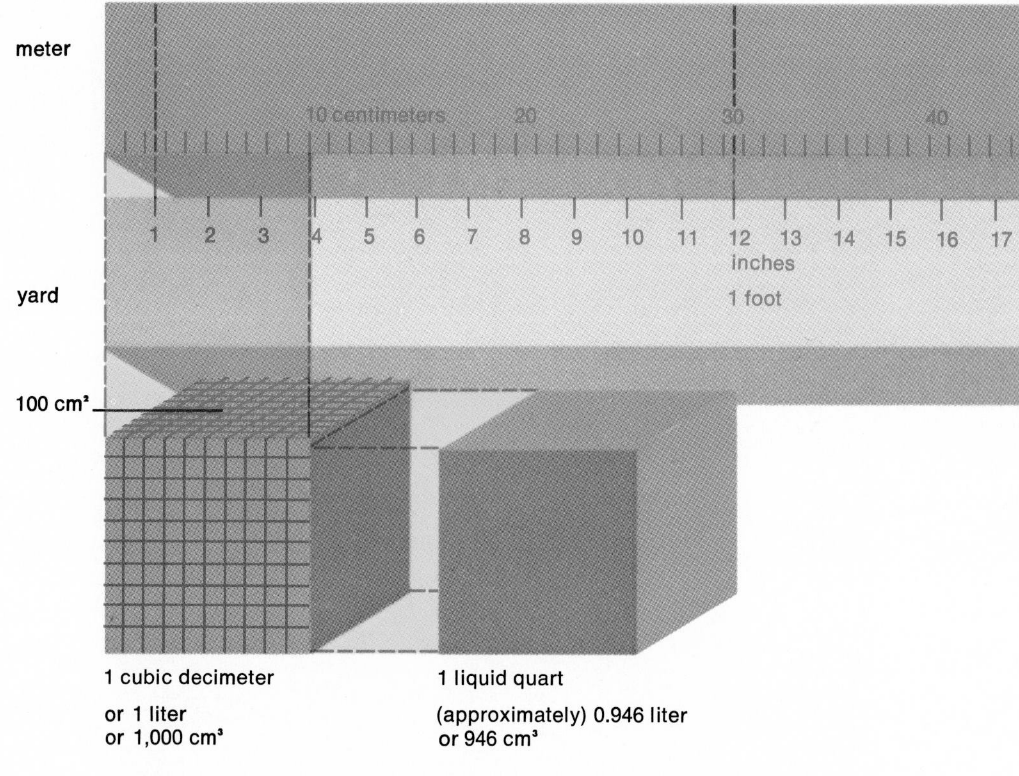


*Relationships between common weights and measures of the metric and U.S. systems are shown graphically. A meter is compared to a yard, a centimeter to an inch, a liter to a liquid quart, and a kilogram to a pound.*

customary and metric. This situation already prevails in technology; scientists and engineers must make these conversions in order to interpret one another's work. It will affect increasingly skilled workers, teachers, and lawmakers. Although the large multinational corporations have demonstrated that they can manufacture U.S.-designed products in metric countries without major difficulties, added costs are incurred, such as two sets of drawings to different dimensions. Workers and small business subcontractors face greater uncertainty in a mixed measurement society because they can predict less easily when they may be called upon to deliver work designed to metric specifications.

The problem of mixed measurement languages cannot be resolved quickly, since a generation must pass before everyone finds the metric language intuitively familiar. In fact there are significant vestiges of imperial units in countries that have been metric for decades, just as the *vara* (a Spanish unit of land measure) persists in Texas and the Southwest. But the implications of metric change for standard designs and dimensions—the hardware aspects of going metric—suggest a more rapid timetable of 10 years for a U.S. changeover.

During the decade of the 1970s, the metric report concludes, there probably will be developed through international engineering organizations a set of metric language industrial standards. These standards are the dictionary of the marketplace in a technologically advanced nation, describing the sizes, test methods, product characteristics, and other conventions for every kind of component and product. Used as

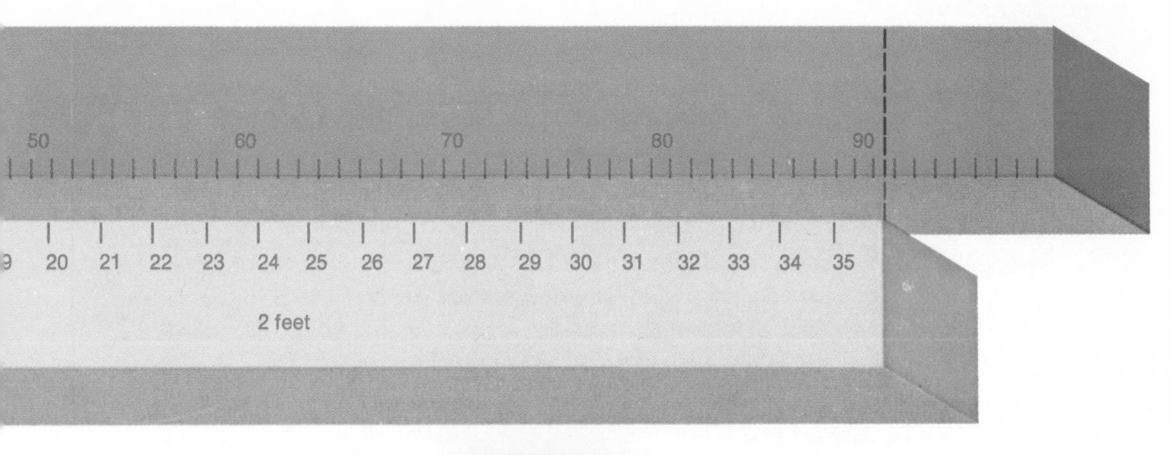


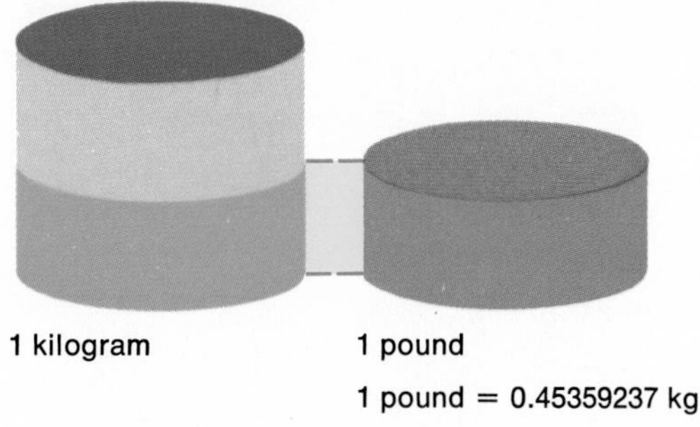


purchase specifications or integrated into product designs, these standards can have great commercial force. The U.S. Metric Study concluded that if U.S. industrial practices and conventions are adopted in these international standards, wherever their technological superiority merits their acceptance, the burden of going metric—that is, of harmonizing national industrial practice to the demands of the international marketplace—will fall on other nations as well as on the United States. For today there are still relatively few international standards, and metric countries use the same language but do not by any means use the same sizes, shapes, and conventions. The United States cannot hope to persuade the more than 60 nations active in the International Standards Organization to adopt its customary measurement language. But we can expect that U.S. determination to participate fully in international negotiations leading to a set of metric language standards for worldwide use—including U.S. domestic use in the future—would remove a significant barrier to world trade while best protecting U.S. commercial interests. If European and Asian powers successfully harmonize their industrial practices through their own international metric standards, without adequate participation by U.S. interests, the result will be to make eventual U.S. conversion to those standards much more costly and difficult.

In summary, it is now, while the metric countries still have divergent standards and practices, that U.S. adoption of metric language and development of metric standards would be most beneficial. However costly the changeover now, it would be more costly in the future.

Keystone

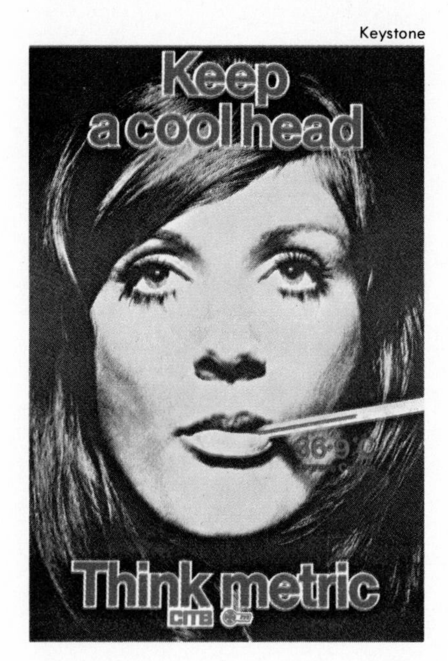


## How conversion would work

How should a 10-year program of conversion be carried out? Secretary of Commerce Stans recommended that while the changeover should be substantially voluntary, it should be accomplished through a coordinated national program. The Congress could establish a suitable central coordinating body to guide the changeover. This organization would be responsible to all sectors of the U.S. society, and would facilitate the generation of detailed plans and timetables. After these plans have been made, under which some sectors might change in only a few years and others might take 20 years or more, Congress would establish a date ten years later by which time the United States would be predominantly, though not exclusively, metric.

Two tasks would need to be undertaken immediately and pressed with some vigor. First, as soon as a new cycle of school textbooks was ready for introduction all public school children would begin to gain a practical mastery of the basic metric dimensions, areas, volumes, weights, and temperature scales. This effort, strongly supported by the major national organizations representing teachers, would be needed to prepare the six-year-olds starting school that year for life in the metric world of the 21st century.

Second, the task of generating new metric language standards for domestic U.S. use, and hopefully for international adoption, would have to begin quickly because conversion of our economy to metric practice would not be possible until the standards exist. This job (like most of the tasks in metric conversion) will fall primarily on the private sector, working through the many voluntary standards-generating bodies in the United States. The American National Standards Institute (ANSI), which arranges for U.S. representation in international standards organizations, would be expected to play a major role in coordinating this effort.

In the process of generating new metric standards, some improvements might be made that could provide substantial savings to the economy once conversion was accomplished. An example is in the area of thread sizes on nuts and bolts. A study by the Industrial Fasteners Institute suggests that the present 59 varieties of U.S. customary thread diameter-pitch combinations, plus the 57 metric sizes in common use in Europe, might be replaced by 25 new metric sizes that could do the job of all 116 old sizes. Substantial inventory savings and design simplifications might result. The British found this aspect of "metrication"—housecleaning of their industrial standards—a major opportunity for productivity improvement.

The standardization of common plumbing will serve to illustrate another area in which savings might result. The present common pipe standards date back to the 1880s and the wall thicknesses reflect the state of metallurgy at that time. Since American pipes are standardized on their inside dimensions, while European pipes are standardized by their outside dimensions, the rationalization and harmonization of pipe standards internationally might provide the opportunity to reduce

**TEMPERATURE** Kelvin-K

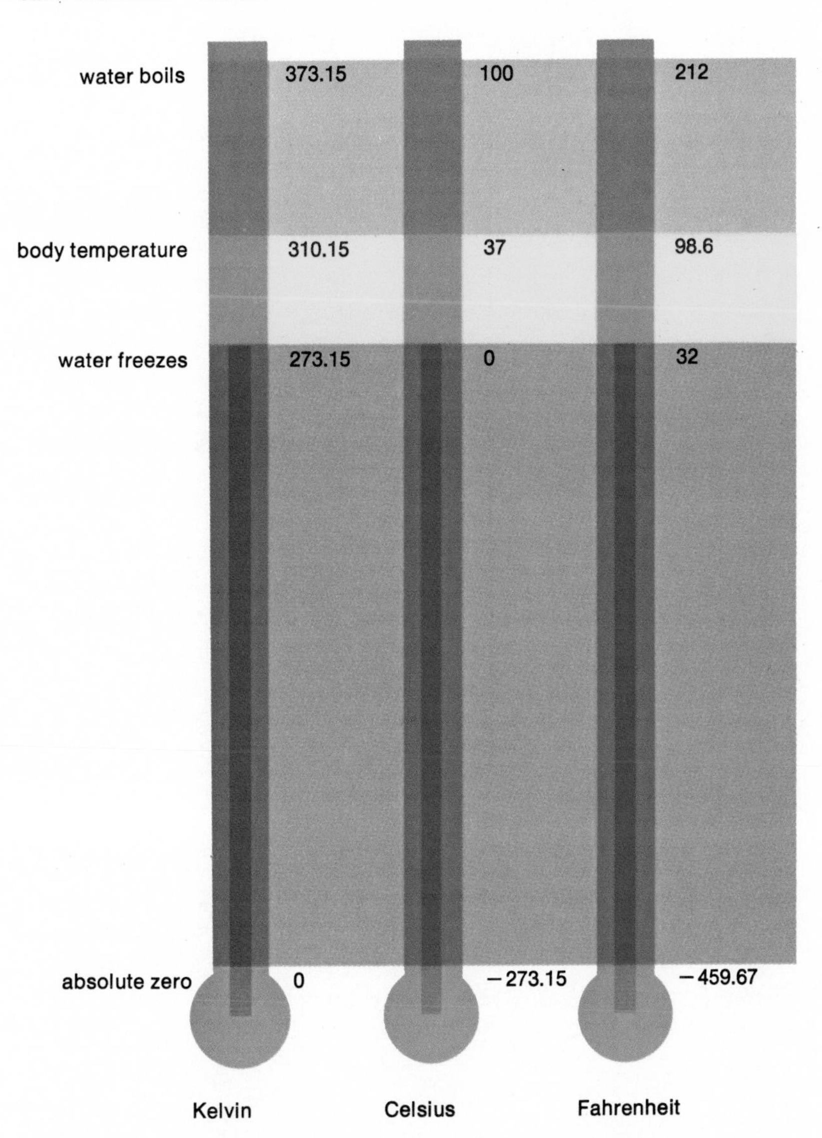


*Three scales on which temperature is measured are the Kelvin, Celsius or centigrade, and Fahrenheit. Celsius is used in the metric system and Fahrenheit in the United States. On the Kelvin scale the unit of measurement equals the Celsius degree, each being 1/100 of the interval between the freezing and boiling points of water. Kelvin is most often used in scientific work. Another of Britain's conversion posters is on the opposite page.*

the pipe wall thickness significantly without any increase in hazard or failure rate. However, it would be a mistake to assume that U.S. pipe standards would have to be changed in the event of a metric change-over simply because the names of pipe sizes are "one-quarter inch," "one-half inch," etc. In fact, nowhere on a piece of one-half-inch threaded pipe does the pipe actually measure one-half inch. The savings in this area thus derive from the use of less metal in pipe production through modernization of the engineering content of the standard.

Following on British experience, and the conclusions of the U.S. metric study, Stans recommended that the costs of metric conversion "lie where they fall." This does not rule out government sharing the cost of its participation in the metric conversion; special consideration might have to be given to the problems of the small businessman and self-employed artisan. The argument against subsidies to defray the

## Origin of our present system of measurement

The origin of American customary units of measurement is lost in antiquity, for it evolved from ancient Roman, Anglo-Saxon, and Norman customs. But England, as a great maritime nation, was an early leader in measurement standardization. King Henry II standardized the yard, and Elizabeth I the avoirdupois pound, both of which are very similar to today's units. The volumetric units are more complicated, since the United States adopted the British wine gallon, about 20% smaller than the Imperial gallon adopted by the British in 1824. Furthermore, American colonists used a different measurement for liquid and dry volume as a matter of convenience because of different types of containers then in use. (John Quincy Adams thought this inconsistency would cause the United States less trouble than other nations because "Wine is an article of importation; an article of luxury; the exactness of the measure by which it is distributed, is not an incident which every day comes home to the interests and necessities of every individual.")

In general, dimensional units of the customary system had an anthropomorphic origin. Limbs of the body were used as conveniently available measurement gauges, the inch being the knuckle of the thumb and the early yard being the distance from King Henry I's nose to the end of his thumb when his arm was fully outstretched.

The metric system derives its anthropomorphic character from the 10 fingers on which many people learned to count. Thomas Jefferson had early seen the advantages of a measurement system based on decimal arithmetic, and proposed the use of a pendulum whose time of swing and length would simultaneously define both length and time.

The metric system was born of the French Revolution, with provisional standards issued in 1795. The definition of the unit of length,

costs of conversion rests on the fact that these costs are difficult to identify and there are many opportunities to reduce them by careful management at the company level. Americans will pay much less for metric conversion as consumers and stockholders than they would as taxpayers.

Even after the planning of timetables for a metric conversion, several years would pass before metric labeled goods would predominate on the grocer's shelves and "kilometers per hour" signs would adorn the roads. It has been estimated that about three years would be required to modify the dial-plates of scales used in retail trade. The changeover of highway signs could be accomplished over several years by concealing metric speeds, distances, and load limits under removable labels in miles per hour, miles, or tons. These two changes, which exemplify encounters that the average citizen would have with metric language the meter, was one ten-millionth of the earth's quadrant from pole to equator. The kilogram, unit of mass, was realized by the water contained in a volume of one liter, that is a cube 10 centimeters on a side.

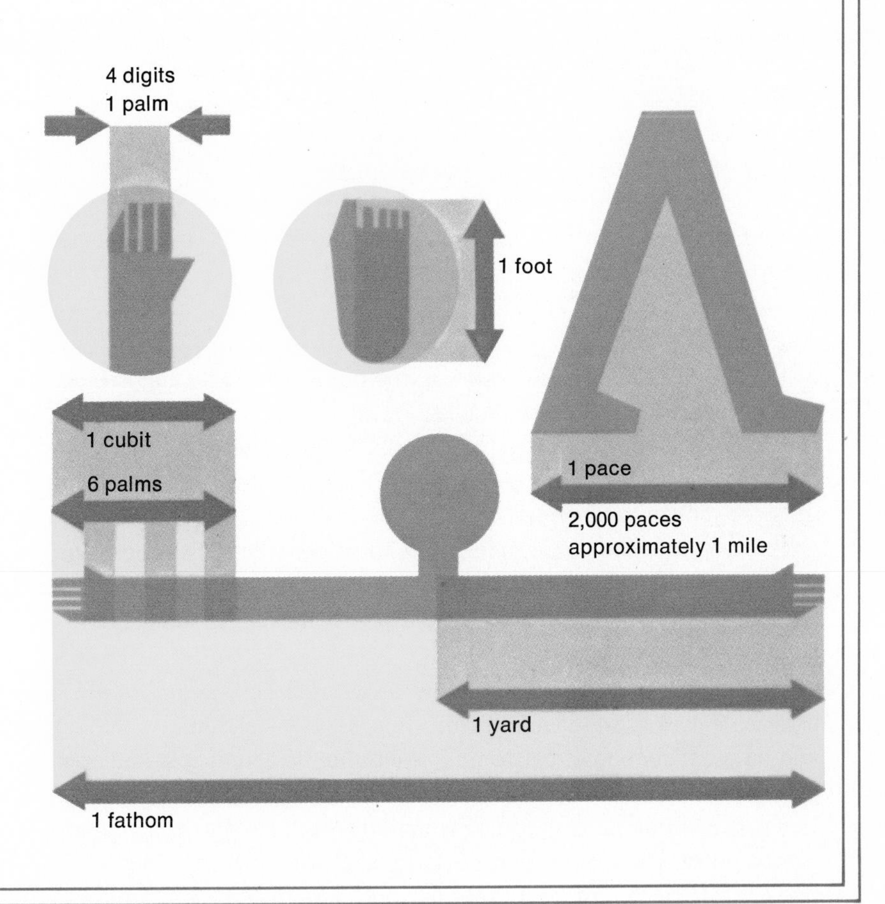


in his everyday life, also illustrate the need to be very practical about arranging the conversion. It probably would be advantageous from the safety point of view to change speed limit signs all at once. But a period of two or three years of dual labeling might prove most helpful to the shopper who could gain familiarity with the new units. After this transitional phase, it could be expected that the states (which regulate the weights and measures used in retail trade) would agree on a date after which metric units must be used in package labeling and retail measures. Customary equivalents also could be given for as long as anyone desires.

It probably would not be necessary for Congress to intrude on the states' authority to set the weights and measures regulations in retail trade, even though it would be quite important that timing of a change to metric be uniform in all 50 states. All the states now coordinate their

Keystone

efforts through a National Conference of Weights and Measures, whose technical recommendations for regulations are adopted by administrative proceedings in most states. Thus, once the conference determined a timetable for conversion, in collaboration with the national coordinating body, it could seek the simultaneous implementation of this schedule in all states. It is, therefore, unlikely that any new mandatory federal authority would be needed to effect an orderly conversion program.

The purchasing power of the federal government would provide a useful market demand for metric products and services in metric language. However, the timing of the government's own actions would have to be dovetailed with timetables developed by the private sector through participation of both sectors in the proposed national coordinating body. Thus, government would not be expected to move first to demand metric products but should set a good example by adhering to the timetable agreed upon by all.

### Reaping the benefits

At the end of the conversion period, the public would be accustomed to metric language predominating in their shops, workplaces, and newspapers. About 5% more milk would appear on the doorstep (the margin by which the liter exceeds the quart), and a meter of cloth would be three inches longer than the yard. But basically the change would not be very noticeable. For many years to come horses, 16 hands high, still would be competing over furlong measures, football fields would remain 100 yards long, skeptics would say, "Give him an inch and he'll take a mile."

Estimating the costs and benefits associated with going metric is elusive. British experience indicates that while the planning is difficult and time-consuming, most companies found the actual cost and complexity of conversion easier than they feared it would be. Conversion costs in general were found reasonable in relation to other uncertainties in ordinary operating costs. Indeed, the "rule of reason" for metric conversion suggests that this should be the case, for no sector should be expected to convert on a timetable that would make the costs unacceptable. Detailed cost surveys, conducted by over 100 companies that volunteered their data to the U.S. Metric Study, showed wide variations from company to company in the same industry on the cost impact of conversion. In extreme cases the estimated cost of conversion provided by each of two companies engaged in the same business differed by as much as 100 to 1. Thus these data reflect a wide variation in experience and perception of metric conversion by different companies. Their problem in making the estimates was compounded, of course, by the fact that the metric standards to which their hypothetical conversion was directed do not in fact exist and had to be imagined. Finally, those U.S. industries that have undertaken a "pay as you go" conversion in recent years report that the process was accomplished with relatively little confusion and that costs, in general, were absorbed in ordinary operating expense.

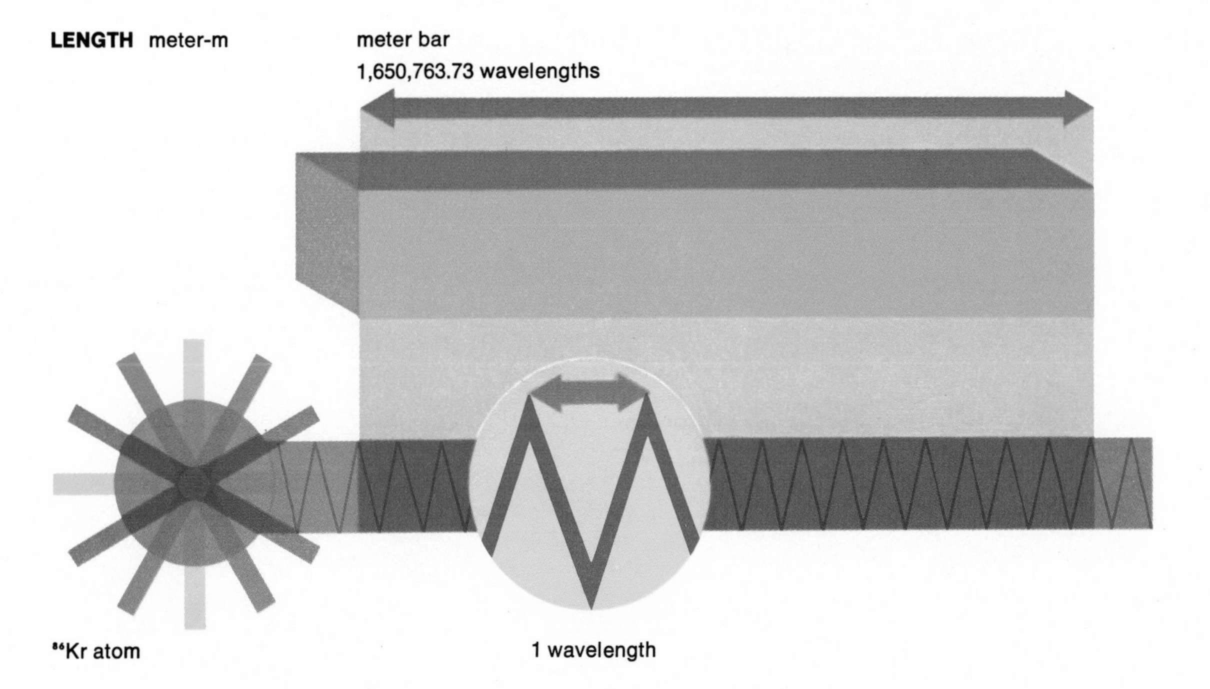


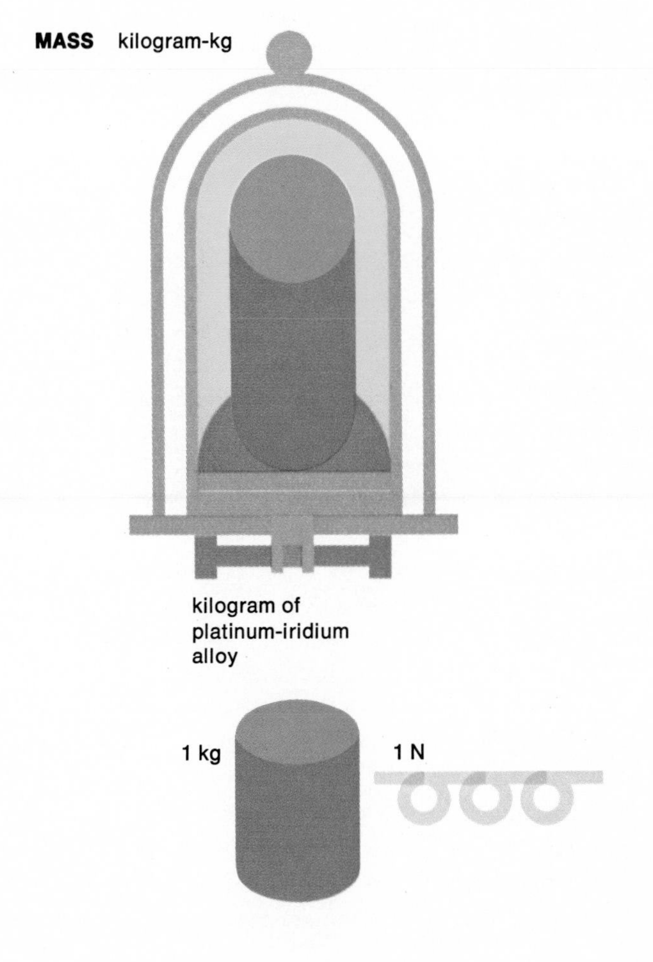


*Common metric units are based on specific physical entities. A meter (above) is 1,650,763.73 wavelengths, in a vacuum, of the orange-red line in the spectrum of krypton-86. A kilogram (left) is the weight of a cylinder of platinum-iridium alloy kept by the International Bureau of Weights and Measures in Paris. The metric unit of force, one newton (1 N), when applied for one second to one kilogram will give it an acceleration of one meter per second per second. On the opposite page is another British conversion poster.*

**LUMINOUS INTENSITY** candela-cd

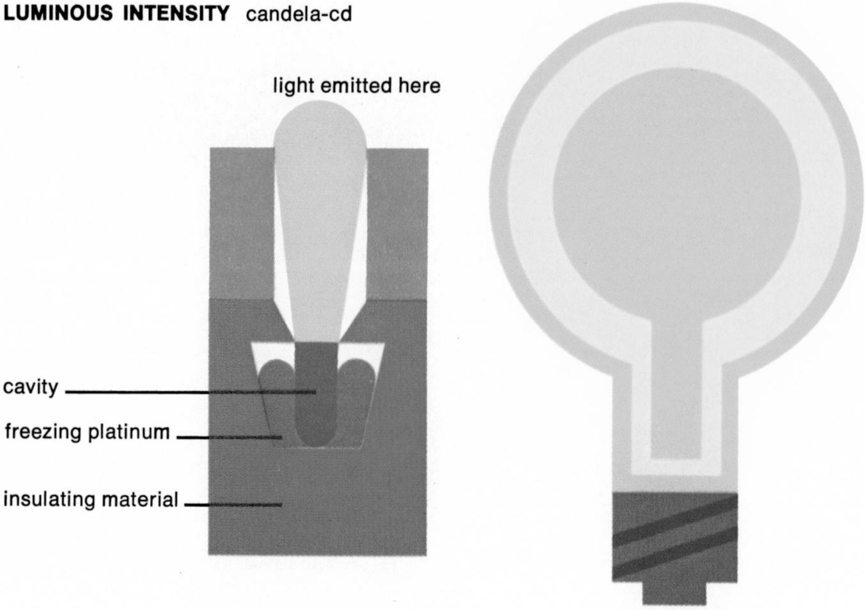

light emitted here

cavity

freezing platinum

insulating material

100-watt light bulb

*The metric candela (above) is 1/60 of the luminous intensity of one square centimeter of a blackbody surface at the temperature of platinum at its solidification point (2,042° K). A source having an intensity of one candela in all directions radiates a light flux of about 12.5 lumens. By comparison a 100-watt light bulb emits about 1,700 lumens. One second (opposite page) is the duration of 9,192,631,770 cycles of the radiation associated with a particular transition of the cesium atom. A second can be realized by tuning an oscillator to the resonance frequency of cesium atoms as they pass through magnets and a resonant cavity into a detector.*

Costs of hardware changes would be determined in large measure by the extent to which the time scale for conversion is commensurate with the normal life of the affected equipment. Machine tools could in many cases be converted from inches to meters by a simple change in the drive gears. Since the inch is internationally defined as 2.54 centimeters, a gear of 127 teeth can be used with a gear of, for example, 50 teeth to produce a metric tool.

On the benefits side, importers and exporters expressed the view that measurement language was not a strong factor in determining the course of international trade. Nonetheless, this survey revealed an estimated net benefit to the U.S. balance of trade of about $600 million a year. About one-fourth of the manufacturers surveyed estimated that savings resulting from conversion would compensate for added costs in a period of time comparable to the conversion process itself. In addition, this survey identified an annual cost of maintaining dual capability for the manufacture of standard parts and materials of about $500 million a year, a cost that will be faced for as long as the United States is making substantial use of both measurement systems. In addition to this cost of "being dual," manufacturers identified a fixed cost of conversion that the study estimated as between $6.2 billion and $14.3 billion for the 10-year period. The study concludes that no matter what the cost of conversion, it would be less costly, and the costs would be recouped sooner, if the nation changes to metric by plan rather than by leaving the change to chance.

All of the analyses in the study were based on examination of today's world. A more relevant question to answer is, "What measurement language will Americans want to be using during the nation's third century?" In view of the extraordinary changes that have swept over

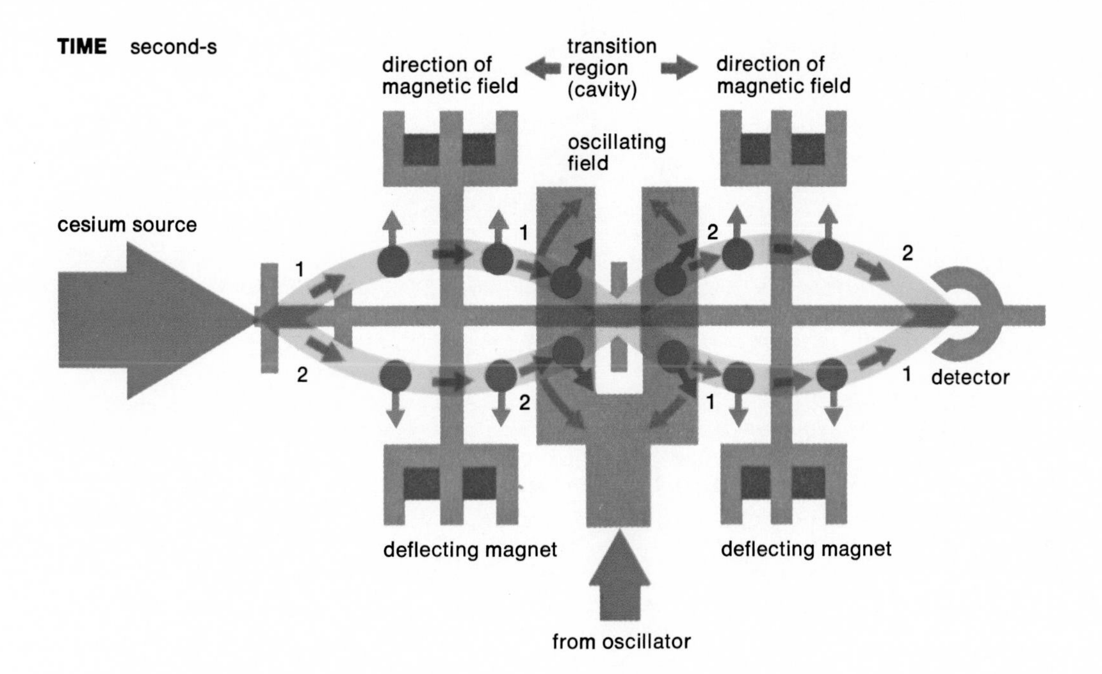


the world in the last 50 years, largely brought on by technology, what can we expect for the year 2000? Technology will continue to make the peoples of all nations more interdependent. Advanced technology will play an ever more important role in U.S. economic life, and we will find ourselves competing with increasingly sophisticated trading partners. Barriers to the transfer of goods, of technology, and of personal communication would become increasingly disadvantageous for the United States. In the light of the needs of future generations of Americans who will live in a metric world, it is time for a decision on this issue which has been too long unresolved.

# The Science Year in Review: A Perspective by John H. Lannan

Skyrocketing research costs, unmet human needs in such areas as health and environment, and the threatened loss of U.S. technological and industrial preeminence combined to place science and technology high on the list of national concerns in 1972. This came about despite—or perhaps as a result of—the "antitechnology" sentiment that had gained momentum in recent years. Concerned by what science and technology were doing to them as well as for them, people had begun to take a more critical look at both.

In part, at least, their concern stemmed from environmental considerations. That concern led, among other things, to a conflict between the antitechnology advocates in the developed nations and the less developed countries, where the pursuit of science remained the mark of societal sophistication and where the resulting technology was a source of jobs and better lives. Nowhere was this dichotomy more evident than at the first UN Conference on the Human Environment, held in Stockholm in June 1972, where an impasse developed between those who sought to protect the world's overall environment and those who felt the world environment had little meaning in the face of poverty and underdevelopment.

In the U.S. strong environmental and consumer pressures brought demands for a "look before we leap" approach to legislation and regulation. Congress reaffirmed the broad regulatory powers it had given to the Environmental Protection Agency, intensified the ongoing attacks on disease, poverty, and functional illiteracy, and took steps toward the creation of an Office of Technology Assessment, designed to evaluate new technologies before they are implemented.

On the executive side, the new emphasis meant additional attention to and direction for the scientific and technological enterprise. Early in 1972 the administration put before Congress the largest science and technology budget ever proposed—$17.8 billion. Significantly, the budget emphasized the civilian or domestic side of the economy as opposed to the military, space, and atomic energy efforts that had long commanded the lion's share of the federal government's research and development funds. Energy research, health care, transportation, education, and the effects of pollutants on health were among the areas designated for significant increases.

The first presidential message on science and technology, promulgated in March, called for broader-based support of fundamental research by government, the channeling of scientific and technological capacity to meet existing needs, and the expansion of international cooperation.

By midyear a search for ways to support basic research was under way throughout the federal establishment. With added funding in sight and a presidential directive calling for new applications of science and technology, the government—which supports more than half the nation's research and development—began moving toward new goals.

On the international front, Pres. Richard Nixon's diplomatic initiatives had important implications for science and technology. The beginnings of political and cultural rapprochement between the U.S. and China opened the way for exchange visits between scientists and physicians of the two countries. On the other side of the world, the series of landmark agreements signed during the president's official visit to the Soviet Union had striking science and technology components.

Of these, the strategic arms limitation agreement —itself made possible by scientific and technological developments stemming from space, geophysical, electromagnetic, and oceanographic research—won the widest public attention, but agreements on space, environment, health, and science and technology all called for joint efforts or cooperation in areas where both countries would benefit. Significantly, the agreements provided for cooperation on a government-to-government basis. Previously, cooperative arrangements had been made between individuals or institutions through the two governments' foreign offices. Furthermore, the new agreements would be monitored not by diplomats and politicians but by joint commissions of industrialists, scientists, technologists, and concerned government agencies.

While the highlights of the year were in the policy area, scientific and technological details were not lacking. The new 200-GeV accelerator near Batavia, Ill., came on line and reached its designed beam power and intensity. The Apollo program, begun as a cold war initiative, approached its conclusion with a series of scientific triumphs. By way of beginnings, the construction of a new ground-based radio-astronomy facility, a Y-shaped interferometer with 13-mi arms, was authorized by Congress.

# Contents

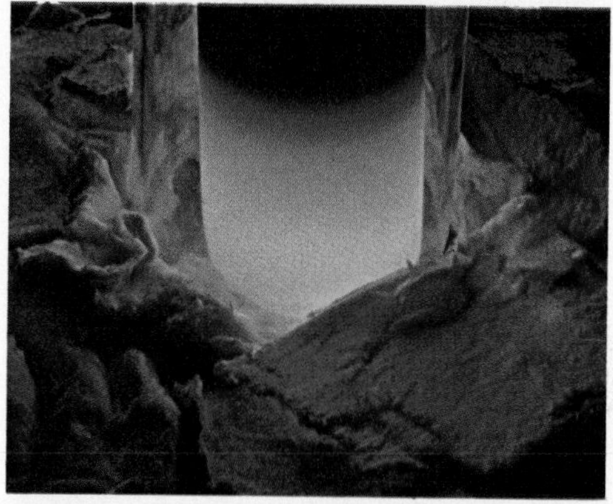

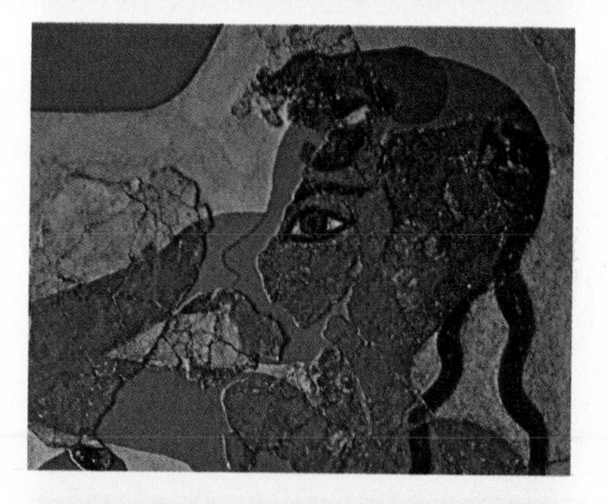

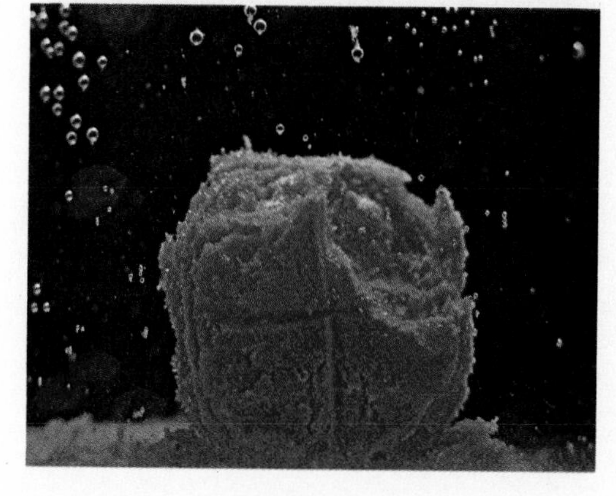

# Agriculture

For more than a century, research pursued by agricultural scientists has been instrumental in making United States agriculture the most productive in the world. Two of the more outstanding research accomplishments of recent months have far-reaching implications in man's search for clues to the causes of human diseases.

Early in 1972, scientists announced that they had isolated a tiny disease-producing particle called a viroid. This discovery opened new paths for research into diseases whose causes have eluded scientists. The viroid is an infectious particle that causes potato spindle tuber disease. Smaller than any known virus, the viroid is a fragment of ribonucleic acid (RNA) with a molecular weight of only approximately 50,000. The smallest known virus is 80 times larger than a viroid. The infectious capabilities of viroids may give scientists new clues to the cause of some human cancers, infectious hepatitis, multiple sclerosis, scrapie of sheep, and other diseases.

A second major scientific breakthrough was the development of the first vaccine found to be effective against a naturally occurring cancer. The vaccine is used in the prevention of Marek's disease, which occurs widely in chicken flocks throughout the world. A highly infectious form of avian leukosis, it has caused losses of an estimated $200 million annually to the U.S. poultry industry. The vaccine proved to be a safe and effective way to control this disease, and its discovery was considered to be among the most important in current cancer research.

**Pest and weed control.** Because recent research has revealed potential dangers in the use of some pesticides, scientists from the Agricultural Research Service (ARS) of the U.S. Department of Agriculture have pioneered the reduction of pesticide use by developing other methods of pest control. Sex attractants may play a more prominent role in future pest management programs than they have in the past. Attention in 1972 was being directed toward their possible use as a direct control rather than as a detection tool. Scientists believed that the attractants might be released in sufficient quantity so that the environment of insect pests would be permeated sufficiently to disorient males in their efforts to find females.

A newly identified sex attractant for houseflies, called muscalure, may reduce the amount of in-

*Romaine lettuce damaged by air pollutants (left) and ozone damage to a leaf of corn (right) dramatically illustrate a problem faced by thousands of U.S. farmers. The Council on Environmental Quality estimated the annual loss to crops and livestock at $500 million.*

Courtesy, Dr. O. C. Taylor, Statewide Air Pollution Research Center, University of California-Riverside secticide needed if it can efficiently attract flies to a small, selected area treated with insecticide. This would eliminate spraying large areas, which often involves a heavy application of insecticide. Scientists also developed a strain of houseflies that produces only male offspring. This could help streamline the procedures used in sterilization programs for insect control. Sterilization has been successfully employed against several kinds of flies but not houseflies. In recent months, however, preliminary tests showed that this approach could be employed in some control programs against this carrier of human diseases. Sterilization programs involve rearing insects in a laboratory, sterilizing the males, and releasing them in quantities sufficient to significantly outnumber other males; as a result, almost all matings would occur between normal females and sterilized males.

ARS scientists were examining ways to manipulate nematodes—tiny parasitic worms—to attack the bark beetle that carries Dutch elm disease from tree to tree. They hoped that the nematodes could sterilize effectively or at least seriously debilitate the beetles.

For the first time, the launching of a full-scale biological attack against the cereal leaf beetle—a serious threat to the production of wheat and other small grains—appeared to be possible. The most promising technique for biological control was the importation, rearing, and release of insects that become parasites on the beetle in its native habitat. In this technique insectaries—places for the rearing of insects—are established in heavily infested grain fields located in wildlife areas or on farms retired from production, where pesticide drift will not be a hazard to newly hatched parasites. During 1971 the ARS and cooperating states released more than 800,000 parasites in 109 locations across the U.S., and nearly 100,000 larval parasites were freed at 11 sites in Indiana, Michigan, Ohio, and West Virginia.

An artificial diet packaged in capsules resembling insect eggs was being used as food to rear large numbers of aphid lions, a beneficial predatory insect. The aphid lion, which is the larva of the green lacewing (*Chrysopa carnea*), eats many kinds of crop insects and their eggs, but most often attacks bollworms and tobacco budworms on cotton plants. In addition, it is resistant to several insecticides, a factor that favors its use in integrated control—a concept in which a minimum amount of insecticide is applied to reduce the number of pests, followed by the release of large numbers of aphid lions to finish the job.

Scientists employing chemical control were using foam experimentally as a better method to control the corn borer, with less chance of environmental pollution. A nozzle was designed and built to generate foam formulations of (1) DDT (used only as the best available standard for comparison) or (2) *Bacillus thuringiensis*, a bacterium that produces protein crystals toxic to some insects, including the corn borer. Each of the two insecticides was combined in a nozzle with water, liquid foaming agent, and air and then forced through a meshed material. The volume of foam and its consistency were controlled by the amounts of insecticide, foaming agent, and air that were used. In the tests *B. thuringiensis* demonstrated potential for reducing pollution by replacing potentially hazardous chemicals.

Weed control is also essential if man, animals, and wildlife are to survive. Scientists discovered that some aquatic weeds can be effectively controlled with the herbicides diquat and paraquat without harming fish. Aquatic weeds obstruct water flow, cause large losses of water through transpiration (evaporation of water from plant parts exposed to air), and prevent proper drainage of land. The weeds may interfere with navigation, prevent fishing and recreation, depress real estate values, and present health hazards. Losses from aquatic weeds amount to tens of millions of dollars annually.

The discovery of an additive that enhances the efficiency of herbicides may lead to effective weed control with fewer applications and at lower herbicide rates per application. Phenylcarbamate herbicides are readily degraded by soil microorganisms, and repeated applications often are needed to control weeds effectively. But research has shown that certain methylcarbamate chemicals (some of which are insecticides), when applied along with certain herbicides, temporarily inhibit the degradative action of the soil microorganisms. Experimenting with pure culture and soil systems, ARS scientists found that p-chlorophenyl methylcarbamate (PCMC) strongly inhibited the biodegradation of the herbicides propham and chlorpropham.

**Treating agricultural wastes.** Dealing with agricultural wastes as a major resource rather than as a necessary evil may hasten the recycling of those materials as one way to help cleanse the environment. Scientists chemically treated barn wastes to increase their digestibility and then blended them into rations for use as feed for sheep. This research demonstrated that barn waste rations do have potential as a forage substitute.

The potential role of nitrogen fertilization in polluting water created renewed interest in ways of suppressing nitrification. Scientists found that small amounts of potassium azide added to anhydrous ammonia fertilizer helped to insure against possible nitrate pollution of groundwater. Main-

Courtesy, Dr. Uheng Khoo and Dr. F. L. Baker, U.S. Department of Agriculture

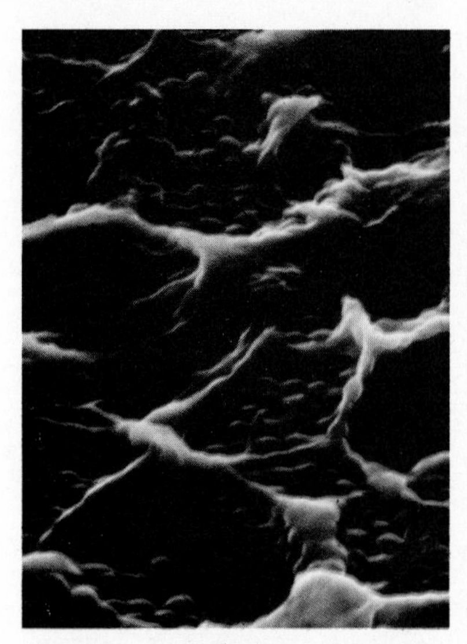

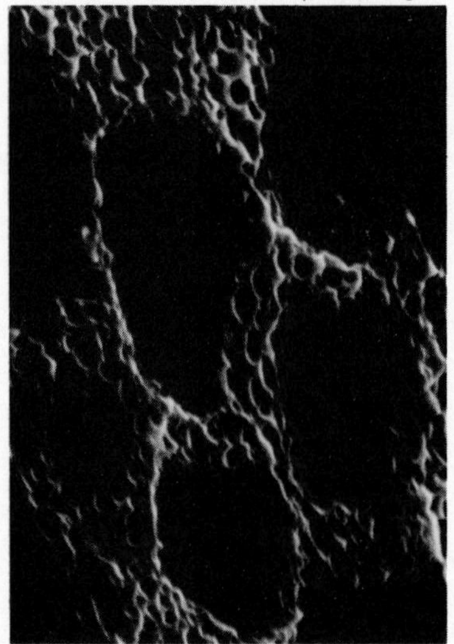

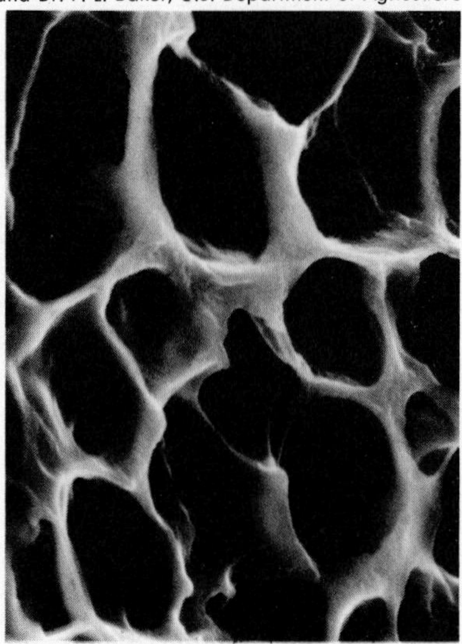

*Normal corn endosperm with starch removed (left) consists of matrix proteins and zein which is deficient in the essential amino acids lysine and tryptophan; center, endosperm after dissolution of zein. Removal of starch from high-lysine "opaque-2" strain (right), which lacks zein, leaves principally the amino acid-rich matrix protein glutelin.*

taining fertilizer nitrogen in the ammonium form with such nitrification inhibitors would not only restrict loss through leaching and runoff but also keep applied nitrogen available to plants for a longer time.

Finding more efficient uses of materials that man cannot eat and that may pose potential pollution problems is another way to aid in cleaning up the environment. Scientists found a unique solution to a part of this problem by feeding old newspapers mixed with molasses to ruminant farm animals. Replacing some forage in the feed ration with newsprint had no adverse effects on the animals, and research showed that newsprint could substitute for at least 8% of the feed ration roughage.

**Animal and plant science.** Although there was considerable emphasis in the past year on environmental studies, research in animal and plant science continued to make important advances. A major breakthrough in the freezing of boar semen promised to bring the full benefits of artificial insemination to the swine industry. Researchers credited the success to their development of a new additive and a special procedure for handling the semen before it is frozen. The use of this new procedure in conjunction with artificial insemination was expected to increase the rate at which genetically superior sires can be used to produce pigs with more lean meat and less fat.

Corn genes with extra food potential were recovered in studies of a primitive South American corn. Scientists discovered that the aleurone, the site of B vitamins and high-quality protein just under the kernel hull, is two to five cells thicker in primitive Coroico corn than in U.S. commercial hybrids. Coroico may serve as parent stock to improve the nutritional value of U.S. hybrids.

A new biochemical technique, called mitochondrial complementation, or MC, promised to shorten the time required for the development of improved crop varieties and hybrids. The technique permitted plant breeders to evaluate the production potential of crosses between proposed parent plants in the laboratory. It would be used to screen out crosses that would produce less vigorous offspring; only plants showing a high performance potential when tested biochemically would need to be crossed and field tested.

**Future outlook.** A hybrid sunflower that could become an important cash crop in the southern U.S. and a replacement for flax in the North was one of the hoped-for results of future research. The outlook for sunflower research was bright because scientists found the missing ingredient for breeding hybrid sunflowers. Called the restorer gene, it restores fertility to male-sterile plants and makes possible the breeding of hybrid sunflowers of consistent high quality with good yield.

Three experimental compounds offered promise as replacements for DDT and dieldrin as mothproofing agents. Because of their chemical nature, they were expected to be much less persistent than either DDT or dieldrin in the discarded solutions.

A highly toxic substance that may occur as a contaminant in a few pesticides synthesized from chlorophenols was under intensive study to learn its chemical and environmental significance. The contaminant, 2,4,7,8-tetrachlorodibenzo-p-dioxin

(TCDD), causes chloracne, a condition characterized by skin eruptions and irritations of the face, arms, and shoulders. TCDD is an important consideration in the recent review of uses of the herbicide 2,4,5-T.

—Robert B. Rathbone

*See also* Feature Article: NUTRITION AND THE AMERICAN DIET.

# Archaeology

During the year, archaeologists became increasingly concerned with the destruction of archaeological sites caused by the burgeoning international market in antiquities. The illicit trade also served to exacerbate the growing political problem of arranging for excavations in foreign countries, although this latter phenomenon appeared to have its roots in rising ethnocentrism, national pride, and resistance to the idea of foreign researchers unraveling the ancient history of one's own region. This, in turn, was leading to many more arrangements for international cooperation in excavations, more training of diggers in the less developed countries, and a somewhat different outlook on archaeological objectives.

**Africa.** The search for ancient man and his progenitors in Africa continued. Bryan Patterson of the Museum of Comparative Zoology in Cambridge, Mass., announced the discovery of a lower jaw with one molar in place, dated at 5.5 million years ago. The specimen was found in deposits on Lothagam Hill in northern Kenya in 1967, but was not announced until several years of research had established its date and identity. Dating of the jaw in the Pliocene period was determined not only by radioactive techniques but through correlation with animal bones of known age from other regions.

Patterson concluded that the bone is from a creature closely related to *Australopithecus*, a hominid ancestral to man, which was first reported from South Africa many years ago. Previously the oldest date for *Australopithecus* had been 3.5 million to 4 million years ago. The discovery of 5.5 million-year-old *Australopithecus* remains fills part of the evolutionary gap between *Ramapithecus*, a "man-ape" dated to 14 million years ago, and the first of the hominids. There was still a question as to when earlier forms developed the upright posture and ground-living habit of *Australopithecus* and the later hominids.

The history of the famed Benin art style, which reached its apogee in the Benin bronze sculptures of Nigeria, was being elucidated by excavations at Owo, 70 mi N of Benin. Some years ago extraordinary sculptured heads found at Ife were dated to the 9th century A.D., indicating that the Benin bronze art had its roots in an ancient African tradition, not in contacts with Europeans in the 16th century as was once supposed. Excavations at Owo were begun by Ekpo Eyo in 1971 after fragments of terra-cotta sculpture in the Ife style were found there by a surveyor. The site turned out to be a former sacred grave used in the worship of the goddess Oronse.

A large amount of sculptured material was found and, although detailed studies had yet to be made, it was clear that the older Ife art tradition did not die out in Owo before at least the 15th century. It now appeared that the Benin tradition was derived from Ife through Owo, and that the art form we know as Benin existed in West Africa for at least a thousand years.

**Near East.** At a site on the Lebanese coast between Tyre and Sidon, presumed to have been the ancient Sarafand referred to in the Bible (I Kings 17), excavations made in 1971 by James Pritchard of the University Museum, Philadelphia, turned up the personal seal of a man from Sarepta, inscribed with the Phoenician letters *s r p t* ("Sarepta" spelled without vowels in the Phoenician mode). Even more significant than this rare circumstance of identifying a site by an inscription was the fact that this was the first Phoenician city ever found on the homeland. These people, who transmitted the

*Leopard in a crouched position with a human leg in its mouth was unearthed at Owo, Nig., in 1971. Authorities hope the Owo finds, when fully analyzed, will shed new light on the evolution of the famed Benin art style.*

Courtesy, Federal Department of Antiquities, Lagos, Nigeria

alphabet to the West, are best known from their colonies, such as Carthage.

The late Minoan site on the island of Santorini (Thera), being excavated by Spyridon Marinatos, continued to make world news, probably because it is theoretically connected with the lost continent of Atlantis. The most significant discovery during the year concerned frescoes pieced together from fragments of wall paintings in buildings buried during the volcanic explosion that destroyed the Minoan town. They were described as the finest frescoes ever discovered in the Mediterranean region, more delicate, free, and rhythmical than the famous frescoes of Knossos on Crete.

**Europe.** Some clue to the facial appearance of men who lived in Europe 200,000 years ago was unearthed during the year in France. This is a period, preceding the age of Neanderthal man (the popular stereotype of the caveman), that had been known primarily from deposits of primitive flint tools. In July 1971, Henry de Lumley and his wife, from the University of Ais-Marseilles, found a largely intact skull with massive teeth in a cave above the village of Tautavel near Perpignan. Previously, two jawbones had been found in the same deposit, but neither of them fit the skull. One jaw is that of a man and the other of a woman. The skull is believed to be that of a youth about 20.

*"Fresco of the Princes" is one of the wall paintings painstakingly reassembled from fragments discovered at the late Minoan site Akrotiri on the island of Thera and installed in the National Museum in Athens.*

Dmitri Kessel

The new discovery shows extreme prognathism (snout-like, with jaws protruding far in front of the upper face) and unusually thick jawbones, both characteristics of ancient hominids. The skull is described as "lantern-jawed" and lacking the chin knob characteristic of modern man. There are sharp structural differences between the male and female jaws.

The cave contains 15 ft of deposit and 20 distinct habitation levels, each laid down during the period of the skull. More than 100,000 pieces of worked stone were found. There were also many bones of rhinoceroses, horses, bears, panthers, elephants, turtles, deer, a big archaic cow, wolves, rabbits, and birds. Broken bones indicate that these animals were eaten. The discovery fills in some of the huge time gap between *Pithecanthropus* (400,000–500,000 years ago) and Neanderthal (90,000 years ago), and suggests that many other links in the chain of hominid evolution will be found. The discoverers were inclined to believe the skull is evidence that Neanderthal man developed independently of his contemporaries in Africa and Asia.

Archaeologists and historians have often raised the question of what motivated the Greeks to colonize southern Italy and Sicily in the 8th century B.C. At least a partial answer emerged from recent discoveries at a new archaic Greek site on the island of Ischia in the Bay of Naples. The buildings at the new site are not simply private dwellings but the remains of a metalworking center. Large and small chunks of raw iron "bloom," fragments of broken implements, and nodules of vitreous slag occur in great abundance. The buildings themselves appear to be designed as metalworking shops. Bronze as well as iron was worked in these shops. Pieces of bronze sheets and wire, lumps of lead, a bronze ingot, and a miscast bronze fibula all attest to the manufacture of metal objects.

There are no ore deposits on the island of Ischia, and the obvious sources of the raw materials found there are the island of Elba and the Etruscan mines in central Italy, near Populonia. Diodorus of Sicily, writing at a later date, describes such a commerce in iron ore, which was carried to emporia, then fashioned into implements and exported throughout much of the known world. Strabo tells us it was men of the Euboean aristocracy who led the Greek colonists westward. It now appears that metal manufacture and trade played a basic role in the

Courtesy, The Institute of Archaeology, Tel Aviv University

*Assyrian-Babylonian cylinder seal of provincial workmanship was recovered at the Tel Beer-sheba site in Israel by an expedition directed by Yohanan Aharoni of Tel Aviv University. The sanctuary of ancient Beer-sheba, a fortified city which divided the land of Israel and Egypt, yielded a rich find of cult objects, all of pagan origin showing strong Egyptian influences. Further excavation was expected to shed more light on questions of early Israelite history and religion.*

first colonization. Work at the site was continuing under the direction of Giorgio Buchner and Jeffrey Klein for the Italian Antiquities Service and the University Museum, Philadelphia.

**Mexico.** In January 1939, in the Olmec country of eastern Mexico, Matthew Stirling of the Smithsonian Institution found a fragment of an inscribed stone monument recording a date in the Mayan calendrical system. The beginning cycle number (Baktun) was missing, but Stirling predicted that it would be a seven, making it the oldest known recorded date in America. During the year another fragment, recently discovered and reposing in a jail cell in the village of Tres Zapotes, Veracruz, was reported by Francisco Beverido and Michael Coe to be a matching piece of Stirling's original discovery. The beginning cycle number is indeed Baktun 7, confirming a date equivalent to 31 B.C. in our own calendar.

The discovery is significant not only because it is the earliest recorded date in America but because it helps to confirm a theory that the Olmecs, rather than the Mayans, invented this particular calendrical system. The earliest known recorded date in the Maya lowland region, found at Tikal in Guatemala, is A.D. 292. Other archaeological evidence also points to Olmec as the earliest civilization in America.

**North America.** Deep, stratified archaeological sites, normal in the Near East where the debris of thousands of years of human settlement are often found at a single site, are rare in North America. At the Koster site in the lower Illinois River Valley, however, excavations in 1971 were carried down to 18 ft and to an occupation level dated at 4200 B.C.; testing indicated that there are four lower levels that should date by the radiocarbon method to about 5000 B.C. Even more remarkable is the fact that the site's 12 cultural horizons, or occupation periods, are neatly separated by deposits of sterile soil washed in during periods when the site was abandoned. Such a separation of levels tends to preserve artifacts and other human and animal remains because they are not disturbed by later occupation. The only other known site in North America with this kind of separated stratigraphy extending over a very long period is at Onion Portage on the Kobuk River in Alaska, where caribou hunters settled repeatedly over a period of about 9,000 years.

Stuart Struever of Northwestern University began excavations at the Koster site in 1969. It is located on a slope in an old stream bed tributary to the Illinois, where loess deposited during periodic floods formed the stratified deposition. Continued excavation posed difficult problems. The depth of the cuts already completed raised the danger of cave-in and problems of earth removal. The two lowest horizons known from soundings are below the groundwater table, so that excavation would require constant pumping. Moreover, this kind of excavation requires a considerable team of researchers, including zoologists and botanists.

In British Columbia, University of Colorado archaeologists under the direction of James J. Hester discovered a site near the village of Namu in Fitz Hugh Sound where the earliest occupation was radiocarbon-dated at approximately 7000–8000 B.C. The significant point about this occupation lies in the fact that the material seems to have no connection with other material in the interior of southern British Columbia, but instead is allied with the earliest occupation at Onion Portage, which dates at approximately the same period.

These occupation levels in British Columbia and Alaska marked the earliest known horizon in the northern regions, though they fell far short of the period assumed for the original settlement of America from Asia. Another interesting aspect of the Namu site was the indication that the artifacts

Courtesy, The University Museum, University of Pennsylvania

*Ancient Greek vessel decorated with an uncomplicated geometric motif was discovered at the Mezzavia site on the island of Ischia. Other finds indicate that the site was a commercial metalworking center.*

reflect a coastal type of culture. This may have a bearing upon the theory that man first moved into North America along the sea front of what was once a land bridge in the Bering Sea area.

**Techniques.** The development of remote sensing devices in aircraft has greatly enlarged the scale of archaeological reconnaissance, and it was expected that further discoveries would result from experiments carried out by Earth Resources Technology Satellites. As an example of the technique, features that look like very straight roadways were traced nearly 15 mi in a northerly direction from Pueblo Alto in Chaco Canyon, New Mexico. Remote sensing from aircraft located a network of five of these so-called roads converging on masonry walls just east of the pueblo. They were first thought to be elements of a water system, but this was disproved by further study, and their nature remains a puzzle.

The discrepancy between radiocarbon ($^{14}C$) dates and certain astronomically fixed dates in the Egyptian calendar was first noted several years ago. Subsequently, comparison of the tree rings of 4,000-year-old bristlecone pine trees in the American Southwest, which could also be radiocarbon dated, indicated that $^{14}C$ dates prior to about 1000 B.C. are too young by several centuries. (See *1971 Britannica Yearbook of Science and the Future,* Feature Article: DATING THE PAST.) The MASCA Laboratory at the University Museum, Philadelphia, and the $^{14}C$ laboratory at the University of California in San Diego have both published systematic correction factors applicable to dates older than 1000 B.C.

Several papers published during the year by Colin Renfrew pointed out that European prehistory is being revolutionized by the corrected $^{14}C$ dates for Neolithic and Bronze Age sites in Europe and the Near East. For many years it has been generally agreed that major cultural advances in ancient Europe were derived from the Near East. With corrected dates, however, megalithic tombs in Western Europe and copper metallurgy in the Balkans proved to be older than their presumed prototypes in the eastern Mediterranean region.

One difficulty that emerged in the Renfrew papers involved the reliability of archaeological dates from Egypt and the Near East. A few of those dates, which can be correlated with actual astronomical phenomena, are certain, but the entire chronology built up from inscriptions recording the reigns of kings and other events is not so reliable. It may well be that much of the accepted Near Eastern archaeological chronology may need correction. In any case, radiocarbon dates around the world were being changed, and the effect on theories of diffusion, centers of origin, and time-relations of human events in different regions was sure to be significant.

The contributions of archaeology to the earth sciences were first noted when radiocarbon dating stations began to observe the increase of carbon dioxide in the atmosphere resulting from the burning of fossil fuels and the rapid dispersal in the earth's atmosphere of the fallout from nuclear tests. During the past year another example came to light. Radiocarbon dates for charcoal from aboriginal fireplaces at Lake Mungo in New South Wales, Austr., provide a record of a brief reversal of the earth's geomagnetic field about 30,000 years ago. The event apparently encompassed no more than 2,500 years. Additional evidence for such controversial events in the earth's history was also coming from Czechoslovakia and Mexico.

—Froelich Rainey

# Architecture and building engineering

The great skyscraper race, abruptly suspended with the completion of the Empire State Building in 1931 and not resumed until the 1960s, continued to accelerate in 1972. As it did so, a new basic architectural concept, tubular design, established itself.

**A new generation of skyscrapers.** The invention of architect-engineer Fazlur Khan, of Skidmore, Owings & Merrill of Chicago, tubular design was only eight years old, but it had been used in four of the world's five tallest buildings. Tubular design permits economical construction because with it the building's perimeter columns resist the lateral force of wind pressure, eliminating the need for expensive internal bracing in the floor system. Khan introduced the concept in a 43-story reinforced-concrete apartment house in Chicago. He then applied it to the design of two other Chicago structures, the 1,105-ft (337-m), steel-skeleton John Hancock Center and the 1,450-ft (442-m) Sears Tower, under construction as the new claimant of the title of world's tallest. The tubular concept was also used by other designers for the 1,350-ft (411-m) towers of the World Trade Center in New York City, and for the 1,136-ft (346-m) Standard Oil Co. of Indiana Building in Chicago.

On the Hancock project, the tubular design was visually marked by diamond-like diagonal truss bracing on the building's exterior, giving it something of the appearance of a bridge standing on end. The unique design saved $15 million and answered the problem of constructing a high-rise building that would have the great flexibility of interior space required to accommodate both office and residential use.

Following the Hancock, Khan used the tubular idea for the One Shell Plaza Building in Houston, Tex., a reinforced-concrete tower that rose 52 stories, setting a new record for all-concrete structures. In it Khan used what he called "tube-in-tube"; the inner tube, the walls surrounding the service core of the building, supported the interior dead loadings, while the closely spaced perimeter columns resisted the wind loads. Khan believed that concrete could be carried much higher by the tubular system and envisioned a 110-story concrete tower with criss-crossed bracing like that of the Hancock. Under construction in Chicago in 1972 was a 63-story combination hotel-office-commercial-residential building of concrete using tubular design. In another modification of his basic design, Khan used steel-and-concrete perimeter columns to take the wind load and a steel interior frame for the floor loadings of three new buildings, a 24-story office tower in Houston, a 36-story structure in Chicago, and a 50-story tower in New Orleans, La.

On the Sears Tower, Khan combined the tube design with a modular system. Nine continuous steel modules, each 75 ft sq in plan, form the tower from the street to the 50th floor. There the northwest and southeast modules stop, with the remaining seven modules continuing up through the next 16 floors. The northeast and southwest modules drop off at the 66th floor, and three others do so at the 90th. The final two continue to create the 20-story upper tower reaching to 1,450 ft (442-m). Common interior columns stiffen the interior of the modular skeleton. The modules create 75 × 75-ft column-free squares on every floor that can be organized freely to meet space requirements. Nine floors contain the building's mechanical operations such as heating and air-conditioning; three sets of two floors each (30th and 31st, 48th and 49th, 64th and 65th) and a single set of three (106th, 107th, and 108th) in the upper tower are clad with louvers, giving a striking exterior appearance of horizontal black bands.

Despite the large floors, the Sears building has an efficient floor-area ratio of 32 to 1, that is, 32 sq ft of building space to each square foot of the site. "We wanted to maintain a decent environment at ground level," said Khan, who planned the building's plaza to be dotted with art pieces. This consideration, rather than a desire to break a record, lay at the basis of the height decision. Sears, Roebuck and Co. required 3.7 million sq ft of office space; to preserve the plaza the building had to rise 110 floors.

If the site—a single square block on the west side of Chicago's Loop—was small for so large a building, it made up for the narrowness of the foundation by its depth. More than 180,000 cu yd of earth were excavated to make room for 240 caissons set on bedrock and hardpan. The hardpan caissons were to support the plaza and lower levels of the building, while the bedrock caissons underpinned the tower, one caisson for each interior and exterior column, which, at 15-ft intervals, made up the nine modules. The bedrock caissons are 6 ft in diameter and were sunk to a depth of 100 ft, some penetrating 9 ft into the rock for secure anchorage. Two 10-ft-diameter caissons in the center module support two columns that rise the full height of the building. The floor of the lowest basement consists of a 5-ft-thick mat that transfers the wind shear forces from above to the bedrock; the caissons take only vertical loadings. The world's largest drill rigs, one of which had a 500,000-ft-lb torque, were used to install the caissons.

Originally the Sears site consisted of two small city blocks separated by a narrow street. Khan and his colleagues considered two slim towers with a crossing over the street, a sort of modern version of the Wrigley Building, a Chicago landmark of another era. They even considered a single building with the street running through it, which would have made use of the tubular design impossible. The Sears Company solved the problem by purchasing the street from the city, but another problem immediately developed. Under the street ran a main sewer line that had to be protected as well as relocated. It was moved under the plaza and given 47 hardpan caissons for support.

Another new Chicago giant, the Standard Oil Building, rising on the south bank of the Chicago River close to the lakefront, used the tubular design to eliminate interior wind bracing. The work of two architectural firms, Edward Durell Stone and Associates of New York and Perkins & Will of Chicago, the building uses a time-saving fabrication and erection system devised by Perkins & Will vice-president E. Alfred Picardi. Folded-plate perimeter columns in the shape of V's and pressed-plate spandrels (exterior panels) were shop-fabricated, eliminating welding at the site. The folded-plate columns were mass-fabricated in three-story sections by automatic cutting and welding techniques and simply bolted in place during erection. The column-spandrel units gave the building a unique appearance during construction, that of a rising forest of crosses.

As in other tubular designs, the core columns of the Standard Oil tower helped take the floor loadings while the perimeter columns handled the wind bracing. The floors act as diaphragms to transfer lateral wind load to the V-shaped columns. The floor trusses, all identical, were also mass-produced. The building had no setbacks. The foundation pad was supported by 56 caissons extending 100 ft to bedrock. The columns were mounted on an "egg crate" concrete grid whose ribs connected the tops of the caissons. Five subplaza floors had reinforced-concrete load-bearing walls and interior columns bearing on caissons in hardpan.

A notable innovation of the building's construction was the use, at the initiative of Standard Oil, of a computerized system of cost-estimating and control developed by the company for use in tanker design and in oil and mineral exploration. The basis of the system is the Monte Carlo simulation technique, which calls for an input of the expected, lowest, and maximum costs of each item. The computer then generates a series of random numbers that are transformed into either the lowest, the highest, or the expected cost of each item, and adds them to arrive at the total cost of a hypothetical project. By repeating the process a thousand times, a realistic low, high, and expected cost figure is reached.

Perkins & Will later developed a second, similar program of its own, in conjunction with Massachusetts Institute of Technology, and was applying it on other projects. "The traditional cost estimate of a building," said Picardi, "has had at best a contingency factor having no really good relation to the numerous variables and conditions that influence the final cost." The Monte Carlo system gave a highly reliable set of estimates (provided of course that the data was sound) and helped alert designers and builders to trouble spots. Picardi pointed out that "this is not a cost-cutting technique; it is a cost-control technique, telling us more accurately what our costs are."

A cost not anticipated by the computer in the Standard Oil and other buildings of the new generation of skyscrapers may be that of solving the television ghost problem. Completion of the World Trade Center in New York brought ghosts (double images) to thousands of TV screens because some of the signals from the antennas atop the Empire State Building were reflected off the new twin towers and so arrived at New Yorkers' sets a split second late. The phenomenon was repeated in Chicago as the Sears and Standard Oil buildings began to rise in the paths of signals sent out from the John Hancock Center and Marina City antennas. Many other cities, both in the U.S. and abroad, were similarly threatened. Relocating transmitting antennas is expensive, and other solutions, such as using signal-absorbing construction materials, were under study. The maturing of cable television (CATV) appeared to be an ultimate solution.

The new skyscraper competition showed signs of stretching across the Atlantic to Europe, where it originated with Alexandre Eiffel's airy 984-ft (300-m) tower built in 1889 for the centennial of the French Revolution and ever since the major landmark of Paris. The tallest building in the world until the Empire State and Chrysler buildings surpassed it in the early 1930s, the Eiffel Tower remained by far the tallest structure in Europe. The 1970s might see it eclipsed, however. A 600-ft (185-m) conventional office building was begun in London in 1972 to house the Westminster Bank. It was a product of Britain's most controversial architect-engineer, Richard Seifert, who was accused of putting commercial considerations ahead of aesthetic ones and of threatening to pollute the low-rise atmosphere of London.

**Opening the Iron Gate.** A long-dreamed-of Eastern European engineering project was finally brought to fruition in 1972 with the completion of the Djerdap Dam over the Danube at the historic Iron Gate, a 70-mi (1,100 km) stretch of swift water, swirling with rapids, between Yugoslavia and Romania. The two countries spent nearly $1 billion to tame the dangerous torrent, with anticipated vast benefits in increased shipping on the river that had been called Eastern Europe's most underdeveloped resource. The project survived the Danube's worst flood of the 20th century in 1970 and stayed on schedule. In 1972 barges moving upriver negotiated the Iron Gate in 31 hr compared with 120 hr required previously, even with the aid of locomotives on shore.

The earthfill dam was 196 ft (60 m) high and

4,100 ft (1,250 m) long, had a two-chamber ship lock on either side, and identical powerhouses that produced 1,050,000 kw for each country. Besides sharing costs and benefits, Romania and Yugoslavia contributed jointly to the engineering design and the labor force. Construction was executed in three stages: (1) cellular cofferdams were built to enclose the lock and powerhouse structures on both sides; (2) sections of the dam were extended out into the river to match completion of powerhouse passages for the water flow; and (3) the final gap in midstream was closed by dumping a mass of concrete blocks. Each powerhouse was equipped with six generators, four of which fed the owner nation's grid, while the other two could be connected to either grid, allowing for varying peak demand in the two countries. In addition, power was to be sold to other Eastern European nations.

One detail of the project furnished a parallel to the rescue of the ancient Abu Simbel temple during construction of Egypt's Aswan High Dam in the 1960s. Yugoslav engineers carefully cut out of the canyon wall a Roman bas-relief and inscription commemorating the road built at the site by the Emperor Trajan nearly 20 centuries before and repositioned it above the new level of the back-up reservoir.

**Pollution control projects.** Air, water, and other types of pollution continued to draw the attention of engineers and public officials in 1972.

The Chicago Metropolitan Sanitary District, for example, began construction of the Salt Creek Water Reclamation Plant, which would pioneer a new two-step aeration process to reduce ammonia to less than 2.5 parts per million (ppm), suspended solids to 5 ppm, and basic oxygen demand (BOD) to 4 ppm. The plant was designed to meet the new, strict standards laid down by the state of Illinois. The plant would be capable of handling 50 million gal of flow per day, and a peak one-hour flow rate of 67 million gal per day. Storm flow was given primary sedimentation and chlorination treatment before being discharged into Salt Creek.

Sweden announced a new incentive plan whereby industries and towns that began to build antipollution systems or install air or water purification equipment could receive government grants

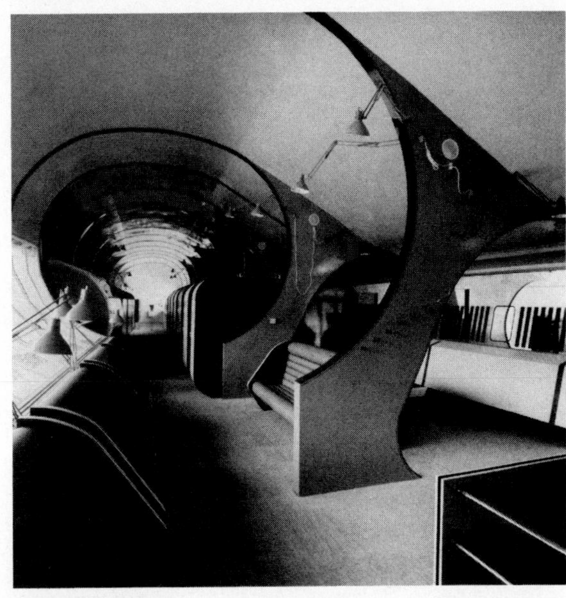

Robert Gray

*A double-airfoil silhouette and entrance portholes (left) are the distinguishing exterior characteristics of an orthopedic clinic in Allentown, Pa. Designer Noel Schaffer also planned the 14,300-sq ft interior (above), using a modular concept that can be subdivided or expanded almost at will to meet changing requirements.*

Courtesy, Standard Oil

*Shop-fabricated column-spandrel units form a series of crosses atop the Standard Oil Building in Chicago during construction in 1972. The building also utilized tubular design to eliminate interior wind bracing.*

covering 75% of the cost. The plan was designed simultaneously to ease Sweden's unemployment and improve its antipollution situation.

Sweden was also the site of the first world meeting to discuss the pollution problem. In June 1972, the United Nations Conference on the Human Environment met in Stockholm seeking to develop environmental control techniques on the international level through monitoring, resource management, and education, including assistance to less-developed countries in finding ways to advance economically while maintaining sound environmental practices. A number of antipollution steps were taken on the international level in Western Europe, where the European Economic Community agreed on continent-wide automobile emission standards, with the British and Swedish governments joining in.

**Tunneling ahead.** One engineering approach to pollution problems may be to dig. Underground construction, definitely on the upswing throughout the world, promised certain advantages in respect to the environment, as Ellis Armstrong, commissioner of the U.S. Bureau of Reclamation, among others, pointed out. Armstrong spoke at a symposium held by the Geological Society of America in Washington, at which another engineer forecast 554 mi (886 km) of transportation tunnels would be constructed in the 1970s, and up to 1,500 mi (2,400 km) in the 1980s.

Transportation tunnels could provide high-speed mass transit without damage to the environment. They could also be beneficial as automobile carriers because the auto emissions might be collected and treated before release to the atmosphere. Tunnels might appear in a new form for one of their oldest functions, that of water transport: a proposal was under study for an offshore pipeline, built as an immersed tube sunk in a trench of the continental shelf, to help solve the water shortage in southern California. Sweden, Norway, and other countries increasingly employed underground chambers for storing petroleum products, including compressed gas cooled to a liquid state.

Behind the growing popularity of tunneling for a variety of purposes was the improved capacity of tunneling machines (moles), especially for tunnelling through hard rock. Within the next decade the giant borers were expected to become capable of 500 ft a day in hard rock and 1,500 ft a day in soft ground. Their progress might soon make economically feasible an important antipollution measure under consideration in many cities: underground storage chambers for storm waters. Such chambers would permit the treatment of all wastewater, including storm water, received by a combined sewer system (virtually the only kind in existence). In the absence of a reservoir, this wastewater must bypass treatment plants whenever a heavy storm flow occurs.

**France's new port.** French engineers completed a major maritime construction project in the Gulf of Fos, 19 mi (30 km) west of Marseilles, a man-made harbor that took advantage of a fine natural harbor site. A long sandspit hooking out from the mouth of the Rhone River on the west side of the gulf half enclosed the inlet. The Marseilles port authority engineers built a 2-mi (3-km) jetty extending from the base of the gulf toward the point of the sandspit, leaving an entranceway between. Along the interior of the jetty four huge berths were built for the new giant oil tankers. To give the jetty a firm foundation, a wide trench was first dug and filled with hard sand from the mouth of the Rhone. Gravel fill and rock fill completed the embankment. Within the harbor three broad channels were cut into the uninhabited marshland to receive ore carriers and container ships. The Fos harbor, added to that at Marseilles proper, made the area Europe's second largest port.

**The Disney World water problem.** Construction of the 30,000-ac Walt Disney World near Orlando, Fla., which opened in October 1971, was made possible by an ingenious system of hydraulic engineering that overcame the table-flat site's two conflicting meteorological phenomena, seasonal droughts and tropical rainstorms. To prevent the water table from dropping and killing vegetation, and to keep the entertainment area unflooded through the year-round tourist season, a network of 50 mi (80 km) of canals with 30 gated control structures, and double-acting levees was built. The gates had to act automatically to cope with the suddenness of Florida downpours. The device employed was not electrical or mechanical, however, but relied simply on a counterweight and a buoyancy compartment; when the counterweight was adjusted for a particular water level, the buoyancy compartment operated the gate unattended.

**The world's largest powerplant.** By mid-1972 work was nearing completion on the foundation and abutment clearing work for the Sayano-Shushenskaya hydroelectric project in the U.S.S.R. When completed in 1977, it was expected to produce a world's record 6,400,000 kw of electric power. The project was part of the development of Siberia's Yenisei-Angara river basin. The arch dam that would block the narrow Yenisei canyon was designed to rise 774 ft (236 m).

Soviet planners hoped to build as many as 15 dams in the project. This would produce enough power to provide the basis for a vast industrial development to exploit the metal-ore riches of the region.

—Joseph Gies

# Astronautics and space exploration

The pace of manned space exploration slowed during the year under review with only one flight to the moon by the United States and none by the Soviet Union. But in unmanned flights considerable activity took place. Highlights included probes to Mars and Venus and the first spacecraft ever to be launched to Jupiter.

## Earth satellites

Earth-oriented satellites utilize the vantage point of space in a number of ways, seeking increased knowledge of the earth, economic benefit, and military security. Also called applications satellites, there are three general classes: communications, earth-survey, and navigation. By 1972 each had developed greatly in performance, capability, sophistication, and promise for the future.

**Communications satellites.** The Communications Satellite Corporation (Comsat), a U.S. firm, operates the global space communications network for an 83-nation consortium. Four fourth-generation Intelsat 4 satellites were in orbit by mid-1972. Each of these 18-ft-tall by 8-ft-diameter satellites provides an average of 6,000 two-way telephone circuits or 12 color television channels or tens of thousands of telephone circuits. The operational lifetime of each is seven years. Thirty-nine countries are now linked through 63 earth stations. By 1979 it was estimated that there would be 90 ground antennae in about 58 countries.

All Comsat satellites are launched into geostationary, or synchronous, orbits. This means that at a certain orbital altitude, 22,300 mi, the satellite velocity matches the angular rate of the earth's rotation and thus remains fixed over one portion of the earth's surface.

When U.S. Pres. Richard Nixon made his historic eight-day trip to China in February 1972, nearly 200 hours of satellite time were used for high-quality color television transmission. After the trip China leased four Comsat circuits for voice, teleprinter, and telephoto service. Earlier in February, the Winter Olympic Games in Sapporo, Jap., were televised by Comsat to Europe and the United States.

The development of satellites for secure communications by U.S. military forces advanced in 1971–72. In November 1971 two high-powered geostationary satellites were launched using a Titan IIIC booster. Four more were planned to replace the smaller, lower-powered, and limited-bandwidth military satellites. In early 1972 this early class of military satellites numbered 21 in operation using 29 earth terminals. The satellites were designed to turn off automatically during the period 1972–74. The development of a tactical communications satellite system for application to small ground terminals and the U.S. fleet continued. Receiving antennae as small as 1 ft in diameter might be feasible in this system.

Several applications were made by Comsat and other U.S. firms for licensing systems to transmit nationwide television by geostationary satellites. The Comsat proposal would employ three satellites larger than Intelsat 4s, each with a capacity of 24 color television channels, and would utilize an initial ground network of 132 stations.

Among the advantages that might be gained by the increased development of communications satellites is the reduced cost of long-distance communications, both for the private consumer and for data transmission by industry. In the realm of television broadcasting, however, such satellites perhaps offer their greatest promise of impact and general benefit. The satellites in operation in 1972 required large, powerful receiving stations that transmit television by microwave or landlines to a local broadcasting station whose own transmitters then beam the television image to home receivers. Eventually, however, scientists may develop satellites capable of broadcasting directly to home receivers. Such a satellite must have high broadcast power and a large antenna. To meet these needs, the U.S. National Aeronautics and Space Administration (NASA) was currently developing the Applications Technology Satellite F (ATS-F). It was to be equipped with a 30-ft-diameter antenna, the largest yet deployed in space.

In 1973 the ATS-F satellite was scheduled to be used by the government of India in a program of mass education. Receiving television signals from an Indian transmitter, the satellite would then rebroadcast from its vantage point in space to approximately 5,000 villages. Each village, according to the plan, would have an antenna and at least one low-cost large-screen television receiver. Initially, the programs were to deal with improvements in agricultural practices and methods of population control. The possible expansion of this experiment within India could lead to direct linkage of all of that nation's approximately 560,000 villages. Since interlinking television landlines do not exist and the country has insufficient funds to build them, the broadcast satellite holds immense potential as the most effective way of educating India's 524 million people.

Meanwhile, Canada was moving ahead to orbit Anik 1, the world's first domestic communications satellite, late in 1972. Three of these satellites were being constructed for Telesat Canada, and each was designed to relay 12 communications channels. Each channel could provide one color television program or as many as 960 voice channels. Anik is an Eskimo word meaning brother.

In Europe development of the Franco-West German Symphonie satellite continued. It was not completed in time for the 1972 Olympic Games at Munich, but a launch of the prototype in early 1974 was thought feasible.

Two more Molniya 1 and the first of the Molniya 2 communications satellites were launched in the U.S.S.R. in 1971. The new design operates on higher frequencies than the Molniya 1, more closely approximating those of Intelsat 4.

The Soviet Union's approach to communications satellite systems remained quite different from that of the U.S. Beginning in 1965, the U.S.S.R. launched a series of Molniya 1 satellites in an orbit inclined about 65° to the equator. The orbital path was not geostationary but eccentric, moving closest to the earth in the Southern Hemisphere and reaching its highest altitude over the Northern Hemisphere. This path permitted a Soviet satellite to travel for a relatively long time, about eight hours, over the portion of the earth visible to U.S.S.R. ground terminals. Because the Soviet craft are higher powered than corresponding U.S. satellites, relatively low-cost ground terminals can be used. Three such satellites, by proper spacing, can achieve 24-hour coverage within the Soviet Union. Ground terminal construction in the U.S.S.R. continued at the rate of 6 to 8 per year, and by late 1972 there were about 50 stations.

*Three simultaneous photographs of a portion of the earth's surface, exposed on (left to right) black and white film with a green filter, near infrared black and white film, and black and white film with a red filter, were taken by a camera system developed for the Earth Resources Technology Satellite, launched in 1972.*

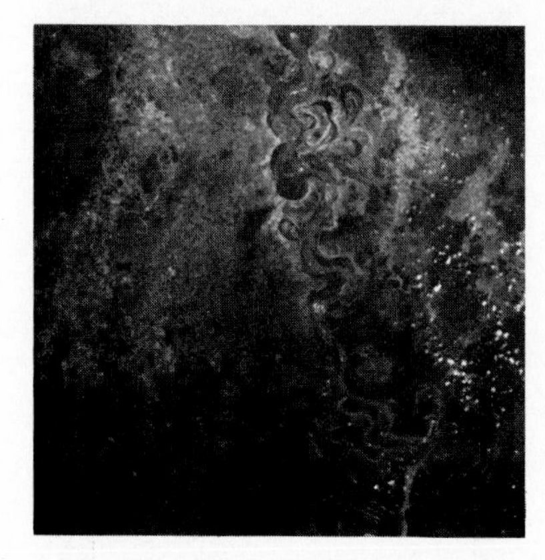

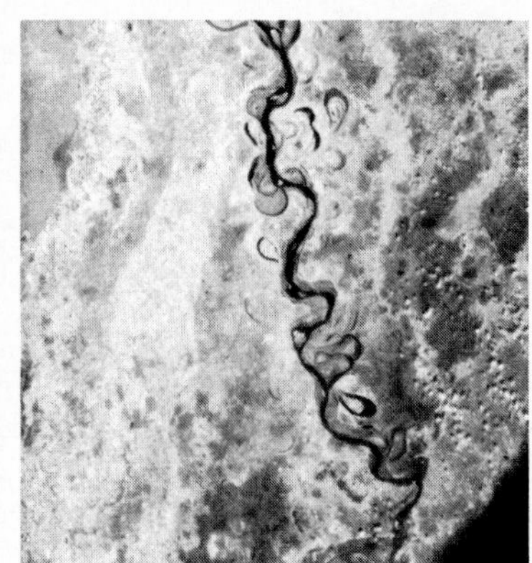

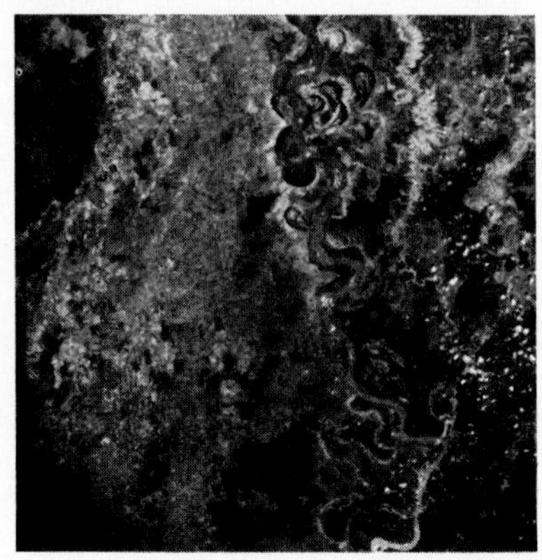

Courtesy, NASA

In late 1971 Cuba joined Intersputnik, the Soviet bloc equivalent of Intelsat. The Cuban ground station was scheduled to become operational in 1973.

**Earth-survey satellites.** In this category are included meteorological (weather), geodetic, earth resources, and reconnaissance satellites. Such satellites are designed to survey the earth with various sensors, both photographic and electronic, in order to obtain and transmit data that could not be gained by other means.

*Weather satellites.* Only the U.S. and the Soviet Union in 1972 had operational satellites with continuing global daily coverage. In the U.S. the National Environmental Satellite Service (NESS) was responsible for weather satellite operations. NESS is a major element of the National Oceanic and Atmospheric Administration of the Department of Commerce. Developmental work on advanced design satellites and all launchings into polar orbit were conducted by NASA.

In early 1971 NESS was controlling four operational meteorological satellites. Two of these spacecraft were primary sources of operational cloud-cover data, and two were backup in case of failure of the primary system. In mid-1971 failure did occur and the two older, standby, spacecraft were "called up." Their sensors and transmitters were activated and they provided data as required.

Two major uses are made of these U.S. satellites. One is to store global cloud-cover data and then "dump" it (or read it out) on command by one of two U.S. command and data stations. This information is then transmitted to central processing centers for operational use on a global scale. The other use for the satellites is the direct readout for immediate local use. This latter worldwide availability of Automatic Picture Transmission (APT) permits direct interrogation of the satellite by more than 500 APT stations located in 94 countries and trust territories. Because the NESS satellites are in polar orbit any point on the earth's surface can have direct access to a picture of regional, cloud-indicated weather at least once a day.

While NESS conducted regular weather satellite operations, NASA continued to develop advanced systems and components. These included Nimbus 4 and two Applications Technology Satellites (ATS-1 and -3). Nimbus 4 carried, among other sensors, an infrared spectrometer, data from which is converted into a description of the vertical temperature distribution in the atmosphere. The ATS satellites were geostationary, and their daytime pictures covered most of the Western Hemisphere. By use of a time-lapse movie sequence technique the development and motion of large- and small-scale cloud systems are studied to estimate low- and upper-level wind speed and direction and in the diagnosis of hurricane and severe local storm situations. National television weather reports now alert regions on a daily basis as to where severe storms may occur.

In 1971 the U.S. operational weather satellite system completed five years of uninterrupted service. During this period great quantities of data on global weather movement were acquired, stored, and studied at the National Meteorological Center, Suitland, Md. They proved useful in forecasting weather conditions that endanger livestock in western ranching regions, in marine advisories, in sea and lake ice forecasts, and in tracking and forecasting the movements and intensities of severe storms and hurricanes.

NESS is conducting a number of studies, including the monitoring of polar ice packs, snow mapping and snow surface temperatures, and the manner in which tropical heat is injected into the middle latitudes. The results of such studies are expected to lead to a better understanding of how weather develops, thus improving short- and long-term forecasting.

The Soviet Union continued to launch its Meteor class of weather satellites, with four orbited successfully in 1971. A cooperative program of rapid exchange of weather photographs was established between the U.S. and the U.S.S.R.

*Geodetic satellites.* By simultaneous observation of a satellite from two distant points on the earth's surface the distance between these two points can be determined with an accuracy not previously possible. The first satellites used for geodesy were launched in the early 1960s. In 1971 a program using laser observation equipment was carried out by the U.S., several Western European countries, Australia, and Japan. Three French and four U.S. satellites were used for observations. Such data can correct maps and properly locate continents as well as measure the small annual continental drift.

*Earth resources satellites.* As a logical outgrowth of growing sophistication of observations of the earth from the vantage point of space, NASA launched an Earth Resources Technology Satellite (ERTS) in mid-1972. Many of the sensors and subsystems were advanced developments of those tested in the ATS series, Nimbus, and other spacecraft. ERTS-A was launched in a near-polar, circular orbit at an altitude of 492 mi. It weighed about 1,800 lb and had an estimated lifetime of one year. The satellite carried three main systems: a television camera, multispectra scanners, and a data collection system.

The multispectra scanners detect the distinctive electromagnetic radiation reflected by every object on earth, including its subsurface strata, its atmosphere, and its oceans. Vegetation, for example, has a distinct reflective "signature" different from water or bare earth. Different crops or trees have different signatures. Moreover, diseased plant life has a different signature from the healthy crop.

ERTS-A circled the earth every 103 minutes, completing almost 14 orbits a day. It thus could see any spot in the U.S. every 18 days. The satellite scanned a 100-sq mi area, relaying overlapping pictures back to earth or storing them to be relayed later. A number of remote automatic data collection platforms were located in the U.S. and its coastal areas. Each platform had as many as eight sensors, which sampled such local environmental conditions as stream flow, snow depth or soil moisture, and air and ground temperature. Information from any platform was available to users in less than 12 hours from time of measurement. Three U.S. data acquisition stations are planned: Fairbanks, Alaska; Mojave, Calif.; and NASA, Greenbelt, Md.

Both ERTS-A and ERTS-B, to be launched in 1973, are essentially concerned with further developing the technology of this type of spacecraft. Many scientists expect these satellites to be the most useful of all applications of space flight. The data that they obtain will be made available to hundreds of scientists in universities, state and local governments, and federal agencies.

*Thirty-foot-diameter antenna was developed for Applications Technology Satellites designed to pioneer new educational television, data relay, air-traffic control, and information transmission systems.*

Courtesy, NASA

Listed below are the hoped-for benefits from earth resources surveys:

In agriculture, the information gathered will aid in land use planning, range management, identification and combatting of crop diseases, and improved irrigation planning.

In geology, scientists envision ERTS data being used in monitoring glaciers and volcanoes, in improving earthquake predictions and warnings, and in identifying terrain features usually associated with oil and mineral deposits.

In hydrology, the ERTS system promises to produce information useful in detecting water-pollution trends, in providing an inventory of lake and reservoir levels and rain and snow levels, in predicting the potential of floods, and in the locating of other water reserves.

In oceanography, the ability of ERTS sensors to detect changes in sea temperature may produce information useful in showing the probable location of schools of fish. ERTS data also promise benefits in aiding maritime commerce by better charting of sea conditions, and in spotting ice fields for iceberg warnings.

In geography, ERTS sensors will produce a constantly updated map showing the various changes in the earth's surface, as the result both of natural events and of building by man. This information promises to be of great value for use in urban development and transportation planning.

From the standpoint of environment it is possible that ERTS will provide a powerful tool to assess and monitor the earth's ecology on a global scale. Infrared scanners, for example, can detect thermal pollution in water—both day and night, while conventional photographs from space can show discoloration and patterns in bodies of water for tracking and controlling pollutants. Ecologists agree generally that only by dealing with the problem on a worldwide basis will man be able to save his oceans and atmosphere from being destroyed by pollution.

*Reconnaissance satellites.* The U.S. and Soviet Union continued their reconnaissance satellite programs, but they continued to be highly classified for security purposes and no details about them were released. It was known, however, that the spacecraft involved in these programs utilized extremely high-resolution photographic and narrow-band electromagnetic radiation surveillance techniques.

Since the early 1960s both countries have launched more than 175 military observation low-orbit, satellites. All payloads were recovered after a few days in orbit.

One additional type of U.S. military reconnaissance satellite that deserves mention is the Vela. Its purpose is to detect nuclear explosions in the atmosphere and in space. Two Vela satellites were placed in orbit in 1970, the sixth pair in a series that began in 1963.

**Navigation satellites.** In the area of navigation satellites, the U.S. Navy Transit system continued to be fully operational and in use by the U.S. fleet throughout the world. Although commercial ships were permitted to use this system, it required a shipborne computer to calculate position from the Doppler shift of the satellite's signal and was thus probably too expensive for widespread public use. Future civilian systems were more likely to depend upon a signal transmitted from a ship via satellite to a shore-based computer, where position data could be determined and broadcast back.

The Soviet Union claimed to have an operational navigation satellite system, but released no details. NASA and the European Space Research Organization were holding joint studies concerning satellites that would be used for communications and for surveillance of aircraft flying in the North Atlantic air routes. This was expected to permit aircraft to fly closer to one another. Thus, while in 1972 airplanes must be separated laterally by distances of about 120 mi, a satellite monitoring system could permit continuing "fixes" of aircraft only a few miles apart. Tests by the U.S. Federal Aviation Administration indicated that such a system could be in operation in the late 1970s.

—F. C. Durant III

## Manned space exploration

During the past year, manned space flight was reduced. The United States successfully completed the Apollo 16 lunar exploration, while the Soviet Union did not launch any manned missions following the tragic Soyuz 11 flight that ended with the death of the three cosmonaut crewmen. In the U.S., the manned space flight program continued to be criticized by some members of the Congress, the science community, and other public leaders. There was a growing feeling that the resources being used for the manned space flight program should be used in the solution of the nation's domestic and welfare problems. Notwithstanding this criticism, U.S. Pres. Richard Nixon announced that he had decided that the U.S. should proceed with the development of a new type of manned space transportation system called the space shuttle. In addition, preliminary planning continued on a joint U.S./Soviet Union international rendezvous and docking mission.

**Apollo.** Apollo 16 was launched from Cape Kennedy on April 16, 1972, after a delay of one month for various technical problems. The commander for the mission was John W. Young. It was his fourth space flight, as he had previously flown on Apollo 10 and Gemini 3 and 10. The lunar module pilot was Charles M. Duke, Jr., and the command module pilot was Thomas K. Mattingly II, both rookies in space. The launch and flight into lunar orbit were accomplished as scheduled except for some minor problems with the S IVB section of the Saturn V booster rocket.

The first major problem encountered in the mission occurred in lunar orbit after separation of the lunar module from the command module prior to the start of lunar landing operations. It was found that a secondary yaw gimbal actuation on the service module propulsion system was oscillating. Tests and analysis showed, however, that the system was still usable and safe. This problem delayed lunar landing approximately six hours.

The landing site for the 11-day mission was 45 mi N of the ancient crater Descartes on the hilly, grooved, and furrowed edge of the Kant Plateau in the central lunar highlands among the highest mountains on the face of the moon. The landing site coordinates were 9°00′01″ S and 15°30′50″ E. This area of the moon was important to geologists because rock and soil samples from this site were expected to provide information about the composition and age of the lunar surface and also increase our knowledge about volcanic activity and the role it played in the evolution of the moon. Near the landing site were two large craters, North Ray and South Ray, which had thrown out huge blocks of the basin fill.

Three lunar surface excursions were taken by Young and Duke. On the first, an electric-powered four-wheeled lunar rover was deployed, an Apollo lunar surface experiments package (ALSEP) was placed on the surface and activated, and geological samples were obtained. The astronauts also deployed the first lunar astronomical observation device, a Far Ultraviolet Camera Spectrograph. This device took photographs of the earth, various galaxies, and the Magellanic Clouds. From these pictures, scientists expected to be able to study the earth's atmosphere and magnetosphere and their interaction with the solar wind. In addition, the spectrograph was designed to help astronomers study the interstellar gas that is present throughout space and the ultraviolet halos that appear around galaxies. This experiment would thus provide an evaluation of the moon as a possible site for future astronomical observations.

On the first excursion by Young and Duke, two lunar exploration stations were made in the vicinity of the Flag and Spook craters. On the second, geological investigations and lunar sampling were conducted at Stone Mountain and at several craters. On the third trip the crew drove to the rim of North Ray Crater and examined this area in detail, obtaining both rock samples and photographs. In total the Apollo 16 astronauts spent 20 hours and 14 minutes in lunar surface exploration, traveling almost 17 mi and collecting about 213 lb of rock and soil samples.

After approximately 71 hours on the moon's surface, the crew initiated lunar ascent and rendezvous and docked with the command module. The mission was terminated one day early. Landing and recovery took place in the Pacific Ocean.

Apollo 17, scheduled to be launched in December 1972, is the last mission in the current U.S. lunar landing exploration program. The landing site selected is Taurus-Littrow in the northeastern portion of the moon at 20°10′N latitude and 30°48′E longitude, and on the southeastern rim of the Sea of Serenity. The area is characterized by material that is believed to be ancient highland crustal blocks emplaced by faulting and uplifting at the time of the formation of the Serenity basin. Both very old and much younger surface material are expected to be present at the site. The objectives of the mission will be to conduct lunar surface geological exploration, to establish another ALSEP

*The reusable space shuttle is designed to make possible a more economical manned space program. After being launched by a reusable booster rocket, the shuttle can deliver scientific payloads into earth orbit through its open hatches (top). When its mission is completed, the shuttle can be landed like an airplane (bottom).*

station, and to conduct lunar surface traverse experiments. In addition, lunar orbital experiments will also be conducted from the command module. The same basic spacecraft and lunar rover vehicle were to be used as in Apollo 15 and 16.

**Skylab.** The fourth U.S. manned space flight program, Skylab, was designed to use existing Apollo spacecraft and launch vehicles with modifications for long-term earth orbital flights. The third stage of the Saturn V, the S IVB, was modified to serve as an orbital workshop. It contained living quarters and the necessary provisions for sleeping, eating, and housekeeping to support three crewmen for one 28-day and two 56-day missions. It also contained medical and other science and engineering experiment equipment. The workshop was 22 ft in diameter, 48 ft long, and had two floors. Its environment was to be maintained at a pressure of five pounds per square inch with an atmosphere of 70% oxygen and 30% nitrogen.

An airlock module, consisting of a special hatch, an airlock compartment, and equipment for supporting space walks by the Skylab crewmen to retrieve science experiments, was to be joined to the orbital workshop. It also contained additional crew working areas to perform earth-resources and space-manufacturing experiments and housed the controls and displays for the Apollo Telescope Mount. A third module, the multiple docking adapter, was designed to provide facilities for docking the Apollo command module, which was to be used to transport the Skylab crewmen to and from the orbital workshop.

During operations, scheduled to begin in the spring of 1973, the unmanned orbital workshop with the airlock module and multiple docking adapter was to be launched by a Saturn V into earth orbit at an altitude of 270 mi. One day after the Skylab launch, three crewmen would be launched in the Apollo command module by a Saturn IB. They would fly the command module to a rendezvous and docking with the orbital workshop. The three crewmen would then activate the orbital workshop and carry out various experiments within it for 28 days, after which they would deactivate the workshop and return to earth in the command module. Sixty days later, another three-man crew would be launched into orbit to revisit the orbital workshop for 56 days. A third three-man crew would be launched 60 days later for another 56-day mission.

In the Skylab program, the crews were scheduled to perform experiments in many technical fields. Medical experiments would be conducted to study the various body systems. From these studies, scientists expected to obtain information on the physiological responses to long-term manned space flight and the crew's ability to adapt to a weightless environment. These medical studies were considered to be of extreme importance to future manned flight programs. Science experiments were expected to include geophysics, physics of the upper atmosphere, astronomy, and studies of the sun. The Apollo Telescope Mount was to contain a complex solar observatory that would permit observations and measurements to be made on the sun that are not possible to make from the ground because of the earth's atmosphere. An earth resources instrument package was to be used to conduct broad surveys of crops, geological formations, weather conditions, and gross vegetational regions. Engineering and technology experiments were to include material processing and manufacturing in the weightless environment.

**Space shuttle.** In the past year, preliminary design concepts were established for a reusable manned spacecraft called the space shuttle. Space engineers conceived it to be an airplanelike vehicle that could be used repeatedly to take satellites into orbit and bring them back for repair and service as well as performing manned orbital missions on its own. It was hoped that the shuttle would reduce the cost of manned and unmanned space flights. The space shuttle was to consist of an orbiter and booster. The orbiter would resemble an airplane and be about the size of a small jet airliner. It would be powered by three hydrogen-oxygen rocket engines and have a cargo bay 15 ft in diameter and 60 ft in length. A two-man crew would fly the orbiter, which would also be able to accommodate 2–14 scientists or technicians. The booster would be an unmanned recoverable solid-fueled rocket.

During operation the orbiter stage would be launched in a "piggybacked" position by the booster. At a predetermined altitude, the orbiter would separate and fly into orbit like a spacecraft. In orbit, the crew and scientist passengers would be able to launch unmanned satellites and space probes, conduct experiments, make observations of the earth and solar system, and retrieve and/or repair satellites previously placed in orbit. When the mission was completed, the orbiter would reenter the earth's atmosphere and land like a conventional airplane. It would then be serviced and made ready for future flights. The first space shuttle flight was scheduled for 1978.

**Soviet manned space flight.** In the past year, the U.S.S.R. did not launch any manned missions. The ill-fated Soyuz 11, launched June 6, 1971, was the last manned Soviet space flight. In this mission, the crew was launched in the Soyuz spacecraft and then rendezvoused with Salyut, a large orbiting space station 66 ft long and 13 ft in diameter.

Courtesy, NASA

*Apollo 16 commander John W. Young works on the Lunar Roving Vehicle (left). Astronaut Charles M. Duke, Jr., piloted the lunar module (right) and astronaut Thomas K. Mattingly II remained with the command and service modules in lunar orbit during NASA's fifth manned voyage to the moon.*

The crew transferred to the Salyut space station and performed biomedical studies similar to some of the planned Skylab medical experiments. Studies in astrophysics and meteorology were also conducted. The crew made observations on the earth's weather conditions, and these were then coordinated by earthbound scientists with photographs transmitted from unmanned weather satellites.

The crewmen terminated the mission 24 days after the launch. They moved back into the Soyuz 11, separated from the Salyut, and started the procedures to reenter the earth's atmosphere. Shortly after the reentry rockets were fired, there was a loss in communications. When the spacecraft landed, ground crewmen found that the three cosmonauts, Georgi T. Dobrovolsky, Vladislav N. Volkov, and Viktor I. Patsayev were dead. A Soviet commission appointed to study the cause of the accident reported that a hatch or valve failure created a rapid decompression of the Soyuz 11 spacecraft, killing the three cosmonauts. The Salyut remained in orbit until Oct. 11, 1971, when retro-rockets were fired and it was destroyed during reentry into the earth's atmosphere.

**International cooperation in space research.** The U.S. and the Soviet Union continued to work together to determine the feasibility of an international docking mission. As envisioned in 1972, three Apollo crewmen would be launched in an Apollo command module to rendezvous and dock with three crewmen aboard the Soyuz spacecraft. The crewmen would move from one spacecraft to the other to perform joint operations. This mission would lay the foundation for international space rescue missions. Engineers and scientists from the two countries exchanged technical design data to establish the required docking interfaces between the two spacecraft. Formal approval of the mission was made during President Nixon's visit to the Soviet Union in May 1972.

In other international cooperative efforts between the U.S. and U.S.S.R., biomedical specialists met in Moscow in the fall of 1971 to start the exchange of medical data from the manned space flight program. A second meeting was held at the Manned Spacecraft Center in Houston, Tex., during May 1972 to exchange medical results from Soyuz 11 and the Apollo program.

—Richard S. Johnston

## Space probes

Despite the landing of still more men on the moon, 1972 was clearly the year of Mars. As the year dawned, three probes were in orbit about the planet, transmitting an incredible amount of information about its topography, geochemistry, and atmospheric physics. In addition to the manned landings by Americans, the moon was also the target for unmanned probes from the U.S.S.R. The year also saw man's first probe launched to Jupiter and another launched to Venus.

**Probing Mars.** With both the United States and the Soviet Union concentrating on studies of Mars, scientists of the world had high hopes for cooperation between the two nations. Indeed, a peaceful "hot line" was operating in November 1971, between the Jet Propulsion Laboratory, Pasadena, Calif., and the Academy of Sciences of the U.S.S.R. in Moscow, to exchange data from the two country's probes. It soon became apparent, however, that the "hot line" was a victim of the cold war. As a scientist at the Jet Propulsion Laboratory remarked, "The big disappointment to some of us is that they haven't responded to our suggestions of some cooperative ventures." Since the Soviet and U.S. spacecraft each had instrumentation that was unique, the lack of a cooperative exchange of data between the two countries was all the more disheartening. However, the "hot line" was not wholly unused. At one point, the Soviets on January 9 detected a 15° rise in temperature at Cereberus, in the Eastern Hemisphere of Mars near the equator. They asked their American counterparts if they too had noted it. A review of data from Mariner 9 did show a 10° rise in the area at the same time. Late in December, the Jet Propulsion Laboratory scientists sent a Mariner 9 photograph of the landing of the Soviet Mars 3 pod to their Soviet counterparts.

Mariner 9 entered orbit about Mars on Nov. 13, 1971. It initially completed one orbit in 12 hours and 34 minutes with a periapsis, or point closest to the planet, of 1,397 km (868 mi) and an apoapsis, or point farthest from it, of 17,927 km (11,135 mi). A maneuver was made two days later that slightly reduced the time of orbit and lowered the apoapsis to 17,048 km (10,655 mi), while changing the spacecraft's angle of inclination to the Martian equator from 64.28° to 64.36°. Since Mariner 9 had arrived at Mars during a violent dust storm, scientists at the Jet Propulsion Laboratory were skeptical of what the photographs from its two television cameras would reveal. Instruments aboard the spacecraft indicated that dust particles stirred up by the storm were thrown as high as 74 km (46 mi) into the planet's extremely thin atmosphere. However, by mid-December the dust storm began abating to such an extent that a decision was made to begin systematic photo-mapping of the planet's surface on Jan. 1, 1972.

The photographs soon revealed a wealth of geological features. Typical of these was a great rille or chasm along the Martian equator about 4,800 km (3,000 mi) long and 96 km (60 mi) wide,

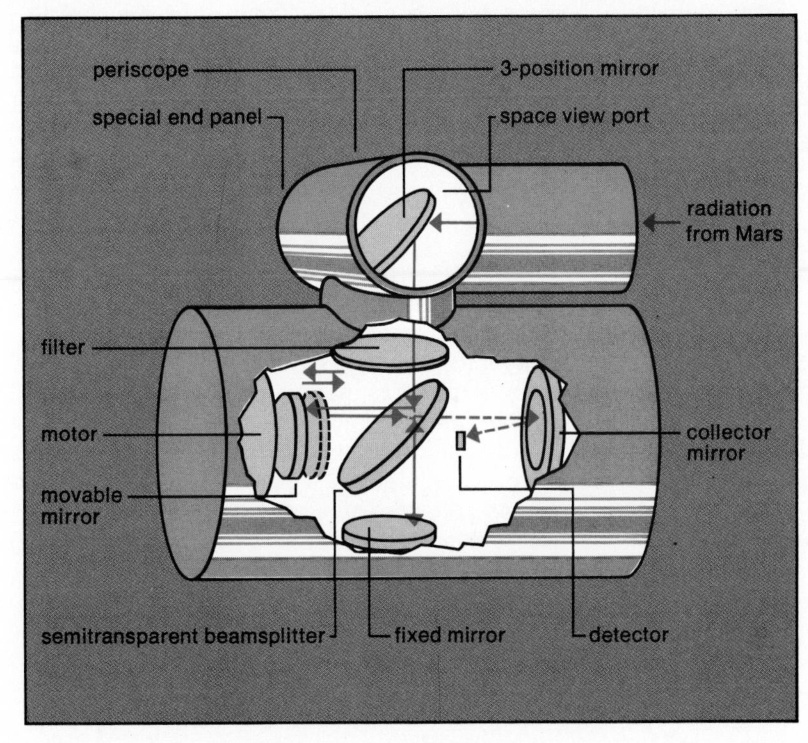


*The infrared spectrometer aboard the Mariner 9 measures atmospheric and surface characteristics of Mars. Radiation from the planet is reflected from a three-position mirror and then passed through a filter that screens out all but a certain range of infrared light. The light waves are then reflected from a fixed mirror to a beamsplitter and a movable mirror. The latter changes the distance that the waves travel, thereby either strengthening or canceling out the waves reflected to the beamsplitter. The detector senses the wave interference caused by the movable mirror, and analysis of the interference patterns identifies gases and particles affecting the infrared radiation that passes through the Martian atmosphere.*

Courtesy, NASA

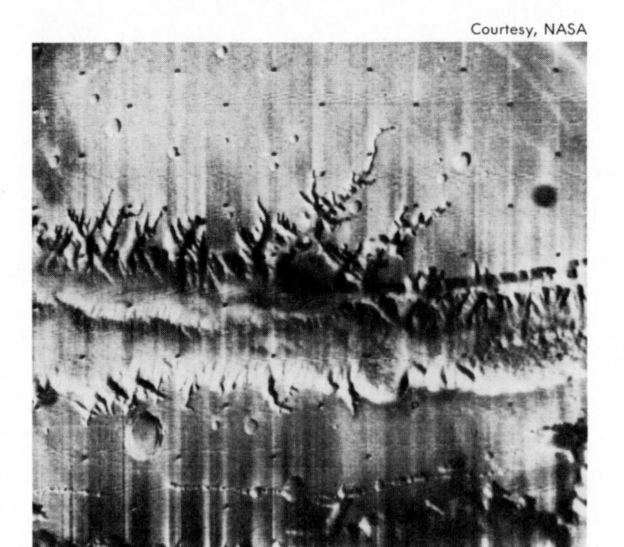

*Wide-angle photograph of Mars taken by Mariner 9 on Jan. 12, 1972, reveals a vast chasm with canyons branching into adjacent plateaus. Crust subsidence and wind erosion may have sculpted this unique landform.*

unobservable from the earth because of its orientation with respect to the sun's rays. Similar features elsewhere found the scientists at Jet Propulsion Laboratory hard put to describe their origin to anything except erosion by running water, although there clearly was no liquid water currently on the planet.

From the pictures a concept of Mars emerged that was radically different from that which scientists held after studying the photographs from Mariners 6 and 7. Harold Masursky, U.S. Geological Survey, said that instead of a dead, primordial planet, "Our photographs show something very, very different indeed. By seeing much more of the planet we can see great volcanic piles and because of the crispness of the edges and the lack of craters, we think these are geologically-young. The only thing we can keep as a primitive body are the two Martian satellites. Those do look like unevolved bodies, but Mars itself turned out to be even more dynamic than we had hoped."

Equally as exciting and important data were returned by the other instruments of Mariner 9. Data derived from the analysis of the refraction of signals from the craft by the Martian atmosphere indicated that the atmospheric pressure ranged between 3 millibars at the highest points on the surface to 8.3 millibars at the lowest point. The average atmospheric pressure was about 5.5 millibars (compared with an average on earth of 1,013 millibars). The same experiment showed that the Martian ionosphere was at an altitude of approximately 149 km (93 mi) compared with a height of only 135 km (84 mi) measured in 1969 by Mariner 7. The dust storm was probably the cause for the difference.

Gravitational variations in the planet caused considerable discussion among scientists. From an analysis of the variations in the velocity of the spacecraft, it appeared that Mars has a bulge of about 1 to 2 km (0.6 to 1.2 mi) on the equator at 110° W and a similar one at 70° E. In between these bulges, at 90-degree intervals, are slight depressions. There was no immediate explanation for this observation.

An infrared spectrometer on Mariner 9 indicated that there was about 0.001 the amount of water vapor in the atmosphere above the Martian south pole as there is in the same region of the earth. It also indicated that the amount of water vapor was greater over the south pole than elsewhere over Mars. The ultraviolet spectrometer provided data which showed that as much as 379,000 l (100,000 gal) of water per day could be lost from the Martian atmosphere. The source of this water was assumed to be volcanic.

Measurements of the Martian surface temperature seemed generally cooler than those reported by previous probes. The daytime high at mid-afternoon was about 240° K (−32.8° F), and it was about 200° K (−104.8° F) at night.

Published information from the Soviet Mars 2 and 3 probes was sketchy at best. Arriving at Mars on Nov. 27, 1971, Mars 2 went into a highly elliptical orbit with a periapsis of 1,376 km (860 mi) and an apoapsis of 25,000 km (15,600 mi). The periapsis was lowered by 149 km (93 mi) a month later. The orbit had an angle of inclination to the Martian equator of 48° 54′ and completed one orbit in 18 hours. Just before the probe entered orbit, it detached a capsule toward the planet. Apparently the capsule was not intended to soft-land.

Mars 3 arrived at the planet on Dec. 2, 1971. It assumed an orbit with a periapsis of 1,484 km (530 mi) and an apoapsis of 192,000 km (120,000 mi), which produced an orbital period of 11 days. Mars 3 also ejected a capsule, which soft-landed on the planet south of its equator in the Simois Strait. Unfortunately, the instrumented pod was damaged on landing, or it sank into a deep layer of Martian dust. Only 20 seconds of video signals were received. The pod was a highly sophisticated scientific package that would have provided a wealth of information on surface conditions had it operated. Among these instruments were sensors to measure the temperature and atmospheric pressure, a mass spectrometer to analyze the chemical composition of the atmosphere, an anemometer for measuring wind velocities at ground level, and instruments to determine the composition of the soil as well as its mechanical properties.

Both the Mars 2 and 3 orbiters had infrared radiometers for mapping surface temperature patterns. Radiofrequency telescopes in the 3-cm (1.2-in.) range probed the properties of the Martian soil. Ultraviolet photometers measured the luminescence of the upper atmosphere to detect atomic hydrogen, oxygen, and argon. In addition, the radiowave signals from both craft were analyzed as were those of Mariner 9 for their diffraction by the Martian atmosphere to determine pressure profiles and altitude of the ionosphere. Each probe also had two cameras, as did Mariner 9, for taking wide-angle and telephoto pictures of the planet. However, the pictures taken by the Soviet cameras were developed on board the orbiters and then transmitted to the earth, while in the Mariner 9 the pictures were converted electronically to signals for dark and light areas and the information was transmitted to the earth for reconstitution as pictures by computers.

Mars 3 reported that temperatures along the planet's equator did not exceed 15° C (59° F). Temperatures along the terminator, or boundary between the dark and sunlit side of the planet, ranged between −80° C and −90° C (−112° F to −130° F). The craft also reported that the maximum daytime temperature in the mid-latitudes was only −15° C to −20° C (5° F to −4° F). Among other data released by the Soviets from their Mars probes were that water vapor was greater in the atmosphere over the Hellas region than at other parts of the planet except the polar regions. Analyses of the orbits of the craft suggested that Mars is almost 50% flatter at the poles than is the earth.

**Exploring the moon.** Throughout July and August 1971, Lunokhod 1 continued to surprise and delight its Soviet developers. The eight-wheeled, self-propelled vehicle moved about its landing site in the Sea of Rains, sampling the soil and taking a variety of panoramic and stereo photographs. Toward the end of its ninth lunar day, it explored a relatively fresh crater approximately 90 m (297 ft) in diameter. By the end of the day it had covered a total distance of 10,237 m (33,782 ft). By Sept. 11, 1971, it had survived 11 lunar days (1 lunar day = about 14 earth days), but its power supplies were weakening drastically. On October 4, Lunokhod 1 ended its mission of active lunar exploration. In 10 months and 17 days, it had travelled 10,540 m (34,782 ft) and taken 200 panoramic pictures and another 20,000 single views of lunar features. It also made mechanical and physical analyses of the lunar soil at 500 different places and chemical analyses at 25 locations. (The chemical analyzer failed after seven months.) The robot was left parked in a position so that its French-built laser reflector was pointed toward the earth, insuring that it could be used for years to come.

On Sept. 2, 1971, the Soviet Union launched Luna 18 from Tyuratam. It entered a lunar orbit on September 7 at an altitude of 96 km (60 mi) at an angle of inclination to the equator of 35°. After making 54 orbits of the moon, Luna 18 crashed into the Sea of Fertility, while trying to make a soft landing.

Luna 19 followed Luna 18, being launched from Tyuratam on September 28. It entered a circular lunar orbit at 140 km (87 mi) altitude on October 3. On November 26 and 28, it made a series of maneuvers that placed it in an elliptical orbit with a perilune of 77 km (48 mi) and an apolune of 385 km (239 mi). The angle of inclination remained unchanged at 40.5°, but the time of one orbit was lengthened from 122 to 131 minutes. By January 1972, the craft had made 1,358 orbits of the moon with no apparent intention of landing. Its major experiment appeared to be the long-term study of the moon's gravitational characteristics. How-

*Seasonal shrinking of the south polar cap of Mars is clearly evidenced by the widening separations and the breaking up of the large detached mass in this series of photographs taken by Mariner 9 over a 25-day period early in 1972.*

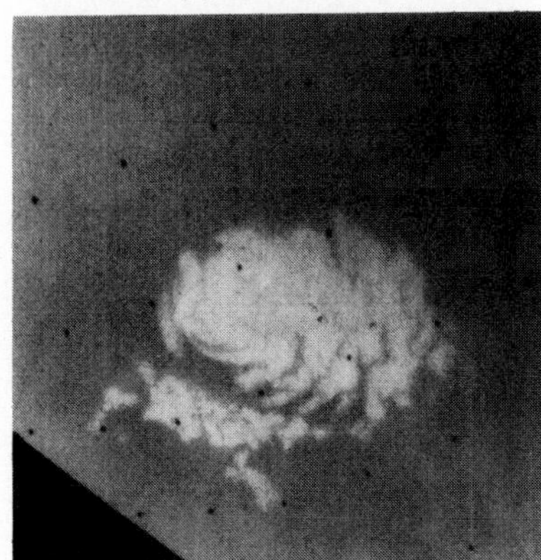
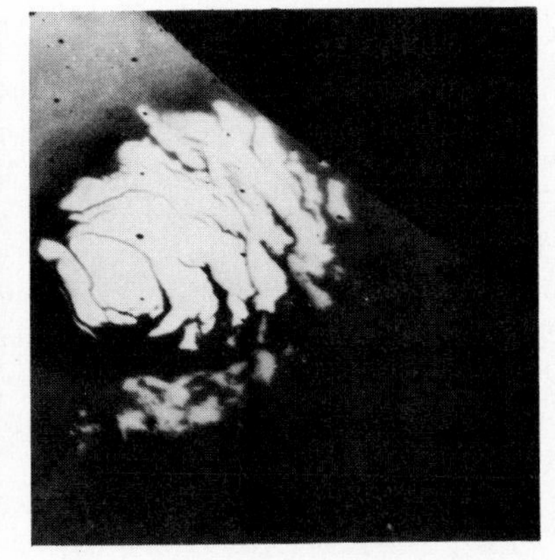

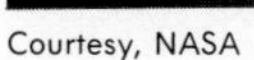

Courtesy, NASA

Courtesy, Tass from Sovfoto

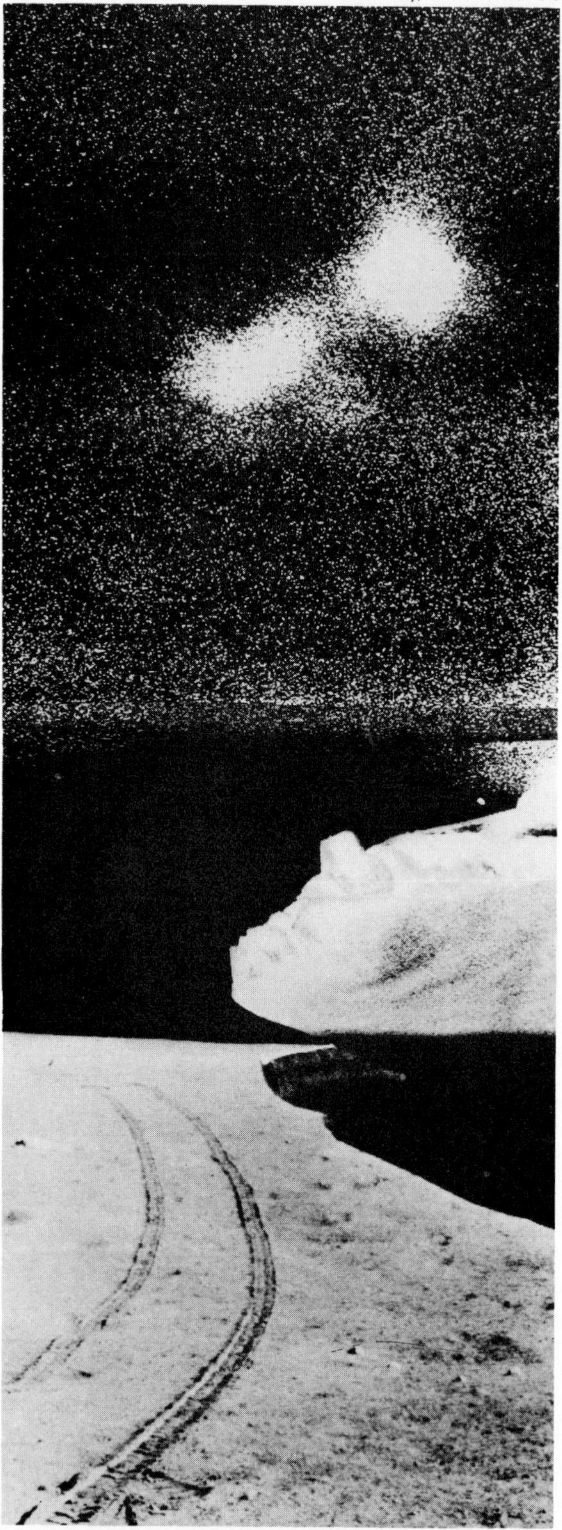

*Portion of a circular panoramic photograph taken by Lunokhod I's astrotelephotometer in 1971 shows portions of the landing stage and the original tracks made by the self-propelled vehicle against a background of the night sky.*

ever, it also investigated the lunar magnetic field and cosmic radiation in the vicinity of the moon.

Luna 20, the back-up probe for Luna 18, was launched from Tyuratam on February 14 and entered a circular orbit four days later. For three days it continued to orbit the moon at 99.2 km (62 mi) at an angle of 65° to the equator. It soft-landed on February 21 in a highland region between the Sea of Fertility and the Sea of Crises about 120 km (72 mi) north of the landing site of Luna 16.

The probe featured a much improved drill that, unlike the wholly automated machinery on Luna 16, was controlled by engineers in the Soviet Union via television. The drilling operation required 30 minutes and reached a depth of approximately 33 cm (13 in.). The drill struck extremely hard material at a depth of 14.5 cm (5.7 in.), and it took seven minutes to remove the drill core from its hole. The return capsule with about 100 g (3.5 oz) of lunar soil lifted off from the moon on February 23 and landed in the U.S.S.R. on February 25.

**A probe to Jupiter.** Pioneer 10 was launched on March 2 from Cape Kennedy, Fla. The 258-kg (570-lb) probe was designed to pass within 144,000 km (90,000 mi) of Jupiter on Dec. 3, 1973, after a voyage of 800 million km (500 million mi). Launched on a direct ascent trajectory, the craft reached a velocity of 51,840 km/hr (32,400 mph), making it the fastest probe ever to be sent into space. A mid-course maneuver was made on March 7 to ensure that the craft was on the desired trajectory to Jupiter.

The probe contained 11 scientific instruments to investigate the interplanetary medium, the asteroid belt, and the environmental and atmospheric properties of Jupiter. These included an imaging photopolarimeter, a helium vector magnetometer, a plasma analyzer, a charged particle counter, a Geiger-tube telescope, a trapped radiation detector, an ultraviolet photometer, an infrared photometer, and an asteroid-meteroid detector.

The trajectory of the probe was planned to take it directly through the asteroid belt between Mars and Jupiter. While the possibility existed that Pioneer 10 could be destroyed by a collision with one of the millions of fragments in the belt, such a probability was believed to be very small. Indications were that it would come no closer than 48 million km (3 million mi) to any of the known asteroids. After swinging past Jupiter and gaining velocity in the process, Pioneer 10 was designed to escape from the sun's gravity and fly into interstellar space, the first man-made object to do so.

**New data from the sun.** OSO 7 (Orbiting Solar Observatory) was launched on Sept. 29, 1971, from

Cape Kennedy. The second stage malfunctioned and the probe went into a slightly elliptical orbit rather than the circular one planned. The spacecraft's perigee was 342 km (214 mi), the apogee was 597 km (373 mi), and the angle of inclination to the earth's equator was 33.1°. The primary mission of OSO 7 was to obtain high-resolution data of the solar corona in special ultraviolet spectra and to investigate the intensity and spectrum of solar and cosmic X rays. By the end of November, scientists had gathered enough data to term the mission a success.

Among the data received from OSO 7 was the fact that the polar regions of the sun were cooler than its equatorial regions, a difference of 1,000,000° C (1,800,032° F) being reported by the probe. It also reported clouds of hot, energized gas erupting from the sun's surface that were 20 to 40 times the size of the earth. The energy involved in these eruptions was equal to the total electrical power requirements for the U.S., at the current rate of consumption, for the next million years.

**Another try at Venus.** On March 26, the Soviet Union launched the 1,180-kg (2,596-lb) Venera 8 on a trajectory toward Venus. Like its predecessors, the probe was designed to soft-land a spheroidal instrument package on Venus to measure temperatures and pressures at ground level.

**A look into the future.** Scientists throughout the world were disappointed with the announcement in 1972 that the U.S. National Aeronautics and Space Administration (NASA) had cancelled its ambitious "Grand Tour" that would have sent a probe to the outer reaches of the solar system in 1977–79. The Grand Tour envisioned a probe that would utilize a lineup of the planets that will not reoccur for about 180 years to propel itself through the solar system by the additional velocity imparted to it by the gravitational fields of the various planets. The program was cancelled because of the decreasing space budget.

As an alternative to the Grand Tour, NASA proposed to launch two Mariner probes in 1977 on flyby missions to Jupiter and Saturn. Plans were

*The path of the U.S. space probe Pioneer 10 will take it within 90,000 mi of the surface of Jupiter in December 1973. Launched on March 2, 1972, the probe must travel 500 million mi to reach its destination, the longest journey yet attempted for a spacecraft.*

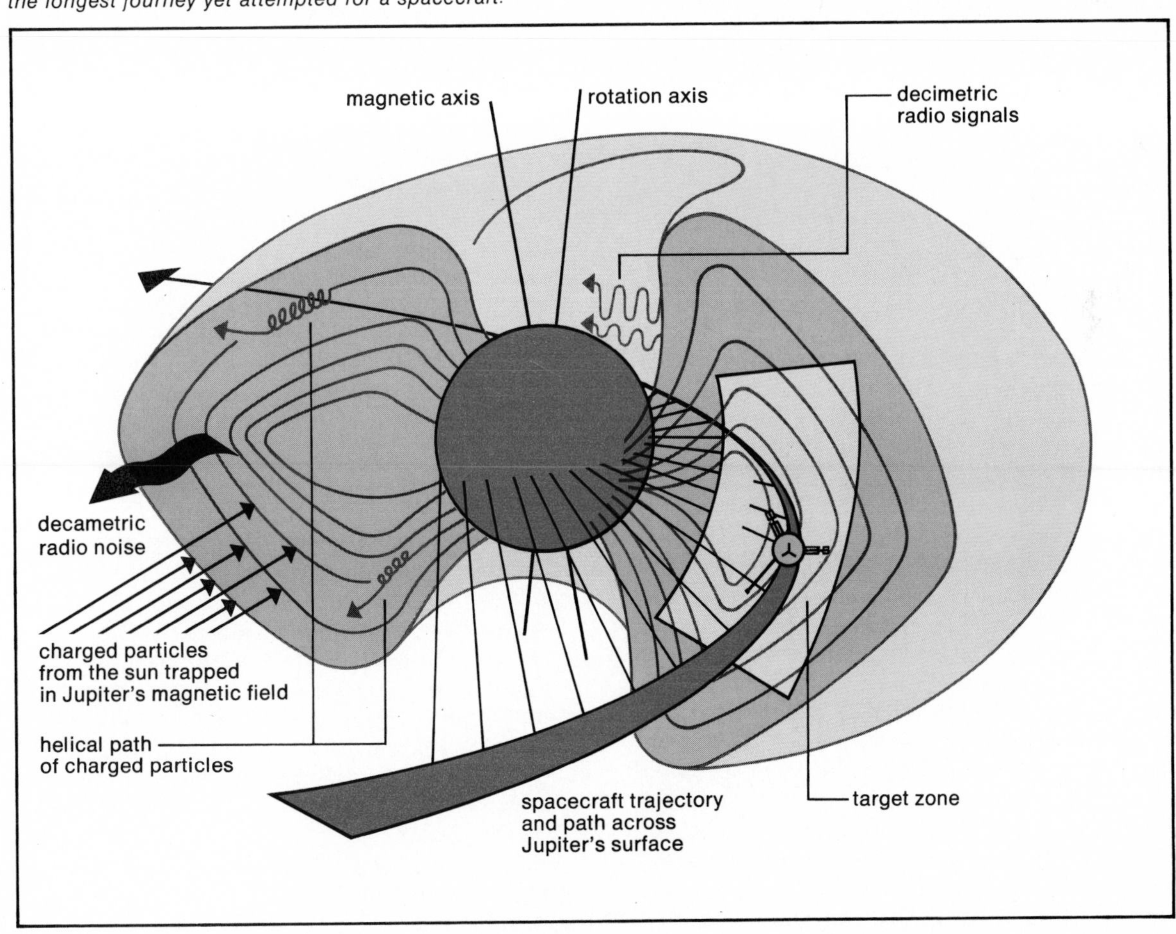

Courtesy, NASA

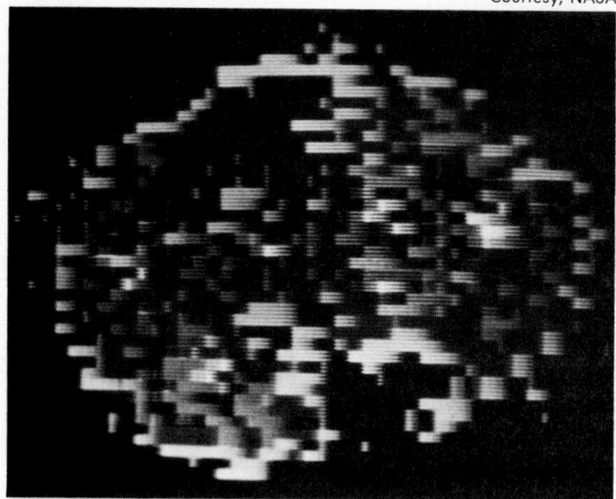

*Map of the solar corona, obtained by the Orbiting Solar Observatory 7, is presented on a computerized display. The white areas are the hottest and contain the greatest quantities of ionized gas.*

also revealed for flyby missions in 1973 to Venus and Mercury.

During 1972, work continued on the Viking, an unmanned probe scheduled to land on Mars in 1976. The 1,200-kg (2,400-lb) probe was expected to be heavily instrumented to analyze the Martian soil for the presence of microbacterial forms of life. However, the wealth of geological data returned by Mariner 9 prompted many scientists to ask the space agency to consider placing more emphasis on geological than biological instrumentation.

—Mitchell R. Sharpe

# Astronomy

During the past year, astronomy of the solar system gained a temporary prominence because of new findings about the moon and Mars. Even so, the underlying trend was a growth in radio astronomy and in high-energy astrophysics, directed toward the problems of stars, the Milky Way, radio sources, galaxies, and the universe.

**New Mexico radio telescope.** In March 1972, the U.S. National Science Foundation announced the selection of a 3,000-ac desert site near Socorro, N.M., as the proposed location for the world's most sensitive radio telescope. This instrument, known as a VLA (very large array), was designed to consist of 27 dish-shaped antennas, each 82 ft in diameter. The antennas are to be movable along each of three 13-mi arms of a Y-shaped layout of railroad track. This array would afford a resolution 10 to 100 times greater than that given by any existing array. The VLA would be managed by the National Radio Astronomy Observatory. If the U.S. Congress appropriates the funds for the $76 million project, the facility could go into partial operation by 1976 and full operation by 1982.

The significance of the New Mexico undertaking is that it would be the first major new U.S. radio astronomy facility in a decade. Its construction would tend to restore the U.S.'s lead in radio astronomy, which had been slipping in favor of the Netherlands, West Germany, and the Soviet Union.

**The moon's interior.** By 1972, the state of lunar science had become much healthier than a decade earlier, as its emphasis changed increasingly from speculation to weighing a great variety of empirical facts. A large amount of this new information was afforded by the successful visits of Apollo 14 to the Fra Mauro region in February 1971, and of Apollo 15 to Hadley Rille in July of that year. Many of the findings from Apollo 14 and some from Apollo 15 were first announced at a lunar science conference held in January 1972, at Houston, Tex. A significant aspect of that conference was the emergence of a fairly detailed picture of the moon's interior, which not so long ago had seemed hopelessly inaccessible.

Seismometers were left behind on the moon by Apollos 11, 12, 14, and 15. While the first of these devices ceased to function after a few weeks, the other three stations registered hundreds of seismic events, including both moonquakes and impacts. To locate the epicenter and depth of a seismic event requires records from a minimum of three suitably spaced stations, which became possible only in August 1971, when the Apollo 15 seismometer was planted.

In new controlled experiments, the impact of the spent third stage of Apollo 15 as it crashed onto the moon's surface was detected by the Apollo 12 and 14 stations, while the lesser impact of the discarded Apollo 15 ascent stage was registered at all three stations. These tests confirmed again that strong oscillations persist for at least an hour when the moon's surface is struck a violent blow. This prolonged vibration time is explained by the presence of a surface layer a few kilometers thick, which scatters seismic waves effectively and tends to confine their energy.

Gary Latham of Columbia University, the leader of an active group of lunar seismologists, was able to distinguish between true moonquakes and meteorite impacts through a study of wave forms (curves representing the conditions of wave-propagating media at particular instants). Despite the extreme sensitivity of the Apollo 15 seismometer, which can detect surface displacements as small as $10^{-8}$ cm, moonquakes are few and weak, even the largest amounting to only 1 or 2 on the Richter scale. (On earth, such quakes are at best barely perceptible without instruments.) Seismically, the moon is much less active than the earth. According to Latham, each square mile of lunar surface releases only $10^{-5}$ as much seismic energy, on the average, as does a square mile on the earth.

Most of this energy release is in the form of periodic moonquakes that tend to occur near the times when the moon is nearest and farthest from the earth (perigee and apogee). These quakes appear to originate only in a dozen or so separate locations. One of these places is responsible for about 80% of the seismic energy. It is located near the crater Bullialdus, at a depth of 500 mi. The seasons of activity (perigee and apogee) are the times when the tides produced in the moon's crust by the earth's gravitational attraction are greatest. Hence, the periodic moonquakes seem to originate in recurrent rock slippages triggered by tidal forces. The existence of a focus 500 mi deep means that even halfway down from the surface to the center the moon's interior is still rigid enough to support appreciable stress. The moon's interior, thus, cannot be molten at that level.

From an analysis of travel times of moonquakes in the Mare Imbrium region, it was found that the velocity of seismic waves increases with depth to about 4 mi/sec at a depth of near 15 mi. The material above 15 mi is thought to be primarily basalt. Then, down to a depth of about 40 mi, the velocity remains nearly constant. This layer is believed to be either eclogite or anorthosite. At about 40 mi, the velocity of compressional waves increases abruptly to about 5½ mi/sec, indicating a discontinuity beneath which the material is probably olivine.

Further information about the moon's interior is furnished by studies of magnetism. The instruments left by Apollo 14 indicated at that site a steady magnetic field for the moon of little more than 50 gammas ($10^{-3}$ of the earth's field). The steady field at the Apollo 12 base was 38 gammas, and at the Apollo 15 site only 5 gammas. Thus, the present general magnetic field of the moon is very weak, as was also shown by magnetometers used aboard orbiting Apollo spacecraft.

Oddly enough, however, most of the rock samples brought back by the astronauts and measured in terrestrial laboratories showed large remnant magnetism. In the remote past when these rocks were formed, they must have solidified in a magnetic field much stronger than the moon now has.

The moon's response to the changing magnetic fields of the earth and the sun has been studied. The results are compatible with a three-layer model of the moon, in which a 60-mi-thick outer crust at 440° K overlies an intermediate zone at 800° and a

Courtesy, NASA

*Tiny green glass fragment was taken from an Apollo 14 core tube sampling. The fragment, constituted mainly of iron and magnesium, has been nicknamed the Genesis bean because of its great age, possibly dating back to the beginnings of the solar system.*

1,100° central core of 1,200-mi diameter. This model assumes that the moon's interior consists primarily of olivine.

A number of lines of evidence pointed to the moon either being solid throughout, or having only a small molten core. One argument is based on the weakness of the steady magnetic field; this field, according to some astronomers, would be much stronger if the moon had a large liquid core, free to move relative to the crust. Another argument is that the deviations of the moon's surface from a sphere are so great that they can persist over long periods of time only if the moon's interior has considerable rigidity.

However, an apparent contradiction is offered by measurements of heat flow outward through the moon's crust. In one of their more difficult tasks while at Hadley Base, Apollo 15 astronauts David Scott and James Irwin drilled two holes into the ground in which they emplaced temperature-sensing probes at depths of 3 ft and 5 ft below the surface. Because of the excellent heat-insulating properties of lunar soil, there is no perceptible day-night temperature at depths of more than 2 ft. Below that, the temperature was observed to increase at a rate of 0.5° K per foot. This implies a heat flow from below the Hadley site about half that in the earth's crust—considerably higher than astronomers had expected. It seemed to indicate very high temperatures in the moon's interior.

Galileo—astronomer, physicist, mathematician—was the first man to use the telescope to study the skies and with it proved that the earth revolves around the sun.

One explanation for this finding was that the Hadley site may not be typical of the moon and that the *average* heat flux could be much smaller. An alternative hypothesis was that lunar radioactivity is concentrated in the moon's outer layers, heating them and thus producing a larger outward heat flux than occurs in the deep interior. This possibility is supported by the high uranium and thorium concentration measured in many of the rock samples returned from the moon's surface.

**New observations of Mars.** An exceptionally favorable opposition of Mars occurred in August 1971, when that planet's distance from the earth became as small as 34.9 million miles. Mars would not be as close again during the rest of the 20th century. With headquarters at Lowell Observatory in Arizona, an international photographic patrol program was mounted at a network of observatories in Europe, Africa, Asia, and North America, whereby thousands of excellent photographs were obtained on a uniform plan. Many other observatories conducted their own photographic or visual programs.

However, these optical observations were severely hampered for several months by a vast dust storm on the surface of Mars, which was first photographed on September 22. It spread rapidly and about two weeks later had enveloped the entire planet in a heavy yellow veil, dimming or obliterating the surface features. This storm did not begin to abate until January 1972.

The U.S. spacecraft Mariner 9 began orbiting Mars on Nov. 13, 1971. Once the planet's atmosphere began to clear, the craft's television cameras obtained highly detailed pictures of the Martian surface. These views provided strong evidence that many features on Mars are of volcanic origin. For example, there are numbers of craters clearly resembling the earth's calderas —depressions made by the collapse of the central parts of volcanoes when magma is drained from beneath them.

A striking Martian example of this is Nix Olympica. Mariner 9 photographs showed this as a 4-mi-high mountain, 300 mi across at its base and topped by a volcanic vent about 40 mi wide. This great mountain is similar to, but even larger than, Mauna Loa in the Hawaiian Islands, which rises from the Pacific floor to a summit volcano.

Other newly discovered Martian surface formations included a long, 75-mi-wide trough cutting through the Tithonius Lacus region, and sinuous valleys that superficially resemble terrestrial arroyos. However, these features are in most cases probably the results of collapses along lines of weakness in the crust, combined with erosion by wind-blown dust. There is too little water on Mars for running water to be an important agency in shaping the surface, although it might have been a factor in the past.

The Mariner pictures gave the general impression of a planet that has been geologically active in the past and whose surface has been greatly altered by volcanism, crustal movements, and erosion. A large proportion of Mars's craters must, however, continue to be regarded as resulting from meteorite impacts.

Close-up Mariner photographs revealed that the Martian satellites Phobos and Deimos are irregularly shaped dark bodies, respectively 14 and 8 mi in diameter. Each is densely cratered, one Phobos crater being 3 mi wide.

Instruments carried by Mariner to measure the planet's ultraviolet and infrared radiations provided interesting findings. For example, a new temperature measurement of the south polar cap confirmed that the cap material is primarily carbon dioxide frost and not frozen water. The observed temperature closely matched that which solid carbon dioxide in equilibrium with its vapor must have at the atmospheric pressure obtaining on Mars. A second result was that the planet-wide obscuration in late 1971 was due to airborne silicate dust, resembling in its optical properties the surface material of Mars's ocher (yellowish) deserts.

Two Soviet spacecraft also orbited Mars beginning in late November and early December 1971. Both Mars 2 and 3 ejected instrumented capsules to make soft landings on the planet. Although the second attempt succeeded, the capsule's radio transmitter failed soon after landing.

Few early results were made public from the instruments operated aboard Mars 2 and 3 while they orbited the planet. A faint airglow of atomic hydrogen extends 500 mi above the Martian surface, and ultraviolet emissions from atomic hydrogen reach much higher. The Soviet observations confirmed the very low abundance of water vapor in the Martian atmosphere, giving it as less than the equivalent of a 0.005-mm layer of liquid water.

Extensive studies of Mars were also made with large earth-based radar telescopes operated by scientific teams at the Massachusetts Institute of Technology (MIT) and the California Institute of Technology (Caltech). A radar observation gave the distance from the telescope to the nearest point on the Martian globe; as the planet rotates, successive observations yield a profile of the vertical relief along an east-west strip of the planet's surface. Such observations had been made since 1963, but by 1971 techniques had become so refined that elevation differences on the order of 200 ft could be measured. Both the MIT and Caltech teams reported that they could detect many individual Martian craters identifiable on Mariner photographs, and that they could determine rim-to-floor depths for the larger craters.

**Occultations by Jupiter.** An astronomical event of extreme rarity occurred on May 13, 1971, when the planet Jupiter passed in front of (occulted) the 2.6-magnitude star Beta Scorpii. It was the brightest star occulted by this planet since observations began with the invention of the telescope in the early 17th century. Moreover, on the same night Jupiter also occulted the 4.9-magnitude compan-

*Man's first close-up views of the Martian satellites Deimos (left) and Phobos (right) were obtained by means of a high resolution television camera aboard Mariner 9. Both moons are irregularly shaped; the dark spot near the bottom of Phobos is a crater about four miles in diameter.*

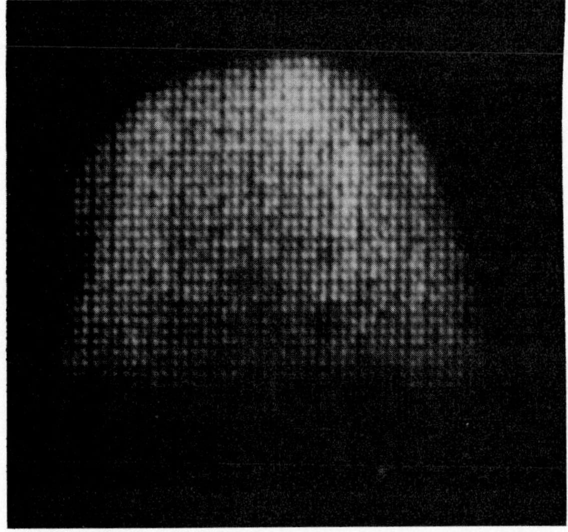

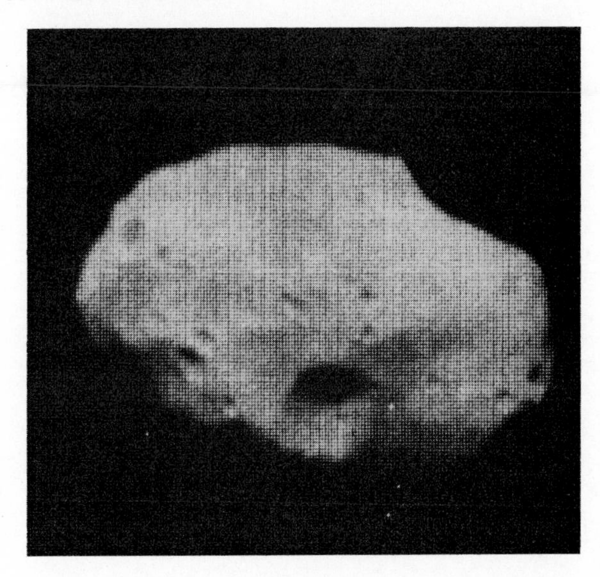

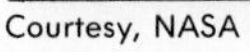
Courtesy, NASA

ion of Beta Scorpii, located 14 seconds of arc to its north. This companion was also occulted by Jupiter's satellite Io.

The occultations by Jupiter were visible from South Africa, India, Australia, and Antarctica. Expeditions were sent from the University of Texas to monitor photoelectrically the gradual fading of the stars as they disappeared behind Jupiter's disk and their reappearances about an hour later. These observations gave an accurate determination of the vertical density gradient in the planet's atmosphere.

A remarkable phenomenon recorded during the fading of each star into occultation was the repeated occurrence of short-lived flashes as the star would momentarily recover to nearly its full brightness. These flashes were conspicuous to visual observers. The photoelectric records showed that the same pattern of flashes was observed from widely separated places on the earth. Astronomers believe that the flashes are associated with a layered structure in the Jovian atmosphere. There are at least three warm layers present at both the eastern and western edges of Jupiter, apparently part of a planet-wide pattern of stratification.

One result from the Beta Scorpii occultations was the most accurate determination yet of the dimensions of Jupiter. According to the analysis by W. B. Hubbard and T. C. Van Flandern, the equatorial radius of the planet is 71,880 km (44,565 mi) and the polar radius is 67,500 km (41,850 mi).

The occultation of the companion of Beta Scorpii by Io was visible about seven hours later from a limited region in the Caribbean Sea and Florida. Both the disappearance and reappearance were sudden, indicating that Io has no appreciable atmosphere. The radius of the satellite was found to be 1,829 ($\pm$3) km. Io, therefore, is about 5% larger than our moon.

**Companion of Sirius.** As seen in fairly large telescopes, the bright star Sirius has a much fainter companion, moving around it in a period of 50 years. This companion, Sirius B, is the most familiar example of the several thousand white dwarf stars now known with more or less certainty. These are tiny, hot stars of extremely high density, comparable to the sun in mass but only as large as planets. Since Sirius B happens to belong to a well-studied binary star system, its distance from us is reliably established as 8.7 light-years and its mass as 1.02 times that of the sun.

According to general relativity theory, the spectrum of a white dwarf such as Sirius B should show a red shift as a result of the star's high surface gravity. Attempts to measure this gravitational red shift were made at Mount Wilson Observatory in the 1920s, when it was mistakenly announced that the prediction was confirmed. But the very difficult spectroscopic observations were in fact inconclusive: the companion's spectrum was badly contaminated by the overpowering glare of Sirius itself, 10,000 times brighter and only a few seconds of arc distant.

Late in 1971 a team of astronomers from Hale Observatories, Pasadena, Calif., announced that this problem had finally been overcome. Using the 200-in. Palomar telescope and taking extreme precautions to avoid stray light from the brilliant primary star, they managed to obtain the first uncontaminated spectrograms of Sirius B. These observations were taken at a time when the companion, in tracing its elongated orbit, was approaching its maximum angular separation from the primary star. Analysis of these observations established that Sirius B is 0.85 the size of the earth, has a surface temperature of 32,000° K, and a density 3,000,000 times that of water. By measuring the wavelength displacement of a hydrogen line in the Sirius B spectrum, the astronomers found a gravitational red shift of 81 km/sec, agreeing well with the predicted value of 83.

**Algol as a radio source.** At the National Radio Astronomy Observatory, Green Bank, W. Va., C. M. Wade and R. M. Hjellming examined various stars as possible radio sources. (The sun, flare stars, and a few novas were already known as sources.) As of early 1972, these astronomers had discovered four optically "normal" stars that were radio emitters: the companion of Antares, the star in Cygnus associated with the X-ray source Cygnus X-1, Beta Lyrae, and Beta Persei (Algol).

Discovery of the latter two stars was announced in February 1972. In both cases, the position of the source, as measured with a large radio interferometer, agreed so exactly with the known optical position of the star that no doubt existed as to the correctness of the identification.

The case of Algol is remarkable. During October and November 1971, it was detectable as a weak radio source at wavelengths of 3.7–11 cm. But in February a new phenomenon appeared: frequent, strong bursts of radio noise, lasting several hours or more, occurring at irregular intervals of a few days. Meanwhile, photographs of the optical spectrum of Algol made at David Dunlap Observatory, Toronto, Can., revealed the unprecedented appearance of bright lines of calcium and other elements.

Both of these anomalous observations were probably connected with the long-known fact that Algol is a close binary system with a period of 2.8673 days, in which occasional burps of gas are ejected from the secondary star. Previously, the most noticeable effect of this gas ejection was to produce minute changes in the period of orbital revolution. Astronomers believe that an unusually marked ejection of gas took place in January or February 1972 and that both the radio bursts and the optical emission lines originated in this transitory gas cloud.

**Future outlook.** In the coming year, an eagerly awaited astronomical event is the total solar eclipse of June 30, 1973, visible from Africa. It will be remarkable for the long duration of totality, more than seven minutes. In late 1972 the Pioneer 10 spacecraft will be traversing the asteroid belt prior to its flyby of Jupiter in December 1973. Astronomers hoped that in 1973 the world's largest reflecting telescope (aperture 236 in.) and largest radio telescope (a stationary dish nearly 2,000 ft across) will go into operation at neighboring sites in the Soviet Union.

—Joseph Ashbrook

*See also* Year in Review: ASTRONAUTICS AND SPACE EXPLORATION, *Space Probes.*

# Atmospheric sciences

As in previous years, atmospheric scientists in many countries were concerned with the same broad problems. The search continued for new techniques for making better observations of the properties of the atmosphere at more places and at more frequent intervals. Greater attention was being devoted to the development of indirect probing techniques to replace the direct sensing instruments that need to be transported on balloons or aircraft. Various radar, laser, and radiometric methods continued to show great promise.

Major efforts were being devoted to scientific programs aimed at improving the accuracy of weather forecasts. This was being accomplished by the development of more realistic mathematical models of atmospheric phenomena and by the use of the most advanced electronic computers.

Although progress appeared to be slow, important advances were made in the study of weather and climate modification. The possible ecological, legal, and other societal consequences of intentional weather modification came under the scrutiny of social scientists as well as of the public at large. On the other side of the coin, inadvertent effects of human activity on the atmosphere also were receiving increasing attention. Scientists and the public were becoming more concerned about the degree to which environmental pollution may be affecting the global climate.

**Atmospheric sciences in the '70s.** In 1971 the Committee on Atmospheric Sciences of the National Academy of Sciences issued a report entitled

Courtesy, Dr. Roger Cheng, Atmospheric Sciences Research Center, State University of New York at Albany

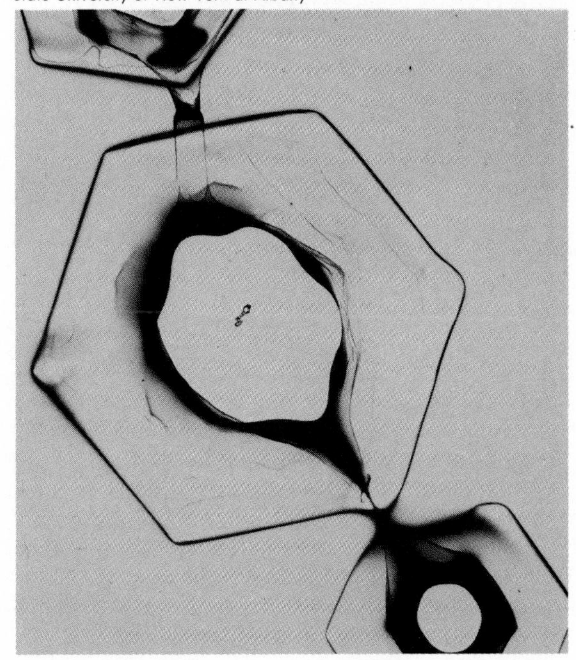

*Electron micrograph shows a lead-nucleated ice crystal formed by the combustion of gasoline in the presence of iodine vapor. Study of the nucleating particles could result in new techniques for cloud seeding.*

*The Atmospheric Sciences and Man's Needs.* It briefly notes the progress made in the 1960s and looks ahead over the next decade. The document focuses on those aspects of the atmospheric sciences that can contribute to important human needs and proposes an ordering of priorities. The problem areas are weather prediction, air quality, weather and climate modification, and weather dangers and disasters. The report notes that these subjects are receiving far from adequate attention.

Greater knowledge in these fields will not only satisfy the never ending hunger to learn but will also benefit society enormously in tangible ways. For example, in the United States alone violent storms do billions of dollars worth of damage and take hundreds of lives each year. In exceptional years and in other countries, destruction and deaths greatly exceed even these figures. The tropical cyclone that hit East Pakistan in November 1970 killed between 200,000 and 300,000 people and damaged some 400,000 homes. Better observations of violent storms, improved forecasts, more effective warnings to the public, and, finally, the development of techniques to weaken the storms would thus provide substantial advantages.

**Observing the atmosphere.** Most observations of the atmosphere are still made in the traditional way—using such instruments as thermometers, barometers, anemometers, and radiosondes. Since the 1950s radar has been used extensively to observe the location, size, and movement of storms. The 1960s saw the introduction of earth-orbiting satellites carrying instruments to observe cloud cover and measure the radiation emitted and reflected from the atmosphere below the satellite. Beginning in about 1970, various types of infrared spectrometers on satellites have supplied information from which cloud top heights could be estimated and the vertical structure of temperature and humidity of the atmosphere could be calculated.

The presence of extensive clouds interferes with the use of infrared radiometer techniques, but apparently this difficulty can be overcome by means of radiometers operating at microwave frequencies. The U.S. National Aeronautics and Space Administration was planning a Nimbus E meteorological satellite which would be equipped to make spectrometer measurements at both infrared and microwave frequencies. The satellite was scheduled for launching in late 1972 or early 1973, and shortly thereafter scientists should be able to tell whether such a combined system can supply reasonably accurate data on the temperature and vapor distributions in the lower atmosphere. Some scientists predicted that these indirect satellite techniques would make the balloon-borne radiosonde as obsolete as the old weather kite, but such predictions seemed premature.

Progress in the use of microwave radar for observing clouds and precipitation was slow during the year. Various promising programs concerned with the development of new pulsed-Doppler radar systems did not advance as rapidly as had been expected. On the other hand, there was enthusiastic interest in the use of lidar, *i.e.*, laser radar, for atmospheric probing. A wide variety of programs in several countries were developing techniques for measuring various properties of the atmosphere, particularly the sizes and numerical concentrations of particulates. An appropriate lidar can monitor the depth of the polluted layer over a city. Lasers were also being used to measure the water vapor content and some of the other gaseous constituents over an extended horizontal path.

Several scientists, notably Gordon C. Little and Donald W. Beran and their colleagues at the Wave Propagation Laboratory of the National Oceanic and Atmospheric Administration, did some exciting work by means of an acoustical echo sounder. Such an instrument is sometimes called an "acoustical radar," but this is not appropriate because the term radar implies the use of radio waves. The echo sounder transmits a sound pulse and detects sound echoes reflected back from the atmosphere. The instrument measures a fantastic amount of detail about the structure of the atmosphere in the lowest 500 m and sometimes higher. Relatively small inhomogeneities of temperature, water-vapor pressure, and wind velocity yield detectable backscattered signals. The echo sounder measures these quantities and allows investigations of low-level wind patterns, wave motions, and turbulence.

*The compounds of sulfur present in the earth's atmosphere have various sources, with man contributing about half as much as nature to the total. In the figure, $XSO_4$ represents sulfates or sulfuric acid formed by the three-body reaction of $SO_2$ with atomic oxygen and a molecule of either oxygen or nitrogen.*

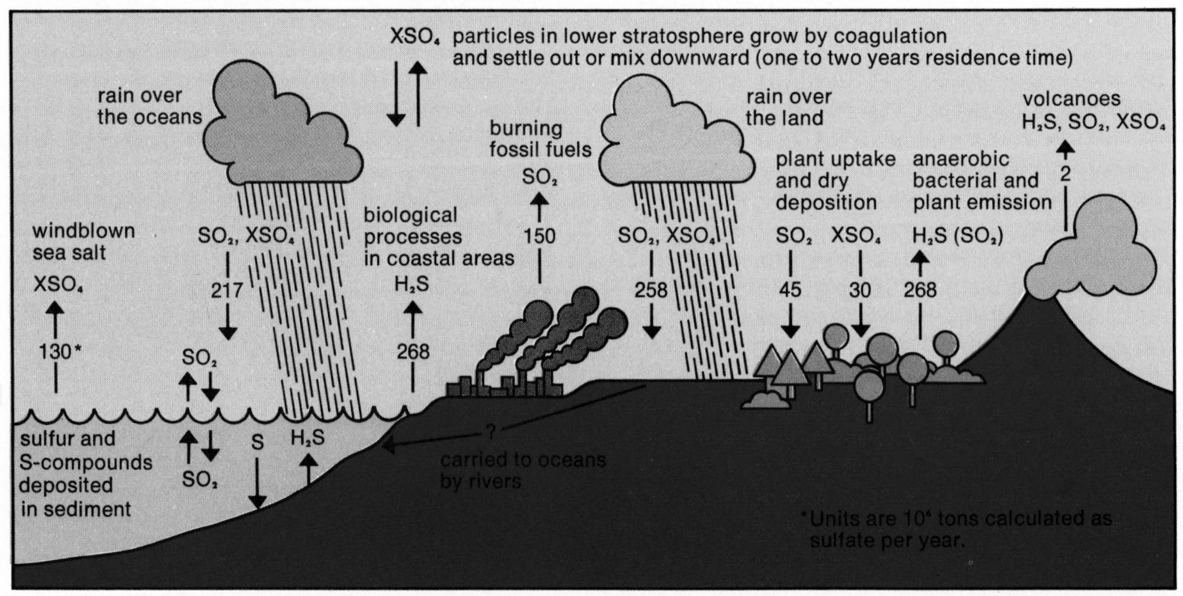


Adapted from SCIENCE, courtesy, National Center for Atmospheric Research

**Mathematical modeling.** In some ways, modern meteorology can be said to have started when electronic computers were first employed to construct mathematical models of the atmosphere by numerical methods. Early work in the 1950s dealt with weather over North America. As computers became bigger, as more was learned about the interaction of the solid and liquid earth and the atmosphere, and as better mathematical techniques were evolved, scientists developed models of the global atmosphere. To a certain extent these theoretical conceptions outstripped the observations. Adequate testing of their validity awaited the collection of worldwide observations. This was one of the major goals of the Global Atmospheric Research Program (GARP).

In recent years many scientists have been working on the development of mathematical models of the global climate. The difficulties involved are great, but the potential value is tremendous. When calculating changes in the global weather, each time step in the computation represents perhaps 10 minutes and the computation is extended for periods of a few days to a few months. When examining global climate, however, the computations extend over periods of hundreds of years. William Sellers at the University of Arizona used time steps of one month in his studies of this problem.

If we are ever to understand man's effect on the worldwide climate and predict what it will be in the future, much more progress must be made in the development of such global climate models. This general problem has been of special interest to Soviet scientists, particularly M. I. Budyko, director of the Main Geophysical Observatory in Leningrad. During August 1971 he hosted a meeting of experts from various parts of the world to examine the questions involved in climate change.

For a decade a number of scientists have been working on the development of numerical models of individual cumulus clouds. Edwin Danielson and his colleagues at the National Center for Atmospheric Research formulated a numerical model of a hailstorm that takes into account the evolution and changes of updrafts and downdrafts. In addition, it includes the initial cloud droplet population, the probability of ice particle formation, the breakup of water drops, and the growth of hail.

In the past relatively little effort has been devoted to the mathematical modeling of mesoscale phenomena, that is, those having horizontal dimensions from perhaps 10 km to a few hundred kilometers. A possible exception to this statement is the hurricane, which for several years has had the attention of many outstanding scientists. Nevertheless, much more research was needed in order to develop techniques for mathematically predicting the future intensification and path of hurricanes. Another mesoscale system that can be treated numerically is one specifying the air pollution concentrations over cities. George Cressman, director of the U.S. National Weather Service, indicated in December 1971 that new techniques for this purpose were being considered.

Courtesy, Roger Cheng, Atmospheric Sciences Research Center, State University of New York at Albany

*Supercooled water drop emits tiny, positively charged microdroplets as it freezes; the parent drop retains a negative charge. This separation mechanism could possibly generate lightning in storm clouds.*

**Man's effect on climate.** In 1971 the Massachusetts Institute of Technology Press issued a volume entitled *Inadvertent Climate Modification: Report of the Study of Man's Impact on Climate.* Familiarly known as the SMIC report, it was a companion to a 1970 MIT volume *Man's Impact on the Global Environment: Study of Critical Environmental Problems* (the SCEP report). The new book is a summary of a conference planned by Carroll L. Wilson of MIT and held in Sweden during the summer of 1970. Some 30 scientists from 14 countries attended. The discussions considered a wide range of ways by which human activities could affect the climate, including changes in atmospheric composition; release of energy into the atmosphere; land-surface alteration; manipulation of surface and underground waters; weather modification;

and the effects of automobiles and aircraft, particularly supersonic aircraft. The conference made it clear that much still had to be learned in these areas.

A subject of more local—but still important—interest was the effect of large industrialized cities on the climate of the cities themselves and of the regions immediately around them. In 1968 Stanley Changnon of the Illinois State Water Survey reported certain anomalies in the weather records from stations downwind of Chicago and Gary, Ind., particularly in the data from the town of La Porte, Ind. There was considerable debate over possible explanations, but the uncertainties might be settled by a program begun in the vicinity of St. Louis, Mo., during the summer of 1971.

The program, called the Metropolitan Meteorological Experiment or METROMEX, was designed by scientists from four institutions, including the Illinois State Water Survey, to investigate changes in weather caused by urban-industrial effects with particular attention to changes in rain and snow. A dense network of observing stations of many kinds was installed around St. Louis. Starting in 1973, scientists of the Environmental Protection Agency would make detailed measurements of the pollutants in the air over St. Louis. The concentration of contaminants still in the atmosphere downwind of that city was to be measured by chemists from the National Center for Atmospheric Research, who had developed techniques for detecting trace quantities of various substances.

**Weather modification.** During the spring and summer of 1971, the southern states of the U.S. from Arizona to Florida experienced abnormally dry weather, and the governors of Florida, Texas, Arizona, and Oklahoma requested the federal government to undertake cloud-seeding operations in an attempt to increase rainfall. Each project lasted for one month or longer. In every case, rain did fall, but there appeared to be little concrete evidence to support claims that it fell because of the seeding or was increased by the seeding. Evaluation procedures were inadequate; the operators tried to seed the maximum number of clouds that appeared favorable, and the design of the programs made it virtually impossible to arrive at a satisfactory determination of their effects.

The year's events clearly indicated that the public and their elected representatives had become more aware of the potential value of cloud seeding. Most atmospheric scientists, however, were of the opinion that the technology is not sufficiently advanced to be used in the uncontrolled way adopted during the summer of 1971. Except in a few circumstances, it still had not been adequately demonstrated that cloud seeding will increase precipitation by substantial amounts. In some seeding programs, in fact, precipitation apparently was decreased.

Several important experimental cloud-seeding programs were in progress during the year. Particularly notable were the so-called San Juan Project in Colorado, in which the Bureau of Reclamation was evaluating the effects of seeding on winter snowfalls; a hail modification program being run by the National Center for Atmospheric Research with support from the National Science Foundation; and Project Stormfury, which was concerned with developing techniques for weakening hurricanes.

Considerable controversy arose around one cloud-seeding project when it was discovered that it had taken place near Rapid City, S.D., several hours before torrential rains caused a devastating flood in that city early in June. Scientists claimed that the seeding was done in clouds that were physically separate from the storm that caused the flood and that any rainfall it might have added to the total would have been comparatively light.

The legal, social, and ecological consequences of weather modification were receiving greater attention, particularly in the United States. At the same time, the international aspects of large-scale weather and climate modification were being recognized by many nations. The National Academy of Sciences report on the atmospheric sciences, acknowledging that weather modification technology could be used in war as well as peace, called on the United States "to present for adoption by the United Nations General Assembly a resolution dedicating all weather-modification efforts to peaceful purposes." Several members of Congress, particularly Sen. Claiborne Pell (Dem., R.I.), were concerned about the use of weather modification techniques in the war in Southeast Asia. Pell called for a treaty to outlaw such practices.

**Global weather programs.** Progress on GARP continued on many fronts. The first global experiment, an attempt to collect detailed observations of the atmosphere and upper layers of the ocean over the entire earth, was scheduled to take place between 1976 and 1980. The data were to be used to develop a better understanding of the atmosphere up to an altitude of 30 km and in constructing a numerical model to forecast the weather some 5 to 10 days in advance.

Beginning in 1967, a series of experiments had been conducted over small oceanic areas. Plans were in progress for a greatly enlarged experiment, during the summer of 1974, over the tropical Atlantic in a zone including parts of Central and South America and Africa and extending from 10° to 20° N. Some 25 countries were contributing ships, airplanes, weather satellites, other equipment, and scientists. The management of the project was in the hands of an international group of scientists headquartered in Bracknell, Eng. This program, called the GARP Atlantic Tropical Experiment, would set the stage for expansion of the research, toward the end of the 1970s, to cover the entire planet.

—Louis J. Battan

# Behavioral sciences

Continuing a trend that had been in evidence for several years, behavioral scientists turned their attention more and more toward the application of their special knowledge to the practical problems facing society. Anthropologists, taking stock of their field, placed special stress on problem-solving. At the same time, psychologists figured prominently in widely publicized arguments over the possibility and desirability of behavior control and over the hypothetical existence of genetic differences affecting the intellectual capabilities of various population groups.

## Anthropology

The human species is now recognized to be some two million years beyond its other-primate ancestry. Its whole career is the subject matter of anthropology. The species carries with it much of its long history; thus it is possible for paleoanthropology and archaeology to make intelligible to modern man the fragments of prehistory for which they continually search. In recent years, the 3,000,000,000 humans living today, their problems and prospects, have also become a major focus of anthropology.

To understand the basic characteristics of the species requires disciplined study of the rich variations in human types and cultures. This has been the task of anthropology for at least a century and a half, but accumulation of such knowledge is slow. Moreover, until this generation anthropology very largely lacked input from scholars of other than European cultural traditions, a flaw that seriously crippled progress in a discipline where scientists—who are also humans partially blinded by their own cultural heritages—are the main tools of cross-cultural research. The body of disciplined knowledge has grown substantially in the past 20 years, but because analysis, publication, and criticism of new field research take time, anthropologists have still been unable to make the contributions required by a world whose problems can only be solved by its people.

In May 1971 the International Union of Anthropological and Ethnological Sciences embarked on a program to expose, synthesize, and summarize anthropological knowledge at its ninth International Congress, scheduled to be held in Chicago, Aug. 28–Sept. 8, 1973.

In August 1971 anthropologists all over the world were asked both to propose their own individual contributions and to suggest symposia and conferences for the congress. By March 1972 over a thousand responses had been received from 75 different nations. They were studied, collated, and synthesized by the organizing committee, which found that they could all be subsumed in four overlapping sets of major themes.

The first two (comprising 60% of the contributions suggested) show the knowledge that we have of the species, the first temporally and developmentally, the second geographically. The other two show the concerns of the disciplines—and particularly how they acquire knowledge—and the intersection of anthropological sciences with problems of the world outside. The whole of anthropology, as it approaches the last quarter of the

*Substandard living conditions, as in this slum in Bogotá, Colombia, typify the lot of the urban poor in many areas of the world. Recently, a number of anthropologists completed studies of the urban poor which contradict the idea of a separate culture of poverty. In important areas, notably employment and family life, the values of slum dwellers correspond to those of the majority culture, but behavior differs because of the absence of opportunity to realize their goals.*

Courtesy, Dr. Thomas Greaves, University Museum, University of Pennsylvania, and EXPEDITION Magazine

Courtesy, Transvaal Museum

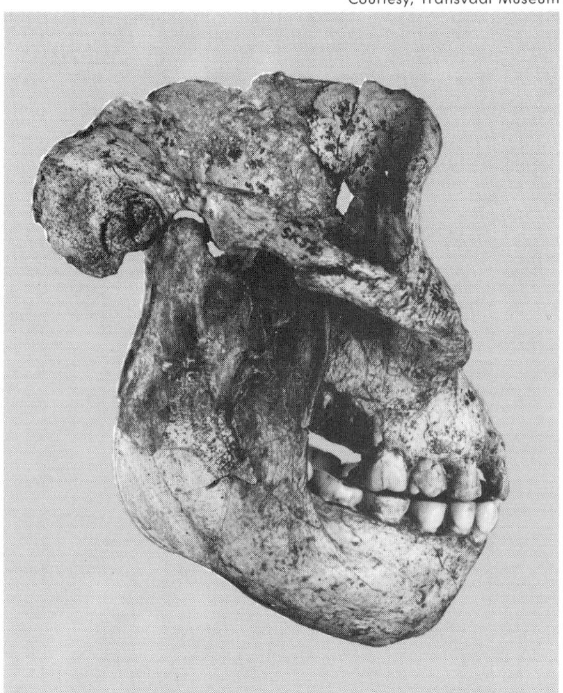

*Reconstructed fragmentary skull was recovered from a limestone quarry at Swartkrans, a fossil-rich site in South Africa. This skull differs in several respects from the* Paranthropus *fossils found at the same site and has been assigned to the genus* Homo.

century, could be summarized in these terms, if one remembers that there is overlapping at every point.

**Nature and development of man.** This first set of themes looks at humankind from its beginnings, in all of its interrelated dimensions; 36.8% of suggested contributions are grouped here, and others almost equally appropriate are included in the geographic and methodological groupings.

*Body and behavior,* the first of three themes under this heading, encompasses the biological problems as they are now seen in three overlapping groups.

• Man and the primates examines human evolution both morphologically and behaviorally, as evidenced by the study of living and fossil genera. The age of the hominid family—going back to the Pliocene and even the Oligocene—requires that the development of humanness be seen in a context far wider than that of the anthropoid apes. Therefore, the search for human ancestry is pushed further back, and our fossils are compared with primates other than apes. Nor can we now tolerate any separation of morphology from behavior. Thus, ethology joins with anatomy and paleontology to provide understanding of human evolution.

• Paleoanthropology: the Pleistocene. As *Homo* developed and differentiated in the Pleistocene, interest shifts from the genus *Homo* to questions concerning the species *Homo sapiens,* and from "behavior" alone to "culture" and "cultures." The evidence shifts from the relatively few human fossils to the important cultural remains. Here, biological or physical anthropology merges with archaeology and prehistory.

• Human differentiation. The richness of evidence from the large post-Pleistocene populations, including the living, permits detailed study of human variability related to both genetic and ecological factors. Emphasis shifts, therefore, to discussion of inheritance and growth within populations and differentiation of groups in different environmental circumstances. Here the simple concept of "race" is falling into disuse. Continental-historic terms such as Africans, Asians, and Europeans are replacing "Negroid," "Mongoloid," and "Caucasoid," and interest in the classification of "races" is giving way to demographic, ecological, developmental, genetic, physiological, and medical studies.

*Mind and culture* encompasses the development of communications, cognition, and science in evolutionary perspective.

• Language and thought are interests that combine the skills of linguists, psychologists, philosophers, and a variety of anthropologists. Sociolinguistics deals with the dialects of social groups and leads into the problems of education and "applied linguistics." Cross-cultural and intensive studies of cognition involve new problems of conceptual categories as well as nonverbal symbolic behavior. When and how symbolic communication began are questions requiring both ethological perspective and study of the functions of the brain.

• Science, technology, and invention are traditional anthropological interests that have recently claimed public attention in the context of "runaway" technology. Tool making is older than was once thought and is not exclusively human. Recent evidence indicates that deliberate scientific thought was already well developed during the Pleistocene, at least in astronomy and mathematics. Anthropology offers new perspectives for general understanding of this subject, as well as for recently emerging social problems.

• Sociological innovation and change are not confined to recent times, and important evidence of deliberate policy changes among even the smallest and most isolated human societies is being provided by field studies. It is apparent that human societies are rarely passive recipients of change. Any contrary impression may be an obvious example of "colonialist" bias.

*Expression in man,* also an area of traditional anthropological interest, can now be treated fully in empirically verifiable developmental and cross-cultural perspectives. The first three overlapping topics under this theme deal with aesthetics broadly considered; the fourth with religious and similar expressions.

• The graphic and plastic arts include all material, forms, and color, two and three-dimensional representations of thought and perception, to which there is affective response.

• The performing arts, music, dance, and theater, are viewed as affective behavior as they recreate human situations to satisfy emotional as well as intellectual needs.

• Folklore includes oral and written literature, seen as a body of information encompassing the study and analysis of myths and legends, fairy tales, riddles, folktales, and creative storytelling, analyzed from the viewpoint of aesthetics, psychology, and anthropology, and in their social and cultural matrix.

• Ritual, cults, and shamanism include studies usually associated with religious behavior; manifestations of a special type of affective responses including supernatural beliefs, worship, ritual, ceremony, and the manipulation of the social and natural world through esoteric means.

**Looking at the world spatially.** This emphasis has a long tradition in anthropology. Twenty-five percent of all respondents' suggestions are classified here under seven headings. Each subject heading includes an area as a whole plus a particular general problem for which the area is central.

*The circumpolar regions,* from the icecap through tundra and taiga. Special problem: man's adaptability to new and difficult environments, including the first peopling of, and dispersion throughout, the Americas.

*The Pacific rim,* from Japan and the northern coasts of mainland Asia through the Aleutians and Alaska and including the mountains south to Tierra del Fuego. Two special problems: (1) maritime anthropology and man's relations to the sea; and (2) communications along a cordillera as compared with cross-relations to the plains on one side and the sea on the other.

*Asian-African hot and cold desert and steppe,* a belt comprising northern and eastern Africa, the Middle East, and central Asia. Special problem: the relations between sedentary and nomadic ways of life in the context of modernization.

*The Indian Ocean areas,* including the South Asian plains, the eastern seaboard of southern Africa, Madagascar, and Southeast Asia through Malaysia. Special problem: comparative study of postcolonial "new nations."

*China to the Antipodes,* including mainland and Oceanic peoples and cultures. The special problem contrasts China as a "mother" culture with island cultures as historic receptors, taking as cases the Philippines and New Guinea.

*Europe.* Special problem: urbanization as a central historic process.

*The Atlantic.* The two coastal areas (Afro-European and American) are treated as virtually separate until A.D. 1500, and then (as the special problem) as they are brought together by northern maritime Europeans on both sides of the Atlantic who developed the international plantation economy, moved populations from east to west, and led to the development of a new Afro-American culture.

**Professional concerns.** It is characteristic of the orientation of anthropology that the smallest number of responses (14%) concerned methodologies and professional problems.

*Theoretical perspectives* can be divided into three sections, the first two of which deal with

*The graveyard of the tiny fishing hamlet Guañape on the coast of Peru illustrates a reality of countless communities around the world in which the graveyard is their most prominent feature. Only seven families remain in Guañape, once a thriving port town.*

Courtesy, Dr. Thomas Greaves, University Museum, University of Pennsylvania

major substantive problems: alternative theoretical orientations required for the analysis of differences in economic and sociopolitical development; the anthropology of complex societies; and current theories, methods, and techniques of research.

*Data storage and retrieval* includes bibliography, museology, cartography, and—what is becoming a major interest—visual-aural anthropology, including, especially, the use of sound motion pictures and television tapes.

*The history and future of the anthropological sciences* subsumes means of overcoming centrifugal tendencies caused by increasingly specialized knowledge; difficulties of communication across disciplines and across linguistic, cultural, and national boundaries; and problems of training and placing professionals and establishing ethical norms on a worldwide scale.

**Social concerns.** Some 24.2% of the responses concerned the social uses of anthropological knowledge.

Three interrelated problems that affect the whole of the species, and whose understanding urgently requires the broadest perspective, are: population and technological increase on a limited planet; colonialism, power abuse, and war; and systems of injustice and discrimination.

An area of major concern involves the fates of indigenous and minority peoples, as seen throughout history and today by historians and anthropologists and by the survivors themselves. The possibility and prospects of cultural pluralism in the industrial age is a major theme of the congress.

Among the problems of modern life are three sets of interrelated problems seen in the context of rapid urbanization: mental and physical illness; malnutrition; and drugs and stimulants.

Social improvement through anthropological understanding includes: the study of reproductive and early childhood behavior basic to the family and to community mental health, keeping in mind that child-rearing practices may become more positive and supportive if they are released from restrictions based on mistaken rules of supposedly universal "human nature"; and the need for education based on recognition of cultural and value differences, and of identification of individuals with different local, class, and culture groups, so that supposed "lack of motivation" can be eliminated and children and adults freed to fulfill themselves.

The future of the species as seen in the context of its entire past is an ultimate goal of the sciences of man and a general theme for the congress.

—Sol Tax

*See also* Feature Article: THE TASADAY—STONE AGE TRIBE OF MINDANAO.

## Psychology

The quiet, steady progression of research findings and theoretical understanding in psychology is ordinarily barely visible except to those most deeply involved, and then not in annual segments, but only over several years. During the year, however, several public events, including reviews of the IQ-race argument and statements advocating deliberate behavior control, attracted public vigilance and excitement that would not easily be extinguished.

**B. F. Skinner.** In his book *Beyond Freedom and Dignity*, B. F. Skinner suggests that since almost all of our major problems involve human behavior, we should turn to the science of behavior for relief. Continuing on our present course, without using a technology of behavior, we will inevitably bypass solutions to these problems, deny the full potential of man, and possibly drive mankind to extinction in the process. Thus he argues that deliberate behavior management is not only desirable but also absolutely essential.

Skinner's primary argument can be summarized easily. Behavior is controlled in any case; the complex relationships that determine behavior encourage what is only an illusion of freedom. And since behavior is already determined, but in most cases by poor and unsystematic environmental arrangements, we should deliberately shape behavior that is desirable to all of us. We should control for the good of man—indeed, for man's survival.

Skinner uses considerable space to demonstrate the varieties of behavior control. He points out that we tend to generalize and to resist any type of control because some forms of it are objectionable. He offers behavior-management procedures that would be agreeable to any subject and suggests ways to institute these procedures piecemeal. Readers are reminded that behavior-management programs do exist. Many have begun in psychotherapy, education, and commerce; they are multiplying, and their coverage and precision are increasing. In many ways Skinner is saying that we should encourage the expansion of behavior-control programs, and we should also consider whether they are based on desirable social policy.

Current psychology has dealt with many issues relating to behavior control, but Skinner widens the debate and strikes at two seemingly fundamental concepts of human nature. In his view, the primary impediments to large-scale employment of behavior technology have nothing to do with knowledge or cost. Rather, it is suggested that the main barriers are existing conceptions of individual freedom and human dignity.

Believers in these conceptions maintain that behavior control would reduce the freedom of any individual, subjugate him to constraints not present now, and, by virtue of considering man as a subject of control, reduce the worth of men. Skinner argues that the "literature of freedom" arose to encourage and justify escape from controls that were seen as oppressive, but that it has grown to regard any form of control as undesirable. Yet, he states, there are alternative kinds of control, involving "positive reinforcement," with which any man would find himself in happy concordance. "The problem," Skinner says, "is to free men not from control, but from certain kinds of control."

Skinner does not urge the immediate adoption of any particular cultural design; rather, he advocates continuous cultural experimentation and piecemeal adoption of those procedures that prove to be successful. The effort "is not to design a world that will be liked by people as they now are but to design one that will be liked by those who live in it." Further, there is no implied distinction between the ruled and ruling classes. A cultural design must answer continuously to the culture itself; those who control are simultaneously those who are controlled.

**Kenneth B. Clark.** Perhaps even more startling was Kenneth B. Clark's call, in his presidential address to the American Psychological Association, for research into a biochemical means that would prevent persons in power from abusing their power. "A few men in the leadership positions in the industrialized nations of the world now have the power to determine among themselves, through collaboration or competition, the survival or extinction of human civilization . . . . The masses of human beings are now required to live and continue to work on . . . [the] hope . . . that their powerful leaders will . . . use their power wisely and morally." This situation requires the development of precise psychological and social sciences; it has made "psychotechnology" imperative.

Citing what he called "many provocative and suggestive findings from neurophysiological, biochemical, psychopharmacological and psychological research," Clark suggested that "we might be on the threshold of that type of scientific biochemical intervention which could stabilize and make dominant the moral and ethical propensities of man and subordinate, if not eliminate, his negative and primitive behavioral tendencies."

Although Clark's target population and methods differ from those of Skinner, he does call for deliberate behavior control and he implies that much of the essential technology is currently available. Clark anticipated certain objections with a

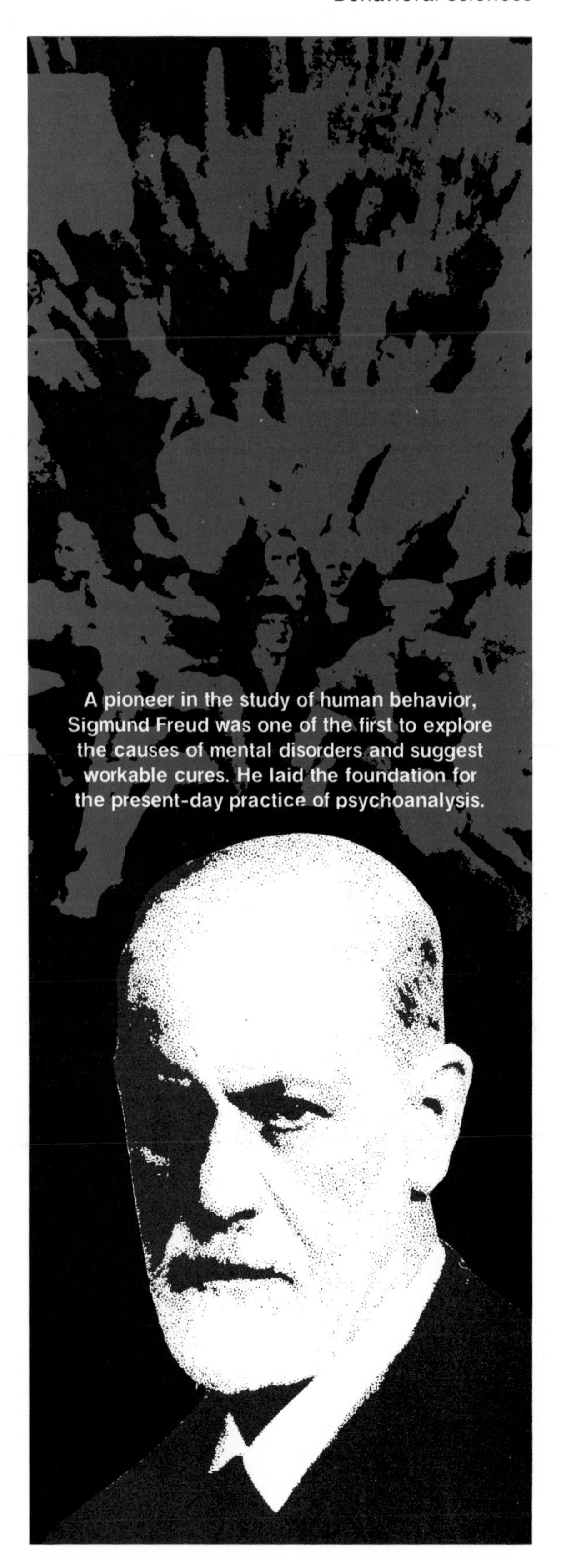

A pioneer in the study of human behavior, Sigmund Freud was one of the first to explore the causes of mental disorders and suggest workable cures. He laid the foundation for the present-day practice of psychoanalysis.

moral argument. He concluded that moral objections to psychotechnology because it is manipulative and "will take away from man his natural right to make errors—even those errors which perpetrate cruelties and destruction upon other human beings, . . . seem mockingly, pathetically immoral."

Professionals reacted quickly to Clark's address. Researchers in the fields Clark singled out were quick to suggest that his appeal was much too simplistic and fantastic. No one denied that considerable behavioral control was now possible, however, and professionals generally agreed that ongoing research of the type proposed by Clark should continue.

Public reaction was even more interesting. Skinner's book and Clark's speech received immediate worldwide attention. *Time* magazine featured Skinner's work in a cover story. *Beyond Freedom and Dignity* led nonfiction bestseller lists for several months and was nominated to receive the National Book Award. Interest was also evident in high places. Shortly after Clark's speech, Rep. C. Gallagher (Dem., N.J.) asked the U.S. General Accounting Office to examine federal funding to the American Psychological Association to determine whether federal money was being spent to implement Clark's proposal. Vice-Pres. Spiro Agnew spoke of Skinner's and Clark's thinking in an address to the Illinois Farm Bureau, calling it "radical surgery on the national psyche." (*See* Year in Review: SCIENCE, GENERAL.)

**IQ again.** The ancient heredity-environment controversy was exhumed by A. R. Jensen in 1969, when he analyzed data on heritability of IQ and observed differences between groups. Jensen hypothesized that social class and racial differences were reflections of significant genetic differences. During the year three reviews of the subject appeared: a book, *The IQ Argument*, by H. J. Eysenck of the University of London; an *Atlantic* magazine article by Harvard's R. J. Herrnstein; and a *Science* magazine article by S. Scarr-Salapatek from the University of Minnesota. In addition, *American Psychologist* carried a debate between Jensen and D. O. Hebb of McGill University.

Based on available data, the genetic argument was strong and consistent, but data comparable to that for the white, middle-class population (on which the argument was based) were lacking for lower-class and nonwhite groups. However, in one series of studies, differing heritability coefficients for different racial and social-class groups were obtained by Scarr-Salapatek.

The dilemma, which could be resolved only with large numbers of measurements, was this: because IQ is a trait standardized for a population, defining the population is critical. A population definition rests on increasingly restrictive specifications, particularly those regarding environments. Within a restricted population, the heritability factor must necessarily increase, but, given such restrictions, the meaningfulness and usefulness of the IQ measure as applied over a more generally defined population must decrease. Clearly, more technical expertise was required for public policy decisions on this issue.

Courtesy, National Institutes of Health

*Two female monkeys play independently in contrast to males, who prefer contact games. Scientists at the Wisconsin Regional Primate Research Center observed that gender is equal to or more important than environment in determining whether the play time of monkeys involves contact or noncontact sports.*

**New theory.** Experimental and theoretical developments in human verbal learning and in conditioning and animal learning, long the central areas of psychology, have progressed largely without regard to one another. Conditioning research generally followed the elemental, so-called "stimulus-response" format, while human learning theorists postulated hypothetical cognitive processes. During the year two conferences, of researchers in human learning and in conditioning, led to similar theoretical formulations. Since these areas give direction to much of psychology, such convergence would have far-reaching consequences.

A conference on conditioning and learning at North Carolina State University heard suggestions, particularly from W. K. Estes of Rockefeller University, in New York City, for the use of a "coding" event transpiring between experimenter-arranged stimulus events and subject-generated responses. The second conference, sponsored by the U.S. Office of Education at Woods Hole, Mass., was specifically devoted to "Coding Processes in Human Memory."

The notion of a coding event is helpful in both areas. Consider, for example, the problem of selective attention. Some subjects choosing between two complex stimulus arrays will base their choice only upon certain features of the array, while other subjects will choose on the basis of other features. Elemental theories suggesting that all the features have an equal potential for eliciting a response are too simple, while cognitive theories are either ad hoc or vague. A coding event that refers to specific aspects of the stimulus and is determined by experimental treatment and historical variables seems a likely and testable component in the information-processing sequence.

Major differences exist concerning the topography of the coding event, however. The original elaborations emphasized behavior in their discussions of coding as mediating responses. On the other hand, cognitive and perceptual theorists generally referred coding to the nervous system. Many of their theories used the early 1960s discovery, by physiologists D. Hubel and T. N. Wiesel at Harvard, of single neurons in the cerebral cortex that are excited by changes in unusual, single aspects of visual stimuli (angular orientation, direction of movement, size, etc.).

Recently N. Weisstein of Loyola University, Chicago, and J. Bisaha of Mundelein College, also in Chicago, obtained visual masking effects in which the apparent contrast for a second visual item is reduced by prior viewing of another visual item. This suggests an analysis involving groups of neurons. Their work is interesting to psychologists because the more complex the coding, the more a response analysis and a neural analysis converge in function.

—Daniel Johnson

*See also* Feature Article: THE PSYCHIC BOOM.

# Botany

Major trends in plant sciences during 1971–72 provided better understanding of the various interactions between plants and animals and heightened perspective on the control of molecular movement in and between plant cells. The organization of cells continued to seem more and more complex as it was better understood and botanists were learning more about the ways in which this complexity governs the relationships between green plants and their environment. One of the most surprising developments was the establishment of the importance of ion movements in setting up action potentials in plants, movements similar to those long known in the nervous systems of animals.

**Recycling of atmospheric ammonia by higher plants.** Nitrogen constitutes about four-fifths of the atmosphere but, although it is an important component of the amino acids that make up all proteins, green plants cannot use it directly from the atmosphere and animals must obtain it from higher plants in the form of organic molecules. The process of utilizing atmospheric nitrogen is controlled mainly by a series of microorganisms that occur either free-living in the soil or in the nodules on the roots of legumes, such as clover and soybeans, the plants often grown to enrich the nitrogen content of soil depleted by other plants.

*Stoma of a cucumber leaf is flanked by guard cells, which control the movement of water and carbon dioxide between the exterior and interior of the leaf. During the year scientists discovered that the guard cells are open only when they have a high concentration of potassium ions.*

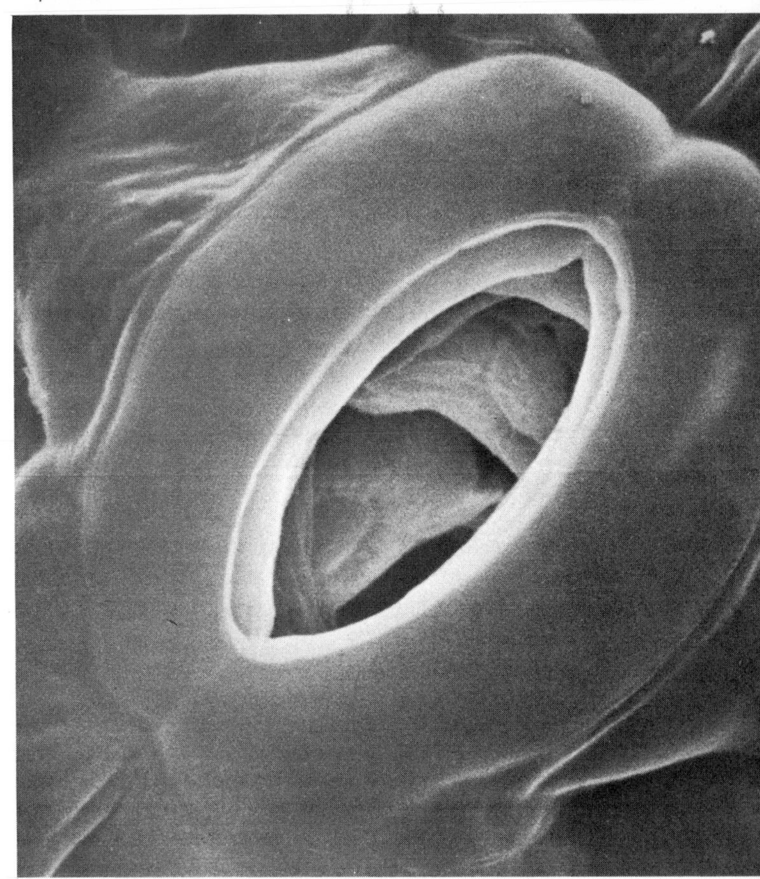

Courtesy, Dr. John Troughton

Most of the ammonia (ammonium ion) in the atmosphere or in the soil is there as a result of organic decay—the breakdown of proteins and amino acids. In the traditional nitrogen cycle, the ammonia is considered to be oxidized to nitrites by the bacteria *Nitrosomonas* and *Nitrosococcus*; the nitrites are in turn oxidized to nitrates by the bacteria *Nitrobacter*. These nitrates can then be taken in by green plants and used in the synthesis of organic molecules.

In early 1972, however, a group of researchers showed that growing plants could satisfy about 10% of their nitrogen requirements by absorbing ammonia, which is a compound of nitrogen and hydrogen, directly from the atmosphere through the leaves. This discovery might be of considerable importance in explaining the balance between plants and their environment, particularly under polluted conditions where ammonia may be abundant.

**Following root growth.** Work by J. Waddington of the Canada Agricultural Research Station in Saskatchewan resulted in the development of a periscope for observing the growth of roots. All of the commonly used techniques had involved considerable disturbance to the roots or the application of radioactive traces that constituted potential radiation hazards. In Waddington's method, transparent glass or plastic tubes could be inserted in the soil near a plant and an inverted periscope constructed with one element transmitting light down to the roots and another transferring the image back to a viewing eyepiece. The whole system could be accommodated in a tube about 2.5 cm in diameter and would cause very little disturbance to the plant when left in place over a period of time. The method could permit observations of patterns of root growth far superior to those possible with any previous method and might provide better knowledge of the relationship between roots and the soil, which would be important for agriculture as well as for theoretical botany.

**Plant-animal interactions.** As the scope of studies into the relationship between green plants and insects and other animals broadened, plants were discovered to employ a surprising variety of methods to ward off predators. Many plants manufacture molecules of alkaloids, cardiac glycosides, cyanide, mustard oils, and the like, which inhibit feeding by most insects. Some of the relatively few insects that are able to feed on these "protected" plants retain these molecules and use them as a defense against their own predators. A familiar example is that of the orange-and-black monarch butterfly (*Danaus plexippus*), which is highly distasteful to birds because it retains cardiac glycosides it took up as a caterpillar from milkweed plants. The viceroy (*Limenitis archippus*) resembles the monarch so closely that it too is avoided by birds, even though it is not distasteful because its larvae did not feed on milkweed.

Some plant defenses were found to be, or were suspected of being, more subtle. Some ferns and gymnosperms, for example, have molecules that resemble ecdysone, the insect molting hormone, which might cause disturbances in the development of the insects that feed on them. Certain bark beetles and bees utilize substances they derive from plants as their own sex attractants. Many of the hairs and spines on plants might constitute almost impenetrable barriers to small insects, and also might regulate water loss from the leaves.

During World War II, in an attempt to discover why bracken fern (*Pteridium aquilinum*) was not eaten by cattle and sheep, it was found that the plant contained substances that inhibit the synthesis of thiamine. Perhaps this also explained why bracken and other ferns are rarely eaten by insects and other herbivores. Recently it was shown that when the leaves of tomato or potato plants are wounded by insect attack, they rapidly accumulate a substance that inhibits digestive enzymes (proteinases). Such wound reactions of plants might eventually be found to discourage insect attack, a role that might have important implications for agriculture.

**The meaning of flower colors.** Recent observations helped to explain why flowers visited for food by hummingbirds and other birds are most frequently red. The energy consumed by a hummingbird hovering at a flower is some 600 times that used by a bumblebee crawling over a cluster of lilac blossoms. Hummingbirds and other animals with high energy requirements, such as hawkmoths, must largely confine their visits to flowers that produce nectar abundant enough and rich enough in sugar to support them. Insects with lower energy requirements will also visit flowers that produce abundant nectar, but they will not normally move from plant to plant and their visits will not result in cross-pollination. The plant that produces enough nectar to support repeated visits by hummingbirds must, therefore, "protect" its nectar from these insects.

The plant appears to accomplish this in three ways. First, its flowers are generally strong and tubular and the nectar is secreted at the base, so that only an animal with a long tongue or beak can reach it. Second, the flowers are odorless. Birds, like men, orient largely by sight, while insects often orient by odor, especially at a distance. Third, the flowers are often red. Although insects are known to detect wavelengths in the ultraviolet that are not visible to men or birds, red, to them, fades into the

Barbara Panessa, New York University

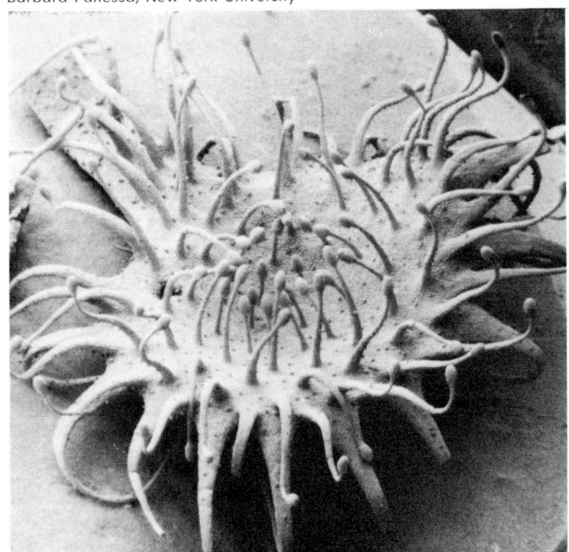

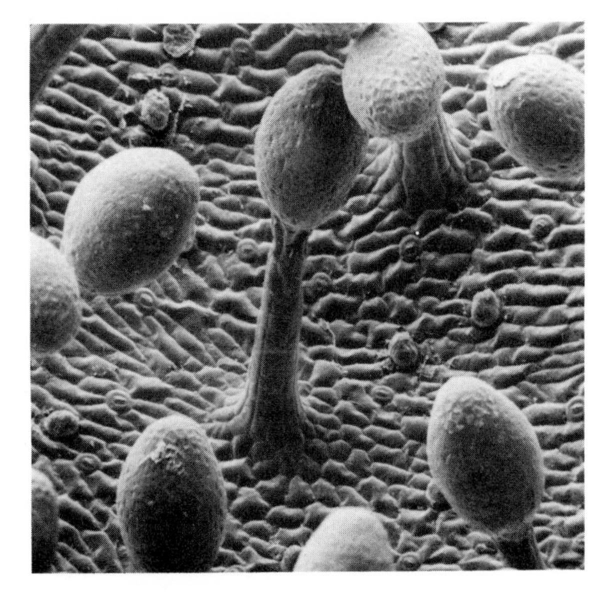

*Scanning electron micrographs of a sundew leaf clearly show the tentacles through which digestion of trapped insects takes place (close-up at right). The sundew may prove valuable in further studies of the role of electrical energy in plant behavior.*

background of green leaves, as it does to a color-blind person. There are few pure red flowers in Europe, where there are no flower-visiting birds. In contrast, the distribution of red flowers in the Americas, where 84% of the South American species known to be bird-pollinated are red, may well depend on the annual migration of hummingbirds.

**Control mechanisms in stomata.** One of the most striking developments in botany was the growing understanding of the mechanisms controlling the movement of water and carbon dioxide between the exterior and the interior of leaves. Water and carbon dioxide are the compounds from which organic compounds are built during photosynthesis. A thin cuticle makes leaf and stem surfaces largely impervious to both of these vital molecules. Water must flow continuously from the soil into the roots, upward through the stems, and out through the leaves. Since there is no mechanism for recirculating water, continuation of plant life depends upon the maintenance of this flow. Carbon dioxide, on the other hand, must enter the green parts of the plant directly from the atmosphere. It does so through the same small openings, called stomata (singular: stoma), that permit the escape of water.

Each stoma is flanked with a pair of specialized cells called guard cells, which, instead of being flat and tablet-shaped like the rest of the surface, or epidermal, cells, are sausage-shaped and thickened on ridges along their inner margins. When the guard cells are turgid, or subject to high internal water pressure, the inner margins spring outward in opposition to one another and the stoma is open. When they are flaccid, or the water pressure is low, the inner margins relax and come together. In many plants additional specialized cells, the subsidiary cells, are arranged in small groups of definite position around the guard cells.

Stomata are often very numerous; there are 12,000 per sq cm on tobacco leaves. In some plants they are confined only to the upper or lower surface of a leaf, while in others they are on both. The stomata are connected directly with a honeycomb of air spaces within the leaf that sometimes makes up nearly half of the total volume of the leaf. The air in these spaces is saturated with water evaporated from the damp surfaces of the leaf's interior cells. Within these cells are the chloroplasts in which photosynthesis occurs. For a plant to continue to live, the flow of water from its leaves and the passage of carbon dioxide into the leaves must be regulated. Under drought conditions, the low water pressure in the leaves often causes the stomata to close. This reduces access to carbon dioxide, and photosynthesis may be very limited.

The strides made during the year, however, dealt with influences other than drought on the regulation of the stomata. The stomata of many plant species were known to close regularly in the evening and open in the morning, even when the supply of water to the plant was constant, presumably because of an increase in carbon dioxide concentration around the plant. It was shown that when stomata are open, the guard cells have high concentrations of potassium ion ($K^+$) and that when they are closed, most of the potassium ions have moved out into the subsidiary cells or into other cells of the leaf. The changes in concentration were sufficiently large to account for other changes associated with the opening of the stomata. Normal opening and closing of the stomata were found to be possible only when the guard cells were attached to the other cells of the leaf or when they floated in a solution containing potassium ions. In darkness, air containing carbon dioxide apparently inhibits the accumulation of potassium ions in guard cells, and thus the stomata do not open.

The ability of stomata to open and close rapidly in response to environmental stimuli is essential to good plant growth, and botanists were considering possible mechanisms whereby the movement of potassium ions might be controlled. Guard cells in higher plants also clearly can take up ions of potassium with a much greater degree of specificity than had been suspected. The way this specificity is controlled was also expected to be studied further, since the considerable degree of specificity found in the movement of ions from the soil into the plant roots remained one of the unsolved questions of botanical science.

**Electrical signals in plants.** As early as 1880, it was known that a leaf of Venus's flytrap (*Dionaea muscipula*) conducts "action potentials" when certain specialized hairs protruding from its surface are touched by an insect. Following the passage of two or more action potentials, the two lobes of the leaf fold together rapidly. If the insect is trapped, small glands secrete digestive enzymes that reduce the hapless creature to its biochemical building blocks, which are rapidly absorbed for the nutrition of the plant. These action potentials, which seemed to be very similar in origin and function to those of animal nerve cells, were long regarded by botanists as freaks of evolution, even though similar action potentials were discovered to mediate the rapid closing of leaves of the sensitive plant *Mimosa pudica*. To some botanists, however, it seemed that such specialized behavior could not have arisen de novo in these peculiar plants, so they looked for electrical signals of a less conspicuous nature under more ordinary circumstances. As a result of their investigations, it appeared that action potentials might be widespread in the plant world, communicating information about the state of the plant's cells or about the external environment.

*Leaves of Venus's flytrap conduct "action potentials" and close rapidly when hairs on their surfaces are touched by an insect. Scientists are currently studying other factors that stimulate electrical signals in plants as well as the roles played by such signals.*

Russ Kinne from Photo Researchers

It was reported, for example, that when pollen lands on the surface of the female part of a lily flower, an action potential passes into the hidden region of the flower where the egg is waiting to be fertilized. The action potential may trigger final stages in the development of the female tissue that aid in the passage of the sperm to the egg and might even aid fertilization itself. In certain families, the female parts of the flower play an active role in securing pollen. The touch of an insect on the stigmatic surface of the female part triggers an action potential that causes the lobes of the stigma to fold around the insect. The insect usually escapes, but any pollen on its body is brushed off and held tightly between the lobes, which form a little chamber within which the pollen develops further. A second action potential then passes to the region of the egg in the ordinary fashion.

When a young seedling, such as that of the garden pea, struggles to emerge from the soil, it is aided by a special sensory system that enables it to judge the depth and compaction of the overlying soil. The detailed mechanism of the system was not understood, but it was clear that when the surface of the stem (technically, of the epicotyl) is rubbed, as by particles of soil, voltage fluctuations that resemble action potentials can be observed almost immediately and continue for some time. A hormone is soon released that causes the plant to develop a stouter stem and keeps the tiny leaves folded over tightly in a bud that will not suffer abrasion as the plant pushes through the soil. When the plant nears the surface, friction between soil particles and the stem diminishes, the voltage fluctuations cease, the release of the hormone decreases, the stem puts on a burst of elongation, the bud unfolds, and the leaves expand in preparation for photosynthesis.

More mysteriously, a variety of spontaneous signals of unknown function were detected when electrodes were applied to the green parts of such plants as the morning glory, garden pea, and cocklebur. These signals sometimes occurred in trains of variable duration. The individual signals might be very brief and rapid in sequence, or they might be longer and separated by fairly long intervals. Isolated individual signals of more erratic character were observed in mushrooms as well as in green plants.

Scientists were trying to learn more about the factors that trigger electrical activity in plants and about the role of electrical signals in plant behavior. A carnivorous plant might well play an important part in these studies. The sundew *(Drosera)* was found not only to have action potentials like those of animal nerves but also to have receptor potentials resembling those of animal receptor organs. Because this versatile plant has cells that are comparatively easy to see and to impale with electrodes, it was thought to be particularly suitable for detailed comparisons of the electrical activity of plants and animals.

—Peter H. Raven

Courtesy, INDUSTRIAL RESEARCH

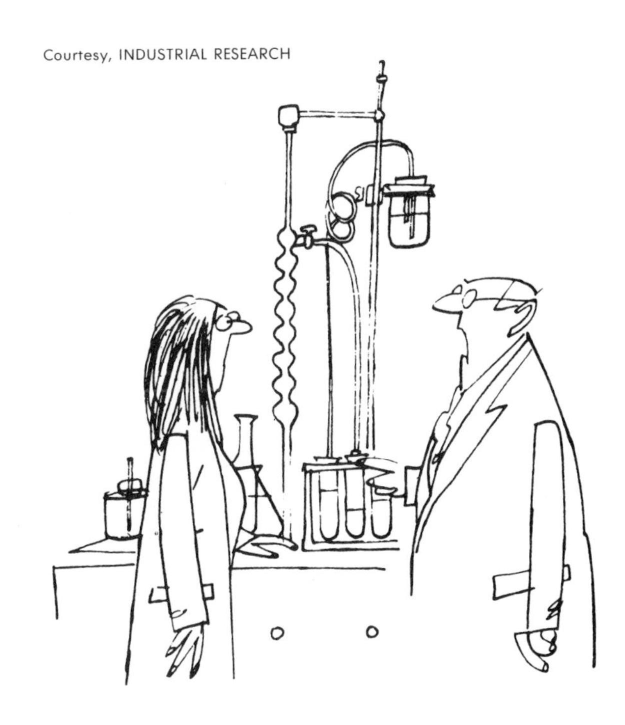

*"We should have a reaction in about 36 hours. . .Call me when it happens."*

# Chemistry

Advances during the year in chemistry ranged from success in limiting the lifetime of plastics to the recording of the two millionth unique chemical structure. Beam collision techniques were used effectively to help discover the nature of chemical reactions, while liquid-liquid chromatography helped determine chemical structures.

## Applied chemistry

"Energy Crisis" was added firmly to the list of public concerns during the past year, and chemists and chemical engineers responded by intensifying their efforts to improve the creation of synthetic natural gas from naphtha and coal. They also suggested hydrogen as a universal fuel, supplementing both fossil fuels and nuclear energy. A successful effort at reducing the persistence of pesticide molecules in the biological environment was also reported.

**Chemistry in the furnace.** About one-third of U.S. energy needs in 1972 were met by natural gas, a remarkable statistic considering that until the end of World War II this fuel was simply flared as a necessary waste in the production of oil. The energy demands of the postwar suburban boom, coupled with the invention of the seamless welded pipeline that made possible long-distance transmission at high pressures, produced natural gas' explosive growth, about 6% a year since 1950.

During the past year, various indications—gas companies cutting back or simply eliminating new orders of gas, for instance—made it evident that shortages of natural gas were developing in the United States. The reasons for the shortage included not only an actual decrease in the supply but also the pricing policies of the Federal Power Commission (FPC) and the apparent reluctance of the gas companies to search out new deposits. There did appear to be ample supplies of natural gas for the next two to three decades if the predictions of the Potential Gas Committee are borne out. The committee estimated proved reserves of natural gas at 260 trillion cu ft (the U.S. now uses about 23 trillion cu ft annually), and potential reserves at 1,178 trillion cu ft. Given the exponential increase in overall energy consumption—a doubling about every 10 years—this meant that the proved reserves would last for 13 years and the potential reserves for 20.

The combination of these limitations on natural gas supplies and a more favorable FPC pricing policy spurred the gas companies into a greater technical effort to make a gas from coal and naphtha (a light hydrocarbon fraction of crude oil) that will be compatible with natural gas. More than a dozen companies recently announced plans to make synthetic natural gas (SNG) from naphtha, while several more planned to make SNG from coal. To the consumer the effect of this will be higher gas prices, perhaps two to three times

Courtesy, ICI Fibres Ltd.

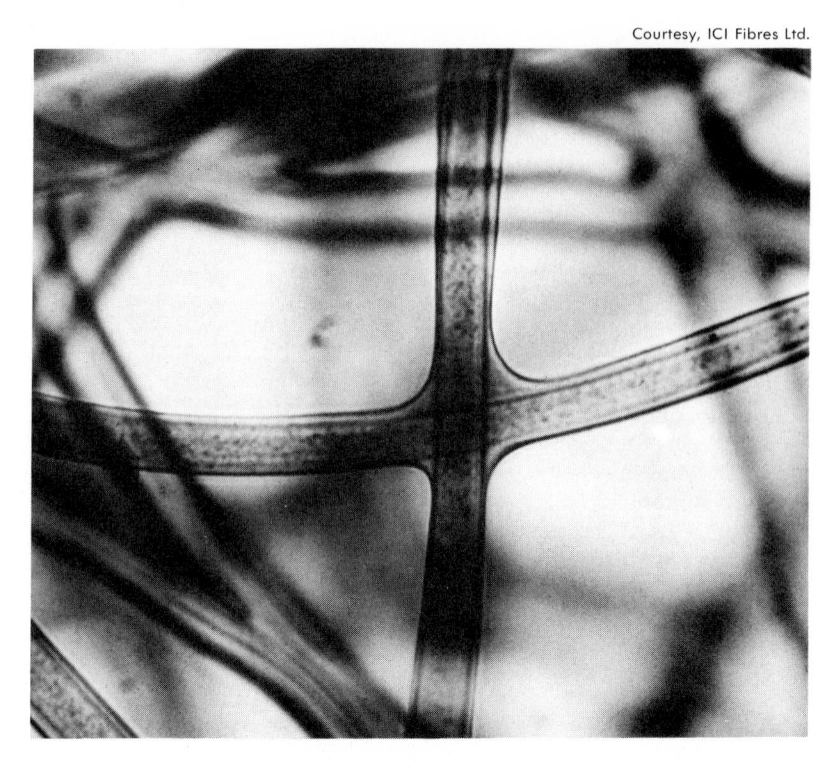

*Heterofil bond is formed when a web containing normal fibers and bicomponent fibers, a fiber with a high melting core and a sheath of a low melting polymer, is heated. The resulting fabric is called a melded fabric.*

above current levels, although the blending of natural with synthetic will somewhat mitigate the price increase.

For scientists and engineers, the effort to create SNG compatible with natural gas creates intense technical pressures. While synthetic gas is not new —town gas made from coal was the chief heat and light fuel in the 19th century—it is quite something else to make a synthetic product as good in its heating qualities and as free of contaminants as natural gas. Town gas, made by simply heating coal, has a heating capacity of about 500 British Thermal Units (BTUs), while natural gas is about twice that, about 1,000 BTUs.

Work on creating SNG from naphtha was well underway in 1972, spurred largely by the initial efforts of the British Gas Council. The largest venture to be announced by mid-year was a 500 million-cu-ft-per-day plant to be built in New Jersey in 1973 jointly by Texas Eastern Transmission Corp. and Consolidated Natural Gas Co. The effort to create SNG from coal is a bit behind naphtha; most experts predict 1977–80 as the time for commercial coal gasification in the U.S. However, in April 1972, the Continental Oil Co., British Gas Council, and Scottish Gas Council jointly announced plans to upgrade an existing coal gasification facility in Scotland to the point where it will produce a gas with the heating capacity and purity of natural gas.

Natural gas is almost exclusively one hydrocarbon, methane ($CH_4$), and the successful creation of SNG consists of converting economically a mixture of many hydrocarbons into methane. The problems are particularly difficult for coal: the hydrogen content is too low and must be upgraded; it must be properly pretreated to avoid caking problems that delayed the start-up of one gasification plant this year; and, there are a host of impurities, including sulfur, that must be removed.

The fuel to be gasified—be it coal or naphtha—is reacted under pressure with air and steam. The product is "synthesis gas," a variable mixture of carbon monoxide, hydrogen, methane, carbon dioxide, and sulfur impurities. A properly catalyzed "shift reaction"

$$CO + H_2O \longrightarrow CO_2 + H_2$$

brings hydrogen and carbon monoxide into proper balance for the subsequent methanation reaction,

$$CO + 3H_2 \longrightarrow H_2O + CH_4$$

Methanation, while successful in the laboratory, has not yet been proven commercially. However, it is necessary if several gasification schemes now proposed are to produce useable SNG.

Coal, as suggested earlier, presents some formidable technical problems as a source of SNG. However, technical experts are optimistic that these problems can be resolved. Coal, that is, carbon (C), is gasified when it reacts with steam ($H_2O$) and air ($O_2$) at about 1,800° F:

$$(1)\quad C + O_2 \longrightarrow CO_2$$
$$(2)\quad C + H_2O \longrightarrow CO + H_2$$
$$(3)\quad C + 2H_2 \longrightarrow CH_4$$

The heat required for these gasification reactions makes them expensive steps in the coal-to-gas process. Heat is supplied in various ways depending on the process: electrical energy, burning of coal, reaction of carbon monoxide with metal oxides, reaction of lime (CaO) with carbon dioxide.

In addition, coal contains various impurities—a typical chemical formula for coal (Illinois #6) is $C_{11}H_{85}S_2N_{1.5}O_{9.5}$—that must be removed. Sulfur is taken out as hydrogen sulfide, nitrogen as ammonia, and oxygen as water.

In 1972 there were about 60 coal gasification plants operating in Europe (none in the U.S.). All were based on the Lurgi Process, which helped keep Germany supplied with synthetic fuels during World War II. It is questionable, because of caking problems, whether bituminous coals found abundantly in the eastern part of the U.S. are suitable for the Lurgi process. However, at least three U.S. plants using the Lurgi process were announced.

There were several new approaches to coal gasification on the planning boards in 1972. The most advanced, in terms of reaching actual production, was the Institute of Gas Technology's HYGAS Process. In this process, hot hydrogen reacts with a mixture of coal and oil in a hydrogasifier. The methane formed is enriched by a subsequent reaction between carbon monoxide and hydrogen. A HYGAS plant was scheduled in October to start producing about 2 million cu ft per day of synthetic gas from lignite (brown coal) or midwestern coal.

The M. W. Kellogg Company's molten salt process may prove the most versatile, both in the kind of fuels it can gasify (ranging from coal to garbage) and in its ability to handle both caking and noncaking coals. The coal is suspended in molten sodium carbonate, which also catalyzes the formation of methane in the gasification steps. One advantage of the molten salt approach, carried out under pressures of 1,200 pounds per square inch, is that combustion can be separated from the synthesis gas, eliminating the need to remove sulfur, carbon oxides, and other impurities.

**Hydrogen as a fuel.** While coal gasification neared the production stage, energy experts were also seriously considering another energy source—hydrogen gas. At a national meeting of the American Chemical Society in April 1972, several speakers considered the prospects for a "hydrogen economy" in which hydrogen would be used as a universal fuel, that is, as an alternative to both nuclear power and fossil fuels, natural or synthetic. Hydrogen fuel has several immediate advantages: it can be made from water, burns easily, is transportable through already existing gas pipelines, and, when burned, leaves only one "ash"—water. Moreover, unlike electricity, which must be made on demand, hydrogen gas can be stored.

The first basic step in developing a hydrogen economy is to use nuclear power to break down water, either electrolytically or by thermal cracking—heating to very high temperatures. However, electrolysis of water to hydrogen and oxygen is a relatively expensive and somewhat inefficient process. Direct thermal breakdown would require temperatures higher than fission (but not fusion) reactors could achieve. However, Cesare Marchetti and G. De Beni of the European Atomic Energy Community proposed a four-step process to make hydrogen that involved a series of recycled chemical intermediates in which the highest temperatures reached would be about 1,350° F. Some problems might arise with this, including possible loss of mercury (one of the chemicals involved) to the environment, corrosive intermediate materials, and an overall efficiency about the same or below that of electrolysis.

**The architecture of pesticides.** The past decades of organic chemistry were marked by the analysis of the structures and subsequent syntheses of often quite complex molecules. By 1972 the emphasis was on coupling the ability to synthesize with the growing knowledge of how chemicals act in living systems, including the human body. The result was molecular architecture; chemicals designed for specific biological applications. One example of molecular architecture of particular satisfaction to environmentalists was provided in the February 1972 issue of *Chemical Technology* by Robert L. Metcalf and his colleagues at the University of Illinois. They designed analogs, or chemically similar materials, to DDT that are reportedly as effective as that insecticide but much more readily broken down by living systems, that is, more biodegradable.

DDT, and its even more stable metabolite DDE, do not break down easily, and this has been blamed for the accumulation of these pesticides in fishes and for the reproductive failures of a variety of birds such as the Bermuda petrel. The basic problem, as Metcalf pointed out, is that both DDT and DDE dissolve much more readily in fat than in water. The consequence is that DDT (or DDE) is poorly excreted and instead accumulates in fatty tissues of fishes, birds, and man. The reason DDT is not broken down in the body to water-soluble products is that enzymes, biological catalysts, cannot attack a certain portion of the DDT molecule. Accordingly, working from clues supplied by a much more biodegradable, if less effective, pesticide called methoxychlor, Metcalf and his colleagues prepared a host of DDT-like compounds in which the enzyme-resistant portions were replaced by other chemical groupings.

The result, reported Metcalf, was a variety of compounds readily biodegradable but just as effective as DDT. In some cases, these products of molecular architecture were even more effective; for example, one designed compound proved effective against a strain of DDT-resistant houseflies.

—Norman Metzger

## Chemical dynamics

Research in chemical dynamics continued to try to shed light on the questions as to how chemical reactions occur, what forces drive them, and what factors determine their rates. Traditionally, the processes involving chemical change have been studied by mixing reactants in the gas phase or in solution and then monitoring the products produced as a function of time. The energy of the reactants was varied by varying the temperature. In such a system the reactants have a spread of kinetic energies corresponding to their thermal distribution. One is limited in the energy range one can cover, and if solutions are involved the matter may be further complicated by effects of the solvents. In order to avoid some of these complications, chemists began to use the so-called beam techniques, in which a beam of ionic or neutral reactants with a narrow spread in kinetic energy is made to collide with the target gas molecules. In such a reaction one studies the energies, the identity, and the angular distribution of the products resulting from a collision between projectile and target.

**Beam collision techniques.** A diagram of a spectrometer used for such studies is shown in figure 1. Ions (charged atoms) are produced in the input mass spectrometer by the electron bombardment of a suitable gas. The ions so produced are then retarded or accelerated to a desired energy and focused into a reaction chamber containing $D_2$. The ionic species ($C^+$ and $CD^+$) that come out of the reaction chamber have their energies determined by the energy analyzer and their identity established by the analyzing mass spectrometer. The ions are then detected by a sensitive detector such as an electron multiplier. An ion-molecule reaction is used here to illustrate this technique since ions can be accelerated, decelerated, and focused. Consequently, they are easier to work with than projectiles without a charge. Similar experiments, however, can be and are being carried out for reactions where both target and projectile are neutral.

When ions are produced by electron bombardment, they are frequently electronically excited; that is, their internal electrons are displaced from their most stable positions. Likewise, a similar situation may exist when an ion reacts with a molecule and produces an ionic product. This product may be electronically excited, or its nuclei may be vibrating more violently than normal (vibrational excitation), or the product may be rotating more rapidly than in its lowest rotational energy state (rotational excitation). Quantum mechanics tells us that these states of excitation are characterized by discrete energies called energy levels. In general, the differences between electronic energy levels are larger than the differences between vibrational energy levels, which in turn are larger than the rotational energy level spacings. In a chemical reaction ideally one would like to know the kinetic, electronic, vibrational, and rotational energies of all the species involved in the reaction. This has been accomplished in only a few favorable cases such as the reactions of deuterium atoms with diatomic halogen molecules in which the various states can be determined by the light emitted during the course of reaction.

*Figure 1. In the beam techniques for studying chemical changes, ions are produced in the input mass spectrometer, focused into the reaction chamber, and, after coming out, analyzed by various instruments.*

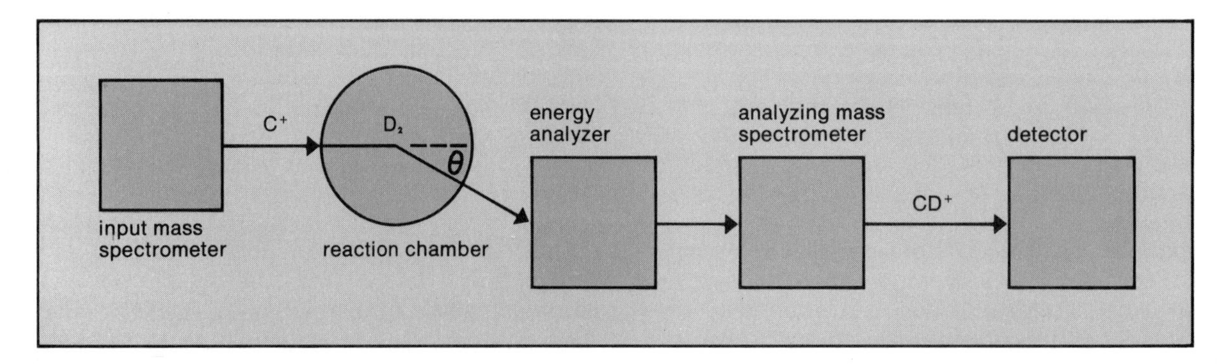

A technique was recently developed for determining how many and what electronic states are present in a projectile ion beam produced by electron bombardment. The ion beam is passed through a chamber containing a suitable gas, the pressure of which can be varied in a quantitative manner. By measuring the diminution in intensity of the ion beam as a function of gas pressure, one can determine if the ions are in a single electronically excited state; if there is more than one electronic state, the fractions of various states present can be evaluated. By studying these fractional abundances as a function of the electron bombarding energy one can determine which electronic states are present. If an ion in an electronically excited state makes a sufficient number of collisions with other gas molecules, it loses its excitation energy and goes into its lowest (ground) state. Using this technique, scientists studied a number of ion-molecule reactions with the projectile ions in their ground electronic state; and by comparing the reactions produced by a projectile beam of ions in mixed electronic states, the role of both the ground electronic state and the excited states could be determined. In the near future one can expect application of these techniques to the identification of states of the product ions.

Another more elegant method of state selection that has been used is photoionization. In this technique one irradiates a molecule such as $H_2$ with a beam of light with a very narrow energy spread but of sufficient energy so that ionization of the molecule occurs. The molecule then can be put into not only a desired vibrational but also into a selected rotational state. This type of refined state selection permits one to investigate the roles that vibrational and rotational energy play in chemical reactions.

A fascinating development is the observation that beam techniques can readily produce rather large ion clusters that are of sufficient stability so that their properties and reactions can be studied in the gas phase. Several studies were reported on ion clusters as $H^+(H_2O)_n$, $NH_4^+(NH_3)_n$, and others. A number of solvated (containing solvent and solute) clusters of both positive and negative ions were reported. In the case of ammonia clusters as many as 20 ammonia molecules were found associated with the $NH_4^+$ ion. These clusters are not just academic oddities. Such solvated ions will probably be found to play a significant role in pollution problems and in the nature of structured water in biological systems.

Studies of the reactions of ion clusters are just beginning; and before quantitative studies can be realized the properties of the excited states of such ions will have to be investigated. Probably the most fascinating implications of these studies is that they will permit the bridging between gas phase and solution chemistry. Chemists for a number of decades have been accustomed to assuming the existence of species such as $H^+(H_2O)_n$ in explaining the migration of ions in solutions, rates of reactions, electrical conductivity of ionic solutions, etc. By 1972 it had become possible to examine the various properties of such species on a molecular basis. This will undoubtedly prove to be a fruitful area for the study not only of aqueous solutions but also of solutions of various organic and inorganic systems.

**Reactions of three-atom systems.** During the past year, investigators continued to report results on the angular and energy distribution for products of relatively simple ion-molecule and neutral-neutral reactions represented by the equation $A + BC \rightarrow AB + C$. Depending on the shape of the angular distribution (the distribution in angle, relative to a specified direction, of the intensity of particles resulting from a reaction), the reaction forms a persistent complex or proceeds by some type of direct mechanism. An example of a complex formation may be a case where A and BC on collision interact with each other for a period of about $10^{-12}$ seconds and then dissociate into products $AB + C$. An extreme example of a direct mechanism is one in which A on moving past BC picks up B to form AB, leaving C behind without any momentum imparted to it. In such a case A may spend a relatively shorter time (less than $10^{-14}$ sec) in the region of BC.

Some investigators believed that the condition necessary for persistent complex formation is a potential "well" in the ABC system that is avail-

*Figure 2. In a schematic of a potential well, reactants overcome a potential barrier (slight rise at left) before forming intermediate complex at lower energy. Energy barrier at right is surmounted to form final products.*

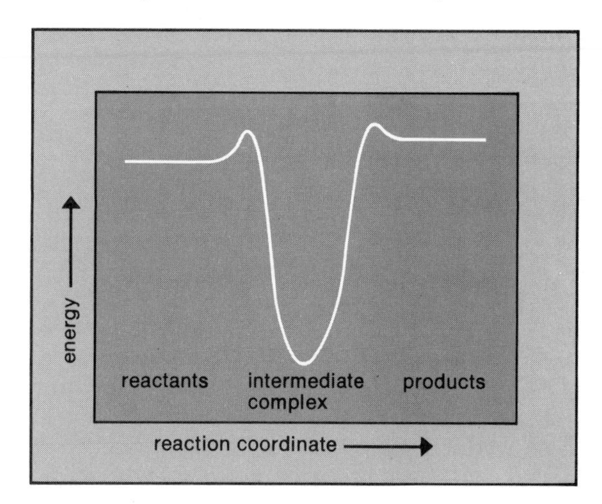

able to the reactants—access is not forbidden by spin, symmetry, or other restrictions. Under such a condition persistent complex formation can be expected if the relative kinetic energy is low enough. Persistent complex formation was observed in sample reactions such as $O_2^+ + D_2 \rightarrow DO_2^+ + D$; and $C^+ + D_2 \rightarrow CD^+ + D$. As the projectile kinetic energies are increased in these reactions, a direct mechanism comes into operation.

A schematic potential well, which may represent the situation in an endothermic reaction such as $C^+ + D_2 \rightarrow CD^+ + D$, is shown in figure 2. In the left-hand side of the diagram, as the projectile approaches the target, it may have to overcome a small potential barrier (the slight rise); but as it gets closer to the target, the potential energy falls strongly. This represents a strong attraction between the reactants, and a complex possibly such as

$$\begin{array}{ccc} & C^+ & \\ / & & \backslash \\ D & & D \end{array}$$

is formed. The kinetic energy of the projectile has been converted in part to vibrational and rotational energy of the complex. The smaller this energy is relative to the depth of the well the longer the complex lives.

After going through a series of internal oscillations, the system surmounts the barrier on the right-hand side of the figure and the products $CD^+ + D$ are formed. The line on the right representing the products is higher than the line on the left that represents the reactants because the reaction is endothermic and requires a certain amount of energy in order to be able to proceed to the final products.

It is clear that much remains to be done before a thorough understanding of these three-atom systems is realized. The role not only of kinetic energy but also of vibrational and rotational energy as driving forces in chemical reactions will have to be investigated.

**Future outlook.** The immediate future outlook for the area of chemical dynamics is bright except for reduced funding for basic research. With the increased availability of tunable lasers and sensitive, high-resolution optical spectroscopic systems, scientists can expect more work on state selection of reactants and the determination of the states of the reaction products. By irradiating suitable target gases with the proper frequency of light obtained from a laser, one will be able to put a target molecule into a particular desired state. The high-resolution optical spectrometers will permit the analysis of light emitted by the product; and in favorable cases, the electronic, vibrational, and rotational states, as well as the geometry of the product, will be determined. This will lead to a deeper understanding of the role of kinetic, vibrational, and rotational energies as driving forces of chemical reactions.

In the coming years, work on solvated ions in the gas phase is expected to increase our understanding of the properties of these ions and, it is hoped, bring us closer to bridging the gap between gas phase reactions and reactions in solution. The question of persistent complex formation *v.* direct mechanisms in chemical reactions will certainly receive attention from a number of research groups, and we will be getting a clearer understanding of exactly what happens in a binary collision between reactants.

—Walter S. Koski

## Chemical synthesis

For modern chemical technology and fundamental studies of molecular behavior, pure chemical compounds of defined constitution must be secured. While naturally occurring materials exhibit great diversity, they are often limited in supply and utility. In fact, their number and structural types pale before those of the conceivable atomic combinations of the elements. In the effort to build new chemical structures, and reproduce or modify complex natural products, chemists rely first upon the vast body of chemical theory to guide them in formulating workable synthetic routes, and then upon the array of selective and efficient reagents for performing each of the individual component transformations. The following discussion presents recent examples of the successful application of this basic procedure.

**Vitamin $B_{12}$.** A clear indicator of the capability of chemical synthesis is the complexity of structure yielding to the chemist's hand. The synthesis of vitamin $B_{12}$ was announced at an international symposium on the chemistry of natural products held at New Delhi, India, in February 1972. Robert B. Woodward delivered the address describing the culmination of a decade of joint activity between his group at Harvard University and a team at the Swiss Federal Institute of Technology, Zürich, under the direction of Albert Eschenmoser. The focusing of talent and resources by these research teams is probably unparalleled in both quality and magnitude, and in the ingenuity required to achieve the goal.

Since the discovery in 1926 that minute quantities of what was later found to be $B_{12}$ are required by liver tissue for the prevention of pernicious anemia, the vitamin yielded to purification (1948) and structure elucidation (1956) only with the greatest difficulty. Vitamin $B_{12}$ (1) is a structural relative of heme, the red blood pigment, and chlorophyll, the green leaf pigment, in that a metal

vitamin $B_{12}$ (1)

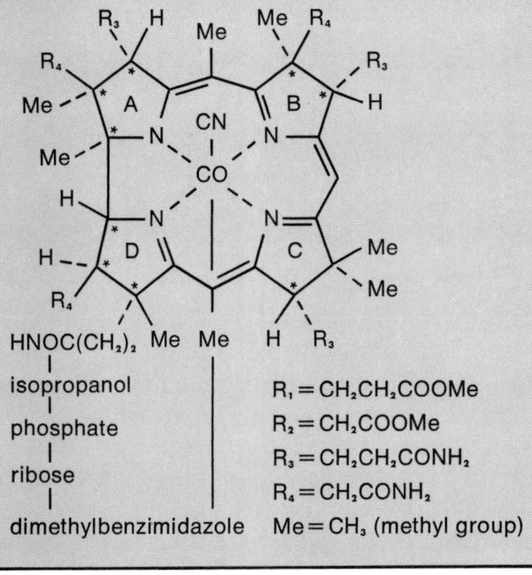

HNOC$(CH_2)_2$ — isopropanol — phosphate — ribose — dimethylbenzimidazole

$R_1 = CH_2CH_2COOMe$
$R_2 = CH_2COOMe$
$R_3 = CH_2CH_2CONH_2$
$R_4 = CH_2CONH_2$
$Me = CH_3$ (methyl group)

tetrodotoxin (2)

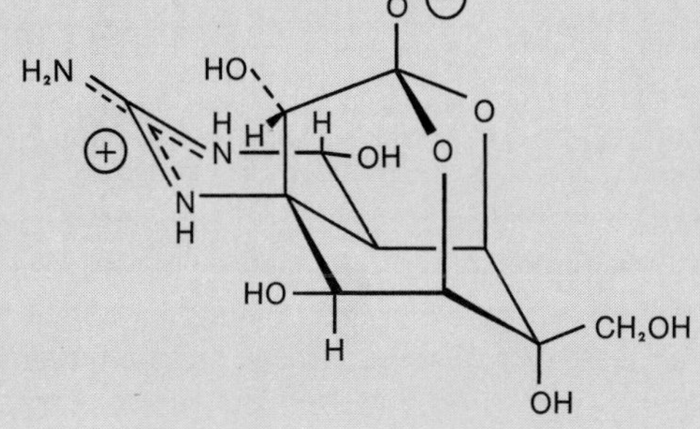

atom (in this instance, cobalt) is embedded within a nearly planar macrocyclic (ring structure of large size) framework of four nitrogen containing rings (A, B, C, and D). The singularity of vitamin $B_{12}$, and the challenge to synthesis, derive from the presence of nine asymmetric centers on the periphery of this large ring. At the starred sites (1) two different substituents project either above (indicated with solid lines) or below (broken lines) the plane. Only the indicated configuration of substituents at each site corresponds to the natural vitamin.

The synthesis of polypeptides (see *1972 Britannica Yearbook of Science and the Future:* CHEMISTRY, *Breakthrough in Chemistry*), because of the repeating amino acid units, could be accomplished automatically with a machine programmed to link the required building blocks. By contrast, the vitamin $B_{12}$ molecule does not have repeating units. Thus, each of the approximately 100 synthetic transformations required individual design. To complete the synthesis a large number of wholly new and ingenious reactions were devised. The broad method of attack involved the construction of two subunits, one comprised of rings A and D, the other of rings B and C. The joining of rings C and D yielded an elongated structure A-D-C-B with many of the necessary features. The joining of A and B required a folded form rather than the extended one preferred by this molecule. Introduction of a metal atom brought the nitrogen atoms into close proximity with the metal, thus bringing rings A and B close enough to be joined by a chemical bond. By following this basic pattern, the researchers synthesized cobyric acid and showed it to be identical with a sample of cobyric acid derived from vitamin $B_{12}$. Some years earlier K. Bernhauer and his colleagues in Stuttgart, W.Ger., had devised a method for converting cobyric acid to vitamin $B_{12}$. Consequently, the synthesis was complete.

Aside from the personal satisfaction gained by these scientists on having surmounted their Everest, what is the significance of their achievement? In earlier days, synthesis was considered the ultimate proof that a structure had been correctly deduced. However, X-ray diffraction analysis is now the prime method for proof of structure. Furthermore, the synthetic route is unlikely to be employed for preparing the vitamin in the future since the cost would be prohibitive. Perhaps this new synthesis will make possible the preparation of some modified form of vitamin $B_{12}$ that will be useful in helping to elucidate the biological role played by the vitamin. Also, the stimulus of achieving their goal inspired the scientists to develop a number of highly ingenious synthetic procedures that will have broad utility in other work. Nonetheless, in retrospect, the argument can be made that the return was too small for the expenditure.

Perhaps the great artistry exemplified by this effort is the final best effort of the golden age of synthesis. As one ponders in what manner these artistic expressions will be passed on to succeeding generations, efforts are beginning to develop in the direction of syntheses designed with the aid of computers. Perhaps a computer can be programmed to duplicate and improve on the ingenuity of man in this area.

**Tetrodotoxin.** Another synthesis of great complexity, described at the New Delhi meeting by Yoshito Kishi and six colleagues of Nagoya University's department of agricultural chemistry, dealt with the elaboration of tetrodotoxin, the poi-

prostaglandin $F_{1\alpha}$ (3)

reaction sequence determined by desired side chain

key intermediate—has the four asymmetric centers (*) in the correct arrangement

sonous principle of the Japanese puffer fish and one of the most potent nerve poisons known. The unusual assembly of rings into a space-enclosing cage-like structure (2) was derived in part from a simple benzene derivative.

**Prostaglandins.** The prostaglandins, a family of physiologically active compounds, were becoming available in research quantities through the efforts of E. J. Corey at Harvard University and many pharmaceutical research teams. Interest in these substances, which occur naturally in only tiny amounts, derives mainly from their unique ability to stimulate muscle contraction at dosages of one-billionth of a gram, and the hope that this action can be safely and reliably used to induce labor and abortion. The prostaglandins' diverse influences upon biological functions have also suggested their utility in the treatment of ulcers and asthma, and as stimulants for the heart. Clinical evaluation and research into all aspects of their pharmacological action are intense.

The structures of the six primary prostaglandins are related to that shown for $PGF_{1\alpha}$ in (3). Two other F-series compounds contain *cis* double bonds at carbons 5,6 or 17,18; the three E-series structures contain a ketone function at carbon 9. All these 20 carbon fatty acids have two long carbon chains on opposite sides (indicated by broken and heavy lines at positions 8 and 12) of a small planar ring. Not only must these and the other substituents be in the proper three-dimensional relation to one another, but the absolute configuration of the whole molecule must be only that shown. That is, the fingers of the right hand are related to each other as are those of the left, but the right and left hands are certainly not the same! This subtle mirror image spatial isomerism of many natural substances presents a particular challenge to the synthetic chemist.

The structural similarity of the several naturally occurring prostaglandins suggested to Corey that, rather than design distinct syntheses of each individual family member, an efficient approach would involve construction of a single key "intermediate" having many crucial features common to all. Further simpler transformations of this intermediate would yield all the desired compounds. The success of this plan and further improvements adaptable to large-scale production were reported.

**Synthesis from free atoms.** A largely neglected area of chemistry was the reactions of atoms. Contrary to popular thought, until recently the chemistry of only a few of the 103 known atoms had been described: hydrogen, nitrogen, oxygen, sulfur, fluorine, chlorine, bromine, and iodine. Chemists generally did not have free atoms for study, but only atoms strongly combined with other atoms (like atoms combined are the elements, unlike atoms combined are compounds). Although in common parlance one makes such statements as water contains oxygen atoms, this statement only means that in principle water can be made from oxygen atoms and hydrogen atoms; there is not even a remote resemblance in chemical behavior between water and free oxygen atoms.

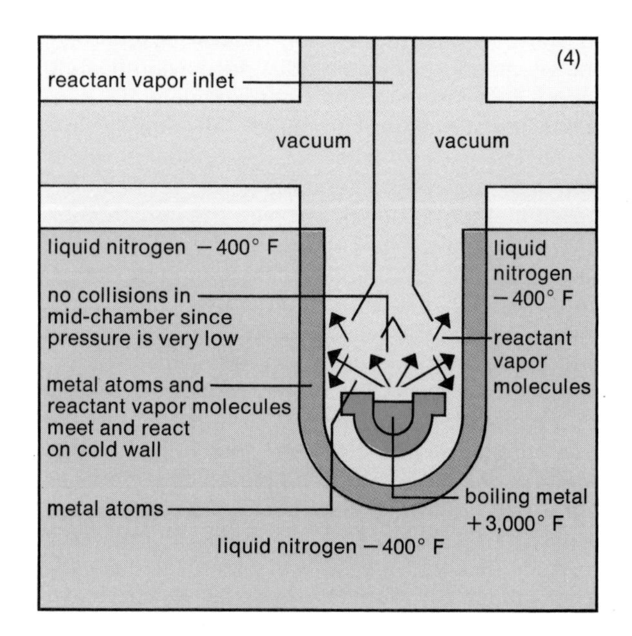


*The boiling metal heated within a highly evacuated flask liberates free atoms which strike a cold surface of reactant vapor molecules. The reactant used may be one of many organic or inorganic compounds. New materials incorporating the atoms form in the solid matrix at the wall.*

At high temperatures, above 3,000° F, most atoms are available in the free state. They had not been studied in this state for want of a method of mixing them with other substances at low temperature. If the high-temperature atoms are cooled in the absence of other substances, they react with one another vigorously to give the usual comparatively unreactive form of the element. Recently, a method was developed by P. S. Skell and coworkers for mixing high-temperature atoms with low-temperature substances. A schematic diagram (4) shows their basic apparatus, which was used to study the reactions of free atoms of many elements. As instances of the difference between bound and free atoms, carbon in its usual form of graphite or diamond is chemically inert, while carbon atoms react at temperatures of −300° to −400° F with almost any substance. Similarly, platinum metal is valued because it is so inert to corrosion and almost all reagents; in contrast, free platinum atoms will react with simple olefins at low temperatures to make organoplatinum compounds.

A simple principle is that the higher the boiling point of the element, the greater will be the difference in chemical properties between the free atom and the usual form. This is a useful principle because the more strongly atoms are bound together the higher will be the element's boiling point. These new reactions of atoms are broadly applicable, and there is promise that many new compounds will be synthesized by application of the technique of carrying out low-temperature reactions with high-temperature substances.

—Robert A. Davidson and Philip S. Skell

## Structural chemistry

A milestone in chemical structure was reached in 1972. In February *Chemical Abstracts*, published by the American Chemical Society, recorded the two millionth unique chemical structure in its files. The substance was 5-amino-3-chloropyrazinecarboxaldehyde oxime. Since compounds are being added at the rate of 300,000 per year at the present time, the three millionth unique chemical structure should be recorded in 1975.

**Structure of atoms.** Structural chemistry is concerned not only with the shape and size of molecules and how the shape affects chemical and physical properties but also with the relation of each individual atom in a molecule and how the atoms are bonded together. This latter concern includes the disposition of the electrons and the shape and size of the individual nuclei. For some time it was thought that the nuclei of atoms were spherical, but in 1971 a group of scientists at Oak Ridge (Tenn.) National Laboratory led by Paul H. Stelson found that the nucleus of the uranium-238 atom was actually elongated in the shape of a football. They were able to demonstrate that the long axis of the atom is about 35% longer than the short axis.

For a long time it was thought that the rare gases were completely inert, but in recent years a number of compounds of these rare gases have been described. In 1971 Roy Gordon at Harvard University was able to show that argon forms a diatomic molecule ($Ar_2$); the bond energy of this molecule, however, is only $^1/_{400}$ of the hydrogen molecule ($H_2$). In addition to this weak binding force, "nonbonded" systems repel each other at short distances. These nonbonding intermolecular forces determine many chemical and physical properties of substances.

In a related study, Stephen Derenzo at the Lawrence Radiation Laboratory, Berkeley, Calif., was able to use some of these properties to build a liquid xenon radiation detector which proved to be about 10 times more accurate than the conventional gas-filled detectors in pinpointing the position of charged particles and the energy of gamma rays. The basis for this detector is that the liquid xenon has properties favorable for electron avalanche, a snowballing effect that greatly amplifies the small electrical signal produced when radiation interacts with a detector. When the avalanche

the LH-RH/FSH-RH hormone

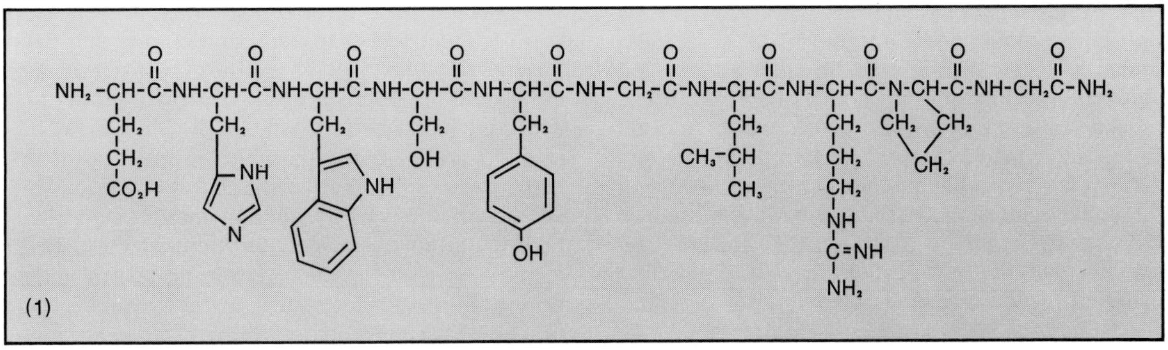

(1)

the antibiotic nisin

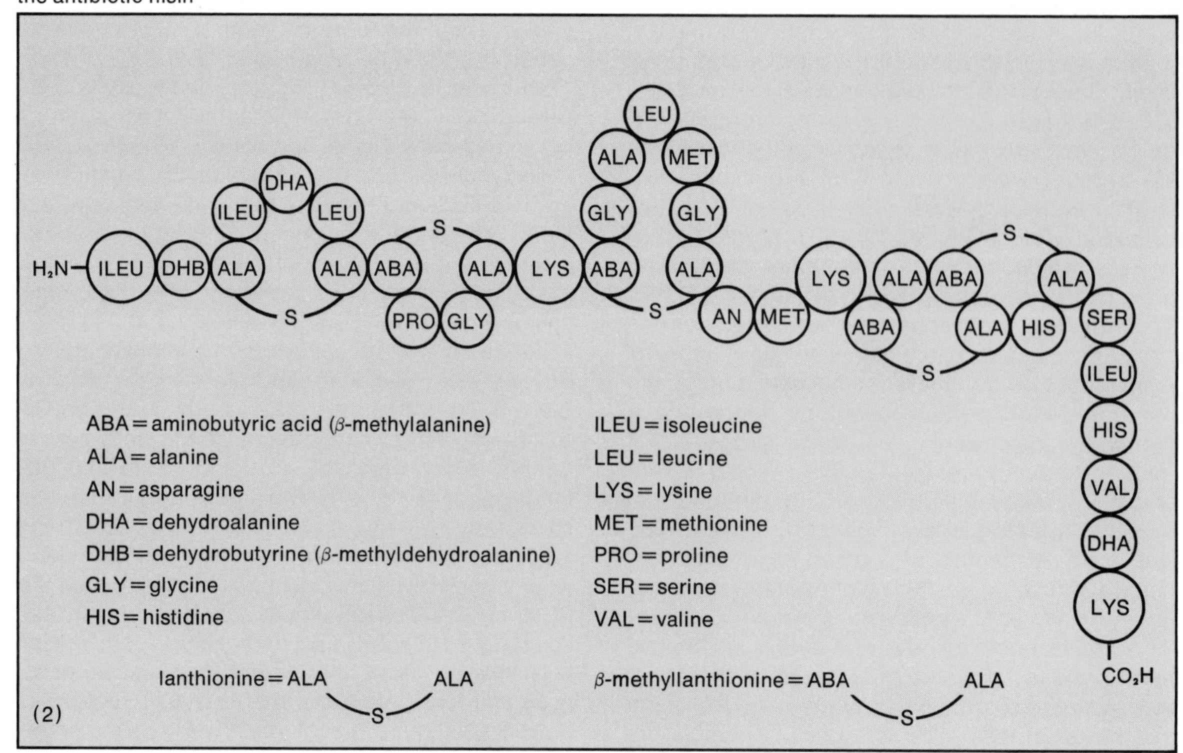

(2)

occurs, a few thousand electrons freed by incident radiation knock loose millions of additional electrons. Because xenon is relatively inert (that is, its outer shell is filled with tightly bound electrons), the electrons, once freed, drift freely in the pure liquid xenon. Very fine positively charged wires criss-crossing the detection chamber pick up these drifting electrons and their electrical signals.

**New tools for structural determination.** The determination of the structure of a complex molecule is often difficult and time-consuming. The first step is the isolation and purification of the material. A new tool for the isolation of compounds of high molecular weight or of those that are heat-sensitive is liquid-liquid chromatography. Robert B. Woodward of Harvard University, who used liquid-liquid chromatography extensively in his recent synthesis of vitamin $B_{12}$, predicted that this method would become in the 1970s what gas-liquid chromatography was to chemistry in the 1960s. (In chromatography a flow of solvent or gas causes the components of a mixture to migrate differentially from a narrow starting point.)

If a material can be obtained in a crystalline form, X-ray crystallography has often been the method of choice to determine its structure. If, however, the material can not be crystallized or if the actual structure or conformation in solution or in a mixture is desired, such a technique is not applicable. For example, it has been widely assumed that enzyme structures are the same in crystals and in solution. However, using a spectral chemical probe with the enzyme carboxypeptidase A, Harvard Medical School chemists Jack T. Johansen and Bert L. Vallee were able to show that a derivative of a tyrosine residue does not complex with the active site of the crystals but does complex with the active site in solution. (The active site of an enzyme is that portion that interacts specifically with the substrate.) Furthermore, the specific activity in the solution is 300 times that of the crystalline enzyme.

Another important problem in which X-ray diffraction is of little use is in the location of the very small hydrogen atom in a hydrogen bond, which is extremely important in the alpha helix of the protein structure (a polypeptide chain folded in a spiral similar to the threads of a screw) and base pairing in nucleic acids. Neutron diffraction can determine the position of the hydrogen atom to an accuracy much better than 0.01 Å, while X-ray diffraction can rarely determine the position of the hydrogen atom to better than 0.2 Å ($1 Å = 10^{-10}$m). With such information one can show that in lysine hydrochloride one hydrogen atom of the $\alpha$-$NH_3^+$ forms a very strong hydrogen bond with a neighboring carboxyl group so that the barrier of rotation is 7.0 kilogram calories (kgcal)/mole. For comparison, the barrier for the amino group, which does not form such a strong hydrogen bond, is only 2.9 kgcal/mole, according to Walter C. Hamilton of the Brookhaven National Laboratory, Upton, N.Y.

A useful tool for examining the shape of noncrystalline large molecules is the transmission electron microscope. Halmet Coppa and Klaus Heinemann, working at the Ames Research Laboratory of the U.S. National Aeronautics and Space Administration, recently improved the resolution of this microscope to about 1 Å from the present limits of 3 Å by use of an angular aperture system.

The molecular formula of a substance can now be determined directly by the use of a high-resolution mass spectrograph that can determine molecular weights accurately within 0.001 of any given molecular weight unit. Because of the structure and natural abundance of various isotopes, the atomic weights of elements are not even integers. For example, if the carbon isotope $^{12}C$, is assigned the value 12.00000, then hydrogen, $^{1}H$, is 1.00783; nitrogen, $^{14}N$, is 14.00307; oxygen, $^{16}O$, is 15.99491. With these values one is able to calculate a molecular formula quickly and accurately from a single determination.

Proton nuclear magnetic resonance has been used for some time to determine the structure of small organic molecules. However, with the advent of the new generation of instruments, such as the 220 and 300 MHz high-resolution spectrometers as compared to the usual 60 MHz instruments, it became possible to examine the structure of various large and complex molecules, such as proteins and synthetic polymers with molecular weights in the millions. Similarly, nuclear magnetic resonance (nmr) is capable of looking at the naturally occurring $^{13}C$ carbon isotope which occurs to the extent of 1.1% together with the very common $^{12}C$ isotope in organic compounds. By the utilization of the newer, more powerful nmr instruments, together with a Fourier analysis in a computer, very accurate information which allows structural assignment can be obtained.

Another useful new technique is electron X-ray spectroscopy, sometimes called ESCA (electron spectroscopy for chemical analysis). This was pioneered by Kai Siegbahn at the University of Uppsala in Sweden. His technique was to bombard a molecule with X rays to eject the tightly held valence electrons, the energy of which is determined by the effective charge on a particular atom and therefore its chemical environment. This method allows one to look at a large variety of chemical elements directly. For example, since there is a good correlation between the effective charge on a sulfur atom and the energy required to eject an electron from the 2p valence orbit (the orbit of third lowest energy level around a nucleus), Siegbahn was able to use this technique to show that the oxidation of sulfur in a molecule of insulin took place to give the thiosulfonate form rather than the disulfoxide form. The spectra showed two peaks for sulfur atoms rather than the single peak that would have been expected if the symmetrical disulfoxide had been produced.

**Structure of inorganic compounds.** In the inorganic field two new isotopes of element 105, for which the name hahnium was proposed, were synthesized with a heavy ion linear accelerator by

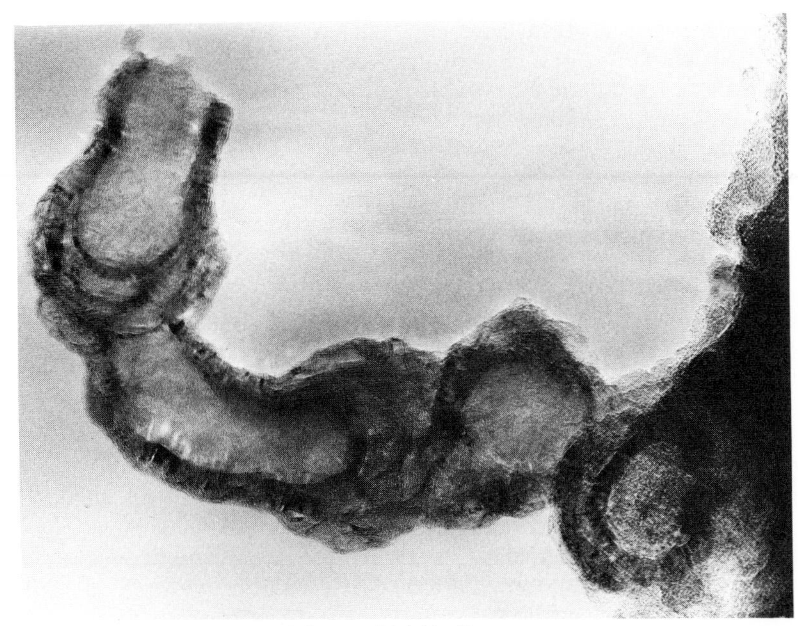

Courtesy, Drs. T. Baird, B. Grant, J. R. Fryer, and J. L. Hutchison

*Electron micrograph shows a carbon fiber produced by propane pyrolysis over nickel. The average direction of the (0002) graphite lattice planes is parallel to the axis of the fiber; i.e., the areas where metal particles are believed to have been (the metal was presumably removed during preparative treatment with concentrated HCl). One recent theory for the formative mechanism of these fibers suggests that atomically dispersed nickel is deposited in each carbon layer, making possible the continuation of catalytic deposition of carbon some distance away from the actual metal surface.*

Albert Ghiorso at Lawrence Radiation Laboratory, Berkeley, Calif. Hahnium-262, which was produced by bombarding berkelium-249 with oxygen-18, has a half-life of about 40 seconds. Hahnium-261, which was produced by bombarding californium-250 with nitrogen-16 or berkelium-249 with oxygen-16, has a half-life of about 1.8 seconds.

Uranium carbonyls, sought unsuccessfully in the 1940s for the separation of uranium isotopes by gaseous diffusion, were successfully prepared at low temperatures by a group of chemists from the University of Florida and Florida State University when a mixture of uranium, carbon monoxide, and argon at 4° K (−269° C) was warmed and then quenched to 4° K. Uranium compounds containing up to six carbonyls (the radical CO) are formed.

**Structure of new polypeptides.** As the structures of more and more proteins were determined, the structures of many different materials were noted to be strikingly similar. For example, Joseph Kraut of the University of California, San Diego, found that two catalytic enzymes, which evolved independently, have identical arrangements of the peptide chains involved in the catalytic mechanism. The two molecules, chymotrypsin from vertebrates and subpilisin from bacteria, are identical at the catalytic site but have peptide chains folded differently in other parts of the two molecules. Kraut noted; "Nature has twice come up with the same solution to a specific problem in molecular engineering."

In a similar vein, Choh Hao Li of the University of California's Hormone Research Laboratory in San Francisco recently determined the structure of the human hormone, chorionic somatomammotropin (HCS), that is present in the placenta of pregnant women. This hormone bears a remarkable similarity to the human growth hormone (HGH) whose structure his group had determined earlier. The two hormones share the same bifunctional physiological action in that they both promote bodily growth and stimulate mammary development and milk secretion, as well as have similar immunological responses. Both hormones are proteins comprised of 190 amino acids, 160 of which are sequentially identical. In addition, both molecules consist of two loops, one large and one small, formed by a disulfide bridge between the cysteine residues at positions 53 and 164 and positions 180 and 188.

Chemists at Ciba-Geigy, Ltd. in Switzerland reported that the previously published structure for the adrenocorticotropic hormone (ACTH) was incorrect. ACTH, which has 39 amino acids, was reassigned a structure that was later confirmed by the synthesis of the full polypeptide molecule.

A nonsteroidal fertility-regulating compound, a hypothalamic hormone that triggers the release of lutinizing hormone and of the follicle-stimulating hormone from the neighboring pituitary gland, and hence called LH-RH/FSH-RH, was recently isolated and its structure determined (1). The structure appears to be simple enough to permit its synthesis on a commercial scale and suggests that analogs will also be available in the near future.

The peptide antibiotic nisin (2), which is active against malarial parasites, also was isolated and its structure determined. Although nisin consists of only 29 amino acids, that number includes 8 of the less common ones; the peptide also contains five cyclic units that may give the antibiotic the capacity to bind ions. Erhard Gross and John Morell of the U.S. National Institutes of Health, Bethesda, Md., speculated that the two double bonds in the molecule play a key part in nisin's biological activity.

—William J. Bailey

# Communications

In charting the growth of the world's population and of the world's telephones over the past 50 years, it becomes quickly apparent that the number of telephones has grown more rapidly than the number of people. While the world's population has approximately doubled since 1921, the number of telephones has grown more than tenfold.

By Jan. 1, 1971, the number of telephones throughout the world had increased to nearly 273 million. And if, as expected, the pattern of growth established over the last decade continued during 1971, nearly 300 million phones should have been in service by the year's end. Also by 1971, 37 countries had more than 500,000 phones. Israel was the latest nation to join this group. Fifteen of the world's cities had more than one million telephones, seven of those within the U.S.

As telephone users, North Americans remained the busiest with each person in the U.S. accounting for 779 calls in 1970, and each Canadian accounting for 739. Sweden ranked third, with 674 telephone conversations for each person during the previous year. Interestingly, Iceland and the Bahamas were not far behind the leaders, with 593 and 512 calls, respectively.

This continuing growth of telephony provided a challenge to scientists and engineers in the field of telecommunications. New ways of transmitting messages had to be invented, and new techniques had to be developed for routing these messages through the complex telephone network to their destination.

**Communications services.** While the basic telephone was being used by more people than ever in 1971, new services and new uses for it were being added. Perhaps the most spectacular of these was the see-while-you-talk PICTURE-PHONE® service, introduced in Pittsburgh during 1970 and in Chicago during 1971. This service permits PICTUREPHONE conversationalists to see and show drawings and objects to each other and, in general, represents a form of communication that is the "next best thing to being there." PICTUREPHONE service also enables the user to dial information from a computer and have it displayed graphically on the screen.

But while a telephone conversation requires about 4,000 Hertz (Hz) of bandwidth for transmission, a PICTUREPHONE signal needs 1,000,000 Hz. Much research was devoted to removing transmission redundancy for areas of the picture that do not change from frame to frame. This would allow the use of a narrower bandwidth, making transmission costs cheaper.

One of the most promising methods for doing this is conditional selective frame replenishment. The concept behind this technique is that when the picture is not changing the only information that must be transmitted to the other screen is just that —"no change." As movement occurs, changing the scene, the particular part of the picture that needs updating is defined and only that area of the picture on the other screen is changed. The amount of frequency bandwidth needed to transmit the no-change condition is quite small, while the bandwidth needed to transmit rapid motion over large areas is quite large—1,000,000 Hz as mentioned earlier. Frame replenishment attempts to average these extremes. The scheme's objective is to process a PICTUREPHONE signal so that it can be transmitted over the equivalent of one-tenth its original bandwidth when the signal is coded in a digital format.

The role of digital coding is important, and digital transmission looms large both in a review of important events in 1971 and in the future of telecommunications. Future systems to carry the world's great variety of communication signals will probably be digital—that is, they will transmit communications as on-off pulses rather than as continuously varying signals.

Digital transmission makes possible the use of another important telecommunication service, digital data. Although digital data transmission service is not new, having been developed in the 1950s, important new advances took place in 1971 that were expected to speed and enlarge its applications. The principal use of digital data is to permit machines—computers, teletypewriters, and

Courtesy, Bell Laboratories

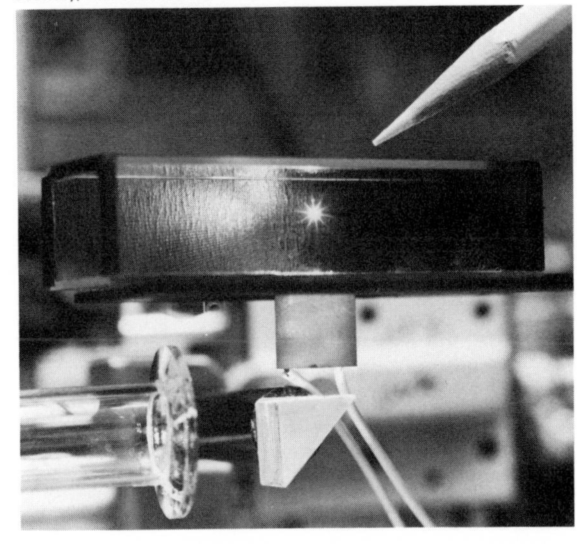

*Tiny starlike particle at the center of the photograph is being held aloft by light. An invisible laser beam passing through a lens at the end of the glass tube is bent by the triangular prism to focus on the transparent glass particle from below.*

telemetering equipment—to communicate with each other. The most efficient means for handling their language—on-off digital signals—is a digital transmission system. The new development achieved during 1971 that would eventually link machines in most cities of the U.S. was a technological breakthrough known as data-undervoice (or DUV). What made DUV so significant and its potential impact so far-reaching was that it could be used on the existing microwave radio relay network, a transmission facility that reaches almost every corner of the U.S. This network in 1972 relayed nearly all television programming and about two-thirds of the long-distance telephone conversations in the U.S. The DUV system, developed at Bell Laboratories, "piggy-backs" digital data signals over the microwave network in frequency bands currently transmitted by the system but below those used for its voice traffic. A field trial of the new DUV system was scheduled for 1972.

DUV's new capability was built around terminal equipment that processes a 1.5 million-bit-per-second (megabit) digital signal so that it can be transmitted over one working channel in the microwave system. In a radio system route with 16 channels, for example, it would be possible to expand digital data service, in steps of 1.5 megabits, from a capacity of 1 channel to 16 1.5-megabit channels, depending on demand for the service.

When transmitting digital data, the error rate of the system is vital. If transmission facilities that carry data distort the signals, the accuracy of the communication will be degraded. The DUV system was expected to achieve a very low error rate. Even if the radio channel carrying the signal should fade severely enough to trigger the system to switch to a protective, back-up channel, the error rate of the digital signal at the time of the switch is expected to be significantly better than one on-off pulse for every 10 million transmitted. When such fading is not a factor, the performance of DUV is expected to make the error rate practically zero.

Another new communications service that was shown to be practical and came a step closer to realization during 1971 was a high-capacity mobile telecommunications system proposed to the U.S. Federal Communications Commission (FCC) by the American Telephone and Telegraph Company (AT & T) in December 1971. The proposed system represented a radical departure from previous mobile radio telephone concepts. By using a honeycomb pattern of small service areas and reusing a limited number of radio frequencies many times, the new system was designed to make it possible to offer mobile telephone service to many more users, and at the same time offer improvements in cost and quality of service.

**Communications devices.** The laser, a device capable of emitting coherent light, seemed likely to be the cornerstone of future high-capacity systems to transmit communications. Optical transmission techniques were being studied by researchers to develop high-capacity systems in which telephone conversations, television, digital data, and other signals would be transmitted by means of visible and infrared light. Lasers were expected to provide potent light sources for these systems.

One of the most unusual discoveries during 1971 was made by research scientists who used radiation pressure from a beam of laser light to raise small transparent glass spheres off a glass surface. The researchers found they were able to lift the spheres and hold them aloft for hours in a stable position, a phenomenon never before observed.

In the experiments, a laser beam was focused upward on a tiny glass sphere about 20 microns (about 0.001 in.) in diameter. The researchers found that radiation pressure from the light not only counteracted gravity and raised the particle, but also trapped the glass sphere and prevented it from slipping out of the beam. In the experiment, successfully conducted both in air and in a partial vacuum, the scientists generated a stable optical trap for holding particles, which they termed an "optical bottle."

Their results demonstrated that light photons have momentum as well as energy. Thus, when a laser is focused on a small transparent particle, the force exerted by its light, though extremely small, is sufficient to lift the sphere off the surface and suspend it.

The experiment was not as simple a procedure as it seemed. The sphere was launched by lifting it off a transparent glass plate with the light beam. Initially, however, the radiation pressure was not sufficient to overcome the molecular attraction between the sphere and the glass plate. This attraction, known as Van der Waals force, is about 10,000 times the force of gravity for a 20-micron sphere. For this experiment, the Van der Waals attraction was broken acoustically by vibrating a ceramic cylinder attached to the plate.

When the attraction was broken, the sphere rose in the light beam and came to rest where the upward pressure caused by the laser was balanced by the earth's gravity. In this position, it was held aloft as long as the light was focused on it. By changing the position of the focus, the researchers caused the trapped sphere to move up and down or sideways very precisely.

In the experiment, the new phenomenon of the trapping forces created by the light was also studied by using a second laser focused on the trapped particle from the side. As the power of the second laser was increased, the particle was displaced within the first laser beam until it was finally driven out and fell.

The experiments may provide research scientists with simple, precise methods for manipulating small particles without mechanical support. This capability could be useful in communications research for measuring scattering loss caused by particles in the atmosphere or in other transmission media. These measurements were expected to help communications researchers develop optical transmission systems for the future.

During 1971, the study of lasers for their communications potential demonstrated some new and unanticipated noncommunications applications for these devices. The studies showed that rapid, precise, on-the-spot identification of pollutant gases in the air might be made possible by using a laser-light technique devised at Bell Laboratories. The equipment used for the tests included a fixed-frequency pump laser (in the present case a carbon monoxide laser), a powerful electromagnet used to "tune" the laser beam to a desired light frequency (by changing its magnetic field), and a rotating shutter to interrupt the beam at regular intervals to cause a pulsing in the light emitted.

The beam of laser light was directed into an optoacoustic absorption cell containing the air sample suspected of being polluted. The cell contained a sensitive microphone with a cylindrical diaphragm that converted changes in air pressure into electrical energy. In the air samples, each type of polluting gas can be identified by the specific frequency of light that it absorbs from the laser light (its infrared absorption spectrum). When the gas absorbs the light energy, it turns it into heat. The heat in turn causes the gas pressure to increase, and it is this increase in pressure that is detected by the sensitive microphone referred to above. A common pollutant, nitric oxide, absorbs infrared light at several wavelengths in the region of 5 to 6 microns.

Infrared absorption measurements are helpful for pollution studies because almost all known pollutants have their fundamental absorption bands in the infrared (nonvisible) portion of the spectrum from about 2–15$\mu$. While many gases may be present in an air sample, each separate gas has its own absorption characteristics, and it is possible to find at least one absorption line for a pollutant that is not overlapped by absorption lines from other gases. One problem has been to distinguish between a pollutant and other overlapping gases. Another difficulty has been that pollutant concentrations are as small as one part per million—too small to be measured directly. The new laser technique solves both problems.

**International communications.** Although international telephone traffic was much smaller than that within nations, international calling was growing in the early 1970s at twice the rate of domestic calls. The international telephone calling rate from the U.S. to other countries continued to grow. In 1969 Americans originated about 20 million international calls and in 1970 about 26 million. Between 1960 and 1970 there was a sevenfold increase.

In 1971 the Intelsat 4 series of satellites was inaugurated with the launch of the first one for the Communications Satellite Corporation in January. A second member of this family was launched in December, followed by the third in January 1972 and the fourth in June. These satellites were designed to provide much-needed relief for overworked channels spanning both the Atlantic and Pacific oceans. Each Intelsat 4 could handle more

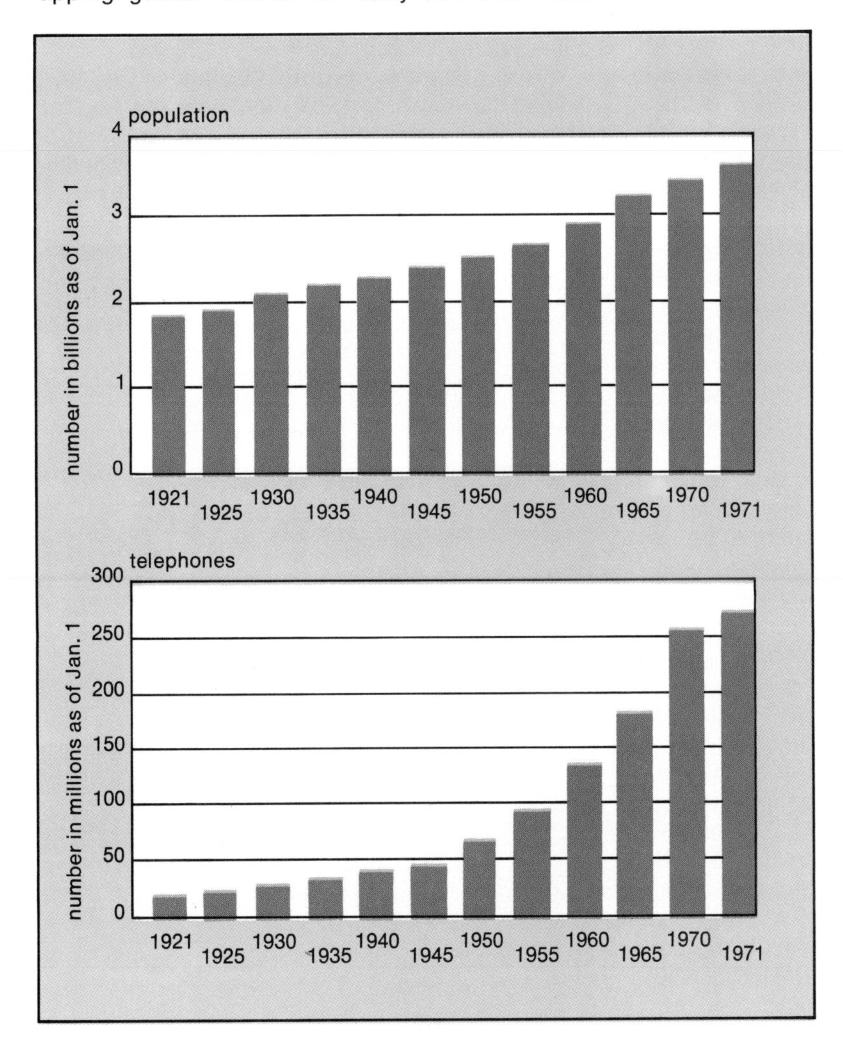


*Growth in the world's population and number of telephones is compared. The number of telephones jumped sharply after 1955 in contrast to the gradual upward curve of population.*

than 6,000 telephone conversations, and its functioning could be modified by commands from the ground as new needs arose.

The Soviet Union also added to its satellite communication system during 1971 with a launch of another in the series of Molniya satellites in July. The satellite system in the U.S.S.R. is used for television and telephony.

In the field of submarine cables, considerable activity took place, although none of the systems terminated in the U.S. Ten cable systems were installed during 1971, ranging in length from 64 naut mi (across the English Channel from Great Britain to Belgium) to 825 naut mi (from Canada to Bermuda). They varied in capacity from as few as 60 channels (Okinawa to Taiwan) to as many as 1,840 (Spain to the Canary Islands).

**Future outlook.** A far-reaching and comprehensive review of the telecommunications field was contained in a report submitted by the National Academy of Engineering (NAE) to the U.S. Department of Housing and Urban Development in June 1971. This report, which summarized the thoughts of many telecommunications planners, discussed four basic future networks: (1) the telephone network for transmitting pictures, voice, and written material between two points; (2) a network, based on existing cable television systems, which can distribute information from central facilities to offices and homes, with a capacity of as many as 30 television channels and a limited call-back capacity for polling or making requests; (3) a broadband communications network carrying up to 30 television channels in both directions that would interconnect major public institutions and large commercial enterprises; (4) a multipurpose city sensing network to collect data on such items as weather, pollution, and traffic.

The NAE report reached the following conclusion: "Many of the cities' problems are caused by high-density living conditions in an era of increasingly rapid change. Communications technology, imaginatively applied, could offset the trend in which the vast majority of Americans today, and more in the future, live on a small percentage of the available land.

"We suggest an exploratory program to examine how broadband communications technology could be applied to business, government, education, health care, and entertainment to stimulate the development of existing small communities, or new communities, in rural areas. As a result, people would have a viable option of settling in either urban or rural America."

—Irwin Welber

*See also* Year in Review: ASTRONAUTICS AND SPACE EXPLORATION, *Earth Satellites.*

# Computers

When viewed from the outside, the computer field in 1972 still appeared to be vigorous and dominated by a "pioneering" psychology—there were so many new applications, so much new technology, and so many pressing national problems requiring intelligent computer support. From the inside, the field seemed less lusty, more deliberate in its choice of problems, more conservative in its methods, and more perfectionist in its certification of software and hardware products. As organizations have become more dependent on computers, they have become more cost-conscious. Competition for the customer's dollar has increased. Overall, there were enormous economic forces on the computer industry to stabilize, to slow its headlong development, to increase its quality control, and to become more conservative in its methods. But there were also enormous technological forces that mitigated against stability and control. The computer field had not begun to use up its reserves of discoveries in physics and electronics, which spawn new components, new design, and new equipment.

**World computer growth.** Estimates in 1972 placed the number of installed computers in the United States at about 80,000. However, most of the newly installed computers were in the small or mini category (rental about $1,500 a month or less). The Japanese had about 8,000 computers installed and were engaged in a national effort to computerize their society. It was estimated that they had

*Experimental electronic chip, shown here on a common window screen, can store 960 bits of computer information. Microscopic circuits on the chip—called charge-coupled devices—store the information by means of minuscule electric charges.*

Authenticated News International

Courtesy, IBM

*Double-exposure of a defective card reader part pinpoints areas of stress that caused the part to fail. By using this new technique, scientists can analyze such failures in as little as two hours, a task that previously took as long as two weeks.*

earmarked $1 billion for revamping primary and junior-high-school education, automated vehicle control, pollution alarm systems, and automated service industries like supermarkets and banking. The government allocated $24 million to subsidize the Japanese computer industry, and from 1972 to 1975 promised computer manufacturers about $110 million to support their research and development programs. Smaller amounts were set aside for the support of software development.

In France there were about 6,500 computers, and, through joint efforts with West German and Dutch firms, the French continued their determined drive to weaken the domination by IBM of computing in Europe. West Germany, which in 1972 had about 8,000 installations, was hoping to reduce IBM's share of the government computing business from 80% to 60%. In Great Britain the government was asked to provide $120 million a year for the support of its computer industry.

The decision of RCA in September to discontinue its computer manufacture and sales was the major business story of the year in the U.S. RCA sustained a $250 million tax loss and laid off 11,000 people. Control Data Corp. took over the support of leased RCA installations. In other business developments IBM delivered more than 10,000 of its System 3 line of small business computers, and the stability of its new System 370 line in its first year already exceeded that of the comparable six-year-old 360 models, even though the 370 had about four times as many components. Digital Equipment Corp. delivered about 13,000 of its diverse minicomputers and continued to be the leading supplier of machines of that size. Overall, in the U.S., the economic slowdown continued to affect the industry: for example, the number of installations in southern California dropped 10% during the year.

**Computer technology.** Very large computer memories were being supplied to government and industry in 1972. Precision Instrument Co. delivered to the U.S. National Aeronautics and Space Administration a memory consisting of one trillion bits, and a three trillion-bit memory was promised for late 1972. (A bit is the basic information symbol in the binary notation system used by digital computers and is represented by either a closed or open circuit.) In the present decade the cost per bit is expected to fall from 0.01 to 0.00001 cents. However, in 1972 the industry was still operating at the expensive end of the spectrum. Thus, a commercial holographic (laser) memory system of 12 million-bit capacity was being delivered at a cost of 0.01 cents per bit. Commercial magnetic spot (or domain) memories were also available at a cost of about five cents per bit, although a slower and smaller version cost about one cent per bit.

By 1972 the age of large-scale integrated (LSI) circuits had arrived. MOS (metal oxide semiconductor) integrated circuits accounted for about 25% of the computer circuit market. Some of this circuitry allowed minicomputers to achieve addition times of 100 nanoseconds (1 nanosecond = 0.000000001 sec). The first "computer on a chip" was announced by Intel Corp. Such a chip, made of silicon and measuring 1/4 × 3/4 in., contained a 4-bit adder, 48-bit program counter and stack, an address incrementer, an 8-bit instruction register and decoder, control logic, and a repertoire of 45 instructions. A computer developed by Intel con-

Harry Seawell

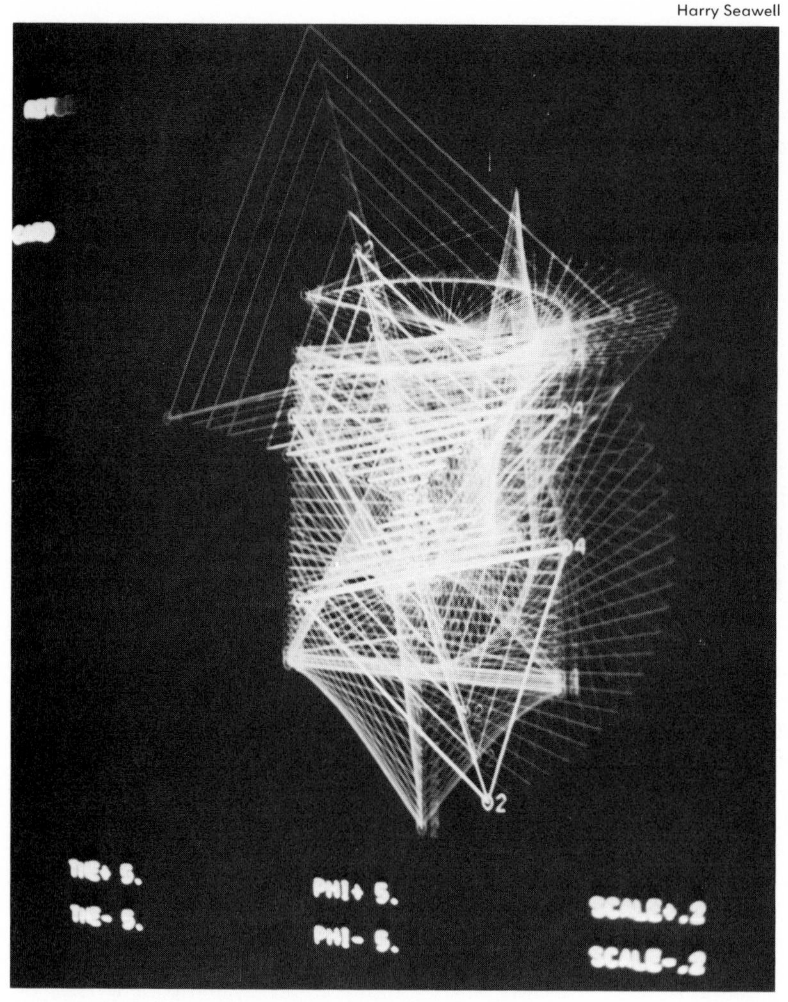


*A graphic design as at the left can be presented and manipulated on a terminal hooked up to a specially programmed computer. The terminal displays mathematical functions and circuit designs. The image can be rotated by pointing a light pen at the area on the terminal labeled "phi"; size changes can be accomplished by pointing the pen at the area labeled "scale."*

tained four such chips, a control memory, a 320-bit memory, and a 10-bit shift register.

Optical character recognition (OCR) machines by 1972 were considered standard commercial items. However, their inclusion in general computer operations had still not been attained. A system being designed for the blind used OCR for putting English text into a computer, utilizing a dictionary to map the words into phonemes, passing these to a speech synthesizer, and then expressing the text in audio form. The system was scheduled to be tested on blind students at the University of Connecticut late in 1972.

Supercomputers were in operation during the year. Goodyear's STARAN IV, under optimal conditions, processed up to 500 million instructions per second, though the average performance rate was more like 16 million instructions per second. The STAR computer of Control Data was somewhat slower but still could process about 15 million instructions per second on data streams.

The network of the Advanced Research Projects Agency (ARPA) of the U.S. Department of Defense in 1972 connected 23 centers and was functioning for a growing population of users. The connected computers included PDP-10, PDP-11, Illiac IV, B5500, IBM 360/91. Each node of the network contained a modified Honeywell DDP 516 computer (IMP), which processed (and forwarded) messages. The connecting links were leased telephone lines operating at 50,000 bits per second. The network was equipped to handle both long and short messages as sequences of 1,000-bit packets which are independently propagated. Routing and reformatting of forwarded messages was routinely handled in the IMPs at the nodes.

The network was used by researchers who wished to have access to special programs and/or large data bases at remote sites. One installation used a remote computer to build and test software for a similar computer that it would later take delivery on. The network had a number of one-of-a-kind computer resources that all subscribers might use, such as special very fast computers, very large memories, and specialized software systems.

Other commercial networks existed, including

the extensive CYBERNET of Control Data Corp. and the INFONET of Computer Sciences Corp., but none was as well organized and as flexible a switching network as ARPA. The U.S. National Science Foundation continued to support various university networks aimed at supplying computing power from central machines to smaller ones at high schools and small colleges. A good example of such a network was one operated by the University of California.

There seemed little doubt that network configurations of computers and terminals would be the dominant mode of future computing. Thus, there was enormous activity in the design and manufacture of terminals and minicomputers that could be connected to networks. Their prices ranged from $800 for the teletype up to $20,000 for a graphics terminal. Their variety was bewildering. Some produced hard copy, and some character graphics only; others provided line-drawing capability.

**Applications.** There seemed to be no slackening in the application of computers to new tasks. At the same time, though it appeared less evident to the casual onlooker, enormous improvement was continually taking place in those applications.

The postal services were prime candidates for computerization. Tests were under way in the Cincinnati, O., post office to sort mail under computer control. A host of problems remained to be solved, and in their solution the postal systems would almost certainly be radically altered. Thus far, the emphasis was on the distribution of mail. However, the sending of mail by optical or electronic means was sure to be actively tried in the next few years.

One of the most successful computer systems was NASDQ, used by stockbrokers. In 1972 it processed about one million calls per day for stock quotations. Few doubted that the automatic stock exchange was near to reality. Many newspapers of medium circulation installed and were using computer-controlled typesetting. Scores of police departments used computer systems for an increasingly sophisticated attack on crime based on the premise that information existing in files can be correlated by computers to narrow the choices that make crime-solving difficult. (*See* Feature Article: CRIME FIGHTERS IN THE LABORATORY.)

Volkswagen installed in its cars a set of sensors whose outputs are connected to one central socket. A computer can be attached to the car through the socket so that a program on the computer can test a number of the critical points in the car. For years many had proposed such a scheme for testing complex devices.

Modeling of complex phenomena is, of course, the area in which the computer is essential. Man is no longer driven by curiosity, but by necessity, to model his society and environment. He has become aware that many of the great forces acting on society are irreversible and that planning—long-range planning—has thus become indispensable. Weather forecasting, in real time, by computers is recognized as an important goal since it has such immediate commercial and social value. But scientists were also beginning to appreciate the need for long-range climate study. Man's handiwork increasingly affects climate, and computer simulations of climatic consequences were increasingly recognized as important contributions to national debates such as those on supersonic transport, the Alaskan pipeline, and urban pollution. An interesting computer simulation of the forces causing the deterioration of Venice is being performed in Italy. Land subsidence, pollution, and floods are taking their toll of that city. What controls will slow this deterioration of an important center of Western culture? So many variables interact, and so many models are possible that the computer is essential for such studies. Similar studies are soon expected

*"You know, I'm beginning to feel guilty about the unemployment situation!"*

Ross, Ben Roth Agency

to be quite common in many U.S. municipalities.

No model provoked more comment than one developed at the Massachusetts Institute of Technology which sought to show that national and international growth of the scale now being sustained would bring irreversible disaster to the world in the next century. The model is international in scope, covers a long time period, and includes a large number of important variables. Its simulation results had a mixed reception. Many outstanding economists publicly derided the approach, the data used, and the conclusions reached. Such has been the fate of most such studies, which have attempted to extrapolate short-term present behavior into the future. It should be noted that simulations can be improved through practice, as witness the enormously improved competence of the election results simulations.

The computer is approaching a combination of speed, memory, size, and cost that allows it to be used in increasingly important control applications. Serious research work was under way to use a computer to signal an undamaged part of a brain to control the performance of motor functions which a damaged part can no longer do. A brain-damaged monkey with one paralyzed arm was taught to flip a switch on a computer which would then transmit signals to its brain to work the paralyzed arm so that it might feed itself.

By putting a scanning video camera and a computer into a car, researchers at Stanford (Calif.) University could study robot-controlled driving. Robots were solving some simple problems, but mostly they worked on the blindly mechanical level indicated by their name. (*See* Feature Article: MECHANICAL SERVANTS FOR MANKIND.)

**The future.** So much happens in one year in the computer field that it is difficult to sense the future. Still, in 1972 some trends were clear. The computer is accelerating the rate at which knowledge gets distributed and the rate at which our systems for performing large tasks get altered. It seems unlikely that there will be less use in the future of computers for processing data about us and our society. Our security and dignity can only be safeguarded by appropriate legal means (legislation and the courts) and, more importantly, by an increasing understanding and appreciation of computers. These machines are not our masters. We are theirs. In the next few years one of the greatest responsibilities of our education systems is to teach our young that computers, and the tasks they perform, can be bent to serve everyone and on a one-by-one basis, and not just the large organizations whose needs and appetites dominate the uses made of computers today.

—Alan J. Perlis

# Earth sciences

The most significant developments during the year were in the history of the ocean basins and in planetary geology. Results of extensive programs over the last decade were yielding a consistent model for the geologic development of planetary bodies of various sizes.

## Geology and geochemistry

As part of the Apollo 15 mission, begun July 26, 1971, astronauts David R. Scott and James B. Irwin collected moon rock samples near the foot of Hadley Delta Mountain, which rises 12,000 ft above the Mare Imbrium plain. The lunar rover traveled three miles from the landing site to the edge of Hadley Rille, a sinuous fissure extending for 70 mi. The rille is probably a collapsed lava tube formed during volcanic eruptions long after the formation of Mare Imbrium. Horizontal layering of a dozen successive lava flows is visible on the sides of the rille.

Included in the 171 lb of rock returned to earth was an important sample of anorthosite (composed of over 99% plagioclase with the remainder pyroxene and ilmenite). It is probably a piece of the lunar crust that existed before the formation of Mare Imbrium and was dated at 4,150,000,000 years, the oldest rock yet collected on the moon. The moon, along with the earth and other planets, probably formed by means of gravitational accretion about 4,600,000,000 years ago, while the Mare Imbrium formed about 4,000,000,000 years ago as a result of a meteorite impact on the moon. Lava flows filling the Mare Imbrium at Hadley Rille are 3,300,000,000 years old.

Instrumentation installed at the Apollo 12 and 14 sites sent back information that trace amounts of water vapor are escaping from the lunar interior. Previously, the moon was thought to be devoid of water. The hydrous mineral goethite ($Fe_2O_3 \cdot nH_2O$) was found in small amounts in the Apollo 14 samples from the Fra Mauro region; this was the first evidence of water incorporated in the mineral structure of any lunar sample. The total amount of water on the moon is minuscule, however. There is no atmosphere to produce precipitation and no indication that enough water ever existed on the moon to produce stream erosion.

In June 1971 the United States exchanged three grams of Apollo 11 samples for three grams of Soviet samples obtained from the Sea of Fertility by the unmanned spacecraft Luna 16. This was the first concrete result of the U.S.-Soviet agreement for cooperation in space.

Photographs from the Mariner 9 spacecraft in

Courtesy, Dr. Fred Mackenzie, Northwestern University, and Dr. Rudi Gees, Dalhousie University

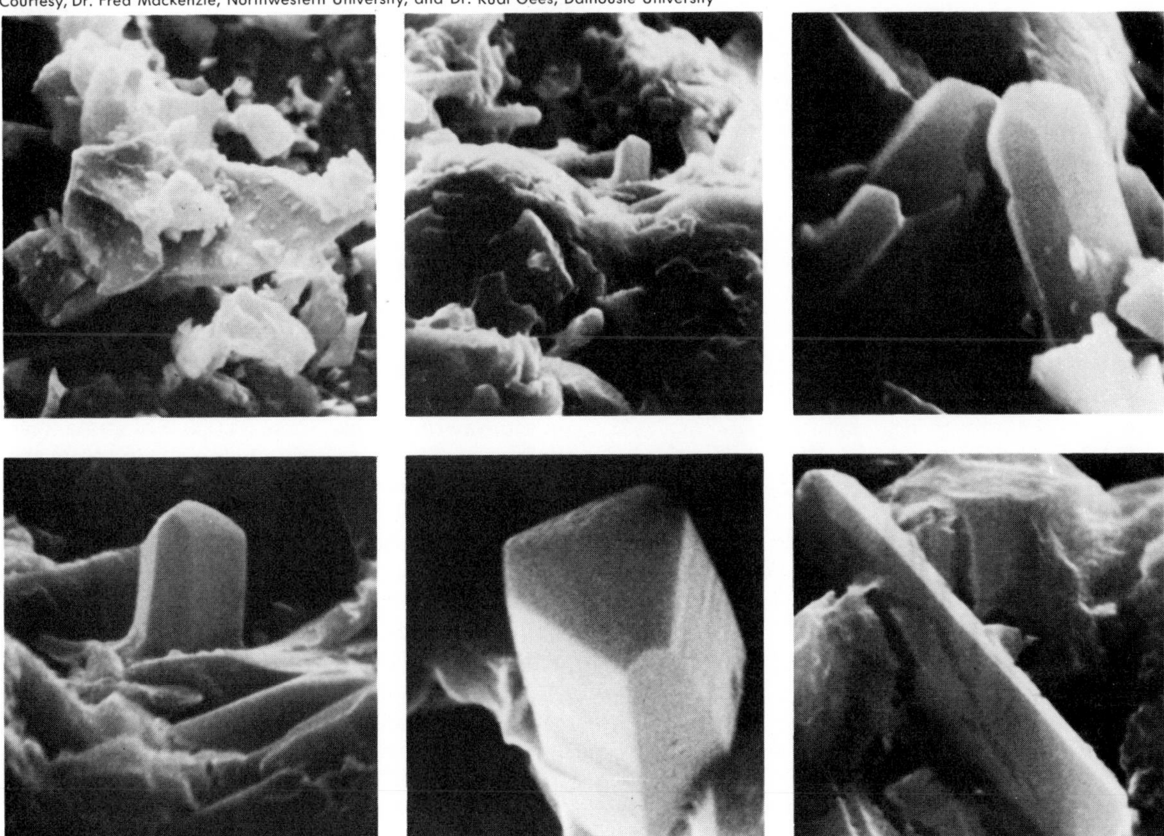

*A series of scanning electron micrographs depict the crystallization of quartz directly from seawater at room temperature. The first figure is the starting material, ground quartz, which exhibits poor sorting, high angularity, and no evidence of external crystal morphology. The remaining figures show the quartz surfaces after immersion in seawater and illustrate the formation of new crystals.*

orbit around Mars revealed a probable volcanic vent at Nix Olympica and three probable volcanic craters in the Tharsis region. Mars is intermediate in geologic behavior between the moon and the earth. Its interior has had more differentiation of igneous rocks than the moon, and it has an atmosphere to produce weathering and wind erosion, phenomena present on the earth but absent on the moon. Impact craters are conspicuous on Mars, because erosion and mountain-building processes have not obliterated early craters, as had happened on the earth.

Radar measurements from the earth were beginning to reveal the topography of Venus, which is invisible to direct observation because of the dense cloud cover of that planet. One mountainous region about 200 mi across averages about 2 mi above the general level of Venus. (*See* Year in Review: ASTRONAUTICS AND SPACE EXPLORATION; ASTRONOMY.)

**Marine geology.** The vessel "Glomar Challenger" completed legs 17, 18, 19, and 20 of the National Science Foundation's Deep Sea Drilling Project, with a total of 37 new holes drilled into the northern Pacific. In leg 20, two drilling records were set at a site 800 mi SE of Tokyo: the "Glomar Challenger's" bit descended through the deepest water yet tested in the drilling project (20,321 ft) and then drilled the greatest rock penetration in the project (1,237 ft). The bottom hole samples were Jurassic rocks over 135 million years old, the oldest yet obtained from the Pacific floor. The findings of the entire project caused a complete rethinking of theories concerning the structure and behavior of rocks beneath the ocean, confirmed continental drift, and revolutionized prospects for long-range mineral exploration.

During the past 125 million years the northern Pacific floor has moved northwestward more than 2,000 mi. Parts of the Pacific that were once under the Equator are now just south of the Aleutian Islands. This motion has not been at a uniform speed, but it came as a total surprise that 70 million to 55 million years ago the motion reversed, so that the Pacific crustal plate moved southward for a time before resuming its northwestward drift. This

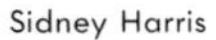
Sidney Harris

*"They did it again—not a word in the weather report about an ice age."*

was the first definite evidence of a complete reversal in motion of a tectonic plate. The reversal might be related to the breaking apart of Australia and Antarctica, which occurred about 65 million years ago. At the present time, the Pacific crust is moving against and downsinking beneath Asia, drifting at the rate of 4 in. per year over the past 10 million years.

The 1970s were designated the International Decade of Ocean Exploration by the UN General Assembly. As part of this effort, the U.S. Geological Survey undertook a six-month cruise to study the sediments and mineral resources beneath the Gulf of Mexico, the Caribbean, and the sea floor adjacent to Liberia, in Africa. The first leg was off the coast of Mexico. Jurassic salt beds seem to underlie the entire Gulf of Mexico and appear to have been formed by a combination of evaporation and restricted circulation as the Gulf was opening up during the beginning of continental drift. Seismic profiles revealed that the sediments along the western margin of the Gulf are sliding toward the deeper portion of the basin, producing folds in the marginal sediments.

The theory of sea-floor spreading implies an upwelling of the suboceanic crust at the site of the Mid-Atlantic Ridge, with a moving of the sea floor away from the ridge at a rate of one to two centimeters a year as new material is brought up. Thus, the axial zone should contain only rocks less than 10 million years old. During the year, however, scientists reported remnant blocks of older rocks in transverse fracture zones left behind as most of the sea floor spread. Ultrabasic (low in silica and high in ferromagnesian materials) igneous rock from St. Paul Rocks is 835 million years old. The crest of the Vema transverse ridge yielded samples of limestone with detrital feldspar and quartz deposited at a time when granitic continental areas were nearby; these probably represent remnant blocks of Jurassic or early Cretaceous sediments deposited as the Atlantic Basin was first spreading open some 120 million years ago.

Scientists at the Scripps Institution of Oceanography, La Jolla, Calif., reported that some of the thickest known ocean sediments in the world are in the Bay of Bengal. Some 54,000 ft of strata have been deposited there after being carried from the Himalayas to the sea by the Ganges River. (*See* Year in Review: MARINE SCIENCES.)

**World geologic map.** The Commission for the Geological Map of the World continued work, under the direction of the International Geological Congress, to develop a geologic map at scales of 1:10,000,000 (1 in. = 160 mi) and 1:30,000,000 (1 in. = 480 mi). The project involved geological representatives from more than 100 countries. The first parts of the finished map were printed in 1972 for presentation at the 24th International Geological Congress in Montreal in August.

**Paleontology.** In 1971 William J. Morris discovered remains from a duck-billed dinosaur over 100 ft long in Baja California. The duck-billed dinosaurs were very prominent in the Late Cretaceous about 75 million years ago. Their average length was about 30 ft, although specimens attaining 50 ft were recently found in Asia.

The extinction of the dinosaurs at the end of the Cretaceous has long puzzled paleontologists. Thomas R. Worsley of the University of Washington proposed a link between dinosaur extinction on the land and events in the ocean basins. During the Mesozoic Era (225 million–65 million years ago), excessive volcanism resulted in a warm, cloudy climate especially favorable to the development of large reptiles. The "Glomar Challenger" cores revealed that about 160 million years ago marine plankton that extracted carbon dioxide from the atmosphere evolved. As these flourished, an ocean-wide layer of calcareous sediment, consisting of the tests or outer coverings of the plankton, formed in the oceans, diminishing the greenhouse effect of the atmosphere through removal of carbon dioxide which was incorporated in the plankton tests. Worsley postulated that the resulting worldwide temperature decrease rendered conditions so unhealthy for dinosaurs that they became extinct. (*See* Year in Review: ARCHAEOLOGY.)

**Geochemistry and geochronology.** Strata with marine fossils associated with very fine-grained quartz (silica) in the form of chert or flint have been known for over a century from many parts of the world. Origin of the silica remains a mystery; such rocks are the only major sediment type that has not been seen forming today somewhere in the earth or in the sea. Geochemical experiments by Fred T. MacKenzie of Northwestern University, Evanston, Ill., and Rudi Gees of Dalhousie University, Halifax, N.S., resulted in overgrowths of silica crystals on finely ground quartz placed in natural seawater from Bermuda. This was the first time identifiable quartz had been formed directly from seawater at surface conditions without aging on amorphous precipitates.

Rubidium-strontium and lead dating of rocks from southwestern Greenland revealed the oldest rock yet discovered on the earth: gneissic granite which crystallized 3,980,000,000 years ago. Previously the oldest known rocks, dated about 3,500,000,000 years ago, were from Minnesota and South Africa. Greenland is a fragment of the ancient continent that ruptured to form the nucleus of North America and Europe as the continents drifted apart. The Greenland sample is considerably younger than the 4,600,000,000 years postulated for the earth-moon system, and its composition shows evidence of a complex earth history of crustal differentiation prior to its formation.

The Precambrian Era, which ended 570 million years ago, comprises the first 85% of the earth's history. A group of scientists from Australia proposed a formal division of the Late Precambrian Erathem, designated the Adelaidean System and defined to range from the beginning of Late Precambrian glaciation in Australia to the beginning of the Cambrian Period (spanning a time roughly 750 million–570 million years ago). This is the first formal subdivision of the Precambrian into periods based on modern dating techniques.

—John M. Dennison

Courtesy, Dr. D. W. Durney, Imperial College, London, and NATURE Magazine

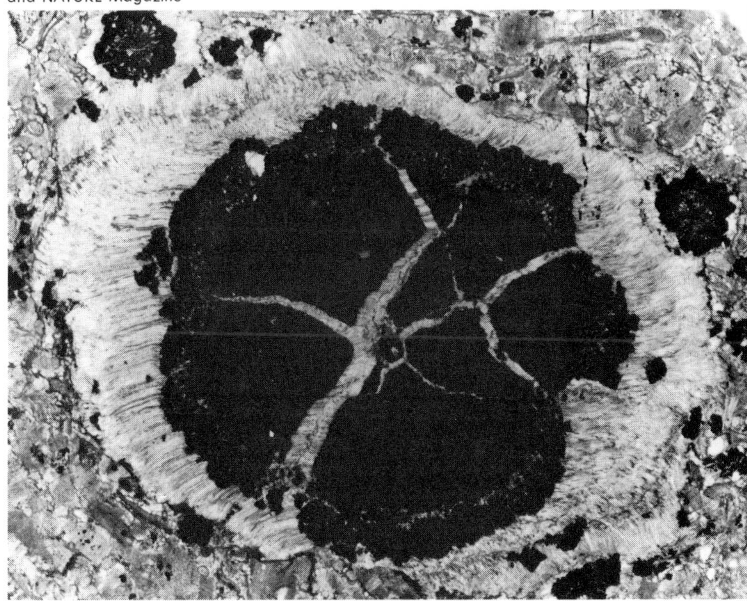

*Pressure shadows and veinlets (light) of symmetrically curved fibrous calcite growths in a pyrite concretion (black) are an example of the redistribution of minerals in nonporous sedimentary rock subjected to stress. This phenomenon was recently explained as arising from the diffusion of ions along grain boundaries.*

## Geophysics

Geophysicists made important contributions to two national policy debates in the U.S. during the year. The Cannikin nuclear test on Amchitka in the Aleutian Islands was hotly disputed, and one of the central points of concern was the possibility that it might trigger a large and damaging earthquake. The second question concerned the ability of seismologists to detect all significant nuclear tests so that a comprehensive test ban treaty could be negotiated without the requirement for on-site inspection. A precise description of the degree of accuracy actually attainable was probed by two symposia and vigorously debated within government circles. Both of these topics were of general interest, since they test the ability of scientists to feed accurate information into the decision-making process in an atmosphere that is not always conducive to scholarly detachment.

**Cannikin.** Amchitka lies in an island arc structure marking a place where a giant crustal plate plunges back into the earth's mantle along a great fault, capable of producing major earthquakes. The island itself is so remote that even a large earthquake probably would not have caused any significant damage by ground shaking, but there was concern that such an earthquake might trigger a seismic sea wave, or tsunami, that could cause high waves at distant sites around the Pacific Basin.

The distribution of small earthquakes located by special networks of seismograph stations revealed that the great thrust fault in the region of Amchitka passed approximately 30 km (1 km = 0.62 mi) below the point of detonation, and studies of ancient shorelines indicated that the island itself was outside the region where large deformations had taken place in past great earthquakes. These observations and the fact that no large tsunamis had been generated in this region in historic time led to the conclusion that the test was not likely to trigger a large, dangerous quake.

The Cannikin nuclear test was detonated on

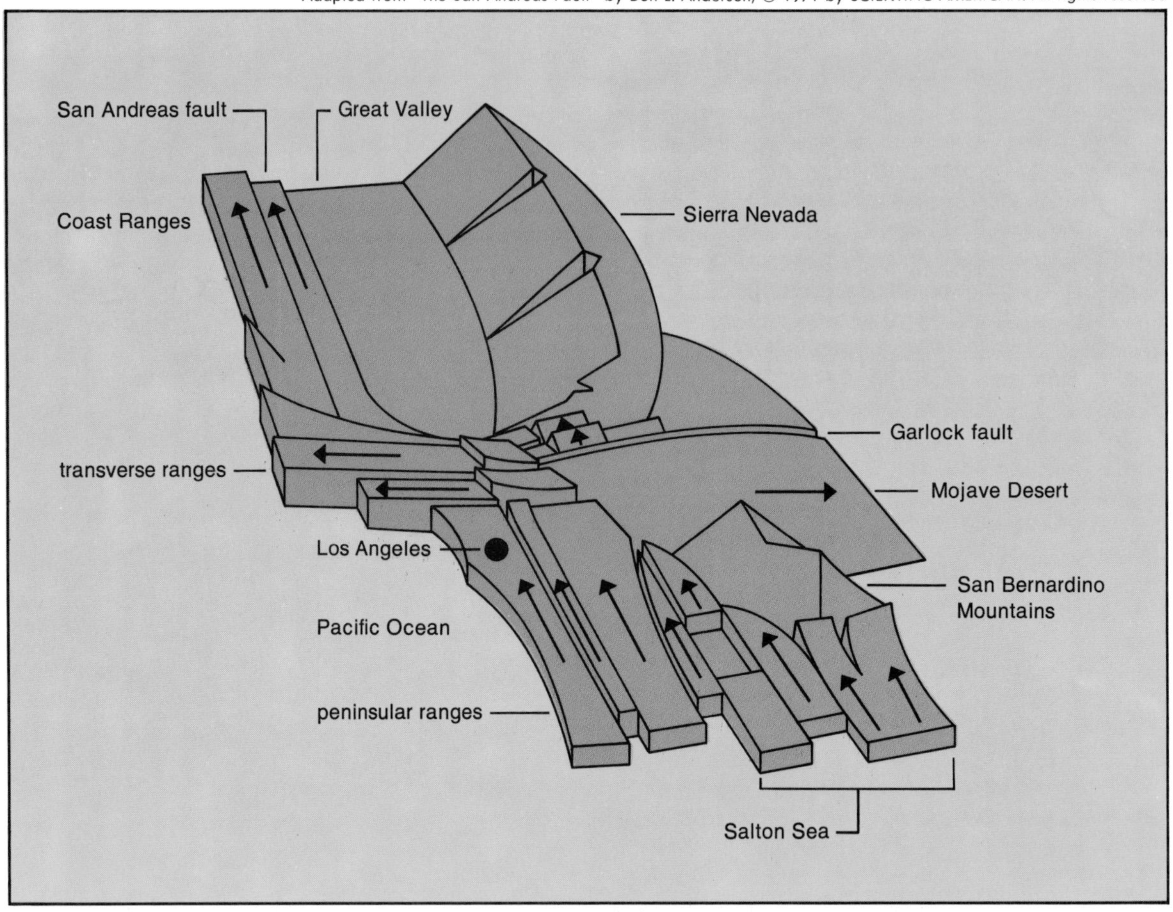


*Movement of the earth's crust in southern California is generally from southeast to northwest along the San Andreas fault. But at the point where the moving crust encounters the deep roots of the Sierra Nevada, the blocks of crust move west, creating the transverse ranges and a bend in the fault.*

Nov. 6, 1971, and the predicted seismologic effects were in agreement with the observations. The test triggered only small earthquakes in the immediate vicinity of the shot. Geophysicists were relieved that their predictions had been borne out. Had knowledge of the earthquake mechanism been more precise, however, it might have been possible to minimize some of the controversy.

**Nuclear test detection.** Negotiation of a comprehensive nuclear test ban treaty, under discussion since the mid-1950s, has been blocked largely because the major nuclear powers are reluctant to permit on-site inspection. The possibility of using the seismic waves generated by tests to detect treaty violations has been recognized, however, and beginning in 1959 a massive research program, sponsored by the U.S. Department of Defense, was undertaken to develop seismic means for detecting and identifying underground nuclear explosions. More than $300 million was spent on this research.

At the beginning of the program, most seismologists believed the task would be relatively easy, but this proved not to be the case. As the research progressed, it was found that complexities in the structure of the earth distorted seismic signals so that it was difficult to obtain accurate locations for some events. Furthermore, it was not always possible to tell the difference between an explosion and an earthquake. Nevertheless, a dramatic improvement was finally achieved. As the number of seismograph stations in the world increased and more data became available for study, corrections could be made for perturbations in the seismic signals caused by structural anomalies. This, in turn, led to an improved capability to detect and locate nuclear explosions and to distinguish them from earthquakes.

A series of studies initiated by Jack F. Evernden, a seismologist then employed by the Department of Defense, led to one of the most effective techniques for identifying nuclear explosions. The first seismic waves to arrive at a seismograph station pass through the deep interior of the earth, where they travel in a mode similar to the propagation of sound waves through the air. These are called body waves. Another class of seismic waves, arriving later, travels over the surface of the earth much as water waves travel over the surface of a body of water. The size, or magnitude, of a seismic event can be estimated by measuring the amplitude of either of these classes of waves at a number of distant seismograph stations and correcting the amplitude measurements for distance and other peculiarities along the propagation path.

Through comparison of the waves from many earthquakes and explosions, it was found that explosions tend to radiate surface waves of lower amplitude than earthquakes of comparable size as determined by the amplitude of the body waves. Continued refinement of the corrections needed to make these determinations and improvements in instrumentation led to an almost perfect set of discriminators to identify nuclear events. Problems with smaller events were gradually reduced by the development of larger arrays of very sensitive instruments, which are theoretically capable of detecting all earthquakes and explosions down to approximately magnitude 4 as measured on the Richter scale.

The seismic signals from underground nuclear explosions depend not only on the size of the test but also on the medium in which the device is detonated. In hard, water-saturated rocks, almost all tests of a yield of one kiloton can be detected. In dry, alluvial rocks, nuclear tests yielding 10 kilotons or more might be fired without detection by seismic means. However, when tests are fired in alluvial rocks, they frequently produce characteristic surface fractures or craters that could be observed from satellites. The exact threshold of detection remained a matter of concern, but it was generally conceded that the seismic methods currently available would severely limit the ability of a country to conduct a secret nuclear test program.

**Earthquake resistant construction.** During the San Fernando, Calif., earthquake on Feb. 9, 1971, an instrument located at the Pacoima Dam measured a ground acceleration of approximately one unit of gravity (G), the highest ever recorded from an earthquake. Engineers and seismologists usually describe ground accelerations as a function of the force of the earth's gravity field. A vertical acceleration of one G would apply a force equal to the weight of an object sitting on the stable surface of the earth. For example, if a 150-lb man was standing at a point that was accelerating at one G, he would feel that his weight had doubled to 300 lb during the upward cycle. If the direction of motion reversed, he would feel as though he had zero weight and was floating through space. If the ground acceleration is greater than one G, objects can be thrown into the air. During great earthquakes rocks have apparently been thrown into the air and come to rest some distance from their original positions. From this, and from some observations of ground failure, seismologists postulated that earthquakes might produce ground accelerations of more than one G, although this had never before been recorded.

The evidence of high accelerations during the San Fernando earthquake, together with the failure of a number of modern structures, indicated that an evaluation of existing building practices was in order. Some engineers believed that the high accelerations measured on Pacoima Dam resulted from unusual local conditions that focused or amplified the seismic waves. However, preliminary results of studies undertaken to determine the nature of this amplification, if any, failed to indicate any local conditions that could account for the Pacoima measurements.

*Portion of the San Andreas fault runs almost due east and west a few miles north of Frazier Park, Calif. The hilly region left of the fault line is moving upward with respect to the plain, causing definite offsets in the courses of the two largest streams in the area.*

Gary Settle, The "New York Times"

This poses a difficult dilemma for engineers and seismologists. If all structures are designed to withstand forces as large as those observed at San Fernando, the cost of construction would be increased substantially. Furthermore, the high accelerations raise serious questions about the safety of some existing structures that were designed in the belief that earthquake forces would not approach one G. On the other hand, accelerations as high as one G are not likely to be experienced by most buildings, even in great earthquakes. Even though the Pacoima Dam observation may not be unique in the sense that such accelerations may be experienced in most earthquakes, the area in which they are experienced and the percentage of buildings subjected to them are likely to be quite small. Eventually, the risk that some structures will be destroyed may have to be accepted if the cost of earthquake-resistant design is to be kept at an acceptable level.

The question of acceptable risk becomes particularly acute for certain critical structures such as nuclear power plants and dams, which have a potential for causing damage far beyond the structures themselves. One possible approach is to identify in advance those regions that might be prone to very large accelerations and to avoid such sites as much as possible. Another is to pay the price of higher costs for buildings located in those areas. It was too early to project all of the implications, but clearly the San Fernando earthquake would have a considerable effect on earthquake-resistant design for many years to come.

**Earthquake prediction.** One of the most attractive possibilities for reducing the danger from a great earthquake is prediction of the approximate time and location in advance. A program of research in earthquake prediction was proposed in 1965 by a panel headed by Frank Press of the Massachusetts Institute of Technology, and a modest level of formal research was undertaken at that time. In response to the San Fernando earthquake and rising public concern about earthquake hazards, proposals for a greatly expanded program were introduced in Congress.

There are numerous reports of phenomena observed prior to great earthquakes, although many are undoubtedly unreliable. Soviet researchers have discovered changes in the velocity of seismic waves propagating through the epicentral zones prior to the occurrence of some large earthquakes. Changes in the ground level and magnetic field have been reported by researchers in Japan and the Soviet Union, and these may be useful. In the United States there have been a few isolated reports of phenomena that might be precursors. However, the U.S. program has concentrated on achieving a better understanding of the fundamental process in the expectation that, if precursory signals do exist, they will be discovered in this way.

One center for this research is the U.S. Geological Survey's National Center for Earthquake Research located in Menlo Park, Calif. This group has deployed dense networks of seismometers and measured strain and tilt in a region extending south from San Francisco to Parkfield, a distance of approximately 200 km. The research of this group and others clearly demonstrates that the San Andreas Fault, the most likely location of future large earthquakes, is currently moving more or less continuously over part of its length. Presumably, where the fault is continuously creeping, at least a portion of the strain is being released by a nondestructive mechanism. Where the fault is locked, particularly in the section north of San Francisco, it is believed that strain is accumulating and the potential for a great earthquake is increasing.

By determining the accurate locations of very small earthquakes (microearthquakes), 4,000 to 5,000 of which occur each year in the region being studied, it is possible to define a three-dimensional picture of the geometry of the fault zone. Further, studies of historical data suggest that no large earthquakes have occurred in California without being preceded by some microearthquake activity. Although not yet proven, the corollary is probably true. That is, if a given region or length of the fault zone does not presently have any microearthquakes at all, that portion of the fault zone is not likely to produce a great earthquake in the future.

Traditional surveying methods, supplemented by the use of an instrument called the geodolite, which employs a laser beam to measure distances between two points, have been used to track the current deformation of large blocks of the earth's crust in California. These measurements confirm the fact that the portion of California west of the San Andreas is moving northward at a rate of one to two inches per year, and an increase in the density of these measurements is revealing details of the pattern of movement and showing how the rate of movement varies from place to place. The measurements also show the pattern of other complexities of the San Andreas fault system.

New research proposals are based on a major extension of these methods. A combination of surface geologic mapping and microearthquake studies would be used to identify the currently active faults. Once identified, these faults would be repeatedly resurveyed to determine the nature of ongoing deformations. Even if the exact time of future earthquakes cannot be predicted, the identification of currently active faults, together with an estimate of their potential for producing large earthquakes, could be of practical value.

**Plumes.** The island of Hawaii is being formed by volcanic activity that is bringing molten rock from a depth of at least 60 km and depositing it on the flanks of great volcanoes. This island is the easternmost and only currently active volcanic island in the Hawaiian chain. Geologists studying the other islands have determined that each has been formed by a similar process.

W. Jason Morgan, associate professor of geological and geophysical sciences at Princeton, suggested a possible mechanism to explain the origin of the islands. He believes they are the result of great vertical upwellings of material, called plumes, rising from deep within the earth in an approximately cylindrical column. According to his theory, the crust of the whole Pacific Ocean is sliding slowly over these fixed plumes at a rate of approximately two centimeters per year, thus accounting for the succession of islands.

—John H. Healy

## Hydrology

The recent rapid pace of research in hydrology and related water-resources fields continued during 1971–72. One index of the pace was the large number of national, regional, and international symposia. Prominent among these were meetings on the following subjects: Man-Made Lakes (Nashville, Tenn.); Mathematical Models in Hydrology (Warsaw); Water Balance in Semi-Closed Sea Bays (Moscow); Surface and Groundwater Pollution (Moscow); Hydraulic Research and Its Impact (Paris); Hydrology of Fractured Rocks (Washington, D.C.); and Stochastic Hydraulics (Pittsburgh, Pa.).

**Organizational changes.** The increased rate of research was matched by changes in the associated institutional structures. In the United States, the past few years witnessed the establishment of the Water Resources Council, the National Water Commission, the Environmental Protection Agency, the National Oceanic and Atmospheric Administration, and the Office of Water Resources Research, together with Water Resources Research institutes in each of the 50 states and Puerto Rico. These, and other federal and state agencies, spurred by the increased awareness of local, regional, and national water problems in the U.S., operated generally within the framework of a report issued by the Committee on Water Resources Research in the Office of Science and Technology. This report, "A Ten-Year Program of Federal Water Resources Research," was issued in 1966. Because of the progress and institutional changes since 1966, the program was being revised. The first step in the revision was a survey of water-research programs in the universities.

The International Hydrological Decade (IHD) in 1972 was in its eighth year. At its latest (1971) session, the IHD Coordinating Council stressed the need to complete documents and publications fulfilling its program by the time the Decade ends in 1974. These reports were prepared to guide hydrological research and education in all countries, particularly developing ones. One of the principal benefits of the IHD program to date has been the increase in the number and functions of regional groups dealing with water problems common to

*Right, after a 44-mi stretch of Crow Creek in Tennessee was rechanneled for flood control, biologists who visited the area found the stream nearly lifeless and termed it an "ecological disaster." Left, an untouched portion of Crow Creek still provides a home for rainbow trout and wood ducks.*

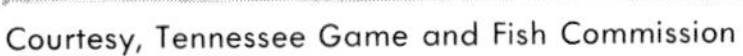
Courtesy, Tennessee Game and Fish Commission

two or more countries. In addition to the Nordic Council, which had been established for several years, there was an International Commission for the Hydrology of the Rhine Basin, an international group to study the water balance of the Danube basin, and an international cooperative program to study the water balance of Baltic waters.

While moving vigorously ahead in research and regional cooperation, hydrology was given another twist in its search for identity as a clearly defined scientific discipline. Hydrology's principal international group, the International Association for Scientific Hydrology (IASH), changed its name and broadened the scope of its interests to become the International Association of Hydrological Sciences (IAHS). The new name reflected the concern of hydrologists with engineering, statistical, mathematical, economic, political, and sociological disciplines where these activities affected man's use of the available water resources.

**Research efforts.** The new name also responded to the increasing concern in many countries for the quality of the environment. This concern particularly focused attention on the need to maintain and improve the quality of atmospheric, surface, and underground waters. Rainfall in Europe, for example, became increasingly corrosive over broader areas between 1956 and 1966. Rain and snow with a pH of 5.5–5.0 fell on parts of the Netherlands, Belgium, eastern England, and the North Sea in 1956. (The pH scale measures acidity and alkalinity, with 7.0 representing pure water at 25° C; lower figures indicate increasing acidity.) By 1966, an area nearly as large received precipitation with a pH of 4.5–4.0—and in spots as low as pH 2.8—while the area receiving a rainfall with a pH of 5.5–5.0 grew to extend from Great Britain on the west to the western border of the Soviet Union on the east, and from southern Norway and Sweden south across West Germany almost to Switzerland. Potentially the effects of the increased pH are multiple and complex, including changing the leaching rates in soils, acidification of lakes and rivers, corrosion of structures, and modification of metabolic systems of organisms. In southern Norway, for example, salmon runs have been eliminated in some streams because the increasing acidity has prevented salmon eggs from developing. Some data indicate that the northeastern United States has been similarly affected.

Increasingly extensive compilations of data show that the pollution of rivers and lakes has continued with little abatement in recent years. Although the problem is common throughout the U.S., it is as yet critical only in selected areas. Among the difficult decisions to be made are those that differentiate between levels of pollution that are "only" undesirable and those that are definitely toxic.

In partial response to this problem, U.S. Pres. Richard M. Nixon and Canadian Prime Minister Pierre Elliott Trudeau signed a joint U.S.-Canadian agreement to clean up the Great Lakes. The treaty, called the Great Lakes Water Quality Agreement, set stringent standards for the quality of the Great Lakes waters and also established a timetable for meeting them. The treaty pinpointed five objectives: to free the waters from substances that enter as a result of human activities and from harmful sludge deposits; to free the waters from floating debris, including oil; to free the waters from undesirable colorations and odors; to free the waters from materials toxic to man or aquatic life; and to free the waters from an excess of nutrients that enter as a result of human activities and result in undesirable growths of weeds and algae.

To provide an improved comprehensive base of hydrological information for the management of the water of Lake Ontario and its basin, the U.S. and Canada launched the International Field Year for the Great Lakes (IFYGL) on March 30, 1972. This program comprised an integrated series of studies and investigations on five major components of the hydrology of Lake Ontario and its basin: the terrestrial water balance, the atmospheric water balance, the heat-energy balance, lake circulation, and the distribution of chemical constituents and life forms. Several governmental agencies and universities on both sides of the border are involved. The IFYGL was expected to accumulate masses of data, some of which would take years to analyze, synthesize, and report. The program was organized, however, to make the actual data available at the earliest possible time through two data-management centers, one in each country.

In a somewhat different category was another proposal to collect large amounts of hydrological data during the last half of 1972. The first U.S. Earth Resources Technology Satellite, ERTS-A, was designed to provide information for several hydrological and water-resources experiments. These included interpretation of ERTS spectral imagery in the preparation of hydrological maps of arid lands; the practicality of using satellites to transmit water data from surface stations to central data-processing facilities; evaluation of hydrological hazards along the Alaska Pipeline corridor; delineation of boundaries of regional wetlands; the evaluation of soil moisture in the Lake Ontario basin; changes in snow cover in the Pacific Northwest; relationships of changes in the quantity of runoff in the Potomac River; and changes in land and water distribution in the Everglades in Florida.

—Leo A. Heindl

# Electronics

The epoch-making visits of U.S. Pres. Richard Nixon to the People's Republic of China and to the Soviet Union threw a spotlight on recent advances in electronics technology's oldest branch, telecommunications. China, through its government's choice, had remained isolated from most of the world for a quarter century. But with Nixon's visit it found itself the center of attention of the mass communications media of the world's technologically most advanced nations. To anticipate the demand, the Chinese government contracted with a number of Western firms for facilities and services to cover Nixon's visit.

An Intelsat 4 communications satellite went into service over the Pacific Ocean, taking up traffic that had been carried by the smaller Intelsat 3. Some 850 circuits between 15 earth stations were transferred to the new satellite. A second ground station was installed in Shanghai, supplementing one previously delivered to Peking. The second station also featured Western-made video and voice circuits, microwave terminals, and other equipment capable of handling television, telephone, and teleprinter transmissions, all of which were being expanded and were intended to remain in service for international communications following the visit. International divisions of four U.S. firms provided television coverage to viewers in America and elsewhere on a rotating basis, in accordance with special authorizations by the U.S. Federal Communications Commission (FCC).

Courtesy, Westinghouse Electric Corporation

*Photograph taken by starlight was made with the aid of a small, self-contained, direct-view image intensifier. The intensifier brightens an image by converting weak incoming light to electrons which are amplified and reconverted to visible light.*

**Mobile communications.** The FCC also figured importantly in several domestic rulings. One concerned mobile communications, especially for moving automobiles and trains. Such service had been limited to 33 channels totaling about 2 MegaHertz (MHz) in bandwidth, around 35, 150, and 450 MHz. This provided a somewhat overloaded system in which the police, fire, taxicab, and other dispatch services in some U.S. metropolitan areas tended to interfere with one another; a move to a higher frequency range, where greater bandwidth (and hence more services) could be accommodated without "crosstalk," had long been indicated. In 1971, the FCC allocated a total of 75 MHz (from 806 to 881 MHz) to mobile communications, opening exciting possibilities for new developments. Among them was a system of particular interest to highly congested areas, such as New York City, in which the mobile radios of several users share a single transmitter, a "community repeater."

In another proposed shared-user technique, computers are used to control a system based on geographical "cells" approximately hexagonal in shape. Three low-power switching units are located in alternate corners of the hexagon and communicate by radiotelephone with mobile units operating within their reach; also several switching units, provided they are sufficiently distant from one another, can use the same frequency, so that a limited band can serve a large number of clients ("frequency reuse"). Selection of the switching unit nearest to the mobile user to be contacted is automated, as is switching to a new frequency when the mobile user moves to an adjacent cell. As the density of users grows, the hexagons can be further subdivided into a more dense honeycomb.

Introduction of such services remained in the future. But some of the individual features were being used on a limited basis in existing systems. For instance, on the new high-speed Metroliner trains between New York City and Washington, D.C., American Telephone and Telegraph Co. (AT & T) as early as 1969 installed pay telephones that automatically switch calls from one radiotelephone channel to another as the train moves from zone to zone.

**Data transmission.** Another important FCC action was the Specialized Carrier Decision, intended to stimulate competition with AT & T's dominant Bell Telephone system in the increasingly important field of computer and other data transmission. Among other things, the decision would provide for the entry of carriers with no interest in the transmission of speech into the lucrative telephone market and force established companies to permit the interconnection of cus-

Courtesy, GTE Sylvania

*Newly developed laser capable of operating five to seven years in a communications satellite is powered by sunlight. Solar energy is collected by a 24-in. mirror and directed into the end of the laser (center). Lenses and mirrors collect and focus rays of the sun, which stimulate material in the laser to produce beams for carrying voice, data, television, and other communications between satellites and earth.*

tomer-owned terminal facilities with existing telephone networks. There was widespread skepticism about the effectiveness of stimulating competition by government measures. The skeptics pointed out that previous attempts in that direction had proved only moderately successful, at least from the customer's viewpoint: even when an alternate service was required to avoid monopoly (as for domestic telegraph and communication-satellite services), it often appeared that Bell could have done the job cheaper. Moreover, potential newcomers were said to be interested in competing only for the most profitable long-haul services, which might ultimately result in higher charges for less profitable local services, overcrowding of existing facilities, and the establishment of incompetent firms.

Representatives of new Bell competitors, among whom Microwave Communications Inc. and Data Transmission Co. (Datran) were the most prominent, retorted that none of these charges was true. Datran spent $10 million over the four years since its formation in 1968 to plan and design a network that would ultimately serve 35 cities and cost $400 million to construct—yet the firm was still two years from earning its first dollar; thus, the cost of competing against the giants would be sufficient reason to keep out get-rich-quick operators, and in any case the FCC would not approve carriers who failed to meet stringent financial requirements. As to jeopardy to local services, the revenues from data transmission were expected to increase tenfold during the 1970s, from $500 million to $5 billion a year. Thus, there would be more than enough business for several carriers and no justification for intrastate rate increases based on the contention that the new entrants were siphoning off the most profitable interstate business.

Meanwhile, the Bell system was advancing its own research and development to meet future demand in data communications. Its Canadian associate was experimenting with an entire digital-data network (known as Project 99). Already linking Toronto, Ottawa, and Calgary, the network was slated to be expanded to Halifax in the east and Vancouver in the west, so that it would become a coast-to-coast facility. The economic constraints in communicating with northern Canada are greater than in more densely populated areas, since there are fewer potential customers separated by larger distances. Therefore, a system that would meet their requirements economically would be sure to be viable in the U.S., Western Europe, and Japan.

**"The New Rural Society."** Thoughtful engineers also looked to electronic communications services to reverse the centuries-old trend of migration from rural areas toward cities. As early as 1962, J. R. Pierce proposed that much of the time now spent in travel, especially to and from work, could be saved by substituting communications for travel through such services as "face-to-face or group-to-group sound and vision, safely encrypted to preclude eavesdropping, documents transmitted by rapid facsimile or by data devices, some control or manipulation at a distance . . . equivalent to operating a computer, turning the pages of a catalog, or moving from department to department in a store while shopping. With cheap enough communication, students could attend classes by television." Now, a decade later, another U.S. engineer, Peter Goldmark (the inventor of the long-playing record and of electronic video recording, or EVR, for the home), suggested that the trend toward urbanization could be reversed by planning for what he dubbed The New Rural Society, which would imaginatively apply modern communications to make rural life more attractive.

A tentative start toward such a society had already been made in Britain, where the government was undertaking a study of the feasibility of dispersing some of its offices out of London. The study was directed toward maximizing the fiscal and social benefits and minimizing what was termed "communications damage" arising from the absence of face-to-face contacts. An interesting approach to the problems involved was to adjust the solution to prescribed amounts of funds available: how much could be accomplished by telecommunications "packages" costing the same as normal telephone services (*T*), four times that amount (4*T*), and sixteen times (16*T*)? Preliminary estimates showed that 4*T* could provide loudspeaking, audio conferences, and transmission of documents. Going to 16*T* would provide two-way television communications of standard broadcast quality. The next step would be to classify existing face-to-face contacts into those which could be carried out by phone, those which would require a 4*T* system, those which would require 16*T*, and those which would still have to be face to face.

Similar planning on an even more ambitious scale was taking place in the U.S., where systems engineers talked about integrating energy flow, waste flow, information flow, and "people" flow into a single, optimized system in which information flow would become a viable alternative to moving people and things and to the consumption of energy. Bell Laboratories announced that it had evolved a method of video conferencing in which two groups of nine conferees each could meet in two separate locations. Three unattended cameras in each room provide uninterrupted coverage without operators or other technicians.

In summary, electronics developments during the past year highlighted an emerging characteristic: a stage at which "hardware" innovations were less prominent than methods of applying them in novel ways that promised to increase communications between city and country and among the nations of the world.

*See also* Year in Review: COMMUNICATIONS.

—Charles Süsskind

Often called the greatest inventor who ever lived, Thomas A. Edison invented the phonograph, electric light bulb, a forerunner of the motion-picture projector, and a host of other devices.

# Environmental sciences

The environmental movement, launched with spectacular bravado and fueled by intense popular enthusiasm, was brought down to earth during the year. Dreams of easy solutions had drifted away. In their place were tough and complex political choices, economic analyses, lawsuits, legislation, technological development, and scientific research.

**Challenge to growth.** The change was graphically illustrated by the appearance of several widely discussed treatises on environmental dilemmas. Two of them, based on a series of computer models of the whole world, made some startlingly gloomy predictions, given current trends in development, environmental degradation, and population growth.

One, *World Dynamics*, was written by Jay W. Forrester, a Massachusetts Institute of Technology engineering professor and pioneer in systems analysis. A similar and more widely publicized book, *The Limits to Growth*, was based on a study sponsored by the Club of Rome and directed by another MIT professor, Dennis L. Meadows. "If the present growth trends in world population, industrialization, pollution, food production, and resource depletion continue unchanged," it warned, "the limits to growth on this planet will be reached sometime within the next 100 years." It stated, ominously, that the "most probable result will be a rather sudden and uncontrollable decline in both population and industrial capacity." *The Limits to Growth* added, however, that man can successfully opt for a state of "equilibrium" by stabilizing population and capital investment.

*Oil-eating bacteria (light globules) devour drops of oil in a test flask. Laboratory tests have identified 62 species of bacteria that eat oil, and scientists envision a day when freeze-dried bacteria can be stockpiled for use on oil spills.*

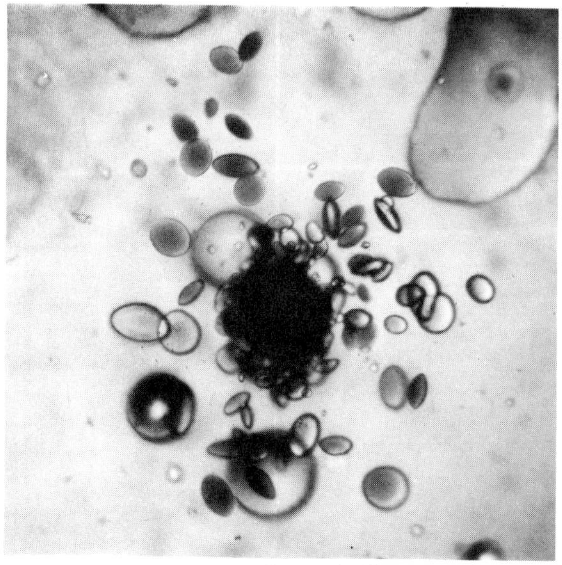

Courtesy, Dr. Thomas O'Neill, U.S. Navy

Concepts and conclusions in *World Dynamics* and *The Limits to Growth* were challenged by a number of experts. Nevertheless, the books raised the level of debate on the complex issues involved, and many who had dismissed ecological warnings as a bird watchers' fad were being forced to reexamine their ideas. (*See* Year in Review: SCIENCE, GENERAL.)

There were other impressive challenges to unharnessed growth and inadequate environmental protection. A long report entitled "A Blueprint for Survival," published in the January 1972 issue of the British magazine *Ecologist*, warned of the "breakdown of society and of the life support systems on this planet." The report was endorsed by 33 of the United Kingdom's most prestigious scientists. In March 1972 the Commission on Population Growth and the American Future, established two years earlier by the U.S. Congress, released its report. The commission concluded that "no substantial benefits would result from continued growth of the nation's population" and called for a federal population-control policy.

Further thrust was added to the environmental movement by Barry Commoner, a biologist and director of Washington University's Center for the Biology of Natural Systems, with his book *The Closing Circle*. In it, Commoner describes in detail how technological change in many industries—particularly during the post-World War II period—has centered on processes that increase pollution and that use a growing amount of electric power. He argues that there has been no compensating increase in the real value of products and services.

Environmental warnings and remedial activities were not unopposed, however. Critics challenged pollution control standards as unnecessarily strict, stressed the high costs to the public, and cited companies that had been forced to close down and jobs that had been lost. The running debate only served to emphasize how little man really knows about pollutants and health, about practical scientific and economic techniques of environmental protection, about the effect of controls on the economy and employment, and about the ramifications of slowing or stopping growth.

**Pollution and health.** A joint study summarizing the known effects of environmental pollutants on health was issued in January 1972 by the U.S. Department of Health, Education, and Welfare (HEW) and the Environmental Protection Agency (EPA). Among the findings:

Courtesy, U.S. Department of Agriculture

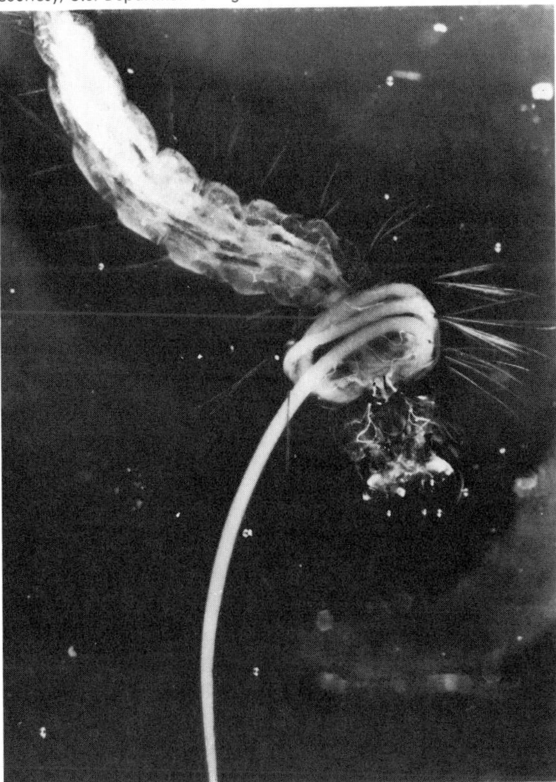

Reesimermis nielseni *nematode coiled inside the thorax of a southern house mosquito kills its host as it emerges to molt, mate, and reproduce. Nematodes have provided good mosquito control in selected areas around the world with fresh, unpolluted water.*

• "Pollution control should diminish the frequency and severity of acute respiratory disease, acute gastroenteric disease, and chemical intoxications of the central nervous system. Environmental aggravation of existing heart and lung diseases should also be substantially reduced."

• Cancer studies demonstrate that "there are striking differences in cancer incidence from place to place in the world . . . and when persons migrate, they appear to adopt the propensity for cancer which is characteristic of their new home." Also, studies of workers in some occupations reveal a number of substances that produce cancers, and cancers have been induced in test animals by the application of many substances. Thus, the study says, there is "strong evidence that environmental factors have a significant role in cancer production."

• Although cigarette smoking explains about 90% of chronic bronchitis and pulmonary emphysema, air pollution is associated with a "demonstrable excess" of chronic respiratory disease. The culprits implicated are sulfur oxides, nitrogen oxides, and particulates.

• "Atmospheric sulfur oxides, particulates and photochemical oxidants aggravate the frequency and severity of asthma symptoms in persons with this disease. . . . Ambient air pollutants have been associated with a significantly increased frequency of acute respiratory disease."

• "Decreased lung function due to air pollution and increased sodium intake from polluted waters can put a strain on the heart and cause exacerbation of cardiovascular disease." There are also suspicions that the carbon monoxide emitted by cars is linked with heart disease.

Concern also mounted over the high incidence of lead poisoning—in some cases leading to mental retardation—among urban children. In addition to the traditional villain, chipping lead-based paint, increasing blame was attached to the toxic exhausts of automobiles burning leaded gasoline. "Some 400,000 inner city children are believed to be suffering from high blood lead levels," said a federal interagency report, "Our Urban Environment," in September 1971.

Significantly, the HEW-EPA study was filled with warnings about man's ignorance of the health effects of pollutants and the need for further research. "The causes of our three major killers—heart disease, cancer, and stroke—have not been identified," the study noted. Without more knowledge, the effect of pollution on these diseases cannot be assessed. The continued introduction and wide distribution of thousands of chemical and physical substances of unknown environmental and health impact constitutes a "hazard of serious proportions." There is need for a more sophisticated, systematic screening of agents entering the environment and for assessment of their "potential harm."

**Clean air standards.** The battle over clean air standards in the U.S. began to take shape as each state drew up a plan to implement the strict provisions of the Clean Air Act Amendments of 1970. These implementation plans were sent to EPA for approval. Under Delaware's plan, for example, Delmarva Power & Light Co. would be required to reduce sulfur dioxide emissions at a plant in Delaware City. The company filed an appeal, whereupon EPA issued a federal notice of violation, making Delmarva the first company so cited under the 1970 law.

Congress had warned that many cities would have to enforce drastic curbs on automobiles in order to attain the air quality standards required for 1975, but the proposed state plans were generally timid on that score. In the District of Columbia, the City Council did propose a $1 per day downtown parking tax on commuters as one way to help meet the standards—as well as to raise revenues. Lusty opposition from a number of business organizations at the council's hearings suggested that it would be a stouthearted local government that would undertake to tell America's spoiled drivers they could not go where they wanted when they wanted. It remained to be seen whether EPA could provide the necessary muscle.

Some encouragement came from Italy where, after a week-long experiment with free bus and streetcar service had brought a dramatic decline in traffic on congested streets, authorities in Rome announced in April 1972 that the free service would be made permanent during rush hours. They added that cars would be banned in stages from the two-square-mile historic and business center of the city. Though traffic problems had sparked the move, the ecological potential for other cities did not go unnoticed.

As challenges to the uncurtailed use of autos increased, so did interest in an important alternative—rapid mass transit. Pressure rose to increase federal financial aid for mass transit systems. Chief target was the enormous Highway Trust Fund, set up in 1956 as the source of federal aid to the states for highway purposes.

Meanwhile, auto manufacturers, insisting they could not produce cars for the 1975 model year that would meet federal emissions standards for carbon monoxide and hydrocarbons, applied for the maximum one-year extension of the deadline to 1976. The major companies continued work on catalytic mufflers or afterburners, designed to cause more complete combustion of pollutants that now escape into the air. One problem was that, as more of the other pollutants are burned up, increasing amounts of nitrogen oxides are formed. Another is that the lead from leaded gasoline chokes the mufflers. The auto industry asked for assurances that lead-free or low-lead gasoline would be available in time for the 1975 models, and on Feb. 23, 1972, EPA took the first step to require that it be sold by refiners and large gas stations.

The auto makers also claimed that cars with pollution-control devices would be expensive, costly to operate, and less responsive than current models. Environmentalists tended to be skeptical about the figures cited, but many of the drawbacks were real enough and served to reinforce the widely held belief that much more drastic measures would be necessary. Auto manufacturers would have to wrench themselves away from the internal-combustion engine or the public would have to wrench itself away from the automobile.

Hundreds of firms around the world were accelerating efforts to develop "clean" cars. Major candidates were cars driven by a gas turbine, by steam, by electricity, by natural gas, by liquefied petroleum gas, or by hybrid engines that would permit the driver to shift from one fuel to another, depending on the circumstances. A gas turbine car made by Williams Research Corp. began undergoing extensive tests in New York City in February 1972. The manufacturer of Mercedes-Benz, which already mass-produced diesel-engine cars with lower emissions, tested a city bus that ran on natural gas. In January a steam-driven bus began experimental operations in the Oakland-Berkeley area of California.

There were several major air pollution episodes during the year, including one in east central New Jersey in September 1971 and another in Birmingham, Ala., two months later. In Birmingham, EPA invoked the emergency provisions of the Clean Air Act for the first time, leading a federal judge to order a two-day shutdown of 23 large industrial plants. EPA sought funds in its fiscal 1973 budget to develop analytical models of urban areas to show the relationships between pollution emissions, atmospheric conditions, and the ambient air quality. In a taste of things to come, Litton Industries was developing an experimental computerized air pollution monitoring system for Los Angeles County.

**The unclean waters.** The U.S. Congress, in effect, conceded failure in the nation's efforts to clean up its waters by moving to completely revamp the basic federal water pollution control law. Proposed legislation would depart from the existing system under which states specify the uses to be made of a body of water, decide what pollutants would be compatible with those uses, and prescribe necessary control measures. Bills passed by both the U.S. Senate and House of Representatives would rely chiefly on a permit program setting effluent limits for specific sources of pollution.

In other respects, however, the Senate and House bills were so divergent that many despaired of any possibility for effective agreement. Among the major differences was the Senate bill's requirement that all pollutant discharges be eliminated by 1981 except when this could not be done "at reasonable cost." The aim was to make all waters clean enough to swim in. The House bill was more lenient.

Major emphasis in water-pollution control continued to be placed on municipal waste-treatment plants and on modifications and processes to remove higher percentages of pollutants. Pointing out that research along these lines "has failed to provide a single significant technological innovation" (in the words of a Ralph Nader group study), many environmentalists were urging that more attention be paid to "closed systems" involving the benign disposal or recycling of sludge and other water wastes as fertilizer, through "spray irrigation" and by filtering the wastes through the ground.

In another controversy involving water pollution, confusion over washday detergents reached new peaks. After phosphates had been implicated in the "death" of lakes and streams, more and more jurisdictions were implementing bans on the sale of high-phosphate detergents. The detergent industry, meanwhile, came up with some highly effective cleaning agents to use as substitutes: nitrilotriacetic acid, or NTA, and caustic soda.

Just as manufacturers were phasing out phosphates, however, studies began to raise the suspicion that NTA and caustic soda were potentially dangerous to humans and perhaps to their environment as well. The industry voluntarily stopped using NTA. Then, in September 1971, the federal government did an about-face and suggested that housewives return to detergents with phosphates as the lesser of two evils. Housewives were understandably perplexed, and some environmentalists suggested a return to soap.

**Solid wastes.** Governments and industries continued to grapple on a more or less piecemeal basis with the myriad problems of solid waste disposal and recycling. Much attention was focused on the use of municipal solid wastes to produce energy. In one of a number of demonstration projects, the Union Electric Co. in St. Louis, Mo., planned to grind up refuse, mix it with pulverized coal, and use the mixture to fire a boiler and generate electricity.

"The United States pays over $3 billion to dispose of the solid waste residue of our affluent society," Richard C. Bailie, professor of chemical engineering at West Virginia University, told the annual meeting of the American Association for the Advancement of Science in December 1971. "This residue has a heating value equivalent to 300,000 tons of coal a day with a sulfur content of 0.2 to 0.3%. Our cities are literally being buried under the thing they desperately need—low-sulfur fuel." He added that processes have been developed that offer the promise of effectively using

Courtesy, National Institute of Mental Health

*Psychologist John Calhoun stands among some of the 2,200 mice bred in a physically optimal environment. As the population, which started with four pairs, increased, the mice began to suffer from a "lack of psychological space." Aberrant behavior, homosexuality, anxiety, and aggression increased, while reproduction and the capacity for reproduction decreased. As the members of this colony die, there will be no replacements for them. From experiments such as this, behavioral scientists believe they can extrapolate insights into such human conditions as urban crowding.*

refuse for at least a fraction of the nation's energy needs. Along the same lines, Hinrich L. Bohn of the University of Arizona, in the December 1971 issue of *Environment* magazine, said that the organic wastes of animal feedlots, which pose an enormous disposal (and water pollution) problem, could be converted to methane gas for a large new fuel supply.

Marketing of recycled paper continued to rise. In an experimental project, the city of Franklin, O., was selling not only recycled paper but also steel cans and other metals to private industry—all of them from the plant that handled the city's solid wastes. Soon it expected to sell recycled glass and aluminum cans as well.

Work continued during the year on various methods for using solid wastes for paving roads. A street in Omaha, Neb., was paved with "glasphalt," made from crushed bottles. The federal government was working on an experimental paving material comprising reclaimed rubber, bottles, garbage, lime, fly ash from incinerators, and calcium sulfate, a steel-processing waste. Michael D. Piburn of Rutgers University suggested in the April 1972 issue of *Natural History* magazine that glass from used bottles be converted back to sand to rebuild eroded beaches.

On the economic and political front, a number of court fights were being waged over statutes requiring deposits on drink containers. In December 1971 a county judge upheld the validity of the first such law, a Bowie, Md., ordinance requiring a five-cent deposit on all beer and soft-drink containers, but the ordinance was not expected to be enforced until further court challenges were settled.

**The pesticide controversy.** Environmentalists had long urged the adoption of biological or "integrated pest control" techniques in place of chemical pesticides. These techniques include the use of natural predators, pest diseases, sterilization, sex attractants, and genetic manipulation—sometimes in combination with a judicious use of chemicals. In his third environmental message, on Feb. 8, 1972, Pres. Richard Nixon said he was directing half a dozen federal agencies to "launch a large-scale integrated pest management research and development program."

Lengthy EPA hearings and court appeals had not laid to rest the thorny question of DDT use.

*A bucket wheel excavator with a digging capacity of 130,000 cu yd per day exposes a layer of lignite during the course of strip mining operations in West Germany. The machine selectively strips off and saves the top layer of fertile loam, which is used for later reclamation.*

Courtesy, Rheinische Braunkohlenwerke A. G.

Despite the imposition of some restrictions, it was still being applied in large quantities by U.S. farmers. EPA announced its intention to ban all use of DDT, but the pesticide industry made an administrative appeal that resulted in the hearings. A federal court declined to suspend sales of DDT pending resolution of the issue. A decision was expected from EPA in the near future, but it was likely to lead to further court challenges. Efforts to ban aldrin and dieldrin—other long-lived chemicals—were going through the same process.

A special advisory panel reported to EPA in September 1971 that even a complete ban on DDT in the U.S. would not solve the problem because of heavy worldwide use and dissemination, but it recommended early curtailment, with a goal of "virtual elimination." Meanwhile, agricultural expert Norman Borlaug, winner of the 1970 Nobel Peace Prize, told a UN meeting in Rome in November 1971 that DDT was indispensable for health and agriculture. Borlaug himself was sharply criticized in return.

With DDT and other persistent pesticides under fire, many were turning to toxic but short-lived chemicals. One widely used pesticide recommended as a safe substitute was carbaryl, sold as Sevin. Some scientists feared carbaryl might turn out to be hazardous, however. It was also one of the many pesticides that are highly toxic to bees. It was an irony of pesticide use that, under federal law, the U.S. Department of Agriculture paid out millions of dollars in indemnities to beekeepers for honeybees lost as a result of nearby spraying, as well as to dairy farmers and manufacturers for milk products removed from the market because they contain excessive residues of approved pesticides.

**The energy crisis, continued.** Long-range energy-environment problems were the subject of intensive study. On several occasions during the year, President Nixon called for a major effort to find new sources of abundant, nonpolluting energy. The administration requested increased funds in fiscal 1973 for research on a variety of potential sources, some of which had received relatively little attention—liquid metal fast breeder nuclear reactors, controlled thermonuclear fusion, coal gasification, solar energy, magnetohydrodynamics, and geothermal power.

Energy problems were exacerbated by the very real shortage of nonpolluting natural gas, the environmental damage caused by strip mining coal, the shortage of low-sulfur coal in many areas, the constant threat of oil spills, and the potential hazards of radiation or nuclear accident.

The federal government was seeking to encourage increased exploration and production of offshore oil and gas, but it received a major setback in December 1971 when a federal judge enjoined the Department of the Interior from selling oil and gas leases in the Gulf of Mexico. Citing the National Environmental Policy Act, which requires preparation of a comprehensive "environmental impact statement" for all federally funded projects, environmentalists argued—and the court agreed—that Interior had not adequately considered alternatives in the search for energy. A complete study was expected to take months. Meanwhile, environmentalists were poised to fight drilling off the East Coast, which was still in the talking stage.

Swan, courtesy, INDUSTRIAL RESEARCH

*"So far, the only really effective smog filter we've found is the human lung."*

Conservation and citizens' groups won a dramatic legal victory in their challenge of the Atomic Energy Commission's procedures for considering environmental factors before licensing nuclear plants. In a case involving the Calvert Cliffs plant in Maryland, a federal court ordered the AEC to revise its rules, and some power plant construction was delayed. Pending were complex challenges of safety and radiation requirements, as well as industry's ability to deal with thermal pollution.

The long-delayed trans-Alaska oil pipeline project was approved by the Department of the Interior, although the impact statement admitted the probability of environmental damage. Actual construction was awaiting action by the courts. Held up, at least temporarily, because of inadequate impact statements were some major highway construction projects and a number of large dam and canal projects, such as the Tennessee-Tombigbee Waterway in Alabama and Mississippi and the Gillham Dam and reservoir in Arkansas. (*See* Year in Review: FUEL AND POWER.)

Courtesy, Royal Aircraft Establishment

*A stained-glass window is a target at the end of the "Blunderbuss," a device designed by British scientists to simulate the sonic booms of the Concorde supersonic transport plane. Preliminary tests indicated that the booms would not affect soundly constructed buildings.*

**The Stockholm conference.** After four years of intensive preparation, the first United Nations Conference on the Human Environment was held in Stockholm in June 1972. Advisory in nature, and with no real power to enforce its recommendations, it nevertheless served to dramatize the growing worldwide awareness of environmental problems.

The conference was attended by delegates from 114 nations, representing some 90% of the world's population. The Soviet Union and its closest allies absented themselves to protest the exclusion of East Germany from full participation, but it was expected that even they would join in support of the conference's aims. These were set forth in 26 principles, embodied in a "Declaration on the Human Environment" which would be submitted to the UN General Assembly for ratification. Also approved were a 200-point "action program" and a recommendation to establish international machinery for coordinating environmental efforts.

The conferees were by no means in total agreement. The developed and less developed countries disagreed on priorities. China objected strongly to the "principle" urging destruction of all nuclear weapons, and there was acrimonious debate on such political matters as the Indochinese war. Despite their political differences, however, the conferees did express a basic unanimity on the need to protect the global environment. It would be a good while before anyone could assess the real effect of the conference, but unquestionably it had forced many nations to begin giving serious attention to their environmental problems.

—Rice Odell

*See also* Feature Articles: SPECIES IN PERIL; THE NOTORIOUS TRIO: MERCURY, LEAD, AND CADMIUM.

# Foods and nutrition

About 25% of the world's total 2,000,000,000-ton food supply is of animal origin—milk, meat, and eggs. To obtain this amount of food from domestic animals, as much food is needed for them in the form of cereals, legumes, and other feedstuffs, as is consumed by humans. Thus, if one counts this as part of the food used to feed mankind, the 1971–72 total is larger than ever before, amounting to roughly 4,000,000,000 tons. This means that there is about one ton of food for every person on earth.

However, there is still a precarious balance between the world's food supply and its population. Millions of persons still do not have enough food or the right kinds of it. Though at present there is theoretically enough food for all, starving and hungry populations can make little or no use of these supplies due to factors largely beyond their own control, such as disease, famine due to drought or insects, war and unfriendly politics, bad social and health standards, ignorance, and infirmities of old age. Bad economic conditions, infertile land, lack of transportation and storage facilities, and overpopulation are also involved.

In the United States approximately $125 billion were spent on food in 1972, with nearly a third of this being spent for meals outside the home. Foods at no cost or at reduced prices were offered to approximately 15 million persons in the U.S. (or about 7% of the population) by means of commodity food programs or food stamps.

More than 40 million school children (about 80% of the total) were offered school food services throughout the year in 90,000 schools. About 60% of these received federal assistance, including funds for eight million children who received free or reduced-price school meals. The experimental school breakfast program continued to grow, with a budget of more than $25 million a year.

In spite of this important government aid, which exists in many countries, malnutrition continued to be a serious problem in the U.S. and throughout the world. Malnutrition in most of the so-called developed and industrialized countries occurs in such forms as obesity, anemia, decayed and weak teeth, diet-related heart disease and diabetes, alcoholism, and borderline nutrient deficiencies. In the developing, less-industrialized countries malnutrition is often more direct, showing up in the form of simple food shortages with resulting deficiencies of protein, energy, vitamins, and minerals. Overeating of the wrong kinds of food was a problem everywhere.

Preliminary estimates of the dollar costs of malnutrition (including all direct and indirect costs) in the U.S. and countries with similar nutrition problems were placed at approximately $1.5 billion per 10 million population (for a total of about $30 billion in the U.S. per year). This amounted to nearly half of all the nation's medical, dental, and health costs, and about a fourth of all food costs.

**Food consumption patterns in the U.S.** Typical of citizens in most modern countries, an average individual in the U.S. normally eats a varied diet based on cereal products, fruit and vegetables, milk and milk products, and meat and legumes. In the U.S. the consumption of red meat and poultry was steadily increasing while milk consumption dropped considerably. Of great significance nutritionally was that the consumption of concentrated energy sources (sugars, alcohol, and pure fats) continued at record high levels. In 1971 more than one-third of the U.S. intake of calories came from these sources, which are completely devoid of most other important nutrients such as protein, minerals, and B vitamins.

A detailed breakdown of the 1971 per capita availability (before cooking) of food in the U.S., based on U.S. Department of Agriculture figures, appears in the Table. The amounts are given on a

Courtesy, U.S. Department of Agriculture

*Navaho nutrition aides prepare a breakfast using USDA-donated foods. A recent agreement giving the Navahos control of the distribution program was expected to achieve a more equitable division of the food among the 83,000 needy individuals.*

**Pounds of available food commodities per capita on a fresh basis**

| | Average 1957-59 | 1971 |
|---|---|---|
| **Meats (carcass weight, including bone and fat)** | 156.6 | 192.0 |
| Beef | 82.1 | 113.3 |
| Veal | 7.1 | 2.7 |
| Lamb and mutton | 4.4 | 3.2 |
| Pork (not including lard) | 63.0 | 72.8 |
| **Fish (edible weight)** | 10.5 | 11.2 |
| **Poultry** | | |
| Eggs (pounds, at rate of 8 eggs/pound) | 44.5 | 40.4 |
| Chicken (ready to cook) | 27.5 | 41.6 |
| Turkey (ready to cook) | 6.0 | 8.6 |
| **Dairy products** | | |
| Cheese | 7.9 | 12.2 |
| Condensed and evaporated milk | 14.8 | 6.8 |
| Fluid milk and cream* | 337.0 | 258.0 |
| Ice cream | 18.4 | 17.7 |
| **Fats and oils** | | |
| Butter | 8.2 | 5.0 |
| Margarine | 8.9 | 11.1 |
| Lard | 9.3 | 4.4 |
| Shortening | 11.4 | 17.0 |
| Other edible fats and oils | 10.8 | 18.6 |
| **Fruits** | | |
| Fresh | | |
| Citrus | 34.0 | 29.2 |
| Apples (commercial) | 21.0 | 16.8 |
| Others (excluding melons) | 40.5 | 35.0 |
| Processed | | |
| Canned fruit | 22.4 | 22.4 |
| Canned juice | 13.5 | 14.9 |
| Frozen (including juices) | 8.6 | 10.4 |
| Dried | 3.3 | 2.9 |
| **Vegetables** | | |
| Fresh (commercial) | 104.1 | 97.3 |
| Canned (excluding potatoes) | 43.3 | 51.3 |
| Frozen (excluding potatoes) | 6.6 | 9.1 |
| Potatoes (fresh equivalent) | 106.9 | 120.9 |
| Sweet potatoes (fresh equivalent) | 8.3 | 5.1 |
| **Grains** | | |
| Cornmeal and flour | 7.4 | 7.4 |
| Corn syrup | 9.4 | 16.6 |
| Corn sugar | 3.6 | 5.0 |
| Wheat flour (including whole) | 120.0 | 111.0 |
| Wheat cereals | 2.8 | 2.9 |
| Rice, milled | 5.4 | 7.6 |
| **Other** | | |
| Coffee (green beans) | 15.7 | 13.6 |
| Tea | 0.6 | 0.8 |
| Cocoa beans | 3.5 | 3.9 |
| Peanuts (shelled) | 4.6 | 5.9 |
| Dry edible beans | 7.7 | 6.0 |
| Melons | 25.1 | 21.7 |
| White sugar, refined | 96.1 | 102.0 |
| Total pounds | 1,386.2 | 1,364.3 |

*Fluid milk equivalent (including dried milk).

fresh basis, thereby including the water content, which ranges from 60 to 90% for most fresh foods. These figures are compared with 1957–59 averages, and show such trends as a decline in availability of milk and an increase in meat.

**Food science and technology.** Probably more than in any previous year, in 1972 the influence of the consumer greatly affected the types of foods made available to the public by the food industry. Of special interest in this connection in the U.S. were the proposals for voluntary labeling of food products by the Food and Drug Administration (FDA). Under these proposals, made as a result of requests of various concerned groups, much more nutritional information about a product could be obtained from the product label (information on the amount of calories, protein, and fat per serving). Detailed information about major vitamins and minerals would also be given if present proposals are approved.

There were many important food technological advances during the past year. Among them was the discovery of many new uses for "texturized proteins" made from soybeans. These products, of lower cost than meat (as low as 30 cents per pound), were added to ground beef or meat-loaf products, for example, in order to raise the levels of protein and decrease costs. They were also used to make imitation meat products, a development that nutritional scientists were studying because the texturized proteins contain very few vitamins and minerals and thus do not provide full nutritive value.

Work continued on the improvement of low-cost vegetable protein sources for human use, includ-

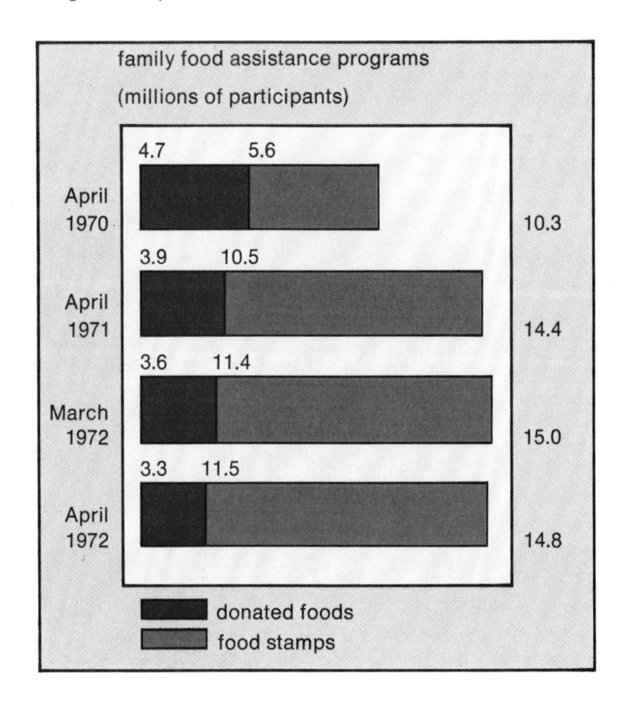

ing those from soybeans, plant leaves, beans, rice, sunflower meal, and rapeseed. Packaging and manufacturing advances resulted in many new products, such as frozen orange slices, frozen meat and cheese specialties, dehydrated batter mixes, frozen ethnic food, dry sauces, new canned foods, specialized "health food" products (many of which are now marketed in large supermarkets), new desserts, and hundreds of other new commercial convenient food items.

The growth of commercial and public food service organizations in the United States was phenomenal, and by 1971 had become a more than $40 billion business. Several private companies produced over 100 million meals per year, and some authorities predicted that within a few years more than half the meals consumed by the American public would be prepared outside the home. Commercial food services for retirement homes and day-care centers in the U.S. were growing at a particularly rapid rate.

The safety of food additives was given more than the usual attention by government agencies in 1972. In the U.S. an entirely new appraisal was being given to those additives designated as "generally recognized as safe," specifically artificial colors and sweetners. The presence of unintentional additives in food, such as synthetic diethylstilbestrol, which is fed currently to most all commercial beef cattle (and which does not remain in meat if removed from the food of the cattle one week before slaughter), and of pesticide residues from treatment of farm crops and animals were under investigation. The widespread publication by the FDA of the legal limitations of "filth" in human foods (such as hair, rodent fecal matter, insects, and worms) attracted much comment from consumers and the food industry alike.

**Nutritional advances.** Basic discoveries in nutrition continued to be made on a very broad front. The trace mineral field continued to be productive. During the past year Klaus Schwarz of the Veterans Administration Hospital at Long Beach, Calif., confirmed that rats require the element tin in their diet and also demonstrated that fluoride deficiency in rats caused a 20% depression in growth. The latter finding, if confirmed, would be a great impetus to the cause of fluoridation of water supplies. Edith Carlisle of the University of California at Los Angeles announced that silicon, the common ingredient of sand, was a dietary essential for the rat.

Also, in the nutrition area, the rapidly growing interest of all citizens in the environment and in the seeming lack of real concern of the food manufacturing and retailing interests in nutrition resulted in much growth of the new "natural" foods movement. Natural-food stores and restaurants expanded at a rapid rate in most large cities, especially around colleges and universities.

Courtesy, Varian Techtron

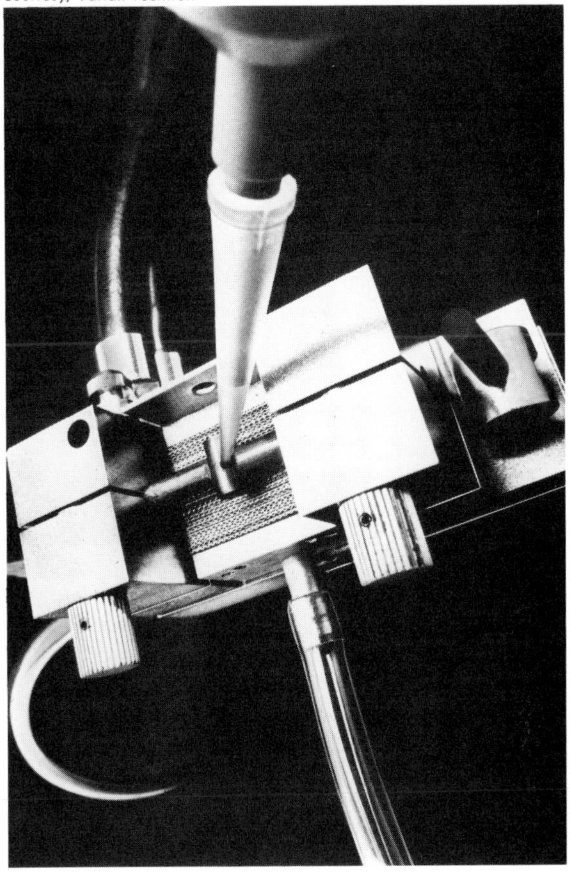

*Newly developed carbon rod work head is used with an atomic absorption spectrophotometer to determine the metal content of food additives. The sample is syringed into the rod, dried, ashed, and atomized for analysis of any or all of 80 metals that might be present.*

Nutritionists applauded the increased concern in developing better eating habits but were troubled over many false and misleading claims made about foods and food supplements and about the unnecessarily high prices paid for these foods. Adherents of health diets generally consumed food of high nutritional value, but in some cases food extremists and fanatics ate nutritionally poor diets for long periods of time. An example was the "seventh stage" of the so-called Zen macrobiotic diet (consisting chiefly of brown rice and a few vegetables), which can result in severe nutritional problems.

In December 1971, the White House Conference on Aging recognized the importance of nutrition in a properly fed aging society by calling for increased research on nutrition problems of the aged. Especially important was the resolution that "the Federal Government assume the responsibility for making adequate nutrition available to all elderly persons of the U.S. and its possessions." Means of doing this, such as by food stamp programs, improved nutrition education, improved food facilities, and availability of food services, and by providing adequate income and housing were spelled out in some detail.

About 10,000 research papers were published in 1972 in the food and nutrition field. Among the areas of most active research were nutrition and behavior, vitamin and mineral metabolism, nutrition as it relates to blood lipids, disease and nutrition interrelationships, measurement of the biological value of new foods for man, psychological aspects of weight control, effects of high sucrose intakes, vitamin D chemistry, and the effect of drugs and alcohol on nutrient levels in the body.

**The future.** Since our most urgent need in the food and nutrition area is the combating of human suffering caused by malnutrition, nutritionists hope that the future will bring a much greater concern about this problem from national leaders, international health agencies, and the food industry. Though the development of technological improvements and the uncovering of basic knowledge in the field is extremely important, these alone are not sufficient to feed the hungry millions. New and attractive low-cost food sources will need to be developed and new educational programs established, the combination of which (with improved economic conditions) is essential if an effective start is to be made in the prevention of malnutrition.

—George M. Briggs

*See also* Feature article: NUTRITION AND THE AMERICAN DIET.

# Fuel and power

Fuel and power matters claimed a growing share of attention during 1971 from political leaders, scientists, environmentalists, and the general public in the United States and the other industrially advanced countries. Adequate supplies of fuels are essential to these complex, highly developed societies because energy is involved in the full range of problems confronting them. These include national economic growth, depletion of resources, security of supply, environmental protection and ecological degradation, conservation and recycling of materials, and changes in living standards and life styles.

Prime evidence of this growing concern was the unprecedented energy message of June 4, 1971, by U.S. Pres. Richard Nixon in which he established national energy goals and outlined a broad range of policies to deal with the nation's growing energy demands and overcome the supply and environmental problems. Nixon stated, "A sufficient supply of clean energy is essential if we are to sustain healthy economic growth and improve the quality of our national life." He noted that existing domestic energy sources and technologies are not sufficient to meet future requirements within acceptable environmental standards and costs. Therefore, Nixon committed the U.S. to a "clean energy" policy and ordered accelerated programs of research and development of three priority technologies—sulfur oxide emissions control, the fast breeder nuclear power reactor, and coal gasification. He also noted the need for greater research and development efforts in nuclear fusion, solar energy, underground electric power transmission, magnetohydrodynamics, and other technologies.

*Drilling rig in west Texas was used to conduct tests of a new technique, abrasive jet drilling. The tests demonstrated drilling rates 4 to 20 times faster with substantially longer bit life when compared with conventional rotary drilling.*

Courtesy, Gulf Oil Corporation

Secretary of the Interior Rogers C. B. Morton announced on July 15, 1971, the essential elements of a proposed national energy policy that was consistent with and supportive of the president's message. These policy elements set forth the objectives, responsibilities, and roles of private enterprise and of local, state, and federal governments in bringing about a more acceptable balance among national energy objectives.

Despite the slowdown of general economic activity during 1971, fuel and power consumption registered healthy gains. In the U.S., energy consumption increased by 2.3% over that of 1970. At the same time, energy demand in all the non-Communist nations was estimated to have increased by more than 5%. Most of the increased consumption occurred in the U.S., Western Europe, Japan, and the other more industrialized and affluent countries.

The less developed countries continued as the principal sources of fuel supplies. Western Europe remained dependent on the oil-rich countries of the Middle East and Africa for nearly two-thirds of its total fuel needs, while Japan obtained about 70% of its fuels from those sources; the U.S. received about 25% of its petroleum requirements from foreign countries, but mostly from Western Hemisphere sources.

As a result of the continuing growth in the demand for fuels in the major consuming nations, the oil-producing and exporting countries were successful in further modifying the terms of concession agreements held by Western oil companies operating in their territories. The exporting countries greatly increased their share of revenues and imposed more stringent controls on the concessionaires. During 1971, oil payments to governments of the oil-producing states in the Persian Gulf and African areas increased by approximately 30 to 40% in accordance with the five-year settlements of early 1971 negotiated in Teheran and Tripoli between oil operators and the Organization of Petroleum Exporting Countries (OPEC). Despite this increase the oil-producing countries pressed new demands for increased revenues and for local government participation in the ownership of the oil companies operating in their territories. These actions seem likely to lead to a greater control of oil exports by the producing countries and further sharp increases in oil prices in the major consuming nations.

One of the most significant steps in nuclear power development in recent years was the announcement on Jan. 14, 1972, by U.S. Atomic Energy Commission (AEC) chairman James R. Schlesinger that the nation's first large-scale nuclear fast breeder electric power plant would be

Novosti from Sovfoto

*Sections of a new pipeline lie on the surface of the Caspian Sea prior to being lowered onto the seabed. The 32-km pipeline will link the offshore Bakhar oil and gas deposits with the Soviet mainland.*

constructed in Tennessee. Since the beginning of the nuclear age, scientists have recognized the need for fast breeder reactors. Such a plant will produce more fissionable fuel than it consumes, in contrast to presently developed nuclear plants that consume approximately twice as much fuel as they produce. President Nixon stated in his energy message that "Our best hope today for meeting the Nation's growing demand for economical clean energy lies with the fast breeder reactor." Not until 1972 had the breeder program progressed enough to warrant a large-scale demonstration plant.

During 1971, environmental concerns remained a major issue in the U.S. and other highly industrialized countries. New environmental protection regulations and measures taken to control pollution increasingly influenced the fuel and power industries, placing constraints on traditional methods of producing, processing, transporting, and using fuels and power.

Scientists in 1972 generally recognized that no fuels or forms of energy were entirely "clean" within existing technologies and that "zero pollution" was not a realistic expectation. The production and use of energy involved many activities that inherently affected the environment adversely, and current technologies were simply not capable of providing the ultimate environmental protection desired. Nevertheless, acceptable pollution limits could be attained through improvements in the production and consumption of fuels. Such improvements would result in an increase in the cost of power to the consumer. Thus, the consumer's need for reliable supplies of adequate and reasonably priced energy had to be balanced with attainable and affordable environmental quality objectives.

As a result of a federal appeals court decision in the Calvert Cliffs (Md.) nuclear power plant case, a decision critical of the plant's environmental protection policies, the AEC issued drastically revised rules for assessing the total environmental impact of nuclear power stations. The AEC assumed full responsibility for evaluating all environmental effects from the nuclear plants and all alternatives for producing electric power prior to issuance of construction permits and operating licenses.

Prior to this decision, the AEC had placed most consideration on radiological effects and had relied on other state and federal agencies for advice on other environmental matters. The new regulations were expected to delay licensing of nuclear plants under construction by as much as one to two years.

Other significant developments affecting fuel and power in the year under review included: (1) world petroleum consumption reached record highs, but a general slowdown in economic expansion limited the gain to about half that of recent years; (2) new large oil discoveries in the North Sea confirmed that area as a major oil province; (3) the discovery of natural gas off Nova Scotia was the first commercial hydrocarbon discovery along the east coast of the North American continental shelf; (4) U.S. crude oil production in 1971 declined for the first time since 1961; (5) international shipments of liquefied natural gas (LNG) expanded rapidly as the U.S. and other highly industrialized countries sought new gas sources; (6) despite a slowdown in gas demand in 1971, the natural gas supply situation in the U.S. continued to worsen; (7) U.S. coal production in 1971 declined by 7% as a result of a six-week labor strike; (8) the number of fatalities and their frequency rate in U.S. bituminous coal mines were at an all-time low in 1971; (9) world electricity demand continued to pace energy demand at an extraordinary rate; and (10) there was a continued strong upsurge throughout the world in new orders for nuclear electric power generation units.

*Engineers weld sections of the Yakutsk-Pavrovsk gas pipeline in the Soviet Union, the northernmost pipeline of its kind in the world. A 148-km (92 mi) sector of the snakelike structure rests on special pile supports aboveground.*

Tass from Sovfoto

## Petroleum and natural gas

Throughout 1971 and early 1972, the world petroleum and natural gas industry was confronted with lower growth rates, new economic and political initiatives by the large oil-exporting countries, growing environmental constraints, and greater challenges to develop resources in difficult operating regions so as to diversify supply sources. In the U.S., domestic oil production declined, petroleum product and natural gas demand growth rates slowed sharply, oil and gas prices increased, and environmental constraints increasingly deterred and restricted petroleum exploration and development activities.

Worldwide oil production averaged almost 48 million bbl daily in 1971, only about 5.4% more than that of 1970 and considerably less than the average growth rate over the past decade of nearly 8%. Crude oil production in the U.S. fell in 1971 by about 1% to a little more than 9.5 million bbl per day. Even so, the U.S. continued as the world leader in crude oil production, although its share of the world total steadily declined from 32% in 1961 to 20% in 1971. The Soviet Union followed as the second largest producer, accounting for about 16% of world supply. Iran and Saudi Arabia ranked third and fourth, respectively.

U.S. consumption of petroleum products averaged just under 15 million bbl daily, an increase of only 2.4% over that of 1970. Consumption in the rest of the world of about 33 million bbl daily was only 5.3% higher than the previous year. These gains were only about half the annual growth rates of recent years. The lower rates of gain in 1971 were attributed to a general slowdown in world economic expansion and to milder weather in the major consuming areas throughout 1971.

In concerted and individual actions the major oil-exporting nations of the world exerted economic and political pressures to obtain greater benefits and control over oil concessionaires operating in their territories. Most negotiations were conducted under the auspices of the Organization of Petroleum Exporting Countries (OPEC), an 11-nation group comprising 6 Persian Gulf area states plus Indonesia, Nigeria, Algeria, Libya, and Venezuela. OPEC controls approximately 90% of the world's oil reserves and exports. .

The international oil and gas industry continued exploration efforts throughout the world and succeeded in finding or extending new reserves. The discovery of three potentially large oil fields in the British sector of the North Sea confirmed this area as a major oil province. The first production from large oil fields in the Norwegian sector of the North Sea was initiated in August 1971. Estimates by industry sources indicated that in 1972 known oil reserves were sufficient to support production of one million barrels daily from the North Sea. The major impediment to development was the lack of adequate drilling equipment to operate in the turbulent sea and drill in water depths up to 400 ft.

Other notable exploration successes were registered in offshore Indonesia, where oil production began in 1971. Large new gas reserves were discovered in the Canadian Arctic. The first commercial discovery off the east coast of North America was the announced natural gas find on Sable Island off Nova Scotia. An oil find in the Mediterranean near Spain, a large new oil discovery in Egypt, another giant oil field found in Saudi Arabia, and more new discoveries in Ecuador were among the more significant events of 1971.

The U.S. petroleum industry waged a losing battle in its efforts to expand domestic producing capacity to meet rising demands. Although domestic production neared capacity, 1971 crude oil production actually declined below that of 1970 by 90,000 bbl per day to an average of 9,545,000 bbl daily. Environmental considerations continued significantly to forestall development of oil and gas resources in the most oil-rich regions of the U.S.—Alaska and the outer continental shelf.

Construction of the trans-Alaska pipeline remained at a standstill as efforts to resolve environmental objections continued. The U.S. Congress passed a bill settling Alaskan native claims, thereby resolving one major problem. As a prerequisite for obtaining a permit for construction of the pipeline, the U.S. Department of the Interior prepared and filed an environmental impact statement for a federal court, in compliance with the National Environmental Policy Act of 1970. Environmental groups were expected to appeal any decisions that would allow issuance of the permit.

A general lease sale of federal outer continental shelf lands offshore of Louisiana, scheduled for Dec. 21, 1971, was cancelled because of court injunctions against the sale obtained through the efforts of three environmental groups. They argued successfully that the required environmental statement was inadequate in its consideration of alternate energy sources.

As a result of these conditions, the U.S. was expected to increase its reliance on foreign oil imports. In 1972 these already provided 25% of the required supply.

Marketed production of natural gas throughout the world amounted to an estimated 40 trillion cu ft in 1971. U.S. production accounted for 57% of this total. Based on gross withdrawals, the world marketed production of natural gas amounted to about

Courtesy, Gazocéan

*Stainless-steel tank in the hold of the tanker "Descartes" is one of six in ship, which can transport 1,100,000,000 cu ft of liquid natural gas at −259° F. Experts predicted that by 1976 such tankers would be bringing natural gas to the U.S. from such remote areas as Siberia.*

70% of producing capacity. The major supply centers were remote from large consuming countries, and inadequate transportation facilities precluded the full utilization of this clean fuel. Few industrial nations had adequate indigenous gas supplies and thus were net importers of natural gas. Pipelines transported most of the gas sold across national boundaries, but LNG continued to grow rapidly as a means of processing and transporting gas in international trade. During 1971, LNG plants were operating in Algeria, Libya, and Alaska, and ten special tankers regularly delivered LNG to European, Japanese, and U.S. markets. The international shipments of LNG in 1971 amounted to 250,000,000,000 cu ft.

In 1971, the U.S. demand for dry natural gas was about 22.1 trillion cu ft, accounting for one-third of the nation's total gross energy consumption. The 1971 demand was only 3.4% more than that of 1970 and was considerably less than the average annual growth rate of 5.8% during the period from 1961 to 1971. Milder weather, a slowdown in the general economy, and limited availability of new gas supplies for some markets were responsible for the sharp decline.

Despite the slowdown in gas demand in 1971, the domestic natural gas supply situation in the U.S. continued to worsen. Consumption continued to exceed the replacement of new gas reserves. Paradoxically, it was reliably estimated that there were large natural gas resources remaining to be discovered and developed in the U.S. According to producers and developers of gas resources, they would remain undiscovered because regulated wellhead prices of gas were inadequate to attract the risk capital necessary for increased natural gas exploration and producing ventures. In an attempt to alleviate this situation the U.S. Federal Power Commission (FPC), the regulator of prices in interstate gas markets, changed its area wellhead pricing rules and took other steps that increased average domestic wellhead prices by 8.4% in 1970 and 6.6% in 1971. But even with the FPC action many producing, gas transmission, and distribution companies maintained that greater price increases were needed.

Expanded domestic natural gas development appeared to be essential to fulfill the expected rising demands of the mid- and late 1970s. If economic incentives prove inadequate, more gas shortages are expected to occur. Alternate sources of supply, such as overland imports from Canada, LNG, and synthetic gas from coal or oil, were not considered adequate supplements to meet fuel demands for gas in the near future.

Nevertheless, the use of synthetic gas as a high-cost supplementary gas supply was growing. Plans were announced for the proposed construction of approximately 20 plants for the conversion of coal, naphtha, and other hydrocarbons to gaseous fuels. The chief detriments to synthetic gas development were that conversion technologies were not perfected and the cost of such gas was very high.

## Coal

Coal continued to be a major energy source in 1971, with worldwide production and demand estimated to be slightly in excess of 3,000,000,000 tons. Production was centered in eight countries that accounted for more than 80% of the total output. These included the Soviet Union, the U.S., China, East and West Germany, Poland, the United Kingdom, and Czechoslovakia. The U.S. continued to account for about one-fifth of the total output.

Production of bituminous coal and lignite in the U.S. reached only 560 million tons in 1971, 7% less than that of 1970. The decline was the result of a six-week strike that shut down most of the U.S. coal industry in the fall. Despite the strike, domestic coal consumption of about 504 million tons was off by only 2.5% from 1970. Coal's most important market, electric power generation, actually increased consumption by about 3.3% to 331 million tons. U.S. coal exports were affected by the strike and by a general slackening of world economic activity. Coal and coke exports declined to an estimated 59 million tons, valued at about $950 million.

Coal in 1972 was the most abundant fossil fuel resource in the U.S., with reserves adequate for centuries at current levels of consumption; however, its combustion was the major source of air pollution from stationary sources. It was essential that new or improved technologies be developed for removal of sulfur dioxide and other pollutants either prior to or after combustion so that coal could make its proper and necessary contribution to future energy needs. Technologies for the removal of sulfur and other harmful contaminants from stack gases, and conversion processes for producing clean synthetic gases or liquids from coal were known but had to be made more efficient and less costly. In his energy message, President Nixon recognized this urgent need and announced high-priority programs and increased funding for research to speed up the time when those technologies become a commercial reality.

The U.S. Bureau of Mines greatly expanded its coal mine inspection and enforcement activities, and the coal industry made considerable progress in complying with the Coal Mine Health and Safety Act of 1969. The number of fatalities and their frequency rate in bituminous coal mines reached an all-time low in 1971. Although there were 177 fatal injuries, this was 78 less than in 1970. The frequency rate of fatal injuries in 1971 was 0.76 per million manhours, a drop of 27% from that of 1970. In addition, four miners lost their lives in anthracite coal and mine accidents in 1971.

## Electric power

World demand for electrical energy continued to grow at an extraordinary rate, nearly double that of energy as a whole. It was estimated that world electricity consumption reached 5 trillion kw-hr in 1971. The U.S. led all nations in power consumption, accounting for more than one-third of the total. In 1971, net electricity generation by the U.S. electric utility industry was an estimated 1,610,000,000,000 kw-hr, a gain of 5.4% over the 1970 output.

Electricity is a secondary form of energy that requires the conversion of a primary source into power by a generation process. In 1972, about 82% of the electricity in the U.S. was generated in fossil-fueled thermal plants; 16% was from falling water in hydroelectric plants; and only about 2% was from nuclear power plants. Insignificant quantities were

*Oil drilling rig stands in the North Sea, where potentially large oil fields were recently discovered. In 1971 production began from wells in the Norwegian sector of the sea in spite of difficulties presented by considerable turbulence and by water depths of 400 ft.*

Courtesy, Standard Oil Co. (N.J.)

generated from geothermal steam, wind, solar, and tidal energy sources. The thermal conversion (of heat energy into electric power) was costly in terms of primary fuels because it was accompanied by a great loss of heat energy. The thermal efficiency of the best power plants in 1972 was only about 40%.

The types of nuclear power plants in use or under construction utilized only 1 to 2% of the thermal energy released from the nuclear fuel. This figure was not expected to improve materially until the so-called fast breeder reactors became a reality. Fast breeder reactors are technically feasible concepts. When perfected for commercial use, each reactor will be capable of producing enough excess fissionable fuel in a period of 7 to 10 years to sustain itself and also fuel another reactor of similar size. In his energy message, Nixon increased the funding for the AEC program to demonstrate the commercial feasibility of a fast breeder reactor by 1980. To achieve this goal, AEC chairman James R. Schlesinger announced on Jan. 14, 1972, a joint industry-government plan to build the nation's first fast breeder prototype power plant.

The breeder demonstration plant was to be built at a site to be selected in Tennessee and would be operated by the publicly owned Tennessee Valley Authority (TVA). The AEC and Commonwealth Edison Co. of Chicago were to manage the project. The U.S. government was to provide $200 million of the anticipated total cost of about $500 million for the plant, the rest to be paid by the major U.S. electric utility companies, which by mid-1972 had pledged $240 million for the project. Construction was expected to begin in early 1973.

Despite the small contribution of nuclear power plants to electricity output to date, the expected growth in nuclear generation was the most important change anticipated in the history of thermal power generation. Assuming that expectations are realized in the construction of reactors of the light water and high-temperature gas types and in the development of fast breeders, nuclear power will become the predominant method of electricity generation before the end of the century. A long-range project that could lead to the production of vast amounts of electric power is the effort to achieve controlled nuclear fusion. (*See* Feature Article: NUCLEAR FUSION, POWER SOURCE OF THE FUTURE?)

At the end of 1971, there were 23 nuclear units having a total generating capacity of 10,000 Mw in commercial operation in the U.S. An additional 54 units were being built, and orders had been placed for 52 more units. All of these units added up to a total generating capacity of 107,000 Mw, or about one-third of the present U.S. generating capacity. The total nuclear capacity and planned construction in all countries outside the U.S. was approximately equivalent to that of the U.S.

## Outlook

The outlook for fuels and power in 1973 is that it will continue to be of major concern and command high-priority attention from the leaders of the world's industrial nations. The matters of greatest interest will involve the reliability and security of fuel supplies, especially those involving petroleum sources; the resolution of environmental and eco-

Courtesy, Battelle-Northwest Photography

*Mock-up shows the Fast Flux Test Facility, an experimental liquid-metal fast-breeder reactor being built in Richland, Wash. Present projections indicate an exhaustion of fossil fuel reserves within several generations; thus, alternate energy sources must be developed. The fast-breeder reactor, when perfected, will be able to provide an essentially inexhaustible, economic supply of energy.*

logical conflicts arising from fuels and power production and usage; and the technological achievements or progress that may ensue from the greatly expanded research and development efforts that have been announced or initiated.

Among the most probable specific developments that can be expected to occur in 1973 are the following: (1) total world energy demand growth rates will rebound in response to improved general economic conditions in the major industrialized countries; (2) U.S. domestic petroleum and natural gas production rates will increase to new highs as demands resume higher growth rates; (3) governmental actions to conform to more strict environmental constraints on fuels and power generation will be implemented, especially in relation to air pollution standards, leasing of federal lands for mineral fuel development, and nuclear plant operations; (4) the environmental review prescribed by the new AEC regulations will delay by one to two years the start of operations of nuclear power plants under construction in the U.S.; (5) the U.S. will begin construction of its first fast breeder nuclear demonstration plant, its most important recent step in the development of nuclear reactor technology; and (6) the first substantial results of expanded research to develop coal conversion technologies should be reported.

—James A. West

# Honors

The following major scientific honors were awarded during the period from July 1, 1971, through June 30, 1972.

## Aeronautics and astronautics

**Robert J. Collier Trophy.** Established in 1912 by the Aero Club of America and presented annually by the National Aeronautic Association, the Collier Award is given in recognition of great achievement in aeronautics or astronautics in America. The 1971 trophy was awarded to the three Apollo 15 astronauts, David R. Scott and James B. Irwin of the team that explored the moon's surface, and Alfred M. Worden, pilot of the command module. A fourth co-recipient was Robert R. Gilruth, director of the Manned Spacecraft Center at Houston, Tex.

**Langley Medal.** In 1908 the Smithsonian Institution established the Langley Medal in memory of Samuel Pierpont Langley, an early pioneer in the physics of flight and secretary of the Smithsonian from 1887 until 1906, for the purpose of honoring a scientist who had conducted especially meritorious investigations in connection with the sciences of aeronautics and astronautics. Previously awarded only 11 times, the honor in 1971 went to Samuel C. Phillips of the United States Air Force, who was chosen for his outstanding contributions as director of the National Aeronautics and Space Administration's (NASA's) Apollo manned spacecraft program from 1964 through 1969.

## Astronomy

**Helen B. Warner Prize.** The American Astronomical Society each year presents the Helen B. Warner Prize in recognition of a significant contribution to astronomy made during the preceding five years by an astronomer under the age of 35. The 1971 award was presented to Kenneth Kellermann, a radio astronomer at the National Radio Astronomy Observatory, Charlottesville, Va. Kellermann was honored for his interpretation of the evolution of radio sources, and for his work on high-resolution radio interferometry and on radio emission from planets and stars.

**Henryk Arctowski Medal.** An award for studies of solar activity, the Henryk Arctowski Medal is bestowed approximately every three years by the National Academy of Sciences and consists of a gold medal and honorarium of $5,000. The 1972 recipient of the award was Francis S. Johnson, director of the Southwest Center for Advanced Studies at the University of Texas at Dallas. Johnson was cited for his pioneering work in the physics of high atmosphere and space. Especially noted was his description of how atmospheric transport processes regulate their response to changes in solar activity as well as to changes in season.

**James Craig Watson Medal.** The National Academy of Sciences presents approximately every three years the James Craig Watson award, consisting of a gold medal and a $2,000 honorarium for achievements in astronomical research. The 1972 medal was presented to André Deprit, a National Research Council postdoctoral resident research associate at the NASA Goddard Space Flight Center, Greenbelt, Md., for his adaptation of modern computing machinery to algebraic rather than arithmetical operations. His work solved in general terms a problem of lunar theory that concerns the motion of the moon around the earth as affected by the gravitational force of the sun.

**Kepler Gold Medal.** Co-sponsored by the geology and geography section of the American Association for the Advancement of Science and the Meteoritical Society, the Kepler Gold Medal awards for 1971 were presented in honor of the 400th anniversary of the birth of Johannes Kepler, the German mathematician and astronomer. Six scientists received the gold medal and certificate for their outstanding contributions to the understanding of the origin of the solar system and the planets: (1) Hannes Alfvén (see *Franklin Medal*, below), visiting professor at the University of California, San Diego, who was cited for past work in the development of magnetohydrodynamics, and recent studies of the role plasmas played in the evolution of the planets. (2) Gerard Kuiper, professor of astronomy and director of the Lunar and Planetary Laboratory at the University of Arizona, who was honored as one of the few 20th-century scientists credited with creating the new discipline of planetary science. Kuiper discovered the atmosphere on Titan, a satellite of Saturn (1944); measured the $CO_2$ atmosphere of Mars; and conducted research on the meteorology of Jupiter. (3) Harold Urey, a professor at the University of California, San Diego, who was cited in particular for his work on the general formulation of the problem of the origin of the planets.

Also receiving the award were: (4) Fred Whipple, director of the Smithsonian Astrophysical Observatory, for his leadership of the observatory and of the High Altitude Rocket Research Panel in 1946, and for his research into the role of meteors and comets in the evolution of the solar system; (5) E. J. Öpik, of the Armagh Observatory in Northern Ireland, in recognition of his study of the interaction of meteorites and the atmosphere, hypersonic collisions in the solar system, the role of dust grains in planetary and interstellar space, the applications of terrestrial physics to the elucidation of the nature of the surfaces and atmospheres of the planets, and the way in which changes in the sun's atmosphere can influence the climate; (6) Boris Levin of the U.S.S.R. Academy of Sciences, for his work as an astronomer, solar-system astrophysicist, and cosmochemist, with the ability to reach beyond the boundaries of the discipline in which he was trained to achieve an understanding of the origin of the planets. Levin's major research covered meteors and the preatmospheric nature of interplanetary particles, comets, and asteroids; the physical and orbital properties of meteorites; and the structure, thermal history, and surface evolution of the moon.

## Biology

**Institute of Life Prize.** Made possible by a donation from the French electricity system, the Institute of Life Prize consists of a $50,000 monetary award which is to be given to a scientist for outstanding contributions to the life sciences. The first prize, presented in 1972, was received by René Dubos, retired microbiologist of Rockefeller University in New York City, who was cited for his work concerning environmental problems. Dubos was quoted as having pledged to use the prize money for furthering his commitment to finding out "what human environment really means."

**Louisa Gross Horwitz Prize.** Established in 1967 by the College of Physicians and Surgeons of New York, the Louisa Gross Horwitz Prize is administered by Columbia University and is awarded annually to recognize outstanding basic research in biology or biochemistry. The 1971 prize of $25,000 was awarded to Hugh E. Huxley, a fellow of Churchill College, Cambridge University, who was cited for his studies that revealed the chemical mechanism of muscle contraction. Huxley's "sliding filament hypothesis" proposed that thick and thin filaments, sliding past one another, could form bridges that provide the basis for the weight-bearing capacity of contracting muscles.

**U.S. Steel Foundation Award in Molecular Biology.** Administered by the National Academy of Sciences, the U.S. Steel Foundation Award may be given annually for a recent discovery in molecular biology by a young scientist. The 1972 award with its honorarium of $5,000 was presented to Howard M. Temin, Wisconsin Alumni Research Foundation Professor of Cancer Research at the University of Wisconsin McArdle Laboratory for Cancer Research in Madison, Wis. Temin was cited for his work in the experimental demonstration that showed some cancer viruses transfer genetic information by a reverse process. This demonstration contradicted the older theory that the transference is only a one-way process, and could lead to a greater understanding of the mechanisms that cause cancer.

## Chemistry

**ACS Award in Pure Chemistry.** Sponsored by Alpha Chi Sigma Fraternity, the $2,000 American Chemical Society's Award in Pure Chemistry is given to recognize and encourage fundamental research in pure chemistry carried out by a U.S. or Canadian citizen. The 1972 award was received by Roy Gerald Gordon, professor of chemistry at Harvard University, who was cited for his many contributions to the field of theoretical chemistry. Gordon has advanced numerous new approaches in quantum mechanics, statistical mechanics, scattering theory, and spectroscopy in order to secure new information concerning the shapes and strengths of intermolecular forces.

**Beilby Medal.** Representing the Royal Institute of Chemistry, the Society of Chemical Industry, and the Institute of Medals, the Sir George Beilby Memorial Fund administrators made an award of the medal and £100 prize from the fund in 1971.

Courtesy, Harvard University News Office

*George Kistiakowsky*

That year's recipient was John Howard Purnell, dean of the Faculty of Science at the University College of Swansea, Wales, for his research in physical chemistry, pertaining particularly to his studies of hydrocarbon pyrolysis, and to the application and development of gas chromatography. Purnell and his associates extended certain findings of their prior academic studies of alkane systems to problems connected with industrial hydrocarbon cracking processes and olefin production. Purnell pioneered in the application of the gas chromatographic method to the measuring of physical properties of molecules, and in the study of solution and liquid surface thermodynamics.

**Garvan Medal.** The American Chemical Society presents each year the Garvan Medal, an award established in 1936 and consisting of a gold medal and a $2,000 honorarium, to a U.S. woman chemist selected for her distinguished service to chemistry. The 1972 winner of the award was Jean'ne M. Shreeve, professor of chemistry at the University of Idaho, for her contributions to the fundamental understanding of the behavior of inorganic fluorine compounds, and to the synthesis of important new fluorochemicals. Among Shreeve's earlier research accomplishments were the discovery of the compound perfluorourea, an oxidizer ingredient, and new methods of synthesizing several important compounds including chlorodifluoroamine and difluorodiazine, both used to synthesize rocket oxidizers.

**Nobel Prize for Chemistry.** The Swedish Royal Academy of Sciences chose Gerhard Herzberg of Ottawa, Ont., as the recipient of the 1971 Nobel Prize for Chemistry. Herzberg, a retired scientist at the National Research Council of Canada, was cited for his contributions to the knowledge of electronic structure and geometry of molecules, particularly fragments of molecules, known as free radicals. The physicist's laboratory was also cited as the foremost molecular spectroscopy center in the world. Herzberg's synthesizing of numerous techniques and theories, which led to the uncovering of structures and characteristics of diatomic and polytomic molecules and of classic free radicals (methyl [$CH_3$] and methylene [$CH_2$]), contributed greatly to the development of chemistry.

**Priestley Medal.** The American Chemical Society's Priestley Medal, established in 1922 and considered the highest honor in American chemistry, is bestowed in recognition of distinguished services to chemistry. The 1972 recipient was George Kistiakowsky, associated with Harvard University for more than 40 years. An authority on explosives, shock waves, thermodynamics, and rates of chemical reactions, he was cited for his more than three decades of service to the U.S. government. Kistiakowsky was head of the explosives division of the National Defense Research Committee (1940), chief of the explosives division of the Los Alamos Scientific Laboratory (1944), and a member of the Science Advisory Board of the Air Force until 1957. From 1959 through 1961 he was science adviser to U.S. Pres. Dwight D. Eisenhower and then served (1962–68) as first chairman of the National Academy of Sciences Committee on Science and Public Policy. From 1965–69 he was vice-president of the Academy.

## Earth sciences

**Alexander Agassiz Medal.** The National Academy of Sciences presents the Alexander Agassiz Medal, usually at three-year intervals, for original contributions to oceanography. The 1972 award of $1,000 and a gold medal was bestowed upon Seiya Uyeda, professor of geophysics at the University of Tokyo's Earthquake Research Institute, for his outstanding contributions to the tectonic and thermal history of the earth. Uyeda made a thorough investigation of heat flow from the earth's interior in Japan and the Pacific Ocean. His work shows how island arcs and the seas were formed through the development of the earth as a whole and supports the new theories of global tectonics.

**Arthur L. Day Prize.** Established by the National Academy of Sciences in the early 1970s, the Arthur L. Day Prize and Lectureship consists of a $10,000 prize and an invitation to deliver four to six lectures later to be published in a monograph or book. The first award, presented in 1972, was given to Hatten

UPI Compix

*Gerhard Herzberg*

Wide World

*Earl W. Sutherland, Jr.*

S. Yoder, Jr., a petrologist and first director of the Geophysical Laboratory of the Carnegie Institution of Washington, D.C., who was cited for developing an apparatus that enabled scientists to investigate a wide range of phenomena within the earth's crust. Yoder's investigations dealt with the properties of minerals and rocks, synthetic and natural, at high pressures and temperatures.

**Charles P. Daly Medal.** The American Geographical Society of New York intermittently presents the Charles P. Daly Medal for valuable or distinguished geographic services or labors. In 1971 the award was made to Gilbert Fowler White, professor of geography and director of the Institute of Behavioral Science at the University of Colorado. Research, administration, and public service all figured in White's career.

In research three fields were of interest to him: arid lands, water resource allocation, and natural hazards. White's public service covered membership on the National Resources Planning Board and work in the United States Bureau of the Budget. He was president of Haverford College, in Pennsylvania for ten years (1946–55) and served as chairman of the American Friends Service Committee from 1963 until 1969.

**Cottrell Award.** Established by the Research-Cottrell Inc., the $5,000 Cottrell Award is to be given annually for outstanding scientific or technical achievement in improving the environment or controlling pollution. First awarded in 1972, it was presented to Arie J. Haagen-Smit, emeritus professor of bio-organic chemistry of the California Institute of Technology, who was cited for his highly innovative studies on the formation of smog, and for his untiring efforts to shape the air pollution control policies of the nation. Haagen-Smit began his studies in the 1940s (long before pollution became a household word) when he undertook the examination of the chemical composition of the polluted atmosphere and its effects on plant life in the Los Angeles basin.

**International Meteorological Organization Prize.** The World Meteorological Organization's 16th prize was awarded in 1971 to Jule G. Charney, professor of meteorology at Massachusetts Institute of Technology. Charney was cited for his "outstanding contribution to dynamical meteorology, his furtherance of research especially in numerical methods of weather prediction, and his services to the cause of international cooperation in the science of the atmosphere."

**Massey Medal.** The Royal Canadian Geographical Society annually presents the Massey Medal, established in 1959, for recognition of outstanding personal achievement in the exploration, development, or description of the geography of Canada. The 1972 silver medal was presented to Isobel Moira Dunbar, geographer with the Canadian Defense Research Board, for her work on Arctic ice studies. Dunbar contributed important new knowledge on ice distribution and the interpretation of ice in photographs, including both satellite and infrared photography.

## Medical sciences

**Albert Lasker Medical Research Awards.** The Albert and Mary Lasker Foundation each year presents a number of prizes for medical research. The 1971 Albert Lasker Clinical Research Award with its $10,000 honorarium was given to Edward D. Freis, senior medical investigator at the Veterans Administration Hospital in Washington, D.C. Fries was honored for a five-year study demonstrating that moderately high blood pressure, though producing no symptoms, could be still dangerous and that proper treatment could reduce greatly the high risk of stroke and heart failure.

Three recipients of the Albert Lasker Basic Medical Research Award shared the $10,000 prize: Seymour Benzer of California Institute of Technology, Sydney Brenner of the University of Cambridge in England, and Charles Yanofsky of Stanford (Calif.) University. The three geneticists pioneered in demonstrating that inherited characteristics in all living things are coded molecule by molecule along strands of deoxyribonucleic acid (DNA) and that messages are "read" to produce specific protein molecules.

**NAMH Research Achievement Award.** Established by the National Association of Mental Health, the Research Achievement Award was to be presented annually at the beginning of May, National Mental Health Month. Chosen in 1972 for the first $10,000 award was Seymour Kety, a psychiatrist at Harvard University. Kety was cited for his contributions to the development of pharmaceutical approaches to the control of various types of mental illnesses, especially schizophrenia.

**Nobel Prize for Physiology or Medicine.** The Swedish Royal Caroline Medico-Chirurgical Institute awarded the 1971 Nobel Prize for Physiology or Medicine, amounting to $90,000, to Earl W. Sutherland, Jr., professor of physiology at Vanderbilt University, Nashville, Tenn. He was honored for his discoveries concerning the mechanisms of the action of hormones, chemical substances secreted by the glands and which influence the activities of various cells, tissues, and organs.

Sutherland's research led to the discovery of cyclic adenosine monophosphate, or cyclic AMP. This intermediary chemical substance proved to play an important role in the mechanisms by which hormones exert their control over various metabolic activities throughout the human body. A later discovery by Sutherland showed that the substances also occurred in bacteria, which opened the door on a whole new biological perspective.

**Procter Medal.** The Philadelphia Drug Exchange sponsors the Procter Medal, which is given in recognition of outstanding achievement that has benefited the health of the public and advanced the progress of the health professions in the pharmaceutical industry. The 1972 medal was bestowed upon Maurice R. Hilleman, director of virus and cell research for the Merck Institute for Therapeutic Research. Hilleman was honored for his contributions in the fields of etiological discovery, classification, laboratory diagnostics, epidemiology, public health, and clinical evaluation.

*Dennis Gabor*

Courtesy, CBS Laboratories

Jay Scarpetti, courtesy, Polaroid Corporation

*Edwin H. Land*

## Physics

**Albert Einstein Award.** Administered by the Lewis and Rosa Strauss Memorial Fund, the Albert Einstein Award is presented approximately every two years for achievement in the mathematical and physical sciences. The 1972 award of a gold medal and $5,000 honorarium was given to Eugene Paul Wigner of Princeton University for his outstanding contribution to the physical sciences.

Wigner, a co-winner of the 1963 Nobel Prize for Physics, was awarded also in 1972 the George Gamow Memorial Lectureship Award presented by the University of Colorado, in recognition for his contributions to the development of physics and to the elucidation of its fundamental concepts.

**Enrico Fermi Award.** The U.S. Atomic Energy Commission presents annually the $25,000 Enrico Fermi Award in recognition of outstanding scientific or technical achievement related to the development, use, or control of nuclear energy, and to stimulate creative work in the development and application of nuclear science. The 1972 award was made to two of the world's leading medical experts on radiation: Stafford L. Warren, emeritus vice-chancellor of health sciences at the University of California at Los Angeles, and Shields Warren, emeritus professor of pathology at Harvard Medical School. The Warrens (unrelated) were cited for having helped make possible the early development of atomic energy so as to assure the protection of man and the environment.

**Franklin Medal.** The Franklin Institute of Philadelphia presents its highest award, the Franklin Medal, annually for recognition of those workers in physical science or technology whose efforts have done most to advance a knowledge of physical science or its application. The 1971 award was presented to Hannes Alfvén of the department of applied physics and information at the University of California, San Diego. Alfvén was cited for his pioneer work in establishing the field of magnetohydrodynamics and his many revolutionary contributions in that field to plasma physics, space physics, and astrophysics.

**Ernest O. Lawrence Memorial Award.** The U.S. Atomic Energy Commission in 1959 established the annual Ernest O. Lawrence Memorial Award to recognize five young scientists who have made recent, especially meritorious contributions to the development, use, or control of nuclear energy in areas of all the sciences relating to nuclear energy, including medicine and engineering. The 1972 awards, consisting of a gold medal and $5,000, were made to: (1) Charles C. Cremer, group leader of the theoretical design division of the Los Alamos (N.M.) Scientific Laboratory; (2) Sidney D. Drell, deputy director and executive head of theoretical physics at the Stanford (Calif.) University Linear Accelerator Center; (3) Paul F. Zweifel, professor of physics and nuclear engineering at Virginia Polytechnic Institute; (4) Marvin Goldman, radiobiologist at the University of California at Davis; and (5) David A. Shirley, chairman of the chemistry department at the University of California at Berkeley.

Cremer was honored for his contributions to the development of weapons design codes; Drell for his theoretical investigations of the range of validity of quantum electrodynamics; and Zweifel for his contributions to the slowing down and thermalizing of neutrons, important to the design and development of water-moderated reactors. Goldman was recognized for his contributions as a radiobiologist, and Shirley was cited for his contributions to the understanding of the chemical nature of matter by his studies of the interactions of the nucleus with its electronic environment.

**Nobel Prize for Physics.** The Royal Swedish Academy of Sciences designated Dennis Gabor, Hungarian-born British scientist, as recipient of the 1971 Nobel Prize for Physics. He received the more than $87,000 award for founding the applied science of holography, a lensless, three-dimensional imagery system. Gabor, emeritus professor of applied electron physics at Imperial College of Science and Technology in London, and a staff scientist at the CBS Laboratories in Stamford, Conn., was an inventor with more than 100 patents, many of them relating to his work in holography.

Gabor began working on his optical system in the late 1940s as he searched for a way to improve the resolution of electron microscopes so that scientists could "see" a single atom. He named his new type of photography hologram for the Greek words *holos* (whole) and *gramma* (letter or message). With the development of laser light, holography became increasingly useful to researchers.

## Science Journalism

**Howard W. Blakeslee Awards.** The American Heart Association presents its annual Blakeslee Awards, named in honor of an Associated Press science editor and founder of the National Association of Science Writers, for outstanding reporting on diseases of the heart and blood vessels. The 1971 awards, consisting of a citation and $500, were made to three reporters and two television stations. The reporters were Jane E. Brody of the *New York Times;* Harold J. Eager of the *Sunday News,* Lancaster, Pa.; and Rachel MacKenzie of *New Yorker* magazine. The TV stations were WHDH-TV of Boston, and WTAR-TV of Norfolk, Va.

**James T. Grady Award.** The American Chemical Society presents annually the James T. Grady Award for Interpreting Chemistry for the Public, with its $2,000 cash prize and a gold medal, to recognize, encourage, and stimulate outstanding reporting directly to the public, which materially increases the public's knowledge and understanding of chemistry, chemical engineering, and related fields. The 1972 award was given to Dan Q. Posin, professor of physical science and chairman of the department of interdisciplinary sciences at San Francisco State College. Posin authored a number of popular science books, including a biography, *Mendeleyev: The Story of a Great Scientist* (1948), and also wrote for newspapers. A noted television and radio personality, he has produced two TV series: "What's New" and "Dr. Posin's Universe."

**Kalinga Prize.** The United Nations Educational, Scientific, and Cultural Organization (UNESCO) annually presents its Kalinga Prize, named for an ancient Indian empire of 2,000 years ago renowned for its policies of peace, for particular merit in the popularization of science in writing. In 1971 UNESCO made the award with its $2,400 stipend to Margaret Mead, the anthropologist, for her writing covering her life's work with the primitive cultures of the South Pacific.

## Miscellaneous

**Founders Medal.** The National Academy of Engineering annually presents the Founders Medal, established in 1965, to honor outstanding contributions by an engineer both to his profession and to society. The 1972 recipient of the medal was Edwin H. Land, president and director of Polaroid Corp., who was cited for his development of cameras, films, and processes for instant photography, and for his invention of synthetic polarizers for light. During the 1940s Land worked on the idea of a hand-held camera that would use apparently dry film and would give a picture instantly. In 1947 he demonstrated the Polaroid Land camera, which became available a year later. One-step color photography followed some ten years afterward.

**E. Harris Harbison Award for Gifted Teaching.** The Danforth Foundation annually presents the E. Harris Harbison Award for Gifted Teaching, estab-

*Howard M. Temin*

Courtesy, University of Wisconsin

lished in the early 1960s, to an outstanding teacher who has taught for at least five years in an accredited U.S. junior college, college, or university. The 1971 recipient of the $10,000 award, was John G. King, professor of physics at Massachusetts Institute of Technology. King, a leader in educational innovation and reform, was cited for his introduction of numerous new methods of teaching into his undergraduate classes, including approaches to concentrated study and "corridor labs."

**Hoover Medal.** The American Society of Mechanical Engineers, with several other professional engineering groups, presents the Hoover Medal, named for its first recipient, former U.S. Pres. Herbert Hoover, in recognition of outstanding engineers who have performed distinguished public service. The 1972 medal was given to Luis A. Ferre, an engineer and governor of Puerto Rico, who was cited for exemplifying an engineer-humanist's achievements in enriching the life of a people.

# Information science and technology

In this era of rapidly expanding knowledge people must be selective in what they hear and read, and must rely more and more on systems that will store this information until it is needed. The problem is particularly acute for scientists and other specialists who have to locate specific information to carry on their work. Information science and technology is concerned with devising means for providing more efficient access to documents and improving the dissemination of information. To accomplish these tasks it is necessary to gather together large quantities of raw information and to process them into a form that is more easily transferable and usable. When these processing procedures are integrated into an efficiently functioning unit, the result is called an information system.

Since the information problem is worldwide, devising such systems was a matter of concern to most nations, as well as to many private organizations. Many countries and organizations provided support during the year for the development of information systems.

**Coordinating information services.** The Congress of the United States in July 1970 created the National Commission on Libraries and Information Science (NCLIS) as an independent agency within the executive branch of the government. In establishing NCLIS, Congress responded to the need for improving information services and for alleviating some of the problems faced by libraries throughout the country. During 1970–71 NCLIS spent most of its energies getting organized. In 1972, however, it undertook its first major study effort, which was to take the form of an overview of significant research affecting the library and information worlds. This study was designed to provide a proper base from which to plan future operations. In addition, NCLIS was expected to help coordinate the information activities of various federal and state agencies.

In the United Kingdom, the Office of Scientific and Technical Information (OSTI) supported various research studies and assessments of library and information needs. These included studies of the uses made by chemists of scientific news periodicals, of the design of library buildings and facilities, and of ways to promote the improvement of public library services.

In Canada, an Office of Library Coordination was organized by the Council of Ontario Universities to develop a wide range of cooperative activities among the provincially assisted university libraries in Ontario. This office was the result of several years of planning. Its aim was to provide a means for achieving close cooperation among the participating public libraries in order to provide better service to students, faculty, and staff.

Many other countries established similar national and regional centers for coordinating the collection and dissemination of scientific and technical information. Clearly, information was being recognized as a national resource, and efforts were being made to control and use it as efficiently as possible.

**A world science information system.** The exchange of information does not stop at national borders, and scientists of different countries have long cooperated in sharing knowledge and seeking the solutions to significant social problems. All of this has been done on a more or less informal basis, and a need exists for a more organized and equitable procedure for coordinating scientific studies and information.

In 1970 studies of the possibility of establishing a world science information system (code-named UNISIST) were begun. In the ensuing two years much progress was made. In October 1971, a UNISIST conference was convened in Paris under the sponsorship of the International Council of Scientific Unions and the United Nations Educational, Scientific and Cultural Organization (UNESCO). The conference was attended by 240 delegates representing 83 member and affiliated nations of UNESCO plus observers sent by many intergovernmental and nongovernmental organizations. On completing its discussions, the conference adopted a broad resolution supporting the creation of UNISIST and laid the foundation for a voluntary world scientific and technical information system to help scientists and engineers gain access to the more than two million articles published annually in approximately 70,000 specialized journals. The plan called for the development of interconnected automated information systems. Thus, a center in Africa would be able to process on its computer information stored on tapes that had been produced in Europe or North America and immediately tell a researcher what material had been published in his field.

Many practical problems remained to be solved, but funds were allocated and plans for implementation were being prepared. Final approval still had to be given by the UNESCO General Conference, scheduled to meet late in 1972.

**Library automation.** One of the great contributions to information science and the dissemination of information was the MARC project (machine-readable cataloging) of the U.S. Library of Congress. This project recorded on computer tape bibliographic information for most current English-language books. Although the value of this project was recognized, not all libraries had access to the computer services necessary to process and retrieve MARC records. In 1972, however, the MARC-Oklahoma Data Base Storage and Retrieval Project went into operation. On request, it searched the complete MARC data base and provided individual libraries with bibliographic and cataloging data on only those books that they had acquired. Thus, smaller libraries were able to take advantage of the savings in time and money that computer processing of library records could provide.

MARC tapes were also being supplied to other countries. The Ontario College Bibliocentre in Canada was evaluating the benefits to be gained by using the tapes to provide a centralized and computerized selection, acquisition, processing, and cataloging service for 20 community libraries.

With the MARC program successfully underway, the Library of Congress began a new project called "Cataloging in Publication." Actually, the basic idea is not new; an experimental project called "Cataloging-in-Source" was undertaken in 1958–60. The results of this experiment demonstrated that although it was possible for the Library of Congress to catalog some books from page proofs before publication, the costs and the disruption of publishing schedules outweighed the benefits. But in 1972, as a result of new techniques, the project was revived.

The goal of "Cataloging in Publication" was to incorporate cataloging data, of the kind now found on library catalog cards, in the printed book at the time of publication. Catalogers at the Library of Congress would receive galley proofs of books from the publishers. They would then prepare cataloging information, including Library of Congress call numbers and Dewey classification numbers, and the publisher would incorporate the complete cataloging data in the book. This new service was expected not only to save processing costs for individual libraries but also to provide faster service for library users and even allow small private libraries and book collectors to arrange their holdings more efficiently.

**Information dissemination systems.** Since scientists generally are most in need of timely information, many of the professional scientific societies have responded by creating specialized information systems. Such systems exist in physics, chemistry, engineering, psychology, education, and medicine. During the year these systems expanded their services and new ones were being formed. The American Geological Institute approved the concept for a geosciences information system that was expected to lead to a more structured pattern of services for the Institute's 17 member societies.

The National Library of Medicine in 1972 introduced a new service called MEDLINE. Previously, hospitals, medical schools, medical libraries, and related research institutes had access to medical literature through an automated system called MEDLARS (Medical Literature Analysis and Retrieval System). In using MEDLARS, a physician or researcher would generally write or telephone a request for a search on a given topic to the National Library of Medicine in Washington. A week or longer would lapse before he would receive the results. Now, through MEDLINE, each user installation would have its own computer terminal connected to a centralized data processing system and would be able to search a large body of the more frequently used medical literature and receive listings of relevant citations in minutes.

The social sciences were also striving to improve access to the literature in their fields. One important study, completed late in 1971, was the compilation of an international registry of *Data Bases in the Social and Behavioral Sciences* under the auspices of the City University of New York Graduate Division. This effort was directed toward helping individual scientists locate files of relevant information outside their immediate specialty.

Frequently, after information dissemination procedures have been proven to be of value in the sciences, they are applied to other more general fields. For example, the Federal Bureau of Investigation (FBI) began operating a computer-based National Crime Information Center (NCIC) in 1967. By 1972 the Center stored more than three million records containing information about wanted persons, stolen vehicles, license plates, and other property. Rigid safeguards were employed to insure the security and confidentiality of these files. The system was being expanded into a nationwide network and was expected to reduce the number of duplicate records now stored in local, state, and other federal agencies and thus save time, money, and manpower. But more importantly, it was hoped that the NCIC would result in more effective law enforcement.

The U.S. Postal Service was also undertaking an extensive automation project, known as the Letter Mail Code Sorting System (LMCSS). The old manual letter-sorting system required postal personnel to read each first-class envelope up to seven times before delivering it to the recipient. Using LMCSS, one would have to scan a letter manually only once, at which time the name and address would be recorded on the envelope in a code that a computer could read. After being coded, letters would be conveyed directly to a letter-sorting machine which would automatically sort them into a storage bin that represents a specified destination. When a letter reaches the post office at its destination, the code would again be used to sort the letters automatically by street and address into the sequence followed by the carrier as he makes his rounds. When fully operational, this system was expected to enable the post office to deliver more than 95% of all first-class mail within two days.

—Harold Borko

*See also* Year in Review: COMPUTERS.

# Marine sciences

Increasing evidence of pollution, even in the deep oceans, was a source of continuing concern to oceanographers during the year. Government support of marine studies continued, and marine scientists obtained valuable data from these programs and also as a by-product of space exploration.

**Pollution and ecology.** The year was an active one on the ecology and pollution fronts. Delaware acted to protect its coastal wetlands. According to Gov. Russell W. Peterson, the people of Delaware "made a choice, and that was to use our valuable bay and coastal areas for recreation and tourism instead of for another purpose for which they are admirably suited, industrial growth." Connecticut's program to save its wetlands was also making progress. Pres. Richard Nixon proposed a wetlands protection measure as part of a general land-use planning proposition that would give priority to coastal areas. A thorough mapping of coastal wetlands was also planned.

Following six years of study, Canada and the United States signed a Great Lakes Water Quality Agreement to highlight an International Field Year for the Great Lakes. The agreement called for a dramatic reduction in the pollution of Lake Erie, Lake Ontario, and the international portion of the St. Lawrence Seaway. With the hope that their initiative would serve as a precedent for worldwide action at the United Nations Conference on the Human Environment in Stockholm in June 1972, 12 European nations signed a voluntary pact on Feb. 14, 1972, at Oslo, Nor., to bar dumping of harmful substances in a wide area of the Atlantic Ocean. The pact would take effect as soon as it was ratified by seven of the signatories.

Scientists were gathering the basic data on many aspects of the dispersion, removal, and recycling of pollutants by rivers, the atmosphere, and the oceans. At the annual meeting of the American Geophysical Union, V. J. Linnenbom and his colleagues reported that the Northern Hemisphere oceans contribute 60,000,000,000 kg of carbon monoxide gas to the atmosphere per year. If the Southern Hemisphere oceans contribute a comparable amount, the total is 140,000,000,000 kg per year, or about 60% of that produced by mankind's activities. Carbon monoxide is removed from the atmosphere in various ways, but these figures are important for a full understanding of the air pollution problem.

At the same meeting, R. A. Carr and J. D. Gassaway described the distribution of mercury in the oceans. Typical open ocean values range from 10 to 60 nanograms per liter (1 nanogram = 0.000,000,001 grams). Sea ice is a reservoir for mercury in the geochemical cycle. Other scientists reported on hydrocarbons in the ocean and their effects on estuaries, on mercury in the Columbia River, and on fluorides in Chesapeake Bay.

E. J. Carpenter and K. L. Smith, Jr., described finding about 3,500 plastic particles per square kilometer of surface in the Sargasso Sea. These particles could be the source of the polychlorinated biphenyls found recently in marine organisms. For example, DDT and polychlorinated biphenyls had been discovered in specimens of plant and animal life throughout the Atlantic Ocean by Woods Hole (Mass.) Oceanographic Institution ships. The full effects on the ocean environment of relatively indestructable plastic pellets, and the polychlorinated biphenyls that leach out of them, had yet to be assessed, however.

Oil, oil spills, cleanup techniques, and the effects of oil on marine life were the subjects of hundreds of studies and papers. At the Offshore Technology Conference in May 1972, for example, 28 papers were devoted to this problem. Ways to prevent spills, to contain spills that have occurred, to clean them up, and to predict their movement were described.

Most of the oil in the marine environment comes from the thousands of ships that clean their fuel tanks by steam and hot water and pump the resulting mixture overboard. The NATO countries agreed to end this practice in the near future but, in general, the outlook for stopping this source of pollution was not bright. The laws that provide penalties against ships discharging their wastes in this way are almost impossible to enforce. The breakup of a ship and the failure of an oil-drilling rig are dramatic examples of oil pollution, but their effects are really small compared with the cumulative results of such everyday practices.

Another practice with similar cumulative effects was identified as the source of 50% of the oil in Long Island Sound. When oil in an automobile is changed, the used oil, which was once collected and reclaimed, is now more often than not poured down a sewer. Oil disposed of in this way in any community near the Sound eventually ends up polluting its waters.

A more encouraging report came from Great Britain. The January 1972 *Marine Pollution Bulletin* reported changes in bird populations on the Thames River since it was cleaned up. Previously, the river had been virtually devoid of life except for hardy scavenging gulls. Now fish are found up-

*Aragonite, an unusually fine, pure form of limestone, is pumped from an ocean-bed site near the Bahamas and moved by conveyor (above) to a ship for delivery to customers (below). The high purity and uniformity of aragonite make it valuable for cement and glass making.*

Ray Fisher, BUSINESS WEEK

Courtesy, Holger Knudsen, Marine Biological Laboratory, Denmark

*Sea creatures that normally move slowly exhibit lively responses when under attack. The cockle* Cardium echinatum, *when touched by the sea star* Asterias rubens, *opens its shell, thrusts out its tubular foot, and, pushing violently against the substrate on which it is resting, jumps as much as 10 cm. It continues this maneuver until the predator is dislodged.*

stream as far as the environs of London and large numbers of migratory birds have returned. (*See* Year in Review: ENVIRONMENTAL SCIENCES.)

**Expanding data and improved theories.** Better data require improved theories to explain them. Improved theories require better data to check them out. The interaction of theory and data was underlined during the year in studies of the constancy of ionic proportions, continental drift, and the tides.

The theory that the major positively and negatively charged ions are in constant proportion in seawater goes back to a study of water samples collected by the "Challenger" oceanographic expedition of 1872–76 and analyzed in 1884 by C. D. Dittmar. As quantitative chemical analysis techniques became more exact, this concept was tested with ever increasing accuracy, sometimes with results that suggested it was not exactly true. R. C. Mangelsdorf, Jr., and T. R. S. Wilson reported on a study of 345 water samples from the Pacific Ocean along 150° W from 63° S to 54° N. Samples were compared with standards established from samples taken from the Sargasso Sea. No significant variations were found in the ratios of $K^+$ and $Mg^{++}$ to $Na^+$, but $Ca^{++}$ to $Na^+$ varied markedly and systematically, with the bottom water having more $Ca^{++}$ relative to surface waters.

With the acceptance of the theory of continental drift, the questions of "which way" and "when" began to arise. After studying data obtained from the Joint Oceanographic Institutions for Deep Earth Sampling (JOIDES) project and other data, W. C. Pittman III tentatively modeled the evolution of the North and South Atlantic. According to this model, Africa and North America began to drift apart 180 million to 200 million years ago. Africa and South America began to drift apart 139 million years ago, while Eurasia and North America started to separate only 80 million years ago.

In 1921 A. T. Doodson published an important analysis of the harmonics of the tides. It took 50 years before better astronomical data became available, but during the year D. E. Cartwright and R. H. Taylor of the National Institute of Oceanography of Great Britain published new computations of the tide-generating potential. As a sign of the times, the basic tables were given in the form of a computer printout. These improved tables were needed for many scientific studies, such as that of Walter Munk and his co-workers, who were working on the problem of the global tides by measuring them in the transition zone between coastal and deep-sea waters.

**Status of oceanographic programs.** In the United States, support for oceanographic research appeared to be holding its own or increasing slightly. The Navy's budget for such research totaled $204 million, compared with $200 million a year earlier, while the National Oceanic and Atmospheric Administration (NOAA) program was increased by almost 20%. Through the National Science Foundation, the United States was helping to support the International Decade of Ocean Exploration (IDOE). The NSF was also suporting a program called Research Applied to National Needs (RANN). Many programs in marine science met the RANN criteria, but it was understood that only $12 million was available for environmental studies and that $300 million to $400 million in proposals had already been submitted.

The IDOE program, recently described by E. M. Davin and F. D. Jennings, involved more than 30 countries in multidisciplinary ocean research programs in the areas of environmental quality, environmental forecasting, and seabed assessment. Other currently active programs included the Mid Ocean Dynamic Experiment (MODE) and a study of how the tectonic plate originating at the mid-Pacific rise moves and is consumed in the Peruvian trench, which encompassed the fields of marine geology, continental drift, and plate tectonics.

The shortage of students entering engineering, including ocean engineering, continued, probably because of the current high unemployment rate in the field. Mrs. Betty Vetter of the Scientific Manpower Commission predicted that there would be a shortage of engineers by 1980. It appeared that overreaction to what was almost certainly a temporary situation might preclude a promising career for many young people.

Sidney Harris, BULLETIN OF THE ATOMIC SCIENTISTS

*"Look at the bright side—employment for thousands trying to clean it up."*

**Space technology and marine science.** The U.S. deep space probe Pioneer 10, launched on March 2, 1972, carried a message that, in a million years or so, might reach intelligent life on some other planet outside our solar system. The message was designed to tell its recipients the location of our star, the sun, to within one star in one thousand, the location of earth in our solar system, what men and women look like and how big they are. It will not, however, tell them that life on the earth is possible because of its oceans, continents, and atmosphere or that life started in the oceans. Carl Sagan, Linda Sagan, and Frank Drake, who devised the message, conceded that "it can be improved upon" and expressed the hope that future spacecraft sent beyond our solar system would carry more detailed messages.

The National Aeronautics and Space Administration (NASA) was developing an Earth Resources Program with numerous marine science applications. To test instruments for possible use on either unmanned or manned spacecraft, NASA flew prototype instruments on aircraft and installed them on platforms. Results of the experiments were described in the first through the fourth annual Earth Resources Program Reviews, published by the Manned Spacecraft Center in Houston, Tex.

These studies reached fruition in the instrumentation of the unmanned Earth Resources Technology Satellite (ERTS), launched in mid-1972. ERTS was planned as the first of many systems designed to study the earth from space with agricultural, geologic, land management, marine resource, oceanographic, and meteorological implications. For marine science, it would provide valuable photographic and other data on coastlines and coastal wetlands that were expected to become available during late 1972 and 1973.

The projected manned orbiting space platform Skylab, along with many other experiments, would carry an Earth Resources Experiment Package consisting of S190, a multispherical photographic facility; S191, an infrared spectrometer; S192, a multispectral scanner; S193, a microwave system; and S194, a radiometer. Skylab was scheduled for launch in May 1973, and would be manned during three periods extending through December 1973.

Of particular interest was S193, a combination scanning pencil beam passive microwave-active radar and radar altimeter. The scanning pencil beam passive microwave-active radar would measure the natural and backscattered radiation from the sea surface at 13.3 GHz (1 gigaHertz = 1,000,000,000 Hertz) over a wide swath to each side of the satellite. Using measurements of this type made from aircraft, it has been demonstrated that winds over the ocean surface can be inferred from these data, while L. M. Druyan showed that the surface atmospheric pressure can be determined from the winds and a few scattered ship reports. These data combined with the infrared atmospheric sounding techniques developed for the Nimbus meteorological satellite should provide greatly improved weather and wave forecasts, especially for the Southern Hemisphere.

R. K. Moore and W. Pierson used the results of studies of the data obtained by the aircraft program and available meteorological and oceanographic theories to predict the usefulness of the data to be obtained by the radar-radiometer mode of S193. They found that the data should provide ideal information on the synoptic-scale (as opposed to local) winds in the planetary boundary layer of the atmosphere.

The altimeter mode of S193 can measure the distance between the spacecraft and the surface of the ocean to within a few meters. Though it may seem strange to laymen, the oceanographers do not know where the surface of the ocean is relative to the center of the earth. The gravity field of the earth is warped because of inhomogeneities of mass in the upper mantle which cause the surface of the ocean to vary by tens of meters from the oblate spheroid that represents an approximation to the shape of the earth. The oceans are not level, and even a level stretch of ocean has bumps and hollows in it.

—Willard Pierson

*See also* Year in Review: ASTRONAUTICS AND SPACE EXPLORATION, *Earth Satellites.*

# Mathematics

Geometry is a branch of mathematics for which a brief definition seems impossible. It is varied and broad. Elements of what most mathematicians would consider to be geometry appear in most areas of modern mathematics, and geometry in its own right is a subject of enormous scope. The most inclusive definition, given by Felix Klein in 1872 and known as the *Erlanger Programm,* states that geometry (roughly) is the study of symmetries of certain structures on sets. By making the structures more and more specialized and detailed, various geometries and subgeometries arise, to be studied in turn. On the other hand, one can specify the symmetries in an abstract manner (they will form a group), proceeding then to study those structures which admit the given group as a group of symmetries. The symmetries are known in general as isomorphism of the structure.

The *Erlanger Programm* is highly successful, even though it does not include all the topics that might be called geometry. In each specialized geometry of the *Erlanger Programm,* there are many questions to consider that are appropriate only to that geometry, but it is generally agreed that a question is "geometrically meaningful" only if it is formulated in a manner harmonious with the symmetries of the geometry being examined.

For example, Euclidean (plane) geometry studies a set called the plane whose elements are called points. The Euclidean structure allows measurement of distances and angles. The symmetries of the Euclidean structure are called congruences. Although several severe crises arose in the 19th century affecting the foundations of Euclidean geometry, they were resolved satisfactorily by 1930 through definitive treatments of the foundations by David Hilbert and his followers. There is little, if any, research of importance being done now in Euclidean geometry, and current trends in education de-emphasize Euclidean geometry in favor of set-theoretic and topological concepts.

Current research in geometry covers too broad an area to be surveyed in a short article. For this reason, this discussion will not deal with algebraic geometry, finite geometries, and coordinatization, all of which are now considered parts of abstract algebra. Of the remaining areas of geometry, the one closest to classical problems is differential geometry, which is the study of geometry by methods of the infinitesimal calculus. Thus, the subject is no older than differential calculus itself, about 300 years. But differential geometry is rooted in the study of the bending and twisting of curves and surfaces, and interest in such problems antedates the discovery of the calculus.

**Differentiable manifolds.** At present, the main effort in differential geometry is directed toward the study of differentiable manifolds and various structures they can carry. Differentiable manifolds are topological spaces; that is, they are sets of points with certain distinguished subsets called open sets (or neighborhoods), two points in an open set being considered in some sense to be "nearby." Furthermore, differentiable manifolds are topological manifolds. Each point of a differentiable manifold is contained in an open set that is required to be homeomorphic (the name for isomorphisms in the topological setting) to an open set in a Euclidean space, and two distinct points can be enclosed in disjoint open sets (Hausdorff property). The special open sets so singled out on the manifold are called coordinate domains, and

Novosti from Sovfoto

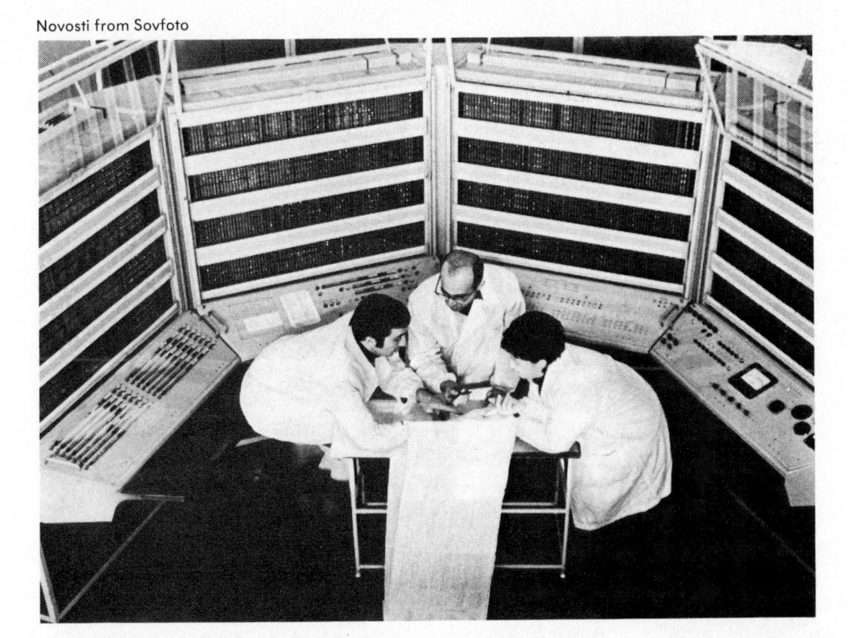

*Scientists at the Applied Mathematics Institute of Tbilisi State University in the U.S.S.R. make calculations on an advanced computer. Soviet-bloc countries are buying computers from U.S. and West European manufacturers, but their goal is self-sufficiency. By 1975 the U.S.S.R. hopes to be producing 12,000 to 15,000 third-generation computers annually. These will equip a network of 4,000 computer centers set up to monitor a predicted national economic growth rate of 37 to 40%.*

the homeomorphic correspondences to Euclidean open sets are called local coordinates.

A topological manifold is covered with coordinate domains in a way resembling an often-repaired automobile inner tube covered with patches. When two of these patches overlap, the open set were they intersect is described by two sets of local coordinates. The topological manifold is a differentiable manifold if the changes of local coordinates from one system to another always have continuous partial derivatives of sufficiently high order. Thus, the manifold will be composed of a certain number of connected pieces, and the dimension of the Euclidean space furnishing the local coordinates will be the same everywhere on a given piece; it is called the dimension of (that piece of) the manifold.

A differentiable manifold is therefore a generalization of the Euclidean space of coordinates. Viewed as a whole, a manifold is a contorted and twisted object, but in the small, thanks to the coordinate patches, it looks like a piece of Euclidean space. Thus, it has an infinitesimal linear structure. The manifold has tangent vectors (corresponding to vectors in a Euclidean space), but they are fixed at a base point and cannot be shifted around without an additional structure, called an affine connection. All tangent vectors with a given base point form a vector space, called the tangent space of the manifold at the point. All the tangent spaces fit together into a new differentiable manifold called the tangent manifold of the original; this has twice the dimension of the original.

If one picks out a tangent vector at each point of the manifold in such a way that the chosen vector varies smoothly as the base point varies, a vector field is obtained. If one supposes the tangent vectors to be hairs, then making the vector fields smooth corresponds to combing the hair down so that there are no cowlicks, whirls, nor, generally, any singularities. On a torus, a differentiable manifold that can be visualized as the surface of an inner tube, many smooth vector fields exist, while on a two-dimensional sphere no smooth vector field can be found.

The first explanation of this came from L. E. J. Brouwer, who used homology theory (a branch of algebraic topology) to examine vector fields on spheres. In 1927, H. Hopf gave a more inclusive explanation. He supposed that the vector field was smooth except at a finite set of points (singularities) and assigned to each of them an integer—the index—which measured its failure to be smoothable. Hopf showed that the sum of all the indexes at all the singularities was a number depending only on the topological nature of the manifold and not upon the particular vector field considered. He calculated this sum, finding its value to be a topological invariant, the Euler characteristic of the manifold. The Euler characteristic of a two-dimensional sphere is 2, and so there must be at least two singularities of index 1 or a singularity of higher index. If there are no singularities, the Euler characteristic of the manifold must be 0; conversely, the manifold has a smooth tangent vector field with no singularities if the characteristic is 0 (for example, the torus).

Recently, M. F. Atiyah and J. L. Dupont generalized Hopf's results. They considered a finite number, r, of smooth tangent vector fields on a differentiable manifold of dimension greater than r. A point is singular if there the vector fields fail to span an r-dimensional vector subspace of the tangent space. Atiyah and Dupont supposed the singularities to occur at a finite number of isolated points and assigned to each an index, which was no longer an integer but an element of a homotopy group. Summing the indexes over all the singularities, they obtain an element of a homotopy group and the vanishing of this element is a necessary and sufficient condition for r vector fields to exist with no singularities.

Besides homotopy theory, other tools from algebraic topology are used, specifically K-theory. This invention of A. Grothendieck's plays an exceedingly important role in current research in topology and is beginning to be used in various parts of differential geometry.

If instead of r tangent vector fields one considers the r-dimensional plane they are to span in each tangent space, the notion of a distribution is obtained. Specifically, a distribution of r-planes on a differentiable manifold selects an r-dimensional subspace from each of the tangent spaces and does so in a smooth fashion from point to point. A distribution of r-planes is called a foliation if the manifold can be filled up with disjoint r-dimensional submanifolds, each of which is tangent to the selected r-plane at each of its points. As a consequence of Hopf's theorem, the two-dimensional sphere can have no foliation by 1-planes. The three-dimensional sphere has a foliation by 2-planes, discovered by G. Reeb. Recently, A. Durfee and I. Tamura discovered a way to foliate any odd-dimensional sphere by planes of one dimension lower (codimension-one). H. Blain Lawson, Jr., and Durfee showed how to obtain codimension-one foliations on a large set of manifolds that are, from the viewpoint of algebraic topology, not too complicated. They obtain obstructions to finding such foliations involving integral 2-torsion of the homology.

**Status of the discipline.** Differential geometry is currently vigorous and growing. There is a wide variety of current research that makes contact with other fields of mathematics, including topology, ordinary and partial differential equations, and algebraic geometry. The benefits flow in both directions across the interfaces, and differential geometry has contributed its full share to the globalization of analysis that has occurred since World War II. As a course in university instruction, differential geometry remains important; for example, the place of geometry in the undergraduate curriculum was recently the subject of a symposium conducted by the Ohio Section of the Mathematical Association of America.

The methods of Élie Cartan (1869–1951), the most profound and original geometer of the 20th century, are beginning to be widely understood, thanks mainly to the unflagging efforts of S. S. Chern, who spent more than three decades explaining and extending Cartan's work, while himself contributing a substantial share of the original work in differential geometry in that time. Early resistance to Cartan's ideas arose in part from his use of infinitesimals, banished from mathematics in the 19th century. The introduction of the notion of fiber bundles in the early 1940s allowed an alternate description of Cartan's ideas within an acceptable context and led to the discovery by Chern of characteristic classes, which now play a prominent role in differential geometry and algebraic topology.

It is interesting to speculate about the subsequent development of differential geometry if in 1940 geometers had been in possession of A. Robinson's nonstandard analysis (1960), which shows how infinitesimals can be restored to mathematical legitimacy in a perfectly rigorous way. The time is ripe for an exposition of Cartan's work based on nonstandard analysis, and students of geometry are hopeful that one will come within the next ten years. One omen for this is the appearance this year of a rigorous beginning calculus book written by H. J. Keisler and based entirely on Robinson's methods.

—Nathaniel Grossman

# Medicine

Significant developments in medicine in the past year are presented below under the title *General Medicine*. The section is followed by a more detailed look at six areas in which there were special developments or accomplishments in the same period. The six are *Paramedical Personnel, Alcoholism, Psychiatry, Death and Dying, Dentistry,* and *Venereal Disease.*

## General medicine

The physician and the delivery of health care continued to come under closer public scrutiny in the U.S. and throughout the rest of the world, with the possible exceptions of the Republic of China and the Soviet Union. The private practice of medicine, so long a cherished tradition by physician as well as patient was under heavy fire for its failure to provide a system of equal health care to all people in the United States. Instead of just coming from the impoverished blacks and browns in urban ghettos, serious questioning of the American medical system was beginning to come from middle class white families, who were finding more and more that health care in the U.S., as good as it was, was simply not sufficiently available. The high and continuously rising cost of medical care was not the sole consideration. The distribution of health services—hospitals, clinics, physicians' private offices—was of equal concern. Physicians not only were fleeing urban America along with middle class whites but tended to establish their private practices in areas where climate and lifestyle were more ideal. California, the southwest, northwest, and Florida were the most popular locations for physicians, both young and old. This left situations in other areas of the country that had even people with the ability to pay for private medical attention unable to find physicians, especially pediatricians, general practitioners, and internists, who could take on new cases.

Other areas in the health care delivery system were locked in struggles. One such battle related to the efficacy of various drugs in preventing death from blockage of the coronary arteries (arteriosclerosis). Blockage of the coronary artery was accounting for 80% of the deaths from heart disease in American males and in 1972 was ranked as the first cause of disability or death in the U.S. A number of drugs had been found to lower blood cholesterol. Over 8,000 U.S. men were participating in a current study of the National Heart and Lung Institute attempting to find the most effective agent. The subjects, ranging in age from 30 to 64, had one thing in common: each had suffered a previous myocardial infarction (heart attack). Final results would not be known until 1974, but medical science already was embroiled in debate, arguing: (1) whether cholesterol was really the key factor to be controlled, (2) whether drugs were superior to prevention via exercise and diet, and (3) whether a massive study of men whose hearts were already "sick" would be valuable to the healthy.

In the interim, clinical investigators in both Scotland and Newcastle-upon-Tyne, Eng., reported that one of the drugs being tested in the U.S., clofibrate, was associated with fewer heart attacks when used by a group of men who suffered previously only from angina—"chronic chest pain" that is related to the oxygen starvation occurring in the heart as arteriosclerosis builds its barriers. The studies were considered the best evidence yet that medicine may find an effective chemical weapon against heart disease.

Another study, conducted in the southern U.S., showed that hard-working black and white sharecroppers had an extremely low incidence of heart disease even though they had high levels of cholesterol and a high incidence of cardiovascular abnormalities.

*continued on page 270*

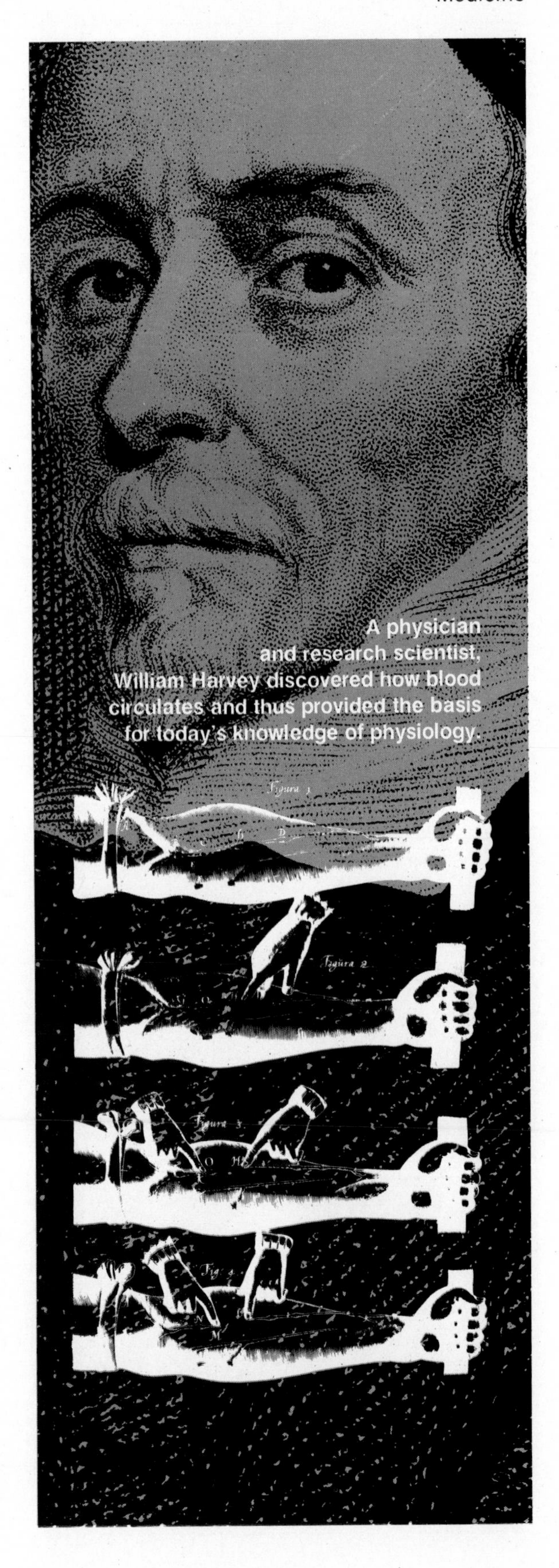

A physician and research scientist, William Harvey discovered how blood circulates and thus provided the basis for today's knowledge of physiology.

## Acupuncture

On April 21, 1972, at Weiss Memorial Hospital in Chicago, Dr. Wei Chi Liu performed a tonsilectomy on a 31-year-old male nurse, Gary Chinn. What distinguished the otherwise commonplace operation was the absence of any of the anesthetics ordinarily used in Western medicine. Instead, the patient was anesthetized by acupuncture, the age-old Chinese practice in which needles are inserted at specified points in the body.

Gary Chinn's tonsilectomy was one of several surgical procedures using acupuncture anesthesia that took place in U.S. hospitals in mid-1972. Only a year or two earlier, American medical men had regarded acupuncture (if they regarded it at all) as a quaint cultural survival, but by 1972 it was the subject of intense interest and study.

The scene for this change had been set in the political arena, as China and the U.S. made tentative advances toward normalizing their relations and Americans were admitted to China for the first time in more than 20 years. Pres. Richard Nixon's much-publicized trip to Peking gave rise to a fad for things Chinese. More to the point, U.S. doctors returned from China with reports of having witnessed major operations in which only slender, vibrating needles were used to deaden pain. The patients remained conscious and were able to converse with the surgeons. They said they felt no pain, although sometimes they could feel their organs being manipulated. In some cases, they sipped tea or ate pieces of orange while the operation was proceeding.

The use of acupuncture as an anesthetic during surgery was relatively new, but acupuncture had been practiced as a form of curative medicine for some 3,000 years. Legend has it that acupuncture originated when an ancient emperor observed that soldiers struck by arrows sometimes obtained relief from ailments in other parts of their bodies. In time, a body of philosophical concepts grew up around it, involving the ideas of yin and yang, the active and passive principles of the universe.

In traditional acupuncture, the body is divided by 12 meridians, the pathways of the body's vital energies. Disease is seen as an imbalance in these energies, and the imbalance is corrected by inserting needles at one or several of 365 points on the head, trunk, and extremities. The practitioner is guided by elaborate charts, the oldest known of which is a bronze model of about A.D. 860. The needles may be of different metals, each with a special significance, although today stainless steel needles are most frequently used. They may be inserted only a short distance or deeply into the body, and sometimes they are rotated or vibrated. Since the points for insertion are based on the meridians, they may bear no visible relationship to the illness being treated.

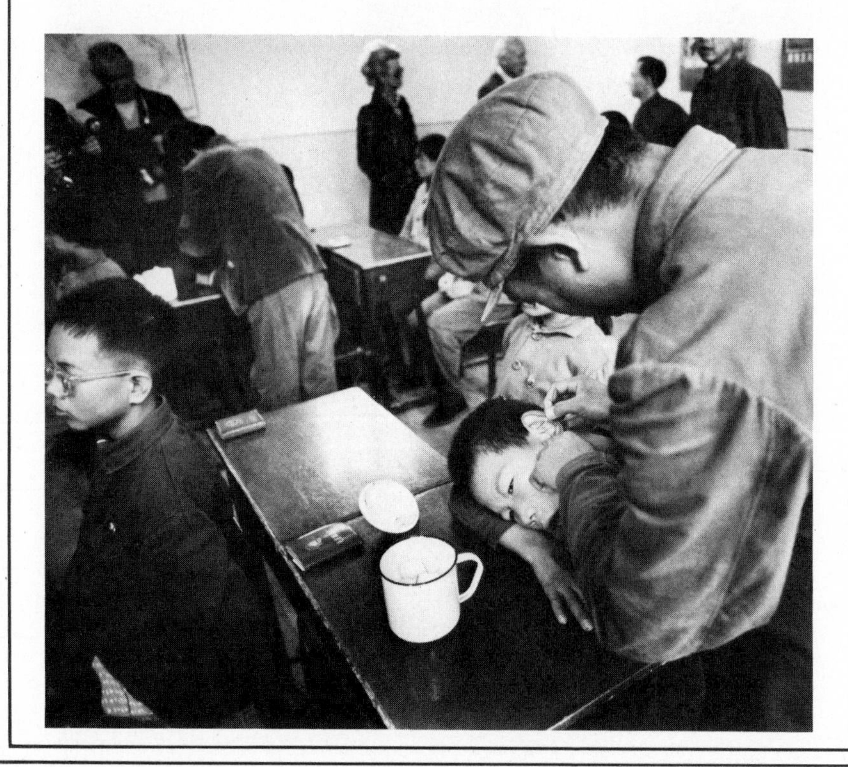

*A deaf child receives acupuncture treatment at the School for the Deaf No. 3 in Peking. The cure is said to last approximately one year.*

Stern from Black Star

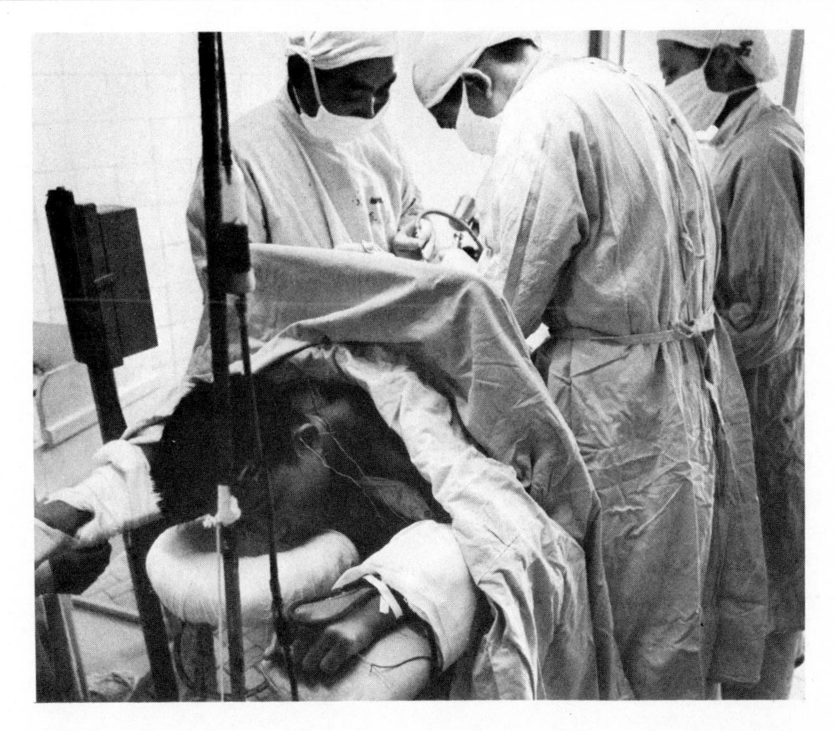

*Chinese surgeons operate on the spinal column of a 20-year-old man anesthetized by acupuncture. The five silver needles inserted into his ear, neck, and hand are periodically vibrated by a six-volt electric current.*

Cures and remissions of many diseases have been claimed for acupuncture, but Western science has no explanation of why—or even whether—it works. Skeptical doctors have postulated a placebo effect—the patient is cured because he believes the cure will work—and note that many of the ailments for which acupuncture seems to succeed have a large psychosomatic component. Others, while discounting yin and yang, believe that the ancient Chinese doctors stumbled into something that works, just as herbalists, by trial and error, built up a pharmacopoeia of folk medicines, many of which have real therapeutic value. The most likely interpretation is that the needles in some way affect the nervous system. One theory is that acupuncture's anesthetic effects result when the sensations caused by the needles block "gateways" in the spinal cord and thalamus through which painful sensations reach the brain.

Europeans have long shown some interest in acupuncture. It reached France via Indochina, and a number of French doctors who practice *acuponcteur* claim it has helped patients who failed to benefit from conventional treatment. The Soviet Union is said to have it under study. In the United States, however, it has been confined largely to Chinese communities.

Even in China, acupuncture was falling into official disfavor prior to the Communist revolution. Mao Tse-tung, however, encouraged a combination of traditional and modern medicine and, following his dicta, traditional practitioners and doctors trained in the Western mode began working together. The trend gained momentum during the Cultural Revolution, when doctors from the cities were sent to work in the countryside. Today, Chinese doctors claim their practice includes the best of both worlds. (*See* Feature Article: SCIENCE IN CHINA.)

It is acupuncture as an anesthetic that has received the most publicity in the West. Partly this is because it is spectacular. Further, although Western anesthesiology is highly developed, a certain small percentage of patients suffer severe reactions or even die from the effects of chemical anesthetics. If acupuncture could be substituted, medicine would have gained a valuable tool.

This does not appear to be in the immediate future, however. The haste with which Western doctors are studying acupuncture stems partly from the fear that, unless the medical profession takes a credible stand one way or the other, it may become a fad and fall into the hands of quacks. Samuel Heller, head of the New York State Board of Medicine, predicts that it will be two to five years before any final judgment can be made. Chinese doctors themselves warn that acupuncture anesthesia is not always suitable and that, even when it is used, it must sometimes be supplemented by chemical anesthetics. Nevertheless, as of 1972 it appeared that acupuncture had emerged from the medical shadows.

*continued from page 267*

**Drugs, health fads, and the FDA.** Meanwhile, more stringent rulings by the U.S. Food and Drug Administration (FDA) resulted in a virtual moratorium on new drugs introduced by the American pharmaceutical industry. By the latest count, no less than three dozen drugs available in Great Britain, France, and elsewhere were not available in the U.S. because FDA regulations relating primarily to safety testing and efficacy prevented their sale. As the result of a National Academy of Sciences review of all prescription drugs, advertisements reaching doctors in the spring of 1972 and thereafter had to carry what the pharmaceutical industry called the "black box." This border-enclosed statement had to show how the drug being sold was rated by the academy. Only if the drug was ruled effective without exception could the manufacturer omit the rating.

The FDA also began a long-term study of nonprescription drugs, which numbered in the thousands. These "over-the-counter" agents ranged from cold remedies to laxatives, and from antiperspirants to suntan lotions. First target of the review was to be the antacids, which had been racing each other to the stomachs of television viewers in growing numbers over recent years. In a related action, the Federal Trade Commission (FTC)—with advice from the FDA—noted that "one aspirin is about as effective as the other" for pain relief. The FTC ordered aspirin manufacturers to spend one-fifth of future advertising budgets in retracting their claims. The order was being contested.

The general effect of such developments was to introduce a sense of uneasiness into the doctor-patient-pharmacist relationship. The surge of self-medication and fads that followed was unmatched by anything since grandmother locked up her roots and snake oil. By far the most popular of recent nostrums was Vitamin E, which supporters claimed can cure everything from dandruff to liver disease. Known to older Americans by its original, mysterious label of Vitamin X, the substance occurs naturally in cereals, nuts, and unrefined vegetable oils. Its proven medical uses were limited and confined to such uncommon diseases as amyotrophic lateral sclerosis (Lou Gehrig's disease). Yet when a Canadian scientist began appearing on television discussing his new book on the wonders of Vitamin E, Americans bought tons of it. Some persons, it was estimated, were taking 30 times the minimum recommended daily requirement. Fortunately, there was no evidence that megadoses of the vitamin do any harm—or any good, for that matter.

Another naturally occurring body substance, discovered about the same time as Vitamin E, observed its golden anniversary. Insulin was discovered in 1921 by two University of Toronto scientists, Frederick Banting and Charles Best, as the

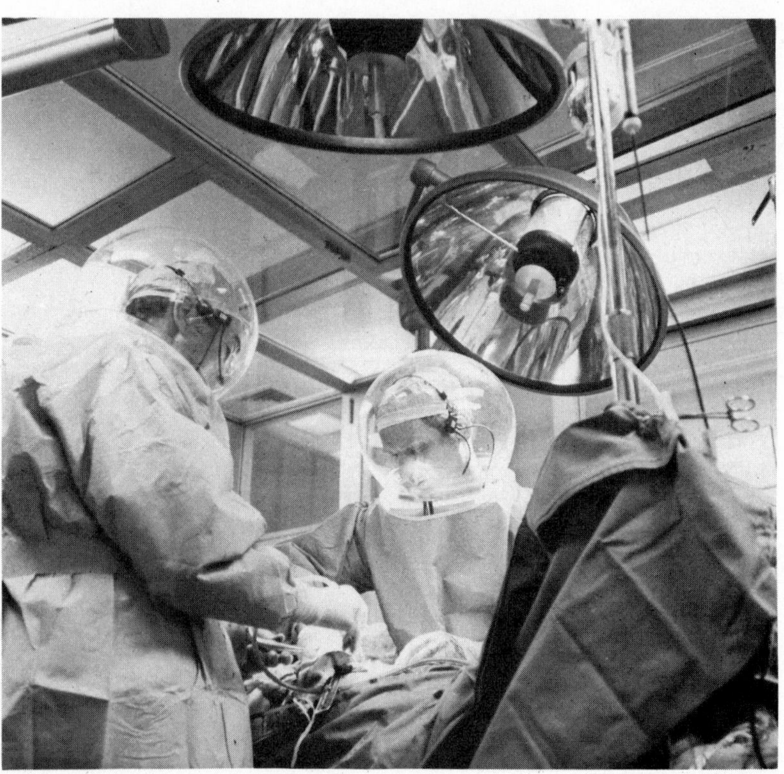

*A surgical team at St. Luke's Hospital in Denver, Colo., performs an operation in a new clean room facility that helps lessen the danger of infection to the patient. During the operation air is forced in a gentle breeze from the rear to the open front of the clean room, which can be folded and stored when not in use. Team members "upwind" of the patient wear helmets and garments that are impermeable to bacteria. The clean room technique for surgery was developed for NASA by Martin Marietta Corp.*

Courtesy, St. Luke's Hospital, Denver

substance that controlled diabetes in their laboratory dogs. Now used by millions of humans in injectable and oral forms, even insulin came under a modern cloud of doubt. Investigators worldwide were becoming concerned with the very long-term survival of diabetics enabled by insulin. Since the disease is probably genetically based, the once relatively rare condition would be carried by close to half of Americans and Europeans by the time the 1970s ended. "Carrier" means only the ability to pass diabetes on to an offspring, but an ever growing number of persons now had the disease itself.

A continuing medical concern was that insulin does not stop the continuing systemic breakdown of diabetes. New attention had been turned to cure by such public groups as the Juvenile Diabetes Foundation. Yet medical researchers agreed that substantially more funds were devoted to manufacturing insulin than to looking for a cure for diabetes. But there were hopeful signs. In New York, a young housewife received a pancreas transplant. Her long-term survival and reversal of symptoms, which had not been achieved in previous transfers of the organ, was credited to the fact that her old pancreas was not removed. Instead, the second one was merely "plugged into" her circulation. As with all transplants, however, the technique was considered too intricate and expensive for general use.

**Surgery, modern and ancient.** Assisted by biomedical engineers, surgeons moved forward on several other fronts. Authorities at the Mayo Clinic and Harvard Medical School announced successful total knee replacements, utilizing a metal-and-plastic prosthesis. Elderly patients with rheumatoid or osteoarthritis could be particularly benefited by the development and wider use of such a technique. A nuclear-powered heart, fueled with plutonium-238, sustained the lives of calves in Boston and Chicago laboratories. With refinements in miniaturization and other factors, such hearts could be available for humans by 1980. A University of Cincinnati, O., surgeon announced reconstruction of the voice box in dogs. If the procedure proved similarly effective in baboons, it would be used to return the powers of speech to human victims of laryngeal cancer.

Surgeons were fiercely debating the development of a new approach to breast cancer that would in many cases make breast removal unnecessary. Some surgeons believed a simple removal of the cancerous breast tumor with appropriate radiological therapy was an effective treatment.

Acupuncture, an ancient Oriental therapy, became a matter of great public and professional interest when China began to allow American visitors. Such distinguished physicians as Paul Dudley White and Samuel Rosen, while visiting and lecturing in China, witnessed acupuncture being administered. *Acupuncteurs,* who had practiced in obscurity if not in secrecy within the Western culture, suddenly found themselves in demand as medical school lecturers. In mid-July 1971 *New York Times* columnist James Reston had his appendix removed in Peking and received acupuncture for postoperative pain. A number of research teams were feverishly trying to discover how this 4,500-year-old treatment really works. The best popular explanations were really guesses; these included self-hypnosis and the possibility of neuroelectrical reactions between the thin metal needles and cellular substances. The president's physician urged that a team of U.S. physicians be sent to China for a lengthy study of acupuncture. (*See* Sidebar: *Acupuncture,* above.)

**Epidemic diseases.** Smallpox, which had been virtually controlled on a global scale, broke out in epidemic proportions following the bloody independence struggle in Bangladesh. More than 700 persons died during the first 10 days of the out-

*Two advanced techniques in hemodialysis are incorporated in the system shown here. The filter contains a hydron-and-albumin-coated activated charcoal which removes uric acid and creatinine much more quickly than conventional dialysis. The system also uses a single-needle cannula that handles both inflow and outflow.*

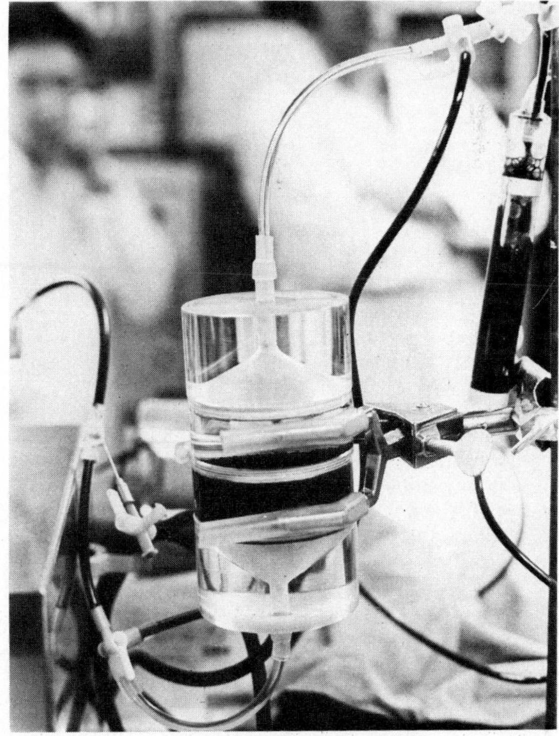

Courtesy, Division of Artificial Organs, University of Utah

break alone. The highly contagious smallpox virus spread quickly to Iran and Pakistan. A Yugoslav Muslim carried the disease back from a religious pilgrimage; a Yugoslav worker imported it into West Germany; and public health officials frantically established mass vaccination programs as modern transportation threatened to send the disease across all of Europe. By spring the outbreak appeared under control and the World Health Organization continued to predict eradication of smallpox by 1976—some 170 years after Edward Jenner introduced his "cowpox" vaccine.

The U.S. Public Health Service (PHS), meanwhile, recommended that compulsory vaccination for smallpox be dropped from the health codes of all states. The infrequent but serious reactions to the vaccine that occurred were a greater risk at this time than smallpox itself, the PHS said. The last U.S. case of the disease was recorded in 1949.

The epidemics of concern to U.S. health planners were venereal disease (see *Venereal Disease*, below) and drug abuse. In a historic and controversial decision, the PHS recommended that private and clinic physicians alike begin routinely testing women patients for gonorrhea. An estimated 650,000 U.S. women were asymptomatic carriers among the 2,250,000 persons who contracted the disease each year. A simple, self-contained test kit for *gonococci* was made available to private practitioners, who were already detecting four of every five VD cases in the nation.

**Resolving drug abuse problems.** Drug abuse was a no less widespread but quite different problem. While considerable federal attention was focused on such "hard drugs" as cocaine and heroin—there were an estimated 650,000 U.S. heroin addicts spending $2 billion yearly to maintain their habit—the prime method of treatment came under fire. Methadone was dispensed by 450 clinics nationwide as a chemical blocker and antagonist of heroin. It was found that some clinics were not requiring the addicts to consume their methadone during their visits, allowing the drug (which produces a mild "high" of its own) to be sold elsewhere to drug abusers. Several persons, many of them children, died of methadone overdoses. More stringent dispensary methods were instituted as the pharmaceutical industry announced a stepped-up campaign to develop a safer narcotic antagonist, perhaps one that would also replace morphine and other strongly addicting substances used in medical practice to lessen extreme pain.

Federal controls were initiated to reduce amphetamine abuse. Overprescription of these drugs by well-meaning doctors and overproduction by drug firms were spotlighted as the leading causes of this kind of abuse as well as of the skyrocketing barbiturate problem. In 1972, U.S. suppliers were being asked to reduce their amphetamine production by about 80%. At the same time, physicians' organizations were urging members to confine their prescriptions of such drugs to those situations where another, safer agent could not do the job, such as narcolepsy (sleep epilepsy) and severe hyperkinesis in children.

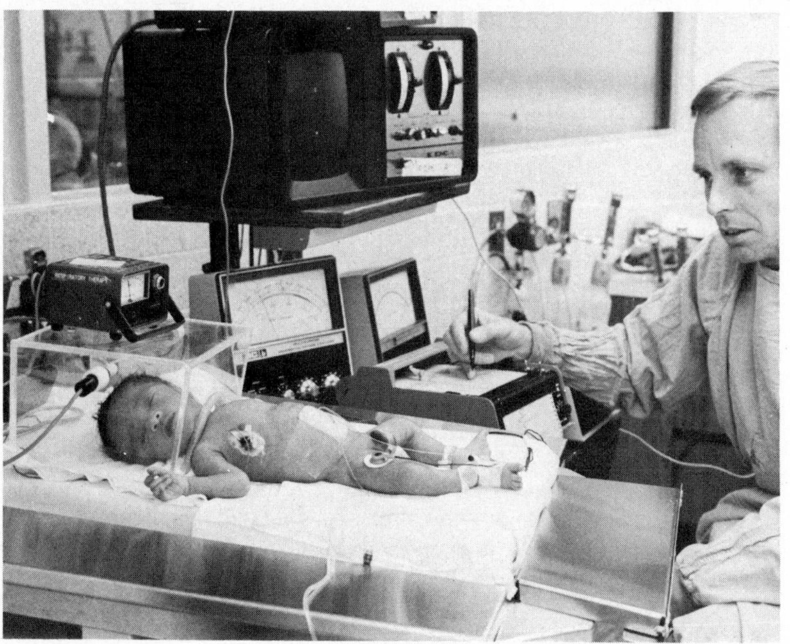

*Blood gases in a premature baby are monitored by an experimental electric microsensor introduced through the umbilical artery into the infant's aorta. University of Arizona pediatrician Thomas Harris and a California firm designed the oxygen-measuring electrode in an effort to avoid missing the signs of distress not detected when blood-gas samples are taken only periodically.*

Dave Davis, MEDICAL WORLD NEWS

A study commissioned by the National Commission on Marihuana and Drug Abuse showed that 24 million Americans have tried marijuana at least once. Of these, 8.4 million became regular users. The commission's chief recommendations in its marijuana report were two: a national policy of discouragement of marijuana use and the decriminalization of marijuana in cases involving use of the drug in the privacy of the home or use in an obvious social situation. The commission recommended continuation of strong penalties for the sale of large amounts of marijuana for profit (as opposed to the sale of one marijuana cigarette to a friend at cost). It also recommended, in keeping with its intent to keep the marijuana problem out of the overloaded criminal justice system, that the contraband theory in law be applied to public possession cases made by law enforcement officers. This would mean that a policeman could seize marijuana from a person in a public place without having to arrest him. The growing of marijuana in a flower box or elsewhere would still be considered a criminal action under the commission's recommendations.

The 13 commissioners, including eminent physicians, psychiatrists, and lawyers, were impressed in their year's study of marijuana use in the U.S. and throughout the world with the fact that incarceration for marijuana offenses may cause far more serious physical and emotional harm to an individual than smoking marijuana. The commission urged that studies of the long-range effects of marijuana use on the body be started immediately. It deplored the lack of such scientific information. At the same time, the commission had to take into consideration the overwhelming evidence from scientific and social inquiry that showed marijuana used in moderation has no serious ill effects on the body or mind.

Within a few months after the commission's report was released, bills were introduced in Congress and in a number of state legislatures to reduce penalties for marijuana use and possession along lines suggested by the commission report. Penalties for simple possession (one ounce or less) of marijuana ranged from seven days in Nebraska to life imprisonment in Texas. The commission was to review its marijuana recommendations in its second report on drug abuse in general due in March 1973.

**Organized medicine.** The public and private lives of doctors were becoming increasingly stressful by the threat of massive changes. National health insurance, stalled in Congress until after the 1972 presidential elections, would surely change the method by which the patient paid his bills. But for the doctor it would mean much more:

Courtesy, Lawrence B. Pinneo, Stanford Research Institute

*A monkey moves its paralyzed right arm by means of a Programmed Brain Stimulator, a computer that sequentially fires electrodes implanted in the brain stem. Such a device may eventually enable persons paralyzed by brain damage to regain use of their limbs.*

audits of his practice procedures, paperwork, and the real possibility that he would no longer control the level of his fees. For these and many other reasons, in recent months U.S. doctors had begun to talk of unionization.

Some physicians preferred to call their organizations guilds or merely associations, but the difference between the new organizations of that title and those that said they were unions seemed purely semantic. All were striving for a "new professionalism" and a collective voice in the socioeconomic struggles ahead. In March 1972, 30 doctors formed the Nevada Physicians Union and became affiliated with a local of the AFL-CIO. In March, over 100 private practitioners met in San Antonio, Tex., to form the American Physicians Union. In April, over 600 San Francisco area physicians formed the Union of American Physicians. Still small and unorganized, the trend to physicians' unions seemed obvious. The key question in the movement was whether a doctor could or would go on strike. Since the idea of these organizations came from the interns and residents who united to stage "job actions" a few years before, the answer might well be affirmative.

The traditional voice of medicine in the U.S. has been the American Medical Association (AMA). In his inaugural speech, the organization's 1971–72 president, Wesley Hall of Nevada, asked for a constitutional convention to reform the organization. While resisting the proposal, the AMA House of Delegates agreed to hold hearings on it over the next year. Because some doctors thought the AMA too conservative, while others condemned it as liberal for proposing its own national health insurance plan, any attempts at change would prove doubly difficult.

Yet another potential spokesman in medical affairs was the newly formed Institute of Medicine of the National Academy of Sciences (NAS). Since the U.S. Civil War, the academy had been a respected adviser to the government in matters of science. Very few physicians were elected to membership, and the medical issues considered by the NAS expert groups were comparatively rare. In 1970, a separate wing of the academy was formed by a mass admission of several score nationally famous medical leaders. At this point, the Institute of Medicine was confining itself to considerations of broad social and ethical issues, such as death and dying. But if the history of the parent academy was a guide, the new body's voice would soon be heard in specific matters of national policy. (See *Death and Dying,* below.)

Of course, medical ethics is a serious concern, and in recent months the public had increased insight of the doctor's day-to-day ethical concerns. The Institute of Society, Ethics, and the Life Sciences, a nonaffiliated private group with members at dozens of medical schools, seminaries, and even in Congress, became the focus for a number of public controversies. The institute's spokesmen condemned a Texas physician who investigated the side effects of oral contraceptives by giving placebos to a group of uneducated, Mexican-American women. Several became pregnant. On another occasion, the *Journal of the American Medical Association* was criticized for publishing Federal Bureau of Investigation wanted notices of criminals with medical problems. The institute argued that this was a clear violation of doctor-patient confidentiality. The AMA Judicial Council declared that it was not.

Even the basic process of educating physicians became a storm center. The National Fund for Medical Education, headed by John S. Millis, proposed that the nation create no more medical

Courtesy, Dr. James G. White, MEDICAL WORLD NEWS

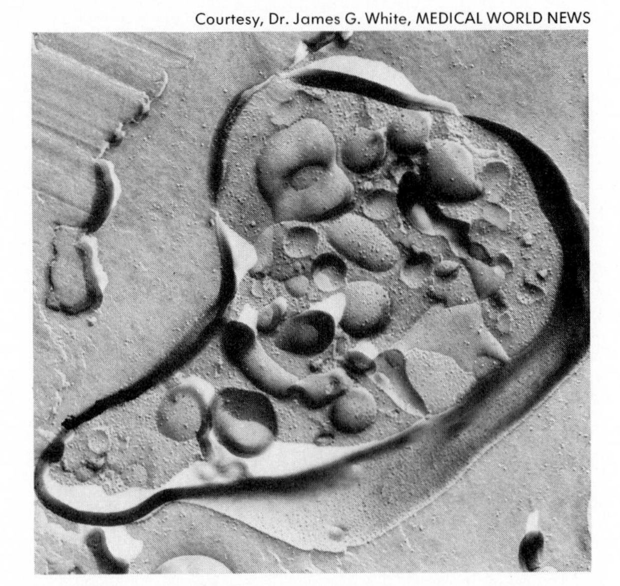

*An electron micrograph of a blood platelet, the particle that aggregates to arrest bleeding, shows the open canal system winding through the organelle's interior. Platelet dysfunction was implicated in numerous diseases in which blood clotting or bleeding is a factor.*

schools. Instead, Millis' report recommended increasing enrollments in existing schools while reducing the training required from four to just two years. This could be done by diversifying and specializing the students' training at an earlier stage on the way to the M.D. degree. Despite these revolutionary proposals, spokesmen for organized medicine praised the study as "the most comprehensive" since the 1909 report by Abraham Flexner, which had set the course for U.S. medical schools to date.

**Research and funding.** Any drastic change in medical education would require an equally impressive change in the pattern of federal funding. Currently, little support was given actual educational programs. The bulk of federal tax moneys going to medical schools was for research and patient care. For the coming months, the "stars" of medical research would include cancer, heart disease, and sickle-cell anemia. Investigators in all three fields received large increases in congressional appropriations for the coming year. Cancer received the first and largest increase (*see* Feature Article: CANCER UNDER ATTACK). Heart disease followed, as both leading killers received nothing less than "Manhattan Project" pledges of federal backing.

Sickle-cell anemia, by contrast, was a significantly less frequent threat to life. It received its name because certain drugs and environmental conditions cause the red blood cells of its victims to collapse into a sickle shape. It occurs predominantly among blacks, and even then, among non-African blacks (an estimated 40 million West Africans carry the trait) the genetically acquired trait occurs far less frequently than such other diseases as diabetes or hypertension. Nonetheless, at least two million blacks in the U.S. carried the trait, and more than 45,000 were estimated to suffer crises of medical significance each year. Those with the trait had to be particularly careful about flying or exerting themselves at relatively high altitudes.

The national sickle-cell program authorized by Congress was to provide $105 million over three years, beginning in 1973. Emphasis would be on treatment, screening, and counseling. Despite the availability of cheap, rapid screening tests, many victims do not discover that they have the disease until the first crisis occurs. Until recently, sickle-cell anemia was a neglected recipient of research funds, averaging less than $1 million annually.

**Mobilized medicine.** Major efforts would be devoted during the next two or three years to detecting breast cancer by means of mobile X-ray units and intensified public education campaigns. The drive was launched by the American Cancer Society, which noted that the incidence of breast cancer had not declined in 35 years. Communities would be asked to push for such projects as the "Cancermobile," sponsored by the Brooklyn chapter of ACS and the Holy Family Hospital. This mobilized detection center utilized refined radiologic techniques, such as mammography and thermography, to pick out breast tumors in their earliest stage.

Ironically, the public health drive that first proved the value of mobile detection units was fading from the American scene. Vans carrying chest X-ray units for tuberculosis detection had done their job, according to the National Tuberculosis and Respiratory Disease Association. TB was now too infrequent to justify continued use of the red-and-white vans, a familiar sight in shopping centers, church parking lots, and other community centers since after World War II.

But the U.S. is basically a culture on wheels, and its concerns in health care were hardly excepted. Major public and professional campaigns were sprouting coast-to-coast to improve systems of prehospital emergency care. In the past, clinicians had derided most ambulance services as "taxis for sick people." But now the technology and the funds were becoming available to give patients better care on their way to the hospital. Three federal programs and uncatalogued scores of community groups in Kansas City, Mo., Akron, O., and Grand Rapids, Mich., were improving emergency systems with the assistance of such national groups as the AMA, American College of Physicians, American Heart Association, and ACT (Advanced Coronary Treatment) Foundation.

Much of the technology and training for better ambulance-to-hospital care was provided by medical interest in mobile coronary care units. Highly sophisticated units in such cities as Seattle, Wash., Miami, Fla., and Columbus, O., featured highly trained paramedical personnel who could give treatments on the spot with the advice of a doctor who communicated with them via radio and telemetry. In the spring of 1972 a University of Virginia mobile coronary care van was credited with bringing critical therapy to a famous heart attack victim—former U.S. President Lyndon B. Johnson. (See *Paramedical Personnel,* below.)

In their efforts to keep healthy and discover more about illness, Americans were spending $75 billion each year—7.4% of the nation's gross national product. Predictably, the expenditure seemed to have little relationship to the effect desired. While the dollars for health had increased steadily, the average citizen remained just as sick. Since 1959, the National Center for Health Statistics reported, average days of disability remained at 14–16 days per year. That included 5–6 days annually at home, sick in bed.

—Byron T. Scott

## Alcoholism

Throughout history, man has used and abused the mood-altering drug ethanol to celebrate his triumphs and to soothe his frustrations. It has long been known that many people, in all walks of life and from many cultures, consume more alcoholic beverages than is good for their health and well-being and for the well-being of those around them. Only recently, however, have scientific and humane measures replaced moralism and punishment as methods of dealing with the problem drinker.

The past five years have witnessed a deepened awareness and understanding of the problems of alcohol abuse and alcoholism. Instead of being relegated to the category of willful misconduct or bad behavior, alcoholism has been recognized as a medical problem, a complex disorder with physical, social, behavioral, and psychiatric aspects. The chronic excessive drinker has come to be regarded with compassion rather than condemnation, as the helpless victim of an illness rather than the deliberate perpetrator of a criminal or moral offense.

Alcohol affects each person differently, and there are almost as many definitions of alcohol abuse and alcoholism as there are brands of whiskey, wine, and beer. In our society, most drinkers keep their consumption of alcohol within socially acceptable and harmless bounds—limited to amounts they can readily handle. Drinking becomes a problem when the use of alcohol causes physical, psychological, or social difficulties to the drinker or to society. Alcohol is being abused when it leads to recurrent episodes of intoxication or heavy drinking that impair health, or when it is consistently used as a mechanism for coping with personal problems, to a point where it seriously interferes with an individual's effectiveness on the job, at home, in the community, or behind the wheel of a car.

**A new force in the war on alcoholism.** During the year a dramatic organizational change took place in the effort to combat alcohol abuse and alcoholism in the United States. Massive federal intervention in this previously neglected area officially began in December 1970, when the Comprehensive Alcohol Abuse and Alcoholism Prevention, Treatment and Rehabilitation Act of 1970 (Public Law 91-616) was passed unanimously by both houses of Congress and signed into law by Pres. Richard Nixon. One of the most important pieces of legislation ever enacted in the field of alcoholism, Public Law 91-616 established the National Institute of Alcohol Abuse and Alcoholism (NIAAA) to serve as the federal government's principal agency for alcohol programs and activities.

Working closely with state and local governments, private voluntary agencies (such as the National Council on Alcoholism and Alcoholics Anonymous), and agencies involved in traffic safety (such as the Department of Transportation, the National Safety Council, and insurance companies), the NIAAA embarked on a broad range of research and service programs to meet the total needs of alcoholics and their families on the community level. NIAAA's project grants totaled $300 million over a three-year period, six times as much as had been available in the three years before its inception.

NIAAA had two main overall goals: to assist in making the best alcoholism treatment and rehabilitation services available at the community level, and the longer-range but even more important goal of developing effective methods of preventing alcoholism and problem drinking. To achieve these objectives, NIAAA aimed to foster, develop, conduct, and support broad programs of research, training, development and coordination of community services, and public education.

**Research and training.** Since alcoholism is the product of a complex interaction of biological, psychological, and social factors, an improved understanding of its causes, natural history, and treatment can only be achieved through a wide range of research. With increased funds made available by NIAAA grants, many highly competent researchers from a variety of disciplines became involved for the first time in research studies on alcohol problems. These studies were being conducted at universities and medical centers located in geographic areas with a high incidence of alcoholism.

One cause of death among alcoholics is cirrhosis of the liver, and investigation into alcohol-related liver disease was given high priority. Attention was also focused on other serious complications of alcoholism, including pancreatitis, gastritis, and withdrawal reactions. Another important research goal was to find ways of speeding the clearance of alcohol from the system. One study showed that alcohol can be removed rapidly from the bloodstream through hemodialysis (blood purification through an artificial kidney). While not currently practical for prolonged treatment, this technique had significant implications for the management of life-threatening intoxications.

In order to match specific treatment methods to particular patients, studies were conducted to differentiate between subtypes of alcoholic persons. One study identified "essential" and "reactive" types of alcoholic persons, while another categorized four subtypes based on the relationship between social factors and the individual. As subtypes of alcoholic individuals were identi-

*"I'll be glad to listen to your troubles—yes. But, first, have you severed your connections with other bartenders to whom you've been relating them in the past?"*

fied and specific treatments devised for each group, a higher success rate could be expected.

New treatments would remain ineffective, however, as long as relatively few physicians and other health personnel were sufficiently trained and experienced to deal with the complex problems of alcohol abuse and alcoholism. To meet this need, programs were initiated during the year for the inclusion of training in alcohol problems in the curricula of schools of medicine, social work, nursing, public health, psychiatry, and psychology. Other programs in progress included those aimed at development of specialists to assume leadership roles in training, research, education, and administration; training for private practitioners and hospital staff; and the preparation of counselors to function in both clinical and coordinating roles.

**Special programs.** NIAAA's formula grant program, the first of its kind to deal with this problem, enabled states to develop comprehensive plans covering all segments of alcohol abuse and alcoholism. Such programs made it possible to involve a wide variety of groups, including private citizens, voluntary agencies, medical professionals, health care and social welfare agencies, and court and enforcement personnel.

NIAAA community assistance programs provided aid in the development of comprehensive community services to meet the needs of alcoholics and their families, to coordinate these services, and assure their continuity. Prior to the creation of these programs, such comprehensive services were largely nonexistent. Most communities did have a core of medical and social services, however, and the comprehensive structures were being built around these existing services. A growing number of community alcoholic treatment centers were being established. Staff was being recruited not only from the ranks of professional health and social workers but from among recovered alcoholics as well.

Marshall Savick

*A Milwaukee, Wis., bartender listens to a customer's problems after completing an experimental course in elementary counseling skills sponsored by the local mental health association.*

NIAAA was also collaborating with other federal agencies, state and local governments, and private organizations in the development of programs targeted at special population groups. One of the most important was a joint effort with the National Highway Traffic Safety Administration of the Department of Transportation aimed at reducing the number of deaths and injuries caused by drunken drivers. This program included a broad public information and education campaign on the dangers of mixing drinking and driving. Special demonstration projects enabled communities to show the appropriateness and effectiveness of providing a public health service for problem drinkers, rather than running them through the revolving door of arrest, fine, loss of driver's license, and jail.

The public inebriate—the chronic drunkenness offender or Skid Row derelict—represents only about 5% of the alcoholic population but accounts for 40% of all arrests for nontraffic offenses. During the year the NIAAA, in cooperation with mental health centers, law enforcement agencies, and other organizations, sought to find more effective and practical alternatives to the present costly and inefficient system of handling these offenders. Several states initiated legislation to remove public intoxication from the criminal statutes and define it as a health problem. Further positive developments in this direction were expected in the near future.

Another collaborative effort was under way on behalf of employed alcoholics, who were estimated to comprise about 5% of the U.S. work force. Based on the experience of industries that had established programs to help these individuals, more than half of alcoholic employees can be rehabilitated. During 1971 the NIAAA, as required by Public Law 91-616, assisted the Civil Service Commission in developing guidelines for the implementation of alcoholism programs in all federal agencies and was fostering similar programs in state and local governments and in industry.

To help meet the needs of the American Indian population, which is troubled by severe problems associated with alcohol abuse, NIAAA inaugurated a collaborative program with the Indian Health Service, the Bureau of Indian Affairs, and other concerned agencies. Twelve treatment and rehabilitation service projects designed and developed by Indians for their people were established during the year.

**Prevention and education.** An illness is never eradicated by treating only the casualties. To be effective, any public health endeavor must also be aimed at prevention. Accordingly, high priority has been placed on public education as a method of reducing and preventing alcohol problems.

The public information and education campaign, conducted through television, radio, the press, and special publications, has several objectives: to develop public recognition of alcoholism as a treatable disease; to encourage the health-care system to regard alcoholism as a medical-social-behavioral problem and to treat the alcoholic patient with the same attention and consideration as any other patient; to inform the public of the properties of alcohol, its effects on the body and on behavior, and its potential for harm; and to produce a new environment concerning the use and misuses of alcoholic beverages, with an eventual reduction in the rate of drunkenness, problem drinking, and alcoholism. No person is asked to abstain from drinking. Rather, the use or nonuse of alcoholic beverages is regarded as a personal decision to be made by each individual.

**A worldwide problem.** Outside the United States alcohol abuse and alcoholism were also being recognized as medical rather than legal problems. After two decades of attempting to enforce antiliquor laws, the Indian government was being forced to recognize that a disease cannot be banned, and that the attempt to do so has resulted in a considerable loss of revenue while encouraging bootlegging, smuggling, and illicit distillation. Surveys in Great Britain and Western Europe revealed a substantial increase in the consumption of alcoholic beverages, particularly among the young. Yugoslavia, Poland, Hungary, and other countries of Eastern Europe were mounting intensive efforts to counteract the growing alcohol abuse problem.

On the international level, the first World Congress on the Prevention of Alcoholism was held in Kabul, Afg., in August 1972. It was followed by another international conference on the increasing problem of alcohol abuse held in Amsterdam, Neth., in September.

**Future developments.** The past year was largely one of planning and organization. The implementation and broadening of the programs described above would be the work of the months and years ahead. Like most legislation, Public Law 91-616 provided a framework for action. Filling in that framework would make a significant contribution to the needs of alcoholic people and to the goal of preventing and controlling alcoholism.

In addition to the benefits to be derived from Public Law 91-616, a uniform law on public intoxication, developed by the Commissioners on Uniform State Laws, was awaiting enactment by the states. The thrust of the law would be to move public intoxication out of the criminal area and into the health care service area where it belongs.

NIAAA's new National Clearinghouse on Alcoholic Information was expected to be in operation shortly. This was to be the mechanism for developing and distributing information on all aspects of the use and misuse of alcohol. In addition, the clearinghouse would distribute proceedings of research workshops and symposia, annotated bibliographies of treatment and rehabilitation literature, guides to federal assistance programs in the area of alcoholism, directories of alcoholic treatment programs, and fact sheets on alcoholism.

—Morris E. Chafetz

*Muscle tissue (left) taken from a subject who consumed 225 gm of ethanol daily for four weeks reveals extensive damage, including swollen interfibrillar spaces clogged with glycogen and lipid droplets. A second specimen (right) taken six months later shows the muscle has returned to normal.*

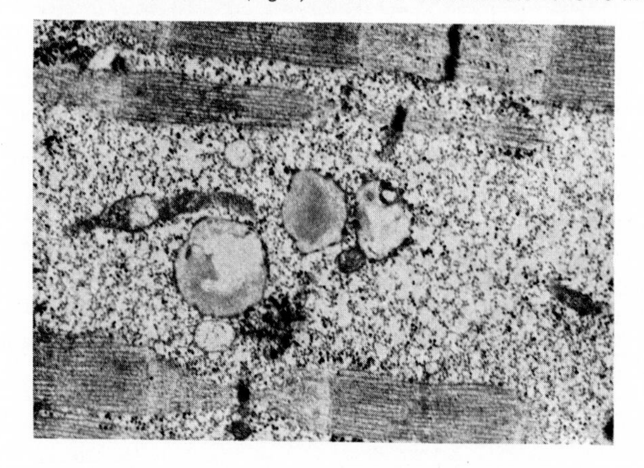

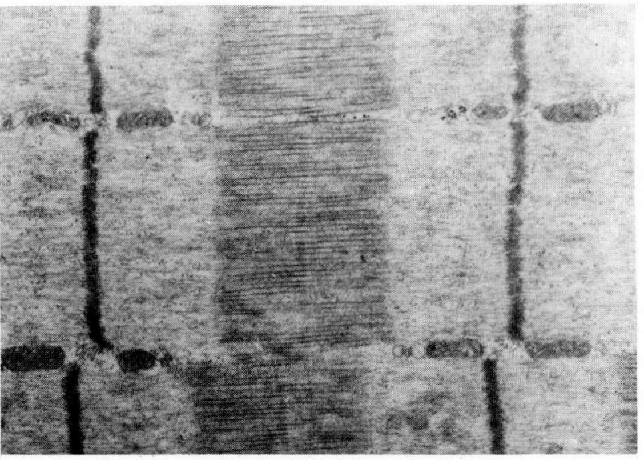

Courtesy, Dr. Emanuel Rubin, Mount Sinai School of Medicine

**Alcoholism today**

The first annual report on alcohol and health, prepared by the National Institute on Alcohol Abuse and Alcoholism, was the most comprehensive report ever published on the subject. It gave ample proof that alcohol is the most abused drug in the United States today. Among the principal findings of the report:

Nine million Americans are alcohol abusers and alcoholics. This constitutes about 5% of the adult population and almost 10% of the nation's work force.

Alcohol plays a major role in half the highway fatalities in the U.S., costing thousands of lives each year.

Alcohol abuse and alcoholism drain the economy of an estimated $15 billion annually.

Public intoxication alone accounts for one-third of all reported arrests.

Among American Indians, alcoholism is at an epidemic level; the rate is at least 10%, or twice as high as the national average.

Alcoholism is not a crime but an illness that requires rehabilitation through a broad range of health and social services tailored to the individual case.

Present programs dealing with alcohol abuse and alcoholism are accorded a low priority and are unrelated to most of the health and social resources within communities. What community services are available are frequently fragmented and fail to take into account changing life-styles or the unique characteristics of various population groups.

Establishment of modern public-health-oriented facilities to deal with intoxicated persons will free law enforcement agencies from being overburdened with a great many ill people.

The criminal law is not an appropriate device for preventing or controlling the health problems brought on by alcoholism. To deal with alcoholic persons as criminals because they appear in public when intoxicated is unproductive and wasteful of human resources.

## Death and dying

In recent months a rash of articles in major newspapers and national magazines dealt with aspects of death and dying that had rarely been written about before. Should a patient dying of widespread cancer be kept going by artificial means only to die several times over and in great discomfort? Should a retarded infant with intestinal obstruction have his life saved surgically only to spend it in a state institution, unproductive and at public expense? Such questions were surfacing more and more frequently and attracting wide public attention. At the opening of the John F. Kennedy Center for the Performing Arts in Washington, D.C., in October 1971, a capacity audience of professional and lay people listened to panels debating these and related ethical issues for two days at the International Symposium on Human Rights, Retardation, and Research.

Several centers for the study of ethical problems in medicine had been established recently in the U.S., notably the Institute of Society, Ethics, and the Life Sciences at Hastings-on-Hudson, N.Y., and the Kennedy Center for Bio-ethics at Georgetown University, Washington, D.C. Programs under government or foundation support were

*The growing percentage of people over the age of 65 presents a set of unique problems for today's physician. Social attitudes toward the elderly and the lack of training in geriatrics and gerontology often hinder decisions regarding the proper care of the aged.*

Joe Baker, MEDICAL WORLD NEWS

exploring these and related problems at Cornell, the University of Virginia, Stanford, Yale, Harvard, Case Western Reserve, the Hershey Medical Center of Pennsylvania State University, and the University of Washington, among others.

**Medical advances.** Until the past few decades the powers provided by medical science to prevent death were relatively feeble. That is, people lived and died largely independently of major intervention. When the time came, the heart stopped, as did breathing, and that was death. But now the human organism can be kept going despite cessation of normal brain, heart, respiratory, or kidney function, and other procedures for replacing lung and liver function by artificial means were in the experimental phase. Several hundred bodies were being maintained at the temperature of liquid nitrogen against the day when medical science will have discovered enough new information to revive them and patch them up.

Thanks to these rapidly developing technologies, dying is no longer a simple matter. Sharp controversy exists as to the use of these powerful tools. What are the proper procedures for deciding who shall be kept alive and who shall be allowed to die? When *is* a person dead? Certainly not when his heart stops; it can often be started again and kept going by an electrical pacer. Is he dead when his brain has irreversibly ceased to function, even though his vital organs otherwise continue with or without assistance? This was spoken of as "brain death," and many felt that it was a more reasonable criterion of death than failure of circulation and respiration, arguing that a human body without a brain, and hence no mind or sensibilities, is no longer a human being. But then the question arises: Must the entire brain cease permanently to function or is the loss of higher cerebral function, that part responsible for conscious thought and feelings, sufficient? Is an individual whose lower brain centers continue to keep the body going, but whose mind is irretrievably gone, no longer a human being? Is he essentially dead? Or should he be allowed to die and if so how? By whose decision?

These were some of the difficult questions being debated. They were raised most urgently by the ever increasing numbers of families who watched the death of a relative by the long-drawn-out and tortuous route made possible by modern medicine—in a respirator, on antibiotics, fed intravenously, with the family excluded from the intensive care unit in which it all took place.

Whatever the criteria for death, it was clear that the doctor could no longer consider it his duty solely to prolong life and fight death to the bitter end as he once could. Part of his charge was the difficult business of considering the quality of life as well as its duration and of accepting the responsibility for helping patients to die with as much dignity and peace as possible at the right time. Public sentiment seemed to be swinging this way, as judged by the recent press and the recent growth of interest in euthanasia.

**Euthanasia and the living will.** Euthanasia, derived from the Greek, meaning "a good death," was once an accepted part of many societies. For example, Eskimos who simply could not survive otherwise let disabled old people die as painlessly as possible when the rations got short. Modern Western societies, on the other hand, generally were strongly oriented toward the protection of individual life at all costs, especially favoring the weak and helpless. Nevertheless, the withholding of extraordinary measures to prolong life was widely condoned in private, and most physicians agreed that when the family consented and the patient was clearly demonstrated to be beyond salvage, resuscitative measures were not required at the end.

Passive, or negative, euthanasia was the next step; that is, the withholding of ordinary life-preserving measures in the hope of shortening painful or useless life indirectly. This also was practiced by many physicians when, in their judgment, it was the proper and humane course of action. Of a group of professors of medicine recently surveyed, 87% favored negative euthanasia.

Active, or positive, euthanasia, that is, the taking of definite measures to terminate or shorten life, is not generally condoned in Western society at present. Sir George Pickering, in his address at the opening of the new medical school in Nottingham, Eng., expressed the prevailing opinion about euthanasia among British physicians when he said, "I reject euthanasia—killing people is not what doctors should, or could, do." At the same time, he strongly endorsed letting old people die with dignity and not trying to prolong the lives of "old men and women stuffed with other people's viscera but with their own senile brains." Various groups, such as some members of the Euthanasia Society, argued that there was little difference ethically between active and passive euthanasia, once it was decided that it would be best for a patient not to go on living. In the survey just mentioned, almost half of the students of medicine and nursing questioned favored positive euthanasia, but this was true of only 15% of the professors of medicine. This perhaps indicated a trend in favor among the young, which society as a whole might well embrace in the future.

The Euthanasia Educational Fund, whose membership had grown from 600 persons in 1969 to

Joe Baker, MEDICAL WORLD NEWS

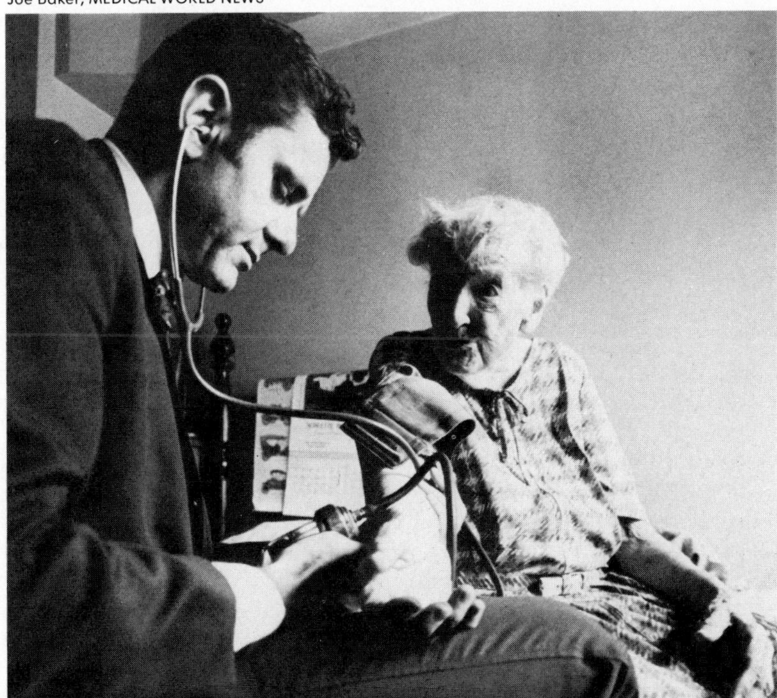

*Though young physicians can learn much from treating the elderly, often the lack of time and patience impels them to concentrate their efforts on younger patients, who respond quickly and provide a greater and more immediate sense of achievement.*

over 5,000 in 1972, had received more than 65,000 requests for living wills, documents designed to help physicians and families make decisions about life-prolonging measures in terminal illnesses where the patient may be unable to communicate his wishes. A person executing a living will indicates in part that "I do not fear death as much as I fear the indignity of deterioration, dependency, and hopeless pain." The signer asks that when any of these conditions are present he be allowed to die "in dignity." It was generally conceded that these documents did not bind the physician or the family legally. They did, however, help the consciences of both when the terminal patient was comatose or non compos mentis.

The legal and medical professions were still quite uncertain as to their respective roles in making the necessary decisions. How much should be left to the doctor, the doctor and the family, a committee of doctors, a committee with lay members, or the courts? Arguments were advanced for all of these, but in general the physician still took the major responsibility, in consultation with the family when possible.

Special problems arose in relation to organ transplantation. This new and very powerful technology depended on the recovery of viable organs from a donor as soon as possible after death. The kidneys of a person who is dead by any of the usual criteria remain viable for a limited time and can be transferred to a patient needing a transplant, but the shorter the time interval the better the chances for success. Obviously, a physician caring for both the donor and the recipient would be in a position of potential conflict over the interests of his two patients. It was, therefore, generally agreed that a different physician or group should be responsible for the donor's care and make decisions quite independent of the needs and wishes of the recipient and his physicians. Also, these decisions were generally made by more than one physician.

**Legal aspects.** Recognizing the need for changes in statutes relating to death, in 1971 the state of Kansas enacted the first law in the U.S. that attempted to redefine death and to take into account problems of organ transplantation. It read in part: "A person will be considered medically and legally dead if, in the opinion of a physician, based on ordinary standards of medical practice, there is the absence of spontaneous brain function; and . . . [that if] during reasonable attempts to either maintain or restore spontaneous circulatory or respiratory function in the absence of aforesaid brain function, it appears that further attempts at resuscitation or supportive maintenance will not succeed, death will have occurred at the time when these conditions first coincide. Death is to be pronounced before artificial means of supporting respiratory and circulatory function are terminated and before any vital organ is removed for purpose of transplantation."

This statute had a mixed reception in both medical and legal circles. Some greeted it as an important advance while others deplored it as both unnecessary and dangerous. In these times of rapidly changing technology, no consensus had developed in the medical world as to the criteria of death or its precise time. In an article in *Science* Robert Morison argued persuasively that death is a process and not an event; in which case, any attempt to fix an exact moment of death is necessarily arbitrary. In the same issue, Leon Kass vigorously defended the classical view that death must be defined and timed precisely as an event.

The courts in several cases upheld the right of an individual to refuse life-prolonging treatment. Thus, the right to be allowed to die had been established under some circumstances, especially, but not exclusively, where religious convictions were involved. On the other hand, other courts had held the reverse, and life saving measures were ordered despite opposition by a patient's family. As of 1972 there had not been enough cases in the courts to establish legal precedents.

**Thanatology: the care of the dying.** In addition to the medical and legal aspects of the care of the dying patient, there are psychological factors of great importance. These were well known to the old-fashioned family doctor who had little but comfort to offer his terminal patient. He usually presided over death in the home, among the patient's family, in circumstances favorable to human contact and psychological support. This pattern of dying now tended to give way, especially in the U.S., to a more common one of death in the hospital. There, in addition to the indignities associated with overenthusiastic prolongation of life, the patient may find himself lonely and virtually abandoned as the end draws near. The team of busy physicians and nurses who visited regularly and were solicitous and attentive early in the terminal illness find it more difficult to face the patient. Too rarely are they willing or able to answer the questions he really wants to pose and to help him understand and accept his impending demise with equanimity and with a hand to hold.

Medical educators were, fortunately, becoming more and more concerned about reintroducing the human touch into modern, scientific medicine. The Society for Medicine and Human Values, with support from the National Endowment for the Humanities, held a national meeting at Williamsburg, Va., in April 1972 where the role of the humanities in medical education was the central theme. In June 1972 another national conference was convened in Tarrytown, N.Y., by the Hastings center, under the directorship of Daniel Callahan, to evaluate the teaching of medical ethics in the nation's medical schools. This center was also serving as a clearinghouse for information on ethical issues in the life sciences and was publishing periodically its own ongoing research into the ethical, social, and legal problems of death and dying. Finally, Robert H. Williams of the University of Washington was engaged in editing discussions on death and dying by a number of distinguished authors.

—Thomas H. Hunter

## Dentistry

For decades the dental profession has advocated that teeth can last a lifetime through proper care. In early 1972 this idea was markedly advanced as dentistry began to move on many fronts toward stepping up the prevention of oral disease. The American Dental Association pointed out that the "profession now has the scientific knowledge to recognize and control dental disease and the technical ability to correct its harmful effects." A special ADA coordinating committee on preventive dentistry had the task of studying ways of implementing a large-scale prevention program that would include utilization of all research achievements and massive public dental education programs.

**Control of dental plaque.** Dental scientists long recognized dental plaque as a major culprit in the development of tooth decay and periodontal disease. The gummy masses of microorganisms develop constantly in the mouth; toothbrushing alone will not remove them. Effective plaque control, according to studies at the National Institute of Dental Research (NIDR), Bethesda, Md., can only be achieved through a variety of steps including proper eating habits, use of dental floss or tape to remove plaque and food debris from spaces between the teeth, use of water sprays and jets, use of solutions to identify areas where plaque is most likely to persist, and most of all, regular dental examinations.

**Tooth decay research.** Scientists at NIDR and at other research centers were studying the development of more effective and practical techniques for destroying existing plaque and blocking its further formation, possibly through the use of enzymes and chemicals. A number of antibacterial agents were discovered to be highly effective but had not been sufficiently tested for human use.

More progress was reported with studies on adhesive sealants that are painted on the occlusal (biting) surfaces of teeth to seal off areas particularly susceptible to decay. The sealants could not be applied to the sides of the teeth and therefore would not protect against smooth-surface or gum-line cavities. The ADA announced provisional recognition of one sealant material as an agent

Courtesy, Dr. Reidar F. Sognnaes, University of California at Los Angeles

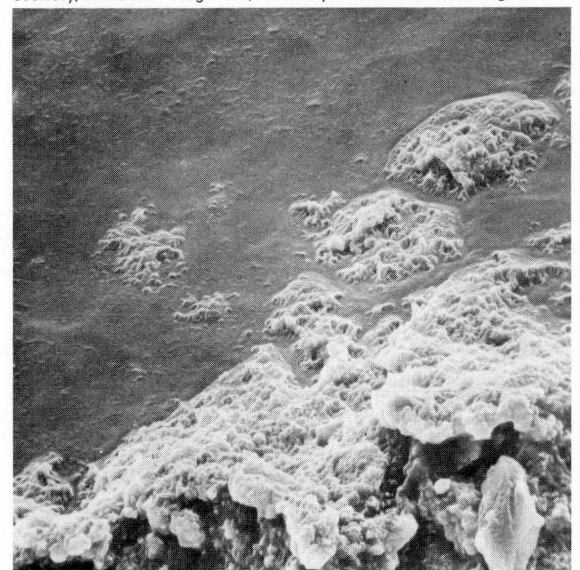

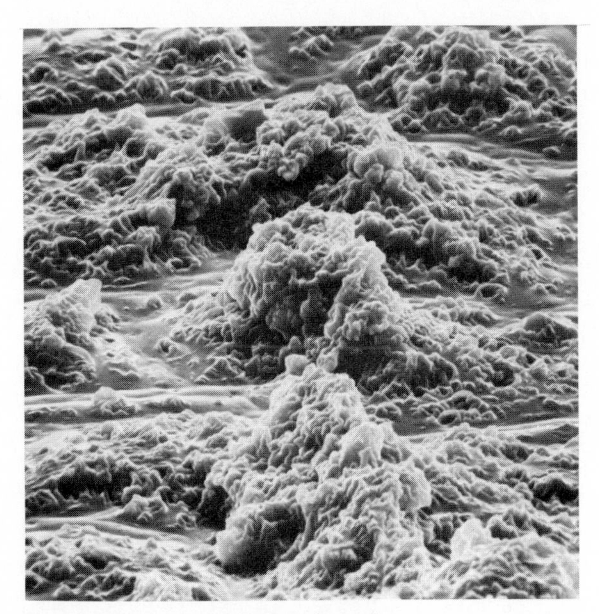

*Precise silicone impressions of living tissue reveal islands of cancerous growth next to normal mouth tissue (left) and a malignant growth under an ill-fitting denture (right) when magnified under a scanning electron microscope. The technique can be used to replicate any easily accessible tissue.*

capable of "restoring or sealing off anatomically deficient regions of the tooth," but cautioned that this recognition in no way pertained to any possible decay-preventing characteristics of the product.

**Emotional stress and gum disease.** Unusual emotional and physical stress may be the cause of a type of periodontal disease in teenagers and young adults known as pericoronitis (an inflammation of the gum tissues surrounding the crowns of partially erupted teeth). In a study conducted at the University of Kentucky dental school, more than 70% of 136 patients suffering from the disorder were under severe emotional or physical pressures. Unerupted third molars or wisdom teeth were most commonly affected. Fatigue, financial worries, and, in women, menstruation were mentioned among major causative factors.

Patients suffering from advanced periodontal disease frequently incur bone loss in the jaw. To correct this, a U.S. Army dentist, Robert G. Schallhorn, developed a unique surgical technique in which he implanted bone grafts from the patient's hip in the defective areas of the jaw.

**Oral cancer prevention.** Smoking was linked to the development of oral cancer and, according to University of California at San Francisco dental scientist Sol Silverman, Jr., even if an oral cancer patient quit smoking he might still develop a second oral malignancy. Such patients, however, might have a better chance of preventing a second malignancy than those continuing to smoke.

At the University of California at Los Angeles, Reidar F. Sognnaes developed an elastic silicone process for making living tissue impressions precise enough to be magnified thousands of times under the powerful scanning electron microscope (SEM), thus helping to detect early oral cancer. The technique provided nondestructive and painless micro-replication of all easily accessible hard and soft tissues, such as those in the mouth. Malignant cancer growths bulging up next to normal tissue were clearly visible in micro-replicas examined under the SEM. Small silicone discs carrying the fluid impression mixture are placed on the surface being examined. In a few minutes the mixture hardens and the replica can then be examined directly or be converted to a positive replica made of a harder resin material. The replica is also coated with a thin film of evaporated gold or other material to make the surface conductive to the beam of the SEM.

—Lou Joseph

## Paramedical personnel

In 1900 the doctor's chief help in caring for the sick came from the patient's family, and most medical services were rendered in the home. At that time there was approximately one nonphysician health worker for every doctor in the United States.

In 1972 almost all medical services were performed in offices, clinics, and hospitals, and there were an estimated 12 nonphysician workers for every doctor, a ratio that continues to rise annually. Of these workers, about half were in nursing and its related professions; about 7% worked as secretaries, receptionists, and in similar positions; slightly more than 6% were in dentistry and its related fields; and slightly less than 6% were various types of environmental control workers. Pharmacists and their helpers, clinical laboratory technicians, physical therapists, various basic scientists, dieticians, and administrators formed the remainder of this large segment of the U.S. work force.

How and for what reason have these professions evolved? Some resulted from the subdivision of already existing professions. Two examples are X-ray technology and laboratory technology. At one time a doctor did his own X rays (radiographs) and his own laboratory work. In the course of time, it became apparent that these tasks could be taught to nonphysician helpers, and that this delegation of work was more efficient. At first, doctors trained their own helpers, but soon schools were created for the purpose.

Other professions emerged in response to new needs. Physical therapy arose out of a post-World War I need for professional persons to care for disabled veterans. Radiation safety, a relatively new field, grew from a need for persons to set up testing and surveillance programs to ensure the safety of patients and workers in hospital radiographic units. These skills were never possessed by more than a very few doctors, and were therefore the basis for truly new professions.

Impetus for this continuing creation of more and more medical professions came from the increasing specialization and institutionalization of medical care. Because of the rapid increase in knowledge and the parallel development of new technical capabilities, new skills became necessary. An example is kidney transplantation. This demands a team to operate the artificial kidney machine while the patient awaits a transplant; an immunologist to determine the best match of donor kidney to recipient patient; and, finally, a team that receives the donor kidney, preserves it, and prepares it for the procedure. All this is in addition to the surgeon who actually performs the transplant, along with his own team of helpers.

**The generalist helper.** While these developments have created a demand for more "specialty" workers in the hospital setting, many segments of the population have not been receiving adequate care for less sophisticated but far more common medical problems, and some members of society have practically no access to medical care. As the magnitude of this problem became apparent to the public and to medical, nursing, and other health professionals, there was a corresponding interest in the creation of still other categories of health workers to deliver needed services to these underserved groups.

*Paramedical students learn laboratory procedures (left) and practice artificial respiration (right) during the course of a six-month training program at the University of New Mexico. The goal of the program is to improve the quality of health care in rural counties, many of which have no doctors.*

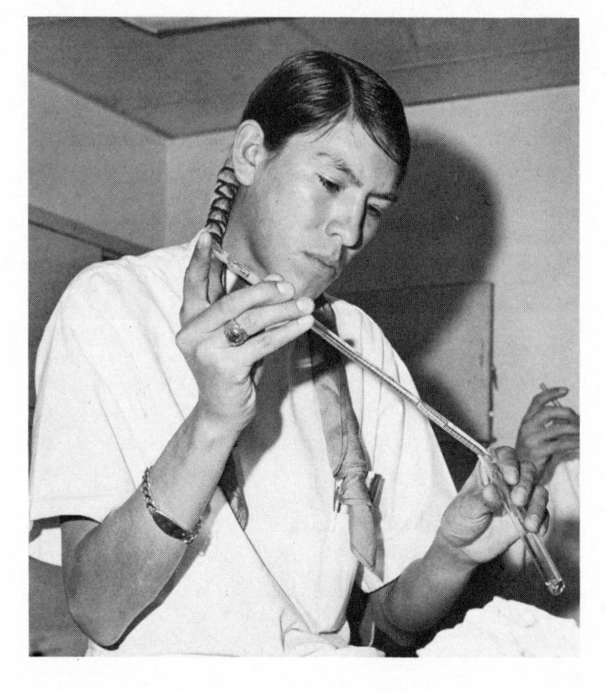

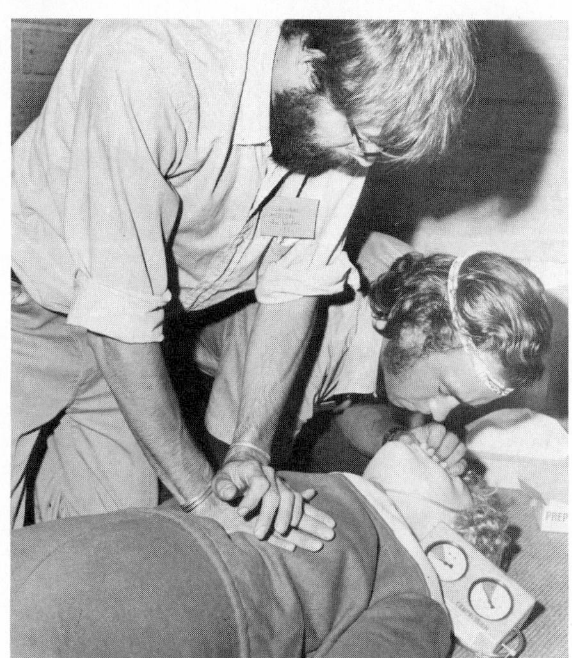

Margie McCurry, University of New Mexico

Joe Baker, MEDICAL WORLD NEWS

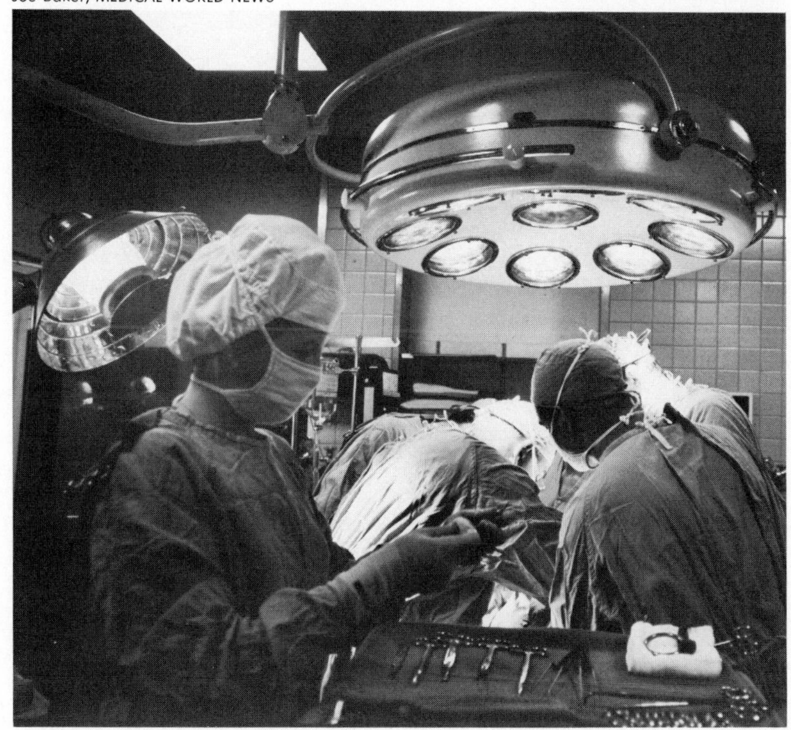

*Operating room nurses at Monmouth Medical Center in Long Branch, N.J., have become better informed about the total patient-care program through a series of regular conferences with physicians and post-operative patients. Many nurses had expressed a feeling of isolation from their patients because of the technological orientation of surgical procedures and the lack of post-operative contact.*

The year 1971 was characterized by the emergence of physician's assistants, nurse practitioners, and other "generalist" assistants as accepted members of the health care team. Although these groups had been on the scene for five or six years, there had been considerable debate regarding their capabilities and, in some cases, serious controversy. During 1971 several developments all but assured the permanence of these new professions. There was a sharp increase in the number of operational training programs. A number of states passed laws recognizing these assistants and providing a legal framework for their activities. The American Medical Association (AMA) adopted a set of minimal educational standards for approved training programs for "assistants to the primary care physician" (as opposed to assistants to various medical specialists). The federal government also recognized the potential importance of such programs, and the creation of new programs was designated in the president's health care message as an area for special emphasis.

These new professionals came from various backgrounds. Physician's assistants usually had previous experience as military corpsmen, but an increasing number were entering the profession from nonmilitary backgrounds. Most were trained in two-year programs in which about half of the curriculum was classroom work, the other half practical work under physician direction. The nurse practitioner was usually a registered nurse who had had six months to a year of added training. The "Medex" was the product of a training program that accepted trainees with previous military or civilian experience and then provided 3 months of intensive instruction followed by 12 months of on-the-job training in a doctor's office.

Relatively few representatives of these new professional groups had been integrated into actual work settings in 1971. According to an AMA survey in October 1971, 96 physician's assistants had graduated (74 from the Duke University program) and Medex programs had graduated 56 persons (22 from the University of Alabama, 19 from the University of North Dakota, and 15 from the University of Washington.)

**Licensing and certification.** These developments highlighted some major problems associated with licensure, certification, and accreditation. The higher levels of medical and related professions are generally licensed by individual states; each state has passed laws defining the role of each profession, setting educational and ethical standards for entry into the profession, and establishing an examination to test skills and knowledge. Such laws exist for doctors, dentists, and nurses, and occasionally for other professionals as well. These laws are designed to protect the public by ensuring a minimum level of competence before entry into practice.

Other health professionals are not licensed, but are certified by a given training program as having

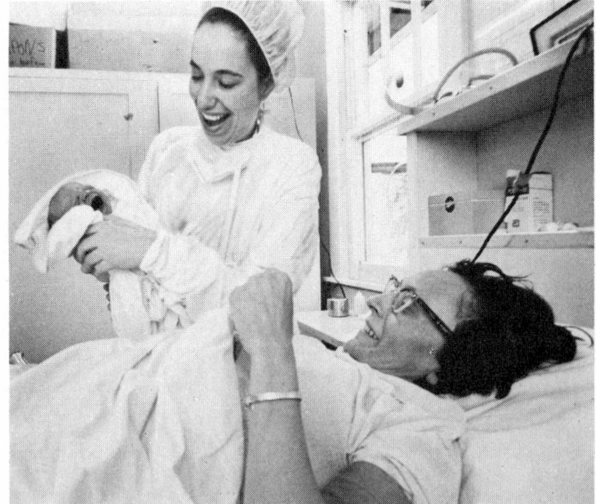

*A nurse-midwife, an R.N. with specialized training to help ease the nationwide doctor shortage, by Kentucky's Frontier Nursing Service presents a newborn baby to its mother shortly after delivery.*

completed a prescribed course of study. Some programs of this type are accredited by various professional bodies as being in conformity with established minimum standards. The AMA, for example, has set up criteria for a number of such programs and accredits those that meet or exceed these criteria. Most certification and accreditation programs, like licensure programs, are established with quality control as the primary objective. However, it is also recognized that such mechanisms can become the means whereby a profession protects its own interests by preventing competition.

In August 1971 the U.S. secretary of health, education, and welfare, Elliot L. Richardson, announced the recommendations of a special committee appointed to study the problem. The states were urged to observe a two-year moratorium on enactment of legislation creating new categories of health-care personnel. The states were also urged to expand existing health practice (licensure) acts, granting broader delegational authority to already licensed professionals. The two-year period was to be utilized to develop equivalency and proficiency examinations for use on a national basis. Inclusion of public (consumer) representatives on health licensing boards was also recommended.

**Community health aides.** Another significant development over the past few years has been the use of community health aides, particularly in low-income, medically deprived areas. These are generally residents of the area to be served, and usually have had no prior formal health training. Often they were employed after efforts to recruit more formally trained personnel had failed, and frequently their employment was opposed or received with misgivings by conventional health-care personnel.

In almost all areas where they have been employed, these workers have more than met expectations. An often encountered comment is that they were able to communicate with residents of the areas where they work when regular professionals could not. Generally, they have been trained in short courses of several weeks' duration, followed by on-the-job supplementation of training. Their jobs vary considerably. Some are trained to educate area residents regarding immunizations and regular health care. Others assist in family planning and in drug abuse programs. Still others assist in direct medical services. In all cases, the empathy, understanding of problems, and ability to communicate have proved highly useful and unique assets, far outweighing the lack of formal instruction.

**Further developments.** Federal support for various paramedical programs received a great boost in 1971 in the form of the Comprehensive Health Manpower Training Act. This act provides added support for training institutions, through special improvement grants and through incentive grants for programs producing physician's assistants and nurse practitioners.

There have also been a number of studies aimed at measuring the improvement in services associated with use of physician's assistants and nurse practitioners. Reports suggest that the use of a physician's assistant augments the volume of medical services produced by 50 to 70%. Analyses of the pediatric nurse practitioner indicate that such a person can effectively handle 80% of the cases once seen by the pediatrician.

Pharmacy faced a number of issues during the year, most of which remained unresolved. The primary issue was should the pharmacist remain as a dispenser of drugs, or should he redefine his role, assuming a larger responsibility in health care? Most observers agreed that the present educational base of the pharmacist should enable him to assume a broader set of functions than merely operating a retail drugstore. One possibility was that he could function within the hospital as an expert on drugs—in charge of dispensing drugs, consulting with doctors on the choice of drugs, and instructing patients regarding them. Another was that he could serve as a primary contact with the patient in areas with a shortage of doctors. In many cases, the latter function would require changes in state laws. It was noteworthy that those who favored changing the role of the pharmacist usually favored the training of a pharmacist's assistant to dispense drugs under his supervision.

Within the nursing profession, a considerable amount of energy was being expended on discussion of the proper stance of nursing vis-à-vis the physician's assistant. Many nurses felt that widespread assumption of such a role by nurses would negate the gains that had been fought for during the past 20 or 30 years in obtaining recognition of nursing as an independent profession. Others believed that a professional colleagueship with the doctor in the care of patients is a satisfying role, in no way threatening to the professional stature of the nurse. The more youthful members of the profession seemed much more inclined toward the latter view.

—E. Harvey Estes

## Psychiatry

During the past year, researchers in the field of psychiatry made a number of significant discoveries. Work was done on the genesis and treatment of such diverse illnesses as schizophrenia, temporal lobe epilepsy, and alcoholism, as well as manic-depressive psychosis.

**Depression.** Lithium carbonate, a drug that had been reported to be effective in the management of manic-depressive psychosis in 1943, was finally cleared by the U.S. Food and Drug Administration for use with manic patients. Numerous papers published during the year attested to the therapeutic usefulness of this compound in the alleviation of mania or hypomania. In addition, recent reports from England and Scandinavia indicated that it may be effective in the treatment of bipolar depression.

Studies by the Scandinavian researcher K. Leonhard indicated that there are two types of depressive illness. One, referred to as unipolar, occurs in patients without any history of a manic attack. The other type, which Leonhard labels bipolar, occurs in patients who have had mania in the past. R. Fieve at Columbia University showed that lithium carbonate is effective in alleviating the symptoms of depression in patients with bipolar illness. Further, lithium carbonate seems to be associated with a decrease in the frequency of recurrent attacks. Lithium does not appear to be effective in the treatment of unipolar depression, however.

This finding supports the concept that patients with bipolar illness may suffer from some form of biochemical abnormality that is corrected by the

*Actors stage "The Concept," a psychodrama on drug abuse, for the personnel of a New York company. Large corporations throughout the U.S. are discovering many drug users among their employees and are sponsoring several types of programs to combat the problem.*

Bill Schropp

administration of lithium carbonate. Further evidence is provided by genetic studies. Several researchers have noted that a large number of bipolar patients have bipolar parents. George Winokur at the University of Iowa showed that, in two families under study, family members who were color-blind were also bipolar patients, indicating a genetic linkage between color blindness and bipolar depression.

**Schizophrenia.** Whether environment or genetic makeup is related to the development of schizophrenia has long been the subject of debate. D. Rosenthal at the National Institute of Mental Health (NIMH), in reviewing case records of the offspring of schizophrenics who were put up for adoption at birth, noted that even though these patients had been adopted by nonschizophrenics, roughly one third of them developed schizophrenia in later life. However, a small number of adoptees from nonschizophrenic families who were raised by schizophrenics also developed schizophrenia. Rosenthal concluded that both genetic and environmental factors are important in the development of the disease.

*A methadone "disket," widely used in heroin addiction therapy, begins to dissolve in water. Properly prescribed and taken, methadone curtails an addict's craving for heroin and stabilizes his condition so that rehabilitative measures have a chance to succeed.*

Leonard Kamsler, MEDICAL WORLD NEWS

Chemotherapy continued to be the primary treatment for schizophrenia. Chlorpromazine was joined by haloperidol as an effective antipsychotic agent. Studies by Larry Stein of Wyeth Laboratories, Radnor, Pa., indicated that the administration of chlorpromazine during the first few months of schizophrenia may prevent the formation of an irreversible brain lesion. Stein hypothesized that schizophrenia in man can be explained on the basis of the destruction of the pleasure center in the septal region of the brain by the formation in the brain of a toxin, 6-hydroxydopamine, which has been shown to destroy selective areas of the brains of rats. Stein identified a metabolite of 6-hydroxydopamine exuded in the sweat of schizophrenics. Since administration of chlorpromazine to rats during the first six months of 6-hydroxydopamine injections prevented the destruction of brain centers, Stein believed that chlorpromazine used sufficiently early in the treatment of the schizophrenic patient could arrest irreversible destruction of the brain's pleasure center. His work had yet to be confirmed, however.

Other researchers believed that the delusions of schizophrenia are produced by a hallucinogen manufactured by the human brain. Under the direction of Julius Axelrod, researchers at NIMH discovered that the human brain converts tryptamine to n-methyl-tryptamine and di-methyl-tryptamine. The latter compound is a hallucinogen, raising the possibility that the delusions of a schizophrenic may be explained on the basis of genetic predisposition to form di-methyl-tryptamine. It is also possible that schizophrenic patients form the compound when undergoing stressful situations. Research was in progress to determine whether schizophrenic patients do in fact synthesize di-methyl-tryptamine in their brains.

**Violent and aggressive behavior.** Many patients who experience episodes of violence and aggressive behavior as a result of a disease known as temporal lobe epilepsy can be treated with tranquilizers, psychic energizers, and antiseizure drugs. In the past, if these compounds were unable to control violent outbursts, such patients were subjected to the surgical removal of part of their temporal lobe. This required quite a large incision, the sacrifice of a large amount of brain cortex that was not diseased, and the removal of a large amount of skull bone.

V. H. Mark and F. R. Ervin of Harvard University perfected a surgical technique in which tiny elec-

Culver Pictures

*An etching by William Hogarth, published in 1735, depicts Londoners amusedly watching the cruel treatment of mental patients in Bedlam. As recently as January 1972, cruelty to patients in a British mental hospital was the subject of an official inquiry. A plan released in 1972 by Britain's Department of Health and Social Security proposed closing all the nation's mental hospitals and replacing them with small psychiatric units in local hospitals and outpatient care facilities.*

trodes are implanted in the brain and used to destroy those few brain cells that act abnormally and that in the past have been associated with the symptoms of seizure and violence. The therapeutic results in patients suffering from unilateral or bilateral temporal lobe epilepsy have been excellent. It must be remembered, however, that this procedure is applicable only to those violent patients whose behavior stems from temporal lobe epilepsy. Mark and Ervin estimated that this disorder occurs in one out of ten people who exhibit excessively violent behavior.

**Community and social psychiatry.** The state of California closed three large state hospitals during the year. This was done in response to the decreasing patient population in these facilities, as well as to the increased availability of community mental health centers. Since chlorpromazine became available for the treatment of mental illness in 1955, the patient population in large state mental hospitals has decreased steadily. At that time there were more than 550,000 patients in state and county mental hospitals; as of 1972 the number had fallen to less than 400,000, despite the increase in the population as a whole. The number of admissions to state mental institutions rose during this period, but the patients were hospitalized for a much shorter time.

**Animal models of psychosis.** One of the factors that has prevented an early understanding of the causes of mental illness has been the lack of an animal model of any psychiatric disturbance. Whereas the infectious diseases, cancer, and a number of metabolic diseases affect animals as they do man, the mental illnesses that afflict man have appeared to lack counterparts in animals.

Working with rhesus monkeys, Harry Harlow and William McKinney at the University of Wisconsin developed an animal model of autistic behavior, involving severe withdrawal and inability to relate. Animals were separated from their mothers early in life and reared in isolation in large metal chambers. Their behavior following confinement in this chamber for two months or longer appeared autistic—the animals had little interest in food or sex, they were afraid of monkeys their own age, and did not enter into the play activities that their normal siblings enjoyed. Whether or not this behavioral state is related to schizophrenia or to any other human psychosis was still in question. Two forms of therapeutic intervention were successful in reversing this autistic behavior. Observations by the group in Wisconsin revealed that if one of the autistic monkeys was placed in a cage with another, younger monkey, this younger and more socially adept "therapist" monkey served to increase the socialization of the autistic animal. Over the course of several weeks the "therapist" monkey was able to engage the autistic animal in play with other animals, and after several months of group interaction with the "therapist" monkey the autistic animal seemed to be highly socialized with respect to the other animals in the group cage. The Wisconsin researchers also noted that the administration of chlorpromazine reversed some of the symptoms of the maternally deprived animals. Thus, both environmental and drug treatments appeared to be effective.

**Future developments.** Rubidium carbonate showed some promise as an antidepressant in pilot studies under Fieve's direction at Columbia University. It has been thought by many researchers that depressed patients may suffer from a functional deficiency of norepinephrine in the brain. Rubidium carbonate increases the rate of release and uptake of norepinephrine in the brains of rats and monkeys, and it has been hypothesized that the compound may be effective in increasing the functional availability of norepinephrine in the brains of depressed patients.

With the federal government placing great emphasis on the treatment of alcoholism and drug abuse, it was not unlikely that major discoveries would occur in this area in the near future. The finding by E. P. Davis in Houston, Tex., that alcohol may lead to the formation of a substance similar to morphine stimulated research to discover whether this morphine derivative is responsible for the addicting effects of alcohol. Several investigators were attempting to isolate this morphine derivative in the spinal fluid of alcoholics.

To date, methadone has proved to be the most effective compound in the treatment of heroin addiction, but its effects are short-lived, requiring frequent dosage. A possible replacement was $\alpha$-acetyl-methadol, a methadone derivative, which was shown to be longer lasting in its antinarcotic effect than methadone. It is also more potent, and therefore can be given in a smaller dose. Another drug treatment was being developed by Sydney Spector at the Roche Institute of Molecular Biology, Nutley, N.J. He produced antibodies to morphine and heroin in animals, and it was possible that the administration of these antibodies to man would eliminate the narcotic effects of heroin.

Studies on the use of nondominant brain hemispheres were expected to reveal some interesting therapeutic results. This work, carried out under the direction of E. Callaway at the University of California in San Francisco, focused on the training of stroke victims to use their nondominant hemisphere. The training involved a feedback technique using electroencephalographic (EEG) tracings taken from scalp electrodes, which were displayed to the patient in a feedback loop. The EEG reading enabled the patient to know when he was using his dominant or nondominant hemisphere in a specific process. The possibility of tapping the potential of the nondominant side of the brain, which normal man has never used, is tremendously exciting and may offer hope for thousands of stroke and cancer victims.

—H. Keith H. Brodie

## Venereal disease

In 1972, venereal diseases continued at pandemic levels in the United States, as an estimated 80,000 new cases of infectious syphilis and 2.2 million new cases of gonorrhea joined the carrier pool. Reported congenital syphilis continued its rise from 300 cases in 1970 to 397 in 1971.

The incidence of venereal diseases in the U.S. varies widely by region and age distribution. Not surprisingly, the greatest incidence occurs in areas of high population density and among the largest age groups within a population. Thus, as the mean age of population decreases, the incidence of venereal disease among young people increases.

Regional variation is more complex: the national rate of infection for gonorrhea is 285 per 100,000 population. Yet in fiscal 1970 (July 1, 1970, through June 30, 1971) gonorrhea occurred at the rate of 2,486.7 per 100,000 in Atlanta, Ga.; 1,262.2 in Chicago, Ill.; and 2,121.4 in San Francisco, Calif. The average rate of gonorrhea infection for the 61 major cities of the U.S. was 640.2 per 100,000. For the same year, syphilis rates ranged comparably from 10.4 cases per 100,000 population nationally to 93.4 cases per 100,000 in Newark, N.J.

**Whence VD?** Venereal disease is an old problem. Gonorrhea has been with us at least 5,500 years, and although the history of syphilis is not as clear, it does extend beyond the memory of modern Western man. The spread of venereal disease is encouraged by its high infectious rate and by its status as a shameful and secret illness. Equally infectious diseases such as smallpox have been conquered, but venereal disease eludes eradication, due in part to a fundamental contradiction between sexual mores and actual practice.

Historically, the chief prophylactic, or preventive, measure offered the general public against venereal disease has been abstinence from sex until the beginning of a monogamous marriage. In fact, however, if one generation had ever practiced virginity until monogamous marriage occurred, no venereal diseases would exist. The fact that gonorrhea and syphilis together outnumber all other communicable diseases attests to the ineffectiveness of such an approach and to the serious need for more realistic preventive measures as well as treatment.

**VD in the twentieth century.** During World War I, venereal disease ranked second only to influenza among infectious diseases. Venereal disease kept men out of the service and, once in service, out of combat. Thus, the military began a massive venereal disease control program which ended at the conclusion of the war. Not until the build-up for World War II was another major program begun, and with the discovery of the effect of penicillin on venereal diseases, early in the 1940s, their eradication seemed in sight.

Unfortunately, that very possibility led to changes in the treatment of venereal disease which virtually assured its survival. With the advent of penicillin, venereal disease treatment was taken out of the hands of the U.S. Public Health Service and shifted into those of the private physician. The American Social Health Association estimated that in 1972 private physicians treated 75% of all venereal disease cases yet reported less than 15% of them, despite mandatory state reporting laws. Because venereal diseases are highly infectious, this failure to report cases and follow up on infectious contacts leaves the diseases free to spread.

Failure to report also accounted for an apparent decline in venereal disease infection, and consequent reduction of venereal disease control programs. Following World War II, tax dollars allocated for venereal disease control were cut from 11 cents per capita to 3 cents per capita, closing down federal, state, and local case-finding programs. By the late 1950s, alarming new rates of syphilis infection had appeared, and in 1961 a federal task force was created to recommend steps to eradicate the disease. 1971 witnessed the formation of a second task force, this time to investigate the epidemic of gonorrhea.

**Why the rise?** A secondary effect of the discovery of penicillin as an effective treatment for venereal disease was public apathy. With a fully effective curative treatment at hand, one could assume that the danger of infection had passed. Many assumed just that, and paid less heed to available preventive measures.

Among the chief preventive measures generally available was the condom. Military experience in World Wars I and II indicated that use of the condom could substantially reduce, if not eliminate, the danger of infection by genital contact. An Army study demonstrated that gonorrhea infection rates could be reduced from the 625 per 1,000 current in the Army at that time to 35 per 1,000.

A second factor in venereal disease control, principally control of gonorrhea, was the use of female contraceptive devices. Although not intended as venereal disease prophylactics, the diaphragm and various contraceptive foams and jellies effectively prevented infection of the cervix by the gonococcus. Use of the birth control pill, which had become widespread by the early 1960s, displaced these devices and opened the way for increased gonorrhea infection.

The pill had one final effect which passed unnoticed for some time. Oral contraceptives cause the normally acid vaginal flora to become alkaline, and increase moisture in the vaginal tract. Gonococci thrive in conditions of moisture and alkalinity, escalating the chances of infection on a single exposure from 40% in the female not taking the pill to 100% for the woman who is.

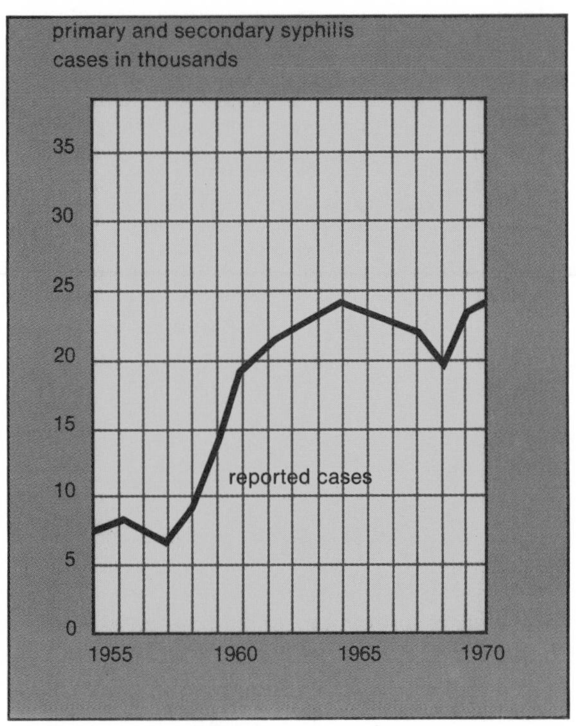


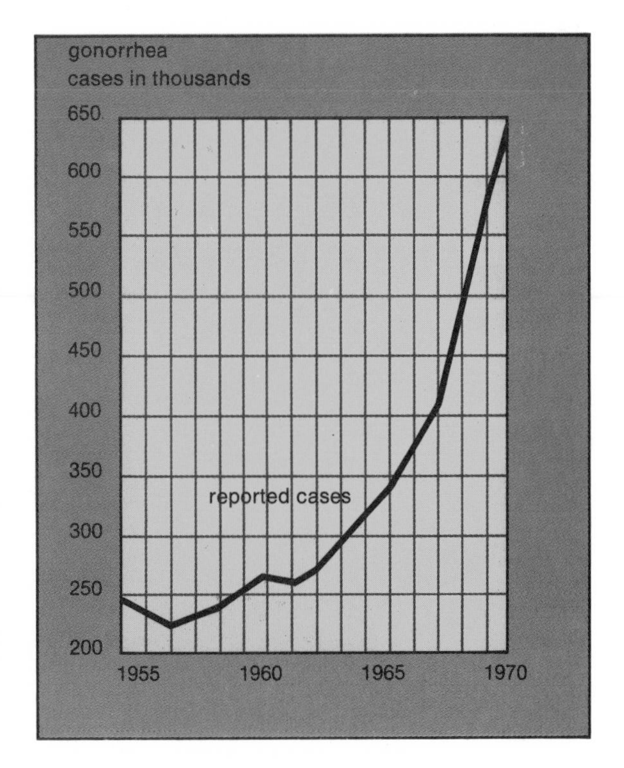

Fenga and Berkovitz, NEWSWEEK

**VD today.** Today, as in the past, venereal disease control depends on five basic procedures: public education; development of effective prophylactics; case finding; treatment; and follow-up. In a sense, these measures are self-evident. They are standard for the conquest of any communicable disease, and have been apparent for many years. With venereal disease, however, shrouded as it is in centuries of unjust prejudice, shame, and myth, conquest depends on the cooperation not only of the public but of the medical profession, drug manufacturers, and government agencies active in funding and carrying out effective programs. Venereal disease education illustrates the interrelationship of these groups.

To be effective, venereal disease education must deal with access to the health care delivery system. It is not enough to inform the public about the symptoms and consequences of venereal diseases. We must also be certain that physicians, hospitals, and the public health services are prepared to handle the huge caseload that would result from effective education. At present this is not so. Many physicians are not adequately trained in detecting and treating venereal diseases; others suffer from the same taboos about venereal disease that afflict the general population. Public health services were admirably designed to tackle case finding, treatment, and public education. However, today, with the barest minimum of funding in relation to the problem, they are unable to provide the services required. Finally, the finest case finding and treatment services will not avail unless financial barriers to care are removed. Even free treatment will fail to solve the problem if unrealistic scheduling or remote locations add indirect costs such as transportation and time off work.

*A poster prepared for the United Givers Fund expresses the growth of venereal disease among teen-agers.*

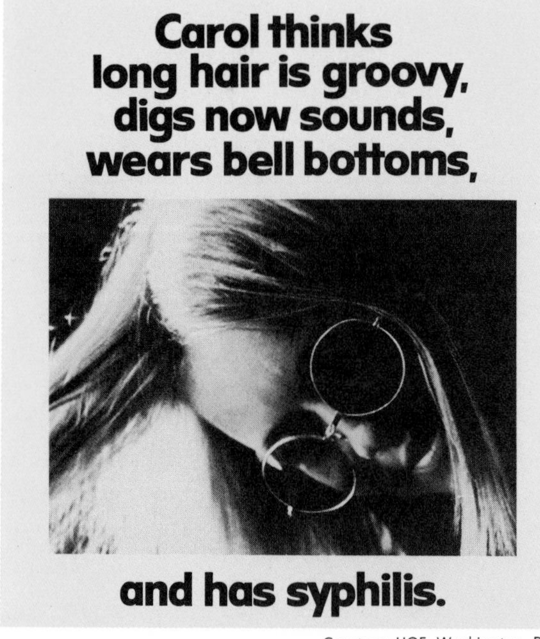


Courtesy, UGF, Washington, D.C.

**National Commission on VD.** In March 1972, the National Commission on Venereal Disease issued its report, detailing the rise of venereal disease infections to their highest point since the introduction of antibiotic drugs 20–30 years earlier. Warning that venereal disease was epidemic and gonorrhea "out of control," the commission offered 40 recommendations aimed at controlling the spread of these diseases. The commission added that even if its recommendations were fully implemented the number of cases could be expected to continue to rise in 1973.

The campaign recommended by the commission would include a voluntary venereal disease education program for school children starting in the seventh grade; enforcement of existing mandatory tests for syphilis and gonorrhea; stricter reporting; and more money for case finding, follow-up, and treatment. Enforcement of mandatory tests alone would result in the screening of 40 million people in fiscal year 1973. The commission recommended a federal investment of almost $300 million over the next five fiscal years, including $46.1 million for the coming year. However, in 1972 the federal budget called for spending only two-thirds of that amount. Commission officials pointed out that while an estimated $43 million is being spent this year on venereal disease control by all groups combined, the total cost of the effects of syphilis and gonorrhea would be $364 million for the same period.

**Beyond the commission.** Broader steps against venereal disease will entail reevaluation of many attitudes toward venereal disease itself and the nature of health care, and a restructuring of the systems that deliver all forms of treatment today. Stricter observation of those firms engaged in testing and manufacturing prophylactics and curative drugs may be required. And recommendations for increased government and medical school funding for venereal disease control programs will require close scrutiny to prevent the creation of a new class of "VD professionals" and assure the delivery of services to the public.

Realistic prevention programs must operate at several levels to identify the massive pool of those carriers who show no symptoms of disease, treat them and their contacts, and reach new cases. The first level of prevention should be case finding, currently impeded by failure to report cases. Mandatory testing should be expanded to screen more potential cases; for example, marriage require-

Paul Kniskern, MEDICAL WORLD NEWS

*Alan Hinman, New York's chief of epidemiology, examines a chart diagramming the sexual contacts of a gonorrhea carrier. New federal grants to state health departments totaling $16 million were to be used for diagnostic laboratories and screening programs to seek out and eradicate the disease in asymptomatic carriers.*

ments could add a gonorrhea test to the syphilis screening already in force. Tests for both syphilis and gonorrhea should be required as part of prenatal care. The U.S. Public Health Service began a pilot screening program for gonorrhea among women having a routine vaginal examination. Other times when a patient routinely passes through the hands of a physician could be developed as opportunities for screening.

All partners of an infected person must be identified, tested, and treated. This second level of prevention, treatment of all exposed persons regardless of test results, must be enforced because of the relative inefficiency of tests and the great increase of gonorrhea that does not present obvious symptoms.

The third level of realistic prevention is the development and use of effective prophylactics, devices that will prevent infection upon exposure. In this area, vaccines are the traditional remedy. But research on vaccines is hampered by the nature of the diseases and the lack to date of a reliable experimental animal model. The potential prophylactic role of contraceptives other than the pill has been largely ignored, despite a report issued by Pfizer Laboratories that indicated that eight such devices have been proven in laboratory studies to prevent gonorrhea, syphilis, and other vaginal tract infections as effectively as they prevent conception.

Realistic prevention depends finally on each of these steps, and upon the equitable distribution of all the services they entail. If treatment, including access to prophylactic drugs, is limited only to those who can afford to pay the direct and indirect cost of service, venereal diseases will continue to spread epidemically.

—Jerry Lama

*See also* Feature Articles: TEST-TUBE BABIES; CANCER UNDER ATTACK; THE NOTORIOUS TRIO: MERCURY, LEAD, AND CADMIUM; NUTRITION AND THE AMERICAN DIET; Year in Review: VETERINARY MEDICINE.

# Microbiology

Advances were made in several areas involving microscopic organisms during 1971–72. As is always the case, these results had implications in other biological and medical sciences.

**Subcellular studies.** Ecologists had long had the problem of how best to measure community activity in a given natural environment. For example, it was difficult to determine whether forms of microbial life observed in a natural environment were metabolically active, dormant, or even dying. Recent studies of the cell's fundamental energy-producing mechanism, however, led to the development of a technique that may be used as an objective criterion of community activity.

Adenosine triphosphate (ATP) is the principal compound involved in a cell's energy-yielding and energy-requiring reactions. Energy is released from ATP by hydrolysis (decomposition by the action of water), and ATP is converted to adenosine diphosphate (ADP) and adenosine monophosphate (AMP). Thus, these three nucleotides couple all of the metabolic reactions of living cells. The amount of metabolically available energy that is

Courtesy, Litton Bionetics, Inc.

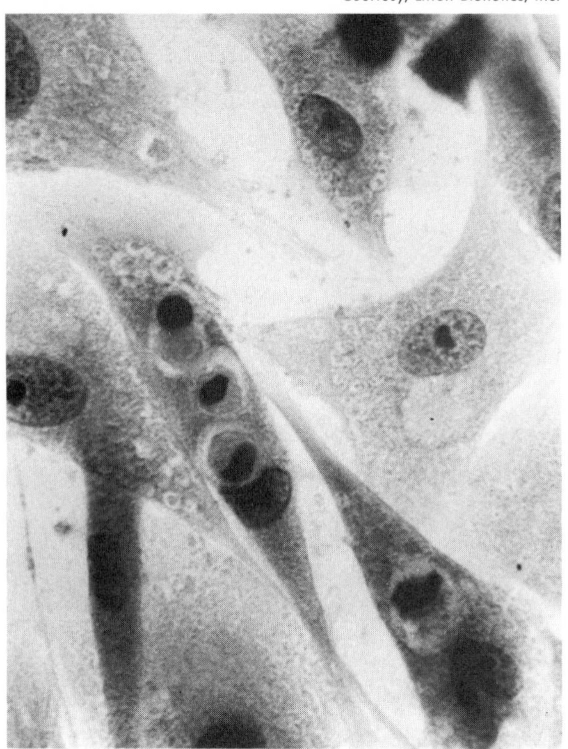

*Degeneration characterized by abnormal structures within the nuclei and by cavities near them was found in lung cells infected with Mason-Pfizer monkey virus. The virus is characteristic of tumor-causing RNA viruses, but its effects on cells had never before been observed.*

momentarily stored in this system is related to the mole fractions, or concentrations, of ATP, ADP, and AMP in the system. This value was being termed the "energy charge." The value of the energy charge during growth of bacterial and mammalian cells is about 0.8. Maintenance of life, but not growth, seems possible at energy charge values between 0.8 and 0.5. Energy charge values below 0.5 appear to be incompatible with maintenance of life, and such cells appear to be in a dying state.

The bacterial cell membrane regulates the transport of nutrients into the cell and wastes out of the cell, contains enzyme systems involved in energy production and in respiration, and serves as the site of synthesis of many cellular components. All biological membranes are composed almost entirely of two classes of compounds, proteins and lipids (fatty subtances). The proteins have enzymatic or transport activities and provide the membrane with its distinctive functional properties. The actual molecular organization of membranes, however, remained poorly understood.

A recent concept visualized a fluid mosaic model for biological membranes. In this model, globular protein molecules are considered to be partially embedded in a matrix of lipid, with the protein molecules being mobile in the plane of the membrane. This fluidity might be important in the transport of substances across the membrane and in other biological functions peculiar to the membrane. Recent experiments provided strong supporting evidence for the fluid mosaic nature of the bacterial cell membrane. Bacteria were induced to synthesize the protein that transports the milk sugar lactose into the bacterial cell. It was found that this transport protein was inserted in all regions of the cell membrane rather than at fixed regions. These experiments also showed that the lactose transport protein molecules were inserted equally well into both previously and newly synthesized portions of the membrane.

In the area of genetics, scientists presented indirect evidence that they had succeeded in introducing an active bacterial gene into human cells. The human cells were derived from a patient with galactosemia, a disease caused by the absence of the gene that directs the synthesis of the enzyme enabling the person to metabolize the sugar galactose. A bacterial virus picked up the specific gene from bacterial cells that could metabolize galactose. When the human cells were exposed to the virus, an active enzyme was produced that enabled the human cells to metabolize galactose. Whether this gene could be replicated by the human cells and passed on to the daughter cells was not determined. It was thought by some scientists that human genetic diseases may be cured some day by such "genetic engineering."

**Disease-causing microbes.** Infection of newborn infants by the bacterium *Staphylococcus aureus* (staph) acquired in hospitals continued to be a serious problem. Although staph is a common human parasite, especially deadly and often drug-resistant forms are found in hospitals. To guard against staph infections, the germicide hexachlorophene long had been used in soaps and medicinal cleansers. Hexachlorophene was also used in certain adult cosmetics. The U.S. Food and Drug Administration (FDA) banned hexachlorophene from cosmetics early in 1972 and advised hospitals not to bathe babies in substances containing it. This action was based on the experimental finding that brain damage occurred in animals that had absorbed high quantities of hexachlorophene through the skin, although no such injuries had been reported for humans.

Following the FDA ban, a sequence of events occurred that illustrated how the beneficial and the potentially harmful effects of a drug must be weighed in order to determine proper use. A rising incidence of outbreaks of staph infections among newborn infants was noted and was attributed to the discontinuance of the use of hexachlorophene. Hospitals may have overreacted to the directive by not using hexachlorophene-containing cleansers at all, even on the hands of nursery personnel, a use that was not banned. The FDA changed its ruling to state that babies may be bathed in hexachlorophene-containing solutions on a temporary basis should there be signs of an impending staph outbreak.

Rabies is a virus considered to be invariably fatal in man. Recently, however, a six-year-old boy was reported to have recovered from the disease, contracted from the bite of an infected bat. The child began the 14-day course of the rabies vaccine four days after his exposure to rabies, but symptoms of the disease began to appear two days after completion of the series. Scientists from the National Communicable Disease Center, Atlanta, Ga., took charge of the case and began a program of symptomatic treatment. There are no drugs known to be effective against the rabies virus (or any other virus), and treatment of the symptoms was the only alternative. The treatment included continuous monitoring of heart and lung functions; prevention of an oxygen deficiency by cutting an opening into the windpipe; and mechanical assistance for the lungs. Further experience with specific treatment measures on rabies-infected laboratory animals and prolonged survivals achieved with other recent human patients resulted in hope that the possibility of recovering from the disease might be increased appreciably.

In contrast to the case of many cancers in animals, no cancer-causing viruses had as yet been isolated from human cancer patients—even though virus-like particles were observed in a variety of human cancer tissues. In late 1971 and early 1972, there were reports of at least three viruses thought to have been isolated from human cancer tissue. One, however, was shown instead to be an animal virus and research was still in progress on the others.

The C-type virus was considered by many scientists to be able to cause cancer. Viruses of this type were thought to be able to exist in unexpressed and unobservable forms in many cells as a part of their genetic makeup. Recently, two chemicals, both of which were structural analogues of a component of the genetic material deoxyribonucleic acid (DNA), were shown to be able to activate C-type virus particles in animal cell lines. These chemicals might be useful as "tools" with which to demonstrate the existence of unexpressed C-type viruses in human cell lines.

New evidence was obtained that Hodgkins' disease, a lymphatic cancer, may be caused by two

Louis Pasteur—chemist and microbiologist—proved that microorganisms cause disease and fermentation and originated vaccines for rabies and anthrax. Pasteurization, a process of sterilizing liquid foods, is based on his work.

viruses working together, one containing ribonucleic acid (RNA) and the other containing DNA. The latter was thought to be related to the herpes family of viruses. There was also speculation that Hodgkins' disease is contagious, most likely because of transmission of the DNA-containing virus.

Although it was discovered in 1874, the bacterium that causes leprosy had never been cultivated in a test tube. Leprosy infects only humans and has not even been transmitted to higher primates. The bacterium prefers the cooler regions of the body for optimal growth, and it had been found to grow very slowly in the foot pads of mice. There was evidence that the armadillo might be a good experimental animal host for leprosy. It has a low skin and body temperature, which the leprosy bacterium requires, and has a long enough life to give scientists time to study the slowly progressive nature of the disease.

Smallpox, an entirely human viral disease, appeared to be almost extinct, so much so that the U.S. Public Health Service no longer advocated routine vaccination against it and most states had ceased to require vaccination of school-age children. Eradication of smallpox was brought about by a worldwide vaccination program. It was thought that the virus no longer existed in those societies in which extensive vaccination programs had been enforced and in which there have been no recently reported cases. (The last reported case in the U.S. occurred in 1949.) There were, however, annual reports of instances of severe complications resulting from the vaccination itself; a few instances even resulted in death. The disease was still residual in a few countries, most notably Ethiopia, India, and Pakistan, and continued surveillance for it would be required, with routine vaccination recommended for persons who might be exposed to it.

A 1971 epidemic in the U.S. of horse sleeping sickness (Venezuelan equine encephalomyelitis, or VEE) was confined to southern Texas by the cooperative action of various state and federal agencies. By a stroke of good fortune a vaccine against VEE had been developed by microbiologists at the U.S. Army Medical Research Institute of Infectious Diseases at Fort Detrick, Md., several years prior to the epidemic. The U.S. Department of Defense provided the vaccine, which was distributed to veterinarians who administered it at federal expense to over two million horses in 19 states and the District of Columbia.

VEE is frequently fatal to horses but is seldom fatal to man, who is also susceptible to it. The vac-

*A new herpes virus, the type responsible for recurring blisters in humans, was found during research on another virus suspected of causing cancer in rhesus monkeys. Disruption of white cells by the virus is evident in the contrast between a normal cell culture (left) and one infected by the new virus (right).*

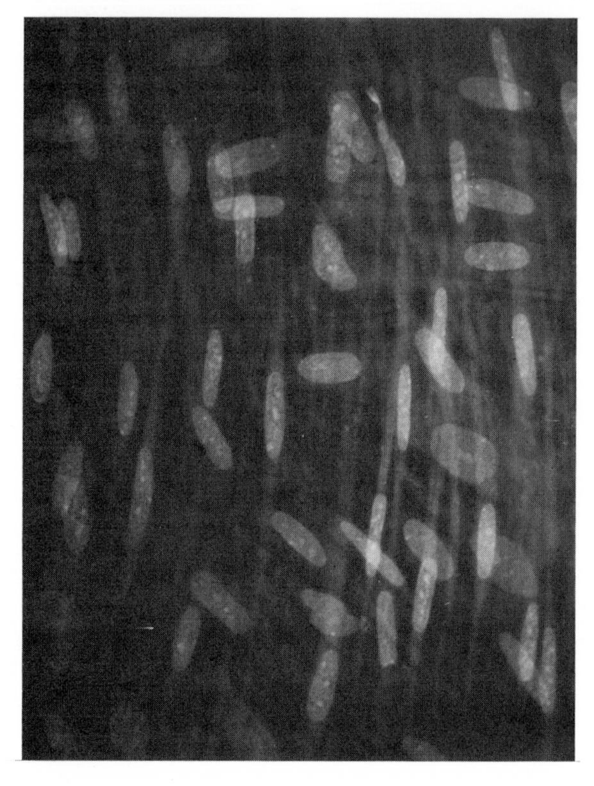

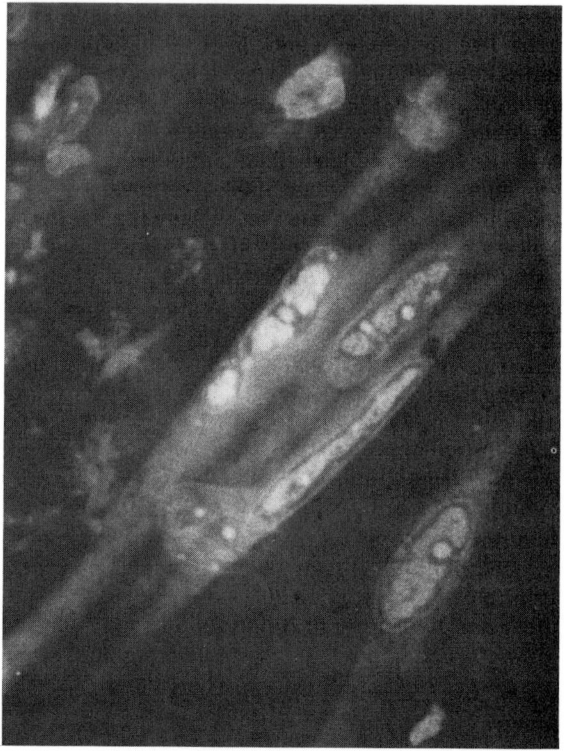

Courtesy, Dr. Paul D. Parkman, National Institutes of Health cine was still considered to be experimental, and it was unlicensed in the U.S. for human use. VEE was first detected in Venezuela in 1935 and spread to southeastern Mexico in 1970. In early 1971 the virus "jumped" into the Tampico area of central Mexico. VEE is usually transmitted through the bites of infected mosquitoes, but it can also be transmitted by other biting insects, or it can even be airborne. Most recently, the VEE virus was also found in an infected vampire bat in Mexico. Vampire bats subsist on a blood diet and frequently feed on horses. They also cohabit with other species of bats, including those that migrate in massive numbers to the southwestern U.S. from Mexico. Thus, there was ample opportunity for the transfer of the VEE virus to areas far from an epidemic area. For these reasons, VEE continued to be a significant public health problem, and a joint U.S.-Mexican effort was under way to vaccinate all horses in northern Mexico to provide a barrier against the northward spread of the disease.

—Robert G. Eagon

Redfern, INDUSTRIAL RESEARCH

*"I've just discovered a new and less expensive method of exciting molecules."*

# Molecular biology

Developments in molecular biology in recent months displayed more clearly than ever the dissolution of distinct areas of study for its subdisciplines. Basic research was more often than not producing information that had immediate, often practical applications in other sciences, especially in medicine.

## Biochemistry

Many developments in biochemistry in the past year provided insight into relationships between the structure and the function of the chemical constituents of living things. Significant advances in our knowledge of carbohydrates, lipids, nucleic acids, and enzymes were reported, but proteins received particular attention. The most noteworthy advances in protein chemistry concerned hemoglobin, the oxygen-carrying protein of red blood cells (erythrocytes); the immunoglobulins, which act to neutralize foreign chemical substances; and the protein hormone insulin.

**Modifying hemoglobins.** Through the efforts of many investigators over the past 20–25 years, it was established that hemoglobin from most vertebrates is composed of two pairs of different polypeptide chains, and that each chain contains a molecule of heme, the iron-containing group to which oxygen molecules are bound during their transport throughout the body. Thus, the protein portion of hemoglobin, called globin, may be designated $\alpha_2\beta_2$, in which $\alpha$ and $\beta$ indicate the two different chains. Each of the chains is folded in a unique, three-dimensional shape, and the four chains are combined together in a precise manner to form the functional molecule.

In man over 100 different variants of hemoglobin are known, most of which occur only in one of the amino acids in either of the two chains. The most widespread variant is hemoglobin S, the major species of hemoglobin in persons with sickle-cell anemia. This molecule contains normal $\alpha$-chains, which are identical to those found in the vast majority of persons, and abnormal $\beta$-chains in which the amino acid valine is found instead of glutamic acid at what is described as the sixth position from the amino-terminus. When oxygen is removed (deoxygenation), hemoglobin S is less soluble than normal hemoglobin and cells composed of it are deformed. The deformed cells have a shorter lifetime than normal and masses of them can block blood flow in capillary networks, causing severe anemia and periodic episodes of pain and fever (a crisis). Although sickle-cell anemia is a genetic disease and an obvious candidate for management in the future by genetic engineering, other, more immediate approaches to its management must be sought.

Recent studies by Anthony Cerami and Charles Manning at Rockefeller University in New York City provided a most promising approach. Other workers had reported that the sickling process was inhibited if high blood levels of urea, the compound that is the chief end product of protein

Courtesy, Dr. Anthony Cerami, Rockefeller University

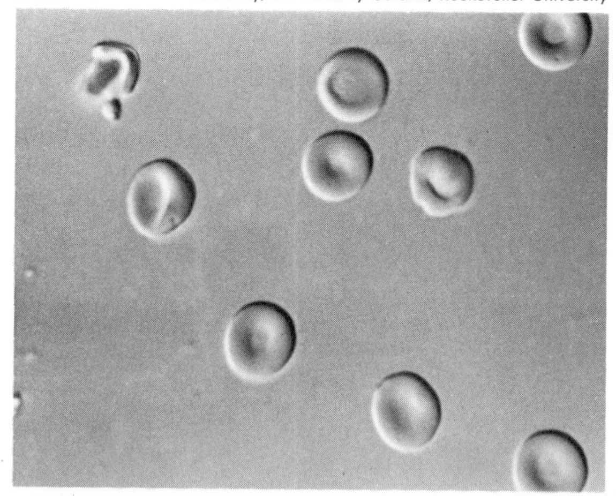

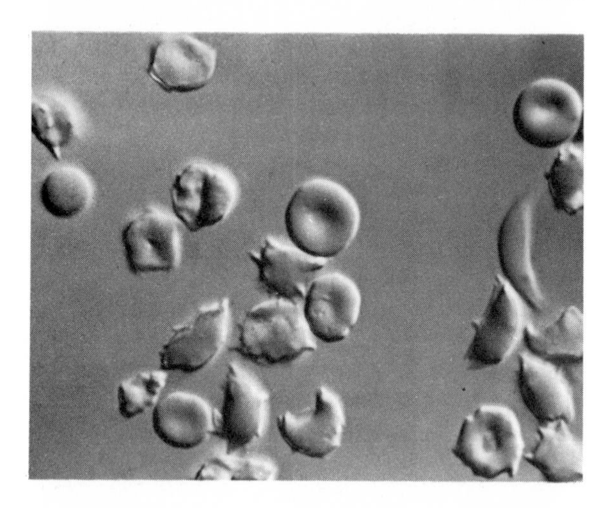

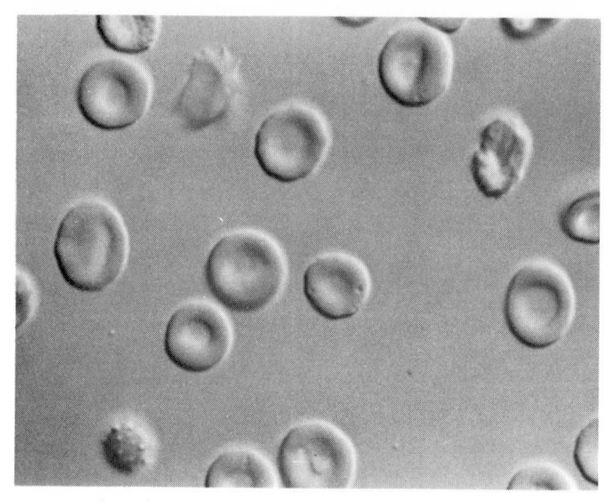

*Red blood cells (erythrocytes) from a patient suffering from sickle-cell anemia show the beneficial effects of treatment with the compound cyanate. Cells carrying oxygen (top) show the normal erythrocyte configuration. Deoxygenated cells (middle) assume the deformed sickle shape. Cells that have reacted with cyanate (bottom) retain their normal shape when oxygen is removed.*

metabolism, could be maintained in patients with the anemia. The concentrations of urea required, however, were many times greater than normal and adverse side effects occurred.

Cerami and Manning were aware that many of the effects of urea on proteins can result from cyanate, a compound that may be derived from urea and is often present in urea solutions. They treated erythrocytes from patients with sickle-cell anemia under very mild conditions with rather low concentrations of cyanate and found that a large majority of the cells retained their normal shape when deoxygenated. In addition, the effects of cyanate were irreversible; treated cells did not sickle even after removal of cyanate and when re-oxygenated and deoxygenated again.

Cerami and Manning also found that no more than two molecules of cyanate were incorporated per molecule of hemoglobin S, and that both cyanate molecules had reacted with the two $\alpha$-amino groups of the valyl residue at the amino-terminus of the two $\alpha$-chains. The modified hemoglobin S also had a near normal solubility, which appeared to account for its lack of sickling. It was of considerable interest that amino groups in other proteins involved were not reactive with cyanate.

These impressive results prompted further studies with experimental animals to determine whether cyanate could be used therapeutically. Blood levels of cyanate sufficient to modify hemoglobin could easily be obtained by injection or feeding under conditions that produced no adverse side effects. In addition, only the hemoglobin had reacted extensively with cyanate. The modified hemoglobin retained its ability to transport oxygen and help maintain the normal pH (acidity) of blood. Early results of further studies, still in progress, indicated that erythrocytes in human subjects partially modified with cyanate were not destroyed at the rapid rate found in sickle-cell anemia, and that there might be a general improvement in the disease.

**Immunoglobulin structure.** Studies over the past decade provided much insight into the structure of the polypeptide chains of the major classes of immunoglobulins and of the G class in particular. Immunoglobulin Gs have a molecular weight of about 150,000 and are comprised of two heavy (H) chains with molecular weights of about 50,000 and two light (L) chains of about half this weight. The four chains are combined to form a molecule thought to be in the shape of the letter Y. The two arms of the Y, called Fab fragments, are each comprised of a light chain and about half of a heavy chain. The leg of the Y, called the Fc fragment, contains the remaining portions of the heavy chains. The heavy chain can be divided into four quarters of about equal length that are designated the $V_H$, $C_H1$, $C_H2$, and $C_H3$ domains, and the light chains can be divided into equal halves designated as the $V_L$ and $C_L$ domains (V = variable region; C = constant region). The amino acid sequences of each of the six domains are very similar. Four domains are present in each Fab fragment, two composed of variable regions ($V_H$ and $V_L$) and two of the constant regions ($C_H1$ and $C_L$). The remaining domains are in the Fc fragment and are formed from two $C_H2$ regions and two $C_H3$ regions.

It is believed that the variable domains in Immunoglobulin G form the antigen binding site, the part of the molecule that combines in a complementary fashion with a specific foreign substance, and that the amino acid sequences forming this site vary from one specific immunoglobulin to another, depending upon the nature of the antigen. The constant domains mediate other important effector functions of immunoglobulins. Although this model of immunoglobulin structure was consistent with available information, the detailed structure of the antigen binding site, as well as the structural relationships among the domains, would be ascertained only with a more exact knowledge of the spatial arrangements of their atoms.

A crystallographic analysis at low resolution of the Fab fragment from a human immunoglobulin was reported in 1972 by R. J. Poljak, L. M. Amzel, and H. P. Avey at Johns Hopkins University, Baltimore, Md., in collaboration with A. Nisonoff at the University of Illinois. Greater resolution was needed to distinguish the exact location of groups and atoms and to see details of the antigen binding site, but the investigators were able to conclude that the Fab fragment contained two globular structures thought to correspond to the variable and constant domains. This was interpreted as good evidence that the similarity in amino acid sequence between regions is reflected in conformational similarities in domains.

*A model of immunoglobulin suggests that it consists of two heavy (H) polypeptide chains and two light (L) chains combined into a Y shape. Each arm (Fab) has two variable (V) and two constant (C) regions, each with an L and part of an H chain. The leg (Fc) contains the rest of the H chains in four C regions.*

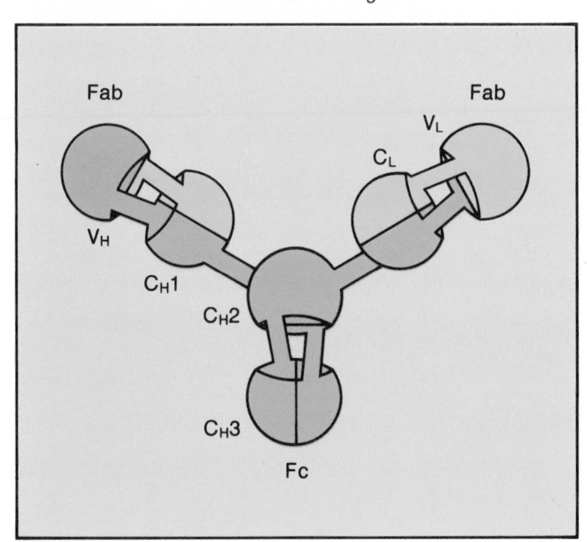


Otto Epp of the University of Graz, Austria, working at the Max Planck Institute in Münich, W.Ger., with W. Palm, H. Fehlhammer, A. Rühlmann, W. Steigeman, P. Schwager, and R. Huber, used X-ray diffraction methods to examine crystals of a Bence-Jones protein, one produced in persons with bone marrow tumors and corresponding to a single type of immunoglobulin light chain. They also found that the molecule is composed of two globular parts, with characteristics most probably corresponding to the $V_L$ and $C_L$ domains.

Another group at the National Institutes of Health in Bethesda, Md. (V. R. Sarma, E. W. Silverton, D. R. Davies, and W. D. Terry) reported crystallographic studies on an intact Immunoglobulin G molecule that provided convincing arguments that the Y-shaped structure for it was indeed correct. Globular regions corresponding to the Fab and Fc fragments were discerned and appeared to be joined by nonglobular stretches of polypeptide chain. Analysis at higher resolution was needed to see more structural detail, including the antigen binding site, but it was encouraging that these analyses confirmed to a substantial degree the model of immunoglobulin structure deduced by other means.

**Insulin binding.** Although insulin had been used for 50 years in controlling diabetes and was the first protein whose complete amino acid sequence was established, its mechanism of action remained poorly understood, primarily because insulin's effects are produced indirectly. It was generally agreed that insulin binds target cells to cause selective amplification of preexisting cellular processes. Recent studies on the nature of the insulin-binding substances in target cells began to shed light on the molecular basis for these events.

Particularly noteworthy were studies by Pedro Cuatrecasas at Johns Hopkins University, who had shown earlier that the metabolic effects of insulin were obtained when it was bound to cells under conditions that did not allow it to penetrate the cell. It was further found that the specific insulin-binding structures were associated with isolated liver membranes. Cuatrecasas now reported that it was possible to solubilize the insulin-binding structures by extraction of either liver or fat cell membranes with detergents. The solubilized preparation specifically and reversibly binds insulin, whereas derivatives of insulin appeared to bind in direct proportion to their biological activity and other polypeptide hormones did not bind. Evidence was also provided that the isolated molecule is a protein with a molecular weight of about 300,000. Although this protein must be closely associated with lipids (fatty substances) in the cell membranes, it did not appear to be a lipoprotein. The dissociation constant for the insulin-receptor protein complex was of the same order of magnitude as antigen-antibody binding and highly suggestive of considerable binding specificity.

G. Vann Bennett and Cuatrecasas also found that the number of insulin receptors in fat cells from rats made diabetic by chemical means was not decreased and that the diminished responsiveness could not be explained by structural alterations in the receptors. How the event of insulin binding to receptors is transmitted to other molecules in the cell and translated into altered, specific chemical processes could only be answered by additional studies.

**Insulin and nerve growth factor.** An interesting and somewhat surprising relationship between insulin and nerve growth factor was reported. NGF is a protein with a molecular weight of about 26,500 that contains two identical polypeptide chains. It was known to enhance the growth of sympathetic ganglia and appeared to be essential for maintaining the sympathetic nervous system. W. A. Frazier, R. H. Angeletti, and R. A. Bradshaw at Washington University in St. Louis, Mo., had established previously the complete amino acid sequence of the 118 residues in the single polypeptide chain of NGF, including three intrachain disulfide bonds. Although at first the sequence did not seem to resemble other proteins, the observation that NGF had many biological effects in common with insulin, prompted a closer comparison of its sequence and those of insulin and its precursor, proinsulin.

Some degree of structural similarity was found, although not as much as is found in other sets of structurally related proteins. The fact that NGF contained 33 more residues than proinsulin did not minimize the resemblance since the additional residues were similar in some respects to the amino-terminal portion of proinsulin. Most significantly, one of the three disulfide bonds in the two molecules was in a closely corresponding position.

On the basis of these apparent structural relationships between NGF and insulin and their analogous physiological effects on their target cells, it was suggested that the two hormones are evolutionarily related. The gene controlling the sequence of NGF may have been derived from an ancestral gene for proinsulin.

The relationships between proinsulin and NGF suggested goals for future studies on the mechanism of action of NGF. For example, it will be interesting to find whether NGF forms a complex on target cells analogous to that of insulin and its receptor protein.

—Robert L. Hill

## Biophysics

X-ray diffraction and refinements of its technique continued to be the mainstay of the biophysicist's application of the principles and tools of the physical sciences to the elucidation of the structure and function of biological systems. Materials that were periodic, or that could be crystallized into periodic structures, including enzymes, muscle proteins, metallo-proteins, and viruses, were studied by this technique. Other techniques were being adapted for use on other biological materials. Nuclear magnetic resonance, for example, was used quite widely for the determination of protein configuration and the mapping of active sites.

**New techniques.** Infrared spectroscopy appeared to hold considerable promise for the study of the orientations of nucleic acids, the compounds found primarily in cell nuclei, usually in combination with proteins. Nucleic acids consist of four bases, or side groups, attached to a backbone of

*Crystals of human immunoglobulin subjected to electron microscopy revealed that the molecule has three very distinct globular regions. This overall shape was consistent with what had been predicted in many other chemical, enzymatic, and microscopic studies.*

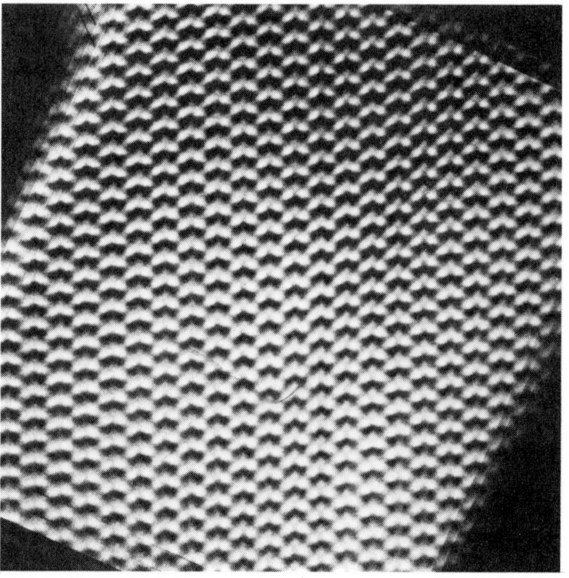

Courtesy, Drs. V. R. Sarma, D. R. Davies, L. W. Labaw, E. W. Silverton, and W. D. Terry, "Cold Spring Harbor Symposia"

alternating sugar and phosphate molecules. Recent studies looked at the transitions of deoxyribonucleic acid (DNA) from the B to the A form. Polarized infrared light was used to measure and compare the differences in absorption of light caused by vibrations of the backbone. From the recorded differences it was possible to estimate angles of the sugar-phosphate groups that were in good agreement with X-ray diffraction studies. Because proteins do not absorb strongly in the infrared region used, this technique should prove ideal for studying nucleic acid structure, and for determining the effects of agents that modify it.

During the year several promising methods were developed for use in biophysical studies. In one of these, the photoelectric effect, a phenomenon long the province of the physicist and more recently of the chemist, was added to the tools of the biophysicist. The photoelectric effect occurs when high-frequency radiation impinging on certain substances causes bound electrons to be given off with a velocity proportional to the frequency of the radiation. Chemists use photoelectron spectroscopy for chemical analysis: a sample is irradiated with ultraviolet radiation and emits photoelectrons (or with X rays and emits Auger electrons) whose energy spectra are analyzed in a high-resolution spectrometer to determine directly the electronic energy levels of the elements present in the sample. This technique can be used to analyze unknown samples for their elements; it is also useful for the study of surface phenomena since in a solid sample the electrons that form the spectrum come from a thin layer near the surface, usually less than 50 Å in depth (one angstrom, Å = $10^{-8}$ cm). By replacing the spectrometer with a phosphor screen and a camera and thus forming a photoelectron microscope, it became possible to map the structure of biological surfaces and membranes.

**Visual excitation studies.** Recent experiments offered intriguing clues about the mechanism by which incident light causes visual excitation of the rods, the achromatic photoreceptor cells in the retina of the eye. In vertebrates, the photosensitive pigment in the rods is the protein rhodopsin. Rhodopsin is located in the membrane of the rod outer segment and accounts for about 80% of its total protein. It consists of the lipoprotein opsin and a group of atoms that gives rise to color, the visual chromophore 11-cis retinal (which is derived from vitamin A). As a result of alterations that can occur in its atomic bonds, retinal can exist in the form of either cis or trans isomers; to combine with opsin it must be in its 11-cis form. Phospholipids in the membrane of the rod outer segments are thought to stabilize this association. Exposure

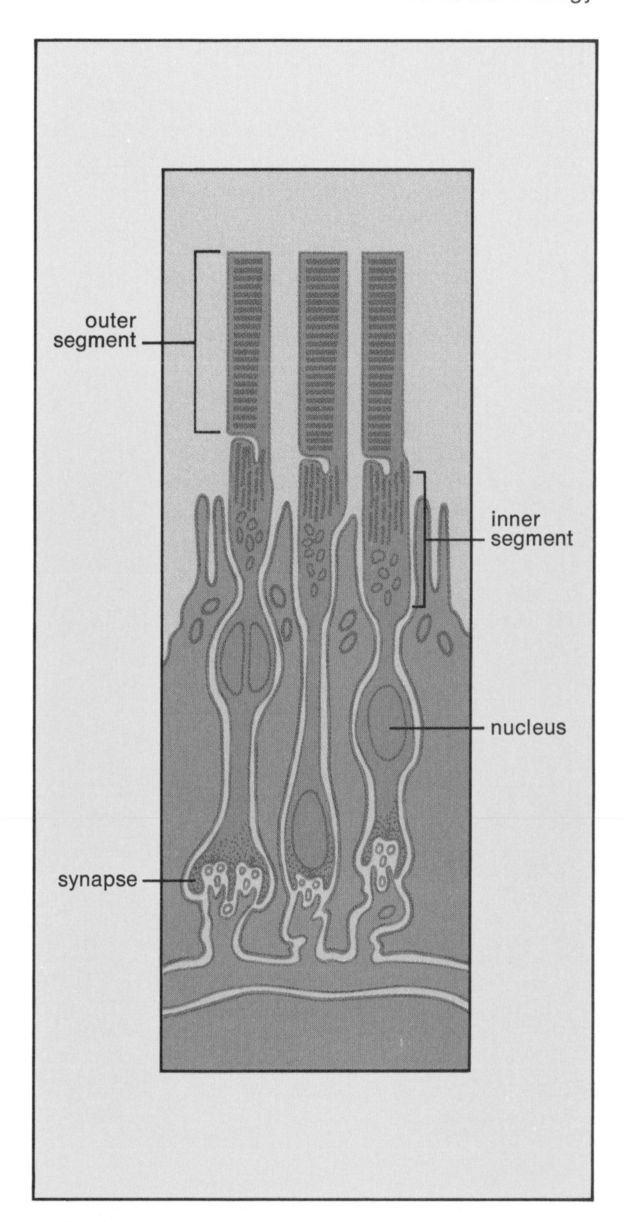


*Rod cells in the retina of the eye have outer segments of thin disks with the pigment rhodopsin attached. The effects of light's action on rhodopsin spread to the inner segment housing the cell's energy supply and then past the nucleus to the synapse, a point of near contact with a nerve cell leading to the brain.*

to light photoisomerizes the retinal and initiates a series of changes in the configuration of opsin as its deep red color is replaced by light yellow (bleaching process), which apparently results in the release of retinal. If during this process retinal is reisomerized to cis form, rhodopsin is regenerated.

Recent studies of frog retina indicated that rhodopsin undergoes free rotational diffusion motion in the visual receptor membrane. Having established the rotational relaxation time of 20 $\mu s$ (one microsecond, $\mu s = 10^{-6}$ seconds) and assuming rhodopsin to be a sphere of about 25 Å radius, researchers determined the effective viscosity (ability to flow) of the membrane medium to be that of a typical light oil, such as olive oil. On the basis of these properties, it was suggested that rhodopsin may function as a diffusional carrier in the membrane transport processes that lead to excitation of the visual receptor.

Other recent experiments suggested that rhodopsin may have a more indirect role in the excitation process. Adenylate cyclase (cyclase), the enzyme that catalyzes the primary energy-producing activity of cells, the conversion of adenosine triphosphate (ATP) to cyclic adenosine 3′,5′-monophosphate (cAMP), was found to be present in the rod outer segments at a specific activity higher than that reported for any other tissue. Furthermore, illumination decreased cyclase activity in direct proportion to the bleaching of rhodopsin, which suggested some sort of direct coupling between the bleaching process and the enzyme. Other investigators reported finding a light-stimulated phosphorylation (conversion into a phosphate) of rhodopsin by ATP that will occur even if ATP is added after the flash of light. Since this reaction would not proceed after the membrane was disrupted, a membrane-bound kinase (enzyme catalyst) appeared to be involved.

It was not clear whether there was any relationship between the inactivation of cyclase by the bleaching of opsin and the photo-induced phosphorylation of rhodopsin, but there was evidence to suggest that the actual excitation of the visual receptor may be mediated by cAMP. Electrophysiological studies using the eye of the horseshoe crab *Limulus* showed that the addition of cAMP mimics the effects of light on the photoreceptor cell. Aminophylline and theophylline, two inhibitors of phosphodiesterase (the enzyme that hydrolyzes cAMP to 5′ AMP), also produce identical effects, presumably by increasing the intracellular concentration of cAMP by inhibiting its breakdown by phosphodiesterase.

It was also reported that the addition of small amounts of extracellular calcium decreases the sodium permeability of visual receptors, again mimicking the effect of light. Since cAMP was known to affect the transport of calcium in heart muscle and intestinal mucosa, it was possible that it also regulates the calcium transport in rod outer segments. Calcium, in turn, may regulate the changes in the permeability of the photoreceptor membrane to sodium and, thus, excitation of the visual receptor.

**Research on membranes.** Cell membranes were the subject of intensive research because of the numerous biochemical processes that occur on or in them. Their dynamic nature was well illustrated in the studies of rhodopsin. The complexity and diversity of their structure was being pursued with a variety of techniques in many laboratories.

From these studies the cell membrane emerged as a structure about 80 Å thick with a bimolecular phospholipid core (two molecules of fatty complexes), large portions of which are covered with proteins, glycoprotein, glycolipids, cholesterol, and other molecules. In some cases these protein molecules span the membrane from inner to outer surface. The ratio of lipid to protein appears to vary with the physiologic function of the membrane, from about 75% lipid for the relatively inactive myelin sheath around nerves to about 75% protein for active bacterial and mitochondrial membranes. Further studies suggested that the conformation of membrane proteins may be related to function. Mitochondrial membranes show a high proportion with spiral shapes, whereas myelin membranes have more of their protein flattened. Recent evidence also suggested that the bimolecular core of the plasma membrane of mammalian erythrocytes is asymmetrical, having two types of phospholipid on the outer half of the bilayer and two different ones on the inner half.

Another very active area of membrane research studied the differences between the surfaces of normal cells and cancer cells (or mammalian cells after viral transformation). After transformation, new antigens (foreign substances) appear; in most cases these are identical to antigens produced by fetal tissue (*see* Feature Article: CANCER UNDER ATTACK). Recent studies indicate that polyoma-transformed cells synthesize more glycoprotein, which is loosely bound to the cell surface. There may also be changes in the sites of transport across the membrane. Finally, the transformed cells lose their sensitivity to contact inhibition, which normally restrains growth.

During the year it was found that transformed cells treated with the cAMP analogue dibutyryl cAMP regain many of the characteristics of normal cells, including sensitivity to contact inhibition. Levels of cAMP were found to be lower in polyoma-transformed cells than in normal cells and this decrease was shown to precede the establishment of the transformed state, suggesting that cAMP is the cause rather than the effect of the properties of the transformed cells. Cyclic AMP, in fact, might control cell growth and morphology in mammalian systems.

**Radiation biology.** It was well known that DNA is implicated as the target of the primary biological lesion induced by ionizing or ultraviolet radiations.

For this reason repair of DNA continued to be the main preoccupation of workers in radiation biology. Three enzymatic mechanisms for the repair of DNA were known to occur in the much-studied bacterium *Escherichia coli (E. coli).* The first mechanism, photoreactivation, acts specifically to split pyrimidine dimers formed in DNA exposed to ultraviolet light into their respective monomers when the cells are subsequently exposed to visible light. The second system excises radiation-induced damage from one strand of DNA and resynthesizes the missing segment by using the opposite, complementary strand of DNA as the template. Since this mode of repair does not require light it is called a dark repair process. The third mechanism is under the control of the genes involved in genetic recombination and is called recombinational, or Rec, repair. It appears to repair single-strand breaks in the sugar-phosphate backbone that are induced primarily by such ionizing radiation as X rays, and by certain chemicals, but that may also be present after DNA replication in ultraviolet-irradiated cells. Rejoining of single-strand breaks by this method requires that the cells be incubated under growth conditions, generally for 10–30 minutes at 37° C.

Evidence for two other repair mechanisms was reported recently. Both of these modes were faster than Rec repair for mending X-ray induced, single-strand breaks in DNA. Ultrafast repair, which occurs in less than one minute at 0° C, rejoins breaks produced under anoxic conditions (severe oxygen deficiency), with the result that there are only about one third as many breaks remaining in the cells irradiated in the absence of oxygen as in cells irradiated in the presence of oxygen. Thus, oxygen acts not by increasing the number of breaks produced but by changing them so that they are no longer reparable by the ultrafast repair process. The second mode of repair recently reported appeared to involve the enzyme DNA polymerase I. Repair of about 80% of the remaining breaks, whether induced by X rays delivered aerobically or anoxically, is completed by this method, Pol repair, in 2–5 minutes at room temperature.

It was thought that these two fast systems repair relatively simple breaks by a simple rejoining of the DNA backbone (ultrafast) or replacing a few bases (Pol repair). Rec repair presumably rejoins more complex breaks using some of the steps involved in genetic recombination.

Certain drugs, namely quinacrine, N-ethyl maleimide, and hydroxyurea, were found to inhibit this ultrafast repair of anoxically produced breaks in *E. coli.* In addition, quinacrine inhibited Rec repair. These results had interesting implica-

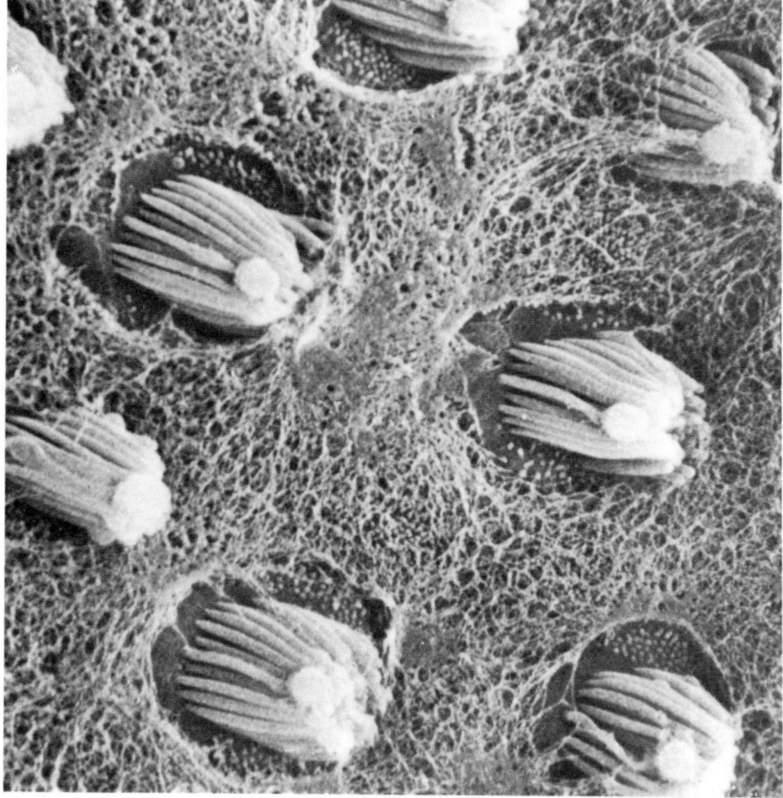

Courtesy, Dr. D. E. Hillman, University of Iowa

*A scanning electron micrograph shows how sensory hairs (cilia) on receptor cells in the saccule of a frog's ear may function to transfer sound waves into messages to the auditory nerve. The cilia can be seen bound to the bulb atop one specialized cilium, which rests on a portion of the receptor membrane lacking the rigid base under the other cilia. Movement of the masses distends the membrane at the base of this cilium; the distension is thought to bring about changes that stimulate nerve activity.*

tions for the treatment of solid cancers, which, because of low oxygen tension in the tumor mass, are generally more resistant to irradiation treatment than surrounding healthy tissue. It would, therefore, be desirable to make these tumors at least as sensitive as the more fully oxygenated surrounding tissue by the use of such drugs.

**Future trends.** After several years of decline and stagnation, the U.S. budget for science appeared to be on the increase. The emphasis, however, would be on applied research, with the National Science Foundation proposing to eliminate its science development grants and further reduce funds for graduate traineeships and institutional grants for science. Basic research funds would increase slightly, but the largest increases would be for mission-oriented research and development. Funds for research on cancer, and probably on diseases of the heart and lungs, would be increased sharply. Thus, human cells and their viruses would be the new *E. coli* as molecular pathology replaced molecular biology.

—Jeremy Baptist

## Genetics

Molecular genetics remained the focus of intensive exploration, with ever more sophisticated techniques, and ever more specific questions being asked about cell mechanisms. The year was marked by no excitements to compare with the discovery the year before of reverse transcriptase, an enzyme responsible for the production of deoxyribonucleic acid (DNA) from a model provided by ribonucleic acid (RNA), a thoroughly unexpected operation for RNA. There was, however, growing appreciation that every seemingly simple mechanism of DNA is involved in increasingly more complex networks of relationships within the cell.

This overall complexity, rather than any profound difference in basic operation, may well have accounted for the puzzling recalcitrance of some underlying problems in the molecular biology of eukaryotic cells (those with distinct nuclei). These problems included the physical structure of the chromosome and the mechanisms of regulation of gene activity—problems that are rather well understood now for the simpler viral and bacterial systems. In responding to these challenges of higher organisms, molecular geneticists began crossing traditional boundaries that once compartmentalized biological research.

**Reverse transcriptase studies.** The main thrust of studies with the enzyme reverse transcriptase was the tracing of viruses newly implicated in cancer in primates and perhaps in man. Research groups led by Sol Spiegelman at Columbia University, David Baltimore at the Massachusetts Institute of Technology (MIT), and J. Ross at the U.S. National Institutes of Health (NIH), Bethesda, Md., also used this enzyme with a sample of messenger RNA (m-RNA) as a template. The m-RNA, isolated from immature red blood cells of rabbits and other mammals, was prepared in a way believed to encode for the globin protein portion of hemoglobin. In a cell-free protein-synthesizing system—containing ribosomes, amino acids linked to transfer RNA (t-RNA) molecules, iron-containing hemin molecules, and the appropriate enzymes—this m-RNA did indeed effect the synthesis of mammalian hemoglobin. The DNA that was synthesized under the influence of the reverse transcriptase may, therefore, be equivalent to the gene for the globin, an identity that remained to be confirmed.

Apart from speculations about its use for the genetic "engineering" of tissue cells, the ability to produce quantities of a specific gene would be technically invaluable. It would be an ideal method of assaying specific m-RNAs in cells at different stages of development or in different states. It would also facilitate the analysis of the nucleotide sequence of RNA through stepwise synthesis of the corresponding DNA by the action of transcriptase. In its natural occurrence, the DNA for a specific gene accounts for less than one millionth of the DNA content in a mammalian cell, which evokes despair about attempts to isolate it directly. However, m-RNAs, which reflect the activity of specific genes, are produced differentially in various tissues, and often amplified manyfold. We can, therefore, be more optimistic about extracting specific RNAs, and then using reverse-transcriptase to produce the DNA corresponding to a specific gene.

Further studies with m-RNAs will be facilitated by the finding, by J. B. Gurdon and his colleagues at the University of Oxford, that living frog eggs can "translate" injected samples of RNA into their corresponding proteins. The eggs are a relatively simple biological system technically and retain their activity for 24 hours or longer, during which each m-RNA molecule can be translated many times. The system can recognize and encode into DNA as little as one billionth of a gram of RNA added to it.

Reverse transcriptase is inactive on pure, single-stranded RNA templates; a short segment of complementary RNA is required as a primer to which DNA can attach and assemble itself. This primer segment functions like the poly-A (adenine)-rich ends found naturally in many RNAs. In the reverse transcriptase experiments just described, the primer was a synthetic poly-T (thymine), which binds to the complementary As on the RNA and helps initiate DNA synthesis.

Similar requirements had been established in 1967 by A. Kornberg's group at Stanford University for the replication of viral DNA. In later studies it was found that the RNA polymerase enzyme probably initiates replication in bacterial cells infected with the bacteriophage (virus) M13. RNA polymerase was now commonly thought of as the agent of translation, the production of m-RNA copies of DNA genes. Short segments of such copies may also be indispensable for the initiation of replication of DNA, which once begun is sustained by DNA polymerase.

**Ligase and the swivelling enzyme.** Ligases, enzymes that repair nicks in one strand of a duplex of DNA, were now a routine part of the repertoire of molecular genetics. V. Sgaramella, H. H. van de Sande, and H. G. Khorana (working at the University of Wisconsin) reported that ligase associated with the bacterial virus T4, besides showing the nick-sealing activity of other ligases, can join free ends of two separate DNA duplexes. The application of this tool to biologically active genes derived from various types of cells was under active study in several laboratories.

Many DNA molecules are found in closed circles, and much thought has been given to the way a double helix might be wrapped into simple circles without excessive twists. When DNA circles were proven to be pure DNA lacking any protein linkers, wrapping was thought to be due to a combined action of nicking enzymes and sealing enzymes. J. C. Wang at the University of California at Berkeley and J. J. Champou and R. Dulbecco at the Salk Institute, San Diego, Calif., have now described a protein (swivelling enzyme) that appears to untwist supercoiled DNA molecules. The reaction appeared to occur at one turn at a time, rather than the simultaneous relief afforded by a nicking enzyme. The molecular mechanism of the swivelling enzyme would be difficult to determine unless intermediate stages could be captured.

**Solving DNA puzzles.** The molecular requirements for DNA recombination were illuminated by studies on a protein coded by T4 and called simply the product of gene 34 of that virus. Because an enormous number of gene-34-protein molecules are produced during viral growth, some struc-

*In the test tube the enzyme reverse transcriptase requires a primer to initiate DNA production from an RNA model, or template. The primer is poly-T (dT), a synthetic of the nucleic acid thymine. When this piece of artificial RNA bonds with its complementary nucleic acid, adenine (A), on the template, the RNA becomes active. DNA then begins assembling with the aid of the enzyme.*

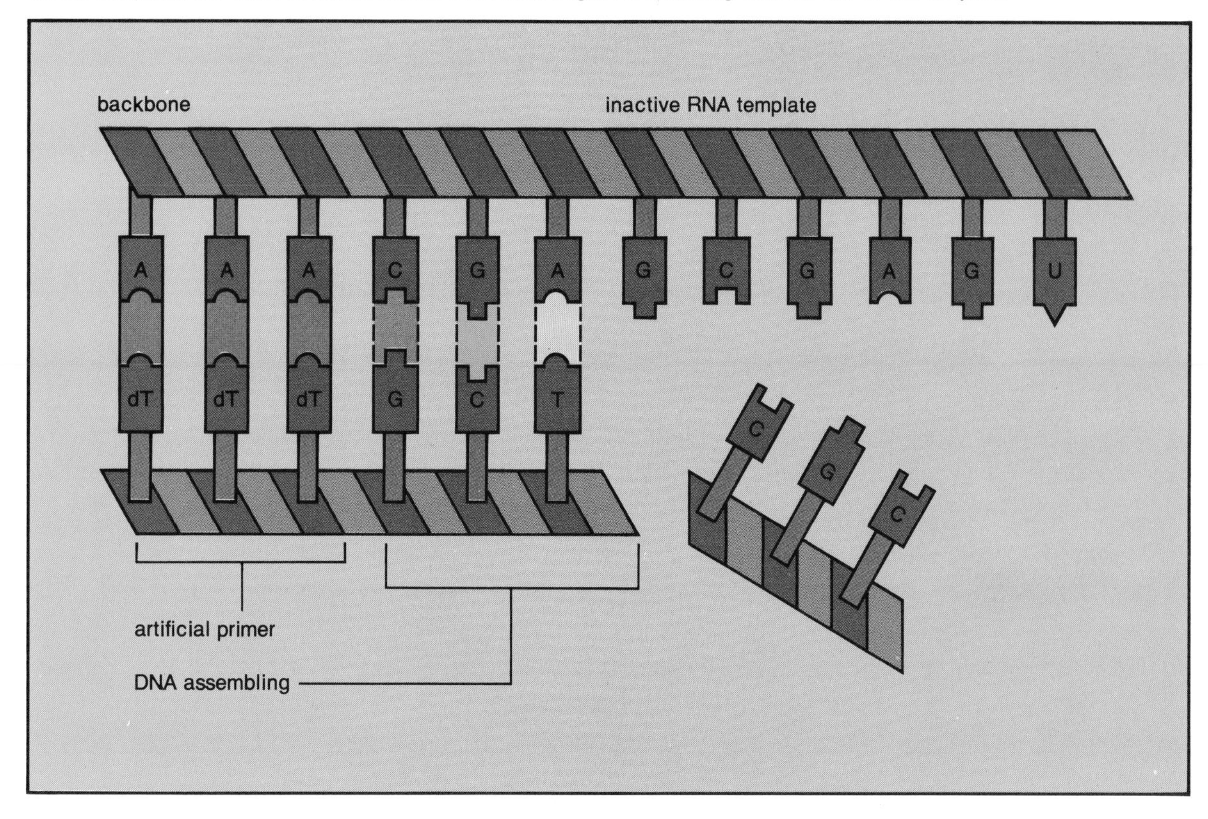

tural rather than catalytic role for the protein had been suspected. Bruce M. Alberts and Linda Frey at Princeton University isolated the protein and showed that it binds to single-stranded DNA in such a way as to minimize self-associations and to encourage the unwinding of duplexes, in effect stretching out sequences to enable their exposure to complementary ones. The addition of gene-34-protein, therefore, facilitates the resynthesis of DNA that has been melted with the strands separated. The protein may be a prime mover in the mechanism of genetic recombination. More recently Y. Hotta and H. Stern at the University of California at San Diego found that this protein, designated by them "meiotic protein," is almost always found in cells in meiosis (reduction division).

Of growing interest was the work by L. S. Lerman at Vanderbilt University, Nashville, Tenn., on another puzzle of DNA—how the extended fibers of the double helix of a phage particle in an infected cell can be condensed and packaged into the tiny space of the phage head. Similar problems of DNA compression in the sperm head and in chromosomes in higher forms had been explained (correctly?) by invoking complexes with basic proteins that neutralize the multiple negative charges of the DNA phosphates and allow tight compaction. Lerman found a remarkable collapse of the extended structure of DNA solutions—but no chemical binding—in the presence of neutral water-soluble polymers. The phage head contains the compact DNA until the phage infects a new host and the DNA springs back into expanded form.

**Cell fusion techniques.** Parallel to the deepening of biochemical probes into basic genetic functions was the explosive development of somatic cell genetics. The underlying theme was the progressive domestication of tissue cells in culture. This consisted primarily of the accumulation of cell lines and techniques to display precise features of chromosome content or of enzymatic function—in all events revealing the potentialities and the regulation of the genetic information in the cell. The most striking advance was the fusion of cells, in many cases facilitated by the use of an enzyme associated with inactivated virus particles.

After cells from different species have been fused, clones of hybrid cells can be found that may contain a sum of the parents' chromosomes, many of which are unfortunately unstable in such com-

*Transcription, the copying of a DNA sequence into messenger RNA, and translation of that information into protein can be seen occurring simultaneously in the bacterium* Escherichia coli. *Special electron microscope techniques developed at Oak Ridge (Tenn.) National Laboratory captured RNA linked by spherical ribosomes attached to strands of DNA magnified 306,000 times (left) and 72,000 times (right).*

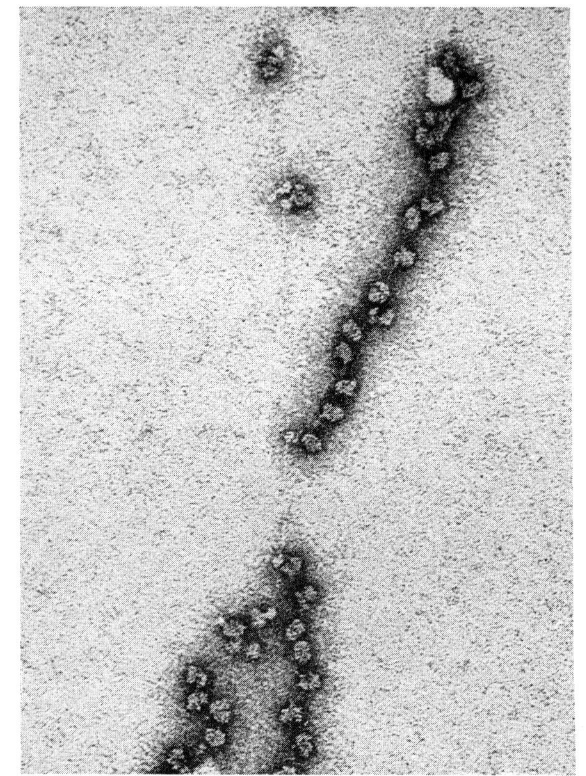

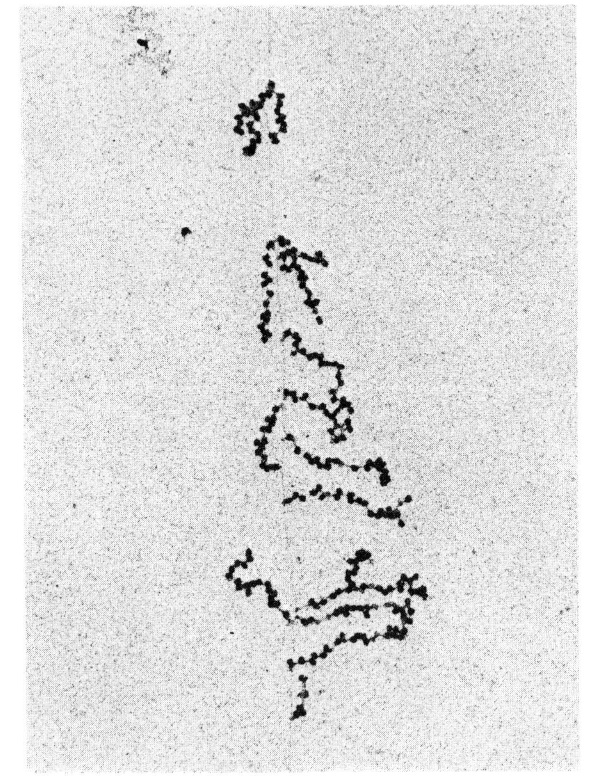

Courtesy, Dr. Barbara A. Hamkalo and Dr. O. L. Miller, Jr., Oak Ridge National Laboratory binations. In the fusion of a mouse cell and a human cell, the human chromosomes are particularly unstable. Such instability is a mixed curse: it interferes with the observation of controlled variations from the norm, but it also produces a rich variety of cell types, with different sets of chromosomes whose effects can then be measured.

The pathologist Henry Harris at Oxford, who introduced virus-mediated fusion for genetic analysis, pursued it with particular vigor to study the genetic basis of malignancy. The results vary considerably depending on the particular cell lines under study; however, overall malignancy is attenuated, or even extinguished, when cancer cells are augmented with normal chromosomes. These observations were the first steps to a formal genetic analysis of the chromosomal basis of malignancy. They would advance much further as techniques were developed for identifying specific chromosomes and for the construction of cells with a normal complement of chromosomes of known origin.

Cell fusion techniques also offered a unique opportunity to study the possible genetic role of particles in the cytoplasm of cells. The occurrence of DNA in mitochondria and other particles suggested that these may be more than mere passive reflectors of the information in chromosomal DNA. In several experiments it appeared that the human chromosomal contribution was overridden by the mouse contribution, arguing for negative control as a common cellular mechanism. Negative control was also inferred from studies on biochemical attributes, including regulatory functions. The conclusions of all such studies indicated: (1) that unexpressed functions in a cell derive from negative control (that is, the active inhibition of certain genes, an inhibition that also regenerates new chromosomes); (2) that these functions can be revived upon extraction of the controlling factors; and (3) that the controlling factors are stably associated with certain chromosomes.

The nature of the stable activation of the controlling chromosome was the central challenge to developmental genetics. One hypothesis had some support from studies of X-chromosomes: stable changes of a chromosome are often deletions of some part of the chromosome. The activation of a controlling chromosome would then result from removing a small regulating unit from it, unleashing the negative control over other specific gene functions. In other cases, such deletions are already known to unleash positive functions.

While these studies were still at an exploratory stage, cell fusion was a well-established method for mapping genetic mutations on various human chromosomes, greatly accelerating work that formerly required the painstaking collection of pedigrees often insufficient or inappropriate to allow firm conclusions. In the course of gene mapping F. H. Ruddle's group at Yale noted the first evidence of a translocation (breakage and new union) between a mouse and a human chromosome, which may reflect the actions of a two-strand-repairing ligase, as noted earlier.

**Testicular feminization.** During the year testicular feminization (Tfm), a fascinating rare genetic disease, was critically reviewed. It is a disturbance whereby an XY embryo, normally male, is switched into a female pattern of development, but is sterile. A similar condition in the mouse is firmly linked to the X chromosome. S. Ohno (City of Hope National Medical Center, Duarte, Calif.) and his colleagues offered an explanation of the defect: the body cells of the Tfm type appear to be unable to respond to testosterone for lack of a primary protein receptor for it. The Tfm "female" thus differs from normal females as well as from males, both of whose tissues do respond to male hormones.

—Joshua Lederberg

# Obituaries

The following persons, all of whom died between July 1, 1971, and June 30, 1972, were noted for distinguished accomplishments in one or more scientific endeavors. Biographies of those whose names are preceded by an asterisk (*) appear in *Encyclopaedia Britannica.*

*__Adams, Roger__ (1889—July 6, 1971). U.S. organic chemist and recipient of the National Medal of Science, Adams was head of the University of Illinois department of chemistry for 28 years. He received an A.B. degree (1909), an A.M. (1910), and a Ph.D. (1912) from Harvard University, and then studied at the Kaiser Wilhelm Institute in Berlin before returning to Harvard as an instructor. In 1916 he went to the University of Illinois as an assistant professor, served as head of the department of chemistry and chemical engineering (1926-54), and became research professor emeritus in 1957. Adams and his students explored the many methods of organic synthesis, determined the structures of synthetic and natural products, and discovered a platinum oxide catalyst for further research in organic chemistry and biochemistry. He investigated (1947) the structure of marijuana, isolating and identifying cannabidiol, and then synthesized a drug that was 70 times stronger.

Among the many honors bestowed upon Adams were the Davy Medal of the Royal Society (London, 1945), the Priestley Medal of the American Chemical Society (1946), and the Perkin Medal of the Society of Chemical Industry (1954). He was awarded the 1964 National Medal of Science for his many years as the recognized leader in organic chemistry.

**Armand, Louis** (Jan. 17, 1905—Aug. 30, 1971). A French engineer and nuclear energy executive, Armand was responsible for the reconstruction of the French railway system after its almost complete destruction during World War II. He had joined the Compagnie des chemins de fer de Paris à Lyon et à la Mediterranée (the PLM) in 1934, and with the Nazi invasion of France in 1940 became a member of the railroad underground intelligence network, and subsequently was captured by the Gestapo. Released in 1944, Armand returned to the railroad and five years later was named director of the nationalized rail system, the Societé Nationale des Chemins de Fer (S.N.C.F.). He was assigned to rebuilding the lines and converting them from steam to electric power.

In 1958 Armand resigned his directorship to serve a two-year term as president of Euratom (the European Atomic Energy Community). He was the author of works on geology and mining, on transportation, and on the construction and unification of Europe. Some of his better known writings include: *Plaidor pour l'avenir* ("Plea for the Future"), published in 1961, and with Michel Drancourt *Le Pari européen* ("The Wager for Europe") published in 1968.

**Babakin, Georgi N.** (1916—Aug. 3, 1971). Soviet space expert, believed to have been involved in the design and production of Soviet automatic space probes, Babakin was credited with having an important role in U.S.S.R. space technology and in the study of the moon and of the planet Venus. He undoubtedly contributed heavily to the construction of the unmanned Lunokhod 1 moon rover and of the Venera 7 space probe.

Babakin began his career in 1930 as a radio technician and by 1949 was actively involved in Soviet aircraft and space technology. Although unlisted, he was presumed to be a corresponding member of the Soviet Academy of Science.

**Bernal, John Desmond** (May 10, 1901—Aug. 30, 1971). British physicist and natural philosopher, Bernal was also a crystallographer and helped establish the foundations of molecular biology. He served as professor of physics at Birkbeck College, University of London, from 1937 until 1963, when he became that institution's first professor of crystallography, a position he held until ill health forced his retirement in 1968.

During his undergraduate days at Cambridge University Bernal wrote a paper that brought him a research post to study crystal chemistry at the Royal Institution. He returned to Cambridge in 1927 and began a series of investigations into a wide range of scientific subjects, including the taking of the first X-ray photographs (in 1933) of a single protein crystal (pepsin). His works include *The Social Function of Science* (1939), *Freedom of Necessity* (1949), *Science in History* (1954), *World Without War* (1958), and *Origin of Life* (1967). Bernal was a fellow of the Royal Society and a recipient of the 1953 Lenin Peace Prize.

**Blegen, Carl W.** (Jan. 27, 1887—Aug. 24, 1971). U.S. archaeologist, Blegen was professor of classical archaeology at the University of Cincinnati from 1927 until 1957, when he became emeritus professor; he also served as head of the classics department from 1950 until 1957. Blegen led many expeditions in search of the site of the ancient city of Troy, and in 1939 discovered the palace of Nestor, King of Pylos. At that site a number of tablet fragments in Linear B (Mycenaean) Script were found.

Blegen was a fellow of the British Academy, a member of the Archaeological Institute of America, and a winner of the gold medal from the Society of Antiquaries of London. He was the author of several books, including six volumes on Troy and the palace of Nestor.

***Bragg, Sir (William) Lawrence** (March 31, 1890—July 1, 1971). British physicist Sir Lawrence Bragg was joint recipient (with his father Sir William) of the 1915 Nobel Prize for Physics, awarded for their discovery of the method of determining the structure of crystals by the use of X rays. Sir Lawrence was Fullerian professor of chemistry at the Royal Institution from 1953 until 1966, and its director from 1954 until 1966. He became a fellow of Trinity College, Cambridge, in 1914, after having produced an original work on the structure of crystals; from 1919 until 1937 he served as Langworthy professor of physics at Victoria University of Manchester, and then as director of the National Physics Laboratory (1938), and Cavendish professor of experimental physics at Cambridge University from 1938 until 1953.

Bragg did fundamental work on the structures of metals and minerals and developed new techniques for the study of X-ray diffraction. From 1954 he concentrated his efforts on the study of the structure of proteins. He became a fellow of the Royal Society in 1921. He was the author of *The Crystalline State* (1934), *Electricity* (1936), and *Atomic Structure of Minerals* (1937). With his father he wrote *X-Rays and Crystal Structure* (1915).

**Buu Hoi, Prince** (1917—Jan. 28, 1972). A Vietnamese physicist, Buu Hoi was head of the National Research Center of the French Radium

Wide World

*Edward C. Kendall*

*Lillian Moller Gilbreth*

Institute (the Curie Foundation), and a pioneer in radiological cancer research. In 1957 he discovered 3-4-9-10 dibenzpyrene, a chemical substance in cigarette smoke that caused cancer when injected into mice.

**Fesenkov, Vasiliy G.** (1889—March 12, 1972). A Soviet astronomer, Fesenkov was an early investigator of zodiacal light, a nebulous glow observed in the west after twilight and in the east before dawn. Fesenkov was chairman of the Soviet Committee on Meteorites and, from 1935, a full member of the Soviet Academy of Sciences, also serving as editor of the academy's astronomical journal.

**Gilbreth, Lillian Moller** (May 24, 1878—Jan. 2, 1972). U.S. industrial engineer who pioneered in time-motion studies, Gilbreth was the inspiration for the popular book *Cheaper by the Dozen*. The book, written by her two oldest children Frank Gilbreth, Jr., and Ernestine Gilbreth Carey and published in 1948, recounted the parents' methods of applying their knowledge of industrial management to the running of the household and family of 12 children.

Gilbreth received her B.A. degree from the University of California in 1900, her M.A. in 1902, and her Ph.D. from Brown University in 1915. In 1924, upon the death of her husband, she took over his consultant firm and continued its successful management. As consultant to the Institute of Rehabilitation Medicine (at the New York University Medical Center) in the early 1950s, she set up a program to augment the capabilities of disabled homemakers, incorporating features to accommodate workers in wheelchairs or on crutches.

***Kendall, Edward Calvin** (March 8, 1886—May 4, 1972). U.S. chemist and Nobel Prize winner, Kendall was visiting professor of chemistry at Princeton University from 1951. Prior to his appointment at Princeton he headed the Mayo Clinic biochemistry section for 37 years, and served as a professor in the graduate school of the University of Minnesota from 1921 until 1951.

While at the Mayo Clinic, Kendall succeeded in isolating the hormone of the thyroid gland (thyroxin), used to correct glandular deficiencies and permit normal growth in human beings. In 1930 he began his extensive studies on the chemical nature, physiological activity, and synthesis of adrenal cortex hormones. By the end of the 1930s he had isolated six of the gland's hormones and named them Compounds A, B, C, D, E, and F—E later became known as cortisone. Further investigations led to the identification of the chemical structure and composition of the compounds, and subsequently to a partial synthesis. Not until 1948, however, was the synthesis of E produced in enough quantity for human experimentation.

For this work on the adrenal cortex compounds and the resulting large-scale production of cortisone and hydrocortisone, Kendall (with P. S. Hench and Tadeusz Reichstein) received the 1950 Nobel Prize for Physiology or Medicine. Other awards bestowed upon Kendall included the Lask-

Wide World

*Maria Goeppert Mayer*

er Award of the American Public Health Association and the Kober Medal of the Association of American Physicians.

**Lynch, William A.** (June 21, 1892—Feb. 14, 1972). U.S. physicist and expert on earthquakes, Lynch was emeritus professor of physics at Fordham University. He taught at New York University a number of years before going to Fordham in 1937. As assistant director of the seismological laboratory at Fordham, Lynch came to be recognized as an authority on deep-focus earthquakes and the analysis of earthquake waves.

**MacLeod, Colin Munro** (Jan. 28, 1909—Feb. 13, 1972). A U.S. microbiologist, MacLeod was a research professor on the faculty of New York University's School of Medicine when, in 1963, he was appointed deputy director of the White House Office of Science and Technology. After leaving Washington in 1966 he returned to New York University and remained there until 1970, when he assumed the presidency of the Oklahoma Medical Research Foundation. MacLeod pioneered in immunology, working in the field of bacterial genetics concerned with phage viruses. His research aided in the discovery of treatment for all forms of pneumonia.

***Mayer, Maria Goeppert** (June 28, 1906—Feb. 20, 1972). German-born U.S. theoretical physicist and Nobel Prize winner, Mayer was a professor of physics at the University of California at San Diego from 1960. Previously she had held teaching positions at Johns Hopkins University (1930–39) and Columbia University (1939–45). During 1945–59 she was a professor at the University of Chicago and at the Enrico Fermi Institute of Nuclear Studies (under Enrico Fermi), at the same time serving as senior physicist at the Argonne National Laboratories, near Chicago.

Early investigations into theories of nuclear structure came into sharp focus for Mayer in 1948 as she discussed her work with Fermi. Her thinking led to the explanation that a measured spin of a nuclear particle could correspond to one of two different orbits, making possible a description of the nucleus in terms of orbits of single particles. Mayer then worked out a "shell model" of the way in which nuclei absorb neutrons in high-energy physics. This work paralleled that of J. Hans D. Jensen, who at that time was working at the University of Heidelberg. The two scientists met in 1950 and collaborated on a book advancing their theories. Their investigations brought Mayer and Jensen the 1963 Nobel Prize for Physics (which they also shared with Eugene P. Wigner).

**Sampsonov, Nikolai Nikolayevich** (1906—1971). A Soviet scientist, Sampsonov's promising career was stifled because of his outspoken criticism of the Stalinist bureaucracy. In the 1930s he worked for the Directorate of the Northern Sea Routes and was in charge of prospecting for valuable arctic mineral deposits. During World War II he was engaged in scientific work, also seeing action on the Leningrad front.

In 1956 Sampsonov published *Thinking Aloud,* an attack on Stalin and a call for the return to Leninist democracy. For this he was arrested and confined in a special psychiatric hospital where he remained for eight years, refusing to plead insanity. After repeated threats to his health and family Sampsonov finally capitulated and was discharged with a pension. In 1950 he and S. A. Poddubny were awarded the Stalin Prize for designing a new type of gravimeter.

**Sarnoff, David** (Feb. 27, 1891—Dec. 12, 1971). U.S. executive and the "father of American television," Sarnoff was president of the Radio Corporation of America (RCA) from 1930 until 1947 when he became chairman of the board, a position he held until retiring in 1970.

General Sarnoff (he was a communications officer during World War II) launched TV into American homes when he made a personal appearance before the television cameras at the New York World's Fair in 1939.

***Tiselius, Arne Wilhelm Kaurin** (Aug. 10, 1902—Oct. 29, 1971). A Swedish chemist and recipient of a Nobel Prize, Tiselius joined the faculty of the University of Uppsala in 1925. He was assigned as an assistant to T. Svedberg of the Institute of Physical Chemistry and in 1930 became a lecturer in chemistry. After several years of study at Princeton University, and work at the Rockefeller Institute in New York City he returned to Uppsala in 1938 as a professor in biochemistry.

Tiselius was awarded the Nobel Prize for Chemistry in 1948 for his discoveries concerning the complex nature of the serum proteins. This work on electrophoresis and adsorption analysis included the development of a system for moving large molecules (such as proteins) through a solution by means of electricity. The two instruments he devised, the "Tiselius apparatus," have been utilized by scientists everywhere to separate proteins into their many complex chemical compounds.

***von Bekesy, Georg** (June 3, 1899—June 14, 1972). Hungarian-born U.S. physicist and Nobel Prize winner, von Bekesy was professor of sensory sciences at the University of Hawaii from 1966. He received a Ph.D. in physics from the University of Budapest in 1923 and became an engineer with the Hungarian telephone system, where he began his research work into the mechanism of hearing. In 1947, after several years at the Royal Caroline Institute in Stockholm, von Bekesy went to the U.S. and joined the faculty at Harvard University. His continuing experiments led to the discovery that the human ear distinguishes the pitch of sounds in the cochlea, the part of the inner ear shaped like a snail shell, and that the cochlea works as a sort of microphone to convert the mechanical energy of sound into electrical impulses received by the brain. For this major research concerning the ear and for his pointing the way for advances in the diagnosis and correction of damaged hearing, von Bekesy was awarded the 1961 Nobel Prize for Physiology or Medicine.

# Photography

The year 1971-72 was one of outstanding progress in photography. In particular, research on the theory of sensitivity, emulsions, sensitizers, and color formers resulted in a tremendous improvement in color films, together with simplification of the processing. Research on black-and-white films also progressed, resulting in the production of recording films of greatly enhanced speed, with low grain and high definition, and direct positive films using a single stage of development.

While amateur photography continued its steady growth of over 5% a year, the largest amount of photographic material was still consumed in the nonamateur fields. Remarkable growth was shown in medical and industrial photography, microfilming, audiovisual uses, and the application of photography to the fields of graphic arts and reprography.

**Silver and nonsilver processes.** Most photographic processes rely on silver; in the U.S. the photographic industry is second only to the U.S. mint in silver consumption. In recent years the price of the metal has undergone extreme fluctuations, ranging from a high of $2.57 an ounce in 1968 to a low of $1.30 in October 1971. As of 1972 it showed a slow upward trend, leading to increased interest in silver-recovery systems. A number of such systems were proposed for photofinishers, radiographers, motion-picture and microfilm processors, and other large producers of waste silver. A further impetus to research in this field stemmed from the general interest in recycling of wastes in order to minimize environmental pollution.

Prompted in part by the high price of silver, considerable investment was being put into research on nonsilver photographic processes. As of 1972, no satisfactory substitute for silver in conventional photographic uses had been found, although nonsilver processes had been adopted for special uses, such as electophotographic systems for document copying, diazo and vesicular systems for microfilm duplicate prints, and photopolymer systems for printing plates and so-called photofabrication.

**Practical photography.** Nine years after the introduction of its highly successful Instamatic cameras, Eastman Kodak Co., in March 1972, introduced the Pocket Instamatics, a series of five $1 \times 2 \times 5$ in. cameras capable of being carried in a pocket or purse. The drop-in cartridges are loaded with Kodak Ektachrome-X and Kodachrome-X films, a Verichrome pan, and a specially developed Kodacolor II film for color negatives. Color prints are $3\frac{1}{2} \times 4\frac{1}{2}$ in., larger than the 3X prints made from normal Kodacolor negatives and of comparable quality. The slides are 30 mm square in plastic mounts. Kodak announced that it would license other manufacturers for production of the new series.

In September 1971 Kodak started to sell its new XL movie cameras and Ektachrome 160 Super 8 film in cassettes, permitting motion-picture photography indoors with existing light (XL). This was made possible by basic changes in camera design, including a specially designed high-aperture lens and an opened-up shutter. With the new film, the system gives 24 times the light-recording power of previous systems, and movies can literally be taken by candlelight.

Considerable work was being devoted to video cassette systems, in which a video cassette player

Courtesy, Köpf, Siemans AG, Munich, Germany

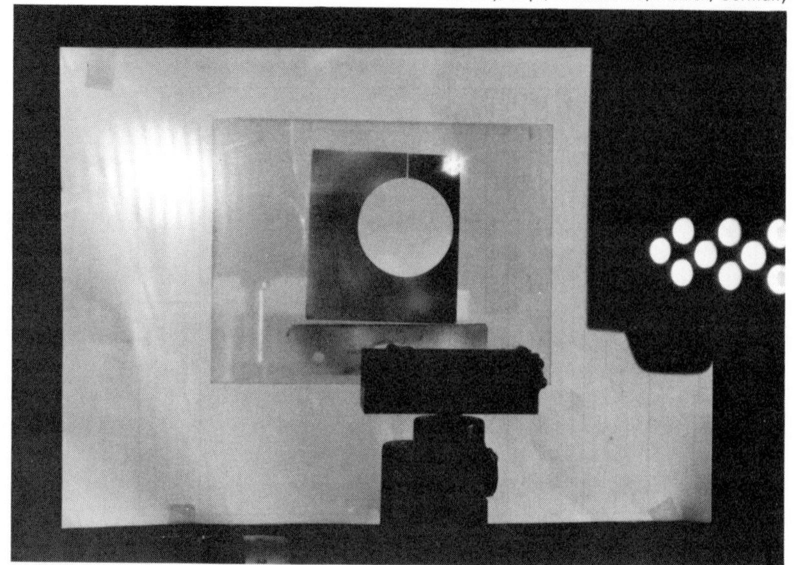

*Device for making speckle interferograms, photographs that accurately measure the tiniest displacements of moving objects, is at the right. Coherent light of a laser is used because it causes speckling when reflected by a dull surface and the amplitude of a displacement can be determined by examining the pattern of the speckles. The patterns below indicate a large (left) and a small (right) displacement.*

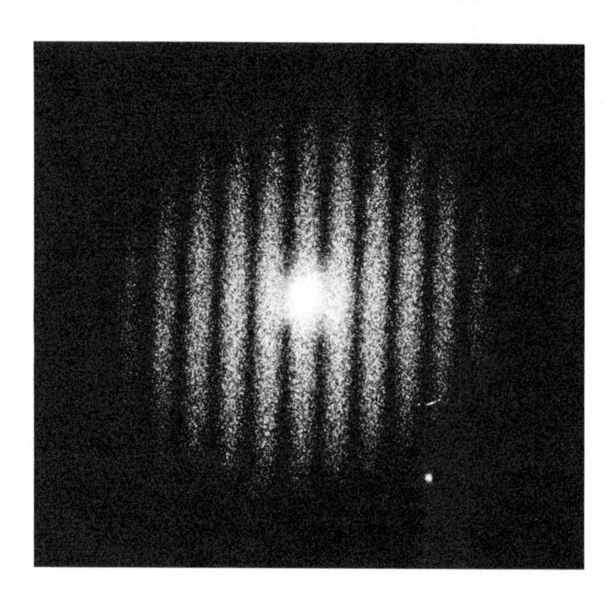

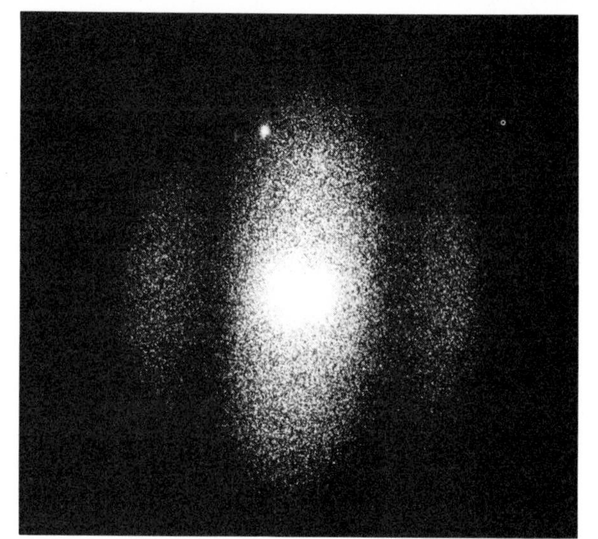

is attached to a conventional television set to give a display on the TV tube. Rapid progress in the field had been hampered by such factors as cost, lack of standardization, copyright and union problems, and programming uncertainties. In 1971, however, Kodak demonstrated a feasibility model of a system using conventional Super 8 film cartridges in a special player attached to the antenna of a TV set, and Fuji announced the CVR (Cine Video Recording) system for its straight-8 movie film.

A number of new Polaroid Land Cameras were introduced in 1971, including a 400 series of folding pack cameras featuring a "Focused Flash" system in which louvers open or shut in front of the flashcube according to the distance of focusing; a camera for making color close-ups; and a camera for taking square pictures. A new Polaroid pocket camera was demonstrated to stockholders at the company's annual meeting in the spring of 1972. It was claimed that the new systems made it unnecessary to stabilize the prints by coating them.

In 1969 Kodak and Polaroid had announced a new agreement whereby Kodak would continue to make color negative film for Polaroid and could make color film packs for Polaroid cameras. In 1971 Kodak announced intensification of research on its own in-camera color system, including new color chemistry and coating techniques and new portable cameras. Meanwhile, Polaroid, which in the past has contracted out much of its manufacturing to other firms, was developing its own chemical-, film-, and apparatus-manufacturing plants.

Use of color photography continued to expand; about 80% of all still pictures were in color and some 70% of these were color prints, while practically all amateur movies were in color. Further, the number of hobbyists making their own color prints was growing. Important advances in color photography, besides the Kodacolor II negative film for the Pocket Instamatic cameras and Ektachrome 160 for the XL movies, included special high-quality color film for photomicrography; high-definition film for high-altitude aerial reconnaissance; Ektachrome duplicating film; and three-solution processing for Ektaprint paper for printing from color negatives and new Ektachrome processing machines, both for photofinishers.

The Kodak Ektaprint three-solution chemical system and processing machines inaugurated a trend toward simplification of color processing, partly by combining bleaching and fixing stages in one solution. Associated with this system was a process that regenerates the bleach-fix (blix) for further repeated use and recovers practically all the silver.

For the professional color finisher, the Vericolor system, using a special color processor and new chemistry, permitted processed and dry color negatives to be produced in 11 minutes, compared with the one-hour wet time previously needed. For color printers, there were attachments that scanned incoming negatives with photosensors and set the printing exposures automatically.

**Current trends.** Active research was taking place in the field of color xerography, and it was expected that a practical system would be available within four years. New instant-picture cameras were anticipated, not only from Polaroid but also from Kodak and probably others. The advanced amateur would probably do more personal darkroom work, especially in color. The 8-mm video cassette systems could well provide an important stimulus to the expected boom in cable TV.

In education, the entire audiovisual field was expected to undergo rapid growth, not only through Super 8 films but also in slide-sound presentation and a variety of other display techniques. In the fields of medical, dental, and industrial radiography, in microforms, especially micropublishing, and in the graphic arts, the systems approach—film, chemistry, processor, and auxiliary equipment regarded as a system—was finding rapid acceptance.

In printing, the trend to phototypesetting continued. There was increased use of scanners to give color-corrected color separations and of photographic methods for producing lithographic printing plates.

—Walter Clark

# Physics

The discovery of the heaviest element occurring in nature, an isotope of plutonium, was among the outstanding developments of the past year in physics. In high-energy physics new facilities at the U.S. National Accelerator Laboratory and the European Nuclear Research Laboratory were being used to probe the world of subatomic particles.

## High-energy physics

**Intersecting storage rings.** A number of important experiments were carried out at the new intersecting storage rings (ISR) of the European Organization for Nuclear Research laboratory at Geneva. The ISR consists of two beams of protons, each with a maximum energy of 26.5 GeV (billion electron volts), circulating in opposite directions. These beams intersect at points about the storage rings. The energy of the head-on collision of such protons is 53 GeV, which is equivalent to the collision between a 1,500 GeV proton and a stationary proton. The ISR thus provides energy 7.5 times higher than the 200 GeV achieved at the National Accelerator Laboratory, Batavia, Ill. (However, large classes of experiments are not accessible to the ISR, since collisions when the two proton beams pass through each other are rare. Even counting all its protons, the ISR beam is an excellent vacuum.)

Two ISR experiments are of special interest. Elastic proton-proton scattering shows an angular distribution which scales to that at lower energy. (A collision is elastic when no energy of motion is lost to other forms, such as the creation of particles.) That is, one looks at this distribution with respect to "momentum transfer" to judge whether there has been a change in apparent particle size compared to that of the protons at low energy. The proton's apparent average diameter was observed not to change with energy. It had been speculated that particle production, through conversion of energy into mass, would lead to an ever-increasing apparent diameter with increasing energy for strongly interacting particles. A classic geometric view of fixed particle size, seems, instead, to be indicated by the experiment.

The second experiment also suggested a simple geometrical view. In this experiment just one particle was observed of the many that generally result from a high-energy collision. If the "Lorentz contraction" of special relativity is taken into account for the initial protons, and special relativity is also taken properly into account for the observed particle, a prediction can be made for the rate of particle production. This prediction was con-

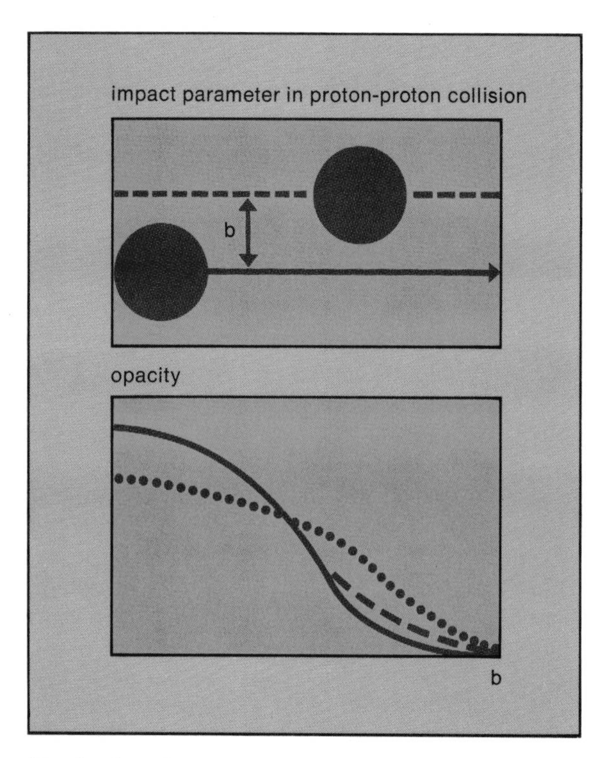


*Distribution of strength of interaction, or opacity, of a proton can be determined during interaction with another proton of high energy. Experiments revealed that the proton's opacity distribution does not expand with increased energy (dotted curve) but is independent (solid curve), though a slight effect (dashed line) might be present.*

firmed in a striking manner by the experiment: the probability for producing a particle of a particular velocity, when expressed in a certain scaling form, was predicted to be independent of the total energy. This was confirmed for a number of energies (quoted for a proton on a stationary proton target) of 12 GeV up to 1,500 GeV.

A number of other new and important results are expected from experiments with the ISR. These include the numbers of particles produced in collisions at various energies and a very extensive "map" of likelihood for producing each of a variety of particles at various velocities.

**Amplitude analysis.** The general technique for discovering properties of nuclear particles is to observe incident and outgoing particles associated with collisions. The general types of information arising from experiments on elastic scattering of particles are the existence of "states" and the properties, such as range, of forces or of distributions of matter. (A state is a spatially localized system that remains compact for some time.)

The tool used to extract detailed information about states and range of forces from collision data is "amplitude analysis." Observables (perceptible events), according to the quantum theory, are always products of two amplitudes, for example $A^2$ or AB. Amplitude analysis involves extracting the amplitude (A or B), which can be difficult. For example, if $A^2 = 4$, the amplitude is ambiguous: $A = \pm 2$. But as improved information has become available, it has gradually become possible to remove these ambiguities for certain cases of scattering.

An exciting development in this subject occurred during the year: the determination of the amplitude for pi-meson nucleon scattering at a high energy (6 GeV) over a range of small angles. This amplitude analysis was not intended to reveal any "states." (States are not expected to be observed at this energy because they may be too broad and overlapping or, perhaps, because there are no more states at this energy.) However, in the case of the two-nucleon problem studied in the 1950s, although no states were found, amplitude analysis had made possible the detailed determination of the two-nucleon potential—the force that holds atomic nuclei together. The high-energy amplitude analysis of the pi-meson nucleon system might have equally powerful implications for the behavior of strongly interacting particles at high energy.

One possibility, about which the present analysis is not yet definitive (complete experiments at other energies are needed), is that the apparent particle size is independent of energy, as also suggested by the ISR experiments discussed above. But a small amount of matter at a large distance from a particle's center may influence collisions at very high energy; thus, although average apparent particle size at high energy does not appear to grow rapidly with increased energy, it is possible that there is a small effect of this kind.

**Deep inelastic scattering of electrons by neutrons.** What is the neutron like internally? Does it differ substantially from the proton? "Deep" inelastic scattering of electrons at the Stanford (Calif.) Linear Accelerator Center—collisions in which the electrons are observed to lose a great deal of their energy of motion to particle production—confirmed that the proton should be thought of as a composite system. That is, the proton is made up of several, or many, parts. The neutron should be similar.

New experiments on the neutron showed that it is not identical to the proton, but the question is whether the difference can be understood in terms of electric charge. As a simple exercise consider a "quark" model for a proton and a neutron such that the proton is made up of three particles, called quarks, with charges $+2/3$, $+2/3$, and $-1/3$ to total one

Courtesy, Physics International Co.

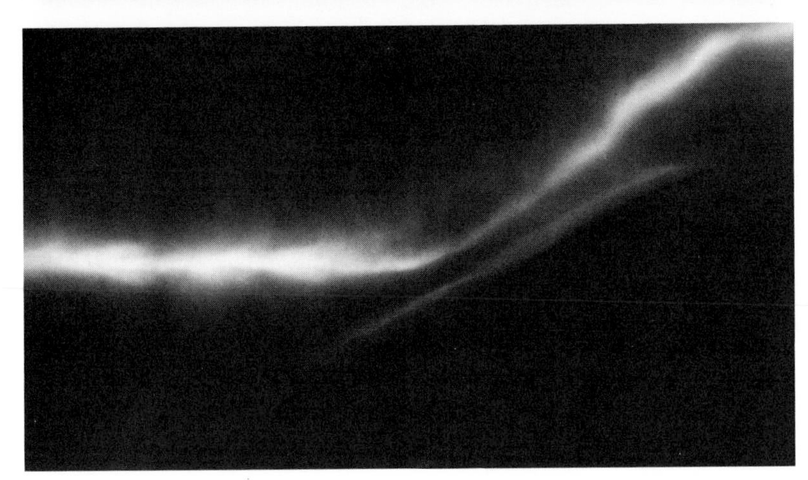

*Very intense beam of electrons moving at about the speed of light is disrupted (top, flare at left) shortly after leaving a vacuum and entering a gas. Electrons collide with the neutral gas atoms and ionize them. The positive ions almost totally neutralize the negative charges of the electron beam, and the magnetic field at the edge of the beam bends the electrons sharply around. Below, a tilted metal plate placed in the path of an intense electron beam causes the beam to appear to "bounce" off without actually striking it. The only illumination in both photographs was that produced by the recombination of ionized gas molecules in the path of the beam.*

unit of positive charge, and the neutron is made up of three quarks with charges $+2/3$, $-1/3$, and $-1/3$, so that it is neutral. The probability of deep inelastic electron scattering would be expected to be proportional to the sum of the squares of the quark charges: $4/9 + 4/9 + 1/9 = 1$ for the proton and $4/9 + 1/9 + 1/9 = 2/3$ for the neutron. Experiments to demonstrate this were exceedingly challenging both to perform and to analyze. The latest results indicated a rate of scattering for the neutron that is lower than the $2/3$ predicted by this model.

**National Accelerator Laboratory.** Important milestones were passed in the construction of the accelerator at the National Accelerator Laboratory during the year. Protons were accelerated to full energy in the "booster ring," and protons were injected and accelerated to nominal full energy (200 GeV) in the "main ring." Despite these achievements, there were serious difficulties, especially with magnets in the main ring. Much work remained to be done before acceleration at full current, or intensity, would be possible. The accelerator, however, was expected to operate effectively at a reduced intensity, thereby making possible a series of interesting experiments. *See* A Gateway to the Future: THE WORLD'S LARGEST NUCLEAR PROBER.

One of the early experiments was scheduled to be a Soviet-U.S. collaboration on elastic proton-proton scattering using a jet of hydrogen gas inside the vacuum chamber of the main ring as the target. This technique had been used successfully at the Serpukhov accelerator near Moscow. Another early experiment was expected to make use of the hydrogen bubble chamber successfully moved from the Argonne (Ill.) National Laboratory. With the bubble chamber, studies were to be made of the number of particles produced in collisions, and of correlations among such properties of the produced particles as velocity.

**Neutron stars.** High-energy physics seldom makes a direct, as contrasted with a philosophical, contribution to other subjects. A recent example of a direct contribution is to the theory of "pulsars," pulsating neutron stars. To understand the state of the condensed nuclear matter in such a neutron star one must understand thoroughly the states of nuclear matter at high energy. Recent speculations on these states, their density, and their "elementary" as opposed to composite, character, were being used by research physicists and astronomers as a basis for constructing theoretical models of the interior of neutron stars.

The key question in the investigation concerned the role of the Pauli exclusion principle. The exclusion principle states that two or more particles of a certain type cannot occupy the same space. If one visualizes a large number of such particles being gradually pressed together, which mutual gravitational attraction would cause to happen in a star, the consequence of the exclusion principle would be a rapid rise in pressure in order to prevent particles from occupying the same space. The exclusion principle does allow a completely different particle to occupy the same space, however. In a neutron star the question concerns the series of excited states of the neutron (particles like the neutron but more massive). If these excited states are fundamentally different particles from the neutron itself, the exclusion principle does not apply and the neutron star would be relatively easily compressed. But if these particles are simple composite systems they would then probably be similar to the neutron, as discussed above in connection with deep inelastic scattering, and a neutron star would, therefore, be relatively incompressible.

—Marc Ross

## Nuclear physics

Some impression of the nuclear domain—how far man has progressed in his exploration of it and how much remains totally unknown—is provided by the accompanying figure. This is a map obtained by plotting the number of protons ($Z$) in each variety of nucleus (nuclear species) against its number of neutrons ($N$). In each atom the number of positively charged protons in the nucleus is balanced by an equal number of orbital electrons, with equal negative charge, so that the atom as a whole is electrically neutral. The number of protons determines which chemical element is involved; different numbers of neutrons associated with a given number of protons result in the different isotopes of this element.

**The valley of stability.** For $N$ or $Z$ of less than 20, the most stable nuclei have $N = Z$; with increasing $Z$, however, the electrostatic repulsion between the protons acts to make the nuclei less stable, and so the most stable, increasingly, are those with more neutrons than protons. If the nuclear domain is considered to have three dimensions, reminiscent of a relief map, the instability of a particular isotope can be represented by its elevation above some selected "sea-level"; therefore, on such a relief map the most stable species lie at the bottom of a deep valley, the valley of stability, which runs diagonally across the map. In the figure the stable species are shown as the black squares; in all of nature there are only approximately 300 of these.

Until late 1971, physicists and chemists always stated, without question, that uranium—the element with $Z = 92$—was the heaviest in nature. This is no longer true. Darleane Hoffman of the Los Alamos (N.M.) Scientific Laboratory and her collaborators succeeded in 1972 in showing that an isotope of plutonium ($Z = 94$) occurs in nature as an extremely rare component of a rare earth mineral, bastnasite, from the Mountain Pass Mine in southern California. This came as a complete surpirse and provided important information that supported those who believe that such heavy nuclear species were formed in a supernova explosion of a giant star somewhere near our galaxy about 5,000,000,000 years ago.

Since the early 1940s nuclear scientists have made and studied approximately 1,300 artificially radioactive isotopes; these fall within the inner solid outline surrounding the stable species in the figure. These 1,300 along with the 300 stable species represented all that physicists knew about the nuclear domain in 1972. Extrapolating from these to describe the characteristics of all possible nuclei is something like attempting to describe the geography of the United States from a detailed study of the floor of the Grand Canyon. Physicists have really only begun to study the very lowest levels of the walls of the valley of stability. Enormous regions remain unexplored.

**The continent of stability.** *Can* nuclear scientists say anything about the unknown regions? One thing they have learned is that adding more and more protons to a nucleus eventually will produce a nuclear species that will disintegrate instantly because of the electrostatic forces of repulsion. If more and more neutrons are added to a nucleus, a limit is eventually reached where the additional neutrons can no longer be bound and they will "drip" out of the nucleus. These two limits can be calculated rather well from current knowledge of nuclear forces, and they establish the upper and lower boundaries of the continent of stability in the figure. In the case of very heavy, proton-rich isotopes, a spontaneous fission process competes with the emission of protons or of alpha particles (two protons and two neutrons very tightly bound to give a helium nucleus, a frequent nuclear decay product).

The continent of stability is a broad and unexplored one. Spanning it are the isotopes of calcium from calcium-31 to calcium-70; each isotope contains 20 protons, but the numbers of neutrons range from 11 to 50. Similarly, uranium extends from uranium-195 to uranium-302 with 92 protons in each isotope and neutron numbers ranging from 103 to 210. Of these 39 calcium and 107 uranium isotopes, which are predicted as stable against instantaneous disintegration, by 1972 only 24 and 14, respectively, had been discovered.

Fortunately, probes for exploring these new nuclear regions were expected to be at hand soon. G. N. Flerov and his associates at the Joint Nuclear Research Institute laboratories at Dubna in the Soviet Union calculated that if a uranium target could be bombarded with a 2,000 MeV (million electron volt) beam of uranium ions, the collision would produce at least 6,000 different nuclear species as compared to the 1,600 now known.

Quite apart from the great scientific interest in exploring this new territory, the discoveries might lead to valuable new applications. For example, many more varieties of radioisotopes would be produced. These radioactive nuclei have already had a revolutionary impact on medicine. In almost every application, however, the radioisotope has been selected for its chemical or biological characteristics—it had to follow and tag a certain chemical species or process, or concentrate in a certain organ, as, for example, iodine in the treatment of human thyroid disorders. Having made this selection, however, the user was then very limited in the radioisotopes of the selected element available to him. At best, he selected the one that came closest to giving the desired results while having minimal unwanted side effects. All too frequently, if the radioisotope had the desired radiation it had the wrong lifetime or dangerous associated radiation. With many more radioisotopes to choose from, physicians could hope to approximate much more closely the ideal isotope characteristics for a given application. The new isotope iodine-131, as an isolated example, was shown to be vastly preferable to the iodine-133 previously used in human thyroid therapy in terms of both its lifetime and its unwanted radiations.

**The islands of stability.** The extent of the continent of stability in the upper right portion of the figure remained unknown except that it shades gradually into the surrounding oceans of instability. Possible outer islands of stability in these oceans were predicted where $Z = 114$ and 126 (and possibly also $Z = 164$) and where, simultaneously, $N = 184$. There are no obvious natural processes whereby the elements in these islands would have been made during the history of the universe so that, even if stable, it is not surprising that they have never been found. Collisions of mas-

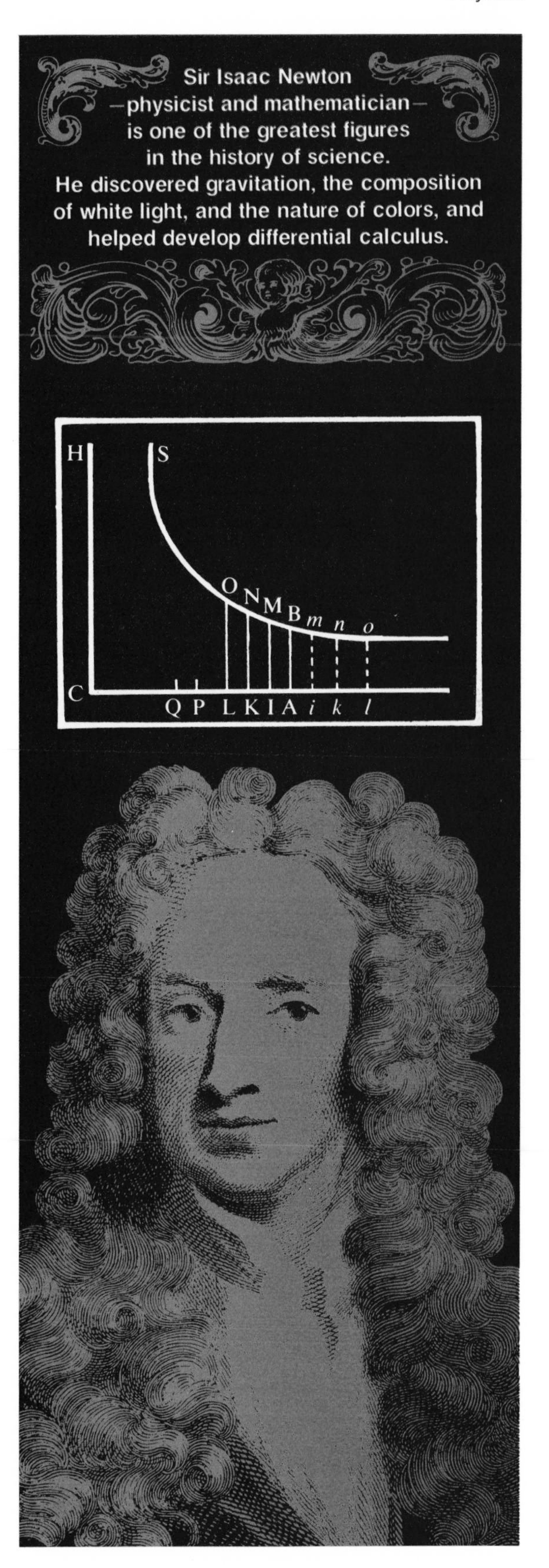


Sir Isaac Newton —physicist and mathematician— is one of the greatest figures in the history of science. He discovered gravitation, the composition of white light, and the nature of colors, and helped develop differential calculus.

sive chunks of nuclear matter, similar to the uranium on uranium example described above, provide a mechanism for the production of these very heavy nuclei, and programs under way at Berkeley, Calif., Dubna, Paris, and at Darmstadt, W. Ger., were directed toward their possible production.

Part of the interest in these potentially stable heavy nuclei stems from the fact that the same calculations that predict the possible long-term stability of these new species also predict that when they fission they might release 10 neutrons as compared to the 3 to 4 in the case of the more familiar fissile materials. This has potentially important consequences in reducing the critical mass required to support a nuclear chain reaction, thereby producing electric power from nuclear energy more easily.

**Excitations above ground level.** Thus far only the actual surface of the continent of stability has been discussed. It is traced out by the location of the most stable, or ground, state of each nuclear species in question. Each of these species, however, has many configurations lying at higher excitations (greater levels of energy) above the ground state; in terms of the figure, these would be in the "atmosphere" above the continent. Until recently, nuclear science concentrated on the structure very close to the ground levels. This reflected both experimental and apparatus limitations that made other studies difficult and also a reluctance to tackle those higher, more tightly packed, states until a better understanding of the lower, and presumably simpler, ones was at hand.

In recent years, however, major progress has been made in moving up from the ground state to more excited levels. In many nuclear regions—around lead ($Z = 82$) for example—it was found that the nuclear structure is much simpler than had been expected, with individual neutrons and protons moving largely independently in well-defined orbits much like those of the more familiar electrons in atoms. Pioneering work on this region was done by Nelson Stein and his collaborators at Yale University. Groups at Copenhagen and at Los Alamos found evidence for pairs of neutrons and protons moving together as single entities and making up whole new classes of states. More recently, Rolf Siemssen and his collaborators at the Argonne National Laboratory in Illinois, Henriette Mathieu-Faraggi and her collaborators at Saclay, in France, and Roy Middleton and a group at the University of Pennsylvania, found evidence for yet another class of structures where four particles—two neutrons and two protons—moved together in so-called quartet configurations. Proceeding in this way researchers were able to disentangle what seemed to be a wildly complex situation into overlapping families of relatively simple nuclear structures. This new work surveyed the "atmosphere" of the nuclear domain up to an elevation of approximately 10 MeV and covered hundreds, if not thousands, of different energy states in many different nuclei.

But what is the situation in the "stratosphere" of the nuclear domain? The question of what happens to a nucleus as energy is continually pumped into it has long been an open one. In the simplest models, such as those that were remarkably successful in the hands of Niels Bohr and John Wheeler in explaining the phenomenon of fission, the nucleus was likened to a liquid drop. When such a droplet has energy added to it—is heated—its components (molecules) move ever more rapidly

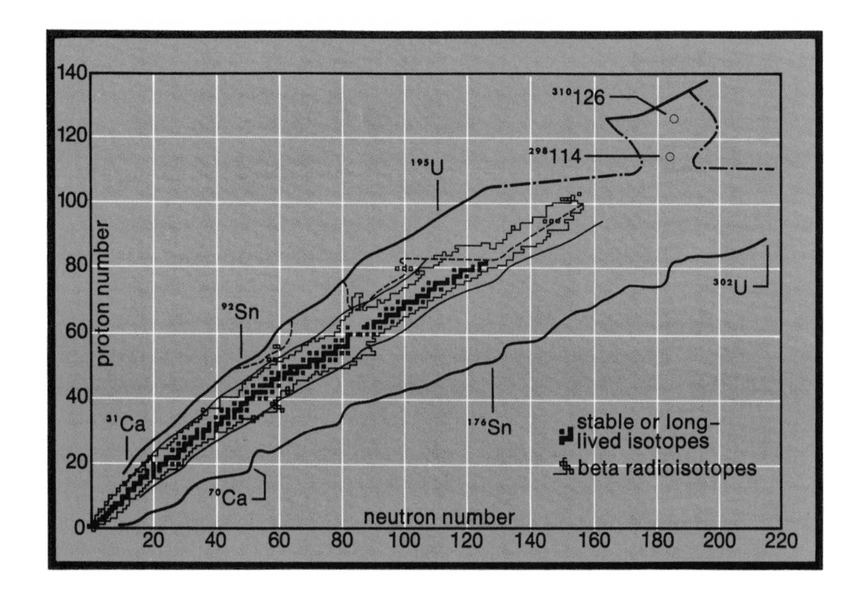


*The "valley of stability" for atomic nuclei can be drawn by plotting the number of protons in each nuclear species against the corresponding number of neutrons. The heavy black lines and dots in the center of the valley represent the most stable species. The solid boundary lines show the limits of stability, beyond which the strong nuclear forces would cause instantaneous decay of a species. The region of heavy man-made elements is defined by the broken line at the upper right; the circles indicate two potential man-made elements, 114 and 126, that would be islands of stability.*

Harry Seawell

*A 2.5 million-v ion accelerator developed by IBM applies the techniques of nuclear physics to analyze materials and make new electronic devices. Beams of ions, giving off glows characteristic of their composition, are directed at high velocities to be embedded in silicon crystals and thereby create new devices.*

under thermal agitation until some, on a statistical basis, achieve enough energy to escape from the drop altogether. They have been evaporated. Similar phenomena have long been observed in nuclear science where neutrons having a familiar evaporation spectrum of energies are frequently observed. If this were all that happened, the nuclear stratosphere would be dull indeed. But recently Adriano Gobbi and his co-workers at Yale obtained evidence for energy states near 40 MeV in silicon-28; these states have a striking molecular structure in which two well-defined carbon nuclei are bound together by the exchange of an alpha particle. Also, preliminary results on the capture of negative K mesons in material such as nickel, obtained by collaborating groups from Carnegie-Mellon University, Pittsburgh, Pa., and the Argonne National Laboratory, provided evidence for the presence of large nuclear subunits (for example, carbon-12 or beryllium-8) orbiting at a large radius from the center of mass of the parent nucleus. At such large radii they can be effectively unhooked from the parent nucleus by the energy release from the decay of the captured K mesons.

**Heavy ion nuclear science.** In a new frontier area of the nuclear sciences, beams of heavy ions (normally any nuclear species heavier than helium, $Z = 2$) are used to initiate nuclear reactions. They have the advantages of permitting the study of reactions when large clusters of neutrons and protons are transferred between projectile and target, of giving access to very-high-spin situations, and of selectively focusing on the surface regions of nuclei rather than upon their interiors.

In the areas of high spin, heavy ions allow the study of states having spins at least 10 times higher than ever before seen. Andrew Sunyar and his collaborators at Brookhaven National Laboratory, Upton, N.Y., found in preliminary heavy ion studies that in dysprosium-156, as the nucleus is spun faster and faster, an abrupt and striking change occurs in the moment of inertia (the sum of the products of each mass particle of a body multiplied by the square of its respective distance from the axis of rotation). The reason for this was unknown, but the phenomenon provided a stringent test for the current understanding of the microscopic bases of nuclear structure.

—D. Allan Bromley

# Science, General

In general, the basis for federal support of the U.S. scientific endeavor has been the belief that, out of its vast array, useful information will surely emerge. Although few specific basic research efforts will produce economically beneficial results, it is necessary to support the overall research program because its value will be greater than the total investment. Although no one could fully document the relationship, it has been a commonly held assumption that the substantial technological superiority of the United States derives from its leadership in almost every field of science. During the year, however, doubts began to emerge.

A business survey published in February 1972 by the Morgan Guaranty Trust Co. of New York reported:

> While the U.S. still enjoys a sizable trade surplus in transactions involving technology-intensive manufactured products—such as chemicals, electronic equipment, machinery, and transport equipment—a threat to that surplus unmistakably exists. This is emphasized by the fact that imports of such products grew more than twice as fast in the 1960s as exports. One of the most visible manifestations of this, of course, has been the

Courtesy, Lawrence Berkeley Laboratory, University of California

*A photoelectron spectrometer at the Lawrence Berkeley Laboratory of the University of California will be used for chemical analysis of atmospheric particulates as part of an interdisciplinary environmental research program coordinated by the laboratory. By mid-1972, 14 projects had been funded by various governmental agencies.*

spectacular progress Japanese industries have made in cracking U.S. markets for electronic products such as calculators and television sets. But such other technology-intensive industries as machinery producers also have found themselves increasingly hard pressed to maintain a competitive stance in domestic and foreign markets.

## New directions for R and D

In the White House the realization was growing that the vaunted U.S. superiority in technology was being challenged—not only by Japan but also by West Germany and other European nations. Furthermore, technological proficiency did not seem to be of great help in solving such severe societal problems as the deterioration of the cities, crime, and the crisis in transportation. The result was a rather fundamental shift in attitudes toward the allocation of federal resources for the support of science and technology.

A decade earlier the scientific community was being asked to identify those fields of science that would most benefit from increased support. Now the question seemed to be: How should the national research and development program be directed in order to solve societal problems and restore the U.S. technological advantage in international trade? In addition, if the research and development effort could provide a boost to the domestic economy, that would be of great interest to an administration concerned about rising prices in a period of rising unemployment.

**Searching for projects.** Few appointments to Washington officialdom aroused as much interest and speculation among professional science-policy watchers as that of William M. Magruder to serve as special consultant to Pres. Richard Nixon, effective Sept. 8, 1971. The initial announcement was made in rather low key by an assistant press officer at a routine White House press briefing. Reporters were told simply that the former director of the supersonic transport program in the Department of Transportation would be working with the White House "in the areas of research and development which hopefully will lead to potential new technology which could help the United States in our domestic and foreign export policy."

Magruder saw his mission as the identification of promising technologies that would have the potential of solving one or more of the nation's pressing problems. In an effort to unearth these new technological opportunities, as he called them, he invited government agencies, private organizations, and industries to send in their pet ideas. Letters went out by the thousands. The following is a typical example outlining the principal concerns of the White House:

Dear Dr.——:

On behalf of the President's Domestic Council, I am writing to solicit the views of your organization, or even the individual views with respect to potential new technology opportunities in the civilian sector and how the Federal Government might stimulate these potential opportunities. Generally, these programs might include efforts (Federal, private, or both) in areas such as desalinization of water, advanced energy systems, low pollution automobiles, solid waste recycling, new teaching methods, new health care methods, new housing construction, oceanographic exploration, means for incurring industrial productivity, transportation systems, new law enforcement methods, and so on. The objective is to stimulate innovation in the civilian sector of the economy directed at basic national problems and/or economic opportunities.

The study should apply at least these basic criteria to each technology project or program before any increased funding is recommended;

1. Importance: Project or program must relate (a) to a significant and urgent national problem (b) to a significant opportunity for economic growth, increased exports, or technological leadership.

2. Pay Off: The cost/benefit ratio must be favorable. What are the social and economic benefits—including impact on balance of trade productivity and employment? What will be the distribution of benefits among different industries and areas of the country?

3. Budget Impact: What are estimated long-term costs over next six years of full systems costs?

4. Non-Federal Support: What is the feasibility and likely maximum extent of cost sharing by industry? Can the project be achieved through steps short of direct Federal funding: regulation or deregulation, standard setting?

5. Potential Problems: Are there potential institutional or economic barriers to acceptance and implementation of the technology and how are they to be overcome? Are there any undesirable side effects from the technology and how are they to be dealt with?

6. Organization and Management: Assuming several programs or projects meet the above criteria, what kind of organization (existing agencies, new agency, or quasi-public corporation) should be created (if any) or redirected to carry out these programs?

A second area wherein I seek your counsel relates to potential stimulus that could be offered to industry in order to create innovation, higher productivity, etc. by revision of tax laws, grants, prizes, etc.

A third area about which I would appreciate your ideas concerns potential changes in anti-trust and patent law practices to encourage innovation, higher productivity, improved export business and greater overall employment . . .

Sincerely,

William M. Magruder
Special Consultant
to the President

As the months wore on and proposals for the new technological initiatives began to pour into the Executive Office Building, visitors returned from the screening sessions with a sense of awe at the size of the input. Piles of paper were said to reach to the ceiling. Projects were coming in by the hundreds, and some of them were accompanied by stacks of supporting documents three feet high. Unfortunately, there appeared to be few good ideas per pound of paper. Proposals from government agencies looked like warmed-over versions of ideas that had been turned down several times by the Office of Management and Budget or by lower echelons of review. Proposals from industry looked as if the sponsors had good reason for not investing their own money.

Courtesy, F. E. Compton Company

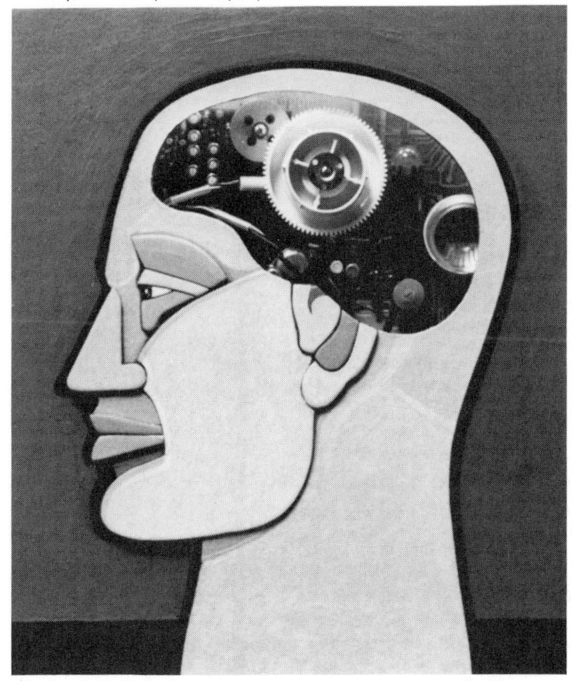

*Can man and his culture survive only if he is conditioned to want what best serves group interests? Harvard psychologist B. F. Skinner* (Beyond Freedom and Dignity) *thinks so, but his critics contend that self-determination is vital and behavioral manipulation violates man's essential humanity.*

Even though Magruder had attempted not to raise expectations too high, the absence of any significant reference to his mission in the president's message on science and technology on March 16, 1972, evoked bitter expressions of disappointment from much of the science press. *Science* magazine called the technology plan "vapid" and described the message as "little more than reshufflings of existing rhetoric and known policies" in "sad contrast to the optimistic hints that emanated from the Administration last summer and fall."

Nevertheless, many of those close to the exercise refused to write it off as a failure. They argued that its primary accomplishments had been, on one hand, to convince top government officials that there was a solid connection between R and D and the quality of American life, and, on the other, to educate them in the intricacies of dealing with such issues. Said one budget official: "The political officers in the Office of Management and Budget began for the first time to understand the complexity of the R and D process—its complicated relationship to such things as balance of trade, productivity, and jobs." Magruder pronounced his own valedictory: "I'm satisfied that we served the top decision makers in at least bringing the conflicts and hard questions out into the open. . . . Beyond that . . . the operation gave the Administration a whole credenza of projects whose time will come sooner or later."

**A budgetary step upward.** One reflection of the administration's growing awareness of the value of a healthy R and D effort appeared in its proposed budget for the 1973 fiscal year. Once again there was a small but significant increase in the overall figures, as contrasted with the abrupt leveling off that had characterized the Johnson administration. The total figure rose 9%, from $16.4 billion to $17.8 billion. The increase was almost precisely divided between the military and civilian sectors.

Responses to the stimuli that had given rise to the Magruder program were also evident in the budget. The high-technology agencies were directed to engage in activities with broader social benefits. Thus the Atomic Energy Commission (AEC) received increased funding to work on electric automobile batteries and cryogenic transmission lines, and the National Aeronautics and Space Administration (NASA) was asked to work with the Department of Transportation on a variety of vehicle-development projects.

Most evident, however, was the effect on the budget of the National Science Foundation (NSF). Although the budgeted figure of $653 million represented only a 5% increase over the previous year, the release of previously impounded NSF funds brought the total available amount to $674.7 million, a far more significant increase of 12%. In a period of belt-tightening, few other agencies gained as much.

Of special interest were three new programs, again related to the new emphasis. One $36 million item, the Experimental R & D Incentives Program, was to be conducted jointly with the National Bureau of Standards. Its goal was to encourage increased investment by the civilian sector and to improve the application of R and D results. Another $2.5 million was requested for a National R & D Assessment Program, to study how science and technology contribute to such national objectives as improvement of the quality of life, economic growth, productivity, and employment. A third program, Institutional Grants for Research Management Improvement, would set aside $4 million to encourage academic scientists and administrators to study and experiment with new arrangements for the management of research.

Once again, the overall NSF budget provided evidence that the administration preferred to spend its funds on targeted research rather than on producing scientists per se. There were almost no funds for the continuing support of science education, and support for graduate students was

Ken Heyman

*Pigeons in B. F. Skinner's laboratory have been central to his experiments in controlled behavior. By setting up "contingencies of reinforcement" under which desired behavior is rewarded with food at critical moments, Skinner has disciplined the pigeons to walk figure eights, dance, and play Ping-Pong.*

reduced from $20 million to $14 million. At the same time, funds allocated to the two-year-old NSF program in Research Applied to National Needs increased 43% to $80 million, mostly under the headings of Advanced Technology Applications and Environmental Systems and Resources. Major emphasis was placed on energy R and D, particularly solar energy. The largest single addition, however, was in basic research: $3.7 million for initial funding of a huge new radio astronomy facility, with a Y-shaped interferometer with 13-mi arms for the placement of movable antennae.

An increase of $736 million in R and D funds, larger than the total NSF budget, was granted to the Department of Defense. The 9.5% increase was justified by the need to catch up on projects deferred because of the Vietnam war and the apprehension that the Soviet Union was outspending the U.S. in this area. NASA's budget remained essentially level at $3.2 billion. A proposed grand tour of the outer planets, dropped as not cost-effective, was to be replaced by individual flights to Jupiter and perhaps Saturn. Additional funds were allocated to two remaining Apollo flights, three Skylab launches, and developmental work on the space shuttle (*see* Year in Review: ASTRONAUTICS AND SPACE EXPLORATION).

The AEC budget showed a slight increase, with R and D emphasis on the development of clean energy sources, such as the liquid metal fast breeder reactor and controlled thermonuclear fusion. Total federal funding allocated to the environment, scattered throughout a number of agencies, rose only slightly—from $1.1 billion to $1.2 billion—despite public interest in the problem. Health-related research was programmed to increase from $1.8 billion to $2 billion, mostly in the National Institutes of Health; cancer research received a substantial increase of $93 million, to $430 million. The Food and Drug Administration was also strengthened, with a jump from $38 million to $56 million in research funds.

## Technology assessment

The year 1972 brought a little closer to reality an attractive governmental concept that had produced more words and fewer concrete results than any similar idea in recent years. This was technology assessment, a term possibly coined and certainly made popular by former Representative Emilio Q. Daddario of Connecticut. (See *1971 Britannica Yearbook of Science and the Future,* Feature Article: PRIORITIES FOR THE FUTURE: GUIDELINES FOR TECHNOLOGY ASSESSMENT, pages 292–311.) For more than five years Daddario had insisted that Congress, whose acts of appropriation make possible the development of such powerful new technologies as atomic energy and supersonic flight, had a corresponding responsibility to discern the future effects of these new technologies on national life.

Despite the obvious difficulties involved, Congress and much of the rest of the national leadership had come to accept the need for some kind of capability to predict the consequences of technological change. A number of programs to prepare and evaluate methodologies for making such predictions sprang up on university campuses, notably at Harvard, Columbia, and Cornell. Of these three, only Cornell remained, along with smaller programs elsewhere. Congress, however, had had the issue brought home to it with increasing intensity. Much of the deliberation about funding the development of the supersonic transport hinged on environmental consequences. The National Environmental Protection Act calls upon government agencies to predict the environmental consequences of their activities. Rep. Mike McCormack (Dem., Wash.) noted that in the 89th Congress 9 bills were introduced to create new committees with a special interest in science, technology, and the environment; in the 90th Congress the number rose to 27, and in the 91st, to 132.

**The OTA proposals.** The principal bills in the 92nd Congress concerned the establishment of an Office of Technology Assessment (OTA). As originally written in the House, the proposal would have provided for a policy-making Technology Assessment Board, consisting of one senator and one

Ken Heyman

*B. F. Skinner demonstrates the laboratory apparatus that he has designed to carry out experiments in controlling animal behavior. The wires are connected to the animals' cages and to the trigger mechanisms that provide rewards to reinforce the desired behavior.*

representative from each major party, four private citizens appointed by the president, plus the comptroller general, the director of the Congressional Research Service, and a director of the OTA, who would be appointed by the board. The principal assignment of the office would be to contract out studies that would provide Congress with early warnings concerning potential consequences of new technologies and with analyses of alternative measures. Its initial operating budget was to be in the neighborhood of $5 million, and it would have a staff of a dozen or so.

When the OTA bill came before the House, there was little argument over the need for such advice, but there was an unexpected reluctance to go outside Congress to find it. Although the framers of the original bill conceived of the board as a device whereby scholars and civic leaders could peer into the crystal ball alongside Congress, House members doggedly refused to admit that they could profit from such assistance. Nor did they want the staff director of the office to enjoy the privilege of assigning studies. Congress has been elected to draw up legislation, they argued, and it alone was capable of deciding what kinds of information it needed to carry out its function.

Since no other body had yet conducted a successful technology assessment study of the magnitude needed by Congress, it could not be argued that Congress would do a worse job. The bill sailed through the House in its amended form. It was expected to pass the Senate easily, and probably in a version very like that approved by the House.

**The Safeguard controversy.** An example of the controversy that can arise when an outside group attempts to assess a technology being promoted by a federal agency occurred during the winter of 1971–72. The groundwork for the furore was laid in a series of Senate hearings in the spring of 1969 dealing with the merits of the Safeguard antiballistic missile system. In those hearings and in various publications, a group of scientists had argued that the supporting evidence for authorizing the Safeguard system was faulty, and that the system itself was neither sufficiently necessary nor reliable enough to justify the funding that would be required.

Much of the argument for and against the Safeguard system fits under the general rubric of operations research or systems analysis. Thus it was that in November 1969 one of the proponents of the Safeguard system, Albert Wohlstetter of the University of Chicago, wrote to the council of the Operations Research Society of America (ORSA), asking it to

> . . . appoint a panel to consider some aspects of professional conduct during the ABM debate this last spring and summer. . . . The conduct of the controversy on the ABM provides many outstanding examples of the violations of professional norms . . . Some basic issues in the public debate concerned matters of analysis. Other issues concerned matters of fact. . . . Many of these questions both of fact and analysis can be answered quite precisely by a competent panel of operations research men. I believe the profession owes this to the public and the Congress.

The ORSA council satisfied itself that Wohlstet-

*"How's this for recycling? We get some heavily polluted air, put it in an aerosol can, and use it as an insecticide."*

Sidney Harris

ter, a member in good standing, had raised an important issue. It agreed to conduct an examination and to accept, by and large, Wohlstetter's formulation of the issues. A committee of inquiry was set up and, in an effort to avoid the appearance of a kangaroo court, the president of ORSA invited George Rathjens of the Massachusetts Institute of Technology (MIT), a chief Safeguard opponent and Wohlstetter target, to designate a competent systems analyst to serve on it. In a joint reply, Rathjens and two other Safeguard opponents, Jerome Wiesner, then provost of MIT and formerly science adviser to Pres. John F. Kennedy, and Steven Weinberg, also of MIT, argued that the inquiry as formulated did not address itself to the principal issues and was, in fact, loaded against them. They also pointed out that, since they were not members of ORSA, that body had no right to evaluate their professional conduct.

Nevertheless, ORSA proceeded with the study and in September 1971 reported its findings. "No significant defects" were found in Wohlstetter's study of Safeguard, and much of it stood "as a model for the professional and constructive conduct of a debate over important technical issues," but those opposing Safeguard were frequently guilty of faulty methodology. Although some shortcomings were found in the testimony of the proponents, the report said, "they nowhere equalled the cumulative mass of inadequacies compiled by the opposition."

The report then addressed itself to the larger and more difficult question of the ambiguous role of scientific advocate:

> There was present, throughout the debate, a complete intermixture and confusion of the roles of the analyst and the advocate. To the extent that advocating analysts attempted to provide convincing arguments supporting the validity of the conclusions and recommendations they favored, the calm orderly presentation of the analysis on which these conclusions were based suffered.

Issuance of the report was followed by a controversy between the original antagonists that often became acrimonious. The fiery exchanges that appeared in the letters columns of major national newspapers and in the *Congressional Record* failed to settle the fundamental questions concerning the usefulness of the Safeguard system, but they did serve to underline in bold strokes that, in the absence of sufficient data, scientific confrontations can be as inconclusive as those in any other field of human endeavor—and as fierce.

## In the arena of public debate

Despite the painful experiences of a number of scientists who had thrust themselves into controversy in public forums, and despite the determined efforts of many scientific societies to avoid politicization, the pressures on scientists to become involved in political issues continued to grow. Many of these pressures were exerted by younger scientists, who shared with some of their generation a growing conviction that the scientific establishment had become, through indifference, an instrument to consolidate the control of an already powerful elite group. But in April 1972, George B. Kistiakowsky, vice-president of the National Academy of Sciences and formerly science adviser to Pres. Dwight Eisenhower, appeared before a meeting of the American Chemical Society and argued for involvement.

Looking to a future of increased population pressures and depleted natural resources, of mounting demands for a cleaner environment, and of rising expectations in the less developed nations, he declared:

> We cannot deal with these problems and meet these obligations without either a large decrease in our standard of living or the development of new technologies for public use and benefit, based on scientific knowledge that is still to come. That is self-evident. The real challenge is to design attacks on specific problems as they arise and bring the resources of the scientific community to bear on them. To meet these challenges, I now urge far greater social and political involvement than heretofore. In the past the practitioners of science have tended to let others in the society decide how to apply their findings. The results, as I in company with many others have emphasized, have not been uniformly good. Together with those especially knowledgeable in the problems of society, we must now try to separate the good and bad potentialities and then lend our weight to the good side.

**Meeting the challenge.** There appeared to be many volunteers to take up the cudgels. The Federation of American Scientists, a Washington-based national organization openly dedicated to lobbying activities, primarily in the field of arms control and disarmament, established a new program called TACTIC (Technical Advisory Committees to Influence Congress). Its announced goal was to create, at the grass roots, a "nationwide network of scientists and engineers" who would press for favorable national policies—"first of all, by providing constituent pressure on their Congressmen, and secondly by alerting the public, thereby generating political pressure for a change." By February 1972 its organizers claimed that 850 participants had signed up in 300 out of the country's 435 congressional districts; their goal was to establish an active working group in each of them.

As the federation set out to pursue its original purposes more vigorously, a number of other scientific groups were following courses of action that had not been envisioned by their founders. Members of the American Physical Society attend-

Stuart Ackerman

*Stanford professor William Shockley confronts a group of demonstrators dressed as Ku Klux Klansmen who disrupted his electrical engineering class. Shockley became a controversial figure in academic circles because of his theories on the relationship between race and intelligence. He requested permission to teach a course based on his thesis that blacks suffer an innate intelligence disability as compared to whites, but was refused after a faculty committee found him unqualified and not sufficiently objective.*

ing its annual meeting in April 1972 found themselves voting on a constitutional amendment to "shun all those activities which are judged to contribute harmfully to the welfare of mankind" and listening to a symposium on the significance to the physical sciences of the conspiracy trial of the "Harrisburg 7" antiwar protestors. The American Chemical Society, considered by many to be the most conservative group of its kind, confronted a choice of presidential candidates that included not only the two routinely selected by a nominating committee but also an aggressively outspoken retired industrial chemist named Alan C. Nixon. Nominated by a discontented group within the membership, Nixon quickly distinguished himself by appointing a campaign manager, stumping for improved working conditions for chemists, and boldly advocating that employed members of ACS be assessed to fund a political arm to lobby in the tradition of the American Medical Association. Nixon further distinguished himself from the other candidates by winning, hands down. He was scheduled to take office Jan. 1, 1973.

Even more flux developed within the engineering societies, whose members were hard hit by cutbacks in specific areas of research, notably space and electronics. Although surveys conducted by the NSF produced far less alarming statistics among the group responding to a questionnaire, a reporter for the *Wall Street Journal* estimated in February 1972 that 35,000 engineers were jobless and as many as 65,000 more were not able to work full time in their profession. Whatever the actual numbers, the technical societies were taking extraordinary steps. Even before Alan Nixon took office, the ACS set up an independent, nonprofit corporation to establish a national pension system that would permit members to take benefits with them when they changed jobs. The American Society of Civil Engineers sought to persuade employers to accept minimum standards for working conditions as well as salary guidelines. With an obvious reference to a burst of organizing activity by major AFL-CIO unions, the executive director of the ASCE said, "We hope employers will voluntarily adopt these guidelines. We're not turning into a union, but we're taking a more realistic role—and maybe we can bring economic improvements through responsible leadership."

**The argument over growth.** Controversy within the scientific community was not limited to institutional problems. Continuing along lines that had been taking shape for several years, scientists began to choose sides on a number of broad policy issues in which science and technology played key roles. Two of the most provocative issue-raisers were an MIT engineer named Jay Forrester and a young colleague named Dennis L. Meadows.

Forrester set the stage for controversy by fitting a pet theory to the announced concern of a floating international body of industrialists, statesmen, and academics calling itself the Club of Rome. The club, organized by an Italian industrialist named Aurelio Peccei, had proclaimed its anxiety over what it termed the "predicament of mankind"—an inexorable slide toward global catastrophe through uncontrolled population growth and industrial pollution. Forrester's theory, which he had espoused in his book *World Dynamics*, was that a global model of the dynamic interactions between world population, industrialization, depletion of natural resources, agricultural productivity, and pollution could be created through the use of an intricate and complicated computer program.

Meadows and a group of young associates took it from there, cranking a variety of numbers and relationships into a computer and reading the depressing tapes that spewed out. To those who questioned the accuracy of the data that went into the machine, Meadows responded that it did not appear to make much difference whether population growth or agricultural productivity or pollution increased according to this projected curve or that one; sooner or later the interrelationship produced a catastrophic dive in the population curve. The only arrangement that forestalled such sudden population decreases within 60–100 years was a tightly managed equilibrium, in which not only population growth but also industrialization and agricultural productivity were flattened out or forced into a negative growth curve.

Meadows' wife, Donella, worked the results of the study into a short book of simple and dramatic statements, *The Limits to Growth.* A Washington firm specializing in attractive social policy issues joined the Club of Rome in promoting it. The result was the most sensational display of publicity ever granted to what purported to be a scientific statement. It was not the publicity, however, but the science in the statement that led to widespread challenges by other scientists. Economists were especially aroused; they charged that the methodology employed by Meadows and his associates was faulty, especially in that no allowances were made for policy adjustments by social and political institutions as the effects of the predicted developments began to be felt. Meadows retorted that his group was simply trying to develop a useful methodology, designed to lessen the possibility that the effects of disastrous policy (or nonpolicy) might be felt too late.

**At issue.** Only slightly less attention was devoted to the theories of the internationally renowned behavioral scientist B. F. Skinner. In his *Beyond Freedom and Dignity,* he argued that human beings are hardly the free spirits of which the poets sing. Instead, they are manipulated throughout their lives by a series of operant conditionings generated in their environment, over which they have little or no control. His fundamental proposal was that since we are all the objects of external manipulation, we should try to order society so that the manipulation is done by individuals equipped to ensure that the greatest benefits ensue.

Skinner was attacked with far greater ferocity than Meadows—primarily because he offended the humanists, who tend to be considerably more articulate than social scientists. As in so many instances in which controversies arise between groups who speak essentially different languages, the argument was rarely joined properly. Skinner was attacked because he dared to suggest that behavioral scientists were uniquely equipped to manipulate human conduct and produce an orderly and placid society. He replied that we are now

Stuart Ackerman

*Officials inspect the damage caused by the explosion of two bombs in the klystron gallery of the Stanford Linear Accelerator Center on Dec. 7, 1971. The blasts damaged $45,000 worth of electronic equipment. Most of the damage was repaired within a week, and tests of the accelerator proceeded on schedule.*

already being manipulated by individuals—knowingly or unknowingly—who cannot always be expected to have the best interests of society at heart. Why not consciously define and strengthen that capability and turn it over to those groups now thought to constitute our ethical and moral leadership—the clergy, the humanists and scholars, and the worldly philosophers?

The arguments raging around the Skinnerian view of life were not notably reduced when the retiring president of the American Psychological Association, in a valedictory address to the annual meeting, suggested that world political leaders be tested for dangerous tendencies in conflict situations and appropriately medicated by psychopharmacologists. (*See* Year in Review: BEHAVIORAL SCIENCES, *Psychology.*)

Two areas of scientific research were so ridden with controversy that some scientists found themselves in the unusual position of suggesting that, in certain fields, too much knowledge is a dangerous thing. One was the artificial development of human beings; the other was the nagging question of whether there are heritable differences in intelligence linked to racial characteristics.

The first dispute arose from the work of Patrick Steptoe and Robert G. Edwards in England. These two researchers had succeeded in carrying the human development process from fertilization of a human ovum through several stages of cell fission to a 32-celled embryo, all within the glass walls of a laboratory test tube. Despite the shock that reverberated through the press when this achievement was announced, Steptoe insisted that he intended to push forward with his research. As the next step, he was seeking willing mothers in whose womb a laboratory-raised embryo would be placed to complete the birth process.

Steptoe faced up to the outcry by pointing out the benefits that might derive from his research, particularly to women who might be able to bear their own children but whose internal environment was hostile to fertilization. But to the more forbidding question—what would be done if the experiment produced living monsters?—the scientific team could only respond by asserting that their proposal involved no more risk than the natural act. (*See* Feature Article: TEST-TUBE BABIES.)

A larger question was posed by Leon Kass, executive secretary of the National Academy of Sciences Committee on Life Sciences and Social Policy. Addressing not only the question of in vitro conception but also such related questions as cloning (producing an embryo from an individual somatic cell of a single parent) and the production of living chimeras (organisms carrying genetic material of two different species), Kass asked whether such avenues of research should be explored at all, given man's present state of wisdom:

> To have developed to the point of introduction such massive powers, with so little deliberation over the desirability of their use, can hardly be regarded as evidence of wisdom. And to deny that questions of desirability, of better and worse, can be the subject of rational deliberation, to deny that rationality might dictate that there are some things that we can do which we must never do—in short, to deny the need for wisdom—can only be regarded as the height of folly.

Contention over the relationship between heredity and intelligence raised even higher waves in academia. Indeed, when physicist William Shockley, Nobel prizewinner, proposed that research be undertaken to determine whether an alleged 15-point deficit in IQ among American blacks was due primarily to heredity or environment, the dispute approached violence. Shockley, whose scientific credentials rested chiefly on his contributions to the development of the transistor, was at the center of the storm. He returned again and again to the assertion that blacks, as an identifiable genetic group, suffered from an inherited deficiency in intelligence as measured by the available standard tests. Others, notably Arthur Jensen of Berkeley and Richard Herrnstein of Harvard, also wrote in this difficult area, but it was Shockley who appeared to move further along the path toward the promulgation of social policies based on his assertions.

On numerous occasions he petitioned the National Academy of Sciences, of which he was a member in good standing, to undertake or foster the required research. Each time, the NAS flatly declined his invitation or pointed out that a majority of its members thought the task to be impossible. Then he appealed to his fellow faculty members at Stanford for approval to teach a special graduate course on "the dysgenic question: new research methodology on human behavior, genetics, and racial differences." In May 1972 the graduate dean at Stanford notified Shockley that the course could not be given. Shockley bowed to the edict, but refused to bend in his beliefs. He retorted publicly, "The flat human equality illusion that thwarts objectivity is, in my opinion, far more threatening to the future of the U.S. than was the flat earth illusion to the future of Italy in Galileo's day."

Even those who quarreled with the methodology of Jay Forrester and Dennis Meadows could agree that there were indeed some limits to growth. The suggestion that there were also limits to knowledge was a far more disturbing thought with which to end the year.

—Howard J. Lewis

# Transportation

A relatively poor year for many transport carriers tended to slow down research and development efforts by private industry but did not have that effect on the U.S. Department of Transportation, which accelerated its efforts in this area. The department continued to concentrate mostly on the passenger field, especially urban mass transport and the development of futuristic high-speed passenger vehicles.

From a technological point of view, a major change in transportation was the strong emphasis being placed on the demands for greater safety and environmental protection. This affected virtually all modes of transport, and caused a growing proportion of research and development to be devoted to such things as antipollution and safety devices, and noise suppressors.

To illustrate the across-the-board nature of this challenge, airlines faced tougher engine noise and smoke standards, as did trucks. Automobiles had to meet new pollution and safety standards. Railroads were confronted with newly imposed federal track standards. Water carriers faced costly penalties for spillages. Even the largely unseen pipelines were facing growing opposition to construction for environmental and safety reasons.

**Aviation.** There were signs that the trend toward progressively larger jumbo jet transports might have reached its peak for the commercial market in the B-747, which experienced both mechanical and load factor problems. Also, the even larger C-5A, built for the U.S. Air Force, continued to have airframe problems, which helped accelerate the cost per aircraft from $28 million to about $56 million. No orders for the L-500 commercial version of the C-5A were reported. In mid-1972 Lufthansa German Airlines introduced the B-747F all-cargo version for service between New York and Frankfurt; Boeing Co., makers of the B-747, hoped that this would help them recapture some of the canceled orders for this version. Another sign that the B-747 might get a firmer hold in the cargo field was an order by World Airways, a U.S. supplemental air carrier, for three B-747Cs, a convertible version that could be switched quickly from passenger to cargo service.

The McDonnell-Douglas DC-10 wide-bodied, tri-jet began commercial service and, over-all, provided "generally trouble-free" service. Major breakthroughs for the 230-to-345 passenger jet

*A radical new design for a supersonic transport incorporating an elliptical pivoting wing was proposed by Robert Jones of NASA's Ames Research Center. He claimed the plane would be relatively quiet, prevent pollution of the stratosphere, and produce no sonic boom under normal atmospheric conditions.*

Courtesy, NASA

**Henry Ford shaped modern industrial society by creating in 1913 the assembly-line method of mass production in his automobile factory.**

included engines that were much quieter than previous ones and that emitted virtually no smoke. Having overcome their serious financial problems through U.S. and British government aid, Lockheed proceeded to build the L-1011 and Rolls-Royce the engines, for this wide-bodied, tri-jet. Flight tests of the plane appeared to substantiate claims by its manufacturers that it would be the "world's quietest jet airliner."

The long-delayed European (France-West Germany) A-300B, twin-engine Airbus finally got the boost needed by receiving its first firm order —from Air France for six aircraft. Deliveries of the 270-passenger, short-haul plane were scheduled to be made in 1974. Other members of the Atlas Consortium (Air France, Alitalia, Lufthansa, and Sabena), along with Iberia of Spain, an associate member, were considering ordering the A-300B after completing a nine-month evaluation of it.

There were no indications that the U.S. would resume its program to develop a commercial supersonic transport (SST). Meanwhile, costs of the Anglo-French Concorde SST continued to rise. The cost per aircraft was expected to be more than $31 million to holders of the initial 74 options to buy. Even at that price, the manufacturers did not expect to recover a major share of the nearly $2 billion committed to research and development on the aircraft. Concern was also expressed that rigorous noise and emission standards in the U.S. might sharply limit operations of the Concorde to and from U.S. airports.

Like the Concorde, the Soviet Union's Tu-144 SST continued to be test-flown. The U.S.S.R. predicted that it would enter regular air service in 1974, although the commercial version was to be designed to have a greater capacity than the three prototypes, from 150 to 180 passengers compared to the 120-passenger version being tested.

Supporters of vertical and short takeoff and landing aircraft (V/STOL) received two setbacks during the year. The U.S. Department of Transportation's Northeast Corridor study of passenger transportation needs through the 1980s concluded that V/STOL aircraft for high-volume, short-haul travel "are not considered practical for application during the 1970s," and that much more research on them would be required for their use in the 1980s. Major roadblocks, according to the study, were community opposition to the noise, air pollution, and the safety hazards that these craft might generate. Also, American Airlines, which conducted extensive operational tests of commercial-type STOL aircraft and which evaluated proposals for propeller-driven STOL aircraft from three U.S. and Canadian firms, terminated its program. The airline concluded that a propeller-driven, lift-augmented STOL aircraft "probably won't work" for a trunk airline and would be "risky financially."

On the plus side for V/STOL aircraft was an agreement by three major West German firms to join in a 12-year development plan, with government backing, to have a competitive prototype vehicle ready by early 1982. Also, a joint Japanese government-industry group was planning to develop a 200-passenger STOL airbus, and the U.S. National Aeronautics and Space Administration announced plans to build two test STOL transports.

**Highway transport.** As the May 12, 1972, deadline for a decision by the U.S. Environmental Protection Agency on imposing rigid statutory emission standards for 1975 autos came closer, the number of opposing views and claims increased. All major automakers stated that they could not meet the deadline and needed another year for further research and tests. Two major studies, by special White House and Academy of Sciences panels, both concluded the extra year was needed. The White House panel criticized both proposed pollution and safety standards for 1975 autos, contending that the public benefits would not justify the higher additional costs, estimated at $873 for 1976 *v.* 1971 models—$350 for emission and $523 for safety devices. The other panel said that compliance with pollution standards, while possible, would be accompanied by the following: a need for more costly low-lead gas, more expensive maintenance, periodic reinstallations of antipollution devices, less engine power, and higher initial costs of about $200 per car.

General Motors began pilot production of 60 gas turbine engines for industrial and large truck and bus use, with installations made on many vehicles for over-the-road tests. Plans called for mass production of the 325-hp engines starting late in 1972, building up to a rate of 50 per month. General Motors estimated that the engines would cost about 20% more than a comparable diesel engine and consume more fuel (getting only 5 *v.* 7 mi/gal), but would otherwise be less costly to maintain and operate.

General Motors also reported "good progress" in its development work on the Wankel rotary engine, but said that reports that it might soon go into mass production were premature. While the Wankel is said to be quieter, smoother, and simpler than the piston engine, and weigh only about one-third as much, it uses as much as 25% more fuel and has an imperfect sealing system. While not much better than the piston engine in preventing air pollution, the Wankel is smaller and thus allows more room for the installation of antipollution devices.

Three of four experimental safety cars were turned over to the U.S. Department of Transportation for further evaluation and for 12 of the best model to be produced at a later date. They were designed to meet demanding goals, including survival in a 50-mph crash, but all proved to be much heavier than present cars; and the cost, if sold on the market, would probably be twice that of a current-model family car, according to a General Motors official.

Courtesy, U.S. Department of Transportation

*Commuters to Washington, D.C., travel unimpeded by rush-hour traffic jams on special highway lanes reserved for buses only. This test and another in New York demonstrated time savings of up to 30 minutes per one-way trip. The U.S. Department of Transportation granted $36 million for a similar project in Los Angeles.*

Regina, Sask. (population 137,759), called its six-month experimental "Dial-a-Bus" service a "remarkable success," and it announced plans to expand it from serving 18,000 residents to 60,000 by the end of 1972 and to the entire city by the end of 1974. In this system collection buses pick up callers within 15 minutes for trips to waiting rooms for regular bus service. Several smaller-scale tests were being financed by the U.S. Department of Transportation, including one in Camden County, N.J.

The U.S. Department of Transportation continued to promote the use of exclusive bus lanes and highways. It approved a $36 million grant toward a $52 million project to build an 11-mi busway in the Los Angeles area for use by buses only during rush hours over a two-year test period. Similar projects involving approaches to New York City and Washington, D.C., proved attractive to commuters, with 35,000 using the former and 20,000 the latter daily—at savings of up to 30 minutes per one-way trip. The Department of Transportation granted an additional $2.2 million for the Washington project for 30 more buses and fringe parking lots.

**Pipelines.** Another major U.S. oil pipeline was completed and being filled for commercial service in 1972. The 1,300-mi Explorer Pipeline moved refined petroleum from the Gulf of Mexico to Chicago via Dallas, Tex., Tulsa, Okla., and St. Louis, Mo.

Despite an unprecedented amount of research and planning to satisfy objections of environmentalists, the Alyeska Pipeline Service Co. was in 1972 still awaiting approval from the U.S. Department of the Interior to proceed with construction of the 789-mi, 48-in. crude-oil pipeline from the Alaskan North Slope south to Valdez, an ice-free port on the Gulf of Alaska. One major roadblock was removed when Congress passed legislation settling century-old land claims of natives of Alaska, thereby providing for rights-of-way for the pipeline. In addition, a comprehensive, nine-volume environmental impact report required by law was submitted, and it did not appear to contain any information that would call for rejection of the project. Even if approved by the secretary of interior, however, the project may well be delayed further because of court action by environmental groups. In the meantime, the estimated cost continues to rise, from an original $1.4 billion to more than $2.5 billion.

A new concept in the movement of solid materials through pipelines was announced during the year by Trans-Southern Pipeline Corp. The system, called TUBEXPRESS, was designed to move either packaged or bulk material through a large pipeline in wheeled capsules that are propelled by a moving column of air at essentially atmospheric pressure. Flow-through pumps move the air, and by-passes separate the vehicles from the air column at loading and unloading sites. Since the system does not require a vacuum, airtight seals are not needed.

The capsules are equipped with flat endplates at the front and rear that have a flexible surface nearly the diameter of the pipeline, thus permitting movement at nearly the velocity of the propelling column of air. The vehicles themselves roll on load-bearing wheels and have side-guide wheels to

Wide World

*Prototype of a monocab, one of several possibilities for a future urban transportation system exhibited at Transpo 72, passes above an old trolley car, one of the older forms of transportation that were also on display.*

stabilize their movement through the pipeline. Two-way traffic can be achieved by using a closed-loop system. Loading, unloading, removing, and inserting the vehicles in the air column can be automated to the extent warranted.

Initial plans called for the use of such a pipeline for the movement of mail and baggage over short hauls, and the U.S. Postal Service is now evaluating the system. Backers claim, however, that the concept has "virtually unlimited" potential as a new mode of transport. A prototype was built near Stockbridge, Ga., and was extensively tested.

**Railways.** To help speed up rail freight service throughout the U.S., railroads introduced more "run-through" trains that by-passed intermediate yards between two distantly separated points. Such service usually required cooperation by at least two roads, especially when the same locomotives were used. These trains carried a variety of commodities, as compared to unit-trains, which carry only one commodity.

The number of products moving in long-distance, shuttle-type, unit-train services continued to grow. One sizeable potential market was the movement of entire trainloads of modular homes, and initial runs were successfully made under a program in which the railroads and builders were cooperating with the U.S. Department of Housing and Urban Development and the General Services Administration. A study of expanded unit-train operations by the steel industry in the Great Lakes area claimed that savings of more than $585 million in capital outlays and $83 million a year in operating costs would be possible if such trains replaced the present rail-ore ship system.

General Motors Corp., which shipped about three million motor vehicles by train in 1971, reported that 10% of them were stolen, vandalized, or damaged. The company foresaw the use of specialized unit-trains as the solution to this problem and continued to work closely with the railroads toward that end. General Motors reported that its Vert-a-pak cars, which load 30 Chevrolet Vegas per enclosed car in a nose-down position, proved successful, with 183 now in use and 134 to be added in 1972. It also said early in 1972 that shipping Cadillacs by railroad in stacked containers had virtually eliminated damage claims, and that it was building 200 more containers and 50 railcars to handle more cars.

The piggyback unit-train, carrying mostly truck trailers but also containers, continued to grow, and by 1972 approximately 115 such trains moved between U.S. cities, mostly in the midwest and northeast. These nonstop trains covered from 400 to 500 mi in 10 to 12 hours, thus averaging about 10 times the mileage of regular-service freight cars.

Because of financial problems the railroads were not able to put into effect their program to computerize the control and utilization of freight cars. The labeling of freight cars was virtually completed, but the number of installed roadside scanners that "read" the labels as the trains moved by them remained small. Although a number of the 62 railroads that planned to use the computerized network had installed their own freight car control systems, the nationwide system was a long way off. Immediate plans called for installing scanners in the high-density Chicago terminal area, with the many railroads serving it sharing the costs. If successful, this may become the model for other rail terminal districts.

The potential of fully automated railroad operations was pointed out by an announced agreement to build such a train in Arizona to haul coal between mines and an electric generating plant. The electric-powered train was being designed to operate without any on-board crew and represented, according to an electrical industry spokesman, "a prototype of a modern electrification concept for high-volume railroads on the North American continent."

While hard-pressed to hold operating losses to a minimum, the U.S. Amtrak passenger-train service did report continued success for its high-speed, electric Metroliner Service between Wash-

*A film of air supports this six-passenger Personal Rapid Transit (PRT) car as it traverses a test guideway in Denver. The vehicles, powered by linear induction motors that utilize the metal strips in the guideway, were demonstrated at Transpo 72.*

Courtesy, Transportation Technology, Inc.

ington, D.C., and New York City. Amtrak reported that annual passenger volume along this route increased from 605,000 in 1969 to 1,635,000 in 1971.

About 40 countries, including the U.S., expressed interest in Britain's entry in the high-speed passenger-train field, the so-called Advanced Passenger Train (APT). The British government expects to put the APT in experimental commercial service in 1975, and, after removing all flaws, into the international market a year or two later. The lightweight trains were designed to operate at speeds up to 155 mph on existing rail routes with minor alteration to the track and signaling. High average speeds could be maintained around curves because of a "unique steering mechanism." Backers claimed that the basic high-speed capability of the APT was due to a new design that sharply reduced the characteristic side-to-side movement of railroad cars when operating at high speeds on tracks.

**Urban mass transit.** The rapidly increasing number of research grants from the U.S. Department of Transportation in the urban mass transit field were beginning to produce useful results. One example was the inauguration into rail transit service in Cleveland, O., of a new car that featured Westinghouse Air Brake Co.'s pulsewidth modulation propulsion. Made possible by a $1.3 million grant from the Department of Transportation, the unique propulsion system assured far faster, yet smoother, starts and stops for improved passenger comfort and safety. It also offered savings in power by returning energy to the electric power lines while braking and by the use of low-cost, more reliable alternating-current motors instead of the usual complex direct-current motors.

Another federal government grant of $7.4 million, to be matched by New York State and private industry funds, was made to develop and test in service eight combination gas-turbine/electric-powered rail commuter cars, which the Department of Transportation foresaw as the "trains of the future for suburban rail commuters." The dual-powered cars were designed to provide greater flexibility by being usable on both electrified and standard track, thus providing commuters with a one-seat, no-change ride between the city and suburbs.

Another milestone in rail rapid transit was passed when an eight-car train built by St. Louis Car for New York City passed a required test of operating 20 hours a day for 30 consecutive days without a breakdown. This cleared the way for delivery of the remaining cars in a 352-car order. They were designed to operate at speeds of 70 mph compared to 50 mph with existing equipment.

Boeing Co. and Bendix Corp. were selected to help build the "Personal Rapid Transit" system in Morgantown, W.Va. It was to consist of fully automated transit cars that would link West Virginia

*New Chevrolet Vegas are shipped from the GM Assembly Division plant at Lordstown, O., by a unique method known as Vert-a-pac. Each Vert-a-pac railway car carries 30 Vegas nose down and is completely enclosed to protect the vehicles against damage during transit.*

Courtesy, Chevrolet Motor Division, General Motors Corporation

University's two campuses, carrying passengers at speeds up to 30 mph. The rubber-tired, electric-propelled cars were to operate on a guideway, mostly elevated. Boeing was to design and construct prototypes of the cars, and Bendix would design and develop the automated control and communication system. The first section of the system was scheduled to start experimental operations by the end of 1972, with completion in 1974.

**Water transport.** While the trend toward construction of super oil tankers continued, even greater interest was being expressed in the need for specialized ships to haul liquefied natural gas (LNG), because of the rapidly declining supply of natural gas in industrialized nations including the U.S. A special Presidential Commission on American Shipbuilding estimated that up to 100 LNG tankers would be needed in the near future for U.S. requirements alone. While 14 such ships, mostly of smaller size, were in service, the new ones were to be huge, with one such ship being able to haul enough gas on a single trip to supply 11,900 homes for a year.

While LNG tankers may help assure an adequate supply of natural gas, they are considered to be the most complicated merchant ship to build and, except for luxury passenger ships, the most expensive. Their cargo tanks must be able to maintain liquefied natural gas at −260° F during long voyages and be strong enough to prevent breakages that would cause the liquid to heat and turn into gas, as its volume would thus expand by more than 600 times. A major supplier, El Paso Natural Gas Co., formally requested a federal government subsidy to help build six LNG tankers, at $72 million each, to haul gas from Algeria to the U.S. In 1972 it was building three similar ships in France.

The first European-built LASH ship started commercial service in February 1972. These specialized ships carry lighters and containers that are loaded and unloaded by on-board cranes. More than 20 were in service or being built. The three even larger SEABEE ships being built for Lykes Bros. Steamship Co. were expected to be in service by late 1972 or early 1973. They were designed to haul large barges that could be floated on and off a stern elevator for lifting and placing on any of three decks. They would have the most power (36,000 shaft horsepower) of any single-screw merchant ship and be the largest general cargo ship afloat.

Litton Industries, Inc., announced plans to build what it called "the largest push-type tug-barge cargo transporter in the world"; it would be 1,000 ft in length when being pushed by its 15,000-hp tug. This unique, self-loading/unloading vessel would be able to carry about 52,000 tons of a variety of dry bulk cargoes, including iron ore, limestone, and coal. Using a novel, on-board rotary elevator —which would be 67½ ft in diameter and look much like a water wheel—that would be linked to special conveyors, the barge could unload at a rate of 10,000 tons per hour. To be operated in the Great Lakes, the tug-barge was expected to be completed and undergoing sea trials and systems tests by the end of 1972.

**Air-cushion and linear-induction motor vehicles.** The U.S. Department of Transportation continued to put a strong research emphasis on the tracked air-cushion vehicle (TACV) as the most probable mode of high-speed surface passenger transport in the future. The department's comprehensive Northeast Corridor Transportation Project concluded that TACV is a "prime alternative" for passenger transport needs in the 1980s, because "it offers the potential for very good reliability, comfort and safety; high capacity; and decreasing passenger-mile costs at high volume."

Three major U.S. aircraft manufacturers were engaged in separate projects for the government. Rohr Corp. and Vought Aeronautics Co., under contracts totaling $2,850,000, were producing designs and mock-ups of a 60-passenger TACV to be powered by an electric linear-induction motor at speeds up to 150 mph. They were to be delivered in 1972, with one selected for possible construction and testing as a prototype for high-speed travel from urban centers to outlying airports and to other cities at distances up to 50 miles. Grumman Aerospace Corp. received the third contract, for $3.5 million, to build a TACV capable of speeds up to 300 mph for longer-distance travel. It was to be tested extensively in 1972.

Other countries were also continuing the development of such vehicles. Bertin & Co., with backing from the French government, had two types of air-cushion "Aero-trains" in operation. A larger version carried 80 passengers at normal speeds of 155 to 160 mph on an 11.2-mi elevated guideway north of Orléans, France. Its lift propulsion and braking (using reverse pitch) were provided by gas turbines. The smaller version carried 40 passengers at speeds of 112 mph. It was being tested on a shorter track at Gometz-la-Ville, France.

**Outlook for 1973.** After two years of traffic and financial setbacks, the airlines were expected to experience a resurgence that would result in the renewal of orders for new jumbo jets such as the B-747, DC-10, and L-1011. Efforts will continue to develop a suitable air transport for high-volume, short-haul trips.

Turbine engine output for large trucks and buses was expected to change from experimental to production-line models, giving a boost to twin-

Courtesy, Los Alamos Scientific Laboratory

*An electrically powered rock-melting drill undergoes preliminary tests at the Los Alamos Scientific Laboratory. Scientists predicted that eventually the drills, called subterrenes, would be used for large-scale projects such as highway and railroad tunnels.*

trailer and intercity bus operations over long, minimum-stop routes. The use of exclusive and/or separate bus lanes for rush-hour commuter traffic will be expanded sharply.

Railroads, with help from the government, will begin to develop their now individualized freight car control systems into a national network. Roadside scanners at key interchange terminals will record car locations for computerized control.

A sharp spurt in orders for supertankers to haul both petroleum and liquid natural gas was expected, along with the development of more offshore terminals to enable the tankers to operate to points that do not have ports that can handle them. The huge SEABEE barge-carrying vessels will start commercial service, and additional variations of deepwater tug-barge vessels will be ordered for construction.

—Frank A. Smith

# Veterinary medicine

Several diseases of livestock and poultry reached epizootic proportions in the U.S. and focused renewed attention upon various eradication efforts. The two most serious outbreaks were those of Venezuelan equine encephalomyelitis (VEE) and of exotic Newcastle disease in poultry and pet birds. That both of these diseases were imported reflected the growing potential for inadvertent introduction of organisms and the need for increased surveillance at ports of entry.

**Venezuelan equine encephalomyelitis.** The first outbreak of VEE in the U.S. began near Brownsville, Tex., in late June 1971 and by the end of July had affected more than 2,000 horses, about half of which died. A national emergency was declared, and immediate steps were taken to control the disease by quarantine, vaccination, and aerial spraying to eradicate the mosquito vector. By the time the last case was confirmed in November, about 3,400 horses had been affected and some 1,400 had died. The virus, which also causes a relatively mild disease in humans, was isolated in 88 persons in Texas; there were no deaths. More than eight million acres in Gulf Coast areas were sprayed with insecticide, and 2.8 million horses in 19 states and the District of Columbia were vaccinated, including 80% of those in high-risk areas.

VEE had been present in South and Central America since 1935, and when a major outbreak erupted in Guatemala in 1969, the U.S. Department of Agriculture (USDA) began a cooperative vaccination program. This was extended to Mexico during 1970–71, when up to 25,000 horses died and perhaps 12,000 persons were affected. The rapid containment of the U.S. outbreak was attributable in large measure to the early release of vaccine that had been stockpiled as a potential biological warfare countermeasure. Though VEE resembles the disease caused by the eastern (EEE) and western (WEE) strains of the encephalomyelitis ("sleeping sickness") virus, both long present in the U.S., the vaccines against these two strains conferred no protection against VEE. Some concern was expressed regarding the safety of the relatively untested VEE vaccine. The results of a USDA study released in March 1972, however, made it clear that the vaccine was safe, and government veterinarians urged that all horses not vaccinated in 1971 be protected. Meanwhile, a commercial vaccine had been produced in ample quantities.

**Poultry diseases.** Newcastle disease of poultry and pet birds had been present in the U.S. and most of the more developed nations for several decades but had been kept under control by vaccination and sanitation. In 1970 an exotic strain of the virus, apparently from the Orient, was found in ornamental birds imported to the U.S., and later in commercial poultry flocks in Texas and New Mexico. Hardest hit was southern California, where, since December 1971, the disease had affected flocks containing some 1.5 million birds. By March 1972 more than 350,000 chickens had died in San Bernardino County alone, and it was estimated that some two million birds would have to be destroyed before vaccination would bring the disease under effective control. In March a state of disaster was declared in this area.

Earlier, a "henocide" program had been urged to bolster egg prices, which were too low to make egg farming profitable, but the disease and the eradication measures being employed against it were expected to make such a program unnecessary. Import regulations affecting some 280,000 pet birds brought into the U.S. annually were extended to require blood tests overseas and again on entry.

Marek's disease, a form of leukosis (leukemia) in chickens and turkeys, had threatened to disrupt if not destroy the poultry industry in the U.S. and Europe until an effective vaccine had become available in 1971. Although the vaccine would not eradicate the disease, tests on two million broilers indicated that each dollar spent on vaccine (no more than two cents per bird) returned $2–$5 in profits, even when losses among unvaccinated birds were only 5%. If all laying hens in the U.S. were vaccinated, 8.1 million fewer chickens could produce the same number of eggs, saving 500,000 tons of feed annually and releasing 100,000 ac of corn land for other uses.

**Toxoplasmosis in cats.** During the year several women's magazines carried well-intended but ill-advised articles indicating that cats presented a major hazard to pregnant women because they transmitted toxoplasmosis. This disease is caused by a protozoan parasite and produces symptoms like those of infectious mononucleosis. If a woman becomes infected during pregnancy, the infant may be born with serious birth defects. But the hazard should be viewed in perspective. About 25% of the U.S. population has toxoplasmosis antibodies and, hence, presumed immunity to further infection. In addition, while toxoplasma eggs can be found in cat feces and can transmit the disease to other animals, there was no instance of infection in man traceable to either source. The American Veterinary Medical Association and the U.S. Public Health Service, without incriminating the cat as a major causal factor in toxoplasmosis, recommended that pregnant women not increase the risk of the disease by acquiring a new cat, nor allow household cats to catch birds or other animals that might transmit the infection. They should also avoid handling cat litter or digging in gardens frequented by cats.

**Drug problems.** In a continuing study of the safety and effectiveness of veterinary medications, the U.S. Food and Drug Administration (as it had with numerous drugs for human use) withdrew its approval of many compounds marketed years before claims for efficacy were closely scrutinized. The FDA, in cooperation with the USDA, was also enforcing more stringent regulations concerning the tolerable residues in meat of the chemical hormonal analogue diethylstilbestrol, fed or implanted in cattle to promote fattening. In late 1971 a random survey of 2,500 liver samples from slaughtered beef cattle turned up 10 with more than the allowable residues, the tolerance being zero, though the test was not sensitive to less than two parts per billion.

Stilbestrol had produced cancers in laboratory mice at six parts per billion, thus warranting rigorous supervision of its use in meat animals. Although no instance was reported of residues in meat causing cancer in man, the period of withdrawal from the drug prior to slaughter was extended from 48 hours to 7 days.

Veterinarians were reporting a few cases of "pot syndrome" in dogs and cats. Both animals appeared to be highly susceptible to marijuana, either by ingestion or by inhaling its smoke.

**Animal disease eradication.** The 10-year hog cholera eradication program begun in 1962 had made rapid strides since the use of vaccine was outlawed in 1969. Forty-one states had been declared cholera-free by late April 1972. In February 1972 the U.S. Department of Agriculture committed itself to a drive to eradicate brucellosis from livestock by the end of 1975, five years ahead of the original schedule. The disease causes abortion in cattle and affects other domestic animals; it can be transmitted to man, causing undulant fever. Twenty-four states were certified as brucellosis-free, but the three making the least progress (Texas, Nebraska, and Mississippi) had about 20% of all U.S. cattle.

Complacency and curtailment of test-and-slaughter programs had permitted swine and poultry to become important reservoirs of the tuberculosis bacterium. In 1970 lesions were found in 1.09% of all swine slaughtered in federally inspected plants, and the avian form was believed to be nearly as prevalent in small farm flocks. Both forms were potentially hazardous to man, but producers did not consider TB an economic problem and eradication would require cooperation with human health agencies.

—J. F. Smithcors

# Zoology

Zoology as a discipline predates Aristotle and has attracted many of the world's greatest minds. In the 1970s the zoologist continued to work along traditional but exciting lines, discovering, describing, and classifying organisms and structures and thereby broadening the base for future research. This kind of work might have seemed unnecessary alongside such contemporary problems as drug abuse, pollution, population growth, and health care until one noticed how much the information gained from organisms could be applied appropriately for the benefit of man.

**General zoology.** The favorable economics of dwarfism, long exploited by plant geneticists in high-yield wheat and rice, was being applied in zoology. Selective breeding of fowl, utilizing a sex-linked dwarfing gene, produced a laying hen of the Leghorn type and a hen that produces broiler chickens; each was only two-thirds normal size. Three of these hens could fit in a standard cage for two, and each could be expected to produce 200 standard-size eggs, compared with 240 for normal birds, for an investment of the same 160 lb of feed per cage. This was a production gain of more than 20% at no increase in feed cost. Broiler chickens offered the same kind of feed economy. Researchers at Hy-Line Poultry Farms, Des Moines, Ia., were examining the possibility of exploiting a similar dwarfing gene in beef cattle, while, at the California Institute of Technology, an injection of minimine—a toxic polypeptide isolated from the bee venom—into fruit fly larvae was found to result in one-quarter-size adults that were normal in every way except that their individual cells were miniature. The new breed of fly lived a normal lifespan and produced regular-size offspring.

The alfalfa leaf-cutting bee, a wild bee, was a highly effective pollinator in Idaho, where after introduction it would often double alfalfa seed yield. But efforts to introduce the bee in California were less than successful mainly because the bee had difficulty in reproducing there. Robbin Thorp of the University of California at Davis said that in California the bee appeared to be the victim of a toxin, a steroid material called saponin, that occurs naturally in California alfalfa varieties. Thorp theorized that the bees, in preparing nests, bite out small circles of alfalfa leaf. Nectar deposited in the circular nest dissolves the saponin, and when the larvae feed on the poisoned food they die. To introduce the bee successfully in California would probably require the selection of nontoxic varieties of alfalfa.

Eldon Reeves and S. V. Amonkar of the University of California at Riverside found two species of algae in California that specifically kill mosquito larvae. The algae, *Cladophora glomerata* and *Chara elegans*, release toxic substances that attack the lining of the larva's alimentary canal and that are especially effective against *Aedes aegypti*, a disease carrier, as well as four other species of mosquito. The insects and fish that live in mosquito habitats remained healthy in the presence of these active compounds at levels that suppressed mosquito development.

Like mayflies, stone flies (order Plecoptera) pass through an aquatic nymphal stage commonly found in streams and rivers. Twenty new Australian species were added to the family list by I. D. McLellan, who, in a revision published in the *Australian Journal of Zoology*, described the new types, which include the new genus and species *Neboissoperla alpina* discovered in a small stream on Mt. Wellington, Victoria.

Because so many aphids were serious pests and because their life cycle was notoriously complicated, it was important to be able to identify the different stages of a particular species. This was often easier said than done. In the case of the woolly aphids, the Adelgidae, specific identification was very difficult or even impossible and in Europe only a dozen or so species of these serious pests of conifer trees had been recorded. Considerable help toward the identification of the winged forms (alatae) of adelgids was given in a new booklet by D. I. Carter of the British Forestry Commission. Three of the species described, *Adelges viridana*, *Pineus pineoides*, and *P. orientalis*, were recorded for the first time in Britain.

Ocelli are simple eyes found in adult insects. They are incapable of perceiving form but are receptive to changes in light intensity. Adult sphinx moths, *Manducta sexta*, lack external ocelli. Microscopic studies revealed tissue inside *M. sexta* that show structural similarities to ocelli of other insects. New electrophysiological studies of this tissue revealed a response to light stimuli similar to that recorded for external ocelli. These studies also showed that illumination of the ocellus inhibits the continuous discharge that occurs in the ocellar nerve during darkness. Further, behavioral studies showed that the ocelli are important in the regulation of the daily rhythm of insects.

It was learned that insects are strongly attracted to a submillimeter radiation source. A theoretical study on the moth antennae, together with the frequency dependence of their response to electromagnetic radiation, strongly suggested that the antennae may be the receptor through which insects are stimulated to respond. The response is very pronounced and has strong potential for traps to control the population of pest insects.

In the central nervous system, the spinal cord and the lower brain were generally viewed as being little more than the passive recipient and the integrator of multiple sensory inputs more or less on a reflex basis. Recent investigations made the reflex hypothesis untenable. Nerve networks located in the central nervous system of both vertebrate and invertebrate animals appeared capable of numerous and varied functions that were essentially independent of sensory controls. Recent work showed that central-nervous-system networks, independent of the senses, could function in a wide variety of ways and give rise to kinds of behavior difficult to explain on the hypothesis that the nervous system is largely a reflex mechanism.

Iridescence caused by regular multilayer structures stacked in a medium of different refractive index was observed in a variety of animal tissues. Corneal iridescence had not received much attention in the scientific literature. It was found that fishes with iridescent corneas often, but not always, possess filters for yellow light at some point in the optical pathway. It was too early to speculate on the role of the yellow optical filters and the iridescent cornea on selective spectral absorption of light from different directions, but all fishes so far known to possess this iridescence lived and fed near the bottom of the sea.

The distribution of marine life in the sea was still difficult to determine with present methods of study. A technique that offered particular promise for defining the locales in which sea creatures dwell was deep-ocean photography. On a recent voyage of the USNS "Lynch," researchers obtained 340 exposures of the Caribbean Sea floor at 17 stations. Among the pictures obtained was the deepest photographic observation yet of a cephalopod (a class of mollusks that includes octopus and squid). The specimen, about 45 cm long, was identified as belonging to the Cirroteuthidae family, that of the octopus. It was about one meter off the sea floor, at a depth of 5,145 m.

The Pogonophora—marine, tubicolous, wormlike animals that live buried in soft sediments often in the abyssal regions of the oceans—were overlooked by biologists until 1955 when the phylum was established by the Soviet scientist A. V. Ivanov. One marine zoologist who made the group her special study was Eve C. Southward of the Marine Biological Laboratory, Plymouth, Eng. In a new report she added three more species —*Diplobrachia floridiensis, Suboglinoides caribbeanus,* and *Siboglinum bayeri*—to a growing list of species recovered by teams from sampling stations along the continental margin of North America. A total of 24 species from depths ranging

**Man's view of himself was fundamentally altered by Charles Darwin. His *Origin of Species* established the theory of evolution by natural selection.**

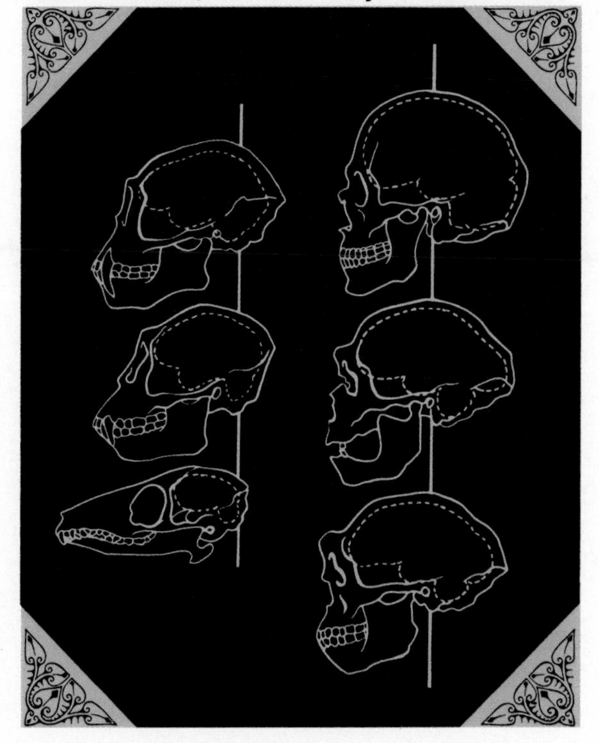

from 40 m to more than 5,000 m had been recovered as of mid-1972 by these teams.

Discussions of the evolution of vertebrate hearing systems tended, for lack of physiological data, to concentrate on changes in middle ear structure. Conspicuous differences are seen among reptiles, birds, and mammals in the shape of the basilar membrane of the inner ear. Use of a new method of frequency analysis indicated that "advanced" reptilian membranes analyze frequencies in a way similar to those of birds and mammals.

The distinctive markings and colors of birds are more than just decoration. They serve as camouflage, mimicry, social signals, and temperature regulators. Certain markings, according to Robert W. Ficken and Paul E. Matthiae of the University of Wisconsin at Milwaukee and Robert Horwich of the Chicago Zoological Society, may have yet another, very specialized, function. Eye lines, marks leading from the eye toward the tip of the bill of some predatory species, may serve as aiming sights in tracking and capturing swiftly moving prey. The researchers found a high correlation between the presence of eye lines in North American songbirds and their tendency to feed on swiftly moving prey.

The basis for the remarkable navigational ability of homing pigeons remained unknown. C. Walcott and M. C. Michener published the results of a major series of experiments designed to change the direction in which pigeons believe their home lofts to be by shifting their internal clocks from five minutes to six hours before releasing them. But no clock-shifted pigeon ever flew to or investigated a false home area. Slowing down the birds' internal clocks also had no effect on their ability to navigate to their home lofts.

Pigeons, like other animals, certainly use the sun as an azimuthal compass reference, but Walcott and Michener found no evidence that they can use it for true navigation. The "map" of the homing bird was, therefore, as great a mystery as ever, even more so in the light of recent evidence that pigeons do not depend on a view of any celestial body and can operate effectively even after many days' imprisonment away from the home loft.

For those who contended that inborn mechanisms other than genes control the timing of biological clocks, a blow was struck by two California Institute of Technology biologists. Hypothesizing that mutation of certain genes might lead to abnormal rhythms, Ronald Konopka and Seymour Benzer exposed *Drosophila* flies to a drug known to cause mutations. It did indeed upset the daily rhythmic cycles of emergence of the adult fly from the pupa and of locomotor activity. The changes in rhythms also carried over into subsequent fly generations. The work was not final proof that the key to the biological clock is internal, but it showed, said Konopka, that "genes play a key role in determining or in specifying rhythms."

**Developmental zoology.** Compared with the successful storage of spermatozoa and other tissue cells at sub-zero temperatures, attempts to preserve frozen mammalian eggs were disappointing. A new technique for obtaining a high proportion of viable embryos after subjecting them to sub-zero temperatures was described. These re-

*Tiny insectlike tardigrade, seen here magnified more than 100 times, exhibits cryptobiosis (below), a dormant, deathlike state of almost complete dehydration. When the animal is moistened, it revives (right). By alternating these states, tardigrades greatly extend their life-spans.*

Courtesy, Dr. John H. Crowe, University of California, Davis sults demonstrated that mouse embryos could be frozen and thawed successfully if a suitable protective agent was added to the medium before freezing.

Having achieved external fertilization, the next step in this area of experimental embryology would be complete in vitro (in the test tube) growth of the embryo. Unfortunately, growth stops when the cells begin to form into specialized organs, in what is called the blastocyst stage, when the developing embryo normally implants itself in the wall of the uterus. Once this stage has been passed, growth in an artificial womb could take place. Therefore, reaching and passing this stage of development was a major research goal. Yu Chih Hsu of Johns Hopkins University, Baltimore, Md., announced that he developed mouse embryos beyond the implantation stage in culture dishes coated with rat tail collagen, a protein substance. Because, as Hsu put it, "all the embryos developed in vitro seem to be defective in one organ or the other with the present method," he concluded that "the supply of nutrients and gases may be inadequate for further embryonic development." Hsu believed that now was the proper time to begin work on an artificial placenta that could supply the necessary nutrients and gases to the embryo.

Independently, a handful of chemical engineers was turning from traditional industrial challenges to the creation of such biomedical devices. One of the more ambitious undertakings was to probe and simulate functions of the placenta in engineering terms. The work was detailed in September 1971 at a national meeting of the American Chemical Society in Washington, D.C. Chemical engineers Daniel Reneau and Eric Guilbeau of Louisiana Polytechnic Institute and their physician-physiologist co-worker, Melvin Knisely of the Medical College of South Carolina, developed a theoretical model of the placenta. They were planning to undertake animal experiments to test their simulation. The model, linked with a computer into which many variables could be fed, would help tell them what they should look for in their experiments. (*See* Feature Article: TEST-TUBE BABIES.)

For all its sophistication, modern biology was not able to explain why certain lower organisms can grow back certain injured or lost parts—the worm its lower body, or the lizard its tail. Nor had it been able to explain why higher animals do not have this capacity. R. Becker of the State University of New York Upstate Medical Center at Syracuse partially regenerated the forelimbs of several dozen rats amputated between what corresponds to the shoulder and the elbow in humans. Minuscule amounts of electric current applied to the severed sites stimulated the limbs to grow down to the elbows. Becker's was the most advanced success yet in regenerating a limb in an animal as high up the evolutionary ladder as the rat. For 15 years Becker has pursued the theory that mammals are not able to regenerate damaged limbs because they have lost the ability to generate enough electricity to stimulate the formation of a new limb bud.

**Ecological studies.** It was found that the large flightless grasshopper from Florida, *Romalea microptera,* probably takes the herbicide 2,4-dichlorophenoxyacetic acid (2,4-D), metabolizes it, then uses it in a kind of chemical warfare to repulse predators. There were many documented instances of insects using chemicals produced by other species, but this was the first report involving a man-made pesticide, although such uses of man-made chemicals may not be all that exceptional.

A report that might surprise many was one that revealed mercury levels in several human tissues declined from the 34.5 parts per million recorded in 1913–19 to about 5.5 ppm currently. At a conference on pollution, J. Ui of the University of Tokyo spoke of the dangers to health that may arise from the release of mercury into the natural waters of densely populated areas. His opinion, endorsed by others, was that mercury pollution was likely only to be a local problem, because the amount of mercury reaching the sea from industrial sources was only a small fraction of that entering by normal geological processes. He considered that a much greater threat to man's health was posed by toxic chlorinated hydrocarbons, such as DDT, and polychlorinated biphenyls (PCBs), which enter the sea by way of the atmosphere and are, therefore, distributed globally. These compounds are concentrated by marine animals at many levels in the food chain. (*See* Feature Article: THE NOTORIOUS TRIO: MERCURY, LEAD, AND CADMIUM.)

DDT and other pesticides rendered some large bird species nearly extinct because they interfered with calcium metabolism and caused eggshell thinning. Such shells break before the embryos mature. L. Z. McFarland and R. L. Garrett of the University of California School of Veterinary Medicine at Davis used a scanning electron microscope to determine that the middle layer of the multilayered eggshell contributes the most toward structural strength, and that it was this layer that was most adversely affected by DDT.

One investigation of how, physiologically, pesticides kill insects was based on the observation that water conservation is critical for insects because of their large surface area relative to their volume and the generally arid environment in which they live. There were indications that the critical water balance is disrupted by poisoning with certain insecticide chemicals, but the physiological basis for this action had not previously been defined. S. Maddrell and J. Casida of the University of Cambridge found that treatment of *Rhodnius* with DDT results in the massive excretion of a large volume of fluid into its rectum. Furthermore, the underlying mechanism involves the release of diuretic hormone, possibly from a bound form in the central nervous system, into the hemolymph (insect "blood"), where it contacts the Malpighian tubules (in the insect's "kidneys") and greatly increases their rate of secretion. The excretion induced by insecticide chemicals was always found to be initiated during the paralytic stage of poisoning. (*See* Feature Article: SPECIES IN PERIL.)

These findings were obviously of potential importance in the use and design of insecticide chemicals. The rapid loss of a large volume of fluid could kill the animal by dehydration. Furthermore, the Malpighian tubules secrete a fluid that is, in general, extremely rich in potassium, high in phosphate, and low in sodium and chloride. This changes the insect's ionic balance drastically, and probably enhances its disruptive action on the nervous system. Finally, these kinds of dehydration changes could be initiated by chemicals that are nonpersistent in the environment. All that needed to be discovered was whether the bug is killed first by paralysis, or by dehydration, or by both.

For those who might favor gentler methods of dealing with pests, J. S. Gill and C. T. Lewis of London's Imperial College of Science and Technology believed that applications of the nontoxic, distasteful chemical substances produced by some plants might deter insects from attacking crops and, thus, offer an alternative to toxic pesticides. A suspension of crushed seeds or leaves from the neem tree, a large East Indian tree with a bitter bark, deterred some insect species from feeding when applied to plant foliage. This compound can be absorbed by the roots of plants.

The ladybird beetle, *Stethorus punctum*, was being used by scientists at Pennsylvania State University and independent apple growers in a program of "integrated control" of mites harmful to apple trees. In this program the grower must arrange a careful balance between protecting the apple crop and the trees from injury by insects while at the same time not harming *Stethorus*, which preys upon the European red mite and the two-spotted spider. If these pests were destroyed by pesticides, the beetle would leave the orchard to search for mites elsewhere. The program had the advantages of eliminating some pesticides from the spray schedule and of reducing amounts of other sprays. This not only lowered the level of environmental pollution but also growers' costs. Another advantage was that *Stethorus* would eat mites that were resistant to the sprays.

The chemical 13-methylhentriacontane was identified in the feces and larvae of the corn-ear worm *Heliothis zea* (Boddie), a pest of considerable economic importance, as the major constituent that triggers the short-range host-seeking response of its parasite, *Microplitis croceipes* (Cresson). Bioassay of closely related compounds indicated that the structural requirements for activity are remarkably specific. This chemical, the first of its kind to be found, could also upgrade efforts to use parasites for insect control as well as lead to a better understanding of insect behavior mechanisms.

In a fascinating study, Sarah Corbet of the University of London reported that when caterpillars of the *Anagosta kuehniella* touch each other they deposit a secretion from their mandibular glands on the substratum. This secretion stimulates oviposition movements in its insect parasite *Venturia canescens* (Grav.) and a larger proportion of the host caterpillars are parasitized in a crowded population than in an uncrowded one. When a female of *V. canescens* oviposits in a larva of *A. kuehniella*, she taps her antennae against the host, unsheaths her ovipositor, and then makes jabbing movements, which, if they penetrate the host's cuticle, may result in the release of an egg. The experiments showed that the parasite responds quantitatively to its host's secretion. The finding of this substance, termed an epideictic pheromone, might have applications in the control of some other butterfly and moth pests, especially if their mandibular glands prove to have a similar function. Work was in progress on the chemical characterization of the components of the secretion, making use of the parasite's ovipositional response as a clever bioassay for its presence.

A very simple agent for controlling insect pests without harming other organisms (including man) was being investigated by three workers with the Agricultural Research Service of the U.S. Department of Agriculture. The agent was light. Larvae of many insects, such as the codling moth and corn borers, go into diapause, a state resembling hibernation, when the days become shorter in the fall. With spring, longer days, and more light, the larvae resume their life cycle and metamorphose into adults. Artificial light might be used to prolong the days so that the larvae would not go into diapause but would metamorphose in the fall and be wiped out by the winter weather. On a test plot where artificial light was supplied by 40-watt fluorescent lamps, more than 70% of the insects failed to go into diapause, continuing instead to develop into adult moths in the fall. Apparently the major drawback of this procedure was the cost of optimum lighting conditions.

**Evolution studies.** The search into the details of the origin of life continued. There was speculation, for example, that the organelles, specialized cellular bodies, might once have been independent organisms. A new symbiosis theory held that the genetic material recently found in chloroplasts and mitochondria demonstrated that these two kinds of organelle were once free-living organisms. The theory proposed that the higher plants are the result of a symbiosis between animal-like hosts and photosynthetic blue-green-algalike guests whose partnership could not have evolved until relatively recent times, when mitosis had been perfected. Proof that the symbiosis theory is correct demanded experiment.

One of the major trends in evolution studies was a decline in subjective typological reasoning involving examination of single features only and a corresponding move toward the use of more objective, comprehensive, and, increasingly, statistical methods analyzing complexes of many traits. Because of the fragmentary condition of most fossil remains, the scope of the trait complex approach was very limited, but application of the statistical concepts of population genetics allowed the investigator to consider his sample, however small, as a part of a larger population, and to make the population, not the individual, the unit of study.

New techniques that detected patterns of fluorescence caused by the binding of the drug guinacrine to specific regions of chromosomes were being used in chromosome analyses of various mammals. It was found that fluorescence of the intensity found on the human Y chromosome was only present in the African great apes and man. Applying this information, the investigators suggested that the absence of intense fluorescence in the orangutan or gibbon makes it likely that these apes evolved from an ancestral stock before the origin of the lineage giving rise to the African great apes and man.

—John G. Lepp

Courtesy, Dr. Clifford O. Berg, Cornell University

*Snails that transmit the debilitating disease schistosomiasis, or snail fever, can be controlled without pesticides. At top a marsh fly larva approaches a snail harboring the tiny flatworm larvae that cause the disease. The larva attacks the snail, pursues it into its shell, and feeds on it. Thus the snails can be destroyed without the harmful effects to the environment caused by pesticides.*

A Pictorial Essay

# Art and Technology

## by Jack W. Burnham

**Modern science has given rise to a wealth of technical gadgetry. Modern artists, in turn, are using this technology for their own purposes.**

*"Day Passage" by Rockne Krebs. Photo, Ed Cornachio.*

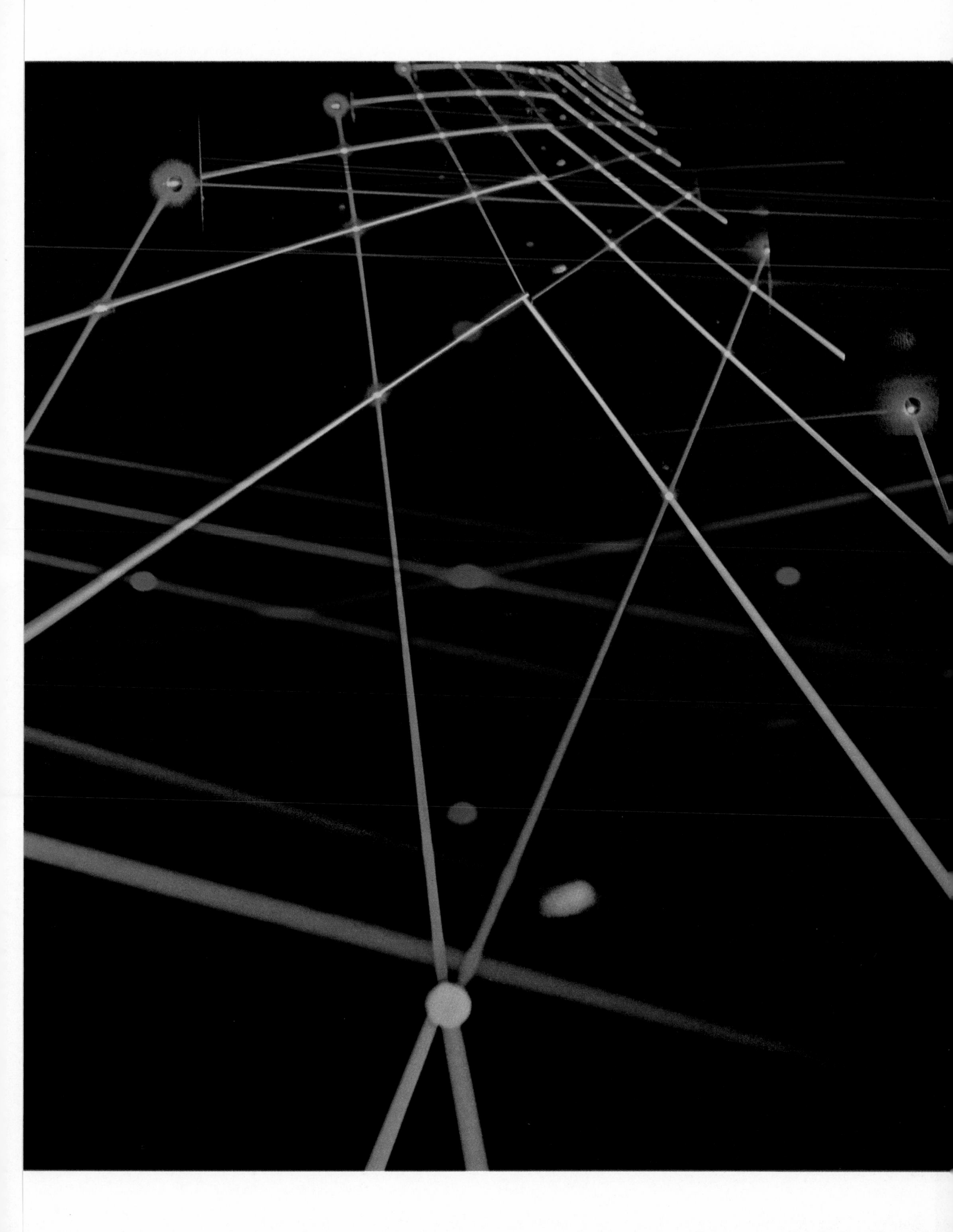

who attended, "9 evenings" appeared to be an unprepared series of Space Age happenings, beset by constant breakdowns and delays. Ironically, though, the year-long preparations for the "9 evenings" demanded a deep emotional involvement from many of the participants. It proved several things. Artists and engineers can work together toward common goals provided they share some degree of technical knowledge. Moreover, it demonstrated beyond doubt that the excitement of technology comes with using it rather than viewing it.

"9 evenings" was masterminded by Bell Laboratories scientist Billy Klüver, who subsequently founded Experiments in Art and Technology, Inc. Originally, Klüver's goal was to provide a parent corporation where artists worldwide could receive technical assistance from scientists and engineers interested in their projects. Soon EAT had several thousand names in its active file and the promised support of several important corporations and foundations. EAT's basic function was to facilitate technicalization of the arts, while also serving as an antidote to the dehumanizing aspects of social innovation.

EAT's most important statement came in the form of an open exhibition at the Brooklyn Museum and the Museum of Modern Art in New York during the winter of 1968–69. Its title, "Some More Beginnings," perhaps gives an indication of the caution that then prevailed. The earlier optimistic belief that technology was the wave of the future had given way to a realization that much technicalized art fails aesthetically and makes little sense financially to the artist, engineer, or gallery. Research costs for unique projects remain relatively high for the artist, as do maintenance expenses for the museum. A unique feature of "Some More Beginnings" was its attempt to give incentive to engineers. Considerable publicity, along with the main prizes, went to engineers rather than to artists. Well over a hundred entries of greatly varying quality were shown.

As the novelty gradually faded from technological art, EAT was confronted with several fundamental difficulties. Such art tends to attract more than its share of bad work. Artists of reputation do not need the sponsorship of EAT to gain help from corporations. Most decisively, few engineers are inclined toward extended collaboration with artists without the incentive of financial help. At present EAT is no longer a force in the art world, and its remaining energies seem to be focused on large-scale urban problems.

Courtesy, Jack W. Burnham

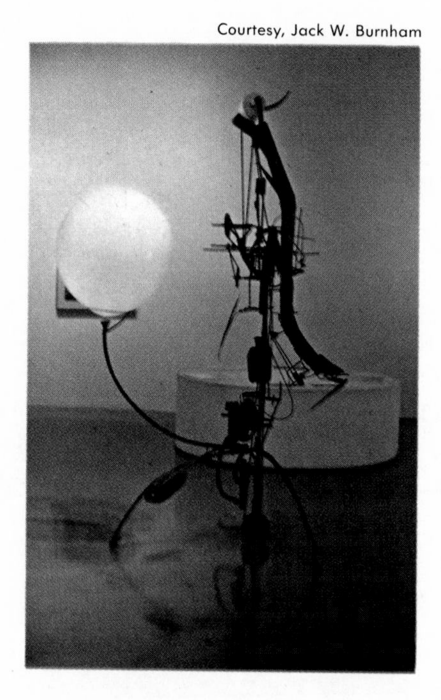

*In "Meta-matic No. 17," 1959, one of a series of art-making machines by the Swiss-French sculptor Jean Tinguely, pens scratch random marks on a moving roll of paper, parodying the abstract expressionist style of the 1950s. Designed to break down, Tinguely's machines represent witty philosophical observations on machine-oriented societies.*

### The programmed art experience

The first international exhibition of "post-machine" art took place at the Institute of Contemporary Arts in London during the summer of 1968. Jasia Reichardt's "Cybernetic Serendipity" attempted to document the creative use of computers and various cybernetic devices. Its book-catalog contains a good layman's account of the historical development of digital computers. "Cybernetic Serendipity" included some strictly scientific experiments, along with works by artists that utilized the principle of feedback in machines responding to both internal and external stimuli. Other exhibits featured printouts (visual diagrams) from computers as used in music analysis, computer graphics and movies, computer-designed choreography, and computer poems and text analysis. Generally speaking, computer graphics and painting have found little favor in the art world. They usually seem to be pale imitations of earlier modernist styles. (There are a few exceptions to the "pretty picture" school of computer graphics. For instance, the English painter Harold Cohen is concerned with a "game theory" of strategies for creating paintings according to very primary assumptions. Cohen devises programs based on theories of graphic creativity.) Unfortunately, much of "Cybernetic Serendipity" was destroyed in transit when it was shipped to the Corcoran Gallery of Art in Washington, D.C.

The 1970 cybernetic exhibition at New York City's Jewish Museum proved to be a more interactional experience for the gallery goer. "Software" had the added complication of using several computers in the museum. It included a programmed catalog where viewers were encouraged to sit down at key-scope terminals and type out questions

*Harold Cohen works with a Data General Nova 1200 computer and a flatbed graphic plotter. This plotter includes a freehand drawing simulator based on the artist's normal style of line drawing. A typical program has two parts. The first provides an "environment" consisting of a given set of digits randomly distributed within the cells of a grid. The program then selects a succession of digits, exhausting one number and then moving on to the next highest, the object being not to cross an existing line. Cohen is interested in problem-solving strategies inasmuch as they simulate a structure common to "art games."*

Courtesy, John Waggaman

about nearby exhibits. "Software" also introduced algebraic and set-theory games played on special display scopes. More than just a display of electronic technology, "Software" focused on the creation of aesthetic realizations through communications media. Such a complete ephemeralization of the art object is connected with the idea of Conceptual Art, which emphasizes that sensory input need not be related to material art objects but only to an "art proposition" in some written or diagrammatic form. Closed-circuit television, audio tape players, environmental sensors, and photography are other forms of information technology organized to produce art experiences.

"Software" was not without its problems. Programs in the main computer, a DEC PDP-8, were short-circuited two days before the opening. The system was not sufficiently reprogrammed and debugged until just a week before the exhibition closed in New York. Because of differences of opinion over editing, five of the introductory film loops were destroyed on the second day of the exhibition and were not replaced for two weeks. Art critics were generally critical of "Software"; commercial trade journals saw it for what it was; some gallery goers construed it as a threat to the traditional art world. If nothing else, technological art seems to generate some very human reactions.

Courtesy, Newton Harrison

*Each of the four water troughs in Newton Harrison's "Salt Water Ecosystem: Brine Shrimp Farm," 1971, has a different degree of salinity and thus a different color, since the amount of salt affects the growth properties of the algae in them. The troughs are seeded with brine shrimp, which grow to maturity in six or seven weeks. After they are harvested, the water evaporates and the salt is collected.*

The Los Angeles County Museum of Art's "Art and Technology" program, as of 1972, was the most ambitious exhibition of its kind. Planned from 1967 by Curator Maurice Tuchman, "Art and Technology" appeared during the summer of 1971. Tuchman was fortunate to have the backing of nearly 40 corporations in the southern California area. In a selection process covering two years, the curator chose about 30 artists from a worldwide group numbering in the hundreds—although the final selection of artists was mainly from New York. Most significant from a planning perspective were "A & T's" exhaustive contractual relationships between artist, company, and museum; its relatively decent budgets for artists; and its attempts to produce a sustained and practical interchange between artists and the corporate-industrial sector. Quite aware of the strong antitechnological sentiments within the art world, Tuchman and his associates wisely chose a limited number of entries from some of the biggest names in the American art world: Roy Lichtenstein, Claes Oldenburg, Robert Rauschenberg, Tony Smith, and Andy Warhol.

There has been a steady escalation of means within technological art, and few significant projects are inexpensive. The proposed outlay for each "A & T" project was $15,000, but it seems certain that several works at least doubled that. Total costs mentioned for the exhibition ran upward to $1 million. For a profit-making investment such as a motion picture this would not seem unreasonable, but in the penurious museum world $50,000 to $100,000 is considered the range for a major show. Although "A & T" produced a fairly successful and entertaining exhibition, many art critics accused it of selling out, Disneyland fashion, to large corporations with interests in profits and war, and of perverting high art "aesthetics."

What was evident from the Los Angeles "A & T" effort was a certain inherent alienation from big technology on the part of the artists. Their projects either threatened viewers' sensory capacities or reflected the ironic vein of Chaplin's 1936 movie *Modern Times.* Much of the exhibition was implicitly antagonistic toward the effects of technology. Below the surface, few of the artists appeared to believe that U.S. corporate interests, controlling the overwhelming portion of the country's technology, have any real sense of social responsibility or direction.

### Is it "art"?

Coming back to the original question, can art and technology be reconciled? Do they have a common meeting ground where both perform according to everyone's expectations? Technological art that simply

Courtesy, Alan Sonfist

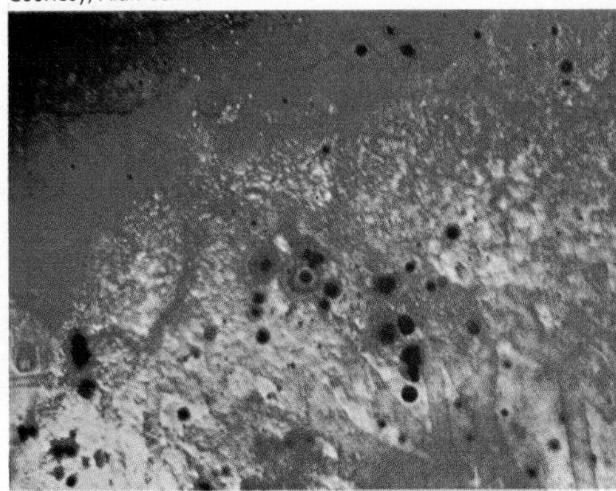

Courtesy, Alan Sonfist

*All Alan Sonfist's works deal with growth, decay, and interdependence. He sees "sculpture" in the ecological exchanges that occur every day and believes that man-nature stability will come only when we have become acutely sensitive to the natural changes around us. "Crystal Globe," 1968 (left), is a sealed hollow glass sphere containing natural mineral crystals. Through the effects of natural heat and light, some of the crystals vaporize into a purplish gas which rises to the top of the globe and crystallizes on its inside surface. "Micro-Organism Window," 1968 (above), appeared in the window of a New York City art gallery. The artist coated the glass with an agar-based substance, and within 48 hours growth patterns began to appear on it. In a sense, this is a way of dramatizing the toxic content of New York air.*

duplicates the results of handcrafted art is in all respects a trivial contribution. Moreover, a great selling point of technological invention, time saving, has little meaning for the real artist. Making art embodies its own psychological reward. The dilemma of the alienated office and industrial worker is still a reality; consequently, it is not surprising that thousands of students enter art schools every year although the chances that they will find work in their chosen field are slim. Why does "art" feed the soul and "work," in a modern industrial society, deaden it?

At a fundamental level, technology serves the same purpose as art. It provides the link between physical need and social fulfillment. And while they are probably not inherently antagonistic, after a thousand years in which they have gradually separated, we have grown to think of production and research as "practical" while art remains merely entertaining and provocative. Many of the justifications for earlier art, *i.e.*, its spirituality, beauty, idealism, and emotional sentiment, no longer hold true for contemporary work. Where art objects were once carried through the streets of towns with adoration, the art gallery today functions as a tame sideshow for an easily jaded public. Perhaps the greatest victims of the Industrial Revolution, many artists are ridiculed as industrial "scavengers," more interested in the up-to-date trickery of modern technology than in its expressive appropriateness.

The artist finds himself in a strange double bind. On the one hand he knows that rewards in the art world usually go to innovators—just as they do in the world of business and production. Yet the craving for "new" effects has reached such a point of absurdity and diminishing returns that the artist is beginning to question the entire meaning of stylistic change. If art was once the essence of social stability and ritual, something has happened during its historical evolution to bring it to its present state of psychic and social ineffectuality.

One aspect seems to be that art fulfills certain inherent psychological needs, which, given man's symbol-making propensity, are biologically unique. Making a completed work of art denotes a kind of cycle. Machines incorporated into art, on the other hand, imply an infinitely linear event; that is, energy tamed to do man's will. Thus it is not surprising that little or no machine-driven kinetic art has found aesthetic acceptance. Machines may be used to *make* art so that the resultant object fits into a historic time slot. In such a way, the work of art is frozen in time. In contrast, many artists have become interested in the cycles of animal and vegetable growth, weather change, and ecological stabilization. Here the temporal dimension is repetitive. Perhaps for very profound reasons, such art appears to embody many of the technological and craft features developed by European Neolithic cultures between 6000 and 2000 B.C.

### Rediscovering wholeness

On one level this represents a rejection of the Industrial Revolution. Or, pessimistically, we could speculate that art is phasing itself out as

(Opposite page) courtesy, Los Angeles County Museum of Art; (below) courtesy, Jack W. Burnham

*Using the research facilities of the Container Corp., Tony Smith designed his massive interior sculpture (opposite page) for the 1971 Los Angeles "Art and Technology" project. The cavelike structure uses thousands of assembled tetrahedral modules. Visually and spatially, it remains one of the most stunning experiences in modern sculpture, although Smith was never satisfied with the structural solution. The intense light source in "Heart Beats Dust" (above), a 1968 work by Jean Dupuy and Harris Hyman, is a 250-w quartz element with shutters and lenses. A very light red pigment, lithol rubine, palpitates on a stretched rubber membrane in time with a continuous loop tape recording of heartbeats played through a coaxial speaker mounted just below it.*

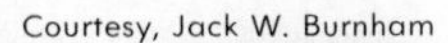

Courtesy, Jack W. Burnham

*The light environment "Tripping" (above) was created by Jack Burnham for the Ravinia Art Festival, held near Chicago in 1969. Two 35-ft tapes, cemented back to back, are suspended the length of a room so that they nearly touch the floor and are illuminated by normal 115-v house current. A phosphorescent glow in materials distributed on the floor is generated by several ultraviolet lamps above the tapes. Another light environment (opposite page), designed by Newton Harrison for the Los Angeles "Art and Technology" project, uses five plastic cylinders, each filled with a different gas. An electric current passed through the cylinders ionizes the gases, causing each to glow with a different color. Changes of form in the light are induced by regulating the gas flow.*

a meaningful activity, due to our ever increasing commitment to a rationalized and systems-oriented life-style. The truth probably lies with neither. The artist may be warning us that many of our technologies are dangerously unbalanced, that our waste disposal practices and use of resources represent massive social pathologies. What is more, unstable technologies have usually generated new methods whose instabilities lie farther in the future. While the concept of a steady-state world remains an impossible dream, many artists intimate that there is no other alternative. They seem to be saying that art properly made—that is, work accepted as "art"—embodies ancient patterns for balanced human behavior.

Some social scientists hold to the possibility that art and ritual are tied to each other through the same structure of relationships. Consequently, if much of the origin of physical sickness is psychosomatic, then the medical and, hence, aesthetic function of the shaman in primitive societies may be critical after all. In his book *Pigs for the Ancestors*, the anthropologist Roy Rappaport reports on a tribe of hill farmers in the Central Highlands of New Guinea. Taking all the aspects of their complex ecological system into consideration, Rappaport decides that their elaborate cycle of rituals functions as a feedback mechanism or a kind of spiritual thermostat controlling animal raising, farming areas, man-land ratios, protein consumption, war-making activities, human procreation, and other vital activities. Traditionally, Western culture has sought purely empirical answers to all of its immediate

Ed Cornachio

Ed Cornachio

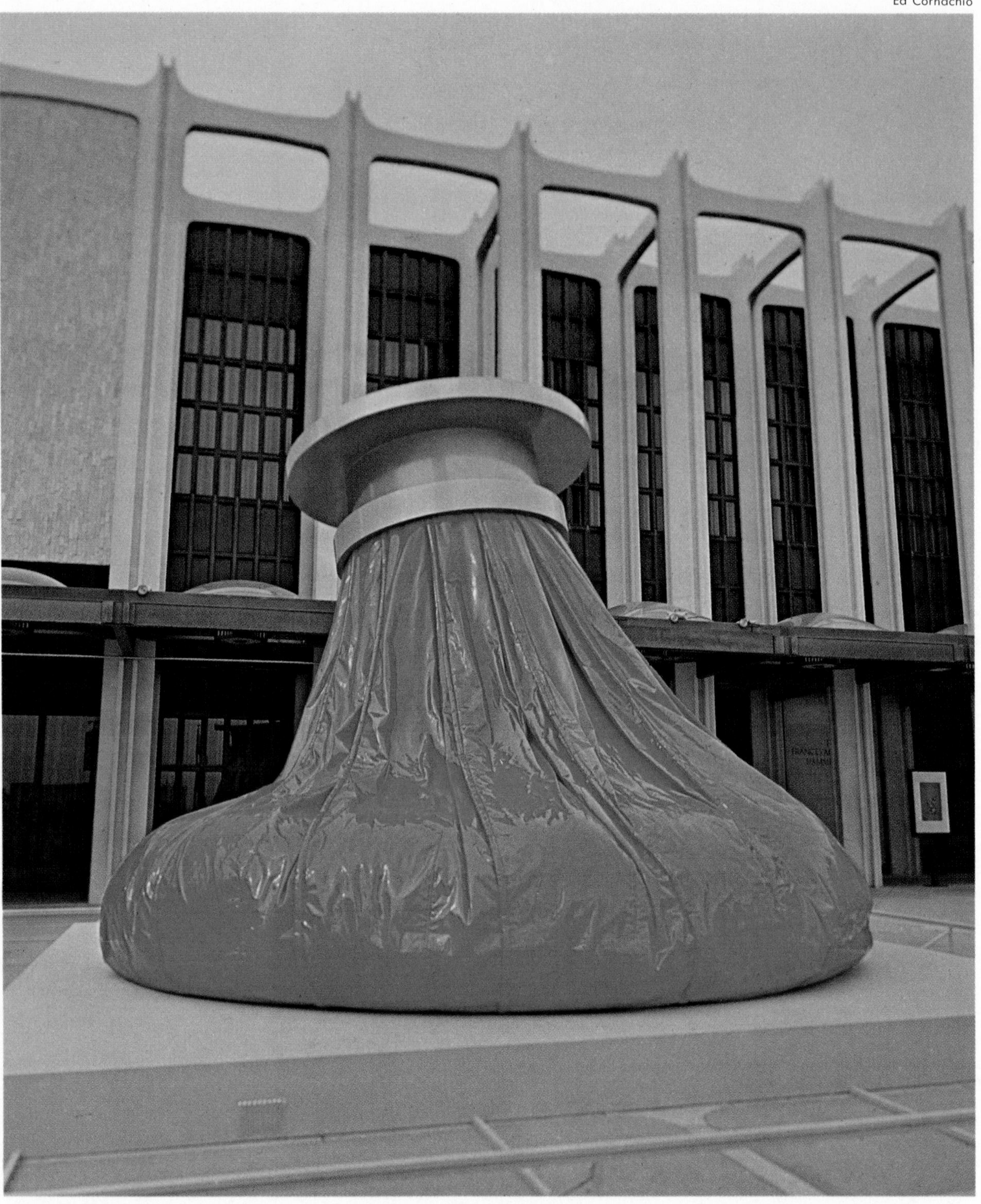

Courtesy, Los Angeles County Museum of Art

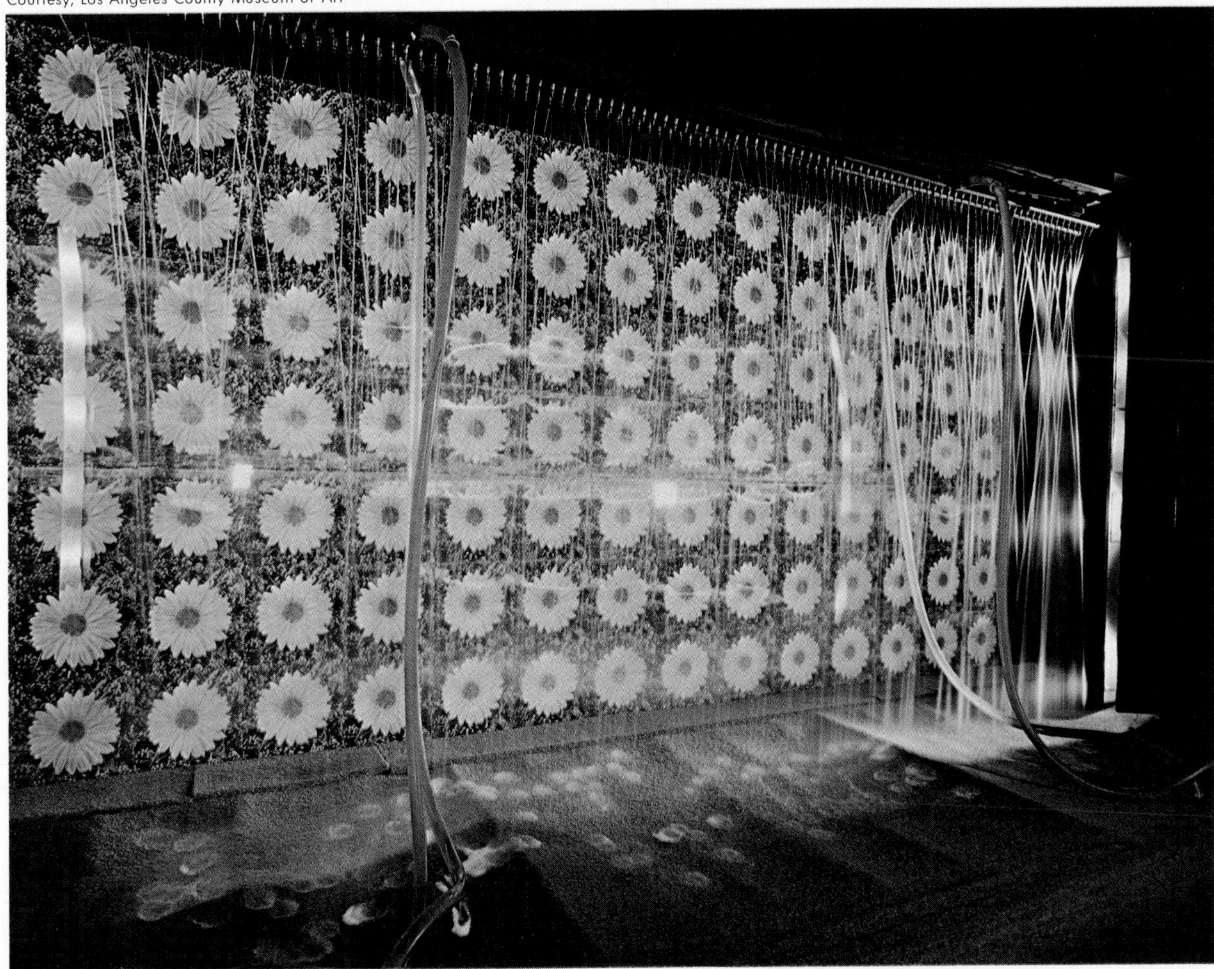

Les Levine

*Claes Oldenburg's "Giant Icebag" (opposite page) rises from 7 to 16 ft by means of a hydraulic mechanism. In this 1968–71 work, Oldenburg, like Tinguely, is making the comment that there is something inherently absurd about both moving sculpture and icebags. Set against a backdrop of three-dimensional plasticized daisy panels, Andy Warhol's 1971 work "Rain Machine" (above) drops "rain" on plastic grass and recycles the water through clearly visible Teflon hoses. "Chain of Command," 1972, by Les Levine (left), involves three stepladders: one with a television monitor, one with a TV camera, and one with a slide projector that shows 80 slides, each with a different view of the three ladders.*

problems. But as we erode our psychically organized activities such as the handcrafts, visual art, and ritual itself, we fall in danger of destroying the human will to act in unison. Given practical material solutions to pressing social problems, we must begin to ask ourselves if such solutions embody an artistic-ritual format, or do they further sever people from their sources of psychic motivation?

The 1970s appears to be a decade in which the artist is beginning to temper his desire to explore new technologies. A mood of discrimination is about. Just as the artist began to look like an ineffective manufacturer or a parody of the lonely research scientist—thereby denying his own unique contribution—we are beginning to rediscover his primal role as the soul's messenger and coordinator. Since the philosophy of aesthetics began in the 17th century, one thing has been

Courtesy, Shunk-Kender

*"Seek," 1969–70, by the MIT Architecture Machine Group (above), deals with small blocks that act as an "environment" for a colony of gerbils. These little animals constantly bump into the blocks, thus disrupting the work of a computer programmed to build the blocks into towers. This produces a mismatch between the computer's memory and the gerbils' randomizing activities. Robert Rauschenberg's "Mud Muse," 1970–71 (opposite page), is a 9 by 12-ft glass-enclosed vat of driller's mud. In collaboration with Teledyne for the Los Angeles "Art and Technology" exhibition, Rauschenberg made recorded tapes of bubbling mud and used them to program a series of air jets, located beneath the tank, which sporadically produce bubbles and eruptions.*

demanded from the artist: that intangible called "quality." Concerning past excellence we have developed adequate perspectives and theories of evaluation. But quality in present art eludes us more than ever. Consequently, we have begun to question the restriction of quality to the fine arts.

There is a deep longing for a wholeness within our technologies and their integration into our life rhythms. And we wonder why no systems analyst has ever been successful in achieving this. Perhaps it lies in the origins of culture itself. Some realize that quality most essentially belongs in our attitudes toward nature and our patterns of daily living. Therefore, a few artists are beginning to reexamine that most complex of all technologies: the human nervous system and its essential relation to the world.

Courtesy, Los Angeles County Museum of Art

# The Notorious Trio: Mercury, Lead, and Cadmium

## by John F. Henahan

**Many refinements of modern industry are the result of man's increasingly sophisticated use of the earth's mineral wealth. The metals mercury, lead, and cadmium have been used with great success and in such quantities that each in its own way now threatens man's health and ultimate well-being.**

At the time the earth was formed some 4,500,000,000 years ago, the metals mercury, lead, and cadmium were more or less safely locked in the minerals that contained them. About a hundred years ago, man discovered he could use these metals to make a better material life for himself, and he began to dig them from the earth. By converting the crude ores to refined metals or metal-containing compounds and by manufacturing new products from them, he gradually introduced millions of pounds of cadmium, lead, and mercury into an environment in which they did not occur naturally. Because all three metals are elements they contain atoms of only one kind and, hence, will not be broken down by chemical forces in the environment. With their concentrations pushed much beyond what had been their natural levels, the three metals began to disturb the balance of nature and by the 1970s their adverse biological properties were being recognized as threats to man's well-being.

Although each of the three metals has increased steadily in the environment in proportion to its value to industrial and technological expansion, each is now seen as a threat to life for different reasons. Lead has become ubiquitous in land, air, sea, plants, and animals in direct proportion to the proliferation of the automobile and the gasoline it uses as a fuel. The deadly effects of mercury, previously considered hazardous only to industrial and laboratory workers, have been magnified ominously with the discovery that by a natural phenomenon mercury is reacting with organisms in the bodies of water in which it is dumped to produce what may well be a common health hazard. Cadmium, which has so many applications that it is used at twice the level of mercury, is now suspected of being a major contributing factor to the increased incidence of high blood pressure and heart disease.

MERCURY—dumped in oceans and lakes for disposal, it has become a toxic compound there.

LEAD—used in gasoline, it is emitted in automobile exhausts and poses a health hazard to those who breathe it.

CADMIUM—emitted in industrial smoke, it causes lung damage and has been linked to high blood pressure and heart disease.

Akhtar Hussein from Woodfin Camp

*The metals that may threaten man's well-being pollute in proportion to their usefulness. Fine particles discharged from forest product industries, such as the paper plant in Mississippi (above), rank second only to those of coal-fired power producers and are believed to include traces of the mercury found in pulp and its bleaching agents. Of all atmospheric emissions from industries, 45% are estimated to come from those processing cadmium-bearing ores, such as the copper being poured into molds at a Nevada mine (opposite page). Metallic runoff directly into water supplies has been curtailed, but little is known of the amounts of metals in particulate emissions and of their eventual dispersal in the environment.*

## Mercury: messenger of death?

Only within the last ten years has man come to realize that the uncounted millions of pounds of mercury wastes previously believed to be safely resting at the bottom of the world's oceans, lakes, and rivers are slowly being converted into a toxic form called methyl mercury that has already had drastic effects on birds, fish, and other animals including man. Methyl mercury is formed by the combination of an inorganic mercury compound with a one-carbon organic molecule, an action that can be precipitated by the organisms in bodies of water. It is seen as a special threat to man because it has a long biological half-life (about 70 days in man) compared to the inorganic forms of mercury, which are usually excreted from the body in a few days. Methyl mercury's bad reputation is fortified by the fact that it easily passes through biological membranes, including the barrier that separates the brain from the rest of the body. The compound is particularly destructive to nerve cells, and early symptoms of methyl mercury poisoning include headache and fatigue. These are followed by sensitivity loss in toes and fingers, visual disorders, poor muscular coordination, speech and hearing difficulties, mental retardation, and death.

Mercury is not normally dangerous in its natural liquid metal form, although scattered deaths have occurred when industrial workers have inhaled its vaporized fumes. Inorganic mercury compounds, such as mercuric chloride, have been used to commit suicide, and workers in felt hat factories have been driven literally "mad as a hatter" by the mercuric nitrate used to improve the felting properties of wool. There have also been many deaths and illnesses recorded in the various industries that mine, manufacture, or use mercury. These cases, however, have never been seen as serious enough to pose an international threat to human health. Industries using the inorganic mercury compounds have felt no qualms about dumping wastes containing them into lakes and rivers, where they were presumed to settle safely.

**JOHN F. HENAHAN,** *a science writer for* Intellectual Digest, *is the author of* Men and Molecules.

Paolo Koch from Rapho-Guillumette

Within the 20th century, about 163 million pounds of mined mercury have been used; according to the U.S. Bureau of Mines, between 1959 and 1969 mercury consumption in the U.S. doubled to more than 6 million pounds annually. The greatest single use of mercury is as a component in the electrodes that prepare high-purity sodium hydroxide and chlorine gas from brine solutions to meet needs of the petroleum, glass, paper, and detergent industries. Mercury is also used in catalysts for the manufacture of the versatile plastic polyvinyl chloride (PVC) and is being used increasingly in specialized electrical applications, such as silent switches, arc rectifiers, high-intensity street lamps, and batteries for transistor radios. Mercury or its compounds is also used for making amalgams for dental fillings, for temperature and pressure control devices, in mildew-resistant paints, in agriculture as a seed sterilant, and in medicine as antibacterial agents (for example, mercurochrome).

Each of these applications is responsible to a greater or lesser extent for the increased presence of mercury in the environment. About one third of all mined mercury cannot be accounted for and is presumed to have been dispersed. Poor pollution control techniques used in smelters and ore-refining plants and in the chloralkali and PVC industries have been major contributing factors to mercury contamination; runoff from farms treated with mercury fungicides has been another; and simple carelessness in industry, dental office, laboratory, and home has also contributed. In addition, the burning of coal, oil, and lignite, which contain traces of mercury, spews a million or so pounds of the metal into the air every year, perhaps more than is released by any other means.

### The methyl mercury crisis

The current concern over methyl mercury pollution can be traced to 1953 when residents of Minamata, a fishing village in Japan, reported

Courtesy, Dr. J. M. Wood, University of Illinois

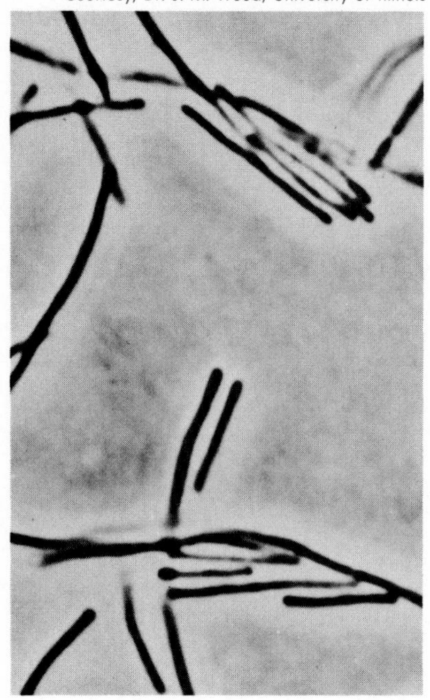

*Poisonous methyl mercury is formed by the chemical reaction of the less harmful inorganic form of mercury with an organic molecule. University of Illinois biochemists established that microorganisms commonly found in bodies of water can accomplish this synthesis. One such organism,* Methanobacterium formicicum, *is shown above as it appears under a phase contrast microscope, which is used to reveal details not visible under a normal microscope.*

strange omens: cats ran screaming into the sea and crows fell from the sky. Soon after, Minamatans began to complain of aching muscles and distorted vision in which they appeared to be seeing through a long tube. Then a father and his young son went mad and died. In the next nine years, another 110 persons up and down the coast of Minamata Bay developed the same symptoms and 36 died. Some survivors were left blind and deaf, and many children were born with what came to be known as "Minamata disease." In 1965, the disease broke out in the Japanese city of Niigata; 50 residents became ill, 16 died, and many babies seemed to be affected by the disease.

In Minamata, physicians began to recognize that the symptoms of the strange disease were very much like methyl mercury poisoning. Their suspicion was strengthened when they learned that a PVC plant nearby was pouring large amounts of mercuric chloride into the bay and that all victims of the disease—including the cats and crows—ate fish from the bay as a major portion of their diet. (A PVC plant was also located in Niigata.) Originally, Japanese scientists speculated that the normally innocuous mercuric chloride had somehow been converted to toxic methyl mercury, perhaps by bacterial action in the water, but they learned later that the electrodes of the chemical plant contained appreciable amounts of methyl mercury.

Sediment near the Minamata plant was shown to contain two million parts per billion (ppb) mercury, while mercury in the bay water itself ranged from 1.6 ppb to 3.6 ppb, compared to natural background levels of from 0.3 ppb to 2.0 ppb. It was also learned that fish and shellfish eaten by the residents of Minamata and Niigata had somehow concentrated the mercury (predominantly in the form of methyl mercury) in their bodies at levels of 5–6 parts per million (ppm), about 100 times the 0.05 ppm the World Health Organization (WHO) considers the irreducible minimum of mercuric residue and 10 times the 0.5 ppm "interim guideline" suggested in 1970 by the U.S. Food and Drug Administration (FDA).

Methyl mercury poisoning cropped up in a different guise in Sweden in the early 1960s when conservationists observed population drops among many seed-eating birds, including pheasants and partridges, and among their predators. The decline was linked to an agricultural seed sterilant (methyl mercury dicyanodiamide), which had been in use in Sweden for more than 20 years. Apparently, the birds fed heavily on treated seeds and were poisoned. Sweden forbade the use of the compound in early 1966, and the bird population began to return to normal levels.

Three years later several members of a family in Alamogordo, N.M., came down with what appeared to be various stages of methyl mercury poisoning after they had eaten the meat of a hog that accidentally had been fed grain (millet seed) treated with the methyl mercury seed sterilant. Although the family was treated quickly with various metal-scavenging drugs, one child went blind and two others developed brain damage. The condition of a fourth child born after the mother had

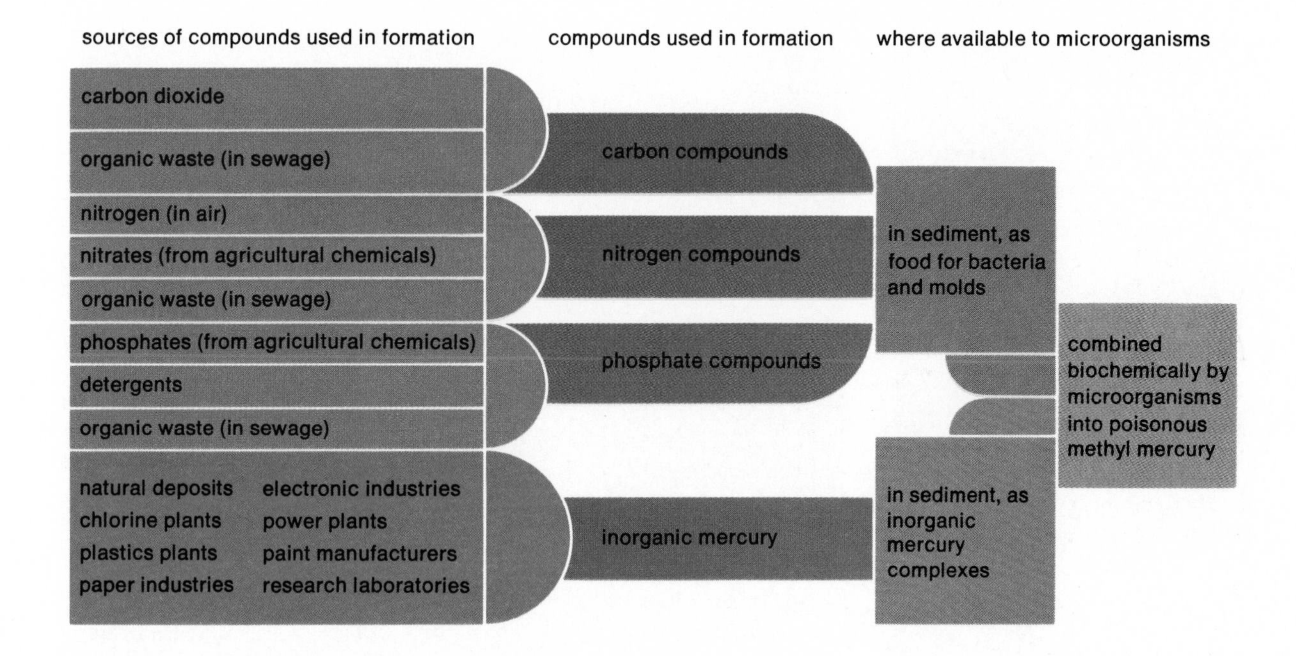


*The potential for widespread methyl mercury contamination is enhanced by many kinds of industrial pollution. Carbon dioxide, organic wastes, nitrates, and phosphates provide food for the organisms that permit transformation of inorganic mercury runoff from industries into methyl mercury. Following alarming reports in 1970, U.S. and Canadian industries virtually eliminated the flow of new mercury into lakes and rivers. Nonetheless, enough remained in some sediment to permit methylation to continue for generations. Some control might be accomplished by limiting the entry of the carbon, nitrogen, and phosphate compounds into the water, which, in turn, would limit growth of the microorganisms.*

eaten the contaminated pork remained uncertain. Shortly after the Alamogordo poisonings, the U.S. Department of Agriculture banned the use of methyl mercury fungicides but the manufacturer obtained a court injunction permitting the chemical to remain in use.

Despite the mercury-linked disasters in Japan, Sweden, and the U.S., response remained minimal until March 1970 when Norvald Fimreite, a chemist at the University of Western Ontario, found that commercial shipments of fish from Lake St. Clair (near Detroit) contained mercury close to the levels found in the fish involved in the Minamata and Niigata episodes. The Canadian government quickly closed its side of the lake to commercial fishing and warned residents against eating fish from the lake. Other chemists reported similarly high levels (about 7 ppm) in fish from many other lakes and rivers along the U.S.-Canadian border, including the Great Lakes, and, subsequently, from the waters of 33 of the 50 U.S. states. Many lakes and rivers were closed to fishing, and localities lost considerable revenues from tourism and license fees. In most cases chloralkali plants or pulp and paper processors were located on or near the polluted waters (the sodium hydroxide used in paper processing contains residual mercury). Prodded by quick action by then U.S. Secretary of the Interior Walter J. Hickel and several suits filed by the U.S. Justice Department, industries cut their mercury flow to a trickle.

The scope of the mercury pollution problem broadened throughout 1970 and 1971: Levels of mercury (primarily other than methyl mercury) in a range of Canadian foodstuffs, including bread, beef, and chicken, were almost invariably determined to be higher than the WHO minimum. The discovery that much tunafish and swordfish sold in the U.S. contained mercury (almost entirely methyl mercury) in excess of the

Jon Brenneis

Chuck Rogers from Black Star

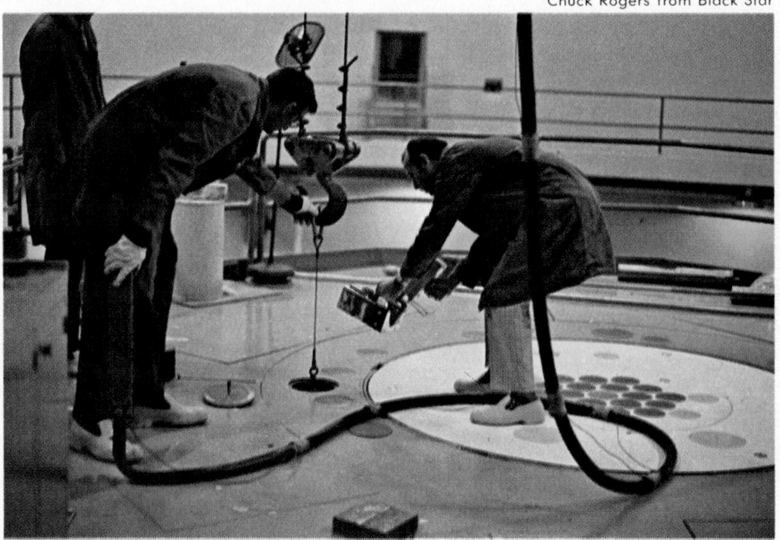

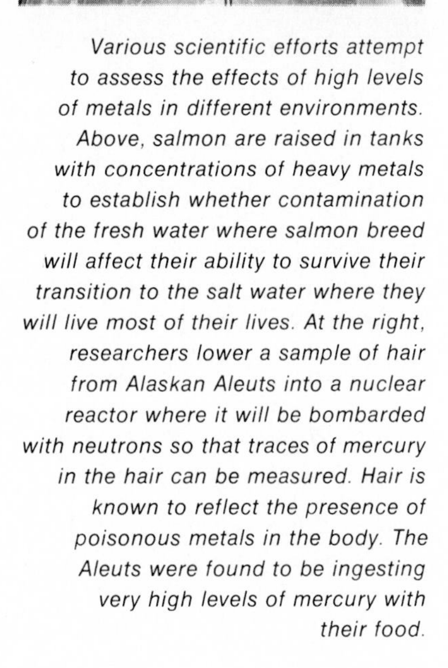

*Various scientific efforts attempt to assess the effects of high levels of metals in different environments. Above, salmon are raised in tanks with concentrations of heavy metals to establish whether contamination of the fresh water where salmon breed will affect their ability to survive their transition to the salt water where they will live most of their lives. At the right, researchers lower a sample of hair from Alaskan Aleuts into a nuclear reactor where it will be bombarded with neutrons so that traces of mercury in the hair can be measured. Hair is known to reflect the presence of poisonous metals in the body. The Aleuts were found to be ingesting very high levels of mercury with their food.*

FDA safety level, resulted in the temporary withdrawal of some brands of canned tunafish and the virtual destruction of the U.S. swordfish industry. A New York physician found that a woman patient on a weight-reducing diet involving heavy intake of swordfish had symptoms typical of methyl mercury poisoning. Aleutian Eskimos were advised to discontinue their traditional practice of eating the livers of Alaskan fur seals and vitamin supplements made from seal liver were forced off the market because of their excessive mercury levels.

The most frightening aspect of mercury pollution, according to many scientists, is that there is an enormous reservoir of the metal on the bottoms of lakes and rivers waiting to be converted by microbial action into the deadly methyl mercury. The methyl mercury is absorbed by phytoplankton, which are eaten by small fish. As the small fish are devoured by larger fish, which are then eaten by man, methyl mercury contamination spreads throughout the food chain. John Wood, a University of Illinois biochemist who was among those who proved that the transformation of inorganic mercury to methyl mercury occurs, estimates that it will take about 5,000 years for microorganisms to eliminate all the 200,000 pounds of mercury now known to be in Lake St. Clair sediments.

A number of researchers have found that preserved fish—some as much as 2,000 years old—have as much mercury in them as today's "contaminated" fish. Some scientists interpret this finding to mean that the hazards of being poisoned because of the current mercury content of fish is overrated. Others maintain that this implies that there may be a vast natural reservoir of mercury, which, when coupled with man's mercury contribution, only means that more methyl mercury may be being formed than is currently believed.

### Problems of mercury control

Many scientists are questioning the security offered by the FDA's 0.5 ppm guideline, preferring the much lower WHO minimum level. Environmentalist Barry Commoner, for example, insists that the FDA guideline is far too narrow when compared with safety margins applied to other toxic materials. Wood argues that the current practice of measuring "total mercury" in foods is pointless because it does not distinguish between inorganic and methyl mercury. Even though Swedish scientists have established that most of the mercury in fish is methyl mercury, Wood maintains it is possible that the high levels of mercury in other foods may be inorganic and, therefore, much less hazardous. The mercury standards also appear to be questionable when eating habits are taken into account. Wood suggests that if Americans and Canadians ate fish as often as the Japanese in Minamata do, methyl mercury poisoning might be rampant in North America.

Safety thresholds for mercury are as vague as they are because it is not known just how much methyl mercury in the blood is required before poisoning occurs. Although death has occurred when mercury levels in blood reached 1.4 ppm, some populations have shown no symptoms of poisoning when mercury levels were as high as those of other populations showing overt symptoms. Neurological damage may occur at levels as low as 0.4–1.0 ppm of mercury, but little is known about what else happens at or below those levels. Treatment for mercury poisoning is even less well-defined. Physicians have used penicillin derivatives and agents for countering the effects of poison gas with uncertain success. In 1971, however, T. W. Clarkson of the University of Rochester, N.Y., reported that ion exchange resins similar to those used in water conditioners increased the rate of mercury excretion in laboratory animals.

Erich Hartmann from Magnum

*An emission spectrograph measures the presence of metals in water. Metal traces are isolated when water in the porcelain tray is vaporized by a spark from a graphite-arc device.*

Several methods, in addition to the better industrial controls, have been suggested to prevent the inorganic mercury–methyl mercury transformation. One obvious method would be to dredge the mercury from lake sediments. But, when that approach was tried in Minamata Bay, the sediments were stirred and inorganic mercury was spread over a wider area. The amount of methyl mercury actually increased. Covering the sediments with nonmercury-containing ores has also been tried on a moderate scale with some success. The best method, suggests Wood, might be to prevent the growth of the microbes needed for methyl mercury formation. This could be done by reducing the amount of phosphates and nitrogen compounds that normally promote microbial growth.

### The lead problem

Man has been poisoning himself with lead for as long as he has known how to extract it from the earth. Romans are known to have died or gone insane from the lead oxide used to reduce the acidity of their wines. In the Massachusetts Bay Colony, settlers were poisoned when they drank rum distilled in lead vessels. In modern times, children are poisoned by nibbling the peeling paint of old buildings and adults can die from the poisoning effects of lead-bearing bullets rather than from the wounds they inflict. Other cases of lead poisoning appear wherever

Courtesy, Dr. Derek J. Cripps, University of Wisconsin

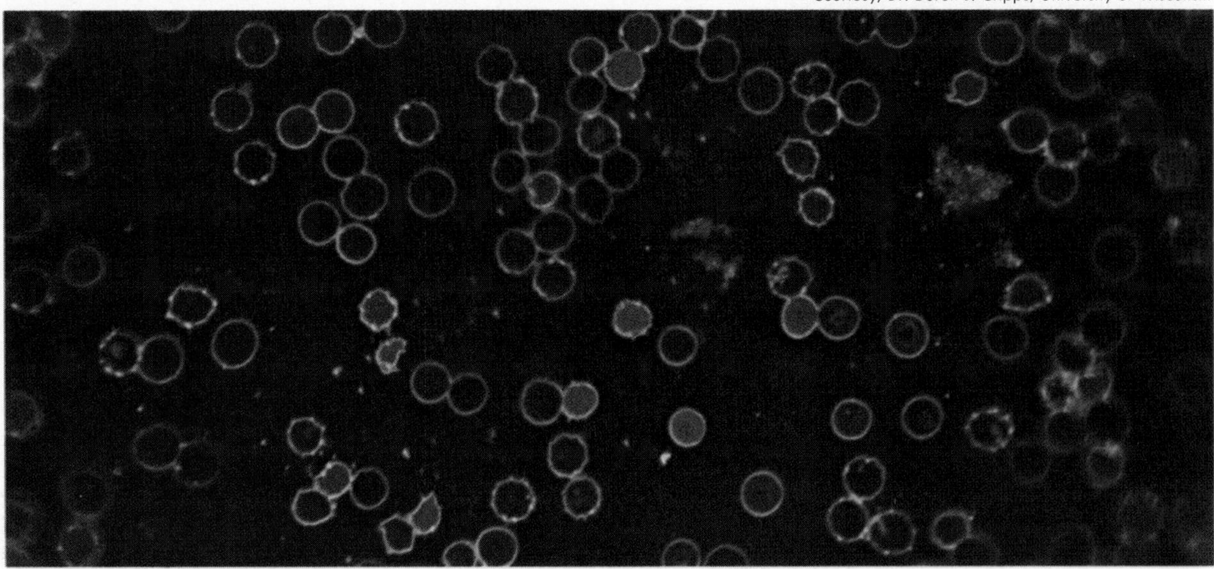

*Industrial activity is responsible for the increased presence of lead in the body as well as in remote parts of the earth. Even exposure to quite low levels of lead produces changes in the synthesis of porphyrins, substances needed for the production of hemoglobin in red blood cells. In certain solutions, porphyrins show an intense red fluorescence, which is captured (above) in a highly specialized type of microphotograph. Far below the entrance to a tunnel driven into the permanent snowfields of Antarctica (opposite page), amounts of lead in ice and snow samples were found to have increased steadily with the passage of time. The largest increase occurred after the introduction of tetraethyl lead into gasoline in 1923.*

lead or some of its more toxic compounds are manufactured, inhaled, ingested, or added to the air.

Lead poisoning is one of the most unpleasant ailments afflicting mankind. Yet, many of its symptoms are deceptively commonplace and often misdiagnosed as lesser problems, such as headache, diarrhea, or just plain irritability. As the lead reaches its as yet poorly defined toxic level in the body, the symptoms of poisoning become more characteristic and lead to wild delirium, coma, convulsions, blindness, mental retardation, brain damage, and death.

Every one million pounds of the earth's crust contains an average of 16 pounds of lead, usually in one of its three major mineral forms, lead sulfide, lead carbonate, or lead sulfate. Man has a great hunger for this lead. In the United States about 590,000 tons were mined in 1969, primarily in Missouri, Idaho, and Utah. Another 500,000 tons of lead scrap were recovered, usually from automobile storage batteries, and 250,000 tons were imported.

Lead ores are usually concentrated by a process called ore flotation and then sent to smelters and refineries to be converted into pig lead. In each part of the process, some lead is lost to the environment. It is spewed into the air through the giant stacks of the smelters or washed into rivers and streams adjacent to mining sites. Later it may be lost into the air, water, and soil in the course of any number of manufacturing processes that use or produce lead-containing materials.

Lead is also dispersed throughout the atmosphere in volcanic eruptions, in sea spray, and in the dust and smoke of occasional forest fires. Such natural sources of lead pollution existed long before man appeared, but it is now clear that the polluting effect of man's use of lead has far outstripped the natural background levels of lead in the world. C. C. Patterson of the California Institute of Technology measured lead levels in ice samples collected from Greenland, with the realization that the deeper the ice was the older it would be. In ice presumed to have

Courtesy, Dr. Clair C. Patterson

been formed about 500 B.C. he found approximately eight grams of lead per ton of ice. Obviously, this lead could only have been introduced by natural processes. The lead levels in ice samples formed in 1950 were found to have increased to 90 grams per ton and to have risen again to 210 grams per ton by 1965. Such sudden increases are hardly the result of natural phenomena. In another study, Henry A. Schroeder and his colleagues at the Dartmouth Medical School, Hanover, N.H., measured the lead levels in the rings of an elm tree growing about 150 feet from a lightly travelled street. They found that the lead content per million parts of wood had increased from about 0.16 in rings dating from the mid-19th century to 3.90 in rings formed in 1960.

As with most of the environmental contaminants that now appear to plague man and threaten his future, lead pollutes more or less in direct proportion to the amount being used, even though different forms of lead pollute in different ways and to different extents. The two major sources of current lead pollution problems are related to the rise of the automobile and the internal combustion engine. In the U.S., most lead goes into the production of storage batteries in automobiles. Lead consumption for this purpose tripled from 1934 to nearly 600 million tons in 1970. Although, fortunately, most of this lead is recycled to make new batteries, battery lead makes up about 10% of the lead now being added to the environment.

The second major use for lead, and by all odds the greatest contributor of lead to the atmosphere, is in tetraethyl lead (TEL), an especially poisonous organic chemical used since 1923 as an antiknock additive to upgrade gasoline. Use of TEL increased phenomenally when it was found that it could enhance gasoline octane sufficiently to power modern high-compression engines. In 1970, 275,506 tons of lead, or anywhere from 2.5 to 4.0 grams of TEL per gallon of high-test gasoline, were added to gasolines in the United States; 50 years ago, essentially no lead was used for this purpose. Unlike battery lead, the lead in TEL cannot be recycled, and is expelled into the atmosphere at the rate of 200,000 tons a year; the rest remains behind in the automobile's lubricating oil or exhaust system, where it prevents buildup of harmful iron oxide particles. In fact, lead from automobile exhausts accounts for 94.8% of the lead in the air we breathe, compared to 0.5% contributed by smelter stacks.

Toxicologists and pathologists are particularly concerned about the lead from automobile exhausts because it enters the air as extremely fine particles averaging 0.25 microns in diameter (one micron = 0.001 millimeter), which can be inhaled easily. When inhaled, about 45% of the lead from auto exhausts is absorbed into the lungs and then distributed throughout the rest of the body via the bloodstream; in contrast, the body retains only about 5–10% of the lead ingested with food.

### Dangers of lead in the atmosphere

Very little is yet known about the effects of low-level exposure to lead, the kind of exposure persons breathing urban air may be experiencing daily, and the kind that poses a long-range threat to man, since lead is known to accumulate in the body, frequently in the bones, with advancing age. There is some evidence that low levels of lead inhibit the formation of an important blood enzyme, delta-aminolevulinic acid dehydrase (ALA-dehydrase), which aids in the synthesis of hemoglobin, the oxygen-carrying molecule in red blood cells. Without better evidence of this or other ill effects of low-level exposure, it remains difficult to establish reliable safety standards for lead in the human body.

There is little doubt, however, that the automobile is adding heavily to the danger of widespread low-level intoxication. In some large U.S. cities, such as Los Angeles and New York, lead concentration in the air is 20 times greater than it is in sparsely populated areas, and 2,000 times greater than it is in the uncluttered atmosphere above the mid-Pacific. In a report in 1971, the National Research Council of the National Academy of Sciences warned that traffic policemen at busy intersections, gasoline service station attendants, mid-city dwellers, and, especially, young children living in cities are potential victims of lead poisoning from the air. Recent surveys of children, and, in some cases, adults, in several large cities have found 40–60 micrograms of lead per gram of blood. This is about half the level known to cause clinical lead poisoning in adults in many well-documented cases. It is also well documented that children are more susceptible to lead intoxication than adults.

### Getting the lead out

In 1970, gasoline and automobile manufacturers responded to the growing public clamor for controls on lead emissions. The gasoline manufacturers came out with low or no-lead grades of fuel, and the automobile makers redesigned the engines used in 1971 models so that they could handle the new gasolines. At the same time, federal, state, and municipal governments enacted legislation demanding decreases in lead levels to 1.7 grams per gallon or less by 1975. The automobile manufacturers were not reluctant to cooperate. They realized, among other factors, that if lead were still being used at 1970 levels in 1975 cars, the catalytic antismog devices they are compelled to develop by that date to meet federal standards will not operate properly.

However, the good and bad aspects of the new low-lead or no-lead gasolines were still being debated in 1972. The new gasolines usually have a lower octane rating (about 91) than leaded gasoline (about 100) and they do not perform well in older cars, some of which will still be on the road in 1980. (Octane rating is an arbitrary measure of a gasoline's resistance to knocking and indirectly of its performance.) If the gasoline industry were to produce lead-free gasolines that would meet the needs of older cars, these highly refined gasolines would contain more hydrocarbons of the olefin and aromatic type, which could cause more eye-smarting smog than leaded gasolines do. The industry estimates that such a conversion would cost it $4 billion–$5 billion and add 4–5 cents to the consumer's per gallon gasoline costs. Manufacturers of TEL (Ethyl Corp. and E. I. du Pont de Nemours & Co.) assert that valves wear out quickly without TEL. Others point out that when lead is removed, mufflers, rings, cylinders, and exhaust control devices work better and longer and result in overall gasoline cost reductions of nearly two cents a gallon.

Although smelters, ore refineries, and other lead-using industries are also responsible for the increase in environmental lead, most of these industries are close-mouthed about how much lead they discharge, how they control it, and how much their abatement measures cost. Currently they use a variety of filters, precipitators, separators, and scrubbers, some of which are as much as 99% effective. Actual performance, however, depends to a great extent on how seriously the individual plant takes its lead control responsibilities. As for lead pollution of water supplies, a recent report prepared by the University of Illinois notes that standard water treatment processes in use in the United States can keep lead in the country's drinking water at a safe level (less than 50 micrograms per liter).

Courtesy, Dr. J. Kobayashi, Okayama University

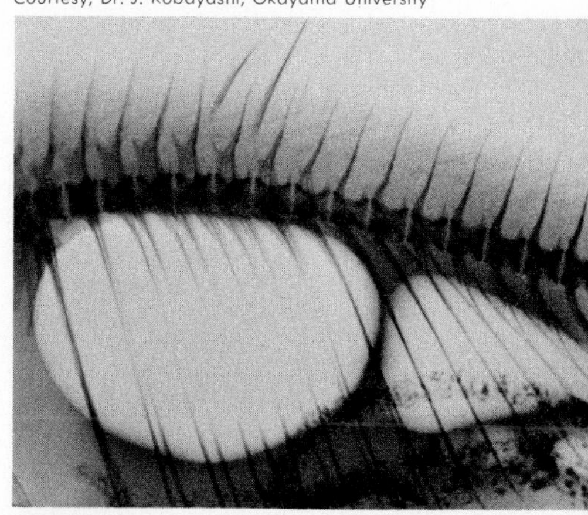

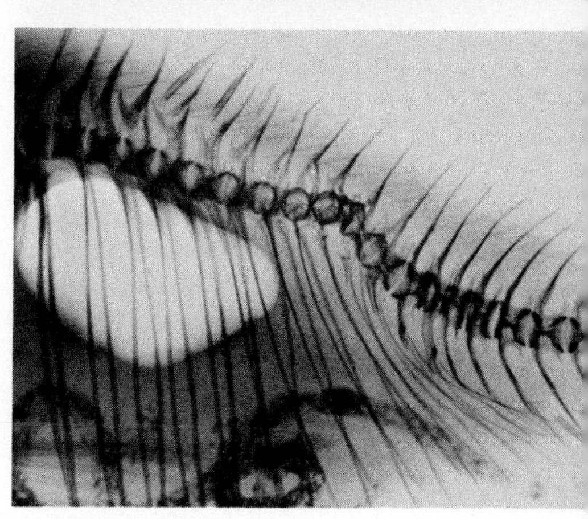

*Chronic cadmium poisoning was blamed for the contrast between the backbone of a normal young carp (top) and that of an abnormal one (bottom), as seen in X-ray photographs. The abnormal fish was raised in water containing a high concentration of cadmium. Its bones show the same damage noted among sufferers of* Itai-Itai *disease in Japan: a loss of calcium leaves the bones transparent and severely softened, so that they collapse under slight pressure.*

### Cadmium: the most lethal metal?

Of the metals that may pose threats to human health, cadmium can be singled out as the most lethal, even though its ill effects have not been studied as painstakingly or completely as those of lead or mercury. Cadmium's deadliness comes from its known ability to build up in the body, especially in the kidneys, and its link, at least statistically, to hypertension (high blood pressure) and heart disease. In animals, even small dosages have produced brain damage, birth defects, damage to the reproductive organs, blood abnormalities, and many other physiological disturbances. In man, inhalation of cadmium dust is known to have caused lung damage resembling emphysema. The Japanese Ministry of Health and Welfare has reported that since 1962 cadmium from mine wastes have contaminated food and water supplies in parts

Courtesy, Dr. J. Kobayashi, Okayama University

*The incidence of* Itai-Itai *disease was limited to inhabitants of low rice fields irrigated by a river carrying the waste waters of a mine producing zinc, lead, and cadmium. The farmers blamed the mine's wastes for damage to their rice crop (above) but did not connect the simultaneous appearance of disease among their families with their use of the river's water in their kitchens (opposite page). After 1955, when a dam began collecting the mine's wastes, crops returned to normal and the number of* Itai-Itai *cases dropped rapidly.*

of northern Japan and led to more than 220 cases of a severe degenerative bone disease, called *Itai-Itai,* or "It hurts! It hurts!" after the cries of patients suffering the severe pain that results. The disease has killed 50 people who were first exposed to cadmium discharges during or after World War II. The cause of the disease was not recognized until 1955 and, by that time, many sufferers had become hopeless cripples.

A study carried out for the National Air Pollution Control Administration (NAPCA) of the Environmental Protection Agency showed that about 13,328,000 pounds of cadmium were used in 1968 and that about 4,600,000 pounds of that were vented into the air. Consumption of cadmium and its discharge into the atmosphere can only have increased since 1968, since the industries that use it and its compounds have expanded. According to the NAPCA report most of the cadmium in the air is vented in the form of cadmium oxides and sulfides from smelters that refine cadmium-containing ores of lead, zinc, and copper. In 1968, about 2.1 million pounds of the cadmium delivered to the air came from ore processing and refining sources. Almost the same amount was produced as a result of the recycling of cadmium-plated iron and steel. Other sources of air-borne cadmium are scrapped automobile radiators, the incineration of plastic wastes, tire wear, combustion of motor oil, compounding of cadmium pigments and plastic stabilizers, preparation of cadmium alloys, and cadmium-containing fertilizers and pesticides. Cities closest to smelters and metallurgical industries usually bear the brunt of cadmium pollution of the air. Their water supplies are probably also polluted by cadmium, but no comprehensive measurements of this type have been made.

Toxicologists are concerned about cadmium as a hazard to man's health not just because of its well known effects on the kidneys and lungs but also because it appears to build up in the kidneys and other organs to concentrations equivalent to those known to shorten life and cause hypertension in rats. Dartmouth's Schroeder found more cause for alarm in the human situation when he measured higher levels of cadmium in the kidneys of patients who died from hypertension than in patients who died from cancer, hepatitis, or diabetes.

Because there is so little irrefutable evidence about the long-range effects of less than obviously toxic exposure to cadmium, health officials have been slow to set up standard levels for it in air, food, and water. A joint FDA/WHO commission has suggested but never established a maximum permissible level in food of 5 ppm (compared to 0.5 ppm for mercury), even though concerned scientists, notably Robert Nilsson of the Swedish Natural Science Research Council, insist that animal studies and the Japanese cadmium poisoning incidents make that level much too high. To date, the United States has established the maximum permissible amount of cadmium in drinking water as 0.01 ppm. Measurements of the metal's presence in rivers and reservoirs throughout the country carried out by the U.S. Geological Survey indicated that 33 out of 720 water samples tested were higher in cadmium than the prescribed level. The implications of these higher levels are not clear, but some scientists give them an ominous interpretation since cadmium has proved to be such a frightening poison.

### The metal crisis and the future

Discovery of the possible hazards evolving from the increasing amounts of mercury, lead, and cadmium in man's environment has triggered investigations of three other likely metallic hazards: nickel, zinc, and arsenic. Nickel, for example, has been found to have a proven ability to induce cancer. But, compared with what is known of the three notorious metals, very little has been established about any current or potential environmental impact of the others.

For mercury, if the most pessimistic predictions are correct, the potential for danger to mankind is growing every day. Even if the most effective methods for controlling further mercury discharges were put into practice overnight, millions of pounds of inorganic mercury would remain in lake and river sediments waiting to be converted to the lethal methyl form. An optimist might decide that the Minamata experience, the Swedish bird crisis, and the New Mexican disaster were only scattered incidents of carelessness and poor planning. If so, and if present levels of mercury contamination are not a long-range threat to the world's health, the optimist will have his day and so will mankind. The question is, how many are willing to gamble their futures on "ifs"?

The solution to the lead problem may be easier. Many authorities believe that if lead is removed from gasoline, the back of the lead pollution problem will be broken. Automobile owners may then have to settle for less powerful cars, but may gain longer lives in exchange.

Cadmium may be the most lethal member of the three notorious metals, but, for the most part, its environmental impact is unknown and befogged by inconclusive data. Perhaps the most that can be said about it at this time is that the enormous amounts of such a deadly metal that are now in the air will probably not improve anyone's health.

Confronted as the deadly trio they are already known to be, mercury, lead, and cadmium represent but one example of how man has altered his relationships with the earth and indirectly impaired his health and well-being. Once, each of the three metals occupied its niche in the balance of nature. Man removed them from the earth in great volume and distributed them unthinkingly into parts of the physical environment where they never really belonged. This may represent a small aspect of the current ecological crisis, but in the case of these three metals, it is certainly an ominous one.

FOR ADDITIONAL READING:

Chisolm, J. Julian, Jr., "Lead Poisoning," *Scientific American* (February 1971).

*Environment*, issue on mercury (May 1969); issue on mercury (May 1971); issue on lead (June 1971).

McCaull, J., "Building a Shorter Life," *Environment* (September 1971).

Schroeder, H. A., "Metals in the Air," *Environment* (October 1971).

# Crime Fighters in the Laboratory

by Alexander Joseph

**Backing up the modern detective is an armory of advanced scientific techniques undreamed of by Arthur Conan Doyle in his wildest flights of fancy.**

> Detection is, or ought to be, an exact science, and should be treated in the same cold and unemotional manner.
> —Sherlock Holmes in *The Sign of Four*

The potential of science and technology in law enforcement was first recognized almost a century ago. Institutionally, it was introduced through the Assay Office of Scotland Yard. Dramatically, it was introduced through Sherlock Holmes in the narratives of Sir Arthur Conan Doyle. By the turn of the century, Holmes was triumphantly the image of the scientific detective—the cerebral man instead of the merely muscular one. In the real world, the Assay Office set the precedent for police departments throughout the world, which were to move from fingerprint identification—then a radically new procedure—and paraffin tests to emission spectrography, analytical gas chromatography, and X-ray diffraction. The "basic" police-lab sciences of chemistry, biology, and physics quickly became allied with the technologies of electronics, wide-spectrum photography, and light- and sound-wave mechanics.

In the last few years, the style, tone, and range of the science and technology of law enforcement have been given an enormous new thrust and a vast new urgency. These were generated in part by appropriations of the U.S. Congress, made in the 1960s to assist further and deeper research into the subject. They were complemented by the marketing mechanisms of U.S. private industry, which developed arcane devices in the 1960s—mostly for use in the Vietnam war—and was later impelled to seek alternative outlets for them.

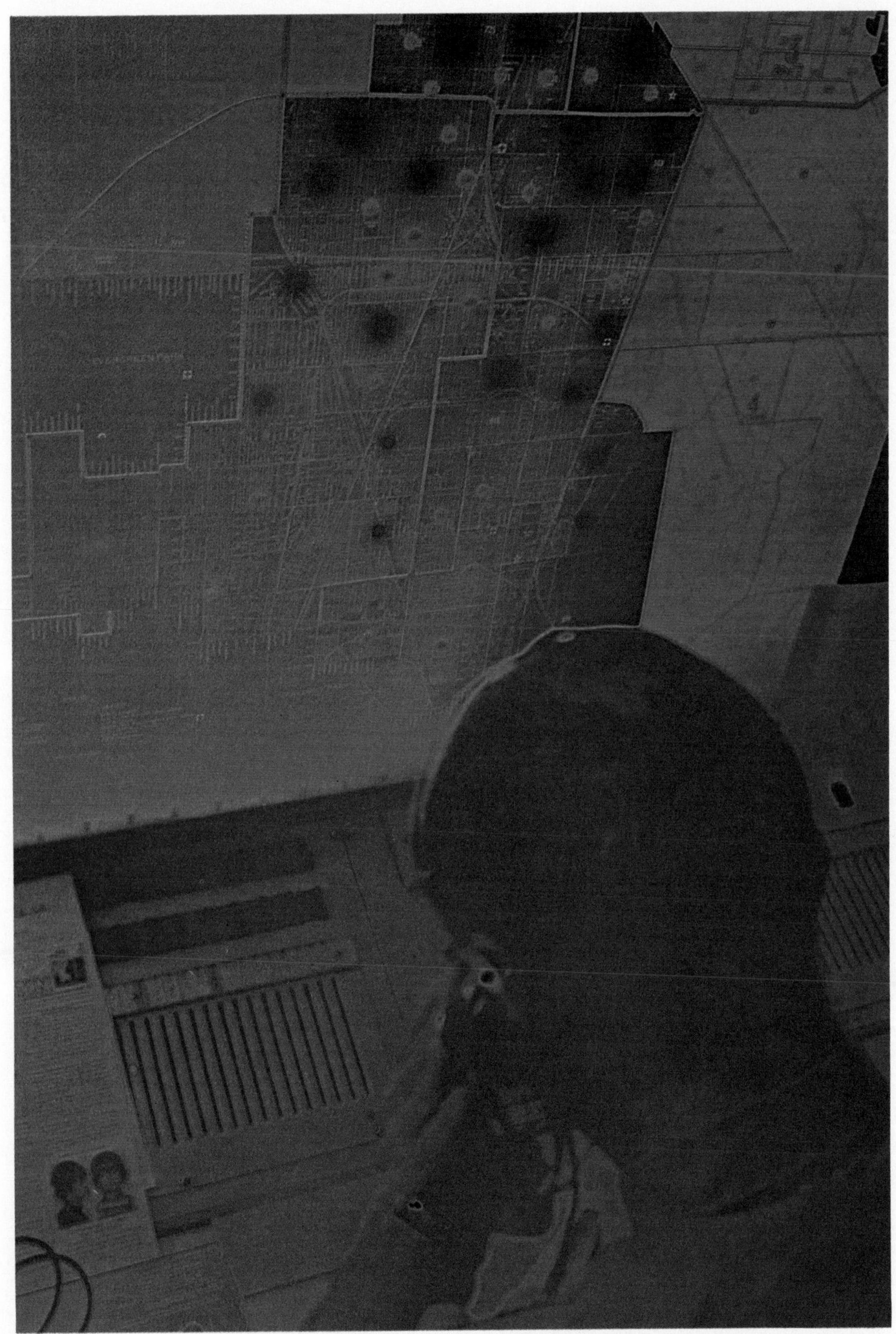

Photos, Dan Morrill, courtesy, Chicago Police Department

Here, for example, are some of the devices and procedures that emerged from the Vietnam war to become part of the science and technology of law enforcement:

• Unmanned, remote-controlled aircraft equipped to pick up signals from ground sensors are being used by the U.S. Border Patrol to fly over isolated stretches of the Mexican-U.S. border. The system was initially designed for an anti-infiltration program in Indochina. The drone picks up signals from hundreds of sensors, adapted from devices used to detect the sounds and vibrations of troops and trucks moving along the Ho Chi Minh Trail. In this way, an enormous stretch of isolated border can be patrolled, without human personnel, in a very short time.

Moreover, the information thus gathered can be relayed instantly to a computer that will sift and collate it. In the past, it took days, weeks, or even months before human patrolmen scouting the same area could get their reports analyzed to see if there was a recognizable pattern of illicit traffic across the border. Now the same work can be done in microseconds. Border patrolmen can be dispatched to a scene while suspected offenders are still in the vicinity. There is one difficulty, however: the system detects movements, but it does not say who, or what, caused them. A wandering burro is as likely as a drug smuggler to send officers of the U.S. Border Patrol rushing to their jeeps.

• A "no-light" technology allows police officers to see suspects and other persons with great clarity at night, no matter how dim the light. The technology is based on a method of amplifying light levels as much as 40,000 times; even the light from a single star can be amplified satisfactorily for use in this system. Perfected to meet the need of the U.S. military for detecting night-fighting guerrillas in Vietnam, the "snooperscope" is now used by some big-city police departments against snipers and in other instances where night vision is required.

A more general adaptation of the technology involves the use of either still photography or closed-circuit television. The latter permits constant surveillance of many locations at night, even from long distances. Such a system is being used to scan the streets of Mount Vernon, N.Y. Its designers say that it is capable of discerning a man-sized object in extreme darkness from more than half a mile away. Thus, it holds promise of relieving the enduring problem of police patrols—that a crime may take place unseen at a location far from the area the squad car is patrolling.

• A "black box" device has been developed that gives radar an X-ray quality, permitting it to suggest who, or what, is behind walls. (Normally, the radar signal bounces back as soon as it encounters an obstruction.) The device, about the size of a cigar box and weighing less than 10 pounds, was originally conceived as a foliage-penetration surveillance system for use in Vietnam. In law enforcement, it can help to determine nighttime occupancy of supposedly "vacant" buildings —such as banks or warehouses—or the potential for ambush in a building that must be entered by police officers.

*__ALEXANDER JOSEPH__ is a professor and director of the Forensic Science Program at the John Jay College of Criminal Justice at the City University of New York.*

To be sure, not all of the new law-enforcement technology is the product of the Vietnam war. For years an enormous stockpile of technological know-how has been building up that is just beginning to be applied to police work. It ranges from aerosol clouds that reflect light from searchlights and can be used to illuminate large areas at night to agents and devices, such as the chemical Mace and rubber bullets, that incapacitate but do not kill, to helicopters.

Some of the technologies have become established parts of the police armamentarium. Others are still controversial. Taken together they are revolutionizing police work.

### Voiceprints: identification through the spoken word

According to the developers of the voiceprint system, no two persons have the same resonant cavities; thus, the voice pattern of each person is highly individual and can be used to identify the person in much the same way that fingerprints are used. Employing a machine called the sound spectrograph, researchers can graph the frequency, loudness, and other characteristics of the human voice. The graph—the voiceprint—is said to reveal the distinctive characteristics of each voice. The status of the voiceprint as legal evidence is still a matter of dispute, although it has been admitted as evidence in some cases. Its potential importance is indicated by those cases in which it has played a part.

One spring night in 1970, for example, police headquarters in St. Paul, Minn., received an anonymous phone call seeking help at a certain address for a woman who was about to give birth. A squad car was rushed to the address, but the policemen found only a darkened building. As they explored the area, a sniper suddenly opened fire and one of the policemen was shot and mortally wounded.

The only clue was the tape recording of the original call for help. The police kept that tape, developed its voiceprint, then began interviewing a number of women who lived in the neighborhood of the shooting. They recorded each response in an effort to get a voiceprint that matched the one from the phone call. Among the first 13 women interviewed, they found a suspect: an 18-year-old girl whose voiceprint exactly matched that on the tape recording. She was subsequently indicted and brought to trial for first-degree murder.

Another case in which a voiceprint proved crucial began as a civil suit. A businessman was sued for breach of contract, and the fundamental evidence against him consisted of an ordinary tape recording. The defendant asserted that he had never made any of the statements or promises that were recorded on the tape. This suggested that the man was lying or that the voice on the tape was an impersonation.

To determine the truth, the defendant was asked to speak into the microphone of a voiceprinter. He agreed readily. The voiceprinter showed his voice pattern was clearly different from that on the original tape, even though the words used in both cases were exactly the same. The result: the civil suit against the defendant was dismissed and a charge of fraud was entered against his accusers.

### X rays leave the laboratory

Improving technology has made it possible to devise portable, low-voltage X-ray equipment that can be taken into the field. One such system, packed into a small box—about the size of an old-fashioned $4 \times 5$ inch press camera—and operated by a low-voltage battery, is virtually revolutionizing the technique of "lifting" latent fingerprints. These are prints taken from a surface that a person has touched. They are called latent because, unlike inked fingerprints taken directly from a person, they are not visible to the naked eye and must be "developed" by various scientific techniques. Historically, the methods used in developing latent prints involved the use of various powders or of chemicals such as iodine, silver nitrate, or Ninhydrin. The prints, when discovered, would be "lifted" off the surface, usually by applying a sticky tape to the chemically treated print. At the police laboratory, a photograph would be made of the tape and the film developed and delivered to a fingerprint division for identification.

This procedure had frustrating limits, however. Prints could not be developed satisfactorily after even a relatively short time had passed. Nor could they be lifted easily off all surfaces. Often the tape could not be applied to—or pulled off of—surfaces with a rough texture or a tendency to adhere or tear when the tape was lifted, such as clothing, paper, or skin. Thus, seven decades after fingerprint identification came into wide use, police departments still relied on prints left on smooth, hard surfaces: glass, polished wood, metal surfaces, and so forth. Not too long ago, New York City police-lab technicians were saying that they got usable fingerprints in only 10% of cases.

The X-ray system has helped overcome these difficulties. It was originally developed in a scientific laboratory in Glasgow, Scot., using full-size X-ray equipment, but recent modifications have brought it into play literally at the scene of the crime. It represents an enormous advance over the older systems. It can be used to pick up latent prints that are many months old. It can develop them from clothing, money, paper documents, "soft" curved surfaces, even the skin of dead bodies. (It was, in fact, used in this way by the Royal Canadian Mounted Police to help solve a quintuple murder in 1969.) And it can produce a usable image of a print within a minute or two of discovery.

The X-ray system involves a "dusting" technique, as in the older system, but the "dust" is made from a dense metal that shows up clearly on X-ray film. The metal is ground down to an extremely fine powder that tends to arrange itself in the pattern of the print. A quick-developing film, similar to that used in home Polaroid cameras, is placed behind the surface that contains the print. The X-ray machine is brought up to the dusted print, the tube is focused on it, the machine is switched on briefly, and an image of the fingerprint is imprinted through the texture onto the Polaroid film mounted behind it. The film is then developed on the spot; in a minute or two the investigators have in their possession a clear image of a fingerprint and can start checking out the identity of the victim or potential suspect.

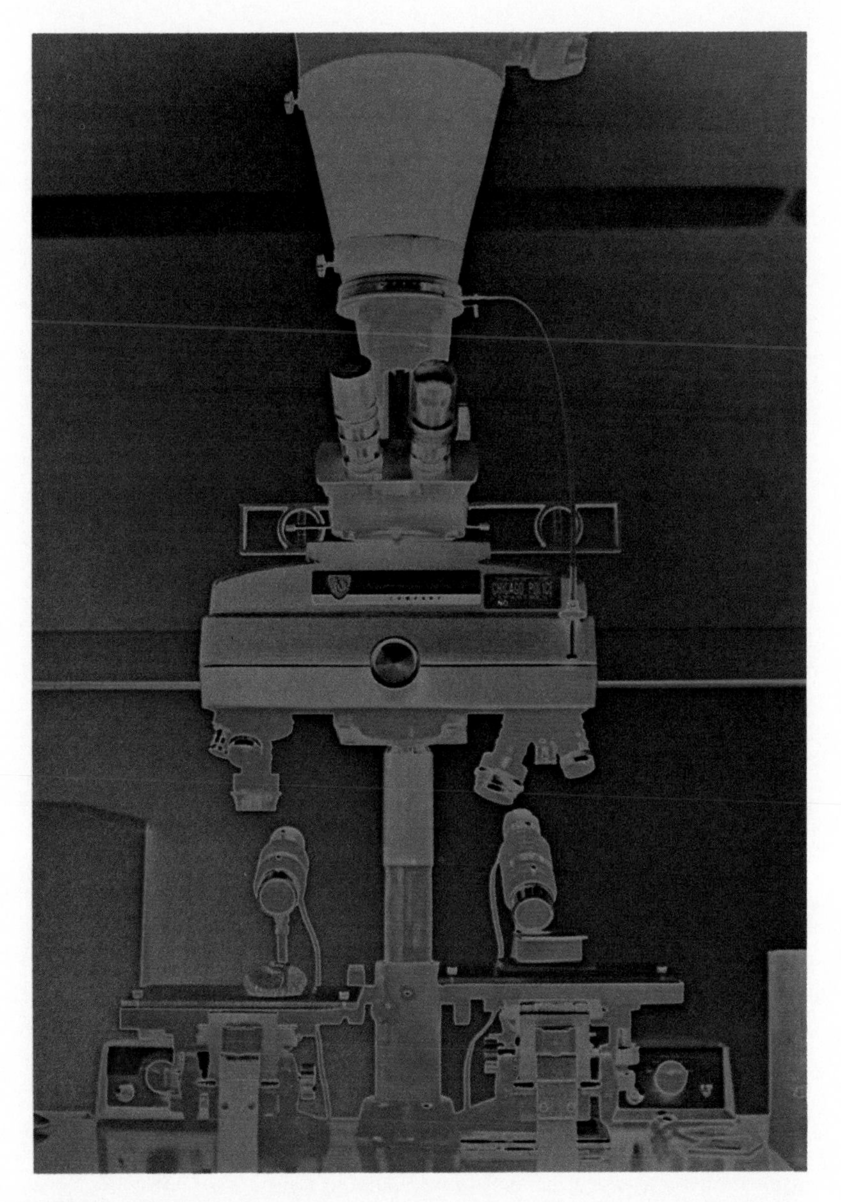

Low-voltage equipment is used in this operation because high-voltage—and more powerful—X-ray units cannot be used casually in areas of high-density human traffic. Thus, magnetometers rather than X rays are used in screening baggage and passengers at airline terminals; otherwise, persons who took a plane more than once or twice a year might receive an overdose of radiation. Nevertheless, portable high-voltage systems are useful in some situations. They make it possible to examine, in isolated areas, those items that cannot possibly be brought to an X-ray laboratory.

*The microscope is a versatile instrument in the hands of the police lab technician. Through comparison of toolmarks, he can identify the tools used in the commission of such crimes as breaking and entering. Another use is in the detection of obliterated gun numbers through microscopic examination of the chemically treated surface.*

For example, in the solitude of the night, U.S. customs agents brought a portable high-voltage X-ray tube to a deserted dockside building in New York City to examine a shipment of canned fish from Spain. Customs had been alerted that a company not normally engaged in canning or fishing was suddenly buying fish cans and—curiously—

lead fishing sinkers. Quick diversification is not a characteristic of most European corporations, so when a shipment from this company arrived in New York, customs officers entered the dockside warehouse where it was stored and began to examine it.

First they weighed sample cans: the weight was exactly as indicated on the label. Then they brought in the X-ray unit and X-rayed a number of cans. Many of the cans did contain fish, but the X rays also showed that some of them contained lead sinkers. When those cans were opened, the customs officers found that they contained—in addition to the lead sinkers—a great deal of heroin. Heroin is lighter than fish, but the combination of heroin and lead sinkers gave the cans a total weight exactly equal to the listed weight of the fish. The discovery led to the seizure of a multimillion dollar cache of heroin and provided an insight into the methods of drug traffickers.

*Fingerprints obtained in the course of an investigation can be transmitted over telephone lines from a precinct station to the photofacsimile recorder (above) at Chicago police headquarters. The recorder produces a photographic copy that is then used to search the records. The computer (opposite page) in the police department's Data Systems Division can provide, almost instantaneously, information ranging from the license numbers of stolen cars to the arrest record of a suspect to overall crime statistics.*

### Still photography and videotapes

That photography is now being used effectively in solving bank robberies—and preventing them, by threatening certain identification—is due directly to advances in film technology. Until fairly recently, this use of photography was impractical because of the low light level in most banks. However, improvements in film speed and wide-aperture lens settings now permit clear photographs to be taken indoors. Indeed, the sharp increase in arrest and conviction rates of bank robbers identified through such photographs may have brought about a change in the robbers' methods of operation. The older bank robbers—who would go to prison for life if they were caught—do not appear on the film. Some law-enforcement officers believe they now act as teachers of apprentice thieves who take the risk and split the loot with the older men, especially since young criminals are turning up with exactly the same techniques long used by the older criminals.

Videotaping has also changed certain criminal methods of operation, particularly when it is combined with other advanced technologies, such as helicopters. For decades, thieves being hemmed in by police approaching at ground level would flee to safety over the roof tops. Now they can be spotted from helicopters and followed across the roofs. They can also be photographed, sometimes from great distances, and this videotaped record constitutes irrefutable proof that the accused man was at the scene of the crime.

One night a year or so ago, a thief who invaded a medical center in Kansas City, Kan., was the target of such technology. The report of a potential threat was radioed both to a patrol car and to a helicopter. The helicopter got there first, with a videotape system geared for night photography ready and running. The camera was on when the suspect appeared on the roof, fled across it, and descended to the ground via a fire escape on a nearby building. The helicopter followed the suspect, now in his car, while radioing reports of his position to the ground patrol. Meanwhile, the cameras recorded his every move, right up to the time of his interception and arrest. There was no alibi he could offer that could contest the videotape record of his flight from the scene of the crime.

It is this element of proof that gives heightened significance to videotaping, even in the most unglamorous areas of law enforcement. Thus the sheriff's police of Bernalillo County, New Mexico—where Albuquerque is located—mounted videotape cameras and recording units in two squad cars used principally in patrolling county roads and in the arrest of drunk drivers. When police officers see a car weaving down the road, they start to take a videotape recording of the car and its movements. When the car is curbed by the police and the driver steps out, his every move is also recorded. If he is falling-down drunk, if he is staggering, if he comes out swinging, the camera has it all on tape. It also has the driver's performance in the usual roadside tests for drunkenness—walking a straight line, picking up coins off the road, touching a finger to the nose.

The camera provides an authentic, irrefutable record of how the driver was behaving and performing at the time he was driving the car, not how he performed 20 or 40 or 60 minutes later when he was booked in the police station. Under the old system, which relied on a blood-alcohol test performed after the suspect was brought in for arrest and booking, the local prosecutor was getting convictions in 66% of drunk-driving cases. After the videotape system was introduced in July 1970, he got 100% convictions in the first 130 cases brought to court.

### Enter the computer

The principal use of computers in law enforcement today is for information retrieval. More sophisticated uses of the computer have been slow in coming, but the speed and thoroughness of the information-retrieval program have been immensely gratifying. In fingerprint identification, for example, one reel of magnetic computer tape can hold 5,000,000 frames of fingerprints and a computer can scan those images at the rate of 2,000 a second. Thus, it is possible to make a fingerprint search of dimensions that would have been all but unthinkable before computers came into use.

When a townhouse on West 11th Street in Manhattan blew up in March 1970, for example, the numerical possibilities in fingerprint identification were astronomical. The investigators did not know how many people had been in the house at the time of the explosion. They did not know how many bodies there were; indeed, the blast had been so violent that they could not find enough remains from any one person to make an identification, much less a "body." They did find a number of finger fragments, but putting them together in the proper combination needed for identification, using pre-computer methods, might have been a job covering several lifetimes. High-speed computers were able to sift through all the possible combinations and produce three positive identifications in a matter of a day or so. At the same time, using an older technology—X-ray diffraction—police were able to analyze traces of dynamite found in the wreckage and compare them with batches from manufacturers' production lines until they got a match. Through sales records, they were able to trace the dynamite to a certain shop in New England, and, by using descriptions developed after fingerprint identification, tie it directly to members of the Weatherman faction of the Students for a Democratic Society who had been using the townhouse as a bomb factory.

The very speed of the computer system has made certain aspects of police work somewhat safer than in the past. Consider the officer cruising a road or street in a police car. In the past, every car that was stopped by the police represented an exceedingly high risk. The police officers simply did not know—unless the suspect came out shooting—whether the car's occupant was dangerous. Today, the police officer can check out the car before he stops it. All he needs to do is radio the license number to headquarters. It can relay the request to a computer where stolen-car records are kept; the computer can scan its records in a matter of microseconds and relay that information back via telephone wires to headquarters, where it is either printed out on a teleprinter or flashed across a television-like screen. Within a minute or so, the police officer on the road can be informed, via his patrol car radio, whether the car is stolen or whether it has been used in the commission of a felony anywhere in the country.

Once the car is stopped, the officer can check the driver's credentials with the computer to find out whether he is wanted or considered to be dangerous. In the past, an officer who suspected such a person had to take him to headquarters. Requests for information about him had to be made from headquarters to a record center by phone, the search had to be made by hand, and frequently the answer was sent by mail. Inevitably, some innocent persons were accused and jailed. In other cases, police officers did nothing for fear of harassing the innocent, and a certain number of wanted and dangerous persons were stopped, then allowed to continue on their way.

But while the most widespread use of the computer in law enforcement is for information retrieval, there are many other possibilities, some of which are only beginning to be investigated. With the aid of a computer, patrol cars can be distributed in such a way as to minimize the elapsed time between receipt of a report of a crime in progress and the arrival of a patrol car at the scene. Computers can be used to aid police in guiding the flow of traffic: changing street lights according to the traffic approaching from various directions; regulating the flow from entrance ramps onto major expressways. In San Francisco a computer was used in a test to collect tolls on the Golden Gate Bridge from certain vehicles, such as buses. In this experiment, the buses were fitted with transponders, radio-crystal transistor devices that can be set up to transmit a code number when activated by another signal. Instead of stopping to pay the toll, the buses cruised through a special lane where the code number was picked up via the transponder and then relayed to a computer that automatically added the toll to the bus company's bill.

## The computer turns highway patrolman

Perhaps the most innovative and farsighted use of the computer in traffic control involves a test being run in Monroe County, Indiana. It was conceived by Kent Joscelyn of the Department of Police Administration at Indiana University, and was intended as a research project that would indicate what drivers do under changing conditions—as dusk comes, as rain or snow falls, during a holiday weekend, as accidents pile up. Initially, some 512 electronic sensors were placed in slots one or two inches wide in a 14.8-mile stretch of Indiana's Highway 37 north and south of Bloomington. Later, all the major roads in the county were seeded with sensors, and the test was extended to provide vital information to the Monroe County sheriff's police.

*At the Chicago Police Department's highly sophisticated communications center, dispatchers seated before computer consoles receive complaints from citizens calling the police emergency number and coordinate police responses. The maps above each console show the boundaries and numbers of patrol car beats and which patrol cars are available.*

The sensors are connected via the regular telephone system to a computer on the Bloomington campus of Indiana University. Each sensor picks up what every car does as it passes over it—as many as 10,000 cars a day—and all of the information on all of the cars passing over all of the sensors in any one instant is relayed to the computer. In millionths of a second the computer assembles, analyzes, and prints out this information. The computer can provide patterns for all of the cars at any one time, or it can provide information on any one car all the time—how long it is, how fast it is going, whether it is speeding up or slowing down and by how much, whether it is weaving from one lane to another. It can spot a Volkswagen playing fendertag with a Greyhound bus at 60 miles per hour. And it can do more: "We can watch a car's behavior and literally *see* it head into an accident," said Joscelyn. This kind of information is vital to police regulation of the highway. Thus forewarned, the police could radio a patrol car or a police helicopter to try and stop the accident before it takes place.

More importantly—if less dramatically—a police computer traffic center could see certain difficulties arising before they reach the critical stage. Even in the present state of the art, it could provide the kind of overall surveillance hitherto reserved for the gigantic TV eye-in-the-sky beloved of science-fiction writers. In certain cases it could do better. Clouds and swirling snow might prevent the TV eye from discerning just how much ice was accumulating in a certain spot, but the computer-linked sensors can detect the very start of ice or snow accumulation. This could be enormously useful to workers in the highway department, since it would enable them to deploy their salt trucks and snowplows with maximum efficiency.

The computer could even help the driver to help the police in regulating traffic. Today's driver moves in an envelope of ignorance when he is on the highway. He knows only what is happening to his car and to those just behind and just in front of him. There could be an accident anywhere from 400 yards to 40 miles ahead and he will continue toward it until he is trapped in a traffic jam backed up behind it. Yet, in many cases, if he had known about the accident he could have taken an alternate route. Police helicopter patrols that relay information on traffic conditions through radio stations are designed to assist such

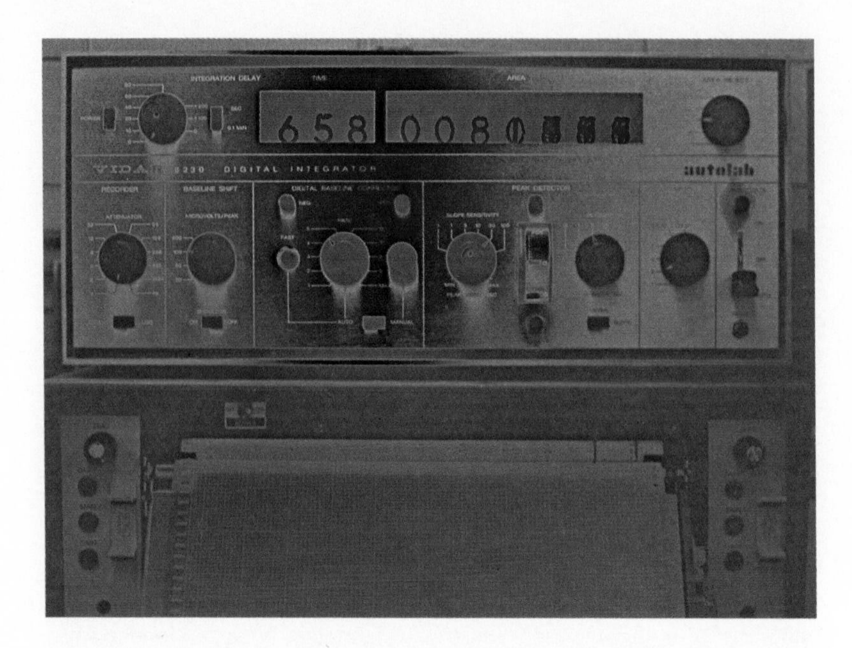


*The gas chromatograph (right) is based on the principle that different complex molecules move at different speeds. Among its uses in police work are the analysis of trace amounts of materials that may constitute evidence, the separation and purification of such substances as narcotic mixtures and volatile solvents suspected of use in arson cases, and the exact determination of the amount of alcohol in the blood of an arrested person. The ultraviolet spectrophotometer (opposite page) has been used to identify certain types of drugs and to identify substances of suspicious origin in assault and homicide cases.*

drivers, but the driver must have a radio, must have it on, and must be within range of the station for this service to be effective. Joscelyn suggests that it would be simple to relay information from the computer to every driver on the road simply by mounting electric signs along the highway. Because of its speed, the computer could flash a trouble signal immediately, while there was still time for the driver to take effective action. Furthermore, this would allow the police to use their energies in the most efficient way: in clearing up the accident, helping its victims, and guiding traffic back onto the main road beyond the accident site. Joscelyn estimates that such a system would cost $3,000 per mile of interstate highway—a negligible sum when compared with the million-dollar-per-mile cost of the superhighways themselves.

### Science wins its badge

The significance of the Indiana experiment lies not merely in the potential of computer-surveyed traffic but in the willingness of a relatively small sheriff's police department to recognize that potential. This represents a notable change in attitude, for the acceptance of science and technology even by big-city police forces in America was slow in coming. At the time of the St. Valentine's Day massacre in 1929, the Chicago Police Department did not even attempt to identify the slugs found in the seven victims. But once the notion caught hold, the most exotic of scientific techniques became routine: the analytical gas chromatograph is used in many police departments for determining the amount of alcohol in the blood of a person accused of drunk driving; infrared and ultraviolet spectrophotometers are used to determine the presence of some narcotics.

More important, in cases that are not routine, the scientific approach can be conclusive. In New York City several years ago, for example, an

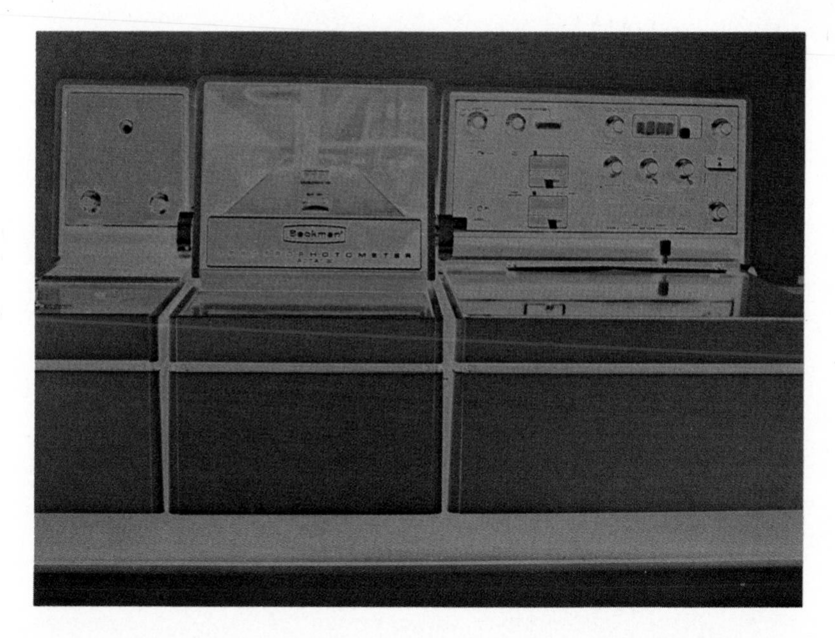


elderly woman was robbed and raped in her home. Her assailant had invaded the home while she was polishing silver, and in the struggle a smudge of the polish came off on his clothing. The police laboratory used X-ray diffraction to determine the crystalline structure of the polish the woman had been using. When a suspect was brought in, his clothing was analyzed by, among other things, X-ray diffraction, and a smudge on it turned out to have exactly the same crystalline structure as the silver polish. Confronted with this evidence, he confessed.

In another case, in a small mountain town in West Virginia in the mid-1960s, a woman we will call Vera Monahan was gradually poisoning her husband with arsenic. Arsenic poisoning is a slow process, however, and one day shortly before Christmas she lost patience and strangled him. Then she burned his remains for several days in a kerosene-fed coal-and-wood fire not far from their home. Eventually the man was missed, his disappearance was investigated, and Mrs. Monahan was arrested on suspicion, but the police lacked two things: a body, and proof that the murder was premeditated. They did suspect where the body had been burned, so they collected several hundred pounds of dirt and ashes from the site and shipped them to the FBI laboratories in Washington, D.C. There, scientists sifted through the debris and collected enough bone and tissue fragments—almost invisible to the naked eye—to fill a quart jar. With the aid of an anthropologist from the Smithsonian Institution, they were able to identify these remains as coming from a human male of the age and build of Monahan. That provided evidence of a body.

Next came the matter of premeditation. To prove this, the laboratory scientists subjected the remains to neutron activation, making the bone and tissue fragments radioactive so that they gave off gamma rays. The various elements present in any radioactive material can be identified precisely by measuring the gamma rays it emits. In this case, the measurement showed an excessive amount of arsenic. The police were able to trace arsenic purchases in the region of the tiny town to Mrs. Monahan, and she was subsequently convicted of first-degree murder.

## Angora fiber and blood alcohol

In a lonely spot in San Francisco one night in the 1960s, a man rose suddenly from the back seat of a parked car occupied by a young couple. He bound, gagged, and blindfolded the young man—a physical education instructor in a local high school—ordered the young woman into the back seat, and then proceeded to abuse her sexually for several hours. Later, neither the young man nor the victim could offer any description of the assailant, but certain aspects of the crime suggested that it had been carefully planned. The assailant carried with him a kit of ropes, chains, and sadistic devices that he was careful to clean up and carry away as he left. The character of the crime and the particular abuse he had heaped on the young woman suggested the working out of a profoundly macabre sadomasochistic drive. Lacking a distinctive description, the police searched through their files of sexual deviates for a man with this particular type of mentality. As it happened, they came up with the wrong man, but justice—and science—was working for him. Using an infrared spectrophotometer, experts in the police lab had found almost microscopic evidence of angora fiber on the victim's bra. At first the investigators hoped the angora had come from the clothing of the assailant, providing irrefutable proof of his presence at the scene of the crime and perhaps a clue to his character as well. Somewhat to their chagrin, it was found that the young woman had been wearing an angora sweater.

*Specimens too minute to be identified with the naked eye may provide important clues to the solution of a crime. Shown as they appear under the microscope are (top to bottom) asbestos, mink, and wool and (opposite page, top to bottom) hemp, kapok, and dog hair.*

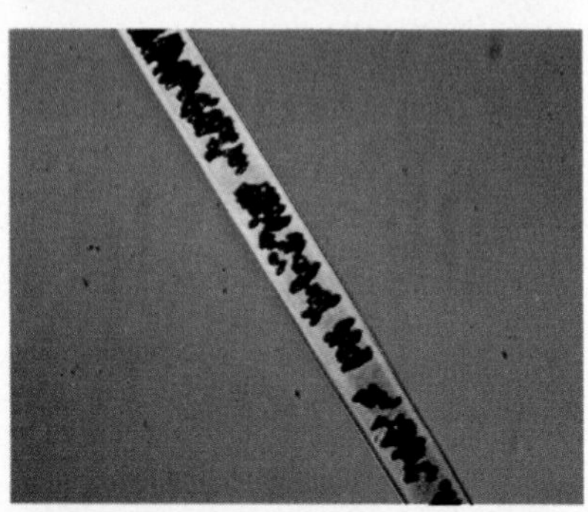

Nevertheless, the discovery was to prove critical. Angora sheds easily, and during the assault it could easily have shed all over the assailant's clothing. But even before investigators could find and test the clothing of the suspect, the laboratory men made an astonishing discovery: under routine infrared spectroscopy, the clothing of another man, arrested for a nonviolent crime, proved to have traces of the same angora fiber as that on the victim's bra.

At first, the police could hardly believe the evidence, for the new suspect—who by some ironic coincidence had been lodged in the cell next to the original suspect—was not quite five feet tall. This fact alone would establish a powerful defense for him. On the face of it, it seemed unlikely that such a tiny man could subdue a healthy, muscular physical education instructor almost a foot taller than he, or that a woman could spend so much time in intimate contact with him without noticing his size. Yet the evidence provided by the laboratory, combined with the further evidence found in his home—including chains used in the assault—proved conclusive. The old suspect was released, and the new one was charged and ultimately convicted of the crime.

This was not the first case in which scientific evidence has been used to free the innocent as well as to convict the guilty. Take the case of a driver in New York City who was taken into custody after his car, weaving back and forth on a crowded city street, sideswiped three cars and finally crashed into a stop sign. Two persons were known to have been in the car, but the passenger escaped before the police arrived. The driver was found slumped in the wreckage. The car reeked of alcohol, and an open bottle of liquor was dribbling the last of its contents into the smashed front seat. The driver was rushed to a hospital and, as a matter of routine, his blood was tested via the analytical gas chromatograph. It showed absolutely no alcohol in his bloodstream. The police wanted another test, but the hospital personnel intervened: their patient had suffered a stroke, which, of course, accounted for his erratic driving. Later, the police discovered that it was the passenger who had been drinking and who had left the open liquor bottle in the wreckage.

A somewhat similar case occurred in Baltimore, Md., where a man was arrested for drunk driving early one morning as he headed toward the Sparrows Point steel mills. When he was stopped by the police, he showed all the signs of drunkenness: slurred speech, very poor coordination, even a strange smell on his breath. But when he was taken to headquarters and given—with his permission—a test for blood-alcohol content, it proved to be negative. Unconvinced, the police made another test, but when it also turned out to be negative they began to investigate further and thus averted a possible tragedy.

The suspect was a diabetic who had been afraid to reveal his illness for fear he would lose his job in the steel mills. That morning he had taken his prescribed dose of insulin but—finding that he was late—he did not eat as he was supposed to do. While he was driving to work, an insulin reaction set in. The symptoms of approaching insulin shock—loss of motor control, nervousness, fear, a flushed face, sweating, and the vague smell of acetone on the breath—are similar to the signs of drunkenness. If the test for blood-alcohol content had not been easily available, he might have been thrown into a cell to "sleep it off," sunk deeply into insulin shock, and died.

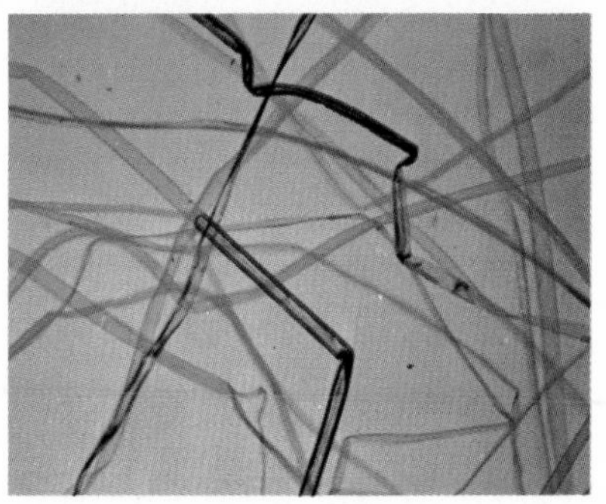

### Overcoming time and space

Some persons—bemused, perhaps, by fictional tales of scientific detection—believe that science and technology have made shoe-leather detective work dispensable. This is not the case: the hard, grinding work of gathering facts and details is more vital than ever. But science and technology have expanded the horizons of detection—and of justice—to an extent that even Sherlock Holmes could never have imagined. Consider what they have made possible.

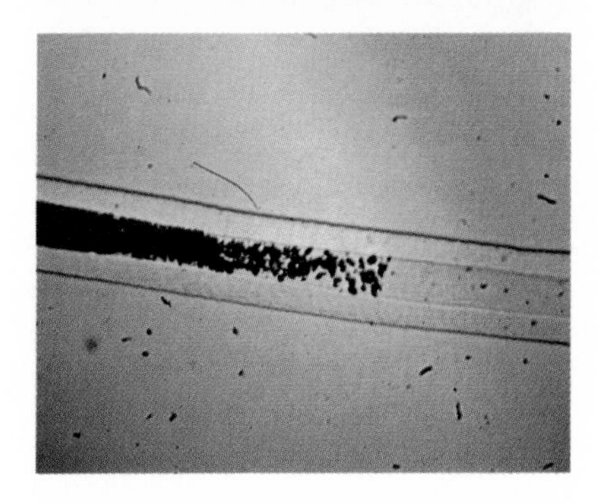

• Overcoming remote evidence: In 1960, the automobile of a suspect in the kidnap-murder of a wealthy brewer, Adolph Coors III, was found rusting in a gutter near Atlantic City, N.J., some 1,800 miles from the scene of the crime in Colorado. Hardened mud scraped from beneath the fenders of the car was subjected to emission spectrography, in which a substance is vaporized and the light-spectrum of the gas it emits is analyzed. After comparison with 421 other soil samples, the mud from the car was found to match the soil from the area in which

## A Gateway to the Future

# The World's Largest Nuclear Prober

by Howard J. Lewis

**Finding the tiniest particle, the most fundamental constituent of all matter, is the goal of the scientists at the U.S. National Accelerator Laboratory, the world's most powerful particle accelerator.**

Photos, courtesy, National Accelerator Laboratory

*Pre-accelerator generates the protons later used in the other accelerators by stripping electrons from hydrogen atoms. Here the protons get their first energy boost, to 750,000 electron volts.*

A billion electron volts is not a great deal of energy. Not when you consider that the equivalent of one billion billion billion electron volts is required to keep a 20-watt light bulb burning for an hour. On the other hand, if you can apply that much energy to particles of matter so small that ten thousand billion of them could crowd into a linear inch, you can give them quite a wallop. There's a problem in doing it, but it's a problem well worth solving if you wish to discover the most fundamental secret of nature—how matter is arranged and held together by a variety of powerful forces we are only dimly beginning to understand.

This problem—of imparting enough energy to atomic particles to break them apart—has occupied scientists since the early years of this century. And since the middle 1940s, when an explosion over the New Mexican desert demonstrated that man could extract the enormous energy locked in the nucleus of the atom, it has also interested governments.

## Pioneering work

Let us start at the beginning. From the work of several French physicists, Henri Becquerel and Pierre and Marie Curie, it was discovered that radium, one of the heavier elements, emitted three different forms of radiation. Two of them could be deflected by magnetic fields, and from this it was determined that one of those two consisted of particles with a positive charge, the other negative. In 1909 a British physicist, Ernest Rutherford, and several colleagues discovered that if a stream of the positively charged particles was directed against an extremely thin sheet of gold foil, almost all of them went through the sheet; a number of these particles, however, were scattered from the sheet at sharp angles. Since it was well known that positively charged particles repel each other, it was clear from the undeflected passage of most of the particles that the atoms in the apparently solid sheet of gold metal were actually far from solid. It also was evident that the sharp deflection of some indicated that almost all the mass of the gold atom was concentrated in a tiny positively charged region, which was called the nucleus.

Later discoveries by others revealed that the basic structure of the atom consisted of a dense nucleus containing one or more positively charged particles called protons, around which orbited, at relatively enormous distances, negatively charged particles called electrons. The key to unlocking the secrets of atomic structure is the proton—specifically the proton of the hydrogen atom, which contains a single electron orbiting around a single proton. Stripped of its satellite, the hydrogen proton becomes an ideal missile with which to bombard the nuclei of more complex atoms. Its relatively large mass, at high speeds, provides for a great deal of energy; its electrical charge makes possible (1) its acceleration, by manipulating the potential of electrical fields through which it passes and (2) its direction, through the use of powerful magnets surrounding its path of flight.

## First accelerators

The fascination that atomic structure holds for scientists can be measured—in part, at least—by the number of Nobel Prizes that have been awarded to scientists working in this field. Besides Becquerel, the Curies, and Lord Rutherford, John D. Cockcroft and Ernest T. S. Walton received one for the design of the first machine that was made specifically to increase the kinetic energy of protons by accelerating their velocity. Although the energy imparted by the Cockcroft-Walton accelerator was measured in the thousands rather than the billions of electron volts, it was adequate to perform an experiment that had been a long-standing dream: the transformation of one element into another. Not lead into gold, as the ancient alchemists had sought, but lithium into helium. More importantly, Cockcroft and Walton demonstrated beyond any question that the accelerator was an instrument whose potential exceeded man's ability even to dream.

The Cockcroft-Walton accelerator was able to impart only one quick boost to the flight of protons. Another Nobel Prize awaited the man who first realized the possibility of bending the flight path of atomic particles into a complete circle, thus providing the opportunity to give an almost limitless series of kicks to speeding particles as they raced around the course. He was Ernest O. Lawrence, of the University of California, the first American to occupy a central role in the development of the accelerator. His machine was called the cyclotron. Its energy was measured in the millions of volts and one of the things that it did was to create a new element, called plutonium.

With such potentialities unfolding, the search began for ways to increase the energy of the accelerator. With the proton synchroton, the Brookhaven National Laboratory, Upton, N.Y., reached 2.3 billion electron volts (GeV) in 1952. Still later developments made it possible for Brookhaven to reach an energy level of 33 GeV. Several European countries combined to build a 28-GeV machine in Switzerland, and early in 1968 the Soviet Union placed in operation a machine operating at a peak level of 76 GeV.

If one thinks of an accelerator as a telescope in reverse, whose energy levels permit it to peer deeper and deeper into inner space, then the question arises: what energies are attainable? By the time the Soviet machine was built at Serpukhov, it was apparent that advances in accelerator theory and technology had reached a stage when man's ignorance was no longer the limiting factor. It was money. When U.S. scientists were asked by their government what kind of a machine should next be built, desire said 1,000 GeV, but a realization of the available funding said 200–300. Two other considerations led to this more conservative decision. Although it was true that available technology made 1,000 GeV possible, scientists also realized that probable advances in technology would make it possible by about 1980 to build such a machine at far less cost than was the case in 1972. Second, since it was apparent that even the largest nations could build only one such accelerator, there was much sentiment for the idea of looking to the day when all the major powers would agree to collaborate in the construction and operation of a single machine at that energy level or better.

***HOWARD J. LEWIS,*** *director of the Office of Information at the U.S. National Academy of Sciences—National Academy of Engineering—National Research Council, is also editorial consultant for the* Britannica Yearbook of Science and the Future.

## Building the national accelerator

For the builders of the U.S. National Accelerator Laboratory at Batavia, Ill., Feb. 29, 1972, was a day of great importance. On that date they wanted to bring their massive new machine to its design level of 200 GeV—before the Joint Committee on Atomic Energy of the U.S. Congress concluded its annual hearings on March 1. It was the powerful Joint Committee that authorized the U.S. Atomic Energy Commission to spend $250 million on this machine.

That the accelerator was so near completion by February 29 was in great part due to its director, Robert R. Wilson. When he was persuaded by the consortium of universities that had contracted to construct and operate the accelerator to leave Cornell University in early 1967 and direct the project, he was given a preliminary design study prepared by the Lawrence Radiation Laboratory at Berkeley, Calif., that called for a seven-year construction program.

Wilson promptly notified all concerned that he would not spend seven years building an accelerator. He announced that his target date was June 30, 1972. Later, he confided to his close associates that his real target date was June 30, 1971—an unheard-of four-year span between the reworking of the original plan and the beginning of operation at 200 GeV.

Wilson and his doughty crew almost made it, mainly by taking a series of audacious shortcuts that brought cries of "Irresponsible!" from some of the most respected accelerator physicists in the country. One shortcut involved the 240 quadrupole magnets that focus the proton beam. Because each must be located relative to its neighbors within an accuracy of 0.0005 inch (1/10th the width of a human hair), the original Berkeley design called for pilings to be sunk down to bedrock. Wilson, however, ordered borings to be taken around the four-mile circle the main accelerator ring was to occupy and decided that the glacial till, lying approximately 20 feet below the surface, had been sufficiently compressed during the ice age to bear up under its new and less demanding load. Nevertheless, after the 36,000 cubic yards of concrete that make up the tunnel for the main ring had been laid down, with its 2,750 tons of reinforcing steel, the ground under

*Aerial view of the National Accelerator Laboratory shows the main ring, four miles in circumference, and the smaller booster ring in the right foreground. After being boosted to 8 GeV in the small ring, proton beams are fed into the main ring, where they are raised to an energy level of 200 GeV.*

the tunnel began to settle at what experts termed a "violent" rate of one-tenth of an inch a month. Again there were bitter outcries that the largest accelerator in the world was too precious an instrument for such shortcuts. By November 1971, however, the settling had smoothed down to a uniform three-hundredths of an inch per month and by mid-1972 posed no problem for the operation of the accelerator.

One other shortcut, however, did hurt. It denied Wilson his dream of a 1971 completion date and worked a considerable hardship on those experimental scientists who had taken academic leave and rushed to bring their instrumentational apparatus to completion by the earlier date. Because the operation of a proton beam inevitably induces radioactivity within the ring enclosure, safety precautions dictate that the ring be buried under 15 feet of earth. When the last concrete section of the Batavia ring was lowered into place, it was midwinter on the Illinois prairie and the 15 feet of dirt that was piled back onto the precast concrete semicylinders was at freezing temperature. The massive concrete envelope remained abnormally cold throughout the spring and when the humid air of midsummer entered the tunnel, condensation swiftly generated a London fog within. Thus it was that when the eager director ordered the first tests to determine whether his personal 1971 date was to be reached, his magnets blew. They had been tested for every environmental factor save one. They had not been tested under water. In order to save time and money, the epoxy resins used to insulate the magnets had been applied at normal atmospheric pressure. Hairline cracks had developed, undetectable by the tests used but sufficiently large to permit permeation by the vapor-laden atmosphere in the ring tunnel. Before energy approached design level, the entire operation had to be shut down to prevent catastrophic damage to the equipment. The staff tried everything, including setting up giant fans to flood the tunnel with superheated air. But eventually, some 350 of the total 1,014 focussing and bending magnets had to be withdrawn from the enclosure and undergo reimpregnation with epoxy in specially built vacuum chambers.

Thus it was that through a succession of daring

*Interior of the linear accelerator tunnel shows several of the nine tanks in which protons, after being received from the pre-accelerator, are boosted to an energy level of 200 MeV.*

forward leaps and an occasional misstep, Wilson found himself at the end of February 1972 hurrying back from the first of two days of hearings in Washington to reach his new self-imposed goal of design energy while the hearings were still in session. The accelerator had already surpassed all previous achievements by reaching 100 GeV more than two weeks earlier, on February 11. When Wilson arrived at Batavia on the evening of February 29, he was told that while the thousands of components were being tuned and adjusted to reach 200 GeV problems had arisen in the magnet ring. A 12-hour shutdown was necessary. Wilson called a full staff meeting the next morning, March 1. He told them of his testimony the previous day and expressed his disappointment that it had not been possible to announce the attainment of 200 GeV. He urged the staff to try to reach the goal before the Congressional committee concluded its hearings that day.

The magnet problems had apparently been cured during the early hours of the morning. While Wilson was exhorting the staff, Francis T. Cole was directing activities in the main control room, communicating constantly with James Griffin some six blocks away. Griffin had the task of turning on the radiofrequency stations that provide the incremental boosts in velocity to the particles as they race around the four-mile circle.

The *Village Crier* of the National Accelerator Laboratory tells the rest of the story:

> At 11:00 (March 1) a steady, stable beam was achieved. By 11:30 beam had passed transition energy (17.4 GeV). According to Ed Gray, synchrotron physicist on duty at the main control console, from then on things just got better and better. A slight jolt to the system—a 'quad bump'—gave the beam further intensity. At 12:30 P.M. the beam reached 167 GeV for the first time. More and more people clustered around the screen in the Control Room lobby, watching the narrow band that typifies the beam as it crept slowly to the right to match the triangular peak representing 200 billion electron volts of energy. By 1:00 the screen was completely hidden by a crowd of forty or more people. The ding-ding-ding of a bell signalled the coming of the 200 GeV pulse in the bi-modal ramp.
>
> At the R. F. (radiofrequency) building, Jim Griffin recalls, all controls were in ideal positions, but the noise of the gathering crowd in the Control Room drowned out the warning bell on the inter-com; his adjustments had to be cautious. Stan Tawzer, working with Griffin, watched the screen. At 1:03, Tawzer noted, 'That one went all the way out,' but they waited for the next pulse to be sure.
>
> In the Control Room on the next pulse, someone in the hushed crowd said, 'There it is!' and a rousing cheer filled the room, at 1:08 P.M.

## Financial problems

What did Wilson build and what did he have to work with? First, he had more than a quarter-century of experience working with protons and accelerators. He had the institutional support of the Universities Research Association, the consortium of 52 universities in the United States and Canada that held the AEC contract to build and operate the accelerator. He had the preliminary design of the Lawrence Radiation Laboratory, made by those who were then generally regarded as the preeminent people in the field. He had a small team of particle physicists and engineers to adapt the Berkeley design and then build the accelerator and experimental buildings. And he had, as a gift from the state of Illinois, 6,800 acres of prairie about 30 miles from Chicago, containing several farm buildings and—more importantly, as it turned out—about 100 small homes, all identical, in the unincorporated village of Weston.

What Wilson did not have was money. Although the U.S. scientific community has come to depend heavily on the federal government for the funding of research, the money is not always easy to obtain. First the budget bureau of the administration in power decides how much the individual executive agencies may ask of Congress for the conduct of their business. Then, normally, one committee in each branch of Congress decides which of the major tasks the agency wants to perform should in fact be authorized. Still other committees decide how much money, if any, should be appropriated to actually carry out the work in any given fiscal year.

Such an arrangement creates an odd variety of problems for the director of a federally funded facility that takes time to build. And it sets a high premium on improvisation. In the case of the National Accelerator Laboratory, for instance, the Bureau of the Budget allocated $50 million for the accelerator, with the promise of additional allocations for experimental equipment. The payroll began in the spring of 1967 and ground was broken in December 1968; final authorization of the full $250 million did not occur until July 11, 1969. The unpredictability of the financing created some of the more unusual physical characteristics of the installation. For example, Wilson had asked for funds to build a 17-story central laboratory building, into which he would move most of the facility's activities from expensive rented facilities in a nearby town. But when the money was not forthcoming and there was almost literally no place to move his now large operation, Wilson delivered a bold stroke indeed. He brought in house-moving equipment and literally picked up scores of the now vacant houses of Weston village and rearranged them to suit the needs of the laboratory. Many of these houses were left as they were, providing individual domiciles for such small-scale operations as public information; but for the central directorate itself 17 of the houses were pushed together and joined as a single structure. When funds did become available to begin the high-rise building, its first floor was already being used as a cafeteria and for office space before the concrete had even been poured for the fourth floor.

*In the bubble chamber area of the neutrino laboratory, the silver building at the lower left contains the 15-foot chamber, and the red building at the right houses the 30-inch device.*

## The accelerator in 1972

But no matter how chancy the venture looked from time to time as authorizations or appropriations were withheld, the construction of the accelerator moved along briskly from one improvisation to the next. By 1972, there were four major sections completed that were used to accelerate protons to the next higher level of speed. A relatively small pre-accelerator, much like the one originally invented by Cockcroft and Walton, generated protons by stripping electrons from hydrogen atoms and then gave them their first kick of 750,000 electron volts. The protons were then fed into a 500-foot-long linear accelerator in which nine similar banks of high-frequency voltage apparatus propel the protons to 200 MeV. From there they enter a miniature version of the accelerator itself, a rapid-cycling synchroton approximately 500 feet in diameter. In this instrument, which works on the same principle as the machines at Brookhaven and Serpukhov, the protons are accelerated around their prescribed circle at close to the speed of light, receiving 750,000-electron-volt boosts with each circling. Upon reaching energies of 8 GeV, the speeding pulses of protons are fed into the main ring. Less than a second is required for this operation.

At this point the full energy of the installation

*Geodesic dome atop a building in the bubble chamber area of the neutrino laboratory consists of reinforced polyester fiberglass panels containing discarded beverage cans.*

is brought to bear on the whirling proton pulses. Contained in a 2 inch × 5 inch vacuum chamber fashioned into a ring four miles in circumference, the protons make 70,000 laps around the course, picking up 2.8 MeV with each revolution. The 280,000-mile trip inside the main ring is completed in 1.6 seconds, within a hair of the speed of light.

When the proton packages—each containing a billion or more—have been accelerated to the desired energy level, they are ejected from the main ring along tangential paths a mile or more long to the experimental areas. There, a variety of instruments will derive data from the protons themselves or from the secondary beams of other particles created when the protons collide with a variety of targets.

The first activity in the experimental area was designed to make use of a 30-inch hydrogen bubble chamber, recently transferred from the Argonne (Ill.) National Laboratory. The bubble chamber, which won a Nobel Prize for its inventor, Donald A. Glaser, is a remarkably simple device. When a volume of liquid hydrogen in a chamber is raised to a superheated condition for only a fraction of a second, the content of the chamber is in a state in which a speeding particle creates a boiling path during its brief penetration of the medium. This path of tiny bubbles, visible only for an instant, is photographed in the brilliant light of a stroboscopic camera. These photographs, in the hundreds of thousands, are then studied and analyzed.

During the summer of 1972, an even larger bubble chamber was placed in operation, permitting a track length of 15 feet. As the protons come speeding into the liquid hydrogen, collisions with their constituent atoms will splatter them into a variety of particles and sub-particles. Work already performed on similar instruments elsewhere revealed the presence of 130 different energy particles within the structure of the atom.

## Accelerator experiments

The search for a simplifying theory in particle physics has preoccupied many physicists since the great variety of so-called "strange particles" began to come to light. Again a Nobel Prize for Physics was awarded for innovative work in this field, this time to Murray Gell-Mann for his theoretical work in postulating a classified structure for these sub-atomic particles. A central postulate of this theory is that there are three *really* elementary particles, which he named "quarks." By assigning certain properties to each of three varieties of quarks, says Gell-Mann, one would be able to construct the entire array of the 130 known particles.

The search for quarks will surely benefit from the extraordinary devices at the National Accelerator Laboratory. Some of the possibilities were explained by Father Timothy E. Toohig, a physicist and a Jesuit priest;

One of the things we shall be able to do with this accelerator is generate very-high-energy muons, which have the special value of being sensitive only to electromagnetic force rather than the nuclear force itself. With these muons, we shall be able to probe the electromagnetic structure of the nucleus and see whether it is a simple thing with a charge on the outside of it, or whether it is actually a collection of very small charges, like a pudding with the charges embedded in it. Then we would be able to study the deeper structure of the nucleus just as Rutherford studied the grosser structure of the atom.

Father Toohig is, however, far more involved with the central focus of his laboratory, which is the study of neutrinos. These particles have negligible mass and no electrical charge. Because they are so small and are not affected by magnetic fields, they pass through objects like the earth as if they were not there. Muons and neutrinos are created simultaneously by the decay of other products of proton bombardment. To separate the two, the neutrino laboratory at Batavia uses, among other more sophisticated devices, 1,000 yards of earth. Almost all the muons, it is hoped, will be stopped by the earthworks as they veer off the main ring; the neutrinos will slip through to the laboratory. There they will encounter an ingenious apparatus that consists primarily of a spark chamber sandwiched by two high-speed scintillators.

The apparatus works as follows: Since neutrinos carry no charge, they are impossible to detect directly. But one can, in a bubble chamber or a spark chamber, detect the products of an impact between a speeding neutrino and another object, such as a proton or neutron. Spark chambers will display the passage of collision products for a thousandth of a second or longer; the scintillators work in millionths of a second. Therefore, if the rear scintillator signals the passage of a charged particle when the front one does not, logic dictates that the charged particles have originated in a neutrino collision. And the millisecond duration of the traced passage in the spark chamber is long enough for high-speed film to record the event.

There are three experimental facilities located about half a mile or more from that point at which the particles they deal with are flung out of the accelerator ring. In addition to the neutrino laboratory, which includes the work with the muons, there are the meson laboratory and the proton laboratory. By 1975, it is expected that approximately 350 researchers from the United States and abroad will be swarming over these experimental laboratories at any given time. Competition for the privilege of utilizing the product of the accelerator for an experiment is as keen as it was among astronomers when the 200-inch telescope of Mount Palomar first went into action.

## Who can use the accelerator?

Who will be allowed to use this unique instrument? Ultimately, that is the decision of director Wilson, but he does not decide alone. With one exception, all the experiments that have been accepted by the laboratory have resulted from proposals submitted by individuals outside the directorate of the facility. All are reviewed by a program advisory committee, chaired in 1972 by Robert G. Sachs, director of the Enrico Fermi Institute of Nuclear Studies at the University of Chicago. Periodically the committee submits its recommendations to the director on which proposals might be accepted and suggests priorities. Once a proposal is accepted, the laboratory staff works closely with the experimental group in arranging scheduling, developing cost estimates, and reviewing logistical needs.

By the spring of 1972, approximately 170 proposals had been received and about 60 had been accepted by the laboratory. Although the directorate of the laboratory sought not to impose its own criteria on the applicants, there is a general strategy of research that Ned Goldwasser described in some detail:

> When you enter a new world, as we shall be doing with this new instrument, you really have two basic choices. You can continue to do much the same kinds of experiments you have been doing, simply trying to attain more precise measurements of known phenomena in this new region of energy; or you can look upon this as an opportunity to explore an entirely new world, to look around to see what is here that wasn't known to exist in the former domain. On the whole, putting together the research program we have approved, we have put the highest priority on the second category. Take the rough look first. In my view, it would be nonsense and trivial to continue what we've done before. When the bevatron was built, it was designed specifically to produce an anti-proton. What they actually found was the whole new world of strange particles that no one had any reason to suspect existed before. Ninety percent of the time of that instrument has been spent in the study of these new particles. If anyone had made out a research program before that accelerator went into operation, he would either have had to have the courage to cancel what they had arranged to do too early or to trudge ahead with a very uninteresting program because of inertia. In actual fact, they were more flexible in those days and were able to respond to the new discoveries that were made. Here we're explicitly making that decision, I hope!

The genius of the men at Batavia has already produced a 200-GeV accelerator at a lower cost and in less time than anyone else had thought possible. But these scientists hold out still greater promises. With modest additions to present equipment, an increased power supply, and increased cooling capacity, they are confident that they can reach 500 GeV, two and a half times the previously rated capacity.

And, in the manner of scientists, they are looking still farther ahead—to the day when it will be possible by the use of jackets of liquid helium to arrange an additional ring of superconducting magnets above the present ring. This addition, they believe, will raise energy levels available at Batavia to an incredible one trillion electron volts—and very probably open up doors to knowledge that no man can now imagine.

# Test-Tube Babies

## by William K. Stuckey

**Does man have the right to interfere with the events of human conception and birth for the purpose of knowing more about such events? Is it better to know than not to know?**

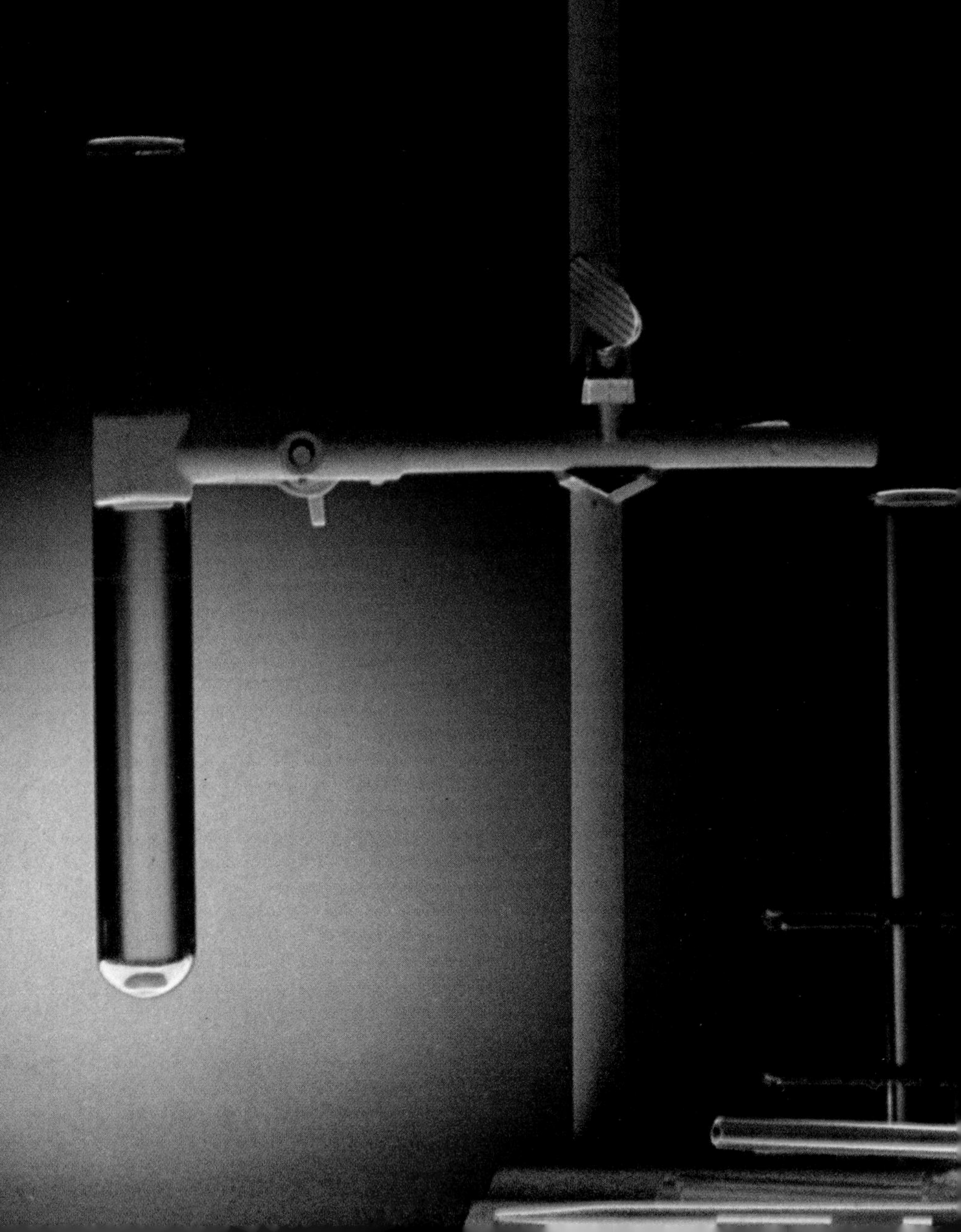

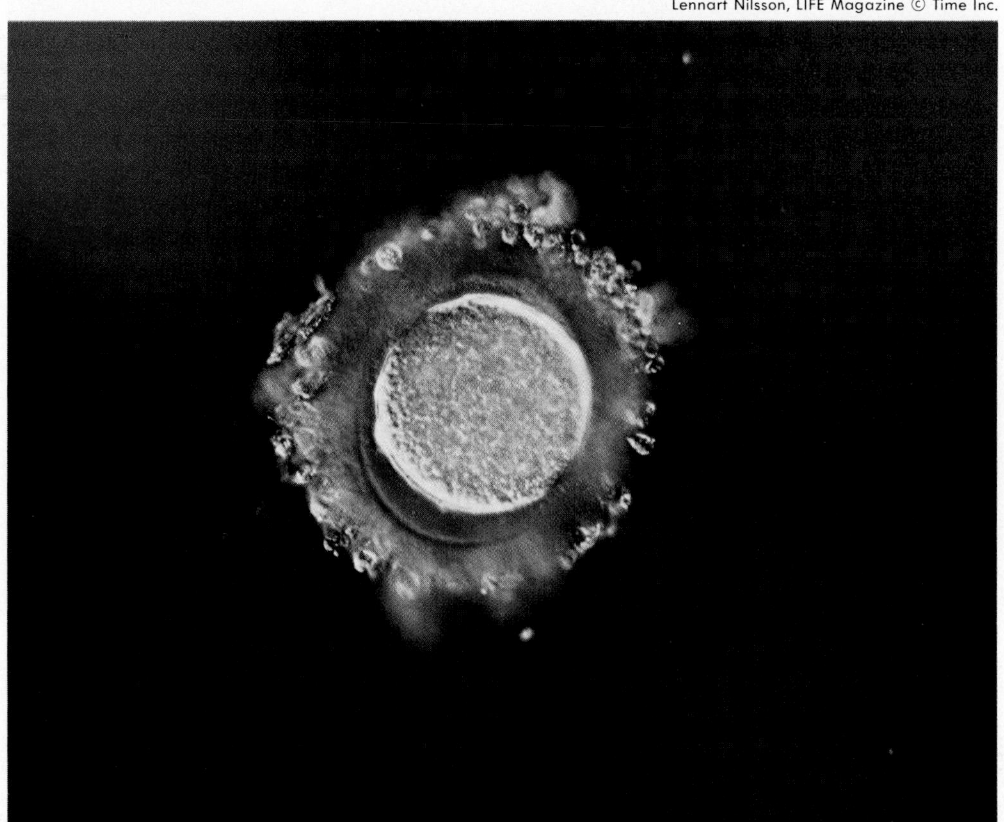

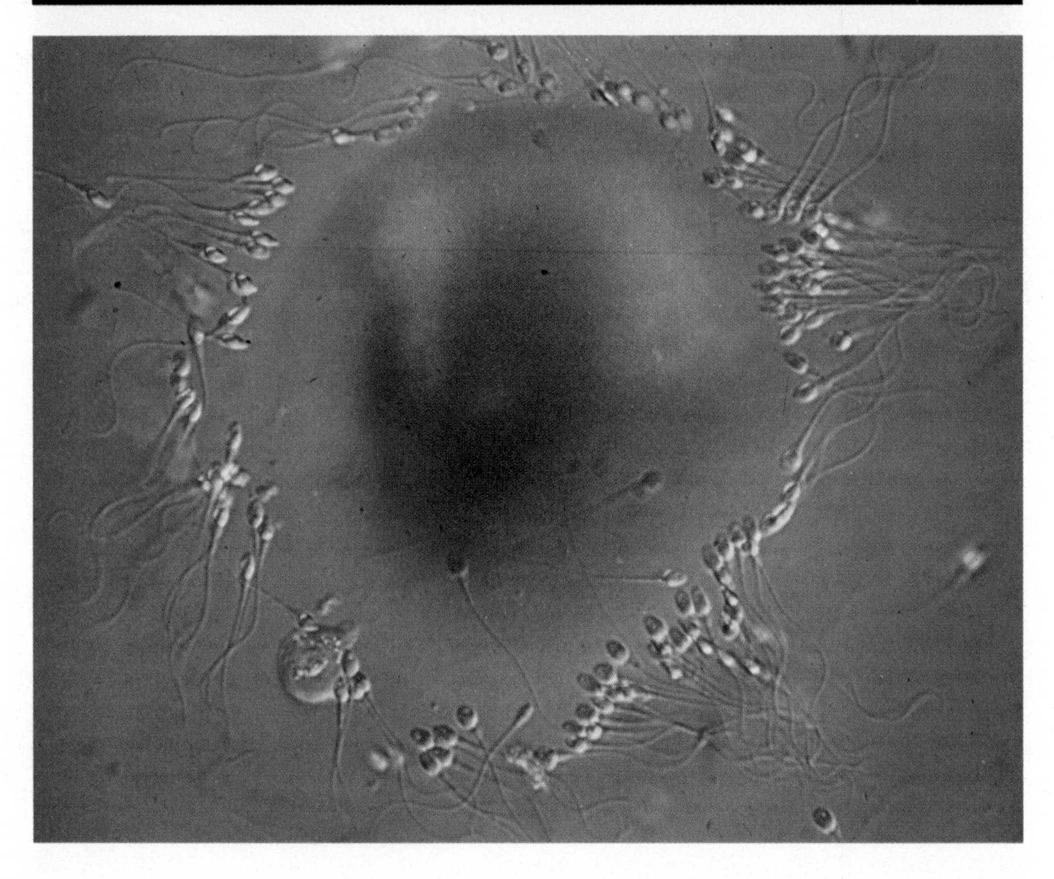

To date, every human being that has ever lived has been conceived in vivo, within a living organism. This seemingly unalterable necessity may soon be dispensed with. A baby—someday soon—will be conceived in a test tube. It will result from a female egg being fertilized by a male sperm cell in laboratory glassware—in vitro. The embryo then will be returned to its mother's womb for the normal nine-month development period. It is possible that such an event will happen this year or the next. Very few births in the history of man will be watched so closely or will create more controversy, fear, apprehension, and, perhaps, hope.

The proposal for such an experiment has come from a group of scientists at the University of Cambridge, in England, led by physiologist Robert G. Edwards.

If Edwards accomplishes such a feat, it will not be without courage, for he has been brought to task for even suggesting so daring an experiment. The language of Edwards' critics is harsh. His work has been described by one outspoken opponent as immoral and a stunt. The tone of this criticism is extreme for the scientific world, but then the issues raised by the possible success of such a venture are profound. The two principal questions raised in the controversy are these:

1. If the technique works (and the protagonists are divided as to whether it will), does the world really need it? Will the way have been opened to the production of a dehumanized, test-tube baby society described by Aldous Huxley in his *Brave New World*?

2. Is the guilt and concern felt by many scientists about past discoveries (such as nuclear weapons and the chemical processes responsible, unwittingly, for much pollution) leading to a rigid new conservatism that may stifle freedom of research? Will scientists be saying, in effect, "It is good to know this and bad to know that"?

It is particularly appropriate that the proposal arose where it did. Indeed, Cambridge was the site also of the crystallization of the idea that suggested the mechanism of transmission of genetic information via the complex deoxyribonucleic acid (DNA) molecule. From such studies emerged the DNA model, deduced by J. D. Watson, Maurice Wilkins, and F. H. C. Crick (see *1969 Britannica Yearbook of Science and the Future:* THE LANGUAGE OF LIFE, pages 122–139), which suggested that DNA operates as precisely and predictably as a machine —and like a machine, it can be adjusted, modified, overhauled. As DNA is manipulable, so is life.

### Tampering with nature

In the two decades since the deduction of the Watson-Crick DNA model, biologists have begun to tamper with nature in ways that earlier generations would have left to evolution. By 1971, when Edwards' studies began to become widely known, the life sciences were already producing startling results.

The genetic code—the arrangement of subunits of DNA and RNA (ribonucleic acid, another important molecule) into informational bits

*The mature human egg (opposite page, top) is the largest cell in the body, typically measuring 0.1 mm in diameter. Magnified over 50 times, it appears as a granular mass hiding its nucleus. The cell is surrounded by the zona pellucida, a protective material. The sperm, seen in large numbers circling an egg (bottom), are among the smallest cells, averaging 0.06 mm in length. Each has a rounded head containing tightly massed chromosomes, a midpiece containing other cellular bodies, and a long tail that gives the cell great mobility. The sperm developed as a result of meiosis, a unique form of cell division. Only one sperm will penetrate and fertilize the egg. The egg is midway through meiosis. It has divided once into a large cell and a very small one (a polar body), which is later discarded. It will divide again in the same way after the sperm has entered.*

***WILLIAM K. STUCKEY*** *is a science writer and former editor of the* M.I.T. Reports on Research.

that call for the chemical construction of cell components—was broken. A gene—a group of these subunits responsible for a particular trait of an organism—was created in the laboratory with off-the-shelf chemicals. It was demonstrated that one type of cell could be converted into another. Laboratory animals were cloned, a procedure that involves taking one animal and producing other animals that are its genetic duplicates. Giant atlases were being prepared to map and identify genes as to their function and their location along the DNA molecule. Methods were being developed to separate and identify sperm cells that, upon fertilization of an egg, could produce progeny of predetermined sex. Fruit flies (long a favorite experimental animal of geneticists) were altered in their DNA composition, and those changes were passed along to future generations. And methods for using harmless viruses to carry alien DNA into cells, incorporating it into the cell's DNA and thereby producing a change (a process referred to as genetic "surgery" or genetic "engineering") were reported.

Ironically, all these events—including the fundamental Watson-Crick work on DNA—were the results of a totally free spirit of scientific inquiry. The guiding maxim behind the work was, in effect: "It is better to know than not to know." Indeed, that has been an idea underpinning science since the Renaissance.

It was within this framework of events and ideas that one of the world's best-known scientists—Watson himself—criticized Edwards' experiments. At a science symposium in the United States, Watson urged Edwards to abandon the idea of reimplanting a test-tube embryo in a uterus because not enough is known to predict whether or not the resulting infant—if it should survive—would be permanently damaged by the procedure.

Late in the fall of 1971, Max Perutz commented on Watson's remarks. Perutz, a senior chemist at the Medical Research Council, the British equivalent of the U.S. National Institutes of Health, and a Nobel laureate, joined with Watson in the belief that scientists must be held responsible for the social ramifications of their otherwise morally neutral discoveries. Perutz' remarks concerning Edwards went even further than those of Watson. Initially, he supported one of Edwards' stated aims for conducting the research—to make visible an otherwise hidden process, to record in fine detail the way a sperm cell penetrates an egg, the way the egg then begins to divide, and the order in which the cells of the baby-to-be are then produced. Perutz felt that an attempt to understand the early development of the human embryo was entirely justified, both scientifically and ethically, but he continued his remarks in a different vein: "His [Edwards'] second aim was to fertilize ova in the test tube and to re-implant these ova into infertile women. I agree with Dr. Watson that this is far too great a risk. Even if only a single abnormal baby is born and has to be kept alive as an invalid for the rest of its life, Dr. Edwards would have a terrible guilt on his shoulders. The idea that this might happen on a larger scale—a new thalidomide catastrophe—is horrifying."

Perutz noted that while he had not read Edwards' scientific papers, such as those published in the widely respected British journal *Nature*, he had read an article prepared for laymen by Edwards and Ruth E. Fowler and published in the December 1970 *Scientific American*. And although he was not familiar with the details of Edwards' experiments, he felt that there was something dangerous here—even to science itself. For if an infant was born that proved to be a freak or monster, public reaction might saddle all of science with the blame. Such backlash would further weaken the already shaky support structure for the basic sciences in a number of countries, Perutz predicted.

By 1972, in the wake of considerable controversy about him, Edwards was remaining silent. Although he had written and spoken extensively since 1965 about his plans to bring an implanted human embryo to term, he was no longer presenting defenses of himself. His last public statement after the controversy erupted was simply, "The work will continue."

**Significance of the work**

Among the defenders of Edwards' work was his administrative superior at the Physiological Laboratory in Cambridge, Colin Austin. "I'm entirely in favor of the work," Austin stated. "Much thought has gone into each step of it. There is much data from similar experiments in animals; none of it suggests that Edwards is taking any particularly unusual risk." And it is a relatively small step from animals to man, he contended. Actually, there are fewer chances for deformities to occur in the embryo in its first five or six days after conception—roughly the period in which it will be in vitro—than in later stages of development. The defects associated with thalidomide, for example, did not develop until much later in pregnancy.

"The tone of the criticism the Edwards work is drawing is not particularly justified—in fact, I'm incensed by it," Austin continued. "A number of women who are donating eggs to the experiments, and who may carry them to full term after the in vitro fertilization, are nurses, doctors, or the wives of doctors. They are fully informed about the risks, yet are willing to go through with it. As to the 'morality' of the experiments, they have considerable humanitarian promise. The Edwards technique not only would permit previously infertile women to have their own children, but also offers the opportunity of separating some types of deformed embryos from normal ones while they are in vitro."

Who is Edwards? Exactly what is he doing? Robert G. Edwards is an intense, articulate man in his forties. He is a transatlantic scholar, having held research positions not only in Great Britain (Cambridge for the past seven years, and universities in Glasgow, Edinburgh, and Wales) but also in the U.S. at the California Institute of Technology, Johns Hopkins University, and the University of North Carolina. He has received U.S. research funds in the form of grants from the Ford Foundation.

His interest in reproductive physiology, the early developmental

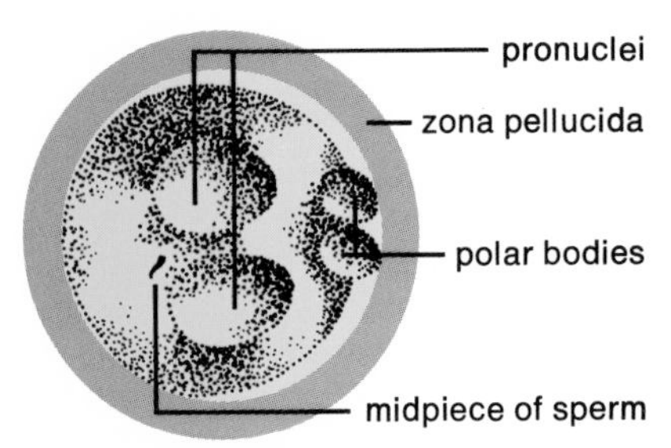


zygote

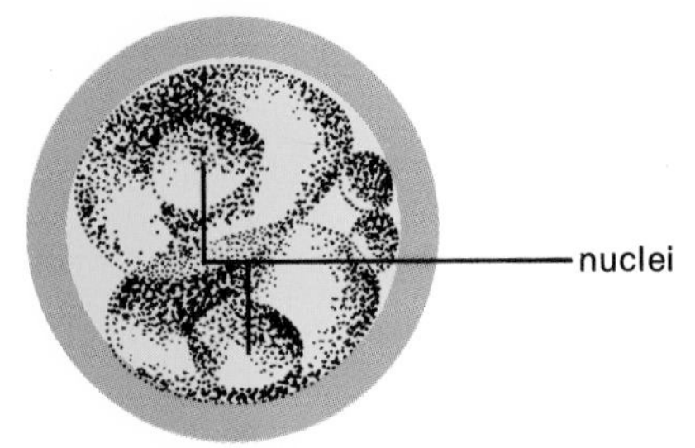


first cleavage

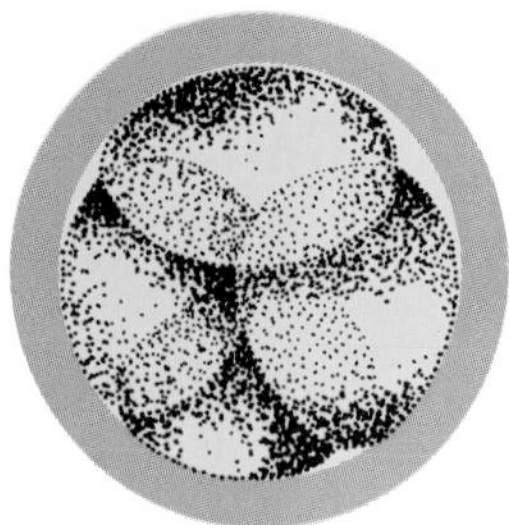

4 cells

*The earliest stages of human growth, diagramed in sequence on these pages, can be duplicated in the test tube. A fertilized egg cell, or zygote, (top) has two polar bodies, showing meiosis is complete; two pronuclei, one the egg's and the other a result of expansion of the sperm's head; and the sperm's discarded midpiece. Before the cell divides, the pronuclei will join, so that in the first cleavage and all cell divisions thereafter all cells will have the full chromosome number. The external form of the egg remains unchanged as the cells, all of equal size, divide within it.*



stages of the embryo, and in certain birth defects was expressed as early as 1957 in experiments on mice, rabbits, and other laboratory animals. By the mid-1960s, he was publishing articles suggesting the value of conducting such research on the human level. Of central interest was the egg, both animal and human.

At birth, each woman possesses in her ovaries a lifetime supply of eggs. At the onset of menstruation, usually only one egg matures at a time and begins its journey through the oviduct to the uterus. If it is not fertilized, it disintegrates and passes out of the body. But if fertilization by sperm occurs—an event that takes place within the oviduct—the egg starts on a long process of division, or cleavage, and initial development of the embryo has begun.

By the time the fertilized egg has entered the uterus, it has divided into 16 or more cells and has been transformed into a structure called a blastocyst. If the blastocyst attaches itself to the wall of the uterus, pregnancy has begun. A placenta is then formed to nourish the embryo, and the nine-month period of development (during which are formed all of the organs and supporting systems necessary for human life outside the womb) is under way.

All of this, of course, is a hidden process. Edwards wants to know what stages and events the egg undergoes between its maturation and implantation in the uterine wall. Are eggs released in any particular order? (Or, as he says, "Does each egg have a number assigned to it, and if so, why?") What hormone systems work to ripen an individual egg? What are the first steps involved in a single cell, the egg, becoming an organism composed of untold trillions of cells—and what might go wrong? What chemical changes does a sperm undergo to enable it to penetrate an egg? And what better way to observe such events than to let them occur in a test tube, subject to full visual and microscopic analysis?

Even before he began such work with humans, Edwards was aware of the controversial aspect of the project as well as of the fears that such tampering with natural processes might draw from the public. In speeches and papers he began to outline his "defense," contending that nothing is worth doing in the laboratory unless it is socially valuable. He considers the socially valuable aspects of his project as the following:

An accurate knowledge of the early events in reproduction can help achieve two goals that at first appear to be contradictory. It could suggest much more effective means of contraception and thereby help to achieve world population control. At the same time, many men and women who want children desperately cannot have them because of physiological barriers. A common barrier in women is a blockage of the oviduct. Some men, on the other hand, generate very low volumes of healthy sperm. To retrieve eggs surgically and to fertilize one under controlled laboratory conditions with what healthy sperm cells can be obtained from the husband would be one way of overcoming such types of infertility.

## The research timetable

In collaboration with Patrick Steptoe, a senior clinician at Oldham General Hospital in Lancashire, Edwards' group began the long transition from animal to human reproductive studies in the mid-1960s. Eventually, a group of about 50 women volunteered for the experiments—most of them in expectation of overcoming infertility. Previous animal experiments indicated that an egg must be at a precise state of maturity before it can be fertilized in vitro. Early work with the volunteers, consequently, was devoted to "gaining control" of the ovulatory cycle (by hormonal injections) so that the eggs could be taken at precisely the right stage of development. A standard surgical technique called laparoscopy—the insertion of a needlelike instrument with an optical viewing attachment through the volunteer's navel into the egg sack—was used to extract the eggs. (The Edwards group plans to implant an in vitro-fertilized egg into a uterus by the same route.)

Several different types of culture media (solutions of salts, proteins, and other nutrients) had already been used successfully in animal in vitro-fertilization experiments, and were tested for their applicability to the human eggs. In September 1970, Edwards and his co-workers published in *Nature* a report stating that a number of the human eggs had not only begun dividing in these test-tube solutions but had been carried as far as the 16-cell stage of development. A *Nature* paper of January 1971 contained evidence that the group had later mastered the in vitro culture media and had indeed gone far beyond the blastocyst stage—to embryos with 110 or more cells—in the test tube.

By late 1971, Edwards was already well into his policy of public silence. Word from his laboratory, however, revealed that no more than a minor alteration or two to the culture media was needed before a laboratory-grown blastocyst would be implanted in its natural mother. When that happens—if it has not already happened—the embryo will be on his or her way to birth.

Both Edwards and his supporters contend that there is far less risk in producing a deformed infant than the critics predict. Edwards, who has been speaking of his work on an international scientific tour circuit, stated that every expert he has met in this general field has assured him he is right—and safe enough—to move ahead to implantation. Furthermore, there are proven techniques for detecting malformations well before birth. Taking samples of the fluid from the sac enveloping the baby, for example, and analyzing the contained cells for genetic or other abnormalities is one such approach.

As Edwards has written: "Bringing such embryos into the light of day provokes various thoughts and opinions: The beginning of test-tube babies, armies of carefully planned robots, playing at God in the laboratory. . . . But every stage of clinical research must be questioned, and care taken to ensure that it is related to social needs and is not stimulated by its own impetus." Edwards states he has no other interests in altering natural embryonic development processes than to overcome

*continued on page 408*

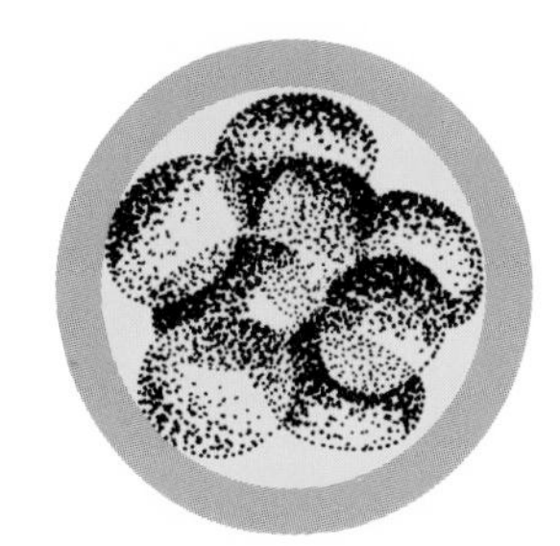

8 cells

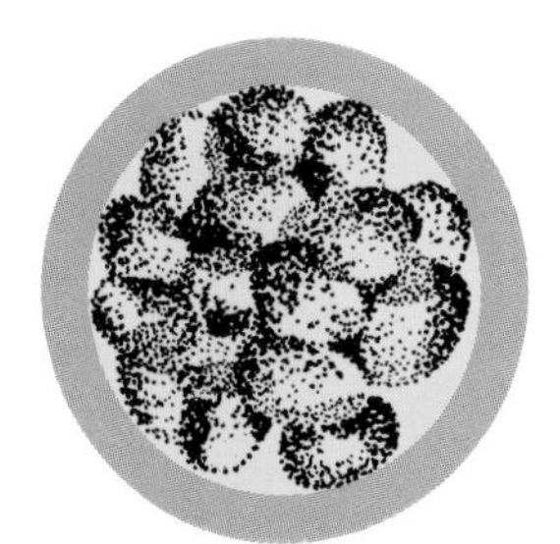

16 cells

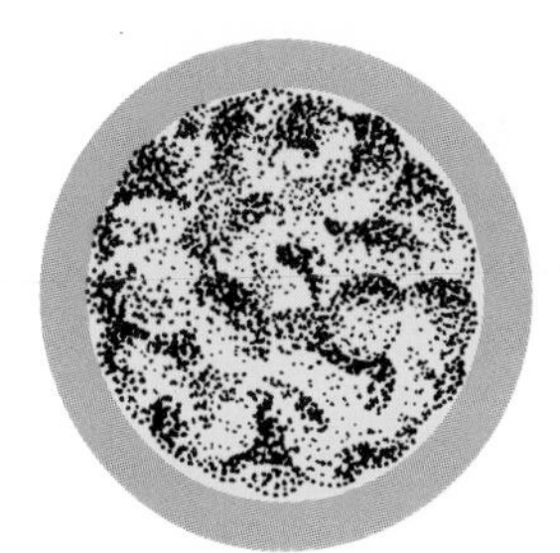

morula

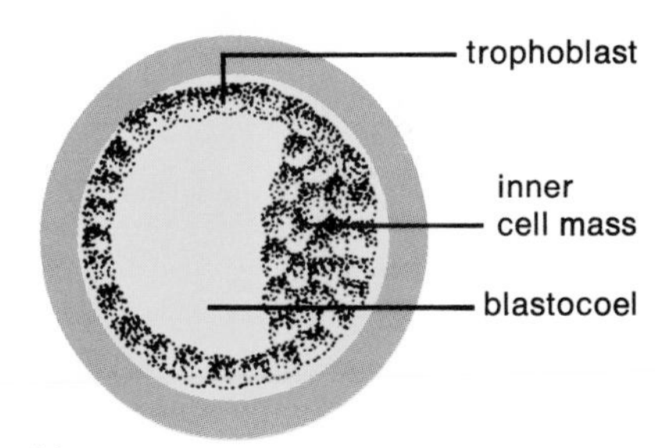


blastocyst

*When the morula is reached, there are about 32 cells and indications of unequal internal cell development. The egg soon becomes a hollow sphere (blastocyst) with a one-cell-thick wall, the trophoblast, which will become the parts of the placenta not contributed by the mother, and, heaped to one side, an inner cell mass, which will become the embryo proper.*

## Pioneering work

The earliest accounts of test-tube conception date back to 1944 when John Rock and M. F. Menkin of Harvard University Medical School achieved this experimental fertilization of human ova outside the body. They obtained ripe eggs from female patients and then exposed them to sperm in a test tube. No special medium was used to incubate the eggs, and success was slight. But these experiments marked the beginning of the elucidation of one of nature's prime phenomena.

In 1952, Landrum B. Shettles, a biologist, obstetrician, and gynecologist in the College of Physicians and Surgeons, Columbia University, began refining the experiments of Rock and Menkin. He demonstrated conclusively that in vitro fertilization of human ova is possible. In the course of performing various operations requiring abdominal incision into the peritoneal cavity of a woman, he pierced the ovarian follicles with a needle, causing some of the eggs in their follicular fluid to aspirate into a syringe. At the same time, without harming the patient, he excised tiny pieces of the tubal fimbrae, the fingerlike projections at the end of the Fallopian tubes that receive the egg when it is released from the ovary, and aspirated some of the mucosa that abounds within the tubes. From all this, he formulated the culture medium in which the eggs could mature and become ready for fertilization in the test tube.

When the egg, bathed in the follicular fluid, was ready for fertilization, it was placed in a sterile petri dish containing another oxygenated culture medium. There the prime ingredient was ovulation mucus taken from the midcervical canal of the woman who provided the egg. Into this Shettles unleashed millions of sperm cells and let them migrate to the egg just as they would in nature. At the right moment, he added the tubal mucosa and tiny bits of fimbrae in order, again, to provide the chemical components that the sperm normally encounter when they pass from the uterus into the tubes.

Thus, Shettles first witnessed the drama of human fertilization. The first report was in the *Federation Proceedings*, 1953, and 29 subsequent publications appeared on in vitro fertilization and the earliest steps of human development. He watched the ovum, glowing like a distant but bright-yellow sun on the microscope as it was pummeled by hundreds of sperm striving for penetration into its interior. The spectacle that was unfolding before him continued for hours while the sperm, their heads firmly against the outer membrane of the ovum, lashed their tails furiously, rotating the egg a full 360° every 15 seconds. At last, one penetrated into the interior of the cell proper, there to merge with its nucleus and thus achieve fertilization.

In the course of these experiments, various criteria were established for successful test-tube fertilization that have since become standard for the researchers in this field. These include the observation of both sperm head and tail within the egg proper, the elimination of excess female chromosomes in the second polar body (a minute cell that separates from the egg during maturation), the pairing of the male and

Courtesy, Dr. Landrum B. Shettles, Columbia University

A human egg cell, still within its zona pellucida, shows two polar bodies released below it, indication that a sperm has entered the egg to begin fertilizing it.

female pronuclei, and the actual division of the fertilized cell, showing that the embryo is growing and developing normally.

All of these criteria were repeatedly satisfied, and several of the fertilized ova were grown to the blastocyst stage, with more than 100 cells. At this time, in a woman's body the dividing egg would normally start to attach itself to the lining of the uterus. When experiments were first begun, however, too little was known to attempt implanting one of the test-tube embryos in the uterus.

Since then, a further step has been taken. Shettles aspirated a nearly mature egg from its follicle in the ovary of a woman undergoing an operation to correct a defect in one of the Fallopian tubes. The egg was matured in vitro, fertilized with sperm from the woman's husband, grown in culture for five days to the blastocyst stage, and then implanted into the uterus of a second individual scheduled for surgery. The menstrual cycles of the two women were synchronized, affording a hospitable environment for the transplanted embryo. Two days after the transplant, the previously scheduled hysterectomy was performed on the recipient. The implanting embryo was then located, with a dissecting microscope, on the upper, posterior lining of the excised uterus. An examination showed that it was implanting properly and consisted of several hundred cells. The objective of the procedure was not to produce a full-term baby but to ascertain whether implantation could occur, given the circumstances of in vitro fertilization, and to provide further information before proceeding to the final transplant.

—L. B. Shettles

Courtesy, Dr. Landrum B. Shettles, Columbia University

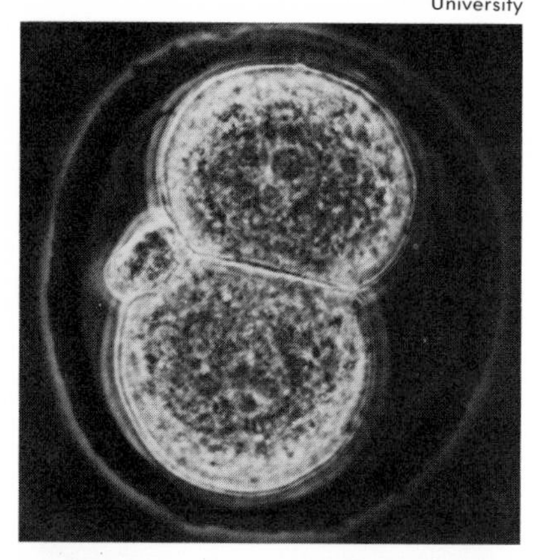

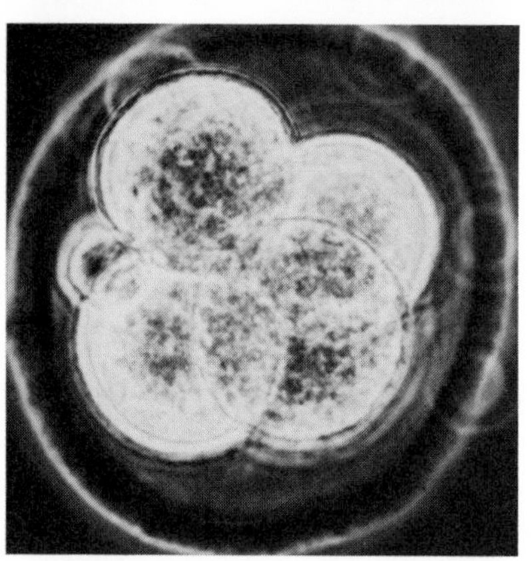

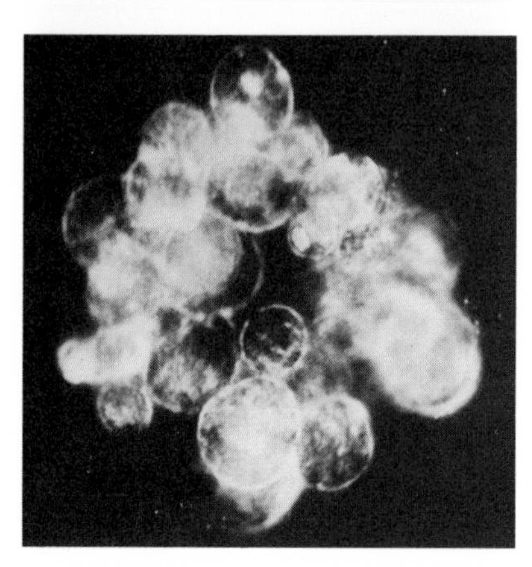

*The first steps of human development can be seen in photomicrographs of an egg fertilized in vitro: top, the two-cell stage, 30 hours after fertilization; middle, a four-cell egg 40 hours after fertilization; and bottom, a forming blastocyst, the state of a $4\frac{1}{2}$-day-old embryo. To develop further, the blastocyst must become attached to the wall of the uterus.*

*continued from page 405*

infertility or to ensure contraception and to perfect means for spotting defective embryos very early in pregnancy. He sees no point to developing human cloning methods—to use one set of "desirable" human genes to produce many living "desirable" duplicates. Also, keeping the baby in vitro for five or six days is quite enough to accomplish his socially valuable purposes. There is no justification for attempting to develop a baby for the full nine-month term in test-tube conditions, he states. (Austin, incidentally, estimates that such an accomplishment, at the present rate of advances in embryology, is still about a decade away. A major barrier currently is lack of knowledge about growing a placenta in vitro for such an embryo.)

### Yet . . . are Edwards' good intentions enough?

Edwards' work is not literally "genetic engineering"—that is, producing changes in DNA in order to alter an undesirable physical trait or induce a desirable one. The prospect of such developments is actually what frightens Watson, Perutz, and many other knowledgeable men: who will decide what is desirable? But the Edwards' in vitro technique could conceivably make it much easier to shape the physical traits and eventually the behavioral characteristics of a baby through genetic or other types of manipulation.

John M. Thoday, chairman of the genetics department at Cambridge, notes that determining the physical characteristics of a baby before birth is a long step toward fixing its behavior as well. Certain physical types could be isolated that are much easier to "condition" than others, he speculates. And sooner or later, such methods might conceivably be employed to satisfy a despotic leader with designs to create and control an entire race of beings.

Are there indeed some things that it is better *not* to know? M. R. Pollock of the University of Edinburgh's molecular biology department writes: "I do not believe there is, has been, or could ever be, a scientific discovery potentially of benefit to humanity that might not be exploited to some degree *against* mankind, if people want to do so."

### Scientific results

The basic scientific advances stemming from the Edwards work to date, according to his supporters, are (1) confirmation of the fact that the early development of the human embryo—from conception to blastocyst stage—is very much like that of the widely used laboratory animals, (2) isolation of the key hormone actions governing the maturation of the human egg and its release into the oviduct and (3) determination of the proper conditions for permitting early embryos to develop outside the womb. As to how such knowledge, and other knowledge that Edwards' experiments might produce, may be used, the debate is wide open. Some possibilities include:

• Coupling the Edwards information with that of the many other geneticists and life scientists for the purpose of remedying widely recognized pathological defects occurring before birth. Perutz supports such an approach, noting that at least 1,000 recorded pathological syndromes are now classed as being associated with genetic errors during the prebirth period. A broad-scale public monitoring of even this goal might be warranted, however, to ensure that "pathological" is not interpreted in a political or racial sense.

• Using Edwards' technique, or something that might succeed it, to clone a genetic blueprint. The usual response to such a possibility is that cloning (1) is no longer interesting for basic scientific reasons since it can be done so readily in animals, (2) is unpredictable, since environmental conditions may lead one intelligent clone into one path of choices and another in a separate direction, and (3) is not necessary if reproducing high intelligence is the goal of cloning (as Edwards says, "There is so much brilliance around today that it is not necessary to copy genetically one form of it").

Speculation can take one much further than such alternatives. For example, would it be better to design people to perform certain jobs —working underwater with lungs that could extract oxygen from its liquid form, or adapting humans to life in other worlds with different gravitational or chemical conditions—or would it be more humane to use contemporary man's considerable technological skill for dealing with such specialized existences? Or, more feasible under the Hippocratic Oath of medical ethics, would it be worthwhile to try to encode DNA to produce brain cells that would regenerate after their normal "life" as do other cells, thus averting loss of memory and senility, which have long created a fear of normal human aging; to produce hearts, kidneys, and livers that could endure more strain, with less damage, than is the case now; or to produce breasts and uteruses that were not so highly prone to cancer?

Such questions will not be resolved by deciding whether Edwards' socially valuable goals are worth as much as his critics' moral ones, dictated as they are by the environments of science and politics today and by the lives of the individuals proclaiming these goals. Perhaps the answers will come into clearer focus when an embryo conceived in vitro is born.

FOR ADDITIONAL READING:

Edwards, R. G., and Fowler, Ruth E., "Human Embryos in the Laboratory," *Scientific American* (December 1970, pp. 44–54).

Edwards, R. G., Purdy, J. M., and Steptoe, P. C., "Fertilization and Cleavage *in vitro* of Preovulator Human Oocytes," *Nature* (Sept. 26, 1970, pp. 1307–09), and "Human Blastocysts Grown in Culture," *Nature* (Jan. 8, 1971, pp. 132–133).

Edwards, R. G., and Sharpe, D. J., "Social Values and Research in Human Embryology," *Nature* (May 14, 1971, pp. 87–91).

Fuller, Watson (ed.), *The Social Impact of Modern Biology* (London: Routledge and Kegan Paul, 1971).

# The Nobel Science Prizes

## by William Kirk

**Although generally considered the most important and prestigious awards in science, the Nobel Prizes have not recognized many significant achievements. How are the winners chosen? Should the system be modified?**

Probably the greatest honor a scientist can receive in his lifetime is a Nobel Prize. And to many scholars and laymen, the Nobel Prize *is* science. There are three science prizes—one for outstanding research in chemistry, one in physics, and one for either physiology or medicine. (Nobel Prizes also are awarded for high achievement in literature and for contributions to world peace, and in 1969 a Nobel Memorial Prize was awarded for research in economics.) No other science prize has been so highly regarded by the world—and particularly by the United States—for so long a time. Since the prizes were first awarded in 1901, winners have returned from the lavish award ceremonies in Stockholm, to find that they are national heroes. Almost everything they ask for in money or equipment for further research is granted by a grateful nation or university. They are sought after as government advisors, as university presidents, as general savants who can make the unknown knowable. André Lwoff, who shared the 1965 prize in physiology/medicine with two other French geneticists, Jacques Monod and François Jacob, summed it up this way: "We have gone from zero to the condition of movie stars."

Not the least of a Nobel laureate's rewards is prize money that may reach $90,000 or more, but the amount of cash associated with the prizes is only one reason for the Nobel prestige. Equally as important is Sweden's reputation as a nation that values objectivity and high intellectual achievement—and the widely held respect for the standards of judgment used by Swedish scientists who select the winners.

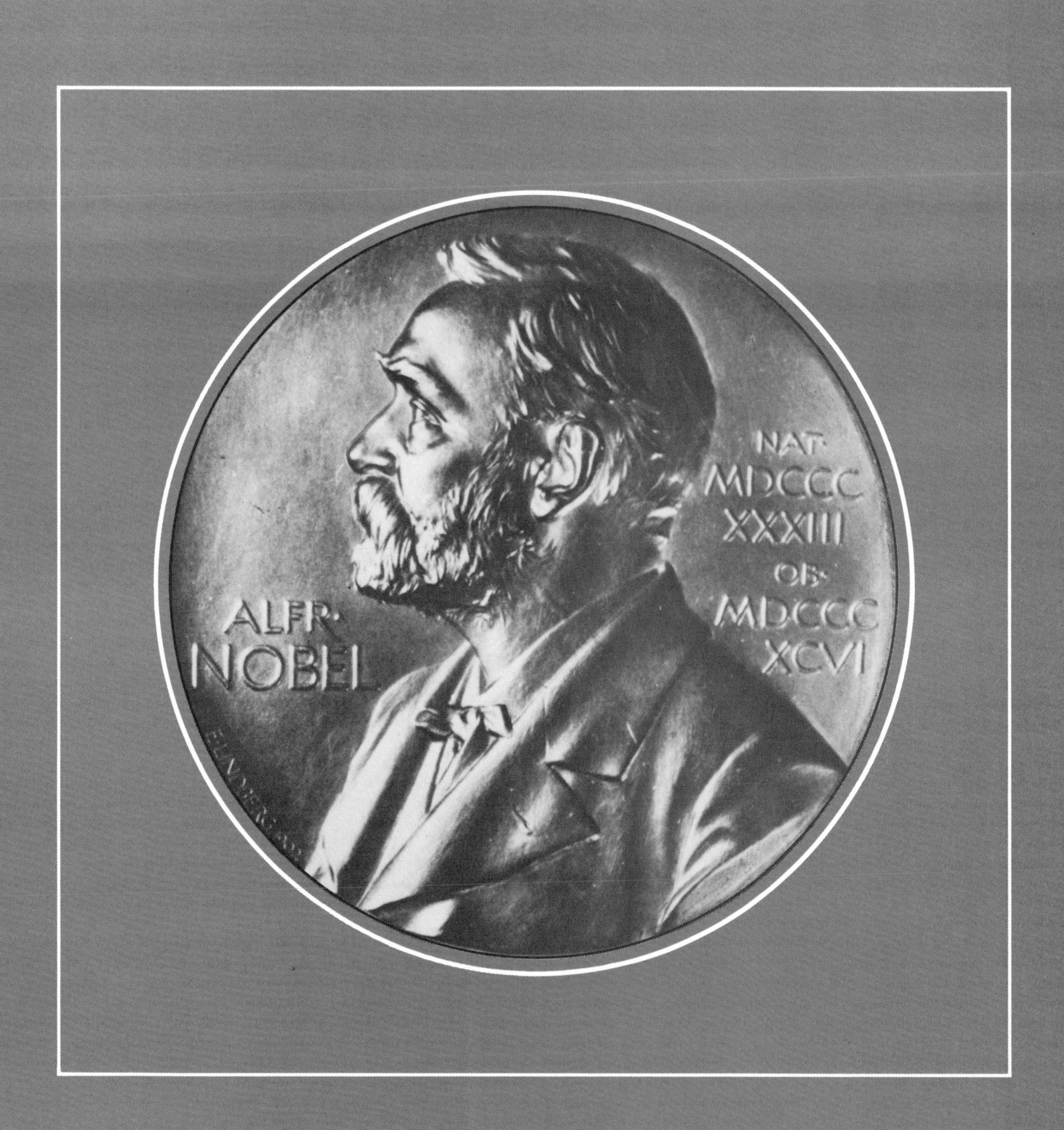
ALFR·
NOBEL
NAT·
MDCCC
XXXIII
OB·
MDCCC
XCVI

For reasons such as these, the casual remarks of a young Swedish geneticist, chatting with a visitor in the summer of 1971, came as a shock. For most of his professional life, the geneticist had been used to the very best in science. He was the protégé of a Nobel Prize judge. It was his privilege to work in the laboratories of one of the world's leading centers for research in life sciences, Sweden's Royal Caroline Medico-Chirurgical Institute with a faculty that not only includes Nobel laureates but which also has the final say each year on who wins the physiology/medicine prize. And in the middle of a conversation about the significance and standards of the prize, he was to say: "Who cares? Nobel prizes aren't all that important. I doubt that it is useful to continue to hand out the prizes. It was all right earlier in the century when there were very few scientists who needed all the encouragement they could get. But much of the best research in science and technology today is not eligible for prize consideration. Very few of the younger scientists here pay any attention to it at all. The prize simply does not affect their lives."

This was one of the more critical remarks encountered during recent interviews with Nobel officials and judges and a cross-section of Swedish governmental and academic figures. Some of these men had equally surprising responses. "So much modern science is excluded from Nobel consideration," remarked Gösta Funke, secretary-general of the Swedish government's two principal organizations for financing research in the life sciences, physics, and mathematics. "It's a pity. There should be several more prizes. In fact, I discussed this need with the Ford Foundation people several years ago, but nothing came of it —our Nobel officials simply said they did not want a Ford prize."

Two of the most influential Nobel judges, Erik Rudberg and the late Arne Tiselius, expressed similar views. "Too often we are regarded as the world's supreme court in science," said Tiselius, a Nobel laureate himself (chemistry, 1948), a former president of the principal administrative and financing body for the prizes, the Nobel Foundation, and at the time the chairman of the chemistry prize committee, "This is unfortunate. We operate on much too narrow a base to reward all of the remarkable people in science today." And Rudberg, who is chairman of the physics prize committee as well as the permanent secretary of the Royal Swedish Academy of Sciences, made this statement: "There is a long list of scientific and technical achievements that will never get the prize, at least as long as it remains in its present form. Such exclusions do raise a legitimate question about the relevance of the prize."

### Exclusions?

If archaeologists 1,000 years from now were attempting to piece together an accurate picture of 20th century science and technology, using only the minutes, reports, and written conclusions of the Nobel prize awarding bodies, what a strange society these records would reveal. These documents would show that 20th century society was

***WILLIAM KIRK*** *is a free-lance science writer.*

unquestionably advanced: that 20th century scientists possessed an excellent basic understanding of how the atomic nucleus was organized; of how genetic molecules encode and pass along the physical traits of all forms of life from generation to generation; of how chemical elements are "fused" within the interior of the sun and other stars to produce vast amounts of thermonuclear heat for the warming of such planets as earth; of how nerve cells exchange information chemically and electrically; of vaccines and vitamins; of lasers, transistors, and wireless communication. Many other discoveries of similar importance would demonstrate that the 20th century scientists were extremely clever in extracting information from a reluctant, secretive nature.

Yet how limited 20th century scientists would appear to be in other areas. The Nobel documents provide no evidence of computers, no indication of the existence of a program or means for exploring space—or even the oceans; nothing showing 20th century knowledge of the physical history of the planet—of how mountains grew, how oceans were formed, or how massive supercontinents sundered and drifted apart to form Antarctica, Europe, Africa, Asia, and the Americas. There is not even a hint that scientists of the era knew how atmospheric pressure systems and fronts shaped the weather; or how to measure the distance to a star; or of the existence of exploding non-stars outside the galaxy; or how animals, including humans, communicate or reproduce. The only medicines mentioned are penicillin, the sulfa drugs, and a handful of antibiotics, and the only synthetic fiber, rayon. And on the disciplines of psychiatry and psychology—nothing, a blank. The evident crudity of 20th century science would be much more heavily underscored, however, by the omissions of the sure signs of an advanced civilization—the development of a body of "pure" mathematics. The Nobel documents give no evidence of the existence of such knowledge.

These future archaeologists also would find no trace of certain seminal figures in 20th century science and technology. There is nothing about Mendeleyev bringing the chaotic science of chemistry into predictable order with a periodic table of chemical elements; no mention of Thomas Edison or electric lights or recording machines; nothing on the Wright brothers and airplanes or on K. E. Tsiolkovski, Robert H. Goddard, and Wernher von Braun and the development of rocketry; and no hint of Henry Ford or the transformation of the 20th century by the development of the techniques of mass production.

And although these future archaeologists would find some names of scientists that would be recognizable to most 20th century schoolchildren, these scientists would not always be cited for their most important accomplishments. Albert Einstein is there, honored for describing how a ray of light releases electrons from the metallic surface it strikes (the photoelectric effect)—good solid physics but certainly not the product of a mind capable of suggesting how gravity might give shape to the universe, or how time itself may in some way be related to the speed of light, or how matter and energy are inter-

The Bettmann Archive

*Albert Einstein, winner of the Nobel Prize for Physics in 1921 for his "services to theoretical physics," leads an informal seminar at the Institute for Advanced Study at Princeton, N.J. Einstein worked at the institute from 1933 until his death in 1955.*

The Bettmann Archive

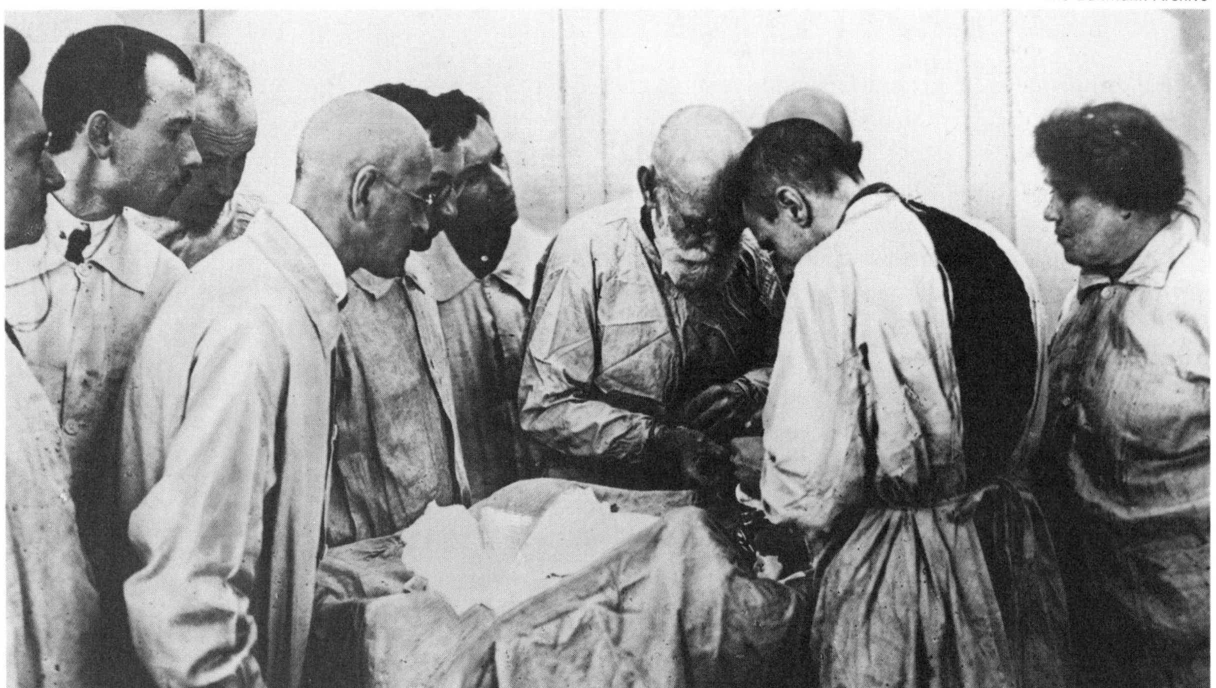

*Ivan Pavlov (center, with beard) conducts an operation on a dog at his institute in Leningrad. A Russian physiologist, Pavlov won the Nobel Prize for Physiology or Medicine in 1904 for his work on the action of the digestive glands, but he became best known for his studies of conditioned reflexes in dogs and other animals.*

changeable. And yes, there is Ivan Pavlov, honored for his studies of gastric juice secretion. Could that possibly be the same man whose discoveries of conditioned reflexes would have profound impact on the development of experimental psychology and the understanding of mental illness? The Nobel documents do not say.

It would also appear that 20th century scientists did not work together in large groups as indicated by the fact that the Nobel documents show that a prize was never awarded to more than three people. Would future archaeologists conclude this to be at least part of the reason for the apparent lapses in 20th century scientific creativity? They might consider this the reason that scientists were unable to send vehicles into space or to harness nuclear and thermonuclear energy, all impossible undertakings without large research and development teams.

There is no doubt, the archaeologists would probably conclude, that 20th century scientific knowledge was very unevenly developed. Based on the evidence provided in the Nobel documents, its gaps were more remarkable than its achievements.

### A defense of prize selection procedures

Almost all of the Nobel judges interviewed were aware, in varying degree, of such criticism and of the resulting charges that the prize is narrow, irrelevant, and obsolete. One Nobel judge, Börje Uvnäs, chairman of Sweden's Royal Caroline Institute's pharmacology department, made this remark: "I'm surprised how highly appraised the prizes are. It is difficult to understand. Of course we need criticism on our choices of winners, but the criticism we get is usually very mild."

But in defense of prize selection procedures the judges offered a number of arguments:

1. *Do not confuse science with technology.* Nobel Prizes are designed to reward discoveries about the organization and function of living and inanimate matter. How such discoveries may be applied to create computers, rockets, or better mouse traps—is outside of our scope. Besides, the men who developed the great technological and industrial inventions such as the telephone, the wide variety of pharmaceuticals, and the like were amply rewarded by the economy and have no need for our prize money.

2. *The will of Alfred Nobel, drafted in 1895, which is the basis for the prizes, excludes several of the established basic sciences and, of course, many of the newer behavioral sciences.* That is why there have been no prizes in astronomy or geology, for example (even though these might be considered to be branches of physics) or in "pure" mathematics. Mathematics is of Nobel prize interest only when it is *applied* in making fundamental discoveries within chemistry, physics, medicine, and physiology.

3. *The statutes of the Nobel Foundation, formed to carry out Nobel's will, limit the number of winners of a single prize to three.* This is why such things as the space program and the computer are not eligible for the prize. Each development required the efforts of thousands of people, and the Nobel committees were unable to single out the three most important individuals.

4. *Sometimes it takes years for a scientific discovery to be confirmed, found useful, and generally accepted by other scientists.* Many of the "gaps" the future archaeologists might find, based on interpretation of Nobel records, are in this category. Some of them will doubtlessly receive future prizes, but more solid evidence of their value is needed first. Unfortunately, by the time such general acceptance comes, some of the scientists will be dead. That is why Mendeleyev, for example, who was nominated several times, received no prize. And for Einstein's theories of general and special relativity still more confirmation is needed, although such confirmations are still gradually being made. And even if the theories are completely upheld, Einstein is dead too.

5. *There are simply too many good scientists today and too few prizes.* Each year some 10 to 20 men are judged by the Nobel committees to be worthy of winning *each* of the three prizes. Again, only a maximum of three scientists may be named for each prize. All of the prizeworthy candidates may be outstanding but may represent scientific fields that are impossible to compare. In medicine, for example, which is more important—a basic discovery in genetics or cell function, a spectacular surgical technique, or an important vaccine? The difficulty of answering such questions makes it a foregone conclusion that much good work will not receive Nobel recognition.

6. *Admittedly, there are inconsistencies and even errors in the prizes awarded over the past 71 years.* The selection committees are human and subject to error. There are also legitimate philosophical

differences among Nobel judges even now on what scientific excellence might be. Even the three Nobel science prize committees have views and procedures that differ from each other.

7. *If the world has an overinflated view of the science prizes, that is not the fault of the Nobel committees or the Nobel Foundation.* Probably the most representative defense of the prize record was given by Carl Gustave Bernhard, also a Nobel judge and a recent president of the Royal Swedish Academy of Sciences. As he writes in the 1971 edition of the Nobel Foundation's official book on the prizes, *Nobel: The Man and His Prizes:* "That the Nobel Prizes, in spite of these restrictions have come to be regarded as the highest scientific distinctions in the whole world, can therefore only mean that *certain well-defined discoveries* (emphasis added) have in fact been of paramount importance to the general progress of science."

### Alfred Nobel's will

Accepting such restrictions, what are the standards used to separate the scientific wheat from the chaff? How are prize candidates found and examined? What *is* a discovery and when is it of benefit to mankind?

Most Nobel judges and officials continually refer to the will, the statutes, and the wishes of Alfred Nobel as explained by his friends' memories of his actions, conversations, and letters in defending prize selection procedure. With those guidelines, the explanation of the Nobel prize standards of excellence in science begins.

Nobel's will was very brief and general. On a single sheet, he wrote only that his vast estate should be invested safely with the resulting interest to be distributed annually "to those who, during the preceding year, shall have conferred the greatest benefit on mankind." His specific instructions were that the interest be divided five ways, with three of these earmarked for physics, chemistry, and physiology or medicine (the other two for literature and peace); that each science prize was to go to the person making "the most important discovery" in the three fields (as well as for "inventions" in physics and "improvements" in chemistry); that the Royal Swedish Academy of Sciences would select the winners of the chemistry and physics awards, while physiology/medicine would be the province of Sweden's Royal Caroline Institute; and that the winners might be of any nationality.

The will created much initial confusion among disappointed relatives and business associates, among the governments of the various nations where Nobel occasionally lived or operated his business empire, and among the membership of the science academy and the Royal Caroline Institute (some of whom feared that administration of the awards would bring political pressures and lobbying into their quiet havens of research). However, influential Swedish academics and industrialists—with the willing assistance of the Swedish King Oscar II—brought about an agreement, established the parent Nobel Foundation, and wrote the statutes.

## The Nobel statutes

The statutes, which are much more detailed than Nobel's will, contained several interesting changes. Whereas Nobel wrote that each prize should go to a "person," presumably meaning one individual, the statutes specified that three—but no more—could share it. Apparently the intent was twofold. The statute writers recognized that an outstanding scientific work was seldom the work of one man, but to divide the prize money more than three ways would produce sums so small that the winners could finance very little future research. The unwitting effect of this provision, however, was forever to exclude discoveries made by large research groups. Also discarded from the will was the directive that the prize be awarded for a scientific achievement made during "the preceding year." The way of science is to treat discoveries very cautiously—to first publish them in a respected professional journal, to allow other scientists to check every detail in search of error or over-enthusiastic claims, to see if they survive the test of time. Decades can roll by before scientific discoveries are accepted as valid. (The most interesting example of this was the case of Francis Peyton Rous of the United States, who received the Nobel Prize in Physiology or Medicine in 1966 for experiments demonstrating that cancer can be transmitted by a virus. The experiments were first reported in 1910!)

The Nobel will says nothing about the cloaking of the prize selection procedure—the sorting out of the winners from the losers—in secrecy. The statutes, however, make selection off limits to outside scrutiny. Minutes of meetings by the three prize committees were directed to be extremely brief and to contain no evidence of disagreement among individual committee members over who should or should not win. Votes are not recorded, the names of all of the scientists considered for the prize in any one year are not released, and the lengthy special reports prepared on the scientific merit of the dozens of nominees are available for inspection only by members of the faculty of the Royal Caroline Institute and the Royal Swedish Academy of Sciences. For some 70 years, then, the nitty-gritty of determining who is "prize-worthy" and who is not, the fine detail of the value judgments rendered on what is good, bad, or indifferent in science has remained a well-kept secret.

The statute makers no doubt were motivated by the best of intentions. To conduct the selection in the open would subject Nobel judges to political influence and lobbying on behalf of particular scientist-candidates, thus destroying the objectivity of the scientific approach, they reasoned. The practice is still defended by Nobel judges who add that it would be unkind to reveal to the world the names of those candidates found to be unqualified for the prize. And the fact that the prizes are still highly regarded throughout the world today in many scientific and governmental circles reflects favorably, the judges believe, not only on the high quality of accomplishment by the prize winners over the past seven decades, but also on the policy of not airing the selection controversies (or the opinions behind value judgments) in public.

The Bettmann Archive

*Alfred Nobel willed the bulk of his vast estate to establish the prizes that bear his name. A Swedish chemist and engineer, Nobel amassed his fortune by the invention and manufacture of dynamite and other explosives and by investments in oil fields.*

Ironically, the Nobel judges themselves suffer from the secrecy mandate in the statutes. Stockholm conversations can be counted upon to produce rumors that such secrecy conceals prejudice or incompetence among the judges. The policy of secrecy also encourages the Swedish political left to suggest frequently in print that Marcus Wallenberg, one of the four members of the Nobel Foundation's governing board and a Rothschild-like figure with great influence in Swedish banking and industrial circles, may also influence prize awards. This is one reason, the left will argue, that the prizes frequently go to scientists of Western anti-Communist nations. (Through 1971, 84 science prizes had been awarded to the United States, 46 to Germany, 50 to Great Britain, 21 to France, and only 9 to the Soviet Union—out of a total of 278 for all nations over the 71 years. Scientists of the People's Republic of China have declined to submit prize nominations on the grounds that they do not believe in prizes awarded on an individual, non-collective basis.) The judges themselves shrug off such attacks as unfounded and generally unworthy of reply. Since the judges cannot produce Nobel records and minutes to defend their choices and actions, however, the rumors continue.

But a far more serious ramification of the secrecy policy is that it might result in the wrong man winning the prize. One old tradition, stemming from the early days of the statutes, is to avoid letting a science nominee know he is being considered for the prize. Yet the judges are committed to conduct as thorough an investigation as possible of the nominee's scientific work, not only to determine its merit as a "discovery" but also to see if he was the first or most important contributor to it. One senior Nobel judge, Sven Gard, commented that the policy forces Nobel investigators to operate so obliquely—trying not to tip off the nominee—that an erroneous assessment of a candidate's work might result.

A case can be made, according to most judges interviewed, that the secrecy policy denies the world's scholars access to civilization's longest and most complete record of science criticism and analysis. Perhaps the most valuable of these records are the lengthy investigations on scores of nominees each year. In a world where the vast majority of people understand little of the methods and goals of science —and increasingly attribute sinister purposes to it—the opening of the Nobel archives might, at the very least, produce a much more accurate picture of just what science can or cannot do.

That the secrecy policy has remained unchallenged for seven decades is particularly bizarre in Sweden. For more than 200 years, Sweden has operated under a government documents act that permits anyone to demand letters, memoranda, reports, and general documents on civil governmental affairs—and to have them produced by the civil servant approached *on the spot.* Swedish universities are nationalized, and since all Nobel judges are university men who are considered government employees, there is a question of whether they could deny access to the archives if such access were requested.

But perhaps the most important element of the statutes was clearly to invest prize decision powers in the hands of university professors specializing not in engineering or the applied sciences but in the fundamental sciences. Their decisions over seven decades have generally favored the basic scientific discovery—which often appears at first to be of no more than intellectual or cultural interest—at the expense of many of the revolutionary technological developments of the century (although occasionally machines or "practical" processes have been honored). Nobel himself was vitally interested in such basic science and wanted to give impetus to its development. Ironically, however, Nobel was also one of the 19th century's supreme industrial technologists.

Finally, the statutes roughly outlined how the judges were to go about canvassing the world for prizeworthy scientists. By 1971, these general instructions had been refined into a set procedure (*see* "How Nobel prizewinners are selected," pages 420–421).

### What constitutes a "discovery?"

Language is full of words like "discovery," the meaning of which everyone is quite sure he understands. However, one often finds that the world is no more agreed on its meaning than it is on "good." There is no doubt that the meaning was clear to Alfred Nobel when he used the word in his will. It is probably very clear in meaning to his successors, the judges, although they are rather vague in discussing its significance with outsiders. When asked about what they considered "discovery" to mean in a Nobel context, the judges responded:

—"Something unexpected."

—"A big step."

—"A development which opens an entire new field of research."

Slightly more detailed is this definition by Nobel judge Gard: "If a scientist has followed a systematic research theme throughout his career that has at last answered a set of complex questions, he has made a discovery." Nobel laureate Ulf von Euler (who is also the son of a Nobel laureate): "It is something that no one believes at first because it departs too sharply from the norm. If it were good average work, it would be believed." And Nobel laureate Axel Hugo Theorell: "A discovery can be the result of a quick, simple experiment that yields an answer so clear and obvious that it is unnecessary for others to check or repeat the experiment."

According to classic scientific practice, the minimum requirement for a discovery is that it be published in a respected scientific journal. A great development kept secret—not presented before a scientific audience for a period of checking and confirmation—is simply not a "discovery." If a considerable number of scientists drop what they are doing to verify a development by another person—and find it is correct and interesting enough to enter that field themselves—it is indeed a valuable discovery.

*continued on page 422*

## How Nobel prizewinners are selected

The Nobel awards complex is to Sweden as the Vatican is to Italy. It is autonomous and international. The judges constitute their own parliament, and preside over an elaborate diplomatic and scientific intelligence network. This Nobel nation begins a new year's activities each fall, and it is also then that the intelligence network is the most active.

The three science prize committees begin the new annual cycle by sending out 2,000 to 3,000 confidential letters. The purpose of the letters is to ask network members for suggestions about top candidates for the prizes. The identity of the letter recipients is not revealed outside of Nobel circles. It is generally known, however, that they are senior academics. Their ranks, of course, include all the members of the Royal Swedish Academy of Sciences and the senior faculty of the Royal Caroline Medico-Chirurgical Institute (each of whom are encouraged to informally contact their own colleagues abroad in search of nominees), senior professors at all the major Scandinavian universities, and leading scientists from two to three universities in each of the scientifically developed European and Asian nations as well as from perhaps a dozen U.S. universities. The principal science academies throughout the world, including the Soviet Academy of Sciences, are contacted, as are the principal officials of major research installations such as the European Organization for Nuclear Research (CERN), a high-energy physics laboratory in Switzerland; Great Britain's Jodrell Bank astrophysical observatory; the U.S. government National Institutes of Health; and the U.S. National Accelerator Laboratory in Batavia, Ill. Final key elements of this intelligence gathering network are the previously chosen Nobel laureates.

UPI Compix

*The members of the Royal Swedish Academy of Sciences meet annually to select the winners of the prizes in physics and chemistry. This 1964 gathering chose Dorothy Crowfoot Hodgkin to receive the chemistry award and Charles H. Townes, Nikolai G. Basov, and Aleksandr M. Prokhorov to share the honor in physics.*

The network then informs Stockholm on the prospects that look best that year. The letters of nomination they return are in themselves fairly detailed investigations of individual candidates that spell out clearly the nature of a scientist's work.

No further nominations are accepted after the following February 1. Then, the prize activities once more become an exclusively Swedish show. A rough weeding out is performed by members of the three committees. Immediately discarded are the names of scientists who nominate themselves (including those names sent by a candidate's loving wife—which actually has happened). The judges live by a kind of Robert's Rules of Order that cherishes merit coupled with modesty and is offended by what appear to be pressure tactics involving lobbying and publicity. An outstanding scientist may prejudice himself in the eyes of the judges by appearing to be trying too hard to win the prize. Finally, the best candidates must have a large body of published scientific work to their credit that can be read and checked in detail by the judges.

By summer, each committee has narrowed the list down to between 10 to 20 candidates. These are selected for the intensive investigations mentioned earlier. Senior judges sometimes are asked to prepare as many as 8 or 10 such reports. On occasion, a young outsider (not a judge or a person affiliated with the awards machinery) is asked to conduct an investigation. In this phase, selected members of the international intelligence network may again be contacted for much more detailed information on a candidate. Judges attend international scientific conferences, inquiring discreetly about this or that candidate but always, of course, denying that anyone actually is a candidate.

In an average year, the name of the winner is usually known by the judges by July. The remainder of the summer is spent in committee meetings and the writing of the final "Committee Report and Recommendations." The reports, containing all the investigations plus a recommendation for a winner, are formally presented to the full membership of the Royal Caroline senior faculty and the Royal Swedish Academy of Sciences in fall meetings (a year after the annual cycle began). On these meeting days, members of the two organizations file into a large auditorium and are presented a golden Nobel medal—or about $25 in cash if they already have enough medals—for attending the debate and vote. Usually, the debate is brief, and the voters endorse the committee recommendations (although judges commented that such recommendations have been altered or ignored on three or four occasions since World War II).

Then, the press is informed of the identity of the winners, and the world is presented its new Siegfrieds of science. All that remains is for the winners to come to Stockholm in December to deliver their Nobel addresses, to be congratulated by the King of Sweden, and to pick up their checks. But by then, the Nobel committees are well into the next cycle of activity, receiving nominations, keeping in touch with their network, pondering the next year's prizewinners.

Culver Pictures

*Marie Curie, a Polish-born French physicist, won Nobel prizes in 1903 and 1911. She shared the first award, in physics, with her husband Pierre and Antoine Becquerel for their pioneering investigations into the nature of radioactivity. The 1911 prize, in chemistry, was for the isolation of pure radium.*

*continued from page 419*

The most commonly mentioned examples of great discoveries, in the opinion of those judges who will discuss the subject, are Wilhelm Röntgen's discovery of X rays and the Antoine Henri Becquerel and Marie and Pierre Curie discovery and exploration of the nature of radioactivity. These were revelations of a fundamental property of matter which took the scientific world by surprise, opened vast new fields of research, and were applied in solving a huge variety of practical and scientific problems. Concerning more recent awards, the entire string of prizes given for genetics research since the 1950s is cited by the judges as examples of key discoveries. In that area, the name most frequently mentioned is that of Marshall W. Nirenberg, who shared the 1968 prize with Robert W. Holley and H. Gobind Khorana for work in deciphering the genetic code.

Erik Rudberg provided a more detailed analysis on what constitutes a Nobel-level discovery. He labeled most of his comments "personal opinions" that may or may not be shared by other judges. The first type of prizeworthy discovery he mentioned was in the class of "an advanced experimental technique or instrument that makes it possible for science to measure a phenomenon for the first time." Examples in physics include the cyclotron developed by Ernest Lawrence of the United States, the 1939 laureate. The cyclotron represented an entirely new approach at that time for probing the atomic nucleus with beams of high-energy electrons. Another example cited is the hydrogen bubble chamber developed by Donald Glaser of the United States (and honored by the physics award in 1960). The device permits a photographic record to be made of fragments broken from an atomic nucleus by a beam of probing particles.

"On the other hand, I would also give special consideration to a Nobel prize candidate who made an important finding with less-than-adequate instruments," Rudberg said. "That would be more impressive than the scientist who made a finding simply by pressing a button on a very sophisticated device." An example in point, he added, was the 1964 award to Charles H. Townes of the United States and to Nikolai G. Basov and Aleksandr M. Prokhorov of the Soviet Union for discovering the principles of the maser, and, indirectly, the laser. The two Soviet scientists worked in much more primitive laboratory conditions than Townes, but still managed to produce valuable and relevant results. (The award, incidentally, was delayed for several years, said Rudberg, because of the difficulty of obtaining information on the Soviet work.)

It is often a frustrating job to determine whether a discovery is something totally new or is merely a continuation of work done by others. One could argue philosophically, for example, that many 20th-century discoveries were simply continuations of work started by Archimedes or Democritus. Consequently, the Nobel judge looks principally for something Rudberg describes as the "sudden jump." Previous findings plotted on a graph, for example, would describe no more than a gently rising curve when compared to the sharp upward rise of the Nobel-level discovery. An example is the 1969 physics award to Murray Gell-Mann of the United States for his theories on how the components of the atomic nucleus are organized. His "symmetry scheme" of nuclear components was later largely confirmed when a research team discovered a particle, the Omega Minus, which his theory had predicted. Another example of the sudden jump was in the field of semiconductors and transistors, which had drawn the attention of many physicists. The contributions of the U.S. team of William Shockley, John Bardeen, and Walter Brattain were sufficiently more precise and dramatic to justify the award of the 1956 physics prize to them, said Rudberg.

One recent award suggests an interesting change in traditional Nobel discovery standards and one that eventually may make findings by large research groups eligible. Luis W. Alvarez of the United States received the 1968 physics prize for his work in directing the accelerator team that confirmed the theories of Gell-Mann. The development of new experimental techniques and instruments was one reason for the award, explained Rudberg, as were the achievements of Alvarez in analyzing the extremely complex information obtained from the experiments. But for probably the first time, the judges also recognized a researcher's ability to administrate, to organize, and to coordinate a large team working on a difficult problem.

The discovery then remains a term of multiple definitions. In most cases, other judges said, when there is disagreement on which of a number of findings is the most legitimate, the contest is usually decided on the basis of which has brought more "benefit to mankind." The vagueness leaves the judges much room for maneuver in making award selections. Do they ever abuse this flexibility and allow prejudice to influence their choices? Rudberg's reply was generally echoed by other judges interviewed. "To my knowledge, very definitely not. I believe that to be the case for all the science prizes. However, in a critical international situation, for example, the wording of the award citation may be dampened a bit for other than purely scientific reasons."

UPI Compix

*King Gustavus V of Sweden (left) congratulates Irène Joliot-Curie, daughter of Pierre and Marie Curie, for winning the 1935 Nobel Prize for Chemistry with her husband Frédéric, standing behind her. The Joliot-Curies won the award for synthesizing artificial radioactive elements.*

## Conclusion

The awarding of the prizes, then, has served high purposes over the past seven decades in stimulating the growth of certain of the basic sciences. The award activity has probably been more fair and objective in the hands of honest and qualified men from a small, neutral country such as Sweden than it might have been under the control of men from more nationalistic and politically ambitious governments. The discoveries honored by the prizes do not, in any sense, represent all of the best that has been produced in science and technology by 20th century man. Errors have been made, but scores of important scientists who might otherwise have been overlooked or mercilessly exploited for financial reasons have been allowed to bask in the glow of world admiration and to develop according to their own lights. That alone—although the prizes appear to many to be entirely too narrow in scope today—may be enough to fulfill Alfred Nobel's good intentions.

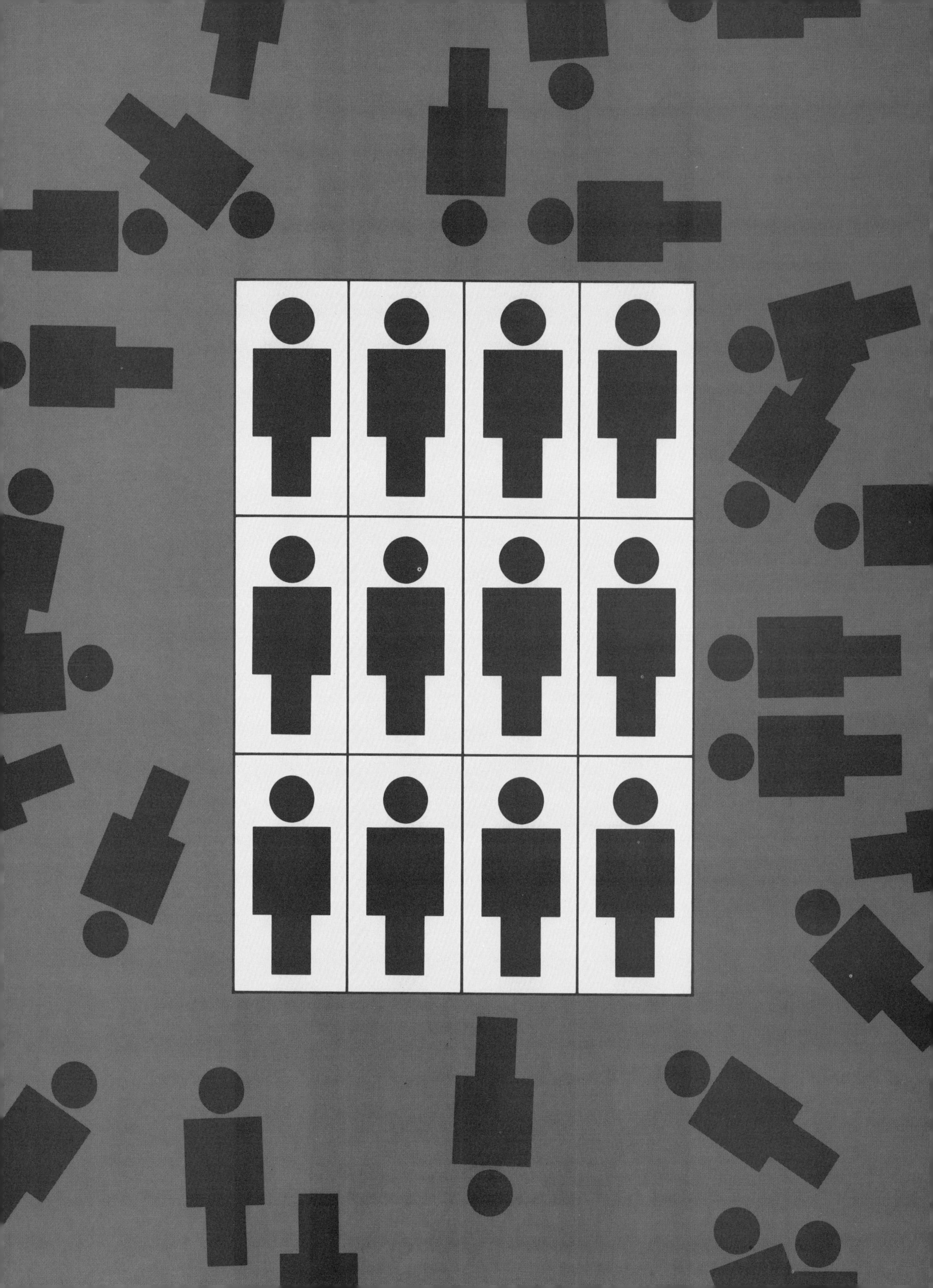

# Too Many Scientists?

## by Charles V. Kidd

**After many years of not having enough manpower in science, the United States now shows signs of having too much. Have we trained too many scientists?**

For nearly 30 years—from the beginning of World War II until the end of the 1960s—the United States faced continuing shortages of all kinds of manpower with advanced training, and particularly scientists and engineers. Only brief episodes in specialized fields disturbed the picture of demand that exceeded supply. All of the signs of shortage were present during this period: salaries rose rapidly; recruitment of both new holders of advanced degrees and older workers ranged from active to frantic; employers offered attractive fringe benefits, perquisites, and amenities to supplement salaries; the output of Ph.D.s rose rapidly in response to high demand; highly-trained foreigners relocated in the United States at a rate sufficient to generate heated debates in the United Nations with charges that the United States was draining from other nations the brains that they needed for economic and social progress; and, finally, mobility was high—virtually always leading to better jobs.

But, over the past several years there was a sharp reversal of the shortages that this nation had grown to accept as facts of life. Newspapers, magazines, and professional journals were full of stories of unemployed scientists and engineers and of the bleak future they faced. What caused this sudden shift? What is the real situation? Is unemployment a passing phenomenon or a warning signal of long-range trouble? Are we overproducing Ph.D.s? What is wise public policy, now and for the future? This essay examines these questions as they relate to the top cut of the nation's scientific and engineering manpower—those with a Ph.D. degree—and it includes social scientists as well as those in the physical and biological sciences.

## Unemployment among scientists and engineers

Unemployment among scientists and engineers is so novel that exaggerated stories have been circulated widely. The actual unemployment rate among scientists in the spring of 1971 was 2.6%, as compared with 1.5% in the spring of 1970. The unemployment rate among the entire labor force for the spring of 1971 was 6.5%. So for scientists as a whole, the unemployment rate remained quite low. The general impression that unemployment was higher probably rested on the existence of pockets of unemployment in particular age groups, geographical areas, disciplines, and levels of training. For example, in the spring of 1971, in Seattle, Wash., a city hard hit by the reduction of work in aerospace companies, 5.6% of all scientists were unemployed. Nationwide, 5.3% of all scientists under the age of 29, 3.9% of the physicists, and 3.8% of the sociologists were unemployed. Most significantly the unemployment rate among those scientists with only a bachelor's degree was 4.1%, while only 1.4% of those with a doctorate were unemployed. The idea that there was widespread unemployment among Ph.D.s in science, or in any field, was contrary to fact. Moreover, even among the recent graduates, unemployment rates for holders of Ph.D.s in science were low. In the early months of 1971, only 0.5% of those who received a Ph.D. degree in science in 1969 and 1.1% of those who received the degree in 1970 were unemployed. The unemployment rates were only slightly higher among those with degrees from the less prestigious institutions. New Ph.D.s had to look harder and longer for jobs, and while they were looking, they were unemployed. After a long search, many of them did not land the job they preferred, but few—only 1.2%—took jobs that did not adequately make use of their graduate training.

Unemployment at this level certainly did not indicate that the country had trained too many scientists and engineers. Changes in a dynamic economy are certain to produce unemployment as emphasis shifts from one field to another. However, those unemployment levels may presage a long-range situation in which a degree of unemployment among scientists and engineers will continue, and in which many scientists and engineers may be forced to take jobs that are neither in universities nor in research and development.

## What happened?

The output of Ph.D.s in science and engineering jumped fantastically in recent years, and then the demand for their services slackened. This produced unemployment. Let us look at each development in some detail.

There were as many Ph.D.s in science and engineering produced in the United States in the decade of the 1960s (112,000) as were graduated in the nation's entire prior history (111,000, *see* Table). This tremendous acceleration of output, which was roughly equal in all fields of science, engineering, and the nonsciences, required a continuing rapid growth of the capacity of the economy to absorb Ph.D.s.

***CHARLES V. KIDD*** *is director of the council on federal relations of the Association of American Universities and the author of* American Universities and Federal Research.

**Ph.D.s graduated in the United States 1880–1974**

| year | total | scientists and engineers | scientists and engineers as % of total |
|---|---|---|---|
| 1880–1929 | 23,000 | 15,000 | 65 |
| 1930–39 | 26,000 | 18,000 | 70 |
| 1940–44 | 15,000 | 10,000 | 66 |
| 1945–49 | 16,000 | 11,000 | 70 |
| 1950–54 | 40,000 | 27,000 | 67 |
| 1955–59 | 45,000 | 30,000 | 66 |
| 1960–64 | 60,000 | 41,000 | 66 |
| 1965–69 | 100,000 | 71,000 | 70 |
| 1970–74 | 160,000 | 101,000 | 63 |

Until 1968, the absorptive capacity did expand at the rate required to provide jobs for the new Ph.D.s. College and university enrollments, including enrollment in graduate science and engineering, doubled from three million to six million between 1960 and 1969. This generated a demand for new faculty members, and full-time faculty in colleges and universities increased from 170,000 in 1960 to 290,000 in 1969. Federal research and development expenditures grew steadily. From 1953 through 1961, the average annual growth rate of federal research and development expenditures was 16.3%, and private research and development expenditures rose by 10.1% a year. The comparable rates from 1961 through 1966 were 8.6 and 9.4%.

But between 1966 and 1968 things began to change. For example, from 1966 through 1971, federal research and development expenditures rose by only 1% a year and in real terms they declined by 3% a year. The aerospace cuts were particularly hard on engineers. Then came the general economic slowdown, with a 3% reduction in industrial research and development expenditures from industrial sources in 1970. These downward pressures were augmented by difficulties in the academic field. Universities, the prime academic employers of Ph.D.s, ran into steadily mounting financial difficulties. Graduate schools were particularly hard hit because of the federal slowdown in research and development expenditures and in their reduced support of graduate students. All of these factors combined explain the painful adjustments in the labor market.

### What does the future hold?

Let us start with an examination of the existing pool of scientists and engineers. The United States now has about 160,000 scientists and engineers with Ph.D. degrees. They are distributed as follows:

| *Employer* | % |
|---|---|
| Universities and colleges | 60 |
| Private industry | 25 |
| Government | 10 |
| Other | 5 |

Therefore, the question is this: How rapidly will that pool expand over the next decade, and will the expanded pool be productively employed —and if so, where? Answers to such questions involve forecasting, and forecasting is now an imperfect art. One problem is that forecasts in the social sciences are not simply efforts to foretell what will happen. They can and often do influence the course of events. For example, a forecast of overproduction of engineers with Ph.D.s can be one of the factors leading to a decrease in enrollment and eventually in the output of Ph.D.s. The problem is further complicated by the long time required to produce a Ph.D. and by the long working life of the average Ph.D. A student who is pondering whether to become a Ph.D. candidate must, so far as job prospects influence him, consider not the current market but the market when he first looks for a job five years later and the market over the succeeding 30 years of his working life. To the extent that people believe manpower forecasts and act upon them, the forecasts are doomed to error unless those who make them predict the effects of their own forecasts. The more widely the forecasts are publicized the greater the extent to which they become an active factor influencing the trend of events.

Many powerful forces will continue to promote a continuing expansion of all graduate study, including that of the sciences and engineering. University science and engineering departments need graduate programs, and they need graduate students. The complexity of technology will continue to generate substantial demands for doctorates in engineering. Substantial support for students will continue, from one source or another, and large amounts of research money will continue to flow into universities. So, all things considered, there is a great deal of momentum in the system. Those students already in graduate school programs probably will push on to the Ph.D. degree in about the same proportions as in the past.

One would think that most students would shy away from enrolling for graduate work in science because of the stories of current unemployment forecasts and the declining support for science students. Rapid increases in graduate enrollment in science and engineering that characterized the 1960s have indeed slowed down. During the 1960s graduate enrollment in the physical sciences and engineering more than doubled; in mathematics and the biological sciences, enrollment almost tripled. But in the fall of 1970 and again in the fall of 1971, first time enrollment of full-time graduate students in science and engineering increased by just 5%. Enrollment in science was up by 6% or so and engineering enrollment was unchanged. Total graduate first-year, full-time enrollment slackened off most rapidly in science and engineering in private institutions. The general public disillusionment with science and engineering and decreases in private and public support for graduate students contributed to this change.

While these negative pressures will probably continue, there is one powerful counter force—a continuing increase in the number of bachelor's degrees, including degrees in science and engineering. In 1965, five years before the 29,000 total doctorates of 1970 (17,300 of whom were scientists and engineers), 530,000 students received bachelor's degrees. In 1975, five years before the doctorate group of 1980, there will be about 960,000 bachelor's degrees. If the same proportion of the 1975 bachelor-degree holders proceed to a Ph.D. degree, as was true of their 1965 predecessors, about 52,000 Ph.D. degrees will be granted in all fields (about 32,000 of whom would be scientists and engineers).

A reasonable estimate of the total available supply of scientists and engineers in 1980 (after taking into account not only production, but death, retirement, emigration, and immigration) would be in the range of 315,000 to 340,000, with about 325,000 as a midpoint. These are the assumptions of the National Science Foundation and they are close to the estimates of others who forecast relatively low numbers of scientists and engineers with Ph.D. degrees.

### Supply and demand

If one accepts the assumption that the existing pool of about 160,000 Ph.D.s in science and engineering will grow to about 325,000 or so by 1980, what is the prospect that the increased number—more than double the 1969 level—can be productively utilized? Assumptions have to be made to answer questions like this: At what rate will the total economy expand? What proportion of the gross national product will go into research and development? At what rates will enrollment in the sciences and engineering increase at undergraduate and graduate levels? What would be the effect of a really large-scale attack on problems of the environment? What will be the ratio of professors to students, and what proportion of faculty will have a Ph.D.? To what extent will Ph.D.s in science and engineering be absorbed in nonacademic, non-research and development jobs that will use their training reasonably well? What will be the long-run effects of the youth rebellion on advanced training in science?

Such a simple list of the questions that have to be answered in forecasting the demand for Ph.D.s in science and engineering suffices to show the impossibility of making firm and precise forecasts. Nevertheless, there is a fairly wide consensus on the dimensions of future demand for Ph.D.s, and the National Science Foundation projections provide a reasonable set of figures. They indicate that there will be suitable jobs for 270,000 to 300,000 scientists and engineers in 1980. Employment substantially above these levels would be generated only by implausible rates of increase in the gross national product devoted to research, in undergraduate and graduate enrollment, in the proportion of all college and university faculty with a Ph.D., and in the absorption of Ph.D.s in nontraditional areas requiring high levels of academic preparation.

The total supply of doctorates in 1980 has been predicted to be 325,000. A likely figure for utilization of doctorates in that year would be 285,000. In this circumstance, there would be a gap of 40,000 between supply and utilization. These Ph.D.s would be for the most part underemployed. These figures are not precise but represent calculations based on plausible but not necessarily precise assumptions. In addition, the figures are for all scientists and engineers and do not indicate differences among fields. With these qualifications, the implications of the forecast are clear.

In 1971 most forecasts were less optimistic than they were in earlier years. For example, the National Science Foundation said of its 1971 projections that this "represents a greater likelihood of a future oversupply than the projections developed two years ago." Allan Cartter, who first pointed out that we should be worried about oversupply rather than shortages in 1980, has noted that the time of trouble has arrived sooner than he had originally thought. As of now, the possibility of oversupply should be viewed as a matter of continuing concern and not as a passing aberration.

The rough overall balance of supply and demand disguises some probable imbalance in specific fields. It appears that there may be substantially more engineering Ph.D.s than will find suitable work, although virtually all of them will have jobs. This may also be the case in mathematics and the social sciences, but the imbalance may not be as severe as in the case of engineering. The outlook is for reasonable balance in the physical sciences.

If there is not to be substantial unemployment among scientists and engineers, many more than the current 10%—as many as 15 or 20%—will have to be employed in nonacademic, non-research and development employment. This includes such fields as management, sales, government administration, consultation, and teaching in junior colleges and high schools. Here one faces the question whether such employment proves that there has been "overproduction." I think not, for the reason that these are worthy and productive careers. Moreover, a consistent line of development in this country, and one which has contributed greatly to social mobility and economic growth, has been a steady elevation of the level of education of those in given lines of work. However, this line of argument cannot be pushed too far because at an extreme it offers a way of dismissing any level of overproduction. That is, those not absorbed in academic work in four-year colleges and universities and in nonacademic research and development are simply assumed to be productively employed in other categories. If the National Science Foundation projections have a weakness it is in the assumption that from 40,000 to 55,000 engineering and science doctorates in the other category would all be suitably employed.

The overall picture of *total* employment also disguises the situation that will be faced by the *new* Ph.D.s. A certain and striking decrease in the proportion of new Ph.D.s who will be employed in universities and colleges is one of the most prominent prospects of the 1970s. This decline will come about as the rate of increase in undergraduate enrollment decreases over the decade and as enrollment actually declines in the 1980s. The proportion of new Ph.D.s in science and engineering who will be employed as faculty in colleges and universities will drop (from 60% during the 1960s) to about 10 to 25% in the last half of the 1970s. This decline will come about even if substantially more Ph.D.s are employed in junior colleges and even if the proportion of faculty with Ph.D.s in all academic institutions increases substantially. On the other hand, the proportion of new Ph.D.s during the 1970s who will be in nonacademic, non-research and development jobs will be 20 to 25%, and the proportion will rise throughout the decade.

Because of the changing job market, a substantial proportion of the Ph.D.s in science and engineering will have to be trained more broadly than is now the case, and with less emphasis on training for academic research. The change in the labor market will generate a debate over the nature of advanced training that will continue for years. If history offers a guide, the necessary adjustments will be made long after the need for them becomes evident.

The bargaining power of Ph.D.s in science and engineering will be less, and the power of employers—academic and nonacademic—will be greater in the present decade than it was during the preceding two decades. Not only jobs but promotions and raises will be harder to come by. To some degree, those with jobs will be in competition with those coming into the market. The average age of faculties will increase, and this will raise acutely the problem of maintaining youthful vigor in research and teaching in fields that will continue to change rapidly. Pressure will be exerted to reduce retirement ages and to modify tenure policies, with resulting financial problems for individuals and institutions.

### Overproduction or underproduction?

Given the fact that supply and demand will rarely be in ideal balance, and that forecasts are virtually certain to be wrong in some respects, what general course of action with respect to training of scientists and engineers is appropriate? Should output be restricted on the ground that supply will apparently outrun demand?

The costs to the nation that would be generated by overproduction of scientists and engineers would be far less than those that would be generated by underproduction. The costs of overproduction are of two kinds. First, there are the economic costs of educating those who are "surplus" to the needs of the economy—defining surplus as the number who are not employed in jobs making reasonably full use of their advanced training. Second, there are the intangible costs to individuals as well as to society when a group of highly intelligent and highly trained people are unable to use their skills fully. However, the costs of overproduction, both to society and the individual, are reduced to a substantial degree by the fact that highly-trained scientists and engineers can perform capably and usefully in many capacities other than those for which they have been specifically trained.

On the other hand, the costs of having too few highly-trained scientists and engineers are extremely high. The inability to staff important segments of the economy with enough highly-trained people will tend to depress the total output of goods and services, retard advances in productivity, decrease the adaptability of the economic system, stifle innovation, and raise prices.

These considerations suggest that when the future is murky, as it generally is, the wisest course is not to take steps designed to achieve a fine balance between current output and anticipated demand, but rather to sustain output at levels somewhat above those that would be indicated if future balance were the goal.

The federal government has overreacted to the short-run situation. From a level of 50,000 graduate students in all fields supported by federal fellowships in 1968, there has been a decline to 30,000 in 1971, and a further decline to 22,000 in 1972 is probable. The cuts in graduate student support in science and engineering have been proportional to the total reduction. This abrupt reversal of federal policy toward support of graduate students threatens to cut the capacity of graduate schools—particularly the best ones—to turn out the highly-trained scientists and engineers needed a decade or more in the future.

If there is a principal lesson to be learned concerning the current changes in the labor market and the responses to them, it is that the U.S. has not yet learned how to combine planning and free choice in a way that will give individuals a reasonable chance to follow their interests while ensuring that the needs of society are reasonably well met.

# Index

Index entries to feature and review articles in this and previous editions of the *Britannica Yearbook of Science and the Future* are set in boldface type, *e.g.*, **Astronomy.** Entries to other subjects are set in lightface type, *e.g.*, Radiation. Additional information on any of these subjects is identified with a subheading and indented under the entry heading. The numbers following headings and subheadings indicate the year (boldface) of the edition and the page number (lightface) on which the information appears.

**Astronomy 73**-184; **72**-187; **71**-133
  Colonizing the Moon **72**-12
  honors **73**-251; **72**-260; **71**-205
  measuring temperature of center of sun **71**-370
  photography
    stars trajectory il. **71**-367
  space probe research **71**-129
  spectrograph **73**-175
  topography of Venus **73**-223

All entry headings, whether consisting of a single word or more, are treated for the purpose of alphabetization as single complete headings and are alphabetized letter by letter up to the punctuation. The abbreviation "il." indicates an illustration.

## A

## B

# C

# D

## G



## H

## J



## K



## L

# M

# N



# O

# Q



# R

# S

# T



# U

# Acknowledgments


| | |
|---|---|
| 2 | Photograph by Ed Cornachio |
| 6 | Photographs by (top, left) Brian Brake from Rapho Guillumette; (top, right) Dan Morrill; (center, left) courtesy, Unimation, Inc.; (bottom, left) *Observer*, Transworld Features; (bottom, right) Carlo Bavignoli, *Life* Magazine © Time Inc. |
| 10–11 | Illustration by Lido Lucchesi; photographs by Richard Keane |
| 31 | Photograph by Marc Riboud from Magnum |
| 66 | Photographed by Bill Arsenault |
| 93 | Illustration by George Suyeoka |
| 98 | Photographed by Richard Fegley |
| 128–143 | Illustrations by Phil Renaud |
| 146–157 | Illustrations by Dave Beckes |
| 159 | Photographs by (from top to bottom) courtesy, Dr. John H. Crowe, University of California, Davis; courtesy, Los Alamos Scientific Laboratory; Dmitri Kessel; Leonard Kamsler, *Medical World News* |
| 179, 183, 190, 206, 207, 209, 210, 211, 212, 217, 226, 242, 291, 299, 301, 305 314, 318 | Illustrations by Dave Beckes |
| 186, 197, 233, 267, 295, 317 330, 339 | Illustrations by Victor F. Seper, Jr. |
| 361 | Photographs by Dan Morrill |
| 365 | Illustration by Dave Beckes |
| 398–399 | Photographed by Bill Arsenault |
| 404–405 | Illustrations by Dave Beckes |

# EMERY AND RIMOIN'S PRINCIPLES AND PRACTICE OF MEDICAL GENETICS AND GENOMICS

## Foundations

Seventh Edition

Edited by

Reed E. Pyeritz
Perelman School of Medicine at the University of Pennsylvania, Philadelphia, PA, United States

Bruce R. Korf
University of Alabama at Birmingham, Birmingham, AL, United States

Wayne W. Grody
UCLA School of Medicine, Los Angeles, CA, United States

Academic Press is an imprint of Elsevier
125 London Wall, London EC2Y 5AS, United Kingdom
525 B Street, Suite 1650, San Diego, CA 92101, United States
50 Hampshire Street, 5th Floor, Cambridge, MA 02139, United States
The Boulevard, Langford Lane, Kidlington, Oxford OX5 1GB, United Kingdom



**Library of Congress Cataloging-in-Publication Data**
A catalog record for this book is available from the Library of Congress

**British Library Cataloguing-in-Publication Data**
A catalogue record for this book is available from the British Library

ISBN: 978-0-12-812537-3

For information on all Academic Press publications visit our website at https://www.elsevier.com/books-and-journals

*Publisher:* Andre Wolff
*Senior Acquisition Editor:* Peter B. Linsley
*Editorial Project Manager:* Pat Gonzalez
*Production Project Manager:* Punithavathy Govindaradjane
*Designer:* Matthew Limbert

Typeset by TNQ Technologies

# CONTENTS

# LIST OF CONTRIBUTORS

**Stylianos E. Antonarakis**
Department of Genetic Medicine and Development, University of Geneva Medical School, Geneva, Switzerland

**T. Mark Beasley**
Department of Biostatistics, School of Public Health, University of Alabama at Birmingham, Birmingham, AL, United States

**Darci T. Butcher**
Genetics and Genome Biology, Research Institute, The Hospital for Sick Children, Toronto, ON, Canada

**Lucas Calais-Ferreira**
Centre for Epidemiology and Biostatistics, Melbourne School of Population and Global Health, University of Melbourne, Melbourne, VIC, Australia; CAPES Foundation, Ministry of Education, Brasilia, Brazil

**Rita M. Cantor**
Department of Human Genetics, David Geffen School of Medicine at UCLA, Los Angeles, CA, United States

**Stephen D. Cederbaum**
Research Professor and Professor Emeritus of Psychiatry, Pediatrics and Human Genetics, University of California, Los Angeles, Los Angeles, CA, United States

**Sanaa Choufani**
Genetics and Genome Biology, Research Institute, The Hospital for Sick Children, Toronto, ON, Canada

**Jackie Cook**
Consultant in Clinical Genetics, Sheffield Clinical Genetics Service, Sheffield Children's NHS Foundation Trust, Sheffield, United Kingdom

**David N. Cooper**
Institute of Medical Genetics, Cardiff University, Cardiff, United Kingdom

**Jeffrey M. Craig**
Deakin University School of Medicine, Geelong, VIC, Australia; Murdoch Children's Research Institute, Royal Children's Hospital, Parkville, VIC, Australia

**Michael R. Crowley**
Department of Human Genetics, Emory University, Atlanta, GA, United States; The Department of Genetics, The University of Alabama at Birmingham, Birmingham, AL, United States

**Cheryl Cytrynbaum**
Genetics and Genome Biology, Research Institute, The Hospital for Sick Children, Toronto, ON, Canada; Clinical and Metabolic Genetics, The Hospital for Sick Children, Toronto, ON, Canada

**Allison Fialkowski**
Department of Biostatistics, School of Public Health, University of Alabama at Birmingham, Birmingham, AL, United States

**Geoffrey S. Ginsburg**
Center for Applied Genomics & Precision Medicine, Duke University School of Medicine, Durham, NC, United States

**Wayne W. Grody**
UCLA School of Medicine, Los Angeles, CA, United States

**Susanne B. Haga**
Center for Applied Genomics & Precision Medicine, Duke University School of Medicine, Durham, NC, United States

**Judith G. Hall**
Departments of Medical Genetics and Pediatrics, British Columbia's Children's Hospital, Vancouver, BC, Canada

**Madhuri R. Hegde**
Department of Human Genetics, Emory University, Atlanta, GA, United States; The Department of Genetics, The University of Alabama at Birmingham, Birmingham, AL, United States

**Fuki M. Hisama**
Division of Medical Genetics, Department of Medicine, University of Washington, Seattle, WA, United States

**H. Richard Johnston**
Department of Human Genetics, Emory University School of Medicine, Atlanta, GA, United States

**Bronya J.B. Keats**
Department of Genetics (Emeritus), Louisiana State University Health Sciences Center, New Orleans, LA, United States

**Bruce R. Korf**
Department of Genetics, University of Alabama at Birmingham, Birmingham, AL, United States

**Marie T. Lott**
Center for Mitochondrial and Epigenomic Medicine, Children's Hospital of Philadelphia, Philadelphia, PA, United States

**George M. Martin**
Department of Pathology, University of Washington, Seattle, WA, United States

**Fady M. Mikhail**
Cytogenetics Laboratory, Department of Genetics, University of Alabama at Birmingham, Birmingham, AL, United States

**Daniel W. Nebert**
Department of Environmental Health and Center for Environmental Genetics, University of Cincinnati School of Medicine, Cincinnati, OH, United States; Department of Pediatrics, University of Cincinnati School of Medicine, Cincinnati, OH, United States; Division of Human Genetics, Cincinnati Children's Hospital Medical Center, Cincinnati, OH, United States

**Junko Oshima**
Department of Pathology, University of Washington, Seattle, WA, United States

**Vincent Procaccio**
Biochemistry and Genetics Department, MitoVasc Institute, UMR CNRS 6015 – INSERM U1083, CHU Angers, Angers, France

**Reed E. Pyeritz**
Perelman School of Medicine at the University of Pennsylvania, Philadelphia, PA, United States

**Katrina J. Scurrah**
Centre for Epidemiology and Biostatistics, Melbourne School of Population and Global Health, University of Melbourne, Melbourne, VIC, Australia

**Stephanie L. Sherman**
Department of Human Genetics, Emory University School of Medicine, Atlanta, GA, United States

**Michelle T. Siu**
Genetics and Genome Biology, Research Institute, The Hospital for Sick Children, Toronto, ON, Canada

**Hemant K. Tiwari**
Department of Biostatistics, School of Public Health, University of Alabama at Birmingham, Birmingham, AL, United States

**Benjamin Tycko**
Division of Genetics & Epigenetics, Hackensack Meridian Health Center for Discovery and Innovation, Nutley, NJ, United States

**Mark P. Umstad**
Department of Maternal-Fetal Medicine, The Royal Women's Hospital, Melbourne, VIC, Australia; University Department of Obstetrics and Gynaecology, University of Melbourne, Melbourne, VIC, Australia

**Douglas C. Wallace**
Center for Mitochondrial and Epigenomic Medicine, Children's Hospital of Philadelphia, Philadelphia, PA, United States; Perelman School of Medicine, University of Pennsylvania, Philadelphia, PA, United States

**Rosanna Weksberg**
Genetics and Genome Biology, Research Institute, The Hospital for Sick Children, Toronto, ON, Canada; Clinical and Metabolic Genetics, The Hospital for Sick Children, Toronto, ON, Canada

**Ge Zhang**
Department of Pediatrics, University of Cincinnati School of Medicine, Cincinnati, OH, United States; Division of Human Genetics, Cincinnati Children's Hospital Medical Center, Cincinnati, OH, United States

# PREFACE TO THE SEVENTH EDITION OF *EMERY AND RIMOIN'S PRINCIPLES AND PRACTICE OF MEDICAL GENETICS AND GENOMICS*

The first edition of *Emery and Rimoin's Principles and Practice of Medical Genetics* appeared in 1983. This was several years prior to the start of the Human Genome Project in the early days of molecular genetic testing, a time when linkage analysis was often performed for diagnostic purposes. Medical genetics was not yet a recognized medical specialty in the United States, or anywhere else in the world. Therapy was mostly limited to a number of biochemical genetic conditions, and the underlying pathophysiology of most genetic disorders was unknown. The first edition was nevertheless published in two volumes, reflecting the fact that genetics was relevant to all areas of medical practice.

Thirty-five years later we are publishing the seventh edition of *Principles and Practice of Medical Genetics and Genomics*. Adding "genomics" to the title recognizes the pivotal role of genomic approaches in medicine, with the human genome sequence now in hand and exome/genome-level diagnostic sequencing becoming increasingly commonplace. Thousands of genetic disorders have been matched with the underlying genes, often illuminating pathophysiological mechanisms and in some cases enabling targeted therapies. Genetic testing is becoming increasingly incorporated into specialty medical care, though applications of adequate family history, genetic risk assessment, and pharmacogenetic testing are only gradually being integrated into routine medical practice. Sadly, this is the first edition of the book to be produced without the guidance of one of the founding coeditors, Dr David Rimoin, who passed away just as the previous edition went to press.

The seventh edition incorporates two major changes from previous editions. The first is publication of the text in 11 separate volumes. Over the years the book had grown from two to three massive volumes, until the electronic version was introduced in the previous edition. The decision to split the book into multiple smaller volumes represents an attempt to divide the content into smaller, more accessible units. Most of these are organized around a unifying theme, for the most part based on specific body systems. This may make the book more useful to specialists who are interested in the application of medical genetics to their area but do not wish to invest in a larger volume that covers all areas of medicine. It also reflects our recognition that genetic concepts and determinants now underpin all medical specialties and subspecialties. The second change might seem on the surface to be a regressive one in today's high-tech world—the publication of the 11 volumes in print rather than strictly electronic form. However, feedback from our readers, as well as the experience of the editors, indicated that access to the web version via a password-protected site was cumbersome, and printing a smaller volume with two-page summaries was not useful. We have therefore returned to a full print version, although an eBook is available for those who prefer an electronic version.

One might ask whether there is a need for a comprehensive text in an era of instantaneous internet searches for virtually any information, including authoritative open sources such as *Online Mendelian Inheritance in Man* and *GeneReviews*. We recognize the value of these and other online resources, but believe that there is still a place for the long-form prose approach of a textbook. Here the authors have the opportunity to tell the story of their area of medical genetics and genomics, including in-depth background about pathophysiology, as well as giving practical advice for medical practice. The willingness of our authors to embrace this approach indicates that there is still enthusiasm for a textbook on medical genetics; we will appreciate feedback from our readers as well.

The realities of editing an 11-volume set have become obvious to the three of us as editors. We are grateful to our authors, many of whom have contributed to multiple past volumes, including some who have updated their contributions from the first or second editions. We are also indebted to staff from Elsevier, particularly Peter Linsley and Pat Gonzalez, who have worked patiently with us in the conception and production of this large project. Finally, we thank our families, who have indulged our occasional disappearances into writing and editing. As always, we look forward to feedback from our readers, as this has played a critical role in shaping the evolution of *Principles and Practice of Medical Genetics and Genomics* in the face of the exponential changes that have occurred in the landscape of our discipline.

# PREFACE TO *FOUNDATIONS*

All previous editions of *Principles and Practice of Medical Genetics* have started with a section on basic principles. Although the primary audience for the book is medical geneticists, who might be expected to have mastered basic genetic principles, in fact the breadth of the field makes it difficult for any individual to keep up in all areas. Moreover, in the current edition we have divided the text into 11 volumes in order to make the book more accessible to practitioners who are not medical geneticists but need a resource on genetic and genomic approaches in their discipline. They might especially find an overview of basic principles to be useful.

This volume begins with a summary of the history of medical genetics, initially written by the late Dr. Victor McKusick, who had a unique vantage point on the origin and development of the discipline. It is impossible to replace that perspective, and the history as he recorded it has not changed, but we are grateful to Dr. Stephen Cederbaum for being willing to add commentary on more recent events. We then provide broad overviews on medicine in a genetic context, the newer concept of precision medicine, and the nature and frequency of genetic disease. Subsequent chapters provide overviews of critical topics in three major themes—the flow of genetic information in the cell, the transmission of genetic traits in families, and the forces that mold allele frequencies in the population. We do not focus on medical applications in this volume. Broad coverage of medical genetic and genomic approaches is provided in Volume 2 and underlies the basis of virtually all the other volumes in this text.

Medical genetics and genomics is the quintessential moving target, advancing at a pace that could never be captured in a textbook. Nevertheless, we hope that this overview volume will provide a comprehensive snapshot of genetic and genomic principles, providing a foundation for the subsequent volumes on medical applications and system-specific genetic and genomic approaches.

# INTRODUCTION[a]

*Stephen D. Cederbaum*

Research Professor and Professor Emeritus of Psychiatry, Pediatrics and Human Genetics, University of California, Los Angeles, Los Angeles, CA, United States

## 1. PREFACE (STEPHEN CEDERBAUM)

History is most often considered to be written in the present to record and analyze events that occurred in the recent or more distant past and makes use of written and oral documentation of the events as they happened. The history of medical genetics is no different. The many outstanding chapters in these volumes are the raw material from which historiographers will, in the future, take this documentation to weave a story of what happened in our time and how it fitted into a larger picture. From that perspective a history that brings us into the present is something of an oxymoron. In this addendum I accept that charge and will attempt to provide a true history and at the same time provide a framework for understanding contemporaneous events that are shaping and influencing our approach to virtually every disorder in our practice and discussed in this text.

In the 1996, third, edition of this book, Victor McKusick published a detailed history of human and medical genetics, with a uniquely personal perspective. When McKusick passed, first Peter Harper and now I have undertaken the unenviable task of bringing this history up to date, reflecting on the progress and changes that have occurred in medical genetics and in medical genetic practice. This is an unenviable task because no one will likely ever achieve the level of authoritative scholarship that is seen in McKusick's essay. Moreover, much of the modern history of medical genetics occurred in his lifetime, and his account of this period has the personal perspective of someone who has lived it and who, to a large extent, helped to shape it.

In Harper's addendum he added a table in which he continued the tradition of highlighting the timeline of major advances in genetics, but without the expanded narrative given by McKusick. He did pay tribute to McKusick and to the other intrepid pioneers who founded medical genetics as a medical specialty and to whom we owe our current exciting and dynamic field. Neither Harper nor I found ourselves in a position to take this unique *tour de force* and rewrite it for the future and for an audience for whom the personal touches would have less meaning.

Thus, for the present edition of this book, as Harper wrote, a compromise has been reached, whereby the chapter remains essentially intact, with Harper's addendum and comprehensive table, but is framed by this introduction and by a postscript outlining some of the relevant topics that bring the story of medical genetics into the brave new world in which we now exist and practice.

## 2. HISTORY OF MEDICAL GENETICS[1] (VICTOR A. McKUSICK)

***Victor A. McKusick***

Formerly University Professor of Medical Genetics, Johns Hopkins University, Physician, Johns Hopkins Hospital, Baltimore, MD, United States

Since the late 1950s, when medical genetics first became a defined field of genetics, and a medical specialty in its own right, it has changed and developed at a remarkable pace. It is thus important to record its history, both its clinical and its laboratory aspects, while most of the original founders are still living, since those

[a] This article is an addendum to the previous edition article by Victor A. McKusick, Peter S. Harper, vol. 1, pp. 1–39, © 2011, Elsevier Ltd.

[1] This article is a revision of the previous edition article by Victor A. McKusick, vol. 1, pp. 3–32, © 2007, Elsevier Ltd.

entering the field and those increasingly using it in wider medical research and practice may have little knowledge of how medical genetics began.

This chapter, largely written by Dr. Victor McKusick, now deceased and himself a central figure in the history of medical genetics, has been extended by Peter Harper. It first outlines the foundations of medical genetics that lay in more general genetics in the first half of the 20th century, before examining the key later scientific discoveries, including the human gene map and the Human Genome Project (HGP), the clinical advances that have increasingly followed from these advances, and the growth of medical genetics as a specialty in both clinical and laboratory medicine.

Finally, the chapter emphasizes the importance and urgency of preserving the written records and other historical material underlying these discoveries, as well as the memories and correspondence of the many scientists and clinical workers worldwide who have been responsible for the development of this remarkable field of science and medicine.

## 2.1 Preface (Peter Harper)

Victor McKusick's chapter "History of Medical Genetics," which first appeared in the third edition of this textbook in 1996, has provided readers with a masterly account of the origins and development of the field, containing a remarkable amount of information in the limited space available. This has been a historical account as seen by somebody who has lived through most of it, and who has perhaps shaped the development of medical genetics more than any other individual. While this may at times have given a personal slant to the areas emphasized, it provides an unrivaled perspective on the field—indeed for many years it has been the only substantive account, no detailed history of medical genetics having appeared until recently [8].

McKusick's death in 2008 has sadly removed the possibility of him providing a revised edition of his chapter, but his continued active attention to, and involvement in, the detailed evolution of medical genetics, up to the very end of his life, seen particularly in his interest in the HGP and in the updating of his catalog OMIM, not to mention his many unique photographs, means that the chapter is still of the greatest value, and that it would be wrong, as well as regrettable, were it to be discarded. At the same time, though, the highly condensed nature of the text makes it difficult, if not impossible, to edit without making radical changes.

Thus, for the present edition of this book, a compromise has been reached, whereby the chapter remains essentially intact, but is framed by this introduction and by a postscript outlining some of the relevant topics that the original chapter does not cover. The Web-based format of this new edition will allow the chapter to evolve progressively over the coming years, in keeping with changes in the field, which have not only seen continuing new advances, but have also shown the urgent need for the history of its recent past to be recorded and preserved.

## 2.2 Introduction

Medical genetics is the science of human biologic variation as it relates to health and disease. Clinical genetics is that part of medical genetics concerned with the health of individual humans and their families. Alternatively, clinical genetics can be defined as the science and practice of diagnosis, prevention, and management of genetic disorders.

Within recent years, medical genetics has become established as a clinical specialty, as the culmination of developments that began in 1956 with the description of the correct chromosome number of the human. With the discovery of specific microscopically visible chromosomal changes associated with clinical disorders, beginning with Down syndrome in January 1959, medical genetics acquired an anatomic base. Medical geneticists now had their specific organ—the genome—just as cardiologists had the heart and neurologists had the nervous system.

The anatomic base of medical genetics was greatly extended with the mapping of genes to chromosomes and specific chromosomal regions, at an ever-accelerating pace, since the late 1980s. Gene mapping has not only enlarged the base for medical genetics but, indeed, as pointed out to me by Charles Scriver (personal communication, 1980), has also provided a neo-Vesalian basis for all of medicine [1]. Medical historians tell us that the anatomy of Vesalius published in 1543 was of pivotal importance in the development of modern medicine. It was the basis of the physiology of William Harvey and the morbid anatomy of Morgagni. Similarly, human gene mapping constitutes an approach to the study of abnormal gene function in all diseases; the gene mapping approach has been adopted by researchers in almost all branches of medicine in the study of their most puzzling disorders. Through mapping, they have sought the basic defect in these disorders, and their clinical colleagues have used mapping information for diagnosis and carrier detection. The ultimate anatomic basis for medical genetics, the DNA sequence, was provided by the HGP.

In this brief history of medical genetics, I trace the foundations of the field that were laid between 1865, when Mendel published his work, and 1956, when the correct chromosome number was reported. I then discuss the events since the late 1960s that have seen the main evolution of the discipline. Finally, I attempt some projections for the future.

## 2.3 Foundations of Medical Genetics Before 1956

Medical genetics in many developed countries is now a recognized specialty. In the United States, for example, the American Board of Medical Genetics and Genomics certifies practitioners in the field, including PhD medical geneticists; in 1991, the American Board of Medical Genetics and Genomics became the 24th organization to join the family of certifying American specialty boards. Medical genetics is a rather unusual branch of clinical medicine; indeed, it may be nearly unique in that it originated out of a basic science. Most specialties started as crafts (or grew out of a technological advance such as radiography) and only subsequently acquired basic science foundations.

The basic science that developed before 1956 and served as the foundation for the developments since the late 1960s included Mendelism, cytogenetics, biochemical genetics, immunogenetics, and statistical, formal, and population genetics.

### 2.3.1 Mendelism

The demonstration of the particulate nature of inheritance was the contribution of Gregor Mendel (1822–84), a monk and later abbot in an Augustinian monastery (Fig. 1.1) in Brünn (now Brno), Moravia (now the Czech Republic). The terms dominant and recessive were his. The delay in recognition of his work has been attributed to various factors, but the most likely is poor timing; in 1865, when Mendel reported his findings and conclusions, the chromosomes had not yet been discovered. Because its physical basis, meiosis had not yet been described, Mendelism had no plausible basis to qualify it over other possible mechanisms of inheritance. such as blending inheritance, which was favored by Francis Galton (1822–1911), one of Mendel's contemporaries [151].

R.A. Fisher (1890–1962) [2] raised a question whether Mendel's results were "too good"; that is, the data agreed too closely with the conclusions (see the discussion of this matter by Novitski [3]). (Mendel's "round" [R] vs. "wrinkled" [r] trait in the garden pea has been shown

**Figure 1.1** *(A) Mendel's monastery as it appeared in 1971. (B) Participants in the 2005 International Workshop on Genetics, Medicine and History, held at Mendel's Abbey, Brno, Czech Republic.*

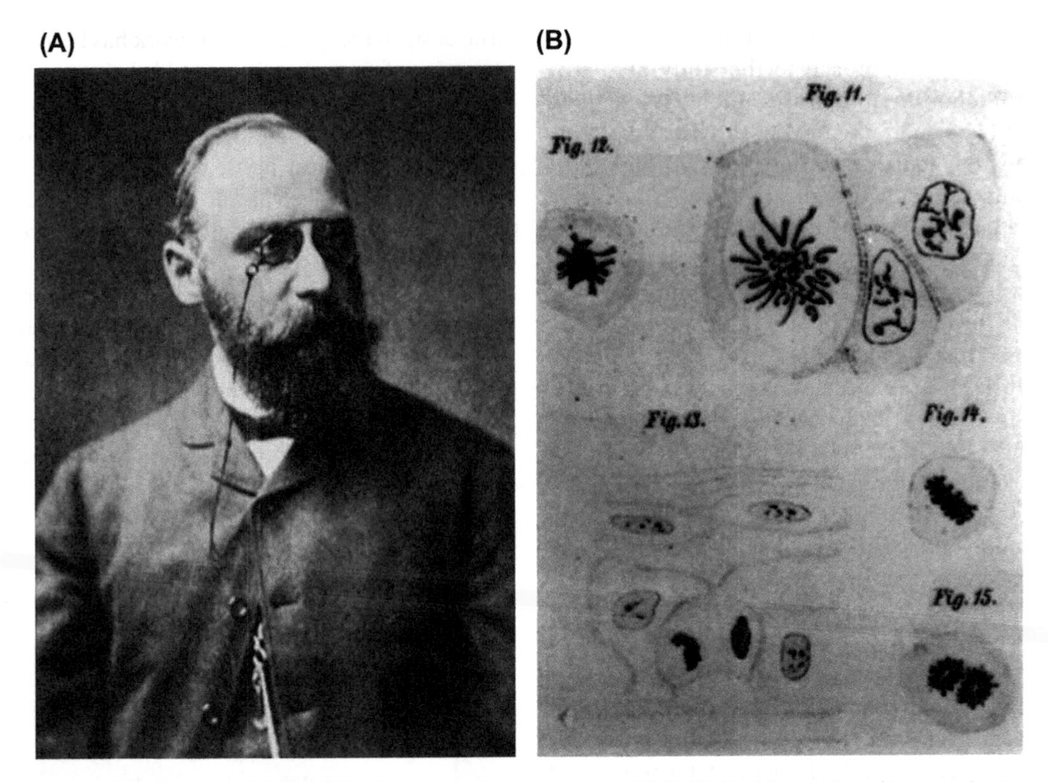


**Figure 1.2** (A) Walther Flemming (1843–1905), discoverer of chromosomes [152]. (B) First illustration of human chromosomes, by Flemming (1882). (Courtesy of the Genetics and Medicine Historical Network.)

to be due to a transposon insertion in the R gene for a starch-branching enzyme. The wrinkled state is due to the lack of an osmotic effect present when the normally functioning enzyme is present. The first demonstration of mutation in the human due to the insertion of a transposon was provided by the Kazazian group [4]: a movable element from chromosome 22 was inserted into the factor VIII gene causing hemophilia A.)

Human chromosomes, visualized in tumor cells in mitosis, were pictured in a paper by Walther Flemming (Fig. 1.2A), professor of anatomy in Kiel, in 1882 (Fig. 1.2B). The term chromosome was introduced by Waldeyer in 1888. Mitosis and meiosis were described in the last quarter of the 19th century. (The term meiosis was introduced in 1905 by Farmer and Moore; the process had been previously referred to as the reduction divisions. The word "miosis," usually taken to mean "reduction in size of the pupil," is from the same root. It is fortunate that the words are spelled differently in the two usages. In fact, Farmer and Moore spelled it "maiosis." They wrote as follows: "We propose to apply the terms Maiosis or Maiotic phase to cover the whole series of nuclear changes included in the two divisions that were designated as Heterotype and Homotype by Flemming.")

During the 1880s, Roux, de Vries, and Weismann developed the theory that the chromosomes carry determinants of heredity and development. The state of cytogenetics before the discovery of Mendel's work was reviewed by E.B. Wilson (Fig. 1.3) in his classic text [5]. In a discussion of the Roux–de Vries–Weismann theory (pp. 182–185), Wilson [5] wrote that "the chromatin is a congeries or colony of invisible self-propagating vital units ..., each of which has the power of determining the development of a particular quality. Weismann conceives these units ... [to be] associated in linear groups to form the ... chromosomes."

In 1900, Mendelism was rediscovered independently by Hugo de Vries (1848–1935) in Amsterdam, The Netherlands; Carl F.J.E. Correns (1864–1933) in Tübingen, Germany; and Erich von Tschermak (1871–1962) in Vienna, Austria. The chromosome theory of

**Figure 1.3** *E.B. Wilson (1856–1939), noted cytologist and author of The Cell in Development and Inheritance (1896).*

Mendelism was put forward about 1903 by Walter S. Sutton (1877–1916) (Fig. 1.4), then a graduate student at Columbia under E.B. Wilson [6,7], and by Theodor Boveri (1862–1915) of Würzburg, Germany, a leading cytologist to whom Wilson dedicated his landmark book (1896). Sutton, a Kansas farm boy, had studied meiosis in the Kansas grasshopper at Kansas University under C.E. McClung. He was in New York in 1902 when William Bateson, the leading English champion for Mendelism, lectured there on the subject. Sutton promptly recognized the behavior of Mendel's so-called factors in transmission from one generation to the next as exactly what one would expect if they were located on the chromosomes. The segregation of alleles and assortment of nonalleles were precisely what one would anticipate given the observed behavior of the chromosomes in meiosis.

William Bateson (1861–1926) had a difficult time converting biometricians to the Mendelian view of heredity. These were the disciples of Francis Galton, including Karl Pearson (1857–1936), who favored

**Figure 1.4** *Walter S. Sutton (1877–1916), codeveloper of the chromosome theory of Mendelism.*

blending inheritance. They were led to this view through experiences with the study of quantitative traits that we would call multifactorial. Their recalcitrance is evident in the following statement [8]:

> *As we have seen in the course of this work, albinism is a graded character, and we have every reason to believe that both in man and dogs separate grades are hereditary ... Mendelism is at present the mode—no other conception of heredity can even obtain a hearing. Yet one of the present writers at least believes that a reaction will shortly set in, and that the views of Galton will again come by their own.*

The views of the biometricians and Mendelists were reconciled by R.A. Fisher in a classic paper published in 1918 [9]. Fisher showed that the Mendelian behavior of multiple genes functioning together can explain the findings of the biometricians in regard to quantitative traits. (In his 1918 paper, Fisher introduced the term variance.)

Bateson made many contributions to genetics, including the introduction of the term. In a letter written in 1905, related to a projected new professorship at Oxford University (which he did not receive), Bateson wrote as follows:

> *If the Quick Fund were used for the foundation of a Professorship relating to Heredity and Variation the best title would, I think, be "The Quick Professorship of the Study of Heredity." No single word in common use quite gives this meaning. Such a word is badly wanted and if it were desirable to coin one "Genetics" might do. Either expression clearly includes Variation and the cognate phenomena.*

Another of Bateson's accomplishments was the discovery (with Punnett) of linkage in the domestic fowl. They misinterpreted the phenomenon and assigned terms that have been perpetuated, however; coupling and repulsion signify whether the two mutant genes of particular interest are on the same chromosome or on opposite chromosomes, respectively. The true significance of linkage and the application of linkage and crossing over to chromosome mapping were contributions of Thomas Hunt Morgan et al. (1866–1945) working in the famous "fly room" at Columbia (Fig. 1.5). Alfred Sturtevant was then a college undergraduate; Calvin Bridges and Hermann Muller were junior associates of Morgan. The unit of genetic map distance, the centimorgan (cM), was named for Morgan. He obtained his PhD in 1892 from Johns Hopkins University, he was the first native-born American to be awarded the Nobel prize in physiology or medicine (1933), given for his contributions to the concept of the gene.

In 1927, Hermann J. Muller (1890–1967) demonstrated that X-irradiation produces an increase in the rate of mutation in *Drosophila*. For this work, he received the Nobel prize in 1946.

### 2.3.2 Cytogenetics

During the 50 years after Flemming's first picturing of human chromosomes in 1882, several attempts were made to determine the chromosome number in the human and to determine the human sex chromosome constitution. A particularly definitive paper appeared to be that of Painter [10], in which the diploid number of 48 was arrived at from study of meiotic chromosomes in testicular material from a hanged criminal in Texas. (There is probably no basis for the favorite suggestion of graduate students that the man may have had the XYY syndrome and therefore appeared to have 24, rather than 23, bivalents.) Painter commented that in some of his best preparations, the diploid number appeared to be 46. As an important contribution of the paper, Painter established the human XY sex chromosome mechanism.

**Figure 1.5** *Thomas Hunt Morgan (1866–1945), student of linkage and the first American-born Nobel laureate in medicine or physiology. Photo c.1910.*

In 1949, Murray Barr, working with Bertram in London, Ontario, discovered the sex chromatin, otherwise known as the Barr body, or X chromatin. The distinctive body was visualized in the nucleus of neural cells of the cat. A neuroanatomist, Barr was studying changes in neural cells with repetitive stimulation of the nerves. Fortunately, he had sufficiently good records that he could establish that the unusual body in the interphase nuclei occurred only in female cats. Ten years later, when the XXY Klinefelter syndrome and the XO Turner syndrome, as well as other sex chromosome aneuploidies, were described, the correlation of the number of Barr bodies with the number of X chromosomes in excess of one was established. The correlation with X-inactivation and the Lyon hypothesis were likewise elaborated. Twenty years after the paper of Barr and Bertram (1949; [11]), the fluorescent (F) body, or

(A)

(B)

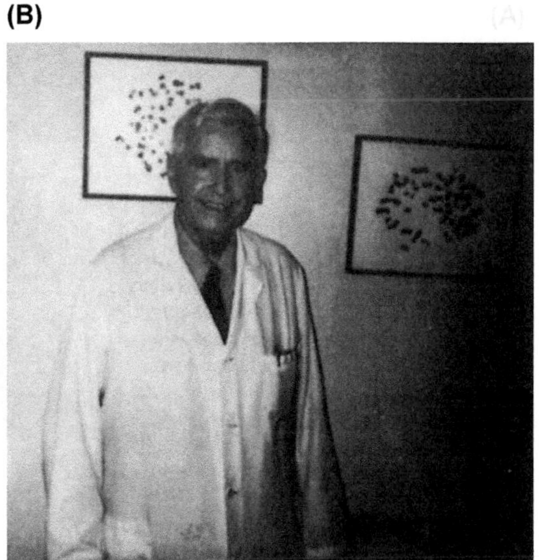

**Figure 1.6** (A) Exhibit by J.H. Tjio at the First World Congress of Human Genetics in Copenhagen (1956). Demonstration of $2N=46$ in mitotic cells. (B) Albert Levan in Lund, Sweden, in 1989. Collaborator of Tjio in work published in 1956.

Y chromatin, was discovered in interphase nuclei by Caspersson et al. [12–14].

Three new techniques developed in the 1950s and early 1960s facilitated the burgeoning of human cytogenetics, including clinical cytogenetics. During the early 1950s, T.C. Hsu (sometimes known to his colleagues and friends as "tissue culture Hsu") was working in Galveston when he accidentally discovered, through an error in the preparation of solutions, that hypotonicity causes nuclei to swell, with dispersion of the mitotic chromosomes, improving visibility. In his paper, Hsu acknowledged that Hughes at Cambridge University had independently made the same discovery. This technical "trick" was used by J.H. Tjio and Albert Levan [15] (Fig. 1.6A and B) in studies of the chromosomes of fetal lung cells in mitosis and by Charles Ford (Fig. 1.7A) and John Hamerton (Fig. 1.7B) in studies of meiotic testicular cells (1956) to determine that the correct chromosome number of the human is $2N=46$.

In the early days of clinical cytogenetics, it was the practice to do bone marrow aspiration to get an adequate number of dividing cells for analysis. This requirement was avoided by the introduction of phytohemagglutinin by Peter Nowell [16–18]. As its name implies, phytohemagglutinin is of plant origin and causes the agglutination of red cells. Its use was introduced by Edwin E. Osgood (1899–1969) of Portland, Oregon, for the purpose of separating white blood cells from red blood cells. It was Nowell's observation that circulating lymphocytes exposed to phytohemagglutinin were stimulated to divide. Thus, it was possible to obtain from a sample of peripheral blood adequate numbers of dividing cells for chromosome studies.

The third technique of particular value to cytogenetics was the use of colchicine to arrest cell division in mitosis. Combined with phytohemagglutinin, it further helped ensure adequate numbers of cells at a stage of division optimal for chromosome identification and enumeration.

### 2.3.3 Biochemical Genetics

The acknowledged father of biochemical genetics is Archibald Garrod (1857–1936), who introduced the concept of inborn errors of metabolism, as well as the term [19]. Alkaptonuria was the first of the disorders he investigated. In his paper on this condition in 1902, after coaching by Bateson, he recognized that its inheritance was probably Mendelian recessive because of the occurrence in both males and females with normal parents who were often consanguineous. In his famous Croonian lectures (delivered in 1908 and published in 1909),

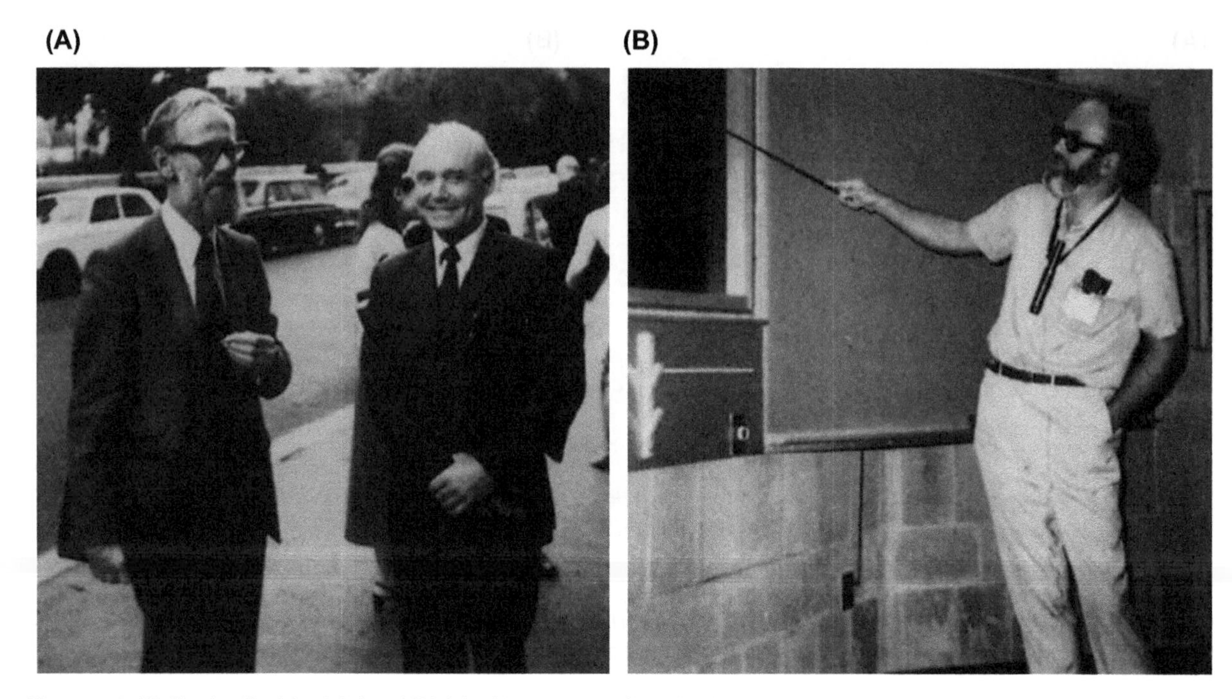

**Figure 1.7** (A) Charles Ford (at right) and (B) John Hamerton confirmed $2N=46$ in meiotic cells in 1956. *Photograph by McKusick (1971 and 1974).*

he formally unveiled the concept of inborn errors of metabolism and discussed three other conditions: pentosuria, albinism, and cystinuria. The nature of the enzymatic defect in all four of these disorders is now known. La Du et al. [20] confirmed Garrod's prediction of an enzyme deficiency in alkaptonuria by demonstrating deficiency of homogentisic acid oxidase in a liver sample. The gene mutant in alkaptonuria has been mapped to chromosome 3 by linkage studies; the gene encoding homogentisic acid oxidase has been cloned and a number of disease-causing mutations identified therein (OMIM 203500). The enzyme deficiency of tyrosinase in classic oculocutaneous albinism has been characterized all the way to the DNA level, and the enzymatic and genetic defects in other forms of albinism have been defined. On the basis of classic biochemistry, the enzyme defect in pentosuria is known to involve xylitol dehydrogenase (L-xylulose reductase); the gene has not been mapped. Cystinuria is not, strictly speaking, a Garrodian inborn error of metabolism but a defect in renal (and intestinal) transport of dibasic amino acids. It is caused by mutation in an amino acid transporter gene located on chromosome 2.

Bateson, Garrod's contemporary and advisor in matters genetical, is credited with a useful piece of advice: "Treasure your exceptions!" He was referring to findings in experimental genetics and perhaps was emphasizing also the value of investigating rare mutants. The same message was conveyed by Archibald Garrod [21]. He quoted a letter written by William Harvey a few months before his death in 1657:

> *Nature is nowhere accustomed more openly to display her secret mysteries than in cases where she shows traces of her workings apart from the beaten path; nor is there any better way to advance the proper practice of medicine than to give our minds to the discovery of the usual law of nature by careful investigation of cases of rarer forms of disease. For it has been found, in almost all things, that what they contain of useful or applicable nature is hardly perceived unless we are deprived of them, or they become deranged in some way.*

Garrod was William Osler's immediate successor as Regius Professor of Medicine at Oxford University. He seems to have been sensitive to the prevailing view that the disorders he studied were unimportant because they were rare, especially in contrast to the infections and nutritional diseases rampant at that time. It was after his retirement from the Regius chair in 1927 that Garrod [22] wrote *The Inborn Factors in Disease*, expanding his earlier *Inborn Errors of Metabolism* [23] to the full-blown theory of biochemical individuality. Indeed, as pointed out by Barton Childs [24], there was prophetic symbolism in the succession from Osler, who emphasized the disease that affected the patient, to Garrod, who focused on the patient affected by the disease.

Alkaptonuria was not the first inborn error of metabolism in which a specific enzyme deficiency was demonstrated. The first was hepatic glucose-6-phosphatase (G6P) deficiency demonstrated in glycogen storage disease type 1 by Carl and Gerty Cori [25]. (Diaphorase deficiency in methemoglobinemia was demonstrated by Gibson in 1948, but perhaps this cannot be considered a Garrodian inborn error of intermediary metabolism.) Deficiency of phenylalanine hydroxylase in phenylketonuria (PKU) was demonstrated soon after by Jervis [26]. Asbjørn Følling in Norway had described PKU in 1934. The molecular nature of a mutation of the gene encoding phenylalanine hydroxylase (MIM 261600) in PKU was first described by Savio Woo et al. in 1986 [27].

Linus Pauling (1901–94) introduced his concept of molecular disease during the late 1940s. It was an outgrowth of his focus on protein structure and specifically his work on sickle cell anemia (with Itano) [28] demonstrating by electrophoresis that the hemoglobin molecule had an abnormal structure. By 1956, Vernon Ingram had narrowed the abnormality of the sickle hemoglobin molecule to a single peptide and by 1957 to a single amino acid difference.

By 1956 also, Oliver Smithies was writing about starch gel electrophoresis, opening up the study of human variation at the protein level. With this and similar methods, Harry Harris (1919–94) at the Galton Laboratory extended the Garrodian concept of biochemical individuality (Fig. 1.8).

**Figure 1.8** *Harry Harris (c.1974).*

As studies of the heredity of biochemical differences were going on, investigations of the chemistry of heredity were under way that would be the basis of what we now call molecular genetics. DNA had been discovered in fish sperm by Miescher in the 19th century. At the Rockefeller Institute for Medical Research in 1944, Avery et al. [29] demonstrated that the transforming factor, which changed the pneumococcus from a rough form to a smooth form, is DNA. In 1953, Watson and Crick suggested a model for the structure of DNA consistent with its X-ray crystallography and with its biologic properties, including the capacity for replication.

### 2.3.4 Immunogenetics

Immunogenetics can be said to have had its start with the discovery of the ABO blood groups. These were demonstrated through the existence of "natural" antibodies (isoantibodies) by K. Landsteiner in 1901. From the distribution of ABO blood types in populations, Felix Bernstein in the early 1920s derived support for the multiple-allele, one-locus explanation rather than the alternative two-locus (A, non-A and B, non-B) hypothesis. Yamamoto et al. [30] defined the molecular differences between the genes for blood groups O and A and between those for A and B.

The next-to-be-discovered blood group was MN, found by Landsteiner and Levine [31]. These workers injected rabbits with different samples of human red cells and absorbed the resulting rabbit immune serum with other red cell samples until they found antibodies that distinguished human blood of the same ABO type.

Yet other blood group systems were discovered on the basis of antibodies (antisera) from mothers immunized to red cell antigens the fetus inherited from the father and antibodies from recipients of mismatched blood transfusions. The Rh blood group system is an example.

As recounted by Race and Sanger [32], Levine and Stetson [33], in a brief but historic paper, described how the mother of a stillborn fetus suffered a severe hemolytic reaction to the transfusion of blood from her husband. The mother's serum agglutinated the cells of her husband and those of 80 of 104 other ABO-compatible donors. The antigen responsible was shown to be independent of the ABO, MN, and P groups. In 1940, Landsteiner and Weiner [34], having immunized rabbits and guinea pigs with the blood of the monkey *Macacus rhesus*, made the surprising discovery that the antibodies agglutinated not only the monkey red cells but also the red cells of about 85% of New York City white people, who were said to be Rh positive. Because the anti-Rh antibody was found in the blood of persons who had suffered a reaction to the transfusion of ABO-matched blood, and the antibody of Levine and Stetson [33] was apparently identical to that raised in rabbits by injection of *rhesus* blood, the system was called rhesus (Rh). Levine et al. [35] showed that erythroblastosis fetalis is the result of Rh incompatibility between mother and child. Years later it became known that rabbit anti-rhesus and human anti-Rh antibodies are in fact not the same, but it was too late to change the name. Instead, the rabbit anti-rhesus antibody was called anti-LW, in honor of Landsteiner and Weiner. The LW antigen was later shown to be encoded by a gene on chromosome 19 that is quite distinct from the *RH* gene on chromosome 1.

Before 1956, blood groups provided some of the clearest examples of the role of Mendelism in the human, as well as some of the most important examples of the application of genetic principles in human health and disease, particularly blood transfusion and maternofetal Rh incompatibility.

### 2.3.5 Statistical, Formal, and Population Genetics

A cornerstone of population genetics is the Hardy–Weinberg principle, named for Godfrey Harold Hardy (1877–1947), distinguished mathematician of Cambridge University, and Wilhelm Weinberg (1862–1937), physician of Stuttgart, Germany, each publishing it independently in 1908. Hardy [36] was stimulated to write a short paper to explain why a dominant gene would not, with the passage of generations, become inevitably and progressively more frequent. He published the paper in the *American Journal of Science*, perhaps because he considered it a trivial contribution and would be embarrassed to publish it in a British journal.

**Figure 1.9** *J.B.S. Haldane with Helen Spurway and Marcello Siniscalco at the Second World Congress of Human Genetics, Rome, 1961.*

R.A. Fisher, J.B.S. Haldane (1892–1964), and Sewall Wright (1889–1988) were the great triumvirate of population genetics. Sewall Wright is noted for the concept and term "random genetic drift." J.B.S. Haldane [37] (Fig. 1.9) made many contributions, including, with Julia Bell [38], the first attempt at the quantitation of linkage of two human traits: color blindness and hemophilia. Fisher proposed a multilocus, closely linked hypothesis for Rh blood groups and worked on methods for correcting for the bias of ascertainment affecting segregation analysis of autosomal recessive traits.

To test the recessive hypothesis for mode of inheritance in a given disorder in humans, the results of different types of matings must be observed as they are found, rather than being set up by design. In those families in which both parents are heterozygous carriers of a rare recessive trait, the presence of the recessive gene is often not recognizable unless a homozygote is included among the offspring. Thus, the ascertained families are a truncated sample of the whole. Furthermore, under the usual social circumstances, families with both parents heterozygous may be more likely to be ascertained if two, three, or four children are affected than they are if only one child is affected. Corrections for these so-called biases of ascertainment were devised by Weinberg (of the Hardy–Weinberg law), Bernstein (of ABO fame), and Fritz Lenz and Lancelot Hogben (whose names are combined in the Lenz–Hogben correction), as well as by Fisher, Norman Bailey, and Newton E. Morton. With the development of methods for identifying the presence of the recessive gene biochemically and ultimately by analysis of the DNA itself, such corrections became less often necessary.

Pre-1956 studies of genetic linkage in the human for the purpose of chromosome mapping are discussed later as part of a review of the history of that aspect of human genetics.

## 2.4 Growth and Development of Medical Genetics: 1956 to the Present

Since the late 1960s, medical genetics has developed through a convergence of Mendelism, cytogenetics, biochemical genetics, immunogenetics, and statistical, formal, and population genetics. The development in each of these areas is traced in the preceding part of this chapter. Since 1956, medical genetics, in building on these foundations, has been blessed with three methodologies more or less specific to the field. These are "chromosomology" (beginning about 1956), somatic cell genetics (beginning about 1966), and molecular genetics (beginning about 1976). As will be indicated later, transgenic mice and all methods for transfer of genes into cultured cells or whole organisms, beginning about 1986, constitute a fourth methodological approach. The gene transfer methods, in combination with directed mutation and gene "knockout," have already proved particularly useful in the analysis of the function of genes, normal and abnormal.

Two further major methodological advances were database searching (research in silico or cybergenomics), as a primary method of genetic and genomic research, and microarray technology for profiling of gene function. These began in the mid- or late 1990s.

### 2.4.1 Chromosomology

Following the lead of Margery Shaw (personal communication, 1971), I divide the history of human cytogenetics into five ages: (1) 1882–1956 (gestation), the period from the first publication on human chromosomes to the reports of the correct chromosome number; (2) 1956–66, a golden age of human clinical cytogenetics; (3) 1966–69 (resting phase), a period when the field seemed to be "in the doldrums," with little progress; (4) 1969–77, the banding era; and (5) 1977 to the present, the era of molecular cytogenetics.

Jerome Lejeune (1926–94) and colleagues (Fig. 1.10) opened up the field of clinical cytogenetics with their report in January 1959 of the extra small chromosome in mongoloid idiocy, as Down syndrome was then called. (A letter in *The Lancet* in 1961, containing a list of 19 signatories resembling a short *Who's Who* in human genetics [39], established the eponymic designation as the preferred one. This is a prime example of the triumph of an eponym.)

The quinacrine fluorescence method was the first of the banding methods developed by Torbjørn Caspersson et al. [12–14] and exploited by Peter Pearson and others. This was followed by the various methods of Giemsa staining following alkali and other treatments for the so-called G-banding and by the method called reverse banding, or R-banding, because the Giemsa-light bands were stained.

The banding techniques permitted the unique identification of each human chromosome. This was immensely useful in experimental situations such as the study of rodent/human somatic cell hybrids and in the precise characterization of chromosomal aberrations. An early result was the demonstration that the smallest autosome is not number 22, but rather number 21, the autosome trisomic in Down syndrome. Jerome Lejeune had thought that Down syndrome was trisomy 21 of the next to the smallest autosome, not the smallest. Furthermore, it was demonstrated by Janet Rowley [40] (Fig. 1.11) that the Philadelphia (Ph) chromosome involves the non–Down syndrome chromosome, number 22, and that it represents a reciprocal translocation (with chromosome 9), not a deletion. The precise delineation of deletions and other aberrations permitted deletion mapping and mapping by dosage effects. An early, perhaps the first, example was the assignment of the locus for red cell acid phosphatase (MIM 171500) to 2p25 by Ferguson-Smith et al. [41]. The refined cytogenetic delineation of aberrations in the chromosomes also permitted the recognition of "new" chromosomal syndromes. Because one could be reasonably certain of having a "pure culture" series of cases with the same anomaly, it was possible to establish karyotype–phenotype correlations. The trisomy 8 syndrome, the 5q− syndrome, the Pallister–Killian syndrome, and the Jacobsen syndrome were examples.

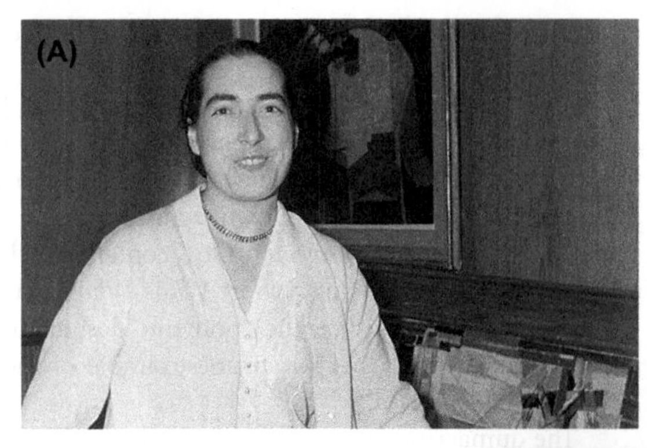

**Figure 1.10** Discoverers of trisomy: (A) Marthe Gautier, (B) Jerome Lejeune, and (C) Raymond Turpin. 1984. (From Harper (2008), courtesy of Oxford University Press.)

**Figure 1.11** *Janet Rowley at the Bar Harbor Course, 1983.*

By 1977, high-resolution cytogenetics involving the banding of extended chromosomes in cells arrested in prophase or prometaphase had been introduced independently by Jorge Yunis [42], Uta Francke [43], and the Manilovs. This improved further the identification of microdeletions in solid tumors such as Wilms' tumor and retinoblastoma, in 11p13 and 13q14, respectively. It revealed specific chromosomal abnormalities in congenital disorders of previously obscure etiology, including Langer–Giedion syndrome, Prader–Willi syndrome, DiGeorge syndrome, and Beckwith–Wiedemann syndrome. It provided the basis for the concept of contiguous gene syndromes put forward by Roy Schmickel [44].

The era of molecular cytogenetics, which persists to this day, began about 1977. With improved methods for studying DNA, the biochemical basis of banding was elucidated. The GC-rich nature of the Giemsa-light bands was demonstrated, and this information was correlated with evidence that these bands are also gene rich. Chromosomal in situ hybridization with radiolabeled DNA probes was first made to work reliably for single-copy genes in 1981, through the work of Mary Harper and Grady Saunders, Cynthia Morton, Malcolm A. Ferguson-Smith (Fig. 1.12), and others. Because of high background noise and perhaps other factors, erroneous

Figure 1.12 *Malcolm A. Ferguson-Smith with Marie G. Ferguson-Smith in Glasgow, 1981.*

results had been previously obtained in experiments that attempted to map single-copy genes. Harper used dextran to create a "wad" of the isotopically labeled probe, thereby achieving a signal at the site of the particular gene that was well above the background level. In the initial paper by Harper et al. [45], the method was used to map the insulin gene to the tip of the short arm of chromosome 11. Fluorescence in situ hybridization, a nonisotopic method, was developed by Ward et al., Landegent et al., and others in about 1985.

The combination of molecular techniques with cytogenetic techniques permitted chromosome mapping of oncogenes and identification of their role in hematologic malignancies associated with translocations. The MYC oncogene on chromosome 8 in Burkitt lymphoma was an early example elucidated by Carlo Croce, Philip Leder (Fig. 1.13), and others; the ABL oncogene on 9q, involved in chronic myeloid leukemia, was worked out by Rowley, Heisterkamp, Grosveld [46], and others. Chromosome sorting with fluorescence-activated devices, followed by analysis of gene content by molecular genetic methods, was developed; for example, Lebo et al. [47] used this method specifically for gene mapping.

Deletions were also found in association with neoplasms, usually solid tumors. The classic example is retinoblastoma. Lionel Penrose et al. [48] were the first to find a deletion in any neoplasm, a deletion in

Figure 1.13 *Philip Leder at the Bar Harbor Course, 1988.*

chromosome 13 in retinoblastoma. (Deletions in retinoblastoma played a role in proof of the Knudson hypothesis [49] and in positional cloning of the RB1 gene.)

Chromosome microdissection, another method of physical mapping, was developed for the collection of DNA from specific regions that could then be subjected to molecular genetic studies. For example, the approach was used by the Horsthemke group [50] to study the gene content of the region of 8q involved in Langer–Giedion syndrome, a contiguous gene syndrome [44].

### 2.4.2 Somatic Cell Genetics

A second large methodological advance in the era of human genetics since 1956 was somatic cell genetics. This has been contributory in several ways. In formal genetic analysis, it permitted the mapping of genes to specific human chromosomes or chromosome regions by the study of interspecies hybrids (e.g., between the human and the mouse). It permitted the differentiation of allelism and nonallelism disorders on the basis of noncomplementation or complementation, respectively, when cells from different patients with a given disorder (e.g., xeroderma pigmentosum) were mixed. In a third place, it permitted the study of the biochemical essence of many inborn errors of metabolism in cultured cells, usually skin fibroblasts. In a fourth, and perhaps its most important, application, somatic cell genetics provided an effective approach to the investigation of that vast category of somatic cell genetic disease—neoplasia.

Somatic cell genetics can be said to have gotten its start in the mid-1960s. The techniques that had been developed for culturing cells during the previous decades and the findings of studies of cultured cells were a useful

**Figure 1.14** *Henrietta Lacks, whose cervical carcinoma was the source of the clonal HeLa cell line.*

background. No cell line has been subjected to more extensive study than the HeLa cell*. This cell line was isolated from the cervical carcinoma of a patient named Henrietta Lacks (Fig. 1.14), who presented to the Johns Hopkins Hospital in early 1951 at the age of 31. Hers was one of some two dozen cervical carcinomas from which George O. Gey (1899–1970) attempted to establish a cell line and the only one yielding a successful result. The fact that it was an unusual cancer, indeed an adenosquamous carcinoma rather than the usual squamous cervical carcinoma, was found on review of the histology by Jones et al. [51]. It had an unusual fungating appearance suggesting a venereal lesion and prompting a dark-field analysis for spirochetes (which were not found). Although there was no evidence of invasion or metastasis at the time she was first seen, and despite radium therapy, Mrs. Lacks was dead in 8 months. The genetic characteristics of the HeLa cell line, including HLA types, were determined by Susan Hsu et al. [52] and compared with the findings in surviving members of her family. That Mrs. Lacks was a heterozygote for G6P dehydrogenase (G6PD) deficiency (G6PD A/B) was established by the fact that she had both G6PD-deficient and G6PD-normal sons. The HeLa cell line is G6PD deficient (G6PD A), indicating its monoclonal origin; this fact was established by Philip Fialkow [53] in studies of the monoclonality of cancers. The vigor of the HeLa cell line is attested to by the extent to which it has contaminated other cell lines in laboratories around the world [51].

Based on the information acquired from studies of cultured cells, the HeLa cell being the prototypic human cell line, cell culture achieved wide use in studies of inborn errors of metabolism in the 1960s. Among the first of such studies, based on the wide enzymatic repertoire of the fibroblast, was that of galactosemia by Bias and Kalckar and by Robert Krooth in the late 1950s and early 1960s. Later in the 1960s, when Seegmiller with Rosenbloom and Kelley was defining the deficiency of hypoxanthine phosphoribosyl transferase in the Lesch–Nyhan syndrome [54], he would refer to making morning rounds on his tissue cultures. He alluded to the fact that cultured skin fibroblasts captured the essence of the patients' inborn errors of metabolism for study. Another notable example of the use of cultured cells in genetic studies was the Goldstein and Brown characterization, in the 1970s, of the low-density lipoprotein receptor and its role in normal cholesterol metabolism and familial hypercholesterolemia.

The development of prenatal diagnosis by amniocentesis was dependent on the fortunate circumstance that the amniocyte for the most part demonstrates the same enzymatic activities (or deficiencies) in cell culture as do other cells and tissues of the patient. Exceptions, however, include conditions such as type 1 glycogen storage disease (von Gierke disease) in which deficiency of G6P had been demonstrated by the Coris. PKU proved to be another exception.

An important application of somatic cell genetics to formal genetics was the clonal proof of the Lyon hypothesis. Ronald Davidson, Harold Nitowsky, and Barton Childs [55] provided the most compelling evidence in the human. These investigators cloned two classes of cells from cultures of females heterozygous for the A/B electrophoretic variant of G6PD, one class of cells being type A and the other type B.

All cancer is genetic disease—somatic cell genetic disease. The chromosome theory of cancer was first

*Editors' Note: The full history of the origin of the HeLa cell line, including ethical issues surrounding the original biopsy, the context and clinical care for Ms. Lacks's disease, and subsequent communications with the family, have been the subject of extensive discussion since publication of the book "The Immortal Life of Henrietta Lacks" (Skloot, Rebecca. The immortal life of Henrietta Lacks. Broadway Books, 2017.). We have left the text intact as originally written by Prof. McKusick, as this is part of the historical record. We refer the reader to Ms. Skloot's book for further information on the history of this cell line, including the author's comments on Prof. McKusick's interactions with the family.

**Figure 1.15** *Guido Pontecorvo (right) with Dirk Bootsma at the Third Human Gene Mapping Workshop, Baltimore, 1975.*

**Figure 1.16** *John Littlefield at the Bar Harbor Course, 1964.*

clearly enunciated by Theodor Boveri in his monograph of 1914 (see Ref. [56] for biographic details).

Somatic cell genetics has played a big role in the proof and clarification of the genetic basis of cancer. It provided the methods by which the clonal nature of cancers was proved. Much of the molecular genetics of cancer, such as the demonstration of oncogenes, has been discovered through somatic cell genetics. The triad of methodologies that have been successively employed beginning in 1956—chromosomology, somatic cell genetics, and molecular genetics—have all played a role.

A particularly substantial contribution of somatic cell genetics has been to gene mapping (see later), specifically physical mapping, that is, the assignment of genes to specific chromosomes and chromosomal regions, as opposed to genetic mapping, the determination of the interval between gene loci on the basis of the amount of crossing over during meiosis. The method substituted interspecies (e.g., mouse and human) differences for interallelic differences used in meiosis-based linkage mapping. Somatic cell hybridization was what Haldane (Fig. 1.9) called a substitute for sex and Pontecorvo (Fig. 1.15) termed a parasexual method.

Somatic cell hybridization was made practical by the discovery of fusigens: first a virus, Sendai, by Y. Okada [57] and later a chemical, propylene glycol, by Guido Pontecorvo [58] (Fig. 1.15). Both damage the plasma membrane of cocultivated rodent and human cells, encouraging fusion. In the second place, the development of selection media allowed the isolation of hybrid cells from among the numerically overwhelming parental cells. The first of the selection media was hypoxanthine–aminopterin–thymidine (HAT), adapted to this use by John Littlefield [59] (Fig. 1.16). The first human gene to be mapped by the method of somatic cell hybridization was that which encodes the cytosolic isozyme of thymidine kinase (TK1; MIM 188300). Mary Weiss and Howard Green [60], who mapped TK1 to a specific chromosome, recognized the power of somatic cell hybridization for chromosome mapping. Barbara Migeon and C.S. Miller [61] identified the TK1-bearing chromosome as belonging to group E, and after the advent of chromosome banding, Miller et al. [62] identified the chromosome as number 17. Through the work of Frank Ruddle (Fig. 1.17), Walter Bodmer (Fig. 1.18), and many others, the application of somatic cell genetics to chromosome mapping was exploited to the fullest. During the 1970s, the somatic hybrid cell method accounted for the largest part of the progress in gene mapping that was collated in the regular Human Gene Mapping Workshops begun in 1973 [63].

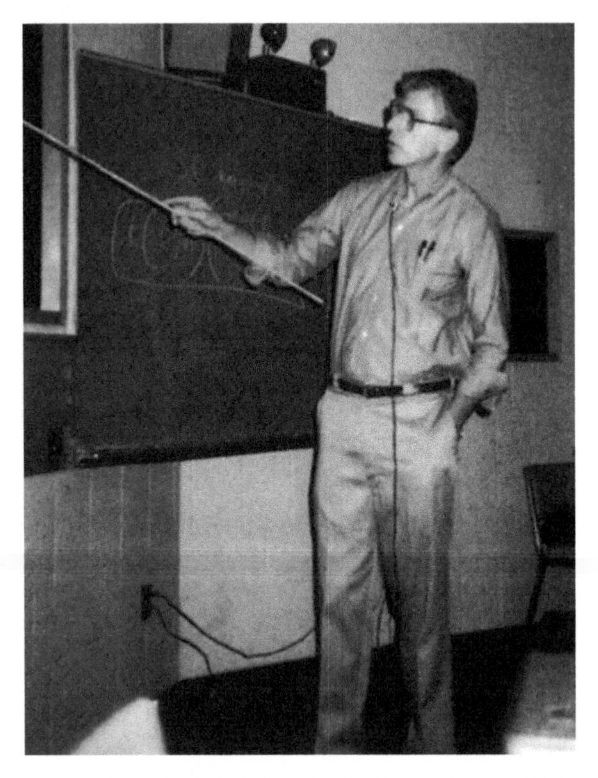

Figure 1.17 *Frank Ruddle at the Bar Harbor Course, 1987.*

Figure 1.18 *Walter Bodmer at the Bar Harbor Course, 1979.*

### 2.4.3 Molecular Genetics

The foundations of molecular genetics were laid in the pre-1956 era by the discovery of DNA, called nuclein, by Miescher in 1867; the demonstration that the

Figure 1.19 *Francis Crick (left) and James Watson at Cavendish Laboratory, University of Cambridge, 1953. (From Photo Researchers, with permission.)*

pneumococcal transforming factor is DNA by Oswald Avery, Colin MacLeod, and Maclyn McCarty (1944); and the solution of the structure of DNA by Watson and Crick [64] (Fig. 1.19). Working out the three-letter code of DNA was begun by Marshall Nirenberg and Heinrich Matthaei, who in 1961 observed the synthesis of a polypeptide composed solely of phenylalanine residues, when they used artificially synthesized RNA consisting only of uracil. Thereafter, the code was "cracked," one trinucleotide at a time, with the complete Rosetta stone of molecular genetics becoming available by 1966.

Restriction enzymes that cut DNA at the site of specific sequences were discovered by Werner Arber and Hamilton O. Smith in about 1970, and Daniel Nathans showed that one can use the enzymes for mapping DNA, the so-called restriction map, first developed for SV40 viral DNA. The Southern blot method for displaying fragments of DNA created by dissection with restriction enzymes was developed by Edwin M. Southern in 1975 [65].

The discovery of specific restriction endonucleases made possible the isolation of discrete molecular fragments of naturally occurring DNA for the first time. This capability was fundamental to the development of molecular cloning [66], which opened up the era of recombinant DNA. The combination of molecular cloning and endonuclease restriction allowed the synthesis and isolation of any naturally occurring DNA that could be cloned into a useful vector and, on the basis of flanking restriction sites, excised from it. The availability of a large variety of restriction enzymes significantly expanded the value of these methods.

The polymerase chain reaction (PCR) was a major addition to the molecular genetics armamentarium. PCR was formally unveiled in full technical detail by Mullis et al. [67] at the same Cold Spring Harbor Symposium on Quantitative Biology, Molecular Biology of *Homo sapiens*, at which the HGP was discussed in a historic rump session. Combined with specific restriction enzyme cleavage at the mutation site, the method had been used in the diagnosis of sickle cell anemia by Saiki et al. [68].

The first human gene to be cloned was chorionic somatomammotropin, by Shine et al. [69]. About the same time, Tom Maniatis cloned two of the smallest human genes, those for the α- and β-chains of hemoglobin. Most genes are at least 20 times bigger, and a few are as much as 1000 times bigger, containing much noncoding DNA (i.e., introns). After the discovery of reverse transcriptase, it was possible to take a shortcut and clone only the coding part of the gene complementary to the processed messenger RNA (mRNA), the so-called complementary DNA (cDNA). The cloning of large segments of DNA (up to a megabase or more) came in 1987 with the invention of YAC (yeast artificial chromosome) cloning by David Burke, Maynard Olson, and others [101].

Two advanced methods for DNA sequencing were reported simultaneously in 1977, by Fred Sanger [70] and Maxam and Gilbert [71]. Sanger and Gilbert shared the Nobel prize in chemistry in 1980. The Sanger dideoxy method for DNA sequencing remained the basic technology upon which the genetics revolution of the last vintage of the 20th century was based, albeit major advances in automation and other modifications were made.

### 2.4.4 Chromosome Mapping

The first assignment of a specific gene to a specific human chromosome can be credited to E.B. Wilson (Fig. 1.3), Columbia University professor and colleague of Thomas Hunt Morgan (Fig. 1.5). Wilson [72] wrote as follows:

> *In the case of color blindness, for example, all the facts seem to follow under this assumption if the male be digametic (as Guyer's observations show to be the case in man). For in fertilization this character will pass with the affected X chromosome from the male into the female, and from the female into half her offspring of both sexes. Color blindness, being a recessive character, should therefore appear in neither daughters nor granddaughters, but in half the grandsons, as seems to be actually the case.*

John Dalton (1766–1844), whose name is memorialized in the unit of molecular mass, had described his own color blindness in the 1790s; the Swiss ophthalmologist Horner had described the pedigree patterns we now recognize as typical of X-linked recessive inheritance. (Even earlier, as indicated by Rushton in 1994 [73], Pliny Earle, a pioneer Philadelphia psychiatrist, in 1845 delineated the inheritance pattern of color blindness on the basis of his own family.) The sex chromosome constitution of the human had been assumed to be either XY or XO in the male and XX in the female, as a result of studies by Wilson's colleague N.M. Stevens.

In a famous paper in 1915, J.B.S. Haldane (1892–1964), with his sister Naomi and fellow student A.D. Sprunt, reported the first example of autosomal genetic linkage in a mammal. This was the linkage of "pink eye" (p) with albinism (c). The authors did not use the term linkage; the title of their paper was "Reduplication in mice." It was published in the early stages of the "War of 1914" (World War I), and in the first paragraph the authors stated: "Owing to the war it has been necessary to publish prematurely, as unfortunately one of us (A. D. S.) has already been killed in France."

The first human genetic map interval to be estimated was that for color blindness and hemophilia. The transmission of hemophilia in the characteristic X-linked pedigree pattern was described in New England families early in the 19th century [74]. J.B.S. Haldane with Julia Bell [38] in 1937 attempted an estimate of the interval on the basis of published pedigrees, and Haldane with C.A.B. Smith revised the estimate in 1947 [75]. They placed the interval in the vicinity of 10 cM; the fact that there are two forms of X-linked hemophilia (A, or classic hemophilia, and B, or Christmas disease) was not found until 1952, when in the Christmas issue of the *British Medical Journal*, Biggs et al. in Oxford, England, reported the case of a 5-year-old boy named Christmas with hemophilia of a distinctive type [76]. It was distinguishable from classic hemophilia by the fact that serum from Master Christmas corrected the clotting defect in the blood of patients with classic hemophilia. Subsequent studies of the linkage between hemophilia A and color blindness showed very tight linkage, and a reanalysis of the published hemophilia/cb data used originally by Haldane et al. led C.A.B. Smith to conclude that the series was a mixture of two forms of hemophilia, one very tightly linked to color blindness and one loosely

Figure 1.20 *Lionel Penrose, 1973.*

Figure 1.21 *Jan Mohr (left) and Victor McKusick with plaque honoring Tage Kemp (1896–1964), Copenhagen, 1979. Mohr, Kemp's successor, was the director of the Institute of Human Genetics in Copenhagen. Kemp was the convener of the First World Congress of Human Genetics (1956).*

linked or not linked at all. We now know that hemophilia B maps to the long arm of the X chromosome, but at a distance proximal to hemophilia A and color blindness, which are at the very tip of the long arm.

During the 1930s and 1940s, considerable thought was given to what approaches might be used to identify genetic linkage, recognizing that it is necessary in the human to use matings as they are found; one cannot, of course, create the matings that are most informative as one can in mouse and fruit fly. Lionel Penrose (1898–1972) (Fig. 1.20) elaborated a sib-pair method in the 1930s [77]. The first autosomal linkage in the human was not identified until the early 1950s, when Jan Mohr (Fig. 1.21) demonstrated linkage of the Lutheran blood group and secretor factor loci as part of his doctoral thesis in the institute headed up by Tage Kemp (Fig. 1.21) in Copenhagen. Mohr was subsequently the director of that institute, and it was in his department that the first-to-be-discovered autosomal linkage group (Lutheran and secretor) was determined to be located on chromosome 19 [78].

The method logarithm of the odds (lod) was developed by C.A.B. Smith in the 1950s and elaborated by Newton Morton. Using these methods, Morton [79] demonstrated linkage of the elliptocytosis and Rh blood group loci (including the demonstration of genetic heterogeneity, as some families did not show linkage). Renwick and others showed linkage of the ABO blood group to the nail–patella syndrome. All three of the early examples of autosomal linkage were established with blood groups (or, in the case of secretor factor, an "honorary" blood group), illustrating the pathetically limited repertoire of markers, a major handicap to gene mapping in that era. By 1951, when the first autosomal linkage was discovered in the human, at least eight autosomal linkage groups had been identified in the mouse. On the other hand, it was about the same time in the early 1950s that the first X-linked locus was identified in the mouse, whereas three dozen or more loci were known to be on the human X chromosome because of X-linked pedigree patterns.

The first gene to be assigned to a specific autosome was the Duffy blood group locus, which was localized to chromosome 1 by Donahue et al. [80]. At that time, R.P. Donahue was a candidate for the doctoral degree in human genetics at Johns Hopkins. As every graduate student in human genetics should probably do, he determined his own karyotype, although his thesis research was not on a cytogenetic problem. He identified a heteromorphism of the chromosome 1 pair. One chromosome 1 showed what looked like an uncoiled region (this was in the prebanding era) in the proximal portion of the long arm. Donahue had both the wit and the gumption to do a linkage study: the wit to sense that this might be a mendelizing character in his family, and the gumption to collect blood samples from widely scattered relatives, to determine

**Figure 1.22** *David Botstein at the Bar Harbor Course, 1987.*

marker traits, and to analyze the data. The analysis suggested linkage of the Duffy blood group locus to the "uncoiler" trait. Although the lod score was far below the value of 3.0 usually accepted as evidence for linkage, confirmation of the assignment came quickly from other workers.

After 1971, abundant information was collected on chromosome mapping by the method of somatic cell hybridization. Radiation by hybrid mapping made use of hybridization with rodent cells to "rescue" human cells in which random fragmentation of chromosomes had been produced by radiation. In situ hybridization of DNA probes to chromosomes, first with radioactively labeled probes and later with fluorescent probes, was a mapping method first made to work reliably for single-copy genes in 1981.

About 1980, molecular genetics entered the chromosome mapping field. It contributed to the field in three ways: (1) it permitted direct identification of the human gene in somatic cell hybrids, so that it was possible to go directly for the gene, rather than requiring expression of the human gene in the hybrid cell; (2) it provided probes for chromosomal in situ hybridization, which was made to work for the first time in a reliable way for single-copy genes in 1981; and (3) it provided DNA polymorphisms as markers for family linkage study, thereby providing a virtually limitless repertoire.

In a seminal paper [81], David Botstein (Fig. 1.22), Ray White (Fig. 1.23), Michael Skolnick, and Ron Davis formally outlined the value of what they called restriction fragment length polymorphisms (RFLPs or "riflips") as linkage markers in family studies. In fact, the first RFLP had been discovered by Kan and Dozy [82], an HpaI variant 3′ to the β-globin locus, and Ellen Solomon and Walter Bodmer [83] had suggested the value of such DNA variants as linkage markers:

**Figure 1.23** *Ray White (left), James Neel (middle), and Joseph Nadeau at Bar Harbor, 1983.*

> *Given the range of available restriction enzymes, one can envisage finding enough markers to cover systematically the whole human genome. Thus, only 200 to 300 suitably selected probes might be needed to provide a genetic marker for, say, every 10% recombination. Such a set of genetic markers could revolutionize our ability to study the genetic determination of complex attributes and to follow the inheritance of traits.*

The late Allan C. Wilson (1934–91) decried the use of the term restriction fragment length polymorphism, arguing that length mutations are one large class of mutations, the other being point mutations, and that it would be better to refer to these simply as restriction fragment polymorphisms. In mice they are referred to as restriction fragment length variants (RFLVs). Indeed this is what they are, differences between inbred strains, and not polymorphisms in the strict sense as defined by E.B. Ford of Oxford University as the occurrence together in the same habitat of two or more distinct forms of a species in such proportions that the rarest of them cannot be maintained by recurrent mutation. Indeed, RFLPs in the human, when first discovered in a small number of samples, could not be known to be polymorphisms in the terms of this definition; 1% is usually taken as the minimal frequency of the rarer allele. However, it turns out that most DNA variants are indeed highly variable and qualify as polymorphisms within the Ford definition.

**Figure 1.24** *Frank Ruddle (left) and Nancy Wexler at Bar Harbor, 1983.*

A second form of DNA variation to be discovered was variable number of tandem repeats (VNTRs), introduced as hypervariable DNA markers for linkage studies, as well as for use in forensic science by Jeffreys et al. [84] and by Nakamura et al. [85]. The third major, and still more variable, type of DNA polymorphism is represented by the dinucleotide repeats, for example $(CA)_n$, the so-called microsatellites. These were discovered and developed by James Weber at the Marshfield Clinic in Marshfield, Wisconsin, and by Michael Litt in Portland, Oregon; they are extensively exploited by many, particularly Jean Weissenbach in Paris, for the creation of genetic maps and the mapping of disease genes.

The first major successful application of RFLPs in linkage study was the mapping of the Huntington's disease (HD) locus to the tip of chromosome 4 through linkage to the so-called GS8 RFLP; this marker at a locus designated D4S10 in the human gene mapping nomenclature was, in turn, mapped to that region of chromosome 4 by in situ hybridization. The initial demonstration of linkage was achieved by James Gusella et al. [86] on samples assembled by Nancy Wexler (Fig. 1.24) from an extraordinarily large and extensively affected kindred with HD in the Lake Maracaibo region of Venezuela.

Haplotyping is the determination of markers close to, and on the same chromosome as, the mutation of interest. The term haplotype was first used in connection with the

**Figure 1.25** *Haig Kazazian at Bar Harbor, 1976.*

**Figure 1.26** *Stylianos Antonarakis at Bar Harbor, 1988.*

major histocompatibility types, HLA types. Haplotyping was developed particularly extensively by Haig Kazazian (Fig. 1.25) and Stylianos Antonarakis (Fig. 1.26) at Johns Hopkins University, in collaboration with Stuart Orkin in Boston and others, for identifying the presence

**Figure 1.27** *Eric Lander at Bar Harbor, 1989.*

of the gene for thalassemia in particular individuals. Because haplotypes represent closely linked markers, it is not unexpected that in families and even in population groups, common forms of thalassemia are associated with the same haplotype. This important application of the linkage principle in diagnosis and in population genetics (for tracing the multiple origins, e.g., of the sickle hemoglobin gene) was extended from the hemoglobinopathies to many inborn errors of metabolism, particularly cystic fibrosis. Because of the many different mutations capable of causing cystic fibrosis and because of the relatively high frequency of the "cystic fibrosis" gene, tracing it in families through the haplotype has been useful.

In some populations, such as the Finns and the Old Order Amish, many usually rare disorders occur in relatively high frequency because of a founder effect. In such populations, most individuals affected with a recessive disorder that is rare in the general population are homozygous for a particular haplotype. Indeed, as was theorized by C.A.B. Smith in 1953 and elaborated more recently by Lander and Botstein [87] (Fig. 1.27), patients with any rare autosomal recessive disorder, if the parents are consanguineous, will be expected to be homozygous for closely linked markers, that is, homozygous for a particular haplotype. This has indeed been shown to be the case in several disorders, including Bloom syndrome, diastrophic dysplasia, and alkaptonuria [88]. Of course, in founder populations, a specific haplotype is likely to be shared in common by persons affected with an autosomal dominant or X-linked recessive as well.

Linkage analysis by the study of PCR-amplified DNA from individual sperm was introduced by Norman Arnheim in 1988 [89]. Essentially, one could fill in the sort of information contained in stick diagrams used to explain linkage in textbooks of genetics. With this approach, it was not necessary to have observations on diploid offspring to identify recombinants among the gametes produced by the parent.

Comparative mapping, most extensively pursued between mouse and human, has been exploited as a clue to the likely location of genes in the human. A great deal of syntenic homology has been found between these two extensively studied species. The X chromosome in the human and in all other placental mammals has essentially the same genetic content, as was pointed out during the 1960s by Ohno [90], who also pointed out that the X chromosome is about the same size in all these organisms. The extensive syntenic homology, in the case of the autosomes, between the human and the mouse came as a surprise to many. What I have chosen to call the Oxford grid was developed by John H. Edwards, professor of genetics at Oxford University, and collaborators, particularly A.G. Searle, at nearby Harwell.

The term synteny, meaning "on the same chromosome," was introduced by James H. Renwick (1926–94) (Fig. 1.28) in about 1970. It proved to be a useful term of distinction from linkage; two loci may be syntenic but not linked in the strict genetic sense because they are so far apart on the same chromosome that they assort independently through crossing over.

Renwick was also responsible for developing an early program for computer-assisted linkage analysis. Jurg Ott [91] also developed programs, as have others. These facilitated widespread analysis of linkage data.

A development of great importance for the advancement of genetic mapping was the creation of the Centre d'Études du Polymorphisme Humain (CEPH) in Paris by Jean Dausset in 1983. CEPH created and maintains a reference panel of family DNAs for mapping by linkage. The three-generation family units consist of the four grandparents, two parents, and a minimum of eight children. The data on markers, as they are determined in these reference families in many laboratories, are collated, and new markers and genetic disorders can be mapped against the reference panel.

Radiation hybrid mapping was first devised by Goss and Harris [92] as a modification of mapping with somatic cell hybrids. It involved fragmentation of the human chromosomes by irradiation of cultured cells and the "rescue" of these cells by fusion with mouse

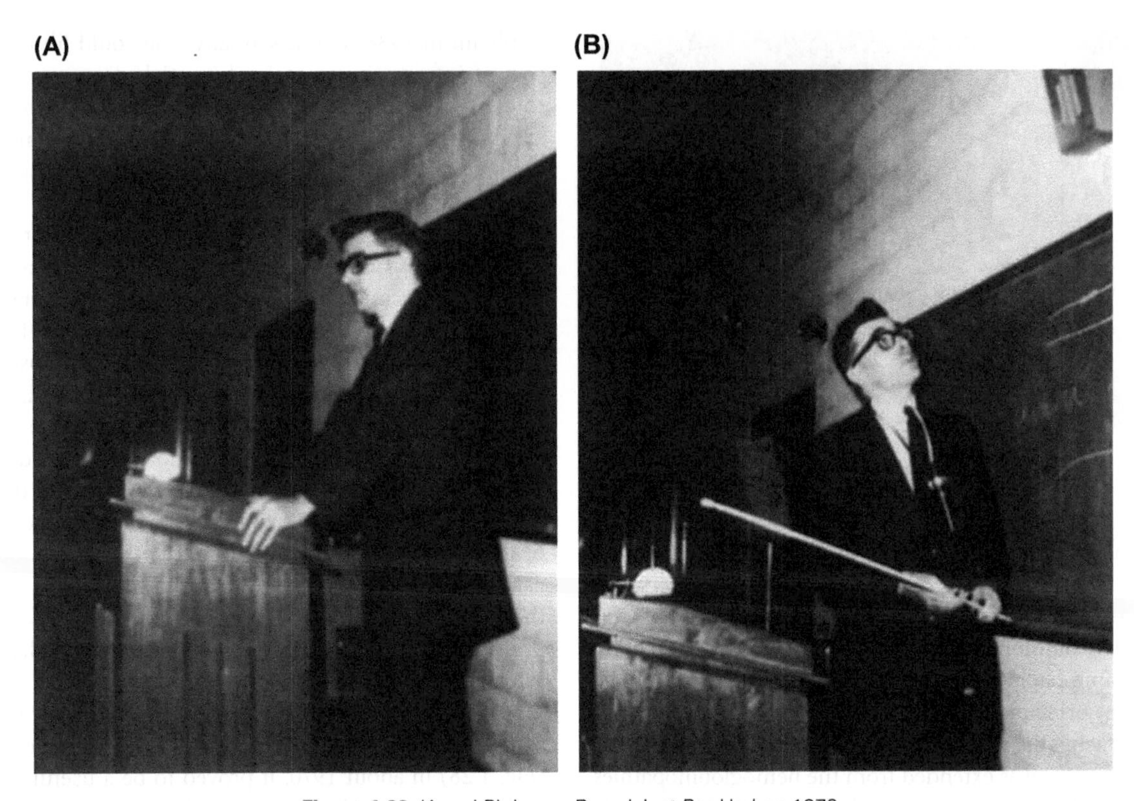

Figure 1.28 (A and B) James Renwick at Bar Harbor, 1972.

cells. The method was revived and revamped about 1990 by Peter Goodfellow of Cambridge University [93] and David Cox of Stanford University [94]. Both created DNA panels from cells with various human chromosomal content.

The creation of libraries of clones, for example, YACs, whose map location was known from radiation hybrid mapping or other methods made it possible to map a newly found sequence by hybridization or by linkage to a marker such as a sequence-tagged site (STS) or gene known to be on the same clone fragment. STSs were proposed by Maynard Olson et al. [95] as useful markers in mapping [96].

### 2.4.5 The Human Genome Project and Genomics

The goal of the HGP is to locate all the genes and sequence the entire haploid. It is primarily a mapping project; even sequencing was determined the ultimate map. In its initial form, however, it was not formally conceived as a mapping project. When proposed by Robert Sinsheimer, Walter Gilbert, Renato Dulbecco, and others in the period 1985–86, the HGP was viewed as a project for the complete sequencing of the genome. Previous progress in gene mapping and the value of the results were apparently unfamiliar to the leading promulgators, who were molecular geneticists. In part, objections to the project raised between 1985, when it was first proposed, and 1990, its official start date, resulted from the impression that it was an enterprise of mindless sequencing without much biology. (A factor also contributing to the unhappiness of many scientists with the project was the price tag. The National Research Council/National Academy of Sciences [NRC/NAS] Committee on Mapping and Sequencing the Human Genome, commissioned in late 1986 and reporting out in February 1988, estimated that the "job" could be done in 15 years for $200 million a year. Journalists in particular advertised the HGP as a $3.0 billion project, which biologists coping with funding problems found disturbing. Of course, the annual budget of the National Institutes of Health [NIH] is approximately 50 times that figure [in 1988 dollars]!)

At a birth defects congress in The Hague in 1969, complete mapping of the genes on the human chromosomes

**Figure 1.29** *Victor McKusick and the human gene map, Bar Harbor Course, 1981.*

**Figure 1.30** *At a Howard Hughes Medical Institute–sponsored NIH conference on the proposed Human Genome Project, July 1986. Left to right: Victor McKusick, L.H. (Holly) Smith, James Watson, Walter Gilbert, Robert Sinsheimer.*

had been proposed as an effective approach to solve problems of congenital malformations and genetic disorders in general [97] (Fig. 1.29). The proposal came close on the heels of the first manned moon landing in July 1969; complete genome mapping was seen as the next moon shot. The potential benefit of complete mapping was frequently emphasized thereafter. For example, it was advanced as an achievable goal for the last "vintage" of the 20th century [98]. The Human Gene Mapping Workshops, initiated by Frank Ruddle in New Haven, Connecticut, in 1973, continued in the same form until 1991, for collation of the world's experience in gene mapping. The function of these workshops was taken over by the Human Genome Organisation (HUGO) (see later).

It is true that the 1969 proposal had little apparent impact even though it was repeatedly restated over the next 15 years [1,98]. In 1969, the methodology that would make the HGP possible was not yet at hand. As outlined elsewhere, recombinant DNA technology and methods of efficient DNA sequencing came in the 1970s; improved methods for both genetic and physical mapping came in the 1980s. As pointed out by R. Cook-Deegan. (Ref. [99], p. 10) in his history of the HGP, whatever revisionist history might be written, the facts are that the HGP was conceived and promoted, not by medical geneticists, but by molecular biologists such as Robert Sinsheimer, Walter Gilbert (Fig. 1.30), and Renato Dulbecco, who worked predominantly on organisms other than the human. And it was conceived and promoted apparently in ignorance, for the most part, of what had been accomplished and proposed in the area of gene mapping. It is true that at the 1986 Cold Spring Harbor Symposium devoted to the human genome, a long review was given of the status of the human gene map [100]. This and the accompanying chromosome-by-chromosome lists of genes then mapped were an eye opener to the molecular geneticists present. Also at that meeting, a rump session led by Gilbert and Paul Berg discussed the pros and cons of the proposed genome project (see Ref. [99] for photographs of Gilbert, Berg, and Botstein at that crucial meeting and for a summary of the discussion).

If the period 1985–90 was a gestational period for the HGP, the NRC/NAS committee (1988) was the midwife. The official birth date for the project in the United States was October 1, 1990, with federal funding jointly through the Department of Energy (DOE) (about one-third) and the NIH (about two-thirds). (The DOE through its national laboratories has long had a mandate to study the biologic effects of radiation. With the leadership of Charles DeLisi, then with the DOE, and the legislative initiative of Senator Pete Domenici of New Mexico, the DOE became involved both intramurally [in its national laboratories] and extramurally in the HGP.)

James G. Watson (Fig. 1.30) of Watson–Crick fame led the NIH program for its first 3 years, bringing to it both his prestige and his wisdom. At the same time he directed the newly created National Center for Human

Genome Research (NCHGR), which functioned in much the same way as the individual institutes, he continued to direct the Cold Spring Harbor Laboratory on Long Island, New York. In 1993, Francis Collins (Fig. 1.31) became the director of the NCHGR and proceeded to develop an intramural program in medical genetics. This reflects his conviction that the HGP is primarily a gene-finding enterprise and is linked to clinical medicine as both an efficient way to find genes and an effective way to realize potential benefits in diagnosis, prevention, and management of genetic disorders. The "center" became an institute, the National Human Genome Research Institute (NHGRI), on January 1, 1995.

**Figure 1.31** *Francis Collins (left) with Leroy Hood, Cold Spring Harbor, 1993.*

In parallel with the HGP in the United States, genome programs were developing in many other countries, particularly the United Kingdom, France, and Japan. In September 1988, a group of 32 scientists from 14 countries met in Montreux, Switzerland, to found HUGO (Fig. 1.32). As a member of the organizing group, Norton Zinder, put it, "HUGO is a UN for the human genome." It is a coordinating organization. Its founder president Victor McKusick was succeeded by Walter Bodmer in 1990, by Thomas Caskey in 1993, by Grant Sutherland in 1996, and thereafter by Gert-Jan van Ommen followed by Lap-Chee Tsui.

The HGP progressed even faster than predicted, mainly because of new methods: PCR [101], YACs (*102*), STSs [95], microsatellite linkage markers [103], and others. The official start date for the HGP in the United States was October 1, 1990. The completion

**Figure 1.32** Founding council for the Human Genome Organisation (HUGO) at Montreux, Switzerland, September 1988 (11 members were absent). First row (left to right): Matsubara, Shows, Tocchini-Valentini, Honjo, Shimizu, McKusick (with Swiss cowbell), Lyon, Gilbert, Cantor, Robson, Karpov (observer). Second row: Hirt, Ruddle, Collins, Zinder, Sutherland, Cavanee, Hinton (staff), Strayer (staff), Tooze, Hood, Frézal, Cahill, Ferguson-Smith. Third row: Pearson, Dulbecco, Philipson, Jacob, Mirzabekov, Goodfellow (observer), Dausset, Watson, Worton, Southern, Grzeschik.

date predicted by the 15-year estimate made by the NRC/NAS Committee on Mapping and Sequencing the Human Genome [104] was September 30, 2005. At the 5-year mark, Francis Collins [105] could report that the HGP was "ahead of schedule and under budget."

A phenomenal speeding up occurred in the last part of the 1990s, with the announcement of a "first draft" of the complete sequence in June 2000. The acceleration can be credited to the development of improved high-throughput sequencers that used capillaries for the dideoxy method developed by Sanger in 1977, and particularly to the application to the human genome of "shotgun" sequencing, which had been successful in the first complete sequencing of a free-living organism, *Haemophilus influenzae*, by the group of Hamilton Smith [106]. Previously, the approach of the publicly funded HGP had been the creation of successive maps: a genetic linkage map of many genetic markers [103], a physical map of overlapping DNA segments (e.g., YACs [102]), and finally the ultimate map, the DNA sequence. As opposed to this top-down approach, a bottom-up strategy, sequencing random fragments with assembly thereafter, was used by Venter et al. in the private genome project of Celera to get the complete genome sequence of microorganisms and subsequently of *Drosophila* [107] and the human.

The Human Mitochondrial Genome Project (HMGP), although never called that, had been completed long before. The HMGP had been done in an order opposite to the way the HGP was conducted. The complete sequencing was done first, by the group of Fred Sanger at Cambridge University [108]. Then all the genes were identified, and finally disorders related to mutations in those genes were characterized.

The term genome appears to have been first used by Winkler in 1920 and to have been created by the elision of GENes and chromosOMEs to signify the set of chromosomes and the genes they contain [63]. Thus, it is an irregular hybrid from two Greek roots. The term genomics is of more recent minting, having been proposed in July 1986 by Thomas H. Roderick of The Jackson Laboratory to designate the field of gene mapping and sequencing and specifically a new journal. The journal's inaugural editorial [63] was entitled "A new discipline, a new name, a new journal." The next 10 years showed a remarkable growth of the field and widespread use of the term genomics, so that in 1996 an editorial was entitled "An established discipline, a commonly used name, a mature journal" [109].

Goodfellow (1997) stated that he "would define genetics as the study of inheritance and genomics as the study of genomes." Adjectives added to the word genomics have indicated specific aspects or applications of genomics. Mapping and sequencing constitute structural genomics. "Functional genomics," continuing with Goodfellow's definitions, "is the attachment of information about function to knowledge of DNA sequence." Pharmacogenomics [110], toxicogenomics, comparative genomics, and physiologic genomics are some of the derivative terms for subfields of genomics [111].

Leading approaches for functional genomics are transgenic methods, which involve manipulation of the particular sequence in transgenic animals. Another is database searching for similar sequences in other organisms where function is already known or can be determined more readily than is possible in the human. The latter is comparative genomics. Database searching has been referred to as research in silico, or cybergenomics.

Already before the HGP was completed, intense attention was directed to the function of the genome in a global sense as reflected by the protein gene products. Proteomics was a term invented about 1995 by Mark Russell who used two-dimensional display methods for peptides [112]. Physiomics, a further derivation, is also called systems analysis or systems biology and refers further to function.

Functional genomics has also involved the study of gene expression as reflected by mRNA. Description of coordinated gene expression in various tissues, at various stages of development and in various physiological states, became possible with the development of microassay methods (chip technology) and other methods such as SAGE (serial analysis of gene expression) in the latter part of the 1990s. These methods for profiling gene expression represent the sixth of the major methodologies that have advanced medical genetics since 1956.

### 2.4.6 Clinical Applications of Gene Mapping

Gene mapping, defined as the location of a gene to a specific chromosome site and/or the identification of markers that are close neighbors, has clinical value in diagnosis by the linkage principle and usefulness in identifying the nature of the basic defect either through positional cloning or through the candidate gene approach. J.B.S. Haldane (Fig. 1.9) suggested in the 1920s that diagnosis by the linkage principle would be both possible and useful. In 1956, after Fuchs had suggested amniocentesis as a method of diagnosis of fetal sex based on the presence or absence of a Barr body in amniocytes, John H. Edwards [113] suggested that with amniocentesis one could do prenatal diagnosis by the linkage principle. The idea was further developed by McKusick [114], and McCurdy [115] demonstrated its practicability in connection with the hemophilia A carrier status, using the very closely linked G6PD marker. Schrott et al. applied the approach in connection with the closely linked myotonic dystrophy and secretor loci, the secretor status of the fetus being determinable in amniotic fluid [116,117]. But the full strength of the linkage approach was not realized until the 1980s, when abundant DNA markers became available. As indicated earlier, haplotypes (i.e., clustered markers around the disease locus) were useful in the thalassemias and in some inborn errors of metabolism. Usually, however, linkage is an arduous, and not always practicable, approach because it requires the availability of multiple family members and heterozygosity of the marker traits in key individuals. Furthermore, even under ideal circumstances, the available markers may be a distance from the disease locus, creating the possibility of a small, but finite chance of recombination, yielding false results. Thus, it is preferable to go for DNA diagnosis, that is, diagnosis based on the precise gene lesion.

In those disorders in which an enzyme deficiency or other protein abnormality has been identified, the wild-type gene can be cloned and defects in the gene identified. In the case of many Mendelian disorders, however, the nature of the biochemical defect was a mystery until the introduction of the mapping approach to identify the basic derangement. The first of the mystery diseases to be mapped by linkage to DNA markers was HD [86]. It was, however, almost exactly 10 years to the week before the nature of the gene defect was reported [118].

Map-based gene discovery involves positional cloning and the candidate gene approach. Positional cloning was referred to initially as reverse genetics [119]. Starting with the phenotype, mapping it, and then going to the gene is, however, the approach of classic genetics; true reverse genetics will be increasingly practiced as fragments of DNA with transcriptional and other characteristics of a gene are investigated to determine their role in the phenotype, by methods such as transgenic, or "knockout," mice. For this reason, Francis Collins [120] (Fig. 1.31) suggested that the approach be called positional cloning, not reverse genetics [119,121]. Table 1.1 lists, in chronologic order of discovery, the disease-producing genes identified by positional cloning (that is, "walking in" on the gene from flanking markers) up to January 1994.

Whereas positional cloning and the candidate gene approach are map-based cloning, functional cloning starts with the functional gene product and uses reverse transcriptase to create a cDNA corresponding to the mRNA encoded by the particular gene. The candidate gene approach involves mapping the disease phenotype to a particular chromosomal location and then scrutinizing that chromosomal region for genes encoding enzymes or other proteins that might plausibly be implicated in the disease in question. Support for the involvement of the candidate gene in the disease is provided by the demonstration of absolute linkage of a RFLP or microsatellite marker within the gene with the disease in question, and the proof is clinched by the demonstration of a specific point mutation or other intragenic lesion. Elucidation of the defect in one form of hypertrophic cardiomyopathy by Seidman et al. [122,123] is an early example.

Many methods for the identification of intragenic lesions have been identified since the early 2000s. By late 1994, disease-producing point mutations had been identified in almost 350 genes, and in many of these (e.g., CFTR, HBB) they numbered in the hundreds.

### 2.4.7 Evolution of Clinical Genetics

Knut Faber [124] attributed to Mendel a major role in shaping our thinking about nosology, particularly the classification of disease and the delineation of distinct disease entities. The advent of the bacteriologic era in the decades immediately after Mendel also had a powerful effect on nosology. Both developments sharpened the focus on etiology: the role of specific microorganisms or the role of specific mutant genes. Until little more than a century ago, jaundice, dropsy, anemia, and other disorders were treated like disease entities in medical textbooks and in medical thinking and practice. Although Mendelian thinking contributed importantly to the general concepts in medicine, genetics did not become involved significantly in clinical medicine until after the acquisition of an anatomic base, beginning in the late 1950s.

One can point to several examples of pre-Mendelian "pedigree genetics." Patterns characteristic of autosomal

TABLE 1.1 **Map-Based Gene Discovery: Positional Cloning (Selected Examples)[a]**

| Disorder | Gene | Location | MIM No. | Year |
|---|---|---|---|---|
| Chronic granulomatous disease, X-linked | *CYBB* | Xp21.1 | 306400 | 1986 |
| Duchenne muscular dystrophy | *DMD* | Xp21.2 | 310200 | 1986 |
| Retinoblastoma | *RB1* | 13q14.1–q14.2 | 180200 | 1986 |
| Cystic fibrosis | *CFTR* | 7q31.25 | 219700 | 1989 |
| Wilms' tumor | *WT1* | 11p13 | 194070 | 1989 |
| Neurofibromatosis type 1 | *NF1* | 17q11.2 | 162200 | 1990 |
| Gonadal dysgenesis, XY female type | *TDF, SRY* | Yp11.3 | 480000 | 1990 |
| Choroideremia | *CHM* | Xq21.2 | 303100 | 1990 |
| Fragile X syndrome | *FMR1* | Xq27.3 | 309550 | 1991 |
| Myotonic dystrophy | *DM* | 19q13.2–q13.3 | 160900 | 1992 |
| X-linked agammaglobulinemia | *AGMX2* | Xp22 | 300310 | 1993 |
| Neurofibromatosis type 2 | *NF2* | 22q12.2 | 101000 | 1993 |
| Huntington's disease | *HD* | 4p16.3 | 143100 | 1993 |
| Multiple endocrine neoplastic type 2 | *RET* | 10q11.2 | 171400 | 1993 |
| Breast cancer, familial, type 1 | *BRCA1* | 17q21 | 113705 | 1994 |
| Polycystic kidney disease 1 | *PKD1* | 16p13.3 | 173900 | 1994 |
| Tuberous sclerosis 2 | *TSC2* | 16p13 | 191092 | 1994 |
| Breast cancer, familial, type 2 | *BRCA2* | 13q | 600185 | 1995 |
| Werner syndrome | *WRN* | 8p | 277700 | 1996 |
| Multiple endocrine neoplasia, type 1 | *MEN1* | 11q | 131100 | 1997 |
| Peutz–Jeghers syndrome | *STK11* | 19p | 175200 | 1998 |
| Rett syndrome | *RTT* | Xq | 312750 | 1999 |
| Ellis–van Creveld syndrome | *EVC* | 4p | 225500 | 2000 |
| Cartilage–hair hypoplasia | *RMRP* | 9p | 250250 | 2001 |
| Alstrom syndrome | *ALMS1* | 2p13 | 203800 | 2002 |
| Hemochromatosis, juvenile | *HJV* | 1q21 | 602390 | 2003 |
| Pernicious anemia, congenital | *GIF* | 11q13 | 261000 | 2004 |
| Roberts syndrome | *ESCO2* | 8p21.1 | 268300 | 2005 |
| Methylmalonic aciduria, cblC type, with homocystinuria | *MMACHC* | 1p34.1 | 277400 | 2006 |

*MIM*, mendelian inheritance in man [144].

[a]In addition to these Mendelian disorders, many genes with somatic mutations involved in neoplasia have also been identified by positional cloning, starting, for example, from translocation breakpoints.

dominant and autosomal recessive inheritance were commented on by Maupertuis in the 1750s, by Adams in 1814 [125], and by Sedgwick in the 1860s. The X-linked recessive pattern of hemophilia was noted in a newspaper account in the 1790s and in medical reports by Otto in 1803 and Hay in 1813 of early New England families [74,126]. Similarly, the X-linked recessive pedigree pattern of color blindness was described clearly by the Swiss ophthalmologist Horner in Zurich in 1876 and even earlier by Earle [127] in Philadelphia.

The relationship of consanguinity to an increased frequency of genetic defects was demonstrated by Bemiss in 1857 in studies of congenital deafness.

Early post-Mendelian examples of pedigree genetics include a report of albinism as a recessive by William E. Castle at Harvard in 1902. Farabee, a graduate student with Castle, described brachydactyly as a Mendelian dominant trait, basing his thesis research on a family in his home town of Old Concord, Pennsylvania. The family was subsequently updated by Haws and McKusick [128]. Harvey Cushing, the neurosurgeon and biographer of William Osler, published a large kindred with symphalangism (his term for anchylosis of the phalanges) in a paper with accompanying foldout pedigree in the first issue of *Genetics* in 1916, which also contained the famous paper by Calvin Bridges on nondisjunction.

The three main principles of clinical genetics (perhaps they should be called the three main phenomena of significance to clinical genetics) are pleiotropism, genetic heterogeneity, and variability. The history of our understanding of each can be traced.

Pleiotropism refers to multiple phenotypic (i.e., clinical) effects of a single mutant gene. This phenomenon is important to clinical medicine because often an external feature that is part of the pleiotropism and that may be in itself benign and insignificant may point to the presence of serious internal disease and/or to the fact that the person is a carrier of the mutant gene. The term was introduced by Plate in 1910. Hadorn developed the concept in considerable detail on the basis of studies in *Drosophila*. In his *Animal Genetics and Medicine* [129], Hans Grüneberg gave numerous examples, particularly from the mouse, and presented what he called "pedigrees of causes" relating all features of the syndrome back to a unitary defect. Analysis of pleiotropism, with the demonstration of plausible pedigrees of causes, was an important aspect of heritable disorders of connective tissue [130]. The evidence for a unitary basic defect was fundamental to research in Marfan syndrome and other disorders. It took more than 35 years for the prediction of a unitary connective tissue defect in Marfan syndrome to be substantiated and particularized [131,132].

Genetic heterogeneity means that any one of several genetic mechanisms can lead to the same or a similar phenotype. The idea was implicit in the work of Johannsen of Copenhagen, who in the first decade of the 20th century distinguished phenotype and genotype. It was he who introduced the two terms (as well as the word "gene") and put forward the concept that the phenotype is no necessary indication of the genotype. Genetic heterogeneity is obviously of practical importance in clinical medicine, as the prognosis, appropriate genetic counseling, and effective treatment may vary among the several genetic forms of a given disorder. Many examples of genetic heterogeneity have been uncovered since the late 1960s, especially through the application of biochemical and molecular methods. A striking example was homocystinuria, which simulates Marfan syndrome closely because of dislocated lenses and skeletal features such as increased height, scoliosis, and deformity of the anterior chest, but homocystinuria has recessive, not dominant, inheritance; has thrombotic, not aortic, complications in the cardiovascular system; and, of course, has a biochemical marker in the form of homocysteine in the urine.

Baur, Lenz, and Fischer, in their textbook published in the 1930s, recognized genetic heterogeneity. William Allen, who is honored by the William Allen Award of the American Society of Human Genetics, wrote about genetic heterogeneity when he pointed out that some disorders, such as Charcot–Marie–Tooth disease, occur in autosomal dominant, autosomal recessive, and X-linked forms. Allen pointed out that as a generalization the autosomal recessive form is most severe, the autosomal dominant form mildest (and most variable), and the X-linked recessive form intermediate in severity. This generalization is sometimes called Lenz's law, after Fritz Lenz of the Baur, Lenz, and Fischer textbook (and father of Widukind Lenz, medical geneticist of Münster, Germany).

Harry Harris [133] (Fig. 1.8) of London emphasized the heterogeneity of apparently simple "characters." One of the problems central to all studies in human genetics arises from the difficulty of knowing whether a particular difference has been characterized in, as it were, a "chemically pure" form. What appears at first sight to be a homogeneous entity readily identifiable by a particular technique, and presumably having a unitary genetical causation, turns out, with the application of newer techniques to the problem, to consist of more than one quite distinct phenomena; the condition known as "cystinuria" provides a simple illustration of this point.

As one aspect of the "darker side" of heredity counseling, F. Clarke Fraser [134] pointed to genetic heterogeneity of clinical entities. He wrote "A lot of difficulty comes from the fact that for many diseases two clinically similar cases may be genetically different, and thus have different genetic prognoses."

Newton E. Morton [79] demonstrated genetic heterogeneity in elliptocytosis when he found that the disorder is linked to the Rh blood group locus in some families and not in others.

Variation in the clinical picture is a characteristic of disease of all etiologies, both genetic and nongenetic. If the clinical picture resulting from a particular etiologic factor were invariant, clinical medicine would be child's play. Learning clinical medicine is, to a large extent, learning how to cope with variation. As well as pointing out the significance of pleiotropism, Hans Grüneberg [129] emphasized that variability can depend on the genetic background of a particular mutation. A frequently cited example from the mouse was provided by the work of L.C. Dunn (1893–1974), who found that the brachyury mutation, usually manifested by a short tail, was accompanied by an almost normal tail in some genetic stocks.

Penetrance and expressivity are aspects of variation. The terms and the concepts they signify were introduced by Vogt in 1926 [135] and were used by Timofeeff–Ressovsky soon thereafter while he was working in Berlin with Vogt. (Timofeeff–Ressovsky, still in Berlin at the time the Russians captured the city in 1944, was sent to a gulag, where he was a fellow prisoner of Solzhenitsyn; see Solzhenitsyn's *The Gulag Archipelago* for references to the evening intellectual interactions by which they helped maintain each other's sanity. Timofeeff–Ressovsky was partially "rehabilitated" toward the end of his life and attended the World Congress of Genetics in Moscow in 1978.)

Penetrance is an all-or-none phenomenon. Expressivity is variation in the severity of a genetic disorder or trait. When expressivity is so low that the disorder cannot be recognized, the gene is said to be nonpenetrant. Obviously, nonpenetrance is to some extent related to the power of the methods for studying the phenotype. The more penetrating the method, the lower the frequency of nonpenetrance.

### 2.4.8 Clinical Armamentarium of Medical Genetics

Part of the reason for the creation of the American Board of Medical Genetics and Genomics and comparable agencies in some other countries is the fact that since 1956 there is so much more that clinical geneticists can do. Thus, a mechanism is needed for oversight of the training and certification of practitioners.

Bradford Hill, the British biostatistician, suggested that the practice of medicine consists of seeking answers to three questions: What is wrong? The answer is diagnosis. What is going to happen? The answer is prognosis. What can be done about it? The answer is treatment. (David Danks of Melbourne suggested to me that the health professional should always keep a fourth question in mind: Why did it happen? The answer is etiology and pathogenesis, on which prevention and treatment can be based.)

Advances in diagnosis have come from both the clinic and the laboratory. Because the individual genetic disorders, of which there are many, are rare, clinical geneticists play a key role in diagnosis. For many of these disorders, there are not yet specific biochemical or other tests. Syndromology and dysmorphology are important aspects of the clinical geneticist's work. Cytogenetic diagnosis is his specific responsibility, at least for nonmalignant disorders. The oncologist makes heavy use of cytogenetic diagnosis, particularly in relation to hematologic malignancies.

Prognosis in clinical genetics has somewhat different implications than prognosis in other parts of clinical medicine. The question—What is going to happen?—relates not only to the person at hand but also to other members of the family, particularly an unborn child. Testing for the carrier status in a condition such as Duchenne muscular dystrophy or hemophilia or for a disease in its presymptomatic stages as in HD is obviously of great importance to prognostication.

Neonatal screening and population-based screening for specific genetic diseases, and prenatal screening for Down syndrome in mothers over 35 years of age, do not fit neatly into the Bradford Hill paradigm of clinical medicine. Although they do address the question—What is going to happen?—the question—What is wrong?—has not been asked. The question is not even—Is something wrong?—as all genetic disorders are not tested for.

These procedures are at the interface between clinical medicine and public health, or at least between clinical medicine and preventive medicine. I would not wish to suggest that I recognize a distinction between preventive medicine and clinical medicine; preventive medicine is an integral and exceedingly important aspect of clinical medicine.

Treatment is generally viewed as the "short suit" of clinical genetics. However, more can be accomplished than is realized, and I would emphasize that management is a better designation for the medical geneticist's role than is treatment, which implies a repertoire of measures almost exclusively pharmaceutical or surgical.

With this overview as a preamble, let me outline the development of the medical geneticist's clinical armamentarium since the 1960s.

An important item in the armamentarium of the clinical geneticist is the command of syndromology and dysmorphology. Genetic diseases include a large number of individually rare disorders, many of which have little basis of diagnosis other than their particular clinical features. Syndromology is the art and science of recognizing distinct genetic entities by characteristic combinations of clinical manifestations. Robert Gorlin of Minneapolis [deceased 2006—Ed.] and John Opitz of Madison, Wisconsin (later of Helena, Montana, and

Salt Lake City), are two American syndromologists from among the many capable ones. Dysmorphology is the term introduced by David W. Smith (1926–81) of Seattle, Washington, as an improvement on clinical teratology. It implies syndromology and as well encompasses considerations of etiology and pathogenesis; for example, the mechanisms by which the several features of malformation syndrome occur together were part of David Smith's focus, and his delineation of the fetal alcohol syndrome illustrated his attention not only to distinctive clinical features but also to causation. *Smith's Recognizable Patterns of Malformation Syndromes* [136] is a classic.

Many malformation syndromes and genetic disorders carry the name of a person, not always the first, who described the condition as a distinct entity. These eponyms have advantages when the basic nature of the disorder is unknown and thus a specific label based thereon is not possible. Usefully, eponyms remind medical geneticists of the roots of their field [137]. Place names (toponyms) likewise link the disorder to the geographic or ethnic setting of the first (or early) description of the disorder or protein (e.g., hemoglobin) variant. Tangier disease and familial Mediterranean fever are examples.

Neither syndromology nor dysmorphology is limited to genetic disorders, but congenital anomalies of all etiologies are the responsibility of clinical geneticists. They must keep in mind all causation, both genetic and nongenetic and an interaction of the two. It was no accident that the teratogenic action of thalidomide was detected by a medical geneticist, Widukind Lenz, in the early 1960s. It was easy for a nongeneticist to consider phocomelia a genetic disorder; a geneticist would be more likely to recognize that the distribution of cases in time, place, and families was not consistent with a genetic basis.

Annual conferences entitled the Clinical Delineation of Birth Defects were initiated in 1968 at Johns Hopkins University with financial support of the March of Dimes. During their heyday, from 1968 to about 1978, these gatherings of the aficionados illustrated the attention given by clinical geneticists to the development of the field of syndromology/dysmorphology. The annual David W. Smith conferences continue in the same tradition.

Several computerized databases have been developed as an aid to the syndromologist: the London Dysmorphology Database of Robin Winter; GenDiag of Ségolène Aymé, Paris; and POSSUM of Agnes Bankier and David Danks of Melbourne.

#### 2.4.9 A Synthesis: 1956–2001

Interestingly, 1956 was not only the year that human chromosomology got off to a firm start but also the date of the First World Congress of Human Genetics, which took place in Copenhagen under the presidency of Tage Kemp. The 10th of these quinquennial congresses was held in 2001. The first congress was a splendid survey of the status of human genetics at the time. The subsequent congresses were milestones measuring progress since 1956 [138].

At the first congress in 1956, Tjio and Levan (Fig. 1.6A and B) and Ford and Hamerton [139] (Fig. 1.7A and B) were getting the chromosome number right. Newton Morton was writing on linkage analysis. Oliver Smithies was beginning to write on starch gel electrophoresis. Vernon Ingram was narrowing down the molecular defect in sickle hemoglobin to a single peptide, and the first edition of my *Heritable Disorders of Connective Tissue* [130] was published.

Behold what happened between the first congress and the eighth held in Washington in 1991. From the anatomy of the chromosomes at the most elementary level of enumeration, we had gone to their dissection by both mechanical and molecular methods. Genetic linkage had enjoyed a phenomenal renaissance, and linkage was by then analyzed on populations of sperm studied individually by direct molecular methods. Rather than studying variation in proteins by electrophoresis, we were examining variation in the DNA itself. From the one example of sickle hemoglobin, the known mutational repertoire of the β-globin gene had expanded to more than 400, for example. At least one disease-related point mutation had been defined in more than 170 different genes, a count that had passed the 1000 mark by the end of 2000 [140].

Among the heritable disorders of connective tissue, clinical delineation had been refined by molecular definition. For example, in the mucopolysaccharidoses, lumped under the Hurler syndrome in 1956, at least 10 enzymatically distinct entities had been defined, and in the disorders of the fibrous elements of connective tissue, precise intragenic lesions had been described in osteogenesis imperfecta, in some of the Ehlers–Danlos syndromes and skeletal dysplasias, and in the Marfan syndrome.

**Figure 1.33** *Mary Lyon (center) with Wesley Whitten at Bar Harbor, 1971.*

By the 1961 congress in Rome, chaired by Luigi Gedda, clinical chromosomology had arrived. From the findings in Turner syndrome and Klinefelter syndrome, the role of the Y chromosome in sex determination was realized; the existence of testis-determining factor (TDF) was deduced. The hypothesis of the single-active X chromosome, advanced by Mary Lyon [141,142] (Fig. 1.33), was the intellectually provocative new concept. Electrophoretic polymorphisms of serum proteins and red cell enzymes were being described. The "Philadelphia chromosome" was found in chronic myelogenous leukemia, one of the first pieces of evidence in humans supporting the chromosome theory of cancer [18].

By 1991, imprinting had taken the place of lyonization. TDF had been cloned and characterized under the label "sex-determining region of Y." The fundamental basis of specific forms of cancer had been traced to specific genes—sometimes multiple, sequentially collaborating genes—and to specific mutations within genes.

The Chicago congress in 1966 was under the presidency of Lionel S. Penrose of London (Fig. 1.20). The genetic code had been completely deciphered that year. Somatic cell genetics had entered the scene for the study of inborn errors of metabolism. For example, it provided the strongest proof of the Lyon hypothesis and, through the study of cultured amniotic cells, opened the way for prenatal diagnosis by amniocentesis. The concept of lysosomal diseases had emerged, and the first edition of *Mendelian Inheritance in Man* (MIM) [143], already computer based, was published earlier that year.

Review of the growth of successive editions of MIM provides an opportunity to engage in some scientometrics. As indicated by the original subtitle *Catalogs of Autosomal Dominant, Autosomal Recessive, and X-linked Phenotypes*, MIM was an encyclopedia of phenotypes—genetic traits and disorders. But early on it was the intent that there should be one entry per gene locus, based largely on the philosophy that, if two genetic diseases or traits result from mutation at different loci, they are fundamentally distinct, however similar in phenotype they may be. Genetics is finding genes—gene hunting. The number of entries in successive editions of MIM can serve as one basis of quantifying progress in our field. There were about 1500 entries in the first edition. In the 1960s, the only way we had to identify separate entries (read "gene loci") was by mendelizing phenotypes, sometimes aided by biochemical or immunologic characteristics or by genetic features such as linkage. In the 1970s, the rate of accession was accelerated by a parasexual method for gene identification and mapping, namely, somatic cell hybridization. By the 1980s, cloning of human genes was practiced. Accordingly, entries in MIM were created for genes when they were cloned, sequenced, and mapped, even though no Mendelian phenotype had been associated. By the 12th edition [144], entries in MIM numbered 8587, an impressive figure but still a long way from the 60,000 to 70,000 expressed genes the human was then thought to have [145]. (By the fall of 2000, the number in the continuously updated online version, OMIM, was approaching 12,000.) Furthermore, by the 1990s, the evolution of the content of MIM as outlined earlier had led to a self-redefinition that prompted a change of the subtitle to *A Catalog of Human Genes and Genetic Disorders* in the 11th edition (1994) and thereafter.

By the Paris congress in 1971, the study of inborn errors of metabolism in cultured cells had paid off. For example, elucidation of the defect in Lesch–Nyhan syndrome and the differentiation of various mucopolysaccharidoses had been accomplished through the study of cultured cells. Chromosome banding had been introduced, and the first genes were being assigned to specific autosomes by linkage to chromosomal heteromorphisms and rearrangements and by interspecies somatic cell hybridization. There were four autosomal gene assignments by 1971.

Plotting the growth of the human gene map since 1971 is another exercise in scientometrics. By June 1976, just before the fifth congress in Mexico City, at least one gene had been assigned to every chromosome. This was largely through the application of somatic cell hybridization. By the 1981 congress in Jerusalem, molecular genetic methods for gene mapping had entered the scene and were responsible for further acceleration of mapping in the 1980s. Molecular genetics provided probes for identification of human genes in rodent/human somatic cell hybrids. It provided probes for in situ hybridization to chromosomes and, importantly, it provided DNA markers (e.g., RFLPs, VNTRs) for family linkage studies, for mapping Mendelian disorders of unknown biochemical basis. By the time of the 1991 congress, a total of at least 2300 genes had been mapped to specific chromosomes and, for most, to specific chromosomal regions.

**Figure 1.34** *Three presidents of the World Congress of Human Genetics (left to right): Arno Motulsky (Berlin, 1986), James V. Neel (Jerusalem, 1981), Victor McKusick (Washington, 1991). (Photo from 1991. Neel died in 2000 and Arno Motulsky died in 2018)*

By the fifth congress in Mexico City in 1976, other advances of note included the concept of receptor diseases—disorders such as familial hypercholesterolemia [146] and androgen insensitivity—and the Knudson hypothesis of hereditary/sporadic tumors [49,147]. The Philadelphia chromosome had been reinterpreted as a reciprocal translocation rather than a deletion.

By the 1981 congress in Jerusalem, under the presidency of James V. Neel (Fig. 1.34), in addition to the advances in the methods and results of gene mapping, human genes were being cloned, the genetic basis of antibody diversity was well on the way to elucidation, human variation was being studied with monoclonal antibodies, and one human chromosome had been completely sequenced—the mitochondrial chromosome.

By 1986 and the congress in Berlin, presided over by Arno Motulsky (Fig. 1.34), HD had been mapped using RFLPs—the first disorder of unknown biochemical basis mapped by this approach. VNTRs were a new class of markers. The previous spring, PCR had been unveiled at the Cold Spring Harbor Symposium, which had been devoted to the human genome. PFGE, DGGE, and CVS were introduced as acronyms for other new techniques or diagnostic methods. Contiguous gene syndromes were conceptualized and the deletions underlying them used for gene mapping and gene isolation. The approach, then called "reverse genetics," had succeeded in the isolation of the gene for chronic granulomatous disease and at the very time of the congress was well on the way to characterization of the large gene that is mutant in Duchenne muscular dystrophy. Transgenic mice expressing human genes had been created. By molecular methods, the Knudson hypothesis had been proved for retinoblastoma; and the specific oncogenic molecular changes in the Philadelphia chromosome of chronic myeloid leukemia had been worked out, as well as those in the translocations underlying Burkitt lymphoma.

By the time of the eighth congress in 1991 (chaired by Victor McKusick; Fig. 1.34), the HGP had been launched. The HGP had been in a stage of debate and planning in 1986. The fruitfulness of positional cloning (alias reverse genetics) had been established for Duchenne muscular dystrophy, cystic fibrosis, neurofibromatoses, polyposis coli, and others. The candidate gene approach had also paid off for retinitis pigmentosa, hypertrophic cardiomyopathy, Marfan syndrome, and malignant hyperthermia. The genetics of common cancers such as colon cancer had been greatly clarified. TDF had been cloned. Specific mutations in the mitochondrial chromosome had been related to specific diseases. Imprinting and uniparental disomy were challenging concepts.

Between the eighth congress and the ninth (1996), in Rio de Janeiro, Brazil, with Newton Morton as president, ESTs (expressed sequence tags) were introduced [148] as a shortcut to the expressed part of the genome, the first free-living organism (*H. influenzae*) was completely sequenced (1995), a detailed linkage reference map using DNA markers was completed, a physical map of YAC contigs was published, radiation hybrid mapping came into wide use, the role of folic acid in birth defects was accepted as established, and folic acid supplementation of food products had been introduced into practice. Preimplantation diagnosis at a four- or eight-cell stage was added to in vitro fertilization as a method for selection of nonmutant concepti.

Between the ninth congress in Rio de Janeiro and the 10th in Vienna in 2001 (G. Utermann, president), database "mining" (research in silico or cybergenomics) came in as a primary method of genomic research. As a result of positional cloning and related approaches to disease gene discovery, well over 1000 genes had been found to have one or more disease-related mutations [140]. Microassay methods (chip technology) had been introduced for profiling gene expression and for gene diagnosis [149]. Yeast became the first eukaryote to be completely sequenced (in the spring of 1996 shortly before the Rio congress). *Caenorhabditis elegans*, a nematode, was the first metazoan completely sequenced (in 1999), followed by *Drosophila* (in 2000). On June 26, 2000, completion of a first draft of the complete human sequence was announced jointly by Venter and Collins, for the private and public projects, respectively, in a historic ceremony in the East Room of the White House, with connection by satellite to 10 Downing Street in London where Prime Minister Blair echoed President Clinton in emphasizing the significance of the event.

During the 1996–2001 interval, rapid progress toward the goal of the HGP was accompanied by the following paradigm shifts:

- from structural genomics to functional genomics;
- from map-based gene discovery to sequence-based gene discovery;
- from genetic disease diagnosis to detection of genetic predisposition of common disorders;
- from etiology (specific causation) to pathogenesis (mechanism);
- from a one-gene-at-a-time approach to the study of systems, pathways, and families of genes;
- from genomics to proteomics.

Another paradigm shift was from medical genetics to genetic medicine. This was not merely a change of name of institutions, although that did occur, as in the program at Johns Hopkins, which began in 1957 as the Division of Medical Genetics in the Department of Internal Medicine, became in 1989 the multidepartmental Center for Medical Genetics, and in 1999 was transformed into the Institute of Genetic Medicine. Beyond a change in name, this shift is a broadening of perspective. Medical genetics tends to imply an exclusive focus on rare Mendelian disorders and chromosomal aberrations; genetic medicine reflects the fact that genetics pervades all parts of clinical medicine, that genetic factors are involved in all disease, and that genetic predisposition in the etiology of common disorders (complex traits) is an important part of the science and practice of medical genetics.

Since 1956, human genetics has become medicalized, to use Motulsky's term, to an enormous extent. It has become subspecialized. Medical genetics has become professionalized through the development of clinical colleges and certifying organizations. During the past decade, human genetics has also become intensely molecularized. Molecular genetics pervades all aspects of human and medical genetics. Human genetics has become commercialized to an extent we might not have predicted. The field has also become democratized and universalized; its implications are felt in all aspects of society. It has become consumerized; consumerism is evident in the role of genetic support groups and foundations for funding of research on single genetic diseases.

### 2.4.10 Forty Years in the History of Medical Genetics

MIM has recorded in detail the advances in medical genetics in the period since it was first initiated in 1960 as a catalog of X-linked traits [126]. The catalog of recessives was undertaken in late 1962 in connection with studies of an inbred group, the Old Order Amish. The dominant catalog was created in 1963. The catalogs were placed on the computer in 1964. The first print edition of all three catalogs (a pioneer in computer-based publication) appeared in 1966, the 12th in 1998. Those 12 editions represent serial cross sections of the field of medical genetics over four decades.

MIM was made publicly available online (under the designation OMIM) in 1987. OMIM (https://www.omim.org) has the virtue of timeliness (it is updated almost daily) and easy searchability. The print version has usefulness in a nonelectronic setting and as an archive.

The evolution of medical genetics since the early 1960s is illustrated in many ways by MIM (and OMIM). The subtitle at the beginning was *Catalogs of Autosomal Dominant, Autosomal Recessive, and X-linked Phenotypes*. Although the gene behind the phenotype was always in mind (e.g., the X-linked catalog was created as an indication of the genetic constitution of the X chromosome), almost the only way to identify a gene was through a mendelizing phenotype, at the beginning. With molecular genetic advances, the subtitle in later print editions became *A Catalog of Human Genes and Genetic Disorders*.

## 2.5 The Future

Since the late 1980s, thought and discussion have been devoted to the future of human genetics to an unprecedented extent—largely growing out of the debate over the HGP and the planning for it. Rarely, if ever, has the future of a scientific field and its implications for society been given such wide and intense attention. An unusual and important feature of the federally funded HGP in the United States was support of research projects in the ethical, legal, and societal implications (ELSI) of genome mapping and sequencing.

ELSI was a proactive effort to avoid misuse and abuse of newfound genomic information. The history of eugenics and its misguided, unethical, and immoral practices based on primitive knowledge of human genetics in the first half of the 20th century was a warning signal. The atrocities in Nazi Germany [150] were only the most flagrant, inhuman, and abominable examples. Furthermore, it was anticipated that the power of information on the full genome of individuals and population groups would raise novel issues for society, medicine, and the law.

Completion of the HGP provides a source book of the human that will be the basis for study in human biology and medicine for a long time to come. When all the genes have been found, we certainly still will not know, for most of them, their function, even in solo, let alone in concert with the rest of the genome. True reverse genetics, in the new genetics according to Walter Gilbert, David Botstein, and others, will involve working from specific DNA sequences of unknown function to the phenotype. Having in the past worked progressively from the phenotype to DNA, we will in the coming years be returning from DNA to the phenotype by determining the function of specific DNA sequences. Just as the function of much of DNA remains to be determined even though the full sequence is known, the worldwide variation in that DNA is largely unknown, and long study will be required of the relationship between DNA variation and variation in function, critical to the understanding of evolution and the genetics of disease susceptibility and performance.

Study of the source book provided by the HGP should be particularly useful in the two great frontiers of human biology: how the mind works and how development is programmed. In the area of health and medicine, great benefit can be expected from the better understanding of genetic factors in multifactorial disorders, which tend to be the common conditions such as hypertension and major mental illness. The source book will surely be very useful to the understanding of somatic cell genetic disease. Gene mapping and related studies have been primarily responsible for appreciation that, in addition to the three classical categories of genetic disease—single-gene disorders, multifactorial disorders, and chromosomal aberrations—somatic cell genetic disease is a large fourth category. This has been most extensively and definitively established through the definition of mutation as the basis of cancer. (Facetiously, it is suggested that medical genetics is taking over oncology.) Somatic cell mutations are likely to occupy us increasingly as the basis of congenital malformations, autoimmune processes, and even aging. (MacFarlane Burnett suggested in about 1960 that somatic cell mutation is the basis of autoimmune processes, and a somatic mutation theory of aging has been entertained for a long time.) The connection between oncogenesis and teratogenesis—between oncogenes and teratogenes, if you will—is already adumbrated by the examples of Wilms' tumor and Greig cephalosyndactyly syndrome, to mention two. It is to be assumed that somatic cell gene therapy not only for inherited diseases but perhaps also for some of these acquired somatic cell genetic diseases, especially neoplasia, will become available during the next decade.

Two scientific and technologic revolutions, the biologic revolution and the information revolution, converged in the human genome initiative. Information is power. Risks can accompany both the political and the scientific changes. Appropriately, the ELSI of the human genome initiative are being examined in many parts of the world.

The methods developed by the HGP will allow the rapid and economical generation of information on the genomes of individuals. In the medical area, this information will widen the gap between what we know how to diagnose and what we know how to treat, between what we can diagnose in the presymptomatic state and what we can prevent. We have had that problem, for example, in the case of HD. The complete map and sequence are also likely to increase the gap between what we think we know and what we really know. But by this second gap, I refer in part to the likelihood that weak associations will be found between particular genomic constitutions and certain presumed characteristics such as criminality or alcoholism or elements of intelligence or performance. Some of these associations are likely to be spurious. Other weak associations may be found to be statistically valid but will be blown out of all proper proportion to the detriment of individuals and of groups. As geneticists, we have a responsibility to avoid unfounded conclusions and overblown interpretations and to inculcate a profound respect for the genetic variability that is the strength of the species and indeed of the individual—referring to the differences in the two genomes each of us has, one from the father and one from the mother.

The mere existence of the complete reference gene map and DNA sequence down to the last nucleotide may lead to the absurdity of reductionism, the misconception that we then know everything it means to be human, or to the absurdity of genetic determinism, that what we are is a direct and inevitable consequence of what our genome is. Our phenotypes are not "hardwired" to our genotypes. Risk figures that state the chance of a given common disorder in an individual based on a genome screen are probabilities, not certainties. They are more analogous to a weather report than to a road map.

Thus, information on the reference gene map and sequence of the human may represent per se a hazard, if it distorts the way we think about ourselves and our fellow human beings. The ability to analyze the genomes of individuals is accompanied by risks of information misuse and abuse. We must be alert to the need to protect the privacy and confidentiality of the information that the HGP will allow to be collected on the genetic constitution of individuals. We must make every effort to avoid the misuse or abuse of such information by third parties, and our governments may need to take measures to ensure these protections under law.

Near the end of his terms of office, President Eisenhower warned against the dangers of the military industrial complex. It is appropriate to warn of a potential hazard of the genetic–commercial complex. The increasing availability of tests for presumed genetic quality or poor quality could lead the commercial sector and the Madison Avenue publicist to bring subtle or not so subtle pressure on couples to make value judgments in the choice of their gametes for reproduction. Autonomy in reproductive choice is a cornerstone of the ethics of genetic counseling. That reproductive choice would not be autonomous if subjected to the Madison Avenue type of pressure. Especially, trivialization of reproductive choices should be avoided.

As human geneticists, we are privileged to work in a scientifically important field and a field of intellectual challenge. Human genetics is a field that holds particular fascination because it involves the most fundamental and pervasive aspects of our own species, an added fascination that the physical sciences or pure mathematics, for example, cannot share. To have combined with this intellectual and anthropocentric fascination the opportunity to contribute to human welfare and to be of service to families and individuals through medical genetics and clinical genetics is a privilege. The privilege carries with it responsibilities to which I have already referred.

## 2.6 Addendum 2013 (Peter S. Harper)

*Peter S. Harper*
Institute of Medical Genetics, School of Medicine, Cardiff University, Heath Park, Cardiff, United Kingdom

### 2.6.1 The Founders of Clinical Genetics

The section headed "Evolution of Clinical Genetics" in McKusick's original chapter tells us much about the advances in the field over the last half of the 20th century, but almost nothing about the people responsible or the pattern of evolution across different countries. Perhaps this is not too surprising, for he would have unavoidably had to give much prominence to himself. Now there need not be any such inhibitions and an account, necessarily brief, of the founders of the field forms an important part of its history.

While there were medically trained researchers in human genetics from the beginning of the 20th century, such as Julia Bell in Britain [105] and Madge Macklin and William Allan in America, clinical genetics as a medical discipline did not take shape in any significant way until the 1950s, when medical workers in several countries began to recognize that the scientific foundations of human genetics were now substantial enough for the systematic development of both clinically oriented research and genetic services.

In North America the initial step can be seen as the founding in 1950 of a medical genetics unit at Montreal Children's Hospital by F. Clarke Fraser, followed in 1957 by academic departments (technically Divisions of Departments of Medicine) in Baltimore and Seattle, by Victor McKusick and Arno Motulsky, respectively. In

Europe, France took an early lead, with pediatrician Maurice Lamy appointed as Professor of Medical Genetics at Hôpital Necker, Paris, in 1953, succeeded by Jean Frézal. In Britain, Lionel Penrose's Galton Laboratory at University College, London, had provided a focal point from 1945 for training and inspiring many of the founders from elsewhere, but did not itself develop as a clinical genetics unit, concentrating on basic human genetics research after Penrose's retirement, while John Fraser Roberts, active from the 1930s and founding the first UK genetic counseling clinic in 1948, also restricted his clinical role to genetic counseling alone, as did his successor Cedric Carter. The two major British founders in the field of clinical genetics were the pediatrician Paul Polani, based at Guy's Hospital, London, and Cyril Clarke, internist and Head of Medicine at the Liverpool Medical School. It is of interest to note that the medical background of these founders in both Britain and North America was at least as much from adult medicine as from pediatrics.

In many countries, development of clinical genetic services was considerably slower; for Germany this is not unexpected, given the legacy of abuses from eugenics under the Third Reich, but it is more surprising in the case of the Scandinavian countries, which had from an early stage been leaders in human genetics research. From the very beginning, though, close international links have been a strong feature of medical genetics, encouraged and reinforced by the numerous visiting workers at Penrose's Galton Laboratory and by the trainees from many different countries with Victor McKusick in Baltimore, most of whom became leaders in their own countries on their return and were strongly influential in determining the pattern and ethos of genetic services which they and their successors established.

### 2.6.2 Victor McKusick and the History of Medical Genetics

The role of Victor McKusick in the history of medical genetics is a threefold one. As a key founder of the field, he and his own work, over a period of more than 50 years [114], form a major strand in its history and development; second, his periodic reviews on the current status of medical genetics chronicle successive advances (Ref. [126]; McKusick, 1989) and many of his other writings have a strongly historical approach. Finally from the beginning of his career he took a keen interest in the history of medicine and published a series of studies on specific physicians whose early reports formed important landmarks in the documentation of inherited disorders [143]. A fuller assessment of McKusick's role as a historian of medical genetics can be found in the chapter by Harper in the book of Dronamraju and Francomano (2012) [153].

Another, more unusual contribution has come from McKusick's habit of taking numerous photographs at conferences, courses, and all other possible occasions,

**TABLE 1.2 History of Human and Medical Genetics Sources and Resources**

**Records of Individual Workers**
American Philosophical Society Genetics Collection (www.amphilsoc.org)
UK Genetics Archive Project (www.genmedhist.org/Records)
Cold Spring Harbor Archive (http://library.cshl.edu/)

**Oral History**
Oral history of human genetics project (www.socgen.ucla.edu/hgp/)
Interviews with human and medical geneticists (http://www.genmedhist.org/Interviews)
Talking of Genetics (Gitschier, 2010)
Witness Seminars. See www.ucl.ac.uk/histmed/publications/wellcome_witnesses_c20th_med for the transcripts of several seminars relevant to medical genetics

**Books**
See "Further reading"; also Harper (2008), Chapter 18 and Appendix 1 for details
Book collections
Human Genetics Historical Library (see http://www.genmedhist.org/HumanHistLib/): a collection of over 3000 books
Electronic Scholarly Publishing (www.esp.org) has digitized a series of books and papers on classical genetics
Cold Spring Harbor Laboratory Press has published a number of important historical books on genetics and eugenics

**TABLE 1.3 A Timeline for Human and Medical Genetics**

| Year | Event |
|---|---|
| 1651 | William Harvey's book *De Generatione Animalium* studies the egg and early embryo in different species and states: "Ex ovo omnium" (all things from the egg) |
| 1677 | Microscopic observations of human sperm (Leeuwenhoek) |
| 1699 | Albinism noted in "Moskito Indians" of Central America (Wafer) |
| 1735 | Linnaeus, *Systema Naturae*; first "natural" classification of plants and animals |
| 1751 | Maupertuis proposes equal contributions of both sexes to inheritance and a "particulate" concept of heredity |
| 1753 | Maupertuis describes polydactyly in Ruhe family; first estimate of likelihood for it being hereditary |
| 1794 | John Dalton: Color blindness described in himself and others; limited to males |
| | Erasmus Darwin publishes *Zoonomia*; progressive evolution from primeval organisms recognized |
| 1803 | Hemophilia in males and its inheritance through females described (Otto) |
| 1809 | Inherited blindness described in multiple generations (Martin) |
| | Lamarck supports evolution, including human, based on inheritance of acquired characteristics |
| 1814 | Joseph Adams, concepts of "predisposition" and "disposition"; "congenital" and "hereditary" |
| 1852 | First clear description of Duchenne muscular dystrophy (Meryon) |
| 1853 | Hemophilic son, Leopold, born to Queen Victoria in England |
| 1858 | Charles Darwin and Alfred Russel Wallace; papers to Linnean Society on natural selection |
| 1859 | Charles Darwin publishes *On the Origin of Species* |
| 1865 | Gregor Mendel's experiments on plant hybridization presented to Brünn Natural History Society |
| 1866 | Mendel's report formally published |
| 1868 | Charles Darwin's "provisional hypothesis of pangenesis" |
| | Charles Darwin collects details of inherited disorders in *Animals and Plants under Domestication* |
| 1871 | Friedrich Miescher isolates and characterizes "nucleic acid" |
| 1872 | George Huntington describes "Huntington's disease" |
| 1882 | First illustration of human chromosomes (Flemming) |
| 1885 | "Continuity of the germ plasm" (August Weismann) |
| 1887 | Boveri shows constancy of chromosomes through successive generations |
| 1888 | Waldeyer coins term "chromosome" |
| | Weismann presents evidence against inheritance of acquired characteristics |
| 1889 | Francis Galton's *Law of Ancestral Inheritance* |
| 1891 | Henking identifies and names "X chromosome" |
| 1894 | Bateson's book *Material for the Study of Variation* |
| 1896 | E.B. Wilson's book *The Cell in Development and Inheritance* |
| 1899 | Archibald Garrod's first paper on alkaptonuria |
| 1900 | Mendel's work rediscovered (de Vries, Correns, and Tschermak) |
| 1901 | Karl Landsteiner discovers ABO blood group system |
| | Archibald Garrod notes occurrence in sibs and consanguinity in alkaptonuria |
| 1902 | Bateson and Saunders' note on alkaptonuria as an autosomal recessive disorder; Bateson and Garrod correspond |
| | Garrod's definitive paper on alkaptonuria an example of "chemical individuality" |
| | Bateson's *Mendel's Principles of Heredity; a Defence* supports Mendelism against attacks of biometricians |
| | Chromosome theory of heredity (Boveri, Sutton) |
| 1903 | American Breeders Association formed; includes section of eugenics from 1909 |
| | Cuénot in France shows Mendelian basis and multiple alleles, for albinism in mice |
| | Castle and Farabee show autosomal recessive inheritance in human albinism |
| | Farabee shows autosomal dominant inheritance in brachydactyly |
| 1905 | Stevens and Wilson separately show inequality of sex chromosomes and involvement in sex determination in insects |
| | Bateson coins term "genetics" |
| 1906 | First International Genetics Congress held in London |

*Continued*

TABLE 1.3 **A Timeline for Human and Medical Genetics—cont'd**

| Year | Event |
|---|---|
| 1908 | Garrod's Croonian lectures on "inborn errors of metabolism" |
| | Royal Society of Medicine, London, "Debate on Heredity and Disease" |
| | Hardy and Weinberg independently show relationship and stability of gene and genotype frequencies (Hardy–Weinberg equilibrium) |
| 1909 | Bateson's book *Mendel's Principles of Heredity* documents a series of human diseases following Mendelian inheritance |
| | Karl Pearson initiates *The Treasury of Human Inheritance* |
| | Wilhelm Johannsen introduces term "gene" |
| 1910 | Thomas Hunt Morgan discovers X-linked "white eye" *Drosophila* mutant |
| | Eugenics Record Office established at Cold Spring Harbor under Charles Davenport |
| 1911 | Wilson's definitive paper on sex determination shows X-linked inheritance for hemophilia and color blindness |
| 1912 | Winiwarter proposes diploid human chromosome number as approximately 47; first satisfactory human chromosome analysis |
| | First International Eugenics congress (London) |
| 1913 | Alfred Sturtevant constructs first genetic map of *Drosophila* X-chromosome loci |
| | American Genetics Society formed as successor to American Breeders Association |
| 1914 | Boveri proposes chromosomal basis for cancer |
| | (Outbreak of World War I) |
| 1915 | Haldane et al. show first mammalian genetic linkage in mouse |
| 1916 | Relationship between frequency of a recessive disease and of consanguinity (F. Lenz) |
| | Calvin Bridges shows nondisjunction in *Drosophila* |
| 1918 | Anticipation first recognized in myotonic dystrophy (Fleischer) |
| | R.A. Fisher shows compatibility of Mendelism and quantitative inheritance |
| 1919 | Hirszfeld and Hirszfeld show ABO blood group differences between populations |
| | Genetical Society founded in UK by William Bateson |
| 1922 | Inherited eye disease volumes of *Treasury of Human Inheritance* (Julia Bell) |
| 1923 | Painter recognizes human Y chromosome; proposes human diploid chromosome number of 48 |
| 1927 | Hermann Muller shows production of mutations by X-irradiation in *Drosophila* |
| | Compulsory sterilization on eugenic grounds upheld by courts in America (Buck v. Bell) |
| 1928 | Stadler shows radiation-induced mutation in maize and barley |
| | Griffiths discovers "transformation" in *Pneumococcus* |
| 1929 | Blakeslee shows effect of chromosomal trisomy in *Datura*, the thornapple |
| 1930 | R.A. Fisher's *Genetical Theory of Natural Selection* |
| | Beginning of major Russian contributions to human cytogenetics |
| | Haldane's book *Enzymes* attempts to keep biochemistry and genetics linked |
| 1931 | Archibald Garrod's second book *Inborn Factors in Disease* |
| | UK Medical Research Council establishes Research Committee on Human Genetics (chairman J.B.S. Haldane) |
| 1933 | Nazi eugenics law enacted in Germany |
| 1934 | Fölling in Norway discovers PKU |
| | *Treasury of Human Inheritance* volume on Huntington's disease (Julia Bell) |
| | O.L. Mohr's book *Genetics and Disease* |
| | Mitochondrial inheritance proposed for Leber's optic atrophy (Imai and Moriwaki, Japan) |
| 1935 | First estimate of mutation rate for a human gene (hemophilia; J.B.S. Haldane); a preliminary estimate had been made by Haldane in 1932 |
| | R.A. Fisher (amongst others) suggests use of linked genetic markers in disease prediction |
| 1937 | First human genetic linkage—hemophilia and color blindness (Bell and Haldane) |
| | Moscow Medical Genetics Institute closed; director Levit and others arrested and later executed; destruction of Russian genetics begins |
| | Seventh International Genetics Congress, Moscow, canceled |
| | Max Perutz begins crystallographic studies of hemoglobin in Cambridge |
| 1938 | Lionel Penrose publishes "Colchester Survey" of genetic basis of mental handicap |

**TABLE 1.3 A Timeline for Human and Medical Genetics—cont'd**

| Year | Event |
|---|---|
| 1939 | Seventh International Genetics Congress held in Edinburgh; "Geneticists' Manifesto" issued |
| | (Outbreak of World War II) |
| | Cold Spring Harbor Eugenics Record Office closed |
| | Rh blood group system discovered (Landsteiner and Wiener) |
| 1941 | Beadle and Tatum produce first nutritional mutants in *Neurospora* and confirm "one gene–one enzyme" principle |
| | Charlotte Auerbach discovers chemical mutagens in Edinburgh (not published until the end of the war) |
| 1943 | Nikolai Vavilov, leader of Russian genetics, dies in Soviet prison camp |
| | First American genetic counseling clinic |
| | Mutation first demonstrated in bacteria (Luria) |
| 1944 | Schrödinger's book *What Is Life?* provides inspiration for the first molecular biologists |
| | Avery shows bacterial transformation is due to DNA, not protein |
| 1945 | Lionel Penrose appointed as head of Galton Laboratory, London; founds modern human genetics as a specific discipline |
| | (Hiroshima and Nagasaki atomic explosions) |
| | Genetic study of effects of radiation initiated on survivors of the atomic explosions (J.V. Neel, director) |
| 1946 | Penrose's inaugural lecture at University College, London, uses PKU as paradigm for human genetics |
| | John Fraser Roberts begins first UK genetic counseling clinic in London |
| | Sexual processes first shown in bacteria (Lederberg) |
| 1948 | Total ban on all genetics (including human genetics) teaching and research in Russia |
| | American Society of Human Genetics founded, H.J. Muller, president; J.B.S. Haldane suggests selective advantage in thalassemia due to malaria |
| 1949 | *American Journal of Human Genetics* begun, Charles Cotterman, first editor |
| | Linus Pauling et al. show sickle cell disease to have a molecular basis; J.V. Neel shows it to be recessively inherited |
| | Barr and Bertram (London, Ontario) discover the sex chromatin body |
| 1950 | Curt Stern's Book *Human Genetics* |
| | Frank Clark Fraser initiates medical genetics at McGill University, Montreal |
| 1951 | Linus Pauling shows triple-helical structure of collagen |
| | HeLa cell line established from cervical cancer tissue of Baltimore patient Henrietta Lacks |
| 1952 | First human inborn error shown to result from enzyme deficiency (glycogen storage disease type 1, Cori and Cori) |
| | Rosalind Franklin's X-ray crystallography shows helical structure of B form of DNA |
| 1953 | Model for structure of DNA as a double helix (Watson and Crick) |
| | Bickel et al. initiate dietary treatment for PKU |
| | Enzymatic basis of PKU established (Jervis) |
| | Specific Chair in Medical Genetics founded (first holder Maurice Lamy, Paris) |
| 1954 | Allison proves selective advantage for sickle cell disease in relation to malaria |
| 1955 | Sheldon Reed's book *Counseling in Medical Genetics* |
| | Oliver Smithies develops starch gel electrophoresis for separation of human proteins |
| | Fine structure analysis of bacteriophage genome (Benzer) |
| 1956 | Tjio and Levan show normal human chromosome number to be 46, not 48 |
| | First International Congress of Human Genetics (Copenhagen) |
| | Amniocentesis first validated for fetal sexing in hemophilia (Fuchs and Riis) |
| 1957 | Ingram shows specific molecular defect in sickle cell disease |
| | Specific medical genetics departments opened in Baltimore (Victor McKusick) and Seattle (Arno Motulsky) |
| 1958 | First HLA antigen detected (Dausset) |
| 1959 | Harry Harris' book *Human Biochemical Genetics* |
| | Perutz completes structure of hemoglobin |
| | First human chromosome abnormalities identified in: |
| | Down's syndrome (Lejeune et al.) |
| | Turner syndrome (Ford et al) |
| | Klinefelter syndrome (Jacobs and Strong) |

Continued

**TABLE 1.3 A Timeline for Human and Medical Genetics—cont'd**

| Year | Event |
|---|---|
| 1960 | Trisomies 13 and 18 identified (Patau et al. and Edwards et al.) |
| | First edition of *Metabolic Basis of Inherited Disease* |
| | Role of mRNA recognized |
| | First specific cytogenetic abnormality in human malignancy (Nowell and Hungerford, Philadelphia chromosome) |
| | Chromosome analysis on peripheral blood allows rapid development of diagnostic clinical cytogenetics (Moorhead et al.) |
| | Denver conference on human cytogenetic nomenclature |
| | First full UK Medical Genetics Institute opened (under Paul Polani, Guy's Hospital, London) |
| | First Bar Harbor Course in Medical Genetics, under Victor McKusick |
| 1961 | Prevention of rhesus hemolytic disease by isoimmunization (Clarke et al., Liverpool) |
| | Mary Lyon proposes X-chromosome inactivation in females |
| | Cultured fibroblasts used to establish biochemical basis of galactosemia (Krooth and Weinberg), establishing value of somatic cell genetics |
| | "Genetic code" linking DNA and protein established (Nirenberg and Matthaei) |
| 1963 | Population screening for PKU in newborns (Guthrie and Susi) |
| 1964 | Ultrasound used in early pregnancy monitoring (Donald) |
| | First journal specifically for medical genetics (*Journal of Medical Genetics*) |
| | Genetics restored as a science in USSR after Nikita Khrushchev dismissed |
| | First HLA workshop (Durham, North Carolina) |
| 1965 | High frequency of chromosome abnormalities found in spontaneous abortions (Carr, London, Ontario) |
| | Human–rodent hybrid cell lines developed (Harris and Watkins) |
| 1966 | First chromosomal prenatal diagnosis (Steele and Breg) |
| | First edition of McKusick's *Mendelian Inheritance in Man* |
| | Recognition of dominantly inherited cancer families (Lynch) |
| 1967 | Application of hybrid cell lines to human gene mapping (Weiss and Green) |
| 1968 | First autosomal human gene assignment to a specific chromosome (Duffy blood group on chromosome 1) by Donahue et al. |
| 1969 | First use of "Bayesian" risk estimation in genetic counseling (Murphy and Mutalik) |
| | First masters degree course in genetic counseling (Sarah Lawrence College, New York) |
| 1970 | Fluorescent chromosome banding allows unique identification of all human chromosomes (Zech, Caspersson et al.) |
| 1971 | "Two-hit" hypothesis for familial tumors, based on retinoblastoma (Knudson) |
| | Giemsa chromosome banding suitable for clinical cytogenetic use (Seabright) |
| | First use of restriction enzymes in molecular genetics (Danna and Nathans) |
| 1972 | Population screening for Tay–Sachs disease (Kaback and Zeiger) |
| 1973 | Prenatal diagnosis of neural tube defects by raised $\alpha$-fetoprotein (Brock) |
| | First Human Gene Mapping Workshop (Yale University) |
| 1975 | DNA hybridization (Southern) on a "Southern blot" |
| 1977 | Human $\beta$-globin gene cloned |
| 1978 | Prenatal diagnosis of sickle cell disease through specific RFLP (Kan and Dozy) |
| | First mutation causing a human inherited disease characterized ($\beta$-thalassemia) |
| | First birth following in vitro fertilization (Steptoe and Edwards) |
| 1979 | Vogel and Motulsky's textbook *Human Genetics, Problems and Approaches* |
| 1980 | Primary prevention of neural tube defects by preconceptional multivitamins (Smithells et al.) |
| | Detailed proposal for mapping the human genome (Botstein et al.) |
| 1981 | Human mitochondrial genome sequenced (Anderson et al.) |
| 1982 | Linkage of DNA markers on X chromosome to Duchenne muscular dystrophy (Murray et al.) |
| 1983 | First autosomal linkage using DNA markers for Huntington's disease (Gusella et al.) |
| 1983 | First general use of chorionic villus sampling in early prenatal diagnosis |

**TABLE 1.3 A Timeline for Human and Medical Genetics—cont'd**

| Year | Event |
|---|---|
| 1984 | DNA fingerprinting discovered (Jeffreys) |
| 1985 | Application of DNA markers in genetic prediction of Huntington's disease |
| | First initiatives toward total sequencing of human genome (US Department of Energy and Cold Spring Harbor meetings) |
| 1986 | PCR for amplifying short DNA sequences (Mullis) |
| 1988 | International Human Genome Organisation established |
| | US congress funds Human Genome Project |
| 1989 | Cystic fibrosis gene isolated |
| | First use of preimplantation genetic diagnosis |
| 1990 | First attempts at gene therapy in immunodeficiencies |
| | Fluorescence in situ hybridization introduced to cytogenetic analysis |
| 1991 | Discovery of unstable DNA and trinucleotide repeat expansion (fragile X) |
| 1992 | Isolation of *PKU* (phenylalanine hydroxylase) gene (Woo et al.) |
| | First complete map of human genome produced by French Généthon initiative (Weissenbach et al.) |
| 1993 | Huntington's disease gene and mutation identified |
| | *BRCA1* gene for hereditary breast–ovarian cancer identified |
| 1996 | "Bermuda Agreement" giving immediate public access to all Human Genome Project data |
| 1997 | First cloned animal (Dolly the sheep), Roslin Institute, Edinburgh |
| 1998 | Total sequence of model organism *Caenorhabditis elegans* |
| | Isolation of embryonic stem cells |
| 1999 | Sequence of first human chromosome [22]. |
| 2000 | "Draft sequence" of human genome announced jointly by International Human Genome Consortium and by Celera |
| | Correction of defect in inherited immune deficiency (SCID) by gene therapy (but subsequent development of leukemia) |
| 2002 | Discovery of microRNAs |
| 2003 | Complete sequence of human genome achieved and published |
| 2005 | Sequencing of chimpanzee genome |
| 2006 | Prenatal detection of free fetal DNA in maternal blood clinically feasible |
| 2007 | First genome-wide association studies giving robust findings for common multifactorial disorders |
| 2008 | First specific individual human genomes sequenced |
| 2010 | Sequencing of Neanderthal genome |
| | Diagnostic use of human exome sequencing |
| 2011 | Modern human genome shown to contain sequence from other ancient hominins (Neanderthal and Denisovan) |

*PCR*, polymerase chain reaction; *PKU*, phenylketonuria; *RFLP*, restriction fragment length polymorphism.
Based on the timeline of the Genetics and Medicine Historical Network (www.genmedhist.org). The original version first appeared in Harper PS. A short history of medical genetics: Oxford University Press; 2008.

both formal and informal. Many of the then young and unknown scientists and clinicians who appear in these informal "snaps" have since risen to fame. This collection must amount to several thousand images over the years and it is greatly to be hoped that they will be carefully preserved and cataloged along with his other personal scientific records at the Johns Hopkins Archive. Only a few have been published, some in this chapter, others in relation to the Bar Harbor "short course" on medical genetics [1].

Altogether medical genetics is fortunate to have had such an assiduous and objective chronicler of its first half-century as Victor McKusick has been.

### 2.6.3 Preserving the History of Medical Genetics

The inevitable loss of the founders of medical genetics reminds us how much of its history we lose when they die. Not all of the key workers have such a keen historical sense as did Victor McKusick; many confine their writing to current scientific or clinical activities.

biology, but the ability of enzymes to treat disorders of amino acid and organic acid metabolism as well as other conditions was less well predicted. Increasingly, parenterally injected different forms of nucleic acids are being developed. Gene therapy trials for disorders of the eye, soluble and secreted enzymes, and hematological disorders, in particular, in vitro and reinjected, are well along and may soon be in common use. In vivo gene correction is not far distant on the horizon, with successful experiments already having taken place in the mouse. To the astonishment of many, injection of mRNA and iRNA is already in clinical use, despite fears that it would be too unstable and short-lived to be clinically useful.

At the time of the last edition of this book, the genetic basis for many cancers had already been established. With the explosive growth in sequencing capacity and the informatics to analyze tumors repeatedly, the field has grown in size and complexity. We now can characterize tumors genetically and identify mutations that may be vulnerable to blockade or stimulation of a specific pathway (precision medicine). Sequencing identifies genetic changes that promote metastasis, and the genetic analysis of cell-free cancer DNA in the circulation allows an alternative approach to site bias during biopsy sampling. To say that genetic analysis is revolutionizing cancer predisposition, diagnosis, treatment, and prognosis would be an understatement. At the same time, the use of cancer panels to look for those with constitutional changes will interdigitate with more general whole-genome sequencing to prevent more cancers, allow their earlier and more successful treatment, and, at the same time, scare many more individuals who might never develop cancer.

## 3.3 Reflections

Charles Dickens begins *A Tale of Two Cities* with the memorable lines, "It was the best of times, it was the worst of times." In many ways this is the situation for medical genetics in the United States, with which I am most familiar, but I am sure elsewhere. On one hand, medical genetics, in the past a stepchild in many institutions, is suddenly getting the respect that we think it deserves. Resources are being invested in hardware, software, and personnel to analyze genomes; we are being incorporated into larger initiatives that emphasize personalized medicine or precision medicine; the demand for our services has never been greater, and through this, our profile and footprint is enlarging. With these advances and this recognition come greater interest by other medical specialties that are larger, have more resources, and have members trying to establish their own professional niches. The future of medical genetics will likely be determined by our ability to remain identified with the special areas that our technology and insights have allowed to grow and thrive. Will we continue to be involved in a meaningful way with cardiac genetics and neurogenetics, two areas in which genetics, either Mendelian or multigenic, plays a role in a large part of their clinical populations? Our ability to maintain a foothold in these areas requires that we add value to the results and analyses that will increasingly become a commodity, interpreted by third-party providers and implemented by clinicians with limited insights into the nuances of genetic analysis. A common refrain among geneticists is that they are stretched thin and faced with too great a workload; off-loading the routine follow-up and treatment of established disorders, the drudge work that keeps our profile up, is a temptation if some other specialty is willing to do this.

Another major change that we face with medical genetics and its disorders is cost. Care of individuals with genetic disorders used to be relatively inexpensive, excepting the costs of supporting the care of affected individuals. A decade ago or so our costs were low, with all of our tests combined in our most extensive evaluations under US$10,000. The widespread use of genomic, and increasingly, metabolomic analysis, now greatly exceeds that. The implementation of "universal" whole-genome or, in the future, whole-epigenome analysis will increase costs exponentially, with analytical and follow-up costs exceeding the costs of the initial sequencing procedure. Treatment for genetic disease, in the past infrequently available and palliative, is becoming more frequent and vastly more expensive, with the cost of drugs now exceeding the costs of follow-up care. The lysosomal disorders were the first to join hemophilia as expensive treatments for genetic diseases, but the list of disorders is growing rapidly and we are just beginning to see the era of nucleic acid therapy, which may influence a great many more disorders. The consequences of this are economic, as well as increasing the interest and the incursion of other specialists into what has been our field. We rarely see patients with hemophilia in a genetic clinic, and other disorders can leave our orbit as well. Are we destined to return to the days of seeing primarily patients with inborn errors of metabolism and developmental delay for which arcane knowledge is required, or will we be involved in the diagnosis and care of all those patients for whom we developed the advanced diagnostic and treatment modalities?

These same forces are determining who and what we will be as a profession in the future. Until recently, medical genetics was a field, save for prenatal diagnosis, that operated primarily in the academic setting. Research was driven by individual initiatives and was often narrowly focused. The newly expanded molecular tools have allowed for an explosion in the description of the molecular basis of many disorders and the definition of new ones. Parallel laboratory studies have confirmed the functional basis of these disorders and have provided the biological basis of genetic markers defined in genome-wide association studies. Now the increasing work force in medical genetics has as one of its primary scholarly endeavors participation in contracted clinical trials, to which they may or may not have significant intellectual input. While some of the participants will become clinical authorities in the field due in part to their role in the trials, it remains to be seen whether these activities risk the quality and independence of critical thinking that has been the hallmark of medical genetics during its creation and in earlier parts of its growth phase.

Medical genetics practitioners are and were trained in the field, and while not all had a strong background in human genetics, the principles of genetics reasoning dominated the field. Gene analysis and pharmaceutical companies have now entered the field and these entities have altered the ethos of medical genetics in several ways. Whole-exome and -genome studies are interpreted by molecular geneticists, with clear formulaic guidelines for addressing the significance of genetic changes. They depend on an ordering clinician for the final interpretation and actions to be taken. Fewer members of this team may be trained geneticists in the future, given that they may specialize in disorders in an organ system, and it is unclear that well-written algorithms can substitute for specific on-site expertise in genetic analysis and interpretation.

We note the many new ethical dilemmas that arise out of these vast changes in the way we practice medical genetics. They are great and some are new. Our practice has always impinged on family members beyond the probands who seek our care and the expanded genetic knowledge that we obtain has an impact in ways we couldn't have guessed and which they and we may not have anticipated. The Aldous Huxley fantasy in *Brave New World* is now beginning to come to pass. These must be addressed and will be in the expanded chapters later in this enterprise.

It is quite clear that the practice of medical genetics is changing in ways that might make it unrecognizable to Victor McKusick observing us from above. These changes are making our profession endlessly exciting and challenging, and they are unlikely to stop soon. Those entering and in the profession are privileged to be there.

## CONTRIBUTOR SECTION

**Stephen D. Cederbaum, MD**

Research Professor and Professor Emeritus of Psychiatry, Pediatrics and Human Genetics, University of California, Los Angeles; Attending Physician, University of California Medical Center, Los Angeles, California.

Stephen Cederbaum was born in Brooklyn, New York, and received his undergraduate degree at Amherst College. He received his medical degree from New York University and trained in medicine at Washington University in St. Louis, and in biochemistry and genetics at Massachusetts General Hospital and the National Institutes of Health in Bethesda. He completed a fellowship in medical genetics with Arno Motulsky in Seattle in 1971. He has been in the Genetics Division at the University of California at Los Angeles, where he is now research professor and Professor Emeritus of Psychiatry, Pediatrics, and Human Genetics.

His major research interests are in inborn errors of metabolism and urea cycle disorders, with special emphasis in the clinical and laboratory aspects of arginase deficiency. He has a passion for all aspects of history and has been teaching the history of human and medical genetics since 1983.

## REFERENCES

[1] McKusick VA. The human genome through the eyes of Mercator and Vesalius. Trans Am Clin Climatol Assoc 1981;42:66–90.

[2] Fisher RA. Has Mendel's work been rediscovered? Ann Sci 1936;1:115–37.

[3] Novitski CE. Another look at some of Mendel's results. J Hered 1995;86:62–6.

1

# Medicine in a Genetic and Genomic Context*

*Reed E. Pyeritz*

Perelman School of Medicine at the University of Pennsylvania, Philadelphia, PA, United States

## 1.1 INTRODUCTION: OUR HISTORY

The history of science is characterized by an exponential rate of expansion [1]. No aspect has escaped, but biology, which is relatively new, has by all accounts exploded. Naturally, these changes are reflected in new principles, new thinking, and new ways of handling new information. Among the problems created is that of making these novelties available to practitioners of science of all kinds.

Among the ways suitable to medicine are massive print volumes that contain detailed summaries of diseases, usually of one class, such as endocrine, gastroenterological, or, as in the case of *Principles and Practice of Medical Genetics and Genomics* (PPMGG), inherited. And of course, the pace of change requires revisions, always characterized by increases that reflect the rate of accumulation. Fission adds volumes whose pages, chapters, contributors, and diseases all do their best to obey the exponential imperative. Each chapter represents one topic more or less, an expansible topic capable of embracing new disorders with each new edition, so the number of chapters is no guide to the number of diseases. Durin the past decade, print volumes are increasingly being supplement, or even replaced, by documents available electronically. In addition to new diseases, new paths of basic science are added, a characteristic of books that mirrors progress in reductionist investigation.

But where is reductionist biology taking us? Clearly, one direction is toward fragmentation; more and more is learned about increasingly restricted fields so that even specialties bifurcate and medicine becomes ever more splintered. But despite such assaults, whatever it is we call "medicine" has at the bottom some integrity, some consistency, and common grounds that are clearly revealed in PPMGG as well as in its sister enterprise, *The Metabolic and Molecular Basis of Inherited Disease* (MMBID) [2]. One such common ground is genetics. And as such a striking variety of disorders of cellular structures and metabolic mechanisms engaging every organ and organ system are included in both of these books, it is easy to imagine that genetic variation is the basis of all diseases. This idea is far from new, having been suggested even in the 18th and 19th centuries, when it took the form of diathesis and idiosyncrasy [3]. Then, in this century, it appears in the shape of a continuity between clear-cut segregating monogenic diseases and varying degrees of familial aggregation of cases that suggest the outcomes of the actions of more than one gene acting in environments favorable for the onset of a disease. But now, with the advent of genomics, which makes possible the study of the genetics of diseases of complex origin in families of patients who have affected relatives, as well as in those who do not, we are learning

*This chapter was authored in earlier editions by Barton Childs, MD, of the Johns Hopkins University School of Medicine. Professor Childs died in 2010.

Emery and Rimoin's Principles and Practice of Medical Genetics and Genomics: Foundations, https://doi.org/10.1016/B978-0-12-812537-3.00001-9

that genetic variation underlies the latter no less than the former. So the continuity of segregating to nonsegregating familial aggregation is extended to include cases where there is neither segregation nor aggregation [4]. Perhaps we should require a disease to be shown not to be associated with any genetic variation, before saying it has no genetic basis. (refer to the chapter on The Genetics of Common Disease).

All professions undergoing rapid change and increasing specialism face the same dilemma. The generalists, who must keep up, find the density of new information daunting, even impossible, to assess and retain. So, books such as the PPMGG are intended to present this information in an orderly way and in relation to specific diseases. But the job is no sooner done than even newer information arrives to change how the various disorders are perceived and, of course, treated. Furthermore, new diseases have been described and must be included. Hence another edition must appear. And that's not all. The various sciences that contribute to our understanding are all changing, too, providing new insights that challenge conventional thinking. Editors respond to this intense pressure by including articles that present not only new information but also new insights, new ways of thinking about groups of diseases or perhaps all, and these usually appear at the beginning, hinting strongly that the reader of any later chapter would do well to read these preliminary ones.

Perhaps focusing first on the principles of chromosomal organization, genomics, and the investigations of diseases of complex origin will assist in understanding the chapters on developmental anomalies, the origins of high blood pressure, or inborn errors. And this may happen, given the effort. But each reader who makes this synthesis for himself or herself is likely to do it in the context of some specific disorder rather than to generalize the principles to all diseases. Indeed, we lack a clearly articulated set of principles of disease as opposed to diseases. That is not to say that medicine lacks principles; the idea of the body as a machine that breaks and needs fixing is one, and the medical history, diagnosis, pathogenesis, treatment, prognosis, and prevention all have a conceptual basis, as do the basic sciences related to medicine. But disease as a concept seems to be taken for granted. However, PPMGG certainly does not suggest that a student of medicine (and we are all students throughout the length of our careers) might take profit in an account of disease as opposed to diseases, including why we have it, who is likely to be affected and how, and when in the lifetime and what forms can it take, as well as what are its constraints. That is, what are the explanatory generalizations that compose a context within which to fit all diseases?

Similarly, definitions of disease have fallen by the wayside. It is true that many such definitions have been offered; there is a sizable literature on the subject [5–8]. Perhaps today's reluctance stems from physicians' perception that we have not had the wherewithal for any but descriptions based on signs and symptoms rather than anything at its core. But today we are satisfied with a definition of a disease when pathogenesis is explained by reference to abnormality of some metabolic or homeostatic system, and we can describe the qualities of the proteins that compose the system. Now, if that is so, why may we not define disease as a consequence of incongruence of a metabolic or homeostatic system with conditions of life? And as all such systems are composed of proteins capable of reflecting the variations of their genetic origins, is it not appropriate to agree with Vogel and Motulsky, who, in the third edition of their book *Human Genetics*, proclaimed genetics as the principal "basic science for medicine" [9]?

If genetics is the basic science for medicine, it should be possible to construct a set of principles that characterize disease in a genetic context—that is, a set of generalizations shared by all diseases and framed in genetic terms. And there should be hierarchies of principles, inclusive and of increasing generality and forming a matrix embracing them all. What follows is one such matrix.

## 1.2 THE PRINCIPLES OF DISEASE

A foundation for developing principles of disease exists in the ideas of Ernst Mayr [10]. Mayr perceived biology as divided into two areas differing in concept and method. One, functional biology, is concerned with the operation and interaction of molecules, systems, and organisms. Causes are proximate, the viewpoint is inward, and questions are commonly preceded by how; how does the organism function? The other area, Mayr calls evolutionary. It is concerned with the history of functional biology, its causes are called ultimate, and its questions are prefaced by why; why in the sense of, what is the history of organisms, what are the conditions of the past that have made it possible to ask for answers to the how questions? The two areas of biology meet, or overlap, at the level of the DNA, so that the functional deals with everything after transcription, whereas the evolutionary centers on the history of the DNA as well as, presumably, with the evolution of the conditions of the environment within which organisms have attained their current state.

Mayr did not include disease in his description of the two biologies, but disease is no less biological than the ideal state, so there should be no difficulty in applying his principles to biological abnormality. Thus, in relation to disease, the proximate causes are the products of the variant genes and the experiences of the environment with which they are incongruent. Ultimate or remote causes are first, the mechanisms of mutation and the causes of fluctuations through time of the elements of the gene pool, including selection, mating systems, founder effects, and drift, and second, the means whereby cultures and social organization evolve. In disease, the variant gene products and the experiences of the environment with which they are incongruent account for characteristic signs and symptoms, but in making available the particular proximate causes assembled by chance in particular patients, it is the remote causes that impart the stamp of individuality to the case.

So the model relates disease to causes, to the gene pool and ultimately to biological evolution, as well as to the evolution of cultures, and to individuality, the latter a consequence of the specificities of both causes. Here there are also elements for constructing a context of principles of disease, always remembering that the word context derives from the Latin word *contexere*, meaning "to weave." That is, the principles must be seen to be related and interdependent so as to form a network of ideas within which to compose one's thoughts about each specific example of each disease.

There is a further feature of Mayr's views on biology, also crucial in its application to disease [11]. It is the state of mind in which to observe patients. In medicine, we tend to think of patients in relation to their disease, that is, as a class of people characterized by the name of the disease. This is what Mayr calls "typological thinking." Although patients do differ somewhat from one another, they all share an essence: the disease. In contrast, Mayr proposes population thinking, in recognition that populations consist not of types but of unique individuals. So, in this context, disease has no essence; its variety is imparted by that of the unique individuals who experience it, each in their own private version, and the name of the disease is a convenience, an acknowledgment of the necessity to group patients for logistical purposes. The fruits of the Human Genome Project (HGP) can be accommodated only with such a population perspective.

Why do we need such principles? Physicians are pragmatic; their way is determined by what they see before them, and students and especially residents are intolerant of anything they can label "philosophical." But the principles are there, explaining the qualities and behaviors of diseases, and they await exposition.

But have we not already discovered them? Medicine is at the peak of success in diagnosis and treatment and moving rapidly to ever new heights of achievement. But all changes may not be equally evident. For example, the analysis of pathogenesis, traditionally a top-down process, is beginning to give way to a bottom-up approach in which discovery of variant genes leads to variant protein products and thence to the same molecular analysis of pathogenesis (refer to the chapter on Pathogenetics). Also, the genetic heterogeneity and individuality of disease are not easily accommodated in traditional thinking. So we are changing how we look at disease, how we define and classify it, and the language we use in describing it. For example, "genomics" and "proteomics" are words that embody ways of thinking new in the past several decades [4,12,13]. These developments are changing our relationships to biology and society. Biologists are expressing interest in the fates of the molecules they discover, and the public is becoming aware of what molecular biology and genetics mean to them, as risk factors, for example [14,15]. So, because this same molecular genetics gives us new insights into the principles that govern—and have always governed—disease, should we not articulate those principles and weave them into our thinking?

Reasons for doing so lie in the need for coherence in medicine, coherence in the face of reductionist dispersion, coherence in bringing new developments to the whole of the medical enterprise and to the public, and coherence in medical education and the thinking that goes into it. No one can possibly know all the information there is, but we all need a context that can supply both a substrate on which to apply the new and a receptacle within which to encompass our own field. The principles of disease bear a relationship to diseases that resemble the relationship of military strategy to integrated systems, responding to influences from adjacent cells, distant organs, and the outside, to maintain the open system in its uncertain relationships in life. They are, therefore, unit steps of homeostasis, and as such are pivotal in concepts of life, development, aging, health, and disease.

Such a list of attainments is banal without explanation and illustration. In the following section, I discuss several ways in which the unit steps of homeostasis fulfill the purposes of the cell. They are called "unit steps" to convey their elemental state as units of pathways and cascades, structural elements, protein machines, transducers in signaling systems, and transporters or receptors of molecules that are going somewhere. The phrase further implies units of integration into systems intended to maintain the organism's steady state; they are the node between nature and nurture, and the phrase has the virtue of being indifferent to whether the specific protein fulfills a useful purpose or is disruptive. And finally, the unit steps have an important historical meaning, representing the central idea of Garrod's inborn error, Beadle and Tatum's one gene–one peptide, and Pauling's molecular disease [21–23].

### 1.4.1 Some Qualities of the Unit Step of Homeostasis

#### 1.4.1.1 As a Unit of History

Clearly, DNA is an instrument of memory, a memory that in preserving the past, gives guidance for the future. That is, the future must always reflect the past, and the means whereby this Janus vision is attained is the protein gene product that repeats its phylogenetic history in its current composition and function and predicts its future in its reincarnation through subsequent generations as itself or in the form of variants. Some of the variants have no future and their incongruence is noted by natural selection. Others are contingent, favorable for some conditions and inappropriate for others. And then there are those proteins that have hardly changed from microbial ancestry and that represent core functions. In human society, political and religious systems have similar capacities for endurance, revealing fundamental unchanged dogma associated with adaptation in ways that promote the cause with little change in the fundamentals. So the proteins that constitute our proteomes descend to us not only from our parents and other human antecedents but also, with variable conservation, from both the ancient and recent past.

#### 1.4.1.2 As Effectors of Gene Intention

We often speak of a gene or genes as being "for" something, by which we indicate some sort of direct relationship to a phenotype. That is, we seem to be saying that the gene's influence is determining. And so it is, if by determining we mean the sequences of bases in mRNA and of amino acids in a protein product. In this sense, the gene is indeed "for" something. But each gene product has in addition, an emergent career of its own, not predicted at all, or only indirectly, by its gene. It assumes a position in the homeostatic device to which it belongs and can now be said to be "for" that system, as the factor VIII gene is both "for" the factor VIII protein and "for" clotting. But it is far from determining clotting; all the other elements are needed, too, or, as we all well know, life-threatening bleeding occurs. We also know that in physiology, system is integrated with system in hierarchical relationships, so that the farther away from the steps of translation and first integration, the more dilute becomes the gene's determining power. No doubt the genes are involved wherever their products are to be found, but indirectly, and any one may have little power to shape the ultimate phenotype. In another sense, the genes appear to be hardly involved at all beyond transcription because it is the quality of the protein product that determines its role in the economy of the cell, a role that is determined by how the protein folds and takes shape, a shape that must accommodate to the shapes of the products of other genes and they with still others. No doubt the protein's folding and shape reflect the information residing in its parent gene, but its gene has no control over the shapes of those other proteins with which it fits, to say nothing of how multiprotein machines work [24]. Here is a question not of genes but of how proteins interact. It is a matter of physiology. Indeed, it is possible that as the fruits of the HGP and the proteomists filter into medicine, we will hear a good deal less about genes and more about proteins [25]. This could be less than ideal were the proteins not perceived to be as closely identified with the concept of variation as the genes. Let us see to it that they are.

#### 1.4.1.3 As a Unit of Development

T. H. Morgan adopted *Drosophila* as an organism suitable for the study of development [26]. But it did not work out that way, and his students led the way to the operational definition of the gene. So it is ironic that technology initially made the fruit fly ideal for the very study that defied Morgan's efforts [27,28].

In development, the genes fulfill their intentions in the ways just described. Their products are the units of developmental change, assuming positions in systems appropriate for their conformation so as to give each organism a matrix, embodying a trajectory of change that is a product of how the embryo, fetus, and infant meet and respond to experiences of intrauterine and external environments. That is, development is a historical process; what the organism is today is built on what it was yesterday and leads to what it will be tomorrow [29]. And because the genes see to the continuity of their products throughout the changes of a lifetime, it is hardly likely that the influences of the past, however, distant, would fail to influence the present. So if we would understand the origins and expressions of disease, it must be in the context of three timescales, all at once: that of phylogeny; that of development maturation and aging; and that of the present [30]. To know what we begin with is to know potential incongruence; to know where development is taking, or has taken, us is to clothe the potential with the probable, one way or the other; and to know where we are at the moment is to know the strengths and weaknesses that we will face tomorrow. There is an increasing interest in the idea that some diseases of middle life have precursors, manifestations dating to early, even intrauterine, life. These expressions may not appear to relate to the disease they are said to characterize. Rather, they may represent subtle changes in trajectories that, if pursued, emerge finally as disease [31]. How else could birth weight be related to type 2 diabetes or heart attack?

#### 1.4.1.4 As a Unit of Individuality

In medicine, patients are seen one at a time. Each one is biologically unique, has different experiences, and tells a different story. These expressions, together with the help of the laboratory and observations over time, are compared with those of the classic case to reach a diagnosis, and, allowing something for variation, treatment or management is devised. This thinking is typological, individuality is usually ignored, and the doctor is in thrall to nosology. The method works well enough, but heterogeneity of proximate cause may be overlooked and patients are likely to be aware when they are being perceived as representative of a class rather than as their unique selves. Now, molecular biology has given us the wherewithal to observe molecular individuality, that is, the capacity to make comparisons between individuals of variations in base pairs in the DNA and differences in amino acid sequences in proteins. The unit of individuality is the unit step of homeostasis, and the expression of uniqueness lies in how the variant proteins affect each its own system and the integrations of the latter with others, as well as how the systems respond to nongenetic proximate causes. Genomic analysis of single nucleotide polymorphisms (SNPs) suggests that the number of polymorphic loci expressed in amino acid substitution in proteins will turn out to be somewhat greater than the 30% we are accustomed to [32,33]. This is the substrate of variability within which additional "private" variants as well as clearly bad mutants express their effects, and all this variation is manifested in how the integrated homeostatic devices are fulfilling their duties. So, if each human being is unique by virtue of the variant proteins in his or her whole physiological apparatus, why should not each such human being express an experience with disease as variously as a career of health?

Variation contributed by variant proteins is far from all. Such variability is compounded by the individuality of the developmental and maturational trajectory characteristic of each person, a path determined no less strongly by the kinds, intensities, and durations of experiences than by the protein gene products with which they interact. But the final arbiter of individuality is the remote causes, which determine the specificity of both genes and experiences. The variation in the parental gene pool is a sample of what is available to the species, but it is necessarily limited, characterized by ethnicity and made local by founder effect, migration, and mating customs. These are all influences that determine the particularity of an individual's genetic endowment. But if genetic individuality is both determined and constrained by the genetic raw material inherited at conception, so is the variety of experiences made possible and limited by the mores of the social and cultural milieux, itself often inherited, which shape our likes and dislikes, our indulgences and restraints, in short, the qualities and quantities of the experiences we encounter. So, in the end, it is the remote causes that confer the specificity of individuality, but the unit steps of homeostasis that supply the substrate. Of course, the idea of variant proteins as units of individuality is not a new one, having been proposed by Archibald Garrod as "chemical individuality" as early as 1902 [22].

#### 1.4.1.5 The Unit Steps as Effectors of Disease

If the gene product is the implement of homeostasis, it follows that it is the effector of disease; certain of its variants are in some degree incongruent with the environment, inside the cell or out. That is, wherever the origins and mechanisms of pathogenesis have been laid bare, there are proteins at the root of it. How could it be otherwise, given that both structures and motivators of the functions of cells are proteins, and disease is a consequence of homeostatic incongruence? A critic might suggest infections as exceptions, but it is the congruence of the microorganism's structures with our unit steps of homeostasis that allows them to attach themselves to cell surfaces and then to release toxin or to gain access to the cell's interior and to reproduce there. It is they who define our strengths as weaknesses and our congruence as incongruence. And they do so by using the human gene products, the human unit steps of homeostasis.

The history of the realization of this role of the unit step in disease is of interest; it paralleled the successive descriptions and definitions of both genes and proteins [5]. We all know that Archibald Garrod was the first to call attention to alkaptonuria as a hereditary alternative form of metabolism because of failure of an enzymatically catalyzed step [22]. He called this, and other such metabolic aberrations, inborn errors to distinguish them from diseases. This was an insight of extraordinary penetration in which he recognized that the differences in protein composition that distinguished species must also differentiate individuals within species [34]. But even by 1909 when his first book, *Inborn Errors of Metabolism*, was published, he could go no further [16,34,35]. Little was known about protein structure and nothing of sequence of amino acids, and it was not even established yet, to everyone's satisfaction, that enzymes were proteins [36].

As for the gene in 1909, it was still defined statistically, and although phenotype and genotype were differentiated in that year, the gene was an unknown entity, perceived by Johannsen as "an accounting or calculating unit." But by 1915 the gene had been defined operationally, so by then Garrod could have proposed the inborn errors as products of mutants of single genes. But he never did. Even in his 1931 book, *Inborn Factors in Disease*, he did not use the word "gene" despite a general recognition that genes were involved somehow in some diseases [19,34]. For example, in 1927 Barker reported that, "No less than 223 heritable anomalies have been described in man already" [37]. And others, not in medicine, recognized a biochemical relationship between genes and phenotypes: Wright in coat colors of guinea pigs and Wheldale in flower pigments [38,39]. Then in the late 1930s and early 1940s, the studies of Ephrussi, Beadle, and Tatum, first in *Drosophila*, then in *Neurospora*, provided a functional definition of the gene that brought gene and protein unequivocally together to clarify Garrod's observations, and capitalizing on rapid advances in biochemistry, to begin in the 1950s an era of biochemical genetics [21]. Biochemical genetics was an ecumenical enterprise. If Garrod was its icon and Harry Harris its chief expositor, there were also contributions of nongeneticists, including Pauling's concept of molecular disease and the elaboration of the enzyme deficiencies in (type I) glycogen storage disease, galactosemia and other disorders, all classic inborn errors, described by biochemists with no primary interest in genetics and who made no reference to Garrod or to Beadle–Tatum [23,40,41]. But whatever the influence, the list of inborn errors expanded rapidly, soon attaining an exponential rate of increase that has barely slackened.

It is worth noting that biochemical genetics flourished before the impact of the discovery of the double helix could be felt. But the later developments led first to Yanofsky's definition of the structural gene with its correspondence to sequences in amino acids in proteins [42] and later to the definition of the gene that includes both transcribed and nontranscribed DNA. And this led, in turn, to the development of genomics as an analytical method. Thus biochemical genetics, whose analysis proceeds from the phenotype to the protein and its gene, met genomics, whose analysis proceeds from the gene toward the phenotype via its protein product [43]. And in time, the glamor passed from biochemical genetics to genomics, perhaps principally because the former had no way to tackle the genetics of complex disorders. Actually both are needed because phenotypes are not necessarily explained on discovery of the gene or genes whose products are acting as proximate causes.

As the focal point in pathogenesis, the protein gene product provides an economical answer to the question of the origins of monogenic diseases. But the question of the moment is how to explain those called complex. The approach includes genomics, by which salient genes can be found and characterized [13]. Further steps involve discovery of their proteins and the homeostatic devices to which they belong, after which the pathophysiology may be elucidated. Additional participation by genetically inclined thinkers lies in sorting out the heterogeneity by means of appropriate family studies, work that must be done before, or together with, efforts to tie treatments to the consequences of particular protein variants.

Today we scoff at such diagnostic "entities" as dropsy and consumption, having begun long since to resolve their heterogeneity. But the sequencing of the HGP will provide the means to show how much more we have to go to characterize distinctive versions of, say, heart attack and stroke. So numerous are the genetic contributors likely to be that a case might be made for everyone having his or her own version of heart attack, stroke, or other multigenic multifactorial disorders. So family studies are vital for deciding which genes play important roles in which versions of the disease. The results will resemble those in the study of monogenic disorders; the heterogeneity will be of both loci and alleles, and the sets thereof will vary from family to family and individual to individual [17]. This kind of genetic thinking, not yet routine in medicine, is crucial to our understanding and represents an important principle of disease.

#### 1.4.1.6 The Protein Product as a Unit of Selection

Neodarwinism is the outcome of a debate in which geneticists agreed that the object of selection must be phenotype, not genes, whereas evolutionary biologists, to whom the phenotype had been that object all along, agreed that both phenotypes, and their variation, originated in the genes [44–46]. If so, although the phenotype remains the unit of selection, it is the variable unit step or steps that cause it to qualify for that fate. In medicine, we are not much concerned with the selection by which species attain their characteristics but with what evolutionary biologists call "purifying" selection, that which removes "undesirable" genes prior to reproduction. So here again the protein product of the gene occupies a central position between two aspects of human biology. And here is yet another example of the cleavage between biology and medicine. The irony in the word purifying is not lost on the physician to whom the protection of life is uppermost, while to biology, with no stake in the individual, the question is purely one of understanding the rise and decline of species. But, in fact, variations in unit steps of homeostasis are no less the stuff of positive selection than negative.

#### 1.4.1.7 As a Hedge Against Genetic Determinism

Institutions change and renew themselves but they usually retain residual signs of their origins. No one would deny that all the genetics of today stems from the concepts elaborated in the fly room at Columbia, or that we continue to use both concepts and language appropriate to the drosophilists' definition of the gene [20]. Theirs was an operational definition in which authority for both heredity and cellular function was accorded to the gene. In his book, *What is Life*, Schroedinger spoke of the gene as "law code and executive power" as well as "architect's plan and builder's craft in one" [47]. So the language of *Drosophila* genetics included such locutions as genes "for" gene–environment interaction, modifiers, penetrance, and pleiotropy, all of which are perceived as properties of the gene, although we know now that they refer to events mediated by the protein unit steps of homeostasis. There is no question of the latter's specification by the genes, but in folding and assuming an appropriate position in a relevant homeostatic device, they become a part of mechanisms that regulate both themselves and the DNA (see chapter on Pathogenetics). Thus, it is not the genes that are penetrant or pleiotropic or that interact with the environment; it is the proteins that do these actions that are removed from the genes' control.

It might be correct to speak of a "gene for," say, an enzyme or even its pathway; for example, there is a "gene for" phenylalanine hydroxylase and "for" phenylalanine degradation. But in their further integration, proteins lose their identity in those of integrated functions, for which any single gene can no longer be perceived to have any authority.

There is another way in which the locution "gene for" is used. When we observe that a disease segregates, we say there is a "gene for" that disease, that one or more mutants act as proximate cause. That is exactly what the drosophilists did for their mutants, unconcerned with their ignorance of how a gene could shorten bristles or deform wings. We continue to use their discourse, even though we know that the protein product is the actual agent of function [20]. But "genes for" is a tricky phrase. When we use it unthinkingly, as in genes for high blood pressure, say, we obscure our own inner view of the reality, whereas when we speak of variant proteins, there springs immediately to mind pathways, cascades, receptors, transducers, and feedback loops such as those that determine blood pressure. Incidentally, it is amusing to imagine that had Archibald Garrod come to alkaptonuria thinking like a geneticist of the time, he would probably have perceived it only as a recessive character, not an inborn error. But he came to it as a biochemist and saw it for what it was: a metabolic alternative due to the absence of an enzyme. He used the genetic evidence expressed in consanguinity to support the idea of heredity, not as evidence of a gene. So, rather than perceiving his lack of interest in genetics as a shortcoming, we should be glad of it because the idea of a "gene for" alkaptonuria could have stood in the way of his biochemical insights. But equally, had he pondered the work of the drosophilists emerging in print from 1905 to 1920, and which included their operational definition of the gene, the second edition of his *Inborn Errors of Metabolism* published in 1926 must surely have anticipated the Beadle–Tatum one gene–one enzyme principle [48].

So, if we human geneticists of today revert occasionally to the drosophiline mentality, how likely are patients, their families, and the public to escape? How are they to know that the words "gene for," say IQ, artistic ability, or criminal behavior obscure the unfathomable complexity of the identity and actions of gene products integrated in hierarchies to compose cells, organs, and whole organisms, all in touch with one another and with the outside? The extremes to which "genes for" can go are summarized in a book called *The DNA Mystique: The Gene as a Cultural Icon* by Nelkin and Lindee [49]. But fortunately, we have our mental image of the products of the genes, the unit steps of homeostasis, with their multifarious behaviors as a bulwark against loose thinking.

#### 1.4.1.8 As the Goal of the HGP

One road to the discovery of new principles of disease has derived from the HGP [50]. Lander has suggested that this bears the same relationship to biology that the periodic table bears to chemistry [13]. So it compels our attention. Furthermore, it is the ultimate identifier of those homeostatic units that lie at the basis of pathogenesis.

About 20,000 or so genes and their products have been identified, and sooner or later, the products' roles in homeostasis will follow, with obvious benefits for investigation of pathogenesis, treatment, and prevention. In addition, definitive samples of gene products, useful in defining disease, will be available for characterizing human biological properties hitherto unknown. A few examples of questions that are being asked are

1. How variable is the human genome? Is it more, or less, than the estimates of Harris and Lewontin? Studies of whole genome sequences suggest more [32,33]. Nothing could be more useful than this answer because it is the common genes that so often act as modifiers and furnish the wherewithal for complex diseases. And we now understand the much of the "junk" DNA encodes various RNA species that exert important control over the expression of genes that encode proteins.
2. Is there an inborn error for every locus? And are all classes of proteins equally involved in diseases? In a comparison of 348 mutant proteins associated with inborn errors (MMBID) and a list of 3000 "core" proteins shared by yeast and *Caenorhabditis elegans*, the distribution of protein types in the two samples was remarkably similar [51]. Although indirect, the suggestion is there that all protein types are involved in inborn errors, but we cannot yet say that there is an inborn error for every locus, however plausible the idea may be.
3. Are diseases characterized by the qualities of the proteins that are their proximate causes? For example, do enzyme deficiencies differ in some systematic way from disorders associated with receptors, transcription factors, or structural proteins?
4. Are conserved genes overrepresented or underrepresented in disease? One might expect them to be overrepresented on the assumption that they fulfill critical functions or underrepresented because their mutants might be so often lethal.
5. What is the role, if any, of developmental constraints in fostering or suppressing disease? These are limitations on the evolution of phenotypic variation expressed in developmental blind alleys. Kirschner and Gerhart have examined ways by which such constraints are loosened to allow new mutation and evolutionary progress. But would some of the latter be disease [52]? And Rutherford showed how, in *Drosophila*, such constraint was exerted by a heat shock protein [53]. When altered by mutation, the constraining force was lifted and the effects of mutants suppressed by the wild-type protein were observed. Some of these effects were developmental anomalies.
6. Are diseases characterized by the evolutionary age of the proteins that lie at their root? That is, we might

suppose that inborn errors of housekeeping genes shared by remotely related species were the oldest. Do they differ in any particular from diseases of the most recent mammalian or human genes?

7. What are the implications for aging? Are some proteins more frequently the object of aging processes, or is it random? Errors in the mitotic machinery that led to multiple abnormalities of regulation of dozens of enzymes increase with age [54]. So, will aging, which has been perceived by some as dishomeostasis, turn out to have the same molecular basis as disease?

Many other questions are being asked, many no doubt not now askable because the contexts in which they are relevant are unknown. As more and more diseases are given molecular definition, we will surely classify them differently, departing from the current anatomical, organ system, age-related rubrics, moving to more molecular designations. As heterogeneity is laid bare, old classes will go and new ones will come, reflecting a sharp revision in how we will see disease itself. In addition, our language will change. It is likely that we will refer less to genes and more to proteins, so our residual drosophiline language is likely to go, too. Of what use are words such as modifier, epistasis, penetrance, pleiotropy, and the like when visualizing the reality as actions and interactions of proteins, in say, multiunit machines, or even in whole systems [13]? This also suggests that we in medicine will be thinking less in units and more in multiunit devices (proteomes, epigenomes, metabolomes, etc.). Linear thinking may be out as complexity moves in. But maybe the most significant change in our thinking will be compelled by the definitive evidence of human variation and individuality. Typological thinking will give way to population thinking. No doubt there will always be use for the former at one level; that of the value of means and the classic case, but only as a preliminary to the population thinking that perceives the extent and impact of variation on human individuality.

#### 1.4.1.9 Social Impact

The unit step of homeostasis is attaining increasing prominence as a risk factor and signal for preventive action, and medicine has been adapting not only to their potential use but also to their impact on their possessors' lives. These concerns are well known to readers of PPMG; they have been the subject of many articles, books, and committee reports and they touch on counseling, ethics, legal matters, and psychological impact [14,15,55–57]. They are mentioned here because of the potential uses of such risk factors in prevention. If the HGP fulfills its promise, there is the possibility to know the protein products of all genes known to participate in pathogenesis. Many scenarios as to the use of these markers have been offered. Only time will tell which, if any, is practical, but we would be wise to continue our study and preparation. How to use information, available at birth, about many variant genes, perhaps dozens in single individuals, and known to be associated with diseases all across the life span is something entirely new in medicine. Is it consonant with good medicine? Is it acceptable to the public? How do we prepare the public to make rational decisions about it? How do we prepare individuals to accept and use constructively such emotionally loaded information? These questions can be answered only in colloquy with the public.

#### 1.4.1.10 As a Source of Coherence

In concentrating on the specificities of the pathogenesis of each disease, reductionist investigation emphasizes the separateness of diseases. It exerts a centrifugal effect that adds to that of our conventional nosology, which divides medicine into specialties across which we interact collaboratively, when at all. But the concept of the unit step of homeostasis as the central focus of all pathogenesis provides a principle of disease that exerts a contrary centripetal force that unifies the thinking about both disease and diseases. It is the difference between analysis and synthesis. Medical thinking, until recently, was mainly synthetic. It dealt with the body as a whole, no doubt because of ignorance of its parts. In contrast, in the thinking of today, the emphasis is on analysis; our attention is directed more to microunitary parts with less attention to the whole. But in acting as units of the mechanisms whereby an open system maintains its adaptation to an indifferent environment, the protein gene product is the effective link between those proximate and remote causes portrayed in the Mayr model. And that link is no less evident when unit step and environment are incongruent than when congruent. In the Mayr model, the proximate causes are consequent on the DNA and pose questions prefaced by how, whereas the remote causes lead up to the DNA and pose questions prefaced by why. In this summary of the role of the protein gene product, we have dealt with "how" questions. In the next section, we examine the "why" questions and the principles they illustrate.

Diabetes Society, and The March of Dimes come readily to mind. In the 1960s and 1970s as more and more inborn errors were described, this principle was also applied to the generation of a multitude of disease-related societies, each dedicated to education, treatment, and prevention of one disease, the latter in the form of reproductive counseling, antenatal diagnosis, and sometimes abortion. Then, as the molecular basis of these disorders was discovered, newborn screening for inborn errors was offered by many state health departments and intensive studies were undertaken of every aspect of this form of preventive medicine including screening, counseling, and issues both ethical and legal [55,65–69]. The question then arose of testing relatives of patients with inborn errors with an eye to reproductive advice, and the triumph of Tay–Sachs testing is one result [14,55]. And now that rapid progress is being made in unraveling the genomics, epigenomics, and proteomics of complex disease, time will give us more risk factors in the way of variant genes and proteins. These developments are reviewed here in this detail to call attention to the movement of the focus of prevention away from populations to individuals, and now to the molecular emphasis in both prevention and treatment. Just as the discovery of genes associated with disease suggests the possibility of cure, so does it suggest prevention by testing of relatives and populations. Indeed, the logic of prevention is even more powerful than that of cure. That is, unlike treatment, which is always after the fact and is occasionally as threatening as the disease it is designed to combat, prevention spares the organism such rigors even while far less disruptive of social and economic life. On the other hand, in keeping with the principle of continuity, the two are sometimes indistinguishable.

So, may we expect miracles of prevention now that we can identify proximate causes? Readers of PPMGG know that we may not [14,15,56,57]. It is a matter of the continuity of the gradient of selective effect. At one end, the virtual elimination of Tay–Sachs disease among Jews and the prevention of a few other inborn errors by the same means represent successes of the high technology promoted by Lewis Thomas [70]. At the other end are healthy centenarians who attribute their robust health to some idiosyncratic behavior. But in between are those genes and their variant products whose virtues are a sometime thing, depending on, on the one hand, the specificities of experiences over the lifetime, and on the other, their support, or reinforcement in failure, by the variant products of other genes. So the same variant gene product may be adequate in one person and fail in another even in the same family. Or it may be within the same person adequate under one circumstance and insufficient under another. Thus, as a predictor, a gene may be of only limited use to an individual even while accepted as a significant risk in a population. This is a frequent problem of epidemiologically designated risk factors; it is not always clear to whom among their possessors the trait is actually risky—to say nothing of gradations in risk. It is the problem also of evidence-based medicine, which, however, valuable in increasing the rigor of diagnosis and treatment provides recommendations suitable for populations, not individuals [71–73]. It is a matter of typological, as opposed to population, thinking. Of course, the HGP has added greatly to the list of our genes and their proteins so that the exact identity of all of the units in pathways and other homeostatic devices will be known, improving thereby the predictive value of various combinations of variants [74]. And, assuming increasing identification of exterior proximate causes, the accuracy and usefulness of preventive predictions may improve remarkably. The necessity for the advancement of knowledge of nongenetic proximate causes cannot be exaggerated. We need projects of similar scope and ambition to that of the HGP. In the meantime, we should do what we can where we can, and for the rest, fall back on an aspect of medicine that may have become unfashionable in modern times but that is perhaps more than ever needed: helping patients to live with uncertainty.

A far less likely, but more effective, means of health promotion is the control of remote causes. The virtue of such an approach is clearly indicated in the Mayr model wherein the relationship of the two kinds of causes and the two kinds of biology is so lucidly stated. To influence by law the distributions of genes is both unconstitutional and in strong opposition to "liberty and the pursuit of happiness," but to influence the organization of society and culture for the betterment of health is not only possible but also already the aim of numerous government agencies, the even more numerous private disease-related societies, as well as physicians who have advised their patients to practice healthy ways. But what has not been emphasized, at least in the United States, is the power of corporate action, of putting the weight of the whole medical profession socially and politically on the side of health promotion. Today in the United States this is impossible, and some will say it is undesirable, but the point to be made here is the logic of the position. On one side is the power of remote causes in the origin of disease (let us think here of the evolution, growth, organization, penetration, and political powers of the tobacco industry) and on the other is the power of societies to organize themselves to influence those remote causes. In the United States, it was public opinion followed by legal action that began the descent of the authority of tobacco. There are other social conditions likely to retard the advance of prevention. One of these is the combination in the public mind of a superficial grasp of progress in biology and genetics and an unreasoning belief in the limitless potential of that progress. But for a recipient to respond realistically to the offer of prevention requires the ability [1] to differentiate between personal and populational probabilities [2], to grasp its potential for success or failure, and [3] to participate constructively with a knowledgeable and sympathetic physician in greeting either success or failure. We do not often think of the evolution of education, or of the public grasp and acceptance of advances in medicine as remote causes of success or failure, but we should.

## 1.6.2 Treatment

The essence of Lewis Thomas' concept of high technology in treatment lay in the discovery of the exact point or points in the machine that were broken and that could be repaired by a single, simple, straightforward maneuver [70]. One of his examples of such a treatment was the use of steroid in the adrenogenital syndrome. And after the 1950s, amid the rapid accumulation of newly described inborn errors, there was optimism that such diseases would be brought under control [75]. But results so far have been something less. In the 1980s, Hayes tabulated successes and failures in 65 inborn errors, a part of a larger randomized sample of monogenic diseases taken from MIM 5 [76]. Table 1.2 shows that for 12%, treatments were successful in rendering the patient normal or essentially so, whereas in 40% there was some improvement, often not very impressive, and about 48% showed no success at all. A further examination of success or failure in treatment of the same 65 diseases 10 years later was reported by Treacy, with results shown in Table 1.2 [77]. There was no increase in the number of very successful treatments, but some of the previously resistant disorders now yielded in some degree, often to such rigorous therapies as tissue and organ transplantation. These, Thomas saw as middle technologies, sometimes effective, but expensive and perhaps hard on the patient. A third look in 1999, this time including 517 inborn errors listed in the seventh edition of MMBID, gave much the same results (see Table 1.2) [78].

**TABLE 1.2 Effectiveness of Treatment of Inborn Errors**

| | 1983 Hayes | 1993 Treacy | 2000 Treacy |
|---|---|---|---|
| Fully beneficial | 12[a] | 12 | 12 |
| Partially beneficial | 40 | 57 | 54 |
| No benefit | 48 | 31 | 34 |

[a]Percentage of total.

Given the qualities of the inborn errors that were the object of treatments, the record is perhaps not surprising. Most are at or near the top of the gradient of selective effect, some are lethal, and some are permanently crippling. And all are heterogeneous, some as to loci, all as to alleles. And we now know that we must expect equal heterogeneity among those shadowy modifiers we presume to exist [17]. So these disorders are simply the most intractable. But farther down the gradient, the diseases are more amenable, not necessarily to cure, but certainly to management. This seems to be telling us that there is a relationship between heritability and success in therapy. When the heritability is high in a population, we expect less of treatment than when it is low (i.e., fewer patients are likely to respond satisfactorily). The history of treatment of rickets with vitamin D is exemplary. When, after the mid-1940s rickets almost disappeared, nearly all that was left were several different kinds of monogenic vitamin D–resistant rickets [19]. This experience seems to furnish medicine, especially preventive medicine, an aim, even a motto. We work to drive the heritability of disease toward 1.0. And we fervently hope that the gene therapists will confound the motto by inventing high-technology treatments that subdue even the most "genetic" and even the most refractory of those disorders that continue to resist every effort to contain them. As for the complex disorders, they are resistant, too, but in a different way. If every individual has his or her own set of proximate causes, the complexity is of a high degree. The problem is one of discovery of the nongenetic causes and trying to eliminate them, as well as discovering which sets of genetic causes pose vulnerability to the threats of those nongenetic influences, both in general and in each affected individual. No easy job, but we are up against a wily opponent. Many years ago, Max Delbrück observed that "any living cell carries with it the experiences of a billion years of experimentation by its ancestors. You cannot expect to explain so wise an old bird in a few simple words" [79]. We are definitely embarked on an effort to expose that wisdom. The next two or three editions of PPMGG should show how wise the old bird is.

Certainly the progress in the past few years in genetically modifying human T-cells to attack various malignancies, the therapy of cystic fibrosis directed at specific mutations in *CFTR*, and the insertion of a normal gene to cure blindness are great examples.

## 1.7 CONCLUSION

The explanatory principles of disease begin with the capacity of the species for genetic variability, a capacity that is required for survival of species and that is experienced randomly resulting in genes whose products have, through time, conferred on their recipients a status of congruence with equally variable environments. But sometimes the result is incongruence, which can lead to disease. It is in the gene products, the protein unit steps of homeostasis, that this species variability is expressed in congruence or incongruence in health or disease. This expression occurs within a biochemical and molecular cellular matrix conditioned by interaction between such protein products and experiences of the environment through development, maturation, and aging. Accordingly, analysis of pathogenesis must be pursued in three timescales all at once: that of phylogeny whence the genes and their products were derived; that of ontogeny, maturation and aging, which condition the ever-changing matrix; and that of the moment representing the impact of today's events. This principle, embracing the three timescales, also incorporates two kinds of causes, proximate and remote, which are expressed in the uniqueness of individuals. The incongruent proximate causes, variable protein unit steps of homeostasis and varying kinds, amounts and durations of experiences of the environment, account for the expressions of disease phenotypes that are subjected to selection and incur the social stigmas that complicate the lives of their victims. Remote causes are composed of the evolution and dynamics of both biological and social milieu that account for the nature and local availability of proximate causes, their unique assembly as genotypes and availability as experiences, to form combinations favorable for disease. And it is this particularity that determines who gets which disease at what time in life. The qualities of diseases are expressions of unique and variable human genomes arranged in a gradient of selective effect, a representation of the removal in early life of those unlikely to reproduce, and in post-reproductive life, a less-intense test of survival in a variable environment. It is in the latter part of the range of the gradient that both prevention and treatment are likely to be most effective; prevention because changes in environment can be effective in avoiding disease, and treatment because the homeostasis can be characterized as inefficient and in need of a boost, rather than broken. The logic of prevention is more powerful than that of treatment, but we need both a more comprehensive knowledge of nongenetic proximate causes and, in time, to learn, understand, and adjust to the social dislocations any sudden spate of preventions could bring. But it is in part in the grasp of the possibility and plausibilities of both prevention and treatment, and in part in understanding the meaning in medicine of individuality and the virtues of population thinking in relation to it, that we may be able at once to pursue the reductionist path we have so successfully traversed and return to embrace the integration, the humanity, of patients who appeal to us for relief of both the consequences of their molecular incongruities and the injury of the disease to that integrated humanity.

## REFERENCES

[1] De Solla Price DJ. Little science big science. New York: Columbia University Press; 1963.
[2] The Online Metabolic and Molecular Basis of Metabolic Disease. https://ommbid.mhmedical.com/book.aspx?-bookID=971.
[3] Rosenberg CE. Banishing risk: or the more things change the more they remain the same. Perspect Biol Med 1995;39:28–42.
[4] Lander ES. Initial impact of the sequencing of the human genome. Nature 2011;470:187–97.
[5] Childs B. The entry of genetics into medicine. J Urban Health 1999;76:497–508.
[6] Cohen H. The evolution of the concept of disease. In: Lush B, editor. Concepts of medicine. Oxford: Oxford University Press; 1960.
[7] King L. Medical thinking. Princeton: Princeton Press; 1982.
[8] Temkin O. The scientific approach to disease: specific entity and individual sickness. In: Crombie AC, editor. Historical studies in intellectual, social and technical

conditions for scientific discovery. New York: Basic Books; 1963. p. 629–47.
[9] Vogel F, Motulsky AG. Human genetics. Preface. 3rd ed. New York: Springer-Verlag; 1998.
[10] Mayr E. Cause and effect in biology. Science 1961;134:1501–6.
[11] Mayr E. The growth of biological thought. Boston: Harvard University Press; 1982. p. 45–7.
[12] Burley SK, Almo SC, Bonnano JB, et al. Structural genomics: beyond the human genome project. Nat Genet 1999;23:151–7.
[13] Lander ES, Weinberg RA. Genomics: journey to the center of biology. Science 2000;287:1777–82.
[14] Andrews LB, Fullarton JE, Holtzman NA, Motulsky AG. Assessing genetic risks. Washington, DC: National Academy Press; 1994.
[15] Motulsky AG. If I had a Gene Test, What would I have and who would I tell? Lancet 1999;354:SI35–7.
[16] Scriver CR. Changing heritability of nutritional disease: another explanation for clustering. In: Simopoulos A, Childs B, editors. Genetic variation and nutrition. New York: Karger; 1989. p. 60–71.
[17] Scriver CR. Monogenic traits are not simple. Trends Genet 1999;15:3–8.
[18] Hill AVS. Genetics and genomics of infectious disease susceptibility. Br Med Bull 1999;55:401–13.
[19] Scriver CR, Childs B. Garrod's inborn factors in disease. New York: Oxford Press; 1989.
[20] Fox-Keller E. Language and science: genetics, embrylogy and the discourse of gene action. In: Fox-Keller E, editor. Refiguring life. New York: Columbia University Press; 1995.
[21] Beadle GW. Genes and chemical reactions in Neurospora. Science 1956;129:1715–9.
[22] Garrod AE. The incidence of alkaptonuria: a study in chemical individuality. Lancet 1902;2:1616–20.
[23] Pauling L, Itano HA, Singer SJ, Wells IC. Sickle cell anemia, a molecular disease. Science 1949;110:543–8.
[24] Alberts B. The cell as a collection of protein machines: preparing the next generation of molecular biologists. Cell 1998;92:291–4.
[25] Gelbert WM. Databases in genomic research. Science 1998;282:659–61.
[26] Kohler RE. Lords of the fly. Chicago: University of Chicago Press; 1994.
[27] Capecchi MR. Hox genes and mammalian develop ment. Cold Spring Harbor Symp Quant Biol 1997;62:273–8.
[28] Kornberg TB, Krosnow MA. The Drosophila genome sequence: implications for biology and medicine. Science 2000;287:2218–20.
[29] Stent GS. Strengths and weaknesses of the genetic approach to the development of the nervous system. In: Cowan WM, editor. Studies in developmental neurobiology. New York: Oxford University Press; 1981.
[30] Waddington CH. The strategy of the genes. London: Allen and Unwin; 1957.
[31] Marmot MG. Early life and adult disorder. Br Med Bull 1997;53:3–9.
[32] Cargill M, Altshuler D, Ireland J, et al. Characterization of single-nucleotide polymorphisms in coding regions of human genes. Nat Genet 1999;22:231–8.
[33] Halushka MK, Fan JB, Bentley K, et al. Patterns of single-nucleotide polymorphisms in candidate genes for blood pressure homeostasis. Nat Genet 1999;22:239–47.
[34] Bearn AG. Archibald Garrod and the individuality of man. Oxford: Oxford University Press; 1993.
[35] Garrod AE. Inborn errors of metabolism. Oxford: Oxford University Press; 1909.
[36] Fruton JS. Molecules and life. New York: Wiley-Interscience; 1972.
[37] Barker LF. Heredity in the clinic. Am J Med Sci 1927;173:597–605.
[38] Scott-Moncrieff R. The classical period in chemical genetics. Notes and records. R Soc London 1981–1983;36–37:125–54.
[39] Wright S. Color inheritance in mammals. J Hered 1917;8:224–35.
[40] Cori GT, Cori CF. Glucose-6-Phosphate of the liver in glycogen storage disease. J Biol Chem 1952;199:661–7.
[41] Isselbacher KJ, Anderson EP, Kurahashi K, Kalckar HM. Congenital galactosemia, a single enzyme block in galactose metabolism. Science 1956;123:635–6.
[42] Yanofsky C. Gene structure and protein structure. Harvey Lect 1965–1966;61:145–67.
[43] Lander ES, Schork NJ. Genetic dissection of complex traits. Science 1994;265:2037–48.
[44] Maynard Smith J, Burian R, Kauffman S, et al. Developmental constraints and evolution. Q Rev Biol 1985;60:265–87.
[45] Mayr E, Provine WB. The evolutionary synthesis. Cambridge: Harvard University Press; 1980.
[46] Rosenberg CR. Toward an ecology of knowledge. In: The organization of knowledge in modern America. Baltimore: The Johns Hopkins Press; 1979. p. 440–55.
[47] Schroedinger E. What is life?. Cambridge: Cambridge University Press; 1967. p. 23.
[48] Garrod AE. Inborn errors of metabolism. 2nd ed. Oxford: Oxford University Press; 1926.
[49] Nelkin D, Lindee MS. The DNA mystique. New York: WH Freeman; 1995.
[50] Collins FS. Shattuck lecture: medical and societal consequences of the human genome project. N Engl J Med 1999;341:28–37.
[51] Jiminez-Sanchez G, Childs B, Valle D. The effect of mendelian disease on human health. In: Scriver CR,

Beaudet AL, Sly WS, Valle D, editors. The metabolic and molecular basis of inherited disease. 8th ed. New York: McGraw-Hill; 2000. p. 167–74.
[52] Kirschner M, Gerhart J. Evolvability. Proc Natl Acad Sci USA 1998;95:8420–7.
[53] Rutherford S, Lindquist S. Hsp 90 as a capacitor for morphological evolution. Nature 1998;396:336–46.
[54] Ly DH, Lockhart DJ, Lerner RA, Shultz PG. Mitotic misregulation and human aging. Science 2000;287:2486–92.
[55] Anon. Genetic screening. Washington, DC: National Academy of Sciences; 1975.
[56] Holtzman NA. Proceed with caution. Baltimore: The Johns Hopkins Press; 1989.
[57] Holtzman NA, Watson MS. Promoting safe and effective genetic testing in the United States. Baltimore: Johns Hopkins University Press; 1998.
[58] Fries JF, Crapo LM. Vitality and aging. New York: WH Freeman; 1981.
[59] Jacobs PA. The role of chromosome abnormalities in reproductive failure. Reprod Nutr Dev 1990;30(Suppl):63s–74s.
[60] Costa T, Scriver CR, Childs B. The effect of mendelian disease on human health: a measurement. Am J Med Genet 1985;21:231–42.
[61] Childs B, Scriver CR. Age at onset and causes of disease. Perspect Biol Med 1986;29:437–60.
[62] Hoffmann AA, Parsons PA. Evolutionary genetics and environmental stress. New York: Oxford University Press; 1991.
[63] Lewontin RC. Gene, organism and environment. In: Bendell DS, editor. Evolution from molecules to men. New York: Columbia University Press; 1983. p. 273–86.
[64] McKeown J. The role of medicine. London: Nuffield Trust; 1976.
[65] Bergsma D. Ethical, social and legal dimensions of screening for human genetic disease. Birth Defects Orig Artic Ser 1974;10(No. 6):1–272.
[66] Burnham JC. America and Medicine's golden age. What happened to it? Science 1982;215:1474–9.
[67] Hsia YE, Hirshhorn K, Silverberg RL, Godmillow L. Counseling in genetics. New York: A.R. Liss; 1979.
[68] Knoppers BM, Labarge CM. Genetic screening: from newborns to DNA typing. Amsterdam: Exerpta Medica; 1990.
[69] Lubs HA, de la Cruz F. Genetic counseling. New York: Raven Press; 1977.
[70] Thomas L. The future impact of science and technology on medicine. Bioscience 1974;24:99–105.
[71] Mant D. Can randomized trials inform clinical decisions about individual patients? Lancet 1999;353:743–6.
[72] Sweeney KG, MacAuley D, Gray DP. Personal significance: the third dimension. Lancet 1998;351:134–7.
[73] Tonelli MR. The philosophical limits of evidence based medicine. Acad Med 1998;73:1234–40.
[74] Van Omenn GJB, Bakker E, den Dunnen JT. The human genome product and the future of diagnostics, treatment and prevention. Lancet 1999;354(Suppl):si5–10.
[75] Scriver CR. Treatment in medical genetics. In: Crow JF, Neel JV, editors. Proceedings of the 3rd international congress on genetics. Baltimore: Johns Hopkins Press; 1967. p. 45.
[76] Hayes A, Costa T, Scriver CR, Childs B. The effect of mendelian disease on human health. II: response to treatments. Am J Med Genet 1985;21:243–55.
[77] Treacy E, Childs B, Scriver CR. Response to treatment in hereditary metabolic disease: 1993 survey and 10 Years comparison. Am J Hum Genet 1995;56:359–67.
[78] Treacy EP, Valle D, Scriver CR. Treatment of genetic disease. In: Scriver CR, Beaudet AL, Sly WS, Valle D, editors. The metabolic and molecular basis of inherited disease. 8th ed. New York: McGraw-Hill; 2000.
[79] Delbruck M. A physicist looks at biology. In: Cairns J, Stent GS, Watson JD, editors. Phage and the origins of molecular biology. Cold Spring Harbor, NY: Cold Spring Harbor Press; 1966. p. 9–22.

2

# Foundations and Application of Precision Medicine

*Geoffrey S. Ginsburg, Susanne B. Haga*

Center for Applied Genomics & Precision Medicine, Duke University School of Medicine, Durham, NC, United States

The goal of precision medicine is to optimize disease prevention, diagnosis, and treatment decision-making based on comprehensive information that incorporates traditional clinical measures, and data with DNA variation (genome), gene expression (RNA or transcriptome), proteins (proteome), metabolites (metabolome), methylation (epigenetics), and/or microbial composition (microbiome). These patient data can provide increased accuracy in assessment of disease risk, diagnosis, prognosis, and drug response. The growth of precision medicine has been possible, in part, due to the convergence of a societal shift toward patient-centered care, transition to electronic medical records (EMRs), and development of decentralized digital health technologies. The integrated use of these technologies, such as linking digital health tools to EMRs, improved patient engagement through online patient portals, and improved portability of medical data, further enables rapid implementation and utilization of precision medicine. In particular, the successful integration of -omics information and technologies into clinical practice is reliant on a robust and secure clinical infrastructure to store and analyze large data sets.

Although much attention has focused on the individual, the field of precision medicine can benefit population (or public) health through the combination of individual-level data and population-based interventions, implementation of genetically targeted approaches such as newborn screening or follow-up testing for some inherited cancers, and application of new technologies to existing public health efforts such as infectious disease surveillance [1]. Preventive medicine can move beyond standard recommendations for health behaviors and work toward tailored lifestyle recommendations to patients' needs and circumstances [2,3]. With the move beyond traditional academia-based initiatives into health systems and exploration of the use of genome technologies in healthy populations, the union of precision medicine and population health stands poised to affect large populations in practical clinical settings [4].

Much work lies ahead to understand the clinical significance of the multiple types of data that can now be generated for a given person, to validate new testing platforms, and to develop evidence-based practice guidelines. We have already witnessed the development and implementation of several new interventions and treatments [5]. In this chapter, we will provide an overview and examples to date that have contributed to the rise of precision medicine and highlight areas where effort is needed to realize the full potential of these applications.

## 2.1 OVERVIEW OF PRECISION MEDICINE

Since the announcement of the Human Genome Project in the 1990s and its completion in 2003, there have been significant advances in medicine, particularly regarding our understanding of risk factors for chronic disease and their molecular basis. Before the sequencing of the human genome and development of a reference human genome sequence [6,7], the process of identifying the genetic basis of disease was a laborious and time-consuming process. The completion of a reference human

Emery and Rimoin's Principles and Practice of Medical Genetics and Genomics: Foundations, https://doi.org/10.1016/B978-0-12-812537-3.00002-0

genome sequence was made possible with substantial developments in sequencing technology, data storage, and bioinformatics and analytical capabilities (which continue to grow to this day [8–10]). The availability of new, rapid, and less costly sequencing and other genomic technologies has led to an ongoing wave of discoveries about the causes of and contributors to rare and complex diseases, respectively, drug targets, and clinical diagnostics [11].

Genomics and precision medicine has focused on the study of large groups or populations with complex, multifactorial conditions aiming to molecularly substratify them, whereas the field of human genetics has primarily focused on families with rare inherited conditions. While one of the major goals of precision medicine is to provide more individualized care, precision medicine may also benefit population health through improved screening interventions and disease prevention for healthy populations [4,12,13]. However, greater efforts are needed to generate comprehensive assessments of the impact of gene–environment interactions and examination of other molecular data sets in order to develop better risk predictors and diagnostic and prognostic markers and to inform treatment decisions [1,14,15]. Large-scale national initiatives such as the All of Us Research Program in the United States [16] and the 100,000 Genomes Project in the UK Project [17] will enable more comprehensive data collection and analysis to gain a better understanding of risk factors as well as to raise public awareness and widespread incorporation of genomics into health and medicine.

## 2.2 PRECISION MEDICINE APPLICATIONS ACROSS THE LIFESPAN AND CLINICAL SPECIALTIES

Precision medicine applications are not targeted to a single disease or patient population but rather can improve well-being and health outcomes for both affected and ill patients across the lifespan (see Fig. 2.1). Complex diseases have long been understood as the culmination of gene and environmental factors. Thus, several types of data sets generated from more than one of these technologies will be essential to understanding disease risk and health outcomes [18,19]. Indeed, the combination of static (genome) and dynamic (RNA, proteins, metabolites, and microbiome) measures has enabled a comprehensive analysis of the interactions between host and microbiome, environmental exposures, and the resulting impact on gene expression and the proteome. The "exposome" or "exposotypes" may only be interpreted through multiple –omic analyses in order to comprehensively evaluate the impact of an exposure on human physiology and health risk [20]. For example, an analysis of several –omics data sets captured over a 14-month period from a single individual highlighted their temporal dynamics associated with changes in environmental exposures (including infection) and lifestyle habits [21]. Another study collected –omics data, clinical tests, and lifestyle data for 108 individuals over a 9-month period and tailored health coaching based on each patient's comprehensive risk assessment [22].

*Preconceptual and Prenatal Screening.* Today, genetic and genomic testing is commonly used in obstetrics and pediatrics. With the rise of in vitro fertilization, preimplantation genetic diagnosis (PGD) provides couples with an option to selectively implant unaffected embryos [23]. Prenatal diagnosis has also been revolutionized with early noninvasive screening of fetal cells isolated from maternal blood for chromosomal anomalies, enabling much earlier results in pregnancy [24,25]. Known as noninvasive prenatal testing (NIPT), the test is recommended for high-risk pregnancies to screen for chromosomal 13, 18, and 21 aneuploidies [26,27]. However, evidence suggest that NIPT performs well in routine (low-risk) pregnancies [24,28], motivating some obstetrics practices to offer NIPT to all women. Although the number of invasive procedures (chorionic villus sampling and amniocentesis) has decreased [29,30], invasive procedures are still necessary to confirm positive NIPT results and in cases with abnormal ultrasonography results to evaluate for other chromosomal abnormalities [31]. In a few cases, NIPT has led to the diagnosis of maternal cancer due to sample contamination with tumor cfDNA [32,33]. Studies are under way to assess the use of fetal cfDNA for single gene analysis, targeted microarray analysis [34], and whole exome or genome sequencing [35,36]. For prenatal diagnosis, chromosomal microarray (CMA) analysis (instead of conventional karyotyping) is recommended for fetuses with a structural abnormality or developmental delay observed by ultrasound examination [37].

*Pediatrics.* Genetic screening begins at birth with newborn screening tests for a suite of conditions whose course can be substantially affected by early interventions. Whole genome sequencing has also been shown

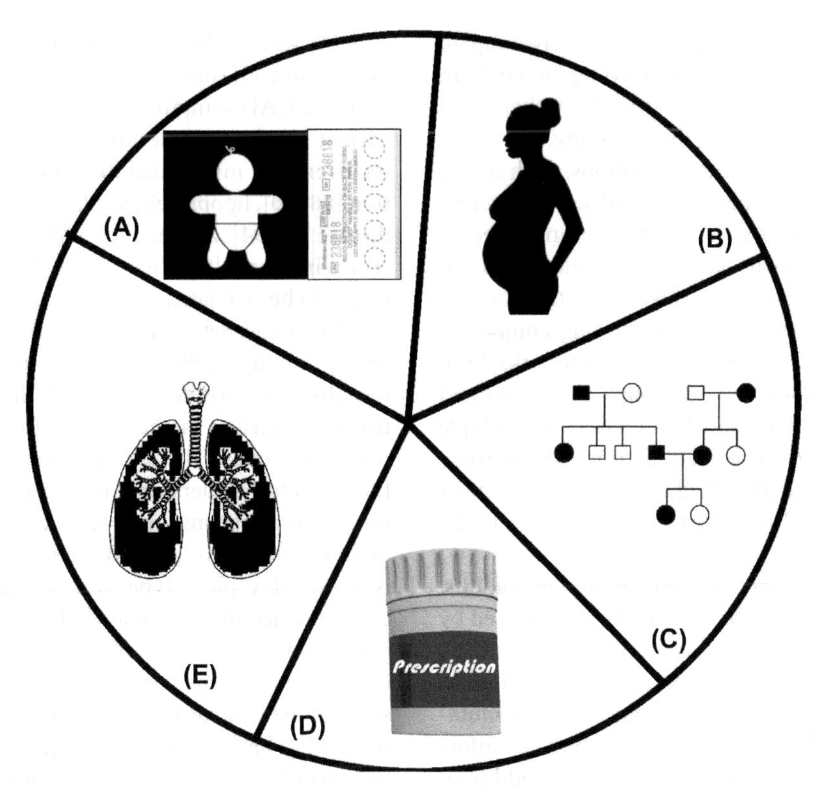


Figure 2.1 Examples of life stages and conditions for which precision medicine applications are used: (A) newborn screening, (B) prenatal screening and diagnosis, (C) family health history, (D) medication use, and (E) lung cancer.

to be useful in establishing a diagnosis for newborns affected with severe congenital malformations or other undiagnosed health issues requiring admission to a neonatal intensive care unit. In these urgent care circumstances, sequencing can be completed in less than 2 days to inform treatment and care decisions [38] and has been shown to be cost effective [39]. For other common conditions that develop during early childhood, CMA is recommended for developmental delay/intellectual disability, congenital anomalies, and syndromic conditions [40–42], and several types of genetic tests have been developed, including panel testing for epilepsy [43–45].

*Risk Assessment and Family Health History.* The clinical benefits of a comprehensive and accurate family health history (FHH) have been well demonstrated, but use of these data has been low and uneven [46]. Barriers to collection and use include time, lack of confidence to interpret, and inexperience with ordering genetic testing if indicated by the FHH [47]. A variety of new tools and approaches to improve FHH collection and utilization are under investigation, including online FHH tools [48] and virtual counseling [49].

Use of online FHH tools can improve the accuracy and comprehensiveness of FHH [50], in turn altering patient risk perceptions and health behaviors [51], improving patient care, and increasing identification of at-risk family members of patients affected with inherited conditions. Cascade screening involves offering testing to family members at risk for a hereditary condition [52]. Cascade screening has often been performed for hereditary cancer syndromes such as Lynch syndrome [53] and hereditary breast and ovarian cancer, familial hypercholesterolemia [54,55], and other rare conditions such as long QT syndrome [56]. However, there is a lack of consensus on cascade screening practices, and other barriers, such as family communication [57] or other strategies to notify family members [58,59], and a low rate of follow-up care of at-risk family members [60], pose challenges to promptly identifying and providing care to at-risk family members.

**TABLE 2.1 List of Genes, Evidence Level, and Type of Information Included in FDA-Approved Package Insert**

| Gene | Drug | CPIC Level | PharmGKB Level of Evidence | PGx on FDA Label |
|---|---|---|---|---|
| *CFTR* | Ivacaftor | A | 1A | Testing required |
| *CYP2C19* | Amitriptyline | A | 1A | |
| *CYP2C19* | Clopidogrel | A | 1A | Actionable PGx |
| *CYP2C19* | Voriconazole | A | 1A | Actionable PGx |
| *CYP2C19* | Citalopram | A | 1A | Actionable PGx |
| *CYP2C19* | Escitalopram | A | 1A | Actionable PGx |
| *CYP2C9* | Phenytoin | A | 1A | Actionable PGx |
| *CYP2C9* | Warfarin | A | 1A | Actionable PGx |
| *CYP2D6* | Amitriptyline | A | 1A | Actionable PGx |
| *CYP2D6* | Codeine | A | 1A | Actionable PGx |
| *CYP2D6* | Fluvoxamine | A | 1A | Actionable PGx |
| *CYP2D6* | Nortriptyline | A | 1A | Actionable PGx |
| *CYP2D6* | Ondansetron | A | 1A | Informative PGx |
| *CYP2D6* | Paroxetine | A | 1A | Informative PGx |
| *CYP2D6* | Tropisetron | A | | |
| *CYP2D6* | Clomipramine | B | 1A | Actionable PGx |
| *CYP2D6* | Desipramine | B | 1A | Actionable PGx |
| *CYP2D6* | Doxepin | B | 1A | Actionable PGx |
| *CYP2D6* | Imipramine | B | 1A | Actionable PGx |
| *CYP2D6* | Trimipramine | B | 1A | Actionable PGx |
| *CYP3A5* | Tacrolimus | A | 1A | |
| *CYP4F2* | Warfarin | A | 1B | |
| *DPYD* | Capecitabine | A | 1A | Actionable PGx |
| *DPYD* | Fluorouracil | A | 1A | Actionable PGx |
| *DPYD* | Tegafur | A | 1A | |
| *G6PD* | Rasburicase | A | 1A | Testing required |
| *HLA-B* | Abacavir | A | 1A | Testing required |
| *HLA-B* | Allopurinol | A | 1A | |
| *HLA-B* | Carbamazepine | A | 1A | Testing required |
| *HLA-B* | Phenytoin | A | 1A | Actionable PGx |
| *IFNL3* | Peginterferon alfa-2a | A | 1A | |
| *IFNL3* | Peginterferon alfa-2b | A | 1A | Actionable PGx |
| *IFNL3* | Ribavirin | A | 1A | |
| *SLCO1B1* | Simvastatin | A | 1A | |
| *TPMT* | Azathioprine | A | 1A | Testing recommended |
| *TPMT* | Mercaptopurine | A | 1A | Testing recommended |
| *TPMT* | Thioguanine | A | 1A | Actionable PGx |
| *UGT1A1* | Atazanavir | A | 1A | |
| *VKORC1* | Warfarin | A | 1A | Actionable PGx |

Data obtained from https://cpicpgx.org/genes-drugs/.

**TABLE 2.2 Approved Oncology Drugs With a Specific Genetic Indication/Target or Known Risk of Adverse Events Associated With a Genetic Variant**

| Indication | Drug (Target Genes) |
|---|---|
| **Approved for Treatment of Single Condition** | |
| Breast cancer | • Abemaciclib (*ESR, ERBB2*)<br>• Ado-trastuzumab emtansine (*ERBB2*)<br>• Anastrozole (*ESR, ERBB2*)<br>• Exemestane (*ESR, PGR*)<br>• Fulvestrant (*ERBB2 ESR, PGR*)<br>• Lapatinib (*ERBB2, ESR, PGR, HLA-DQA1, HLA-DRB1*)<br>• Letrozole (*ESR, PGR*)<br>• Neratinib (*ERBB2, ESR, PGR*)<br>• Palbociclib (*ESR, ERBB2*)<br>• Ribociclib (*ESR, PGR, ERBB2*)<br>• Pertuzumab (*ERBB2, ESR, PGR*)<br>• Tamoxifen (*ESR, PGR, F5, F2*) |
| Non–small cell lung cancer (NSCLC) | • Afatinib (*EGFR*)<br>• Alectinib (*ALK*)<br>• Brigatinib (*ALK*)<br>• Ceritinib (*ALK*)<br>• Crizotinib (*ALK, ROS1*)<br>• Erlotinib (*EGFR*)<br>• Gefitinib (*EGFR*)<br>• Osimertinib (*EGFR*) |
| Acute promyelocytic leukemia | • Arsenic trioxide (*PML-RARA*)<br>• Tretinoin (*PML-RARA*) |
| Acute lymphoblastic leukemia (ALL) | • Blinatumomab (*BCR-ABL1*)<br>• Mercaptopurine (*TMPT*)<br>• Inotuzumab ozogamicin (*BCR-ABL1*) |
| AML | • Enasidenib (*IDH2*)<br>• Thioguanine (*TPMT*) |
| Cutaneous T-cell lymphoma | • Denileukin diftitox (*IL2RA* [CD25 antigen]) |
| Peripheral T-cell lymphoma | • Belinostat (*UGT1A1*) |
| Colon/colorectal cancer | • Irinotecan (*UGT1A1*)<br>• Panitumumab (*EGFR, RAS*) |
| Chronic myelogenous leukemia (CML) | • Bosutinib (*BCR-ABL1*)<br>• Busulfan (*BCR-ABL1*)<br>• Nilotinib (*BCR-ABL1*)<br>• Omacetaxine (*BCR-ABL1*) |
| Chronic lymphocytic leukemia (CLL) | • Venetoclax (*17p*) |
| Hyperuricemia associated with malignancy | • Rasburicase (*G6PD, CYB5R*) |
| Melanoma | • Cobimetinib (*BRAF*) |
| Neuroblastoma | • Dinutuximab (*MYCN*) |
| Ovarian cancer | • Rucaparib (*BRCA, CYP2D6, CYP1A2*) |
| Soft tissue sarcoma | • Olaratumab (*PDGFRA*) |
| **Approved for Treatment of Multiple Conditions** | |
| • -Urothelial carcinoma<br>• NSCLC | Atezolizumab (*CD274* [PD-L1]) |
| • -Merkel cell carcinoma<br>• Urothelial carcinoma | Avelumab (*CD274* [PD-L1]) |

*Continued*

**TABLE 2.2 Approved Oncology Drugs With a Specific Genetic Indication/Target or Known Risk of Adverse Events Associated With a Genetic Variant—cont'd**

| Indication | Drug (Target Genes) |
|---|---|
| • Anaplastic large cell lymphoma<br>• Hodgkin lymphoma<br>• Mycosis fungoides | Brentuximab vedotin (*ALK*) |
| • Renal cell carcinoma<br>• Thyroid cancer | Cabozantinib (*RET*) |
| • Breast cancer<br>• Colorectal cancer | Capecitabine (*DPYD*) |
| • Colorectal cancer<br>• Head and neck cancer | Cetuximab (*EGFR RAS*) |
| • Bladder cancer<br>• Ovarian cancer<br>• Testicular cancer | Cisplatin (*TPMT*) |
| • Melanoma<br>• NSCLC<br>• Thyroid cancer | Dabrafenib (*BRAF, G6PD, RAS*) |
| • ALL<br>• CML | Dasatinib (*BCR-ABL1*) |
| • NSCLC<br>• Urothelial carcinoma | Durvalumab (*CD274* [PD-L1]) |
| • Breast cancer<br>• Neuroendocrine tumors<br>• Renal cell carcinoma | Everolimus (*ERBB2, ESR*) |
| • Breast cancer<br>• Colon and rectal cancer<br>• Gastric cancer<br>• Pancreatic cancer | Fluorouracil (*DPYD*) |
| • Chronic graft-versus-host disease (refractory)<br>• CLL/small lymphocytic lymphoma<br>• Mantle cell lymphoma (MZL)<br>• Waldenström macroglobulinemia | Ibrutinib (*17p, 11q*) |
| • ALL<br>• CML<br>• Gastrointestinal stromal tumors | Imatinib (*KIT, BCR-ABL1, PDGFRB, FIP1L1-PDGFRA*) |
| • AML<br>• Mast cell leukemia | Midostaurin (*FLT3, NPM1, KIT*) |
| • Ovarian, fallopian tube, or primary peritoneal cancer | Niraparib (*BRCA*) |
| • Colorectal cancer<br>• Head and neck cancer<br>• Hepatocellular carcinoma<br>• Hodgkin lymphoma<br>• Melanoma<br>• NSCLC<br>• Renal cell cancer | Nivolumab (*BRAF, CD274* [PD-L1], microsatellite instability, mismatch repair) |
| • CLL<br>• Follicular lymphoma | Obinutuzumab (*MS4A1* [CD20 antigen]) |
| • Breast cancer<br>• Ovarian cancer | Olaparib (*BRCA*) |

Continued

**TABLE 2.2 Approved Oncology Drugs With a Specific Genetic Indication/Target or Known Risk of Adverse Events Associated With a Genetic Variant—cont'd**

| Indication | Drug (Target Genes) |
|---|---|
| • Renal cell carcinoma<br>• Soft tissue sarcoma | Pazopanib (*UGT1A1, HLA-B*) |
| • Gastric cancer<br>• Head and neck cancer<br>• Hodgkin lymphoma<br>• Melanoma<br>• NSCLC<br>• Urothelial carcinoma | Pembrolizumab (*BRAF, CD274* [PD-L1], microsatellite instability, mismatch repair) |
| • ALL<br>• CML | Ponatinib (*BCR-ABL1*) |
| • CLL,<br>• Non-Hodgkin lymphomas | Rituximab (*MS4A1* [CD20 antigen]) |
| • Melanoma<br>• NSCLC<br>• Thyroid cancer | Trametinib (*BRAF, G6PD, RAS*) |
| • Breast cancer<br>• Gastric cancer | Trastuzumab (*ERBB2, ESR, PGR*) |
| • Melanoma<br>• Erdheim-Chester disease | Vemurafenib (*BRAF, RAS*) |

Data source: https://www.fda.gov/Drugs/ScienceResearch/ucm572698.htm.

gene panels [142] or whole genome sequencing of paired tumor/normal tissue [143] has been explored and developed to inform treatment decision and prognosis in cancer patients. Yet, the options for genetic testing in oncology are numerous, creating challenges for providers, patients, and insurers to determine which is most appropriate for a given patient [144]. In 2018, the first comprehensive test panel (FoundationOne CDx) was approved to inform treatment decisions for multiple types of cancer.

Along with targeted drugs, promising advances with biologics, such as immunotherapeutics (immune checkpoint inhibitors) [145], chimeric antigen receptors T-cell (CAR-T) therapy [146], and gene therapy [147] and gene editing, have yielded a new suite of tools and interventions. In 2017, two CAR-T interventions (for acute lymphocytic leukemia and diffuse large B-cell lymphoma), in addition to the first adeno-associated viral (AAV) gene therapy for inherited retinal dystrophy, were approved by the FDA. Also in 2017, the FDA approved the first gene therapy that delivers copies of the *RPE65* gene through a viral vector in patients with a rare genetic disorder, retinal blindness. Continued development and evaluation of engineered interventions focus on improving the delivery methods (viral type, ex vivo vs. in vivo), target specificity, reducing systemic toxicities, and expanding applications to other diseases [148].

## 2.4 PRECISION MEDICINE RESEARCH

Several technologies are under intense investigation, adding layers of analysis of various molecular entities and expanding understanding of development, gene—environment interactions, and risk markers. With the layered data sets for a given individual, the complexity of data analysis is increased, warranting new tools and algorithms. Although much emphasis has focused on the study of DNA-based data and technologies, other developments are under investigation to gain a fuller assessment of various biomolecular entities and response to internal and external factors.

*The Epigenome.* The field of epigenetics has also garnered substantial attention due to the development of new technologies (whole genome bisulfite sequencing) and focus on the study of regulatory elements of the genome [149–151]. DNA methylation is recognized as a critical factor in numerous cellular processes including development and differentiation, cell proliferation, and tumorigenesis. Epigenetics and the importance of methylation status in gene regulation were limited to gene-by-gene analysis of methylation patterns. Today, whole genome analysis can generate comprehensive snapshots of methylation status ("methylome") in a range of tissues, developmental stages, and diseases [152]. Large-scale studies have characterized methylation patterns in pregnancy [153], gametogenesis [154], and diseases such as cancer [155,156] and diabetes [157]. The combined analysis of DNA methylation and transcriptome from a single cell can enable correlative analysis [158].

*The Microbiome.* With the development of new sequencing technologies, it became feasible to characterize microbial communities residing on various human tissues, which was not possible with traditional culturing methods. In 2008, the US National Institutes of Health launched the Human Microbiome Project (HMP), with the goal of characterizing the microbiome in 300 healthy individuals across multiple tissues (https://hmpdacc.org/hmp/) [159–161]. With a national initiative and newly developed sequencing and analytical capabilities, many studies have been conducted of microbiota on a range of human tissues, from healthy and affected patients, and from samples collected at regular intervals. The association of microbiota composition (and potential perturbations) to human development and disease has become an area of intense study for a wide range of diseases, including cancer, dermatological conditions [162], obesity, gastroenterological conditions [163], and diabetes [164]. The impacts of microbiome manipulation through diet (prebiotics and probiotics) [165], antibiotics [166], drug response [167,168], and fecal transplantation [169] are under investigation to better understand the role of local microbial communities to disease susceptibility, onset, outcome, and treatment. Related analyses, such as the mycobiome, are also under way [170,171].

*Data Sciences.* While the singular use of various -omics technologies can provide great insight into particular level of cellular operations, a more comprehensive picture can be achieved through analysis of multiple data sets generated from a single individual. In addition to the sheer size of combined data sets, the interpretation of these large data sets will remain a key challenge. For example, the combination of the microbiome and metabolome can begin to elucidate the physiological impacts of certain microbiota compositions and perturbation sot the microbiome [172,173]. The unprecedentedly large multi-omic data sets generated per individual patient, referred to as "panomics" [174], multidimensional integrative genomics [175], or integrated -omics [176], will greatly benefit from sophisticated analytical capabilities to assess temporal changes in the data as well as other changes associated with aging, lifestyle, or environmental exposures. With the availability of large heterogeneous data sets, including medical records and genomic data, machine learning approaches have begun to be used to develop predictive or diagnostic algorithms [177–179]. Substantial effort has been devoted to artificial intelligence (AI) or machine learning of large, heterogeneous patient data sets to assess phenotypes [180] and identify pathogenic variants [181], drug response [182], and risk or outcome predictors.

## 2.5 THE PRACTICE OF PRECISION MEDICINE

The implementation and integration of precision health applications into routine clinical practice are the subjects of numerous research programs supported by the National Human Genome Research Institute of the NIH. In particular, Newborn Sequencing in Genomic Medicine and Public Health (NSIGHT; https://www.genome.gov/27558493/newborn-sequencing-in-genomic-medicine-and-public-health-nsight/), Clinical Sequencing Evidence-Generating Research (CSER; https://www.genome.gov/27546194/clinical-sequencing-exploratory-research-cser/), the Electronic Medical Records and Genomics (eMERGE; https://www.genome.gov/27540473/) network, and the Implementing GeNomics In PracTicE (IGNITE; https://www.genome.gov/27554264/) network are programs whose goals are to develop the insights and tools required for the use of diverse genome-based precision medicine technologies in day to day health care in diverse settings.

In addition to demonstrating the effectiveness of new interventions, the practice of precision medicine faces several obstacles that will require provider and patient support including educational resources, an EMR system that can provide adequate storage and facilitate easy look-up of laboratory reports, and positive coverage determinations. The impact of personal genomic risk data continues to be under investigation in order to assess patients' informational needs, delivery approaches, and likelihood of adopting healthy behaviors or complying with screening or treatment recommendations [183,184]. Coinciding with other developments in precision medicine is greater patient engagement through online patient portals and self-collection of health information through mobile health technologies. Such barriers may limit access or cause uneven access to these interventions [185,186]. We highlight a few of these areas next.

*Digital Health Tools.* With the expanding use of digital health technologies such as wearables to measure and record health-related behaviors, health monitoring has become increasingly convenient and easy [187]. Particularly for patients with chronic conditions, evidence has demonstrated greater patient engagement, symptom monitoring, and/or adherence, and behavioral changes between clinic visits with use of mobile health applications [188–194]. The quality of data from consumer wearables may be inconsistent between brands and vary between features [195], though they may still sufficiently motivate healthy behaviors and patient engagement [196]. Other digital health applications include wearable biosensors for vital sign monitoring [197], ingestible biosensors for medication compliance monitoring [198], and smart homes [199]. Patients may also increase the likelihood of establishing and maintaining healthy behaviors with a variety of support from health coaches, support groups or social networks, and/or daily reminders through mobile apps and text messaging. Future patient medical records will likely include data collected from self-monitored measures to provide a more comprehensive picture of a patient's lifestyle and environment.

*Clinical Decision Support.* With the transition to EMRs, the provision of point-of-care clinical decision support is now feasible, improving the ability to interpret and generate evidence-based recommendations based on self-reported data, FHH, laboratory testing, and other clinical information. One example of the application of digital health tools is the online collection and analysis of FHH. Long captured through paper-based formats with reportedly limited use, FHH has been revitalized with the development of online FHH tools [200]. Provider use of FHH is limited by time for collection and interpretation and beliefs that detailed FHH (three generations) fall outside the scope of general practice [47]. Similarly, several patient barriers exist to the comprehensive collection of FHH, understanding about FHH, knowledge about family members' health, family dynamics, cultural norms, and privacy concerns [201–207]. Online collection of FHH can help overcome challenges of timely patient reporting and education as well as immediate generation of recommendations based on clinical guidelines that may be reviewed during the office visit. In some cases, a virtual provider can promote family history-taking [49].

MeTree is one example of a patient-facing Web-based FHH driven risk assessment application [48]. The app is integrated into clinical practices and provides clinical decision support to patients and their primary care providers about risk level for 30 different conditions and evidence-based recommendations for how to manage that risk. Such electronic FHH collection tools may greatly improve the comprehensiveness of information collected from patients [208], spur family health information-sharing [209], particularly if educational support is available to help patients understand what type of information to report and about which family members [210], and the ability to update it outside of a clinic visit. The impact of MeTree on patients and providers and the type of care patients has been piloted [211] and is currently being assessed in a large multi-institutional study of patients in five national healthcare settings [212]. For providers that have not implemented electronic FHH tools, software for specific conditions has been developed to generate risk assessments compatible with publicly available FHH tools [213].

*Coverage and Reimbursement.* Routine use of new clinical applications is unlikely to occur without positive coverage determinations based in large part on evidence of clinical utility and cost-effectiveness. While several studies have assessed the cost-effectiveness of a wide range of tests [214–217], evidence from large clinical trials is lacking. Furthermore, clinical guidelines are lacking regarding the use of many genetic and genomic tests, with the one exception being PGx testing [218]. Insurers and government review bodies

YH, Romblad D, Ruhfel B, Scott R, Sitter C, Smallwood M, Stewart E, Strong R, Suh E, Thomas R, Tint NN, Tse S, Vech C, Wang G, Wetter J, Williams S, Williams M, Windsor S, Winn-Deen E, Wolfe K, Zaveri J, Zaveri K, Abril JF, Guigo R, Campbell MJ, Sjolander KV, Karlak B, Kejariwal A, Mi H, Lazareva B, Hatton T, Narechania A, Diemer K, Muruganujan A, Guo N, Sato S, Bafna V, Istrail S, Lippert R, Schwartz R, Walenz B, Yooseph S, Allen D, Basu A, Baxendale J, Blick L, Caminha M, Carnes-Stine J, Caulk P, Chiang YH, Coyne M, Dahlke C, Mays A, Dombroski M, Donnelly M, Ely D, Esparham S, Fosler C, Gire H, Glanowski S, Glasser K, Glodek A, Gorokhov M, Graham K, Gropman B, Harris M, Heil J, Henderson S, Hoover J, Jennings D, Jordan C, Jordan J, Kasha J, Kagan L, Kraft C, Levitsky A, Lewis M, Liu X, Lopez J, Ma D, Majoros W, McDaniel J, Murphy S, Newman M, Nguyen T, Nguyen N, Nodell M, Pan S, Peck J, Peterson M, Rowe W, Sanders R, Scott J, Simpson M, Smith T, Sprague A, Stockwell T, Turner R, Venter E, Wang M, Wen M, Wu D, Wu M, Xia A, Zandieh A, Zhu X. The sequence of the human genome. Science 2001;291:1304–51.

[8] Levy SE, Myers RM. Advancements in next-generation sequencing. Annu Rev Genom Hum Genet 2016;17:95–115.

[9] Organick L, Ang SD, Chen YJ, Lopez R, Yekhanin S, Makarychev K, Racz MZ, Kamath G, Gopalan P, Nguyen B, Takahashi CN, Newman S, Parker HY, Rashtchian C, Stewart K, Gupta G, Carlson R, Mulligan J, Carmean D, Seelig G, Ceze L, Strauss K. Random access in large-scale DNA data storage. Nat Biotechnol 2018;36:242–8.

[10] Roy S, LaFramboise WA, Nikiforov YE, Nikiforova MN, Routbort MJ, Pfeifer J, Nagarajan R, Carter AB, Pantanowitz L. Next-generation sequencing informatics: challenges and strategies for implementation in a clinical environment. Arch Pathol Lab Med 2016;140:958–75.

[11] van Dijk EL, Auger H, Jaszczyszyn Y, Thermes C. Ten years of next-generation sequencing technology. Trends Genet 2014;30:418–26.

[12] Ganguli M, Albanese E, Seshadri S, Bennett DA, Lyketsos C, Kukull WA, Skoog I, Hendrie HC. Population neuroscience: dementia epidemiology serving precision medicine and population health. Alzheimer Dis Assoc Disord 2018;32:1–9.

[13] Vaithinathan AG, Asokan V. Public health and precision medicine share a goal. J Evid Based Med 2017;10:76–80.

[14] Belsky DW, Moffitt TE, Caspi A. Genetics in population health science: strategies and opportunities. Am J Public Health 2013;103(Suppl. 1):S73–83.

[15] Meagher KM, McGowan ML, Settersten Jr RA, Fishman JR, Juengst ET. Precisely where are we going? Charting the new terrain of precision prevention. Annu Rev Genom Hum Genet 2017;18:369–87.

[16] Collins FS, Varmus H. A new initiative on precision medicine. N Engl J Med 2015;372:793–5.

[17] Turnbull C. Introducing whole genome sequencing into routine cancer care: the genomics England 100,000 genomes project. Ann Oncol 2018.

[18] Cuypers B, Berg M, Imamura H, Dumetz F, De Muylder G, Domagalska MA, Rijal S, Bhattarai NR, Maes I, Sanders M, Cotton JA, Meysman P, Laukens K, Dujardin JC. Integrated genomic and metabolomic profiling of ISC1, an emerging Leishmania donovani population in the Indian subcontinent. Infect Genet Evol 2018.

[19] Grunert M, Dorn C, Cui H, Dunkel I, Schulz K, Schoenhals S, Sun W, Berger F, Chen W, Sperling SR. Comparative DNA methylation and gene expression analysis identifies novel genes for structural congenital heart diseases. Cardiovasc Res 2016;112:464–77.

[20] Rattray NJW, Deziel NC, Wallach JD, Khan SA, Vasiliou V, Ioannidis JPA, Johnson CH. Beyond genomics: understanding exposotypes through metabolomics. Hum Genom 2018;12:4.

[21] Chen R, Mias GI, Li-Pook-Than J, Jiang L, Lam HY, Miriami E, Karczewski KJ, Hariharan M, Dewey FE, Cheng Y, Clark MJ, Im H, Habegger L, Balasubramanian S, O'Huallachain M, Dudley JT, Hillenmeyer S, Haraksingh R, Sharon D, Euskirchen G, Lacroute P, Bettinger K, Boyle AP, Kasowski M, Grubert F, Seki S, Garcia M, Whirl-Carrillo M, Gallardo M, Blasco MA, Greenberg PL, Snyder P, Klein TE, Altman RB, Butte AJ, Ashley EA, Gerstein M, Nadeau KC, Tang H, Snyder M. Personal omics profiling reveals dynamic molecular and medical phenotypes. Cell 2012;148(6):1293–307.

[22] Price ND, Magis AT, Earls JC, Glusman G, Levy R, Lausted C, McDonald DT, Kusebauch U, Moss CL, Zhou Y, Qin S, Moritz RL, Brogaard K, Omenn GS, Lovejoy JC, Hood L. A wellness study of 108 individuals using personal, dense, dynamic data clouds. Nat Biotechnol 2017;35:747–56.

[23] Rechitsky S, Pakhalchuk T, San Ramos G, Goodman A, Zlatopolsky Z, Kuliev A. First systematic experience of preimplantation genetic diagnosis for single-gene disorders, and/or preimplantation human leukocyte antigen typing, combined with 24-chromosome aneuploidy testing. Fertil Steril 2015;103:503–12.

[24] Iwarsson E, Jacobsson B, Dagerhamn J, Davidson T, Bernabe E, Heibert Arnlind M. Analysis of cell-free fetal DNA in maternal blood for detection of trisomy 21, 18 and 13 in a general pregnant population and in a high risk population - a systematic review and meta-analysis. Acta Obstet Gynecol Scand 2017;96(1):7–18.

[25] Renga B. Non invasive prenatal diagnosis of fetal aneuploidy using cell free fetal DNA. Eur J Obstet Gynecol Reprod Biol 2018;225:5–8.

[26] Committee opinion No. 640: cell-free DNA screening for fetal aneuploidy. Obstet Gynecol 2015;126:e31–37.
[27] Chen F, Liu P, Gu Y, Zhu Z, Nanisetti A, Lan Z, Huang Z, Liu SJ, Kang X, Deng Y, Luo L, Jiang D, Qiu Y, Pan J, Xia J, Xiong K, Liu C, Xie L, Shi Q, Li J, Zhang X, Wang W, Drmanac S, Jiang H, Drmanac R, Xu X. Isolation and Whole Genome Sequencing of fetal cells from maternal blood towards the ultimate non-invasive prenatal testing. Prenat Diagn 2017.
[28] McLennan A, Palma-Dias R, da Silva Costa F, Meagher S, Nisbet DL, Scott F. Noninvasive prenatal testing in routine clinical practice–an audit of NIPT and combined first-trimester screening in an unselected Australian population. Aust N Z J Obstet Gynaecol 2016;56:22–8.
[29] Bjerregaard L, Stenbakken AB, Andersen CS, Kristensen L, Jensen CV, Skovbo P, Sorensen AN. The rate of invasive testing for trisomy 21 is reduced after implementation of NIPT. Dan Med J 2017;64.
[30] Johnson K, Kelley J, Saxton V, Walker SP, Hui L. Declining invasive prenatal diagnostic procedures: a comparison of tertiary hospital and national data from 2012 to 2015. Aust N Z J Obstet Gynaecol 2017;57:152–6.
[31] Kane SC, Reidy KL, Norris F, Nisbet DL, Kornman LH, Palma-Dias R. Chorionic villus sampling in the cell-free DNA aneuploidy screening era: careful selection criteria can maximise the clinical utility of screening and invasive testing. Prenat Diagn 2017;37:399–408.
[32] Bianchi DW, Chudova D, Sehnert AJ, Bhatt S, Murray K, Prosen TL, Garber JE, Wilkins-Haug L, Vora NL, Warsof S, Goldberg J, Ziainia T, Halks-Miller M. Noninvasive prenatal testing and incidental detection of occult maternal malignancies. J Am Med Assoc 2015;314:162–9.
[33] Cohen PA, Flowers N, Tong S, Hannan N, Pertile MD, Hui L. Abnormal plasma DNA profiles in early ovarian cancer using a non-invasive prenatal testing platform: implications for cancer screening. BMC Med 2016;14:126.
[34] Schmid M, Wang E, Bogard PE, Bevilacqua E, Hacker C, Wang S, Doshi J, White K, Kaplan J, Sparks A, Jani JC, Stokowski R. Prenatal screening for 22q11.2 deletion using a targeted microarray-based cell-free DNA test. Fetal Diagn Ther 2017.
[35] Best S, Wou K, Vora N, Van der Veyver IB, Wapner R, Chitty LS. Promises, pitfalls and practicalities of prenatal whole exome sequencing. Prenat Diagn 2018;38:10–9.
[36] Hayward J, Chitty LS. Beyond screening for chromosomal abnormalities: advances in non-invasive diagnosis of single gene disorders and fetal exome sequencing. Semin Fetal Neonatal Med 2018;23:94–101.
[37] Committee opinion No.682: microarrays and next-generation sequencing technology: the use of advanced genetic diagnostic tools in obstetrics and gynecology. Obstet Gynecol 2016;128:e262–8.
[38] Fukami M, Miyado M. Next generation sequencing and array-based comparative genomic hybridization for molecular diagnosis of pediatric endocrine disorders. Ann Pediatr Endocrinol Metab 2017;22:90–4.
[39] Farnaes L, Hildreth A, Sweeney NM, Clark MM, Chowdhury S, Nahas S, Cakici JA, Benson W, Kaplan RH, Kronick R, Bainbridge MN, Friedman J, Gold JJ, Ding Y, Veeraraghavan N, Dimmock D, Kingsmore SF. Rapid whole-genome sequencing decreases infant morbidity and cost of hospitalization. NPJ Genom Med 2018;3:10.
[40] Bartnik M, Wisniowiecka-Kowalnik B, Nowakowska B, Smyk M, Kedzior M, Sobecka K, Kutkowska-Kazmierczak A, Klapecki J, Szczaluba K, Castaneda J, Wlasienko P, Bezniakow N, Obersztyn E, Bocian E. The usefulness of array comparative genomic hybridization in clinical diagnostics of intellectual disability in children. Dev Period Med 2014;18:307–17.
[41] Manning M, Hudgins L. Array-based technology and recommendations for utilization in medical genetics practice for detection of chromosomal abnormalities. Genet Med 2010;12:742–5.
[42] South ST, Lee C, Lamb AN, Higgins AW, Kearney HM. ACMG Standards and Guidelines for constitutional cytogenomic microarray analysis, including postnatal and prenatal applications: revision 2013. Genet Med 2013;15:901–9.
[43] Bevilacqua J, Hesse A, Cormier B, Davey J, Patel D, Shankar K, Reddi HV. Clinical utility of a 377 gene custom next-generation sequencing epilepsy panel. J Genet 2017;96:681–5.
[44] Butler KM, da Silva C, Alexander JJ, Hegde M, Escayg A. Diagnostic yield from 339 epilepsy patients screened on a clinical gene panel. Pediatr Neurol 2017;77:61–6.
[45] Chambers C, Jansen LA, Dhamija R. Review of commercially available epilepsy genetic panels. J Genet Couns 2016;25:213–7.
[46] Berg AO, Baird MA, Botkin JR, Driscoll DA, Fishman PA, Guarino PD, Hiatt RA, Jarvik GP, Millon-Underwood S, Morgan TM, Mulvihill JJ, Pollin TI, Schimmel SR, Stefanek ME, Vollmer WM, Williams JK. National Institutes of health state-of-the-science conference statement: family history and improving health. Ann Intern Med 2009;151(12):872–7.
[47] Saul RA, Trotter T, Sease K, Tarini B. Survey of family history taking and genetic testing in pediatric practice. J Community Genet 2017;8:109–15.
[48] Orlando LA, Buchanan AH, Hahn SE, Christianson CA, Powell KP, Skinner CS, Chesnut B, Blach C, Due B, Ginsburg GS, Henrich VC. Development and validation of a primary care-based family health history and decision support program (MeTree). N C Med J 2013;74:287–96.

[125] Beckman RA, Antonijevic Z, Kalamegham R, Chen C. Adaptive design for a confirmatory basket trial in multiple tumor types based on a putative predictive biomarker. Clin Pharmacol Ther 2016;100:617–25.
[126] Park JW, Liu MC, Yee D, Yau C, van 't Veer LJ, Symmans WF, Paoloni M, Perlmutter J, Hylton NM, Hogarth M, DeMichele A, Buxton MB, Chien AJ, Wallace AM, Boughey JC, Haddad TC, Chui SY, Kemmer KA, Kaplan HG, Isaacs C, Nanda R, Tripathy D, Albain KS, Edmiston KK, Elias AD, Northfelt DW, Pusztai L, Moulder SL, Lang JE, Viscusi RK, Euhus DM, Haley BB, Khan QJ, Wood WC, Melisko M, Schwab R, Helsten T, Lyandres J, Davis SE, Hirst GL, Sanil A, Esserman LJ, Berry DA. Adaptive randomization of neratinib in early breast cancer. N Engl J Med 2016;375:11–22.
[127] Patel JN. Cancer pharmacogenomics, challenges in implementation, and patient-focused perspectives. Pharmgenom Pers Med 2016;9:65–77.
[128] Robson M, Im SA, Senkus E, Xu B, Domchek SM, Masuda N, Delaloge S, Li W, Tung N, Armstrong A, Wu W, Goessl C, Runswick S, Conte P. Olaparib for metastatic breast cancer in patients with a germline BRCA mutation. N Engl J Med 2017;377:523–33.
[129] Perl AE. The role of targeted therapy in the management of patients with AML. Blood Adv 2017;1:2281–94.
[130] Hainsworth JD, Meric-Bernstam F, Swanton C, Hurwitz H, Spigel DR, Sweeney C, Burris H, Bose R, Yoo B, Stein A, Beattie M, Kurzrock R. Targeted therapy for advanced solid tumors on the basis of molecular profiles: results from MyPathway, an open-label, phase IIa multiple basket study. J Clin Oncol 2018;36:536–42.
[131] Xu MJ, Johnson DE, Grandis JR. EGFR-targeted therapies in the post-genomic era. Cancer Metastasis Rev 2017;36:463–73.
[132] Schuette W, Schirmacher P, Eberhardt WE, Fischer JR, von der Schulenburg JM, Mezger J, Schumann C, Serke M, Zaun S, Dietel M, Thomas M. EGFR mutation status and first-line treatment in patients with stage III/IV non-small cell lung cancer in Germany: an observational study. Cancer Epidemiol Biomarkers Prev 2015;24(8):1254–61.
[133] Sharma SV, Bell DW, Settleman J, Haber DA. Epidermal growth factor receptor mutations in lung cancer. Nat Rev Cancer 2007;7:169–81.
[134] Tufman A, Kahnert K, Duell T, Kauffmann-Guerrero D, Milger K, Schneider C, Stump J, Syunyaeva Z, Huber RM, Reu S. Frequency and clinical relevance of EGFR mutations and EML4-ALK translocations in octogenarians with non-small cell lung cancer. OncoTargets Ther 2017;10:5179–86.
[135] Lemery S, Keegan P, Pazdur R. First FDA approval agnostic of cancer site - when a biomarker defines the indication. N Engl J Med 2017;377:1409–12.
[136] Afghahi A, Kurian AW. The changing landscape of genetic testing for inherited breast cancer predisposition. Curr Treat Options Oncol 2017;18:27.
[137] Robinson JG, Farnier M, Krempf M, Bergeron J, Luc G, Averna M, Stroes ES, Langslet G, Raal FJ, El Shahawy M, Koren MJ, Lepor NE, Lorenzato C, Pordy R, Chaudhari U, Kastelein JJ. Efficacy and safety of alirocumab in reducing lipids and cardiovascular events. N Engl J Med 2015;372:1489–99.
[138] Sabatine MS, Giugliano RP, Wiviott SD, Raal FJ, Blom DJ, Robinson J, Ballantyne CM, Somaratne R, Legg J, Wasserman SM, Scott R, Koren MJ, Stein EA. Efficacy and safety of evolocumab in reducing lipids and cardiovascular events. N Engl J Med 2015;372:1500–9.
[139] Reiss AB, Shah N, Muhieddine D, Zhen J, Yudkevich J, Kasselman LJ, DeLeon J. PCSK9 in cholesterol metabolism: from bench to bedside. Clin Sci (Lond) 2018;132:1135–53.
[140] Dijkstra KK, Voabil P, Schumacher TN, Voest EE. Genomics- and transcriptomics-based patient selection for cancer treatment with immune checkpoint inhibitors: a review. JAMA Oncol 2016;2:1490–5.
[141] Rizvi NA, Hellmann MD, Snyder A, Kvistborg P, Makarov V, Havel JJ, Lee W, Yuan J, Wong P, Ho TS, Miller ML, Rekhtman N, Moreira AL, Ibrahim F, Bruggeman C, Gasmi B, Zappasodi R, Maeda Y, Sander C, Garon EB, Merghoub T, Wolchok JD, Schumacher TN, Chan TA. Cancer immunology. Mutational landscape determines sensitivity to PD-1 blockade in non-small cell lung cancer. Science 2015;348:124–8.
[142] Campesato LF, Barroso-Sousa R, Jimenez L, Correa BR, Sabbaga J, Hoff PM, Reis LF, Galante PA, Camargo AA. Comprehensive cancer-gene panels can be used to estimate mutational load and predict clinical benefit to PD-1 blockade in clinical practice. Oncotarget 2015;6:34221–7.
[143] Rennert H, Eng K, Zhang T, Tan A, Xiang J, Romanel A, Kim R, Tam W, Liu YC, Bhinder B, Cyrta J, Beltran H, Robinson B, Mosquera JM, Fernandes H, Demichelis F, Sboner A, Kluk M, Rubin MA, Elemento O. Development and validation of a whole-exome sequencing test for simultaneous detection of point mutations, indels and copy-number alterations for precision cancer care. NPJ Genom Med 2016;1.
[144] Borad MJ, LoRusso PM. Twenty-first century precision medicine in oncology: genomic profiling in patients with cancer. Mayo Clin Proc 2017;92:1583–91.
[145] Atkins MB, Larkin J. Immunotherapy combined or sequenced with targeted therapy in the treatment of solid tumors: current perspectives. J Natl Cancer Inst 2016;108:djv414.
[146] Sadelain M, Riviere I, Riddell S. Therapeutic T cell engineering. Nature 2017;545:423–31.

[147] Dunbar CE, High KA, Joung JK, Kohn DB, Ozawa K, Sadelain M. Gene therapy comes of age. Science 2018;359.
[148] Ellebrecht CT, Bhoj VG, Nace A, Choi EJ, Mao X, Cho MJ, Di Zenzo G, Lanzavecchia A, Seykora JT, Cotsarelis G, Milone MC, Payne AS. Reengineering chimeric antigen receptor T cells for targeted therapy of autoimmune disease. Science 2016;353:179–84.
[149] An integrated encyclopedia of DNA elements in the human genome. Nature 2012;489:57–74.
[150] Kellis M, Wold B, Snyder MP, Bernstein BE, Kundaje A, Marinov GK, Ward LD, Birney E, Crawford GE, Dekker J, Dunham I, Elnitski LL, Farnham PJ, Feingold EA, Gerstein M, Giddings MC, Gilbert DM, Gingeras TR, Green ED, Guigo R, Hubbard T, Kent J, Lieb JD, Myers RM, Pazin MJ, Ren B, Stamatoyannopoulos JA, Weng Z, White KP, Hardison RC. Defining functional DNA elements in the human genome. Proc Natl Acad Sci USA 2014;111:6131–8.
[151] Siggens L, Ekwall K. Epigenetics, chromatin and genome organization: recent advances from the ENCODE project. J Intern Med 2014;276:201–14.
[152] Fouse SD, Nagarajan RO, Costello JF. Genome-scale DNA methylation analysis. Epigenomics 2010;2:105–17.
[153] Lapato DM, Moyer S, Olivares E, Amstadter AB, Kinser PA, Latendresse SJ, Jackson-Cook C, Roberson-Nay R, Strauss JF, York TP. Prospective longitudinal study of the pregnancy DNA methylome: the US pregnancy, race, environment, genes (PREG) study. BMJ Open 2018;8:e019721.
[154] Castillo J, Jodar M, Oliva R. The contribution of human sperm proteins to the development and epigenome of the preimplantation embryo. Hum Reprod Update 2018.
[155] Beekman R, Chapaprieta V, Russinol N, Vilarrasa-Blasi R, Verdaguer-Dot N, Martens JHA, Duran-Ferrer M, Kulis M, Serra F, Javierre BM, Wingett SW, Clot G, Queiros AC, Castellano G, Blanc J, Gut M, Merkel A, Heath S, Vlasova A, Ullrich S, Palumbo E, Enjuanes A, Martin-Garcia D, Bea S, Pinyol M, Aymerich M, Royo R, Puiggros M, Torrents D, Datta A, Lowy E, Kostadima M, Roller M, Clarke L, Flicek P, Agirre X, Prosper F, Baumann T, Delgado J, Lopez-Guillermo A, Fraser P, Yaspo ML, Guigo R, Siebert R, Marti-Renom MA, Puente XS, Lopez-Otin C, Gut I, Stunnenberg HG, Campo E, Martin-Subero JI. The reference epigenome and regulatory chromatin landscape of chronic lymphocytic leukemia. Nat Med 2018.
[156] Lee EJ, Luo J, Wilson JM, Shi H. Analyzing the cancer methylome through targeted bisulfite sequencing. Cancer Lett 2013;340:171–8.
[157] Chen Z, Miao F, Paterson AD, Lachin JM, Zhang L, Schones DE, Wu X, Wang J, Tompkins JD, Genuth S, Braffett BH, Riggs AD, Natarajan R. Epigenomic profiling reveals an association between persistence of DNA methylation and metabolic memory in the DCCT/EDIC type 1 diabetes cohort. Proc Natl Acad Sci USA 2016;113:E3002–11.
[158] Hu Y, Huang K, An Q, Du G, Hu G, Xue J, Zhu X, Wang CY, Xue Z, Fan G. Simultaneous profiling of transcriptome and DNA methylome from a single cell. Genome Biol 2016;17:88.
[159] Structure, function and diversity of the healthy human microbiome. Nature 2012;486:207–14.
[160] Aagaard K, Petrosino J, Keitel W, Watson M, Katancik J, Garcia N, Patel S, Cutting M, Madden T, Hamilton H, Harris E, Gevers D, Simone G, McInnes P, Versalovic J. The human microbiome project strategy for comprehensive sampling of the human microbiome and why it matters. Faseb J 2013;27:1012–22.
[161] Nelson KE, Weinstock GM, Highlander SK, Worley KC, Creasy HH, Wortman JR, Rusch DB, Mitreva M, Sodergren E, Chinwalla AT, Feldgarden M, Gevers D, Haas BJ, Madupu R, Ward DV, Birren BW, Gibbs RA, Methe B, Petrosino JF, Strausberg RL, Sutton GG, White OR, Wilson RK, Durkin S, Giglio MG, Gujja S, Howarth C, Kodira CD, Kyrpides N, Mehta T, Muzny DM, Pearson M, Pepin K, Pati A, Qin X, Yandava C, Zeng Q, Zhang L, Berlin AM, Chen L, Hepburn TA, Johnson J, McCorrison J, Miller J, Minx P, Nusbaum C, Russ C, Sykes SM, Tomlinson CM, Young S, Warren WC, Badger J, Crabtree J, Markowitz VM, Orvis J, Cree A, Ferriera S, Fulton LL, Fulton RS, Gillis M, Hemphill LD, Joshi V, Kovar C, Torralba M, Wetterstrand KA, Abouellleil A, Wollam AM, Buhay CJ, Ding Y, Dugan S, FitzGerald MG, Holder M, Hostetler J, Clifton SW, Allen-Vercoe E, Earl AM, Farmer CN, Liolios K, Surette MG, Xu Q, Pohl C, Wilczek-Boney K, Zhu D. A catalog of reference genomes from the human microbiome. Science 2010;328(5981):994–9.
[162] Yamazaki Y, Nakamura Y, Nunez G. Role of the microbiota in skin immunity and atopic dermatitis. Allergol Int 2017;66:539–44.
[163] Goulet O. Potential role of the intestinal microbiota in programming health and disease. Nutr Rev 2015;73(Suppl. 1):32–40.
[164] Sohail MU, Althani A, Anwar H, Rizzi R, Marei HE. Role of the gastrointestinal tract microbiome in the pathophysiology of diabetes mellitus. J Diabetes Res 2017;2017:9631435.
[165] Tankou SK, Regev K, Healy BC, Tjon E, Laghi L, Cox LM, Kivisakk P, Pierre IV, Lokhande H, Gandhi R, Cook S, Glanz B, Stankiewicz J, Weiner HL. A probiotic modulates the microbiome and immunity in multiple sclerosis. Ann Neurol 2018.
[166] Becattini S, Taur Y, Pamer EG. Antibiotic-induced changes in the intestinal microbiota and disease. Trends Mol Med 2016;22:458–78.

[167] Doestzada M, Vila AV, Zhernakova A, Koonen DPY, Weersma RK, Touw DJ, Kuipers F, Wijmenga C, Fu J. Pharmacomicrobiomics: a novel route towards personalized medicine? Protein Cell 2018;9:432–45.

[168] Swanson HI. Drug metabolism by the host and gut microbiota: a partnership or rivalry? Drug Metab Dispos 2015;43:1499–504.

[169] Mintz M, Khair S, Grewal S, LaComb JF, Park J, Channer B, Rajapakse R, Bucobo JC, Buscaglia JM, Monzur F, Chawla A, Yang J, Robertson CE, Frank DN, Li E. Longitudinal microbiome analysis of single donor fecal microbiota transplantation in patients with recurrent *Clostridium difficile* infection and/or ulcerative colitis. PLoS One 2018;13:e0190997.

[170] Hoarau G, Mukherjee PK, Gower-Rousseau C, Hager C, Chandra J, Retuerto MA, Neut C, Vermeire S, Clemente J, Colombel JF, Fujioka H, Poulain D, Sendid B, Ghannoum MA. Bacteriome and mycobiome interactions underscore microbial dysbiosis in familial Crohn's disease. mBio 2016;7.

[171] Nash AK, Auchtung TA, Wong MC, Smith DP, Gesell JR, Ross MC, Stewart CJ, Metcalf GA, Muzny DM, Gibbs RA, Ajami NJ, Petrosino JF. The gut mycobiome of the human microbiome project healthy cohort. Microbiome 2017;5:153.

[172] Aw W, Fukuda S. An integrated outlook on the metagenome and metabolome of intestinal diseases. Diseases 2015;3:341–59.

[173] Shaffer M, Armstrong AJS, Phelan VV, Reisdorph N, Lozupone CA. Microbiome and metabolome data integration provides insight into health and disease. Transl Res 2017;189:51–64.

[174] Sandhu C, Qureshi A, Emili A. Panomics for precision medicine. Trends Mol Med 2017.

[175] Arneson D, Shu L, Tsai B, Barrere-Cain R, Sun C, Yang X. Multidimensional integrative genomics approaches to dissecting cardiovascular disease. Front Cardiovasc Med 2017;4:8.

[176] Karczewski KJ, Snyder MP. Integrative omics for health and disease. Nat Rev Genet 2018;19:299–310.

[177] Beam AL, Kohane IS. Big data and machine learning in health care. J Am Med Assoc 2018.

[178] Diao J, Kohane IS, Manrai AK. Biomedical informatics and machine learning for clinical genomics. Hum Mol Genet 2018.

[179] Lin E, Lane HY. Machine learning and systems genomics approaches for multi-omics data. Biomark Res 2017;5:2.

[180] Basile AO, Ritchie MD. Informatics and machine learning to define the phenotype. Expert Rev Mol Diagn 2018;18:219–26.

[181] Quang D, Chen Y, Xie X. DANN: a deep learning approach for annotating the pathogenicity of genetic variants. Bioinformatics 2015;31:761–3.

[182] Way GP, Sanchez-Vega F, La K, Armenia J, Chatila WK, Luna A, Sander C, Cherniack AD, Mina M, Ciriello G, Schultz N, Sanchez Y, Greene CS. Machine learning detects Pan-cancer ras pathway activation in the cancer genome atlas. Cell Rep 2018;23:172–80.

[183] Hay JL, Berwick M, Zielaskowski K, White KA, Rodriguez VM, Robers E, Guest DD, Sussman A, Talamantes Y, Schwartz MR, Greb J, Bigney J, Kaphingst KA, Hunley K, Buller DB. Implementing an internet-delivered skin cancer genetic testing intervention to improve sun protection behavior in a diverse population: protocol for a randomized controlled trial. JMIR Res Protoc 2017;6:e52.

[184] Smit AK, Espinoza D, Newson AJ, Morton RL, Fenton G, Freeman L, Dunlop K, Butow PN, Law MH, Kimlin MG, Keogh LA, Dobbinson SJ, Kirk J, Kanetsky PA, Mann GJ, Cust AE. A pilot randomized controlled trial of the feasibility, acceptability, and impact of giving information on personalized genomic risk of melanoma to the public. Cancer Epidemiol Biomarkers Prev 2017;26:212–21.

[185] Childers KK, Maggard-Gibbons M, Macinko J, Childers CP. National distribution of cancer genetic testing in the United States: evidence for a gender disparity in hereditary breast and ovarian cancer. JAMA Oncol 2018;4:876–9.

[186] Dellefave-Castillo LM, Puckelwartz MJ, McNally EM. Reducing racial/ethnic disparities in cardiovascular genetic testing. JAMA Cardiol 2018;3:277–9.

[187] Piwek L, Ellis DA, Andrews S, Joinson A. The rise of consumer health wearables: promises and barriers. PLoS Med 2016;13:e1001953.

[188] Buller DB, Berwick M, Lantz K, Buller MK, Shane J, Kane I, Liu X. Smartphone mobile application delivering personalized, real-time sun protection advice: a randomized clinical trial. JAMA Dermatol 2015;151:497–504.

[189] Lee JA, Choi M, Lee SA, Jiang N. Effective behavioral intervention strategies using mobile health applications for chronic disease management: a systematic review. BMC Med Inform Decis Mak 2018;18:12.

[190] Low CA, Dey AK, Ferreira D, Kamarck T, Sun W, Bae S, Doryab A. Estimation of symptom severity during chemotherapy from passively sensed data: exploratory study. J Med Internet Res 2017;19(12):e420.

[191] Murphy J, Holmes J, Brooks C. Measurements of daily energy intake and total energy expenditure in people with dementia in care homes: the use of wearable technology. J Nutr Health Aging 2017;21:927–32.

[192] Rathbone AL, Prescott J. The use of mobile apps and SMS messaging as physical and mental health interventions: systematic review. J Med Internet Res 2017;19:e295.
[193] Whitehead L, Seaton P. The effectiveness of self-management mobile phone and tablet apps in long-term condition management: a systematic review. J Med Internet Res 2016;18:e97.
[194] Zhan A, Mohan S, Tarolli C, et al. Using smartphones and machine learning to quantify Parkinson disease severity: the mobile Parkinson disease score. JAMA Neurol 2018;75(7):876–80.
[195] Shcherbina A, Mattsson CM, Waggott D, Salisbury H, Christle JW, Hastie T, Wheeler MT, Ashley EA. Accuracy in wrist-worn, sensor-based measurements of heart rate and energy expenditure in a diverse cohort. J Pers Med 2017;7.
[196] Mercer K, Giangregorio L, Schneider E, Chilana P, Li M, Grindrod K. Acceptance of commercially available wearable activity trackers among adults aged over 50 and with chronic illness: a mixed-methods evaluation. JMIR Mhealth Uhealth 2016;4:e7.
[197] Ajami S, Teimouri F. Features and application of wearable biosensors in medical care. J Res Med Sci 2015;20(12):1208–15.
[198] Chai PR, Rosen RK, Boyer EW. Ingestible biosensors for real-time medical adherence monitoring: MyTMed. Proc Annu Hawaii Int Conf Syst Sci 2016;2016:3416–23.
[199] Karunanithi M, Zhang Q. An innovative technology to support independent living: the smarter safer homes platform. Stud Health Technol Inform 2018;246:102–10.
[200] Welch BM, Wiley K, Pflieger L, Achiangia R, Baker K, Hughes-Halbert C, Morrison H, Schiffman J, Doerr M. Review and comparison of electronic patient-facing family health history tools. J Genet Counsel 2018.
[201] Ashida S, Kaphingst KA, Goodman M, Schafer EJ. Family health history communication networks of older adults: importance of social relationships and disease perceptions. Health Educ Behav 2013;40:612–9.
[202] Ashida S, Schafer EJ. Family health information sharing among older adults: reaching more family members. J Community Genet 2015;6:17–27.
[203] Chen LS, Li M, Talwar D, Xu L, Zhao M. Chinese Americans' views and use of family health history: a qualitative study. PLoS One 2016;11:e0162706.
[204] Hughes Halbert C, Welch B, Lynch C, Magwood G, Rice L, Jefferson M, Riley J. Social determinants of family health history collection. J Community Genet 2016;7:57–64.
[205] Koehly LM, Ashida S, Goergen AF, Skapinsky KF, Hadley DW, Wilkinson AV. Willingness of Mexican-American adults to share family health history with healthcare providers. Am J Prev Med 2011;40:633–6.
[206] Thompson T, Seo J, Griffith J, Baxter M, James A, Kaphingst KA. The context of collecting family health history: examining definitions of family and family communication about health among African American women. J Health Commun 2015;20:416–23.
[207] Underwood SM, Kelber S. Enhancing the collection, discussion and use of family health history by consumers, nurses and other health care providers: because family health history matters. Nurs Clin N Am 2015;50:509–29.
[208] Conway-Pearson LS, Christensen KD, Savage SK, Huntington NL, Weitzman ER, Ziniel SI, Bacon P, Cacioppo CN, Green RC, Holm IA. Family health history reporting is sensitive to small changes in wording. Genet Med 2016;18:1308–11.
[209] Wang C, Sen A, Plegue M, Ruffin MTt, O'Neill SM, Rubinstein WS, Acheson LS. Impact of family history assessment on communication with family members and health care providers: a report from the Family Healthware Impact Trial (FHITr). Prev Med 2015;77:28–34.
[210] Beadles CA, Ryanne Wu R, Himmel T, Buchanan AH, Powell KP, Hauser E, Henrich VC, Ginsburg GS, Orlando LA. Providing patient education: impact on quantity and quality of family health history collection. Fam Cancer 2014;13:325–32.
[211] Wu RR, Orlando LA, Himmel TL, Buchanan AH, Powell KP, Hauser ER, Agbaje AB, Henrich VC, Ginsburg GS. Patient and primary care provider experience using a family health history collection, risk stratification, and clinical decision support tool: a type 2 hybrid controlled implementation-effectiveness trial. BMC Fam Pract 2013;14:111.
[212] Wu RR, Myers RA, McCarty CA, Dimmock D, Farrell M, Cross D, Chinevere TD, Ginsburg GS, Orlando LA, Family Health History N. Protocol for the "Implementation, adoption, and utility of family history in diverse care settings" study. Implement Sci 2015;10:163.
[213] Feero WG, Facio FM, Glogowski EA, Hampel HL, Stopfer JE, Eidem H, Pizzino AM, Barton DK, Biesecker LG. Preliminary validation of a consumer-oriented colorectal cancer risk assessment tool compatible with the US Surgeon General's my family health portrait. Genet Med 2015;17:753–6.
[214] Eccleston A, Bentley A, Dyer M, Strydom A, Vereecken W, George A, Rahman N. A cost-effectiveness evaluation of germline BRCA1 and BRCA2 testing in UK women with ovarian cancer. Value Health 2017;20(4):567–76.
[215] Kazi DS, Garber AM, Shah RU, Dudley RA, Mell MW, Rhee C, Moshkevich S, Boothroyd DB, Owens DK, Hlatky MA. Cost-effectiveness of genotype-guided and dual antiplatelet therapies in acute coronary syndrome. Ann Intern Med 2014;160:221–32.

[216] Li Y, Arellano AR, Bare LA, Bender RA, Strom CM, Devlin JJ. A multigene test could cost-effectively help extend life expectancy for women at risk of hereditary breast cancer. Value Health 2017;20:547–55.

[217] Yuen T, Carter MT, Szatmari P, Ungar WJ. Cost-effectiveness of genome and exome sequencing in children diagnosed with autism spectrum disorder. Appl Health Econ Health Policy 2018.

[218] Caudle KE, Rettie AE, Whirl-Carrillo M, Smith LH, Mintzer S, Lee MT, Klein TE, Callaghan JT, Clinical Pharmacogenetics Implementation, C. Clinical pharmacogenetics implementation consortium guidelines for CYP2C9 and HLA-B genotypes and phenytoin dosing. Clin Pharmacol Ther 2014;96:542–8.

[219] Plumpton CO, Alfirevic A, Pirmohamed M, Hughes DA. Cost effectiveness analysis of HLA-B*58:01 genotyping prior to initiation of allopurinol for gout. Rheumatology 2017;56:1729–39.

[220] Alfares AA, Kelly MA, McDermott G, Funke BH, Lebo MS, Baxter SB, Shen J, McLaughlin HM, Clark EH, Babb LJ, Cox SW, DePalma SR, Ho CY, Seidman JG, Seidman CE, Rehm HL. Results of clinical genetic testing of 2,912 probands with hypertrophic cardiomyopathy: expanded panels offer limited additional sensitivity. Genet Med 2015;17(11):880–8.

[221] Dong OM, Li A, Suzuki O, Oni-Orisan A, Gonzalez R, Stouffer GA, Lee CR, Wiltshire T. Projected impact of a multigene pharmacogenetic test to optimize medication prescribing in cardiovascular patients. Pharmacogenomics 2018.

[222] Karakaya M, Storbeck M, Strathmann EA, Vedove AD, Holker I, Altmueller J, Naghiyeva L, Schmitz-Steinkruger L, Vezyroglou K, Motameny S, Alawbathani S, Thiele H, Polat AI, Okur D, Boostani R, Karimiani EG, Wunderlich G, Ardicli D, Topaloglu H, Kirschner J, Schrank B, Maroofian R, Magnusson O, Yis U, Nurnberg P, Heller R, Wirth B. Targeted sequencing with expanded gene profile enables high diagnostic yield in non-5q-spinal muscular atrophies. Hum Mutat 2018.

[223] Carroll JC, Makuwaza T, Manca DP, Sopcak N, Permaul JA, O'Brien MA, Heisey R, Eisenhauer EA, Easley J, Krzyzanowska MK, Miedema B, Pruthi S, Sawka C, Schneider N, Sussman J, Urquhart R, Versaevel C, Grunfeld E. Primary care providers' experiences with and perceptions of personalized genomic medicine. Can Fam Physician 2016;62:e626–35.

[224] Chan WV, Johnson JA, Wilson RD, Metcalfe A. Obstetrical provider knowledge and attitudes towards cell-free DNA screening: results of a cross-sectional national survey. BMC Pregnancy Childbirth 2018;18:40.

[225] Hamilton JG, Abdiwahab E, Edwards HM, Fang ML, Jdayani A, Breslau ES. Primary care providers' cancer genetic testing-related knowledge, attitudes, and communication behaviors: a systematic review and research agenda. J Gen Intern Med 2017;32:315–24.

[226] Johnson LM, Valdez JM, Quinn EA, Sykes AD, McGee RB, Nuccio R, Hines-Dowell SJ, Baker JN, Kesserwan C, Nichols KE, Mandrell BN. Integrating next-generation sequencing into pediatric oncology practice: an assessment of physician confidence and understanding of clinical genomics. Cancer 2017;123:2352–9.

[227] Schully SD, Lam TK, Dotson WD, Chang CQ, Aronson N, Birkeland ML, Brewster SJ, Boccia S, Buchanan AH, Calonge N, Calzone K, Djulbegovic B, Goddard KA, Klein RD, Klein TE, Lau J, Long R, Lyman GH, Morgan RL, Palmer CG, Relling MV, Rubinstein WS, Swen JJ, Terry SF, Williams MS, Khoury MJ. Evidence synthesis and guideline development in genomic medicine: current status and future prospects. Genet Med 2014.

[228] Sperber NR, Carpenter JS, Cavallari LHLJD, Cooper-DeHoff RM, Denny JC, Ginsburg GS, Guan Y, Horowitz CR, Levy KD, Levy MA, Madden EB, Matheny ME, Pollin TI, Pratt VM, Rosenman M, Voils CIKWW, Wilke RA, Ryanne Wu R, Orlando LA. Challenges and strategies for implementing genomic services in diverse settings: experiences from the Implementing GeNomics in pracTicE (IGNITE) network. BMC Med Genom 2017;10:35.

[229] Gammon BL, Kraft SA, Michie M, Allyse M. "I think we've got too many tests!": prenatal providers' reflections on ethical and clinical challenges in the practice integration of cell-free DNA screening. Ethics Med Public Health 2016;2:334–42.

[230] Filipski KK, Pacanowski MA, Ramamoorthy A, Feero WG, Freedman AN. Dosing recommendations for pharmacogenetic interactions related to drug metabolism. Pharmacogenet Genom 2016;26:334–9.

[231] Hartzler A, McCarty CA, Rasmussen LV, Williams MS, Brilliant M, Bowton EA, Clayton EW, Faucett WA, Ferryman K, Field JR, Fullerton SM, Horowitz CR, Koenig BA, McCormick JB, Ralston JD, Sanderson SC, Smith ME, Trinidad SB. Stakeholder engagement: a key component of integrating genomic information into electronic health records. Genet Med 2013;15:792–801.

[232] Sitapati A, Kim H, Berkovich B, Marmor R, Singh S, El-Kareh R, Clay B, Ohno-Machado L. Integrated precision medicine: the role of electronic health records in delivering personalized treatment. Wiley Interdiscip Rev Syst Biol Med 2017;9.

[233] Spanakis EG, Santana S, Tsiknakis M, Marias K, Sakkalis V, Teixeira A, Janssen JH, de Jong H, Tziraki C. Technology-based innovations to foster personalized healthy lifestyles and well-being: a targeted review. J Med Internet Res 2016;18:e128.
[234] Aronson SJ, Clark EH, Babb LJ, Baxter S, Farwell LM, Funke BH, Hernandez AL, Joshi VA, Lyon E, Parthum AR, Russell FJ, Varugheese M, Venman TC, Rehm HL. The GeneInsight suite: a platform to support laboratory and provider use of DNA-based genetic testing. Hum Mutat 2011;32:532–6.
[235] Caraballo PJ, Bielinski SJ, St Sauver JL, Weinshilboum RM. Electronic medical record-integrated pharmacogenomics and related clinical decision support concepts. Clin Pharmacol Ther 2017;102:254–64.
[236] Freimuth RR, Formea CM, Hoffman JM, Matey E, Peterson JF, Boyce RD. Implementing genomic clinical decision support for drug-based precision medicine. CPT Pharmacometrics Syst Pharmacol 2017;6(3):153–5.
[237] Melton BL, Zillich AJ, Saleem J, Russ AL, Tisdale JE, Overholser BR. Iterative development and evaluation of a pharmacogenomic-guided clinical decision support system for warfarin dosing. Appl Clin Inform 2016;7:1088–106.
[238] Rohrer Vitek CR, Abul-Husn NS, Connolly JJ, Hartzler AL, Kitchner T, Peterson JF, Rasmussen LV, Smith ME, Stallings S, Williams MS, Wolf WA, Prows CA. Healthcare provider education to support integration of pharmacogenomics in practice: the eMERGE Network experience. Pharmacogenomics 2017;18:1013–25.
[239] Vermeulen E, Henneman L, van El CG, Cornel MC. Public attitudes towards preventive genomics and personal interest in genetic testing to prevent disease: a survey study. Eur J Public Health 2014;24:768–75.
[240] Chapman R, Likhanov M, Selita F, Zakharov I, Smith-Woolley E, Kovas Y. New literacy challenge for the twenty-first century: genetic knowledge is poor even among well educated. J Community Genet 2018.
[241] Haga SB, Barry WT, Mills R, Ginsburg GS, Svetkey L, Sullivan J, Willard HF. Public knowledge of and attitudes toward genetics and genetic testing. Genet Test Mol Biomarkers 2013;17:327–35.
[242] Frost CJ, Andrulis IL, Buys SS, Hopper JL, John EM, Terry MB, Bradbury A, Chung WK, Colbath K, Quintana N, Gamarra E, Egleston B, Galpern N, Bealin L, Glendon G, Miller LP, Daly MB. Assessing patient readiness for personalized genomic medicine. J Community Genet 2018.
[243] Buchanan AH, Rahm AK, Williams JL. Alternate service delivery models in cancer genetic counseling: a mini-review. Front Oncol 2016;6:120.
[244] Vrecar I, Hristovski D, Peterlin B. Telegenetics: an update on availability and use of telemedicine in clinical genetics service. J Med Syst 2017;41:21.
[245] Haga SB, Mills R, Pollak KI, Rehder C, Buchanan AH, Lipkus IM, Crow JH, Datto M. Developing patient-friendly genetic and genomic test reports: formats to promote patient engagement and understanding. Genome Med 2014;6.
[246] Gammon BL, Otto L, Wick M, Borowski K, Allyse M. Implementing group prenatal counseling for expanded noninvasive screening options. J Genet Counsel 2017;27(4):894–901.

3

# Nature and Frequency of Genetic Disease

*Bruce R. Korf[1], Reed E. Pyeritz[2], Wayne W. Grody[3]*

[1]Department of Genetics, University of Alabama at Birmingham, Birmingham, AL, United States
[2]Perelman School of Medicine at the University of Pennsylvania, Philadelphia, PA, United States
[3]UCLA School of Medicine, Los Angeles, CA, United States

## 3.1 INTRODUCTION

Genes are major determinants of human variation. Genome sequencing studies have shown that an individual genome contains millions of differences from the reference genome, some of which occur in coding or regulatory sequences that may have a phenotypic effect. Any individual is heterozygous for 50–100 variants that have been associated with genetic disorders [1]. The de novo mutation rate in the human genome is on the order of $10^{-8}$ per nucleotide per generation [2–4]. The phenotypic effect of genetically determined characteristics spans a continuum and may be nil at one extreme or lethal at the other. In determining a trait, one or two alleles may be all important, but more commonly genes interact with one another and with environmental factors. This interaction between genetic and environmental factors is now apparent for numerous conditions and includes many previously believed to have a purely environmental etiology. For example, genetically determined susceptibility has now been identified for several infections [5], for many drug-induced effects, and for several carcinogens (e.g., bladder cancer in aniline dye workers who are slow acetylators [6]). We suspect that such interactions are commonplace and that relatively few conditions are solely environmental in causation. In addition, to the environment, interactions with the microbiome, epigenetic modifications resulting in altered gene expression, and chance all influence how the genome translates into a phenotype.

## 3.2 FREQUENCY OF GENETIC DISEASE

For definitions of the types and frequency of genetic disorders, we use the currently available information, with the proviso that these are all the minimum frequencies and based on imperfect categorization. Interpretation of what constitutes a genetic disorder will depend on the situation (e.g., red-green color blindness may be a serious disability to the hunter-gatherer) or on the public perception (e.g., persons with albinism are considered blessed in some populations). Hence, the distinction between normal variation and disease is blurred and variable criteria have been used in different surveys. The situation is further complicated by the continued delineation of new phenotypic subtypes and by population variations in the frequencies of different genetic disorders [7]. Thus, most of the available data pertain to specific conditions either in a sample of the general population or in specific populations (e.g., the learning disabled); extrapolation to the overall population is difficult.

### 3.2.1 Chromosomal Disorders

A chromosomal disorder is classically defined as the phenotype resulting from visible alteration in the number or structure of the chromosomes. Using routine

Emery and Rimoin's Principles and Practice of Medical Genetics and Genomics: Foundations, https://doi.org/10.1016/B978-0-12-812537-3.00003-2

light microscopy and a moderate level of chromosome banding, the frequency of balanced and unbalanced structural rearrangements in newborns has been estimated at about 9.2:1000 [8]. Some of those with unbalanced rearrangements will have congenital anomalies and/or intellectual disabilities. A proportion of those with balanced changes will, in adult life, be at increased risk of either miscarriage or having a disabled child. The incidence of aneuploidy in newborns is about 3:1000, but the frequency increases dramatically among stillbirths or in spontaneous abortions [9]. It is estimated that one in two conceptuses has a chromosome abnormality, usually resulting in miscarriage [10]. Different types of chromosomal abnormalities predominate in spontaneous abortions compared with live-born infants. For example, trisomy 16 is the most common autosomal trisomy in abortions [11], whereas trisomies for chromosomes 21, 18, and 13 are the only autosomal trisomies occurring at appreciable frequencies in liveborn infants. Monosomy for the X chromosome (45,X) occurs in about 1% of all conceptions, but 98% of those affected do not reach term. Triploidy is also frequent in abortions but is exceptional in newborns. The high frequency of chromosomally abnormal conceptions is mirrored by results of chromosome analysis in gametes, which reveal an approximate abnormality rate of 4%–5% in sperm [12] and 12%–15% in oocytes [13]. When a family experiences three or more miscarriages, one parent is identified with an autosomal chromosome abnormality in 8.5% of analyses [14].

Routine light microscopy cannot resolve small amounts of missing or additional material (less than 4 Mb of DNA). The advent of cytogenomic microarray analysis has revealed a high frequency of submicroscopic deletions and duplications and other copy number variations, including both apparently benign and pathological changes [15]. Multiple microdeletion and microduplication disorders have been defined in recent years, and undoubtedly more await discovery. Such microdeletions, which epistemologically link "chromosomal disorders" with single-gene disorders, account for a proportion of currently unexplained learning disability and multiple malformation syndromes [16,17].

### 3.2.2 Single-Gene Disorders

By definition, single-gene disorders arise as a result of variation in one or both alleles of a gene on an autosome or sex chromosome or in a mitochondrial gene. There have been many investigations into the overall frequency of single-gene disorders. Many early estimates were misleadingly low due to underascertainment, especially of late-onset disorders (e.g., familial hypercholesterolemia, adult polycystic kidney disease, and Huntington disease). Carter [18] reviewed the earlier literature and estimated an overall incidence of autosomal dominant traits of 7.0 in 1000 live births, of autosomal recessive traits of 2.5 in 1000 live births, and of X-linked disorders of 0.5 in 1000 live births. This gave a combined frequency of 10 in 1000 live births (1%). At that time, approximately 2500 single-gene disorders had been delineated. The number of recognized Mendelian phenotypes has since more than doubled, and these new entities include several particularly common conditions (e.g., familial breast cancer syndromes, with a combined estimated frequency of five in 1000; hereditary nonpolyposis colon cancer syndromes, with a combined frequency of five in 1000). In addition, new technologies for DNA analysis have revealed a higher-than-expected frequency of generally asymptomatic people with one or two variant alleles at a locus [1]. For example, up to 1% of the population has an allele for von Willebrand factor, but many of these people have few or no symptoms, so again there is the problem of the imprecise and variable boundary between a harmless variant and a clinically important one. Furthermore, DNA analysis has shown that for some disorders, such as fragile X syndrome, relatives of an affected individual may harbor a premutation that has the potential for expansion to a full deleterious mutation in an offspring. The prevalence of such premutation carriers may be as high as one in 178 females for fragile X syndrome [19], and female heterozygotes may present with premature ovarian failure.

The frequencies of many single-gene disorders show population variation. Geographic variation may be explained by selection, by founder effects, or attributed to random genetic drift. Selection has resulted in a carrier frequency of one in three for sickle cell anemia in parts of equatorial Africa, and the Afrikaners of South Africa have a high frequency of variegate porphyria and familial hypercholesterolemia due to a founder effect. The carrier frequency for mutations in *HFE*, one of the genes responsible for hemochromatosis, is one in 10 in individuals of Celtic ancestry. As another example, heterozygosity for a mutation in *F11* associated with deficiency of coagulation factor 11 causes a bleeding disorder of variable severity. The prevalence of heterozygous mutations varies markedly among populations, with Ashkenazi Jews having 9%–12%, East Asians having 0.45%, and Finns having 0.03% [20]. Undoubtedly, more single-gene disorders are going to be delineated. In theory, at least one per locus will eventually be recognized (about 20,000) minus those with no or a mild phenotype and minus those incompatible with establishment/continuance of a pregnancy. Increasing the total are those loci for which different mutations cause entirely different phenotypes. For example, mutations in *LMNA* can result in at least 13 distinct disorders [21]. There is also overlap with the multifactorial category. For example, many patients with acute intermittent porphyria are asymptomatic in the absence of an environmental trigger, and epistatic involvement of other genes is believed to contribute to intrafamilial phenotypic variation for patients with the same mutation. As more gene–environment and gene–gene interactions are identified, the boundary between single-gene disorders and multifactorial disorders will become further blurred. Among children diagnosed with a developmental disorder, 40% had a confirmed genetic diagnosis. Consanguineous parents are more likely to have affected children (34% vs. 22%) [22].

### 3.2.3 Mitochondrial Disorders

Metabolic defects in the respiratory chain can be due to mutations in autosomal or X-linked genes or to mutations in the genes encoded by the mitochondrial chromosome (mtDNA). Because mitochondria are inherited from females, all their offspring are at risk for the defect in oxygenation present in mother. If all of mother's mtDNA carries the mutation (homoplasmy), then all of the offspring will as well. More frequently, only a fraction of female mtDNA carries the mutation (heteroplasmy), so the offspring will inherited variable proportions of mutant mtDNA and their clinical features will vary in severity [23].

### 3.2.4 Multifactorial Disorders

Multifactorial disorders result from an interaction of one or more genes with one or more environmental factors. Thus, in effect, the genetic contribution predisposes the individual to the actions of environmental agents. Such an interaction is suspected when conditions show an increased recurrence risk within families that does not reach the level of risk or pattern seen for single-gene disorders and when identical twin concordance exceeds that for nonidentical twins but is less than 100%. For most multifactorial disorders, however, the nature of the environmental agent(s) and the genetic predisposition are currently unclear and are the subject of intensive research efforts. The ability to conduct genome-wide association studies (GWAS) has accelerated progress in this area [24]. A relevant example is asthma.

Multifactorial disorders are believed to account for approximately one-half of all congenital malformations and to be relevant to many common chronic disorders of adulthood, including hypertension, rheumatoid arthritis, psychoses, and atherosclerosis (complex common disorders). The former group had an estimated frequency of 46.4 per 1000 in the British Columbia Health Surveillance Registry [25]. In addition, a multifactorial etiology is suspected for many common psychological disorders of childhood, including dyslexia (5%–10% of the population), specific language impairment (5% of children), and attention-deficit/hyperactivity disorder (4%–10% of children). Hence the multifactorial disorder category represents the most common type of genetic disorder in both children and adults. Spontaneous preterm birth is contributed to by maternal genetic predispositions, whose deleterious effects can be stimulated by external factors [26]. Multifactorial disorders also show considerable ethnic and geographic variation. For example, talipes equinovarus is about six times more common among Maoris than among Europeans [27], and neural tube defects were once 10 times more frequent in Ireland than in North America [28].

Often ignored, both intellectually and in research, are genotypes that reduce susceptibility to potentially harmful environmental factors. An understanding of alleles that provide protection from disease or increase longevity will yield insight into pathogenesis as well as novel approaches to therapy and prevention. Genome-wide association study (GWAS) is one approach to deciphering the genetic contributions to overt diseases as well as benign variations [29].

### 3.2.5 Somatic Cell Genetic Disorders

Somatic cell mutation is a natural developmental process in the immune system, but it is also responsible for a significant burden of genetic disease. This includes somatic or germline mosaicism for single-gene disorders [30], as well as mutations that give rise to cancer. Cancer cells tend to have accumulated multiple mutations; the first step in the cascade of mutations may be inherited (i.e., involving germ cells and all somatic cells). Carcinogens are important causes of noninherited mutations, and genetic susceptibility is suspected to account for individual variation in risk on exposure. Somatic cell genetic disorders might also be involved in other clinical conditions, such as autoimmune disorders and the aging process.

## 3.3 MORBIDITY AND MORTALITY DUE TO GENETIC DISEASE

The same general difficulties that pertain to the frequency estimates for genetic disorders also apply to estimates of the contribution of the various types of genetic disorders to morbidity and mortality during pregnancy, in childhood, and in adulthood. Hence these figures should be taken as minimum estimates.

### 3.3.1 Conception and Pregnancy

One in 15 recognized pregnancies spontaneously miscarry, and a higher percentage (up to 50%) of conceptions are lost before recognition of the pregnancy [10]. The majority of these losses are caused by numerical chromosomal abnormalities.

### 3.3.2 Childhood

Since the turn of the century in many developed countries, advances in medicine and public health have resulted in a gradual decline in the contribution of environmental factors to childhood morbidity and mortality. The result of these changes has been to raise genetic disorders to greater prominence. By contrast, in developing countries, nongenetic causes of childhood mortality continue to predominate. An idea of the contribution of genetic disease to morbidity can be judged from the prevalence of such diseases among pediatric inpatients. In reviewing 4115 inpatients, Hall and colleagues [31] found multifactorial disease in 22.1%, a single-gene disorder in 3.9%, and a chromosomal disorder in 0.6%. Thus, more than one in four pediatric inpatients has a genetic disorder in one of these categories, compared with the general population frequency estimate of one in 20 by age 25 years. This does not include morbidity that does not lead to inpatient admission. An updated survey by McCandless and colleagues [32] revealed that 71% of children admitted to the hospital had a disorder with a significant genetic component. Stevenson and Carey [33] found that 34.4% of deaths among children hospitalized in a tertiary care center could be attributed to congenital anomalies, of which 16.7% were due to chromosomal abnormality and 11.7% were due to a recognized malformation syndrome.

### 3.3.3 Adulthood

In developed countries, the most common causes of death are cancer and cardiovascular disease. All cancers are now known to have a cumulative somatic cell genetic basis, and there is evidence for a major genetic contribution to cardiovascular disease. Single-gene disorders causing diabetes or high blood pressure are relatively uncommon, but multifactorial inheritance accounts for a large proportion of patients with premature vascular disease and systemic hypertension. Similarly, there is a growing recognition of the importance of multifactorial inheritance for many other common disorders of adulthood responsible for both morbidity and mortality.

## REFERENCES


[1] Genomes Project C, Abecasis GR, Auton A, Brooks LD, DePristo MA, Durbin RM, Handsaker RE, Kang HM, Marth GT, McVean GA. An integrated map of genetic variation from 1,092 human genomes. Nature 2012;491:56–65.

[2] de Ligt J, Veltman JA, Vissers LE. Point mutations as a source of de novo genetic disease. Curr Opin Genet Dev 2013;23:257–63.

[3] Genomes Project C, Abecasis GR, Altshuler D, Auton A, Brooks LD, Durbin RM, Gibbs RA, Hurles ME, McVean GA. A map of human genome variation from population-scale sequencing. Nature 2010;467:1061–73.

[4] Veltman JA, Brunner HG. De novo mutations in human genetic disease. Nat Rev Genet 2012;13:565–75.

[5] Mozzi A, Pontremoli C, Sironi M. Genetic susceptibility to infectious diseases: current status and future perspectives from genome-wide approaches. Infect Genet Evol 2017. Sep 22. pii: S1567-1348(17):30334-9. https://doi.org/10.1016/j.meegid.2017.09.028. [Epub ahead of print]..

[6] An Y, Li H, Wang KJ, Liu XH, Qiu MX, Liao Y, Huang JL, Wang XS. Meta-analysis of the relationship between slow acetylation of N-acetyl transferase 2 and the risk of bladder cancer. Genet Mol Res 2015;14:16896–904.

[7] Verma IC, Puri RD. Global burden of genetic disease and the role of genetic screening. Semin Fetal Neonatal Med 2015;20:354–63.

[8] Jacobs PA, Browne C, Gregson N, Joyce C, White H. Estimates of the frequency of chromosome abnormalities detectable in unselected newborns using moderate levels of banding. J Med Genet 1992;29:103–8.

[9] Hassold T, Hunt P. To err (meiotically) is human: the genesis of human aneuploidy. Nat Rev Genet 2001;2:280–91.
[10] Boue J, Boue A, Lazar P. Retrospective and prospective epidemiological studies of 1500 karyotyped spontaneous human abortions. 1975. Birth Defects Res A Clin Mol Teratol 2013;97:471–86.
[11] Boue JG, Boue A. Chromosomal anomalies in early spontaneous abortion. (Their consequences on early embryogenesis and in vitro growth of embryonic cells). Curr Top Pathol 1976;62:193–208.
[12] Templado C, Vidal F, Estop A. Aneuploidy in human spermatozoa. Cytogenet Genome Res 2011;133:91–9.
[13] Rosenbusch B. The incidence of aneuploidy in human oocytes assessed by conventional cytogenetic analysis. Hereditas 2004;141:97–105.
[14] Ayed W, Messaoudi I, Belghith ZHammami W, Chemkhi I, Abidi N, Guermani H, Obay R, Amouri A. Chromosomal abnormalities in 163 Tunisian couples with recurrent miscarriages. Pan Afr Med J 2017;28:99–104.
[15] Miller DT, Adam MP, Aradhya S, Biesecker LG, Brothman AR, Carter NP, Church DM, Crolla JA, Eichler EE, Epstein CJ, Faucett WA, Feuk L, Friedman JM, Hamosh A, Jackson L, Kaminsky EB, Kok K, Krantz ID, Kuhn RM, Lee C, Ostell JM, Rosenberg C, Scherer SW, Spinner NB, Stavropoulos DJ, Tepperberg JH, Thorland EC, Vermeesch JR, Waggoner DJ, Watson MS, Martin CL, Ledbetter DH. Consensus statement: chromosomal microarray is a first-tier clinical diagnostic test for individuals with developmental disabilities or congenital anomalies. Am J Hum Genet 2010;86:749–64.
[16] Kirov G. CNVs in neuropsychiatric disorders. Hum Mol Genet 2015;24:R45–9.
[17] Mikhail FM. Copy number variations and human genetic disease. Curr Opin Pediatr 2014;26:646–52.
[18] Carter CO. Monogenic disorders. J Med Genet 1977;14:316–20.
[19] Hantash FM, Goos DM, Crossley B, Anderson B, Zhang K, Sun W, Strom CM. FMR1 premutation Carrier frequency in patients undergoing routine population-based Carrier screening: insights into the prevalence of fragile X syndrome, fragile X-associated tremor/ataxia syndrome, and fragile X-associated primary ovarian insufficiency in the United States. Genet Med 2011;13:39–45.
[20] Asselta R, Paraboschi EM, Rimoldi V, Menegatti M, Peyvandi F, Salomon O, Duga S. Exploring the global landscape of genetic variation in coagulation factor XI deficiency. Blood 2017;130:e1–6.
[21] Capell BC, Collins FS. Human laminopathies: nuclei gone genetically awry. Nat Rev Genet 2006;7:940–52.
[22] Best S, Rosser E, Bajaj M. Fifteen years of genetic testing from a London developmental clinic. Arch Dis Child 2017;102:1014–8.
[23] Davis RL, Liang C, Sue CM. Mitochondrial disease. Hanb Clin Neurol 2018;147:125–41.
[24] O'Rielly DD, Rahman P. Clinical genetic research 2: genetic epidemiology of complex phenotypes. Methods Mol Biol 2015;1281:349–67.
[25] Baird PA, Anderson TW, Newcombe HB, Lowry RB. Genetic disorders in children and young adults: a population study. Am J Hum Genet 1988;42:677–93.
[26] Strauss 3rd JF, Romero R, Gomez-Lopez N, Haymond-Thornburg H, Modi BP, Teves ME, Pearson LN, York TP, Schenkein HA. Spontaneous preterm birth: advances toward the discovery of genetic predisposition. Am J Obstet Gynecol 2017:32484–5. S0002-9378(17).
[27] Cartlidge I. Observations on the epidemiology of club foot in Polynesian and Caucasian populations. J Med Genet 1984;21:290–2.
[28] McDonnell RJ, Johnson Z, Delaney V, Dack P. East Ireland 1980-1994: epidemiology of neural tube defects. J Epidemiol Community Health 1999;53:782–788.
[29] Timpson NJ, Greenwood CMT, Soranzo N, Lawson DJ, Richards JB. Genetic architecture: the shape of the genetic contribution to human traits and disease. Nat Rev Genet 2018;19(2):110–124. https://doi.org/10.1038/nrg.2017.101. [Epub 2017 Dec 11. Review].
[30] Youssoufian H, Pyeritz RE. Mechanisms and consequences of somatic mosaicism in humans. Nat Rev Genet 2002;3:748–58.
[31] Hall JG, Powers EK, McLlvaine RT, Ean VH. The frequency and financial burden of genetic disease in a pediatric hospital. Am J Med Genet 1978;1:417–36.
[32] McCandless SE, Brunger JW, Cassidy SB. The burden of genetic disease on inpatient care in a children's hospital. Am J Hum Genet 2004;74:121–7.
[33] Stevenson DA, Carey JC. Contribution of malformations and genetic disorders to mortality in a children's hospital. Am J Med Genet A 2004;126A:393–7.

4

# Genome and Gene Structure*

*Madhuri R. Hegde*[1,2], *Michael R. Crowley*[1,2]

[1]Department of Human Genetics, Emory University, Atlanta, GA, United States
[2]The Department of Genetics, The University of Alabama at Birmingham, Birmingham, AL, United States

The past decade of biological research has focused heavily on the human genome, and the Human Genome Project has had a significant impact on biomedical research. Our genetic material is encoded in two genomes: nuclear and mitochondrial. Both genomes reflect the molecular evolution of humans, which started about 4.5 billion years ago. The function of the human genome is to transfer information reliably from parent to daughter cells and from one generation to the next. At particular developmental times and in specific tissues, the transcriptional machinery initiates programmed patterns of gene expression that are dictated by chromatin structure and the activities of transcriptional regulatory factors. Gene expression is followed by the processes of splicing, translation, and protein localization, ultimately leading to synthesis of the protein and RNA molecules that mediate cellular function. Variability in genome structure, including single nucleotide differences and larger-scale variations at the genome level between humans, dictates the traits we manifest as well as the diseases to which individuals are predisposed.

## 4.1 INTRODUCTION: COMPOSITION OF THE NUCLEAR HUMAN GENOME

The present assembly of human DNA sequence contains approximately 3.1 billion bp, which covers most of the nonheterochromatic portions of the genome and contains about 250 gaps. The 3.2 Mbp are packed into 22 pairs of chromosomes and two sex chromosomes, X and Y. The human chromosomes are not equal sizes; the smallest, chromosome 21, is 54 Mbp long, and the largest, chromosome 1, is about 249 Mbp long (Fig. 4.1 and Table 4.1). From a functional point of view, the genomic sequences are distinguished by genes, pseudogenes, and noncoding DNA, and only a minute fraction of the sequences code for proteins (approximately 2%). There are many pseudogenes (0.5%) but most of the genome consists of introns and intergenic DNA. Almost half of intergenic sequences consist of transposons. Gene clusters have evolved from several duplication events in deep evolutionary time, and these include clusters such as the *HOX* and *globin* clusters. The chemical structure of genetic material as well as the storage, processing, and transfer of genetic information from one generation to the next are similar in all living organisms. Thus, it was expected that the complexity of the human phenotype would be explained by a significantly higher number of genes in humans compared with simpler organisms. Surprisingly, instead of the predicted 50,000–150,000 human genes, the sequence of the human genome revealed about 20,000–25,000 genes, similar to the number of genes in many other organisms. However, analysis of the genome and its products has revealed complexity in the form of exquisite temporal and spatial regulation of gene alternative transcripts,

*This article is a revision of the previous edition article by David H. Cohen, vol. 1, pp. 61–80, © 2007, Elsevier Ltd.

Emery and Rimoin's Principles and Practice of Medical Genetics and Genomics: Foundations, https://doi.org/10.1016/B978-0-12-812537-3.00004-4

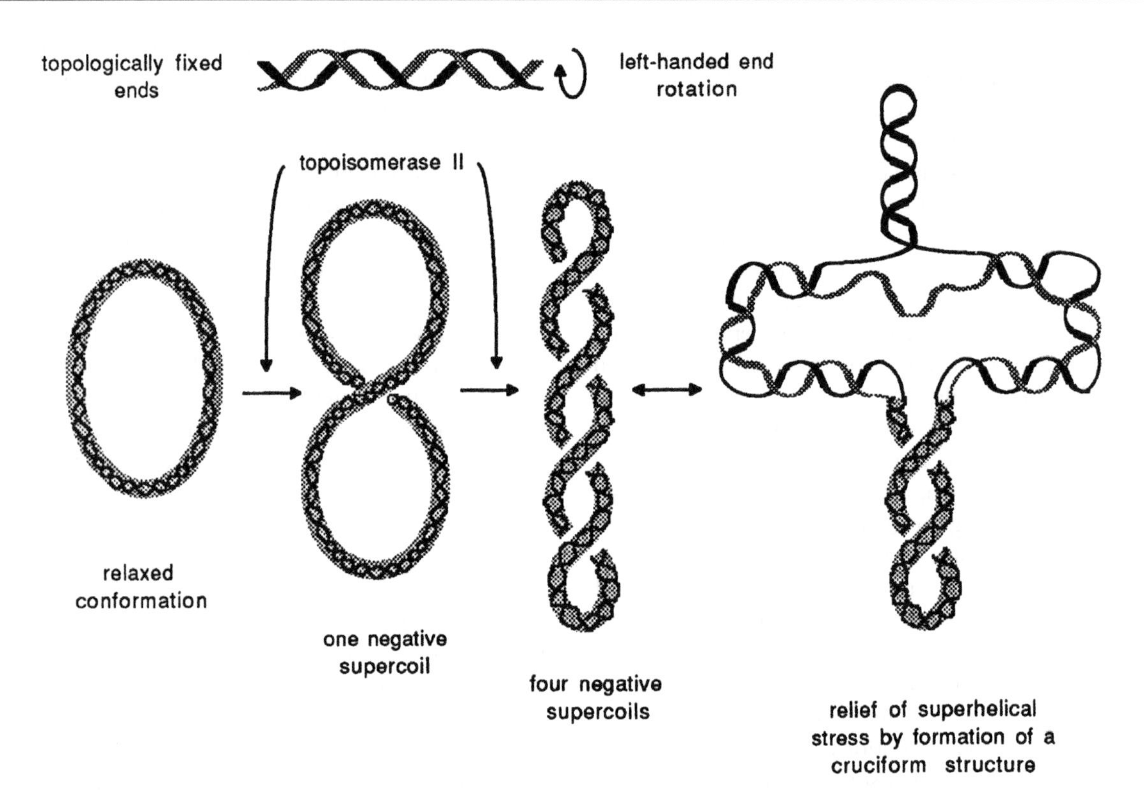


Figure 4.5 Superhelical turns in DNA.

nucleus, is achieved by coiling and folding the double helix into a series of progressively shorter and thicker structures (Fig. 4.5). Proteins that bind to DNA help direct and organize this folding, and the folded complex of DNA and protein is referred to as chromatin.

### 4.3.1 Nucleosomes and Higher Order Chromatin Structure

In addition to compacting the genetic material to fit into the nucleus, chromatin condensation can regulate accessibility of the DNA for transcription and other processes. The simplest level of chromatin structure is the organization of DNA and histones into nucleosomes [2]. Each nucleosome is a 147-bp-long segment of DNA tightly wrapped almost two times around an octamer histone core. This octamer core contains two molecules each of the histones H2A, H2B, H3, and H4. Nucleosomes are the fundamental feature of all eukaryotic DNA, and the sequences of the core histones are well conserved among even, distantly related species. A fifth histone, H1, binds to a particular modification on the core histone H3 at lysine 9 and its sequence is less well conserved. A region of linker DNA, about 60 bp in length in humans, usually separates adjacent nucleosomes, so that nucleosomes are for the most part regularly spaced.

Nucleosomes represent the first level in the packaging of naked DNA into chromatin and appear in the electron microscope as strings of 11-nm "beads." Previous models have the next level in packaging as the coiling of the nucleosomes to form a 30-nm structure with a solenoid conformation. However, recent data suggest the 30-nm solenoid structure is an artifact of X-ray diffraction studies of chromatin in vitro and does not reflect chromatin organization in the nucleus. Indeed, studies by Maeshima and colleagues [3] failed to find the 30-nm structure using small-angle X-ray scattering on purified nuclei or chromosomes in the absence of contaminating ribosomes. More recently, a novel method of decorating the chromatin with photoactivatable long polymers of diaminobenzidine (DAB), staining with $OsO_4$, and visualization by multi-tilt scanning electron microscopy tomography reveals in situ chromatin to be disordered polymers of 5 and 24 nm with no evidence of the 30-nm fibers (Fig. 4.6) [4]. Epigenetic modifications,

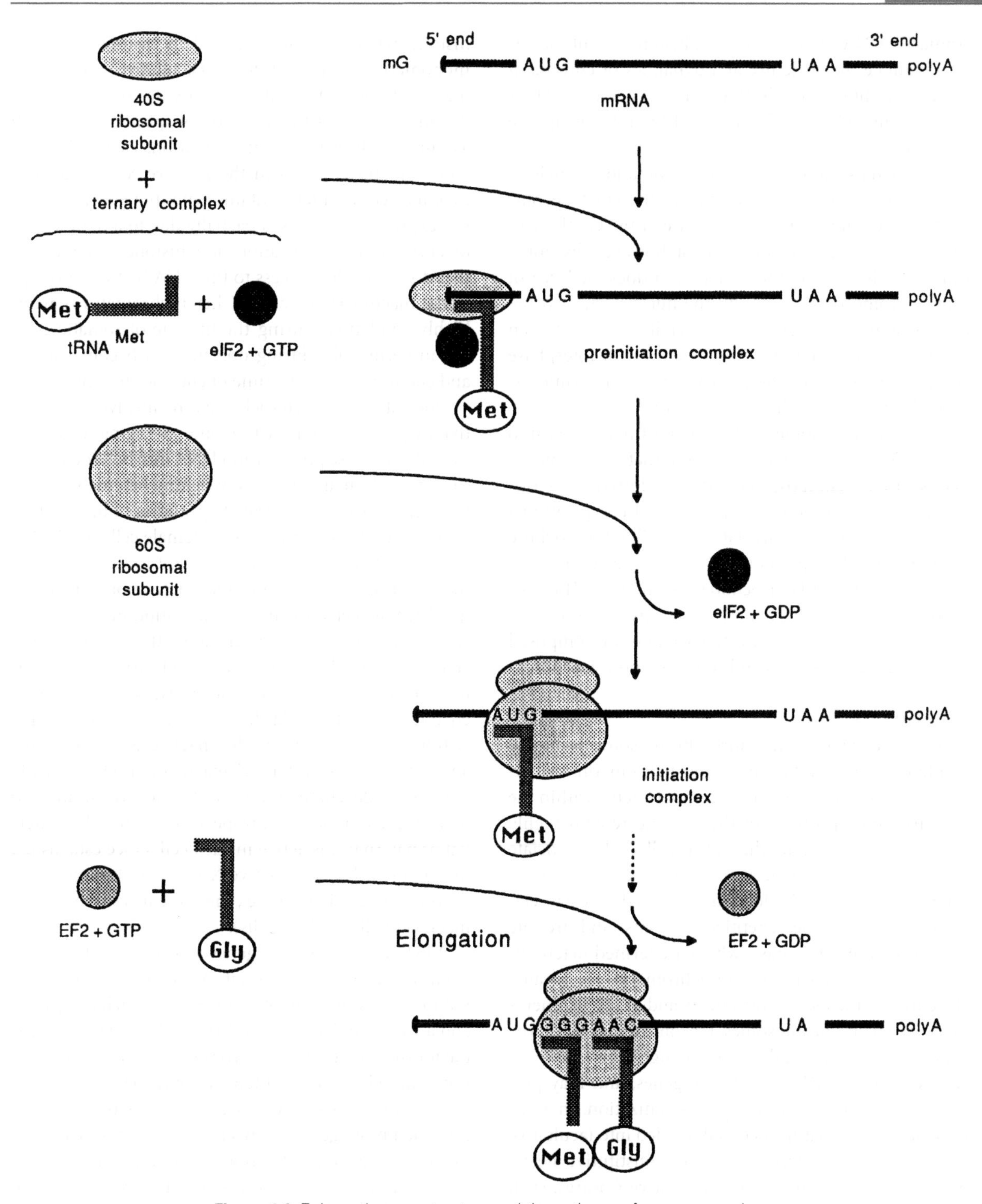


**Figure 4.6** Eukaryotic gene structure and the pathway of gene expression.

DNA replication, and reinserted into the genome in a new location. These transposons use a cut-and-paste mechanism for moving through genomes.

**Pseudogenes (full and partial)** are regions of DNA with many sequence elements of a potential transcriptional unit (e.g., promoter, protein-coding region, splice junctions, etc.), yet do not code for a functional product. They can originate after gene duplication when the duplicated sequence acquires a mutation that prevents its expression. For example, a member of the α-globin gene family, ψζ, has all the sequence characteristics of a functional globin gene, but the protein-coding region contains a point mutation that prevents the expression of a full-length globin [12]. A second way in which pseudogenes originate is via the pathway of reverse transcription and integration. If the mRNA of a cellular gene is converted into complementary DNA by reverse transcriptase, a duplex DNA molecule can be formed that lacks introns and contains a poly(A) tract. Pseudogenes with this pattern are commonly found in genomic DNA, showing that cellular mRNAs are occasional substrates for reverse transcriptase and that the DNA products can integrate back into the genome.

**Large segmental duplications** [13] are especially enriched in pericentromeric and subtelomeric regions of chromosomes. These intrachromosomal or interchromosomal duplications range between 1 and 300 kb, are >90% identical at the sequence level, and are much more common in humans than in yeast, flies, or worms, suggesting a relatively recent origin for these genomic elements. Segmental duplications have been demonstrated to serve as templates for the production of copy number variants (CNVs) within the genome through different mechanisms (see later).

**Highly homologous sequences** can reside within the CDS of the same gene (intragenic homology; same gene), have homology to functional genes (different genes), and have homology to nonfunctional pseudogenes or homology to sequence regions that are still poorly annotated.

Satellite DNA consists of arrays of simple tandem repeats; microsatellites are composed of repeats primarily of 4 bp or less dispersed throughout the genome [14]. These cover about 0.5% of the genome and exist in the form of dinucleotide repeat CA/TG. Microsatellites are well known for their causative roles in as many as 40 neurological diseases. Certain specific triplet repeats can be unstable, expanding and contracting during meiosis and/or mitosis. If the repeat becomes excessively long, it can cause diseases such as Huntington disease or spinocerebellar ataxia, both of which are caused by the expanded repeat CAG in the coding region of a gene and result in a long polyglutamine tract within the gene product. Some expanded repeats occur in 5′- and 3′-UTRs and result in a disease due to an inhibitory effect on gene expression (Fig. 4.7).

Effects of microsatellites can occur at different levels of RNA, which includes alternative splicing, structural changes, and working as microRNAs (Fig. 4.7). Minisatellites are tandemly repeated sequences of DNA lengths from 1 to 15 kbp. The telomeric DNA sequences contain 10–15 kb of hexanucleotide repeats—TTAGGG. Macrosatellites are very long arrays of sequences up to hundreds of kilobases of tandemly repeated DNA. The α-satellite DNA constitutes the bulk of the centromeric heterochromatin on all chromosomes.

### 4.3.5 Gene Families

Genes belong to a family of closely related DNA sequences, which cluster together as families because of similarity in their nucleotide sequence or amino acid sequences. Gene families consist of structurally (and usually functionally) related genes with a common evolutionary origin. Multiple levels of hierarchical subfamily structure are common. A well-studied and extensively described example due to its clinical significance is the gene family consisting of genes that code for α- and β-globin gene clusters, which are assumed to have arisen from a gene duplication event 500 million years ago. This gene family is an example of duplication events due to retrotransposition. These two gene clusters also code for globin chains that are expressed during stages of human developmental stages. Another example of gene family with extensive diversity is the immunoglobin superfamily.

Some gene family members, such as collagens, are dispersed among different chromosomal locations, but many others, such as the κ or λ variable region genes, are physically linked. Among gene families that exhibit linkage, members are usually oriented in the same direction. In some cases, linkage is thought to be an evolutionary footprint without functional significance, suggesting that evolutionary divergence of the gene family occurred through successive rounds of duplication in tandem arrays. However, in other gene families, such as the immunoglobulins, linkage has been conserved

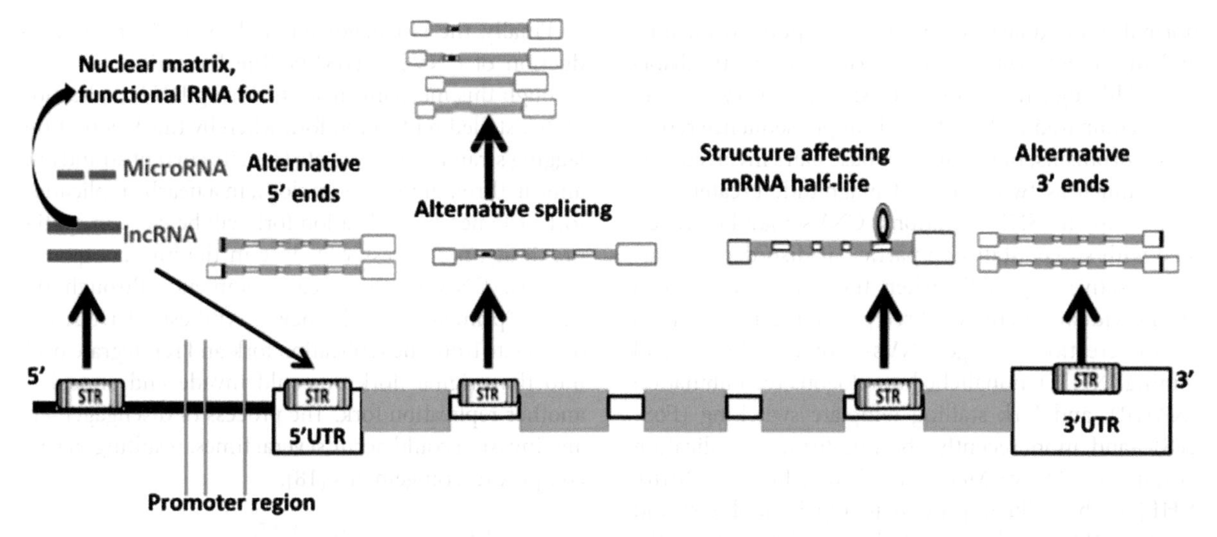


**Figure 4.7** Effects of microsatellites at the level of RNA. Long noncoding RNAs (lncRNAs) predominantly consisting of microsatellites have been observed to function in the nuclear matrix and to aggregate into nuclear foci with indications of functional significance. They also may associate with DNA microsatellites (short tandem repeat [STR]) in both UTRs. Microsatellite-dominated microRNAs have been observed, but their function is not yet known. Intronic microsatellites can regulate splicing efficiency that can lead to exon skipping, intron inclusion or new splice site selection. STRs located in the UTRs can influence the locations of the start and end sites of transcription. Microsatellites transcribed can also affect the mRNA half-life, which may be due to formation of secondary structures such as hairpins.

during evolution because it provides a mechanism for coordinated or regulated control of gene expression. Even when dispersed, coordinated expression of members of gene families can be regulated by similar control mechanisms by carrying similar response elements in their 5′ regulatory regions.

### 4.3.6 Interindividual Variations in the Human Genome

The Human Haplotype Map (HapMap) project is a key component in understanding the genetic potential of the Human Genome Project. Sequencing of the human genome has exposed multiple interindividual variations. These variants include both simple-sequence repeat polymorphisms and single nucleotide polymorphisms (SNPs). Repeat polymorphisms can represent di- (mostly CA), tri-, or tetranucleotide repeats, and they form the basis of the genetic map of the human genome. These repeat markers are multiallelic; the alleles differ in the number of repeat units and thus can be used to identify the maternal and paternal alleles of individuals as well as to define recombinations between marker loci. The high degree of length polymorphism among simple-sequence repeats within the human population is due to frequent slippage by DNA polymerase during replication. These repeats comprise about 3% of the human genome, and there is approximately one such repeat per 2 kb of genomic sequence. The large number and wide distribution of these repeats have facilitated mapping and identification of many genes associated with inherited human disorders solely based on their chromosomal position.

SNPs are also nonrandomly distributed in the human genome. Millions of SNPs have been identified in the genome sequence but only a small fraction is predicted to affect the protein sequence. This limits the extent to which such genetic variations contribute to the structural diversity of human polypeptides, but regulatory effects on gene expression may cause or result in susceptibility to a variety of human phenotypes.

Copy number variation (CNV) describes the variation identified within the population or associated with human diseases in genomic segments larger than the SNP and simple-sequence repeat polymorphisms but smaller than cytogenetically visible chromosomal abnormalities [15]. An appreciation of the number and diversity of such variants has primarily been a product of comparative genomic hybridization studies in normal individuals (copy number polymorphisms) and in patients with a wide variety of genetic disorders. Although the number of sites that vary is small when compared with SNPs and simple-sequence repeat polymorphisms, the number of base pairs involved may be as much as two orders of magnitude greater [16]. Similar to the SNP variations, CNVs may be associated with susceptibility to particular disorders or may be causative, especially when they arise de novo in an individual. There are three prevailing mechanisms for the creation of large CNVs—nonhomologous end joining (NHEJ), nonallelic homologous recombination (NAHR), and fork stalling template switching (FoSTeS)—and, more recently, aberrant firing of replication origins [17] (Maya-Mendoza, Nature. June 27, 2018). NHEJ is the cellular process to repair double-strand breaks (DBSs) and proceeds in four basic steps; the DBS is recognized, bridging of both broken DNA ends, preparation of the ends for ligation, and finally, ligation of the two DNA strands. NHEJ can result in small alterations within the genome such as microdeletions at the repair site, which may or may not have an effect on the organism. Moreover, NHEJ can lead to larger rearrangements such as chromosomal translocations, which have been associated with different types of cancer. NHEJ events are nonrecurrent types of CNVs and can, but not always, be found within low CNRs (SDs) (LCRs/SDs). NAHR events, on the other hand, are associated with LCRs/SDs and can be recurrent due to the presence of similar alterations in different individuals between the same LCR/SD. NAHR can occur during meiosis or mitosis and will result in either constitutive deletions/duplications that are associated with genetic abnormalities or sporadic deletion/duplication events that are mosaic or observed in cancers, respectively. Normal homologous recombination occurs between sister chromatids during meiosis to exchange genetic information or between sister chromosomes during mitosis. If the pairing of the different chromosomes occurs through allelic segments of the genome the resulting recombination has no phenotypic effect. If, however, the pairing occurs between two nonallelic segments of the genome, the result can be duplications or deletions if the LCRs/SDs are in a direct orientation or an inversion if the LCRs/SDs are in opposite orientations. Translocations can also occur if the recombination occurs between LRCs/SDs on different chromosomes.

Finally, the last major method involved in the production of CNVs is FoSTeS. The formation of CNVs through this mechanism starts during DNA replication with a stalled replication fork whereby the 3′ end of the lagging strand can dissociate from its current strand and anneal, through microhomology, in a nearby replication fork. The nearby replication fork will be close in spatial proximity, not necessarily close in the linear sequence context. DNA synthesis can commence through the newly "primed" site. The newly synthesized DNA can dissociate from the replication fork and reintegrate back into the original fork or could invade and anneal to another replication fork. The process of disengagement and invasion could occur several times, resulting in very complex rearrangements [18].

### 4.3.7 DNA Looping and TADs

It has become increasing clear in the past several years that the DNA in the nucleus has a higher order structure in the form of loops. As stated earlier, enhancer elements that lie either upstream or downstream of a gene's promoter can influence developmental timing of gene expression or tissue or cell type–specific gene expression. Enhancer sequences work in a position- and orientation-independent manner to affect gene transcription. Some early studies had suggested that the enhancer and the promoter form a loop in the DNA, bringing the proteins that bind both elements in close proximity to regulate transcription [19]. The locus control region (LCR) of the globin locus is a notable example where the LCR resides 50 kb upstream of the globin genes (α, β, and γ) themselves and regulates the switch between fetal hemoglobin and adult hemoglobin. More recently with the advent of next-generation sequencing (see later), it has become possible to assess contacts between regions of the genome in *cis* and in *trans* on a genome-wide scale. Indeed, through the use of a proximity-based ligation assay termed HiC (a process whereby DNA and proteins are crosslinked), the DNA is digested with a restriction enzyme, and the DNA within the complex is ligated, bringing together normally distant sequences. The ligation products are eventually sequenced with high throughput DNA sequencing); it has been shown that regions of chromatin that are in close proximity to each other have a higher probability to interact with each other, whereas more distant regions of chromatin have a lower probability of interaction. However, there are many regions of the genome that are separated by great distances on the linear chromatin fiber that preferentially interact with each other. These segments form large DNA loops. The prevailing hypothesis of how the loops form is through the loop extrusion model whereby the ring protein cohesin binds to and draws the DNA through itself until the complex encounters the protein CTCF, which is also bound to the DNA. These proteins demarcate most, although not all DNA loops. They also act as boundary elements restricting access to the loop. The loops have been found to persist between cell types and throughout vertebrate evolution and have been termed topologically associated domains (TADs) [20,21]. One example of how TADs are important for regulating gene transcription was the discovery that individuals with copy number alterations at 2q35-36, which caused malformation of the hands and feet, were associated with the deletion of boundary elements and not with mutation of individual genes [22]. The location of 2q35-36 contains the genes *WNT6*, *IHH* (*Indian hedgehog*), *EPHA4*, and *PAX3* separated into TADs with boundaries between the *WNT6/IHH* and *EPHA4* and between the *EPHA4* and PAX3 TADs. In other words, a *WNT6/IHH* TAD, an *EPHA4* TAD, and a *PAX3* TAD, each TAD separated by binding sites for the boundary proteins CTCF and cohesin. Importantly, the chromosome structure is conserved between mouse and humans at this locus. Modeling the human CNVs in the mouse resulted in similar phenotypes in the paws of the affected mice. Deletion of the boundary element, and hence the binding sites for the CTCF/cohesin complex, was responsible for the observed phenotypes in both humans and mice [22]. These results provide evidence that alterations in chromatin architecture can lead to phenotypic changes in the organism without mutations to the genes that reside within the loops.

### 4.3.8 Analysis of Genomes

During the past 10 years there has been an explosion of genomic data due primarily to the use of next-generation sequencing (NextGen or NGS). NGS, at its most basic level, is the process of sequence analysis of DNA (or cDNA) in a massively parallel configuration [23–25]. There are several competing concepts for NGS, but the most prevalent technology in use today is the Illumina Sequencing by Synthesis method. All NGS assays begin with the creation of a sequencing library. The process is basically the same for genomic sequencing and begins with shearing the DNA in some fashion, either by mechanical means (sonication) or enzymatically (tagmentation with a DNA transposon). Following fragmentation, the ends of the DNA are repaired to blunt ends and phosphorylated, and an adenosine residue attached to provide a more efficient substrate for ligation of platform-specific adaptors. (These steps are not necessary during tagmentation as oligomers are inserted into the DNA by the transposase.) Following adaptor ligation, the library is amplified by PCR and is ready for sequencing [23–25].

Sequencing by synthesis on the Illumina platform begins by creating individual sequencing reactions, known as clusters, on the surface of a flow cell. The prepared DNA libraries are denatured and flowed across the surface of the flow cell in order for the adaptor sequences to bind their complementary sequences that are physically attached to the flow cell surface. Once attached the molecules are amplified in situ by bridge amplification, and the resulting clusters are ready for sequencing. In the Illumina system, nucleotides, with a specific fluorescence for each base and a 3′ block, are added to the flow cell with the polymerase and one base is incorporated into the growing strand. The base is "read" by its unique color tag and the nucleotide is recorded. The 3′ block is removed from nucleotide and the cycle begins again. This is repeated from 50 cycles up to 600 cycles in either one direction (single end sequencing) or in both directions (paired end sequencing) depending on the instrument and length of chemistry. Once the sequencing is finished, the raw data file is converted to a usable format, the FASTQ file. The FASTQ file format is similar to a FASTA except it has the addition of run quality metrics for each base position within the read. The FASTQ files are used in downstream bioinformatics analysis of the sequencing run.

## 4.4 STRUCTURE OF GENES (TRANSCRIPTIONAL UNITS): EXONS AND mRNA

Sequences coding for a single eukaryotic mRNA molecule are typically separated by noncoding sequences into noncontiguous segments along the chromosomal DNA strand (Fig. 4.6). The segments that are retained in the mature mRNA are referred to as exons. During transcription, the exons are spliced together from a larger precursor RNA that contains, in addition to the exons, interspersed noncoding segments referred to as introns.

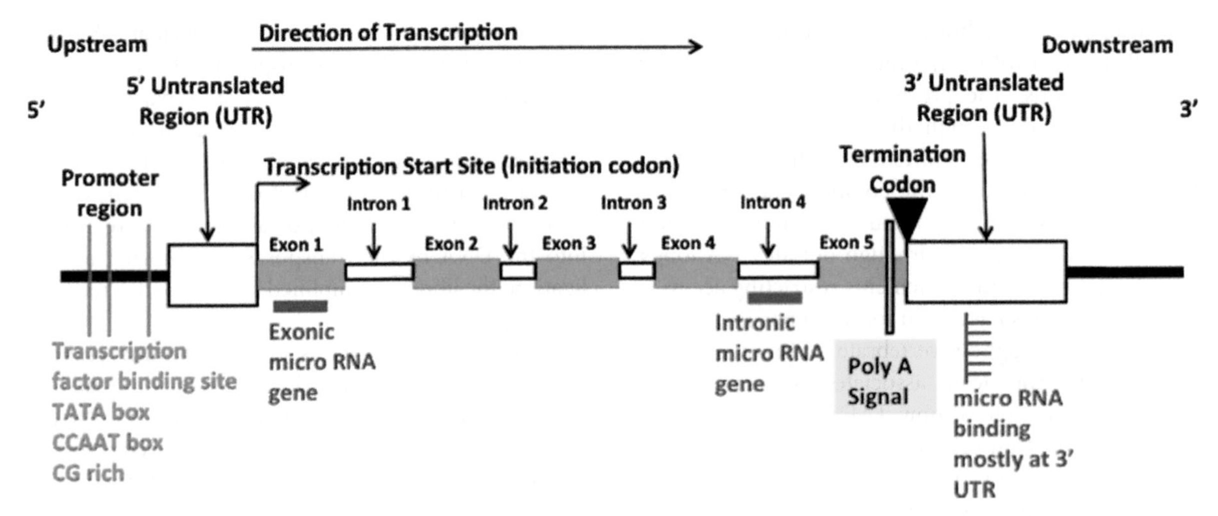


**Figure 4.8** Organization of a human gene. At the 5′ end (upstream) of each gene lies a promoter region that includes sequences (*red lines*) responsible for the proper initiation of transcription (TATA, CCAAT), including regulatory elements. At the 5′ and 3′ end of the gene is the untranslated region (5′-UTR and 3′-UTR; *white boxes*). The 3′UTR contains a signal for the addition of polyA tail (yellow) to the end of the mature mRNA. Gene expression can be regulated by binding of specific microRNAs (purple) in microRNA recognition sequences mostly present in the 3′-UTR. Genes (open reading frames) expressing untranslated microRNAs can also be embedded within exons (exonic microRNA; purple) or introns (intronic microRNA) such as that in immunoglobulin lambda variable region gene family (Das S. Mol Biol Evol 2009;26(5):1179–89.). The nucleotide sequences adjacent after the 5′-UTR or before the 3′-UTR provide the molecular "start" (Initiation codon) and "stop" (Termination codon, *black arrow*) signals respectively for the mRNA synthesis from the gene. Exons (light brown) and intervening introns (white) are the coding and noncoding parts of the gene.

The number of exons coding for a single mRNA molecule depends on the gene and the organism, but ranges from one to more than 100 (Fig. 4.8). The noncoding mRNA sequences are spliced out during mRNA maturation. Human genes tend to have small exons, with a median value of only 167 bp and mean equal to 216 bp. The shortest exon is only 12 bp while the longest is 6609 bp. The exons are separated by introns, which can be less than 100 bp, but can also exceed 10 kb. The size distribution of exons and introns of human genes based on the analyzed sequence information and comparison to worm and fly sequences are provided in Table 4.2. It is important to note that some introns carry significant information and even code for other complete genes (nested genes).

**TABLE 4.2 Physical Sizes of Human Chromosomes**

| | Size |
|---|---|
| Median exon | 167 bp |
| Longest exon | 6609 bp |
| Smallest exon | 12 bp |
| Exon number | 9 |
| Introns | 3300 bp |
| 3′-UTR | 770 bp |
| 5′-UTR | 300 bp |
| Average coding sequence | 1340 bp |
| Polypeptide | 447 aa |
| Overall size | 27 kb |

*aa*, amino acids; *bp*, base pairs; *kb*, kilobase pairs; *UTR*, untranslated region.

Individual exons may correspond to structural and/or functional domains of the proteins for which they code, such as the signal peptide of secreted polypeptides or the heme-binding domain of globin. For some complex proteins, domains encoded by single exons often appear in apparently unrelated proteins, suggesting that the evolution of these proteins may have been facilitated by the ability to bring together different protein subdomains by exon shuffling. The origin of intron/exon structure is thought to be extremely ancient and to predate the divergence of eukaryotes and prokaryotes. However, prokaryotes and small eukaryotes

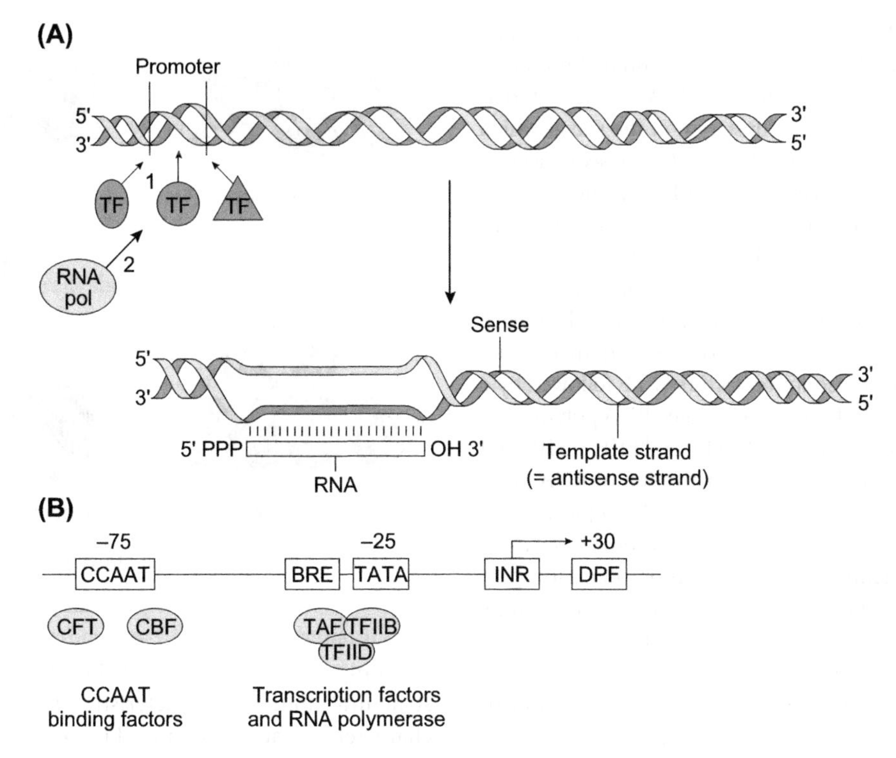


**Figure 4.9** (A) Transcription of eukaryotic genes. (B) Basic promoter elements in eukaryotes.

(e.g., yeast) have lost their introns during evolution, perhaps because of the strong selective pressure on these organisms to retain a small genome size. Therefore, exons can be classified as follows, 5′-UTR exons, coding exons, 3′-UTR exons, and all possible combinations of those three main components, including single exons that cover the whole mRNA.

## 4.4.1 Gene Expression

The expression of individual genes can be regulated at multiple levels. Before a gene sequence gets translated into a polypeptide sequence, multiple events take place: activation of the local DNA structure, initiation and completion of transcription, processing of the primary transcript, transport of the mature transcript to the cytoplasm, and translation of the mRNA. All these steps can be the target of regulation and thus are potential control points for altering gene expression. Some genes are needed in all cell and tissue types as they encode a crucial gene product. Such genes are often referred to as "housekeeping genes." However, numerous human and mammalian genes show highly restricted cell- or tissue-specific expression patterns, and this spatial and/or temporal restriction of gene expression can also be regulated at multiple levels.

## 4.4.2 Transcription

Initiation of transcription happens when the compact DNA structure is loosened and short sequence elements in the 5′ end of the gene guide and activate RNA polymerase (Fig. 4.9). A group of such sequences is often clustered upstream of the transcription initiation site to form the promoter. The promoter is a region of DNA at the 5′ end of the genes that bind RNA polymerase.

There are different types of promoters for RNA polymerases I, II, and III. RNA polymerases I and III are dedicated to transcribing genes encoding RNA molecules (rRNA and tRNA), which assist in the translation of the polypeptide-coding genes. All RNA polymerases are large proteins and appear as aggregates consisting of 8–14 subunits. Significant amounts of information exist on promoter sequences specific for these polymerases. The basal apparatus, the generic minimal promoter sequence that is sufficient to initiate transcription of any protein-coding gene, contains an RNA polymerase II recognition signal

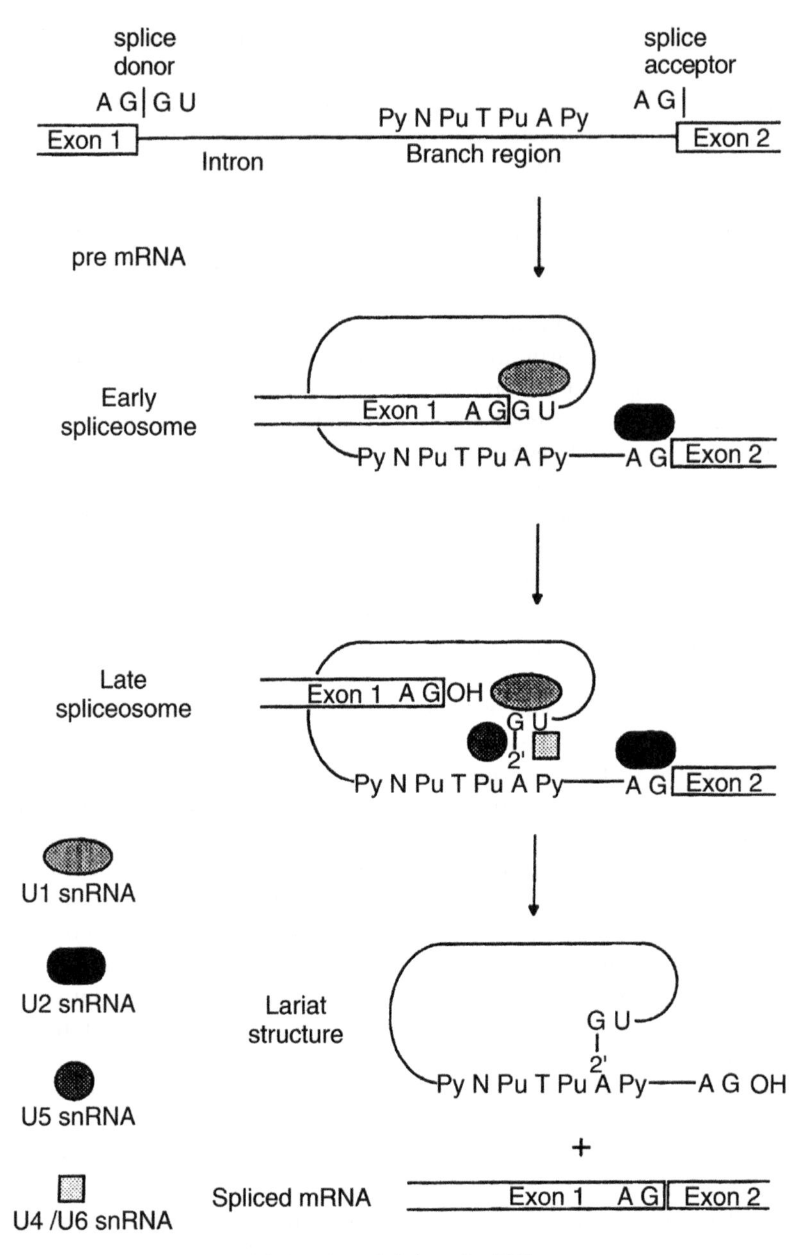


**Figure 4.11** Splicing of mRNA.

(snRNPs), mediates the splicing of the large number of pre-mRNA transcripts, collectively referred to as heterogeneous nuclear RNA (hnRNA). Spliceosomes are multienzyme complexes that both catalyze the splicing reaction and stabilize the intermediates in the splicing process. The snRNPs composing the spliceosome consist of a set of five integral snRNA molecules (U1, U2, U4, U5, and U6) tightly associated with a large number of proteins [34]. RNA molecules in the snRNPs are among the most highly evolutionarily conserved sequences among eukaryotes. An initial intermediate of the splicing reaction is formed when the 5′ guanylate end of an intron (the splice donor) is joined to an adenylate residue near the 3′ end of the intron (the branch point) through a 2′–5′

phosphodiester linkage (Fig. 4.11). After the completion of exon–exon fusion, the excised intron is released as a "lariat structure" by cleavage at the splice acceptor.

The genes encoding rRNA and tRNA also contain exons and introns but are spliced by different mechanisms than those required for mRNA splicing. Self-splicing of RNA without any protein factors is known to happen in prokaryotes, which suggests that introns have an extremely ancient evolutionary origin, predating not only the eukaryote/prokaryote divergence but also perhaps the origin of proteins as well.

### 4.4.6 3′-Untranslated Sequences and Transcriptional Termination

The 3′ ends of primary transcripts are determined by transcriptional termination signals located downstream of the ends of each coding region. However, the 3′ ends of mature mRNA molecules are created by cleavage of each primary precursor RNA and the addition of a several hundred nucleotide polyadenylate [poly(A)] tails (see Fig. 4.6). The cleavage site is marked by the sequence 5′-AAUAAA-3′ located 15–20 nucleotides upstream of the poly(A) site and by additional GU-rich sequences 10–30 nucleotides downstream. Histone mRNAs, which do not have poly(A) tails, have stem-loop structures instead with cleavage of the primary transcript mediated by a distinct protein complex that includes the U7 snRNP [35].

Some complex transcriptional units contain several potential polyadenylation and/or transcription termination sites. It is often difficult to distinguish the latter from the former as the product available for analysis (mRNA) has lost the portion of the 3′-terminus originally transcribed by RNA polymerase. Alternative polyadenylation (or termination) sites can determine final protein structure if the longer precursor RNA contains an exon not found in the shorter precursor RNA. In a simple case, two proteins with different carboxyl termini are formed. But if alternative exon splice sites are made available in the longer precursor RNA, proteins with entirely different sequences can be produced.

The region from the translation termination codon to the poly(A) addition site may contain up to several hundred nucleotides of a 3′-UTR, which includes signals that affect mRNA processing and stability. Many mRNAs that are known to have a very short half-life contain AU-rich elements, 50- to 150-nucleotide sequences containing AUUUA motifs that regulate mRNA stability [36]. Other, less well-characterized sequences can have similar effects. Removal or alteration of these sequences can prolong the half-life of mRNA, indicating that such elements represent a general regulatory feature of mRNAs whose level of expression can be rapidly altered.

## 4.5 TRANSLATION OF RNA INTO PROTEIN

### 4.5.1 Genetic Code

After intron sequences are spliced out of the primary RNA transcript and the 3′-terminus is generated [in most cases, by the addition of a poly(A) tail], the mature mRNA is transported from the nucleus to the cytoplasm, where it is translated into a polypeptide chain. In the cytoplasm, tRNA molecules provide a bridge between mRNA and free amino acids (see Fig. 4.3). Adjacent groups of three nucleotide sequences in the mRNA (codons) each bind to complementary three nucleotide sequences in tRNA (anticodons). Unlike most other nucleic acids, tRNA molecules have rigid tertiary structures. All tRNAs are L-shaped, with the anticodon located at one end and the amino acid binding site at the other end. Modified nucleotides, such as methylguanosine (mG) and pseudouridine ($\psi$), are common in tRNA and help determine the specific three-dimensional characteristics of tRNA molecules. Aminoacyl tRNA synthetases specifically recognize different tRNAs and attach each tRNA to the correct amino acid. The last base in each codon is followed by the first base in the next, and thus the first codon in an mRNA molecule determines the reading frame for all subsequent codons.

The relationship between codon and amino acid sequence is referred to as the genetic code (Fig. 4.12). Different tertiary structures of each tRNA are specifically recognized by the proper tRNA synthetase, ensuring the accuracy of the code. As the anticodon sequence itself does not determine tRNA tertiary structure, each amino acid may have several possible codons recognized by tRNAs with different anticodons but similar tertiary structures; that is, they are recognized by the same tRNA synthetase. For example, 5′-AAA-3′ $tRNA^{Phe}$ (the tRNA coding for phenylalanine with the anticodon 5′-AAA-3′) has the same tertiary structure and is charged by the same tRNA synthetase as 5′-GAA-3′ $tRNA^{Phe}$. Thus, both codons 5′-UUU-3′ and 5′-UUC-3′ code for phenylalanine using different tRNAs but the same tRNA synthetase. Additional redundancy in the genetic code arises because the third base in each codon–anticodon duplex (which is the first base from the 5′ end of the anticodon)

| | | | |
|---|---|---|---|
| UUU, UUC – Phe<br>UUA, UUG – Leu | UCU, UCC, UCA, UCG – Ser | UAU, UAC – Tyr<br>UAA - Stop<br>UAG - Stop | UGU, UGC – Cys<br>UGA - Stop<br>UGG - Trp |
| CUU, CUC, CUA, CUG – Leu | CCU, CCC, CCA, CCG – Pro | CAU, CAC – His<br>CAA, CAG – Gln | CGU, CGC, CGA, CGG – Arg |
| AUU, AUC, AUA – Ile<br>AUG - Met | ACU, ACC, ACA, ACG – Thr | AAU, AAC – Asn<br>AAA, AAG – Lys | AGU, AGC – Ser<br>AGA, AGG – Arg |
| GUU, GUC, GUA, GUG – Val | GCU, GCC, GCA, GCG – Ala | GAU, GAC – Asp<br>GAA, GAG – Glu | GGU, GGC, GGA, GGG – Gly |

**Figure 4.12** The genetic code.

can be flexible according to the rules of Watson–Crick base pairing. In particular, G:U or U:G base pairs are often found in the third position of a codon–anticodon duplex, and the guanine analog inosine, found only in tRNA, can pair or wobble with A, C, or U in the codon. Despite the redundancy of the genetic code, synonymous codons are not used with equal frequency, and the pattern of codon usage (codon bias) may vary tremendously among different species and between nuclear and mitochondrial mRNAs.

The AUG codon, which codes for methionine, nearly always begins the protein-coding portion of each mRNA molecule. Therefore, the vast majority of newly synthesized peptides begin with methionine. The $tRNA^{Met}$ for the initiator AUG codon has a different tertiary structure from all other tRNAs, including the $tRNA^{Met}$ that functions in elongation. Translation of most mRNAs generally begins with the first AUG from the 5′ end, which is typically embedded within a Kozak consensus sequence (5′-CCA/GCC**AUG**G-3′) and establishes the reading frame [37]. The UAA, UAG, and UGA codons are stop codons and have no cognate tRNAs. Thus, recognition of any one of these codons by the protein synthesis machinery terminates the protein-coding portion of every mRNA molecule.

Mutations that change a codon into a different codon and would therefore encode a different amino acid result in a protein with an amino acid substitution, and these are described as missense mutations. However, the UAA, UAG, and UGA codons do not code for an amino acid but instead serve as a signal to terminate protein synthesis. Mutations that produce one of these codons in the middle of a normal reading frame cause truncation of the newly synthesized protein during protein synthesis and are referred to as nonsense mutations. Frequently transcripts with a premature termination codon are degraded by a process called nonsense-mediated decay, so that no protein product is synthesized from the mutant allele, resulting in haploinsufficiency for the gene product.

### 4.5.2 Protein Synthesis

The biochemistry of protein synthesis (Fig. 4.13) can be divided into the stages of initiation, elongation, and termination. All three processes occur on ribosomes, cytoplasmic particles of protein and rRNA that align the different substrates of each reaction. When inactive, ribosomes exist as separate pools of the two ribosome subunits, described by their size or sedimentation coefficient (S value). The small 40S ribosomal subunit contains 18S rRNA and approximately 33 different proteins, and the large 60S subunit contains 28S rRNA, 5.8S rRNA, 5S rRNA, and approximately 50 different proteins. Beyond these structural components, there is a wealth of additional factors, including both proteins and functional RNA molecules, that are required for ribosome biogenesis [38]. Translation begins with the formation of a preinitiation complex that contains the 40S ribosomal subunit, initiator $tRNA^{Met}$, GTP, and several protein initiation factors. An mRNA molecule initially binds to the preinitiation complex in conjunction with several initiation factors that interact with the 5′ cap structure. The canonical model for identification of the AUG start codon involves scanning the mRNA in a 5′–3′ direction until the consensus sequence 5′-CCA/GCC**AUG**G-3′ is reached. However, internal ribosome entry sites (IRES) mediate ribosome recruitment and translational initiation for uncapped mRNA molecules and for translation when cap-dependent processes are inhibited [39]. Binding of the 60S ribosomal subunit and dissociation of several initiation factors generate a complex of proteins and subcellular particles poised to begin synthesis of the first peptide bond.

A ribosome contains room for two tRNAs and their respective amino acids (see Fig. 4.13). One tRNA at the peptidyl or P site is attached to the amino acid that has just been incorporated into a nascent peptide chain, and another tRNA at the aminoacyl or A site is attached to its cognate amino acid and is ready to participate in protein synthesis. During elongation, a peptide bond

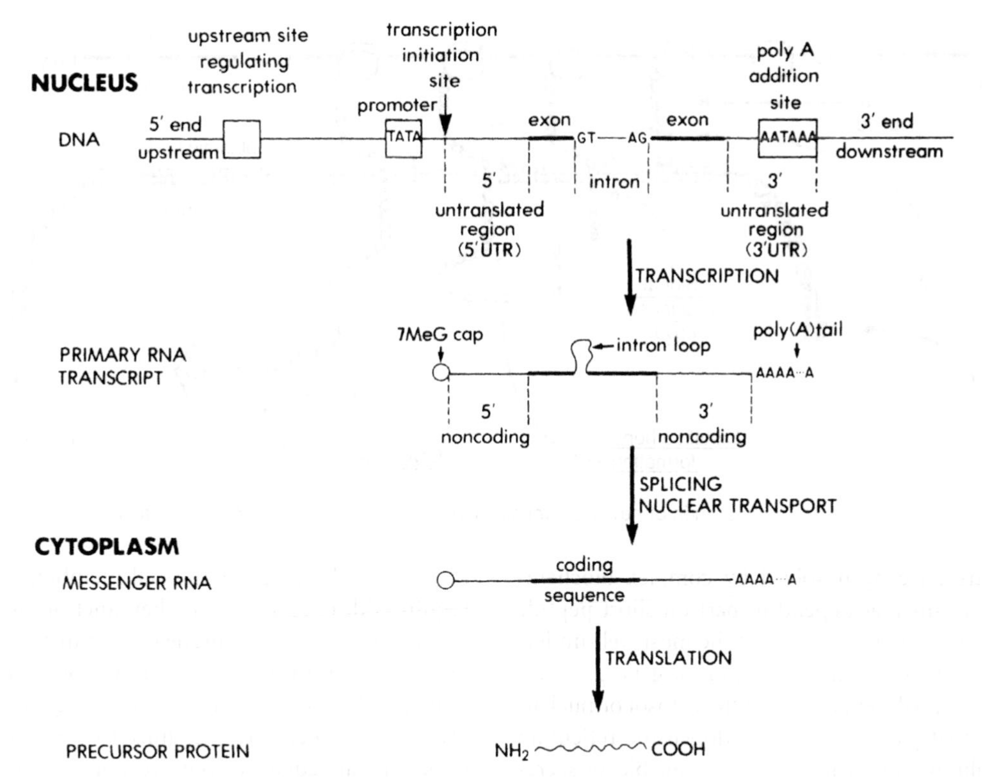


**Figure 4.13** Translation of mRNA into protein.

is formed between the two adjacent amino acids, the ribosome moves to the next codon in the mRNA, the tRNA at the P site dissociates from the nascent peptide chain, and the tRNA at the A site is translocated to the P site. This series of reactions is dependent on elongation factor 1 (EF1), which binds to free charged tRNAs, and EF2, which facilitates translocation from the A site to the P site. A single mRNA can be simultaneously translated by several active ribosomes, forming a polysome that can contain as many as 50 ribosomes.

When a codon signifying termination of protein synthesis (UAA, UAG, or UGA) is reached, the completed polypeptide separates from the tRNA at the P site, and the ribosome dissociates. In bacteria and yeast, a unique group of suppressor tRNA mutations is caused by changes in an anticodon that then permit binding of a charged tRNA to a termination codon. Point mutations in an mRNA that would normally lead to premature termination codon (e.g., by changing a UUA codon into a UAA codon) are then partially suppressed by the mutant tRNA, allowing synthesis of a full-length protein with a missense change at the position of the mutant codon. The principle of suppression of nonsense codons, which represent about 30% of disease-causing mutations in humans, can be mimicked by treatment with aminoglycoside antibiotics and other pharmaceutical compounds [40]. Used as a therapeutic approach, such treatment has the potential to affect therapy in a wide variety of genetic disorders.

### 4.5.3 Protein Localization

Gene products function in particular cellular compartments. For example, histones, tubulin, glycosyltransferases, peptide hormone receptors, and collagen are specifically localized to the nucleus, cytosol, Golgi apparatus, cell membrane, and extracellular space, respectively. Although many membranes contain pores large enough to accommodate a linear polypeptide chain, completely folded proteins are generally too large to fit through these pores. In addition, to the problem of translocating soluble proteins across membranes, proteins that remain attached to the membrane must be placed and oriented in specific ways. These problems—protein sorting, translocation, and membrane

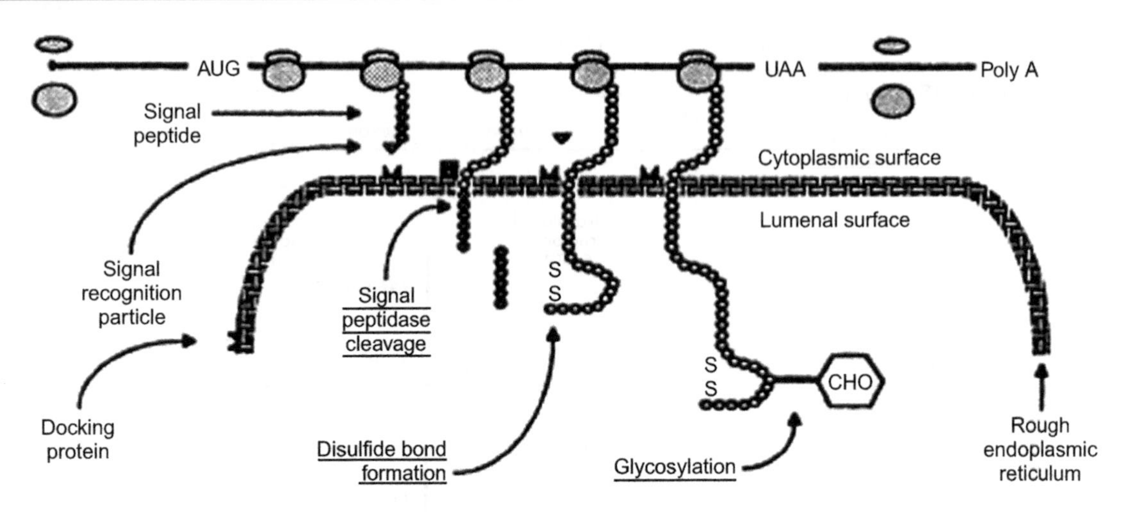


**Figure 4.14** Translocation of newly synthesized proteins across the endoplasmic reticulum.

orientation—have been solved by complex biochemical mechanisms that depend in part on short peptide sequences in each protein. One of the most well-understood pathways is the initial sorting of gene products into those that will remain inside the cytosol or nucleus and those that pass across the endoplasmic reticulum (ER) membrane, and which are then available for secretion into the extracellular space. This initial sorting is determined early in the translation of proteins destined to cross the ER by the presence of a specialized hydrophobic signal sequence of 20–30 amino acids [41] usually located at the N-terminus (Fig. 4.14).

The signal sequence is first recognized when about 25 amino acids of the growing polypeptide have emerged from the ribosome and bind to a protein–RNA complex called the signal recognition particle (SRP). The SRP stops further translation until bound to a docking protein complex, the translocon, which is located on the surface of the ER and forms a hydrophilic membrane pore [42]. The signal peptide passes through the pore, translation recommences, and the growing polypeptide crosses the membrane cotranslationally. After the protein has passed into the lumen of the ER, signal peptidase, a protein on the luminal surface of the ER, cleaves the signal peptide to complete the initial phase of protein sorting.

Proteins extruded from the RER pass through the Golgi apparatus into secretory vesicles for transport to the cell surface. Proteins destined for the extracellular matrix are secreted from the cell when the vesicles fuse with the plasma membrane. Insertion and orientation of proteins destined for the cell membrane, however, require additional sequences that function either to stop transfer across the membrane (stop transfer sequences) or to initiate transfer of an internal loop of the nascent polypeptide chain (start transfer sequences). Start transfer sequences are recognized by SRP-like N-terminal signal sequences but are not cleaved from the protein after translocation. The number, order, and orientation of start transfer and stop transfer sequences determines the conformation of complex integral membrane proteins that span the membrane multiple times. Some soluble proteins that contain the short peptide sequence Lys–Asp–Glu–Leu (KDEL) remain in the lumen of the RER, such as binding protein (BiP) and protein disulfide isomerase (PDI). Both BiP and PDI facilitate the folding of newly synthesized proteins in the RER. PDI catalyzes the rearrangement of Cys–Cys disulfide bonds; BiP is a so-called chaperone that binds temporarily to portions of other proteins normally not exposed to the surface and, in doing so, prevents partially folded proteins from misfolding and/or aggregating. Soluble proteins destined for specialized compartments inside the cell, such as lysosomes or peroxisomes, use a signal sequence to gain access to the ER lumen but require additional mechanisms for proper subcellular localization. Lysosomal sorting depends on amino acid sequences that specify posttranslational addition of a mannose 6-phosphate residue. Proteins containing this modification are selectively transferred from the Golgi apparatus to the lysosomal interior. Failure to modify proteins destined for the lysosome in this way is responsible for the inherited disease mucolipidosis II (I-cell disease).

There are more thn 1000 mitochondrial proteins, the great majority of which are encoded in the nucleus and synthesized in the cytosol. Similar to the proteins destined for the ER, transport of proteins across the mitochondrial membrane also depends on a signal sequence and uses several translocator complexes [43,44]. The targeting sequence is usually located at the N-terminus and is cleaved on import, but sometimes is internal to the protein and is therefore not cleaved. Unlike RER proteins, for which translation and translocation are codependent, mitochondrial proteins are first translated completely, released into the cytosol, and then translocated into the mitochondrial membranes, intermembrane space or matrix. The potential problem of translocating a completely folded protein across a membrane pore is solved for mitochondria by complexes that include chaperone proteins of the Hsp70 family, which bind to proteins destined for the mitochondria and stabilize them in the unfolded state until after they have passed across the mitochondrial membrane, after which they assume their folded conformation.

### 4.5.4 Posttranslational Modification

Alterations to protein structure that occur after translation include the formation of disulfide bonds, hydroxylation, glycosylation, proteolytic cleavage, and phosphorylation. Phosphorylation of serine, tyrosine, and threonine residues is a common reversible modification that alters protein–protein interactions or controls enzymatic activity, mostly of intracellular proteins. The formation of disulfide bonds, hydroxylation, glycosylation, and proteolytic cleavage are generally not reversible and mostly involve extracellular proteins.

Intramolecular disulfide bond formation can begin cotranslationally as the growing polypeptide chain enters the lumen of the ER. Some proteins, such as immunoglobulin light chains, have a sequential pattern of intrachain disulfide bonds (e.g., between the first and second cysteines or third and fourth cysteines). Other proteins, such as proinsulin, have a more complicated pattern. Protein folding and establishment of the correct arrangement of disulfide bonds are critical steps in synthesizing a three-dimensional protein structure. Glycosylation of newly synthesized proteins may be O-linked via serine, threonine, or hydroxylysine residues, or N-linked via asparagine residues. O-linked glycosylation is catalyzed by glycosyltransferases located on the luminal surface of the Golgi apparatus. N-linked glycosylation begins with transfer of a 14-residue oligosaccharide from a lipid molecule (dolichol) embedded in the RER membrane to the asparagine residue of a growing polypeptide chain. At some sites, the oligosaccharide is highly modified by removal of some carbohydrates and addition of other carbohydrates to form a complex glycoprotein modification. Other sites are less modified and contain the original high mannose composition of the dolichol intermediate. Many glycoprotein modifications help determine the specificity of extracellular protein–protein interactions, such as antigen–antibody binding or attachment of cells to the extracellular matrix.

Proteoglycans are a specialized class of extensively glycosylated proteins that contain a protein core with long disaccharide chains branching off at regular intervals and can contain as much as 95% carbohydrate by weight. Proteoglycans are extremely hydrophilic and form hydrated gels that provide structural integrity to the extracellular space. During growth and development, extracellular remodeling is accompanied by endocytosis and degradation of proteoglycans by lysosomal enzymes specific for different disaccharide chains; absence of these lysosomal enzymes produces mucopolysaccharidoses, such as Hunter syndrome or Hurler syndrome.

### 4.5.5 Expression of Housekeeping and Tissue-Specific Genes

Many proteins that operate in basic metabolic functions such as energy generation or nutrient transport are found in all cells, and the genes that encode these proteins are described as housekeeping genes. They are characteristically expressed at a relatively constant level in all cells. More specialized genes that are not housekeeping are used only at specific times and places during development or in one or a limited set of tissues. The sequence of the human genome has revealed that the most common genes in our genome are those encoding TFs and nucleic acid binding proteins, which together encode 26.8% of the proteins with known or putative function. Other highly represented genes encode receptors, transferases, signaling molecules, and transporters [45]. These and other housekeeping genes usually account for 90% or more of the transcripts expressed in any particular cell type.

# 5 Epigenetics

*Rosanna Weksberg[1,2], Darci T. Butcher[1], Cheryl Cytrynbaum[1,2], Michelle T. Siu[1], Sanaa Choufani[1], Benjamin Tycko[3]*

[1]Genetics and Genome Biology, Research Institute, The Hospital for Sick Children, Toronto, ON, Canada
[2]Clinical and Metabolic Genetics, The Hospital for Sick Children, Toronto, ON, Canada
[3]Division of Genetics & Epigenetics, Hackensack Meridian Health Center for Discovery and Innovation, Nutley, NJ, United States

## 5.1 INTRODUCTION

Despite the tremendous advances in human genetics enabled by the original genome projects and brought to fruition with high-throughput genotyping and massively parallel DNA sequencing, many aspects of human biology still cannot be explained by genetics alone. Over the years a largely parallel line of research has implicated epigenetic dysregulation in human diseases. With advancing technologies, the two areas of genomics and epigenomics are starting to come together to yield fundamental insights. Epigenetics is defined as modifications of DNA and its associated proteins and ribonucleoproteins that do not involve alterations in the DNA sequence itself [1]. Normal human development requires the specification of a multitude of cell types and tissues that depend on transcriptional programs that are regulated by epigenetic mechanisms. These mechanisms enforce the acquisition and maintenance of the characteristic gene expression profiles of specific cell types in developing and adult tissues. Enzymes and other proteins that act as epigenetic regulators can be categorized based on their specific functions as "writers," "erasers," or "readers" of epigenetic modifications. Writers enzymatically deposit specific substrates (including acetyl, methyl, and phosphate groups) to modify DNA. Erasers catalyze the removal of these posttranslational modifications. Readers have specific domains that recognize individual epigenetic modifications, often in a unique context, which can then recruit other epigenetic machinery to these sites and fine-tune transcriptional activity. Disruption of such control mechanisms is associated with a wide variety of human disorders with behavioral, endocrine, and/or neurologic manifestations, and quite strikingly with disorders of tissue growth, including cancer. While epigenetic factors affecting disease susceptibility and progression have been studied for decades, this topic has attracted much more attention in our current "postgenomic" era. Ongoing research is focused on understanding the functions of epigenetic regulators that modify or read epigenetic marks, characterizing *cis*- and *trans*-acting influences of the genetic background on epigenetic marks, delineating cell type/tissue–specific epigenetic marks in human health and disease, and studying the interactions between epigenetic marks and the environment, especially with respect to fetal programming and risks for common adult-onset disorders. Further, modulation of adverse epigenetic states by drug-based and nutritional therapies has become an important field of investigation. These efforts have been facilitated by large-scale "epigenome projects" that have defined epigenetic patterns across multiple human cell types and throughout different periods of development, producing whole-epigenome maps that are being integrated with parallel genomic data via publicly accessible platforms such as the Human Genome Browser at the University of California at Santa Cruz (UCSC) (see list of websites).

Emery and Rimoin's Principles and Practice of Medical Genetics and Genomics: Foundations, https://doi.org/10.1016/B978-0-12-812537-3.00005-6

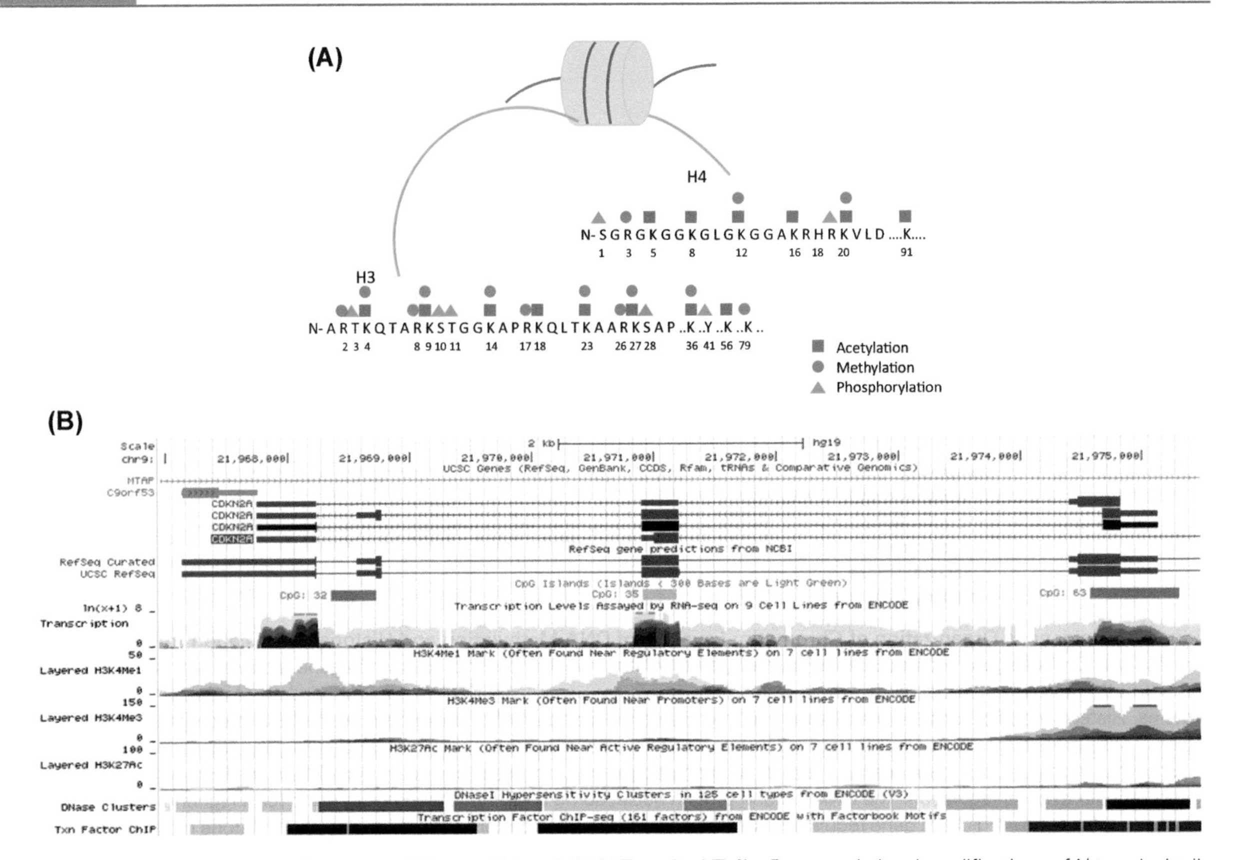


**Figure 5.2** **(A) Histone Modifications of Histone H3 and H4 N-Terminal Tails.** Posttranslational modifications of N-terminal tails (these can also occur in the C-terminal domain but are not shown here) can occur in combination and are read by the appropriate protein to establish local and global open or closed chromatin states. **(B) Snapshot From the University of California at Santa Cruz (UCSC) Genome Browser Demonstrating the Layers of Epigenetic Regulation in the Promoter of the Tumor-Suppressor Gene *CDKN2A*.** This diagram is an example of epigenetic data available in UCSC genome browser. The description of each genomic feature is shown on the left. Two isoforms of the *CDKN2A* gene are shown in *black* and *blue*; here we focus on a shorter (*black*) isoform. Enrichment for the regulatory histone mark H3K4me1, the active histone H3K4me3, and the active histone H3K27Ac is shown by multiple colors for different cell lines. The peaks for each of these marks coincide with the transcription start site of *CDKN2A* overlapping a CpG island (*green*), as well as transcription factor binding sites (TxN factor CHIP) and DNase 1 clusters, which are indicators of open chromatin. All these marks—H3K4me3, transcription factor binding, and DNase1 clusters—indicate that *CDKN2A* is actively transcribed in these cell lines (shown in the transcription track). Information about other histone marks and DNA methylation levels is available from the UCSC genome browser under multiple tracks from the Regulation tab. This image was downloaded from the UCSC genome browser hg19: http://genome.ucsc.edu/ [5,6]. The ENCODE regulation data are from Ref. [7–9].

talk between these modifications occurs through protein networks that write and read these patterns [24]. These two central types of epigenetic modifications are interdependent; histone marks are more labile and DNA methylation marks more stable [24–29], so DNA methylation can act to "lock in" epigenetic states. However, regulating metastable states of gene expression is so crucial in development and tissue homeostasis that other mechanisms, in addition to histone modifications and DNA methylation, come into play to establish and maintain epigenetic states.

Regulatory noncoding RNAs, including small interfering RNAs (siRNAs), microRNAs (miRNAs), and long noncoding RNAs (lncRNAs), play important roles in gene expression regulation at several levels, including transcription mRNA degradation, splicing, transport, and translation [30]. While in plants small RNAs are important for targeting specific loci for cytosine methylation [18], in mammals the main function of siRNAs and miRNAs is posttranscriptional regulation [31]. However, lncRNAs, defined as RNA species >200 nucleotides in length (and often much larger), are increasingly recognized as playing a key role in modulating the epigenetic states of some highly regulated mammalian genes. LncRNAs are transcribed from various genomic locations in relation to their target protein-coding genes. They may be antisense, intronic, intergenic, or promoter or enhancer associated and can regulate transcription both *in cis* and *in trans* by a number of different mechanisms [2,32]. Specifically, certain lncRNAs have been shown to establish specialized nuclear compartments devoid of RNA polymerase II, in which the chromatin-associated polycomb repressive complex-2 (PRC2) catalyzes the formation of a repressive constellation of histone marks [2,30,32,33]. A particularly well-studied example is the *XIST* lncRNA that is essential for initiating the silencing *in cis* of one X-chromosome in female cells [2,34], discussed in greater detail in Section 5.4. LncRNAs are also found in imprinted gene clusters, where monoallelically expressed lncRNAs mediate the repression of one allele via the propagation of repressive chromatin marks such as DNA methylation and histone (H3K27 and H3K9) methylation on that allele *in cis* [35]. In contrast to gene silencing mediated *in cis* by *XIST* and imprinted lncRNAs, the nonimprinted lncRNA *HOTAIR* expressed from the *HOXC* gene recruits PRC2 and the histone demethylase LSD1, resulting in acquisition *in trans* of silencing histone marks at genes within the *HOXD* gene cluster located on another chromosome [36]. It has been estimated that 20% of lncRNAs expressed in human cells are bound by polycomb group proteins, suggesting a shared biochemical mechanism for their role in epigenetic silencing [37,38]. The roles of lncRNAs are diverse; some of them, exemplified by the *H19* untranslated RNA, which is abundant in fetal tissues, are host transcripts that give rise to specific miRNAs, while others appear to function as "molecular sponges" to sequester miRNAs in the cytoplasm [2].

In addition to the "fifth base," 5mC, mammalian DNA contains a related modified base, 5-hydroxymethylcytosine (5hmC). This "sixth base" has taken center stage with the demonstration that proteins in the Ten-Eleven-Translocation (TET) family of oxygenases catalyze the conversion of 5mC to 5hmC, and that the gene for one of these enzymes, *TET2*, is sometimes mutated in human cancers, notably in cases of acute myeloid leukemia and myelodysplastic syndrome (MDS). The formation of 5hmC can, in principle, lead to demethylation of DNA, either by a passive process (failure of remethylation of these sites in S phase) or by an active mechanism (direct demethylation of 5hmC or base excision of a further modified product such as carboxy-mC), so it has been suggested that TET family oxidases probably contribute to the dynamics of DNA methylation [18]. 5hmC has been found in many cell types and tissues, with particularly high levels in the brain and in embryonic stem (ES) cells. As discussed in the next section, on epigenetic reprogramming, TET1 has been shown to be important for self-renewal and maintenance of ES cells.

## 5.3 EPIGENETIC REPROGRAMMING

Global epigenetic reprogramming occurs first in cell populations destined to become germ cells. Epigenetic marks (including DNA methylation) are erased, following which the egg and sperm acquire their highly specialized and divergent epigenetic marks [39]. The second wave of epigenetic reprogramming occurs after fertilization [40]. A small number of genomic regions are protected from this zygotic wave of epigenetic reprogramming; these regions retain an epigenetic "memory" of their parent of origin and thus are crucially involved in the non-Mendelian phenomenon of genomic imprinting [41].

Genomic imprinting (also known as parental imprinting) has been well studied at the whole-genome level and for specific imprinted genes. In most areas of the diploid genome, epigenetic marks on the paternal and maternal alleles are equalized during preimplantation development, with strong allelic asymmetries persisting mainly at imprinted loci [42] and at certain other loci with haplotype-dependent allele-specific methylation. This equalization of the two alleles at most autosomal loci is essential for classical Mendelian transmission of human genetic disorders and involves early postzygotic reductions in DNA methylation. The reduction seems to result from both active demethylation, via mechanisms that are still being investigated experimentally, and passive demethylation, in which CpG methylation is diluted through early rounds of DNA replication in the presence of low-maintenance methylase activity.

Genomic levels of 5hmC change during development, probably as a function of the activity of TET family enzymes, which are particularly highly expressed in ES cells [43–45]. In the mouse genome, 5hmC is widely distributed throughout nonrepetitive regions, whereas satellite repeats (which are located in heterochromatin) are highly enriched for 5mC but substantially less so for 5hmC [46]. Distinct 5hmC and 5mC patterns are observed at CpG islands overlapping gene promoters: while 5mC is depleted from these regions, 5hmC is well represented. Interestingly, the presence of 5hmC and depletion of 5mC at CpG island promoters is associated with increased transcription in ES cells [46]. Consistent with these observations, active histone marks H3K4me3 are enriched in promoters with high 5hmC, suggesting that enriched 5hmC at CpG island promoters is positively correlated with active transcription. Decline of TET oxidase levels during differentiation of ES cells is accompanied by reduced promoter 5hmC and increased 5mC levels, correlating with silencing of certain key developmental regulator genes [46,47]. Thus, one current hypothesis is that hydroxymethylation and the TET proteins could play a role in erasing methylation marks from promoters of pluripotency-related genes during differentiation [46,48]. Hydroxymethylation may also play a role in epigenetic reprogramming of primordial germ cells (paternal genome) and early embryos [49,50].

## 5.4 EPIGENETIC REGULATION OF X INACTIVATION

Inactivation of one of the two X chromosomes in female cells was first described by Mary Lyon in 1961 [51] and since then remains the prototypical example of chromosomal epigenetic silencing. The process of X inactivation is regulated in *cis* by a small control region known as the X-inactivation center (XIC). X-chromosome inactivation (XCI) has evolved in placental mammals to achieve dosage compensation of X-linked genes between females, who have two X chromosomes, and males, with only one X and one Y chromosome. It is hypothesized that the X and Y chromosomes evolved from a pair of autosomes, coinciding in time with the evolution of the placenta and driven by acquisition of the Sex-Determining Region Y (*SRY*) gene. In the course of mammalian evolution, the *SRY*-carrying Y chromosome has been significantly reduced in size and lost most of its active genes, retaining, in addition to *SRY*, a few other genes playing a role in male reproduction, whereas the X chromosome has acquired additional genetic material through translocations from autosomes. As a result, homology between the X and the Y chromosomes is now limited to two small pseudoautosomal regions (PAR1 and PAR2) (reviewed in Ref. [52,53]). It is estimated that the human Y chromosome contains ~45 expressed genes, whereas the X chromosome has ~1300 [54]. The resulting major bias in copy number for X-linked genes between males and females is, for the most part, transcriptionally compensated via XCI.

The XIC is critical for the mechanism of XCI. X-chromosome deletion and translocation mapping in mouse and human defined a critical region for the XIC covering ~1 Mb and mapping to chromosome band Xq13 in humans [55,74]. A major breakthrough in understanding the mechanism of XCI followed the discovery of an lncRNA within this region, the X-inactive-specific transcript *XIST*, which is expressed from the XIC at high levels only on the X chromosome destined for inactivation, and which quickly spreads to coat the entire inactive X (Xi) in female cells [56–58]. The majority of the work on mechanisms of X inactivation was performed using mouse preimplantation embryos and ES cells. Before turning to the details of this mechanism, it should be noted that although mouse models have been very valuable in unraveling XCI, there are a number of substantive differences in the XCI process between mice and humans. In mice, there are two waves of XCI; the first is imprinted inactivation of the paternal X, which is initiated shortly after fertilization. This pattern of paternal X-chromosome silencing is maintained in extraembryonic tissues, but is erased and reestablished in a random manner in the inner cell mass (ICM) of blastocysts, which gives rise to the embryo proper [59,60].

In human embryos, XCI is initiated at the blastocyst stage and occurs at random in both the ICM and the trophoblast [61]. Random XCI is a multistep process, which can be divided into three steps: initiation, spreading, and maintenance. Initiation involves counting the number of X chromosomes per cell so that one X remains active per diploid number of autosomes. In other words, XY males and XO females keep their single X chromosome active, whereas XX, XXX, and XXXX females and XXY males inactivate all but one X, upregulating *XIST* RNA on all the X chromosomes destined to become inactive. The spreading of inactivation is achieved through sequential acquisition of epigenetic marks starting with *XIST* RNA, followed by PRC2 recruitment, a shift to late replication timing, enrichment of histone macro H2A, and silencing of chromatin marks, such as histone H3 and H4 hypoacetylation, H3 lysine 27 methylation, and finally

DNA methylation of CpG-rich promoters [62,63]. Once established early in embryonic development, the inactive state of an X chromosome is maintained through somatic cell divisions utilizing epigenetic modifications of DNA and histones. As DNA methylation is sufficient to lock in the inactive state, *XIST* expression is no longer required [64].

Regulation of *XIST/Xist* expression itself is a complex process involving multiple *cis*- and *trans*-acting factors [2]. In mice, the pluripotency transcription factors Oct4 and Nanog are negative regulators of *Xist* expression, such that a decrease in their expression early in post-zygotic development coincides with cell differentiation, upregulation of *Xist*, and onset of XCI [65–67]. Several subregions within the mouse Xic are involved in *Xist* regulation: *Tsix*, an antisense transcript of *Xist*, is a negative regulator of its expression and it protects the active X (Xa) from inactivation [68]; the noncoding RNA *Jpx* [69], X-pairing region [59,70], and protein-coding *Rnf12* [71,72] are positive regulators of *Xist* expression. There are a number of differences in the organization of the mouse and human XIC/Xic and there is a little sequence conservation between mouse and human [73,74]. Based on comparative analysis of Xic in several mammalian species, it seems that Xic is an evolutionarily labile locus and the orthologs across mammalian species act via multiple diverse strategies [61]. For example, in mice, *Jpx* is located 9 kb upstream of *Xist*, whereas in human it is separated by 90 kb [73,74]. Furthermore, human *TSIX* shows little conservation with mouse and is not transcribed through the entire *XIST*, as in mouse [75]. Also, in human fetal cells *XIST* and *TSIX* are coexpressed from the Xi [76], suggesting that *TSIX*-mediated downregulation of *XIST* might be less functional in humans.

More research is required to understand the regulation of XCI in human embryonic development. Despite the elegant picture that has emerged, some steps of the XCI process are still not completely understood. Many intriguing questions remain, such as how the cell counts the number of X chromosomes and decides how many to inactivate, how X chromosomes communicate with each other to retain one Xa and avoid a lethal state with two active or inactive X chromosomes, how *XIST* RNA recruits repressive chromatin markings, and how spreading of the inactivated state occurs along the Xi chromosome. These are active areas of current research [2,60].

### 5.4.1 Special Aspects of X Inactivation Relevant to Human Genetic Diseases

Not all X-linked genes are subject to X inactivation; some genes are robustly expressed from both the Xa and the Xi. It is estimated that 3% and 15% of genes escape XCI in mouse and human, respectively [77]. Again, there are fundamental differences between human and mouse genes that escape X inactivation. In mouse, genes escaping X inactivation are randomly distributed along X, whereas in humans they tend to cluster together. Only six of these genes overlap between mouse and human [74]. Genes that escape XCI are frequently expressed at lower levels from Xi than from Xa [78]. Further, based on DNA methylation analysis of X-linked promoters it is estimated that 12% of the X-linked genes show variable inactivation status among different somatic tissues [79]. Most, but not all, genes that escape XCI have a Y-linked homologue either within or outside of a PAR; some are functional, whereas others are pseudogenes. On the human X chromosome, the locations of genes that escape XCI are seemingly nonrandom, as the majority are clustered within regions of X that have the highest degree of homology to the Y chromosome [78].

Genes that escape XCI are important potential contributors to phenotypes of X-chromosome aneuploidy, both in the relatively common situation of X chromosome deficiency (females with Turner syndrome, karyotype 45,X; often abbreviated XO) and in supernumerary X-chromosome syndromes. Less than 1% of XO conceptuses survive to birth [80], with surviving individuals having a female phenotype and manifesting Turner syndrome, characterized by short stature, ovarian dysfunction, and a variety of somatic abnormalities such as webbed neck, high arched palate, increased carrying angle of elbows, aortic coarctation, renal malformations, and cognitive problems with visual–spatial perception and social interactions [81]. These problems are presumed to be due to haploinsufficiency of the few genes that normally escape X inactivation (i.e., are present in two active copies in normal XX females). Most of these genes are in the PARs: one of the genes, *SHOX*, has been implicated in the short stature phenotype of females with Turner syndrome, and short stature is observed in individuals with subchromosomal deletions encompassing this region on either the X or the Y chromosome [82]. Individuals carrying supernumerary X chromosomes, of low-copy-number repeats in the 15q11–q13 region, which increases susceptibility to nonhomologous recombination. These interstitial deletions involve the entire imprinting domain and several nonimprinted genes (class I and II deletions) [129,134]. In PWS, deletions occur on the paternal chromosome, whereas in AS deletions occur on the maternal chromosome. In addition to deletions of the 15q11–q13 region, other mechanisms that result in PWS and AS include UPD 15q11–q13 and imprinting defects. Specifically, 25%–30% of PWS cases result from maternal UPD for this chromosomal region, and conversely, 2%–5% of AS results from paternal UPD [129]. Imprinting defects have been reported in 1%–3% of PWS and 2%–4% AS patients. This type of defect results from an acquisition of maternal-type imprint (gain of DNA methylation) on the paternal chromosome in PWS and, conversely, loss of the maternal imprint (loss of DNA methylation) on the maternal chromosome in AS. Imprinting defects typically result from microdeletions in the 15q11–q13 region. The location of microdeletions is distinct in AS versus PWS cases; the smallest region of overlap (SRO) of the microdeletions in PWS is 4.3 kb in size and is located within the *SNURF/SNRPN* exon 1/promoter [135], while the AS SRO is 880 bp and is located more centromerically; ~35 kb upstream of exon 1 of *SNURF/SNRPN* [136]. For both syndromes, a microdeletion can be de novo or inherited from an unaffected carrier parent, through the male germline in PWS and through the female germline in AS [129]. Rarely, IC methylation defects occur in individuals with AS or PWS; these can result from failure of imprint erasure/acquisition or maintenance. No mutations in single genes have been demonstrated to date in PWS, but in 10% of AS cases mutations of the maternal copy of *UBE3A* have been documented [137,138]. This gene encodes an E3 ubiquitin ligase, a protein that functions in protein degradation. About 80% of mutations occur de novo, and about 20% are inherited from unaffected mothers [139]. By a mechanism that is not yet fully understood, maternal expression of *UBE3A* is observed only in the brain and not in other tissues, accounting for the neurobehavioral AS phenotype of deficient expression [140,141].

Last, in contrast to PWS, where most cases can be explained by genetic and/or epigenetic alterations at the 15q11-q13 imprinted cluster, 10%–15% of suspected AS cases have no identifiable molecular alteration [129].

### 5.5.3 Diagnostic Testing and Recurrence Risk in Imprinting Disorders

Identifying specific molecular defects in imprinting disorders provides important information for patient management and for estimating recurrence risk. Molecular diagnosis, specifically testing for abnormal DNA methylation in the relevant imprinted domains, is usually the first line of investigation for many imprinting disorders, including PWS/AS (domain in chromosome band 15q11–q13) and BWS/RSS (domains in chromosome band 11p15.5). The majority of molecular alterations within imprinting domains (UPD, IC epimutations, and microdeletions) can be diagnosed by assaying DNA methylation in the respective IC. Techniques such as Southern blot or PCR-based assays for DNA methylation [33,142] are useful, in addition to the more recently developed method, methylation-sensitive multiplex ligation-dependent probe amplification (MS-MLPA) [143]. The benefit of using MS-MLPA is that it can assess both methylation levels and copy number across several sites within the imprinted cluster. There are advantages and limitations of DNA methylation analysis for the various imprinting disorders. For PWS and AS, methylation analysis is sufficient to establish a clinical diagnosis, but cannot usually elucidate the underlying genetic mechanism, i.e., it cannot distinguish between methylation abnormalities due to a deletion, UPD, or IC mutation. In some cases, MS-MLPA can identify microdeletions within the IC in PWS and AS; however, typically further testing is required to identify the specific mechanism, which is critical to provide accurate recurrence risk information. This is particularly important as the recurrence risk can range from less than 1% for de novo deletions or UPD to 50% for some IC defects. For BWS, MS-MLPA can detect methylation abnormalities and strongly suggest 11p15 UPD when simultaneous gain of DNA methylation at H19 DMR and loss of DNA methylation at *KCNQ1* intronic DMR is identified. Additional testing for chromosome 11p15 UPD using either a PCR-based dosage assay or microsatellite genotyping should be undertaken if this molecular change is suspected, especially given the high frequency of somatic mosaicism. For RSS, DNA methylation is usually performed for the distal (IC1) IC on chromosome 11p15.5 and the imprinted genomic regions at 7p13 and 7q32 [126]. For AS and BWS, if no loss of DNA methylation at the respective ICs is detected, sequencing of *UBE3A* or *CDKN1C*, respectively, should be performed.

If no molecular defects are identified by methylation and mutation screening, comparative genome hybridization arrays to identify small atypical deletions or duplications could be pursued. It should be kept in mind that identification of small deletions or duplications is dependent on the resolution of microarrays, which can vary significantly among different diagnostic laboratories. Although chromosome translocations, inversions, and duplication infrequently cause imprinting disorders, their presence is associated with a significant risk of recurrence. A chromosome abnormality associated with an imprinting disorder may or may not have an associated methylation defect. For BWS, whether or not a methylation defect is present, high-resolution banding of the critical chromosomal region(s) should be considered for individuals who do not have a positive molecular diagnosis.

Individuals with imprinting disorders and UPD have not, as of this writing, been reported to transmit the molecular alteration or imprinting disorder to the next generation. In fact, theoretically this is very unlikely. Recurrence risk is usually low (<1%) in individuals with IC epimutations and imprinting disorders. This low risk implies that, in most cases, the IC defect can be rectified by the normal germline reprogramming mechanism; however, there are a small percentage of cases with IC defects that are heritable. These can be recognized from a positive family history or an associated genomic alteration. When inherited genetic alterations are present, such as IC microdeletions in AS and PWS, *UBE3A* mutation in AS, or *CDKN1C* mutation in BWS, the associated recurrence risk rises to 50%. Such genetic alterations segregate in a Mendelian fashion, but the penetrance of the imprinting disorder depends on which parent transmits the mutation. For example, a parent carrying a mutation in *CDKN1C* (BWS) or *UBE3A* (AS) has a 50% chance of transmitting the mutation, but the imprinting disorder is expressed only if the mother transmits the mutation, since the paternally transmitted gene, whether it carries a mutation or not, is normally silenced in the male germline.

## 5.6 GENETIC DISORDERS CAUSED BY MUTATIONS IN EPIGENES

Within the last decade, an increasing number of Mendelian disorders have been recognized to be caused by mutations in genes that are important for maintaining normal epigenetic regulation, or "epigenes" [144]; [145–147]. Among these disorders, neurologic function, in particular intellectual disability, and growth dysregulation appear to be common features. Loss-of-function mutations in epigenes can disrupt normal establishment, maintenance, or reading of epigenetic marks, thereby resulting in altered chromatin structure and gene expression. Here we will provide examples of disorders caused by pathogenic mutations in four categories of epigenes: writers, erasers, readers, and chromatin remodelers. For a more comprehensive list see Table 5.1.

**TABLE 5.1 Genetic Disorders Caused by Mutations in Epigenes[a]**

| Disorder | Gene | Function | Locus | OMIM |
|---|---|---|---|---|
| Sotos syndrome | *NSD1* | Histone methyltransferase (writer) | 5q35.3 | 117550 |
| Weaver syndrome | *EZH2* | Histone methyltransferase (writer) | 7q36.1 | 277590 |
| Cohen–Gibson syndrome | *EED* | Histone methyltransferase (writer) | 11q14.2 | 617561 |
| Wiedemann–Steiner syndrome | *KMT2A* | Histone methyltransferase (writer) | 11q23.3 | 605130 |
| Tatton–Brown–Rahman syndrome | *DNMT3A* | DNA methyltransferase (writer) | 2p23.3 | 615879 |
| Rubinstein–Taybi syndrome | *CREBBP* | Histone acetyltransferase (writer) | 16p13.3 | 180849 |
| | *EP300* | Histone acetyltransferase (writer) | 22q13.2 | 613684 |
| Hereditary sensory neuropathy type 1E | *DNMT1* | DNA methyltransferase (writer) | 19p13.2 | 614116 |
| Cerebellar ataxia, deafness, and narcolepsy, autosomal dominant | *DNMT1* | DNA methyltransferase (writer) | 19p13.2 | 604121 |
| Kleefstra syndrome | *EHMT1* | DNA methyltransferase (writer) | 9q34.3 | 610253 |
| Kabuki syndrome | *KMT2D* | Histone methyltransferase (writer) | 12q13.12 | 147920 |
| | *KDM6A* | Histone demethylase (eraser) | Xp11.3 | 300867 |

*Continued*

TABLE 5.1 Genetic Disorders Caused by Mutations in Epigenes[a]—cont'd

| Disorder | Gene | Function | Locus | OMIM |
|---|---|---|---|---|
| X-linked syndromic mental retardation; Claes–Jensen type | *KDM5C* | Histone demethylase (eraser) | Xp11.22 | 300534 |
| Brachydactyly–mental retardation syndrome | *HDAC4* | Histone deacetylase (eraser) | 2q37.3 | 600430 |
| Cornelia de Lange syndrome | *HDAC8* | Histone deacetylase (eraser) | Xq13.1 | 300882 |
| | *NIPBL* | Cohesin complex (reader) | 5p13.2 | 122470 |
| | *RAD21* | Cohesin complex (reader) | 8q24.11 | 614701 |
| | *SMC1A* | Cohesin complex (reader) | Xp11.22 | 300590 |
| | *SMC3* | Cohesin complex (reader) | 10q25.2 | 610759 |
| Rett syndrome | *MECP2* | Methyl-binding domain (reader) | Xq28 | 312750 |
| CHARGE syndrome | *CHD7* | Chromatin remodeling (remodeler) | 8q12.2 | 214800 |
| Coffin–Siris syndrome | *ARID1B* | Chromatin remodeling (remodeler) | 6q25.3 | 135900 |
| | *ARID1A* | Chromatin remodeling (remodeler | 1p36.11 | 614607 |
| | *SMARCA4* | Chromatin remodeling (remodeler | 19p13.2 | 135900 |
| | *SMARCB1* | Chromatin remodeling (remodeler) | 22q11.23 | 135900 |
| | *SMARCE1* | Chromatin remodeling (remodeler) | 17q21.2 | 135900 |
| Nicolaides–Baraitser syndrome | *SMARCA2* | Chromatin remodeling (remodeler) | 9p24.3 | 601358 |
| α-Thalassemia mental retardation syndrome, X-linked | *ATRX* | Chromatin remodeling (remodeler) | Xq21.1 | 301040 |
| Floating–Harbor syndrome | *SRCAP* | Chromatin remodeling (remodeler) | 16p11.2 | 136140 |
| Epileptic encephalopathy | *CHD2* | Chromatin remodeling (remodeler | 15q26.1 | 615369 |
| Autism spectrum disorder susceptibility | *CHD8* | Chromatin remodeling (remodeler) | 14q11.2 | 615032 |

[a]There are over 55 known epigenes—readers, writer, erasers, and chromatin remodelers—that are associated with Mendelian disorders. This table includes examples for each type of epigene. For a more detailed list see Kleefstra et al. Neuropharmacology 2014;80:83–94 and Bjornsson. Genome Res 2015;25:1473–81.

## 5.6.1 Human Disorders due to Mutations in Writers of Epigenetic Marks

Writers of epigenetic marks place specific modifications on either the DNA or its associated histone proteins, which constitutes one of the mechanisms that control gene transcription. Hemizygous, typically loss-of-function, mutations in a variety of epigenes encoding writers, including histone methyltransferases, histone acetyltransferases, and DNA methyltransferases, have been associated with well-known Mendelian disorders.

Sotos syndrome (*NSD1*): Sotos syndrome is caused by hemizygous mutations (including deletions) of the *NSD1* gene, which encodes a histone methyltransferase. Sotos syndrome is an overgrowth condition associated with macrocephaly, facial dysmorphology, advanced bone age, and learning difficulties or intellectual disability [148]. *NSD1*, which has a catalytic lysine methyltransferase Su(var)3-9, Enhancer-of-zeste and Trithorax (SET) domain and four zinc-binding plant homeodomain (PHD domains), functions primarily to mono- and dimethylate H3K36 [149]. The role of H3K36 methylation is not completely understood; in model organisms it has been found within gene bodies of expressed genes and is associated with suppression of intragenic transcriptional initiation [150]. Chromatin immunoprecipitation (ChIP)–ChIP experiments using promoter microarrays have shown that NSD1 binds to promoters of genes, playing a role in regulating various cellular processes, such as cell growth/cancer, keratin biology, and bone morphogenesis [151]. In addition, four of the *NSD1* PHD domains normally bind histone H3 methylated at K4 and K9; this binding is disrupted by most of the point mutations found in Sotos syndrome [152].

Weaver syndrome (*EZH2*): Weaver syndrome is characterized by pre- and postnatal overgrowth, accelerated osseous maturation, characteristic craniofacial appearance, and developmental delay. Mutations in *EZH2* (*Enhancer of Zeste, Drosophila, homolog 2*) are the primary cause of this syndrome [153,154]. *EZH2* encodes the catalytic component of PRC2, which regulates chromatin structure and gene expression through trimethylation of lysine 27 of histone H3. The core complex of PCR2 is made up of three components: *EZH2, EED*, and *SUZ12*. It has been established that mutations in all three components of the PCR2 complex cause Weaver or a Weaver-like syndrome [155,156]. Further studies are needed to determine if the phenotypes associated with mutations in these different proteins represent a single clinical entity—Weaver syndrome—or diverse syndromes with overlapping clinical features.

Kabuki syndrome (*KMT2D*): Kabuki syndrome is characterized by typical facial features, skeletal and dermatoglyphic anomalies, postnatal growth deficiency, and mild to moderate intellectual disability [157]. Mutations in *KMT2D* constitute the primary cause of Kabuki syndrome and are identified in about 80% of individuals. A second gene, *KDM6A*, accounts for about 6% of cases (see next section on erasers). *KMT2D* belongs to the SET1 family of histone H3K4 methyltransferases. It has a catalytic SET domain, five PHD domains, and an HMG-I binding motif [158]. *KMT2D* is a part of a multiprotein complex (which includes *KDM6A*) that catalyzes mono-, di-, and trimethylation of H3K4 [159] and regulates the expression of a wide range of downstream genes. H3K4 trimethylation is associated with active transcription [150], and the reduction of *KMT2D* in human HeLa cells results in downregulation of a number of genes involved in cell adhesion, cytoskeleton organization, transcriptional regulation, and development [159]. Interestingly, in a mouse model, *Kmt2d* has been shown to be crucial for the epigenetic reprogramming that takes place before fertilization in oocytes, such that reduced trimethylation of H3K4 due to deficiency of Kmt2d results in anovulation [160].

### 5.6.2 Human Disorders Due to Mutations in Erasers of Epigenetic Marks

Erasers of epigenetic marks remove specific modifications from either the DNA or its associated histone proteins and constitute another mechanism that controls gene transcription. Hemizygous mutations in several epigenetic erasers, including histone demethylases and HDACs, have been associated with well-known Mendelian disorders.

Kabuki syndrome (*KDM6A*): Kabuki syndrome, as described in the writer section, can be caused by loss-of-function mutations in both a writer, *KMT2D* (80%), and an eraser, *KDM6A* (6%). *KDM6A* encodes a histone H3 lysine 27 (H3K27) demethylase that is important for general chromatin remodeling, creating a closed chromatin mark. KDM6A interacts with KMT2D in a conserved SET1-like complex that targets H3K4, and they are both trithorax complex-like proteins. Therefore, the overall effect of mutations in either gene is predicted to be the same: regulating chromatin state and transcriptional activity at a specific set of target genes [161].

X-linked mental retardation (*KDM5C*): Mutations in the X-linked gene *KDM5C*, encoding a histone demethylase, cause a spectrum of phenotypes, ranging from syndromic to nonsyndromic intellectual disability. The clinical features in males with *KDM5C* mutations include mild to severe intellectual disability, epilepsy, short stature, aggressive behaviors, and microcephaly [162–167]. *KDM5C* escapes X inactivation and has a functional Y-linked homologue, *KDM5D*. Therefore, female heterozygous mutation carriers are usually unaffected; some demonstrate mild intellectual disability or learning difficulties [84]. KDM5C has several conserved functional domains, including the Bright/ARID domain responsible for DNA binding, the catalytic JmjC domain, and two PHD domains, responsible for histone binding [168,169]. KDM5C can bind the repressive histone mark H3K9me3 and removes the active mark H3K4me3/2, thus establishing a repressive chromatin state [170,171]. *KDM5C* point mutations identified in humans can suppress demethylase activity and/or H3K9me3 binding in vitro, depending on the location of the mutation [171]. ChIP in cell lines showed that KDM5C colocalizes with REST, a transcriptional repressor acting via neuron-restrictive silencing elements in the promoters of target genes such as *BDNF* and *SCN2A*, suggesting that downregulation of KDM5C activity impairs REST-mediated neuronal gene regulation [172].

### 5.6.3 Human Disorders Due to Abnormal Readers of Epigenetic Marks

Epigenetic regulators can affect gene expression by encoding proteins that read epigenetic marks in a temporal and spatial-specific manner.

Rett syndrome (*MECP2*): Heterozygous mutations of the X-linked gene *MECP2* cause Rett syndrome, a condition that occurs almost exclusively in girls. Rett syndrome is characterized by developmental arrest between 5 and 18 months of age, followed by regression of acquired skills, loss of speech, stereotypical movements, microcephaly, seizures, and severe intellectual disability [173]. The function of MECP2 has been studied extensively but the mechanism by which its deficiency results in the phenotypes of Rett syndrome remains incompletely understood. Initially, MECP2 was identified as a protein capable of binding methylated DNA [174]. It was found to have abundant binding sites distributed throughout the genome and was demonstrated to function in repression of transcription [175–177]. The best established mechanism by which MECP2 downregulates gene expression is through recruitment of HDACs, which transform specific regions of chromatin into a repressive state by removing acetyl groups from histones H3 and H4 [175,177,178]. There is growing evidence that the role of MECP2 in transcription regulation is more complex; for example, in mice it was shown to bind to the transcriptional activator CREB to activate transcription of a large number of genes in the hypothalamus [179]. Further, MECP2 deficiency is associated with dysregulation of specific genes, such as *BDNF*, which has been shown to have MECP2 binding sites. Reduction of Bdnf in a mouse model mimics some features of the *Mecp2*-null mouse phenotype [180] and Bdnf overexpression in these mice can partially rescue the phenotype by improving locomotor function, extending life span [181], and rescuing synaptic dysfunction [182]. These data suggest that *Bdnf* is indeed an important and clinically relevant Mecp2 transcriptional target. Findings suggest that Mecp2 is almost as abundant as histone H1 in mouse neurons but not in glia [183], so Mecp2 function in neurons might affect genome-wide chromatin remodeling in addition to regulating the expression of specific genes. Moreover, targeted deletion of *Mecp2* in mice results in increased expression of repetitive elements in neurons [183], prompting investigators to suggest a role for this protein in limiting overall transcriptional noise in neurons and the capacity of neurons to respond to environmental signals [184].

### 5.6.4 Human Disorders Due to Mutations in Chromatin Remodelers

Chromatin remodeling is an important epigenetic mechanism critical for nucleosome mobility and transcriptional control. Chromatin remodelers often function as part of large macromolecular protein complexes that can reorganize nucleosome structure in an ATP-dependent fashion [185]. Many such genes have been implicated in Mendelian disorders.

CHARGE syndrome (*CHD7*) and other disorders due to mutations in chromodomain helicase enzymes: CHARGE syndrome is characterized by coloboma, heart defects, atresia of the choanae, retardation of growth and development, genital hypoplasia, and ear abnormalities, including deafness and vestibular disorders [186,187]. Nonsense or missense mutations and deletions that result in haploinsufficiency of the *CHD7* gene cause the majority of CHARGE syndrome cases [187,188]. CHD7 is an ATP-dependent chromodomain helicase chromatin-remodeling protein involved in the formation of several large protein complexes that regulate the movement of nucleosomes along DNA, thereby affecting the activity of numerous signaling pathways during embryonic development. The *CHD7* gene is expressed in ES cells, and its expression becomes restricted to specific tissues, including the brain, eye, heart, and ear, during differentiation. CHD7-containing protein complexes bind to DNA at specific sites, the majority of which constitute regulatory elements such as gene promoters or enhancers [189]. The epigenetic effects of CHD7 on chromatin and gene regulation appear to vary both temporally and spatially, depending largely upon the function of the protein complex with which it interacts. Other genes in the Chromodomain Helicase DNA-binding protein family, including *CHD2* and *CHD8*, have been associated with epileptic encephalopathy and susceptibility to autism, respectively.

α-Thalassemia/mental retardation syndrome, X-linked (ATR-X) (*ATRX*): Mutations in an X-linked gene, *ATRX*, cause ATR-X, which is characterized by severe intellectual disability, facial dysmorphology, urogenital anomalies, and α-thalassemia. As the *ATRX* gene normally undergoes X inactivation, affected individuals are almost exclusively males, while females usually are unaffected, due to preferential X inactivation of the chromosome with the *ATRX* mutation [190]. The ATRX protein is involved in epigenetic regulation through two functional domains: an ATP/helicase domain and an ATRX–Dnmt3–Dnmt3L (ADD) domain that shares homology with de novo methyltransferases. The ATP/helicase domain is proposed to be involved in nucleosome repositioning and making DNA more accessible for protein binding [191], while the ADD domain has been shown to bind histone H3 tails with the silencing mark H3K9me3 but not the active mark H3K4me3/2 [192]. In terms of genomic targets, ATRX has been shown to localize to the nucleus in heterochromatin, telomeric/subtelomeric chromosomal regions, ribosomal DNA (rDNA) and promyelocytic leukemia bodies [193–195]. Furthermore, peripheral blood cells of ATR-X patients exhibit changes in DNA methylation of rDNA, subtelomeric repeats, and Y-chromosome-specific satellites [193]. By ChIP-sequencing, it was established that in erythroid cells ATRX binds to CpG-rich tandem repeat sequences clustered at subtelomeric regions, thereby affecting the expression of associated genes, including α-globin, which accounts for the α-thalassemia phenotype of ATR-X syndrome [196].

## 5.7 METHODS FOR STUDYING EPIGENETIC MARKS

### 5.7.1 Mapping DNA Methylation

One of the first methods to score DNA methylation at specific loci was Southern blotting of genomic DNA digested with methylation-sensitive restriction enzymes [197]. Certain restriction enzymes (e.g., *Hpa*II, *Sma*I, *Not*I) that contain a CpG as part of their recognition sequence do not cut at methylated sites. Therefore, failure to cleave by a methyl-sensitive restriction enzyme is evidence of DNA methylation at that site. Restriction enzymes can also be used in combination with microarray platforms to evaluate genome-wide DNA methylation patterns, including promoter methylation and allele-specific methylation [198,199]. Methylation-specific PCR, based on predigestion of genomic DNAs with such restriction enzymes, provides a semiquantitative measurement of DNA methylation levels. While such methods have generally been supplanted by approaches based on bisulfite conversion of DNA (see later), assays with methylation-sensitive restriction enzymes can still be useful for independently validating the results from bisulfite-based methods.

Chemical conversion of DNA via sodium bisulfite is the gold standard for DNA methylation analysis, allowing for quantitative analysis of CpG methylation. Sodium bisulfite deaminates nonmethylated dC's to dU residues. During subsequent PCR amplification, the dU's are paired with T's and amplified as A/T base pairs; however, if a C is methylated, the DNA sequence does not change, and a C will be paired with a G [200]. Pyrosequencing after bisulfite conversion can determine average DNA methylation levels at individual CpG sites across a short genomic region. A more informative approach involves amplification of bisulfite PCR products followed by cloning and Sanger sequencing or deep massively parallel sequencing. This approach reveals DNA methylation levels of a larger number of individual CpG sites and can be useful for detection of heterogeneous and allele-specific patterns of methylation [201].

By combining sodium bisulfite conversion and microarrays or massively parallel bisulfite sequencing, net and allele-specific DNA methylation patterns are being mapped genome-wide in an ever-increasing number of cell types from individuals with a variety of diseases as well as controls (e.g., [116,202]). These methods to characterize the methylation status of all or many of the ~28 million genomic CpG sites can be broadly classified into two categories: microarray and next-generation sequencing (NGS) based. The Illumina Methylation BeadChip assays (450K and EPIC arrays) are the most commonly used methods for epigenome-wide association studies (EWAS) and related types of studies [203,204]. The BeadChips are limited by a fixed number of probes, but are widely used because of their low cost, low DNA input requirement, and significantly reduced sample processing time. NGS-based approaches such as whole-genome bisulfite sequencing are generally regarded as the gold standard genome-wide methods because they provide the broadest coverage at single-base resolution. So far, this definitive approach has been less widely used, since the read depth required, data storage requirements, and computational processing time for the resulting terabyte-scale data have remained cost prohibitive for many researchers [205]. Two alternative NGS approaches, reduced representation bisulfite sequencing (RRBS) [206] and methyl-capture sequencing (MC-Seq), aim to enrich for specific regulatory genomic regions known to be targets for epigenetic dysregulation (i.e., CpG islands, CpG shores, and CpG shelves, close to transcriptional start sites and promoters) [207]. RRBS uses a restriction enzyme that recognizes CpG's to enrich for CpG-rich regions, while MC-Seq uses target-specific bait sequences. Going forward, efforts are under way to adopt single-molecule sequencing technologies for the direct detection of multiple types of DNA methylation in unamplified DNA, with a view to analyzing various sample types, including low-abundance specimens [208].

### 5.8.3 Abnormalities of Histones and Histone Modifications in Cancer

Epigenetic alterations in cancer are not restricted to DNA methylation. Genome-wide mapping of histone marks has demonstrated global changes in histone modifications in most cancer types. For example, acetylated H4 lysine 16 and H4 lysine 20 trimethylation are commonly identified alterations in many tumors correlating with repression of gene expression [269]. HDACs have been found to be overexpressed in a number of cancer types, and in some cancers there can be dysregulation of histone acetyltransferases due to translocations resulting in deleterious gene fusion products [270–272]. Aberrant histone methylation of H3K9 and H3K27 also results in gene silencing in many cancers [26,273,274]. *EZH2*, a histone methyltransferase of H3K27, is frequently overexpressed in breast and prostate tumors, in addition to other tumors [275,276]. Loss-of-function mutations in *EZH2* have also been seen in myeloid malignancies and T cell leukemias, demonstrating the importance of H3K27 methylation [277,278]. As is true for aberrant DNA methylation, most abnormalities in histone modifications in cancer are not yet explained by a single genetic lesion. However, with high-coverage sequencing technologies an increasing number of chromatin-modifying enzymes, such as CREB, JARID1C, EZH2, and the SWI/SNF family proteins hSNF5/INI1 and PBRM1, are now being found mutated in specific types of human cancers [235,277,279–282]. Mutations have been found not only in histone-modifying proteins but also in histones themselves. A recurrent mutation, p.K36M of histone H3 variants, which disrupts H3K36 methylation, has been identified in a number of tumor types, including glioblastomas and head and neck tumors [283,284]. In diffuse intrinsic pontine glioma, a fatal pediatric brain tumor, this mutation has been shown to be a driver of oncogenesis [285].

### 5.8.4 Aberrant miRNA and lncRNA Expression in Cancer

Comparisons of tumor tissues and corresponding normal tissues have revealed global changes in miRNA expression during tumorigenesis [286]. Upregulation of *miR-21*, which targets *PTEN*, occurs in glioblastomas [287], and in chronic lymphocytic leukemia *miR-15* and *-16*, which target the antiapoptotic gene *BCL2*, are downregulated [288]. Similarly, *let-7*, which targets the oncogene *RAS* is downregulated in lung cancer [288]. These alterations in miRNA expression may occur through a number of mechanisms, including chromosomal abnormalities, transcription factor binding, and epigenetic alterations [289]. Silencing of miRNA expression has been shown to occur by aberrant hypermethylation in a number of cancers [290,291] and, like other epigenetic factors discussed earlier, the role of miRNA dysregulation in cancer has been validated genetically by findings of DNA deletions encompassing miRNA genes, e.g., on chromosome 13 in chronic lymphocytic leukemias [292]. The possibility of introducing miRNA as a therapy using a variety of methods is being explored in vitro and also in in vivo models of many different cancers [293]. In an early study restoration of let-7 in a mouse model of lung cancer resulted in decreased tumor growth [294].

LncRNAs have also been shown to be dysregulated in cancer [295,296]. Overexpression of HOTAIR in early-stage breast cancer is predictive of progression to metastatic disease as well as overall survival [297]. Further studies have shown that aberrant expression of HOTAIR is associated with cancer progression in a number of tumor types [298]. MALAT1 overexpression predicts metastatic progression in early-stage non-small-cell lung cancer (NSCLC) [299]. Decreased expression of Malat1 in a mouse model reduces metastasis of lung carcinomas [300]. Overexpression of MALAT1 has also been reported in other cancer types, including breast cancer [301]. These studies suggest that lncRNAs may provide targets for therapeutics that could be applicable to a wide range of tumor types [296].

### 5.8.5 Therapies Targeting Epigenetic Modifications

A number of classes of medications that target epigenetic pathways that are disrupted in cancer have been approved by the US Food and Drug Administration (FDA), either as a treatment for cancer or for clinical trials [240,302]. The DNA methylation inhibitors 5-azacytidine (azacytidine) and 5-aza-2′-deoxycytidine (decitabine), are nucleoside analogues that get incorporated into the genomes of growing tumor cells and inhibit DNA methyltransferase enzymes, leading to progressive loss of DNA methylation with each S phase of the cell cycle. These medications have been approved for use in the treatment of MDS and have shown some promise for treating acute lymphoblastic leukemia and other hematological malignancies. [303–305]. HDAC inhibitors (HDACi), which have been approved by the FDA (including vorinostat, suberoylanilide hydroxamic acid), are being used with good results in treating patients with cutaneous T cell lymphoma in the United States [306]. There have been a number of clinical trials in solid tumors with HDACi with little success, due to adverse events [307]. This problem can be addressed by using an HDACi in combination with other chemotherapeutics, specifically for NSCLC and ovarian cancer [302,308].

There are several recurrent cancer-associated somatic mutations, for example, activating mutations in *IDH1/2* or *EZH2*, which result in the alteration of histone marks and DNA methylation [240]. For both of these genes, drugs that have been designed to inhibit the aberrant protein are in clinical trials [240]. *EZH2* is activated by mutations in lymphomas; the use of an inhibitor has been shown to induce apoptosis or differentiation in cell lines, and this drug is now in clinical trials [309].

Beyond treatment for cancers, such medications are being used and developed to treat a wider spectrum of diseases. Resveratrol, a natural compound in red wine, which inhibits sirtuins (a family of HDACs), is being evaluated as a treatment for T2D and metabolic syndrome [310]. Valproic acid (VPA), also an HDAC inhibitor, is used to treat seizures and mood disorders [311,312], and VPA in combination with other medications has been shown to inhibit cancers in vitro [313]. The wide array of epigenetic alterations identified in human disease could present valuable targets for repurposing approved medications and developing novel agents. For example, small molecules that inhibit bromodomain and extraterminal proteins, which are important epigenetic readers of acetylated lysine residues in histones, are showing promising preclinical results in diseases ranging from cancers to organ fibrosis and osteoporosis [314–319].

## 5.9 ENVIRONMENTAL INFLUENCES ON THE EPIGENOME

The concept of environmental influence on the genome encompassing transgenerational epigenetic inheritance is of great importance to our current understanding of the underlying molecular mechanisms of health and disease. These environmental influences can be transgenerational, i.e., they can affect not only the individuals exposed but also future generations. The effects of environment on our epigenome have been documented for two tragic historical events, the Dutch Hunger Winter (1944) and the Great Chinese Famine (1958–61). Studies of children born following these periods reported them to be small for gestational age at birth and to have an increased risk for schizophrenia, effects that were found to last two generations following the famine [320,321]. Children who were conceived during the Dutch famine demonstrated long-term epigenetic effects, in that changes in DNA methylation were observed six decades later in the *IGF2* gene, which encodes a growth hormone critical for normal embryonic/fetal development [322]. These data constitute the first concrete evidence that in humans the maternal diet early in pregnancy can directly affect epigenetic programming early in utero, impacting growth, metabolism, and neurodevelopment throughout life. Further, the impact is carried forward to future generations. The "developmental origins of health and disease" (DOHAD) hypothesis, pioneered by David Barker [323], has predicted, among other things, that maternal stress during pregnancy (dietary inadequacy, toxic exposures, and perhaps psychological stress) might lead to persistent epigenetic changes in the fetus, which could play a role in modulating the subsequent onset of adult cardiovascular, metabolic, and psychiatric diseases. Other maternal gestational environmental factors, such as gestational diabetes, autoimmune diseases, and infections, have also been associated with adverse pregnancy outcomes, including an increased risk for neuropsychiatric disorders [324–331]. The transgenerational effects of these exposures are not yet documented.

For many of the diseases and health outcomes discussed here, the underlying etiology has not been fully elucidated. It is likely, for example, in the case of phenotypically heterogeneous diseases such as schizophrenia and autism spectrum disorder (ASD), that there are multifactorial etiologic factors to consider. These encompass genetics and environment (often referred to as "G × E" effects), where epigenetics can be used as a powerful tool to reflect their interactions. We are only beginning to understand which environmental signals can alter epigenetic marks, most commonly DNA methylation, of specific genes/regions or the genome. Data are emerging from the exploration of other stable epigenetic marks such as miRNAs, which appear to be a promising epigenetic and molecular mark for studying the origins of brain disorders, given their stability [332–335], their abundance in the brain, and their role in regulating neuronal plasticity and development [336,337]. We will discuss these data further in the context of exogenous, endogenous, or social environmental exposures.

### 5.9.1 Exogenous Exposures

Epigenetic marks can be altered by a broad range of exogenous exposures. Theoretically, it is possible to detect and quantify specific epigenetic alterations for many types of exposures. Whether the observed epigenetic dysregulation is directly linked to adverse health outcomes resulting from these exposures has yet to be confirmed, especially given that the tissues assessed in such studies are not necessarily the primary tissue impacted by the exposure. For example, environmental exposure to compounds such as cadmium [338] and arsenic [339] may predispose to epigenetic instability, aging, and cancer. Lifestyle factors such as diet, cigarette smoking, and alcohol have also been shown to alter DNA methylation in the individual exposed, as well as in the offspring, if exposure occurs during pregnancy [340,341]. Gestational exposure to cigarette smoking and alcohol has been associated with growth dysregulation and an increase in the rate of certain NDDs (e.g., fetal alcohol spectrum disorder, attention deficit hyperactivity disorder) and cancer. DNA methylation and histone methylation alterations have been associated with smoking or alcohol exposure in both human and animal models, across various tissues [240,342–344], demonstrating the utility of epigenetic marks as potential biomarkers of environmental exposures and/or disease.

Certain medications can also have direct or indirect effects on DNA methylation and histone acetylation, as demonstrated in both animal and human studies [324,340,345,346]. The commonly prescribed antiepileptic drug VPA, if taken during pregnancy, can lead to adverse birth outcomes, including teratogenic effects and impaired postnatal cognitive development [347,348]. VPA inhibits HDACs and has been shown to alter DNA methylation and DNA methyltransferase activity in various tissues in both the mother and the exposed embryo [347,349]. Other drugs with epigenetic targets have been used widely in the treatment of certain cancers and neurodegenerative diseases [350–352], but their therapeutic potential may extend beyond these diseases. Further, the detection and quantitation of epigenetic alterations may be useful as a predictive biomarker to identify individuals who may be more amenable to specific interventions and also as an endpoint for evaluating therapeutic efficacy.

Alterations to the one-carbon metabolism (OCM) cycle can have a significant impact on the cell's ability to methylate DNA, and thus can disrupt normal epigenetic regulation [353]. Both folate, which is obtained in part through our diet, and the enzyme methyltetrahydrofolate reductase (MTHFR) are components of OCM, and are important for DNA synthesis and methylation. Mice lacking the enzyme Mthfr have been shown to have decreased global DNA methylation. In humans, MTHFR enzyme activity depends on an individual's genotype for the functional polymorphism MTHFR 1298A→C, which correlates positively with the level of global DNA methylation [354]. Further, humans on a folate-depleted diet demonstrate decreased global DNA methylation [355]. Conversely, in a study of adult males on hemodialysis, adding an exogenous source of folate led to an increase in both global and locus-specific DNA methylation, including *H19*, *IGF2*, and *SYDL1* [356]. Maternal folate and/or related B vitamin status has been shown to influence the cognitive outcome of offspring, although the evidence for this remains conflicting and the direct mediating role of DNA methylation remains unclear [357,358]. Last, dietary supplementation with folate and B vitamins has been clearly shown to modify tumor incidence in mouse models [359,360]. Interest in investigating the impact and therapeutic potential of folate and related vitamins in brain disorders in which epigenetic dysregulation has been implicated in the molecular etiology (e.g., ASD, neurodegenerative diseases) has increased in recent years [361–363], but is still in the early stages of research.

### 5.9.2 Impact of Endogenous Gestational Environment and Assisted Reproductive Technologies on Epigenetic Programming

Infertility and assisted reproductive technologies (ART) may have an impact on epigenetic reprogramming. In mice and humans, oocytes retrieved following hormonal induction or embryos studied after in vitro culture have shown DNA methylation and/or expression anomalies in several imprinted genes [364–367]. Studies of human oocytes harvested after medical hormonal induction showed loss of methylation at the maternal *MEST/PEG1* DMR on chromosome band 7q33 [368] and gain of methylation at the maternal

*H19* DMR (IC1) on chromosome band 11p15.5 [367]. Increasing attention has been focused on reports of increased rates of epigenetic errors in humans following infertility/ART, particularly given that ART are administered during a sensitive developmental period when epigenetic reprogramming is occurring. The use of ART, as of this writing, accounts for ~6% of live births in North America, with continually increasing rates [369]; [370]. In particular, two rare epigenetic disorders, BWS and AS, exhibited an increased incidence in retrospective studies (odds ratios of 6–17 and 6–12, respectively) in children born following infertility/ART [371–376]. The data are especially compelling in that the increased incidence is attributable to an increase in specific epigenetic errors at chromosome regions 11p15 (BWS) and 15q11–q13 (AS), with both locations being affected by abnormal imprinting on the maternal (oocyte-derived) alleles. Furthermore, in ART-conceived AS and BWS patients, loss of maternal methylation at their respective DMRs occurs 8 and 1.9 times more often, respectively, than in individuals born from spontaneous conceptions [377–379]. Such evidence supports the hypothesis that ART-conceived children have an increased rate of epigenetic errors over that in the general population.

In humans, it is still unclear whether maternal loss of methylation observed in children post-ART is the result of the procedure itself or of an underlying infertility with oocyte abnormalities in the couple seeking ART interventions, or both. Idiopathic male infertility is also associated with aberrant methylation at both maternal and paternal alleles, suggesting that male germ cells represent another potential source for methylation defects in children conceived via ART [380,381]. Ovarian stimulation (part of subfertility/infertility treatment) has been linked to perturbed genomic imprinting at both maternally and paternally expressed genes. These data demonstrate that superovulation has dual effects during oogenesis: disruption of imprint acquisition in growing oocytes and disruption of a maternal-effect gene product subsequently required for imprint maintenance during preimplantation development [382]. The use of ART itself is associated with an increased risk for several adverse birth outcomes, including preterm labor, multiple births, and low birth weight, that, independent of ART, also confer an increased risk for adverse health outcomes, including NDDs such as ASD [383–385].

### 5.9.3 Social Environmental Exposures and Their Impact on Epigenetic Outcomes

In both mice and humans, maternal behavior in the neonatal period may correlate with epigenetic programming of the offspring's adult behavior [386,387]. Studies in mice have indicated that mothers showing strong nurturing behavior toward their pups, by frequently licking and grooming their offspring, produce alterations in the patterns of DNA methylation, for example, in the promoter of the *glucocorticoid receptor* gene (*GR*), in the hippocampus of their pups [388]. This area of research has been steadily growing and it will be important to follow this work over the next several years. The link between altered epigenetic marks and postnatal health outcomes is still unclear. The literature on this topic has not been well replicated, and further, there is not yet a clear causative link between the genes and biological pathways affected by altered epigenetic mechanisms and the health outcomes discussed.

As an extension of these observations on gestational exposures, early life adversity and early childhood environment are well known to be associated with long-term health outcomes, like neuropsychiatric problems and cancer [389–391]. One of the mechanisms hypothesized to be mediating these outcomes involves the perturbation of epigenetic marks such as DNA methylation. It has been suggested that specific biological pathways are particularly affected by epigenetic changes as a result of these environmental conditions. In both mouse and human models, it has been shown that early life environment, including the quality of maternal care, can influence hypothalamic–pituitary–adrenal (HPA) axis function [392,393]. Alterations in DNA methylation of the aforementioned *GR* gene, a critical component regulating the HPA axis and stress response, have been well studied in the context of the early life social environment [394]. Although results can be conflicting, this is largely attributable to the variation in experimental designs, models, tissue types, and gene regions examined. There is now growing evidence to corroborate the role of the methylation status of several important players in the HPA axis (e.g., *FKBP5* in humans, *Pomc* in mice) [395,396]. Interestingly, the involvement of the methylation status of *GR* and other components of the HPA axis has also been observed in the social environmental effects on epigenetic mechanisms

W, Waterland RA, Zhang MQ, Chadwick LH, Bernstein BE, Costello JF, Ecker JR, Hirst M, Meissner A, Milosavljevic A, Ren B, Stamatoyannopoulos JA, Wang T, Kellis M. Integrative analysis of 111 reference human epigenomes. Nature 2015;518:317–30.
[12] Stunnenberg HG, International Human Epigenome C, Hirst M. The international human epigenome consortium: a blueprint for Scientific collaboration and discovery. Cell 2016;167:1897.
[13] Tilghman S.M. The sins of the fathers and mothers: genomic imprinting in mammalian development. Cell 1999;96:185–193.
[14] Yoder J.A., Walsh C.P., Bestor T.H. Cytosine methylation and the ecology of intragenomic parasites. Trends Genet 1997;13:335–340.
[15] Pradhan S., Bacolla A., Wells R.D., Roberts R.J. Recombinant human DNA (cytosine-5) methyltransferase. I. Expression, purification, and comparison of de novo and maintenance methylation. J Biol Chem 1999;274:33002–33010.
[16] Bourc'his D., Xu G.L., Lin C.S., Bollman B., Bestor T.H. Dnmt3L and the establishment of maternal genomic imprints. Science 2001;294:2536–2539.
[17] Law J.A., Jacobsen S.E. Establishing, maintaining and modifying DNA methylation patterns in plants and animals. Nat Rev Genet 2010;11:204–220.
[18] Wu X., Zhang Y. TET-mediated active DNA demethylation: mechanism, function and beyond. Nat Rev Genet 2017;18:517–534.
[19] Wolffe A.P., Matzke M.A. Epigenetics: regulation through repression. Science 1999;286:481–486.
[20] Costello J.F., Plass C. Methylation matters. J Med Genet 2001;38:285–303.
[21] Straussman R., Nejman D., Roberts D., Steinfeld I., Blum B., Benvenisty N., Simon I., Yakhini Z., Cedar H. Developmental programming of CpG island methylation profiles in the human genome. Nat Struct Mol Biol 2009;16:564–571.
[22] Jones P.A., Baylin S.B. The fundamental role of epigenetic events in cancer. Nat Rev Genet 2002;3:415–428.
[23] Portela A., Esteller M. Epigenetic modifications and human disease. Nat Biotechnol 2010;28:1057–1068.
[24] Cedar H., Bergman Y. Linking DNA methylation and histone modification: patterns and paradigms. Nat Rev Genet 2009;10:295–304.
[25] Epsztejn-Litman S., Feldman N., Abu-Remaileh M., Shufaro Y., Gerson A., Ueda J., Deplus R., Fuks F., Shinkai Y., Cedar H., Bergman Y. De novo DNA methylation promoted by G9a prevents reprogramming of embryonically silenced genes. Nat Struct Mol Biol 2008;15:1176–83.
[26] Kondo Y. Epigenetic cross-talk between DNA methylation and histone modifications in human cancers. Yonsei Med J 2009;50:455–463.
[27] Lehnertz B., Ueda Y., Derijck A.A., Braunschweig U., Perez-Burgos L., Kubicek S., Chen T., Li E., Jenuwein T., Peters A.H. Suv39h-mediated histone H3 lysine 9 methylation directs DNA methylation to major satellite repeats at pericentric heterochromatin. Curr Biol 2003;13:1192–1200.
[28] Ooi S.K., Qiu C., Bernstein E., Li K., Jia D., Yang Z., Erdjument-Bromage H., Tempst P., Lin S.P., Allis C.D., Cheng X., Bestor T.H. DNMT3L connects unmethylated lysine 4 of histone H3 to de novo methylation of DNA. Nature 2007;448:714–717.
[29] Weber M., Hellmann I., Stadler M.B., Ramos L., Paabo S., Rebhan M., Schubeler D. Distribution, silencing potential and evolutionary impact of promoter DNA methylation in the human genome. Nat Genet 2007;39:457–466.
[30] Kaikkonen M.U., Lam M.T., Glass C.K. Non-coding RNAs as regulators of gene expression and epigenetics. Cardiovasc Res 2011;90:430–440.
[31] Kim D.H., Saetrom P., Snove Jr. O., Rossi J.J. MicroRNA-directed transcriptional gene silencing in mammalian cells. Proc Natl Acad Sci U S A 2008;105:16230–16235.
[32] Ponting C.P., Oliver P.L., Reik W. Evolution and functions of long noncoding RNAs. Cell 2009;136:629–641.
[33] Dittrich B., Robinson W.P., Knoblauch H., Buiting K., Schmidt K., Gillessen-Kaesbach G., Horsthemke B. Molecular diagnosis of the Prader-Willi and Angelman syndromes by detection of parent-of-origin specific DNA methylation in 15q11-13. Hum Genet 1992;90:313–315.
[34] Zhao J., Sun B.K., Erwin J.A., Song J.J., Lee J.T. Polycomb proteins targeted by a short repeat RNA to the mouse X chromosome. Science 2008;322:750–6.
[35] Nagano T, Mitchell JA, Sanz LA, Pauler FM, Ferguson-Smith A.C., Feil R., Fraser P. The Air noncoding RNA epigenetically silences transcription by targeting G9a to chromatin. Science 2008;322:1717–1720.
[36] Tsai M.C., Manor O., Wan Y., Mosammaparast N., Wang J.K., Lan F., Shi Y., Segal E., Chang H.Y. Long noncoding RNA as modular scaffold of histone modification complexes. Science 2010;329:689–693.
[37] Davidovich C., Cech T.R. The recruitment of chromatin modifiers by long noncoding RNAs: lessons from PRC2. RNA 2015;21:2007–2022.
[38] Khalil A.M., Guttman M., Huarte M., Garber M., Raj A., Rivea Morales D., Thomas K., Presser A., Bernstein B.E., van Oudenaarden A., Regev A., Lander E.S., Rinn J.L. Many human large intergenic noncoding

RNAs associate with chromatin-modifying complexes and affect gene expression. Proc Natl Acad Sci U S A 2009;106:11667–11672.
[39] Mackay D.J.G., Temple I.K. Human imprinting disorders: principles, practice, problems and progress. Eur J Med Genet 2017;60:618–626.
[40] Arand J., Wossidlo M., Lepikhov K., Peat J.R., Reik W., Walter J. Selective impairment of methylation maintenance is the major cause of DNA methylation reprogramming in the early embryo. Epigenet Chromatin 2015;8:1.
[41] Barlow D.P., Bartolomei M.S. Genomic imprinting in mammals. Cold Spring Harb Perspect Biol 2014;6.
[42] Feng S., Jacobsen S.E., Reik W. Epigenetic reprogramming in plant and animal development. Science 2010;330:622–627.
[43] Ito S., D'Alessio A.C., Taranova O.V., Hong K., Sowers L.C., Zhang Y. Role of Tet proteins in 5mC to 5hmC conversion, ES-cell self-renewal and inner cell mass specification. Nature 2010;466:1129–1133.
[44] Kriaucionis S., Heintz N. The nuclear DNA base 5-hydroxymethylcytosine is present in Purkinje neurons and the brain. Science 2009;324:929–930.
[45] Tahiliani M., Koh K.P., Shen Y., Pastor W.A., Bandukwala H., Brudno Y., Agarwal S., Iyer L.M., Liu D.R., Aravind L., Rao A. Conversion of 5-methylcytosine to 5-hydroxymethylcytosine in mammalian DNA by MLL partner TET1. Science 2009;324:930–5.
[46] Ficz G, Branco MR, Seisenberger S, Santos F, Krueger F, Hore TA, Marques CJ, Andrews S, Reik W. Dynamic regulation of 5-hydroxymethylcytosine in mouse ES cells and during differentiation. Nature 2011.
[47] Koh K.P., Yabuuchi A., Rao S., Huang Y., Cunniff K., Nardone J., Laiho A., Tahiliani M., Sommer C.A., Mostoslavsky G., Lahesmaa R., Orkin S.H., Rodig S.J., Daley G.Q., Rao A. Tet1 and Tet2 regulate 5-hydroxymethylcytosine production and cell lineage specification in mouse embryonic stem cells. Cell Stem Cell 2011;8:200–213.
[48] Bhutani N., Brady J.J., Damian M., Sacco A., Corbel S.Y., Blau H.M. Reprogramming towards pluripotency requires AID-dependent DNA demethylation. Nature 2010;463:1042–1047.
[49] Iqbal K., Jin S.G., Pfeifer G.P., Szabo P.E. Reprogramming of the paternal genome upon fertilization involves genome-wide oxidation of 5-methylcytosine. Proc Natl Acad Sci U S A 2011;108:3642–3647.
[50] Wossidlo M., Nakamura T., Lepikhov K., Marques C.J., Zakhartchenko V., Boiani M., Arand J., Nakano T., Reik W., Walter J. 5-Hydroxymethylcytosine in the mammalian zygote is linked with epigenetic reprogramming. Nat Commun 2011;2:241.
[51] Lyon M.F. Gene action in the X-chromosome of the mouse (Mus musculus L.). Nature 1961;190: 372–373.
[52] Graves J.A. Review: Sex chromosome evolution and the expression of sex-specific genes in the placenta. Placenta 2010;31(Suppl.):S27–S32.
[53] Vicoso B., Charlesworth B. Evolution on the X chromosome: unusual patterns and processes. Nat Rev Genet 2006;7:645–653.
[54] Graves J.A. Sex chromosome specialization and degeneration in mammals. Cell 2006;124:901–914.
[55] Heard E., Avner P. Role play in X-inactivation. Hum Mol Genet 1994;(3 Spec No):1481–1485.
[56] Brown CJ, Ballabio A, Rupert JL, Lafreniere RG, Grompe M, Tonlorenzi R, Willard HF. A gene from the region of the human X inactivation centre is expressed exclusively from the inactive X chromosome. Nature 1991;349:38–44.
[57] Brown C.J., Hendrich B.D., Rupert J.L., Lafreniere R.G., Xing Y., Lawrence J., Willard H.F. The human XIST gene: analysis of a 17 kb inactive X-specific RNA that contains conserved repeats and is highly localized within the nucleus. Cell 1992;71:527–542.
[58] Clemson C.M., Chow J.C., Brown C.J., Lawrence J.B. Stabilization and localization of Xist RNA are controlled by separate mechanisms and are not sufficient for X inactivation. J Cell Biol 1998;142:13–23.
[59] Augui S., Filion G.J., Huart S., Nora E., Guggiari M., Maresca M., Stewart A.F., Heard E. Sensing X chromosome pairs before X inactivation via a novel X-pairing region of the Xic. Science 2007;318:1632–1636.
[60] Augui S., Nora E.P., Heard E. Regulation of X-chromosome inactivation by the X-inactivation centre. Nat Rev Genet 2011;12:429–442.
[61] Okamoto I., Patrat C., Thepot D., Peynot N., Fauque P., Daniel N., Diabangouaya P., Wolf J.P., Renard J.P., Duranthon V., Heard E. Eutherian mammals use diverse strategies to initiate X-chromosome inactivation during development. Nature 2011;472:370–374.
[62] Morey C., Avner P. Genetics and epigenetics of the X chromosome. Ann N Y Acad Sci 2010;1214:E18–E33.
[63] Okamoto I., Heard E. Lessons from comparative analysis of X-chromosome inactivation in mammals. Chromosome Res 2009;17:659–669.
[64] Brown C.J., Willard H.F. The human X-inactivation centre is not required for maintenance of X-chromosome inactivation. Nature 1994;368:154–156.
[65] Donohoe M.E., Silva S.S., Pinter S.F., Xu N., Lee J.T. The pluripotency factor Oct4 interacts with Ctcf and also controls X-chromosome pairing and counting. Nature 2009;460:128–132.

[66] Guo G., Yang J., Nichols J., Hall J.S., Eyres I., Mansfield W., Smith A. Klf4 reverts developmentally programmed restriction of ground state pluripotency. Development 2009;136:1063–9.
[67] Navarro P, Oldfield A, Legoupi J, Festuccia N, Dubois A., Attia M., Schoorlemmer J., Rougeulle C., Chambers I., Avner P. Molecular coupling of Tsix regulation and pluripotency. Nature 2010;468:457–460.
[68] Lee J.T., Davidow L.S., Warshawsky D. Tsix, a gene antisense to Xist at the X-inactivation centre. Nat Genet 1999;21:400–404.
[69] Tian D., Sun S., Lee J.T. The long noncoding RNA, Jpx, is a molecular switch for X chromosome inactivation. Cell 2010;143:390–403.
[70] Xu N., Tsai C.L., Lee J.T. Transient homologous chromosome pairing marks the onset of X inactivation. Science 2006;311:1149–1152.
[71] Jonkers I., Barakat T.S., Achame E.M., Monkhorst K., Kenter A., Rentmeester E., Grosveld F., Grootegoed J.A., Gribnau J. RNF12 is an X-encoded dose-dependent activator of X chromosome inactivation. Cell 2009;139:999–1011.
[72] Shin J., Bossenz M., Chung Y., Ma H., Byron M., Taniguchi-Ishigaki N., Zhu X., Jiao B., Hall L.L., Green M.R., Jones S.N., Hermans-Borgmeyer I., Lawrence J.B., Bach I. Maternal Rnf12/RLIM is required for imprinted X-chromosome inactivation in mice. Nature 2010;467:977–981.
[73] Chureau C., Prissette M., Bourdet A., Barbe V., Cattolico L., Jones L., Eggen A., Avner P., Duret L. Comparative sequence analysis of the X-inactivation center region in mouse, human, and bovine. Genome Res 2002;12:894–908.
[74] Yang C., Chapman A.G., Kelsey A.D., Minks J., Cotton A.M., Brown C.J. X-chromosome inactivation: molecular mechanisms from the human perspective. Hum Genet 2011.
[75] Migeon B.R., Chowdhury A.K., Dunston J.A., McIntosh I. Identification of TSIX, encoding an RNA antisense to human XIST, reveals differences from its murine counterpart: implications for X inactivation. Am J Hum Genet 2001;69:951–60.
[76] Migeon BR, Lee CH, Chowdhury AK, Carpenter H. Species differences in TSIX/Tsix reveal the roles of these genes in X-chromosome inactivation. Am J Hum Genet 2002;71:286–293.
[77] Berletch J.B., Yang F., Xu J., Carrel L., Disteche C.M. Genes that escape from X inactivation. Hum Genet 2011;130:237–245.
[78] Carrel L., Willard H.F. X-inactivation profile reveals extensive variability in X-linked gene expression in females. Nature 2005;434:400–404.
[79] Cotton A.M., Lam L., Affleck J.G., Wilson I.M., Penaherrera M.S., McFadden D.E., Kobor M.S., Lam W.L., Robinson W.P., Brown C.J. Chromosome-wide DNA methylation analysis predicts human tissue-specific X inactivation. Hum Genet 2011;130(2):187–201.
[80] Nussbaum R.L., McInnes R.R., Willard H.F. Thompson&Thonpson genetics in medicine. 6th ed. 2001.
[81] Ross J., Zinn A., McCauley E. Neurodevelopmental and psychosocial aspects of Turner syndrome. Ment Retard Dev Disabil Res Rev2000;6:135–141 .
[82] Rao E., Weiss B., Fukami M., Rump A., Niesler B., Mertz A., Muroya K., Binder G., Kirsch S., Winkelmann M., Nordsiek G., Heinrich U., Breuning M.H., Ranke M.B., Rosenthal A., Ogata T., Rappold G.A. Pseudoautosomal deletions encompassing a novel homeobox gene cause growth failure in idiopathic short stature and Turner syndrome. Nat Genet 1997;16:54–63.
[83] Chiurazzi P., Schwartz C.E., Gecz J., Neri G. XLMR genes: update 2007. Eur J Hum Genet 2008;16:422–434.
[84] Tarpey P.S., Smith R., Pleasance E., Whibley A., Edkins S., Hardy C., O'Meara S., Latimer C., Dicks E., Menzies A., Stephens P., Blow M., Greenman C., Xue Y., Tyler-Smith C., Thompson D., Gray K., Andrews J., Barthorpe S., Buck G., Cole J., Dunmore R., Jones D., Maddison M., Mironenko T., Turner R., Turrell K., Varian J., West S., Widaa S., Wray P., Teague J., Butler A., Jenkinson A., Jia M., Richardson D., Shepherd R., Wooster R., Tejada M.I., Martinez F., Carvill G., Goliath R., de Brouwer A.P., van Bokhoven H., Van Esch H., Chelly J., Raynaud M, Ropers HH, Abidi FE, Srivastava AK, Cox J, Luo Y, Mallya U, Moon J, Parnau J, Mohammed S, Tolmie JL, Shoubridge C, Corbett M, Gardner A, Haan E, Rujirabanjerd S, Shaw M, Vandeleur L, Fullston T, Easton DF, Boyle J, Partington M, Hackett A, Field M, Skinner C, Stevenson RE, Bobrow M, Turner G, Schwartz CE, Gecz J, Raymond FL, Futreal PA, Stratton MR. A systematic, large-scale resequencing screen of X-chromosome coding exons in mental retardation. Nat Genet 2009;41:535–43.
[85] Orstavik K.H. X chromosome inactivation in clinical practice. Hum Genet 2009;126:363–373.
[86] Burn J., Povey S., Boyd Y., Munro E.A., West L., Harper K., Thomas D. Duchenne muscular dystrophy in one of monozygotic twin girls. J Med Genet 1986;23:494–500.
[87] Maier E.M., Kammerer S., Muntau A.C., Wichers M., Braun A., Roscher A.A. Symptoms in carriers of adrenoleukodystrophy relate to skewed X inactivation. Ann Neurol 2002;52:683–688.
[88] Chung B.H., Drmic I., Marshall C.R., Grafodatskaya D., Carter M., Fernandez B.A., Weksberg R., Roberts W., Scherer S.W. Phenotypic spectrum associated with duplication of Xp11.22-p11.23 includes autism spectrum disorder. Eur J Med Genet 2011;54(5):e516–520.

[89] Giorda R., Bonaglia M.C., Beri S., Fichera M., Novara F., Magini P., Urquhart J., Sharkey F.H., Zucca C., Grasso R., Marelli S., Castiglia L., Di Benedetto D., Musumeci S.A., Vitello G.A., Failla P., Reitano S., Avola E., Bisulli F., Tinuper P., Mastrangelo M., Fiocchi I., Spaccini L., Torniero C., Fontana E., Lynch S.A., Clayton-Smith J., Black G., Jonveaux P., Leheup B., Seri M., Romano C., dalla Bernardina B., Zuffardi O. Complex segmental duplications mediate a recurrent dup(X)(p11.22-p11.23) associated with mental retardation, speech delay, and EEG anomalies in males and females. Am J Hum Genet 2009;85:394–400.

[90] Twigg SR, Kan R, Babbs C, Bochukova EG, Robertson SP, Wall SA, Morriss-Kay GM, Wilkie AO. Mutations of ephrin-B1 (EFNB1), a marker of tissue boundary formation, cause craniofrontonasal syndrome. Proc Natl Acad Sci U S A 2004;101:8652–8657.

[91] Ryan S.G., Chance P.F., Zou C.H., Spinner N.B., Golden J.A., Smietana S. Epilepsy and mental retardation limited to females: an X-linked dominant disorder with male sparing. Nat Genet 1997;17:92–95.

[92] Trappe R., Laccone F., Cobilanschi J., Meins M., Huppke P., Hanefeld F., Engel W. MECP2 mutations in sporadic cases of Rett syndrome are almost exclusively of paternal origin. Am J Hum Genet 2001;68:1093–1101.

[93] Van den Veyver I.B. Skewed X inactivation in X-linked disorders. Semin Reprod Med 2001;19:183–191.

[94] Huppke P., Maier E.M., Warnke A., Brendel C., Laccone F., Gartner J. Very mild cases of Rett syndrome with skewed X inactivation. J Med Genet 2006;43:814–816.

[95] Wan M., Lee S.S., Zhang X., Houwink-Manville I., Song H.R., Amir R.E., Budden S., Naidu S., Pereira J.L., Lo I.F., Zoghbi H.Y., Schanen N.C., Francke U. Rett syndrome and beyond: recurrent spontaneous and familial MECP2 mutations at CpG hotspots. Am J Hum Genet 1999;65:1520–1529.

[96] Allen R.C., Zoghbi H.Y., Moseley A.B., Rosenblatt H.M., Belmont J.W. Methylation of HpaII and HhaI sites near the polymorphic CAG repeat in the human androgen-receptor gene correlates with X chromosome inactivation. Am J Hum Genet 1992;51:1229–1239.

[97] Hunter P. The silence of genes. Is genomic imprinting the software of evolution or just a battleground for gender conflict? EMBO Rep 2007;8:441–443.

[98] McGrath J., Solter D. Nuclear and cytoplasmic transfer in mammalian embryos. Dev Biol 1986;4:37–55.

[99] Surani M.A., Barton S.C., Norris M.L. Nuclear transplantation in the mouse: heritable differences between parental genomes after activation of the embryonic genome. Cell 1986;45:127–136.

[100] Tycko B., Morison I.M. Physiological functions of imprinted genes. J Cell Physiol 2002;192:245–58.

[101] Murphy SK, Jirtle RL. Imprinting evolution and the price of silence. Bioessays 2003;25:577–588.

[102] Weksberg R. Imprinted genes and human disease. Am J Med Genet C Semin Med Genet 2010;154C:317–320.

[103] Monk D. Genomic imprinting in the human placenta. Am J Obstet Gynecol 2015;213:S152–S162.

[104] Reik W., Walter J. Genomic imprinting: parental influence on the genome. Nat Rev Genet 2001;2:21–32.

[105] Tycko B. Imprinted genes in placental growth and obstetric disorders. Cytogenet Genome Res 2006;113: 271–278.

[106] Peters J. The role of genomic imprinting in biology and disease: an expanding view. Nat Rev Genet 2014;15:517–530.

[107] Roberts D.J., Mutter G.L. Advances in the molecular biology of gestational trophoblastic disease. J Reprod Med 1994;39:201–208.

[108] Van den Veyver I.B., Al-Hussaini T.K. Biparental hydatidiform moles: a maternal effect mutation affecting imprinting in the offspring. Hum Reprod Update 2006;12:233–242.

[109] Monk D., Sanchez-Delgado M., Fisher R. NLRPs, the subcortical maternal complex and genomic imprinting. Reproduction 2017;154:R161–R170.

[110] Murdoch S., Djuric U., Mazhar B., Seoud M., Khan R., Kuick R., Bagga R., Kircheisen R., Ao A., Ratti B., Hanash S., Rouleau G.A., Slim R. Mutations in NALP7 cause recurrent hydatidiform moles and reproductive wastage in humans. Nat Genet 2006;38:300–302.

[111] Parry D.A., Logan C.V., Hayward B.E., Shires M., Landolsi H., Diggle C., Carr I., Rittore C., Touitou I., Philibert L., Fisher R.A., Fallahian M., Huntriss J.D., Picton H.M., Malik S., Taylor G.R., Johnson C.A., Bonthron D.T., Sheridan E.G. Mutations causing familial biparental hydatidiform mole implicate c6orf221 as a possible regulator of genomic imprinting in the human oocyte. Am J Hum Genet 2011;89:451–458.

[112] Sanchez-Delgado M., Martin-Trujillo A., Tayama C., Vidal E., Esteller M., Iglesias-Platas I., Deo N., Barney O., Maclean K., Hata K., Nakabayashi K., Fisher R., Monk D. Absence of maternal methylation in biparental hydatidiform moles from women with NLRP7 maternal-effect mutations reveals widespread placenta-specific imprinting. PLoS Genet 2015;11:e1005644.

[113] de Grouchy J. Human parthenogenesis: a fascinating single event. Biomedicine 1980;32:51–53.

[114] Mutter G.L. Teratoma genetics and stem cells: a review. Obstet Gynecol Surv 1987;42:661–670.

[115] Mutter G.L. Role of imprinting in abnormal human development. Mutat Res 1997;396:141–147.

[116] Choufani S., Shapiro J.S., Susiarjo M., Butcher D.T., Grafodatskaya D., Lou Y., Ferreira J.C., Pinto D.,

Scherer S.W., Shaffer L.G., Coullin P., Caniggia I., Beyene J., Slim R., Bartolomei M.S., Weksberg R. A novel approach identifies new differentially methylated regions (DMRs) associated with imprinted genes. Genome Res 2011;21:465–476.

[117] Pettenati M.J., Haines J.L., Higgins R.R., Wappner R.S., Palmer C.G., Weaver D.D. Wiedemann-Beckwith syndrome: presentation of clinical and cytogenetic data on 22 new cases and review of the literature. Hum Genet 1986;74:143–154.

[118] Weksberg R., Shuman C., Smith A.C. Beckwith-Wiedemann syndrome. Am J Med Genet C Semin Med Genet 2005;137C:12–23.

[119] Weksberg R., Smith A.C., Squire J., Sadowski P. Beckwith-Wiedemann syndrome demonstrates a role for epigenetic control of normal development. Hum Mol Genet 2003;12(Spec No 1):R61–R68.

[120] Bens S., Kolarova J., Beygo J., Buiting K., Caliebe A., Eggermann T., Gillessen-Kaesbach G., Prawitt D., Thiele-Schmitz S., Begemann M., Enklaar T., Gutwein J., Haake A., Paul U., Richter J., Soellner L., Vater I., Monk D., Horsthemke B., Ammerpohl O., Siebert R. Phenotypic spectrum and extent of DNA methylation defects associated with multilocus imprinting disturbances. Epigenomics 2016;8:801–816.

[121] Eggermann T., Elbracht M., Schroder C., Reutter H., Soellner L., Spengler S., Begemann M. Congenital imprinting disorders: a novel mechanism linking seemingly unrelated disorders. J Pediatr 2013;163:1202–1207.

[122] Meyer E, Lim D, Pasha S, Tee LJ, Rahman F, Yates JR, Woods CG, Reik W, Maher ER. Germline mutation in NLRP2 (NALP2) in a familial imprinting disorder (Beckwith-Wiedemann Syndrome). PLoS Genet 2009;5:e1000423.

[123] Choufani S., Shuman C., Weksberg R. Molecular findings in Beckwith-Wiedemann syndrome. Am J Med Genet C Semin Med Genet 2013;163C:131–140.

[124] Brzezinski J., Shuman C., Choufani S., Ray P., Stavropoulos D.J., Basran R., Steele L., Parkinson N., Grant R., Thorner P., Lorenzo A., Weksberg R. Wilms tumour in Beckwith-Wiedemann Syndrome and loss of methylation at imprinting centre 2: revisiting tumour surveillance guidelines. Eur J Hum Genet 2017;25:1031–1039.

[125] Bliek J., Gicquel C., Maas S., Gaston V., Le Bouc Y., Mannens M. Epigenotyping as a tool for the prediction of tumor risk and tumor type in patients with Beckwith-Wiedemann syndrome (BWS). J Pediatr 2004;145:796–799.

[126] Eggermann T. Russell-Silver syndrome. Am J Med Genet C Semin Med Genet 2010;154C:355–364.

[127] Gucev Z.S., Saranac L., Jancevska A., Tasic V. The degree of H19 hypomethylation in children with Silver-Russel syndrome (SRS) is not associated with the severity of phenotype and the clinical severity score (CSS). Prilozi 2013;34:79–83.

[128] Cytrynbaum C., Chong K., Hannig V., Choufani S., Shuman C., Steele L., Morgan T., Scherer S.W., Stavropoulos D.J., Basran R.K., Weksberg R. Genomic imbalance in the centromeric 11p15 imprinting center in three families: further evidence of a role for IC2 as a cause of Russell-Silver syndrome. Am J Med Genet 2016;170:2731–2739.

[129] Buiting K. Prader-Willi syndrome and Angelman syndrome. Am J Med Genet C Semin Med Genet 2010;154C:365–376.

[130] Holm V.A., Cassidy S.B., Butler M.G., Hanchett J.M., Greenswag L.R., Whitman B.Y., Greenberg F. Prader-Willi syndrome: consensus diagnostic criteria. Pediatrics 1993;91:398–402.

[131] Gunay-Aygun M, Schwartz S, Heeger S, O'Riordan MA, Cassidy SB. The changing purpose of Prader-Willi syndrome clinical diagnostic criteria and proposed revised criteria. Pediatrics 2001;108:E92.

[132] Williams C.A., Beaudet A.L., Clayton-Smith J., Knoll J.H., Kyllerman M., Laan L.A., Magenis R.E., Moncla A., Schinzel A.A., Summers J.A., Wagstaff J. Angelman syndrome 2005: updated consensus for diagnostic criteria. Am J Med Genet 2006;140:413–418.

[133] Hogart A., Patzel K.A., LaSalle J.M. Gender influences monoallelic expression of ATP10A in human brain. Hum Genet 2008;124:235–242.

[134] Nicholls R.D., Knepper J.L. Genome organization, function, and imprinting in Prader-Willi and Angelman syndromes. Annu Rev Genomics Hum Genet 2001;2:153–175.

[135] Ohta T., Gray T.A., Rogan P.K., Buiting K., Gabriel J.M., Saitoh S., Muralidhar B., Bilienska B., Krajewska-Walasek M., Driscoll D.J., Horsthemke B., Butler M.G., Nicholls R.D. Imprinting-mutation mechanisms in Prader-Willi syndrome. Am J Hum Genet 1999;64:397–413.

[136] Buiting K., Lich C., Cottrell S., Barnicoat A., Horsthemke B. A 5-kb imprinting center deletion in a family with Angelman syndrome reduces the shortest region of deletion overlap to 880 bp. Hum Genet 1999;105:665–666.

[137] Kishino T., Lalande M., Wagstaff J. UBE3A/E6-AP mutations cause Angelman syndrome. Nat Genet 1997;15:70–73.

[138] Matsuura T., Sutcliffe J.S., Fang P., Galjaard R.J., Jiang Y.H., Benton C.S., Rommens J.M., Beaudet A.L. De novo truncating mutations in E6-AP ubiquitin-protein ligase gene (UBE3A) in Angelman syndrome. Nat Genet 1997;15:74–77.

[139] Clayton-Smith J., Laan L. Angelman syndrome: a review of the clinical and genetic aspects. J Med Genet 2003;40:87–95.
[140] Rougeulle C., Glatt H., Lalande M. The Angelman syndrome candidate gene, UBE3A/E6-AP, is imprinted in brain. Nat Genet 1997;17:14–5.
[141] Vu TH, Hoffman AR. Imprinting of the Angelman syndrome gene, UBE3A, is restricted to brain. Nat Genet 1997;17:12–13.
[142] Velinov M., Jenkins E.C. PCR-based strategies for the diagnosis of Prader-Willi/Angelman syndromes. Methods Mol Biol 2003;217:209–216.
[143] Bittel D.C., Kibiryeva N., Butler M.G. Methylation-specific multiplex ligation-dependent probe amplification analysis of subjects with chromosome 15 abnormalities. Genet Test 2007;11:467–475.
[144] Bjornsson H.T. The Mendelian disorders of the epigenetic machinery. Genome Res 2015;25:1473–1481.
[145] Fahrner J.A., Bjornsson H.T. Mendelian disorders of the epigenetic machinery: tipping the balance of chromatin states. Annu Rev Genomics Hum Genet 2014;15:269–293.
[146] Kleefstra T., Schenck A., Kramer J.M., van Bokhoven H. The genetics of cognitive epigenetics. Neuropharmacology 2014;80:83–94.
[147] Tatton-Brown K., Loveday C., Yost S., Clarke M., Ramsay E., Zachariou A., Elliott A., Wylie H., Ardissone A., Rittinger O., Stewart F., Temple I.K., Cole T., Mahamdallie S., Seal S., Ruark E., Rahman N. Mutations in epigenetic regulation genes are a major cause of overgrowth with intellectual disability. Am J Hum Genet 2017;100:725–736.
[148] Baujat G., Cormier-Daire V. Sotos syndrome. Orphanet J Rare Dis 2007;2:36.
[149] Li Y., Trojer P., Xu C.F., Cheung P., Kuo A., Drury 3rd W.J., Qiao Q., Neubert T.A., Xu R.M., Gozani O., Reinberg D. The target of the NSD family of histone lysine methyltransferases depends on the nature of the substrate. J Biol Chem 2009;284: 34283–34295.
[150] Bannister A.J., Kouzarides T. Regulation of chromatin by histone modifications. Cell Res 2011;21:381–395.
[151] Lucio-Eterovic A.K., Singh M.M., Gardner J.E., Veerappan C.S., Rice J.C., Carpenter P.B. Role for the nuclear receptor-binding SET domain protein 1 (NSD1) methyltransferase in coordinating lysine 36 methylation at histone 3 with RNA polymerase II function. Proc Natl Acad Sci U S A 2010;107:16952–7.
[152] Pasillas MP, Shah M, Kamps MP. NSD1 PHD domains bind methylated H3K4 and H3K9 using interactions disrupted by point mutations in human sotos syndrome. Hum Mutat 2011;32:292–298.
[153] Gibson W.T., Hood R.L., Zhan S.H., Bulman D.E., Fejes A.P., Moore R., Mungall A.J., Eydoux P., Babul-Hirji R., An J., Marra M.A., Consortium F.C., Chitayat D., Boycott K.M., Weaver D.D., Jones S.J. Mutations in EZH2 cause Weaver syndrome. Am J Hum Genet 2012;90:110–118.
[154] Tatton-Brown K., Hanks S., Ruark E., Zachariou A., Duarte Sdel V., Ramsay E., Snape K., Murray A., Perdeaux E.R., Seal S., Loveday C., Banka S., Clericuzio C., Flinter F., Magee A., McConnell V., Patton M., Raith W., Rankin J., Splitt M., Strenger V., Taylor C., Wheeler P., Temple K.I., Cole T., Childhood Overgrowth C., Douglas J., Rahman N. Germline mutations in the oncogene EZH2 cause Weaver syndrome and increased human height. Oncotarget 2011;2:1127–1133.
[155] Cohen A.S., Tuysuz B., Shen Y., Bhalla S.K., Jones S.J., Gibson W.T. A novel mutation in EED associated with overgrowth. J Hum Genet 2015;60:339–342.
[156] Imagawa E., Higashimoto K., Sakai Y., Numakura C., Okamoto N., Matsunaga S., Ryo A., Sato Y., Sanefuji M., Ihara K., Takada Y., Nishimura G., Saitsu H., Mizuguchi T., Miyatake S., Nakashima M., Miyake N., Soejima H., Matsumoto N. Mutations in genes encoding polycomb repressive complex 2 subunits cause Weaver syndrome. Hum Mutat 2017;38:637–648.
[157] Ng S.B., Bigham A.W., Buckingham K.J., Hannibal M.C., McMillin M.J., Gildersleeve H.I., Beck A.E., Tabor H.K., Cooper G.M., Mefford H.C., Lee C., Turner E.H., Smith J.D., Rieder M.J., Yoshiura K., Matsumoto N., Ohta T., Niikawa N., Nickerson D.A., Bamshad M.J., Shendure J. Exome sequencing identifies MLL2 mutations as a cause of Kabuki syndrome. Nat Genet 2010;42:790–793.
[158] Prasad R, Zhadanov AB, Sedkov Y, Bullrich F, Druck T, Rallapalli R, Yano T, Alder H, Croce CM, Huebner K, Mazo A, Canaani E. Structure and expression pattern of human ALR, a novel gene with strong homology to ALL-1 involved in acute leukemia and to Drosophila trithorax. Oncogene 1997;15:549–560.
[159] Issaeva I., Zonis Y., Rozovskaia T., Orlovsky K., Croce C.M., Nakamura T., Mazo A., Eisenbach L., Canaani E. Knockdown of ALR (MLL2) reveals ALR target genes and leads to alterations in cell adhesion and growth. Mol Cell Biol 2007;27:1889–1903.
[160] Andreu-Vieyra C.V., Chen R., Agno J.E., Glaser S., Anastassiadis K., Stewart A.F., Matzuk M.M. MLL2 is required in oocytes for bulk histone 3 lysine 4 trimethylation and transcriptional silencing. PLoS Biol 2010;8.
[161] Lederer D., Grisart B., Digilio M.C., Benoit V., Crespin M., Ghariani S.C., Maystadt I., Dallapiccola B., Verellen-Dumoulin C. Deletion of KDM6A, a histone demethylase interacting with MLL2, in three patients with Kabuki syndrome. Am J Hum Genet 2012;90:119–124.

[244] Dobrovic A., Simpfendorfer D. Methylation of the BRCA1 gene in sporadic breast cancer. Cancer Res 1997;57:3347–3350.
[245] Esteller M., Levine R., Baylin S.B., Ellenson L.H., Herman J.G. MLH1 promoter hypermethylation is associated with the microsatellite instability phenotype in sporadic endometrial carcinomas. Oncogene 1998;17:2413–2417.
[246] Herman J.G., Latif F., Weng Y., Lerman M.I., Zbar B., Liu S., Samid D., Duan D.S., Gnarra J.R., Linehan W.M., et al. Silencing of the VHL tumor-suppressor gene by DNA methylation in renal carcinoma. Proc Natl Acad Sci U S A 1994;91:9700–4.
[247] Mancini DN, Rodenhiser DI, Ainsworth PJ, O'Malley FP, Singh SM, Xing W, Archer TK. CpG methylation within the 5' regulatory region of the BRCA1 gene is tumor specific and includes a putative CREB binding site. Oncogene 1998;16:1161–1169.
[248] Veigl M.L., Kasturi L., Olechnowicz J., Ma A.H., Lutterbaugh J.D., Periyasamy S., Li G.M., Drummond J., Modrich P.L., Sedwick W.D., Markowitz S.D. Biallelic inactivation of hMLH1 by epigenetic gene silencing, a novel mechanism causing human MSI cancers. Proc Natl Acad Sci U S A 1998;95:8698–8702.
[249] Gossage L., Eisen T. Alterations in VHL as potential biomarkers in renal-cell carcinoma. Nat Rev Clin Oncol 2010;7:277–288.
[250] Belinsky S.A., Palmisano W.A., Gilliland F.D., Crooks L.A., Divine K.K., Winters S.A., Grimes M.J., Harms H.J., Tellez C.S., Smith T.M., Moots P.P., Lechner J.F., Stidley C.A., Crowell R.E. Aberrant promoter methylation in bronchial epithelium and sputum from current and former smokers. Cancer Res 2002;62:2370–2377.
[251] Klump B., Hsieh C.J., Holzmann K., Gregor M., Porschen R. Hypermethylation of the CDKN2/p16 promoter during neoplastic progression in Barrett's esophagus. Gastroenterology 1998;115:1381–1386.
[252] Wong D.J., Barrett M.T., Stoger R., Emond M.J., Reid B.J. p16INK4a promoter is hypermethylated at a high frequency in esophageal adenocarcinomas. Cancer Res 1997;57:2619–2622.
[253] Enokida H., Shiina H., Urakami S., Igawa M., Ogishima T., Li L.C., Kawahara M., Nakagawa M., Kane C.J., Carroll P.R., Dahiya R. Multigene methylation analysis for detection and staging of prostate cancer. Clin Cancer Res 2005;11:6582–6588.
[254] Hoque M.O., Begum S., Topaloglu O., Chatterjee A., Rosenbaum E., Van Criekinge W., Westra WH, Schoenberg M, Zahurak M, Goodman SN, Sidransky D. Quantitation of promoter methylation of multiple genes in urine DNA and bladder cancer detection. J Natl Cancer Inst 2006;98:996–1004.
[255] Belinsky SA, Liechty KC, Gentry FD, Wolf HJ, Rogers J, Vu K, Haney J., Kennedy T.C., Hirsch F.R., Miller Y., Franklin W.A., Herman J.G., Baylin S.B., Bunn P.A., Byers T. Promoter hypermethylation of multiple genes in sputum precedes lung cancer incidence in a high-risk cohort. Cancer Res 2006;66:3338–3344.
[256] Chen W.D., Han Z.J., Skoletsky J., Olson J., Sah J., Myeroff L., Platzer P., Lu S., Dawson D., Willis J., Pretlow T.P., Lutterbaugh J., Kasturi L., Willson J.K., Rao J.S., Shuber A., Markowitz S.D. Detection in fecal DNA of colon cancer-specific methylation of the nonexpressed vimentin gene. J Natl Cancer Inst 2005;97:1124–1132.
[257] Rasmussen S.L., Krarup H.B., Sunesen K.G., Pedersen I.S., Madsen P.H., Thorlacius-Ussing O. Hypermethylated DNA as a biomarker for colorectal cancer: a systematic review. Colorectal Dis 2016;18:549–561.
[258] Lapeyre J.N., Becker F.F. 5-Methylcytosine content of nuclear DNA during chemical hepatocarcinogenesis and in carcinomas which result. Biochem Biophys Res Commun 1979;87:698–705.
[259] Feinberg A.P., Gehrke C.W., Kuo K.C., Ehrlich M. Reduced genomic 5-methylcytosine content in human colonic neoplasia. Cancer Res 1988;48:1159–1161.
[260] Ehrlich M. DNA methylation in cancer: too much, but also too little. Oncogene 2002;21:5400–5413.
[261] Chen R.Z., Pettersson U., Beard C., Jackson-Grusby L., Jaenisch R. DNA hypomethylation leads to elevated mutation rates. Nature 1998;395:89–93.
[262] Madakashira B.P., Sadler K.C. DNA methylation, nuclear organization, and cancer. Front Genet 2017;8:76.
[263] Mudbhary R., Hoshida Y., Chernyavskaya Y., Jacob V., Villanueva A., Fiel M.I., Chen X., Kojima K, Thung S, Bronson RT, Lachenmayer A, Revill K, Alsinet C, Sachidanandam R, Desai A, SenBanerjee S, Ukomadu C, Llovet JM, Sadler KC. UHRF1 overexpression drives DNA hypomethylation and hepatocellular carcinoma. Cancer Cell 2014;25:196–209.
[264] Berman BP, Weisenberger DJ, Aman JF, Hinoue T, Ramjan Z, Liu Y, Noushmehr H, Lange CPE, van Dijk CM, Tollenaar R.A.E.M., Van Den Berg D., Laird P.W. Regions of focal DNA hypermethylation and long-range hypomethylation in colorectal cancer coincide with nuclear lamina-associated domains. Nat Genet 2012;44:40–46.
[265] Timp W., Bravo H.C., McDonald O.G., Goggins M., Umbricht C., Zeiger M., Feinberg A.P., Irizarry R.A. Large hypomethylated blocks as a universal defining epigenetic alteration in human solid tumors. Genome Med 2014;6:61.
[266] Gao F., Shi L., Russin J., Zeng L., Chang X., He S., Chen T.C., Giannotta S.L., Weisenberger D.J., Zada G., Mack W.J., Wang K. DNA methylation in the

malignant transformation of meningiomas. PLoS One 2013;8:e54114.
[267] Kim Y.I., Giuliano A., Hatch K.D., Schneider A., Nour M.A., Dallal G.E., Selhub J., Mason J.B. Global DNA hypomethylation increases progressively in cervical dysplasia and carcinoma. Cancer 1994;74:893–899.
[268] Baer C., Claus R., Plass C. Genome-wide epigenetic regulation of miRNAs in cancer. Cancer Res 2013;73:473–477.
[269] Fraga M.F., Ballestar E., Villar-Garea A., Boix-Chornet M., Espada J., Schotta G., Bonaldi T., Haydon C., Ropero S., Petrie K., Iyer N.G., Perez-Rosado A., Calvo E., Lopez J.A., Cano A., Calasanz M.J., Colomer D., Piris M.A., Ahn N., Imhof A., Caldas C., Jenuwein T., Esteller M. Loss of acetylation at Lys16 and trimethylation at Lys20 of histone H4 is a common hallmark of human cancer. Nat Genet 2005;37:391–400.
[270] Halkidou K., Gaughan L., Cook S., Leung H.Y., Neal D.E., Robson C.N. Upregulation and nuclear recruitment of HDAC1 in hormone refractory prostate cancer. Prostate 2004;59:177–189.
[271] Song J., Noh J.H., Lee J.H., Eun J.W., Ahn Y.M, Kim SY, Lee SH, Park WS, Yoo NJ, Lee JY, Nam SW. Increased expression of histone deacetylase 2 is found in human gastric cancer. APMIS 2005;113:264–8.
[272] Yang XJ. The diverse superfamily of lysine acetyltransferases and their roles in leukemia and other diseases. Nucleic Acids Res 2004;32:959–76.
[273] Morin RD, Mendez-Lago M, Mungall AJ, Goya R, Mungall KL, Corbett RD, Johnson NA, Severson TM, Chiu R, Field M., Jackman S., Krzywinski M., Scott D.W., Trinh D.L., Tamura-Wells J., Li S., Firme M.R., Rogic S., Griffith M., Chan S., Yakovenko O., Meyer I.M., Zhao E.Y., Smailus D., Moksa M., Chittaranjan S., Rimsza L., Brooks-Wilson A., Spinelli J.J., Ben-Neriah S., Meissner B., Woolcock B., Boyle M., McDonald H., Tam A., Zhao Y., Delaney A., Zeng T., Tse K., Butterfield Y., Birol I., Holt R., Schein J., Horsman D.E., Moore R., Jones S.J., Connors J.M., Hirst M., Gascoyne R.D., Marra M.A. Frequent mutation of histone-modifying genes in non-Hodgkin lymphoma. Nature 2011;476:298–303.
[274] Nguyen C.T., Weisenberger D.J., Velicescu M., Gonzales F.A., Lin J.C., Liang G., Jones P.A. Histone H3-lysine 9 methylation is associated with aberrant gene silencing in cancer cells and is rapidly reversed by 5-aza-2'-deoxycytidine. Cancer Res 2002;62:6456–6461.
[275] Kleer C.G., Cao Q., Varambally S., Shen R., Ota I., Tomlins S.A., Ghosh D., Sewalt R.G., Otte A.P., Hayes D.F., Sabel M.S., Livant D., Weiss S.J., Rubin M.A., Chinnaiyan A.M. EZH2 is a marker of aggressive breast cancer and promotes neoplastic transformation of breast epithelial cells. Proc Natl Acad Sci U S A 2003;100:11606–11611.
[276] Varambally S., Dhanasekaran S.M., Zhou M., Barrette T.R., Kumar-Sinha C., Sanda M.G., Ghosh D., Pienta K.J., Sewalt R.G., Otte A.P., Rubin M.A., Chinnaiyan A.M. The polycomb group protein EZH2 is involved in progression of prostate cancer. Nature 2002;419:624–629.
[277] Ernst T., Chase A.J., Score J., Hidalgo-Curtis C.E., Bryant C., Jones A.V., Waghorn K., Zoi K., Ross F.M., Reiter A., Hochhaus A., Drexler H.G., Duncombe A., Cervantes F., Oscier D., Boultwood J, Grand FH, Cross NC. Inactivating mutations of the histone methyltransferase gene EZH2 in myeloid disorders. Nat Genet 2010;42:722–6.
[278] Ntziachristos P, Tsirigos A, Van Vlierberghe P, Nedjic J, Trimarchi T, Flaherty MS, Ferres-Marco D, da Ros V, Tang Z, Siegle J, Asp P, Hadler M, Rigo I, De Keersmaecker K, Patel J, Huynh T, Utro F, Poglio S, Samon J.B., Paietta E., Racevskis J., Rowe J.M., Rabadan R., Levine R.L., Brown S., Pflumio F., Dominguez M., Ferrando A., Aifantis I. Genetic inactivation of the polycomb repressive complex 2 in T cell acute lymphoblastic leukemia. Nat Med 2012;18:298–301.
[279] Dalgliesh G.L., Furge K., Greenman C., Chen L., Bignell G., Butler A., Davies H., Edkins S., Hardy C., Latimer C., Teague J., Andrews J., Barthorpe S., Beare D., Buck G., Campbell P.J., Forbes S., Jia M., Jones D., Knott H., Kok C.Y., Lau K.W., Leroy C., Lin M.L., McBride D.J., Maddison M., Maguire S., McLay K., Menzies A., Mironenko T., Mulderrig L., Mudie L., O'Meara S., Pleasance E., Rajasingham A., Shepherd R., Smith R., Stebbings L., Stephens P., Tang G., Tarpey P.S., Turrell K., Dykema K.J., Khoo S.K., Petillo D., Wondergem B., Anema J., Kahnoski R.J., Teh B.T., Stratton M.R., Futreal P.A. Systematic sequencing of renal carcinoma reveals inactivation of histone modifying genes. Nature 2010;463:360–363.
[280] Pasqualucci L., Dominguez-Sola D., Chiarenza A., Fabbri G., Grunn A., Trifonov V., Kasper L.H., Lerach S., Tang H., Ma J., Rossi D., Chadburn A., Murty V.V., Mullighan C.G., Gaidano G., Rabadan R., Brindle P.K., Dalla-Favera R. Inactivating mutations of acetyltransferase genes in B-cell lymphoma. Nature 2011;471:189–195.
[281] Varela I., Tarpey P., Raine K., Huang D., Ong C.K., Stephens P., Davies H., Jones D., Lin M.L., Teague J., Bignell G., Butler A., Cho J., Dalgliesh G.L., Galappaththige D., Greenman C., Hardy C., Jia M., Latimer C., Lau K.W., Marshall J., McLaren S., Menzies A., Mudie L., Stebbings L., Largaespada D.A., Wessels L.F.,

[327] Krakowiak P, Walker CK, Tancredi D, Hertz-Picciotto I, Van de Water J. Autism-specific maternal anti-fetal brain autoantibodies are associated with metabolic conditions. Autism Res 2016;10(1):89–98.

[328] Lee B.K., Magnusson C., Gardner R.M., Blomstrom A., Newschaffer C.J., Burstyn I., Karlsson H., Dalman C. Maternal hospitalization with infection during pregnancy and risk of autism spectrum disorders. Brain Behav Immun 2015;44:100–105.

[329] Li M., Fallin M.D., Riley A., Landa R., Walker S.O., Silverstein M., Caruso D., Pearson C., Kiang S., Dahm J.L., Hong X., Wang G., Wang M.C., Zuckerman B., Wang X. The association of maternal obesity and diabetes with autism and other developmental disabilities. Pediatrics 2016;137:e20152206.

[330] Li Y.M., Ou J.J., Liu L., Zhang D., Zhao J.P., Tang S.Y. Association between maternal obesity and autism spectrum disorder in offspring: a meta-analysis. J Autism Dev Disord 2016;46:95–102.

[331] Xiang A.H., Wang X., Martinez M.P., Walthall J.C., Curry E.S., Page K., Buchanan T.A., Coleman K.J., Getahun D. Association of maternal diabetes with autism in offspring. J Am Med Assoc 2015;313:1425–1434.

[332] Fregeac J., Colleaux L., Nguyen L.S. The emerging roles of MicroRNAs in autism spectrum disorders. Neurosci Biobehav Rev 2016;71:729–738.

[333] Mundalil Vasu M., Anitha A., Thanseem I., Suzuki K., Yamada K., Takahashi T., Wakuda T., Iwata K., Tsujii M., Sugiyama T., Mori N. Serum microRNA profiles in children with autism. Mol Autism 2014;5:40.

[334] Parikshak N.N., Swarup V., Belgard T.G., Irimia M., Ramaswami G., Gandal M.J., Hartl C., Leppa V., Ubieta L.T., Huang J., Lowe J.K., Blencowe B.J., Horvath S., Geschwind D.H. Genome-wide changes in lncRNA, splicing, and regional gene expression patterns in autism. Nature 2016;540:423–427.

[335] Wang Y., Zhao X., Ju W., Flory M., Zhong J., Jiang S., Wang P., Dong X., Tao X., Chen Q., Shen C., Zhong M., Yu Y., Brown W.T., Zhong N. Genome-wide differential expression of synaptic long noncoding RNAs in autism spectrum disorder. Transl Psychiatry 2015;5:e660.

[336] Cortez MA, Calin GA. MicroRNA identification in plasma and serum: a new tool to diagnose and monitor diseases. Expert Opin Biol Ther 2009;9:703–711.

[337] Kosik K.S. The neuronal microRNA system. Nat Rev Neurosci 2006;7:911–920.

[338] Poirier L.A., Vlasova T.I. The prospective role of abnormal methyl metabolism in cadmium toxicity. Environ Health Perspect 2002;110(Suppl. 5):793–795.

[339] Pilsner J.R., Liu X., Ahsan H., Ilievski V., Slavkovich V., Levy D., Factor-Litvak P., Graziano J.H., Gamble M.V. Genomic methylation of peripheral blood leukocyte DNA: influences of arsenic and folate in Bangladeshi adults. Am J Clin Nutr 2007;86:1179–1186.

[340] Feil R., Fraga M.F. Epigenetics and the environment: emerging patterns and implications. Nat Rev Genet 2012;13:97–109.

[341] Vaiserman A. Epidemiologic evidence for association between adverse environmental exposures in early life and epigenetic variation: a potential link to disease susceptibility? Clin Epigenet 2015;7:96.

[342] Chater-Diehl E.J., Laufer B.I., Castellani C.A., Alberry B.L., Singh S.M. Alteration of gene expression, DNA methylation, and histone methylation in free radical scavenging networks in adult mouse Hippocampus following fetal alcohol exposure. PLoS One 2016;11:e0154836.

[343] Joubert B., London S. Epigenomics and maternal smoking, with Bonnie Joubert and Stephanie London by Ashley Ahearn. Environ Health Perspect 2012;120.

[344] Lee K.W., Richmond R., Hu P., French L., Shin J., Bourdon C., Reischl E., Waldenberger M., Zeilinger S., Gaunt T., McArdle W., Ring S., Woodward G., Bouchard L., Gaudet D., Smith G.D., Relton C., Paus T., Pausova Z. Prenatal exposure to maternal cigarette smoking and DNA methylation: epigenome-wide association in a discovery sample of adolescents and replication in an independent cohort at birth through 17 years of age. Environ Health Perspect 2015;123:193–9.

[345] Houtepen LC, van Bergen AH, Vinkers CH, Boks MP. DNA methylation signatures of mood stabilizers and antipsychotics in bipolar disorder. Epigenomics 2016;8:197–208.

[346] Zahnow CA, Topper M, Stone M, Murray-Stewart T, Li H, Baylin SB, Casero Jr RA. Inhibitors of DNA methylation, histone deacetylation, and histone demethylation: a perfect combination for cancer therapy. Adv Cancer Res 2016;130:55–111.

[347] Roullet FI, Lai JK, Foster JA. In utero exposure to valproic acid and autism–a current review of clinical and animal studies. Neurotoxicol Teratol 2013;36:47–56.

[348] Tomson T, Battino D, Perucca E. The remarkable story of valproic acid. Lancet Neurol 2016;15:141.

[349] Detich N, Theberge J, Szyf M. Promoter-specific activation and demethylation by MBD2/demethylase. J Biol Chem 2002;277:35791–4.

[350] Abel T, Zukin RS. Epigenetic targets of HDAC inhibition in neurodegenerative and psychiatric disorders. Curr Opin Pharmacol 2008;8:57–64.

[351] Kelly TK, De Carvalho DD, Jones PA. Epigenetic modifications as therapeutic targets. Nat Biotechnol 2010;28:1069–78.

[352] Schneider A, Chatterjee S, Bousiges O, Selvi BR, Swaminathan A, Cassel R, Blanc F, Kundu TK, Boutillier AL. Acetyltransferases (HATs) as targets for neurological therapeutics. Neurotherapeutics 2013;10:568–88.
[353] Friso S, Udali S, De Santis D, Choi SW. One-carbon metabolism and epigenetics. Mol Aspects Med 2017;54:28–36.
[354] Friso S, Girelli D, Trabetti E, Olivieri O, Guarini P, Pignatti PF, Corrocher R, Choi SW. The MTHFR 1298A>C polymorphism and genomic DNA methylation in human lymphocytes. Cancer Epidemiol Biomarkers Prev 2005;14:938–43.
[355] Rampersaud GC, Kauwell GP, Hutson AD, Cerda JJ, Bailey LB. Genomic DNA methylation decreases in response to moderate folate depletion in elderly women. Am J Clin Nutr 2000;72:998–1003.
[356] Ingrosso D, Cimmino A, Perna AF, Masella L, De Santo NG, De Bonis ML, Vacca M, D'Esposito M, D'Urso M, Galletti P, Zappia V. Folate treatment and unbalanced methylation and changes of allelic expression induced by hyperhomocysteinaemia in patients with uraemia. Lancet 2003;361:1693–9.
[357] Caramaschi D, Sharp GC, Nohr EA, Berryman K, Lewis SJ, Davey Smith G, Relton CL. Exploring a causal role of DNA methylation in the relationship between maternal vitamin B12 during pregnancy and child's IQ at age 8, cognitive performance and educational attainment: a two-step Mendelian randomization study. Hum Mol Genet 2017;26:3001–13.
[358] Irwin RE, Pentieva K, Cassidy T, Lees-Murdock DJ, McLaughlin M, Prasad G, McNulty H, Walsh CP. The interplay between DNA methylation, folate and neurocognitive development. Epigenomics 2016;8:863–79.
[359] Song J, Medline A, Mason JB, Gallinger S, Kim YI. Effects of dietary folate on intestinal tumorigenesis in the apcMin mouse. Cancer Res 2000;60:5434–40.
[360] Trasler J, Deng L, Melnyk S, Pogribny I, Hiou-Tim F, Sibani S, Oakes C, Li E, James SJ, Rozen R. Impact of Dnmt1 deficiency, with and without low folate diets, on tumor numbers and DNA methylation in Min mice. Carcinogenesis 2003;24:39–45.
[361] Bjork M, Riedel B, Spigset O, Veiby G, Kolstad E, Daltveit AK, Gilhus NE. Association of folic acid supplementation during pregnancy with the risk of autistic traits in children exposed to antiepileptic drugs in utero. JAMA Neurol 2018;75:160–8.
[362] Castro K, Klein Lda S, Baronio D, Gottfried C, Riesgo R, Perry IS. Folic acid and autism: what do we know? Nutr Neurosci 2016;19:310–7.
[363] McGarel C, Pentieva K, Strain JJ, McNulty H. Emerging roles for folate and related B-vitamins in brain health across the lifecycle. Proc Nutr Soc 2015;74:46–55.
[364] Li T, Vu TH, Ulaner GA, Littman E, Ling JQ, Chen HL, Hu JF, Behr B, Giudice L, Hoffman AR. IVF results in de novo DNA methylation and histone methylation at an Igf2-H19 imprinting epigenetic switch. Mol Hum Reprod 2005;11:631–40.
[365] Mann MR, Chung YG, Nolen LD, Verona RI, Latham KE, Bartolomei MS. Disruption of imprinted gene methylation and expression in cloned preimplantation stage mouse embryos. Biol Reprod 2003;69:902–14.
[366] Mann MR, Lee SS, Doherty AS, Verona RI, Nolen LD, Schultz RM, Bartolomei MS. Selective loss of imprinting in the placenta following preimplantation development in culture. Development 2004;131:3727–35.
[367] Sato A, Otsu E, Negishi H, Utsunomiya T, Arima T. Aberrant DNA methylation of imprinted loci in superovulated oocytes. Hum Reprod 2007;22:26–35.
[368] Kobayashi H, Sato A, Otsu E, Hiura H, Tomatsu C, Utsunomiya T, Sasaki H, Yaegashi N, Arima T. Aberrant DNA methylation of imprinted loci in sperm from oligospermic patients. Hum Mol Genet 2007;16:2542–51.
[369] CDC, Centers for Disease Control and Prevention.
[370] Schieve LA, Devine O, Boyle CA, Petrini JR, Warner L. Estimation of the contribution of non-assisted reproductive technology ovulation stimulation fertility treatments to US singleton and multiple births. Am J Epidemiol 2009;170:1396–407.
[371] Chang AS, Moley KH, Wangler M, Feinberg AP, Debaun MR. Association between Beckwith-Wiedemann syndrome and assisted reproductive technology: a case series of 19 patients. Fertil Steril 2005;83:349–54.
[372] Cox GF, Burger J, Lip V, Mau UA, Sperling K, Wu BL, Horsthemke B. Intracytoplasmic sperm injection may increase the risk of imprinting defects. Am J Hum Genet 2002;71:162–4.
[373] Halliday J, Oke K, Breheny S, Algar E, D JA. Beckwith-Wiedemann syndrome and IVF: a case-control study. Am J Hum Genet 2004;75:526–8.
[374] Maher ER, Brueton LA, Bowdin SC, Luharia A, Cooper W, Cole TR, Macdonald F, Sampson JR, Barratt CL, Reik W, Hawkins MM. Beckwith-Wiedemann syndrome and assisted reproduction technology (ART). J Med Genet 2003;40:62–4.
[375] Orstavik KH, Eiklid K, van der Hagen CB, Spetalen S, Kierulf K, Skjeldal O, Buiting K. Another case of imprinting defect in a girl with Angelman syndrome who was conceived by intracytoplasmic semen injection. Am J Hum Genet 2003;72:218–9.
[376] Sutcliffe AG, Peters CJ, Bowdin S, Temple K, Reardon W, Wilson L, Clayton-Smith J, Brueton LA, Bannister W, Maher ER. Assisted reproductive therapies and imprinting disorders–a preliminary British survey. Hum Reprod 2006;21:1009–11.

[377] DeBaun MR, Niemitz EL, Feinberg AP. Association of in vitro fertilization with Beckwith-Wiedemann syndrome and epigenetic alterations of LIT1 and H19. Am J Hum Genet 2003;72:156–60.

[378] Gicquel C, Gaston V, Mandelbaum J, Siffroi JP, Flahault A, Le Bouc Y. In vitro fertilization may increase the risk of Beckwith-Wiedemann syndrome related to the abnormal imprinting of the KCN1OT gene. Am J Hum Genet 2003;72:1338–41.

[379] Ludwig M, Katalinic A, Gross S, Sutcliffe A, Varon R, Horsthemke B. Increased prevalence of imprinting defects in patients with Angelman syndrome born to subfertile couples. J Med Genet 2005;42:289–91.

[380] Kobayashi H, Hiura H, John RM, Sato A, Otsu E, Kobayashi N, Suzuki R, Suzuki F, Hayashi C, Utsunomiya T, Yaegashi N, Arima T. DNA methylation errors at imprinted loci after assisted conception originate in the parental sperm. Eur J Hum Genet 2009;17:1582–91.

[381] Poplinski A, Tuttelmann F, Kanber D, Horsthemke B, Gromoll J. Idiopathic male infertility is strongly associated with aberrant methylation of MEST and IGF2/H19 ICR1. Int J Androl 2009.

[382] Market-Velker BA, Zhang L, Magri LS, Bonvissuto AC, Mann MR. Dual effects of superovulation: loss of maternal and paternal imprinted methylation in a dose-dependent manner. Hum Mol Genet 2010;19:36–51.

[383] Gardener H, Spiegelman D, Buka SL. Perinatal and neonatal risk factors for autism: a comprehensive meta-analysis. Pediatrics 2011;128:344–55.

[384] Grafodatskaya D, Cytrynbaum C, Weksberg R. The health risks of ART. EMBO Rep 2013;14:129–35.

[385] Savage T, Peek J, Hofman PL, Cutfield WS. Childhood outcomes of assisted reproductive technology. Hum Reprod 2011;26:2392–400.

[386] Szyf M, Weaver I, Meaney M. Maternal care, the epigenome and phenotypic differences in behavior. Reprod Toxicol 2007;24:9–19.

[387] Weaver IC, Cervoni N, Champagne FA, D'Alessio AC, Sharma S, Seckl JR, Dymov S, Szyf M, Meaney MJ. Epigenetic programming by maternal behavior. Nat Neurosci 2004;7:847–54.

[388] Fish EW, Shahrokh D, Bagot R, Caldji C, Bredy T, Szyf M, Meaney MJ. Epigenetic programming of stress responses through variations in maternal care. Ann N Y Acad Sci 2004;1036:167–80.

[389] Heim C, Binder EB. Current research trends in early life stress and depression: review of human studies on sensitive periods, gene-environment interactions, and epigenetics. Exp Neurol 2012;233:102–11.

[390] Holman DM, Ports KA, Buchanan ND, Hawkins NA, Merrick MT, Metzler M, Trivers KF. The association between adverse childhood experiences and risk of cancer in adulthood: a systematic review of the literature. Pediatrics 2016;138:S81–91.

[391] McGowan PO, Szyf M. The epigenetics of social adversity in early life: implications for mental health outcomes. Neurobiol Dis 2010;39:66–72.

[392] Levine AB, Lockwood CJ, Chitkara U, Berkowitz RL. Maternal renal artery Doppler velocimetry in normotensive pregnancies and pregnancies complicated by hypertensive disorders. Obstet Gynecol 1992;79:264–7.

[393] Meaney MJ. Maternal care, gene expression, and the transmission of individual differences in stress reactivity across generations. Annu Rev Neurosci 2001;24:1161–92.

[394] Turecki G, Meaney MJ. Effects of the social environment and stress on glucocorticoid receptor gene methylation: a systematic review. Biol Psychiatr 2016;79:87–96.

[395] Klengel T, Mehta D, Anacker C, Rex-Haffner M, Pruessner JC, Pariante CM, Pace TW, Mercer KB, Mayberg HS, Bradley B, Nemeroff CB, Holsboer F, Heim CM, Ressler KJ, Rein T, Binder EB. Allele-specific FKBP5 DNA demethylation mediates gene-childhood trauma interactions. Nat Neurosci 2013;16:33–41.

[396] Wu Y, Patchev AV, Daniel G, Almeida OF, Spengler D. Early-life stress reduces DNA methylation of the Pomc gene in male mice. Endocrinology 2014;155:1751–62.

[397] Vukojevic V, Kolassa IT, Fastenrath M, Gschwind L, Spalek K, Milnik A, Heck A, Vogler C, Wilker S, Demougin P, Peter F, Atucha E, Stetak A, Roozendaal B, Elbert T, Papassotiropoulos A, de Quervain DJ. Epigenetic modification of the glucocorticoid receptor gene is linked to traumatic memory and post-traumatic stress disorder risk in genocide survivors. J Neurosci 2014;34:10274–84.

[398] Zannas AS, Provencal N, Binder EB. Epigenetics of posttraumatic stress disorder: current evidence, challenges, and future directions. Biol Psychiatr 2015;78:327–35.

[399] Do C, Xing Z, Yu YE, Tycko B. Trans-acting epigenetic effects of chromosomal aneuploidies: lessons from Down syndrome and mouse models. Epigenomics 2017;9:189–207.

[400] Strong E, Butcher DT, Singhania R, Mervis CB, Morris CA, De Carvalho D, Weksberg R, Osborne LR. Symmetrical dose-dependent DNA-methylation profiles in children with deletion or duplication of 7q11.23. Am J Hum Genet 2015;97:216–27.

[401] Aref-Eshghi E, Rodenhiser DI, Schenkel LC, Lin H, Skinner C, Ainsworth P, Pare G, Hood RL, Bulman DE, Kernohan KD, Boycott KM, Campeau PM, Schwartz C, Sadikovic B. Genomic DNA methylation signatures enable concurrent diagnosis and clinical genetic variant classification in neurodevelopmental syndromes. Am J Hum Genet 2018;102:156–74.

[402] Aref-Eshghi E, Schenkel LC, Lin H, Skinner C, Ainsworth P, Pare G, Rodenhiser D, Schwartz C, Sadikovic B. The defining DNA methylation signature of Kabuki syndrome enables functional assessment of genetic variants of unknown clinical significance. Epigenetics 2017;12:923–33.
[403] Aref-Eshghi E, Schenkel LC, Lin H, Skinner C, Ainsworth P, Pare G, Siu V, Rodenhiser D, Schwartz C, Sadikovic B. Clinical validation of a genome-wide DNA methylation assay for molecular diagnosis of imprinting disorders. J Mol Diagn 2017;19:848–56.
[404] Butcher DT, Cytrynbaum C, Turinsky AL, Siu MT, Inbar-Feigenberg M, Mendoza-Londono R, Chitayat D, Walker S, Machado J, Caluseriu O, Dupuis L, Grafodatskaya D, Reardon W, Gilbert-Dussardier B, Verloes A, Bilan F, Milunsky JM, Basran R, Papsin B, Stockley TL, Scherer SW, Choufani S, Brudno M, Weksberg R. CHARGE and Kabuki syndromes: gene-specific DNA methylation signatures identify epigenetic mechanisms linking these clinically overlapping conditions. Am J Hum Genet 2017;100:773–88.
[405] Hood RL, Schenkel LC, Nikkel SM, Ainsworth PJ, Pare G, Boycott KM, Bulman DE, Sadikovic B. The defining DNA methylation signature of Floating-Harbor Syndrome. Sci Rep 2016;6:38803.
[406] Kernohan KD, Cigana Schenkel L, Huang L, Smith A, Pare G, Ainsworth P, Boycott KM, Warman-Chardon J, Sadikovic B. Identification of a methylation profile for DNMT1-associated autosomal dominant cerebellar ataxia, deafness, and narcolepsy. Clin Epigenet 2016;8:91.
[407] Schenkel LC, Aref-Eshghi E, Skinner C, Ainsworth P, Lin H, Pare G, Rodenhiser DI, Schwartz C, Sadikovic B. Peripheral blood epi-signature of Claes-Jensen syndrome enables sensitive and specific identification of patients and healthy carriers with pathogenic mutations in KDM5C. Clin Epigenet 2018;10:21.
[408] Schenkel LC, Kernohan KD, McBride A, Reina D, Hodge A, Ainsworth PJ, Rodenhiser DI, Pare G, Berube NG, Skinner C, Boycott KM, Schwartz C, Sadikovic B. Identification of epigenetic signature associated with alpha thalassemia/mental retardation X-linked syndrome. Epigenet Chromatin 2017;10:10.

**Websites**

GeneImprint database: http://www.geneimprint.com/.
ENCODE at the UCSC Genome Browser: https://genome.ucsc.edu/encode/.
NIH Roadmap Epigenomics Project: http://www.roadmapepigenomics.org/.
International Human Epigenome Consortium (IHEC): http://ihec-epigenomes.org/.
FANTOM: fantom.gsc.riken.jp/.
Regulome DB: http://regulomedb.org/.
GRASP (Genome-Wide Repository of Associations Between SNPs and Phenotypes): https://grasp.nhlbi.nih.gov/Overview.aspx.

## FURTHER READING

Augui S, Nora EP, Heard E. Regulation of X-chromosome inactivation by the X-inactivation centre. Nat Rev Genet 2011;12:429–42.

# Human Genomic Variants and Inherited Disease: Molecular Mechanisms and Clinical Consequences

*Stylianos E. Antonarakis*[1], *David N. Cooper*[2]

[1]Department of Genetic Medicine and Development, University of Geneva Medical School, Geneva, Switzerland
[2]Institute of Medical Genetics, Cardiff University, Cardiff, United Kingdom

## 6.1 INTRODUCTION

Each individual human genome is unique and varies from the reference genome at millions of sites. This genomic individuality contributes considerably to the biological identity of each person; a very small fraction of the genomic variation is pathogenic and contributes to the various disease phenotypes. The recent advances in rapid and relatively inexpensive DNA sequencing, the development of computational tools for data analysis, the functional exploration of the genome, the creation of databases of genomic variants from hundreds of thousands of individuals, the use of model organisms for the functional characterization of variants, have had a major impact on the development of genomic medicine. Genomic medicine, which is based on genomic variation, is at the heart of understanding the molecular pathophysiology of human disorders. This chapter provides examples of the different varieties of pathogenic variants in the different functional elements of the human genome.

A major aid in recognizing a high-impact pathogenic variant in the sea of neutral or slightly deleterious variants is the existence of databases of variants linked to phenotypes. The existing databases, however, are still rather small, and a major challenge in the coming years is to create, maintain and update large, accessible, and high-quality international databases [1].

A wide variety of different types of pathogenic variants occur in the human genome, with many diverse mechanisms being responsible for their generation: single base-pair substitutions in coding, regulatory and splicing-relevant regions of human genes (67.4%), as well as microdeletions (14.7%), microinsertions (6.2%), gross insertions and duplications (1.8%), repeat expansions (0.2%), combined microinsertions/deletions ("indels") (1.4%), gross deletions (7.4%), gross insertions (1.8%), inversions, and other complex rearrangements (0.9%).

Characterized genomic variants occur not only in coding sequences but also in promoter regions, splice junctions, and within introns and untranslated regions, and noncoding RNAs. Different types of human gene variants may vary in size, from structural variants to single base-pair substitutions, but what they all have in common is that their nature, size and location are often determined either by specific characteristics of the local DNA sequence environment or by higher order features of the genomic architecture. A major goal of genomic medicine is to be able to predict the nature of the clinical phenotype through ascertainment of the genotype. The extent to which this is feasible in medical genetics is very much disease, gene, and mutation dependent. The study of variants in human genes is nevertheless of paramount importance for understanding the pathophysiology of inherited disorders, optimizing diagnostic testing and

Emery and Rimoin's Principles and Practice of Medical Genetics and Genomics: Foundations, https://doi.org/10.1016/B978-0-12-812537-3.00006-8

guiding the design of emergent therapies. A major goal of molecular genetic medicine is to be able to predict the nature of the clinical phenotype through ascertainment of the genotype. The extent to which this is feasible in medical genetics is very much disease, gene, and variant dependent.

The first description of the precise molecular defect in a human disease (the sickle cell variant, a Glu to Val substitution at the sixth codon of the β-globin [*HBB*] gene) was identified by Ingram in 1956 [2], who found that the difference between hemoglobin A and hemoglobin S lies in a single tryptic peptide. His analysis was made possible by the methods developed by Sanger for determining the structure of insulin and by Edman to effect the stepwise degradation of peptides. This was followed, 40 years ago, by the characterization of the first heritable pathogenic variants in a human gene at the DNA level: gross deletions of the human α-globin (*HBA*) and *HBB* gene clusters giving rise to α- and β-thalassemia [3] and a single base-pair substitution (Lys17Term) in *HBB* gene causing β-thalassemia [4]. Since then, continuous technical advances have enabled the identification of numerous disease-related genes and the discovery of thousands of underlying pathological lesions [5]. Single base-pair substitutions (67%) and microdeletions (15.6%) are the most frequently encountered variants in the human genome, the remainder comprising an assortment of microinsertions (6.5%), indels (1.5%), gross deletions (6.6%), gross insertions and duplications (1.4%), inversions, repeat expansions (0.3%), and complex rearrangements (1.0%).

The vast majority of single nucleotide variants listed in Human Gene Mutation Database (HGMD) reside within the coding region (84%), the remainder being located in either intronic (13%) or regulatory (3%, promoter, untranslated, or flanking regions) sequences. Variants may interfere with any stage in the pathway of expression, from gene activation to synthesis and secretion of the mature protein product. The question of the proportion of possible variants within human disease genes that are likely to be of pathological significance is one that is difficult to address because it is dependent not only on the type and location of the variant but also on the functionality of the nucleotides involved (itself dependent in part upon the amino acid residues that they encode), which is often hard to assess [6–11]. In addition, some types of variants are likely to be much more comprehensively ascertained than others, making observational comparisons between mutation types an inherently hazardous undertaking.

Different types of human gene variants may vary in size, from structural variants (SVs) to single base-pair substitutions, but what they all have in common is that their nature, size, and location are often determined either by specific characteristics of the local DNA sequence environment or by higher order features of the genomic architecture [12]. This chapter attempts to provide an overview of the nature of variants causing human genetic disease and then considers their consequences for the clinical phenotype. Three online databases, which interested readers may consult, contain information on known disease-related (pathogenic) variants: the *HGMD* (hgmd.org), *Mendelian Inheritance in Man* (omim.org/), and *ClinVar* (ncbi.nlm.nih.gov/clinvar/). The *HGMD* contains 225,000 likely pathogenic variants; *ClinVar* includes 431,000 variants with interpretation, a substantial fraction of which are not pathogenic; *OMIM* contains only representative pathogenic variants per gene in 4150 protein-coding genes.

The advances in DNA sequencing technologies, the computational analysis of the data, the development of public databases of genomic variants of thousands of individuals and the international exchanges of prepublication data, as well as guidelines and criteria for assessing the pathogenicity of genomic variants have greatly advanced the discovery of gene–disease links, primarily on mendelian phenotypes, and has tremendously expanded the pool of pathogenic and likely pathogenic variants [13–18]. The criteria for assessing pathogenicity from the American College of Medical Genetics (ACMG) are particularly useful and are implemented in the diagnostic services worldwide [17]. The ACMG criteria categorize the variants in five classes regarding pathogenicity: benign, likely benign, variant of unknown significance, likely pathogenic, and pathogenic.

An excellent discussion on the origins, determinants, and consequences of human mutations has been recently published [19] and is recommended to the reader.

## 6.2 MOLECULAR MECHANISMS OF VARIANTS CAUSING HUMAN INHERITED DISEASE

### 6.2.1 "Neutral Variation"/DNA Polymorphisms

The term *polymorphism* has been defined [20] as a "Mendelian trait that exists in the population in at least two phenotypes, neither of which occurs at a frequency of less than 1%." Polymorphisms are not therefore rare. Indeed, there is enormous variation in the DNA sequences of any two randomly chosen human haploid genomes. Clearly, not all variations within a gene result in the abnormal expression of protein products. Indeed, single nucleotide substitutions/polymorphisms (SNPs) occur in 1:~600–1200 nucleotides in intervening sequences and flanking DNA (2005; [21–25]). These substitutions represent the most common form of DNA polymorphism that can be used as markers for specific regions of the human genome. Similarly, some single nucleotide substitutions in the coding regions of genes may also be normal (nonpathogenic) polymorphic variants even if they result in nonsynonymous substitutions of the polypeptide product [26]. For example, there are three common forms of *HBB* gene on chromosome 11p; these forms differ at five nucleotides, one of which lies within the first exon of the gene and results in a synonymous codon. The average human gene contains >120 biallelic polymorphisms, 46 of which occur with a frequency >5%, with five occurring within the coding region [27].

Some polymorphisms entail the alteration of an encoded amino acid, for example, the Lewis *Le* alleles of the *FUT3* gene [28], whereas others may introduce a stop codon that serves to inactivate the gene in question—for example, the secretor *se* allele of the *FUT2* gene present in 20% of the population [29]. However, not all polymorphisms are SNPs. Examples of other types of gene-associated polymorphisms in the human genome include triplet repeat copy number (e.g., in the *FMR1* gene; see 9.2.1.3), gross gene deletion (e.g., *GSTM1* and *GSTT1* [30]), gene duplication (e.g., *HBG2* [31]), intragenic duplication (e.g., *IVL* [32]), microinsertion/deletion (e.g., *PAI1* [33]), indel (e.g., *APOE* [34]), gross insertion (e.g., the inserted *Alu* sequence in intron 16 of the *ACE* gene [35]), inversion (e.g., the 48-kb Xq28 inversion involving the *EMD* and *FLN1* genes [36]), and gene fusion (e.g., between the *RCP* and *GCP* visual pigment genes [37]). Functional polymorphisms may occur within the coding region [38] or regulatory regions [39] of a gene or may impact on pre-messenger RNA (mRNA) splicing [40] and therefore can have consequences for protein structure/function, gene expression, or mRNA splicing. It can be seen that the spectrum of polymorphisms in the human genome is qualitatively different than the variants underlying human disease; they may vary in terms of location and frequency but otherwise they display remarkable similarities indicative of the same underlying mutational mechanisms.

It is likely that some SNPs, whether frequent or rare, alter the risk of common complex human phenotypes ("functional SNPs"). A public SNP database now contains >660 million entries (dbSNP; ncbi.nlm.nih.gov-/SNP/snp_summary.cgi). An international project termed the "HapMap project" [41–43] had the objective to define the patterns of common SNP genetic variation in a sample of 270 DNAs from individuals of European, African, Chinese, and Japanese origin (hapmap.org). The data obtained from this project constitute ~2.8 million SNPs and are publicly available. The results of this and other similar projects are contributing significantly to our understanding of both common and rare human genetic disorders and traits. Furthermore, recent advances in high-throughput sequencing have led to the discovery of a large number of individually rare polymorphic variants in samples from the 1000 Genomes [44] and other projects. The protein-coding regions and splice junctions of a typical human genome (also known as the exome) contain 9000–11,000 nonsynonymous variants, ~100 nonsense codons, and 35 splice variants [45]. Analysis of 54 human genomes, sequenced by Complete Genomics (completegenomics.com), revealed 3,700,000–4,700,000 single nucleotide variants per genome; the frequency of these variants was not identical in various fractions of the genome and is related to regional evolutionary constraints. Fig. 6.1 illustrates that the protein-coding fraction of the genome, which is under evolutionary pressure, contains the smallest number of variants per kilobase (kb); by contrast, the repeat fraction of the genome and the intergenic regions, which presumably evolve neutrally, contain almost double the number of variants per kb.

The genome of each individual contains a number of likely damaging alleles for the encoded protein, and not all of them contribute to recognizable phenotypes. The sequence of exomes of 3222 British Pakistani-heritage adults from consanguineous marriages has revealed 1111 homozygous rare variants with predicted loss of function in 781 genes. On average, there were 1.6 homozygous LOF variants per individual; remarkably, these homozygous variants were found in apparently healthy people [46]. One of the first studies to assess the number of predicted deleterious variants in normal individuals has examined low-coverage whole-genome sequences from 179 individuals. Each individual carried 281–515 missense substitutions, 40–85 of which were homozygous, predicted to be highly damaging. They also carried

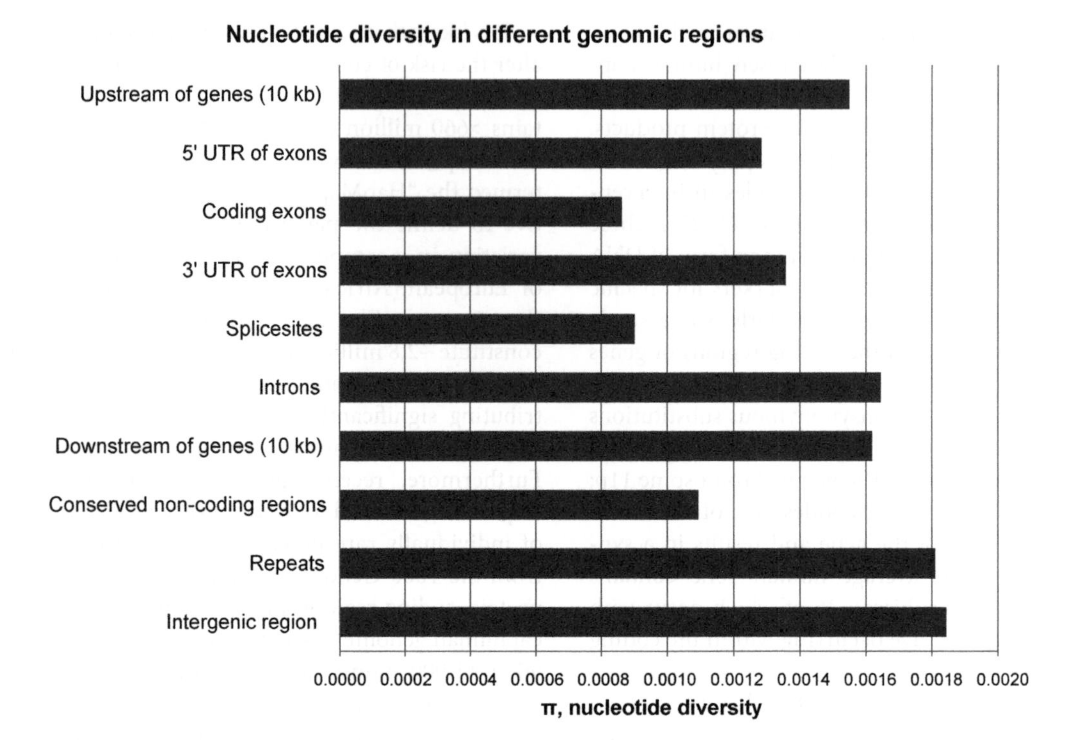


• Genomic variants from CompleteGenomics, 54 unrelated individuals from different ethnic groups (www.completegenomics.com)
• Genomic regions drawn from UCSC genome browser (http://genome.ucsc.edu)

**Figure 6.1** Nucleotide diversity (equivalent to frequency of polymorphic variants) in different genomic regions. The genomic variants analyzed are from the whole genome sequences of 54 unrelated human genomes (see text).

40–110 variants classified by the Human Gene Mutation Database (HGMD) as disease-causing variants, 3–24 variants in the homozygous state [47].

Another form of polymorphic variation in our genome is the presence of variable numbers of tandem repeats. The repeat unit can be 10–60 nucleotides in length and many different alleles may exist at a given locus [48,49]. The combination of a VNTR and single nucleotide substitutions within the repeat unit results in an extremely high level of polymorphic variability, which can be used as a unique bar code to distinguish different individuals [50]. The introduction of the polymerase chain reaction (PCR) [51] permitted the rapid detection and analysis of variation in short sequence repeats (SSRs), for example $(GT)_n$ repeats [52,53]. These are common polymorphisms that occur on average once for every 50 kb of genomic DNA. The SSRs also display many alleles and the repeat unit can be two, three, four, five, or more nucleotides. Poly(A) tracts may also be polymorphic, exhibiting variation in the number of A residues [54]; many of these polymorphisms are localized at the ends of *Alu* repetitive elements. Another kind of polymorphism in the human genome involves the presence or absence of retrotransposons (i.e., *Alu* or LINE repetitive elements or pseudogenes) at specific locations [55,56]. Duplicational polymorphisms in some human genes, such as *HBA1*, *PRB1-4*, *HBZ*, and *CYP21/C4A/C4B*, have been known for some time [56,57]. The use of comparative genomic hybridization against BAC or oligonucleotide arrays has revealed extensive copy number polymorphism/variation (CNP or CNV) of sizeable genomic regions [58–60]. Details of many thousands of such genomic variants may be found in the following databases: *CNV Project*, http://www.sanger.ac.uk/humgen/cnv, and *Database of Genomic Variants*, http://projects.tcag.ca/variation. A first CNV map of the human genome of the 270 "HapMap" individuals revealed a total of 1440 CNV regions covering ~360 megabases

(Mb) (12% of the genome) [61]. High-resolution tiling oligonucleotide microarrays have been used to generate comprehensive genomic maps of >10,000 CNVs [62,63]. The functional significance, if any, of most of these polymorphic variants is, however, unknown. To understand the prevailing mutational mechanisms responsible for human genome structural variation, a total of 1054 large structural variants have been sequenced (589 deletions, 384 insertions, and 81 inversions from 17 human genomes). The prevailing mechanisms for these structural variations were: i/microhomology-mediated processes involving short (2–20 bp) sequences (28%), nonallelic homologous recombination (22%), and L1 retrotransposition (19%) [64].

It is clear that no single individual genome contains the full complement of functional genes [65], a paradigm shift that strikes at the heart of the concept of a "reference genome" [66].

Deletion polymorphisms are also remarkably frequent in the human genome: a typical individual has been estimated to be hemizygous for some 30–50 deletions >5 kb, spanning >550 kb in total, and encompassing >250 known or predicted genes [67,68]. Because such deletions appear to be in linkage disequilibrium with neighboring SNPs, we may surmise that they share a common evolutionary history [69].

Human DNA polymorphisms have proven extremely useful in developing linkage maps, for mapping monogenic and polygenic complex disorders, for determining the origin of aneuploidies and chromosomal abnormalities, for distinguishing normal from mutant chromosomes in genetic diagnoses, for performing forensic, paternity, and transplantation studies, for studying the evolution of the genome, the loss of heterozygosity in certain malignancies, the detection of uniparental disomy, the instability of the genome in certain tumors, recombination at the level of the genome, the study of allelic expression imbalance, and the development of haplotype maps of the genome. In studying the role of a candidate gene in a given disorder, it is imperative to distinguish between pathogenic variants that cause a clinical phenotype and the polymorphic variability of the normal genome.

### 6.2.2 Nonsense SNPs

The loss of a particular gene/allele is not invariably associated with a readily discernible clinical phenotype [70,71]. This assertion is supported by the identification of more than 1000 putative nonsense SNPs (i.e., nonsense variants that have attained polymorphic frequencies) in human populations [72,73]. About half of these nonsense SNPs have been validated by dbSNP (ncbi.nlm.nih.gov/projects/SNP), a process that involves the exclusion of variants in pseudogenes and of artifacts caused by sequencing errors. Bona fide nonsense SNPs are expected either to lead to the synthesis of a truncated protein product or alternatively to the greatly reduced synthesis of the truncated protein product (if the mRNA bearing them is subject to nonsense-mediated mRNA decay [NMD]). Based on the relative locations of the nonsense SNPs and the exon–intron structures of the affected genes, Yamaguchi-Kabata et al. [74] concluded that 49% of nonsense SNPs would be predicted to elicit NMD, whereas 51% would be predicted to yield truncated proteins. Some of these nonsense SNPs have been found to occur in the homozygous state in normal populations [73], attesting to the likely functional redundancy of the corresponding genes. At the very least, genes harboring nonsense SNPs may be assumed to be only under weak selection [72].

It should be appreciated that nonsense SNPs may even occur in "essential" genes, yet still fail to come to clinical attention (or give rise to a detectable phenotype) if these genes are subject to CNV (see CNVs later) that masks any deleterious consequences by ensuring an adequate level of gene expression from additional wild-type copies either in *cis* or in *trans*. Thus, CNV might serve to "rescue" the full or partial loss of gene function brought about by the nonsense variants, thereby accounting for the occurrence of the latter at polymorphic frequencies. Consistent with this postulate [72], it was reported [72] that ~30% of nonsense SNPs occur in genes residing within segmental duplications, a proportion some threefold larger than that noted for synonymous SNPs. Genes harboring nonsense SNPs were also found to belong to gene families of higher than average size [72], suggesting that some functional redundancy may exist between paralogous human genes. In support of this idea, Hsiao and Vitkup [75] reported that those human genes that have a homolog with ≥90% sequence similarity are approximately three times less likely [76] to harbor disease-causing variants than are genes with less closely related homologs. They interpreted their findings in terms of "genetic robustness" against null variants, with the duplicated sequences providing "back-up" by potentiating the functional compensation/complementation of homologous genes in the event that they acquire deleterious variants.

## 6.3 DISEASE-CAUSING VARIANTS

### 6.3.1 The Nature of Genomic Variants

Fig. 6.2 depicts the frequencies of the various genomic variant types responsible for molecularly characterized human genetic disorders, as recorded in HGMD (http://www.hgmd.org) and studies [77–80]. HGMD records each variant *once*, regardless of the number of independent occurrences of that lesion. Fig. 6.2 shows the frequency of the first variant per disease recorded in MIM (omim.org/). As of July 2018, HGMD contained some 225,000 different disease-causing variants and disease-associated/functional polymorphisms in 8784 human genes (Fig. 6.2), whereas MIM contained selected examples of allelic variants in 4150 human genes associated with a specific phenotype.

### 6.3.2 Nucleotide Substitutions

Single nucleotide substitutions are the most frequent pathological variants in the human genome (Fig. 6.2). Most of these alterations occur during DNA replication, which is an accurate yet error-prone multistep process. The accuracy of DNA replication depends on the fidelity of the replicative step and the efficiency of the subsequent error correction mechanisms [81]. Analysis of more than 7000 missense and nonsense variants associated with human disease has indicated that the most common nucleotide substitution for T (thymine) is to C (cytosine), for C it is to T, for A (adenine) it is to G (guanine), and for G it is to A [82]. Transitions are, therefore, much more common than transversions. Some 61% of the missense and nonsense variants currently logged in

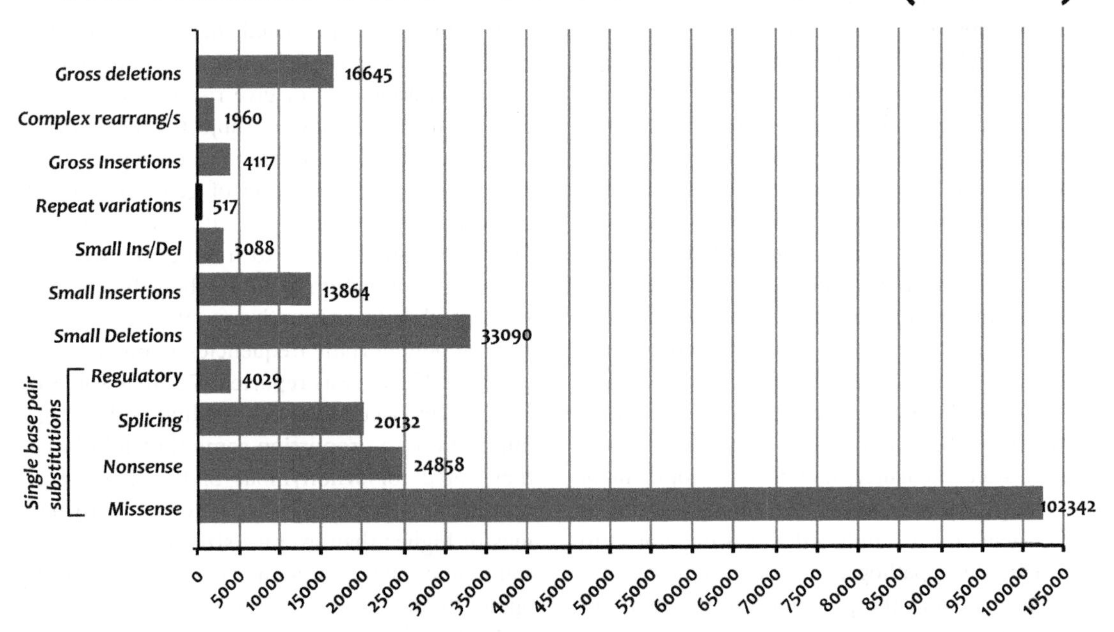


**Figure 6.2** Spectrum of different types of human disease-causing mutations and disease-associated/functional polymorphisms logged in the HGMD as of July 2018.

HGMD are transitions (T to C, C to T, A to G, G to A) while 39% are transversions (T to A or G, A to T or C, G to C or T, C to G or A).

Among single nucleotide substitutions, there is one that clearly predominates and represents the most common type of mutational lesion in the human genome: CpG dinucleotides mutate to TpG at a frequency that is about five times higher than mutations in all other dinucleotides [82–85]. This substitution, which when it occurs on one DNA strand generates TG, and on the other, CA (the "CG to TG or CA rule") represents a major cause of human genetic disease. This phenomenon was first observed in the factor VIII (*F8*) gene in cases of hemophilia A [85], but it was soon noted in the studies of many other genes [86]. In hemophilia A, CG to TG or CA mutations account for 46% of single nucleotide variants in unrelated patients [87]. In the HGMD (www.hgmd.org), such variants currently account for ~18% of the total number of missense and nonsense variants [76]. Among CpG dinucleotide mutations, transitions to TG or CA account for ~90% of substitutions. The mechanism of this common type of mutation appears to be methylation-mediated deamination of 5-methylcytosine (5mC). In eukaryotic genomes, 5mC occurs predominantly in CpG dinucleotides, most of which appear to be methylated (see [88] for review). 5mC then undergoes spontaneous nonenzymatic deamination to form thymine (Fig. 6.3). There is a bias in terms of the origin of CpG to TpG mutations: most occur in male germ cells (the male:female ratio is 7:1). One reason for this may be that sperm DNA is heavily methylated, whereas oocyte DNA is comparatively undermethylated [89]. Another reason may be the considerably higher number of germline cell divisions in males as compared to females [90].

Cytosine methylation also occurs in the context of CpNpG sites [91]. If we assume not only that CpNpG methylation occurs in the germline but also that 5mC deamination can occur within a CpNpG context, then it follows that methylated CpHpG sites are also very likely to constitute mutation hotspots causing human inherited disease. Initial evidence that this might indeed be the case came from the observation that disproportionately high numbers of C>T and G>A transitions occur at CpNpG sites in studies of the human genes, *NF1* [92] and *BRCA1* [93]. Further, ~9.9% of 54,625 missense and nonsense variants from 2113 genes causing inherited disease (HGMD) are C>T and G>A transitions located within CpNpG trinucleotides, approximately twofold higher proportion than would have been expected by chance alone [76]. Some 5% of missense or nonsense variants causing human inherited disease may, therefore, be attributable to methylation-mediated deamination of 5mC within a CpNpG context.

In a recent analysis, the average direct estimate of the combined rate of all mutations was $1.8 \times 10^{-8}$ per nucleotide per generation [94]. Single nucleotide substitutions were found to be ~25 times more common than all other mutations, while deletions were ~3 times more common than insertions; complex mutations were very rare and the CpG context was found to increase substitution rates by an order of magnitude [94]. Rates of different kinds of mutations were also found to be strongly correlated across different loci [94].

It has been estimated that ~20% of new missense variants in humans result in a loss of function, whereas 53% have mildly deleterious effects, and 27% are effectively neutral with respect to phenotype [95]. These estimates have received independent support, at least qualitatively, from a study of human coding SNPs by

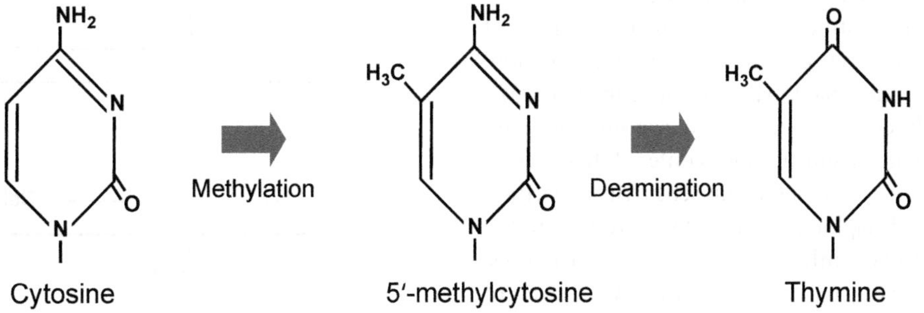

**Figure 6.3** Schematic representation of cytosine, 5′-methylcytosine, and thymine, and the chemical events involved in the mutational transformation of cytosine to thymine.

Boyko et al. [96], who predicted that 27%–29% of missense variants would be neutral or near neutral, 30%–42% would be moderately deleterious, with most of the rest (i.e., 29%–43%) being highly deleterious or lethal, and by Eyre-Walker et al. [97] who estimated that >50% of variants would be likely to exert only a mild effect on the phenotype.

A recent study using human osteosarcoma cell lines has shown that noncanonical (non-B) DNA conformations are capable of increasing the overall spectrum of single base-pair substitutions in a reporter gene in *cis* by exposing those DNA sequences to oxidative damage [98]. In this study, the spectrum of single base-pair substitutions was shown to be indistinguishable from that induced by other conditions known to lead to a hyperoxidative state (such as WRN deficiency and lung tumorigenesis), an observation that lends support to a model whereby DNA bases become oxidized, followed by the transfer of their oxidized state ("hole migration") to target neighboring bases. If these observations are eventually found to be relevant in the context of "natural" chromatin during meiosis, then the impact of non-B DNA conformations on human inherited disease, both with respect to single base-pair substitutions and gross rearrangements (see Section 6.3.6), could be quite significant. Further, because non-B DNA structures can interfere with DNA replication and repair, and may serve to increase mutation frequencies in generalized fashion, they have the potential to serve as a unifying concept in studies of mutational mechanisms underlying human inherited disease.

### 6.3.3 Synonymous Nucleotide Substitutions

Synonymous ("silent") variants, although not altering the amino acid sequence of the encoded protein directly, can still influence splicing accuracy or efficiency [99–104]. It has become increasingly clear that apparently silent SNPs may also become distinctly "audible" in the context of mRNA stability or even protein structure and function. Thus, three common haplotypes of the human *COMT* gene, which differ in terms of two synonymous and one nonsynonymous substitution, confer differences in COMT enzymatic activity and pain sensitivity [105,106]. The major *COMT* haplotypes differed with respect to the stability of the *COMT* mRNA local stem-loop structures, the most stable being associated with the lowest levels of COMT protein and enzymatic activity [106]. In a similar vein, synonymous SNPs in the *ABCB1* gene have been shown to alter ABCB1 protein structure and activity [107], possibly by changing the timing of protein folding following extended ribosomal pause times at rare codons [108]. Finally, it should be understood that although the deleteriousness of the average synonymous variants is always likely to be less than that of a nonsynonymous (missense) variants [96], the higher prevalence of synonymous variant means that they may actually make a significantly greater contribution to the phenotype than nonsynonymous variants [109].

### 6.3.4 Microdeletions and Microinsertions

Deletions or insertions of a few nucleotides are also fairly common as a cause of human inherited disease. Most of these are less than 20 bp in length. Indeed, the majority of microdeletions involve <5 nucleotides. In HGMD, the deletion of 1 bp accounts for 48% of small deletions while an additional 30% involve two or three nucleotides. The majority of microdeletions recorded (78%) result in an alteration of the reading frame. Most microdeletions occur in regions that contain direct repeats of 2 bp or more. The most common length of direct repeat is 3 bp (48% of direct repeats associated with short deletions [83]. The most plausible mechanism for small deletions mediated by the presence of direct repeats is the slipped mispairing model [110] (Fig. 6.4). In addition, deletions of one or a few nucleotides frequently occur in runs of the same nucleotide, for example, a poly(T) region [111]. Finally, inverted repeats and "symmetric elements" are also frequently found in the immediate vicinity of microdeletions [112,113]. Krawczak

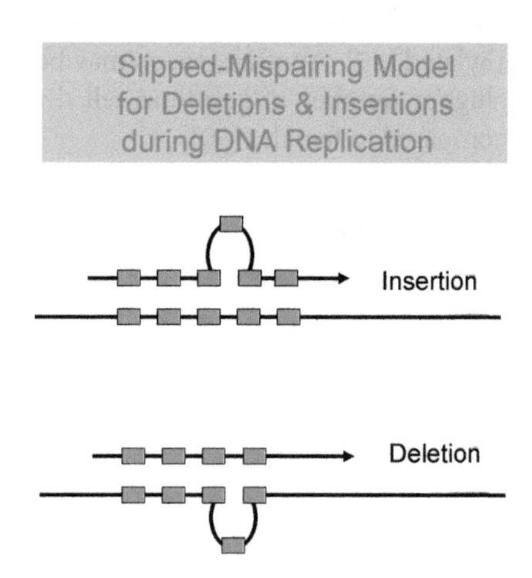


**Figure 6.4** Schematic representation of the slipped mispairing model for deletions and insertions during DNA replication.

and Cooper [114] identified a consensus sequence—TG(A/G)(A/G)(G/T)(A/C)—which they claimed to represent a deletion hotspot.

Microinsertions (again up to 20 nucleotides) are rarer than microdeletions; thus, in HGMD there are three times as many microdeletions as microinsertions (Fig. 6.2). Nearly half of these involve the insertion of only one nucleotide (Fig. 6.5). As is the case with microdeletions, most microinsertions lead to alterations of the reading frame and are located in regions containing direct or inverted repeats or runs of the same nucleotide. Details of possible mechanisms of generation during replication can be found in [115]; however, there are as-yet insufficient data available to estimate the frequency ratio of microinsertions or microdeletions in male or female germ cells. In the case of such lesions in factor VIII (*F8*) gene, 56% of microdeletions/insertions have been reported to occur in DNA regions harboring direct repeats or runs of the same nucleotide [87].

HGMD data (3767 microdeletions and 1960 microinsertions) were used to perform a meta-analysis of microdeletions and microinsertions causing inherited disease, both defined as involving ≤20 bp DNA [116]. A positive correlation was noted between the microdeletion and microinsertion frequencies for 564 genes in which both microdeletions and microinsertions have

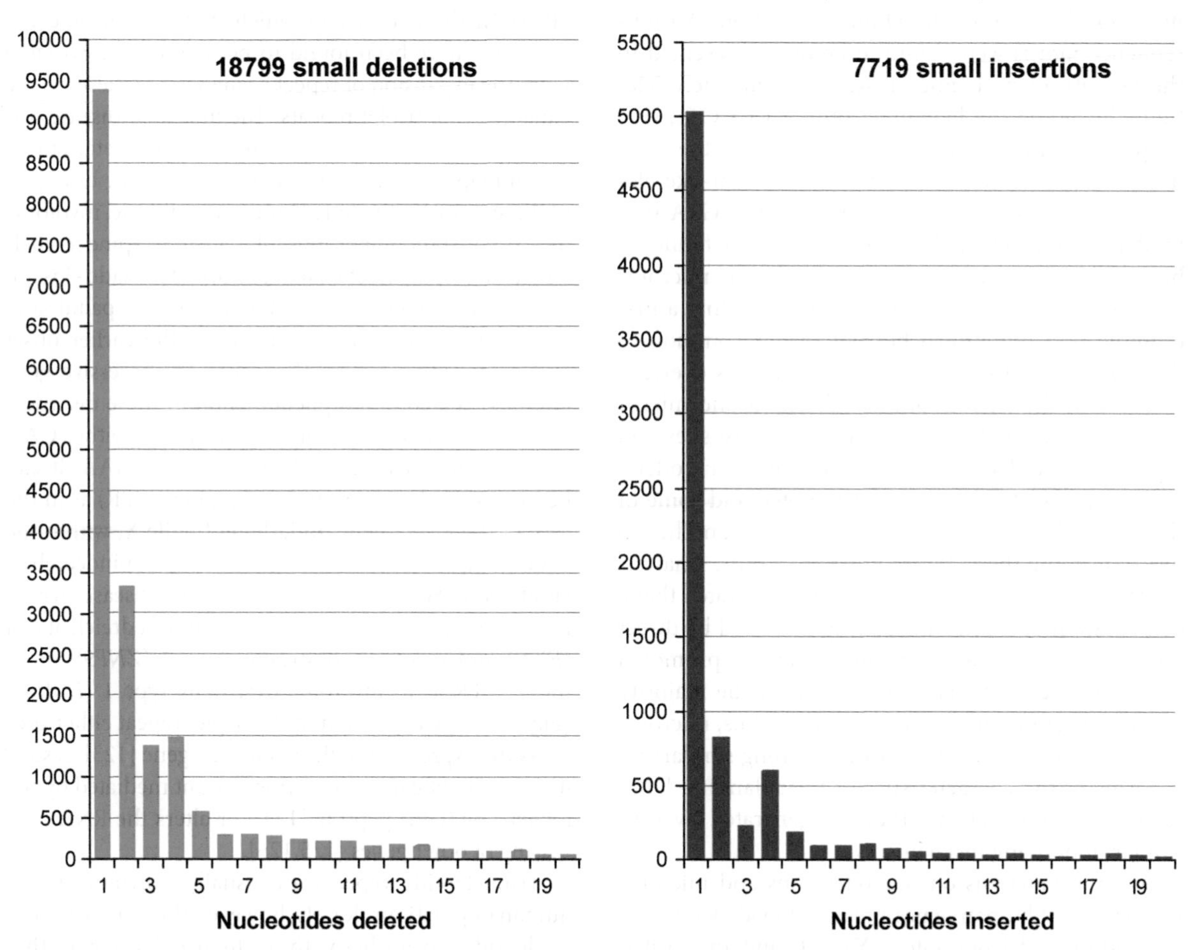


**Figure 6.5** Size distribution of short (<20 bp) pathogenic human gene deletions and insertions (HGMD; http://www.hgmd.org; January 5, 2007).

been reported. This is consistent with the view that the propensity of a given gene/sequence to undergo microdeletion is related to its propensity to undergo microinsertion. While microdeletions and microinsertions of 1 bp constitute, respectively, 48% and 66% of the corresponding totals, the relative frequency of the remaining lesions correlates negatively with the length of the DNA sequence deleted or inserted. Many microdeletions and microinsertions of >1 bp are potentially explicable in terms of slippage mutagenesis, involving the addition or removal of one copy of a mono-, di-, or trinucleotide tandem repeat. The frequency of in-frame 3-bp and 6-bp microinsertions and microdeletions was, however, found to be significantly lower than that of variants of other length, suggesting that some of these in-frame lesions may not have come to clinical attention. Various sequence motifs were found to be overrepresented in the vicinity of both microinsertions and microdeletions, including the heptanucleotide CCCCCTG that shares homology with the complement of the 8-bp human minisatellite conserved sequence/chi-like element (GCWGGWGG). The "indel hotspot" GTAAGT (and its complement ACTTAC) were also found to be overrepresented in the vicinity of both microinsertions and microdeletions, thereby providing a first example of a mutational hotspot that is common to different types of gene lesions. Other motifs overrepresented in the vicinity of microdeletions and microinsertions included DNA polymerase pause sites and topoisomerase cleavage sites. Several novel microdeletion/microinsertion hotspots were noted and some of these exhibited sufficient similarity to one another to justify terming them "super-hotspot" motifs. Analysis of DNA sequence complexity also demonstrated that a combination of slipped mispairing mediated by direct repeats, and secondary structure formation promoted by symmetric elements, can account for the majority of microdeletions and microinsertions. Thus, microinsertions and microdeletions exhibit strong similarities in terms of the characteristics of their flanking DNA sequences, implying that they are generated by very similar underlying mechanisms.

A similar analysis on microdeletions and microinsertions in 19 human genes presented evidence for an elevated microdeletion rate at YYYTG and an elevated microinsertion rate at TACCRC and ATMMGCC [117]. These authors also found that ~45% of microdeletions led to the removal of a repeated sequence, an event they termed "deduplication" in order to highlight the identity of the deleted sequence and the sequence abutting the site of deletion.

Another mutational mechanism, DNA triplex formation followed by DNA repair, has been proposed to explain ~ 5% of microdeletions and microinsertions at mirror repeats [118].

### 6.3.5 Expansion/CNV of Trinucleotide (and Other) Repeat Sequences

Another mechanism of human gene variation causing hereditary disease is the instability of repeat (mainly trinucleotide) sequences and their expansion in affected genes [119–121]. A growing number of repeat expansion disorders (in excess of 66 are now recorded in HGMD), the majority of which involve neuromuscular tissue, have been found to be due to, or associated with, the expansion of repeat sequences; of these, 53 are expansions of triplet repeats. The first such disease was fragile X, a common cause of male mental retardation, which mapped to chromosome Xq27.3. Some examples of these disorders include Huntington disease, myotonic dystrophy, spinobulbar muscular atrophy, spinocerebellar ataxia 1, spinocerebellar ataxia 3, or Machado–Joseph disease, the fragile E site, and dentatorubral pallidoluysian atrophy. Genetic "anticipation" (the earlier onset and increasingly severe phenotype in successive generations) is a common phenomenon in these disorders [122]. The trinucleotide involved is usually either CAG or CGG but occasionally CTG, GCG, or GAA. It can be located in the 5′ untranslated region (UTR) as in the case of the *FMR1* gene underlying fragile X, within the coding region (as in Huntington disease, spinocerebellar ataxia 1 [*SCA1*], *SCA3*, and Kennedy disease) where it encodes poly(Gln), in an intron as in Friedreich ataxia (*FXN*) and myotonic dystrophy type 2 (*ZNF9*), or in the 3′ UTR as in myotonic dystrophy type 1 (*DMPK*) (Fig. 6.6). The expansion of the triplet repeat either prevents the expression of the associated gene [123], results in a dominant gain-of-function variant mediated by the longer poly(Gln) peptide [124], or alters the RNA processing of other genes [125,126].

Trinucleotide repeats are usually polymorphic in human populations. Rarely, however, the number of trinucleotide repeats lies within a high-risk category that is termed a "premutation." In such a case, the premutation exhibits a high probability of further expansion

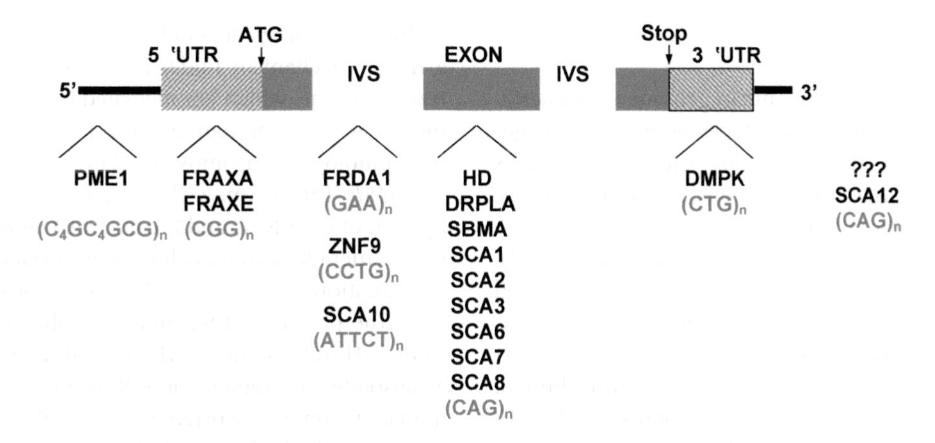


**Figure 6.6** Location of the repeat expansion in selected human disorders.

(instability) to yield disease-related alleles ("full mutation"). In fragile X, for example, the normal polymorphic alleles of the CGG repeat contain between 10 and 50 triplets, the premutation between 50 and 200 triplets, and the full mutation more than 200 triplets [127]. Expansion of premutations to full mutations only occurs during female meiotic transmission. The probability of repeat expansion correlates with repeat copy number in the premutated allele. Because the premutation must precede the appearance of a full mutation, all mothers of affected children carry either a full mutation or a premutation [127]. Premutation alleles may also be associated with late-onset movement disorders and premature ovarian failure [128,129].

The precise mechanism of repeat expansion is unclear, although it is known that DNA polymerase progression is blocked by CTG and CGG repeats and the resultant idling of the polymerase could serve to catalyze slippage leading to repeat expansion [130]. In the case of SCA1, interruption of the CAG repeat with a CAT unit is associated with more stable trinucleotide repeat [131]. More details of these "dynamic mutations" can be found in the appropriate sections covering individual disorders, and in [132]. Short expansions of GCG trinucleotide codons encoding Ala have been observed in the *HOXD13* gene causing dominant polydactyly, and in the *PABP2* gene causing oculopharyngeal muscular dystrophy [133,134]. These mutations may be due to unequal crossing-over rather than polymerase slippage. Generally speaking, it is likely that repeat instability is a consequence of the resolution of unusual secondary structure intermediates during DNA replication, repair, and recombination [135].

A repeat expansion of 12 nucleotides (CCCCGC-CCCGCG) in the 5′ flanking region of the *CSTB* gene causes one form of the recessive progressive myoclonus epilepsy type 1 (EPM1) [136]. This indicates that repeat sequences other than trinucleotides can expand and cause human disorders. This particular expansion silences the *CSTB* gene, probably because it alters the spacing of transcription factor binding sites from each other and/or the transcriptional initiation site [137].

A tetranucleotide repeat expansion $(CCTG)_n$ in intron 1 of the *ZNF9* gene causes myotonic dystrophy type 2 [125]. This expansion can be between 75 and 11,000 repeats in length. The expansion of the pentanucleotide repeat $(ATTCT)_n$ is responsible for the phenotype of spinocerebellar ataxia 10 (SCA10). The expansion occurs in intron 9 of the *SCA10* gene and can be up to 22.5 kb in length [138]. Expansions of even longer repeats have been reported. In Usher syndrome type 1C, for example, there is an expansion of a 45-bp VNTR in intron 5 of the *USH1C* gene (nine tandem repeats instead of the usual less than six such repeats); this expansion has been predicted to inhibit the transcription of the gene [139]. There are also cases in which a large repeat expansion is not associated with a particular phenotype, for example, the expansion of an AT-rich 33-mer repeat in the dictamycin-sensitive fragile site 16B [140].

### 6.3.6 Mechanisms of Gross Genomic Rearrangement

Structural variation in the human genome is characterized by a number of different types of gross rearrangements including deletions, duplications, insertions (termed CNVs), as well as inversions, and translocations. Four major mutational mechanisms account for these SVs: nonallelic homologous recombination (NAHR), nonhomologous end joining (NHEJ), replication-based mechanisms, and L1-retrotransposition [62–64].

*NAHR*: Sequence analysis of the breakpoints of 1054 SVs identified in the genomes of 17 healthy human individuals revealed that NAHR accounts for 22% of insertions and deletions as well as 69% of inversions [64]. The majority of these SVs are likely to represent neutral polymorphisms but ~1% may be disease associated. Some apparently neutral SVs appear to predispose to further structural rearrangements, such as deletions and duplications, which in turn give rise to disease [141–145]. Thus, for example, heterozygosity for the ~970-kb inversion polymorphism of the *MAPT* locus at 17q21.3 predisposes to the NAHR events that underlie the 17q21.31 microdeletion syndrome [146,147]. It may be that inversion heterozygosity perturbs the pairing of homologous chromosomes during meiosis, which then promotes interchromosomal NAHR between the inversion-flanking low copy repeats (LCRs), thereby giving rise to the 17q21.3 microdeletion.

During meiosis, NAHR between sequences that are nonallelic (i.e., paralogous) can result in recurrent deletions and duplications that cause specific genomic disorders. Liu et al. [148] studied two patient cohorts with reciprocal genomic disorders localized to chromosome 17p11.2: the deletion-associated Smith–Magenis syndrome and the duplication-associated Potocki–Lupski syndrome. They reported that complex rearrangements (those with more than one breakpoint) were more prevalent in copy number gains (17.7%) than in copy number losses (2.3%), an observation which supports a role for replicative mechanisms in the formation of complex rearrangements. With respect to the NAHR-mediated recurrent rearrangements, the crossover frequency was found to be positively associated with the flanking LCR length and inversely influenced by the inter-LCR distance. It would, therefore, appear that the probability of ectopic chromosome synapsis increases with increasing LCR length, with ectopic synapsis being a prerequisite for ectopic crossing-over.

Recent findings also indicate that NAHR represents a major mechanism underlying unbalanced recurrent translocations, which are mediated by either interchromosomal LCRs or segmental duplications located on nonhomologous chromosomes [149].

*NHEJ*: The defining characteristic of NHEJ is the ligation of double-strand break (DSB) ends without the requirement for extensive homology, in stark contrast to the situation pertaining with homologous recombination. The presence of terminal microhomologies (typically 1–3 bp) facilitates NHEJ, but this appears not to be an absolute requirement; only 30%–50% of all SVs in the human genome have originated through microhomology-mediated NHEJ events [62,150].

Although some NHEJ events would have resulted from the repair of DSBs that originated quasi-randomly, there are also many well-documented cases in which the locations of the NHEJ-initiating DSBs appear to be highly dependent on the local DNA sequence environment. The role of the local DNA sequence context in generating NHEJ-mediated germline mutations is exemplified by the constitutional t(11;22)(q23;q11), the most common type of recurrent non-Robertsonian translocation in humans [151,152]. The breakpoint sequences of both chromosomes are characterized by several hundred base pairs of inverted AT-rich repeats; similar sequences have also been identified at the breakpoints of other nonrecurrent translocations [153]. It would appear that the NHEJ of two ends from different DSBs requires those ends to be physically located in the immediate vicinity. Indeed, DSBs tend to undergo translocations with those chromosomes with which they share nuclear space [154]. This provides strong support for the "contact-first" hypothesis, which proposes that interactions between different DSBs can only take place if they are colocalized at the time of DNA damage [155]. Consistent with this hypothesis, close spatial proximity has been observed between several frequent translocation partners [156,157].

A number of recombination-predisposing motifs and non-B DNA-forming sequences have been found to be overrepresented at NHEJ breakpoints, indicative of the sequence-directed nature of many NHEJ-mediated rearrangements [158,159]. It has also been observed that at least one of the breakpoints of NHEJ-mediated rearrangements is often located within repetitive elements (such as LTRs, LINE or *Alu* elements) and sequence motifs capable of causing DSBs have been frequently identified in the vicinity of the breakpoints of these NHEJ-mediated rearrangements [160]. The breakpoints of many nonrecurrent CNVs mediated by NHEJ map to LCRs, suggesting that LCRs can promote genomic instability by inducing certain chromatin secondary structures.

*Replication-based mechanisms*: Replication slippage or template switching during replication account for both small and large deletions and duplications with terminal microhomologies. Recently, relevant replication-based models including serial replication slippage (SRS) [161–163], fork stalling and template switching (FoSTes) [164], and microhomology-mediated break-induced replication [165] (MMBIR), which were collectively termed microhomology-mediated replication-dependent recombination by Chen et al. [166], have been used to explain the generation of a diverse range of complex genomic rearrangements [164,167].

DNA replication stalling-induced chromosome breakage has also been found to be an important mechanism causing deletions at chromosomal ends. Different types of telomeric deletions have been described [168]: type A terminal deletions are formed by chromosomal ends that are stabilized by the capture of a telomere from another source, whereas type B deletions are actually interstitial deletions toward the chromosomal ends. By contrast, type C deletions describe the process by which chromosomal ends are stabilized by telomere healing, namely the telomerase-dependent de novo addition of telomeres at nontelomeric sites. Terminal deletions associated with inverted duplications [169] can be classified as either type A or type C. Hannes et al. [170] cloned the breakpoints of nine chromosome 4p terminal deletions. All nine cases were shown to be type C terminal deletions. Bioinformatic analysis of the breakpoint-flanking regions involved in these nine cases, together with 12 previously fully characterized type C terminal deletions, led to the realization that there is an enrichment in secondary structure–forming sequences and replication stalling site motifs in these regions compared with a randomly selected sequence dataset [170].

Certain sequence features, such as microsatellites and transposon-rich regions, can serve to induce replication stalling, thereby acting as potential sources of genome instability [171,172]. On this basis, Koszul et al. [173] proposed a two-step mechanism to account for the generation of large segmental duplications: "First, a replication fork pauses and collapses generating a chromosome breakage. Second, the double-strand break can be processed into a new replication fork either intra- or inter-molecularly by a break-induced replication-like mechanism that does not necessarily need a long sequence homology." It was this "microhomology-dependent BIR" model that was subsequently deployed to explain disease-causing CNVs. In MMBIR, replication ends with the engagement of a misaligned template instead of reannealing to its original template; the synthesis of the second strand then follows the synthesis of the first (see review [166]). In practice, mutations due to SRS/FoSTes are often indistinguishable from those due to MMBIR. Indeed, the two terms have sometimes been used interchangeably [174,175].

Ankala et al. [176] have recently proposed another mechanism, the aberrant firing of replication origins, to explain a number of complex lesions in a series of 62 intragenic nonrecurrent rearrangements within various genes (mainly in the *DMD* gene). While repetitive sequence elements were noted in only four individual cases, microhomologies (2–10 bp) were observed at breakpoint junctions in 56% of the cases studied; further, insertions ranging from 1 to 48 bp were noted in 16 of the 62 cases. The sequence proximal to the breakpoints in six individual Duchenne muscular dystrophy (DMD) cases was characterized by tandem repetitions of short segments (5–20 bp). The repeated replication of template sequences proximal to the mapped deletion breakpoints was taken as evidence of attempts by the replication machinery to bypass a stalled replication fork. This mutational mechanism, based on the replication rescue model originally suggested by Doksani et al. [177], constitutes a novel type of template slippage event. Indeed, it can be seen that microhomologies at CNV breakpoints may be attributed to microhomology-mediated end joining (MMEJ), a replication repair mechanism, rather than to a recombination-based mechanism.

A recent review on the mechanisms underlying structural variant formation in genomic disorders [178] is also recommended to the reader.

### 6.3.7 Gross Deletions

Gross deletions are common causes of certain disorders and rare in others. In most of the X-linked disorders, for example, large deletions account for ~5% of molecular defects. In other disorders, however, such as steroid sulfatase deficiency, large deletions of the *STS* gene account

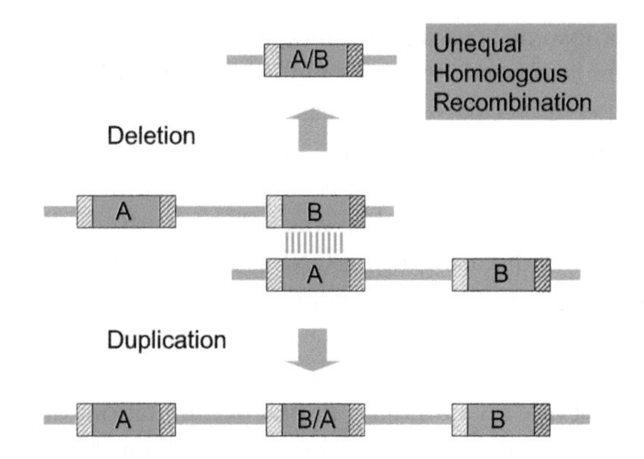


**Figure 6.7** Homologous unequal recombination between similar regions of sequences A and B. The recombination events cause either deletions or duplications. In the case of a deletion, a hybrid sequence is generated with the first part from sequence A and the second from sequence B. The middle sequence in the duplication product is also a hybrid sequence; the first part is from sequence B and the second from sequence A.

for 84% of patients [179]. The same is true for disorders such as DMD, growth hormone deficiency, and α-thalassemia [180–182].

A considerable number of large deletions probably are generated by mispairing of homologous sequences and unequal recombination (Fig. 6.7). One of the best examples of homologous unequal recombination is the case of α-globin genes on chromosome 16p. As a result of a recent evolutionary duplication of the α-globin genes, extensive regions of sequence homology exist between the two closely linked α-genes. Unequal crossover results in either the deletion of one α-gene or the creation of a fusion hybrid gene [183]. The reciprocal product chromosomes carry three α-genes and are not associated with a clinical phenotype [184]. Another example of a fusion gene resulting from an unequal crossover is the case of Hemoglobin Lepore, characterized by a hybrid gene between the δ- and β-globin genes on chromosome 11p [185]. In the case of steroid sulfatase deficiency, the deletion can be as large as 1 Mb [186]. In Kallmann syndrome, translocation can occur as a result of unequal mispairing of X and Y homologous sequences [187].

Several genetic disorders are due to large deletions (or duplications) caused by unequal crossing-over of homologous sequences. Fig. 6.8 depicts various examples that include a 1.5 Mb deletion of 17p12 in hereditary neuropathy with liability to pressure palsies (HNPP) [188], deletion of 1.5 Mb of 17q11.2 in neurofibromatosis type 1 [189], deletion of 1.6 Mb of 7q11.23 in Williams syndrome [190], deletion of 5 Mb of 17p11.2 in Smith–Magenis syndrome [191], deletion of either 3 Mb or more rarely 1.5 Mb of 22q11 in DiGeorge and velo-cardio-facial syndromes [192,193], and 4 Mb deletions of 15q in Prader–Willi and Angelman syndromes [194]. A recurrent deletion of ~0.5 Mb of 17q21.3, which may be mediated by a common inversion polymorphism, has also been described [195–198]. For a review of chromosomal "duplicons," the LCRs that mediate deletions and duplications, see Ref. [199]. It has been estimated that ~5% of the human genome is duplicated either intra- or interchromosomally [200]. The large deletions or duplications (see below) due to duplicon crossover are also termed "genomic disorders." A review of such genomic disorders may be found in Ref. [201].

In many cases of large deletion, homologous unequal crossover occurs between repetitive elements such as *Alu* sequences [202]. The *Alu* repeat is the most abundant repetitive element, with $\sim 1.5 \times 10^6$ copies in the human genome [202,203]. The element is ~300 bp in length and consists of two similar regions separated by a short A-rich region. Unequal crossover can occur between *Alu* sequences oriented in either the opposite or the same direction. In addition, unequal crossing over events have been noted between *Alu* elements and nonrepetitive DNA sequences without homology to *Alus*. The best examples of *Alu-Alu* recombination occur in the genes encoding the low-density lipoprotein receptor (*LDLR*) which underlies familial hypercholesterolemia and complement component 1 inhibitor (*C1I*; [204,205]). All but one of the breakpoints associated with *LDLR* gene deletions occur within *Alu* repeats. By contrast, deletions in other *Alu*-rich genes (e.g., *GLA1*) do not necessarily involve *Alu* repetitive elements [206]. This notwithstanding, *Alu*-mediated recombination between nonallelic *Alu* sequences is a fairly frequent cause of gene deletion causing human genetic disease [207–216]. It should be appreciated that the importance of *Alu* sequences in the context of mediating genomic deletions does not lie simply with their sheer abundance; *Alu* elements also possess inherent recombination-predisposing properties [217].

Nonhomologous (illegitimate) recombination occurs between two DNA sites that share minimal sequence homology of a few base-pairs. This type of recombination during meiosis or alternatively, slipped mispairing during DNA replication mediated by short (2–8)

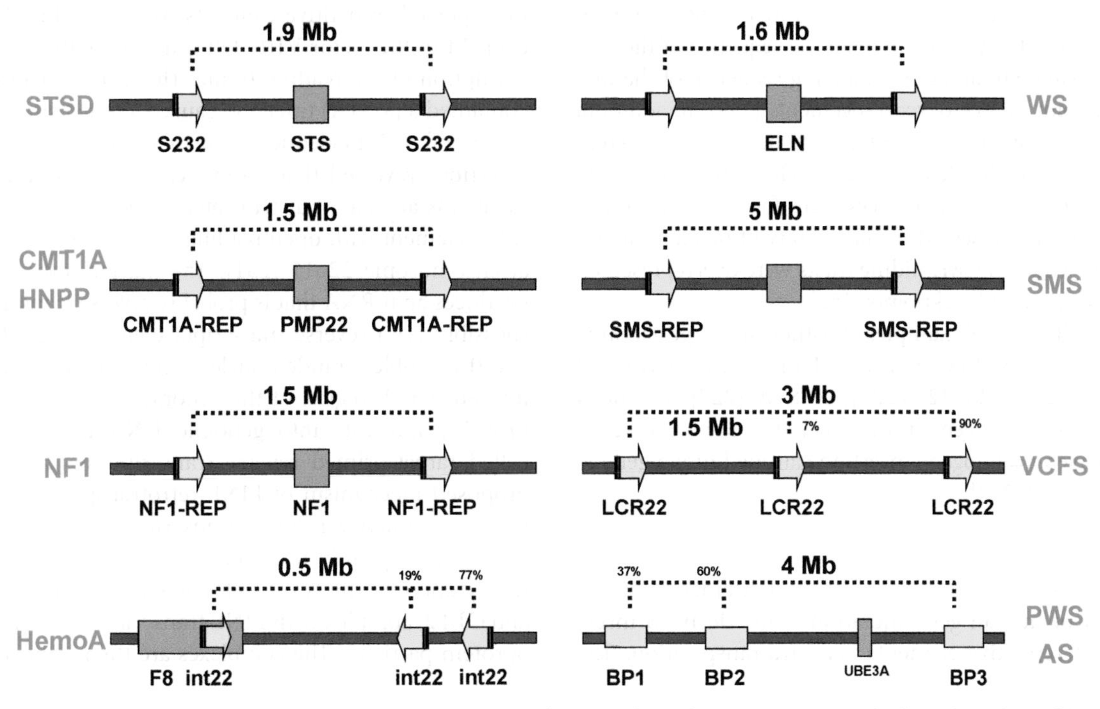


**Figure 6.8** Genes, duplicons, and diseases. Unequal crossover between homologous sequences (duplicons) produce either deletions or duplications of the DNA between the duplicons. The duplicons are shown by *arrows* or by *clear boxes*. Genes included in the duplications/deletions are shown as *dark boxes*. *AS*, Angelman syndrome; *CMTA1*, Charcot-Marie-Tooth type A1; *HemoA*, hemophilia A; *HNPP*, hereditary neuropathy with liability to pressure palsies; *NF1*, neurofibromatosis 1; *PWS*, Prader–Willi syndrome; *SMS*, Smith–Magenis syndrome; *STSD*, steroid sulfatase deficiency; *VCFS*, velo-cardio-facial syndrome; *WS*, Williams syndrome.

nucleotide direct repeats flanking the deletions, is a common finding in many instances of large gene deletions [218]. Such deletions have been studied, for example, in hemophilia A; a compilation of 46 junctions from large deletions revealed that ~50% shared 2- to 6-bp homology at the breakpoint junction, as compared with only 17% in which the deletion was due to *Alu–Alu* recombination [219]. Similar results have been reported from the intron 7 deletion hotspot in the *DMD* gene; 8/9 deletion breakpoints examined were found to be flanked by DNA sequences with minimal homology [220].

It has also been proposed that alternative DNA conformations may trigger genomic rearrangements through recombination–repair activities. Distance measurements have indicated the significant proximity of alternating purine-pyrimidine and oligo(purine-pyrimidine) tracts to breakpoint junctions in 222 gross deletions and translocations, respectively, involved in human diseases. In 11 deletions analyzed, breakpoints were explicable by non-B DNA structure formation [221].

The Gross Rearrangement Breakpoint Database (GRaBD; uwcm.ac.uk/uwcm/mg/grabd/) was established primarily for the analysis of the sequence context of translocation and deletion breakpoints in a search for characteristics that might have rendered these sequences prone to rearrangement [222]. GRaBD, which contains 397 germline and somatic DNA breakpoint junction sequences derived from 219 different rearrangements underlying human inherited disease and cancer, represents a large but not comprehensive collection of sequenced gross gene rearrangement breakpoint junctions. Analysis of these breakpoints has extended our understanding of illegitimate recombination by highlighting the importance of secondary structure formation between single-stranded

DNA ends at breakpoint junctions. For example, potential secondary structure was noted between the 5′ flanking sequence of the first breakpoint and the 3′ flanking sequence of the second breakpoint in 49% of rearrangements, and between the 5′ flanking sequence of the second breakpoint and the 3′ flanking sequence of the first breakpoint in 36% of rearrangements [159]. In addition, deletion breakpoints were found to be AT-rich, whereas translocation breakpoints were GC-rich. Alternating purine-pyrimidine sequences were found to be significantly overrepresented in the vicinity of deletion breakpoints while polypyrimidine tracts were overrepresented at translocation breakpoints [158].

Finally, several examples of pathogenic large genomic deletions caused by the prior L1-mediated insertion of L1 [223,224], *Alu* [225,226], or SVA [227] insertions (see later) have been reported, as well as the first cases of L1-driven pseudogene insertion causing human genetic disease [228,229].

### 6.3.8 Large Retrotranspositional Insertions

A less common but nevertheless still fascinating mechanism of human gene mutation is the de novo insertion of repetitive elements via retrotransposition. The phenomenon was first observed in humans in the factor VIII (*F8*) gene in two unrelated de novo cases of severe hemophilia A [230]. Truncated LINE (long interspersed) repetitive elements were introduced into exon 14 of the factor VIII (*F8*) gene where they caused disruption of the reading frame. The inserted elements contained a poly(A) tract and caused a target site duplication of >12 nucleotides. Further analysis of these insertions revealed that, in one case, the inserted element was an exact but truncated copy of a full-length LINE element with open reading frames (ORFs) found at chromosome 22q11 [231]. The master source gene produces an mRNA that is probably reverse transcribed (possibly via a reverse transcriptase encoded by itself) and the double-stranded nucleic acid is then reinserted into an A-rich region of the genome (Fig. 6.9). LINEs probably integrate into genomic DNA by a process called target-primed reverse transcription [232]. The proposed mechanism of LINE retrotransposition is as follows: An active LINE is transcribed in the nucleus and is subsequently transported to, and translated in, the cytoplasm. The two LINE-encoded proteins, ORF1 and ORF2, complex with LINE transcripts in ribonucleoprotein particles. The complexes are then transported

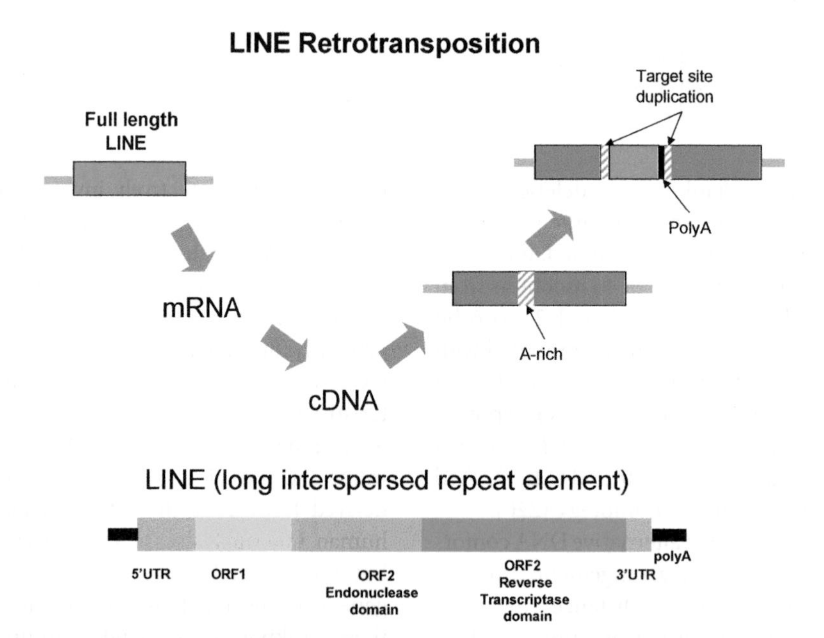


**Figure 6.9** Schematic representation of LINE retrotransposition. A master retrotransposon (full length LINE) from one chromosomal location is transcribed to mRNA; then reverse transcribed to double-stranded DNA and inserted into an adenine-rich region of another chromosomal location. The transposon has a poly(A) tail and produces a target site duplication.

to recipient DNA sequences where target-primed reverse transcription occurs. The new, integrated LINE copy is usually truncated at its 5′ end. Over evolutionary time, L1s have shaped mammalian genomes through a number of different mechanisms. First, they have greatly expanded the genome both by their own retrotransposition and by providing the machinery necessary for the retrotransposition of other mobile elements, such as *Alu* sequences or SVA elements [163]. Second, they have shuffled non-L1 sequence throughout the genome by a process termed transduction. Accidents of retrotransposition can cause disease, and a number of such insertions have been reported to date [232,233]. It is noteworthy that insertions of these elements within introns of genes or flanking regions are probably not associated with disease, but instead represent rare, private polymorphisms [234].

Similar retrotranspositions that involve members of the *Alu* sequence family have also been reported in several genes (examples include *Alu* insertions into the *NF1* gene causing type 1 neurofibromatosis, into the factor IX (*F9*) gene causing hemophilia B, and into the cholinesterase (*BCHE*) gene in a case of acholinesterasemia; [235–237]). It is likely that LINEs provide the molecular machinery necessary for the retrotransposition of *Alus*. One study using mutation analysis of the *F9* gene has estimated the frequency of retrotransposition to be such that it occurs somewhere in the genome of ~1 in every 17 children born [238].

Some 17% of a collection of gross insertions, all ≥276 bp in length, were due to LINE-1 (L1) retrotransposition involving different types of elements (L1 *trans*-driven *Alu*, L1 direct, and L1 *trans*-driven SVA) [163]. A meta-analysis of 48 recent L1-mediated retrotranspositional events known to have caused human genetic disease revealed that 26 were L1 *trans*-driven *Alu* insertions, 15 were direct L1 insertions, 4 were L1 *trans*-driven SVA insertions, and 3 were associated with simple poly(A) insertions [239]. The systematic study of these lesions, when combined with previous in vitro and genome-wide analyses, allowed several conclusions regarding L1-mediated retrotransposition to be drawn: (a) ~25% of L1 insertions are associated with the 3′ transduction of adjacent genomic sequences, (b) ~25% of the new L1 inserts are full-length, (c) poly(A) tail length correlates inversely with the age of the element, and (d) the length of target site duplication in vivo is rarely longer than 20 bp. This analysis also suggested that some 10% of L1-mediated retrotranspositional events are associated with significant genomic deletions in humans. Chen et al. [240] reported an indel in the *CFTR* gene that involved the insertion of a short 41-bp sequence with partial homology to a retrotranspositionally competent LINE-1 element. These authors dubbed such insertions of ultra-short LINE-1 elements "hyphen elements."

Several instances of the clustering of pathogenic L1-mediated insertion events have also been observed. Thus, three independent *Alu* insertions have been found to be integrated into a 104 bp region of the *FGFR2* gene [241,242], two independent L1 insertions have been reported to have inserted into an 89 bp region of exon 44 of the dystrophin (*DMD*) gene [243,244], while six different insertions were found in a 1.5-kb region of the *NF1* gene [245]. It should also be noted that independent L1-retrotransposition elements can integrate at precisely the same chromosomal sites. Thus, two markedly different *Alu* Ya5a2 elements became integrated at precisely the same site in the *F9* gene causing severe hemophilia B [236,246], whereas an SVA element and an *Alu* sequence were inserted at the same site within the coding region of the *BTK* gene [247]. These observations are consistent with some genomic locations being exquisitely prone to L1-retrotransposition [239].

### 6.3.9 Large Insertion of Repetitive and Other Elements

The insertion of nonretrotransposons, namely β-satellite repeats, has been observed in the human genome. The insertion of 18 copies of the 68-bp monomer of the β-satellite repeat in exon 11 of the *TMPRSS3* gene on chromosome 21 caused one form of recessive nonsyndromic deafness DFNB10 [248]. This may have been mediated by invasion of the genomic DNA by a small polydispersed circular DNA.

A patient with a sporadic case of Pallister–Hall syndrome has been shown to have experienced a de novo nucleic acid transfer from the mitochondrial to the nuclear genome. This variant, a 72-bp insertion into exon 14 of the *GLI3* gene, creates a premature stop codon and predicts a truncated protein product. Both the mechanism and the cause of the mitochondrial-nuclear transfer are however, unknown [249]. Further examples of pathological mitochondrial-nuclear sequence transfers have been subsequently identified in the *USH1C* gene [163] and the *PAFAH1B1* gene [250].

Gross insertions (>20 bp) comprise <1% of disease-causing variants. In an attempt to study these insertions in a systematic way, 158 gross insertions ranging in size between 21 bp and ~10 kb were identified from the HGMD; their study has revealed extensive diversity in terms of the nature of the inserted DNA sequence and has provided new insights into the underlying mutational mechanisms [163]. Some 70% of gross insertions were found to represent sequence duplications of different types (tandem, partial tandem, or complex). In the context of a 26-bp insertion into the *ERCC6* gene, the authors also speculated as to whether they had found evidence for another mechanism of human genetic disease, involving the possible capture of DNA oligonucleotides [163].

### 6.3.10 Inversions

The most common inversion found to date is that associated with the factor VIII (*F8*) gene, which occurs via intrachromosomal recombination mediated by a 9.5-kb sequence that is repeated three times in the last megabase of Xqter; once in intron 22 of the *F8* gene and twice ~400 kb telomeric to the first [251,252] (Fig. 6.10). Most inversions, which are high-frequency independent recurring events, involve the distal sequence. The vast majority of inversions occur in male germ cells [253], perhaps because intrachromosomal recombination is inhibited by the presence of homologous X chromosomes (the male:female ratio was estimated to be ~300:1). Almost all mothers of inversion hemophilia A cases are carriers of the abnormality. DNA diagnosis of the molecular lesion in severe hemophilia A has been greatly facilitated by the frequent occurrence of this common inversion of the *F8* gene (45% of individuals with severe hemophilia A). The frequency of de novo *F8* gene inversion has been estimated to be $7.2 \times 10^{-6}$ per gamete per generation. Another example of inversion has been described in the *IDS* gene (also on Xq) in ~13% of cases of Hunter syndrome [254]. Inversions of DNA sequences have also been reported in the β-globin gene cluster on 11p and in the *APOA1-APOC3-APOA4* gene cluster on 11q [255,256].

A meta-analysis of inversions of ≥5 bp but <1 kb has been performed by Chen et al. [162]. Of the 21 mutations studied, 19 were found to be compatible with a model of intrachromosomal serial replication slippage in *trans* (SRStrans) mediated by short inverted repeats. Eighteen (one simple inversion, six inversions involving sequence replacement by upstream or downstream sequence [DSS],

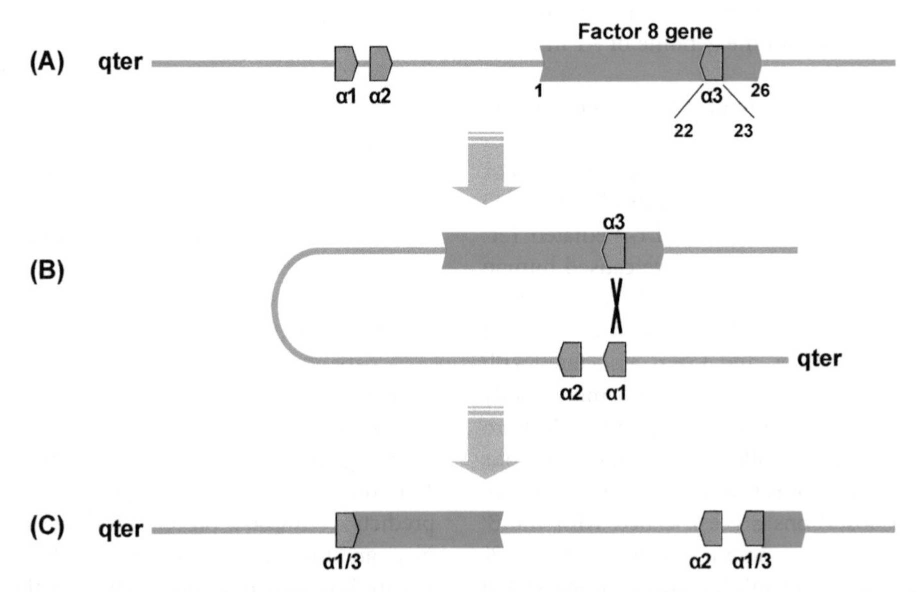


**Figure 6.10** Common inversion of the factor VIII (*F8*) gene in severe hemophilia A. (A) Schematic representation of the most distal 1 Mb of Xq. Regions $\alpha_1$, $\alpha_2$, and $\alpha_3$ represent 9.5-kb highly homologous DNA elements. The orientations of these sequences are shown by *arrows*. (B) Intrachromosomal recombination between elements $\alpha_1$ and $\alpha_3$. (C) The crossover results in the inversion of exons 1–22 of the *F8* gene.

five inversions involving the partial reinsertion of removed sequence, and six inversions that occurred in a more complicated context) of these were found to be consistent with either two steps of intrachromosomal SRStrans or a combination of replication slippage in *cis* plus intrachromosomal SRStrans. The remaining lesion, a 31-kb segmental duplication associated with a small inversion in the *SLC3A1* gene, was explicable in terms of a modified SRS model incorporating the concept of "break-induced replication." This study has, therefore, lent broad support to the idea that intrachromosomal SRStrans can account for a variety of complex gene rearrangements involving inversions.

### 6.3.11 Duplications

Duplications of whole genes or exons have contributed very significantly to the evolution of the human genome [56]. Indeed, most gene clusters (e.g., β-globin, growth hormone, Hox) owe their origin to gene duplications that have occurred during vertebrate evolution. Furthermore, the presence of similar domains in proteins (e.g., immunoglobulin-like domains in many transmembrane proteins) is due to duplications of certain exons.

Occasionally, however, duplications may also be the cause of genetic disorders. The most frequent mechanism of duplication is homologous unequal crossover as described for large deletions. In fact, most large duplications generated as the reciprocal product of a deletion resulting from homologous unequal crossover. Duplications are less common, however, than their theoretically reciprocal deletions (e.g., see Ref. [257], for the *DMD* gene). This may be due to the nonpathogenicity of a duplication (e.g., α-globin genes [184]), elimination of duplications as is the case for the *HPRT1* gene, or to the fact that not all mechanisms that lead to deletions also produce duplications. A large and common duplication has been identified in cases of Charcot–Marie–Tooth disease type 1A [258]. This duplication involves 1.5 Mb of DNA on chromosome 17p containing the peripheral myelin protein 22 (*PMP22*) gene. It results from homologous unequal crossover events between 24-kb repeats that flank the duplicated region. The reciprocal deletion product of this recombination event is responsible for a completely different clinical phenotype: HNPP (Fig. 6.8). Another notable duplication of at least 500 kb that includes the *PLP1* gene is a frequent cause of Pelizaeus–Merzbacher disease [259]. The pathogenetic mechanism of these duplications involves unequal crossing-over in meiosis mediated by "duplicons" in the genome [199].

The molecular defect in the majority of cases with ectrodactyly type SHFM3 on chromosome 10q24, is an approximately 0.5-Mb tandem duplication. The precise pathogenetic mechanism of this duplication is unknown [260]. Additional gene duplications causing recognizable syndromes include the *APP* duplication causing early-onset Alzheimer disease [261], the *SNCA* duplication and Parkinson disease [262], and the triplication of ~605-kb segment containing the *PRSS1* gene in families with hereditary pancreatitis [263].

### 6.3.12 CNV in Association With Disease

CNVs are a form of genomic diversity that involves DNA sequences ≥1 kb in length that are present in the human genome in a variable number of copies. CNVs may be recurrent (which arise by NAHR) or nonrecurrent (which arise by nonhomologous end joining and replication-based mechanisms) [264]. Such gross duplications/deletions not only are rather abundant but also often occur at polymorphic frequencies. Conrad et al. [62] generated a comprehensive map of >8500 validated CNVs >500 bp (detected in 41 Europeans/West Africans) that together cover a total of 112.7 Mb (3.7% of the genome). These authors estimated that 39% of the validated CNVs overlapped 13% of RefSeq genes (NCBI mRNA reference sequence collection). Further, they concluded that the CNVs detected resulted in the "unambiguous loss of function" of alleles for 267 different genes.

It has been estimated that on average, 73–87 genes vary in terms of their copy number between any two individuals [265]. This high degree of interindividual variability with regard to gene copy number has challenged traditional definitions of wild type and "normality" and even the very concept of a "reference genome" itself. High-resolution breakpoint mapping is a prerequisite for the accurate assessment of CNV size, the identification of the genes and regulatory elements affected, and hence, for the determination of the consequences of CNV for gene expression [266] and the phenotypic sequelae [267–270]. This notwithstanding, it is already becoming clear that these consequences may go far beyond the physical bounds of a given CNV. For example, a CNV involving the human *HBA* gene has a dramatic influence on the expression of the *NME4* gene some 300 kb distant [271]. In addition, a 5.5-kb microduplication of a conserved noncoding sequence with demonstrated enhancer function, ~110 kb downstream of the bone morphogenic protein 2 (*BMP2*) gene, has been found to cause brachydactyly type 2A in two families [272].

It may well be that the precise extent and/or location of many CNVs will vary between individuals, thereby further increasing both the mutational and phenotypic heterogeneity. The extent to which CNVs are likely to contribute to the diversity of human phenotypes, including "single gene defects," genomic disorders, and complex disease, is increasingly being recognized. Thus, CNV of the *FCGR3B* genes is a determinant of susceptibility to immunologically mediated glomerulonephritis [273]. CNVs in the *CCL3L1* and *DEFB4* genes have also been found to be associated with increased susceptibility to HIV infection and Crohn disease, respectively [274,275], whereas rare CNVs associated with various complex phenotypes have been identified in studies of schizophrenia [276], epilepsy, and severe early-onset obesity [277]. CNVs are now being widely recruited to genome-wide association studies (GWAS) with the aim of assessing their influence on human disease causation/susceptibility [267,278,279].

To date, several dozen human disease conditions have been identified, which are either caused by CNVs or whose relative risk is increased by CNVs [267,280]. Remarkably, an excess of both rare and de novo CNVs has been identified in patients with psychiatric disorders and obesity [276,277,281–285]. These findings point to genetic heterogeneity in these conditions, thereby illustrating the likely complexity inherent in identifying all disease-causing CNVs. Shlien et al. [286] reported a highly significant increase in CNV number among patients with Li–Fraumeni syndrome, carriers of inherited *TP53* variants. Hence, it would appear that heritable genetic variants have the potential to modulate the rate of germline CNV formation.

It is already clear that the disease relevance of CNVs represents a continuum, stretching from "neutral" polymorphisms on the one hand to directly pathogenic copy number changes on the other [267]. Between these two extremes may lie those CNVs that are capable of acting as predisposing (or protective) factors in relation to complex disease [287,288]. Thus, for example, a 117-kb deletion encompassing the *UGT2B17* gene has been found to be associated with an increased risk of osteoporosis [289]. Some germline CNVs appear to predispose to disease even although no known genes reside within their boundaries [290,291]. Importantly, a 520-kb microdeletion has been identified at 16p12.1, which predisposes to various neuropsychiatric phenotypes as a single copy number variant and aggravates neurodevelopmental disorders if it co-occurs together with other large deletions and duplications [288]. It remains to be seen whether "CNV equivalents," <1 kb in size, which actually occur more frequently than true CNVs (>1 kb) [62], will also be relevant to disease. What is already clear is that, over the coming years, an increasing number of important CNV-disease associations are going to come to light [270].

### 6.3.13 Gene Conversion

Gene conversion is the modification of one of two alleles by the other. It involves the nonreciprocal correction of an "acceptor" gene or DNA sequence by a "donor" sequence, which itself remains physically unchanged. In most known instances of gene conversion as a cause of human genetic disease, the functional gene has been wholly or partially converted to the sequence of a highly homologous and closely linked pseudogene, which therefore acts as the donor sequence [292]. Probable examples include the genes for steroid 21-hydroxylase (*CYP21* [293]), polycystic kidney disease (*PKD1* [294]), neutrophil cytosolic factor p47-*phox* (*NCF1* [295]), immunoglobulin λ-like polypeptide 1 (*IGLL1* [296]), glucocerebrosidase (*GBA* [297]), von Willebrand factor (*VWF* [298]), and phosphomannomutase (*PMM2* [299]). These gene/pseudogene pairs are all closely linked with the exception of the *VWF* gene (12p13) and its pseudogene (22q11–q13) and the *PMM2* gene (16p13) and its pseudogene (18p). Together, these two exceptions would seem to establish a precedent for the occasional occurrence of gene conversion between unlinked loci in the human genome.

An in silico analysis of the DNA sequence tracts involved in 27 well-characterized nonoverlapping gene conversion events in 19 different genes reported in the context of inherited disease was recently performed [300]. It was noted that gene conversion events tended to occur within (C+G)- and CpG-rich regions and that sequences with the potential to form non-B DNA structures (and which might be involved in the generation of double-strand breaks that could, in turn, serve to promote gene conversion) occurred disproportionately within maximal converted tracts and/or short flanking regions. Maximal converted tracts were also found to be enriched in a truncated version of the chi-element (a TGGTGG motif), immunoglobulin heavy chain class switch repeats, translin target sites, and several novel motifs including (or overlapping) the classic meiotic recombination hotspot, CCTCCCCT [300]. Finally, it was found that gene conversions tended to occur in genomic regions that had the potential to fold into stable hairpin conformations. Taken together, these findings support the concept that recombination-inducing motifs, in association with alternative (non-B DNA) conformations, can promote recombination in the human genome.

The large number of duplicated gene sequences in the human genome implies that a considerable number of disease-associated variants could originate via interlocus gene conversion. A genome-wide computational approach to identify disease-associated variants derived from interlocus gene conversion events recently revealed hundreds of known pathological variants that could have been caused by interlocus gene conversion [301]. In addition, several dozen high-confidence cases of inherited disease variants resulting from interlocus gene conversion were identified in ~1% of all genes analyzed. About half of the donor sequences associated with such variants were functional paralogous genes, suggesting that epistatic interactions or differential expression patterns would determine the impact upon fitness of a single amino acid substitution between duplicated genes. In addition, Casola et al. [301] identified thousands of hitherto undescribed deleterious variants that could potentially arise via interlocus gene conversion. It would therefore appear that the impact of interlocus gene conversion upon the spectrum of human inherited disease may be considerably greater than has hitherto been appreciated.

Although variants that are detrimental to the fitness of individuals are expected to be rapidly purged from the population by natural selection, some pathological variants are nevertheless retained at high frequencies in human populations. Several hypotheses have been proposed to account for this apparent paradox (high new mutation rate, genetic drift, overdominance, or recent changes in selective pressure). However, there is an additional process that appears to contribute to the spreading of deleterious variants: GC-biased gene conversion (gBGC), a process associated with recombination that tends to favor the transmission of GC-alleles over AT-alleles. Necsulea et al. [302] have shown that the spectrum of amino acid–altering polymorphisms in human populations exhibits the footprints of gBGC. This pattern is not explicable in terms of selection and is evident with all nonsynonymous variants, including those predicted to be detrimental to protein structure and function as well as those that have been implicated in the causation of human genetic disease. These results indicate that gBGC meiotic drive contributes to the spreading of deleterious variants in human populations.

### 6.3.14 Insertion–Deletions (Indels)

A relatively rare type of mutation causing human genetic disease is the *indel*, a complex lesion that appears to represent a combination of microdeletion and microinsertion. One example is the nine deleted base pairs encoding codons 39–41 of the α2-globin (*HBA2*) gene that were replaced by eight inserted bases that serve to duplicate the adjacent downstream sequence (DSS) [303]. Indels constitute a fairly infrequent type of lesion causing human genetic disease; ~1.5% of lesions in HGMD fall into this category.

Several indel hotspots have been noted in a meta-analysis of HGMD data on 211 different indels underlying genetic disease [304]. A GTAAGT motif was found to be significantly overrepresented in the vicinity of the indels studied. The change in complexity consequent to a mutation was also found to be indicative of the type of repeat sequence involved in mediating the event, thereby providing clues as to the underlying mutational mechanism. The majority of indels (>90%) were explicable in terms of a two-step process involving established mutational mechanisms. Indels equivalent to double base-pair substitutions (22% of the total) were found to be mechanistically indistinguishable from the remainder and may therefore be regarded as a special type of indel.

### 6.3.15 Other Types of Complex Rearrangement

Complex mutational events that involve combined gross duplications, deletions, and/or insertions of DNA sequence have been not infrequently observed and together constitute ~1% of entries in HGMD. One example of this type of gene defect is a 10.9-kb deletion coupled with a 95-bp inversion in the factor IX (*F9*) gene causing hemophilia B [305]. The molecular characterization of this type of lesion is often extremely complicated, and in most cases, the underlying mutational mechanisms could not be readily inferred.

Recently, however, a meta-analysis of 21 complex gene rearrangements derived from the HGMD revealed that all but one could be accounted for by a model of SRS, involving twin or multiple rounds of replication slippage [161]. Thus, of the 20 complex gene rearrangements, 19 (seven simple double deletions, one triple deletion, two double mutational events comprising a simple deletion and a simple insertion, six simple indels that may constitute a novel and noncanonical class of gene conversion, and three complex indels) were compatible with the model of SRS in *cis*; by contrast, the remaining indel in the *MECP2* gene appears to have arisen via interchromosomal replication slippage in *trans*.

A novel type of complex genomic rearrangement, comprising intermixed duplications and triplications of genomic segments, has recently been described at both the *MECP2* and *PLP1* loci [306]. These complex rearrangements share a common genomic organization, viz., a duplication-inverted triplication-duplication (DUP-TRP/INV-DUP), in which the triplicated segment is inverted and located between directly oriented duplicated genomic segments. The DUP-TRP/INV-DUP structures appear to be mediated by inverted repeats, up to >300 kb apart.

### 6.3.16 Multiple Simultaneous Mutations

Transient hypermutability is a general mutational mechanism with the potential to generate multiple synchronous mutations, a phenomenon probably best exemplified by "closely spaced multiple mutations" (CSMMs). From a collection of human inherited disease-causing multiple mutations, Chen et al. [307] retrospectively identified numerous potential examples of pathogenic CSMMs that exhibited marked similarities to the CSMMs reported in other systems. These examples included (1) eight multiple mutations, each comprising three or more components within a sequence tract of <100 bp (*CBS*, *MPZ*, *OPN1LW*, and *STK1* genes), (2) three possible instances of "mutation showers" in the *PTCH1*, *FANCA*, and *KNG1* genes, respectively, and (3) numerous highly informative "homocoordinate" mutations (multiple mutations involving the same mutation type).

Recently, a remarkable phenomenon has been reported in multiple cancer samples: The presence of tens to hundreds of genomic rearrangements involving spatially localized genomic regions [308]. These complex rearrangements primarily affected a single chromosome, although in some cases, multiple apparently concomitant alterations affected several different chromosomes. Stephens et al. convincingly argued that these massive, yet spatially localized, genomic rearrangements must have resulted from a single catastrophic event (which they termed "chromothripsis") rather than from a series of progressive and hence independent alterations. Chromothripsis also appears to be capable of explaining the generation of some complex de novo structural rearrangements in the germline [309]. An illustrative example pertains to a highly complex chromosomal rearrangement, identified in a child with severe congenital abnormalities, which comprised at least 12 de novo breakpoint junctions and involved chromosomes 1, 4, and 10. These breakpoints were clustered in small genomic regions of up to 3.5 Mb in size on each chromosome. Reconstruction of the derivative chromosomes indicated that the breakpoints formed concordantly oriented pairs on the reference genome. Both intra- and interchromosomal junction sequence features were compatible with those commonly associated with NHEJ [309]. The insights generated from the seminal work of Stephens and his colleagues may well help us to understand the mutational mechanisms underlying some previously reported germline complex rearrangements [310]. For example, with respect to the de novo mutational event on chromosome 2 in a patient with Waardenburg syndrome and other congenital defects, the original authors suggested that all five breaks might have occurred simultaneously but were unable to explain why the breakpoints had occurred within a single chromosome [311]. In the light of our emerging knowledge of chromothripsis, the idea that this complex rearrangement could have been generated in such a way as to be compatible with the NHEJ repair of simultaneously generated DSBs becomes quite attractive. Liu and colleagues subsequently investigated 17 subjects with various development abnormalities by means of high-resolution genome analysis [312]. Constitutional multiple copy number changes, including deletions, duplications, and/or triplications, as well as inversions were observed in all cases. Strikingly, in each case, all rearrangements occurred within a single chromosome; in 15 of the 17 cases, the rearrangements were localized to the distal half of the affected chromosomal arms. FISH and breakpoint junction data indicated that all additional copies of the duplicated and triplicated segments appear to be randomly joined, forming a large "breakpoint junction cluster" on 9q21. By analogy with the phenomenon of chromothripsis, the observation of these extremely complex rearrangements in a single chromosome was also described as a chromosome catastrophe event [312]. However, this kind of chromosomal change cannot be easily explained by the previously described NHEJ repair of simultaneously generated DSBs. Instead, Liu and colleagues envisaged the involvement of a replicative mechanism in the generation of this complex chromosome catastrophe event comprising multiple duplications and/or triplications; they regarded MMBIR as the most likely underlying mechanism. They further suggested that a potential replication fork collapse at 9q21 could account for the breakpoint clustering therein [312]. A catalogue of published cases of germline chromothripsis has been provided in ref [313]. Most affected individuals present with developmental delay and dysmorphic features.

### 6.3.17 Molecular Misreading

Long runs of adenines (and perhaps other mononucleotides or dinucleotides) promote a phenomenon termed "molecular misreading" by which DNA replication/RNA transcription and/or translation result in erroneous products with different numbers of (A)s derived from the original DNA sequence [314]. In a family with hypobetalipoproteinemia, a deletion of one C in the $A_5CA_3$ coding sequence of the *APOB* gene results in a run of $(A)_8$. The affected individual, however, did not have severe disease, because some ApoB protein was made. This was the result of molecular misreading in which ~10% of the resulting mRNAs contained $(A)_9$ instead of the expected $(A)_8$; this partially restored the reading frame thereby templating the synthesis of low amounts of normal ApoB [315]. Similarly, a family with mild to moderately severe hemophilia A with a deletion of one T within the coding $A_8TA_2$ sequence of the *F8* gene has been reported. The partial "correction" of the phenotype was due to restoration of the reading frame because of molecular misreading in which ~5% of the resulting RNAs contained $(A)_{11}$ instead of the expected $(A)_{10}$. In this family, there was also evidence for ribosomal frameshifting during translation of the mutant RNA [316].

Another example of this phenomenon was observed in the *APC* gene. A T-to-A transversion is present in the coding $A_3TA_4$ sequence of the *APC* gene in 6% of Ashkenazi Jews, and in ~28% of Ashkenazim with a family history of colorectal cancer. This variant creates a small hypermutable region, indirectly causing cancer predisposition because there are many somatic cells in which stretches of $(A)_9$ occur instead of the expected $(A)_8$; the $(A)_9$ results in frameshifting and a truncated dysfunctional APC [317]. Interestingly, in the neurofibrillary tangles, neuritic plaques, and neuropil threads in the cerebral cortex of Alzheimer disease and Down syndrome, abnormal forms of β-amyloid precursor protein and ubiquitin B have been observed. These aberrant proteins were produced because of +1 frameshifting that resulted from a deletion of AG in a sequence GAGAG that occurred in the coding regions of both genes (*APP* and *UBB*, respectively). This dinucleotide deletion was again the result of molecular misreading during transcription or posttranscriptional editing of RNA [318]. This mechanism is likely to yield a considerable quantity of abnormal RNA molecules and protein products in somatic cells [319].

### 6.3.18 Germline Epimutations

*Epimutations* are modifications of DNA that constitute clonally heritable (yet potentially reversible) alterations in the transcriptional status of a gene that lead to the abnormal silencing of that gene. Epimutations are not mutations *sensu stricto* because they do not alter the gene's nucleotide sequence; however, germline epimutations of the *MLH1* gene have been reported in individuals with multiple cancers [320] and in the *MLH1* and *MSH2* genes in hereditary nonpolyposis colorectal cancer [321]. These heritable inactivating epimutations are characterized by monoallelic hypermethylation of the *MLH1* or *MSH2* genes and, to all intents and purposes, are functionally equivalent to conventional mutations. A maternal epimutation in the *GNAS* gene has also been reported as a cause of Albright osteodystrophy and parathyroid hormone resistance [322]. With the determination of the human methylome [91] and the recent recognition that DNA sequence polymorphisms can exert an effect on gene function via allele-specific methylation in *cis* [323], the number of recognized epimutations should rise quite significantly in the coming years. If eventually shown to be both of pathological significance and heritable, some examples of histone modification [324,325] or RNA editing [326,327] could also turn out to represent "honorary mutations."

### 6.3.19 Frequency of Disease-Producing Variants

*Mutation frequency within genes*: The frequency of different molecular defects is not the same for every gene and every disorder. Indeed, human disease genes exhibit very considerable allelic heterogeneity in terms of their mutational spectra; for some genes, a few predominant disease alleles predominate whereas for others, there is a wide range of disease alleles, each relatively rare [328]. The mutational spectrum depends very largely on the DNA sequence characteristics of the gene in question (e.g., the presence of repeat units or homologous sequences), and the function of, and evolutionary constraints experienced by, its encoded protein [329]. For some genes, deletions predominate; for others, one particular type of lesion such as an inversion may be especially common. Some genes exhibit mainly frameshifts and stop codons associated with a specific disorder, whereas others manifest mainly missense variants for a given phenotype, or expansions of trinucleotide repeats.

Disease variants are nonuniformly distributed within genes [330]. Such variants were found to be statistically overrepresented in conserved domains, and underrepresented in variable regions, even after allowing for the amino acid site variability of domains over long-term evolutionary history. This finding suggests that there is a nonadditive influence of amino acid site conservation on the observed intragenic distribution of disease variants.

*Mutation frequency within human populations*: Population genetic considerations are also likely to be very important in determining why some variants occur frequently, either within a patient cohort or in the population at large (see Frequency of Inherited Disorders Database, http://archive.uwcm.ac.uk/uwcm/mg/fidd/; FINDbase, http://www.findbase.org/). Selection, migration, and genetic drift are all likely to play a role as well as the mutation rate [331–333]. Thus, the mutational spectrum of the *PAH* gene underlying phenylketonuria appears to result from a range of different factors including founder effect, range expansion and migration, genetic drift, and possibly also heterozygote advantage [334]. Selection can also serve to maintain deleterious variants at high frequencies in particular populations by overdominant selection (heterozygote advantage). Good examples of this phenomenon are provided by a reduction in risk of severe malaria associated with female heterozygotes and male hemizygotes for variants in the X-linked *G6PD* gene [335,336], for individuals heterozygous for the β-globin (*HBB*) sickle cell variant, Glu6Val [337], and for individuals heterozygous and homozygous for $\alpha^+$-thalassemia [338]. Intriguingly, however, the protection against malaria afforded by sickle cell disease and $\alpha^+$-thalassemia when inherited individually is lost when the two conditions are coinherited [339]. Other possible examples of heterozygote advantage include an elevated cortisol response in heterozygous carriers of *CYP21A* variants [340], higher values for hemoglobin, serum iron, and transferrin saturation in women heterozygous for *HFE* gene variants [341], resistance to prion infection conferred by a common prion protein (*PRNP*) polymorphism [342], resistance to severe sepsis in heterozygous carriers of the factor V Leiden polymorphism, Arg506Gln [343], and increased keratinocyte cell survival in individuals heterozygous for *GJB2* gene variants [344]. Resistance to cholera toxin [345], protection against bronchial asthma [346], and resistance to *Pseudomonas aeruginosa* infection [347] have all been mooted as possible bases for overdominant selection in heterozygous carriers of *CFTR* gene variants. However, cystic fibrosis heterozygotes have been shown to secrete chloride at the same rate as individuals lacking *CFTR* gene variants [348].

A number of genetic diseases are known to be particularly prevalent in Jewish populations [349,350]. The presence of four distinct lysosomal storage diseases at significant frequencies among Ashkenazi Jews has often been considered as providing evidence for a selective advantage accruing to heterozygotes in this population. However, evidence in support of the idea of genetic drift appears to be more compelling [351,352].

Selection may also act at an extremely early stage to boost the frequency of some variants that are deleterious at a later stage in development. For example, gain-of-function missense variants in the fibroblast growth factor receptor 2 (*FGFR2*) gene responsible for Apert syndrome have been shown to confer a selective advantage on spermatogonial cells by promoting the clonal expansion of mutant cells [353,354].

### 6.3.20 Functional Characteristics of Human Disease Genes

Human disease genes appear to be distinguishable from "nondisease genes" (in reality, the latter can only be defined as genes that are not yet known to cause inherited disease) in terms of a range of features including gene structure, gene expression, physicochemical properties, protein structure, and evolutionary conservation [38,329,355–360]. Thus, human disease genes are characterized by the greater length of their encoded amino acid sequences, a larger number of longer introns, a broader range of tissue expression, and a wider phylogenetic distribution [360,361]. Human disease genes are also known to be unevenly distributed between the human chromosomes [362,363]. Further, synonymous nucleotide substitutions appear to occur at a higher rate in human disease genes, a finding that may reflect increased mutation rates in the chromosomal regions in which disease genes are found [363]. It may be that disease genes are more prevalent in genomic regions that experience elevated rates of mutation [364]. Another possible explanation is that the disease gene set may contain a disproportionately lower number of genes expressed in the germline [363]. This is because variants in such genes might be expected to be more effectively repaired by transcription-coupled repair (transcription-coupled repair in the germline appears to account for the strand asymmetry that the human genome exhibits in terms of inherited variants; [365,366]). Strand asymmetries with respect to the mutation rate may, however, also arise through the influence of DNA replication origins [367], recombination [368,369], and strand-biased repair [370].

### 6.3.21 Mutation/Variant Nomenclature

Some consistency in the way in which variants are described is essential for the accurate and unambiguous reporting and curation of genomic variation data. Most guidelines on how to describe mutational changes in human genes are to be found in [371,372] and on the Human Genome Variation Society (HGVS) Website (hgvs.org/mutnomen/). A program, *Mutalyzer*, is available to check sequence variation nomenclature (lovd.nl/mutalyzer) using a human genome reference sequence and following the current recommendations of the HGVS; *Mutalyzer* is capable of handling most types of variants, including nucleotide substitutions, deletions, duplications, insertions, indels, and splice-site changes [373].

### 6.3.22 Mutations in Gene Evolution

Mutations in human gene pathology and evolution represent two sides of the same coin in that those same mutational mechanisms that have frequently been implicated in human pathology have also been involved in potentiating evolutionary change [56]. Regardless of whether they are advantageous, disadvantageous, or neutral, these mutational changes and their putative underlying causal mechanisms are very similar. It is now clear that the gene has often been a dynamic entity over evolutionary time, not a static one. Indeed, during vertebrate evolution, many genes have undergone gross rearrangement as a result of the action of a variety of mutational processes including insertion, inversion, duplication, repeat expansion, translocation, or deletion. What links pathology and evolution is the underlying genomic architecture with its hitherto largely unexplored vocabulary of structural elements, and different types and patterns of repetitive DNA sequences [201]. It can thus be seen that the mutational spectra of germline mutations responsible for inherited disease, somatic mutations underlying tumorigenesis, polymorphisms (either neutral or functionally significant), and differences between orthologous gene sequences, exhibit remarkable similarities implying that they are very likely to have causal mechanisms in common.

## 6.4 CONSEQUENCES OF MUTATIONS

### 6.4.1 Variants Affecting the Amino Acid Sequence of the Predicted Protein but Not Gene Expression

Many missense variants (i.e., nucleotide substitutions that result in an amino acid substitution) cause hereditary disease in humans. Missense variants are of importance, in understanding the structure or function of a protein because they usually occur in amino acid residues of structural or functional significance [11]. Occasionally, however, not only is the mutated residue not conserved in mouse, but also the substituting residue in humans is identical to its wild-type counterpart in the orthologous (e.g., murine) gene [374]. It is thought that the most likely explanation for the majority of these cases of fixation of disease variants in mice is *compensatory variant.* The chimpanzee genome has been found to harbor a number of examples of potentially compensated mutations (PCMs), defined as human disease-causing or disease-associated missense variants for which the substituting amino acid is identical to the wild-type amino acid residue at the orthologous position in chimpanzee [375]. The absence of strongly deleterious consequences of a specific PCM in chimpanzee would be explicable either by virtue of the very different (simian) environment or by dint of hitherto unidentified variants ("compensatory variants") in the chimpanzee genome that have served to epistatically buffer the PCM [375]. It should be noted, however, that the PCM may only have become seriously disadvantageous in the human lineage, either as a consequence of other lineage-specific genetic changes or due to changes in the human environment and/or lifestyle [376,377]. In this case, it would not have been necessary for the chimpanzee PCM to be compensated for at any time, which is why the qualifier "potentially" is important when PCMs are defined on the basis of current genetic and clinical data.

It is sometimes difficult to establish a causative link between a missense variant and a disease phenotype [378]. The absence of the variant in a large sample (usually 200 individuals) from the same ethnic group as the patient serves to exclude the possibility of a common polymorphism. Amino acid substitutions in evolutionarily conserved residues can also be good candidates for true pathogenicity [11]. If the function of the protein is known, assessment of the effect of the missense variant can be performed by in vitro mutagenesis and functional assay. Finally, the introduction of the variant into an entire organism (e.g., transgenic mice) and the study of its systemic effects provide one of the best means to assess its contribution to a particular clinical phenotype. Amino acid substitutions can be shown to reduce or abolish the physiological function of a protein; for example, missense variants have been identified in factor VIII that abolish thrombin cleavage, which is necessary for its activation [379], interfere with binding to other proteins such as vWF [380], or create or abolish *N*-glycosylation sites [381]. In other proteins, variants have been identified, for example in DNA binding domains, catalytic domains, transmembrane domains, ATP-binding regions, receptor-ligand contact sites, phosphorylation, or other chemical modification sites. Missense variants may also affect protein folding causing a dramatic change in secondary and tertiary structures such that the protein can no longer fulfill its physiological function.

A classic example of a missense variant in the active site of an enzyme is provided by $\alpha_1$-antitrypsin, Pittsburgh, found in an individual with a fatal bleeding disorder [382]. The underlying variant in the $\alpha_1$-antitrypsin (*SERPINA1*) gene is Met358Arg within the active site of the molecule. The substitution by Arg alters the substrate specificity of $\alpha_1$-antitrypsin by converting its "bait loop" (which is specific for elastase) to one that was specific for thrombin. In effect, the molecule lost its anti-elastase activity and became a serine protease inhibitor capable of inhibiting thrombin and factor Xa.

Variants involving gains of glycosylation have generally been considered rare, and the pathogenic role of the new carbohydrate chains has never been formally established [383]; however, the three children identified with Mendelian susceptibility to mycobacterial disease were homozygous with respect to a missense variant in the *IFNGR2* gene that created a new *N*-glycosylation site in the interferon (IFN)$\gamma$R2 chain. The resulting additional carbohydrate moiety was found to be both necessary and sufficient to abolish the cellular response to IFN$\gamma$. From 10,047 HGMD variants in 577 genes encoding proteins trafficked through the secretory pathway, 142 candidate missense variants (~1.4%) in 77 genes (~13.3%) for potential gain of *N*-glycosylation were identified. Six mutant proteins were shown to bear new *N*-linked carbohydrate moieties. Thus, it may be that an unexpectedly, high proportion of variants causing human genetic disease do so via the creation of new *N*-glycosylation sites. Indeed, the pathogenic effects of these variants may be a direct consequence of the addition of *N*-linked carbohydrate.

Missense variants can result in disease by (1) elimination or reduction of the physiological activity/role of the protein; (2) gain of function by which the amino acid substitution creates new functional capabilities of the protein in biochemical and developmental processes in which the protein either does not participate or has a different role; (3) change of the target function of another protein as in the case of the variant in the protein C cleavage site at Arg 506 of coagulation factor V, which is associated with thrombophilia [384], or in the case of a variant in the thrombin cleavage site of factor VIII that eliminates normal activation of factor VIII [379], or in the case of severe obesity from childhood and R236G in the human pro-opiomelanocortin (*POMC*) gene that disrupts the dibasic cleavage site between $\beta$-melanocyte–stimulating hormone ($\beta$-MSH) and $\beta$-endorphin [385]; and (4) participation of the mutant polypeptide in protein complexes, which renders the entire complex abnormal or nonfunctional, as in the case of the triple helical structure of certain collagens in which incorporation of one abnormal collagen chain results in "protein suicide" or an abnormal structure that degrades rapidly [386].

Missense variants have a multitude of different effects on protein structure and function including (1) introduction of larger residues within the hydrophobic protein core leading to adverse interactions between residues [387,388], (2) introduction of buried charged residues [387–389], (3) disruption of protein–protein interactions [390], (4) disruption of hydrogen bonding [388,389], (5) interference with DNA binding [388], (6) breakage of disulfide covalent linkages [388], (7) variant of catalytic residues [389,391], (viii) perturbation of metal binding [388], (9) loss of post-translational modification sites [392], (10) gain of intrinsic disorder [387,393], (11) loss of stability [394,395], and (xii) disruption of quaternary structure [388,396].

Without in-depth analytical studies, missense variants are often difficult to distinguish from polymorphisms with little or no clinical significance, either in the context of candidate gene sequencing studies [397] or in the context of exome sequencing studies [398]. In the "genomic era," a substantial amount of human genetic variation will become amenable to high-throughput analysis in the form of SNPs, and many of these SNPs will influence directly the structure, function, or expression of genes and the RNAs/proteins they encode. Prior knowledge as to which SNPs are most likely to be clinically relevant would greatly enhance the power of studies that aim to identify disease genes through the genotypic screening of patients in both families and populations. Inclusion of structural/functional information could be especially important in the elucidation of multifactorial disease, where genetic heterogeneity and complex interactions between genes and environment have so far limited the success of genetic epidemiological studies [399]. Recently, several predictive models have been developed that use a number of different biophysical parameters to estimate the likely functional impact of an amino acid substitution on the structure and function of a protein [8,394,400,401]. These models have been used to distinguish reasonably and successfully between pathological substitutions, functional polymorphisms, and neutral polymorphisms. Vitkup et al. [402] have claimed that variants at arginine and glycine residues are together responsible for ~30% of cases of genetic disease, whereas random variants at tryptophan and cysteine have the highest probability of causing disease.

### 6.4.2 Variants Affecting Gene Expression

Variants that do not result in amino acid substitution invariably affect gene expression, that is, transcription, RNA processing, and maturation, translation, or protein stability. Total or partial gene deletions, insertions, inversions, and other gross rearrangements obviously result in the loss of gene expression. These types of variants are usually less frequent unless the genomic sequence environment of specific genes (e.g., presence of repeats) predisposes to such lesions. Disorders with high frequencies of gross rearrangements include α-thalassemia, DMD, steroid sulfatase deficiency, and hemophilia A. Some partial gene deletions that eliminate one or a few exons in-frame result in milder clinical phenotypes because gene expression is not totally eliminated; the resulting protein may lack an amino acid domain that is not critical for its function [403].

### 6.4.3 Promoter (Transcription Regulatory) Variants

Microlesions within proximal gene regulatory regions currently comprise only ~1.7% of known variants causing or associated with human inherited disease (see HGMD). Their relative rarity may be in part because not all regulatory elements occur immediately 5′ to the genes that they regulate. Indeed, many such elements are located within the first exon, within introns [404] or within 5′ or 3′ UTRs. In the same vein, upstream ORFs (uORFs), present in 50% of human genes, often impact on the expression of the primary ORFs; indeed, both pathogenic and common variants have been reported within uORFs that can modulate or even abolish the expression of the downstream gene [405,406].

Variants in known promoter motifs usually lead to reduced (or occasionally increased) mRNA levels. Such variants have been studied in the TATA box of the β-globin (*HBB*) gene [407]. Other disease-associated nucleotide substitutions occurring within DNA motifs that bind transcription factors include those located in the CACCC motif of the β-globin (*HBB*) gene influencing transcription factor EKLF binding [408,409], several motifs in the γ-globin (*HBG*) genes [410], the CCAAT motif of the *F9* gene influencing C/EBP binding [411], the SP1 motif of the *LDLR* gene promoter [412], the HNF-1 binding site in the *PROC* gene [413], and the binding site for the transcription factor Oct-1 in the lipoprotein lipase (*LPL*) gene [414]. These few examples are only representatives of a total of over 2200 known

**Mammalian Splice Site Consensus Sequences**

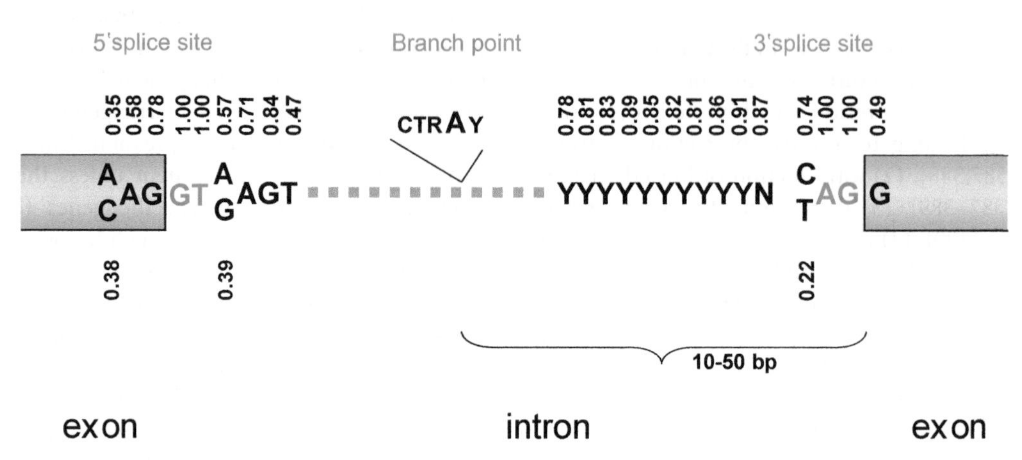


**Figure 6.11** Consensus sequences for the donor (5′ splice) and acceptor (3′ splice) sites and the branchpoint. Numbers above or below the nucleotides correspond to frequencies of a given nucleotide in a large number of mammalian splice-site sequences. Note that the dinucleotides GT and AG (in red) at the beginning and end of the intron are invariant.

promoter variants listed in HGMD and causing human genetic disease. The importance of these mutants lies in the specific DNA sequences thereby implicated in binding to transcription factors. Although most of the known variants reduce the levels of mRNA production, some substitutions actually increase it. Examples include various lesions in the promoters of the Gγ- and Aγ-globin (*HBG1* and *HBG2*) genes that cause hereditary persistence of fetal hemoglobin (HPFH) due to the inappropriate continuation of δ-globin (*HBD*) gene expression into adult life [415] and a gain-of-function (creates a GATA1 binding site) regulatory SNP, which is located in a nongenic region between the α-globin genes and their upstream regulatory elements [416]. An increase in the distance of promoter elements from the transcriptional start site may also result in gene silencing. Such an example has been found in the promoter elements of the *CSTB* gene in EPM1 [137]. Variants that alter the transcriptional regulation of gene expression have been reviewed in [417].

The concomitant change in local DNA sequence complexity surrounding a substituted nucleotide is directly related to the likelihood of a regulatory variant coming to clinical attention [77]. This finding is consistent with the view that DNA sequence complexity is a critical determinant of gene regulatory function and may reflect the internal axial symmetry that frequently characterizes transcription factor binding sites. Polymorphisms in the promoter region that are associated with differential levels of gene expression may predispose to common disorders. For example, a G>An SNP, at nucleotide −6 relative to the transcriptional initiation site of the angiotensin (*AGT*) gene, influences the basal level of transcription and may predispose to essential hypertension [418]. In excess of 400 disease-associated promoter polymorphisms are listed in HGMD plus >700 functional promoter polymorphisms that significantly increase or decrease promoter activity but which have not yet been associated with a clinical phenotype.

### 6.4.4 mRNA Splicing Mutants

Single base-pair substitutions in splice junctions constitute at least 10% of all variants causing human inherited disease. There are, however, a wide variety of variants within both introns and exons that can affect normal RNA splicing (see Ref. [419] for review). The different mechanisms by which disruption of pre-mRNA splicing play a role in human disease have been reviewed in Ref. [420]. The most commonly found variants occur in the dinucleotides GT and AG found at the beginning and end of the donor (5′) and acceptor (3′) consensus splice sequences (see Fig. 6.11 for the consensus splice elements and Fig. 6.12 for the different kinds of

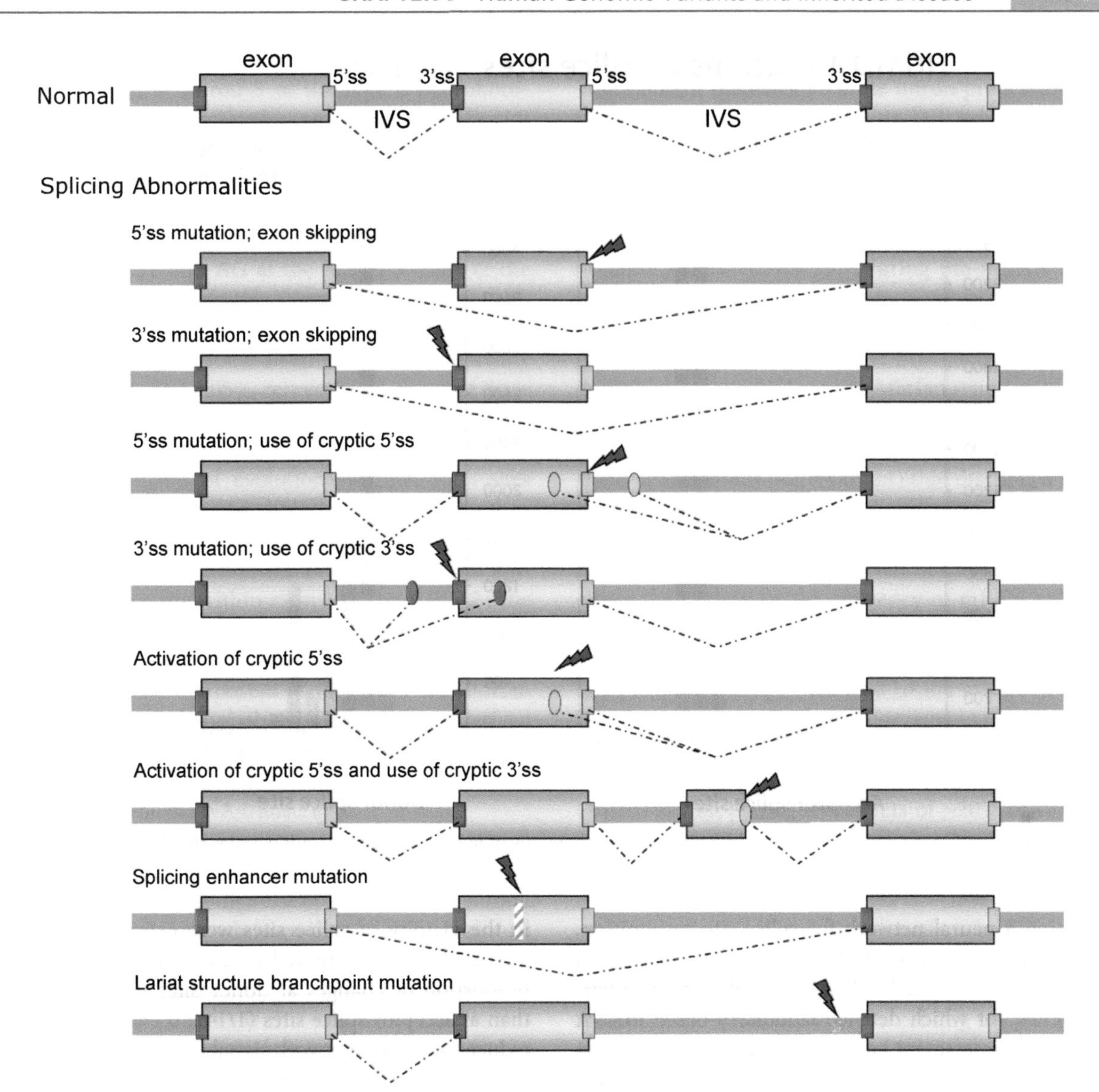


**Figure 6.12** Examples of splicing abnormalities in introns of human genes. Exons are shown as *blue boxes*; introns as *lines* between exons. Green squares denote the normal 5′ (donor) splice sites; red squares represent the normal 3′ (acceptor) splice sites. *Green and red circles* denote cryptic 5′ and 3′ splice sites, respectively. The *broken blue wedge* represents the site of mutation.

RNA splicing abnormalities). Almost all of these variants cause either exon skipping or cryptic splice-site utilization resulting in the severe reduction or absence of normally spliced mRNA. In addition, variants in nucleotides +3, +4, +5, +6, −1, and −2 of the consensus donor splice site have also been observed (Fig. 6.13), with variable severity of the RNA splicing defect. Similarly, likely pathogenic variants in positions −3 and the polypyrimidine tract of the consensus acceptor splice site have been noted (Fig. 6.13). In the majority of these cases, some normal splicing occurs and the defect is not severe. Utilization of cryptic splice sites leads to the production of abnormal mature mRNA with premature stop codons or to the inclusion of additional amino acids after translation (see Ref. [83] for examples and references cited therein).

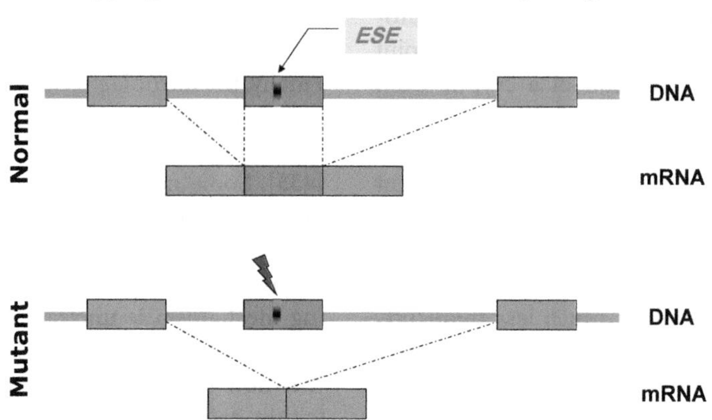


**Figure 6.14** Exon skipping due to nonsense, missense, and silent mutations in enhancer splicing elements (ESE). This element is shown as a darkened segment of the middle exon.

5′ splice site. This activated an upstream cryptic 3′ splice site [the poly(T) complement of the *Alu* poly(A) tail followed by an AG dinucleotide] and a new "exon," containing the majority of the right arm of the *Alu* sequence, was recognized by the splicing apparatus and incorporated into the mRNA. The splice-mediated insertion of an *Alu* sequence in reverse orientation has also been reported in the *COL4A3* gene causing Alport syndrome [444]. Deep intronic variants, located at some considerable distance from splice sites and known splicing-related sequence elements, generally appear to comprise <1% of known splicing variants [445–447]. Such lesions often create novel splice sites thereby activating cryptic exons ("pseudoexons"). It should be appreciated that the <1% figure is very likely to be an underestimate owing to the inherent difficulty in detecting splicing variants located outside of (and distant from) exon–intron splice junctions. Thus, for example, when the *NF1* gene was methodically screened for variants that altered splicing, 5% of the identified lesions that altered splicing were deep intronic variants [448]. Among disease-causing lesions, inclusion of a pseudoexon as a consequence of cryptic splice-site activation appears to be the most common consequence of deep intronic variant [449]. If we also consider the deep intronic polymorphic variants that have the potential to confer susceptibility to disease [450–452], it is very likely that splicing-relevant intronic variation will have been seriously underascertained thus far. A recent review contains comprehensive tables listing the majority of the known deep intronic pathogenic variants [453].

### 6.4.5 RNA Cleavage-Polyadenylation Mutants

A number of examples of RNA cleavage-polyadenylation variants have now been described [454]. Those reported occur in the sequence AAUAAA, which is 10–30 nucleotides upstream of the polyadenylation site and is important for the endonucleolytic cleavage and polyadenylation of the mRNA. A Variant in this sequence of the β-globin (*HBB*) gene results in mild thalassemia [455]. In these cases, normal polyadenylation and cleavage occurs at a level ~10% of normal. Alternative AAUAAA sites downstream of the mutated ones are used, resulting in larger mRNAs that are highly unstable. Other variants near the poly(A) cleavage sequence may result in mRNA destabilization; one such variant has been described 12 bp upstream of the AAUAAA sequence of the *HBB* gene in a patient with β-thalassemia [456].

The G>A mutation at the 3′-terminal nucleotide of the 3′ UTR of the *F2* (prothrombin) gene mRNA gives rise to an elevated prothrombin plasma level and represents a common genetic risk factor for the occurrence of thromboembolic events. This variant creates an inefficient 3′ end cleavage signal and represents a gain-of-function variant, causing increased cleavage site recognition, increased 3′ end processing and increased mRNA accumulation and protein synthesis [457,458].

### 6.4.6 Variants in MicroRNA-Binding Sites

MicroRNAs (miRNAs) posttranscriptionally downregulate gene expression by binding to complementary sequences on the 3′ UTRs of their cognate mRNAs, thereby inducing either mRNA degradation or translational repression. Over 700 human miRNAs have so far been identified but many more probably still remain to be discovered. These miRNAs are each likely to downregulate a large number of different target mRNAs.

The first reported pathological variant in an miRNA-binding site was a G→A transition in a binding site for miR-189 within the 3′ UTR of the *SLITRK1* gene of two apparently unrelated Tourette syndrome patients [459]. Experimental confirmation of the functional effect of this variant came from the demonstration that, in the presence of miRNA-189, in vitro constructs bearing the 3′ UTR variant served to increase repression of a reporter gene by comparison with the wild type. A further example of a functional miRNA-target site variation involves an SNP in the 3′ UTR of the human *AGTR1* gene; although the variant allele is not downregulated by miR155, it has been associated with hypertension in numerous studies [460]. An increasing number of genetic variants located in microRNA target sites are being reported, either causing or associated with an increased risk of inherited disease [461–465].

### 6.4.7 Variants in Non–Protein-Coding Genes

In contrast to the plethora of variants identified in protein-coding genes, the identification of variants in non–protein-coding genes is still very much in its infancy [466]. A number of disease-causing or disease-associated variants have already been reported in various small nucleolar RNA genes [467,468] and miRNA genes [469–472]. In addition, pathogenic variants have also been documented in the longer noncoding RNA genes *XIST*, *TERC*, *H19*, and *RMRP* (see HGMD for details). Numerous pathogenic variants numerous have been described in the untranslated RMRP long noncoding RNA gene that cosegregate with the cartilage hair hypoplasia phenotype [473].

A putative pathological variant has been described in a "gene" encoding a paternally expressed antisense transcript of the GNAS complex locus (*GNASAS*) [474], whereas a functional polymorphism has been reported within an enhancer at the 30 end of the *CDKN2BAS* "gene," which encodes an antisense RNA transcript [475]. A *CRYGEP1* pseudogene-reactivating variant associated with hereditary cataract formation [476] probably also falls into this category. The above examples are likely to comprise only the tip of a fairly large iceberg that still remains essentially unexplored. Thus, for example, both SNP and CNV are likely to impact significantly on miRNA gene expression with myriad potential pathological consequences [461].

Pathogenic variants in microRNAs have also been observed. A pathogenic single nucleotide variant in the seed sequence of the human miR-96 gene causes nonsyndromic autosomal dominant progressive hearing loss (DFNA50) [470]. A mouse mutant from ENU mutagenesis (Dmdo mouse) has been identified with a pathogenic variant in the seed sequence of the mouse miR-96 gene. Heterozygous mice show progressive hearing loss, while homozygous have no cochlear responses [477]. Another example is the pathogenic variant in the seed region of miR-184 which is responsible for familial severe keratoconus combined with early-onset anterior polar cataract. The mutant form fails to compete with miR-205 for overlapping target sites on the 3′ UTRs of INPPL1 and ITGB4 [478]. Furthermore, germline hemizygous deletions of MIR17HG, encoding the miR-17~92 polycistronic miRNA cluster, cause microcephaly, short stature, and digital abnormalities [479].

Pathogenic variants that cause Mendelian phenotypes have also been observed in small nuclear RNAs (snRNAs). The gene encoding the U4atac snRNA, a component of the minor U12-dependent spliceosome, is mutated in individuals with microcephalic osteodysplastic primordial dwarfism type I (MOPD I), a severe developmental disorder characterized by extreme intrauterine growth retardation and multiple organ abnormalities [480,481].

A recent review [482] provides the majority of the known cases of pathogenic variants in the noncoding genes.

### 6.4.8 Variants in Noncoding Regions of Functional Significance

By adopting a gene-centric view, we have until now largely ignored the extensive non–protein-coding portion of the human genome in our quest for variants of pathological significance. As a consequence, we have not only seriously underestimated the extent of the functional component of the genome but may also have overlooked many variants within this genomic "dark matter" [483]. In both the human and the mouse genomes, many noncoding regions exhibit a similar level of evolutionary conservation to that evident in protein-coding regions [484,485]. As yet, however, little is known of the effect that variants in these regions might have on either the phenotype or on overall fitness. Studies of the most evolutionarily conserved noncoding regions have yielded results that are consistent with the view that most variants in noncoding regions are only slightly deleterious [485,486].

To obtain a first, necessarily rather crude, estimate of the contribution of variation in human noncoding sequences to phenotypic and/or disease traits, Visel et al. [487] performed a meta-analysis of ~1200 SNPs that have been identified as the most significantly associated variants in published GWAS. They found that, in 40% of cases, neither the SNP in question nor its associated haplotype block overlapped with any known exons. These authors, therefore, concluded that in at least one-third of detected disease associations, variation in noncoding sequence rather than coding sequence could have causally contributed to the trait in question. We suspect that this could be because the common disease-common variant hypothesis [488] may be much more likely to apply to noncoding sequence than to coding sequence, owing to the selection constraints impacting on sufficiently frequent functional variation in the latter. In similar vein, others have also estimated that 39%–43% of trait/disease-associated SNPs in GWAS are located within intergenic regions [489,490]. This notwithstanding, it should be appreciated that any given variant apparently detected within a noncoding region may actually reside within a hitherto undiscovered exon [491]. We should, however, also be aware that rare variants, in *cis* to those found to be associated with a given disease or trait in GWAS studies, may simply by chance give rise to "synthetic associations" that are then attributed to much more common variants [492].

In the context of identifying genetic variants responsible for human inherited disease, Cooper et al. [5] have argued that it will become increasingly important to consider functional elements in the genome (the "functionome") rather than simply genes per se. We usey the term "functionome" here to describe the totality of the biologically functional nucleotide sequences in the human genome, irrespective of whether they are associated with genes. Because conserved noncoding sequences in the human genome appear to be ~10-fold more abundant than known genes [493], it is likely that (1) currently known variants within coding regions are unlikely to be fully representative of the universe of pathological variants and (2) a whole new grouping of disease-causing variants may await identification and characterization. Once again, a paradigm shift in our thinking may well be required if we are to maximize the potential of the emerging high-throughput technology to detect new (hitherto latent) types of human genomic pathogenic variants.

The above notwithstanding, it is rather unlikely that the functional non–protein-coding portion of the human genome will prove to be quite as mutation-dense as the protein-coding portion. For most inherited disorders, the mutation detection rate is already fairly high (49%), although this success rate is often achieved by combining different mutation detection methodologies, for example, to screen for exon deletions and CNVs as well as more subtle lesions [494]. At least some of the "missing lesions" may nevertheless be found by screening extragenic functional elements.

A recent review on the noncoding genetic variants in human disease is also recommended to the reader [495]. Another recent review further discusses the noncoding pathogenic variation and particularly structural variations such as deletions, inversions or duplications that have the potential to disturb normal chromatin folding [496]. These abnormalities may lead to the repositioning or disruption of the chromatin topological associating domains (TADs [497]) and the relocation of enhancer elements with altered expression of the corresponding genes. Several recent studies highlight this as important disease mechanisms in developmental disorders. Therefore, the regulatory landscape of the genome has to be taken into consideration when investigating the molecular pathophysiology of human disease. Lupianez et al. [498] presented evidence that human limb malformations are caused by deletions, inversions or duplications that alter the structure of TADs. This in turn resulted in misplacement of limb enhancers of the *EPHA4* gene that drive ectopic gene expression. In addition, the creation of new TADs (neo-TADs) due to genomic duplications, results in altered gene expression and limb malformation phenotype [499].

### 6.4.9 Cap Site Variants

Transcription of the mRNA is initiated at the so-called cap site, which is protected from exonucleolytic degradation by the addition of α-methylguanine. An A-to-C transversion at the cap site of the β-globin (*HBB*) gene was found in a patient with β-thalassemia [500]. It is not, however, clear if this variant causes reduced transcription or abnormal initiation of transcription because C is found in 6% of transcriptional initiation sites [501] (the most common nucleotide [76%] at position +1 is A). A functional (C/A) polymorphism of the transcriptional initiation site has been noted in the *APOH* gene; the rarer A allele displayed a carrier frequency of 0.12 and was associated with markedly reduced plasma β2 glycoprotein I [502].

### 6.4.10 Variants in 5′ UTRs

Sequence motifs in the 5′ UTRs of genes are thought to play a role in controlling the translation of the encoding mRNA. The phenotypic effects of lesions in 5′ UTRs and their clinical consequences have been reviewed [503]. Pathogenic variants in the iron response element (IRE) in the 5′ UTR of the ferritin (*FTH1*) gene interfere with the post-transcriptional regulation of ferritin synthesis by decreasing the affinity of IRE for IRE-binding protein [504]. In contrast, decreases in the steady-state level of β-globin (*HBB*) mRNA have been noted in association with a single base deletion at position +10, a G-to-A substitution at position +22, a C-to-G transversion at position +33, and a 4-bp deletion (AAAC) at position +(40–43) in the *HBB* 5′ UTR [505–507].

### 6.4.11 Variants in 3′ Regulatory Regions

Sequences in the 3′ regulatory regions (3′ RRs) of genes are known to be involved in controlling mRNA cleavage/polyadenylation and determining mRNA stability, nuclear export, intracellular localization, and translational efficiency. Although such regions are rich in regulatory elements, relatively few pathological variants have been reported [454,508]. Although only ~0.2% of likely pathogenic variants currently logged in HGMD are located within 3′ RRs, this is likely to represent a rather conservative estimate of their actual prevalence. A typical example is the G→A transition 69 nucleotides downstream of the polyadenylation site of the δ-globin (*HBD*) gene causing δ-thalassemia [509]; the variant occurs within a GATA motif and serves to increase the binding affinity of the sequence for erythroid-specific DNA binding protein.

In an attempt to study 3′ RR variants systematically, Chen et al. [454,510] performed a systematic analysis of disease-associated variants in the 3′ RRs of human protein-coding genes. A total of 121 3′ RR variants in 94 human genes were collated including 17 variants in the upstream core polyadenylation signal sequence (UCPAS), 79 in the upstream sequence (USS) between the translational termination codon and the UCPAS, 6 in the left arm of the "spacer" sequence between the UCPAS and the pre-mRNA cleavage site, 3 in the right arm of the "spacer" sequence or downstream core polyadenylation signal sequence, and 7 in the DSS of the 3′-flanking region. All the UCPAS variants and the rather unusual cases of *DMPK*, *SCA8*, *FCMD*, and *GLA* variants were found to exert a significant effect on the mRNA phenotype and the majority cause monogenic disease. By contrast, most of the remaining variants were polymorphisms, were found to exert a comparatively minor influence on mRNA expression, but may predispose to, protect from, or modify complex clinical phenotypes. The systematic study of these lesions permitted the identification of consistent patterns of secondary structural change that promise to allow the discrimination of nonfunctional USS variants from their functional counterparts.

### 6.4.12 Translational Initiation Codon Variants

Pathogenic variants in the ATG translational initiation codon have been reported in quite a wide variety of disorders (e.g., Ref. [511]). Instances of substitutions in all three nucleotides have been observed in β-thalassemia, Norrie disease, albinism, phenylketonuria, McArdle disease, and Albright osteodystrophy among others. Indeed, a total of 405 likely pathogenic variants within ATG translational initiation codons are recorded in HGMD representing ~0.7% of the total number of reported coding sequence variants causing human inherited disease. Almost invariably, the variant leads to severe reduction of steady-state mRNA levels similar to that associated with nonsense variants. The mutant mRNA is presumably not translated. The first AUG codon occurs in the context of the so-called Kozak consensus sequence GCCA/GCCAUGG, which is thought to be recognized by the 40S ribosomal subunit [512]. Variants at the initiator methionine ATG may completely abolish translation; however, there are alternative possibilities, namely utilization of the mutant ATG with much reduced efficiency or translational initiation at the next available ATG codon. A C/T polymorphism immediately 5′ to the ATG codon within the Kozak sequence of the *CD40* gene is thought to influence translation efficiency [513].

Some diseases are caused by variants that perturb the initiation step of translation by changing the context around the start AUG codon or introducing upstream

AUG codons (see Ref. [514] for review). The scanning mechanism provides a framework for understanding the effects of these changes in mRNAs. The scanning mechanism refers to the entry of the small ribosomal subunit at the (usually capped) 5′ end of the mRNA and linear migration until an AUG codon is encountered. Mutational mechanisms such as (1) reinitiation at an internal start codon (e.g., thrombopoietin, *TPO*) and (2) leaky scanning (as in the case of the *Rx/rax* gene underlying the mouse eyeless variant) probably account for such cases.

Naturally occurring variants in the GCCA/GCCAUGG motif include (for the numbering of the mutant nucleotide, the A of the AUG codon is +1; see references in Ref. [514]): (1) +4 G-to-A in the androgen receptor (*AR*) gene in a family with partial androgen insufficiency; (2) −1 C-to-T transition in the α-tocopherol transfer protein (*TTPA*) gene in a family with vitamin E deficiency; (3) a 2 nt deletion causes an A-to-C change at position −3 of the α-globin gene (*HBA*) in a patient with α-thalassemia; (4) −3 A-to-T transversion in the mouse *Pax6* gene causes defects in eye development; and (5) −3 G-to-C somatic variant in the *BRCA1* gene in one case of highly aggressive sporadic breast cancer. It is not surprising that most of the naturally occurring variants involve positions −3 and +4, the positions wherein experimentally induced mutations have the strongest effect.

A meta-analysis of 405 unique (HGMD-derived) single base-pair substitutions, located within the ATG translation initiation codons of 255 different genes, reported to cause human genetic disease has been performed [515]. Although these lesions comprised only 0.7% of coding sequence variants in HGMD, they nevertheless were 3.4-fold overrepresented as compared to other missense variants. The distance between a translation initiation codon and the next downstream in-frame ATG codon was significantly greater for genes harboring ATG codon variants than for the remainder of genes in HGMD (control genes). This suggests that the absence of an alternative ATG codon in the vicinity of an ATG translation initiation codon increases the likelihood that a given ATG mutation will come to clinical attention. An additional 42 single base-pair substitutions in 37 different genes were identified in the vicinity of ATG translation initiation codons (positions −6 to +4, comprising the Kozak consensus sequence). These substitutions were, however, not evenly distributed, being significantly more abundant at position +4. Finally, contrary to the authors' initial expectation, the match between the original translation initiation codon and the Kozak consensus sequence was significantly better (rather than worse) for genes harboring ATG codon mutations than for the HGMD control genes [515].

### 6.4.13 Termination Codon ("Nonstop") Variants

"Nonstop" variants are single base-pair substitutions that occur within translational termination (stop) codons, which can lead to the continued and inappropriate translation of the mRNA into the 3′-UTR. The classic example of a termination codon mutant is the case of the $\alpha_2$-globin Constant Spring, with a variant in the normal stop codon; this substitution leads to incorporation of an additional 31 amino acid residues in the $\alpha_2$-globin polypeptide chain [516]. The resulting protein is unstable and does not interact properly with the β-globin chains of hemoglobin. Some 119 variants within Term codons (in 87 different genes) have been recorded in HGMD, representing ~0.2% of all missense/nonsense variants.

A meta-analysis of these 119 nonstop variants noted a paucity of alternative in-frame stop codons in the immediate vicinity (0–49 nucleotides downstream) of the mutated stop codons as compared with their control counterparts [517]. This implies that at least some nonstop variants with alternative stop codons in close proximity will not have come to clinical attention, possibly because they will have given rise to stable mRNAs (not subject to nonstop mRNA decay) that are translatable into proteins of near-normal length and biological function. A significant excess of downstream in-frame stop codons was, however, noted in the range 150–199 nucleotides from the mutated stop codon [517]. The authors speculated that recruitment of an alternative stop codon at greater distance from the mutated stop codon might trigger nonstop mRNA decay, thereby decreasing the amount of protein product and yielding a readily discernible clinical phenotype.

### 6.4.14 Frameshift Variants

A large number of frameshift variants have been described in numerous disease-related genes. All lead to altered translational termination with abnormal polypeptide chains after the frameshifts; severe phenotypes are usually seen [518]. Frameshifts occur with microdeletions or microinsertions and exon skipping. The mechanisms underlying these variants were discussed earlier in this chapter.

### 6.4.15 Nonsense Variants

Nonsense variants give rise to premature termination of translation and truncated polypeptides. They account for ~11% of all described gene lesions causing human inherited disease and ~20% of disease-associated single base-pair substitutions affecting gene coding regions [519]. Pathological nonsense variants resulting in TGA (38.5%), TAG (40.4%), and TAA (21.1%) occur in different proportions to naturally occurring stop codons [519]. Of the 23 different nucleotide substitutions giving rise to nonsense variants, the most frequent are CGA→TGA (21%; resulting from methylation-mediated deamination) and CAG→TAG (19%) [519]. The differing nonsense mutation frequencies are largely explicable in terms of variable nucleotide substitution rates such that it is unnecessary to invoke differential translational termination efficiency or differential codon usage. Nonsense variants are usually associated with a reduction in the steady-state level of cytoplasmic mRNA [520]. This mechanism of "NMD" is responsible for the degradation of mRNAs that contain a premature termination codon at a position at least 50 nt upstream of an exon–exon boundary [521] but it is not universal [522]. A recent study of 4584 protein-truncating variants has introduced a better predictor of NMD than the 50 bp rule [523]. One or more parameters could be affected: the transcription rate, the efficiency of mRNA processing or transport to the cytoplasm, or mRNA stability.

In the majority of described instances of nonsense variants, the resulting disorders are recessive in nature as a consequence of the haploinsufficiency resulting from the NMD-induced absence of the truncated proteins (which ensures that such polypeptides do not interfere with the function of the wild-type protein). Nonsense variants that do not elicit NMD can, however, give rise to a dominant negative condition (e.g., variants in the *SOX10* gene causing Waardenburg Shah syndrome [524]). Because for NMD to be activated, the nonsense variant must reside at least 50–55 nt upstream of an exon–exon boundary, it follows that the precise location of the nonsense variant could be an important factor in predicting the pathogenicity of that lesion. By way of example, nonsense variants within the last exon of the human β-globin (*HBB*) gene do not elicit NMD. As a consequence, the truncated β-globin product has near-normal abundance, fails to associate properly with α-globin, and hence gives rise to a dominantly inherited form of α-thalassemia [31]. Different nonsense variants within the same gene may thus be associated with different clinical phenotypes depending on whether or not NMD is activated. Another example of this is provided by a nonsense variant (Q37X) in the *DAX1* gene of an adrenal hypoplasia congenita patient; this lesion is associated with a milder-than-expected clinical phenotype on account of the expression of a partially functional, amino terminal-truncated DAX1 protein synthesized from an alternative in-frame translational start site at Met83 [525]. In a recent meta-analysis, the proportion of known disease-causing nonsense variants predicted to elicit NMD was found to be significantly higher than among nonobserved (potential) nonsense variants, implying that nonsense variants that elicit NMD are more likely to come to clinical attention [519]. In practical terms, the observation of greatly reduced or absent cytoplasmic mRNA associated with nonsense variants has important implications for mutation screening. Thus, attempts to obtain mRNA for RT-PCR and mutation detection may result in amplification of nucleic acid from only the non–nonsense variant-bearing allele. Nonsense variants in the factor VIII (*F8*) gene (hemophilia A) and fibrillin (*FBN1*) gene (Marfan syndrome) have been associated with the skipping of exons containing these variants [252,434] and this observation has now been extended to other genes; exon skipping is either complete or partial. The mechanism underlying this phenomenon is unknown although a number of intriguing models have been proposed [526].

Some genes are characterized by numerous nonsense variants but relatively few if any missense variants (e.g., *CHM*), whereas other genes exhibit many missense variants but few if any nonsense variants (e.g., *PSEN1*). Genes in the latter category have a tendency to encode proteins characterized by multimer formation [519]. Consistent with the operation of a clinical selection bias, genes exhibiting an excess of nonsense variants are also likely to display an excess of frameshift variants [519]. Recently, an example of the spontaneous read-through of a premature termination codon was reported in a patient who was a compound heterozygote for two nonsense variants in the *LAMA3* gene (R943X/R1159X) [527]. The patient, who presented with junctional epidermolysis bullosa, was expected to die as a consequence of harboring these nonsense variants but was "rescued" by spontaneous read-through of the R943X-bearing allele. This patient's full-length R943X-bearing *LAMA3* mRNA escaped nonsense-mediated decay, thereby ensuring near-normal *LAMA3* mRNA and laminin-α3 protein levels. The genetic context of the *LAMA3* variant R943X was found to be close to a hypothetical consensus sequence for optimal premature termination codon read-through.

### 6.4.16 Unstable Protein Mutants

Missense (nonsynonymous) variants can cause abnormal protein folding and are, therefore, associated with reduced expression owing to instability of the protein. Reviews of variants that affect protein stability can be found in Refs. [528,529]. For proteins that circulate in body fluids, most variants are associated with "CRM negative" status in which the amount of protein correlates with the amount of activity or "CRM reduced" status in which the amount of activity is still lower than the amount of protein produced. Many such variants have been seen in factor VIII causing mild/moderate hemophilia A [87].

The nature of the biophysical properties of amino acid substitutions in p53 that increase their likelihood of coming to clinical attention has been explored [530]; these include solvent inaccessibility, the number of adverse steric interactions introduced, and a reduction in H-bond number. This study was extended by modeling in silico all amino acid replacements that could potentially have arisen from an inherited single base-pair substitution in five human genes encoding arylsulfatase A (*ARSA*), antithrombin III (*SERPINC1*), protein C (*PROC*), phenylalanine hydroxylase (*PAH*), and transthyretin (*TTR*) [401]. A total of 9795 possible mutant structures were modeled and 20 different biophysical parameters assessed. Comparison with the HGMD-derived spectra of 469 clinically detected missense variants indicated that several types of mutation-associated change affected protein function, including the energy difference between wild-type and mutant structures, solvent accessibility of the mutated residue, and distance from the binding/active site. These parameters are considered to be important in protein folding which adds support to the view that many missense variants come to clinical attention by virtue of their consequences for protein folding and stability [531,532].

### 6.4.17 Variants in Remote Gene Regulatory Elements

In the β-globin gene cluster, a regulatory region ~10 kb upstream of the ε-globin (*HBE*) gene has been identified that is capable of directing a high level of position-independent β-globin gene expression [533]. This region, termed the locus control region (LCR), is thought to organize the entire 60-kb β-globin gene cluster into an active chromatin domain and to enhance the transcription of individual globin genes [534]. A similar LCR is also present in the α-globin gene cluster and other gene clusters [535]. Deletions of the LCR in the β-globin gene cluster result in silencing of the β-globin (*HBB*) and other genes of the cluster, even although the coding regions of these genes are still intact [415]. A particular 25-kb deletion, known as Hispanic γδβ-thalassemia, which deletes sequences 9.5–39-kb upstream of the ε-globin (*HBE*) gene including the LCR renders the *HBB* gene 60 kb downstream of the deletion nonfunctional [536]. This extraordinary effect of the deletion of the LCR is thought to be due to an altered (DNase I-resistant) state of chromatin associated with nonfunctional genes. Several other examples of similar deletions in the LCR of the α-globin gene cluster have been reported [537].

Several other examples of remote regulatory elements have come to attention as a consequence of their ablation by gross deletions located at some considerable distance (from 10 kb to several megabases) from the genes whose expression they disrupt [160]. For instance, a 960-kb deletion of noncoding sequence, lying between 1.477 Mb and 517 kb upstream of the *SOX9* gene gives rise to the acampomelic form of campomelic dysplasia [538]. Such pathological deletions, however, are not necessarily always so large. Indeed, a 7.4-kb deletion, located 283 kb upstream of the *FOXL2* gene, has been identified as a cause of blepharophimosis syndrome; it disrupts a long noncoding RNA (*PISRT1*) as well as eight conserved noncoding sequences [539]. For some conditions, such lesions may actually occur quite frequently, as in the case of the *SHOX* gene where ~22% of Leri–Weill syndrome patients and ~1% of individuals with idiopathic short stature harbor a microdeletion spanning the upstream enhancer region that leaves the coding region of the *SHOX* gene intact [540].

During the past few years, a number of other examples of likely pathogenic variants in remote promoter elements have been reported. These include a total of nine variants within a 1-kb region (termed the long-range or limb-specific enhancer) ~979 kb 5′ to the transcriptional initiation site of the sonic hedgehog (*SHH*) gene [541] and a T>C transition 1.44 Mb upstream of the *SOX9* gene associated with cleft palate/Pierre Robin sequence [542]. Far upstream polymorphic variants that influence gene expression and that are relevant to disease are also beginning to be documented. Thus, for example, the C>T functional SNP 14.5 kb upstream of the *IRF6* gene, associated with cleft palate, alters the binding of transcription factor AP-2α [543]. Similarly, a functional SNP ~6 kb upstream of the α-globin–like *HBM* gene serves to create a binding site for the erythroid-specific transcription factor GATA1 and interferes with the activation of the downstream α-globin genes [416]. A functional SNP ~335 kb upstream of the *MYC* gene increases the risk of colorectal and prostate cancer by increasing the expression of the *MYC* gene by altering the binding strength of transcription factors TCF4 and/or TCF7L2 to a transcriptional enhancer [544–547]. Finally, in the context of pointing out the shortcomings of the gene-centric approach to mutation detection, we should be aware that functional SNP rs4988235, located 13.9 kb upstream of the lactase (*LCT*) gene and associated with adult-type hypolactasia, actually resides deep within intron 13 (c.19171+326C>T) of the *MCM6* gene [548–550]. Given that up to 5% of quantitative trait loci for gene expression lie >20 kb upstream of transcriptional initiation sites [551], many more far upstream polymorphic variants that influence gene expression are likely to be identified in the coming years.

Rather fewer pathological variants are known to be located at a considerable distance downstream of human genes. One example is the C>G transversion 2528 nt 3′ to the term codon of the *CDK5R1* gene, which has been postulated to play a role in nonspecific intellectual disability [552]. Perhaps more dramatic is the A>G SNP (rs2943641), 565,981 bp 3′ to the Term codon of the *IRS1* gene, which is associated with type 2 diabetes, insulin resistance, and hyperinsulinemia; the G allele was found to be associated with a reduced basal level of IRS1 protein [553].

In the light of the above, it can be seen that the under-ascertainment of disease-associated variants within regulatory regions is likely to be quite substantial but can potentially be rectified by emerging high-throughput entire genome sequencing protocols.

### 6.4.18 Cellular Consequences of Trinucleotide Repeat Expansions

Trinucleotide repeat expansion has been discussed earlier. In the case of fragile X, the $(CGG)_n$ repeat is located in the 5′ UTR of the *FMR1* gene, and its expansion to full mutation results in hypermethylation of the promoter region, loss of transcription, and hence silencing of the gene [554]. Loss of the encoded protein, fragile X mental retardation protein (FMRP), which is thought to play a role in dendritic mRNA transport and translation, is responsible for the classic fragile X syndrome phenotype. Gene inactivation can also be caused by altering the spacing of promoter elements from the transcriptional start site as in the case of the 12-mer repeat expansion in the *CSTB* gene [137].

When the trinucleotide repeat lies within the gene-coding region as in Huntington disease, its expansion results in an abnormal protein with a gain of function due to the enlargement of the polyglutamine tract. Mutant huntingtin exerts its pathological effects via abnormal protein aggregation, transcriptional dysregulation, mitochondrial dysfunction, excitotoxicity, and abnormal cellular trafficking, leading to neuronal loss particularly in the dorsal substratum [555].

Another example of a gain-of-function mutation is provided by the expansion of the CTG repeat in the 3′ UTR of the *DMPK* gene causing type 1 myotonic dystrophy (DM1). This does not abolish transcription but rather causes nuclear retention of RNA transcripts leading to the transcriptional dysregulation of other genes [556]. CTG expansion appears to lead to the sequestration of cellular RNA-binding proteins, which in turn gives rise to the abnormal splicing of multiple transcripts [557]. DM1 thus exemplifies a disease whose mechanistic basis lies at the RNA level.

### 6.4.19 Variants That Give Rise to Inappropriate Gene Expression

HPFH and hereditary persistence of α-fetoprotein (HPAFP) are two clinical conditions that are prototypes for the inappropriate expression of γ-globin (*HBG1* and *HBG2*) and α-fetoprotein (*AFP*) genes, respectively. Normally, the levels of fetal hemoglobin (HbF; α2γ2) in adult life are very low, as there is a switch from fetal to adult hemoglobin during the perinatal period. Similarly, AFP is produced at high level in fetal liver but declines rapidly after birth. In HPFH and HPAFP, however, the levels of HbF and

AFP, respectively, are inappropriately high in adult life. This is often due to single nucleotide substitutions in the promoter regions of the *HBG2*, *HBG1*, or *AFP* genes. A considerable number of variants that occur in the region −114 to −202 of the γ-globin genes have been characterized and presumably cause persistent expression of their corresponding genes [415]. A similar situation has been observed with a −119 variant in the *AFP* gene [558]. These variants occur within DNA binding motifs for transcriptional regulators. A very interesting mutational mechanism has been proposed for facioscapulohumeral muscular dystrophy (FSHD), a common autosomal dominant myopathy associated with a typical pattern of muscle weakness. Most FSHD patients carry a large deletion of a 4q35-located polymorphic D4Z4 macrosatellite repeat array and present with fewer than 11 repeats whereas normal individuals possess between 11 and 150 repeats [559]. An almost identical D4Z4 repeat array is present at 10q26 [560] and the high sequence homology between these two arrays can cause difficulties in molecular diagnosis. Each 3.3-kb D4Z4 repeat contains a *DUX4* (double homeobox 4) gene that, among others, is activated on contraction of the 4q35 D4Z4 array due to the induction of chromatin remodeling of the 4qter region. An increasing number of 4q subtelomeric sequence variants are now recognized, although FSHD only occurs in association with three "permissive" haplotypes, each of which are associated with a polyadenylation signal located immediately distal of the last D4Z4 repeat [561]. This poly(A) signal stabilizes any *DUX4* mRNAs transcribed from this most distal D4Z4 repeat in FSHD muscle cells. Synthesis of both the *DUX4* transcripts and the protein in FSHD muscle cells induces significant cell toxicity. DUX4 is a transcription factor that targets several genes, which results in a deregulation cascade that inhibits myogenesis, sensitizes cells to oxidative stress, and induces muscle atrophy, thereby recapitulating many of the key molecular features of FSHD [562].

### 6.4.20 Position Effect in Human Disorders

In several instances, a DNA alteration is found well outside the putative gene that is primarily involved with a disease. Variants acting by "positional effect" are those in which the transcription unit and minimal promoter of the gene remain intact, but there is a nearby alteration that influences gene expression [563]. These positional effect DNA lesions may involve distal promoter regions, enhancer/silencer elements, or changes in the local chromatin environment. The positional effect could be up to several megabases away from the gene of interest. The examples of the LCR in the β-globin gene cluster and the transcriptional repressor D4Z4 in FSHD are provided elsewhere in this chapter. Most of the position effects are due to chromosomal rearrangements that frequently lead to alteration of the chromatin environment of the gene. Possible mechanisms which may lead to a positional effect include the following: (1) The rearrangement separates the transcription unit from distant *cis*-regulatory elements (enhancer removal results in gene silencing, whereas silencer removal results in inappropriate gene activation); (2) juxtaposition of the gene with an enhancer element from another part of the genome; (3) removal of an insulator or boundary element may also lead to inappropriate gene silencing; (4) enhancer competition of DNA sequences that were juxtaposed to the gene; (5) positional effect variegation in which the chromosomal rearrangement causes the juxtaposition of an euchromatic gene with a region of heterochromatin.

Some examples of positional effect mutations due to translocation breakpoints include genes *PAX6* in aniridia [564], *SOX9* in campomelic dysplasia [565,566], *POU3F4* in X-linked deafness [567], *HOXD* complex in mesomelic dysplasia [568], *FOXL2* in blepharophimosis/ptosis/epicanthus inversus syndrome (*BPES*) [569,570], and the *SHH* gene in preaxial polydactyly [571]. In these cases, the translocation breakpoints may be in excess of a megabase away from the inappropriately expressed/silenced gene. Indeed, in one example of campomelic dysplasia, the breakpoint maps ~1.3 Mb downstream of the *SOX9* gene, making this the longest-range position effect so far found [566]. For a recent review of position effect mutations, see [572].

It is likely that in the majority of cases, the position effect involves a highly conserved *cis*-acting regulatory element. These *conserved noncoding elements* (CNCs; also termed multiple-species conserved sequences; conserved nongenic sequences; the most highly conserved are also called ultraconserved elements) comprise approximately 1%–2% of the human genome and represent potential targets for pathogenic variants [573–577]. An example of such a lesion is provided by the 52-kb deletion of a large noncoding region downstream of the sclerostin (*SOST*) gene in patients with van Buchem disease, leading to altered expression of the *SOST* gene [578]. The deletion disrupts a bone-specific enhancer element that drives *SOST* gene expression.

Pathogenic variants may also occur in nonconserved elements that could become functional after the introduction of the mutant sequence. This pathogenetic mechanism has been described underlying a variant form of α-thalassemia. Affected individuals from Melanesia have a gain-of-function regulatory single nucleotide polymorphism (rSNP) in a nongenic region between the α-globin genes and their upstream regulatory elements. The rSNP creates a new promoter-like element that interferes with the normal activation of all downstream α-like globin genes [416].

### 6.4.21 Position Effect by an Antisense RNA

An individual with an inherited α-thalassemia has been described who has a deletion that results in a truncated, widely expressed gene (*LUC7L*) becoming juxtaposed to a structurally normal α-globin (*HBA2*) gene. Although it retained all of its local and remote *cis*-regulatory elements, expression of the *HBA2* gene was nevertheless silenced and its CpG island became completely methylated at an early stage during development. The antisense RNA of the *LUC7L* gene appears to have been responsible for the silencing of the *HBA2* gene [579].

### 6.4.22 Abnormal Proteins Due to Fusion of Two Different Genes

The translation of fusion genes results in novel proteins with different or abnormal properties from their parent polypeptides. Fusion genes are either the result of homologous unequal crossing-over or the junction sequences at breakpoints of chromosomal translocations. Hemoglobin Lepore, a fusion of δ- and β-globin genes, is the prime example of the first mechanism. Other examples of abnormal fusion genes due to unequal crossover include the case of glucocorticoid-suppressible hyperaldosteronism (GSH), an autosomal dominant form of hypertension, caused by oversecretion of aldosterone [580]; some GSH patients have hybrid genes between *CYP11B1* and *CYP11B2*, two highly homologous cytochrome P450 genes on 8q22. The hybrid gene contains the regulatory elements of *CYP11B1*, expressed in the adrenal gland, and the 3′ coding region of *CYP11B2*, which is essential for aldosterone synthesis. Another example is the case of abnormalities of color vision resulting from fusion of the green and red color pigment (*RCP*, *GCP*) genes [581]. Recombination between the Kallmann gene on Xp22.3 (*KALX*) and its homolog (*KALY*) at Yp11.21 results in a fusion gene that is transcriptionally inactive and is associated with Kallmann syndrome secondary to an X; Y translocation. Finally, Francis et al. [582] identified a large atypical hemolytic uremic syndrome family in whom a deletion occurred through MMEJ rather than by NAHR. The deletion resulted in the formation of a *CFH/CFHR3* hybrid gene. The protein product of this gene, a 24 short consensus repeat protein, was found to be secreted at slightly lower levels than wild-type factor H, but the decay accelerating and cofactor activities of this protein were significantly impaired. A growing number of hematological malignancies are associated with abnormal fusion proteins, the genes of which are found at the breakpoints of chromosomal translocations. One of the first reported examples was the case of fusion of the *BCR* and *ABL* genes in the *t*(9;22) known as Philadelphia (Ph) chromosome in chronic myelogenous leukemia. The *BCR* gene is on chromosome 22 and the *ABL* gene is on chromosome 9; after the translocation junction, a fusion gene is created with the promoter elements of the *ABL* gene and the 3′ half of the *BCR* gene [583]. A new abnormal protein is detected in the leukemia cells, the abnormal function of which probably contributes to the malignant phenotype. Another example is the case of Ewing sarcoma (a solid tumor of bone) in which an 11;22 translocation results in a fusion of the *FLI1* gene on 11q24 with the *EWS* gene on 22q12 [584]; for a classic review, see [585]. Fusion genes can be readily identified by PCR and can serve either as diagnostic indicators for relapse in the disorders concerned or as indicators of the need for an alternative therapeutic regimen.

### 6.4.23 Mutations in Genes Involved in Mismatch Repair Associated With Genomic Instability in the Soma

The study of somatic mutation is extremely important both for the study of cancer [586] and for other diseases such as paroxysmal nocturnal hemoglobinuria [587]. Variants that lead to abnormal or abolished function of genes encoding for proteins involved in DNA mismatch repair are of particular importance because they lead to accumulation of variants throughout the genome. For example, some forms of hereditary nonpolyposis colon cancer (HNPCC), which may account for up to 10% of colon carcinoma, are due to variants in genes such as heterozygotes [622]. If this is not so, and the homozygote is more seriously affected, then the respective alleles may be regarded as *semidominant* [623].

In general, most recessive alleles are loss-of-function alleles and include gross gene deletions and rearrangements, frameshift variants, nonsense variants, and so on. By contrast, dominant alleles are often associated with gain of function, either due to dominant negative variants (which interfere with and hence abrogate the function of the wild-type allele) or dominant positive variants (which confer increased, constitutive, novel, or toxic activity on the mutant protein). Examples of dominant negative variants are to be found in the *GH1* [624] and *KIT* [625] genes, while dominant positive variants have been reported in the *PMP22* [626], *GNAS1* [627], *DMPK* [628], and *SERPINA1* [382] genes. It should be noted that loss-of-function variants (e.g., *TERT* [629] and *RUNX2* [630]) can also be associated with dominantly inherited conditions in cases where a 50% reduction in the level of the protein product is sufficient to impede function.

For X-linked diseases, it is probably inappropriate to use the terms "dominant" and "recessive" because males are hemizygous and females often display variable expressivity of their heterozygous variants due to skewed X-inactivation or clonal expansion [631].

### 6.4.28 Genetic Architecture of Complex Diseases

The study of the genetic architecture of complex disease has revolved around the discussion of two apparently opposing models: the common disease–common variant (CD/CV) hypothesis and the multiple rare variant or common disease–rare variant (CD/RV) hypothesis [488]. Because the CD/CV model conceptually underpinned the HapMap Project, GWAS that have used HapMap data have tended to interrogate the association of common SNPs (MAF >5%) with complex diseases and traits. Initial GWAS data, therefore, strongly supported the involvement of common variants, especially common SNPs in complex phenotypes [490]. However, such studies have succeeded in explaining only a small fraction of the heritability of complex phenotypes [632] and this "missing heritability" has tended to challenge the validity of the CD/CV hypothesis. Perhaps not surprisingly, more recent data are revealing contributions from both common and rare variants to complex phenotypes. Thus, although common SNPs can explain a greater proportion of the heritability than was initially appreciated [633], support for a role for rare variants has also been accumulating from studies of rare SNPs [634,635] and rare CNVs [277,636]. This suggests that the genetic architecture of complex phenotypes is likely to comprise both common and rare variants.

## 6.5 GENERAL PRINCIPLES OF GENOTYPE–PHENOTYPE CORRELATIONS

Given knowledge of a specific clinical phenotype, to what extent can the underlying causal genotype be inferred? Conversely, given knowledge of a specific genotype, to what extent is it possible to infer the likely clinical phenotypic consequences (in terms, for example, of the penetrance, age of onset, and severity of the disease)? The study of the genotype–phenotype relationship is essentially an exploration of the actual correspondence between the genotype and the phenotype where any particular genotype usually corresponds to multiple phenotypes whereas many different genotypes can often correspond to a given phenotype. Several general principles have emerged as a result of the intensive study of causative variants in genetic disorders. The following discussion highlights some of these principles. The reader is encouraged to use the Online Mendelian Inheritance in Man (OMIM) at http://www3.ncbi.nlm.nih.gov/Omim for further information or for specific genes and clinical phenotypes. It is likely that the phenotypic consequences of a given variant will depend on other genetic variants present in the same gene or in the same genome [637]. The review of Wolf [638] provides an excellent guide to the complex issues inherent in the study of the relationship between the mutant genotype and the clinical phenotype.

Cooper et al. reviewed the different proposed molecular mechanisms of penetrance. This could be a function of the specific mutation(s) involved or of allele dosage, differential allelic expression, CNV or the modulating influence of additional genetic variants in *cis* or in *trans* [639].

### 6.5.1 Variants in the Same Gene May Be Responsible for More than One Disorder

There are many examples to illustrate the principle that variants in a single gene can cause different and distinct clinical phenotypes ("allelic heterogeneity"). Historically, the first example is that of the β-globin (*HBB*) gene on 11pter. Mutations of this gene cause β-thalassemia, sickle cell disease, and methemoglobinemia. The *L1CAM* gene on Xq28 has been shown to be mutated in hydrocephalus and stenosis of aqueduct of Sylvius, MASA syndrome (*m*ental retardation, *a*phasia, *s*huffing gait, and *a*dducted thumbs), and spastic paraplegia 1. The *COL1A2* gene on 7q21–q22 is involved in four different clinical forms of osteogenesis imperfecta (types II, III, IV, and atypical) as well as Ehlers–Danlos syndrome type VII B. The fibroblast growth factor receptor 2 (*FGFR2*) gene is mutated in three different craniosynostosis syndromes, namely Pfeiffer, Crouzon, and Jackson–Weiss. The*COL2A1* gene is implicated in Stickler syndrome type 1, SED congenita, Kneist dysplasia, achondrogenesis-hypochondrogenesis type 2, precocious osteoarthritis, Wagner syndrome type 2, and SMED Strudwick type. In a survey of 1014 genes causing disorders in OMIM, 165 genes were associated with two disorders, 52 genes with three disorders, 24 genes with four disorders, and 19 genes with five or more disorders [80].

### 6.5.2 One Disorder May Be Caused by Variants in More than One Gene

There are a plethora of similar clinical phenotypes due to mutations in different genes. This observation, also known as "nonallelic" or "locus" heterogeneity, is well understood thanks to linkage analyses for genetic disorders and the search for mutations in different genes. Thus, tuberous sclerosis, a relatively common autosomal dominant disorder, is caused by lesions in at least two different loci: *TSC1* on 9q34 and *TSC2* on 16p13.3. Approximately 60% of TSC families show linkage to the *TSC2* locus and 40% to the *TSC1* locus. HNPCC has been associated with pathogenic variants in five different genes: *MLH1* on 3p, *MSH2* on 2p16, *PMS1* on 2q31–q33, *PMS2* on 7p22, and *MSH6* on 2p16. Retinitis pigmentosa has so far been associated with a total of 23 different genes and the list is still growing. We expect that disorders of complex or polygenic phenotypes, such as hypertension, atherosclerosis, diabetes, schizophrenia, and manic-depressive illness, will be associated with a considerable number of genes scattered throughout the genome.

### 6.5.3 One and the Same Variant May Give Rise to Different Clinical Phenotypes ("Polypheny")

The clinical phenotype does not only depend on the one variant in the responsible gene; it can be modified by the action of any of the other ~20,000 protein coding genes and other functional genomic elements in the genome. The environment can also play an important role in the full development of the clinical phenotype. The classic sickle cell disease variant in the β-globin (*HBB*) gene (Glu6Val) may be associated with severe or mild sickle cell disease. The amelioration of the severe clinical phenotype in this case can be attributed to the increased expression of γ-globin genes and the presence of high levels of HbF. The genomic environment of the β-globin gene cluster may, therefore, modify the severity of sickle cell disease as may genetic variation originating from other loci, for example, the α-globin genes [112]. Another example of this phenomenon has recently been provided by studies of certain craniosynostoses. Both Pfeiffer and Crouzon syndromes can be associated with the same C342Y or C342R variants in the *FGFR2* gene.

The clinical phenotype associated with the D178N missense variant in the prion protein (*PRNP*) gene is critically dependent on the presence of the Met or Val129 polymorphic allele to which it is coupled. When D178N lies in *cis* to the Met129 allele, fatal familial insomnia (FFI) results, whereas D178N coupled to the Val129 allele is associated with Creutzfeldt–Jakob disease [640]. The Met/Val129 polymorphism also exerts an effect in *trans* through the normal allele because FFI is more severe and of longer duration in patients homozygous for either the Met or the Val allele.

One of the best examples of the contribution of the environment to the clinical phenotype of single gene disorders is that of phenylketonuria due to PAH deficiency. Individual homozygous or compound heterozygous for pathogenic variants in the *PAH* gene develop severe mental handicap if fed a normal diet. However, the cognitive status remains normal if these individuals are fed with a special, "phenylalanine-free" diet.

### 6.5.4 Variants in More than One Gene May Be Required for a Given Clinical Phenotype (Digenic Inheritance; Triallelic Inheritance)

Digenic inheritance refers to clinical phenotypes caused by the coinheritance of variants in two unlinked genes. Thus, one form of retinitis pigmentosa is due to the coinheritance of variants in the *RDS* gene on 6p and the *ROM* gene on 11q [641]. Individuals with either one or the other variant do not suffer from the disease. In similar vein, digenic inheritance of variants in the *MITF* and *TYR* genes has been reported as a cause of Waardenburg

[19] Shendure J, Akey JM. The origins, determinants, and consequences of human mutations. Science 2015;349:1478–83.
[20] Vogel F, Motulsky A. Human genetics. Verlag Berlin: Springer; 1986.
[21] Antonarakis SE, Kazazian Jr HH, Orkin SH. DNA polymorphism and molecular pathology of the human globin gene clusters. Hum Genet 1985;69:1–14.
[22] Cooper DN, Smith BA, Cooke HJ, Niemann S, Schmidtke J. An estimate of unique DNA sequence heterozygosity in the human genome. Hum Genet 1985;69:201–5.
[23] Nickerson DA, Taylor SL, Weiss KM, Clark AG, Hutchinson RG, Stengard J, Salomaa V, Vartiainen E, Boerwinkle E, Sing CF. DNA sequence diversity in a 9.7-kb region of the human lipoprotein lipase gene. Nat Genet 1998;19:233–40.
[24] Sachidanandam R, Weissman D, Schmidt SC, Kakol JM, Stein LD, Marth G, Sherry S, Mullikin JC, Mortimore BJ, Willey DL, Hunt SE, Cole CG, Coggill PC, Rice CM, Ning Z, Rogers J, Bentley DR, Kwok PY, Mardis ER, Yeh RT, Schultz B, Cook L, Davenport R, Dante M, Fulton L, Hillier L, Waterston RH, McPherson JD, Gilman B, Schaffner S, Van Etten WJ, Reich D, Higgins J, Daly MJ, Blumenstiel B, Baldwin J, Stange-Thomann N, Zody MC, Linton L, Lander ES, Altshuler D. A map of human genome sequence variation containing 1.42 million single nucleotide polymorphisms. Nature 2001;409:928–33.
[25] Wang DG, Fan JB, Siao CJ, Berno A, Young P, Sapolsky R, Ghandour G, Perkins N, Winchester E, Spencer J, Kruglyak L, Stein L, Hsie L, Topaloglou T, Hubbell E, Robinson E, Mittmann M, Morris MS, Shen N, Kilburn D, Rioux J, Nusbaum C, Rozen S, Hudson TJ, Lipshutz R, Chee M, Lander ES. Large-scale identification, mapping, and genotyping of single-nucleotide polymorphisms in the human genome. Science 1998;280:1077–82.
[26] Orkin SH, Kazazian Jr HH, Antonarakis SE, Ostrer H, Goff SC, Sexton JP. Abnormal RNA processing due to the exon mutation of beta E-globin gene. Nature 1982;300:768–9.
[27] Crawford DC, Akey DT, Nickerson DA. The patterns of natural variation in human genes. Annu Rev Genomics Hum Genet 2005;6:287–312.
[28] Nishihara S, Narimatsu H, Iwasaki H, Yazawa S, Akamatsu S, Ando T, Seno T, Narimatsu I. Molecular genetic analysis of the human Lewis histo-blood group system. J Biol Chem 1994;269:29271–8.
[29] Kelly RJ, Rouquier S, Giorgi D, Lennon GG, Lowe JB. Sequence and expression of a candidate for the human Secretor blood group alpha(1,2)fucosyltransferase gene (FUT2). Homozygosity for an enzyme-inactivating nonsense mutation commonly correlates with the non-secretor phenotype. J Biol Chem 1995;270:4640–9.
[30] Rebbeck TR. Molecular epidemiology of the human glutathione S-transferase genotypes GSTM1 and GSTT1 in cancer susceptibility. Cancer Epidemiol Biomark Prev 1997;6:733–43.
[31] Thein SL. Genetic insights into the clinical diversity of beta thalassaemia. Br J Haematol 2004;124:264–74.
[32] Green H, Djian P. Consecutive actions of different gene-altering mechanisms in the evolution of involucrin. Mol Biol Evol 1992;9:977–1017.
[33] Dawson SJ, Wiman B, Hamsten A, Green F, Humphries S, Henney AM. The two allele sequences of a common polymorphism in the promoter of the plasminogen activator inhibitor-1 (PAI-1) gene respond differently to interleukin-1 in HepG2 cells. J Biol Chem 1993;268:10739–45.
[34] Fullerton SM, Clark AG, Weiss KM, Nickerson DA, Taylor SL, Stengard JH, Salomaa V, Vartiainen E, Perola M, Boerwinkle E, Sing CF. Apolipoprotein E variation at the sequence haplotype level: implications for the origin and maintenance of a major human polymorphism. Am J Hum Genet 2000;67:881–900.
[35] Rigat B, Hubert C, Alhenc-Gelas F, Cambien F, Corvol P, Soubrier F. An insertion/deletion polymorphism in the angiotensin I-converting enzyme gene accounting for half the variance of serum enzyme levels. J Clin Invest 1990;86:1343–6.
[36] Small K, Iber J, Warren ST. Emerin deletion reveals a common X-chromosome inversion mediated by inverted repeats. Nat Genet 1997;16:96–9.
[37] Neitz M, Neitz J, Grishok A. Polymorphism in the number of genes encoding long-wavelength-sensitive cone pigments among males with normal color vision. Vision Res 1995;35:2395–407.
[38] Ng PC, Henikoff S. Predicting the effects of amino acid substitutions on protein function. Annu Rev Genomics Hum Genet 2006;7:61–80.
[39] Pastinen T, Ge B, Hudson TJ. Influence of human genome polymorphism on gene expression. Hum Mol Genet 2006;15(Spec No 1):R9–16.
[40] ElSharawy A, Hundrieser B, Brosch M, Wittig M, Huse K, Platzer M, Becker A, Simon M, Rosenstiel P, Schreiber S, Krawczak M, Hampe J. Systematic evaluation of the effect of common SNPs on pre-mRNA splicing. Hum Mutat 2009;30:625–32.
[41] Altshuler D, Brooks LD, Chakravarti A, Collins FS, Daly MJ, Donnelly P. A haplotype map of the human genome. Nature 2005;437:1299–320.
[42] Consortium TIH. The International HapMap project. Nature 2003;426:789–96.
[43] Hinds DA, Stuve LL, Nilsen GB, Halperin E, Eskin E, Ballinger DG, Frazer KA, Cox DR. Whole-genome patterns of common DNA variation in three human populations. Science 2005;307:1072–9.

[44] 1000 Genomes Project Consortium. A map of human genome variation from population-scale sequencing. Nature 2010;467:1061–73.
[45] Bamshad MJ, Ng SB, Bigham AW, Tabor HK, Emond MJ, Nickerson DA, Shendure J. Exome sequencing as a tool for Mendelian disease gene discovery. Nat Rev Genet 2011;12:745–55.
[46] Narasimhan VM, Hunt KA, Mason D, Baker CL, Karczewski KJ, Barnes MR, Barnett AH, Bates C, Bellary S, Bockett NA, Giorda K, Griffiths CJ, Hemingway H, Jia Z, Kelly MA, Khawaja HA, Lek M, McCarthy S, McEachan R, O'Donnell-Luria A, Paigen K, Parisinos CA, Sheridan E, Southgate L, Tee L, Thomas M, Xue Y, Schnall-Levin M, Petkov PM, Tyler-Smith C, Maher ER, Trembath RC, MacArthur DG, Wright J, Durbin R, van Heel DA. Health and population effects of rare gene knockouts in adult humans with related parents. Science 2016;352:474–7.
[47] Xue Y, Chen Y, Ayub Q, Huang N, Ball EV, Mort M, Phillips AD, Shaw K, Stenson PD, Cooper DN, Tyler-Smith C, Genomes Project C. Deleterious- and disease-allele prevalence in healthy individuals: insights from current predictions, mutation databases, and population-scale resequencing. Am J Hum Genet 2012;91:1022–32.
[48] Jeffreys AJ, Wilson V, Thein SL. Hypervariable 'minisatellite' regions in human DNA. Nature 1985;314:67–73.
[49] Wyman AR, White R. A highly polymorphic locus in human DNA. Proc Natl Acad Sci U S A 1980;77:6754–8.
[50] Jeffreys AJ, Neumann R, Wilson V. Repeat unit sequence variation in minisatellites: a novel source of DNA polymorphism for studying variation and mutation by single molecule analysis. Cell 1990;60:473–85.
[51] Saiki RK, Scharf S, Faloona F, Mullis KB, Horn GT, Erlich HA, Arnheim N. Enzymatic amplification of beta-globin genomic sequences and restriction site analysis for diagnosis of sickle cell anemia. Science 1985;230:1350–4.
[52] Litt M, Luty JA. A hypervariable microsatellite revealed by in vitro amplification of a dinucleotide repeat within the cardiac muscle actin gene. Am J Hum Genet 1989;44:397–401.
[53] Weber JL, May PE. Abundant class of human DNA polymorphisms which can be typed using the polymerase chain reaction. Am J Hum Genet 1989;44:388–96.
[54] Economou EP, Bergen AW, Warren AC, Antonarakis SE. The polydeoxyadenylate tract of Alu repetitive elements is polymorphic in the human genome. Proc Natl Acad Sci U S A 1990;87:2951–4.
[55] Anagnou NP, O'Brien SJ, Shimada T, Nash WG, Chen MJ, Nienhuis AW. Chromosomal organization of the human dihydrofolate reductase genes: dispersion, selective amplification, and a novel form of polymorphism. Proc Natl Acad Sci U S A 1984;81:5170–4.
[56] Cooper DN. Human gene evolution. Oxford: Bios Scientific; 1999.
[57] Buckland PR. Polymorphically duplicated genes: their relevance to phenotypic variation in humans. Ann Med 2003;35:308–15.
[58] Iafrate AJ, Feuk L, Rivera MN, Listewnik ML, Donahoe PK, Qi Y, Scherer SW, Lee C. Detection of large-scale variation in the human genome. Nat Genet 2004;36: 949–51.
[59] Sebat J, Lakshmi B, Troge J, Alexander J, Young J, Lundin P, Maner S, Massa H, Walker M, Chi M, Navin N, Lucito R, Healy J, Hicks J, Ye K, Reiner A, Gilliam TC, Trask B, Patterson N, Zetterberg A, Wigler M. Large-scale copy number polymorphism in the human genome. Science 2004;305:525–8.
[60] Sharp AJ, Locke DP, McGrath SD, Cheng Z, Bailey JA, Vallente RU, Pertz LM, Clark RA, Schwartz S, Segraves R, Oseroff VV, Albertson DG, Pinkel D, Eichler EE. Segmental duplications and copy-number variation in the human genome. Am J Hum Genet 2005;77:78–88.
[61] Redon R, Ishikawa S, Fitch KR, Feuk L, Perry GH, Andrews TD, Fiegler H, Shapero MH, Carson AR, Chen W, Cho EK, Dallaire S, Freeman JL, Gonzalez JR, Gratacos M, Huang J, Kalaitzopoulos D, Komura D, MacDonald JR, Marshall CR, Mei R, Montgomery L, Nishimura K, Okamura K, Shen F, Somerville MJ, Tchinda J, Valsesia A, Woodwark C, Yang F, Zhang J, Zerjal T, Zhang J, Armengol L, Conrad DF, Estivill X, Tyler-Smith C, Carter NP, Aburatani H, Lee C, Jones KW, Scherer SW, Hurles ME. Global variation in copy number in the human genome. Nature 2006;444:444–54.
[62] Conrad DF, Pinto D, Redon R, Feuk L, Gokcumen O, Zhang Y, Aerts J, Andrews TD, Barnes C, Campbell P, Fitzgerald T, Hu M, Ihm CH, Kristiansson K, Macarthur DG, Macdonald JR, Onyiah I, Pang AW, Robson S, Stirrups K, Valsesia A, Walter K, Wei J, Tyler-Smith C, Carter NP, Lee C, Scherer SW, Hurles ME. Origins and functional impact of copy number variation in the human genome. Nature 2010;464:704–12.
[63] Mills RE, Walter K, Stewart C, Handsaker RE, Chen K, Alkan C, Abyzov A, Yoon SC, Ye K, Cheetham RK, Chinwalla A, Conrad DF, Fu Y, Grubert F, Hajirasouliha I, Hormozdiari F, Iakoucheva LM, Iqbal Z, Kang S, Kidd JM, Konkel MK, Korn J, Khurana E, Kural D, Lam HY, Leng J, Li R, Li Y, Lin CY, Luo R, Mu XJ, Nemesh J, Peckham HE, Rausch T, Scally A, Shi X, Stromberg MP, Stutz AM, Urban AE, Walker JA, Wu J, Zhang Y, Zhang ZD, Batzer MA, Ding L, Marth GT, McVean G, Sebat J, Snyder M, Wang J, Eichler EE, Gerstein MB, Hurles ME, Lee C, McCarroll SA, Korbel JO. Mapping copy number variation by population-scale genome sequencing. Nature 2011;470:59–65.

[64] Kidd JM, Graves T, Newman TL, Fulton R, Hayden HS, Malig M, Kallicki J, Kaul R, Wilson RK, Eichler EE. A human genome structural variation sequencing resource reveals insights into mutational mechanisms. Cell 2010;143:837–47.
[65] Balasubramanian S, Habegger L, Frankish A, MacArthur DG, Harte R, Tyler-Smith C, Harrow J, Gerstein M. Gene inactivation and its implications for annotation in the era of personal genomics. Genes Dev 2011;25:1–10.
[66] Dear PH. Copy-number variation: the end of the human genome? Trends Biotechnol 2009;27:448–54.
[67] Conrad DF, Andrews TD, Carter NP, Hurles ME, Pritchard JK. A high-resolution survey of deletion polymorphism in the human genome. Nat Genet 2006;38:75–81.
[68] McCarroll SA, Hadnott TN, Perry GH, Sabeti PC, Zody MC, Barrett JC, Dallaire S, Gabriel SB, Lee C, Daly MJ, Altshuler DM. Common deletion polymorphisms in the human genome. Nat Genet 2006;38:86–92.
[69] Hinds DA, Kloek AP, Jen M, Chen X, Frazer KA. Common deletions and SNPs are in linkage disequilibrium in the human genome. Nat Genet 2006;38:82–5.
[70] MacArthur DG, Tyler-Smith C. Loss-of-function variants in the genomes of healthy humans. Hum Mol Genet 2010;19:R125–30.
[71] Waalen J, Beutler E. Genetic screening for low-penetrance variants in protein-coding genes. Annu Rev Genomics Hum Genet 2009;10:431–50.
[72] Ng PC, Levy S, Huang J, Stockwell TB, Walenz BP, Li K, Axelrod N, Busam DA, Strausberg RL, Venter JC. Genetic variation in an individual human exome. PLoS Genet 2008;4:e1000160.
[73] Yngvadottir B, Xue Y, Searle S, Hunt S, Delgado M, Morrison J, Whittaker P, Deloukas P, Tyler-Smith C. A genome-wide survey of the prevalence and evolutionary forces acting on human nonsense SNPs. Am J Hum Genet 2009;84:224–34.
[74] Yamaguchi-Kabata Y, Shimada MK, Hayakawa Y, Minoshima S, Chakraborty R, Gojobori T, Imanishi T. Distribution and effects of nonsense polymorphisms in human genes. PLoS One 2008;3:e3393.
[75] Hsiao TL, Vitkup D. Role of duplicate genes in robustness against deleterious human mutations. PLoS Genet 2008;4:e1000014.
[76] Cooper DN, Mort M, Stenson PD, Ball EV, Chuzhanova NA. Methylation-mediated deamination of 5-methylcytosine appears to give rise to mutations causing human inherited disease in CpNpG trinucleotides, as well as in CpG dinucleotides. Hum Genomics 2010;4:406–10.
[77] Krawczak M, Chuzhanova NA, Stenson PD, Johansen BN, Ball EV, Cooper DN. Changes in primary DNA sequence complexity influence the phenotypic consequences of mutations in human gene regulatory regions. Hum Genet 2000;107:362–5.
[78] Stenson PD, Ball EV, Mort M, Phillips AD, Shiel JA, Thomas NS, Abeysinghe S, Krawczak M, Cooper DN. Human gene mutation database (HGMD): 2003 update. Hum Mutat 2003;21:577–81.
[79] Stenson PD, Mort M, Ball EV, Evans K, Hayden M, Heywood S, Hussain M, Phillips AD, Cooper DN. The Human Gene Mutation Database: towards a comprehensive repository of inherited mutation data for medical research, genetic diagnosis and next-generation sequencing studies. Hum Genet 2017;136:665–77.
[80] Antonarakis SE, McKusick VA. OMIM passes the 1,000-disease-gene mark. Nat Genet 2000;25:11.
[81] Loeb LA, Kunkel TA. Fidelity of DNA synthesis. Annu Rev Biochem 1982;51:429–57.
[82] Krawczak M, Ball EV, Cooper DN. Neighboring-nucleotide effects on the rates of germ-line single-base-pair substitution in human genes. Am J Hum Genet 1998;63:474–88.
[83] Antonarakis SE, Krawczak M, Cooper DN. The nature and mechanisms of human gene mutation. In: Scriver CR, Beaudet AL, Valle D, et al., editors. The metabolic and molecular bases of inherited disease. New York: McGraw-Hill; 2001. p. 343–77.
[84] Youssoufian H, Antonarakis SE, Bell W, Griffin AM, Kazazian Jr HH. Nonsense and missense mutations in hemophilia A: estimate of the relative mutation rate at CG dinucleotides. Am J Hum Genet 1988;42:718–25.
[85] Youssoufian H, Kazazian Jr HH, Phillips DG, Aronis S, Tsiftis G, Brown VA, Antonarakis SE. Recurrent mutations in haemophilia A give evidence for CpG mutation hotspots. Nature 1986;324:380–2.
[86] Cooper DN, Youssoufian H. The CpG dinucleotide and human genetic disease. Hum Genet 1988;78:151–5.
[87] Antonarakis SE, Kazazian HH, Tuddenham EG. Molecular etiology of factor VIII deficiency in hemophilia A. Hum Mutat 1995;5:1–22.
[88] Cooper DN. Eukaryotic DNA methylation. Hum Genet 1983;64:315–33.
[89] Driscoll DJ, Migeon BR. Sex difference in methylation of single-copy genes in human meiotic germ cells: implications for X chromosome inactivation, parental imprinting, and origin of CpG mutations. Somat Cell Mol Genet 1990;16:267–82.
[90] Hurst LD, Ellegren H. Sex biases in the mutation rate. Trends Genet 1998;14:446–52.
[91] Lister R, Pelizzola M, Dowen RH, Hawkins RD, Hon G, Tonti-Filippini J, Nery JR, Lee L, Ye Z, Ngo QM, Edsall L, Antosiewicz-Bourget J, Stewart R, Ruotti V, Millar AH, Thomson JA, Ren B, Ecker JR. Human DNA methylomes at base resolution show widespread epigenomic differences. Nature 2009;462:315–22.

[92] Rodenhiser DI, Andrews JD, Mancini DN, Jung JH, Singh SM. Homonucleotide tracts, short repeats and CpG/CpNpG motifs are frequent sites for heterogeneous mutations in the neurofibromatosis type 1 (NF1) tumour-suppressor gene. Mutat Res 1997;373:185–95.
[93] Cheung LW, Lee YF, Ng TW, Ching WK, Khoo US, Ng MK, Wong AS. CpG/CpNpG motifs in the coding region are preferred sites for mutagenesis in the breast cancer susceptibility genes. FEBS Lett 2007;581:4668–74.
[94] Kondrashov AS. Direct estimates of human per nucleotide mutation rates at 20 loci causing Mendelian diseases. Hum Mutat 2003;21:12–27.
[95] Kryukov GV, Pennacchio LA, Sunyaev SR. Most rare missense alleles are deleterious in humans: implications for complex disease and association studies. Am J Hum Genet 2007;80:727–39.
[96] Boyko AR, Williamson SH, Indap AR, Degenhardt JD, Hernandez RD, Lohmueller KE, Adams MD, Schmidt S, Sninsky JJ, Sunyaev SR, White TJ, Nielsen R, Clark AG, Bustamante CD. Assessing the evolutionary impact of amino acid mutations in the human genome. PLoS Genet 2008;4:e1000083.
[97] Eyre-Walker A, Woolfit M, Phelps T. The distribution of fitness effects of new deleterious amino acid mutations in humans. Genetics 2006;173:891–900.
[98] Bacolla A, Wang G, Jain A, Chuzhanova NA, Cer RZ, Collins JR, Cooper DN, Bohr VA, Vasquez KM. Non-B DNA-forming sequences and WRN deficiency independently increase the frequency of base substitution in human cells. J Biol Chem 2011.
[99] Cartegni L, Chew SL, Krainer AR. Listening to silence and understanding nonsense: exonic mutations that affect splicing. Nat Rev Genet 2002;3:285–98.
[100] Gorlov IP, Kimmel M, Amos CI. Strength of the purifying selection against different categories of the point mutations in the coding regions of the human genome. Hum Mol Genet 2006;15:1143–50.
[101] Hunt R, Sauna ZE, Ambudkar SV, Gottesman MM, Kimchi-Sarfaty C. Silent (synonymous) SNPs: should we care about them? Methods Mol Biol 2009;578:23–39.
[102] Sanford JR, Wang X, Mort M, Vanduyn N, Cooper DN, Mooney SD, Edenberg HJ, Liu Y. Splicing factor SFRS1 recognizes a functionally diverse landscape of RNA transcripts. Genome Res 2009;19:381–94.
[103] Sauna ZE, Okunji C, Hunt RC, Gupta T, Allen CE, Plum E, Blaisdell A, Grigoryan V, Geetha S, Fathke R, Soejima K, Kimchi-Sarfaty C. Characterization of conformation-sensitive antibodies to ADAMTS13, the von Willebrand cleavage protease. PLoS One 2009;4:e6506.
[104] Wang GS, Cooper TA. Splicing in disease: disruption of the splicing code and the decoding machinery. Nat Rev Genet 2007;8:749–61.
[105] Nackley AG, Shabalina SA, Lambert JE, Conrad MS, Gibson DG, Spiridonov AN, Satterfield SK, Diatchenko L. Low enzymatic activity haplotypes of the human catechol-O-methyltransferase gene: enrichment for marker SNPs. PLoS One 2009;4:e5237.
[106] Nackley AG, Shabalina SA, Tchivileva IE, Satterfield K, Korchynskyi O, Makarov SS, Maixner W, Diatchenko L. Human catechol-O-methyltransferase haplotypes modulate protein expression by altering mRNA secondary structure. Science 2006;314:1930–3.
[107] Kimchi-Sarfaty C, Oh JM, Kim IW, Sauna ZE, Calcagno AM, Ambudkar SV, Gottesman MM. A "silent" polymorphism in the MDR1 gene changes substrate specificity. Science 2007;315:525–8.
[108] Tsai CJ, Sauna ZE, Kimchi-Sarfaty C, Ambudkar SV, Gottesman MM, Nussinov R. Synonymous mutations and ribosome stalling can lead to altered folding pathways and distinct minima. J Mol Biol 2008;383:281–91.
[109] Goode DL, Cooper GM, Schmutz J, Dickson M, Gonzales E, Tsai M, Karra K, Davydov E, Batzoglou S, Myers RM, Sidow A. Evolutionary constraint facilitates interpretation of genetic variation in resequenced human genomes. Genome Res 2010;20:301–10.
[110] Efstratiadis A, Posakony JW, Maniatis T, Lawn RM, O'Connell C, Spritz RA, DeRiel JK, Forget BG, Weissman SM, Slightom JL, Blechl AE, Smithies O, Baralle FE, Shoulders CC, Proudfoot NJ. The structure and evolution of the human beta-globin gene family. Cell 1980;21:653–68.
[111] Kunkel TA. The mutational specificity of DNA polymerases-alpha and -gamma during in vitro DNA synthesis. J Biol Chem 1985;260:12866–74.
[112] Cooper DNaKM. Human gene mutation. Oxford: Bios Scientific; 1993.
[113] Schmucker B, Krawczak M. Meiotic microdeletion breakpoints in the BRCA1 gene are significantly associated with symmetric DNA-sequence elements. Am J Hum Genet 1997;61:1454–6.
[114] Krawczak M, Cooper DN. Gene deletions causing human genetic disease: mechanisms of mutagenesis and the role of the local DNA sequence environment. Hum Genet 1991;86:425–41.
[115] Cooper DN, Krawczak M. Mechanisms of insertional mutagenesis in human genes causing genetic disease. Hum Genet 1991;87:409–15.
[116] Ball EV, Stenson PD, Abeysinghe SS, Krawczak M, Cooper DN, Chuzhanova NA. Microdeletions and microinsertions causing human genetic disease: common mechanisms of mutagenesis and the role of local DNA sequence complexity. Hum Mutat 2005;26:205–13.
[117] Kondrashov AS, Rogozin IB. Context of deletions and insertions in human coding sequences. Hum Mutat 2004;23:177–85.

[118] Kamat MA, Bacolla A, Cooper DN, Chuzhanova N. A role for non-B DNA forming sequences in mediating microlesions causing human inherited disease. Hum Mutat 2016;37:65–73.
[119] Caskey CT, Pizzuti A, Fu YH, Fenwick Jr RG, Nelson DL. Triplet repeat mutations in human disease. Science 1992;256:784–9.
[120] Mandel JL. Questions of expansion. Nat Genet 1993;4:8–9.
[121] Rousseau F., Heitz D., Mandel J.L.. The unstable and methylatable mutations causing the fragile X syndrome. Hum Mutat 1992;1:91–6.
[122] Harper PS, Harley HG, Reardon W, Shaw DJ. Anticipation in myotonic dystrophy: new light on an old problem. Am J Hum Genet 1992;51:10–6.
[123] Van Esch H. The Fragile X premutation: new insights and clinical consequences. Eur J Med Genet 2006;49:1–8.
[124] Housman D. Gain of glutamines, gain of function? Nat Genet 1995;10:3–4.
[125] Liquori CL, Ricker K, Moseley ML, Jacobsen JF, Kress W, Naylor SL, Day JW, Ranum LP. Myotonic dystrophy type 2 caused by a CCTG expansion in intron 1 of ZNF9. Science 2001;293:864–7.
[126] Savkur RS, Philips AV, Cooper TA, Dalton JC, Moseley ML, Ranum LP, Day JW. Insulin receptor splicing alteration in myotonic dystrophy type 2. Am J Hum Genet 2004;74:1309–13.
[127] Fu Y-H, Kuhl D, Pizzuti A, Pieretti M, Sutcliffe JS, Richards CS, Verkerk AJMH, Holden J, Fenwick RJ, Warren ST, Oostra BA, Nelson DL, Caskey CT. Variation of the CGG repeat at the Fragile X site results in genetic instability: resolution of the Sherman paradox. Cell 1991;67:1047–58.
[128] Conway GS, Hettiarachchi S, Murray A, Jacobs PA. Fragile X premutations in familial premature ovarian failure. Lancet 1995;346:309–10.
[129] Jacquemont S, Hagerman RJ, Leehey M, Grigsby J, Zhang L, Brunberg JA, Greco C, Des Portes V, Jardini T, Levine R, Berry-Kravis E, Brown WT, Schaeffer S, Kissel J, Tassone F, Hagerman PJ. Fragile X premutation tremor/ataxia syndrome: molecular, clinical, and neuroimaging correlates. Am J Hum Genet 2003;72:869–78.
[130] Kang S., Ohshima K., Jaworski A., Wells R.D.. CTG triplet repeats from the myotonic dystrophy gene are expanded in *Escherichia coli* distal to the replication origin as a single large event. J Mol Biol 1996;258:543–7.
[131] Chung MY, Ranum LP, Duvick LA, Servadio A, Zoghbi HY, Orr HT. Evidence for a mechanism predisposing to intergenerational CAG repeat instability in spinocerebellar ataxia type I. Nat Genet 1993;5:254–8.
[132] Wells RD, Warren AC. Genetic instabilities and hereditary neurological disorders. San Diego: Academic Press; 1998.
[133] Brais B, Bouchard JP, Xie YG, Rochefort DL, Chretien N, Tome FM, Lafreniere RG, Rommens JM, Uyama E, Nohira O, Blumen S, Korczyn AD, Heutink P, Mathieu J, Duranceau A, Codere F, Fardeau M, Rouleau GA. Short GCG expansions in the PABP2 gene cause oculopharyngeal muscular dystrophy. Nat Genet 1998;18:164–7.
[134] Muragaki Y, Mundlos S, Upton J, Olsen BR. Altered growth and branching patterns in synpolydactyly caused by mutations in HOXD13. Science 1996;272:548–51.
[135] Pearson C.E., Nichol Edamura K., Cleary J.D.. Repeat instability: mechanisms of dynamic mutations. Nat Rev Genet 2005;6:729–42.
[136] Lalioti MD, Scott HS, Buresi C, Rossier C, Bottani A, Morris MA, Malafosse A, Antonarakis SE. Dodecamer repeat expansion in cystatin B gene in progressive myoclonus epilepsy. Nature 1997;386:847–51.
[137] Lalioti MD, Scott HS, Antonarakis SE. Altered spacing of promoter elements due to the dodecamer repeat expansion contributes to reduced expression of the cystatin B gene in EPM1. Hum Mol Genet 1999;8:1791–8.
[138] Matsuura T, Yamagata T, Burgess DL, Rasmussen A, Grewal RP, Watase K, Khajavi M, McCall AE, Davis CF, Zu L, Achari M, Pulst SM, Alonso E, Noebels JL, Nelson DL, Zoghbi HY, Ashizawa T. Large expansion of the ATTCT pentanucleotide repeat in spinocerebellar ataxia type 10. Nat Genet 2000;26:191–4.
[139] Verpy E., Leibovici M., Zwaenepoel I., Liu X.Z., Gal A., Salem N, Mansour A, Blanchard S, Kobayashi I, Keats BJ, Slim R, Petit C. A defect in harmonin, a PDZ domain-containing protein expressed in the inner ear sensory hair cells, underlies Usher syndrome type 1C. Nat Genet 2000;26:51–5.
[140] Yu S, Mangelsdorf M, Hewett D, Hobson L, Baker E, Eyre HJ, Lapsys N, Le Paslier D, Doggett NA, Sutherland GR, Richards RI. Human chromosomal fragile site FRA16B is an amplified AT-rich minisatellite repeat. Cell 1997;88:367–74.
[141] Antonacci F, Kidd JM, Marques-Bonet T, Teague B, Ventura M, Girirajan S, Alkan C, Campbell CD, Vives L, Malig M, Rosenfeld JA, Ballif BC, Shaffer LG, Graves TA, Wilson RK, Schwartz DC, Eichler EE. A large and complex structural polymorphism at 16p12.1 underlies microdeletion disease risk. Nat Genet 2010;42:745–50.
[142] Ciccone R, Mattina T, Giorda R, Bonaglia MC, Rocchi M, Pramparo T, Zuffardi O. Inversion polymorphisms

and non-contiguous terminal deletions: the cause and the (unpredicted) effect of our genome architecture. J Med Genet 2006;43:e19.
[143] Gimelli G, Pujana MA, Patricelli MG, Russo S, Giardino D, Larizza L, Cheung J, Armengol L, Schinzel A, Estivill X, Zuffardi O. Genomic inversions of human chromosome 15q11-q13 in mothers of Angelman syndrome patients with class II (BP2/3) deletions. Hum Mol Genet 2003;12:849–58.
[144] Hobart HH, Morris CA, Mervis CB, Pani AM, Kistler DJ, Rios CM, Kimberley KW, Gregg RG, Bray-Ward P. Inversion of the Williams syndrome region is a common polymorphism found more frequently in parents of children with Williams syndrome. Am J Med Genet C Semin Med Genet 2010;154C:220–8.
[145] Visser R, Shimokawa O, Harada N, Kinoshita A, Ohta T, Niikawa N, Matsumoto N. Identification of a 3.0-kb major recombination hotspot in patients with Sotos syndrome who carry a common 1.9-Mb microdeletion. Am J Hum Genet 2005;76:52–67.
[146] Donnelly MP, Paschou P, Grigorenko E, Gurwitz D, Mehdi SQ, Kajuna SL, Barta C, Kungulilo S, Karoma NJ, Lu RB, Zhukova OV, Kim JJ, Comas D, Siniscalco M, New M, Li P, Li H, Manolopoulos VG, Speed WC, Rajeevan H, Pakstis AJ, Kidd JR, Kidd KK. The distribution and most recent common ancestor of the 17q21 inversion in humans. Am J Hum Genet 2010;86:161–71.
[147] Rao PN, Li W, Vissers LE, Veltman JA, Ophoff RA. Recurrent inversion events at 17q21.31 microdeletion locus are linked to the MAPT H2 haplotype. Cytogenet Genome Res 2010;129:275–9.
[148] Liu P, Lacaria M, Zhang F, Withers M, Hastings PJ, Lupski JR. Frequency of nonallelic homologous recombination is correlated with length of homology: evidence that ectopic synapsis precedes ectopic crossing-over. Am J Hum Genet 2011;89:580–8.
[149] Ou Z, Stankiewicz P, Xia Z, Breman AM, Dawson B, Wiszniewska J, Szafranski P, Cooper ML, Rao M, Shao L, South ST, Coleman K, Fernhoff PM, Deray MJ, Rosengren S, Roeder ER, Enciso VB, Chinault AC, Patel A, Kang SH, Shaw CA, Lupski JR, Cheung SW. Observation and prediction of recurrent human translocations mediated by NAHR between nonhomologous chromosomes. Genome Res 2011;21:33–46.
[150] Kidd JM, Sampas N, Antonacci F, Graves T, Fulton R, Hayden HS, Alkan C, Malig M, Ventura M, Giannuzzi G, Kallicki J, Anderson P, Tsalenko A, Yamada NA, Tsang P, Kaul R, Wilson RK, Bruhn L, Eichler EE. Characterization of missing human genome sequences and copy-number polymorphic insertions. Nat Methods 2010;7:365–71.
[151] Ashley T, Gaeth AP, Inagaki H, Seftel A, Cohen MM, Anderson LK, Kurahashi H, Emanuel BS. Meiotic recombination and spatial proximity in the etiology of the recurrent t(11;22). Am J Hum Genet 2006;79:524–38.
[152] Edelmann L, Spiteri E, Koren K, Pulijaal V, Bialer MG, Shanske A, Goldberg R, Morrow BE. AT-rich palindromes mediate the constitutional t(11;22) translocation. Am J Hum Genet 2001;68:1–13.
[153] Kurahashi H, Inagaki H, Ohye T, Kogo H, Tsutsumi M, Kato T, Tong M, Emanuel BS. The constitutional t(11;22): implications for a novel mechanism responsible for gross chromosomal rearrangements. Clin Genet 2010;78:299–309.
[154] Soutoglou E, Dorn JF, Sengupta K, Jasin M, Nussenzweig A, Ried T, Danuser G, Misteli T. Positional stability of single double-strand breaks in mammalian cells. Nat Cell Biol 2007;9:675–82.
[155] Nikiforova MN, Stringer JR, Blough R, Medvedovic M, Fagin JA, Nikiforov YE. Proximity of chromosomal loci that participate in radiation-induced rearrangements in human cells. Science 2000;290:138–41.
[156] Meaburn KJ, Misteli T, Soutoglou E. Spatial genome organization in the formation of chromosomal translocations. Semin Cancer Biol 2007;17:80–90.
[157] Wijchers PJ, de Laat W. Genome organization influences partner selection for chromosomal rearrangements. Trends Genet 2011;27:63–71.
[158] Abeysinghe SS, Chuzhanova N, Krawczak M, Ball EV, Cooper DN. Translocation and gross deletion breakpoints in human inherited disease and cancer I: nucleotide composition and recombination-associated motifs. Hum Mutat 2003;22:229–44.
[159] Chuzhanova N, Abeysinghe SS, Krawczak M, Cooper DN. Translocation and gross deletion breakpoints in human inherited disease and cancer II: potential involvement of repetitive sequence elements in secondary structure formation between DNA ends. Hum Mutat 2003;22:245–51.
[160] Cooper DN, Bacolla A, Ferec C, Vasquez KM, Kehrer-Sawatzki H, Chen JM. On the sequence-directed nature of human gene mutation: the role of genomic architecture and the local DNA sequence environment in mediating gene mutations underlying human inherited disease. Hum Mutat 2011;32:1075–99.
[161] Chen JM, Chuzhanova N, Stenson PD, Ferec C, Cooper DN. Complex gene rearrangements caused by serial replication slippage. Hum Mutat 2005a;26:125–34.
[162] Chen JM, Chuzhanova N, Stenson PD, Ferec C, Cooper DN. Intrachromosomal serial replication slippage in trans gives rise to diverse genomic rearrangements involving inversions. Hum Mutat 2005b;26:362–73.

[163] Chen JM, Chuzhanova N, Stenson PD, Ferec C, Cooper DN. Meta-analysis of gross insertions causing human genetic disease: novel mutational mechanisms and the role of replication slippage. Hum Mutat 2005c;25:207–21.
[164] Lee JA, Carvalho CM, Lupski JR. A DNA replication mechanism for generating nonrecurrent rearrangements associated with genomic disorders. Cell 2007;131:1235–47.
[165] Hastings PJ, Ira G, Lupski JR. A microhomology-mediated break-induced replication model for the origin of human copy number variation. PLoS Genet 2009;5:e1000327.
[166] Chen JM, Cooper DN, Ferec C, Kehrer-Sawatzki H, Patrinos GP. Genomic rearrangements in inherited disease and cancer. Semin Cancer Biol 2010;20:222–33.
[167] Bauters M, Van Esch H, Friez MJ, Boespflug-Tanguy O, Zenker M, Vianna-Morgante AM, Rosenberg C, Ignatius J, Raynaud M, Hollanders K, Govaerts K, Vandenreijt K, Niel F, Blanc P, Stevenson RE, Fryns JP, Marynen P, Schwartz CE, Froyen G. Nonrecurrent MECP2 duplications mediated by genomic architecture-driven DNA breaks and break-induced replication repair. Genome Res 2008;18:847–58.
[168] Kulikowski LD, Yoshimoto M, da Silva Bellucco FT, Belangero SI, Christofolini DM, Pacanaro AN, Bortolai A, Smith Mde A, Squire JA, Melaragno MI. Cytogenetic molecular delineation of a terminal 18q deletion suggesting neo-telomere formation. Eur J Med Genet 2010;53:404–7.
[169] Zuffardi O, Bonaglia M, Ciccone R, Giorda R. Inverted duplications deletions: underdiagnosed rearrangements? Clin Genet 2009;75:505–13.
[170] Hannes F, Van Houdt J, Quarrell OW, Poot M, Hochstenbach R, Fryns JP, Vermeesch JR. Telomere healing following DNA polymerase arrest-induced breakages is likely the main mechanism generating chromosome 4p terminal deletions. Hum Mutat 2010;31:1343–51.
[171] Cha RS, Kleckner N. ATR homolog Mec1 promotes fork progression, thus averting breaks in replication slow zones. Science 2002;297:602–6.
[172] Pelletier R, Krasilnikova MM, Samadashwily GM, Lahue R, Mirkin SM. Replication and expansion of trinucleotide repeats in yeast. Mol Cell Biol 2003;23:1349–57.
[173] Koszul R, Caburet S, Dujon B, Fischer G. Eucaryotic genome evolution through the spontaneous duplication of large chromosomal segments. EMBO J 2004;23:234–43.
[174] Choi BO, Kim NK, Park SW, Hyun YS, Jeon HJ, Hwang JH, Chung KW. Inheritance of Charcot-Marie-Tooth disease 1A with rare nonrecurrent genomic rearrangement. Neurogenetics 2010.
[175] Zhang F, Khajavi M, Connolly AM, Towne CF, Batish SD, Lupski JR. The DNA replication FoSTeS/MMBIR mechanism can generate genomic, genic and exonic complex rearrangements in humans. Nat Genet 2009;41:849–53.
[176] Ankala A, Kohn JN, Hegde A, Meka A, Ephrem CL, Askree SH, Bhide S, Hegde MR. Aberrant firing of replication origins potentially explains intragenic nonrecurrent rearrangements within genes, including the human DMD gene. Genome Res 2011.
[177] Doksani Y, Bermejo R, Fiorani S, Haber JE, Foiani M. Replicon dynamics, dormant origin firing, and terminal fork integrity after double-strand break formation. Cell 2009;137:247–58.
[178] Carvalho CM, Lupski JR. Mechanisms underlying structural variant formation in genomic disorders. Nat Rev Genet 2016;17:224–38.
[179] Ballabio A, Carrozzo R, Parenti G, Gil A, Zollo M, Persico MG, Gillard E, Affara N, Yates J, Ferguson-Smith MA, et al. Molecular heterogeneity of steroid sulfatase deficiency: a multicenter study on 57 unrelated patients, at DNA and protein levels. Genomics 1989;4:36–40.
[180] den Dunnen JT, Bakker E, Breteler EG, Pearson PL, van Ommen GJ. Direct detection of more than 50% of the Duchenne muscular dystrophy mutations by field inversion gels. Nature 1987;329:640–2.
[181] Nicholls RD, Fischel-Ghodsian N, Higgs DR. Recombination at the human alpha-globin gene cluster: sequence features and topological constraints. Cell 1987;49:369–78.
[182] Vnencak-Jones CL, Phillips 3rd JA. Hot spots for growth hormone gene deletions in homologous regions outside of Alu repeats. Science 1990;250:1745–8.
[183] Embury SH, Miller JA, Dozy AM, Kan YW, Chan V, Todd D. Two different molecular organizations account for the single alpha-globin gene of the alpha-thalassemia-2 genotype. J Clin Invest 1980;66:1319–25.
[184] Goossens M, Dozy AM, Embury SH, Zachariades Z, Hadjiminas MG, Stamatoyannopoulos G, Kan YW. Triplicated alpha-globin loci in humans. Proc Natl Acad Sci U S A 1980;77:518–21.
[185] Baglioni C. The fusion of two peptide chains in hemoglobin Lepore and its interpretation as a genetic deletion. Proc Natl Acad Sci U S A 1962;48:1880–6.
[186] Shapiro LJ, Yen P, Pomerantz D, Martin E, Rolewic L, Mohandas T. Molecular studies of deletions at the human steroid sulfatase locus. Proc Natl Acad Sci U S A 1989;86:8477–81.
[187] Guioli S, Incerti B, Zanaria E, Bardoni B, Franco B, Taylor K, Ballabio A, Camerino G. Kallmann syndrome due to a translocation resulting in an X/Y fusion gene. Nat Genet 1992;1:337–40.

[188] Reiter LT, Hastings PJ, Nelis E, De Jonghe P, Van Broeckhoven C, Lupski JR. Human meiotic recombination products revealed by sequencing a hotspot for homologous strand exchange in multiple HNPP deletion patients. Am J Hum Genet 1998;62:1023–33.

[189] Dorschner MO, Sybert VP, Weaver M, Pletcher BA, Stephens K. NF1 microdeletion breakpoints are clustered at flanking repetitive sequences. Hum Mol Genet 2000;9:35–46.

[190] Francke U. Williams-Beuren syndrome: genes and mechanisms. Hum Mol Genet 1999;8:1947–54.

[191] Juyal RC, Figuera LE, Hauge X, Elsea SH, Lupski JR, Greenberg F, Baldini A, Patel PI. Molecular analyses of 17p11.2 deletions in 62 Smith-Magenis syndrome patients. Am J Hum Genet 1996;58:998–1007.

[192] Edelmann L, Pandita RK, Morrow BE. Low-copy repeats mediate the common 3-Mb deletion in patients with velo-cardio-facial syndrome. Am J Hum Genet 1999;64:1076–86.

[193] Shaikh TH, Kurahashi H, Saitta SC, O'Hare AM, Hu P, Roe BA, Driscoll DA, McDonald-McGinn DM, Zackai EH, Budarf ML, Emanuel BS. Chromosome 22-specific low copy repeats and the 22q11.2 deletion syndrome: genomic organization and deletion endpoint analysis. Hum Mol Genet 2000;9:489–501.

[194] Christian SL, Fantes JA, Mewborn SK, Huang B, Ledbetter DH. Large genomic duplicons map to sites of instability in the Prader-Willi/Angelman syndrome chromosome region (15q11-q13). Hum Mol Genet 1999;8:1025–37.

[195] Koolen DA, Vissers LE, Pfundt R, de Leeuw N, Knight SJ, Regan R, Kooy RF, Reyniers E, Romano C, Fichera M, Schinzel A, Baumer A, Anderlid BM, Schoumans J, Knoers NV, van Kessel AG, Sistermans EA, Veltman JA, Brunner HG, de Vries BB. A new chromosome 17q21.31 microdeletion syndrome associated with a common inversion polymorphism. Nat Genet 2006;38:999–1001.

[196] Sharp AJ, Hansen S, Selzer RR, Cheng Z, Regan R, Hurst JA, Stewart H, Price SM, Blair E, Hennekam RC, Fitzpatrick CA, Segraves R, Richmond TA, Guiver C, Albertson DG, Pinkel D, Eis PS, Schwartz S, Knight SJ, Eichler EE. Discovery of previously unidentified genomic disorders from the duplication architecture of the human genome. Nat Genet 2006;38:1038–42.

[197] Shaw-Smith C, Pittman AM, Willatt L, Martin H, Rickman L, Gribble S, Curley R, Cumming S, Dunn C, Kalaitzopoulos D, Porter K, Prigmore E, Krepischi-Santos AC, Varela MC, Koiffmann CP, Lees AJ, Rosenberg C, Firth HV, de Silva R, Carter NP. Microdeletion encompassing MAPT at chromosome 17q21.3 is associated with developmental delay and learning disability. Nat Genet 2006;38:1032–7.

[198] Stefansson H, Helgason A, Thorleifsson G, Steinthorsdottir V, Masson G, Barnard J, Baker A, Jonasdottir A, Ingason A, Gudnadottir VG, Desnica N, Hicks A, Gylfason A, Gudbjartsson DF, Jonsdottir GM, Sainz J, Agnarsson K, Birgisdottir B, Ghosh S, Olafsdottir A, Cazier JB, Kristjansson K, Frigge ML, Thorgeirsson TE, Gulcher JR, Kong A, Stefansson K. A common inversion under selection in Europeans. Nat Genet 2005;37:129–37.

[199] Ji Y, Eichler EE, Schwartz S, Nicholls RD. Structure of chromosomal duplicons and their role in mediating human genomic disorders. Genome Res 2000;10:597–610.

[200] Bailey JA, Gu Z, Clark RA, Reinert K, Samonte RV, Schwartz S, Adams MD, Myers EW, Li PW, Eichler EE. Recent segmental duplications in the human genome. Science 2002;297:1003–7.

[201] Shaw CJ, Lupski JR. Implications of human genome architecture for rearrangement-based disorders: the genomic basis of disease. Hum Mol Genet 2004;13(Spec No 1):R57–64.

[202] Deininger PL, Batzer MA. Alu repeats and human disease. Mol Genet Metab 1999;67:183–93.

[203] Lander ES, Linton LM, Birren B, Nusbaum C, Zody MC, Baldwin J, Devon K, Dewar K, Doyle M, FitzHugh W, Funke R, Gage D, Harris K, Heaford A, Howland J, Kann L, Lehoczky J, LeVine R, McEwan P, McKernan K, Meldrim J, Mesirov JP, Miranda C, Morris W, Naylor J, Raymond C, Rosetti M, Santos R, Sheridan A, Sougnez C, Stange-Thomann N, Stojanovic N, Subramanian A, Wyman D, Rogers J, Sulston J, Ainscough R, Beck S, Bentley D, Burton J, Clee C, Carter N, Coulson A, Deadman R, Deloukas P, Dunham A, Dunham I, Durbin R, French L, Grafham D, Gregory S, Hubbard T, Humphray S, Hunt A, Jones M, Lloyd C, McMurray A, Matthews L, Mercer S, Milne S, Mullikin JC, Mungall A, Plumb R, Ross M, Shownkeen R, Sims S, Waterston RH, Wilson RK, Hillier LW, McPherson JD, Marra MA, Mardis ER, Fulton LA, Chinwalla AT, Pepin KH, Gish WR, Chissoe SL, Wendl MC, Delehaunty KD, Miner TL, Delehaunty A, Kramer JB, Cook LL, Fulton RS, Johnson DL, Minx PJ, Clifton SW, Hawkins T, Branscomb E, Predki P, Richardson P, Wenning S, Slezak T, Doggett N, Cheng JF, Olsen A, Lucas S, Elkin C, Uberbacher E, Frazier M, Gibbs RA, Muzny DM, Scherer SE, Bouck JB, Sodergren EJ, Worley KC, Rives CM, Gorrell JH, Metzker ML, Naylor SL, Kucherlapati RS, Nelson DL, Weinstock GM, Sakaki Y, Fujiyama A, Hattori M, Yada T, Toyoda A, Itoh T, Kawagoe C, Watanabe H, Totoki Y, Taylor T, Weissen-

bach J, Heilig R, Saurin W, Artiguenave F, Brottier P, Bruls T, Pelletier E, Robert C, Wincker P, Smith DR, Doucette_Stamm L, Rubenfield M, Weinstock K, Lee HM, Dubois J, Rosenthal A, Platzer M, Nyakatura G, Taudien S, Rump A, Yang H, Yu J, Wang J, Huang G, Gu J, Hood L, Rowen L, Madan A, Qin S, Davis RW, Federspiel NA, Abola AP, Proctor MJ, Myers RM, Schmutz J, Dickson M, Grimwood J, Cox DR, Olson MV, Kaul R, Shimizu N, Kawasaki K, Minoshima S, Evans GA, Athanasiou M, Schultz R, Roe BA, Chen F, Pan H, Ramser J, Lehrach H, Reinhardt R, McCombie WR, de_la_Bastide M, Dedhia N, Blocker H, Hornischer K, Nordsiek G, Agarwala R, Aravind L, Bailey JA, Bateman A, Batzoglou S, Birney E, Bork P, Brown DG, Burge CB, Cerutti L, Chen HC, Church D, Clamp M, Copley RR, Doerks T, Eddy SR, Eichler EE. Initial sequencing and analysis of the human genome. Nature 2001;409:860–921.

[204] Lehrman MA, Goldstein JL, Russell DW, Brown MS. Duplication of seven exons in LDL receptor gene caused by Alu-Alu recombination in a subject with familial hypercholesterolemia. Cell 1987;48:827–35.

[205] Stoppa-Lyonnet D, Duponchel C, Meo T, Laurent J, Carter PE, Arala-Chaves M, Cohen JH, Dewald G, Goetz J, Hauptmann G, et al. Recombinational biases in the rearranged C1-inhibitor genes of hereditary angioedema patients. Am J Hum Genet 1991;49:1055–62.

[206] Kornreich R, Bishop DF, Desnick RJ. a-galactosidase A gene rearrangements causing Fabry disease. J Biol Chem 1990;265:9319–26.

[207] Abo-Dalo B, Kutsche K, Mautner V, Kluwe L. Large intragenic deletions of the NF2 gene: breakpoints and associated phenotypes. Genes Chromosomes Cancer 2010;49:171–5.

[208] Champion KJ, Basehore MJ, Wood T, Destree A, Vannuffel P, Maystadt I. Identification and characterization of a novel homozygous deletion in the alpha-N-acetylglucosaminidase gene in a patient with Sanfilippo type B syndrome (mucopolysaccharidosis IIIB). Mol Genet Metab 2010;100:51–6.

[209] Cozar M, Bembi B, Dominissini S, Zampieri S, Vilageliu L, Grinberg D, Dardis A. Molecular characterization of a new deletion of the GBA1 gene due to an inter Alu recombination event. Mol Genet Metab 2010.

[210] Gentsch M, Kaczmarczyk A, van Leeuwen K, de Boer M, Kaus-Drobek M, Dagher MC, Kaiser P, Arkwright PD, Gahr M, Rosen-Wolff A, Bochtler M, Secord E, Britto-Williams P, Saifi GM, Maddalena A, Dbaibo G, Bustamante J, Casanova JL, Roos D, Roesler J. Alu-repeat-induced deletions within the NCF2 gene causing p67-phox-deficient chronic granulomatous disease (CGD). Hum Mutat 2010;31:151–8.

[211] Goldmann R, Tichy L, Freiberger T, Zapletalova P, Letocha O, Soska V, Fajkus J, Fajkusova L. Genomic characterization of large rearrangements of the LDLR gene in Czech patients with familial hypercholesterolemia. BMC Med Genet 2010;11:115.

[212] Resta N, Giorda R, Bagnulo R, Beri S, Della Mina E, Stella A, Piglionica M, Susca FC, Guanti G, Zuffardi O, Ciccone R. Breakpoint determination of 15 large deletions in Peutz-Jeghers subjects. Hum Genet 2010;128:373–82.

[213] Shlien A, Baskin B, Achatz MI, Stavropoulos DJ, Nichols KE, Hudgins L, Morel CF, Adam MP, Zhukova N, Rotin L, Novokmet A, Druker H, Shago M, Ray PN, Hainaut P, Malkin D. A common molecular mechanism underlies two phenotypically distinct 17p13.1 microdeletion syndromes. Am J Hum Genet 2010;87:631–42.

[214] Tuohy TM, Done MW, Lewandowski MS, Shires PM, Saraiya DS, Huang SC, Neklason DW, Burt RW. Large intron 14 rearrangement in APC results in splice defect and attenuated FAP. Hum Genet 2010;127:359–69.

[215] Yang Z, Funke BH, Cripe LH, Vick 3rd GW, Mancini-Dinardo D, Pena LS, Kanter RJ, Wong B, Westerfield BH, Varela JJ, Fan Y, Towbin JA, Vatta M. LAMP2 microdeletions in patients with Danon disease. Circ Cardiovasc Genet 2010;3:129–37.

[216] Zhang F, Seeman P, Liu P, Weterman MA, Gonzaga-Jauregui C, Towne CF, Batish SD, De Vriendt E, De Jonghe P, Rautenstrauss B, Krause KH, Khajavi M, Posadka J, Vandenberghe A, Palau F, Van Maldergem L, Baas F, Timmerman V, Lupski JR. Mechanisms for nonrecurrent genomic rearrangements associated with CMT1A or HNPP: rare CNVs as a cause for missing heritability. Am J Hum Genet 2010;86:892–903.

[217] Rudiger NS, Gregersen N, Kielland-Brandt MC. One short well conserved region of Alu-sequences is involved in human gene rearrangements and has homology with prokaryotic chi. Nucleic Acids Res 1995;23:256–60.

[218] Roth DB, Wilson JH. Nonhomologous recombination in mammalian cells: role for short sequence homologies in the joining reaction. Mol Cell Biol 1986;6:4295–304.

[219] Woods-Samuels P, Kazazian Jr HH, Antonarakis SE. Nonhomologous recombination in the human genome: deletions in the human factor VIII gene. Genomics 1991;10:94–101.

[220] McNaughton JC, Cockburn DJ, Hughes G, Jones WA, Laing NG, Ray PN, Stockwell PA, Petersen GB. Is gene deletion in eukaryotes sequence-dependent? A study of nine deletion junctions and nineteen other deletion breakpoints in intron 7 of the human dystrophin gene. Gene 1998;222:41–51.

[221] Bacolla A, Jaworski A, Larson JE, Jakupciak JP, Chuzhanova N, Abeysinghe SS, O'Connell CD, Cooper DN, Wells RD. Breakpoints of gross deletions coincide with non-B DNA conformations. Proc Natl Acad Sci U S A 2004;101:14162–7.
[222] Abeysinghe SS, Stenson PD, Krawczak M, Cooper DN. Gross rearrangement breakpoint database (GRaBD). Hum Mutat 2004;23:219–21.
[223] Mine M, Chen JM, Brivet M, Desguerre I, Marchant D, de Lonlay P, Bernard A, Ferec C, Abitbol M, Ricquier D, Marsac C. A large genomic deletion in the PDHX gene caused by the retrotranspositional insertion of a full-length LINE-1 element. Hum Mutat 2007;28:137–42.
[224] Morisada N, Rendtorff ND, Nozu K, Morishita T, Miyakawa T, Matsumoto T, Hisano S, Iijima K, Tranebjaerg L, Shirahata A, Matsuo M, Kusuhara K. Branchio-oto-renal syndrome caused by partial EYA1 deletion due to LINE-1 insertion. Pediatr Nephrol 2010;25:1343–8.
[225] Okubo M, Horinishi A, Saito M, Ebara T, Endo Y, Kaku K, Murase T, Eto M. A novel complex deletion-insertion mutation mediated by Alu repetitive elements leads to lipoprotein lipase deficiency. Mol Genet Metab 2007;92:229–33.
[226] Schollen E, Keldermans L, Foulquier F, Briones P, Chabas A, Sanchez-Valverde F, Adamowicz M, Pronicka E, Wevers R, Matthijs G. Characterization of two unusual truncating PMM2 mutations in two CDG-Ia patients. Mol Genet Metab 2007;90:408–13.
[227] Takasu M, Hayashi R, Maruya E, Ota M, Imura K, Kougo K, Kobayashi C, Saji H, Ishikawa Y, Asai T, Tokunaga K. Deletion of entire HLA-A gene accompanied by an insertion of a retrotransposon. Tissue Antigens 2007;70:144–50.
[228] Awano H, Malueka RG, Yagi M, Okizuka Y, Takeshima Y, Matsuo M. Contemporary retrotransposition of a novel non-coding gene induces exon-skipping in dystrophin mRNA. J Hum Genet 2010.
[229] Tabata A, Sheng JS, Ushikai M, Song YZ, Gao HZ, Lu YB, Okumura F, Iijima M, Mutoh K, Kishida S, Saheki T, Kobayashi K. Identification of 13 novel mutations including a retrotransposal insertion in SLC25A13 gene and frequency of 30 mutations found in patients with citrin deficiency. J Hum Genet 2008;53:534–45.
[230] Kazazian Jr HH, Wong C, Youssoufian H, Scott AF, Phillips DG, Antonarakis SE. Haemophilia A resulting from *de novo* insertion of L1 sequences represents a novel mechanism for mutation in man. Nature 1988;332:164–6.
[231] Dombroski BA, Mathias SL, Nanthakumar E, Scott AF, Kazazian Jr HH. Isolation of an active human transposable element. Science 1991;254:1805–8.
[232] Ostertag EM, Kazazian Jr HH. Biology of mammalian L1 retrotransposons. Annu Rev Genet 2001;35:501–38.
[233] Kazazian Jr HH. Mobile elements and disease. Curr Opin Genet Dev 1998;8:343–50.
[234] Woods-Samuels P, Wong C, Mathias SL, Scott AF, Kazazian Jr HH, Antonarakis SE. Characterization of a nondeleterious L1 insertion in an intron of the human factor VIII gene and further evidence of open reading frames in functional L1 elements. Genomics 1989;4:290–6.
[235] Muratani K, Hada T, Yamamoto Y, Kaneko T, Shigeto Y, Ohue T, Furuyama J, Higashino K. Inactivation of the cholinesterase gene by Alu insertion: possible mechanism for human gene transposition. Proc Natl Acad Sci U S A 1991;88:11315–9.
[236] Vidaud D, Vidaud M, Bahnak BR, Siguret V, Gispert Sanchez S, Laurian Y, Meyer D, Goossens M, Lavergne JM. Haemophilia B due to a de novo insertion of a human-specific Alu subfamily member within the coding region of the factor IX gene. Eur J Hum Genet 1993;1:30–6.
[237] Wallace MR, Andersen LB, Saulino AM, Gregory PE, Glover TW, Collins FS. A de novo Alu insertion results in neurofibromatosis type 1. Nature 1991;353:864–6.
[238] Li X, Scaringe WA, Hill KA, Roberts S, Mengos A, Careri D, Pinto MT, Kasper CK, Sommer SS. Frequency of recent retrotransposition events in the human factor IX gene. Hum Mutat 2001;17:511–9.
[239] Chen JM, Stenson PD, Cooper DN, Ferec C. A systematic analysis of LINE-1 endonuclease-dependent retrotranspositional events causing human genetic disease. Hum Genet 2005;117:411–27.
[240] Audrezet MP, Chen JM, Raguenes O, Chuzhanova N, Giteau K, Le Marechal C, Quere I, Cooper DN, Ferec C. Genomic rearrangements in the CFTR gene: extensive allelic heterogeneity and diverse mutational mechanisms. Hum Mutat 2004;23:343–57.
[241] Bochukova EG, Roscioli T, Hedges DJ, Taylor IB, Johnson D, David DJ, Deininger PL, Wilkie AO. Rare mutations of FGFR2 causing apert syndrome: identification of the first partial gene deletion, and an Alu element insertion from a new subfamily. Hum Mutat 2009;30:204–11.
[242] Oldridge M, Zackai EH, McDonald-McGinn DM, Iseki S, Morriss-Kay GM, Twigg SR, Johnson D, Wall SA, Jiang W, Theda C, Jabs EW, Wilkie AO. De novo alu-element insertions in FGFR2 identify a distinct pathological basis for Apert syndrome. Am J Hum Genet 1999;64:446–61.
[243] Musova Z, Hedvicakova P, Mohrmann M, Tesarova M, Krepelova A, Zeman J, Sedlacek Z. A novel insertion of a rearranged L1 element in exon 44 of the dystrophin

gene: further evidence for possible bias in retroposon integration. Biochem Biophys Res Commun 2006;347:145–9.

[244] Narita N, Nishio H, Kitoh Y, Ishikawa Y, Minami R, Nakamura H, Matsuo M. Insertion of a 5′ truncated L1 element into the 3′ end of exon 44 of the dystrophin gene resulted in skipping of the exon during splicing in a case of Duchenne muscular dystrophy. J Clin Invest 1993;91:1862–7.

[245] Wimmer K, Callens T, Wernstedt A, Messiaen L. The NF1 gene contains hotspots for L1 endonuclease-dependent de novo insertion. PLoS Genet 2011;7:e1002371.

[246] Wulff K, Gazda H, Schroder W, Robicka-Milewska R, Herrmann FH. Identification of a novel large F9 gene mutation-an insertion of an Alu repeated DNA element in exon e of the factor 9 gene. Hum Mutat 2000;15:299.

[247] Conley ME, Partain JD, Norland SM, Shurtleff SA, Kazazian Jr HH. Two independent retrotransposon insertions at the same site within the coding region of BTK. Hum Mutat 2005;25:324–5.

[248] Scott HS, Kudoh J, Wattenhofer M, Shibuya K, Berry A, Chrast R, Guipponi M, Wang J, Kawasaki K, Asakawa S, Minoshima S, Younus F, Mehdi SQ, Radhakrishna U, Papasavvas MP, Gehrig C, Rossier C, Korostishevsky M, Gal A, Shimizu N, Bonne-Tamir B, Antonarakis SE. Insertion of beta-satellite repeats identifies a transmembrane protease causing both congenital and childhood onset autosomal recessive deafness. Nat Genet 2001;27:59–63.

[249] Turner C, Killoran C, Thomas NS, Rosenberg M, Chuzhanova NA, Johnston J, Kemel Y, Cooper DN, Biesecker LG. Human genetic disease caused by de novo mitochondrial-nuclear DNA transfer. Hum Genet 2003;112:303–9.

[250] Millar DS, Tysoe C, Lazarou LP, Pilz DT, Mohammed S, Anderson K, Chuzhanova N, Cooper DN, Butler R. An isolated case of lissencephaly caused by the insertion of a mitochondrial genome-derived DNA sequence into the 5′ untranslated region of the PAFAH1B1 (LIS1) gene. Hum Genomics 2010;4:384–93.

[251] Lakich D, Kazazian Jr HH, Antonarakis SE, Gitschier J. Inversions disrupting the factor VIII gene are a common cause of severe haemophilia A. Nat Genet 1993;5:236–41.

[252] Naylor JA, Green PM, Rizza CR, Giannelli F. Analysis of factor VIII mRNA reveals defects in everyone of 28 haemophilia A patients. Hum Mol Genet 1993;2:11–7.

[253] Rossiter JP, Young M, Kimberland ML, Hutter P, Ketterling RP, Gitschier J, Horst J, Morris MA, Schaid DJ, de Moerloose P, Sommer SS, Kazazian HH, Antonarakis SE. Factor VIII gene inversions causing severe haemophilia A originate almost exclusively in male germ cells. Hum Mol Genet 1994;3:1035–9.

[254] Bondeson ML, Dahl N, Malmgren H, Kleijer WJ, Tonnesen T, Carlberg BM, Pettersson U. Inversion of the IDS gene resulting from recombination with IDS-related sequences is a common cause of the Hunter syndrome. Hum Mol Genet 1995;4:615–21.

[255] Jennings MW, Jones RW, Wood WG, Weatherall DJ. Analysis of an inversion within the human beta globin gene cluster. Nucleic Acids Res 1985;13:2897–907.

[256] Karathanasis SK, Ferris E, Haddad IA. DNA inversion within the apolipoproteins AI/CIII/AIV-encoding gene cluster of certain patients with premature atherosclerosis. Proc Natl Acad Sci U S A 1987;84:7198–202.

[257] Hu XY, Ray PN, Murphy EG, Thompson MW, Worton RG. Duplicational mutation at the Duchenne muscular dystrophy locus: its frequency, distribution, origin, and phenotypegenotype correlation. Am J Hum Genet 1990;46:682–95.

[258] Pentao L, Wise CA, Chinault AC, Patel PI, Lupski JR. Charcot-Marie-Tooth type 1A duplication appears to arise from recombination at repeat sequences flanking the 1.5 Mb monomer unit. Nat Genet 1992;2:292–300.

[259] Woodward K, Kendall E, Vetrie D, Malcolm S. Pelizaeus-Merzbacher disease: identification of Xq22 proteolipid-protein duplications and characterization of breakpoints by interphase FISH. Am J Hum Genet 1998;63:207–17.

[260] de Mollerat XJ, Gurrieri F, Morgan CT, Sangiorgi E, Everman DB, Gaspari P, Amiel J, Bamshad MJ, Lyle R, Blouin JL, Allanson JE, Le Marec B, Wilson M, Braverman NE, Radhakrishna U, Delozier-Blanchet C, Abbott A, Elghouzzi V, Antonarakis S, Stevenson RE, Munnich A, Neri G, Schwartz CE. A genomic rearrangement resulting in a tandem duplication is associated with split hand-split foot malformation 3 (SHFM3) at 10q24. Hum Mol Genet 2003;12: 1959–71.

[261] Rovelet-Lecrux A, Hannequin D, Raux G, Le Meur N, Laquerriere A, Vital A, Dumanchin C, Feuillette S, Brice A, Vercelletto M, Dubas F, Frebourg T, Campion D. APP locus duplication causes autosomal dominant early-onset Alzheimer disease with cerebral amyloid angiopathy. Nat Genet 2006;38:24–6.

[262] Singleton AB, Farrer M, Johnson J, Singleton A, Hague S, Kachergus J, Hulihan M, Peuralinna T, Dutra A, Nussbaum R, Lincoln S, Crawley A, Hanson M, Maraganore D, Adler C, Cookson MR, Muenter M, Baptista M, Miller D, Blancato J, Hardy J, Gwinn-Hardy K. alpha-Synuclein locus triplication causes Parkinson's disease. Science 2003;302:841.

[263] Le Marechal C, Masson E, Chen JM, Morel F, Ruszniewski P, Levy P, Ferec C. Hereditary pancreatitis caused by triplication of the trypsinogen locus. Nat Genet 2006;38:1372–4.
[264] Fu W, Zhang F, Wang Y, Gu X, Jin L. Identification of copy number variation hotspots in human populations. Am J Hum Genet 2010;87:494–504.
[265] Alkan C, Kidd JM, Marques-Bonet T, Aksay G, Antonacci F, Hormozdiari F, Kitzman JO, Baker C, Malig M, Mutlu O, Sahinalp SC, Gibbs RA, Eichler EE. Personalized copy number and segmental duplication maps using next-generation sequencing. Nat Genet 2009;41:1061–7.
[266] Wang RT, Ahn S, Park CC, Khan AH, Lange K, Smith DJ. Effects of genome-wide copy number variation on expression in mammalian cells. BMC Genomics 2011;12:562.
[267] Beckmann JS, Sharp AJ, Antonarakis SE. CNVs and genetic medicine (excitement and consequences of a rediscovery). Cytogenet Genome Res 2008;123:7–16.
[268] de Smith AJ, Walters RG, Coin LJ, Steinfeld I, Yakhini Z, Sladek R, Froguel P, Blakemore AI. Small deletion variants have stable breakpoints commonly associated with alu elements. PLoS One 2008;3:e3104.
[269] Henrichsen CN, Chaignat E, Reymond A. Copy number variants, diseases and gene expression. Hum Mol Genet 2009;18:R1–8.
[270] Stankiewicz P, Lupski JR. Structural variation in the human genome and its role in disease. Annu Rev Med 2010;61:437–55.
[271] Lower KM, Hughes JR, De Gobbi M, Henderson S, Viprakasit V, Fisher C, Goriely A, Ayyub H, Sloane-Stanley J, Vernimmen D, Langford C, Garrick D, Gibbons RJ, Higgs DR. Adventitious changes in long-range gene expression caused by polymorphic structural variation and promoter competition. Proc Natl Acad Sci U S A 2009;106:21771–6.
[272] Dathe K, Kjaer KW, Brehm A, Meinecke P, Nurnberg P, Neto JC, Brunoni D, Tommerup N, Ott CE, Klopocki E, Seemann P, Mundlos S. Duplications involving a conserved regulatory element downstream of BMP2 are associated with brachydactyly type A2. Am J Hum Genet 2009;84:483–92.
[273] Aitman TJ, Dong R, Vyse TJ, Norsworthy PJ, Johnson MD, Smith J, Mangion J, Roberton-Lowe C, Marshall AJ, Petretto E, Hodges MD, Bhangal G, Patel SG, Sheehan-Rooney K, Duda M, Cook PR, Evans DJ, Domin J, Flint J, Boyle JJ, Pusey CD, Cook HT. Copy number polymorphism in Fcgr3 predisposes to glomerulonephritis in rats and humans. Nature 2006;439:851–5.
[274] Fellermann K, Stange DE, Schaeffeler E, Schmalzl H, Wehkamp J, Bevins CL, Reinisch W, Teml A, Schwab M, Lichter P, Radlwimmer B, Stange EF. A chromosome 8 gene-cluster polymorphism with low human beta-defensin 2 gene copy number predisposes to Crohn disease of the colon. Am J Hum Genet 2006;79:439–48.
[275] Gonzalez E, Kulkarni H, Bolivar H, Mangano A, Sanchez R, Catano G, Nibbs RJ, Freedman BI, Quinones MP, Bamshad MJ, Murthy KK, Rovin BH, Bradley W, Clark RA, Anderson SA, O'Connell RJ, Agan BK, Ahuja SS, Bologna R, Sen L, Dolan MJ, Ahuja SK. The influence of CCL3L1 gene-containing segmental duplications on HIV-1/AIDS susceptibility. Science 2005;307:1434–40.
[276] Walsh T, McClellan JM, McCarthy SE, Addington AM, Pierce SB, Cooper GM, Nord AS, Kusenda M, Malhotra D, Bhandari A, Stray SM, Rippey CF, Roccanova P, Makarov V, Lakshmi B, Findling RL, Sikich L, Stromberg T, Merriman B, Gogtay N, Butler P, Eckstrand K, Noory L, Gochman P, Long R, Chen Z, Davis S, Baker C, Eichler EE, Meltzer PS, Nelson SF, Singleton AB, Lee MK, Rapoport JL, King MC, Sebat J. Rare structural variants disrupt multiple genes in neurodevelopmental pathways in schizophrenia. Science 2008;320:539–43.
[277] Bochukova EG, Huang N, Keogh J, Henning E, Purmann C, Blaszczyk K, Saeed S, Hamilton-Shield J, Clayton-Smith J, O'Rahilly S, Hurles ME, Farooqi IS. Large, rare chromosomal deletions associated with severe early-onset obesity. Nature 2010;463:666–70.
[278] McCarroll SA. Extending genome-wide association studies to copy-number variation. Hum Mol Genet 2008;17:R135–42.
[279] Merikangas AK, Corvin AP, Gallagher L. Copy-number variants in neurodevelopmental disorders: promises and challenges. Trends Genet 2009;25:536–44.
[280] Lee C, Scherer SW. The clinical context of copy number variation in the human genome. Expert Rev Mol Med 2010;12:e8.
[281] Elia J, Gai X, Xie HM, Perin JC, Geiger E, Glessner JT, D'Arcy M, deBerardinis R, Frackelton E, Kim C, Lantieri F, Muganga BM, Wang L, Takeda T, Rappaport EF, Grant SF, Berrettini W, Devoto M, Shaikh TH, Hakonarson H, White PS. Rare structural variants found in attention-deficit hyperactivity disorder are preferentially associated with neurodevelopmental genes. Mol Psychiatry 2010;15:637–46.
[282] Glessner JT, Wang K, Cai G, Korvatska O, Kim CE, Wood S, Zhang H, Estes A, Brune CW, Bradfield JP, Imielinski M, Frackelton EC, Reichert J, Crawford EL, Munson J, Sleiman PM, Chiavacci R, Annaiah K, Thomas K, Hou C, Glaberson W, Flory J, Otieno F, Garris M, Soorya L, Klei L, Piven J, Meyer

replication mechanisms generating complex genomic rearrangements. Cell 2011;146:889–903.
[313] Fukami M, Shima H, Suzuki E, Ogata T, Matsubara K, Kamimaki T. Catastrophic cellular events leading to complex chromosomal rearrangements in the germline. Clin Genet 2017;91:653–60.
[314] van Leeuwen FW, Kros JM, Kamphorst W, van Schravendijk C, de Vos RA. Molecular misreading: the occurrence of frameshift proteins in different diseases. Biochem Soc Trans 2006;34:738–42.
[315] Linton MF, Pierotti V, Young SG. Reading-frame restoration with an apolipoprotein B gene frameshift mutation. Proc Natl Acad Sci U S A 1992;89:11431–5.
[316] Young M, Inaba H, Hoyer LW, Higuchi M, Kazazian Jr HH, Antonarakis SE. Partial correction of a severe molecular defect in hemophilia A, because of errors during expression of the factor VIII gene. Am J Hum Genet 1997;60:565–73.
[317] Laken SJ, Petersen GM, Gruber SB, Oddoux C, Ostrer H, Giardiello FM, Hamilton SR, Hampel H, Markowitz A, Klimstra D, Jhanwar S, Winawer S, Offit K, Luce MC, Kinzler KW, Vogelstein B. Familial colorectal cancer in Ashkenazim due to a hypermutable tract in APC. Nat Genet 1997;17:79–83.
[318] van Leeuwen FW, de Kleijn DP, van den Hurk HH, Neubauer A, Sonnemans MA, Sluijs JA, Koycu S, Ramdjielal RD, Salehi A, Martens GJ, Grosveld FG, Peter J, Burbach H, Hol EM. Frameshift mutants of beta amyloid precursor protein and ubiquitin-B in Alzheimer's and Down patients. Science 1998;279:242–7.
[319] Paoloni-Giacobino A, Rossier C, Papasavvas MP, Antonarakis SE. Frequency of replication/transcription errors in (A)/(T) runs of human genes. Hum Genet 2001;109:40–7.
[320] Suter CM, Martin DI, Ward RL. Germline epimutation of MLH1 in individuals with multiple cancers. Nat Genet 2004;36:497–501.
[321] Hitchins M, Williams R, Cheong K, Halani N, Lin VA, Packham D, Ku S, Buckle A, Hawkins N, Burn J, Gallinger S, Goldblatt J, Kirk J, Tomlinson I, Scott R, Spigelman A, Suter C, Martin D, Suthers G, Ward R. MLH1 germline epimutations as a factor in hereditary nonpolyposis colorectal cancer. Gastroenterology 2005;129:1392–9.
[322] Mariot V, Maupetit-Mehouas S, Sinding C, Kottler ML, Linglart A. A maternal epimutation of GNAS leads to Albright osteodystrophy and parathyroid hormone resistance. J Clin Endocrinol Metab 2008;93:661–5.
[323] Schalkwyk LC, Meaburn EL, Smith R, Dempster EL, Jeffries AR, Davies MN, Plomin R, Mill J. Allelic skewing of DNA methylation is widespread across the genome. Am J Hum Genet 2010;86:196–212.
[324] Luco RF, Pan Q, Tominaga K, Blencowe BJ, Pereira-Smith OM, Misteli T. Regulation of alternative splicing by histone modifications. Science 2010;327:996–1000.
[325] Wang Z, Zang C, Rosenfeld JA, Schones DE, Barski A, Cuddapah S, Cui K, Roh TY, Peng W, Zhang MQ, Zhao K. Combinatorial patterns of histone acetylations and methylations in the human genome. Nat Genet 2008;40:897–903.
[326] Li M, Wang IX, Li Y, Bruzel A, Richards AL, Toung JM, Cheung VG. Widespread RNA and DNA sequence differences in the human transcriptome. Science 2011;333:53–8.
[327] Lualdi S, Tappino B, Di Duca M, Dardis A, Anderson CJ, Biassoni R, Thompson PW, Corsolini F, Di Rocco M, Bembi B, Regis S, Cooper DN, Filocamo M. Enigmatic in vivo iduronate-2-sulfatase (IDS) mutant transcript correction to wild-type in Hunter syndrome. Hum Mutat 2010;31:E1261–85.
[328] Reich DE, Cargill M, Bolk S, Ireland J, Sabeti PC, Richter DJ, Lavery T, Kouyoumjian R, Farhadian SF, Ward R, Lander ES. Linkage disequilibrium in the human genome. Nature 2001;411:199–204.
[329] Subramanian S, Kumar S. Evolutionary anatomies of positions and types of disease-associated and neutral amino acid mutations in the human genome. BMC Genomics 2006;7:306.
[330] Miller MP, Parker JD, Rissing SW, Kumar S. Quantifying the intragenic distribution of human disease mutations. Ann Hum Genet 2003;67:567–79.
[331] Flint J, Harding RM, Clegg JB, Boyce AJ. Why are some genetic diseases common? Distinguishing selection from other processes by molecular analysis of globin gene variants. Hum Genet 1993;91:91–117.
[332] Tishkoff SA, Verrelli BC. Patterns of human genetic diversity: implications for human evolutionary history and disease. Annu Rev Genomics Hum Genet 2003;4:293–340.
[333] Zlotogora J. High frequencies of human genetic diseases: founder effect with genetic drift or selection? Am J Med Genet 1994;49:10–3.
[334] Zschocke J. Phenylketonuria mutations in Europe. Hum Mutat 2003;21:345–56.
[335] Mockenhaupt FP, Mandelkow J, Till H, Ehrhardt S, Eggelte TA, Bienzle U. Reduced prevalence of Plasmodium falciparum infection and of concomitant anaemia in pregnant women with heterozygous G6PD deficiency. Trop Med Int Health 2003;8:118–24.
[336] Ruwende C, Khoo SC, Snow RW, Yates SN, Kwiatkowski D, Gupta S, Warn P, Allsopp CE, Gilbert SC, Peschu N, et al. Natural selection of hemi- and heterozygotes for G6PD deficiency in Africa by resistance to severe malaria. Nature 1995;376:246–9.

[337] Aidoo M, Terlouw DJ, Kolczak MS, McElroy PD, ter Kuile FO, Kariuki S, Nahlen BL, Lal AA, Udhayakumar V. Protective effects of the sickle cell gene against malaria morbidity and mortality. Lancet 2002;359:1311–2.
[338] Williams TN, Wambua S, Uyoga S, Macharia A, Mwacharo JK, Newton CR, Maitland K. Both heterozygous and homozygous alpha+ thalassemias protect against severe and fatal Plasmodium falciparum malaria on the coast of Kenya. Blood 2005;106:368–71.
[339] Williams TN, Mwangi TW, Wambua S, Peto TE, Weatherall DJ, Gupta S, Recker M, Penman BS, Uyoga S, Macharia A, Mwacharo JK, Snow RW, Marsh K. Negative epistasis between the malaria-protective effects of alpha+-thalassemia and the sickle cell trait. Nat Genet 2005;37:1253–7.
[340] Witchel SF, Lee PA, Suda-Hartman M, Trucco M, Hoffman EP. Evidence for a heterozygote advantage in congenital adrenal hyperplasia due to 21-hydroxylase deficiency. J Clin Endocrinol Metab 1997;82:2097–101.
[341] Datz C, Haas T, Rinner H, Sandhofer F, Patsch W, Paulweber B. Heterozygosity for the C282Y mutation in the hemochromatosis gene is associated with increased serum iron, transferrin saturation, and hemoglobin in young women: a protective role against iron deficiency? Clin Chem 1998;44:2429–32.
[342] Mead S, Stumpf MP, Whitfield J, Beck JA, Poulter M, Campbell T, Uphill JB, Goldstein D, Alpers M, Fisher EM, Collinge J. Balancing selection at the prion protein gene consistent with prehistoric kurulike epidemics. Science 2003;300:640–3.
[343] Kerlin BA, Yan SB, Isermann BH, Brandt JT, Sood R, Basson BR, Joyce DE, Weiler H, Dhainaut JF. Survival advantage associated with heterozygous factor V Leiden mutation in patients with severe sepsis and in mouse endotoxemia. Blood 2003;102:3085–92.
[344] Common JE, Di WL, Davies D, Kelsell DP. Further evidence for heterozygote advantage of GJB2 deafness mutations: a link with cell survival. J Med Genet 2004;41:573–5.
[345] Gabriel SE, Brigman KN, Koller BH, Boucher RC, Stutts MJ. Cystic fibrosis heterozygote resistance to cholera toxin in the cystic fibrosis mouse model. Science 1994;266:107–9.
[346] Schroeder SA, Gaughan DM, Swift M. Protection against bronchial asthma by CFTR delta F508 mutation: a heterozygote advantage in cystic fibrosis. Nat Med 1995;1:703–5.
[347] Pier GB. Role of the cystic fibrosis transmembrane conductance regulator in innate immunity to *Pseudomonas aeruginosa* infections. Proc Natl Acad Sci U S A 2000;97:8822–8.
[348] Hogenauer C, Santa Ana CA, Porter JL, Millard M, Gelfand A, Rosenblatt RL, Prestidge CB, Fordtran JS. Active intestinal chloride secretion in human carriers of cystic fibrosis mutations: an evaluation of the hypothesis that heterozygotes have subnormal active intestinal chloride secretion. Am J Hum Genet 2000;67:1422–7.
[349] Motulsky AG. Jewish diseases and origins. Nat Genet 1995;9:99–101.
[350] Ostrer H. A genetic profile of contemporary Jewish populations. Nat Rev Genet 2001;2:891–8.
[351] Frisch A, Colombo R, Michaelovsky E, Karpati M, Goldman B, Peleg L. Origin and spread of the 1278insTATC mutation causing Tay-Sachs disease in Ashkenazi Jews: genetic drift as a robust and parsimonious hypothesis. Hum Genet 2004;114:366–76.
[352] Risch N, Tang H, Katzenstein H, Ekstein J. Geographic distribution of disease mutations in the Ashkenazi Jewish population supports genetic drift over selection. Am J Hum Genet 2003;72:812–22.
[353] Goriely A, McVean GA, Rojmyr M, Ingemarsson B, Wilkie AO. Evidence for selective advantage of pathogenic FGFR2 mutations in the male germ line. Science 2003;301:643–6.
[354] Goriely A, McVean GA, van Pelt AM, O'Rourke AW, Wall SA, de Rooij DG, Wilkie AO. Gain-of-function amino acid substitutions drive positive selection of FGFR2 mutations in human spermatogonia. Proc Natl Acad Sci U S A 2005;102:6051–6.
[355] Aerts S, Lambrechts D, Maity S, Van Loo P, Coessens B, De Smet F, Tranchevent LC, De Moor B, Marynen P, Hassan B, Carmeliet P, Moreau Y. Gene prioritization through genomic data fusion. Nat Biotechnol 2006;24:537–44.
[356] Cai JJ, Borenstein E, Chen R, Petrov DA. Similarly strong purifying selection acts on human disease genes of all evolutionary ages. Genome Biol Evol 2009;1:131–44.
[357] Domazet-Loso T, Tautz D. An ancient evolutionary origin of genes associated with human genetic diseases. Mol Biol Evol 2008;25:2699–707.
[358] Jimenez-Sanchez G, Childs B, Valle D. Human disease genes. Nature 2001;409:853–5.
[359] Lage K, Hansen NT, Karlberg EO, Eklund AC, Roque FS, Donahoe PK, Szallasi Z, Jensen TS, Brunak S. A large-scale analysis of tissue-specific pathology and gene expression of human disease genes and complexes. Proc Natl Acad Sci U S A 2008;105:20870–5.
[360] Lopez-Bigas N, Ouzounis CA. Genome-wide identification of genes likely to be involved in human genetic disease. Nucleic Acids Res 2004;32:3108–14.

[361] Kondrashov FA, Ogurtsov AY, Kondrashov AS. Bioinformatical assay of human gene morbidity. Nucleic Acids Res 2004;32:1731–7.
[362] Chelala C, Auffray C. Sex-linked recombination variation and distribution of disease-related genes. Gene 2005;346:29–39.
[363] Huang H, Winter EE, Wang H, Weinstock KG, Xing H, Goodstadt L, Stenson PD, Cooper DN, Smith D, Alba MM, Ponting CP, Fechtel K. Evolutionary conservation and selection of human disease gene orthologs in the rat and mouse genomes. Genome Biol 2004;5:R47.
[364] Chuang JH, Li H. Functional bias and spatial organization of genes in mutational hot and cold regions in the human genome. PLoS Biol 2004;2:E29.
[365] Green P, Ewing B, Miller W, Thomas PJ, Green ED. Transcription-associated mutational asymmetry in mammalian evolution. Nat Genet 2003;33:514–7.
[366] Majewski J. Dependence of mutational asymmetry on gene-expression levels in the human genome. Am J Hum Genet 2003;73:688–92.
[367] Touchon M, Nicolay S, Audit B, Brodie of Brodie EB, d'Aubenton-Carafa Y, Arneodo A, Thermes C. Replication-associated strand asymmetries in mammalian genomes: toward detection of replication origins. Proc Natl Acad Sci U S A 2005;102:9836–41.
[368] Hurles M. How homologous recombination generates a mutable genome. Hum Genomics 2005;2:179–86.
[369] Reich DE, Schaffner SF, Daly MJ, McVean G, Mullikin JC, Higgins JM, Richter DJ, Lander ES, Altshuler D. Human genome sequence variation and the influence of gene history, mutation and recombination. Nat Genet 2002;32:135–42.
[370] Chen CL, Duquenne L, Audit B, Guilbaud G, Rappailles A, Baker A, Huvet M, d'Aubenton-Carafa Y, Hyrien O, Arneodo A, Thermes C. Replication-associated mutational asymmetry in the human genome. Mol Biol Evol 2011;28:2327–37.
[371] den Dunnen JT, Antonarakis SE. Nomenclature for the description of human sequence variations. Hum Genet 2001;109:121–4.
[372] den Dunnen JT, Dalgleish R, Maglott DR, Hart RK, Greenblatt MS, McGowan-Jordan J, Roux AF, Smith T, Antonarakis SE, Taschner PE. HGVS recommendations for the description of sequence variants: 2016 update. Hum Mutat 2016;37:564–9.
[373] Wildeman M, van Ophuizen E, den Dunnen JT, Taschner PE. Improving sequence variant descriptions in mutation databases and literature using the Mutalyzer sequence variation nomenclature checker. Hum Mutat 2008;29:6–13.
[374] Gao L, Zhang J. Why are some human disease-associated mutations fixed in mice? Trends Genet 2003;19:678–81.
[375] Azevedo L, Suriano G, van Asch B, Harding RM, Amorim A. Epistatic interactions: how strong in disease and evolution? Trends Genet 2006;22:581–5.
[376] Corona E, Dudley JT, Butte AJ. Extreme evolutionary disparities seen in positive selection across seven complex diseases. PLoS One 2010;5:e12236.
[377] Di Rienzo A, Hudson RR. An evolutionary framework for common diseases: the ancestral-susceptibility model. Trends Genet 2005;21:596–601.
[378] Cotton RG, Scriver CR. Proof of "disease causing" mutation. Hum Mutat 1998;12:1–3.
[379] Arai M, Inaba H, Higuchi M, Antonarakis SE, Kazazian Jr HH, Fujimaki M, Hoyer LW. Direct characterization of factor VIII in plasma: detection of a mutation altering a thrombin cleavage site (arginine-372---histidine). Proc Natl Acad Sci U S A 1989;86:4277–81.
[380] Higuchi M, Wong C, Kochhan L, Olek K, Aronis S, Kasper CK, Kazazian Jr HH, Antonarakis SE. Characterization of mutations in the factor VIII gene by direct sequencing of amplified genomic DNA. Genomics 1990;6:65–71.
[381] Aly AM, Higuchi M, Kasper CK, Kazazian Jr HH, Antonarakis SE, Hoyer LW. Hemophilia A due to mutations that create new N-glycosylation sites. Proc Natl Acad Sci U S A 1992;89:4933–7.
[382] Owen MC, Brennan SO, Lewis JH, Carrell RW. Mutation of antitrypsin to antithrombin. alpha 1-antitrypsin Pittsburgh (358 Met leads to Arg), a fatal bleeding disorder. N Engl J Med 1983;309:694–8.
[383] Vogt G, Chapgier A, Yang K, Chuzhanova N, Feinberg J, Fieschi C, Boisson-Dupuis S, Alcais A, Filipe-Santos O, Bustamante J, de Beaucoudrey L, Al-Mohsen I, Al-Hajjar S, Al-Ghonaium A, Adimi P, Mirsaeidi M, Khalilzadeh S, Rosenzweig S, de la Calle Martin O, Bauer TR, Puck JM, Ochs HD, Furthner D, Engelhorn C, Belohradsky B, Mansouri D, Holland SM, Schreiber RD, Abel L, Cooper DN, Soudais C, Casanova JL. Gains of glycosylation comprise an unexpectedly large group of pathogenic mutations. Nat Genet 2005;37:692–700.
[384] Bertina RM, Koeleman BP, Koster T, Rosendaal FR, Dirven RJ, de Ronde H, van der Velden PA, Reitsma PH. Mutation in blood coagulation factor V associated with resistance to activated protein C. Nature 1994;369:64–7.
[385] Challis BG, Pritchard LE, Creemers JW, Delplanque J, Keogh JM, Luan J, Wareham NJ, Yeo GS, Bhattacharyya S, Froguel P, White A, Farooqi IS, O'Rahilly S. A missense mutation disrupting a dibasic prohormone processing site in pro-opiomelanocortin (POMC) increases susceptibility to early-onset obesity through a novel molecular mechanism. Hum Mol Genet 2002;11:1997–2004.

[386] Byers P. Disorders of collagen biosynthesis and structure. In: Scriver CR, Beaudet AL, Valle D, et al., editors. The metabolic and molecular bases of inherited disease. New York: McGraw-Hill; 2001. p. 5241–86.
[387] Mort M, Evani US, Krishnan VG, Kamati KK, Baenziger PH, Bagchi A, Peters BJ, Sathyesh R, Li B, Sun Y, Xue B, Shah NH, Kann MG, Cooper DN, Radivojac P, Mooney SD. In silico functional profiling of human disease-associated and polymorphic amino acid substitutions. Hum Mutat 2010;31:335–46.
[388] Steward RE, MacArthur MW, Laskowski RA, Thornton JM. Molecular basis of inherited diseases: a structural perspective. Trends Genet 2003;19:505–13.
[389] Gong S, Blundell TL. Structural and functional restraints on the occurrence of single amino acid variations in human proteins. PLoS One 2010;5:e9186.
[390] Schuster-Bockler B, Bateman A. Protein interactions in human genetic diseases. Genome Biol 2008;9:R9.
[391] Xin F, Myers S, Li YF, Cooper DN, Mooney SD, Radivojac P. Structure-based kernels for the prediction of catalytic residues and their involvement in human inherited disease. Bioinformatics 2010;26:1975–82.
[392] Li S, Iakoucheva LM, Mooney SD, Radivojac P. Loss of post-translational modification sites in disease. Pac Symp Biocomput 2010:337–47.
[393] Midic U, Oldfield CJ, Dunker AK, Obradovic Z, Uversky VN. Protein disorder in the human diseasome: unfoldomics of human genetic diseases. BMC Genomics 2009;10(Suppl. 1):S12.
[394] Wang Z, Moult J. SNPs, protein structure, and disease. Hum Mutat 2001;17:263–70.
[395] Yue P, Li Z, Moult J. Loss of protein structure stability as a major causative factor in monogenic disease. J Mol Biol 2005;353:459–73.
[396] Laskowski RA, Thornton JM. Understanding the molecular machinery of genetics through 3D structures. Nat Rev Genet 2008;9:141–51.
[397] Thusberg J, Vihinen M. Pathogenic or not? And if so, then how? Studying the effects of missense mutations using bioinformatics methods. Hum Mutat 2009;30:703–14.
[398] Cooper GM, Shendure J. Needles in stacks of needles: finding disease-causal variants in a wealth of genomic data. Nat Rev Genet 2011;12:628–40.
[399] Hirschhorn JN, Lohmueller K, Byrne E, Hirschhorn K. A comprehensive review of genetic association studies. Genet Med 2002;4:45–61.
[400] Sunyaev S, Ramensky V, Koch I, Lathe 3rd W, Kondrashov AS, Bork P. Prediction of deleterious human alleles. Hum Mol Genet 2001;10:591–7.
[401] Terp BN, Cooper DN, Christensen IT, Jorgensen FS, Bross P, Gregersen N, Krawczak M. Assessing the relative importance of the biophysical properties of amino acid substitutions associated with human genetic disease. Hum Mutat 2002;20:98–109.
[402] Vitkup D, Sander C, Church GM. The amino-acid mutational spectrum of human genetic disease. Genome Biol 2003;4:R72.
[403] Youssoufian H, Antonarakis SE, Aronis S, Tsiftis G, Phillips D.G., Kazazian Jr. H.H.. Characterization of five partial deletions of the factor VIII gene. Proc Natl Acad Sci U S A 1987;84:3772–6.
[404] Cecchini KR, Raja Banerjee A, Kim TH. Towards a genome-wide reconstruction of cis-regulatory networks in the human genome. Semin Cell Dev Biol 2009;20:842–8.
[405] Calvo SE, Pagliarini DJ, Mootha VK. Upstream open reading frames cause widespread reduction of protein expression and are polymorphic among humans. Proc Natl Acad Sci U S A 2009;106:7507–12.
[406] Wen Y, Liu Y, Xu Y, Zhao Y, Hua R, Wang K, Sun M, Li Y, Yang S, Zhang XJ, Kruse R, Cichon S, Betz RC, Nothen MM, van Steensel MA, van Geel M, Steijlen PM, Hohl D, Huber M, Dunnill GS, Kennedy C, Messenger A, Munro CS, Terrinoni A, Hovnanian A, Bodemer C, de Prost Y, Paller AS, Irvine AD, Sinclair R, Green J, Shang D, Liu Q, Luo Y, Jiang L, Chen HD, Lo WH, McLean WH, He CD, Zhang X. Loss-of-function mutations of an inhibitory upstream ORF in the human hairless transcript cause Marie Unna hereditary hypotrichosis. Nat Genet 2009;41:228–33.
[407] Antonarakis SE, Irkin SH, Cheng TC, Scott AF, Sexton JP, Trusko SP, Charache S, Kazazian Jr HH. Beta-thalassemia in American Blacks: novel mutations in the "TATA" box and an acceptor splice site. Proc Natl Acad Sci U S A 1984;81:1154–8.
[408] Orkin SH, Antonarakis SE, Kazazian Jr HH. Base substitution at position -88 in a beta-thalassemic globin gene. Further evidence for the role of distal promoter element ACACCC. J Biol Chem 1984;259:8679–81.
[409] Perkins AC, Sharpe AH, Orkin SH. Lethal beta-thalassaemia in mice lacking the erythroid CACCC-transcription factor EKLF. Nature 1995;375:318–22.
[410] Collins FS, Stoeckert Jr CJ, Serjeant GR, Forget BG, Weissman SM. G gamma beta+ hereditary persistence of fetal hemoglobin: cosmid cloning and identification of a specific mutation 5′ to the G gamma gene. Proc Natl Acad Sci U S A 1984;81:4894–8.
[411] Crossley M, Brownlee GG. Disruption of a C/EBP binding site in the factor IX promoter is associated with haemophilia B. Nature 1990;345:444–6.
[412] Koivisto UM, Palvimo JJ, Janne OA, Kontula K. A single-base substitution in the proximal Sp1 site of the human low density lipoprotein receptor promoter as a

cause of heterozygous familial hypercholesterolemia. Proc Natl Acad Sci U S A 1994;91:10526–30.

[413] Berg LP, Scopes DA, Alhaq A, Kakkar VV, Cooper DN. Disruption of a binding site for hepatocyte nuclear factor 1 in the protein C gene promoter is associated with hereditary thrombophilia. Hum Mol Genet 1994;3:2147–52.

[414] Yang WS, Nevin DN, Peng R, Brunzell JD, Deeb SS. A mutation in the promoter of the lipoprotein lipase (LPL) gene in a patient with familial combined hyperlipidemia and low LPL activity. Proc Natl Acad Sci U S A 1995;92:4462–6.

[415] Weatherall DJ, Clegg JB, Higgs DR, Wood WG. The hemoglobinopathies. In: Scriver CR, Beaudet AL, Valle D, et al., editors. The metabolic and molecular bases of inherited disease. New York: McGraw-Hill; 2001. p. 4571–636.

[416] De Gobbi M, Viprakasit V, Hughes JR, Fisher C, Buckle VJ, Ayyub H, Gibbons RJ, Vernimmen D, Yoshinaga Y, de Jong P, Cheng JF, Rubin EM, Wood WG, Bowden D, Higgs DR. A regulatory SNP causes a human genetic disease by creating a new transcriptional promoter. Science 2006;312:1215–7.

[417] Semenza GL. Transcriptional regulation of gene expression: mechanisms and pathophysiology. Hum Mutat 1994;3:180–99.

[418] Inoue I, Nakajima T, Williams CS, Quackenbush J, Puryear R, Powers M, Cheng T, Ludwig EH, Sharma AM, Hata A, Jeunemaitre X, Lalouel JM. A nucleotide substitution in the promoter of human angiotensinogen is associated with essential hypertension and affects basal transcription in vitro. J Clin Invest 1997;99:1786–97.

[419] Krawczak M, Reiss J, Cooper DN. The mutational spectrum of single base-pair substitutions in mRNA splice junctions of human genes: causes and consequences. Hum Genet 1992;90:41–54.

[420] Faustino NA, Cooper TA. Pre-mRNA splicing and human disease. Genes Dev 2003;17:419–37.

[421] Krawczak M, Thomas NS, Hundrieser B, Mort M, Wittig M, Hampe J, Cooper DN. Single base-pair substitutions in exon-intron junctions of human genes: nature, distribution, and consequences for mRNA splicing. Hum Mutat 2006;28:150–8.

[422] Treisman R, Orkin SH, Maniatis T. Specific transcription and RNA splicing defects in five cloned beta-thalassaemia genes. Nature 1983;302:591–6.

[423] Sharp PA. Splicing of messenger RNA precursors. Science 1987;235:766–71.

[424] Rosenthal A, Jouet M, Kenwrick S. Aberrant splicing of neural cell adhesion molecule L1 mRNA in a family with X-linked hydrocephalus. Nat Genet 1992;2:107–12.

[425] De Klein A, Riegman PH, Bijlsma EK, Heldoorn A, Muijtjens M, den Bakker MA, Avezaat CJ, Zwarthoff EC. A G→A transition creates a branch point sequence and activation of a cryptic exon, resulting in the hereditary disorder neurofibromatosis 2. Hum Mol Genet 1998;7:393–8.

[426] Burset M, Seledtsov IA, Solovyev VV. Analysis of canonical and non-canonical splice sites in mammalian genomes. Nucleic Acids Res 2000;28:4364–75.

[427] Shaw MA, Brunetti-Pierri N, Kadasi L, Kovacova V, Van Maldergem L, De Brasi D, Salerno M, Gecz J. Identification of three novel SEDL mutations, including mutation in the rare, non-canonical splice site of exon 4. Clin Genet 2003;64:235–42.

[428] Bradley KJ, Cavaco BM, Bowl MR, Harding B, Young A, Thakker RV. Utilisation of a cryptic non-canonical donor splice site of the gene encoding PARAFIBROMIN is associated with familial isolated primary hyperparathyroidism. J Med Genet 2005;42:e51.

[429] Cogan JD, Prince MA, Lekhakula S, Bundey S, Futrakul A, McCarthy EM, Phillips 3rd JA. A novel mechanism of aberrant pre-mRNA splicing in humans. Hum Mol Genet 1997;6:909–12.

[430] Santisteban I, Arredondo-Vega FX, Kelly S, Loubser M, Meydan N, Roifman C, Howell PL, Bowen T, Weinberg KI, Schroeder ML, et al. Three new adenosine deaminase mutations that define a splicing enhancer and cause severe and partial phenotypes: implications for evolution of a CpG hotspot and expression of a transduced ADA cDNA. Hum Mol Genet 1995;4:2081–7.

[431] Hutton M, Lendon CL, Rizzu P, Baker M, Froelich S, Houlden H, Pickering-Brown S, Chakraverty S, Isaacs A, Grover A, Hackett J, Adamson J, Lincoln S, Dickson D, Davies P, Petersen RC, Stevens M, de Graaff E, Wauters E, van Baren J, Hillebrand M, Joosse M, Kwon JM, Nowotny P, Che LK, Norton J, Morris JC, Reed LA, Trojanowski J, Basun H, Lannfelt L, Neystat M, Fahn S, Dark F, Tannenberg T, Dodd PR, Hayward N, Kwok JB, Schofield PR, Andreadis A, Snowden J, Craufurd D, Neary D, Owen F, Oostra BA, Hardy J, Goate A, van Swieten J, Mann D, Lynch T, Heutink P. Association of missense and 5′-splice-site mutations in tau with the inherited dementia FTDP-17. Nature 1998;393:702–5.

[432] Pagani F, Buratti E, Stuani C, Bendix R, Dork T, Baralle FE. A new type of mutation causes a splicing defect in ATM. Nat Genet 2002;30:426–9.

[433] von Ahsen N, Oellerich M. The intronic prothrombin 19911A>G polymorphism influences splicing efficiency and modulates effects of the 20210G>A polymorphism on mRNA amount and expression in a stable reporter gene assay system. Blood 2004;103:586–93.

[434] Dietz HC, Valle D, Francomano CA, Kendzior Jr RJ, Pyeritz RE, Cutting GR. The skipping of constitutive exons in vivo induced by nonsense mutations. Science 1993;259:680–3.

[435] Chao HK, Hsiao KJ, Su TS. A silent mutation induces exon skipping in the phenylalanine hydroxylase gene in phenylketonuria. Hum Genet 2001;108:14–9.
[436] Liu HX, Cartegni L, Zhang MQ, Krainer AR. A mechanism for exon skipping caused by nonsense or missense mutations in BRCA1 and other genes. Nat Genet 2001;27:55–8.
[437] Blencowe BJ. Exonic splicing enhancers: mechanism of action, diversity and role in human genetic diseases. Trends Biochem Sci 2000;25:106–10.
[438] Pagani F, Raponi M, Baralle FE. Synonymous mutations in CFTR exon 12 affect splicing and are not neutral in evolution. Proc Natl Acad Sci U S A 2005;102:6368–72.
[439] Pagani F, Baralle FE. Genomic variants in exons and introns: identifying the splicing spoilers. Nat Rev Genet 2004;5:389–96.
[440] Gorlov IP, Gorlova OY, Frazier ML, Amos CI. Missense mutations in hMLH1 and hMSH2 are associated with exonic splicing enhancers. Am J Hum Genet 2003;73:1157–61.
[441] Lim KH, Ferraris L, Filloux ME, Raphael BJ, Fairbrother WG. Using positional distribution to identify splicing elements and predict pre-mRNA processing defects in human genes. Proc Natl Acad Sci U S A 2011;108:11093–8.
[442] Sterne-Weiler T, Howard J, Mort M, Cooper DN, Sanford JR. Loss of exon identity is a common mechanism of human inherited disease. Genome Res 2011;21:1563–71.
[443] Mitchell GA, Labuda D, Fontaine G, Saudubray JM, Bonnefont JP, Lyonnet S, Brody LC, Steel G, Obie C, Valle D. Splice-mediated insertion of an Alu sequence inactivates ornithine delta-aminotransferase: a role for Alu elements in human mutation. Proc Natl Acad Sci U S A 1991;88:815–9.
[444] Knebelmann B, Forestier L, Drouot L, Quinones S, Chuet C, Benessy F, Saus J, Antignac C. Splice-mediated insertion of an Alu sequence in the COL4A3 mRNA causing autosomal recessive Alport syndrome. Hum Mol Genet 1995;4:675–9.
[445] Coutinho G, Xie J, Du L, Brusco A, Krainer AR, Gatti RA. Functional significance of a deep intronic mutation in the ATM gene and evidence for an alternative exon 28a. Hum Mutat 2005;25:118–24.
[446] Harland M, Mistry S, Bishop DT, Bishop JA. A deep intronic mutation in CDKN2A is associated with disease in a subset of melanoma pedigrees. Hum Mol Genet 2001;10:2679–86.
[447] Tuffery-Giraud S, Saquet C, Chambert S, Claustres M. Pseudoexon activation in the DMD gene as a novel mechanism for Becker muscular dystrophy. Hum Mutat 2003;21:608–14.
[448] Pros E, Gomez C, Martin T, Fabregas P, Serra E, Lazaro C. Nature and mRNA effect of 282 different NF1 point mutations: focus on splicing alterations. Hum Mutat 2008;29:E173–93.
[449] Dhir A, Buratti E. Alternative splicing: role of pseudoexons in human disease and potential therapeutic strategies. FEBS J 2010;277:841–55.
[450] Choi JW, Park CS, Hwang M, Nam HY, Chang HS, Park SG, Han BG, Kimm K, Kim HL, Oh B, Kim Y. A common intronic variant of CXCR3 is functionally associated with gene expression levels and the polymorphic immune cell responses to stimuli. J Allergy Clin Immunol 2008;122:1119–26. e1117.
[451] Fraser HB, Xie X. Common polymorphic transcript variation in human disease. Genome Res 2009;19:567–75.
[452] Susa S, Daimon M, Sakabe J, Sato H, Oizumi T, Karasawa S, Wada K, Jimbu Y, Kameda W, Emi M, Muramatsu M, Kato T. A functional polymorphism of the TNF-alpha gene that is associated with type 2 DM. Biochem Biophys Res Commun 2008;369:943–7.
[453] Vaz-Drago R, Custódio N, Carmo-Fonseca M, PMID:28497172, DOI:10.1007/s00439-017-1809-4.
[454] Chen JM, Ferec C, Cooper DN. A systematic analysis of disease-associated variants in the 3′ regulatory regions of human protein-coding genes II: the importance of mRNA secondary structure in assessing the functionality of 3′ UTR variants. Hum Genet 2006b;120:301–33.
[455] Orkin SH, Cheng TC, Antonarakis SE, Kazazian Jr HH. Thalassemia due to a mutation in the cleavage-polyadenylation signal of the human beta-globin gene. EMBO J 1985;4:453–6.
[456] Cai SP, Eng B, Francombe WH, Olivieri NF, Kendall AG, Waye JS, Chui DH. Two novel beta-thalassemia mutations in the 5′ and 3′ noncoding regions of the beta-globin gene. Blood 1992;79:1342–6.
[457] Gehring NH, Frede U, Neu-Yilik G, Hundsdoerfer P, Vetter B, Hentze MW, Kulozik AE. Increased efficiency of mRNA 3′ end formation: a new genetic mechanism contributing to hereditary thrombophilia. Nat Genet 2001;28:389–92.
[458] Poort SR, Rosendaal FR, Reitsma PH, Bertina RM. A common genetic variation in the 3′-untranslated region of the prothrombin gene is associated with elevated plasma prothrombin levels and an increase in venous thrombosis. Blood 1996;88:3698–703.
[459] Abelson JF, Kwan KY, O'Roak BJ, Baek DY, Stillman AA, Morgan TM, Mathews CA, Pauls DL, Rasin MR, Gunel M, Davis NR, Ercan-Sencicek AG, Guez DH, Spertus JA, Leckman JF, Dure LSt, Kurlan R, Singer HS, Gilbert DL, Farhi A, Louvi A, Lifton RP, Sestan N, State MW. Sequence variants in SLITRK1 are associated with Tourette's syndrome. Science 2005;310:317–20.

[460] Sethupathy P, Borel C, Gagnebin M, Grant G.R, Deutsch S., et al. Am J Hum, Gent. 2006;81:405–13.
[461] Bandiera S, Hatem E, Lyonnet S, Henrion-Caude A. microRNAs in diseases: from candidate to modifier genes. Clin Genet 2010;77:306–13.
[462] Martin MM, Buckenberger JA, Jiang J, Malana GE, Nuovo GJ, Chotani M, Feldman DS, Schmittgen TD, Elton TS. The human angiotensin II type 1 receptor +1166 A/C polymorphism attenuates microrna-155 binding. J Biol Chem 2007;282:24262–9.
[463] Rademakers R, Eriksen JL, Baker M, Robinson T, Ahmed Z, Lincoln SJ, Finch N, Rutherford NJ, Crook RJ, Josephs KA, Boeve BF, Knopman DS, Petersen RC, Parisi JE, Caselli RJ, Wszolek ZK, Uitti RJ, Feldman H, Hutton ML, Mackenzie IR, Graff-Radford NR, Dickson DW. Common variation in the miR-659 binding-site of GRN is a major risk factor for TDP43-positive frontotemporal dementia. Hum Mol Genet 2008;17:3631–42.
[464] Sethupathy P, Borel C, Gagnebin M, Grant GR, Deutsch S, Elton TS, Hatzigeorgiou AG, Antonarakis SE. Human microRNA-155 on chromosome 21 differentially interacts with its polymorphic target in the AGTR1 3′ untranslated region: a mechanism for functional single-nucleotide polymorphisms related to phenotypes. Am J Hum Genet 2007;81:405–13.
[465] Simon D, Laloo B, Barillot M, Barnetche T, Blanchard C, Rooryck C, Marche M, Burgelin I, Coupry I, Chassaing N, Gilbert-Dussardier B, Lacombe D, Grosset C, Arveiler B. A mutation in the 3′-UTR of the HDAC6 gene abolishing the post-transcriptional regulation mediated by hsa-miR-433 is linked to a new form of dominant X-linked chondrodysplasia. Hum Mol Genet 2010;19:2015–27.
[466] Esteller M. Non-coding RNAs in human disease. Nat Rev Genet 2011;12:861–74.
[467] Dong XY, Rodriguez C, Guo P, Sun X, Talbot JT, Zhou W, Petros J, Li Q, Vessella RL, Kibel AS, Stevens VL, Calle EE, Dong JT. SnoRNA U50 is a candidate tumor-suppressor gene at 6q14.3 with a mutation associated with clinically significant prostate cancer. Hum Mol Genet 2008;17:1031–42.
[468] Sahoo T, del Gaudio D, German JR, Shinawi M, Peters SU, Person RE, Garnica A, Cheung SW, Beaudet AL. Prader-Willi phenotype caused by paternal deficiency for the HBII-85 C/D box small nucleolar RNA cluster. Nat Genet 2008;40:719–21.
[469] Li H, Xie H, Liu W, Hu R, Huang B, Tan YF, Xu K, Sheng ZF, Zhou HD, Wu XP, Luo XH. A novel microRNA targeting HDAC5 regulates osteoblast differentiation in mice and contributes to primary osteoporosis in humans. J Clin Invest 2009;119:3666–77.
[470] Mencia A, Modamio-Hoybjor S, Redshaw N, Morin M, Mayo-Merino F, Olavarrieta L, Aguirre LA, del Castillo I, Steel KP, Dalmay T, Moreno F, Moreno-Pelayo MA. Mutations in the seed region of human miR-96 are responsible for nonsyndromic progressive hearing loss. Nat Genet 2009;41:609–13.
[471] Saus E, Soria V, Escaramis G, Vivarelli F, Crespo JM, Kagerbauer B, Menchon JM, Urretavizcaya M, Gratacos M, Estivill X. Genetic variants and abnormal processing of pre-miR-182, a circadian clock modulator, in major depression patients with late insomnia. Hum Mol Genet 2010;19:4017–25.
[472] Sun G, Yan J, Noltner K, Feng J, Li H, Sarkis DA, Sommer SS, Rossi JJ. SNPs in human miRNA genes affect biogenesis and function. RNA 2009;15:1640–51.
[473] Ridanpaa M, van Eenennaam H, Pelin K, Chadwick R, Johnson C, Yuan B, vanVenrooij W, Pruijn G, Salmela R, Rockas S, Makitie O, Kaitila I, de la Chapelle A. Mutations in the RNA component of RNase MRP cause a pleiotropic human disease, cartilage-hair hypoplasia. Cell 2001;104:195–203.
[474] Bastepe M, Frohlich LF, Linglart A, Abu-Zahra HS, Tojo K, Ward LM, Juppner H. Deletion of the NESP55 differentially methylated region causes loss of maternal GNAS imprints and pseudohypoparathyroidism type Ib. Nat Genet 2005;37:25–7.
[475] Jarinova O, Stewart AF, Roberts R, Wells G, Lau P, Naing T, Buerki C, McLean BW, Cook RC, Parker JS, McPherson R. Functional analysis of the chromosome 9p21.3 coronary artery disease risk locus. Arterioscler Thromb Vasc Biol 2009;29:1671–7.
[476] Brakenhoff RH, Henskens HA, van Rossum MW, Lubsen NH, Schoenmakers JG. Activation of the gamma E-crystallin pseudogene in the human hereditary Coppock-like cataract. Hum Mol Genet 1994;3:279–83.
[477] Lewis MA, Quint E, Glazier AM, Fuchs H, De Angelis MH, Langford C, van Dongen S, Abreu-Goodger C, Piipari M, Redshaw N, Dalmay T, Moreno-Pelayo MA, Enright AJ, Steel KP. An ENU-induced mutation of miR-96 associated with progressive hearing loss in mice. Nat Genet 2009;41:614–8.
[478] Hughes AE, Bradley DT, Campbell M, Lechner J, Dash DP, Simpson DA, Willoughby CE. Mutation altering the miR-184 seed region causes familial keratoconus with cataract. Am J Hum Genet 2011;89:628–33.
[479] de Pontual L, Yao E, Callier P, Faivre L, Drouin V, Cariou S, Van Haeringen A, Genevieve D, Goldenberg A, Oufadem M, Manouvrier S, Munnich A, Vidigal JA, Vekemans M, Lyonnet S, Henrion-Caude A, Ventura A, Amiel J. Germline deletion of the miR-17 approximately 92 cluster causes skeletal and growth defects in humans. Nat Genet 2011;43:1026–30.
[480] Abdel-Salam GM, Abdel-Hamid MS, Issa M, Magdy A, El-Kotoury A, Amr K. Expanding the phenotypic and mutational spectrum in microcephalic osteodys-

plastic primordial dwarfism type I. Am J Med Genet A 2012;158A:1455–61.
[481] He H, Liyanarachchi S, Akagi K, Nagy R, Li J, Dietrich RC, Li W, Sebastian N, Wen B, Xin B, Singh J, Yan P, Alder H, Haan E, Wieczorek D, Albrecht B, Puffenberger E, Wang H, Westman JA, Padgett RA, Symer DE, de la Chapelle A. Mutations in U4atac snRNA, a component of the minor spliceosome, in the developmental disorder MOPD I. Science 2011;332:238–40.
[482] Makrythanasis P, Antonarakis SE. Pathogenic variants in non-protein-coding sequences. Clin Genet 2013;84:422–8.
[483] Collins LJ, Penny D. The RNA infrastructure: dark matter of the eukaryotic cell? Trends Genet 2009;25:120–8.
[484] Asthana S, Noble WS, Kryukov G, Grant CE, Sunyaev S, Stamatoyannopoulos JA. Widely distributed noncoding purifying selection in the human genome. Proc Natl Acad Sci U S A 2007;104:12410–5.
[485] Kryukov GV, Schmidt S, Sunyaev S. Small fitness effect of mutations in highly conserved non-coding regions. Hum Mol Genet 2005;14:2221–9.
[486] Chen CT, Wang JC, Cohen BA. The strength of selection on ultraconserved elements in the human genome. Am J Hum Genet 2007;80:692–704.
[487] Visel A, Rubin EM, Pennacchio LA. Genomic views of distant-acting enhancers. Nature 2009;461:199–205.
[488] Schork NJ, Murray SS, Frazer KA, Topol EJ. Common vs. rare allele hypotheses for complex diseases. Curr Opin Genet Dev 2009;19:212–9.
[489] Glinskii AB, Ma J, Ma S, Grant D, Lim CU, Sell S, Glinsky GV. Identification of intergenic trans-regulatory RNAs containing a disease-linked SNP sequence and targeting cell cycle progression/differentiation pathways in multiple common human disorders. Cell Cycle 2009;8:3925–42.
[490] Hindorff LA, Sethupathy P, Junkins HA, Ramos EM, Mehta JP, Collins FS, Manolio TA. Potential etiologic and functional implications of genome-wide association loci for human diseases and traits. Proc Natl Acad Sci U S A 2009;106:9362–7.
[491] Denoeud F, Kapranov P, Ucla C, Frankish A, Castelo R, Drenkow J, Lagarde J, Alioto T, Manzano C, Chrast J, Dike S, Wyss C, Henrichsen CN, Holroyd N, Dickson MC, Taylor R, Hance Z, Foissac S, Myers RM, Rogers J, Hubbard T, Harrow J, Guigo R, Gingeras TR, Antonarakis SE, Reymond A. Prominent use of distal 5′ transcription start sites and discovery of a large number of additional exons in ENCODE regions. Genome Res 2007;17:746–59.
[492] Dickson SP, Wang K, Krantz I, Hakonarson H, Goldstein DB. Rare variants create synthetic genome-wide associations. PLoS Biol 2010;8:e1000294.
[493] Attanasio C, Reymond A, Humbert R, Lyle R, Kuehn MS, Neph S, Sabo PJ, Goldy J, Weaver M, Haydock A, Lee K, Dorschner M, Dermitzakis ET, Antonarakis SE, Stamatoyannopoulos JA. Assaying the regulatory potential of mammalian conserved non-coding sequences in human cells. Genome Biol 2008;9:R168.
[494] Quemener S, Chen JM, Chuzhanova N, Benech C, Casals T, Macek Jr M, Bienvenu T, McDevitt T, Farrell PM, Loumi O, Messaoud T, Cuppens H, Cutting GR, Stenson PD, Giteau K, Audrezet MP, Cooper DN, Ferec C. Complete ascertainment of intragenic copy number mutations (CNMs) in the CFTR gene and its implications for CNM formation at other autosomal loci. Hum Mutat 2010;31:421–8.
[495] Zhang F, Lupski JR. Non-coding genetic variants in human disease. Hum Mol Genet 2015;24:R102–10.
[496] Spielmann M, Mundlos S. Looking beyond the genes: the role of non-coding variants in human disease. Hum Mol Genet 2016;25:R157–65.
[497] Dixon JR, Selvaraj S, Yue F, Kim A, Li Y, Shen Y, Hu M, Liu JS, Ren B. Topological domains in mammalian genomes identified by analysis of chromatin interactions. Nature 2012;485:376–80.
[498] Lupianez DG, Kraft K, Heinrich V, Krawitz P, Brancati F, Klopocki E, Horn D, Kayserili H, Opitz JM, Laxova R, Santos-Simarro F, Gilbert-Dussardier B, Wittler L, Borschiwer M, Haas SA, Osterwalder M, Franke M, Timmermann B, Hecht J, Spielmann M, Visel A, Mundlos S. Disruptions of topological chromatin domains cause pathogenic rewiring of gene-enhancer interactions. Cell 2015;161:1012–25.
[499] Franke M, Ibrahim DM, Andrey G, Schwarzer W, Heinrich V, Schopflin R, Kraft K, Kempfer R, Jerkovic I, Chan WL, Spielmann M, Timmermann B, Wittler L, Kurth I, Cambiaso P, Zuffardi O, Houge G, Lambie L, Brancati F, Pombo A, Vingron M, Spitz F, Mundlos S. Formation of new chromatin domains determines pathogenicity of genomic duplications. Nature 2016;538:265–9.
[500] Wong C, Dowling CE, Saiki RK, Higuchi RG, Erlich HA, Kazazian Jr HH. Characterization of beta-thalassaemia mutations using direct genomic sequencing of amplified single copy DNA. Nature 1987;330:384–6.
[501] Kozak M. Compilation and analysis of sequences upstream from the translational start site in eukaryotic mRNAs. Nucleic Acids Res 1984;12:857–72.
[502] Mehdi H, Manzi S, Desai P, Chen Q, Nestlerode C, Bontempo F, Strom SC, Zarnegar R, Kamboh MI. A functional polymorphism at the transcriptional initiation site in beta2-glycoprotein I (apolipoprotein H) associated with reduced gene expression and lower plasma levels of beta2-glycoprotein I. Eur J Biochem 2003;270:230–8.

[503] Cazzola M, Skoda RC. Translational pathophysiology: a novel molecular mechanism of human disease. Blood 2000;95:3280–8.
[504] Girelli D, Corrocher R, Bisceglia L, Olivieri O, De Franceschi L, Zelante L, Gasparini P. Molecular basis for the recently described hereditary hyperferritinemia-cataract syndrome: a mutation in the iron-responsive element of ferritin L-subunit gene (the "Verona mutation"). Blood 1995;86:4050–3.
[505] Athanassiadou A, Papachatzopoulou A, Zoumbos N, Maniatis GM, Gibbs R. A novel beta-thalassaemia mutation in the 5′ untranslated region of the beta-globin gene. Br J Haematol 1994;88:307–10.
[506] Ho PJ, Rochette J, Fisher CA, Wonke B, Jarvis MK, Yardumian A, Thein SL. Moderate reduction of beta-globin gene transcript by a novel mutation in the 5′ untranslated region: a study of its interaction with other genotypes in two families. Blood 1996;87:1170–8.
[507] Sgourou A, Routledge S, Antoniou M, Papachatzopoulou A, Psiouri L, Athanassiadou A. Thalassaemia mutations within the 5′UTR of the human beta-globin gene disrupt transcription. Br J Haematol 2004;124:828–35.
[508] Conne B, Stutz A, Vassalli JD. The 3′ untranslated region of messenger RNA: a molecular 'hotspot' for pathology? Nat Med 2000;6:637–41.
[509] Moi P, Loudianos G, Lavinha J, Murru S, Cossu P, Casu R, Oggiano L, Longinotti M, Cao A, Pirastu M. Delta-thalassemia due to a mutation in an erythroid-specific binding protein sequence 3′ to the delta-globin gene. Blood 1992;79:512–6.
[510] Chen JM, Ferec C, Cooper DN. A systematic analysis of disease-associated variants in the 3′ regulatory regions of human protein-coding genes I: general principles and overview. Hum Genet 2006a;120:1–21.
[511] Pirastu M, Saglio G, Chang JC, Cao A, Kan YW. Initiation codon mutation as a cause of alpha thalassemia. J Biol Chem 1984;259:12315–7.
[512] Kozak M. Structural features in eukaryotic mRNAs that modulate the initiation of translation. J Biol Chem 1991;266:19867–70.
[513] Jacobson EM, Concepcion E, Oashi T, Tomer Y. A Graves' disease-associated Kozak sequence single-nucleotide polymorphism enhances the efficiency of CD40 gene translation: a case for translational pathophysiology. Endocrinology 2005;146:2684–91.
[514] Kozak M. Emerging links between initiation of translation and human diseases. Mamm Genome 2002;13:401–10.
[515] Wolf A, Caliebe A, Thomas NS, Ball EV, Mort M, Stenson PD, Krawczak M, Cooper DN. Single base-pair substitutions at the translation initiation sites of human genes as a cause of inherited disease. Hum Mutat 2011;32:1137–43.
[516] Clegg JB, Weatherall DJ, Milner PF. Haemoglobin Constant Spring–a chain termination mutant? Nature 1971;234:337–40.
[517] Hamby SE, Thomas NS, Cooper DN, Chuzhanova N. A meta-analysis of single base-pair substitutions in translational termination codons ('nonstop' mutations) that cause human inherited disease. Hum Genomics 2011;5:241–64.
[518] Zia A, Moses AM. Ranking insertion, deletion and nonsense mutations based on their effect on genetic information. BMC Bioinformatics 2011;12:299.
[519] Mort M, Ivanov D, Cooper DN, Chuzhanova NA. A meta-analysis of nonsense mutations causing human genetic disease. Hum Mutat 2008;29:1037–47.
[520] Benz EJ, Forget BG, Hillman DG, Cohen-Solal M, Pritchard J, Cavallesco C, Prensky W, Housman D. Variability in the amount of beta-globin mRNA in beta0 thalassemia. Cell 1978;14:299–312.
[521] Maquat LE. Nonsense-mediated mRNA decay: splicing, translation and mRNP dynamics. Nat Rev Mol Cell Biol 2004;5:89–99.
[522] Inacio A, Silva AL, Pinto J, Ji X, Morgado A, Almeida F, Faustino P, Lavinha J, Liebhaber SA, Romao L. Nonsense mutations in close proximity to the initiation codon fail to trigger full nonsense-mediated mRNA decay. J Biol Chem 2004;279:32170–80.
[523] Rivas MA, Pirinen M, Conrad DF, Lek M, Tsang EK, Karczewski KJ, Maller JB, Kukurba KR, DeLuca DS, Fromer M, Ferreira PG, Smith KS, Zhang R, Zhao F, Banks E, Poplin R, Ruderfer DM, Purcell SM, Tukiainen T, Minikel EV, Stenson PD, Cooper DN, Huang KH, Sullivan TJ, Nedzel J, Consortium GT, Geuvadis C, Bustamante CD, Li JB, Daly MJ, Guigo R, Donnelly P, Ardlie K, Sammeth M, Dermitzakis ET, McCarthy MI, Montgomery SB, Lappalainen T, MacArthur DG. Human genomics. Effect of predicted protein-truncating genetic variants on the human transcriptome. Science 2015;348:666–9.
[524] Inoue K, Khajavi M, Ohyama T, Hirabayashi S, Wilson J, Reggin JD, Mancias P, Butler IJ, Wilkinson MF, Wegner M, Lupski JR. Molecular mechanism for distinct neurological phenotypes conveyed by allelic truncating mutations. Nat Genet 2004;36:361–9.
[525] Ozisik G, Mantovani G, Achermann JC, Persani L, Spada A, Weiss J, Beck-Peccoz P, Jameson JL. An alternate translation initiation site circumvents an amino-terminal DAX1 nonsense mutation leading to a mild form of X-linked adrenal hypoplasia congenita. J Clin Endocrinol Metab 2003;88:417–23.
[526] Frischmeyer PA, Dietz HC. Nonsense-mediated mRNA decay in health and disease. Hum Mol Genet 1999;8:1893–900.
[527] Pacho F, Zambruno G, Calabresi V, Kiritsi D, Schneider H. Efficiency of translation termination in humans

is highly dependent upon nucleotides in the neighbourhood of a (premature) termination codon. J Med Genet 2011;48:640–4.
[528] Alber T. Mutational effects on protein stability. Annu Rev Biochem 1989;58:765–98.
[529] Pakula AA, Sauer RT. Genetic analysis of protein stability and function. Annu Rev Genet 1989;23:289–310.
[530] Wacey AI, Cooper DN, Liney D, Hovig E, Krawczak M. Disentangling the perturbational effects of amino acid substitutions in the DNA-binding domain of p53. Hum Genet 1999;104:15–22.
[531] Bross P, Corydon TJ, Andresen BS, Jorgensen MM, Bolund L, Gregersen N. Protein misfolding and degradation in genetic diseases. Hum Mutat 1999;14:186–98.
[532] Gregersen N, Bross P, Jorgensen MM, Corydon TJ, Andresen BS. Defective folding and rapid degradation of mutant proteins is a common disease mechanism in genetic disorders. J Inherit Metab Dis 2000;23:441–7.
[533] Grosveld F, van Assendelft GB, Greaves DR, Kollias G. Position-independent, high-level expression of the human beta-globin gene in transgenic mice. Cell 1987;51:975–85.
[534] Stamatoyannopoulos G. Human hemoglobin switching. Science 1991;252:383.
[535] Vyas P, Vickers MA, Simmons DL, Ayyub H, Craddock CF, Higgs DR. Cis-acting sequences regulating expression of the human alpha-globin cluster lie within constitutively open chromatin. Cell 1992;69:781–93.
[536] Driscoll MC, Dobkin CS, Alter BP. Gamma delta beta-thalassemia due to a de novo mutation deleting the 5′ beta-globin gene activation-region hypersensitive sites. Proc Natl Acad Sci U S A 1989;86:7470–4.
[537] Liebhaber SA, Griese EU, Weiss I, Cash FE, Ayyub H, Higgs DR, Horst J. Inactivation of human alpha-globin gene expression by a de novo deletion located upstream of the alpha-globin gene cluster. Proc Natl Acad Sci U S A 1990;87:9431–5.
[538] Lecointre R, Lima S, Varlet MN, Combe C. Immunoglobulin treatment for neonatal hemochromatosis: a case report in a context of immunoglobulin delivery quotas. Ann Pharm Fr 2009;67:304–9.
[539] D'Haene B, Attanasio C, Beysen D, Dostie J, Lemire E, Bouchard P, Field M, Jones K, Lorenz B, Menten B, Buysse K, Pattyn F, Friedli M, Ucla C, Rossier C, Wyss C, Speleman F, De Paepe A, Dekker J, Antonarakis SE, De Baere E. Disease-causing 7.4 kb cis-regulatory deletion disrupting conserved non-coding sequences and their interaction with the FOXL2 promotor: implications for mutation screening. PLoS Genet 2009;5:e1000522.
[540] Chen J, Wildhardt G, Zhong Z, Roth R, Weiss B, Steinberger D, Decker J, Blum WF, Rappold G. Enhancer deletions of the SHOX gene as a frequent cause of short stature: the essential role of a 250 kb downstream regulatory domain. J Med Genet 2009;46:834–9.
[541] Gordon CT, Tan TY, Benko S, Fitzpatrick D, Lyonnet S, Farlie PG. Long-range regulation at the SOX9 locus in development and disease. J Med Genet 2009;46:649–56.
[542] Benko S, Fantes JA, Amiel J, Kleinjan DJ, Thomas S, Ramsay J, Jamshidi N, Essafi A, Heaney S, Gordon CT, McBride D, Golzio C, Fisher M, Perry P, Abadie V, Ayuso C, Holder-Espinasse M, Kilpatrick N, Lees MM, Picard A, Temple IK, Thomas P, Vazquez MP, Vekemans M, Roest Crollius H, Hastie ND, Munnich A, Etchevers HC, Pelet A, Farlie PG, Fitzpatrick DR, Lyonnet S. Highly conserved non-coding elements on either side of SOX9 associated with Pierre Robin sequence. Nat Genet 2009;41:359–64.
[543] Rahimov F, Marazita ML, Visel A, Cooper ME, Hitchler MJ, Rubini M, Domann FE, Govil M, Christensen K, Bille C, Melbye M, Jugessur A, Lie RT, Wilcox AJ, Fitzpatrick DR, Green ED, Mossey PA, Little J, Steegers-Theunissen RP, Pennacchio LA, Schutte BC, Murray JC. Disruption of an AP-2alpha binding site in an IRF6 enhancer is associated with cleft lip. Nat Genet 2008;40:1341–7.
[544] Haiman CA, Le Marchand L, Yamamato J, Stram DO, Sheng X, Kolonel LN, Wu AH, Reich D, Henderson BE. A common genetic risk factor for colorectal and prostate cancer. Nat Genet 2007;39:954–6.
[545] Pomerantz MM, Ahmadiyeh N, Jia L, Herman P, Verzi MP, Doddapaneni H, Beckwith CA, Chan JA, Hills A, Davis M, Yao K, Kehoe SM, Lenz HJ, Haiman CA, Yan C, Henderson BE, Frenkel B, Barretina J, Bass A, Tabernero J, Baselga J, Regan MM, Manak JR, Shivdasani R, Coetzee GA, Freedman ML. The 8q24 cancer risk variant rs6983267 shows long-range interaction with MYC in colorectal cancer. Nat Genet 2009;41:882–4.
[546] Tuupanen S, Turunen M, Lehtonen R, Hallikas O, Vanharanta S, Kivioja T, Bjorklund M, Wei G, Yan J, Niittymaki I, Mecklin JP, Jarvinen H, Ristimaki A, Di-Bernardo M, East P, Carvajal-Carmona L, Houlston RS, Tomlinson I, Palin K, Ukkonen E, Karhu A, Taipale J, Aaltonen LA. The common colorectal cancer predisposition SNP rs6983267 at chromosome 8q24 confers potential to enhanced Wnt signaling. Nat Genet 2009;41:885–90.
[547] Wright JB, Brown SJ, Cole MD. Upregulation of c-MYC in cis through a large chromatin loop linked to a cancer risk-associated single-nucleotide polymorphism in colorectal cancer cells. Mol Cell Biol 2010;30:1411–20.
[548] Enattah NS, Sahi T, Savilahti E, Terwilliger JD, Peltonen L, Jarvela I. Identification of a variant associated with adult-type hypolactasia. Nat Genet 2002;30:233–7.

[549] Lewinsky RH, Jensen TG, Moller J, Stensballe A, Olsen J, Troelsen JT. T-13910 DNA variant associated with lactase persistence interacts with Oct-1 and stimulates lactase promoter activity in vitro. Hum Mol Genet 2005;14:3945–53.

[550] Olds LC, Sibley E. Lactase persistence DNA variant enhances lactase promoter activity in vitro: functional role as a cis regulatory element. Hum Mol Genet 2003;12:2333–40.

[551] Veyrieras JB, Kudaravalli S, Kim SY, Dermitzakis ET, Gilad Y, Stephens M, Pritchard JK. High-resolution mapping of expression-QTLs yields insight into human gene regulation. PLoS Genet 2008;4:e1000214.

[552] Venturin M, Moncini S, Villa V, Russo S, Bonati MT, Larizza L, Riva P. Mutations and novel polymorphisms in coding regions and UTRs of CDK5R1 and OMG genes in patients with non-syndromic mental retardation. Neurogenetics 2006;7:59–66.

[553] Rung J, Cauchi S, Albrechtsen A, Shen L, Rocheleau G, Cavalcanti-Proenca C, Bacot F, Balkau B, Belisle A, Borch-Johnsen K, Charpentier G, Dina C, Durand E, Elliott P, Hadjadj S, Jarvelin MR, Laitinen J, Lauritzen T, Marre M, Mazur A, Meyre D, Montpetit A, Pisinger C, Posner B, Poulsen P, Pouta A, Prentki M, Ribel-Madsen R, Ruokonen A, Sandbaek A, Serre D, Tichet J, Vaxillaire M, Wojtaszewski JF, Vaag A, Hansen T, Polychronakos C, Pedersen O, Froguel P, Sladek R. Genetic variant near IRS1 is associated with type 2 diabetes, insulin resistance and hyperinsulinemia. Nat Genet 2009;41:1110–5.

[554] Warren ST, Nelson DL. Advances in molecular analysis of fragile X syndrome. JAMA 1994;271:536–42.

[555] Borrell-Pages M, Zala D, Humbert S, Saudou F. Huntington's disease: from huntingtin function and dysfunction to therapeutic strategies. Cell Mol Life Sci 2006;63:2642–60.

[556] Davis BM, McCurrach ME, Taneja KL, Singer RH, Housman DE. Expansion of a CUG trinucleotide repeat in the 3′ untranslated region of myotonic dystrophy protein kinase transcripts results in nuclear retention of transcripts. Proc Natl Acad Sci U S A 1997;94:7388–93.

[557] Day JW, Ranum LP. Genetics and molecular pathogenesis of the myotonic dystrophies. Curr Neurol Neurosci Rep 2005;5:55–9.

[558] McVey JH, Michaelides K, Hansen LP, Ferguson-Smith M, Tilghman S, Krumlauf R, Tuddenham EG. A G-->A substitution in an HNF I binding site in the human alpha-fetoprotein gene is associated with hereditary persistence of alpha-fetoprotein (HPAFP). Hum Mol Genet 1993;2:379–84.

[559] Wijmenga C, Hewitt JE, Sandkuijl LA, Clark LN, Wright TJ, Dauwerse HG, Gruter AM, Hofker MH, Moerer P, Williamson R, et al. Chromosome 4q DNA rearrangements associated with facioscapulohumeral muscular dystrophy. Nat Genet 1992;2:26–30.

[560] Deidda G, Cacurri S, Grisanti P, Vigneti E, Piazzo N, Felicetti L. Physical mapping evidence for a duplicated region on chromosome 10qter showing high homology with the facioscapulohumeral muscular dystrophy locus on chromosome 4qter. Eur J Hum Genet 1995;3:155–67.

[561] Lemmers RJ, van der Vliet PJ, Klooster R, Sacconi S, Camano P, Dauwerse JG, Snider L, Straasheijm KR, van Ommen GJ, Padberg GW, Miller DG, Tapscott SJ, Tawil R, Frants RR, van der Maarel SM. A unifying genetic model for facioscapulohumeral muscular dystrophy. Science 2010;329:1650–3.

[562] Richards M, Coppee F, Thomas N, Belayew A, Upadhyaya M. Facioscapulohumeral muscular dystrophy (FSHD): an enigma unravelled? Hum Genet 2011.

[563] Kleinjan DJ, van Heyningen V. Position effect in human genetic disease. Hum Mol Genet 1998;7:1611–8.

[564] Fantes J, Redeker B, Breen M, Boyle S, Brown J, Fletcher J, Jones S, Bickmore W, Fukushima Y, Mannens M, Danes S, van Heyningen V, Hanson I. Aniridia-associated cytogenetic rearrangements suggest that a position effect may cause the mutant phenotype. Hum Mol Genet 1995;4:415–22.

[565] Pfeifer D, Kist R, Dewar K, Devon K, Lander ES, Birren B, Korniszewski L, Back E, Scherer G. Campomelic dysplasia translocation breakpoints are scattered over 1 Mb proximal to SOX9: evidence for an extended control region. Am J Hum Genet 1999;65:111–24.

[566] Velagaleti GV, Bien-Willner GA, Northup JK, Lockhart LH, Hawkins JC, Jalal SM, Withers M, Lupski JR, Stankiewicz P. Position effects due to chromosome breakpoints that map approximately 900 Kb upstream and approximately 1.3 Mb downstream of SOX9 in two patients with campomelic dysplasia. Am J Hum Genet 2005;76:652–62.

[567] de Kok YJ, Vossenaar ER, Cremers CW, Dahl N, Laporte J, Hu LJ, Lacombe D, Fischel-Ghodsian N, Friedman RA, Parnes LS, Thorpe P, Bitner-Glindzicz M, Pander HJ, Heilbronner H, Graveline J, den Dunnen JT, Brunner HG, Ropers HH, Cremers FP. Identification of a hot spot for microdeletions in patients with X-linked deafness type 3 (DFN3) 900 kb proximal to the DFN3 gene POU3F4. Hum Mol Genet 1996;5:1229–35.

[568] Spitz F, Montavon T, Monso-Hinard C, Morris M, Ventruto ML, Antonarakis S, Ventruto V, Duboule D. A t(2;8) balanced translocation with breakpoints near the human HOXD complex causes mesomelic dysplasia and vertebral defects. Genomics 2002;79: 493–8.

[569] Beysen D, Raes J, Leroy BP, Lucassen A, Yates JR, Clayton-Smith J, Ilyina H, Brooks SS, Christin-Maitre S, Fellous M, Fryns JP, Kim JR, Lapunzina P, Lemyre E, Meire F, Messiaen LM, Oley C, Splitt M, Thomson J, Peer YV, Veitia RA, De Paepe A, De Baere E. Deletions involving long-range conserved nongenic sequences upstream and downstream of FOXL2 as a novel disease-causing mechanism in blepharophimosis syndrome. Am J Hum Genet 2005;77:205–18.
[570] Crisponi L, Deiana M, Loi A, Chiappe F, Uda M, Amati P, Bisceglia L, Zelante L, Nagaraja R, Porcu S, Ristaldi MS, Marzella R, Rocchi M, Nicolino M, Lienhardt-Roussie A, Nivelon A, Verloes A, Schlessinger D, Gasparini P, Bonneau D, Cao A, Pilia G. The putative forkhead transcription factor FOXL2 is mutated in blepharophimosis/ptosis/epicanthus inversus syndrome. Nat Genet 2001;27:159–66.
[571] Lettice LA, Horikoshi T, Heaney SJ, van_Baren MJ, van_der_Linde HC, Breedveld GJ, Joosse M, Akarsu N, Oostra BA, Endo N, Shibata M, Suzuki M, Takahashi E, Shinka T, Nakahori Y, Ayusawa D, Nakabayashi K, Scherer SW, Heutink P, Hill RE, Noji S. Disruption of a long-range cis-acting regulator for Shh causes preaxial polydactyly. Proc Natl Acad Sci U S A 2002;99:7548–53.
[572] Kleinjan DA, van Heyningen V. Long-range control of gene expression: emerging mechanisms and disruption in disease. Am J Hum Genet 2005;76:8–32.
[573] Bejerano G, Pheasant M, Makunin I, Stephen S, Kent WJ, Mattick JS, Haussler D. Ultraconserved elements in the human genome. Science 2004;304:1321–5.
[574] Boffelli D, Nobrega MA, Rubin EM. Comparative genomics at the vertebrate extremes. Nat Rev Genet 2004;5:456–65.
[575] Dermitzakis ET, Reymond A, Antonarakis SE. Conserved non-genic sequences - an unexpected feature of mammalian genomes. Nat Rev Genet 2005;6:151–7.
[576] Dermitzakis ET, Reymond A, Lyle R, Scamuffa N, Ucla C, Deutsch S, Stevenson BJ, Flegel V, Bucher P, Jongeneel CV, Antonarakis SE. Numerous potentially functional but non-genic conserved sequences on human chromosome 21. Nature 2002;420:578–82.
[577] Thomas JW, Touchman JW, Blakesley RW, Bouffard GG, Beckstrom-Sternberg SM, Margulies EH, Blanchette M, Siepel AC, Thomas PJ, McDowell JC, Maskeri B, Hansen NF, Schwartz MS, Weber RJ, Kent WJ, Karolchik D, Bruen TC, Bevan R, Cutler DJ, Schwartz S, Elnitski L, Idol JR, Prasad AB, Lee-Lin SQ, Maduro VV, Summers TJ, Portnoy ME, Dietrich NL, Akhter N, Ayele K, Benjamin B, Cariaga K, Brinkley CP, Brooks SY, Granite S, Guan X, Gupta J, Haghighi P, Ho SL, Huang MC, Karlins E, Laric PL, Legaspi R, Lim MJ, Maduro QL, Masiello CA, Mastrian SD, McCloskey JC, Pearson R, Stantripop S, Tiongson EE, Tran JT, Tsurgeon C, Vogt JL, Walker MA, Wetherby KD, Wiggins LS, Young AC, Zhang LH, Osoegawa K, Zhu B, Zhao B, Shu CL, De Jong PJ, Lawrence CE, Smit AF, Chakravarti A, Haussler D, Green P, Miller W, Green ED. Comparative analyses of multi-species sequences from targeted genomic regions. Nature 2003;424:788–93.
[578] Loots GG, Kneissel M, Keller H, Baptist M, Chang J, Collette NM, Ovcharenko D, Plajzer-Frick I, Rubin EM. Genomic deletion of a long-range bone enhancer misregulates sclerostin in Van Buchem disease. Genome Res 2005;15:928–35.
[579] Tufarelli C, Stanley JA, Garrick D, Sharpe JA, Ayyub H, Wood WG, Higgs DR. Transcription of antisense RNA leading to gene silencing and methylation as a novel cause of human genetic disease. Nat Genet 2003;34:157–65.
[580] Pascoe L, Jeunemaitre X, Lebrethon MC, Curnow KM, Gomez-Sanchez CE, Gasc JM, Saez JM, Corvol P. Glucocorticoid-suppressible hyperaldosteronism and adrenal tumors occurring in a single French pedigree. J Clin Invest 1995;96:2236–46.
[581] Nathans J, Piantanida TP, Eddy RL, Shows TB, Hogness DS. Molecular genetics of inherited variation in human color vision. Science 1986;232:203–10.
[582] Francis NJ, McNicholas B, Awan A, Waldron M, Reddan D, Sadlier D, Kavanagh D, Strain L, Marchbank KJ, Harris CL, Goodship TH. A novel hybrid CFH/CFHR3 gene generated by a microhomology-mediated deletion in familial atypical hemolytic uremic syndrome. Blood 2011.
[583] Bartram CR, de Klein A, Hagemeijer A, van Agthoven T, Geurts van Kessel A, Bootsma D, Grosveld G, Ferguson-Smith MA, Davies T, Stone M, et al. Translocation of c-ab1 oncogene correlates with the presence of a Philadelphia chromosome in chronic myelocytic leukaemia. Nature 1983;306:277–80.
[584] Delattre O, Zucman J, Plougastel B, Desmaze C, Melot T, Peter M, Kovar H, Joubert I, de Jong P, Rouleau G, et al. Gene fusion with an ETS DNA-binding domain caused by chromosome translocation in human tumours. Nature 1992;359:162–5.
[585] Rabbitts TH. Chromosomal translocations in human cancer. Nature 1994;372:143–9.
[586] Frank SA, Nowak MA. Problems of somatic mutation and cancer. Bioessays 2004;26:291–9.
[587] Erickson RP. Somatic gene mutation and human disease other than cancer. Mutat Res 2003;543:125–36.
[588] Fishel R, Lescoe MK, Rao MR, Copeland NG, Jenkins NA, Garber J, Kane M, Kolodner R. The human mutator gene homolog MSH2 and its association with hereditary nonpolyposis colon cancer. Cell 1993;75:1027–38.
[589] Leach FS, Nicolaides NC, Papadopoulos N, Liu B, Jen J, Parsons R, Peltomaki P, Sistonen P, Aaltonen LA, Nystrom-Lahti M, et al. Mutations of a mutS homolog

in hereditary nonpolyposis colorectal cancer. Cell 1993;75:1215–25.
[590] Papadopoulos N, Nicolaides NC, Wei YF, Ruben SM, Carter KC, Rosen CA, Haseltine WA, Fleischmann RD, Fraser CM, Adams MD, et al. Mutation of a mutL homolog in hereditary colon cancer. Science 1994;263:1625–9.
[591] Ionov Y, Peinado MA, Malkhosyan S, Shibata D, Perucho M. Ubiquitous somatic mutations in simple repeated sequences reveal a new mechanism for colonic carcinogenesis. Nature 1993;363:558–61.
[592] Markowitz S, Wang J, Myeroff L, Parsons R, Sun L, Lutterbaugh J, Fan RS, Zborowska E, Kinzler KW, Vogelstein B, et al. Inactivation of the type II TGF-beta receptor in colon cancer cells with microsatellite instability. Science 1995;268:1336–8.
[593] Schmutte C, Jones PA. Involvement of DNA methylation in human carcinogenesis. Biol Chem 1998;379:377–88.
[594] Upadhyaya M, Han S, Consoli C, Majounie E, Horan M, Thomas NS, Potts C, Griffiths S, Ruggieri M, von Deimling A, Cooper DN. Characterization of the somatic mutational spectrum of the neurofibromatosis type 1 (NF1) gene in neurofibromatosis patients with benign and malignant tumors. Hum Mutat 2004;23:134–46.
[595] Greenblatt MS, Grollman AP, Harris CC. Deletions and insertions in the p53 tumor suppressor gene in human cancers: confirmation of the DNA polymerase slippage/misalignment model. Cancer Res 1996;56:2130–6.
[596] Jego N, Thomas G, Hamelin R. Short direct repeats flanking deletions, and duplicating insertions in p53 gene in human cancers. Oncogene 1993;8:209–13.
[597] Kolomietz E, Meyn MS, Pandita A, Squire JA. The role of Alu repeat clusters as mediators of recurrent chromosomal aberrations in tumors. Genes Chromosomes Cancer 2002;35:97–112.
[598] Oldenburg J, Rost S, El-Maarri O, Leuer M, Olek K, Muller CR, Schwaab R. De novo factor VIII gene intron 22 inversion in a female carrier presents as a somatic mosaicism. Blood 2000;96:2905–6.
[599] Ivanov D, Hamby SE, Stenson PD, Phillips AD, Kehrer-Sawatzki H, Cooper DN, Chuzhanova N. Comparative analysis of germline and somatic microlesion mutational spectra in 17 human tumor suppressor genes. Hum Mutat 2011;32:620–32.
[600] Zlotogora J. Germ line mosaicism. Hum Genet 1998;102:381–6.
[601] Hall JG. Review and hypotheses: somatic mosaicism: observations related to clinical genetics. Am J Hum Genet 1988;43:355–63.
[602] Kehrer-Sawatzki H, Cooper DN. Mosaicism in sporadic neurofibromatosis type 1: variations on a theme common to other hereditary cancer syndromes? J Med Genet 2008;45:622–31.
[603] Campbell IM, Shaw CA, Stankiewicz P, Lupski JR. Somatic mosaicism: implications for disease and transmission genetics. Trends Genet 2015;31:382–92.
[604] Pham J, Shaw C, Pursley A, Hixson P, Sampath S, Roney E, Gambin T, Kang SH, Bi W, Lalani S, Bacino C, Lupski JR, Stankiewicz P, Patel A, Cheung SW. Somatic mosaicism detected by exon-targeted, high-resolution aCGH in 10,362 consecutive cases. Eur J Hum Genet 2014;22:969–78.
[605] Lindhurst MJ, Sapp JC, Teer JK, Johnston JJ, Finn EM, Peters K, Turner J, Cannons JL, Bick D, Blakemore L, Blumhorst C, Brockmann K, Calder P, Cherman N, Deardorff MA, Everman DB, Golas G, Greenstein RM, Kato BM, Keppler-Noreuil KM, Kuznetsov SA, Miyamoto RT, Newman K, Ng D, O'Brien K, Rothenberg S, Schwartzentruber DJ, Singhal V, Tirabosco R, Upton J, Wientroub S, Zackai EH, Hoag K, Whitewood-Neal T, Robey PG, Schwartzberg PL, Darling TN, Tosi LL, Mullikin JC, Biesecker LG. A mosaic activating mutation in AKT1 associated with the Proteus syndrome. N Engl J Med 2011;365:611–9.
[606] Lindhurst MJ, Parker VE, Payne F, Sapp JC, Rudge S, Harris J, Witkowski AM, Zhang Q, Groeneveld MP, Scott CE, Daly A, Huson SM, Tosi LL, Cunningham ML, Darling TN, Geer J, Gucev Z, Sutton VR, Tziotzios C, Dixon AK, Helliwell T, O'Rahilly S, Savage DB, Wakelam MJ, Barroso I, Biesecker LG, Semple RK. Mosaic overgrowth with fibroadipose hyperplasia is caused by somatic activating mutations in PIK3CA. Nat Genet 2012;44:928–33.
[607] Lee JH, Huynh M, Silhavy JL, Kim S, Dixon-Salazar T, Heiberg A, Scott E, Bafna V, Hill KJ, Collazo A, Funari V, Russ C, Gabriel SB, Mathern GW, Gleeson JG. De novo somatic mutations in components of the PI3K-AKT3-mTOR pathway cause hemimegalencephaly. Nat Genet 2012;44:941–5.
[608] Lynch M. Rate, molecular spectrum, and consequences of human mutation. Proc Natl Acad Sci U S A 2010;107:961–8.
[609] Roach JC, Glusman G, Smit AF, Huff CD, Hubley R, Shannon PT, Rowen L, Pant KP, Goodman N, Bamshad M, Shendure J, Drmanac R, Jorde LB, Hood L, Galas DJ. Analysis of genetic inheritance in a family quartet by whole-genome sequencing. Science 2010;328:636–9.
[610] Itsara A, Wu H, Smith JD, Nickerson DA, Romieu I, London SJ, Eichler EE. De novo rates and selection of large copy number variation. Genome Res 2010;20:1469–81.
[611] Crow JF. The origins, patterns and implications of human spontaneous mutation. Nat Rev Genet 2000;1:40–7.

[612] Becker J, Schwaab R, Moller-Taube A, Schwaab U, Schmidt W, Brackmann HH, Grimm T, Olek K, Oldenburg J. Characterization of the factor VIII defect in 147 patients with sporadic hemophilia A: family studies indicate a mutation type-dependent sex ratio of mutation frequencies. Am J Hum Genet 1996;58:657–70.
[613] Grimm T, Meng G, Liechti-Gallati S, Bettecken T, Muller CR, Muller B. On the origin of deletions and point mutations in Duchenne muscular dystrophy: most deletions arise in oogenesis and most point mutations result from events in spermatogenesis. J Med Genet 1994;31:183–6.
[614] Sayres MA, Venditti C, Pagel M, Makova KD. Do variations in substitution rates and male mutation bias correlate with life-history traits? A study of 32 mammalian genomes. Evolution 2011;65:2800–15.
[615] Conrad DF, Keebler JE, DePristo MA, Lindsay SJ, Zhang Y, Casals F, Idaghdour Y, Hartl CL, Torroja C, Garimella KV, Zilversmit M, Cartwright R, Rouleau GA, Daly M, Stone EA, Hurles ME, Awadalla P. Variation in genome-wide mutation rates within and between human families. Nat Genet 2011;43:712–4.
[616] Kong A, Frigge ML, Masson G, Besenbacher S, Sulem P, Magnusson G, Gudjonsson SA, Sigurdsson A, Jonasdottir A, Jonasdottir A, Wong WS, Sigurdsson G, Walters GB, Steinberg S, Helgason H, Thorleifsson G, Gudbjartsson DF, Helgason A, Magnusson OT, Thorsteinsdottir U, Stefansson K. Rate of de novo mutations and the importance of father's age to disease risk. Nature 2012;488:471–5.
[617] Francioli LC, Polak PP, Koren A, Menelaou A, Chun S, Renkens I, Genome of the Netherlands C, van Duijn CM, Swertz M, Wijmenga C, van Ommen G, Slagboom PE, Boomsma DI, Ye K, Guryev V, Arndt PF, Kloosterman WP, de Bakker PIW, Sunyaev SR. Genome-wide patterns and properties of de novo mutations in humans. Nat Genet 2015;47:822–6.
[618] Rahbari R, Wuster A, Lindsay SJ, Hardwick RJ, Alexandrov LB, Turki SA, Dominiczak A, Morris A, Porteous D, Smith B, Stratton MR, Consortium UK, Hurles ME. Timing, rates and spectra of human germline mutation. Nat Genet 2016;48:126–33.
[619] Alexandrov LB, Nik-Zainal S, Wedge DC, Aparicio SA, Behjati S, Biankin AV, Bignell GR, Bolli N, Borg A, Borresen-Dale AL, Boyault S, Burkhardt B, Butler AP, Caldas C, Davies HR, Desmedt C, Eils R, Eyfjord JE, Foekens JA, Greaves M, Hosoda F, Hutter B, Ilicic T, Imbeaud S, Imielinski M, Jager N, Jones DT, Jones D, Knappskog S, Kool M, Lakhani SR, Lopez-Otin C, Martin S, Munshi NC, Nakamura H, Northcott PA, Pajic M, Papaemmanuil E, Paradiso A, Pearson JV, Puente XS, Raine K, Ramakrishna M, Richardson AL, Richter J, Rosenstiel P, Schlesner M, Schumacher TN, Span PN, Teague JW, Totoki Y, Tutt AN, Valdes-Mas R, van Buuren MM, van 't Veer L, Vincent-Salomon A, Waddell N, Yates LR, Australian Pancreatic Cancer Genome I, Consortium, I.B.C., Consortium, I.M.-S., PedBrain I, Zucman-Rossi J, Futreal PA, McDermott U, Lichter P, Meyerson M, Grimmond SM, Siebert R, Campo E, Shibata T, Pfister SM, Campbell PJ, Stratton MR. Signatures of mutational processes in human cancer. Nature 2013;500:415–21.
[620] Acuna-Hidalgo R, Bo T, Kwint MP, van de Vorst M, Pinelli M, Veltman JA, Hoischen A, Vissers LE, Gilissen C. Post-zygotic point mutations are an underrecognized source of de novo genomic variation. Am J Hum Genet 2015;97:67–74.
[621] Goldmann JM, Wong WS, Pinelli M, Farrah T, Bodian D, Stittrich AB, Glusman G, Vissers LE, Hoischen A, Roach JC, Vockley JG, Veltman JA, Solomon BD, Gilissen C, Niederhuber JE. Parent-of-origin-specific signatures of de novo mutations. Nat Genet 2016;48:935–9.
[622] Wexler NS, Young AB, Tanzi RE, Travers H, Starosta-Rubinstein S, Penney JB, Snodgrass SR, Shoulson I, Gomez F, Ramos Arroyo MA, et al. Homozygotes for Huntington's disease. Nature 1987;326:194–7.
[623] Zlotogora J. Dominance and homozygosity. Am J Med Genet 1997;68:412–6.
[624] Cogan JD, Phillips 3rd JA, Schenkman SS, Milner RD, Sakati N. Familial growth hormone deficiency: a model of dominant and recessive mutations affecting a monomeric protein. J Clin Endocrinol Metab 1994;79:1261–5.
[625] Spritz RA, Giebel LB, Holmes SA. Dominant negative and loss of function mutations of the c-kit (mast/stem cell growth factor receptor) proto-oncogene in human piebaldism. Am J Hum Genet 1992;50:261–9.
[626] Patel PI, Roa BB, Welcher AA, Schoener-Scott R, Trask BJ, Pentao L, Snipes GJ, Garcia CA, Francke U, Shooter EM, Lupski JR, Suter U. The gene for the peripheral myelin protein PMP-22 is a candidate for Charcot-Marie-Tooth disease type 1A. Nat Genet 1992;1:159–65.
[627] Aldred MA, Trembath RC. Activating and inactivating mutations in the human GNAS1 gene. Hum Mutat 2000;16:183–9.
[628] Mooers BH, Logue JS, Berglund JA. The structural basis of myotonic dystrophy from the crystal structure of CUG repeats. Proc Natl Acad Sci U S A 2005;102:16626–31.
[629] Armanios M, Chen JL, Chang YP, Brodsky RA, Hawkins A, Griffin CA, Eshleman JR, Cohen AR, Chakravarti A, Hamosh A, Greider CW. Haploinsufficiency of telomerase reverse transcriptase leads to anticipation in autosomal dominant dyskeratosis congenita. Proc Natl Acad Sci U S A 2005;102:15960–4.

[630] Kim HJ, Nam SH, Kim HJ, Park HS, Ryoo HM, Kim SY, Cho TJ, Kim SG, Bae SC, Kim IS, Stein JL, van Wijnen AJ, Stein GS, Lian JB, Choi JY. Four novel RUNX2 mutations including a splice donor site result in the cleidocranial dysplasia phenotype. J Cell Physiol 2006;207:114–22.

[631] Dobyns WB, Filauro A, Tomson BN, Chan AS, Ho AW, Ting NT, Oosterwijk JC, Ober C. Inheritance of most X-linked traits is not dominant or recessive, just X-linked. Am J Med Genet A 2004;129:136–43.

[632] Manolio TA, Collins FS, Cox NJ, Goldstein DB, Hindorff LA, Hunter DJ, McCarthy MI, Ramos EM, Cardon LR, Chakravarti A, Cho JH, Guttmacher AE, Kong A, Kruglyak L, Mardis E, Rotimi CN, Slatkin M, Valle D, Whittemore AS, Boehnke M, Clark AG, Eichler EE, Gibson G, Haines JL, Mackay TF, McCarroll SA, Visscher PM. Finding the missing heritability of complex diseases. Nature 2009;461:747–53.

[633] Park JH, Wacholder S, Gail MH, Peters U, Jacobs KB, Chanock SJ, Chatterjee N. Estimation of effect size distribution from genome-wide association studies and implications for future discoveries. Nat Genet 2010;42:570–5.

[634] Bodmer W, Bonilla C. Common and rare variants in multifactorial susceptibility to common diseases. Nat Genet 2008;40:695–701.

[635] Bodmer W, Tomlinson I. Rare genetic variants and the risk of cancer. Curr Opin Genet Dev 2010;20:262–7.

[636] Mefford HC, Shafer N, Antonacci F, Tsai JM, Park SS, Hing AV, Rieder MJ, Smyth MD, Speltz ML, Eichler EE, Cunningham ML. Copy number variation analysis in single-suture craniosynostosis: multiple rare variants including RUNX2 duplication in two cousins with metopic craniosynostosis. Am J Med Genet A 2010;152A:2203–10.

[637] Lehner B. Molecular mechanisms of epistasis within and between genes. Trends Genet 2011;27:323–31.

[638] Wolf U. Identical mutations and phenotypic variation. Hum Genet 1997;100:305–21.

[639] Cooper DN, Krawczak M, Polychronakos C, Tyler-Smith C, Kehrer-Sawatzki H. Where genotype is not predictive of phenotype: towards an understanding of the molecular basis of reduced penetrance in human inherited disease. Hum Genet 2013;132:1077–130.

[640] Parchi P, Petersen RB, Chen SG, Autilio-Gambetti L, Capellari S, Monari L, Cortelli P, Montagna P, Lugaresi E, Gambetti P. Molecular pathology of fatal familial insomnia. Brain Pathol 1998;8:539–48.

[641] Kajiwara K, Berson EL, Dryja TP. Digenic retinitis pigmentosa due to mutations at the unlinked peripherin/RDS and ROM1 loci. Science 1994;264:1604–8.

[642] Morell R, Spritz RA, Ho L, Pierpont J, Guo W, Friedman T.B., Asher Jr. J.H.. Apparent digenic inheritance of Waardenburg syndrome type 2 (WS2) and autosomal recessive ocular albinism (AROA). Hum Mol Genet 1997;6:659–64.

[643] Katsanis N, Ansley SJ, Badano JL, Eichers ER, Lewis RA, Hoskins BE, Scambler PJ, Davidson WS, Beales PL, Lupski JR. Triallelic inheritance in Bardet-Biedl syndrome, a Mendelian recessive disorder. Science 2001;293:2256–9.

[644] Nichols WC, Ginsburg D. von Willebrand disease. Medicine (Baltimore) 1997;76:1–20.

[645] Dreszer TR, Karolchik D, Zweig AS, Hinrichs AS, Raney BJ, Kuhn RM, Meyer LR, Wong M, Sloan CA, Rosenbloom KR, Roe G, Rhead B, Pohl A, Malladi VS, Li CH, Learned K, Kirkup V, Hsu F, Harte RA, Guruvadoo L, Goldman M, Giardine BM, Fujita PA, Diekhans M, Cline MS, Clawson H, Barber GP, Haussler D, James Kent W. The UCSC Genome Browser database: extensions and updates 2011. Nucleic Acids Res 2011.

[646] Harrow J, Nagy A, Reymond A, Alioto T, Patthy L, Antonarakis SE, Guigo R. Identifying protein-coding genes in genomic sequences. Genome Biol 2009; 10:201.

[647] Myers RM, Stamatoyannopoulos J, Snyder M, Dunham I, Hardison RC, Bernstein BE, Gingeras TR, Kent WJ, Birney E, Wold B, Crawford GE. A user's guide to the encyclopedia of DNA elements (ENCODE). PLoS Biol 2011;9:e1001046.

[648] Ying H, Huttley G. Exploiting CpG hypermutability to identify phenotypically significant variation within human protein-coding genes. Genome Biol Evol 2011;3:938–49.

[649] Millar DS, Lewis MD, Horan M, Newsway V, Easter TE, Gregory JW, Fryklund L, Norin M, Crowne EC, Davies SJ, Edwards P, Kirk J, Waldron K, Smith PJ, Phillips 3rd JA, Scanlon MF, Krawczak M, Cooper DN, Procter AM. Novel mutations of the growth hormone 1 (GH1) gene disclosed by modulation of the clinical selection criteria for individuals with short stature. Hum Mutat 2003;21:424–40.

[650] Botstein D, Risch N. Discovering genotypes underlying human phenotypes: past successes for mendelian disease, future approaches for complex disease. Nat Genet 2003;33(Suppl.):228–37.

# 7 Genes in Families

*Jackie Cook*

**Consultant in Clinical Genetics, Sheffield Clinical Genetics Service, Sheffield Children's NHS Foundation Trust, Sheffield, United Kingdom**

## 7.1 INTRODUCTION

This chapter provides an overview of genetic conditions within families and how to approach and analyze a family history of genetic disease. Many of the topics touched upon are covered in more detail elsewhere in this volume. The principles of Mendelian inheritance for single nuclear gene disorders have been known for many years, but new insights are being constantly gained from new technologies, particularly whole-genome sequencing. An increasing understanding of nontraditional inheritance patterns and the study of the genome rather than single genes are adding to our knowledge and enabling clinicians to provide information to a growing number of families. Most of our current understanding of genetic disease is related to pathogenic variations affecting the exomes or coding regions of the genes. It is likely that, with the advent of whole-genome sequencing, we will start to identify a whole new group of genetic disorders and/or explain disorders of unknown etiology that are caused by alterations in the non-protein-coding parts of the genome, including sequences involved in the regulation of gene expression both in time and at the tissue site.

Genetic counseling is the process by which patients and relatives at risk of a disorder that may be hereditary are advised of the consequences of the disorder, the probability of developing and transmitting it, and the ways in which this may be prevented, avoided, or ameliorated. To achieve these aims, an accurate diagnosis and detailed information regarding the family history are essential. The basis for establishing a diagnosis depends on medical history, examination, and investigation. The diagnostic information is combined with information obtained from the family pedigree to determine the mode of inheritance of the disorder and to calculate the risk of recurrence, so that family members can be appropriately counseled.

A family history of a genetic condition may be due to:

1. a pathogenic variant within a single nuclear gene;
2. a pathogenic variant within a mitochondrial gene;
3. a contiguous gene deletion or duplication involving one or many genes;
4. a chromosomal rearrangement resulting in unbalanced products at meiosis;
5. polygenic and multifactorial inheritance;
6. multiple genetic conditions contributing to the overall phenotype.

## 7.2 PEDIGREE CONSTRUCTION

Accurate documentation of the family history is an essential part of genetic assessment, and the best method of recording this information is by constructing a family pedigree. Pedigrees are universally used in patients' genetic records, journal articles, and textbooks as the means of relaying information in an easily interpretable visual format. Pedigrees also provide the basis for calculations required for both recurrence risk estimation in individual families and linkage analysis in gene-mapping studies.

**Emery and Rimoin's Principles and Practice of Medical Genetics and Genomics: Foundations, https://doi.org/10.1016/B978-0-12-812537-3.00007-X**

affected with Huntington disease. Such individuals do not appear different phenotypically from heterozygotes for the disorder.

#### 7.4.1.1 Incomplete Dominance

If the phenotype of the heterozygous state, AB, is intermediate between the phenotypes of AA and BB, allele A is said to be incompletely dominant or semidominant to allele B. The skeletal dysplasia achondroplasia causes rhizomelic shortening of the limbs, a characteristic facies with midface hypoplasia, exaggerated lumbar lordosis, limitation of hip and elbow extension, genu varum, and trident hand. It was conventionally thought to be due to a dominant allele, but homozygotes for the variant gene have a much more severe skeletal dysplasia, resulting in early death from respiratory obstruction due to a small thoracic cage and neurologic deficit due to hydrocephalus. Therefore, achondroplasia is an example of incomplete or semidominance. Homozygotes for most dominant alleles causing human genetic disorders occur so rarely that it is not known whether they exhibit complete or incomplete dominance.

#### 7.4.1.2 Codominance

If the phenotype of AB displays the phenotypic features of both the homozygotic states, then alleles A and B are said to be codominant. The human ABO blood group system exhibits codominance. The system consists of three alleles A, B, and O. Both A and B are dominant in relation to O, and therefore blood group A can have the genotype AA or AO. Blood group B can have the genotype BB or BO. However, neither A nor B shows dominance over the other, and therefore individuals with the genotype AB have the phenotypic characteristics of both blood group A and blood group B.

### 7.4.2 Mechanisms of Dominance

Most pathogenic variants result in an allele that is recessive to the wild-type allele; the phenotype is therefore expressed only in the homozygous state. This is because most pathogenic variants result in an inactive gene product, but the reduced level of activity due to the remaining wild-type allele is sufficient to achieve the effects of that gene product. An example is a gene for an enzyme that is required only in small amounts as a catalyst for a metabolic pathway. Although in some recessive inborn errors of metabolism it is possible to identify heterozygotes biochemically, more often than not, the only way to identify carriers is by direct mutation analysis, using molecular genetic techniques.

There are several mechanisms by which a pathogenic variant can lead to a dominant allele whose phenotype is expressed in the heterozygous state [2,3].

#### 7.4.2.1 Loss-of-Function Variants

For most variant alleles, loss of function will usually exhibit recessive behavior. Where a reduced amount or reduced activity of the gene product results in the phenotypic features, this is termed haploinsufficiency (e.g., in a critical rate-limiting step of a metabolic pathway). Haploinsufficiency can be due to a number of different mechanisms, including the following:

1. deletion of a whole allele;
2. a pathogenic variant causing gene inactivation;
3. a regulatory variant resulting in failure of transcription;
4. incorrect translational control;
5. decreased mRNA or protein stability; for example, a premature termination codon can lead to nonsense-mediated mRNA decay.

Any reduction in the amount of gene product will result in that pathway not being able to function at full activity. The same appears to apply to regulatory genes that could have a threshold level of activity. *PAX3* is a gene coding for a DNA-binding protein, and single-nucleotide variants in this gene result in Waardenburg syndrome type 1, characterized by deafness and pigmentary disturbances. Certain variants in *PAX3* have been shown to abolish all protein function of that allele, so the phenotype must be due to a dosage effect as it manifests in the heterozygous state.

Other examples in which the quantitative amount of a gene product is important are genes that produce proteins in large quantities. An example is the *C1NH* gene for C1 esterase inhibitor, in which pathogenic variants cause the disorder hereditary angioneurotic edema. C1 esterase inhibitor is removed rapidly from the circulation at a rate independent of its concentration. Therefore, although heterozygotes produce 50% of the normal amount, they have only 15%–20% of the normal amount in the circulation, leading to the clinical manifestations of the disorder.

#### 7.4.2.2 Gain-of-Function Variants

*7.4.2.2.1 Increased gene dosage.* This mechanism involves an excess of gene product leading to a disease phenotype. Although the gene dosage of critical regions or genes has been invoked as the cause of the phenotypic features associated with the autosomal trisomies, there are few examples involving single-gene disorders. One example involves the *PMP22* gene, which codes for the peripheral myelin protein 22. Duplication of the DNA sequence of one allele is associated with hereditary motor and sensory neuropathy type 1A.

*7.4.2.2.2 Ectopic or temporally altered messenger RNA expression.* Genes can be expressed or turned off at different times throughout development and the life of the individual and are also differentially expressed in different tissues. Ectopic or temporally altered messenger RNA is expressed when a pathogenic variant occurs that affects the time or place of gene expression and usually involves a regulatory part of the gene. For example, during development in erythroid precursor cells, there is a switch from the production of γ-globin to the production of δ- and β-globin. This switch is controlled, at least in part, by the binding of transcription factors to the γ-globin promoter. Single-nucleotide variants in the globin promoter region prevent the normal switch, resulting in the disorder of hereditary persistence of fetal hemoglobin.

*7.4.2.2.3 Increased protein activity.* Pathogenic variants can lead to proteins with a prolonged half-life or proteins that have lost their normal constitutive inhibitory regulatory activity. If a pathogenic variant occurs in a part of a gene that codes for the protein sequence acting as the recognition site for proteolytic degradation, this will not take place, with the protein remaining active. Many proteins possess domains that allow their activity to be reversibly inhibited. For example, skeletal muscle sodium channels undergo voltage-sensitive regulation, and variants in the gene *SCN4A*, which codes for the α subunit of the sodium channels, result in the disorder hyperkalemic periodic paralysis, characterized by muscle myotonia and paralysis due to loss of regulatory inactivation of the sodium channel.

*7.4.2.2.4 Dominant–negative variants.* If a variant allele interferes with the wild-type allele, this is termed a dominant–negative variant. This could occur in a multimeric protein in which an abnormal subunit has an intact binding domain but altered catalytic activity, affecting the function of the entire multimer. If a protein is a dimer, one variant and one wild-type allele would result in only 25% normal dimers, with up to a 75% reduction in activity.

Many structural proteins are multimers (e.g., the various types of collagen proteins). Each of the collagen subunit genes has a central portion coding for repeating tripeptide units that are essential for the assembly of the collagen molecule. The disorder osteogenesis imperfecta is caused by single-nucleotide variants in the central portion of one of the collagen subunit genes *COL1A1* or *COL1A2* leading to a structural deformation that causes disruption of the whole collagen protein.

*7.4.2.2.5 Toxic protein alterations.* Toxic protein alterations are pathogenic variants that cause structural alterations in proteins, thus disrupting normal function and leading to toxic products that poison the cell. An example is hereditary amyloidosis, in which variants in the transthyretin gene *TTR* lead to resistance to proteolysis, with resultant increased stability of the protein. The protein then undergoes multimerization and accumulates in the cell as fibrils, causing disruption of the cell.

*7.4.2.2.6 New protein functions.* Some variants have been found to confer a new function on a gene product. For example, a fatal bleeding disorder was found to be caused by a missense pathogenic variant in the α1-antitrypsin gene, in which methionine was replaced by arginine at position 358, the effect of which was to convert α1-antitrypsin, normally an inhibitor of elastase, into an inhibitor of thrombin. This thrombin-inhibitory activity was not compensated for by an increase in endogenous coagulant production, resulting in a severe bleeding disorder [4].

#### 7.4.2.3 Recessive Variants With Dominant Effects

The mechanisms described so far show how variants can cause dominant effects at a cellular level by the effects on the proteins produced. It is possible to have variants that show a dominant pattern of inheritance in families, yet are recessive at the cellular or molecular level; that is, the gene is inactivated but has no other effect. The classic example of this is the retinoblastoma gene *RB1*, inactivation of which can lead to the formation of the developmental eye tumor retinoblastoma. Families can show a dominant mode of inheritance for this disorder, yet cells heterozygous for the variant are completely normal, the variant itself being recessive.

The dominant pattern of inheritance of familial retinoblastoma is the result of transmission of a first variant with a second somatic variant occurring in the normal allele of at least one retinal cell during a critical period of development, leading to the formation of a retinoblastoma—the "two-hit" hypothesis [5]. There are several ways in which the normal allele in somatic cells can be inactivated. These include single-nucleotide variants, deletions, translocations, and mitotic nondisjunction, resulting in the loss of a whole chromosome. It is now known that the two-hit hypothesis applies to most of the dominantly inherited familial cancer syndromes in which the germline variant in a tumor suppressor gene is recessive and an acquired variant in a somatic cell in the corresponding allele leads to the development of a tumor.

## 7.5 AUTOSOMAL DOMINANT INHERITANCE

Autosomal dominant inheritance refers to disorders caused by genes located on the autosomes, thereby affecting both males and females. The variant alleles are dominant to the wild-type alleles, so the disorder is manifest in the heterozygote (i.e., an individual who possesses both the wild-type and the variant allele). The necessary characteristics to be certain a disorder is inherited in an autosomal dominant manner are listed in Table 7.1 and depicted in Fig. 7.3.

| TABLE 7.1 **Characteristics of an Autosomal Dominant Inherited Disorder** |
|---|
| Successive or multiple generations in a family are affected |
| Males and females are both affected in approximately equal proportions |
| Males and females can both be responsible for transmission |
| There is at least one instance of male-to-male transmission |

### 7.5.1 Recurrence Risks

Because individuals with autosomal dominant disorders are heterozygous for a variant and a normal allele, there is a 1 in 2 (50%) chance a gamete will carry the normal allele and a 1 in 2 (50%) chance a gamete will carry the variant allele. Assuming that the individual's partner will contribute a normal allele, there is a 1 in 2 (50%) chance that the offspring, regardless of sex, will inherit the disorder with each pregnancy (Fig. 7.4).

### 7.5.2 Penetrance

There can be marked variability in the clinical manifestations of autosomal dominant disorders, and they can demonstrate reduced penetrance (i.e., not every person with the variant allele shows features of the disorder). The penetrance of a disorder is an index of the proportion of individuals with a variant allele who manifest the disorder. An allele is said to be nonpenetrant if an individual known to be heterozygous for the allele, either by pedigree analysis or by molecular investigation, shows no signs of the disorder when subjected to appropriate clinical investigation. The penetrance of some genes is dependent on the age of the individual, as in Huntington disease, in which the penetrance is age dependent or is said to show delayed penetrance.

The penetrance of a disorder is usually expressed as the proportion or percentage of the individuals carrying the gene who develop the disorder. If the penetrance is known for a particular condition, the risk to the offspring of an apparently unaffected individual can be calculated. In practice, the risk is usually less than 10%, as an unaffected relative is not likely to carry

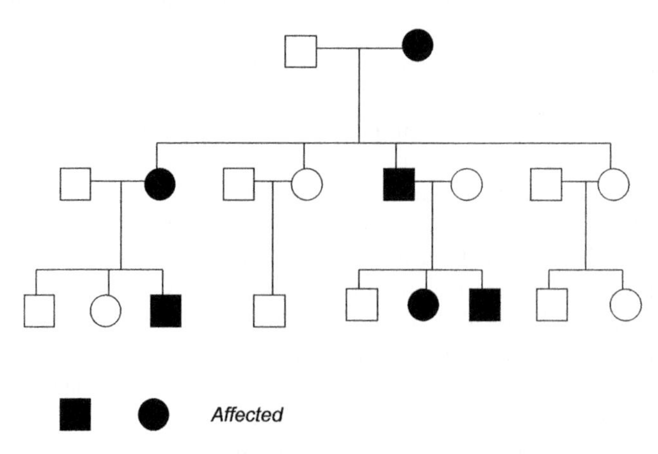


**Figure 7.3** Pedigree consistent with autosomal dominant inheritance.

the gene if the penetrance is high, and a gene carrier is not likely to develop the disorder if the penetrance is low.

### 7.5.3 Expressivity

The expressivity of a gene is the degree to which a particular phenotype is expressed in an individual. Many autosomal dominant disorders show variable expressivity such that individuals in the same family who carry an identical variant can vary considerably in the severity of their disorder. For example, in the autosomal dominant disorder neurofibromatosis type 1, the number of neurofibromas that an individual develops can vary dramatically from a few to many hundreds even within the same family. The variability seen in autosomal dominant disorders may present as both inter- and intrafamilial differences. Intrafamilial variability may reflect the action of modifying genes, but interfamilial variability can also be due to allelic heterogeneity at a single locus. A problem encountered in genetic counseling is that a mildly affected individual, such as a parent with only skin manifestations of tuberous sclerosis, may have a severely affected child. This situation is seen in many autosomal dominant disorders, in that the mildly affected individuals are more likely to reproduce than the severely affected individuals.

### 7.5.4 Anticipation

A disorder is said to demonstrate anticipation if the phenotype of the variant allele increases in severity as it is passed down the generations. An example of a disorder that demonstrates anticipation is myotonic dystrophy. A typical three-generation family with myotonic dystrophy showing anticipation is shown in Fig. 7.5. Another example of anticipation is seen in Huntington disease, in which the onset of symptoms is often seen to occur earlier with each succeeding generation.

### 7.5.5 Sex Influence

Sex influence involves the expression of an autosomal allele that occurs more frequently in one sex than the other. An example in humans is early pattern baldness, with males affected more frequently than females, an effect probably mediated by hormonal differences.

### 7.5.6 Sex Limitation

Some traits are manifested only in individuals of one sex, an extreme situation known as sex limitation. This can occur when a gene affects an organ possessed by only one of the sexes (e.g., unicornuate uterus or ovarian cancer).

### 7.5.7 Pleiotropy

Pleiotropy refers to the phenomenon in which a single gene is responsible for a number of distinct and seemingly unrelated phenotypic effects. For example, the allele causing neurofibromatosis type 1 can produce abnormalities of skin pigmentation, neurofibromas of the peripheral nerves, short stature, macrocephaly, skeletal abnormalities, and fits. Each of the pleiotropic effects of an allele can show reduced or nonpenetrance and variable expressivity.

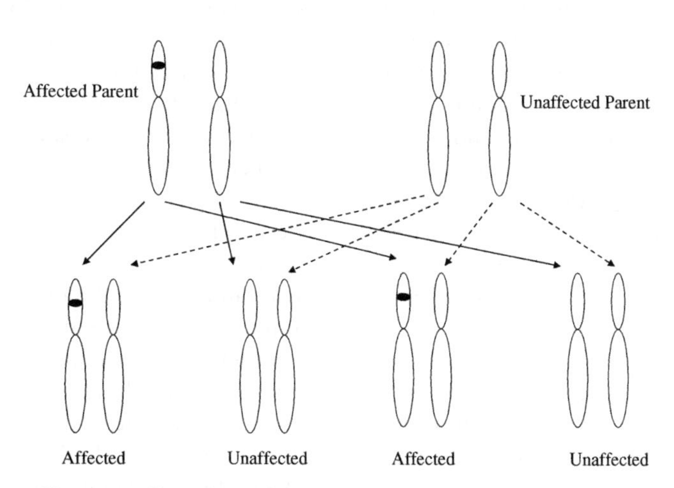


**Figure 7.4** Recurrence risks in autosomal dominant inheritance.

The frequency of gonadal mosaicism differs between disorders. As of this writing, it is unclear why it is a frequent finding in some conditions and not others. Studies on the recurrence of lethal osteogenesis imperfecta show genetic heterogeneity, with some cases due to recessive inheritance and some due to dominant parental mosaicism. The rate of parental mosaicism in families in which a dominant variant was identified in the first affected child has been reported in one study as 16% with a recurrence rate of 1.3% [9]. Recurrence of achondroplasia to normal stature parents, where there is a high new germline mutation rate, has been reported in only a few rare cases. This is an important point to remember when providing recurrence risk advice in genetic counseling. Counseling in cases of gonadal mosaicism is complex because the recurrence risk is dependent on the percentage of cells carrying the variant in the gonads, which can be difficult to determine.

Somatic mosaicism may also be present in an individual who manifests the phenotype of an autosomal dominant disorder. A study looking at individuals who were the first members of their families to develop the signs of neurofibromatosis type 2, a disorder characterized by bilateral vestibular schwannomas, came up with a mosaicism rate of 33% for classical neurofibromatosis type 2 with bilateral tumors and 60% for those presenting with unilateral tumors, with the variant often detected in tumor material and not in lymphocyte DNA [10]. Again, the risk of passing the variant on to the next generation is determined by whether the variant is present in the gonads, and this information is not straightforward to ascertain.

## 7.6 AUTOSOMAL RECESSIVE INHERITANCE

Autosomal recessive inheritance refers to disorders due to genes located on the autosomes, but in which the variant alleles are recessive to the wild-type alleles and are therefore not evident in the heterozygous state, being manifest only in the homozygous state. The necessary characteristics to be certain that a disorder is inherited in an autosomal recessive manner are listed in Table 7.2 and depicted in Fig. 7.6. The parents of an individual with an autosomal recessive disorder are heterozygous for the variant allele and are usually referred to as being carriers for the disorder.

| TABLE 7.2 **Characteristics of an Autosomal Recessive Inherited Disorder** |
|---|
| Both males and females are affected |
| The disorder normally occurs in only one generation, usually within a single sibship |
| The parents can be consanguineous |

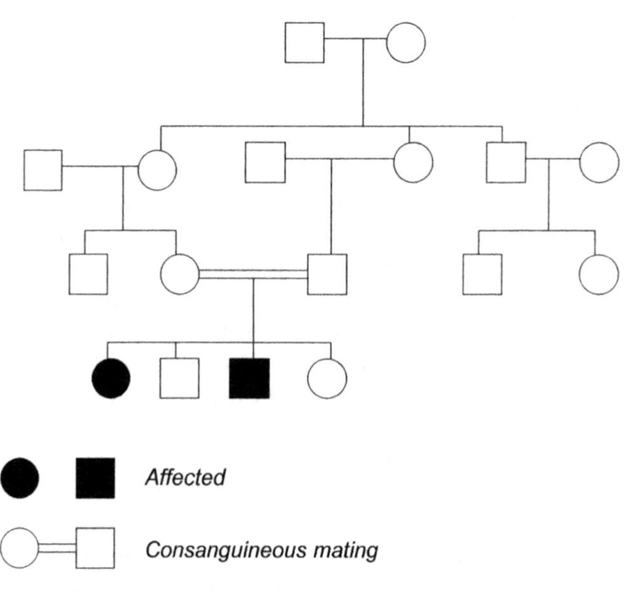


**Figure 7.6** Pedigree consistent with autosomal recessive inheritance.

### 7.6.1 Consanguinity

If a couple are consanguineous, they have at least one ancestor in common in the preceding few generations. First cousins share approximately 1/8 of their alleles in common. This means that they are more likely to carry identical alleles inherited from their common ancestor and could both transmit an identical allele to their offspring, who would then be homozygous for that allele. A consanguineous couple has an increased risk that their offspring will be affected with a recessive disorder. The rarer a particular disorder is in a population, the more likely the parents are to be consanguineous. For example, cystic fibrosis is a common autosomal recessive disorder in whites in Western Europe, with an incidence of approximately 1 in 2000. The incidence of consanguinity in the parents of children with cystic fibrosis is not appreciably greater than that in the general population. By contrast, with very rare autosomal recessive disorders such as alkaptonuria, 8 of the first 19 families originally described by Garrod were consanguineous [11].

### 7.6.2 Recurrence Risks

When two parents, who each carry a variant allele for a genetic disorder, reproduce, there is an equal chance that the gametes will contain the variant or the wild-type allele. There are four possible combinations of these gametes, resulting in a 1 in 4 (25%) chance of having a homozygous affected offspring, a 1 in 2 (50%) chance of having a heterozygous unaffected carrier offspring, and a 1 in 4 (25%) chance of having a homozygous unaffected offspring (Fig. 7.7).

When an individual with an autosomal recessive disorder has children, they will produce only gametes containing the variant allele. Since it is most likely that their partner will be homozygous for the wild-type allele, the partner will always contribute a wild-type allele and therefore all the children will be heterozygous carriers and unaffected (Fig. 7.8). If, however, an affected individual has children with a partner who happens to be heterozygous for the variant allele, there will be a 50% chance of transmitting the disorder, depending on whether the partner contributes a variant or a wild-type allele (Fig. 7.9). Such a pedigree is said to exhibit pseudodominance (Fig. 7.10).

In autosomal recessive disorders, the difficulty lies not with risk estimation, but in determining the underlying mode of inheritance, as these disorders usually present as isolated cases with little contributory information to be gleaned from the pedigree. Carrier risks to other relatives can be calculated from the pedigree, and carrier testing may be appropriate for disorders with a high variant allele frequency or when consanguineous marriages are planned. The risk to members of the extended family for having an affected child will depend on their own risk calculated from the pedigree data and the population-based carrier risk appropriate to their partner. Higher risks in consanguineous partnerships are dependent on the degree of relationship between the partners. In disorders of unknown genetic etiology, consanguinity suggests, but does not prove, an autosomal recessive mode of inheritance.

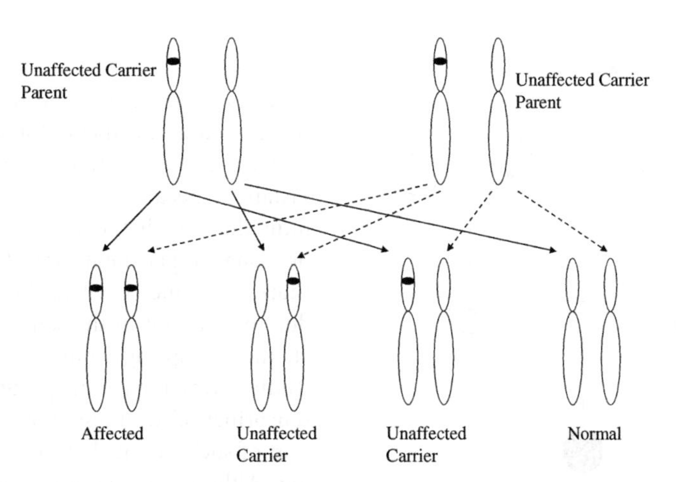


**Figure 7.7** Recurrence risks in autosomal recessive inheritance.

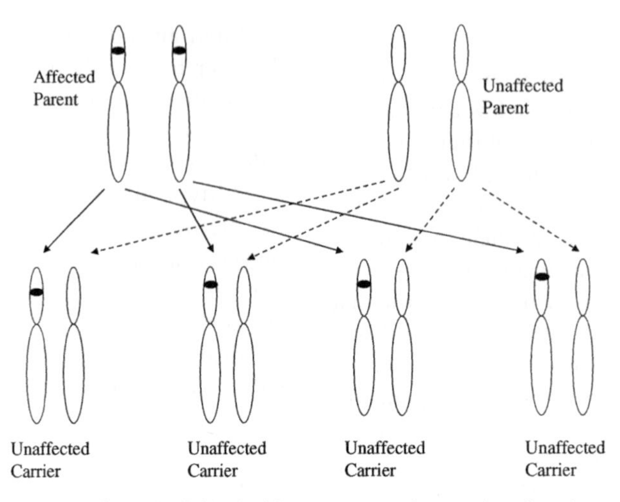


**Figure 7.8** Recurrence risks for an individual with an autosomal recessive disorder and a normal partner.

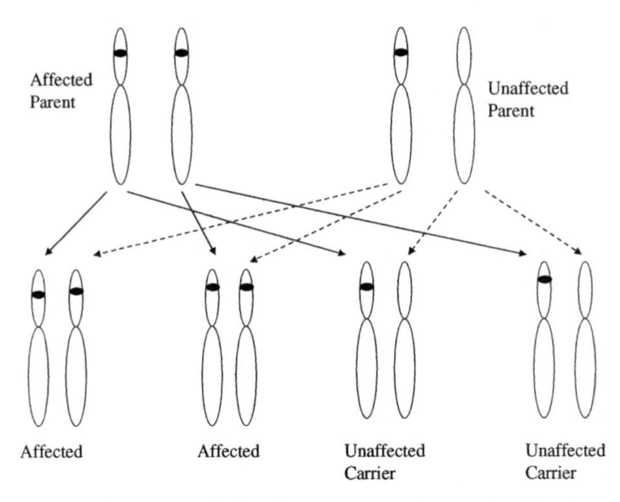


**Figure 7.9** Recurrence risks for an individual with an autosomal recessive disorder and a carrier partner.

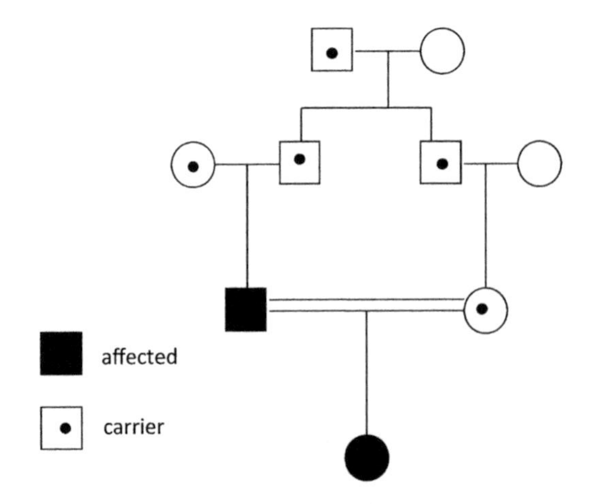


**Figure 7.10** Pedigree of an autosomal recessive disorder showing pseudodominance.

### 7.6.3 Genetic Heterogeneity

It is not unusual in some recessive disorders, such as sensorineural deafness, for two affected individuals to have children. Assortative mating occurs in such instances because of social circumstances in which individuals with the same disability, such as deafness or visual impairment, are often educated together or share the same social facilities. If their disorder were due to a variant in the same autosomal recessive gene, all their offspring would be affected. In a number of studies involving the offspring of parents with inherited sensorineural deafness, however, a significant proportion of such unions led to offspring with normal hearing. Although in some instances this could be due to other causes (e.g., acquired causes being mistaken for inherited deafness), in most instances the gene causing the deafness in the two parents is different, a phenomenon known as genetic heterogeneity. Each parent will transmit the variant allele for their own deafness, but a wild-type allele of the gene involved in their partner's deafness. Therefore, the child is heterozygous for the two variant alleles, referred to as double heterozygosity. This type of genetic heterogeneity involving different genes is known as locus heterogeneity.

Different modes of inheritance have also been documented for a number of clinically defined disorders with similar phenotypes. For example, autosomal dominant, autosomal recessive, and X-linked recessive inheritance have all been documented in hereditary spastic paraplegia, hereditary motor and sensory neuropathy, and retinitis pigmentosa, depending on the causative gene. Without information from genetic testing, locus heterogeneity makes it difficult to determine the risks of recurrence for phenotypes that follow both dominant and recessive inheritance unless the mode of inheritance is clearly defined by the family pedigree. Next-generation sequencing and the use of gene panels have revolutionized testing in these circumstances. It is now possible to analyze very large numbers of genes causing a particular phenotype in a single test, enabling accurate risk analysis.

Heterogeneity can also exist at the same locus; thus, an individual affected with a recessively inherited disorder can have two different variants in the two alleles of the gene and is often called a compound heterozygote. Most individuals with recessive disorders are compound heterozygotes unless a specific variant is especially prevalent in a particular population or the affected individual is the offspring of a consanguineous relationship, in which case the allele is likely to be identical by descent. This is known as allelic or mutational heterogeneity. The specific variants that an affected individual possesses can, in fact, determine the severity of the disorder, as in cystic fibrosis, in which individuals who are homozygous for the most common variant in the cystic fibrosis gene, phe508del, have a higher incidence of pancreatic insufficiency. As the variants underlying disease are identified, it is becoming increasingly apparent that in many cases the exact nature of the variant will determine the phenotype—a phenomenon known as genotype–phenotype correlation.

In some cases different variants in a particular gene may act in a dominant or in a recessive manner. For example, a wide variety of phenotypes and inheritance patterns have been associated with variants in the Lamin A/C gene [12]. These include variants acting in an autosomal dominant fashion, causing limb girdle muscular dystrophy type 1B, dilated cardiomyopathy, familial partial lipodystrophy, and Hutchinson–Gilford progeria syndrome, as well as variants acting in an autosomal recessive fashion, such as type 2B1 axonal neuropathy, and mandibuloacral dysplasia.

### 7.6.4 Uniparental Disomy

Uniparental disomy (UPD) refers to the presence of both homologues of a chromosome pair or chromosomal region in a diploid offspring being derived from a single parent. If the two homologues are identical due to an error in meiosis II, this is known as uniparental isodisomy, while if the two homologues are different but still from the same parent due to a meiosis I error, this is known as uniparental heterodisomy. UPD has been reported as a rare cause of the autosomal recessive disorder cystic fibrosis in the offspring of a couple in whom only one parent was a heterozygous carrier of the variant allele. The affected offspring received both chromosome 7 homologues with the variant allele from that parent. The recurrence risk in this situation would be negligible.

### 7.6.5 New Variants

An autosomal recessive disorder may potentially be due to the inheritance of a variant from one parent, with a de novo variant occurring at the same locus on the chromosome inherited from the other parent. This is likely to be a very rare phenomenon, but accounts for some cases of spinal muscular atrophy because of a predisposition to generate deletions in the 5q13 region involving the *SMN1* gene [13]. The recurrence risk in this situation would relate to the risk of a repeated de novo mutation, and is therefore negligible.

## 7.7 SEX-LINKED INHERITANCE

Strictly speaking, sex-linked inheritance refers to the inheritance patterns shown by genes on the sex chromosomes. If the gene is on the X chromosome, it is said to show X-linked inheritance, and if on the Y chromosome, Y-linked or holandric inheritance.

### 7.7.1 X-Linked Recessive Inheritance

This form of inheritance is conventionally referred to as sex-linked inheritance. It refers to phenotypes due to recessive genes on the X chromosome. Males have a single X chromosome and are therefore hemizygous for most of the alleles on the X chromosome, so that, if they have a variant allele, they will manifest the disorder. Females, on the other hand, will usually manifest the disorder only if they are homozygous for the variant allele, and if heterozygous will usually be unaffected. Since it is rare for females to be homozygous for a variant allele, X-linked recessive disorders usually affect only males.

The necessary characteristics to be certain a disorder is inherited in an X-linked recessive manner are listed in Table 7.3 and portrayed in Fig. 7.11.

### 7.7.1.1 Recurrence Risks

If a male affected with an X-linked recessive disorder survives to reproduce, he will always transmit his X chromosome with the variant allele to his daughters, who will be obligate carriers. An affected male will always transmit his Y chromosome to a son, and therefore none of his sons will be affected (Fig. 7.12). A carrier female has one X chromosome with the wild-type allele and one X chromosome with the variant allele; therefore, her sons have a 1 in 2 (50%) chance of being affected, while her daughters have a 1 in 2 (50%) chance of being carriers (Fig. 7.13).

| TABLE 7.3 Characteristics of an X-Linked Recessive Inherited Disorder |
|---|
| Males are affected almost exclusively |
| Transmission occurs through unaffected or carrier females to their sons |
| Male-to-male transmission is not observed |
| Affected males are at risk of transmitting the disorder to their grandsons through their obligate carrier daughters |

For a female to be affected with an X-linked recessive disorder, her mother would have to be a carrier and her father affected with the disorder. Obviously this situation is encountered only very rarely. Another possibility is that her mother is a carrier and the X chromosome transmitted by her father undergoes a new mutation. A female can also be affected by an X-linked recessive disorder if she has a single X chromosome (i.e., Turner syndrome), in which case she will be hemizygous for alleles on the X chromosome, like a male.

### 7.7.1.2 X Inactivation

Early in embryonic development, one of the X chromosomes in females is inactivated in each cell, with the result that the female, like the male, has a single functional X chromosome. This phenomenon was first described by Mary Lyon is 1961 and is sometimes referred to as lyonization. In individuals with X-chromosome aneuploidies, all but one of the X chromosomes are inactivated in each cell. The process of X inactivation is controlled by a region of the X chromosome, the X-inactivation center, situated on the proximal portion of the long arm. X inactivation usually occurs as a random process, such that in approximately 50% of the cells in a female the maternally derived X chromosome is active, and in the other 50% the paternally derived X chromosome is active. In females who are carriers of an X-linked recessive variant, approximately one-half of the cells will actively express the variant allele. Occasionally, this can

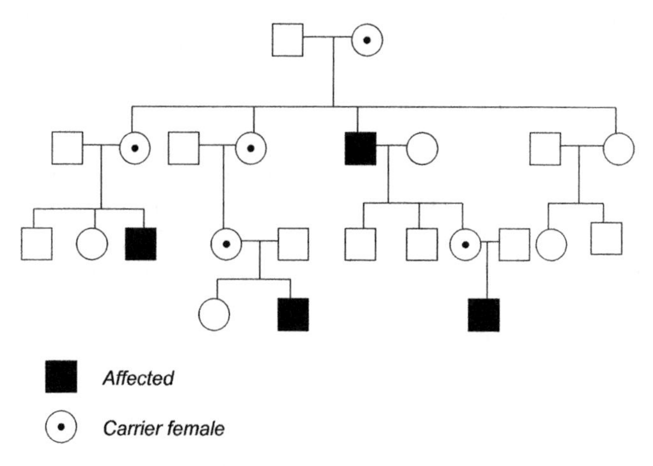


**Figure 7.11** Pedigree consistent with X-linked recessive inheritance.

be demonstrated clinically. In X-linked retinitis pigmentosa, for example, careful fundoscopic examination of a female carrier can show a mosaic pattern of pigmentation.

### 7.7.1.3 Manifesting Female Carriers of X-Linked Recessive Disorders

Although female carriers of X-linked recessive disorders are usually asymptomatic, they can manifest signs of the disorder. They are usually much less severely affected than males, however. There are a number of different mechanisms by which a female heterozygote can manifest signs of an X-linked recessive disorder, but the underlying cause for each one is nonrandom or skewed X inactivation. In this situation, there is a departure from the normal random process of X inactivation, with a greater proportion of one X chromosome being inactivated than the other. If, in most cells, the active X chromosome is the one with the variant allele, a female may manifest the disorder.

***7.7.1.3.1 Mechanisms of nonrandom X inactivation.*** A number of mechanisms can lead to nonrandom X inactivation:

1. *Chance*: Skewed X inactivation can occur by chance.
2. *Monozygotic twinning*: There have been several reports of monozygotic female twins, both heterozygous for a dystrophin gene deletion, one of whom was a manifesting carrier for Duchenne muscular dystrophy and the other an unaffected carrier. In some cases it has been demonstrated that, in the

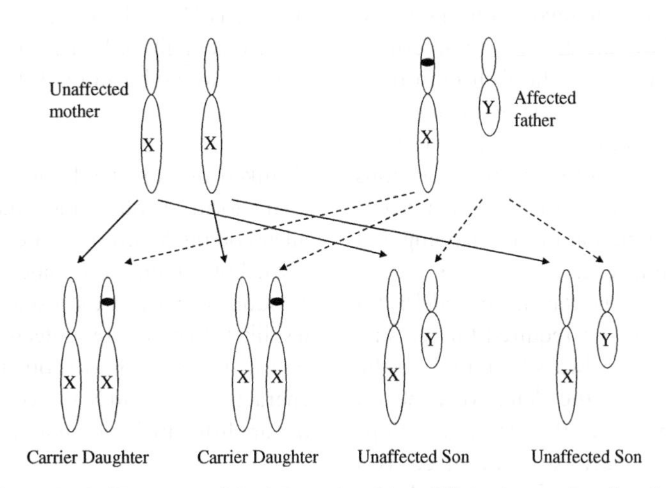


Figure 7.12 Recurrence risks for a male with an X-linked recessive disorder.

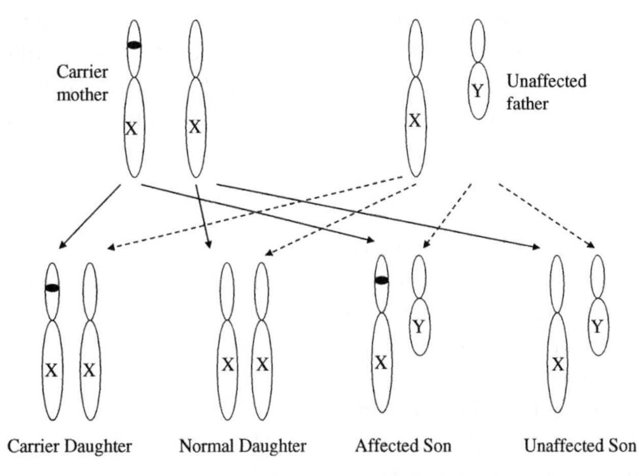


Figure 7.13 Recurrence risks for a female carrier of an X-linked recessive disorder.

lymphocytes and fibroblasts of the affected twin, the majority of active X chromosomes had the deletion, while in the unaffected twin, most active X chromosomes possessed the intact gene or there was random X inactivation. A number of hypotheses have been suggested to explain this observation [14]. One hypothesis is that random X inactivation produces two clusters of cells in the initial cell mass with opposite X-inactivation patterns, which stimulate the monozygotic twinning event. Another hypothesis is that the twinning event leads to unequal allocation of cells, leading to catch-up growth and skewed X inactivation in the twin with fewer cells
3. *Cytogenetic abnormalities*: In females with an X–autosome translocation, the normal X chromosome will be preferentially inactivated, maintaining the diploid state for the autosome involved in the translocation. If the translocation disrupts or interferes with the expression of a gene on the X chromosome, or if the X chromosome involved in the translocation carries a variant allele, that female will manifest the disorder. The finding of X–autosome translocations in manifesting female carriers of Duchenne muscular dystrophy was instrumental in the mapping and cloning of the dystrophin gene.
4. *Elimination of cells expressing the variant allele*: If a gene on the X chromosome is required for cell survival, the normal gene will always be found on the active X chromosome and the defective gene on the inactive X chromosome in the mature cell population, even though X inactivation occurred as a random process. This has been demonstrated in a number of disorders, including X-linked severe combined immunodeficiency, and can be used in the determination of carrier status for females at risk for that condition, although carrier status is now more easily identified by direct molecular testing for the variant.

#### 7.7.1.4 Gonadal Mosaicism

As with autosomal dominant disorders, gonadal mosaicism is an important phenomenon in X-linked recessive disorders, occurring particularly frequently in Duchenne muscular dystrophy, in which it has been shown to occur in both male and female gametogenesis. It is important to take this into account when advising mothers of apparently sporadically affected males with Duchenne muscular dystrophy of recurrence risks. Even though the mother does not carry the variant in lymphocytes, the risk to another son who inherits the same X as his affected brother can be up to 20% due to gonadal mosaicism.

#### 7.7.1.5 New Variants

Variants arising during meiosis may occur in male or female germ cells. If a particular variant arises largely in male germ cells, as in Lesch–Nyhan syndrome [15] and hemophilia A [16], the majority of mothers of affected boys will be carriers, and risks to sisters will be 50% regardless of how many unaffected brothers they have. In Duchenne muscular dystrophy, the overall mutation rate appears to be equal in males and females, but it has been suggested that most single-nucleotide variants occur in spermatogenesis, while most deletions arise in oogenesis [17]. Therefore, in isolated cases, the recurrence risk might be higher in nondeletion cases, as these mothers would be more likely to be carriers.

### 7.7.2 X-Linked Dominant Inheritance

X-linked dominant inheritance is a relatively less frequent form of inheritance and is caused by dominant alleles on the X chromosome. The phenotype will manifest in both hemizygous males and heterozygous females. Random X inactivation usually means that females are less likely to be severely affected than hemizygous males, unless they are homozygous for the variant allele. The characteristics of an X-linked dominant inherited disorder are listed in Table 7.4 and depicted in Fig. 7.14.

#### 7.7.2.1 Recurrence Risks

The offspring of either sex have a 1 in 2 (50%) chance of inheriting the disorder from affected females (Fig. 7.15). The situation is different for males affected by X-linked dominant disorders, whose daughters will always inherit the gene and whose sons cannot inherit

| TABLE 7.4 Characteristics of an X-Linked Dominant Inherited Disorder |
|---|
| Daughters of affected males always inherit the disorder |
| Sons of affected males never inherit the disorder |
| Affected females can transmit the disorder to offspring of both sexes |
| An excess of affected females exists in pedigrees for the disorder |

the gene (Fig. 7.16). An example of an X-linked dominant disorder is vitamin D–resistant rickets. X inactivation results in females with this disorder having less severe skeletal changes than those that occur in affected males.

An exception to the rule that females are less severely affected than males is seen in craniofrontonasal dysplasia, in which heterozygous females have a coronal craniosynostosis and affected hemizygous males do not have a craniosynostosis. The condition is caused by variants in the ephrin B1 gene *EFNB1*, and it has been proposed that in heterozygous females, patchwork loss of ephrin B1 disturbs tissue boundary formation at the developing coronal suture, whereas in males deficient in ephrin B1, an alternative mechanism maintains the normal boundary [18].

#### 7.7.2.2 X-Linked Dominant Lethal Alleles

In some disorders due to variant alleles of genes on the X chromosome, affected males are never or very rarely seen (e.g., incontinentia pigmenti and Goltz syndrome). This is thought to be due to a lethal effect of the variant allele in the hemizygous male, resulting in nonviability of the conceptus during early embryonic development. As a consequence, if an affected female were to have children, one would expect a sex ratio of 2:1, female to male, in the offspring and that one-half of the females would be affected, while none of the male offspring would be affected (Fig. 7.17). The majority of the mothers of females with these X-linked dominant lethal disorders are unaffected, and the variant alleles are therefore thought to arise as new mutations.

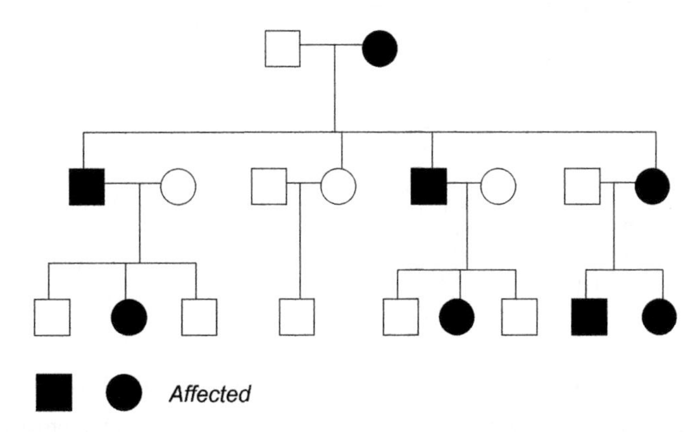


**Figure 7.14** Pedigree consistent with X-linked dominant inheritance.

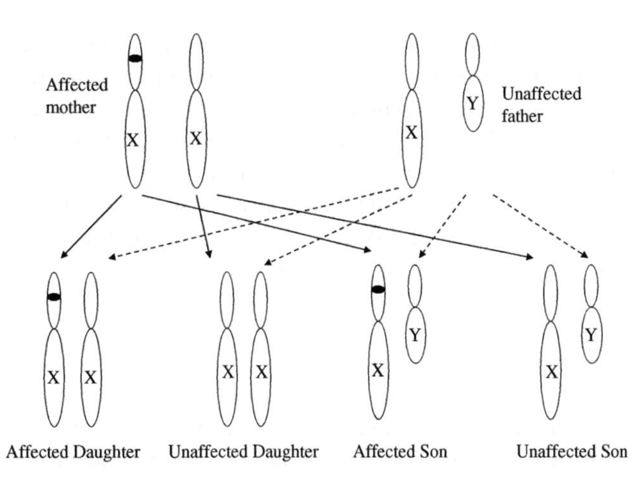


**Figure 7.15** Recurrence risks for a female affected with an X-linked dominant disorder.

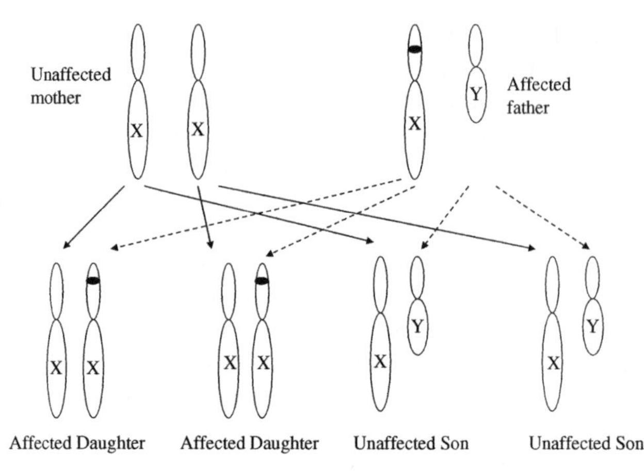


**Figure 7.16** Recurrence risks for a male affected with an X-linked dominant disorder.

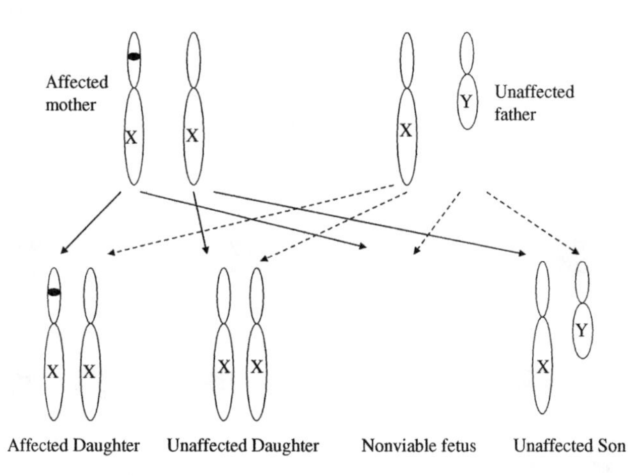


**Figure 7.17** Recurrence risks for a female affected with an X-linked dominant disorder lethal in males.

Rett syndrome is an X-linked dominant condition caused by variants in the *MECP2* gene. Girls with classical Rett syndrome have severe mental retardation developing after a period of relatively normal development. The variants that cause classical Rett syndrome are lethal in males; however, other variants in the gene can give a pattern of neuroencephalopathy and profound intellectual impairment in males that behaves as an X-linked recessive condition—another example of genotype–phenotype correlation.

### 7.7.3 Y-Linked (Holandric) Inheritance

Y-linked, or holandric, inheritance refers to genes carried on the Y chromosome. They therefore will be present only in males, and the disorder would be passed on to all their sons but never their daughters (Figs. 7.18 and 7.19). Genes involved in spermatogenesis have been mapped to the Y chromosome, but a male with a variant in a Y-linked gene involved in spermatogenesis would probably be infertile or hypofertile, making it difficult to demonstrate Y-linked inheritance. This situation may well change with the use of techniques such as intracytoplasmic sperm injection to treat male infertility, which will result in the transmission of the infertility to male offspring.

## 7.8 PARTIAL SEX LINKAGE

A small region of sequence identity exists between the X and the Y chromosomes, located at the tips of the long and short arms, known as the pseudoautosomal regions of the sex chromosomes. A high rate of recombination

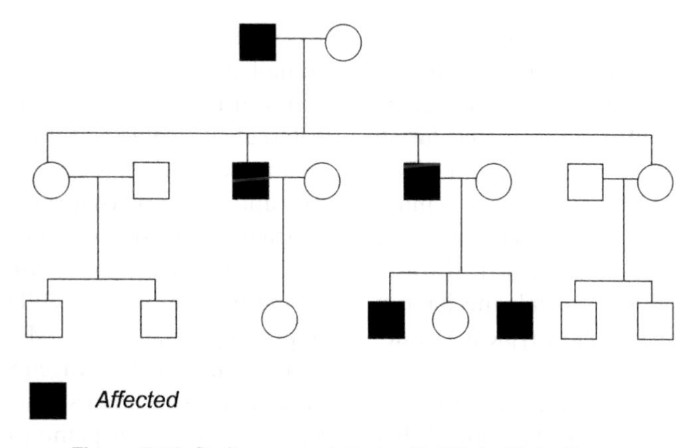


**Figure 7.18** Pedigree consistent with Y-linked inheritance.

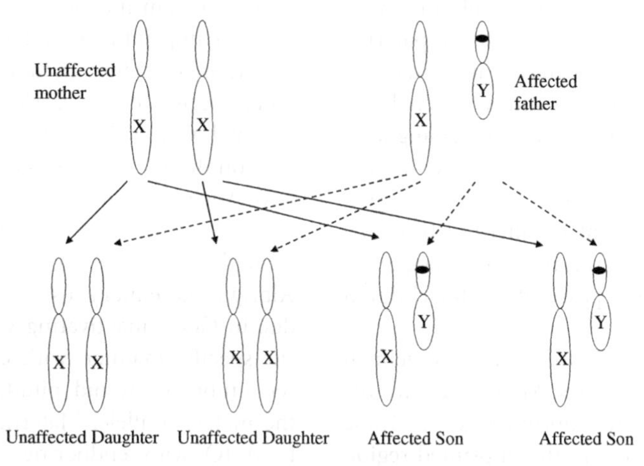


**Figure 7.19** Recurrence risks for a male affected with a Y-linked disorder.

at the telomeres of the short arms is thought to be obligatory for normal meiosis of these chromosomes. The genes within these regions, known as pseudoautosomal genes, escape X inactivation in the female; therefore, both sexes have two active alleles at these loci. The pseudoautosomal gene *SHOX* has been postulated to account for some of the features seen in the numerical sex chromosome disorder Turner syndrome. As a result of the high recombination frequency, variant genes located within the pseudoautosomal region can be transferred from the Y chromosome to the X chromosome and vice versa. Haploinsufficiency of the *SHOX* gene is the cause of Leri–Weill dyschondrosteosis and, in one study, about half of the segregations investigated showed a transfer of the *SHOX* variant to the alternate sex chromosome. Therefore, the condition can be inherited as either an X-linked dominant condition or, more rarely, a Y-linked condition. Affected men can transmit the variant to a son as well as to a daughter [19].

## 7.9 NONTRADITIONAL INHERITANCE

New techniques in molecular genetics are starting to reveal mechanisms that result in unexpected patterns of inheritance.

### 7.9.1 Genomic Imprinting and Epigenetic Mechanisms

Genomic imprinting is a phenomenon in which gene expression depends on parental origin. Imprinting is due to an epigenetic mechanism, in which the primary DNA sequence of the gene is not changed, but transcriptional regulation is affected by mechanisms such as DNA methylation, histone acetylation, and histone methylation. The imprint is erased during the early development of the male and female germ cells and reset prior to germ cell maturation so, for example, the imprint of a paternally imprinted gene is reset during oogenesis to a maternal imprint. In contrast to the biallelic expression of most genes, imprinted genes demonstrate monoallelic expression. This process modifies the transmission and expression of certain genetic diseases and should be borne in mind as a possible mechanism underlying disorders that do not follow typical Mendelian inheritance.

Many disorders due to defects affecting imprinted genes arise de novo, but they can also be familial. For example, the familial paraganglioma syndrome is due to variants in the succinate dehydrogenase subunits B, C, and D. *SDHD* is an imprinted gene, with the paternal allele active in each cell and the maternal allele inactive. The pathogenic variant is dominant and an affected parent (male or female) has a 1 in 2 chance of passing it on to each child but the child is at increased risk to develop paragangliomas only if the variant is inherited from the father (Fig. 7.20). If the variant is inherited from the mother, the risk to develop paraganglioma is negligible but, if her son inherits it and passes it on, his children are at increased risk.

Prader–Willi syndrome (PWS) and Angelman syndrome are probably the best-known syndromic examples of disorders due to imprinted genes. These disorders affect different genes in the imprinted region at 15q11–q13. The first finding was that deletions within the region on the paternally derived chromosome led to PWS, while deletions of the maternally derived chromosome led to Angelman syndrome. This is because the genes in this region that cause PWS are active on the paternal allele only and the gene that causes Angelman syndrome, *UBE3A*, is active only on the maternal allele. The underlying mechanism is often a de novo deletion or single-nucleotide variant within the imprinted gene itself, but in some cases disorders are due to epigenetic mechanisms such as UPD or the consequence of an imprinting center defect. Maternal UPD results in PWS and paternal UPD in Angelman syndrome. Some familial cases of Angelman syndrome are due to variants in the gene *UBE3A*. A pathogenic variant inherited from a mother will cause Angelman syndrome but, when inherited from the father, the condition will not manifest. The imprint is reset during male and female gametogenesis. Therefore, in familial Angelman syndrome, if an unaffected female carries a variant in *UBE3A* on her paternal allele it is reset at gametogenesis, so that it is now on the active allele and children who inherit it will have Angelman syndrome.

Some genes show tissue-specific imprinting, for example, the *GNAS* gene, which is associated with Albright hereditary osteodystrophy (AHO). AHO is due to G(s)α inactivating variants, imprinted in a tissue-specific manner, with expression in the proximal renal tubules, thyroid, pituitary, and ovaries being from the maternal allele. Maternally inherited variants lead to AHO with endocrine involvement (pseudohypoparathyroidism type 1A), whereas paternally inherited variants lead to AHO alone. Pseudohypoparathyroidism

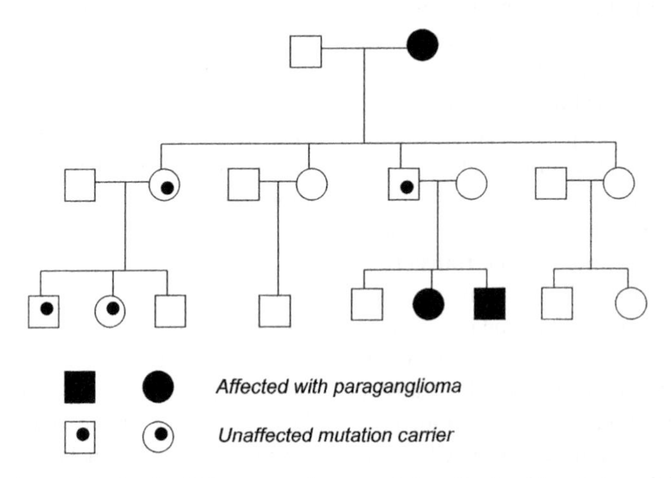


**Figure 7.20** Pedigree showing familial paragangliomas and the effects of imprinting of the *SDHD* gene.

type 1B (parathormone resistance without AHO) can be caused by a deletion of the imprinted promoter regions of the gene that control gene expression [20].

### 7.9.2 Digenic Inheritance

Genetic or locus heterogeneity by which different genes can cause clinically identical disorders has been discussed previously in this chapter. However, these cases are considered to be monogenic in that, in any one family, only one locus is thought to harbor pathogenic variants. Digenic inheritance is the simplest form of inheritance in complex genetic disease and occurs when the phenotype is due to the coinheritance of variants in two distinct genes, i.e., the affected individuals are double heterozygotes [21].

Retinitis pigmentosa was the first inherited condition in which digenic inheritance was convincingly described [22]. Although the families described were initially thought to be compatible with autosomal dominant retinitis pigmentosa with reduced penetrance, the families showed a number of unusual features:

1. In each family, the disorder originated in the offspring of unaffected individuals.
2. Affected individuals transmitted the disorder to statistically significantly fewer than 50% of their offspring.

On molecular testing, it was found that both affected and unaffected individuals carried a variant in the *peripherin/RDS* gene. Affected individuals were also heterozygous for a variant in the *ROM1* gene. These genes encode two of the polypeptide subunits of an oligomeric transmembrane protein complex present at the photoreceptor outer segment disc rims. Abnormal peripherin/RDS protein can assemble with wild-type *ROM1* protein to form structurally normal complexes but cannot assemble with abnormal ROM1 protein. Therefore, only the combination of the two heterozygous variants is pathogenic [23].

Another example of digenic inheritance is seen in facioscapulohumeral muscular dystrophy type 2 (FSHD2). The more common FSHD1 is a dominant disorder associated with a contraction of the D4Z4 array on chromosome 4 from the normal 11-100 repeats down to 1-10 D4Z4 units, when that array is on a specific, permissive chromosome 4 haplotype. The permissive chromosome 4 haplotype contains a single-nucleotide polymorphism in the D4Z4-adjacent sequence at the distal end of the array and is found in about 50% of the general population. Each D4Z4 unit contains a copy of the *DUX4* gene, a transcription factor gene, and contraction of the D4Z4 array on the permissive haplotype leads to chromatin relaxation and overexpression of *DUX4*. Overexpression of *DUX4* is toxic to regenerating and developing muscle cells. FSHD2 occurs in individuals who inherit a loss-of-function variant in the *SMCHD1* gene on chromosome 18 and a normal-sized D4Z4 array on a chromosome 4 haplotype permissive for *DUX4* expression. Reduction of *SMCHD1* gene expression results in D4Z4 CpG hypomethylation, presumably resulting in overexpression of *DUX4*. Both the *SMCHD1* variant and the chromosome 4 permissive haplotype need to be present for disease expression [24].

### 7.9.3 Mitochondrial Inheritance

The nuclear chromosomes are not the only source of coding DNA sequences within the cell. Mitochondria possess their own DNA, which, as well as coding for mitochondrial transfer RNA and ribosomal RNA, also carries the genes for 13 structural proteins that are all mitochondrial enzyme subunits. Variants within these genes have been shown to cause disease (e.g., Leber hereditary optic neuropathy). The inheritance pattern of mitochondrial DNA (mtDNA) is, however, very different from that of nuclear DNA, as mitochondria are exclusively maternally inherited. Therefore, mitochondrial variants can be transmitted only through females, although they can affect both sexes equally. Genetic assessment of mitochondrial disorders is complicated by the great variability of these disorders and a pedigree that is seldom conclusive of maternal transmission. That is because there are a number of different genetic mechanisms that can underlie mitochondrial disorders [25]. These include:

1. variants of the mtDNA, including single-nucleotide variants, deletions, and duplications;
2. dominant variants in nuclear genes involved in mitochondrial structure, function, and mtDNA maintenance;
3. recessive variants in nuclear genes involved in mitochondrial structure, function, and mtDNA maintenance.

In Leber hereditary optic neuropathy, the pattern of maternal inheritance is well documented. The commonest mtDNA variant is a single-nucleotide variant in base pair 1178 of the ND4 gene of complex I of the respiratory chain. Two other common variants have also been at chromosome 22q11 in some cases of nonsyndromic congenital heart disease, is important in providing genetic advice appropriate to a particular case. Hirschsprung disease provides a good example of a polygenic disorder in which involvement of several single loci has now been identified in a proportion of families. Data from initial family studies suggested sex-modified polygenic inheritance, and empirical recurrence risks have been produced for genetic counseling based on the sex of the index case and relatives and the length of the aganglionic segment involved; however, a mode of inheritance compatible with an incompletely penetrant autosomal dominant gene was suggested in some families. Linkage to a gene on chromosome 10 was subsequently demonstrated and variants in the *RET* oncogene were demonstrated. Variants in other genes, notably, the endothelin receptor type B gene *EDNRB*, have also been implicated. Thus in a subset of families with Hirschsprung disease, there is a major unifactorial predisposition, a situation that is seen in many other multifactorial disorders, e.g., Alzheimer and Parkinson disease.

## 7.12 ISOLATED CASES

In the days before widespread molecular genetic testing, isolated, presumed genetic, cases within a family posed problems in terms of genetic counseling unless the condition had a well-recognized inheritance pattern. Nowadays modern molecular genetics techniques are identifying the causes of many of these conditions, although they are still not giving the answers in all cases. Many studies of whole-exome and -genome sequencing are coming up with a similar diagnosis rate of about one-third. Therefore, there is clearly a long way to go in understanding all the mechanisms of genetic disease in humans.

We now understand that there may be many causes of an isolated case within a family. These include the following:

1. The disorder may be due to an autosomal dominant gene variant, arising by a new mutation, transmitted through a nonpenetrant or very mildly affected parent or by a clinically unaffected parent who carries a mosaic germline variant. The situation may also represent misattributed paternity. Risk to siblings varies between 0% and 50%, depending on the origin of the variant, while risk to the offspring of the affected individual would be 50%.
2. The disorder may be caused by an autosomal recessive gene variant with a 25% recurrence risk for siblings unless due to UPD or a de novo variant. If carrier state can be confirmed by molecular or biochemical analysis, cascade screening of other family members may be appropriate when the population carrier frequency is high or consanguineous marriages are planned.
3. The disorder may be due to an X-linked gene, usually presenting in a hemizygous male but affecting females if the X-inactivation pattern is skewed or the gene acts dominantly. Isolated cases may represent de novo variants, which are frequent in lethal X-linked recessive disorders, or may be transmitted by asymptomatic mothers who are carriers or who are gonadal mosaics for the variant.
4. The disorder may be due to two or more independently inherited monogenic or chromosomal defects.
5. The disorder may be due to an mtDNA variant representing a sporadic case or maternal transmission.
6. The disorder may be due to a chromosomal abnormality. Many of these, including the common trisomies due to nondisjunction, have a low risk of recurrence, but unbalanced karyotypes due to familial chromosomal rearrangements may carry high risks of recurrence, and investigation of relatives is required.
7. The disorder may be polygenic, and recurrence risks depend on the disorder. These are based on empirical data derived from family studies.
8. The disorder may have a nongenetic etiology with no increase in the risk of recurrence, unless due to a teratogenic agent to which further pregnancies will also be exposed.

This list will increase with new technologies and new understanding. The use of whole-genome sequencing is likely to uncover a significant number of genetic conditions that are due to gene regulation rather than the gene itself. It is hoped that the future will provide more and more answers to families seeking an explanation for the genetic disease within their family.

## REFERENCES

[1] Peters J. Classic papers in genetics. London: Prentice-Hall; 1959.
[2] Wilkie A. The molecular basis of genetic dominance. J Med Genet Genomics 1994:3189–98.
[3] Veitia RA, Caburet S, Birchler JA. Mechanisms of Mendelian dominance. Clin Genet 2017. https://doi.org/10.1111/cge13107.

[4] Sheffield WP, Bhakta V. The M358R variant of α (1)-proteinase inhibitor inhibits coagulation factor VIIA. Biochem Biophys Res Commun 2016;470(3):710–3.
[5] Knudson A. Mutation and cancer: statistical study of retinoblastoma. Proc Natl Acad Sci U S A 1971. 168820–168823.
[6] Harris P, Rossetti S. Determinants of renal disease variability in ADPKD. Adv Chron Kidney Dis 2010;17(2):131–9.
[7] Acuna-Hidalgo R, Veltman JA, Hoischen A. New insights into the generation and role of de novo mutations in health and disease. Genome Biol 2016;17:241.
[8] Penrose L. Parental age and mutation. Lancet 1955:312–3.
[9] Pyott SM, Pepin MG, Schwarze U, Yang K, Smith G, Byers P. Recurrence of perinatal lethal osteogenesis imperfecta in sibships: parsing the risk between parental mosaicism for dominant mutations and autosomal recessive inheritance. Genet Med 2011;13(2):125–30.
[10] Evans DGR, Ramsden RT, Shenton A, Gokhale C, Bowers NL, Huson SM, Pichert G, Wallace A. Mosaicism in neurofibromatosis type 2: an update of risk based on uni/bilaterality of vestibular schwannoma at presentation and sensitive mutation analysis including multiple ligation-dependent probe amplification. J Med Genet 2007;44:424–8.
[11] Garrod AE. The incidence of alkaptonuria: a study in chemical individuality. Lancet 1902:1616–20.
[12] Rankin J, Ellard S. The laminopathies: a clinical review. Clin Genet 2006;70(4):261–74.
[13] Rodrigues NR, Owen N, Talbot K. Deletions in the survival motor neuron gene on 5q13 in autosomal recessive spinal muscular atrophy. Hum Mol Genet 1995:4631–4.
[14] Machin G. Some causes of genotypic and phenotypic discordance in monozygotic twin pairs. Am J Med Genet 1996;61:216–28.
[15] Francke U, Felsenstein J, Gartler SM, Migeon BR, Dancis J, Seegmiller JK, Bakay F, Nyhan WL. The occurrence of new mutants in the X-linked recessive Lesch–Nyhan disease. Am J Hum Genet 1976:28123–37.
[16] Bröcker-Vriends AHJT, Rosendaal FR, van Houwelingen JC, Bakker E, van Ommen GJ, van de Kamp JJ, Briet E. Sex ratio of the mutation frequencies in haemophilia A: coagulation assays and RFLP analysis. J Med Genet 1991:28672–80.
[17] Lee T, Takeshima Y, Kusonoki N, Awano H, Yagi M, Matsuo M, Iijima K. Differences in Carrier frequency between mothers of Duchenne and Becker muscular dystrophy patients. J Hum Genet 2014;59:46–50.
[18] Twigg SR, Kan R, Babbs C, Bochukova EG, Robertson SP, Wall SA, Morriss-Kay GM, Wilkie AO. Mutations of Ephrin-B1 (EFNB1), a marker of tissue boundary formation, cause craniofrontonasal syndrome. Proc Natl Acad Sci U S A 2004:1018652–7.
[19] Kant SG, van der Kamp HJ, Kriek M, Bakker E, Bakker B, Hoffer MJ, van Bunderen P, Losekoot M, Maas SM, Wit JM. The jumping SHOX gene-crossover in the pseudoautosomal region resulting in unusual inheritance of Leri–Weill dyschondrosteosis. J Clin Endocrinol Metab 2011;96(2):E356–9.
[20] Liu J, Che,n M, Deng C, Bourc'his D, Nealon JG, Erlichman B, Bestor TH, Weinstein LS. Identification of the control region for tissue-specific imprinting of the stimulatory G protein alpha-subunit. Proc Natl Acad Sci U S A 2005;102(15):5513–8.
[21] Scaffer AA. Digenic inheritance in medical genetics. J Med Genet 2013;50:641–52.
[22] Kajiwara K, Berson E, Dryja T. Digenic retinitis pigmentosa due to mutations at the unlinked peripherin/RDS and ROM1 loci. Science 1994:2641604–7.
[23] Goldberg AF, Molday RS. Defective subunit assembly underlies a digenic form of retinitis pigmentosa linked to mutations in peripherin/RDS and ROM-1. Proc Natl Acad Sci U S A 1996;93:13726–30.
[24] Lemmers RJLF, Tawil R, Petek LM, Balog J, Block GJ, Santen GWE, Amell AM, van der Vliet PJ, Almomani R, Straasheijm KR, Krom YD, Klooster R, Sun Y, den Dunnen JT, Helmer Q, Donlin-Smith CM, Padberg GW, van Engelen BGM, de Greef JC, Aartsma-Rus AM, Frants RR, de Visser M, Desnuelle C, Sacconi S, Filippova GN, Bakker B, Bamshad MJ, Tapscott SJ, Miller DG, van der Maarel SM. Digenic inheritance of an SMCHD1 mutation and an FSHD-permissive D4Z4 allele causes fascioscapulohumeral muscular dystrophy type 2. Nat Genet 2012;44:1370–4.
[25] Chinnery P. Mitochondrial disease in adults: what's old and what's new? EMBO Mol Med 2015;7:1503–12.
[26] Balci TB, Hartley T, Xi Y, Beaulieu CL, Bernier FP, Dupuis L, Horvath GA, Mendoza-Londono R, Prasad C, Richer J, Yang X-R, Armour CM, Bareke E, Fernandez BA, McMillan HJ, Lamont RE, Majewski J, Parboosingh JS, Prasad AN, Rupar CA, Schwartzentruber J, Smith AC, Tetreault M, FORGE Canada Consortium, Care4Rare Canada Consortium, Innes AM, Boycott KM. Debunking Occam's razor: diagnosing multiple genetic disease in families by whole-exome sequencing. Clin Genet 2017;92:281–9.

8

# Analysis of Genetic Linkage

*Rita M. Cantor*

Department of Human Genetics, David Geffen School of Medicine at UCLA, Los Angeles, CA, United States

## 8.1 INTRODUCTION TO LINKAGE ANALYSIS

Linkage analysis is a well-established genetic method used to map the genes for heritable traits to their chromosome locations. It is part of a larger process that has been referred to as "reverse genetics," because the approach works in the reverse order from our model of how genes operate, biologically. That is, while genes act in a forward fashion to produce a trait, reverse genetics starts with the trait and uses linkage analysis along with other analytic methods to identify the predisposing genes. Reverse genetics became feasible in the 1990s, when a very extensive panel of multiallelic markers that spanned the human genome was established. Since 2008, the genome-wide markers in use have evolved from multiallelic to biallelic single-nucleotide polymorphisms (SNPs), where their spacing is much denser. The whole genome is analyzed and the approach is referred to as a full-genome linkage scan.

The genome-wide markers are genotyped and tested in a study sample of pedigrees, and those showing the strongest statistical evidence of linkage exceeding a predetermined threshold localize the trait gene to the chromosome segment where the markers reside. The resolution at the locus is usually quite poor, as many genes will reside within a linked region. Nevertheless, a statistically significant linkage result limits the search for the predisposing gene to those in the linked region, thus reducing cost and follow-up time. Once a trait gene is mapped by linkage, other strategies such as fine mapping, linkage analysis with additional markers, targeted association analysis, and sequencing of the chromosome region can be used to identify the gene of interest. Using the advances made by the Human Genome Project, reverse genetics has been very effective in identifying genes causing rare Mendelian disorders that result from fully penetrant single genes.

Linkage analysis has two requirements. First, one must identify and study families containing individuals who exhibit a heritable trait of interest. Then, a substantial number of family members, both affected and unaffected, must be genotyped for genetically informative markers. A critical step is to establish that the trait of interest is heritable and to assess its mode of inheritance. The hallmark of a heritable trait is that it is shared to a greater degree by genetically close relatives compared with distant relatives. Comparison of trait concordance rates in monozygotic and dizygotic twin pairs is the classic approach to assess heritability, but other designs have also been used. The initial genetic markers were multiallelic single tandem repeats, which are very informative. However, these are now genotyped rarely, and less-informative biallelic SNPs genotyped on arrays have replaced them.

In a linkage analysis, the families can vary in size from nuclear families or sibships with at least two genotyped children to larger pedigrees with complicated structures in which a substantial percentage of the members are genotyped and measured for the trait. The trait can be binary, having only two values, such as the absence or presence of a disease, or quantitative, continuous, and possibly normally distributed, such as height. Statistical algorithms are applied to the family

Emery and Rimoin's Principles and Practice of Medical Genetics and Genomics: Foundations, https://doi.org/10.1016/B978-0-12-812537-3.00008-1

number of environmental triggers. The analyses of both traits are conducted using the same model-free analytic methods. In addition, there has been an interest in the genetics of quantitative traits that may be important on their own or correlate with binary traits of interest. Methods and computer programs have been developed to identify the genetics of these traits as well.

### 8.4.1 Model-Free Linkage Analysis

Model-free methods test if the allele-sharing patterns in families are consistent with linkage in a very general way. That is, those in the pedigree who have the trait should display evidence of marker allele sharing to a greater degree than one would expect by chance alone in the linked regions. Most model-free tests include only individuals with the trait of interest. However, genotypes of their relatives are important to make more precise estimates of allele sharing in those exhibiting the trait. Using a study sample that includes only those individuals with the trait reduces the risk of including individuals who do not have the trait but share alleles with those who do, because of reduced penetrance. Including these individuals in the analysis would mask evidence of linkage. The most common model-free test statistics are based on allele sharing in sibling pairs. Limiting the analysis to sibling pairs has been effective because it is usually difficult to ascertain larger pedigrees with multiple members having the trait if it is complex. In addition, with multiple common risk alleles, members of large families may develop the trait as the result of several genetic etiologies, thus introducing within-pedigree heterogeneity, for which we do not have appropriate analytic approaches. Sibling pairs with a trait are usually available for study and are more likely to have the trait because of a shared genetic etiology. However, nuclear families contain less information about linkage than a large pedigree, and thus a larger number of these families will have to be studied to find linkage.

The sib-pair allele sharing linkage methods are based on a simple expectation. At a marker, the sibling pairs can inherit four possible alleles from their parents. If there is no gene predisposing to the trait in the region of the marker, the sibling pairs should share no alleles 25% of the time, one allele 50% of the time, and both alleles 25% of the time for that marker. The expected proportion of allele sharing is 50%. A statistically significant deviation from these theoretical values at a marker provides evidence of linkage of the trait to that marker.

For model-free analyses, the recombination fraction between the marker and the trait gene is assumed to be zero. Thus, an estimate of $r$ is not included in the analysis. For several linkage statistics, the degree of allele sharing for each sibling pair is assessed, and the allele sharing estimates are averaged over the pairs. These individual estimates would be 0, ½, or 1 for each pair if the genotypes were fully informative and all parents were completely genotyped. When data are missing or the parental genotypes do not allow for an unambiguous assignment of their inheritance, the marker allele frequencies are used in the estimate of allele sharing. The allele sharing estimates will usually be different from 0, ½, or 1 to reflect this. In the simplest case, the average allele sharing value is tested against its theoretical value of ½. An additional consideration is that if there are more than two siblings in a sibship, the pairs will not provide statistically independent allele sharing estimates. A weighting scheme to account for this can be employed. For example, three sibs in a sibship contribute information from three pairs, and since the information for the third pair can be derived from the first two pairs, allele sharing for these three pairs can be weighted by a factor of ⅔. A subset of the programs listed in Table 8.1 implement model-free allele sharing linkage methods for binary traits in nuclear families. They are the SIBPAL program of SAGE, the MLS and NPL options of GENEHUNTER, and the MERLIN REGRESS program. These software packages allow for generation of the allele sharing test statistics at evenly spaced intervals along the entire chromosome, given the marker map, and a multipoint analysis is conducted for each chromosome. As with binary linkage, multipoint analyses require that the markers are independent, and the SNPs must be pruned to satisfy that assumption. The packages provide plots of the statistical significance of the tests across the chromosomes so that the loci with significant evidence for linkage can be identified easily.

### 8.4.2 Linkage Analysis of Quantitative Traits

A quantitative trait can be dichotomized into one that is binary for a parametric linkage analysis. For example, a systolic blood pressure of greater than 140 is used to classify an individual as hypertensive, and hypertension has often been the trait tested for linkage. In fact, the quantitative trait itself may be better suited to linkage analysis than the dichotomized trait, as the extent of variation can be assessed in all family members, thus making it likely to provide increased statistical power to detect genes. In addition, the etiologic heterogeneity of genetically complex traits may be reduced by the analysis of a correlated quantitative trait that reflects only one feature of the complex disorder. For example, cardiovascular disease may be better addressed by studying a particular lipid trait, such as cholesterol, rather than the binary trait of a myocardial infarction, which is very likely to have a much broader etiology. Programs to analyze quantitative traits in large pedigrees usually require the assumption of trait normality, while those for smaller pedigrees do not require rigorous adherence to these assumptions [7]. Programs and their features are given in Table 8.2.

Variation in a quantitative trait usually results from the contributions of multiple genes with small effects modified by environmental influences. If none of the genes contributes a substantial amount to this quantitative variation, loci can be difficult to detect using linkage analysis. However, a gene contributing to a relatively large proportion of the variance of the trait, a major gene, is a good candidate for localization by quantitative trait linkage analysis. When conducting a quantitative trait linkage analysis, it is important to select families for which the trait exhibits marked variation within the pedigree. Those families having multiple members with extreme values of the trait are most likely to provide support for a major gene. Regions that are identified by the linkage analysis of a quantitative trait are usually referred to as quantitative trait loci or QTL.

A list of computer software commonly used for the linkage analysis of quantitative traits is given in Table 8.2. A variance component analysis of a quantitative trait, such as that conducted by the SOLAR software, identifies linked chromosome regions by decomposing the variance of the trait into the components that contribute to it [8]. The log of the ratio of the likelihood of the data with a major gene is compared with the likelihood of the data when there is no major gene modeled at that location. Normality of the trait distribution is an important assumption and if the trait is not normally distributed, transformation to normality is critical to a successful analysis. For nonnormal traits in smaller pedigrees, GENEHUNTER/MAPMAKER/SIBS, MERLIN/REGRESS, and SAGE/SIBPAL provide good alternatives [9]. GENEHUNTER/NPL, which uses a nonparametric method that ranks the trait values, is robust against nonnormality. The ordered subset analysis approach implemented in OSA can be used to identify QTL for quantitative traits that are correlated with binary traits. The families in the analysis are ordered according to their scores on a quantitative correlated value and the evidence for linkage is assessed as the ordered families are sequentially included in the analysis.

## 8.5 LINKAGE ANALYSIS: FUTURE DIRECTIONS

Linkage analysis has been used to identify chromosome locations of human trait genes for more than 50 years. Its well-developed tools include families, markers, and statistical methods of analysis. The genes for many Mendelian disorders have been identified with these tools. However, as the number of genetic markers increased from 30 to 10 million, our ability and enthusiasm to localize and identify genes for traits of increasing genetic complexity have grown proportionally. Such marker density allows us to capitalize on linkage disequilibrium to identify trait genes, and consequently, since the early 2000s, there have been few modifications to the well-established linkage methods. The primary approach to gene identification has quickly transitioned to genome-wide association studies (GWAS), which capitalizes on the linkage disequilibrium among closely spaced markers in populations [10]. In some sense, analyzing linkage disequilibrium in populations is similar to analyzing linkage in families. To clarify, in the current generation, within a population, SNP alleles are in linkage disequilibrium because they are too close to each other on the chromosomes to have undergone significant recombination over the generations. Thus, one can view GWAS as linkage analyses of a large pedigree that is the current population under analysis. All mapping information is in the current generation, and we test for association of specific marker alleles and the trait of interest.

GWAS are based on association tests of common SNP variants whose frequencies are larger than 1%. The genotyped SNPs have been selected to tag trait variants that may not be tested directly. Although consistent gene associations have been identified for a substantial number of complex traits, their effects have been surprisingly small. Successful GWAS have required very large samples, and the detected effect sizes indicate that the risk is raised at most by about 10%–20% compared with that of the background genotype. Consequently, the genetics literature has expressed concern that GWAS have not revealed the etiologies of traits with significant heritabilities.

Recently, concern about the small effect sizes of common variants, as well as the dramatic reduction in the cost of whole-exome sequencing with "next-generation" methods, has refocused the interest of many of those studying complex disorders toward the detection of rare genetic variants via sequencing. These variants are expected to have a frequency less than 1%, and may even consist of private mutations. They are also expected to exhibit greater penetrance values and effect sizes than the common variants. To sort through the many variants likely to be uncovered, there has been a renewed interest in the study of large pedigrees. If they exist for a complex trait, analyses will be focused on identifying those pedigrees that segregate the trait where it is acting in a quasi-Mendelian fashion. As with Mendelian disorders, model-based linkage analysis that allows for reduced penetrance and phenocopies may reveal important loci. Targeted sequencing in linked regions could reveal the genes with the segregating rare variants. With this approach, parametric linkage analysis is likely to again become a natural first step in gene-finding efforts over the next few years.

## REFERENCES


[1] Ott J. Analysis of human genetic linkage. Baltimore: Johns Hopkins University Press; 1999.

[2] Morton NE. Sequential tests for the detection of linkage. Am J Hum Genet 1955;7:277–318.

[3] Nyholt DR. Invited Editorial: all LODs are not created equal. Am J Hum Genet 2000;67:282–8.

[4] Cantor RM. Model-based linkage analysis of a binary trait. Methods Mol Biol 2012:285–300.

[5] Risch N. Linkage strategies for genetically complex traits. I. Multilocus models. Am. J. Hum. Genet. 1990;46:222–8.

[6] Risch N. Linkage strategies for genetically complex traits. II. The power of affected relative pairs. Am J Hum Genet 1990;46:229–41.

[7] Haseman JK, Elston RC. The investigation of linkage between a quantitative trait and a marker locus. Behav Genet 1972;2:3–19.

[8] Almasy L, Blangero J. Multipoint quantitative-trait linkage analysis in general pedigrees. Am J Hum Genet 1998;62:1198–211.

[9] Kruglyak L, Lander ES. Complete multipoint sib-pair analysis of qualitative and quantitative traits. Am J Hum Genet 1995;57:439–54.

[10] Hirschhorn JN, Daly MJ. Genome-wide association studies for common diseases and complex traits. Nat Rev Genet 2005;6:95–108.


## BIBLIOGRAPHY

The following are Web pages for linkage analysis software mentioned in the text and listed in Tables 8.1 and 8.2.

GENEHUNTER: http://www.broad.mit.edu/ftp/distribution/software/genehunter/.

LINKAGE: ftp://linkage.rockefeller.edu/software/linkage.

LOKI: http://www.stat.washington.edu/thompson/Genepi/Loki.shtml.

MENDEL: http://www.genetics.ucla.edu/software/.

MERLIN: http://www.sph.umich.edu/csg/abecasis/Merlin.

OSA: http://wwwchg.duhs.duke.edu/research/aplosa.html.

SAGE: http://darwin.cwru.edu/sage/.

SIMWALK: http://www.genetics.ucla.edu/software/simwalk.

SOLAR: http://www.sfbr.org/Departments/genetics_detail.aspx?p=37.

9

# Chromosomal Basis of Inheritance*

*Fady M. Mikhail*

Cytogenetics Laboratory, Department of Genetics, University of Alabama at Birmingham, Birmingham, AL, United States

## 9.1 INTRODUCTION

The human genome is packaged into a set of chromosomes as in other eukaryotes. Chromosomes are thus the vehicles of inheritance as they contain virtually the entire cellular DNA, with the exception of the small fraction present in the mitochondria. The structure, function, and behavior of chromosomes are therefore of much interest and importance. Chromosomes are derived in equal numbers from the mother and the father. Each ovum and sperm contains a set of 23 different chromosomes, which is the haploid number ($n$) of chromosomes in humans. The diploid fertilized egg and virtually every cell of the body arising from it has two haploid sets of chromosomes, resulting in the diploid human chromosome number ($2n$) of 46. The human karyotype consists of 22 pairs of autosomes and a pair of sex chromosomes. The correct chromosome number in humans was determined and confirmed in 1956 [1,2].

The behavior of chromosomes during meiotic cell division provides the basis for the Mendelian laws of inheritance, whereas their abnormal behavior in cell division leads to abnormalities of chromosome number. In this chapter, we examine the current understanding of the structure, molecular organization, and behavior of human chromosomes and explore how these features contribute to chromosomal diseases.

## 9.2 CHROMOSOME STRUCTURE

Although the structure of human and other eukaryotic chromosomes is not understood in full detail, recent investigations have provided insights into several aspects of chromosome structure at the molecular level. The haploid human genome consists of about $3 \times 10^9$ base pairs (bp) of DNA. Since 3000 bp of naked DNA are ~1 μm long, the total length of the diploid human genome is about 2 m. As the cell nucleus is no more than 10 μm in diameter, it is necessary to fold and compact this DNA, which is accomplished by packaging it in a hierarchy of levels into chromosomes of manageable size (Fig. 9.1). Organization of the DNA into chromosomes also maintains the linear order of genes and facilitates faithful replication and segregation of genetic material during cell division. The first level of this packaging, and thus the fundamental unit of chromosome organization, is a regularly repeating protein–DNA complex called the nucleosome. The basic structural features of the nucleosome were established in the early 1970s and have been further confirmed by high-resolution analysis of its crystal structure [3]. The nucleosome has the same design in all eukaryotes and consists of a cylindrical core about 11 nm in diameter and 6 nm in height made up of two molecules each of the four core histones (H2A, H2B, H3, and H4) with 147 bp of DNA wrapped around it. A "linker" DNA connects adjacent nucleosomes. Each nucleosome is also associated with a molecule of histone H1, which changes the path of the DNA as it exits from the nucleosome, and plays a role in further condensation of chromosomal DNA. Formation of the nucleosomes achieves a sevenfold compaction of the

*This chapter is a revision of the previous edition chapter by Julie R. Korenberg and T.K. Mohandas, vol. 1, pp. 167–190, © 2007, Elsevier Ltd.

Emery and Rimoin's Principles and Practice of Medical Genetics and Genomics: Foundations, https://doi.org/10.1016/B978-0-12-812537-3.00009-3

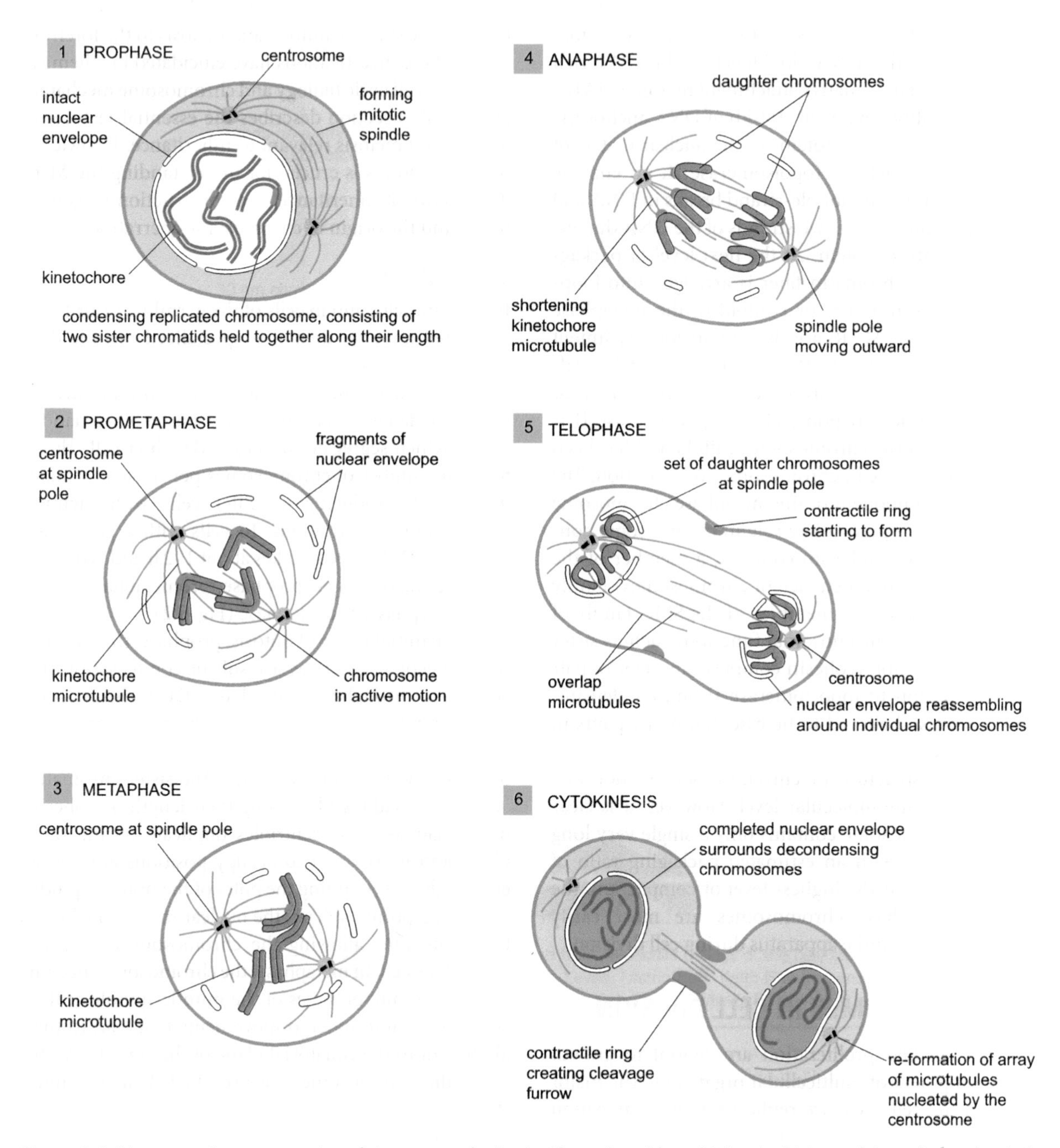


**Figure 9.2** Diagrammatic representation of the stages of mitosis. (Reproduced from Molecular biology of the cell, fourth ed., by Bruce Alberts, et al., Copyright © 2002 by Bruce Alberts, Alexander Johnson, Julian Lewis, Martin Raff, Keith Roberts, and Peter Walter. (c) 1983, 1989, 1994 by Bruce Alberts, Dennis Bray, Julian Lewis, Martin Raff, Keith Roberts, and James D. Watson. Used by permission of W. W. Norton & Company, Inc.)

and the action of motor proteins. The two sister chromatids of a chromosome are held together following chromosome replication by cohesins. This is a multisubunit protein complex that ensures correct segregation of daughter chromosomes at anaphase. At the beginning of anaphase, the cohesin complex is cleaved by a protease called separase, allowing separation of the sister chromatids [16].

At telophase, each set of daughter chromosomes arrives at the centriole at one of the two ends of the mitotic spindle, and reconstitution of the nuclear membrane begins. Cytokinesis, the division of the cytoplasm, follows telophase and leads to the formation of two genetically identical daughter cells.

### 9.3.2 Meiosis

Meiosis is a specialized cell division in germ cells that generates gametes with the haploid set of 23 chromosomes. The final gametic set includes single representatives of each of the 23 chromosome pairs selected at random. The details of meiosis and gamete formation are somewhat different in males and females, but the basic features are the same in both and are of fundamental importance. Meiosis accounts for the major principles of Mendelian genetics: segregation, independent assortment, and recombination of linked genes. Recombination or crossing over is the exchange of genetic material between homologous nonsister chromatids, a process that adds to genetic diversity by generating new combinations of genes.

Meiosis consists of two cell divisions (meiosis I and II) and is distinguished from mitosis by the following:

1. Homologous pairing: Maternal and paternal homologues of each chromosome are replicated and then undergo exact pairing along their lengths during prophase of meiosis I. Such a paired unit is called a "bivalent" because there are only two centromeres, although it is composed of four chromatids.
2. Recombination (crossing over): Crossing over occurs at the four-strand stage between nonsister chromatids, that is, chromatids from each of the pair of homologous chromosomes. The probability of recombination increases with the physical distance between two chromosomal sites and therefore provides a basis for the genetic map.
3. Segregation of maternal and paternal homologues: Centromeres do not divide at the first meiotic division (meiosis I). Instead, the members of a homologous pair go to opposite poles at anaphase of the first meiotic division. This accounts for Mendel's first law, the segregation of homologous genetic units. The segregation of maternal and paternal homologues in each bivalent chromosome pair occurs independent of the segregation in all the other bivalents. That is, the segregation of chromosome 1 homologues is independent of that of chromosome 2 homologues and so on. This accounts for Mendel's second law of independent assortment of genes. Meiosis I also leads to a reduction in the number of chromosomes from the diploid number ($2n=46$) to the haploid number ($n=23$) in the gametes.
4. Division of the haploid set with centromere division: The second meiotic division (meiosis II) occurs without a preceding round of DNA synthesis and chromosome duplication. In meiosis II, the two chromatids of a chromosome move to opposite daughter cells.

Meiotic prophase I is rather prolonged and can be subdivided into five stages on the basis of condensation of chromosomes and the extent of homologous pairing: leptotene, zygotene, pachytene, diplotene, and diakinesis. In leptotene, the chromosomes start to condense and are visible as long threads, but the homologues are still not paired. Chromosomal condensation continues and pairing of homologues (synapsis) begins at zygotene and is completed at pachytene, the stage at which recombination occurs (Fig. 9.3). By the pachytene stage, synapsis between homologues is completed and crossing over between nonsister chromatids occurs, during which homologous regions of DNA are exchanged. Synapsis is thought to be mediated and stabilized by the formation of the synaptonemal complex (SC) between homologous chromosomes. The SC is a protein-rich ladderlike structure that has a central element flanked by two lateral elements. The lateral elements and the central element are held together by transverse filaments. The lateral elements are formed of the axial elements of sister chromatids of the paired homologues, and the bulk of the chromosomal DNA is found in chromatin loops emanating from the outer sides of the two lateral elements. Initially, the SC was characterized by ultrastructural analysis. More recent studies have identified several protein components of the SC, although the functions of many of these are unknown. A constituent of the lateral elements is cohesin, consistent with the fact that the sister chromatids of each of the homologues are held together at pachytene. Another interesting feature

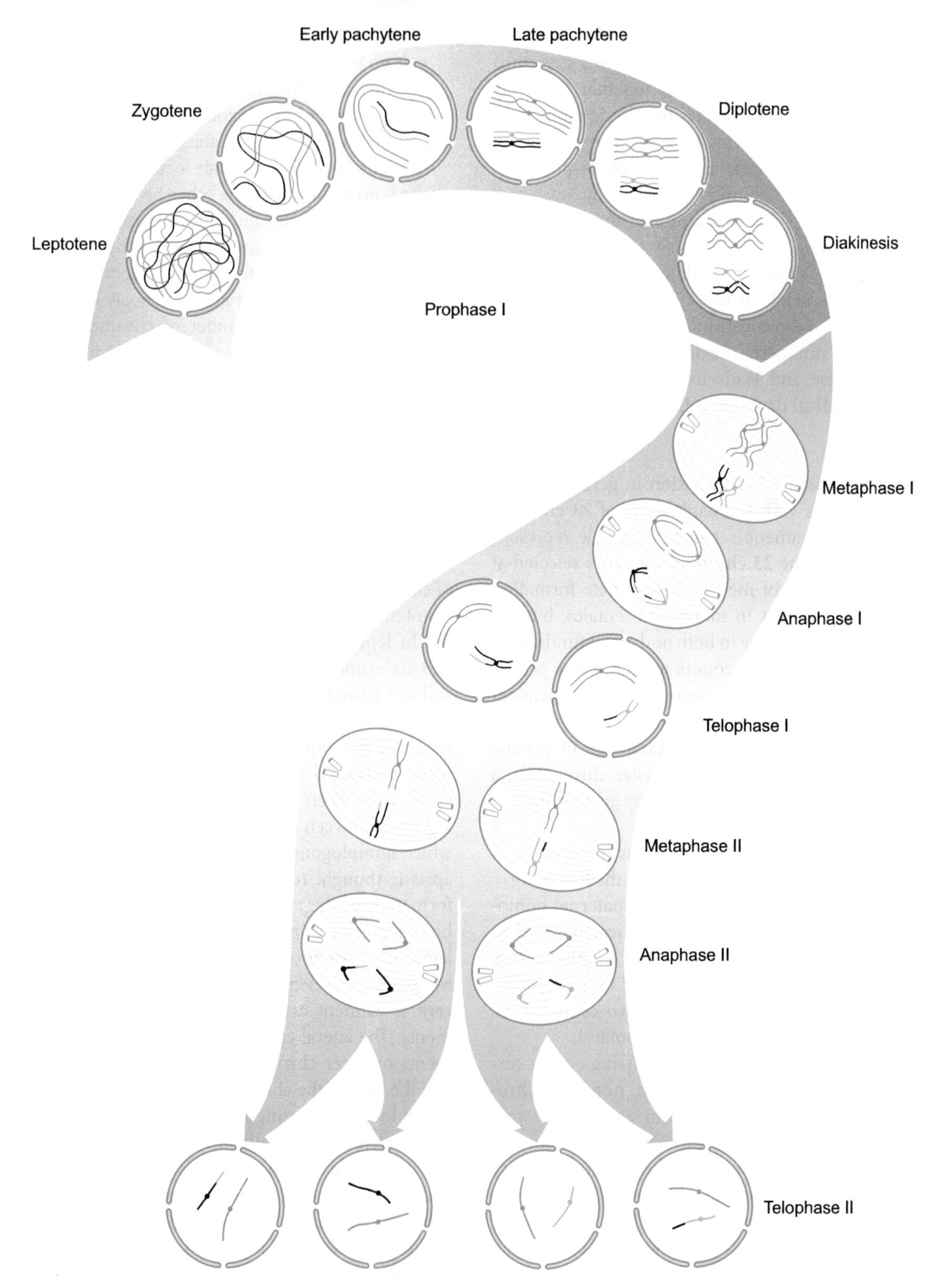


**Figure 9.3** Diagrammatic representation of the stages of meiosis. (Reproduced from Turnpenny PD, Ellard S. Chapter 3: chromosomes and cell division, Figure 3.19: stages of meiosis. In: Emery's elements of medical genetics, 12th ed. Churchill Livingstone: Elsevier; 2005.)

of SC is the presence of recombination nodules along its length. These are thought to be enzyme complexes that mediate genetic recombination via DNA breakage and repair. Chiasmata, or cruciform structures, become visible at the more condensed diplotene stage as cohesion is lost along the chromosome arms except at each chiasma, the point of recombination between homologous chromosomes.

Chiasmata are still visible at diakinesis, the stage of maximal condensation, and can be used to determine the frequency as well as the location of recombination. Chiasmata, like their underlying recombination events, play an important role in the normal segregation (disjunction) of homologues, and each pair of homologues has at least one chiasma per chromosome arm. Moreover, failure of chiasma formation predisposes to nondisjunction of homologues.

During prophase I in males, pairing and crossing over between the X and the Y chromosomes are possible because of a small region of homology at the terminal ends of their arms (i.e.,pseudoautosomal regions). The two chromosomes pair and cross over in these regions during prophase I.

In meiotic metaphase I, the nuclear membrane disappears and the chromosomes become aligned on the equatorial plane of the cell where they have become attached to the spindle, as in metaphase of mitosis. Then in anaphase I, the chromosomes now separate to opposite poles of the cell as the spindle contracts. In telophase I, each set of haploid chromosomes has now separated completely to opposite ends of the cell, which cleaves into two daughter gametes, so-called secondary spermatocytes or oocytes.

The meiosis II division resembles an ordinary mitotic division, except for the presence of a single set of 23 duplicated chromosomes, each with two chromatids held together at their centromere. Also, meiosis II is not strictly a genetically equal division as the two chromatids of a chromosome may not be identical as a result of genetic exchange(s) with a nonsister chromatid. At the end of the two meiotic divisions, each primary spermatocyte or oocyte has given rise to four haploid products (see Fig. 9.3). Their fatesare rather different in males and females, as discussed later.

### 9.3.3 Spermatogenesis and Oogenesis

In the human male, the production of sperms begins at puberty and continues throughout life. Undifferentiated stem cells of the germline, the spermatogonia, are abundant in the seminiferous tubules of the testis and show a high rate of mitotic activity throughout the adult life of a normal male. Of the two types of spermatogonia, only type A can differentiate into primary spermatocytes that enter meiosis, whereas type B are the long-lived progenitors that divide to generate daughter cells of both types A and B. In meiosis, each diploid spermatocyte gives rise to four haploid cells, each of which differentiates into a functional sperm. The entire process, from spermatogonium to sperm, takes about 70 days. The rate of sperm production may be as high as 50–100 million per day over many years, and thus the parental spermatogonia undergo many successive mitoses. It is estimated that the number of mitoses before sperm production in a 20-year-old male is about 200, while in a 45-year-old it is about 800 [17]. This provides the opportunity for the occurrence of more adverse genetic change with age in males, which is reflected in an increased mutation rate for certain inherited diseases.

The behavior of germline cells in the female is quite different from that in the male. By about the fourth month of prenatal development, about 7 million oogonia have begun to develop into primary oocytes and to enter meiosis. Primary oocytes proceed only as far as prophase of meiosis I by the time of birth, in which they remain until ovulation. This suspended stage of prophase occurs after pachytene, is referred to as dictyotene, and lasts from birth until after puberty, when small cohorts of the germ cells progress further into meiosis. The first meiotic division is stimulated by ovulation and is an unequal division in that most of the cytoplasm remains in the ovum and very little is pinched off to enter the first polar body, containing one set of homologues. Sperm penetration of the ovum stimulates the second meiotic division, leading to formation of the second polar body that contains a haploid set of chromosomes. On average, one oocyte per ovarian cycle completes the first meiotic division and proceeds to metaphase of the second meiotic division; if fertilized by a sperm, it completes the second division and embryonic development ensues. Thus, over the approximate 30-year reproductive lifetime of a female, only a few hundred oocytes complete the first meiotic division and few—if any—complete the second [18].

It is of interest to note that the frequency of point mutations and structural chromosomal changes is in general higher in male gametes and increases with age.

This increased mutation rate in males is attributed to the much larger number of cell divisions in the male germline. In contrast, changes in chromosome number increase with age in female gametes. Errors of disjunction seen with advanced maternal age appear to be related to the 13–50 years the oocytes spend in prophase before chromosome segregation. Genetic mapping studies indicate that the number as well as the positioning of crossover events influences meiotic segregation of chromosomes [19,20]. However, the molecular causes underlying age-dependent nondisjunction are still poorly understood [18].

## 9.4 METHODS FOR STUDYING HUMAN CHROMOSOMES

Technical innovations since the 1960s have revolutionized the study of human chromosomes. Chromosomes are normally visible only during cell division as they become condensed in preparation for orderly division. Therefore, chromosomes can be studied only in cells that are dividing in vivo or in vitro. Dividing cells are sufficiently common in some tissues in vivo to permit the direct study of chromosomes. This is true of meiotic divisions in the testis and embryonic ovary, and of mitotic divisions in the bone marrow, some epithelia, and tumors. However, cell culture methods have greatly extended the range of tissue and cell types from which dividing cells can be obtained in vitro. These include blood lymphocytes, fibroblasts from skin and other tissues, and cells from amniotic fluid or chorionic villi. Viable cells can even be obtained for a number of hours after the death of an individual or spontaneous abortion of an embryo. It is thus possible to carry out chromosome studies in a wide range of clinical situations.

The introduction of a short-term peripheral blood culture technique provided a reliable way to obtain human chromosome preparations of good quality for human cytogenetic investigations and for clinical diagnosis [21]. In this widely used technique, Tlymphocytes from a small sample of peripheral blood are stimulated to divide in culture with a mitogen, such as phytohemagglutinin. The blood culture is initiated in a suitable culture medium at 37°C, and within 3 days the stimulated lymphocytes provide very large numbers of dividing cells. These are blocked in metaphase by adding a mitotic spindle poison, such as colchicine, to the culture for a few minutes. Treatment of the cells with a hypotonic solution swells the cells and allows spreading of the chromosomes, which are then fixed and mounted on a glass slide. This makes it possible to prepare well-spread, flattened metaphase chromosome preparations on slides suitable for microscopic analysis using chromosome banding methods and molecular cytogenetic techniques.

### 9.4.1 Human Chromosome Identification

In the early days of human cytogenetic investigations, chromosomes were stained with Giemsa or a similar dye, yielding uniform staining along their lengths. Based on these studies, human chromosomes were classified according to their size and morphology (Figs. 9.4 and 9.5). The primary constriction represents the centromere, the chromosomal locus responsible for proper segregation of chromosomes to daughter cells during cell division. Based on the position of their centromere, human chromosomes are classified as metacentric, in which the centromere is at or near the middle of the chromosome; submetacentric, in which the centromere is located significantly off center; or acrocentric, in which the centromere is very close to one end. For all categories, the short arm of the chromosome is referred to as "p" for petite, and the long arm as "q". In addition to the centromere or primary constriction, five pairs of the acrocentric chromosomes (numbers 13, 14, 15, 21, and 22) may exhibit secondary constrictions on their short arms (Fig. 9.6). These mark the site of each cluster of ribosomal RNA (rRNA) genes and are called nucleolar organizing regions (NORs) because, at telophase, nucleoli are formed at a subset of these sites that are transcriptionally active. The rRNA genes remain in a moderately extended state at metaphase, reflecting the late shutoff of these genes in prophase and the rapid reinitiation of their transcription after the anaphase separation of sister chromatids. Originally, the chromosomes were assigned to groups A through G according to their general size and the position of the centromere (group A, 1–3;group B, 4 and 5;group C, 6–12 and X;group D, 13–15;group E, 16–18;group F, 19 and 20;group G, 21, 22, and Y) (Fig. 9.4).

### 9.4.2 Chromosome Banding

The conventional (Giemsa) staining without pretreatment does not permit precise identification of each chromosome in the human complement. A major technical innovation in human cytogenetics came in 1970, when Caspersson and colleagues discovered that human

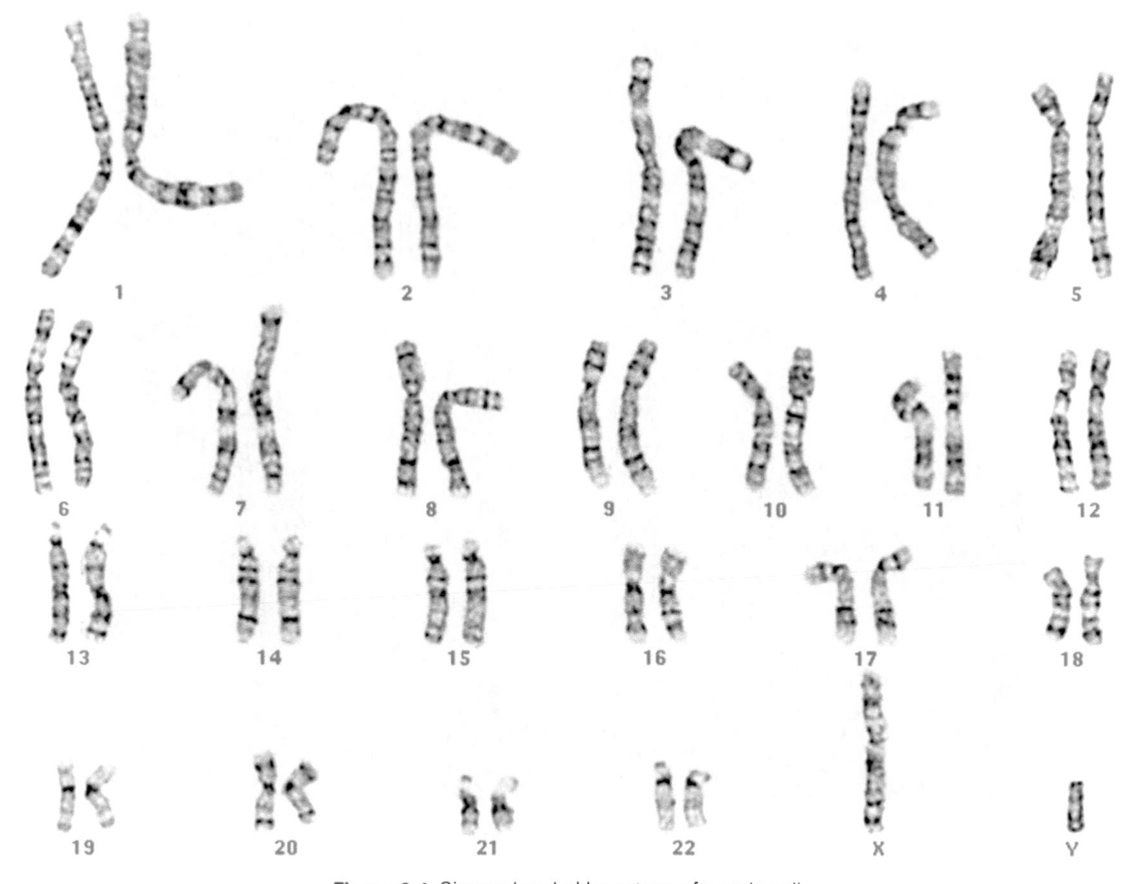


**Figure 9.4** Giemsa-banded karyotype of a male cell.

chromosomes stained with quinacrine mustard, a fluorescent DNA-binding compound, and examined under ultraviolet light show characteristic variation of fluorescence intensity along the length of each chromosome, producing a banded appearance [22]. Each chromosome could then be identified by its characteristic quinacrine (Q)-banding pattern (see Fig. 9.5). Subsequently, several techniques were developed that reveal banding patterns reflecting the underlying structural features of chromosomes. Techniques such as Giemsa (G)-banding and reverse (R)-banding produce the full range of bands along each chromosome, allowing identification of individual human chromosomes. Other banding techniques produce much more restricted staining of specific subsets of chromosome bands and include centromere (C)-banding and NOR-banding. A technique that differentially stains the two sister chromatids of a chromosome is also of particular interest. Chromosome banding methods of special interest are discussed in the following paragraphs.

Although Q-banding was the method first employed for human chromosome identification, it is rarely used today for routine chromosome analysis in clinical cytogenetics laboratories, as simpler methods have become available that do not require the use of a fluorescence microscope. A banding pattern that is almost identical to the Q-banding pattern can be produced by treatment of chromosomes with a denaturing agent or a proteolysis enzyme, prior to staining them with Giemsa. In the most consistent and commonly used version of this technique, chromosomes are treated with a dilute solution of trypsin followed by staining with Giemsa [23]. The resulting G-banding is the most widely used technique for human chromosome identification in clinical cytogenetics laboratories today (see Fig. 9.4). The G-banding patterns are also readily captured and analyzed by

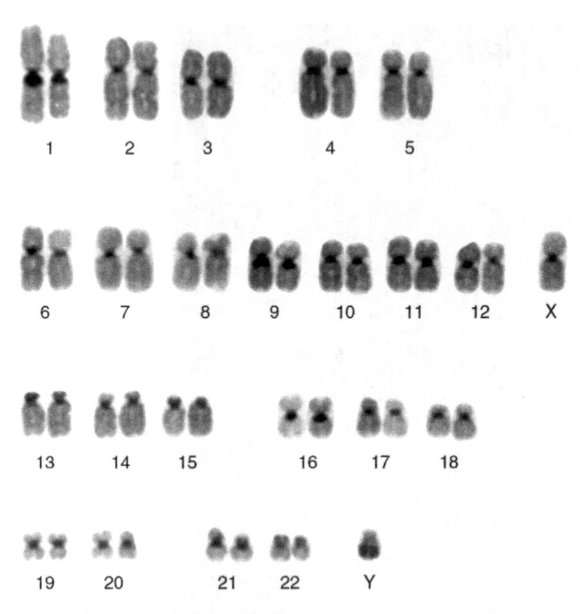


**Figure 9.8** C-banded karyotype of a male cell.

The banding pattern of each chromosome is specific and can be shown in the form of a continuous series of bands. A standardized map of banded chromosomes is known as an "idiogram." Subsequent to the development of banded human karyotypes, a standardized nomenclature for the bands was established by the International Standing Committee on Human Cytogenetic Nomenclature. Updated regularly, this standardized system allows the precise description of chromosome abnormalities [26].

The C-banding method selectively stains the areas located around the centromeres of all chromosomes and on the distal long arm of the Y chromosome [27]. The largest C-bands usually occur on chromosomes 1, 9, 16, and Y in regions that contain highly repetitive, nontranscribed DNA. To elicit C-bands, metaphase chromosome preparations are treated with sodium hydroxide or barium hydroxide followed by Giemsa staining (Fig. 9.8). The size of the C-band on a given chromosome is usually constant in all the cells of an individual but is highly variable from person to person, reflecting variations in the amount of heterochromatic DNA present at the centromeric regions. Such C-band heteromorphisms on chromosomes are transmitted from parent to offspring as simple Mendelian dominant traits. These variations in chromosome morphology are not associated with any known phenotypic effects and are referred to as chromosome polymorphisms. They are, however, useful as heritable chromosome markers in various clinical and epidemiologic studies of chromosome abnormalities.

Silver NOR (AgNOR) staining uses a silver nitrate solution to selectively stain the sites of transcriptionally active rRNA genes, which are located in the stalk regions on the short arms of human acrocentric chromosomes [28]. Silver staining regions are usually present on 6–8 of the 10 acrocentric chromosomes, 13, 14, 15, 21, and 22 (Fig. 9.6), although they may be seen on as few as 3 or as many as all 10 of these chromosomes. The sizes of the AgNORs are highly variable in the human population, although the size of each AgNOR in the cells of one individual is quite consistent and usually remains unchanged from one generation to the next. AgNOR staining is useful in characterizing rearrangements involving human acrocentric chromosomes. The mechanism of AgNOR staining is based on the oxidation of nucleolar nonhistone proteins with silver nitrate, by which Ag is reduced to black native silver. Interestingly, the acrocentric chromosomes show association of their satellite stalk regions even in metaphase chromosome preparations, reflecting the functional association of these sites in the formation of the nucleolus in the interphase nucleus. This association of the NORs is considered to be a factor responsible for the high incidence of Robertsonian translocations involving the short arms of acrocentric chromosomes.

Sister chromatid exchange (SCE) is an extension of the replication banding technique using BrdU incorporation to produce differential staining of the two sister chromatids of the metaphase chromosome. This requires incorporation of the thymidine (T) analogue BrdU (B) into DNA during two successive rounds of DNA replication. At the end of the first round of DNA replication, the two newly synthesized strands of DNA in the double-stranded helix will contain B, but not the two template strands. At the end of the second round of DNA replication, two new double-stranded helices will be produced, of which one will have B incorporated on both strands (BB) and the other will have B substitution in only one strand of the DNA double helix (TB). When the chromosomes containing singly (TB) and doubly (BB) substituted chromatids are stained with the DNA-binding fluorochrome Hoechst 33258, and exposed to ultraviolet light, they show differential sister chromatid staining, with the bifilarly substituted chromatid exhibiting paler fluorescence [29]. Staining of these B-incorporated chromosomes with Giemsa produces darkly stained (TB) and lightly stained (BB) sister chromatids [30] (Fig. 9.9).

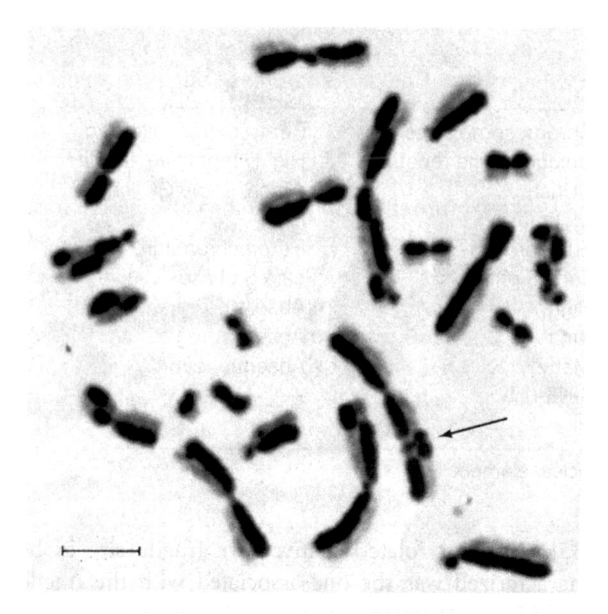

**Figure 9.9** Sister chromatid exchanges shown in Chinese hamster ovary cells.

Therefore, exchanges of material between sister chromatids are readily visible at high resolution following this staining protocol. The differential sister chromatid staining observed following the SCE protocol is a remarkable cytologic demonstration of the semiconservative replication of DNA. It also demonstrates that each chromosome is composed of a single very long duplex of DNA. Further, it shows that exchanges between the two sister chromatids take place in somatic cells that could potentially have mutagenic effects. SCE is used to diagnose diseases associated with chromosomal instability in clinical cytogenetics laboratories. For example, SCE analysis is a diagnostic test for Bloom syndrome, a rare autosomal recessive disease caused by mutations in a DNA helicase of the RecQ family that catalyze the unwinding of duplex nucleic acid molecules [31]. It is characterized by growth deficiency, predisposition to neoplasia, and chromosomal instability in somatic cells. The frequency of spontaneous SCEs in cells from patients with Bloom syndrome is markedly increased. SCE analysis is also used to monitor the effects of potentially mutagenic or carcinogenic agents that enhance the rate of SCEs.

### 9.4.3 Chromosome Banding Reveals Genome Sequence Organization

Quinacrine associates directly with DNA by intercalating between base pairs. Although quinacrine binds equally well to DNA of any base composition, its fluorescence is enhanced in regions containing uninterrupted runs of AT base pairs, and is quenched in regions with more frequent GC base pairs. In the Q-banding pattern of human chromosomes (see Fig. 9.5), the intensity of fluorescence is generally proportional to the ATrichness of the DNA [32]. However, the highly AT-rich satellite DNA that is concentrated at the C-bands of chromosomes 1, 9, and 16 has interspersed GC base pairs and usually fails to show bright Q-banding. That on the Y, in contrast, has no such GC pairs and is intensely fluorescent. Thus Q-banding is related to both base composition and base interspersion,which result in the differential fluorescence or quenching of signals produced by the fluorescent dye. DNA–protein interactions may also be important in the generation of Q-bands.

G-banding is produced most commonly by treatment of chromosomal preparations with the proteolytic enzyme trypsin. Giemsa stains DNA primarily by intercalating between adjacent base pairs in double-stranded regions. G-bands result from the degradation of chromosomal proteins by trypsin, which modifies the interaction of chromosomal DNA with the Giemsa dyes. Since the fixative used in standard chromosome preparation methods, methanol:acetic acid (3:1), removes some of the histone proteins, it is the degradation of the nonhistone proteins that appears to be critical for the production of G-bands. The DNA–protein interactions at the G-band-positive regions apparently render these sites resistant to denaturation by the enzyme.

The commonly used method to generate R-bands is to subject chromosome preparations to moderate heat (~85°C in the presence of high salt) before staining them with Giemsa. The heat pretreatment is thought to selectively denature the more AT-rich DNA sequences, which have a lower thermal stability than GC base pairs, and to result in altered DNA structure on renaturation. Therefore, after chromosomes are exposed to moderate heat, Giemsa stains the unaffected GC-rich double-stranded DNA regions, producing R-banding. R-bands can also be produced by the replication banding technique, which demonstrates that R-band-positive regions contain early-replicating DNA. It also follows that G-band- and Q-band-positive regions contain AT-rich DNA that replicates relatively late in the cell cycle [32].

C-band-positive regions have been found by in situ hybridization and DNA sequencing to consist of α-satellite (discussed later) sequences at the

**TABLE 9.1 Characteristics of Chromosome Bands**

| Characteristic | Q- or G-Bands | R-Bands | C-Bands |
|---|---|---|---|
| Location | Chromosome arms | Chromosome arms | Centromeres, distal Yq |
| Type of DNA sequence | Repetitive, some unique | Unique, some repetitive | Highly repetitive satellite |
| Base composition | ATrich | GCrich | ATrich, some GCrich |
| 5-Methylcytosine content | Low | Moderate | High |
| Type of chromatin | Heterochromatin | Euchromatin | Heterochromatin |
| Replication | Middle to late S phase | Early S phase | Late S phase |
| Transcription | Low | High | Absent |
| Gene density | Low | High | Absent |
| CpG-rich islands | Few | Many | Absent |
| Repeats | LINE-rich | SINE-rich | — |
| Acetylated histones | Low | High | Absent |

*LINE*, long interspersed nuclear element; *SINE*, short interspersed nuclear element.

centromeres of human chromosomes and of different families of simple-sequence satellite DNAs at the large pericentromeric C-band blocks on chromosomes 1, 9, and 16 and distal Yq. Analyses of the completed human genome sequence have defined further families of repetitive DNA [33], but these have not yet been associated with functional or structural landmarks of chromosomes. In contrast, studies employing in situ hybridization as well as in silico analyses of the genome sequence have revealed that the human genome also includes highly homologous duplications of DNA ranging in size from 1 to more than 500 kb. These repeats, called segmental duplications, are located mainly in the pericentromeric and subtelomeric regions of chromosomes, although they are also present as interspersed repeats along the length of the chromosome [33,34]. While some of these segmental duplications are known to predispose to genomic deletions and duplications, their significance for chromosomal function is otherwise unknown. A comparison of the characteristics of Q-/G-, R-, and C-bands is presented in Table 9.1.

Also related to simple sequences are chromosomal regions called fragile sites that remain stretched at metaphase after various treatments that limit DNA replication [35]. Fragile sites are classified as rare (inherited) or common (constitutional) and are further subdivided according to the conditions under which they are induced (e.g., folate or aphidicolin sensitive). Several fragile sites have now been cloned and sequenced. These studies have shown that the expression of rare, inherited fragile sites is associated with repeat expansions [35]. The first folate-sensitive rare fragile site to be characterized was the one associated with the fragile X syndrome (FMR1), which was shown to result from the expansion and methylation of a CGG trinucleotide repeat in the 5′untranslated region of the *FMR1* gene. Other folate-sensitive fragile sites characterized thus far also result from the expansion of trinucleotide repeats [36]. A distamycin-sensitive rare fragile site on chromosome 16 has been shown to involve the expansion of a 33-bp AT-rich minisatellite[36]. In contrast, sequencing of constitutional fragile sites has not revealed any characteristic DNA sequences at these sites [37].

### 9.4.4 Molecular Cytogenetics

The gap between light microscope resolution of chromosome structure and the gene was bridged by the introduction of several molecular cytogenetic techniques. Fluorescence in situ hybridization (FISH) involves hybridizing a fluorescently labeled single-stranded DNA probe to denatured chromosomal DNA on a microscope slide preparation of metaphase chromosomes and/or interphase nuclei prepared from the patient's sample. After overnight hybridization, the slide is washed and counterstained with a nucleic acid dye (e.g., DAPI), allowing the region where hybridization has occurred to be visualized using a fluorescence microscope [38]. FISH is now widely used for clinical diagnostic purposes. There are different types of FISH probes, including locus-specific probes, centromeric probes (CEPs), and whole-chromosome paint probes. Locus-specific probes are specific for a specific single locus. They are particularly useful for identifying subtle submicroscopic

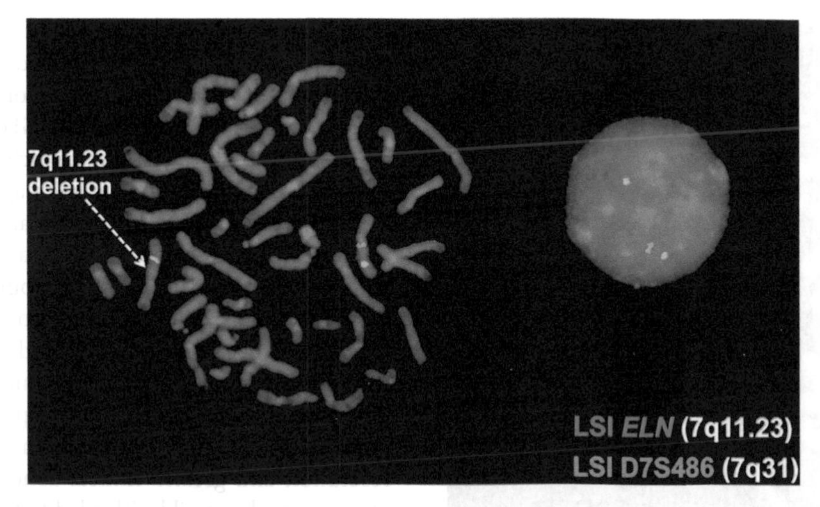


**Figure 9.10** Metaphase and interphase fluorescence in situ hybridization analysis in a patient with William syndrome due to deletion on chromosome 7 band q11.23. Note the deletion of the *ELN* gene probe labeled in red.

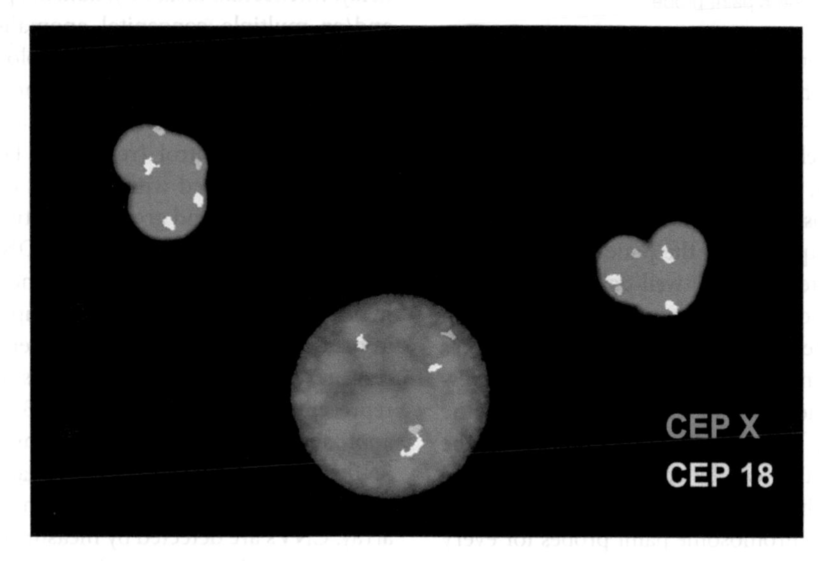


**Figure 9.11** Interphase fluorescence in situ hybridization analysis in a patient with trisomy of chromosome 18. Note the three copies of the chromosome 18 centromeric probe labeled in *aqua*.

deletions and duplications (Fig. 9.10). CEPs are specific for unique repetitive DNA sequences (e.g.,α-satellite sequences) in the centromere of a specific chromosome. They are suitable for making a rapid diagnosis of one of the common aneuploidy syndromes (trisomies 13, 18, and 21, and sex chromosome aneuploidies) using non-dividing interphase nuclei. This is particularly useful in a prenatal setting using amniotic fluid or chorionic villi samples (Fig. 9.11). Whole-chromosome paint probes consist of a cocktail of probes obtained from different regions of a particular chromosome. When this cocktail mixture is used in a single hybridization, the entire relevant chromosome fluoresces (is "painted") (Fig. 9.12). Whole-chromosome paints are useful for characterizing complex chromosomal rearrangements, and for identifying the origin of additional chromosomal material such as small marker or ring chromosomes.

FISH using locus-specific probes has been extremely useful in the detection of "microdeletion syndromes" resulting from deletions of multiple contiguous genes.

in this case requires a separate chromosome-specific confirmatory testing. ROHs can harbor homozygous mutations in autosomal recessive genes, and follow-up sequencing is usually indicated when a recessive condition is suspected. SNP analysis can also allow the detection of polyploidy (triploidy and tetraploidy) in prenatal and postnatal neonatal settings, which is an advantage over the use of array CGH. In recent years, SNP analysis has been added to array CGH platforms.

CMA technologies have two main limitations, namely, their inability to detect balanced chromosomal rearrangements and low-level mosaicism. Several CMA studies have reported finding CNVs at the breakpoints of some de novo apparently balanced rearrangements detected by karyotyping [96,97]. In recent years, next-generation sequencing technologies have been rapidly emerging. Whole-genome sequencing (WGS) is a high-resolution methodology that has the potential to eventually replace some cytogenetic techniques. In addition to the detection of sequence variants, WGS is capable of detecting structural chromosomal abnormalities, including CNVs and balanced chromosomal rearrangements. Many groups are evaluating the accuracy of WGS for detecting CNVs of varying sizes and are developing new analysis algorithms based on read depth of coverage to enhance CNV calling capabilities. Assessment of these next-generation sequencing-based technologies compared with CMA-based technologies for CNV detection will provide an opportunity to evaluate which approach can provide the most accurate high-resolution data for routine clinical testing.

## 9.5 FUNCTIONAL ORGANIZATION OF CHROMOSOMES

Chromatin is classified into euchromatin and heterochromatin. Euchromatin consists of active genes; however, not all genes in euchromatic regions are active at any given time. Therefore, localization in euchromatin is currently thought to be necessary but not sufficient for gene activity. Euchromatin is dispersed in the interphase nucleus and replicates its DNA early in the S phase of the cell cycle. Heterochromatin consists predominantly of inactive genetic material, replicates its DNA late in the S phase, and is condensed in the interphase nucleus. Heterochromatin is further classified into constitutive heterochromatin and facultative heterochromatin. Constitutive heterochromatin consists of highly repetitious simple-sequence DNA, remains transcriptionally inactive, and is located at specific regions of the chromosomes such as the centromere and the distal long arm of the human Y chromosome. Facultative heterochromatin also remains condensed in the interphase nucleus, replicates its DNA late in the S phase, and is largely transcriptionally inactive;however, it is not inactive permanently, does not consist exclusively of repetitious DNA, and can become transcriptionally active. The inactive X chromosome in the human female is a good example of facultative heterochromatin. However, localization in facultative heterochromatin does not exclude transcription altogether, as several genes on the inactive X chromosome are expressed (see later). As already noted, the R-band-positive regions of human chromosomes have characteristics of euchromatin in that they replicate their DNA in early S phase and have high transcriptional activity due in part to high gene density (see Table 9.1). The G-band-positive regions, on the other hand, are more heterochromatic, as they replicate their DNA in late S phase and are low in transcriptional activity associated with low gene density. Integration of the whole human genome sequence with the cytogenetic map shows a lower density of genes in G-positive bands [45]. The C-band-positive regions consist of constitutive heterochromatin with no known functional genes. The facultative heterochromatin of the inactive X chromosome replicates its DNA in late S phase, and forms the condensed Barr body in the interphase nucleus. Consequently, there is a general relationship of functional properties (time of replication during the S phase and transcriptional status or gene density) with chromosome band classes characterized by differential condensation and staining characteristics [46].

Investigations have provided insights into the molecular organization of two specialized structures on chromosomes, the centromere and the telomere, which are summarized below.

### 9.5.1 The Centromere

As already noted, each chromosome has a primary constriction, the centromere, where the sister chromatids of a replicated chromosome are held together until the anaphase stage of cell division. A subdomain of the centromere is the kinetochore, a protein–DNA complex that serves as the attachment site for the spindle fibers essential for chromosome movement and segregation during mitosis and meiosis. The structure of the centromere has been a focus of molecular cytogenetic investigations in recent years. The best characterized eukaryotic centromere is that of the budding yeast *Saccharomyces cerevisiae*. In this organism, a short sequence of about 125 bp specifies the centromere of each of the chromosomes. The nucleotide sequence and organization of this centromere DNA are conserved among the different chromosomes in the budding yeast. The search for a similar specific sequence in the larger and more complex centromeres of higher eukaryotes has not been successful. Rather, the centromeres in these organisms consist of large arrays of repeated α-satellite DNA sequences. In human centromeres, the arrays consist of tandem, head-to-tail repeats of a 171-bp monomer that is further organized into higher order repeats [47]. The centromeric chromatin of human chromosomes spans from 0.1 to 4.0 Mb. The sequence of the basic 171-bp unit is sufficiently divergent among human chromosomes that, with very few exceptions, centromere-specific α-satellite DNA probes can generate fluorescent signals on specific chromosomes in a FISH assay. This is useful from a practical standpoint for identifying and determining the copy number for specific human chromosomes in interphase cells.

Several lines of evidence implicate a critical role for α-satellite DNA in centromere function. Although there are other repeated sequences in the centromeric heterochromatin, α-satellite is the only one localized to the centromeres of all normal human chromosomes. Moreover, studies have shown that human artificial chromosome constructs containing α-satellite DNA are able to form functional centromeres [48]. However, independent evidence from rearranged chromosomes suggests that the presence of α-satellite DNA alone is not sufficient for the formation of an active centromere. Many cases of rearranged human chromosomes containing two centromeric regions have been identified. A true dicentric chromosome with two primary constrictions would be unstable during cell division as spindle fiber attachment occurs independently at the two centromeres, if these are sufficiently far apart. The two centromeres on a single chromatid could then be pulled toward opposite poles of the spindle, breaking the chromosome. However, many dicentric chromosomes with two blocks of α-satellite DNA and C-band regions are stable and show only one primary constriction, indicating that only one of the two centromeres is active. Such stable dicentric chromosomes, referred to as pseudodicentrics, indicate that the presence of α-satellite DNA alone is not sufficient for the formation of an active centromere. In addition, several human marker chromosomes have been characterized that originate from normal human chromosomes but lack α-satellite DNA sequences. These functional centromeres lacking α-satellite DNA are called neocentromeres[49]. As these chromosomes are mitotically stable, the presence of α-satellite DNA is not an absolute requirement for functional centromeres. Thus, although normal human centromeres are composed of α-satellite DNA, it appears to be neither necessary nor sufficient for centromere formation.

Investigations have identified several proteins associated with centromeres that have contributed to our understanding of centromere structure and function [50,51]. A group of these proteins are constitutively associated with centromeres, while others are associated with centromeres only during a part of the cell cycle and are involved in chromosome movement during cell division. The major constitutive centromere proteins identified are CENP-A, CENP-B, and CENP-C. The localization of these proteins at centromeres has been determined by immunofluorescence microscopy using antibodies specific for these proteins. CENP-A is a 17-kDa histone H3–like protein that participates in producing centromere-specific nucleosomes (in place of histone H3) and altered chromatin structure. CENP-A is detected at all functional centromeres, including the neocentromeres. CENP-B is an 80-kDa protein that binds to a specific 17-bp sequence, the CENP-B box, in α-satellite DNA and is found, as expected, even at the inactive centromere of pseudodicentric chromosomes. CENP-C, a 140-kDa protein, is also found at active centromeres, where it is located in the proteinaceous kinetochore. CENP-C shares homology with a domain of the Mif2 protein of yeast that is essential for normal chromosome segregation. In addition to the CENP-A, -B, and -C proteins that associate with centromeres constitutively, many more that associate transiently during cell division have been identified. An example of the latter class of proteins is CENP-E, a 275-kDa kinesin-related protein that is associated with centromeres and the mitotic spindle during mitosis and plays a role in chromosome movement.

### 9.5.2 The Telomere

Telomeres are special DNA–protein structures that are present at the ends of linear chromosomes and prevent origin-dependent portion of a chromosome region. In the case of PWS, about 70% of the patients have a deletion in the proximal q arm of the paternally inherited chromosome 15. In normal individuals, the PWS critical gene(s) is transcribed only from the paternal homologue. Therefore, with the deletion of the PWS critical region on the paternal 15, PWS patients are completely deficient for the products of these imprinted genes. The remaining 30% of PWS patients have two chromosomes 15 derived from their mother and none from their father. In the absence of a paternal 15, these patients also lack the expression of the PWS critical gene(s). A likely mechanism for the origin of this UPD is the conception of a fetus with trisomy for chromosome 15 with two chromosomes from the mother and one from the father. Trisomy 15 is usually lethal and will lead to miscarriage. However, the loss of a chromosome 15 in an occasional cell during early embryogenesis will allow that cell line to proliferate and result in a viable fetus. If the sole paternal chromosome is the one that is lost in this trisomy rescue, the resulting infant will have maternal UPD and PWS. Alternatively, UPD could arise from the rescue of a monosomic conceptus, by duplication of the single homologue. Maternal and paternal UPDs for many of the human chromosomes have now been identified. Several of these result in a normal phenotype, presumably because the chromosome does not harbor any imprinted gene(s) [87]. However, these individuals may be at risk for being homozygous for recessive genes. The possible role of UPD, a unique form of chromosomal inheritance, in disease states of unknown etiology is being investigated.

## 9.8 CHROMOSOME ABNORMALITIES

Human cytogenetics has advanced since the late 1970s because of continuing technical advances and the high incidence of chromosome abnormalities in the human population. It is estimated that the frequency of significant chromosome abnormalities among live births is about 1 in 150. It is well documented that about 50% of first-trimester pregnancy losses are due to chromosome abnormalities, mostly numerical anomalies. Thus, chromosome aberrations have a significant impact as causes of pregnancy wastage, congenital malformations, mental retardation, abnormalities of sex differentiation, and behavior problems. Acquired chromosomal changes play a significant role in carcinogenesis and in tumor progression.

Most chromosomal abnormalities exert their phenotypic effects by increasing or decreasing the quantity of genetic material. Chromosomal abnormalities can be divided into numerical and structural abnormalities. Structural changes such as translocations and inversions pose a much more serious familial recurrence risk for chromosome abnormalities. This is due to aberrant segregation of chromosomes during meiosis in clinically normal carriers of these balanced rearrangements.

### 9.8.1 Numerical Chromosome Abnormalities

The most straightforward of chromosomal abnormalities are alterations of chromosome number. Deviation from the normal diploid complement of 46 chromosomes is referred to as "aneuploidy"; an extra chromosome results in "trisomy," whereas a missing chromosome results in "monosomy." Although all the possible chromosomal trisomies have been observed in spontaneous abortions, trisomies 13, 18, and 21 are the only autosomal trisomies to be observed in a nonmosaic state in live-born infants, and are discussed in detail in later volumes. All autosomal monosomies are lethal. The only viable monosomy involves the X chromosome (45, X resulting in Turner syndrome). Abnormalities associated with sex chromosomes are discussed in detail in later volumes. Aneuploidy results from nondisjunction, in which two copies of a chromosome go to the same daughter cell during meiosis or mitosis. Nondisjunction occurs most often in the first meiotic division in the maternal germline. In meiosis I nondisjunction, both homologues of a chromosome move to the same pole during anaphase I instead of moving to opposite poles, giving rise to one daughter cell with two copies of the chromosome and the other with none. The latter product is never recovered because of lethality associated with monosomy. In the case of meiosis II nondisjunction, the two sister chromatids of a homologue move to the same pole, again giving rise to one daughter cell with two copies of the chromosome and the other with none. Mitotic nondisjunction results in the presence of an aneuploid and a normal cell line—a condition referred to as "mosaicism." The causes of nondisjunction are unknown. The only well-documented risk factor is advanced maternal age.

The term "polyploidy," on the other hand, refers to the presence of a complete extra set of chromosomes; "triploidy" represents three sets with 69 chromosomes, whereas "tetraploidy" represents four sets with 92 chromosomes. Rarely, a triploid fetus will be liveborn, but in general polyploidy is lethal. In a few instances, however, mosaicism for a diploid and a triploid cell line producing congenital anomalies has been compatible with long-term survival.

### 9.8.2 Structural Chromosome Abnormalities

Structural chromosomal rearrangements result from chromosome breakage with subsequent reunion in a different configuration. They can be balanced or unbalanced. In balanced rearrangements, the chromosome complement is complete with no loss or gain of genetic material. Consequently, balanced rearrangements are generally harmless, with the exception of rare cases in which one of the breakpoints disrupts an important functional gene. Carriers of balanced rearrangements are often at risk of having children with an unbalanced chromosome complement. When a chromosome rearrangement is unbalanced, the chromosome complement contains an incorrect amount of genetic material, usually with serious clinical effects.

A deletion involves the loss of part of a chromosome and results in monosomy for that segment of the chromosome, whereas duplication represents the doubling of part of a chromosome, resulting in trisomy for that segment. The result is either decrease or increase in gene dosage. In general, duplications appear to be less harmful than deletions. Very large deletions usually are incompatible with survival to term. Deletions or duplications larger than ~5 Mb in size can be visualized under the microscope using G-banded chromosome analysis. Genomic disorders resulting from submicroscopic deletions and duplications (i.e., microdeletions and microduplications) with a size <5 Mb have been identified with the help of molecular cytogenetic techniques. In these syndromes, groups of contiguous genes are either deleted or duplicated, resulting in a defined set of congenital anomalies.

The use of CMA to analyze the genomes of normal humans led to the discovery of extensive genomic benign CNVs with the majority <500 kb in size [88,89]. Benign CNVs have been proposed to be a major factor responsible for human diversity [90]. Through genomic rearrangement of rearrangement-prone regions as a result of the genomic architecture, pathogenic CNVs, on the other hand, can cause genomic disorders due to the loss or gain of a dosage-sensitive gene(s) resulting in a clinical phenotype [91]. These pathogenic CNVs include recurrent microdeletions and microduplications with common size and breakpoint clustering, which are flanked by segmental duplications (also called low-copy repeats) and mediated by nonallelic homologous recombination, as well as nonrecurrent deletions and duplications with different sizes and variable breakpoints for each CNV, which vary in size from a few hundred kilobases to a few megabases, and are mediated by other molecular mechanisms [92,93]. Well-known genomic disorders can be phenotypically heterogeneous and variable due to incomplete penetrance or variable expression. Clinical variability could also be explained in part by other genetic or environmental determinants, modifying factors of other genes, multigenic inheritance, imprinting, and unmasking of recessive genes. For many years, genomic disorders due to microdeletions and microduplications that are clinically recognizable by their typical constellation of clinical features were tested for by FISH analysis. The advances in CMA technologies since 2008 have allowed their widespread use as a clinical diagnostic modality in a wide variety of human genetic diseases. High-resolution CMA allows the detection of pathogenic CNVs in nearly 17%–19% of patients with developmental delay and/or intellectual disability who had a normal G-banded karyotype [94]. Because of its high diagnostic yield, CMA was recommended in 2010 by the American College of Medical Genetics and Genomics as the preferred first-tier clinical diagnostic test for individuals with developmental delay, intellectual disability, and/or multiple congenital anomalies [94,95].

Translocations involve the exchange of genetic material between chromosomes. In a balanced reciprocal translocation the exchange is equal, with no loss or gain of genetic material, though it is possible for a gene to be disrupted at one of the breakpoints. More often, the carrier of a balanced translocation is free of clinical signs or symptoms but is at risk for having offspring with unbalanced chromosomes. The phenotype usually is a complex mixture as a result of the loss or gain of at least two chromosome segments and therefore can be difficult to predict. One specific type of translocation that is relatively common is the "Robertsonian translocation." This results from a fusion of two acrocentric chromosomes at the centromere. Carriers of a Robertsonian translocation have 45 chromosomes and are clinically unaffected. The most common clinically significant outcome is trisomy 21, in which a carrier for a Robertsonian translocation involving chromosome 21 produces a gamete with both the translocation chromosome and a normal 21, resulting in trisomy 21 after fertilization.

Inversions occur when there are two breaks in a chromosome and the intervening material flips 180 degrees. Inversions that span the centromere are referred to as "pericentric," whereas those that do not are called "paracentric." Inversions generally do not result in added or lost genetic material, and therefore usually are viewed as neutral changes. Disruption of a gene at one of the breakpoints, however, could change the function of that gene. Also, alteration of the gene order at the borders of the inversion could affect the function of blocks of genes that are coordinately regulated ("position effect"). If a crossover occurs in the inverted segment of a pericentric inversion during meiosis, two recombinant chromosomes result, one with duplication of one end and deletion of the other end, and the other having the opposite arrangement. Such a crossover event in a paracentric inversion results in dicentric or acentric chromosomes that tend to be unstable.

An insertion occurs when a segment of one chromosome becomes inserted into another chromosome. Because these changes require three chromosomal breakpoints, they are relatively rare. Abnormal segregation in a balanced insertion carrier can produce offspring with either duplication or deletion of the inserted segment, as well as balanced carriers and normal offspring.

A "marker" chromosome is a rearranged chromosome whose genetic origin is unknown based on its G-banded chromosome morphology. Usually they are present in addition to the normal chromosome complement and are thus called supernumerary marker chromosomes. Two-thirds of de novo marker chromosomes can be associated with an abnormal outcome, whereas inherited ones can be passed from generation to generation without apparent clinical effects. Larger markers with more genetically active material are more likely to be of clinical significance. FISH and CMA have proved very helpful in the precise identification of the genetic origin of supernumerary marker chromosomes. Ring chromosomes are formed when a chromosome undergoes two breaks and the broken ends reunite in a ring structure. Rings encounter difficulties in mitosis and are unstable, resulting in some cells that lose the ring and are therefore monosomic for the chromosome, and others that have multiple copies of the ring. An "isochromosome" is a chromosome in which one arm is missing and the other duplicated in a mirror-image fashion. The most commonly encountered isochromosome is that which consists of two long arms of the X chromosome. This accounts for ~15% of all cases of Turner syndrome [98].

## 9.9 CONCLUDING REMARKS

In this chapter, we have outlined the structural, functional, and behavioral aspects of human chromosomes and their relationship to disease states. Although investigations have provided insights into several aspects of chromosome structure, the details of the higher order structure of chromosomes are not well understood at the molecular level in full detail.

We have begun to understand the DNA sequences that are associated with chromosomal landmarks, such as centromeres, telomeres, chromosome bands, and fragile site; however, we still do not understand the role of such sequences in producing the associated functional correlates. For example, which sequences are critical for centromere function, and how do dicentric chromosomes decide which will function? Most enigmatic, what is the origin of chromosome bands and what is the molecular organization of a band border?

The availability of the finished human genome sequence and CMA has allowed the detection of genomic CNVs on a global scale. It is now appreciated that the underlying genomic architecture plays a crucial role in the origin of these genomic rearrangements in rearrangement-prone regions. Segmental duplications have arisen in the primate genome, driving the process of chromosome evolution. In addition to creating a dynamic, evolvable genome, these segmental duplications result in instability, genomic rearrangement, and disease. We have begun to understand the organization of segmental duplications, which predisposes the chromosomes that carry them to germline genomic rearrangements such as deletions, duplications, and inversions; however, we do not understand the mechanisms that initially led to the formation of segmental duplications, nor the sequences or structures responsible for their continued instability.

Many new insights have come from understanding the structure and function of the human X chromosome and genomic imprinting; however, we do not know to what extent the remainder of the genome may contain imprinted or partially imprinted genes whose parental origin in part determines tissue-specific expression. Might such "epigenetic" phenomena provide another mechanism for both normal human variation and disease susceptibility?

Much remains to be learned about the molecular aspects of chromosome structure, function, and behavior. It is anticipated that the human genome sequence and its functional characterization will provide the tools with which to approach these problems and define a new frontier for the role of chromosomes in human disease.

## REFERENCES

[1] Ford CE, Hamerton JL. The chromosomes of man. Nature 1956;178:1020–3.
[2] Tjio JH, Levan A. The chromosome number of man. Hereditas 1956;42:1–6.
[3] Kornberg RD, Lorch Y. Twenty-five years of the nucleosome, fundamental particle of the eukaryote chromosome. Cell 1999;98:285–94.
[4] Finch JT, Klug A. Solenoidal model for superstructure in chromatin. Proc Natl Acad Sci USA 1976;73:1897–901.
[5] Woodcock CL, Frado LL, Rattner JB. The higher-order structure of chromatin: evidence for a helical ribbon arrangement. J Cell Biol 1984;99:42–52.
[6] Bassett A, Cooper S, Wu C, Travers A. The folding and unfolding of eukaryotic chromatin. Curr Opin Genet Dev 2009;19:159–65.
[7] Dorigo B, Schalch T, Kulangara A, Duda S, Schroeder RR, Richmond TJ. Nucleosome arrays reveal the two-start organization of the chromatin fiber. Science 2004;306:1571–3.
[8] Schalch T, Duda S, Sargent DF, Richmond TJ. X-ray structure of a tetranucleosome and its implications for the chromatin fibre. Nature 2005;436:138–41.
[9] Grigoryev SA, Arya G, Correll S, Woodcock CL, Schlick T. Evidence for heteromorphic chromatin fibers from analysis of nucleosome interactions. Proc Natl Acad Sci USA 2009;106:13317–22.
[10] Hart CM, Laemmli UK. Facilitation of chromatin dynamics by SARs. Curr Opin Genet Dev 1998;8: 519–25.
[11] Hirano T. At the heart of the chromosome: SMC proteins in action. Nat Rev Mol Cell Biol 2006;7:311–22.
[12] Maeshima K, Hihara S, Eltsov M. Chromatin structure: does the 30-nm fibre exist in vivo? Curr Opin Cell Biol 2010;22:291–7.
[13] Maeshima K, Eltsov M. Packaging the genome: the structure of mitotic chromosomes. J Biochem 2008;143:145–53.
[14] Nasmyth K. A prize for proliferation. Cell 2001;107:689–701.
[15] Nurse P. The incredible life and times of biological cells. Science 2000;289:1711–6.
[16] Nasmyth K. Segregating sister genomes: the molecular biology of chromosome separation. Science 2002;297:559–65.
[17] Crow JF. The high spontaneous mutation rate: is it a health risk? Proc Natl Acad Sci USA 1997;94:8380–6.
[18] Hassold T, Hunt P. Maternal age and chromosomally abnormal pregnancies: what we know and what we wish we knew. Curr Opin Pediatr 2009;21:703–8.
[19] Hassold T, Sherman S. Down syndrome: genetic recombination and the origin of the extra chromosome 21. Clin Genet 2000;57:95–100.
[20] Lamb NE, Hassold TJ. Nondisjunction—a view from ringside. N Engl J Med 2004;351:1931–4.
[21] Moorhead PS, Nowell PC, Mellman WJ, et al. Chromosome preparations of leukocytes cultured from peripheral blood. Exp Cell Res 1960;20:613–6.
[22] Caspersson T, Zech L, Johansson C, Modest EJ. Identification of human chromosomes by DNA-binding fluorescent reagents. Chromosoma 1970;30:215–27.
[23] Seabright M. Rapid banding technique for human chromosomes. Lancet 1971;2:971–2.
[24] Yunis JJ. High resolution of human chromosomes. Science 1976;191:1268–70.
[25] Dutrillaux B, Laurent C, Couturier J, Lejeune J. Coloration par l'acridine orange de chromosomes prealablementtraites par le 5-bromadeoxyuridine (BUDR). CR Acad Sci D Sci Naturelles (Paris) 1973;276(24):3179–81.
[26] Shaffer LG, Slovak ML, Campbell LJ, editors. An international system for human cytogenetic nomenclature (ISCN). Basel: S. Karger; 2009.
[27] Arrighi FE, Hsu TC. Localization of heterochromatin in human chromosomes. Cytogenetics 1971;10:81–6.
[28] Goodpasture C, Bloom SE, Hsu TC, Arrighi FE. Human nucleolus organizers: the satellites or the stalks? Am J Hum Genet 1976;28:559–66.
[29] Latt SA. Microfluorometric detection of deoxyribonucleic acid replication in human metaphase chromosomes. Proc Natl Acad Sci USA 1973;70:3395–9.
[30] Korenberg JR, Friedlander EF. Giemsa technique for the detection of sister chromatid exchanges. Chromosoma 1974;48:355–60.
[31] van Brabant AJ, Stan R, Ellis NA. DNA helicases, genomic instability and human genetic disease. Ann Rev Genomics Hum Genet 2000;1:409–59.
[32] Korenberg JR, Engels WR. Base ratio, DNA content, and quinacrine-brightness of human chromosomes. Proc Natl Acad Sci USA 1978;75:3382–6.
[33] Lander ES, Linton LM, Birren B, et al. Initial sequencing and analysis of the human genome. Nature 2001;409:860–921.

10

# Mitochondrial Biology and Medicine

*Douglas C. Wallace*[1,2], *Marie T. Lott*[1],
*Vincent Procaccio*[3]

[1]Center for Mitochondrial and Epigenomic Medicine, Children's Hospital of Philadelphia, Philadelphia, PA, United States
[2]Perelman School of Medicine, University of Pennsylvania, Philadelphia, PA, United States
[3]Biochemistry and Genetics Department, MitoVasc Institute, UMR CNRS 6015 – INSERM U1083, CHU Angers, Angers, France

## 10.1 INTRODUCTION

Western medicine is organized around anatomy, yet life is the interplay between structure (anatomy), energy (vital force), and information. Consequently, the anatomical paradigm of Western medicine has largely overlooked the central role of bioenergetics in health and disease. This may be a critical factor in our inability to understand and develop effective therapies for the common metabolic and degenerative diseases and aging. While bioenergetics has been neglected by Western medicine, it is central to Eastern medicine with the concept of Qi, which can be loosely translated as "vital force."

The dichotomy between anatomy and energy for our cells harkens back to the origin of the eukaryotic cell about 2.5 billion years ago. In a unique single event, two co-equal micro-organisms formed a symbiosis that set the stage for all multicellular plants and animals. The original microorganisms were an archaebacterium that gave rise to the eukaryotic cell nucleus and cytosol and an oxidative eubacterium that gave rise to the cytoplasmic mitochondria. While the original archaebacterium and eubacterium had similar-size genomes, most of the eubacterial mitochondrial genes were transferred to the archaebacterial nuclear genome in association with the proliferation of the mitochondria within the cytoplasm. Because most of a bacterium's energy is used in replicating its DNA, transcribing its RNA, and translating its proteins, by transferring the mitochondrial genes into the nucleus, the number of gene copies for each gene could be reduced from hundreds to thousands down to two, with hundreds of -fold savings of energy. The excess mitochondrial energy could then be used to sustain a much larger nuclear genome with extra genes allocated for multicellularity and organogenesis.

By this process, the residual mitochondrial DNA (mtDNA) of multicellular animals and humans has been reduced to about 13 polypeptide genes plus the rRNA and tRNA genes for their translation on mitochondria-specific ribosomes. The mitochondrial ribosomes retain several features of bacterial translation including chloramphenicol and aminoglycoside antibiotic sensitivity and polypeptide initiation with a formyl-methionine. While few in number, the mtDNA polypeptide genes are essential components of the mitochondrial energy-generating process, oxidative phosphorylation (OXPHOS). In essence, the mtDNA codes for the wiring diagram of the cellular power plant while the nuclear DNA (nDNA) contains the blueprints for building the power plants.

Emery and Rimoin's Principles and Practice of Medical Genetics and Genomics: Foundations, https://doi.org/10.1016/B978-0-12-812537-3.00010-X

Interest in mitochondrial medicine has been increasing rapidly during the past decade, with the current annual number of mitochondria-related biomedical papers exceeding that of genomics papers [1]. Furthermore, the massive efforts to sequence nDNAs to identify the common genetic variants that cause the common metabolic and degenerative disease have been disappointing. This suggests that the classic Mendelian paradigm of genetics is inadequate for a coherent understanding of human genetics. Many of the seemingly puzzling features of the genetics of common diseases such as variable penetrance, delayed onset and progressive course, and multiorgan involvement are naturally explained by mitochondrial genetics. Thus, by combining Mendelian genetics with mitochondrial genetics we can arrive at a synthetic paradigm that can explain many of the novel features of clinical genetics.

The first report that mitochondrial dysfunction could be associated with a clinical phenotype came with the report of a woman with hypermetabolism, abnormal muscle mitochondria (mitochondrial myopathy), and uncoupled mitochondrial OXPHOS [2]. Subsequent studies revealed a variety of clinical phenotypes associated with mitochondrial myopathy and OXPHOS dysfunction resulting in a proliferation of clinical descriptors such as chronic progressive external ophthalmoplegia (CPEO), mitochondrial encephalomyopathy, lactic acidosis and stroke-like episodes (MELAS), and myoclonic epilepsy and ragged red fiber (MERRF) disease [3,4]. However, family studies showed considerable variability in phenotypes leading to the debate as to whether "mitochondrial diseases" should be split into subphenotypes or lumped into larger categories ("splitters" versus "lumpers"). This ambiguity was resolved by the demonstration that mtDNA mutations could cause disease [5–8] and the realization that cells can have thousands of copies of the mtDNA. The high mtDNA ploidy could then accommodate different percentages of mutant resulting in variable biochemical defects and diverse phenotypes [9]. It is now clear that pathogenic mtDNA mutations are not rare, but common. The prevalence of pathogenic mtDNA mutation cases has been estimated at 1:4300 [10–13], but greater than 1:200 newborn cord bloods harbor one of the 10 most common pathogenic mtDNA mutations [14].

Current estimates of the prevalence of mitochondrial disease are based on "primary mitochondrial diseases," those that are caused by nDNA or mtDNA mitochondrial gene mutations that are severe enough to cause a clinically relevant disease by themselves. The phenotypes and mtDNA and nDNA mutations associated with primary mitochondrial diseases were reported in our previous chapter in this series [15], with a current listing of the nDNA and mtDNA genes implicated provided in MITOMAP [16].

However, even in the early days of clinical mitochondrial genetics, it was clear that nDNA gene mutations could alter mtDNA structures, which alter mitochondrial functions and which generate mitochondrial disease. It is now clear that mitochondrial bioenergetic dysfunction, often resulting from faulty nDNA–mtDNA interactions, is prevalent and associated with the etiology of a wide range of metabolic and degenerative diseases and aging. Thus, the characterization of mitochondrial medical genetics during the past 30 years is restructuring the way we understand human genetics and reorienting the way we investigate the etiology of common diseases.

## 10.2 MITOCHONDRIAL BIOCHEMISTRY

Each cell harbors hundreds to thousands of double-membrane mitochondria. The outer mitochondrial membrane is perhaps the remnant of the phagocytic vesicle and the inner membrane that of the eubacterial plasma membrane. The matrix is the bacterial cytoplasm, and the intermembrane space separates the two organisms.

The highly invaginated inner membrane harbors the enzymes of OXPHOS on in-foldings called cristae. The cristae are closed at the junction with the intermembrane space to create "cristae lumens" [17]. The mitochondria inner membrane has a unique composition that includes cardiolipin, and the mitochondrion has an independent bacteria-like lipid biosynthesis system that produces the essential lipoic acid [18]. The mitochondria are the metabolic hub of the cell. Central metabolic pathways include the tricarboxylic cycle (TCA, Krebs cycle), fatty acid β-oxidation, amino acid metabolism, and cholesterol and heme synthesis. Pyruvate from glycolysis enters the mitochondrion via the pyruvate carrier [19] and is then cleaved by pyruvate dehydrogenase (PDH) to generate acetyl-CoA and reduced nicotinamide dinucleotide (NADH). Acetyl-CoA is also generated by β-oxidation, and the mitochondrial acetyl-CoA is condensed with oxaloacetate to generate citrate. Citrate can

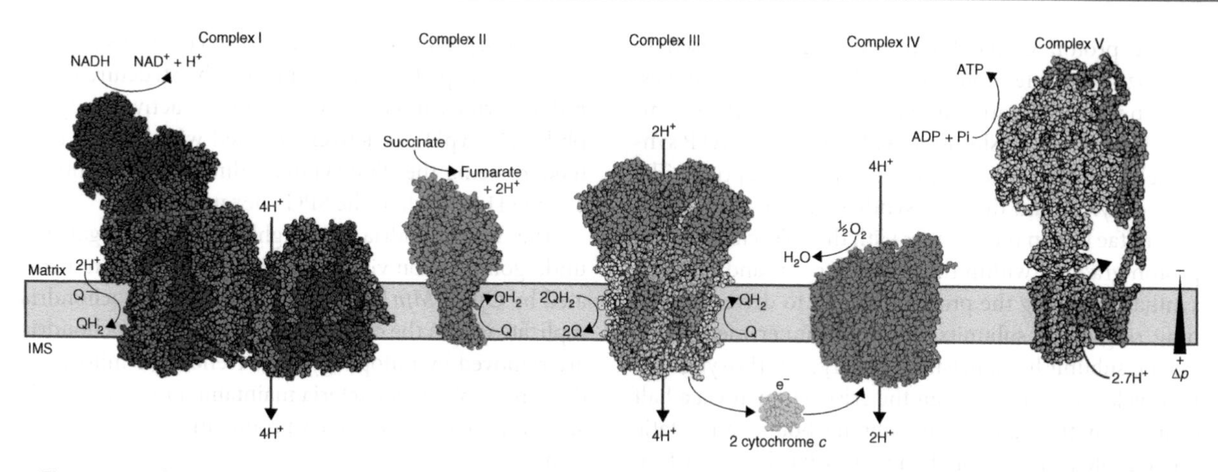


**Figure 10.1** Overview of the OXPHOS complexes I–V showing the molecular structures of the five OXPHOS complexes [36].

be exported into the cytosol where it is cleaved by ATP-citrate-lyase to acetyl-CoA and oxaloacetate. Cytosolic acetyl-CoA can be used for fatty acid synthesis or as substrates for protein and histone acetylation [20]. Within the mitochondrion, citrate is metabolized to isocitrate, aconitate, α-ketoglutarate, succinyl-CoA, succinate, malate, and oxaloacetate via the TCA cycle. Isocitrate dehydrogenase (IDH) has three isoforms—two in the mitochondrion, an $NADP^+$-linked IDH2 and an $NAD^+$-linked IDH3, and one in the cytosol, an $NADP^+$-linked form. IDH1, IDH1, and IDH2 are important in cancer genetics [21,22]. PDH, α-ketoglutarate dehydrogenase, and branched chain-keto acid dehydrogenase (BCKDH) are multiple polypeptide complexes that share common subunits and cofactors, including lipoic acid, and are modulated by $Ca^{2+}$ [23]. The integrity of the mitochondrial inner membrane requires special transport systems to move organic molecules in and out of the mitochondrion. This is accomplished in part by the 53 members of the solute carrier family 25 (SLC25), also called the mitochondrial carrier family, which transport carboxylates, amino acids, nucleotides, and cofactors across the inner mitochondrial membrane [24–26]. Mitochondria have also been proposed to contain an adenylyl cyclase [27] and a nitric oxide (NO) synthase [28].

Each mitochondrion is a capacitor composed of a proton gradient across the mitochondria inner membrane generated by the electron transport chain (ETC). The ETC burns dietary calories (reducing equivalents) with oxygen in a stepwise process that transmits electrons from reduced to oxidized starting with NADH and succinate. NADH is oxidized by complex I (NADH:CoQ oxidoreductase or NADH dehydrogenase) and succinate is oxidized by complex II (succcinate:CoQ oxidoreductase, or succinate dehydrogenase). Complexes I and II then transfer the electrons to the lipid carrier, coenzyme $Q_{10}$ (CoQ), which transfers them to complex III. Complex III transfers the electrons to cytochrome c and cytochrome c transfers the electrons to complex IV (cytochrome c oxidase or COX). Complex IV combines four electrons with $O_2$ to generate $2H_2O$. As the electrons traverse complex I, III, and IV, the energy released is used to pump protons across the mitochondrial inner membrane, four through complex I and one each through complexes III and IV (Fig. 10.1A).

Each of the respiratory complexes is composed of multiple polypeptides. The molecular structures of complex I [29,30], complex II [31], complex III [32], and complex IV [33,34] have been established. Complex I is composed of 45 polypeptides, complex II of four, complex III of 11, and complex IV of 13 (Fig. 10.1). Each complex also contains various prosthetic groups that conduct electrons. Complex I harbors an FMN site to collect electrons from NADH, eight iron-sulfur (FeS) centers, and a CoQ binding site. Complex II harbors an FAD, an FeS, a cytochrome b, and a CoQ binding site. Complex III has the heme-based cytochromes b and c1, the Rieske FeS center, and two CoQ binding sites and the cytochrome c contains heme c. Complex IV encompasses cytochromes a + $a_3$, two Cu centers, and the oxygen reaction center [35]. The respiratory chain is also assembled into super-complexes, the physiological function of which remains to be determined [36–39].

The proton gradient ($\Delta P = \Delta\Psi + \Delta\mu^{H+}$) is used to energize multiple mitochondrial functions, the best known is the generation of ATP from ADP + Pi by complex V (proton-translocating ATP synthase or ATP synthase). Complex V is composed of 18 polypeptides. The ETC enzymes and the ATP synthase are arrayed within the cristae lumen membranes with the ETC charging the proton gradient within the cristae lumens and the ATP synthase utilizing the proton gradient to drive its spinning ring of "c" subunits within the inner membrane. Each c subunit has a negatively charged carboxyl group that picks up a proton from the cristae lumen via a half channel in the abutting membrane bound and static ATP6 protein. The c ring then rotates 360 degrees within the plane of the inner membrane until it comes back to the ATP6 subunit. The ATP6 protein has a second half proton channel open to the matrix through which the proton is released. The c-ring wheel has a γ subunit axial that protrudes inside the F1 (3α:3β) barrel. The barrel is fixed to the ATP6 inner membrane protein and is also static [40]. The wheel spins at 300 Hz [41], and the spinning axial contacts the three β subunits sequentially, causing conformational changes to condense ADP + Pi to make ATP. The ATP synthase is an offset dimer [42], which is aggregated around the edges of the cristae lumens.

ATP generated in the matrix is exported to the intermembrane space by the adenine nucleotide translocators (ANTs) which belonging to the mitochondrial carrier protein family. There are four ANT isoforms in humans, ANT1 being expressed in the heart and muscle, with ANT1 mutations having been identified in human CPEO and mitochondrial myopathy and cardiomyopathy [43–45]. The intermembrane space communicates with the cytosol via the voltage-dependent anion channels (VDAC, porin).

Besides ATP synthesis, the proton gradient is used for multiple other processes. The best characterized of these is the uptake of cytosolic $Ca^{2+}$ though the mitochondria $Ca^{2+}$ uniporter (MCU) complex [46,47]. The cristae lumens are closed at the intermembrane space by the MICOS + Opal complex [48,49]. Opal can be cleaved by OMA1 resulting in the opening of the cristae and the release of cytochrome c and protons into the cytoplasm initiating apoptosis. This is associated with the activation of the inner membrane mitochondrial permeability transition pore (mtPTP) located within the inner membrane and resulting in depolarization of the proton gradient. This combination of events initiates the intrinsic pathway of apoptosis. The structure of the mtPTP, which interacts with the $Ca^{2+}$-activated cyclophilin D (cypD), is actively debated with current contenders being the ATP synthase dimers, ATP synthase c-ring [41,50–54], or the SPG7 hexamer [55].

The mitochondria are highly dynamic organelles undergoing fission via activated *Drp1* and fusion mediated by *Opa1, Mfn1,* and *Mnf2* [56]. The mitochondria replicate within the cytosol and the excess mitochondria are removed by mitophagy [57]. Hence, the mitochondria are a colony of bacteria maintained in a metastable state via a balance between proliferation and degradation.

## 10.3 MITOCHONDRIAL GENETICS

The cellular mitochondrial genome is composed of hundreds to thousands of copies of the mtDNA plus between 1 and 2000 nDNA coded mitochondrial genes. While the Mendelian rules for nDNA gene inheritance were well established by the mid-twentieth century, the principles of human mtDNA genetics were elucidated much more recently.

### 10.3.1 Genetics of mtDNA Genes

The first evidence that the human mtDNAs could code for inheritable traits was obtained by showing that chloramphenicol resistance could be transferred from cell to cell via fusion of cytoplasmic fragments in the absence of the nucleus (transmitochondrial cybrids) [58]. Subsequently, somatic cells were shown to harbor many copies of the mtDNA [59], which can encompass mixtures of different proportions of mutant and normal mtDNAs (heteroplasmy). During cellular replication and cytokinesis, the proportion of mutant and normal mtDNAs can segregate (replicative segregation) and the proportion of mutant mtDNAs determines the degree of the cellular and tissue bioenergetics defect. Each cell, tissue, organ, and individual has a minimal mitochondrial bioenergetic capacity required for normal function. As the proportion of mutant mtDNAs increases, cellular energetics declines until it crosses this minimum bioenergetic threshold and results in phenotypic manifestations (threshold effect) [60–63]. Finally, the human mtDNA is exclusively maternally inherited [64,65].

The mtDNA codes for 13 of the most important polypeptides of OXPHOS: the *ND1, ND2, ND3, ND4, ND4L,*

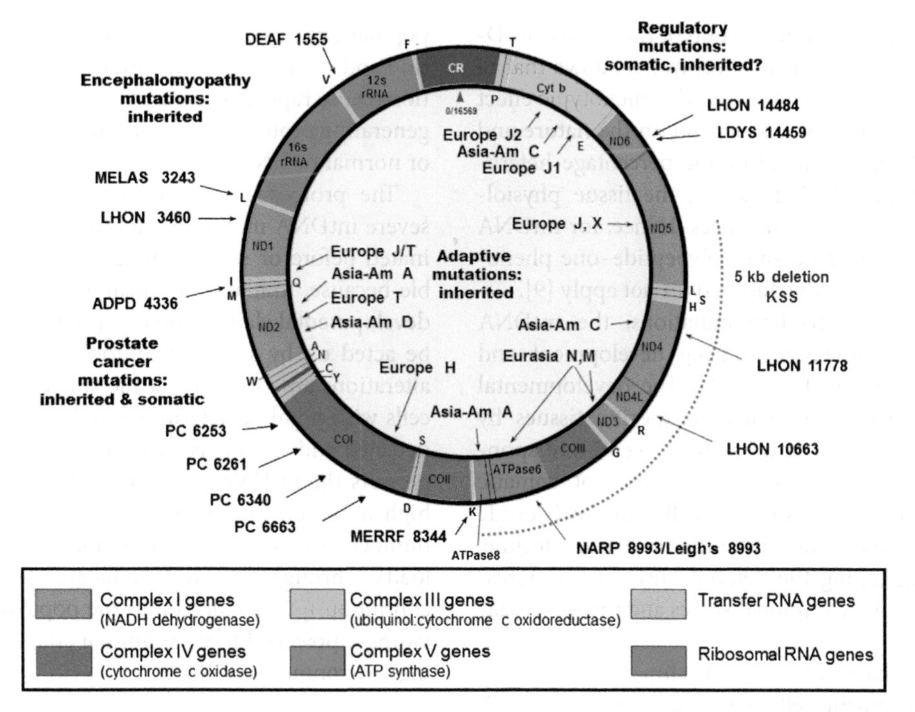


**Figure 10.2** The Human mtDNA Map. The gene locations are shown between the concentric lines. Examples of common mtDNA pathogenic mutations are presented on the outside of the circle while the positions of haplogroup specific markers are located within the circle. Letters stand for the amino acid tRNAs. *CR*, control region. Clinical phenotypes are *Deaf*, deafness; *LDYS*, Leber and dystonia; *LHON*, Leber hereditary optic neuropathy; *MELAS*, mitochondrial encephalomyopathy, lactic acidosis, and stroke-like episodes; *MERRF*, myoclonic epilepsy and ragged red fibers; *NARP*, neurogenic muscle weakness, ataxia, and retinitis pigmentosa; *PC*, prostate cancer. (Figure reproduced from MITOMAP. A Human Mitochondrial Genome Database, 2018. http://www.mitomap.org.)

*ND5*, and *ND6* genes of complexes I; the *cytochrome b* gene of complex III; the *COI*, *COII*, and *COIII* genes of complex IV; and the *ATP6* and *ATP8* genes of complex V. In addition, the mtDNA codes for the 22 tRNAs and two rRNAs for mitochondrial protein synthesis and contains an approximately 1000-nucleotide "control region" that regulates mtDNA transcription and replication [15] (Fig. 10.2).

In addition to the major OXPHOS genes, recent studies have revealed additional polypeptide open reading frames embedded within the mtDNA rRNA genes. Two of these, Humanin and MOTS-c, are thought to generate mRNAs that are exported from the mitochondrion where they are translated on cytosolic ribosomes to generate diffusible peptide hormones [66,67]. The mtDNA may code for additional functional elements such as regulatory RNAs that have yet to be delineated.

Because the mtDNA codes for key OXPHOS polypeptides, genetic alterations in the mtDNA will affect energy metabolism. However, the central role of the mitochondrial membrane potential and mitochondrial intermediary metabolism means that the physiological effects of mtDNA variation can impinge of virtually every cellular and tissue function.

Because it is exclusively maternally inherited, there is virtually no physical interaction between maternal and paternal mtDNAs. Hence, there is little if any recombination and the mtDNA sequence can change only by the sequential accumulation of mutations along radiating maternal lineages. Thus, mtDNA sequence variants remain in total linkage disequilibrium so the functional effects of individual mtDNA nucleotide variants cannot be analyzed in isolation but must be considered within the context of all of the other variants within that mtDNA haplotype.

The mtDNA has a very high mutation rate [68–70]. For a mutant mtDNA to have a phenotypic effect, it must become enriched within the cell from a single mutant mtDNA to a significant proportion of the cellular mtDNAs so that its biochemical effect overshadows that of the residual nonmutant mtDNAs. The phenotypic effect of an mtDNA mutation will depend on the nature and severity of the gene alteration, the percentage heteroplasmy, nDNA-modifying factors, the tissue physiology, and environmental influences. Hence, for mtDNA mutations, the one gene–one polypeptide–one phenotype model of Beadle and Tatum does not apply [9].

In addition to germline mutations, the mtDNA can accumulate mutations during development and in somatic tissues with age [71]. The developmental mutants can become enriched in specific tissues by replicative segregation and result in seemingly spontaneous disease [72]. The accumulation of somatic mtDNA mutations in postmitotic cells can slowly erode mitochondrial energetics resulting in tissue decline, potentially explaining the delayed onset and progressive course of adult common diseases and the molecular basis of aging.

The factors that result in the enrichment of a mtDNA mutation in a somatic cell are still unknown. In CPEO, caused by single mtDNA deletions, the deleted mtDNA becomes clonally and regionally enriched along the skeletal muscle fibers [73]. Also, individual mtDNA mutations accumulate in individual neurons in neurodegenerative diseases [74,75] and in cardiomyocytes during aging [76]. One possible mechanism for the preferential accumulation of the mutant mtDNAs is that an individual mtDNA mutation that affects OXPHOS could change the local redox state. The altered redox state could signal the nucleus to upregulate mitochondrial biogenesis to increase oxidation of the excess reducing equivalents. The increased replication and turnover of mtDNAs within a cell could, with a certain probability, favor a mutant mtDNA, which would become progressively enriched and lead to a functional mitochondrial defect.

Enrichment of mutant mtDNAs within the mtDNA female germline has a different mechanism. The mammalian oocyte contains several hundred thousand mtDNAs. After fertilization and up to the blastocyst stage, the mtDNAs do not actively replicate but become distributed into the blastocyst cells. Hence, the resulting primordial germ cells contain only a very few mtDNA, estimates ranging from a couple to a couple hundred. Subsequent mtDNA replication in the derived oogonia leads to proto-oocytes with reexpanded mtDNA populations of several thousand mtDNAs. The contraction and expansion of the intracellular mtDNA populations cause rapid genetic drift of heteroplasmic mtDNAs generating proto-oocytes enriched for either the mutant or normal mtDNAs [77].

The proto-oocytes and/or oocytes with the most severe mtDNA mutations can then be selectively eliminated before or soon after fertilization. This is possible because, unlike anatomical alterations that require developmental elaboration of structures before they can be acted on by selection, mitochondrial physiological alterations are expressed at the single-cell level. Hence, cells with highly deleterious mtDNA mutations can be detected and eliminated within the ovary [78–80]. This permits the mtDNAs of mammalian species to have a high mutation rate without the accumulation of large numbers of deleterious mutations (excessive genetic load). Through this system bioenergetic variation is continuously introduced into the population, thus providing a powerful tool for animal adaptation to changing environments [81].

### 10.3.2 Genetics of nDNA Mitochondrial Genes

The 1 to 2000 nDNA genes of the mitochondrial genome encompass all of the polypeptide genes for mitochondrial replication, transcription, and translation; mitochondrial intermediate metabolism; mitochondrial dynamics and mitophagy; and regulation [82]. These proteins must be synthesized on cytoplasmic ribosomes and the polypeptides imported into the mitochondrion. Several import systems have been identified to target cytosolic ribosome synthesized polypeptides into the mitochondrion. The best characterized systems are the transport through the outer mitochondrial membrane (TOM) complex, the transport through the inner mitochondrial membrane (TIM22 and TIM23) complexes, and the Mia40–Erv1 intermembrane space complex. The TOM–TIM23 complexes mediate the import of polypeptides having an N-terminal mitochondrial targeting peptide, the TOM–TIM22 complexes integrate the carrier proteins including the ANTs into the inner membrane, and the Mia40–Erv1 complex facilitates the import for disulfide containing proteins [83].

Concurrent with elaboration of the rules of mtDNA genetics and their relevance to disease in the 1980s, efforts were underway to clone and characterize essential nDNA OXPHOS genes such as the *ANT*s and the β subunit of complex V [69,70,84–86]. As knowledge of

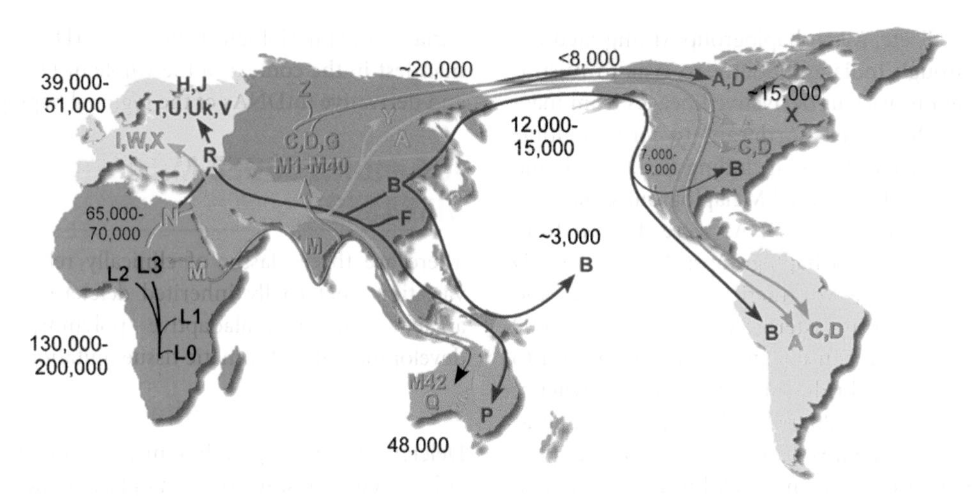


**Figure 10.3** Regional radiation of human mtDNAs from their origin in Africa and colonization of Eurasia and the Americas implies that environmental selection constrained regional mtDNA variation. All African mtDNAs are subsumed under macrohaplogroup L and coalesce to a single origin about 130,000–200,000 YBP. African haplogroup L0 is the most ancient mtDNA lineage found in the Koi-San peoples, L1 and L2 in Pygmy populations. The M and N mtDNA lineages emerged from Sub-Saharan African L3 in northeastern Africa, and only derivatives of M and N mtDNAs successfully left Africa, giving rise to macrohaplogroups M and N. N haplogroups radiated into European and Asian indigenous populations, while M haplogroups were confined to Asia. Haplogroups A, C, and D became enriched in northeastern Siberia and were positioned to migrate across the Bering land bridge 20,000 YBP to found Native Americans. Additional Eurasian migrations brought to the Americas haplogroups B and X. Finally, haplogroup B colonized the Pacific Islands. (Figure reproduced from MITOMAP. A Human Mitochondrial Genome Database, 2018. http://www.mitomap.org.)

the nDNA mitochondrial genes accumulated, it became possible to identify nDNA mutations associated with mitochondrial disease [87] and then diseases resulting from faulty interactions of mtDNA and nDNA mitochondrial gene variants [88].

## 10.4 mtDNA AND HUMAN ORIGINS

The human mtDNA sequence is highly polymorphic, and a survey of mtDNA variation among indigenous populations revealed that different populations have population-specific mtDNA variants [89,90]. Because of its high mutation rate and maternal transmission, it followed that by characterizing the mtDNA sequence variation of indigenous populations around the world and incorporating the mtDNA changes into a sequential mutational tree, it would be possible to determine the genetic relationship of all human populations through the maternal lineage. The result is the reconstruction of the origins and ancient migrations of women [90–93].

The detailed characterization of regional population mtDNAs led to the discovery of regional clusters of mtDNA haplotypes, designated haplogroups. These, in turn, were shown to be founded by one or more functional mtDNA mutations that altered energy metabolism in ways that were adaptive for the regional environment and thus were enriched by natural selection. This created regionally localized clusters of related haplotypes, haplogroups [94–96] (Fig. 10.3).

The global survey of mtDNA variation revealed that the greatest mtDNA variation was found in Africans [91] leading to the conclusion that human mtDNAs originated in Africa approximately 150,000–200,000 years before the present (YBP) [91,97,98]. Because all of the African haplogroups were derived from a common origin, they have been clustered together into macrohaplogroup L [99]. Of all of the African mtDNA variations, only two mtDNA lineages, which arose in Ethiopia [100], left Africa about 65,000 YBP to colonize the rest of the world founding the Eurasian lineages macrohaplogroups M and N [90,94]. The N lineage moved directly north and westward to found all of the European mtDNA haplogroups (H, I, J, Uk, T, U, V, W, and X [101]) and eastward to establish Asian haplogroups

**TABLE 10.1 Confirmed Mitochondrial DNA Mutations—cont'd**

**PANEL A: CODING REGION MUTATIONS**

| Locus | Disease Presentations | Mutation | Amino Acid Change | Homoplasmy Reported | Heteroplasmy Reported | References |
|---|---|---|---|---|---|---|
| *MT–ND6* | LHON | m.14495A>G | L>S | – | + | hh |
| *MT–ND6* | LHON | m.14568C>T | G>S | + | – | ii |
| *MT-CYB* | EXIT/septo-optic dysplasia | m.14849T>C | S>P | – | + | jj |
| *MT-CYB* | MELAS | m.14864T>C | C>R | – | + | kk |

**PANEL B: tRNA AND rRNA MUTATIONS**

| Locus | Disease Presentations | Mutation | RNA | Homoplasmy Reported | Heteroplasmy Reported | References |
|---|---|---|---|---|---|---|
| *MT-TF* | MELAS/ MM + EXIT | m.583G>A | tRNA Phe | – | + | ll |
| *MT-RNR1* | DEAF | m.1494C>T | 12S rRNA | + | – | mm |
| *MT-RNR1* | DEAF | m.1555A>G | 12S rRNA | + | – | nn |
| *MT-TV* | AMDF | m.1606G>A | tRNA Val | – | + | oo |
| *MT-TV* | LS/HCM/ MELAS | m.1644G>A | tRNA Val | – | + | pp |
| *MT-TL1* | MELAS/LS/ DMDF/MIDD/ SNHL/CPEO/ MM/others | m.3243A>G | tRNA Leu (UUR) | – | + | qq |
| *MT-TL1* | MM/MELAS/ SNHL/CPEO | m.3243A>T | tRNA Leu (UUR) | – | + | rr |
| *MT-TL1* | MELAS | m.3256C>T | tRNA Leu (UUR) | – | + | ss |
| *MT-TL1* | MELAS/ myopathy | m.3258T>C | tRNA Leu (UUR) | – | + | tt |
| *MT-TL1* | MMC/MELAS | m.3260A>G | tRNA Leu (UUR) | – | + | uu |
| *MT-TL1* | MELAS/DM | m.3271T>C | tRNA Leu (UUR) | – | + | vv |
| *MT-TL1* | PEM | m.3271T>del | tRNA Leu (UUR) | – | + | ww |
| *MT-TL1* | Myopathy | m.3280A>G | tRNA Leu (UUR) | – | + | xx |
| *MT-TL1* | MELAS/ myopathy/ deafness + cognitive impairment | m.3291T>C | tRNA Leu (UUR) | – | + | yy |

TABLE 10.1 **Confirmed Mitochondrial DNA Mutations—cont'd**

**PANEL B: tRNA AND rRNA MUTATIONS**

| Locus | Disease Presentations | Mutation | RNA | Homoplasmy Reported | Heteroplasmy Reported | References |
|---|---|---|---|---|---|---|
| *MT-TL1* | MM | m.3302A>G | tRNA Leu (UUR) | – | + | zz |
| *MT-TL1* | MMC | m.3303C>T | tRNA Leu (UUR) | + | + | aaa |
| *MT-TI* | CPEO/MS | m.4298G>A | tRNA Ile | – | + | bbb |
| *MT-TI* | MICM | m.4300A>G | tRNA Ile | + | + | ccc |
| *MT-TI* | CPEO | m.4308G>A | tRNA Ile | – | + | ddd |
| *MT-TQ* | Encephalopathy/MELAS | m.4332G>A | tRNA Gln | – | + | eee |
| *MT-TW* | LS | m.5537A>AT | tRNA Trp | – | + | fff |
| *MT-TA* | Myopathy | m.5650G>A | tRNA Ala | – | + | ggg |
| *MT–TN* | CPEO + ptosis + proximal myopathy | m.5690A>G | tRNA Asn | – | + | hhh |
| *MT–TN* | CPEO/MM | m.5703G>A | tRNA Asn | – | + | iii |
| *MT-TS1* | PEM/AMDF/ motor neuron disease-like | m.7471C>CC | tRNA Ser (UCN) | + | + | jjj |
| *MT-TS1* | MM/EXIT | m.7497G>A | tRNA Ser (UCN) | + | + | kkk |
| *MT-TS1* | SNHL | m.7510T>C | tRNA Ser (UCN) | – | + | lll |
| *MT-TS1* | SNHL | m.7511T>C | tRNA Ser (UCN) | + | + | mmm |
| *MT-TK* | MERRF | m.8344A>G | tRNA Lys | – | + | nnn |
| *MT-TK* | MERRF | m.8356T>C | tRNA Lys | – | + | ooo |
| *MT-TK* | MICM + DEAF/ MERRF/ autism/LS/ ataxia + lipomas | m.8363G>A | tRNA Lys | – | + | ppp |
| *MT-TG* | PEM | m.10010T>C | tRNA Gly | – | + | qqq |
| *MT-TH* | MERRF-MELAS/encephalopathy | m.12147G>A | tRNA His | – | + | rrr |
| *MT-TL2* | CPEO | m.12276G>A | tRNA Leu (CUN) | – | + | sss |
| *MT-TL2* | CPEO/KSS | m.12315G>A | tRNA Leu (CUN) | – | + | ttt |
| *MT-TL2* | CPEO | m.12316G>A | tRNA Leu (CUN) | – | + | uuu |

*Continued*

**TABLE 10.1 Confirmed Mitochondrial DNA Mutations—cont'd**

**PANEL B: tRNA AND rRNA MUTATIONS**

| Locus | Disease Presentations | Mutation | RNA | Homoplasmy Reported | Heteroplasmy Reported | References |
|---|---|---|---|---|---|---|
| *MT-TE* | Reversible COX deficiency myopathy | m.14674T>C | tRNA Glu | + | – | vvv |
| *MT-TE* | MM + DMDF/ encephalo- myopathy/ dementia + dia- betes + oph- thalmoplegia | m.14709T>C | tRNA Glu | + | + | www |

See http://www.mitomap.org/MITOMAP/MutationsCodingControl and http://www.mitomap.org/MITOMAP/MutationsRNA for additional reports and phenotypes. *ADPD*, Alzheimer disease and Parkinson disease; *AMDF*, ataxia, myopathy, and deafness; *COX*, cytochrome c oxidase; *CPEO*, chronic progressive ophthalmoplegia; *DEAF/SNHL*, deafness/sensorineural hearing loss; *DEMCHO*, dementia and chorea; *DM*, diabetes mellitus; *DMDF*, diabetes mellitus and deafness; *ESOC*, epilepsy, strokes, optic atrophy, and cognitive decline; *EXIT*, exercise intolerance; *FBSN*, familial bilateral striatal necrosis; *FSGS*, focal segmental glomerulosclerosis; *GER*, gastrointestinal reflux; *HCM*, hypertrophic cardiomyopathy; *LDYT*, LHON + dystonia; *LHON*, Leber hereditary optic neuropathy; *LS*, Leigh syndrome; *MELAS*, mitochondrial encephalomyopathy, lactic acidosis and stroke-like episodes; *MERRF*, myoclonic epilepsy and ragged red fiber disease; *MICM*, maternally inherited cardiomyopathy; *MIDD*, maternally inherited diabetes and deafness; *MILS*, maternally inherited Leigh syndrome; *MM*, mitochondrial myopathy; *MMC*, mitochondrial myopathy and cardiomyopathy; *MNGIE*, mitochondrial neurogastrointestinal encephalopathy; *MS*, multiple sclerosis; *NARP*, neurogenic muscle weakness, ataxia, and retinitis pigmentosa; *PEM*, progressive encephalomyopathy; *SNHL*, sensorineural hearing loss, *+*, reported; *–*, not reported.

[a]Blakely EL, de Silva R, King A, Schwarzer V, Harrower T, Dawidek G, Turnbull DM, Taylor RW. LHON/MELAS overlap syndrome associated with a mitochondrial MTND1 gene mutation. Eur J Hum Genet 2005;13:623–27.

[b]Howell N, Bindoff LA, McCullough DA, Kubacka I, Poulton J, Mackey D, Taylor L, Turnbull DM. Leber hereditary optic neuropathy: identification of the same mitochondrial ND1 mutation in six pedigrees. Am J Hum Genet 1991a;49:939–50; Huoponen K, Vilkki J, Aula P, Nikoskelainen EK, Savontaus ML. A new mtDNA mutation associated with Leber hereditary optic neuroretinopathy. Am J Hum Genet 1991;48:1147–53.

[c]Brown MD, Zhadanov S, Allen JC, Hosseini S, Newman NJ, Atamonov VV, Mikhailovskaya IE, Sukernik RI, Wallace DC. Novel mtDNA mutations and oxidative phosphorylation dysfunction in Russion LHON families. Hum Genet 2001;109:33–9.

[d]Kirby DM, McFarland R, Ohtake A, Dunning C, Ryan MT, Wilson C, Ketteridge D, Turnbull DM, Thorburn DR, Taylor RW. Mutations of the mitochondrial ND1 gene as a cause of MELAS. J Med Genet 2004;41:784–9.

[e]Achilli A, Iommarini L, Olivieri A, Pala M, Kashani BH, Reynier P, La Morgia C, Valentino ML, Liguori R, Pizza F, Barboni P, Sadun F, De Negri A, Zeviani M, Dollfus H, Moulignier A, Ducos G, Orssaud C, Bonneau D, Procaccio V, Leo-Kottler B, Fauser S, Wissinger B, Amati-Bonneau P, Torroni A, Carelli V. Rare primary mitochondrial DNA mutations and synergistic variants in Leber's Hereditary Optic Neuropathy. PLoS One 2012;7:e42242, Fauser S, Luberichs J, Besch D, Leo-Kottler B. Sequence analysis of the complete mitochondrial genome in patients with Leber's hereditary optic neuropathy lacking the three most common pathogenic DNA mutations. Biochem Biophys Res Commun 2002b;295:342–7.

[f]Achilli A, Iommarini L, Olivieri A, Pala M, Kashani BH, Reynier P, La Morgia C, Valentino ML, Liguori R, Pizza F, Barboni P, Sadun F, De Negri A, Zeviani M, Dollfus H, Moulignier A, Ducos G, Orssaud C, Bonneau D, Procaccio V, Leo-Kottler B, Fauser S, Wissinger B, Amati-BonneauP, Torroni A, Carelli V. Rare primary mitochondrial DNA mutations and synergistic variants in Leber's Hereditary Optic Neuropathy. PLoS One 2012;7:e42242, Valentino ML, Barboni P, Ghelli A, Bucchi L, Rengo C, Achilli A, Torroni A, Lugaresi A, Lodi R, Barbiroli B, Dotti M, Federico A, Baruzzi A, Carelli V. The ND1 gene of complex I is a mutational hot spot for Leber's hereditary optic neuropathy. Ann Neurol 2004;56:631–41.

[g]Caporali L, Ghelli AM, Iommarini L, Maresca A, Valentino ML, La Morgia C, Liguori R, Zanna C, Barboni P, De Nardo V, Martinuzzi A, Rizzo G, Tonon C, Lodi R, Calvaruso MA, Cappelletti M, Porcelli AM, Achilli A, Pala M, Torroni A, Carelli V. Cybrid studies establish the causal link between the mtDNA m.3890G>A/MT–ND1 mutation and optic atrophy with bilateral brainstem lesions. Biochim Biophys Acta 2013;1832:445–52.

[h]Blakely EL, Rennie KJ, Jones L, Elstner M, Chrzanowska-Lightowlers ZM, White CB, Shield JP, Pilz DT, Turnbull DM, Poulton J, Taylor RW. Sporadic intragenic inversion of the mitochondrial DNA MTND1 gene causing fatal infantile lactic acidosis. Pediatric

Research 2006;59:440–4, Musumeci O, Andreu AL, Shanske S, Bresolin N, Comi GP, Rothstein R, Schon EA, DiMauro S. Intragenic inversion of mtDNA: a new type of pathogenic mutation in a patient with mitochondrial myopathy. Am J Hum Genet 2000;66:1900–4.
[i]Kim JY, Hwang JM, Park SS. Mitochondrial DNA C4171A/ND1 is a novel primary causative mutation of Leber's hereditary optic neuropathy with a good prognosis. Ann Neurol 2002;51:630–4.
[j]Reid FM, Vernham GA, Jacobs HT. A novel mitochondrial point mutation in a maternal pedigree with sensorineural deafness. Human Mutat 1994;3:243–7.
[k]Ware SM, El-Hassan N, Kahler SG, Zhang Q, Ma YW, Miller E, Wong B, Spicer RL, Craigen WJ, Kozel BA, Grange DK, Wong LJ. Infantile cardiomyopathy caused by a mutation in the overlapping region of mitochondrial ATPase six and eight genes. J Med Genet 2009;46:308–14.
[l]De Vries DD, Van Engelen BG, Gabreels FJ, Ruitenbeek W, Van Oost BA. A second missense mutation in the mitochondrial ATPase six gene in Leigh's syndrome. Ann Neurol 1993;34:410–2.
[m]Harding, AE Holt IJ, Sweeney MG, BrockingtonM, Davis MB. Prenatal diagnosis of mitochondrial DNA8993 T-G disease. Am J Hum Genet 1992;50:629–33, Holt IJ, Harding AE, PettyRK, Morgan-Hughes JA. A new mitochondrial disease associated with mitochondrial DNA heteroplasmy. Am J Hum Genet 1990;46:428–33.
[n]Pfeffer, G Blakely, EL Alston, CL Hassani A, Boggild M, Horvath R, Samuels DC, Taylor RW, Chinnery PF. Adult-onset spinocerebellar ataxia syndromes due to MTATP6 mutations J Neurol Neurosurg Psychiatry 2012;83:883–6, Sikorska M, Sandhu JK, Simon DK, Pathiraja V, Sodja C, Li Y, Ribecco-Lutkiewicz M, Lanthier P, Borowy-Borowski H, Upton A, Raha S, Pulst SM. Tarnopolsky MA Identification of ataxia-associated mtDNA mutations (m.4452T>C and m.9035T>C) and evaluation of their pathogenicity in transmitochondrial cybrids. Muscle Nerve 2009;40:381–94.
[o]Thyagarajan D, Shanske S, Vazquez-Memije M, De Vivo D, DiMauro S. A novel mitochondrial ATPase six point mutation in familial bilateral striatal necrosis. Ann Neurol 1995;38:468–72, Verny C, Guegen N, Desquiret V, Chevrollier A, Prundean A, Dubas F, Cassereau J, Ferre M, Amati-Bonneau P, Bonneau D, Reynier P, Procaccio V. Hereditary spastic paraplegia-like disorder due to a mitochondrial ATP6 gene point mutation. Mitochondrion 2011;11:70–5.
[p]Carrozzo R, Murray J, Santorelli FM, Capaldi RA. The T9176G mutation of human mtDNA gives a fully assembled but inactive ATP synthase when modeled in *Escherichia coli*. FEBS Letters 2000;486:297–9.
[q]Moslemi AR, Darin N, Tulinius M, Oldfors A, Holme E. Two new mutations in the MTATP6 gene associated with Leigh syndrome. Neuropediatrics 2005;36:314–8.
[r]Temperley RJ, Seneca SH, Tonska K, Bartnik E, Bindoff LA, Lightowlers RN, Chrzanowska-Lightowlers ZM. Investigation of a pathogenic mtDNA microdeletion reveals a translation-dependent deadenylation decay pathway in human mitochondria. Hum Mol Genet 2003;12:2341–8.
[s]Crimi M, Papadimitriou A, Galbiati S, Palamidou P, Fortunato F, Bordoni A, Papandreou U, Papadimitriou D, Hadjigeorgiou GM, Drogari E, Bresolin N, Comi GP. A new mitochondrial DNA mutation in ND3 gene causing severe Leigh Syndrome with early lethality. Pediatric Research 2004;55:842–6, Lebon S, Chol M, Benit P, Mugnier C, Chretien D, Giurgea I, Kern I, Girardin E, Hertz-Pannier L, de Lonlay P, Rotig A, Rustin P, Munnich A. Recurrent de novo mitochondrial DNA mutations in respiratory chain deficiency. J Med Genet 2003;40:896–9, McFarland R, Kirby DM, Fowler KJ, Ohtake A, Ryan MT, Amor DJ, Fletcher JM, Dixon JW, Collins FA, Turnbull DM, Taylor RW, Thorburn DR. De novo mutations in the mitochondrial ND3 gene as a cause of infantile mitochondrial encephalopathy and complex I deficiency. Ann Neurol 2004;55:58–64.
[t]Taylor RW, Singh-Kler R, Hayes CM, Smith PE, Turnbull DM. Progressive mitochondrial disease resulting from a novel missense mutation in the mitochondrial DNA ND3 gene. Ann Neurol 2001;50:104–7.
[u]Kirby DM, McFarland R, Ohtake A, Dunning C, Ryan MT, Wilson C, Ketteridge D, Turnbull DM, Thorburn DR, Taylor RW. Mutations of the mitochondrial ND1 gene as a cause of MELAS. Journal of Medical Genetics 2004;41:784–9, Sarzi E, Brown M, Lebon S, Chretien D, Munnich A, Rotig A, Procaccio V. A novel recurrent mitochondrial DNA mutation in ND3 gene is associated with isolated complex I deficiency causing Leigh syndrome and dystonia. American Journal of Medical Genetics 2007;143A:33–41.
[v]Deschauer M, Bamberg C, Claus D, Zierz S, Turnbull DM, Taylor RW. Late-onset encephalopathy associated with a C11777A mutation of mitochondrial DNA. Neurology 2003;60:1357–9, Komaki H, Akanuma J, Iwata H, Takahashi T, Mashima Y, Nonaka I, Goto Y. A novel mtDNA C11777A mutation in Leigh syndrome. Mitochondrion 2003;2:293–304.
[w]Wallace DC, Singh G, Lott MT, Hodge JA, Schurr TG, Lezza AM, Elsas LJ, Nikoskelainen EK. Mitochondrial DNA mutation associated with Leber's hereditary optic neuropathy. Science 1988a;242:1427–30.
[x]Taylor RW, Morris AA, Hutchinson M, Turnbull DM. Leigh disease associated with a novel mitochondrial DNA ND5 mutation. Eur J Hum Genet 2002;10:141–4.
[y]Naini AB, Lu J, Kaufmann P, Bernstein RA, Mancuso M, Bonilla E, Hirano M, DiMauro S. Novel mitochondrial DNA ND5 mutation in a patient with clinical features of MELAS and MERRF. Arch Neurol 2005;62:473–6, Valentino ML, Barboni P, Rengo C, Achilli A, Torroni A, Lodi R, Tonon C, Barbiroli B, Fortuna F, Montagna P, Baruzzi A, Carelli V. The 13,042G–>A/ND5 mutation in mtDNA is pathogenic and can be associated also with a prevalent ocular phenotype. J Med Genet 2006;43:e38.
[z]Dombi E, Diot A, Morten K, Carver J, Lodge T, Fratter C, Ng YS, Liao C, Muir R, Blakely EL, Hargreaves I, Al-Dosary M, Sarkar G, Hickman SJ, Downes SM, Jayawant S, Yu-Wai-Man P, Taylor RW, Poulton J. The m.13,051G>A mitochondrial DNA mutation results in variable neurology and activated mitophagy. Neurology 2016;86:1921–23, Howell N, Oostra RJ, Bolhuis PA, Spruijt L, Clarke LA,

Mackey DA, Preston G, Herrnstadt C. Sequence analysis of the mitochondrial genomes from Dutch pedigrees with Leber hereditary optic neuropathy. Am J Hum Genet 2003;72:1460–9.
[aa]Santorelli FM, Tanji K, Kulikova R, Shanske S, Vilarinho L, Hays AP, DiMauro S. Identification of a novel mutation in the mtDNA ND5 gene associated with MELAS. Biochem Biophys Res Commun 1997a;238:326–328.
[bb]Corona P, Antozzi C, Carrara F, D'Incerti L, Lamantea E, Tiranti V, Zeviani M. A novel mtDNA mutation in the ND5 subunit of complex I in two MELAS patients. Ann Neurol 2001;49:106–10.
[cc]Jun AS, Brown MD, Wallace DC. A mitochondrial DNA mutation at np 14,459 of the ND6 gene associated with maternally inherited Leber's hereditary optic neuropathy and dystonia. Proc Natl Acad Sci USA 1994;91:6206–10, Kirby DM, Kahler SG, Freckmann ML, Reddihough D, Thorburn DR. Leigh disease caused by the mitochondrial DNA G14459A mutation in unrelated families. Ann Neurol 2000;48:102–4.
[dd]Achilli A, Iommarini L, Olivieri A, Pala M, Kashani BH, Reynier P, La Morgia C, Valentino ML, Liguori R, Pizza F, Barboni P, Sadun F, De Negri A, Zeviani M, Dollfus H, Moulignier A, Ducos G, Orssaud C, Bonneau D, Procaccio V, Leo-Kottler B, Fauser S, Wissinger B, Amati-Bonneau P, Torroni A, Carelli V. Rare primary mitochondrial DNA mutations and synergistic variants in Leber's Hereditary Optic Neuropathy. PLoS One 2012;7:e42242, Valentino ML, Avoni P, Barboni P, Pallotti F, Rengo C, Torroni A, Bellan M, Baruzzi A, Carelli V. Mitochondrial DNA nucleotide changes C14482G and C14482A in the ND6 gene are pathogenic for Leber's hereditary optic neuropathy. Ann Neurol 2002;51:774–8.
[ee]Howell N, Bogolin C, Jamieson R, Marenda DR, Mackey DA. mtDNA mutations that cause optic neuropathy: how do we know? Am J Hum Genet 1998;62:196–202.
[ff]Brown MD, Voljavec AS, Lott MT, MacDonald I, Wallace DC. Leber's hereditary optic neuropathy: a model for mitochondrial neurodegenerative diseases. FASEB J 1992;6:2791–9, Howell N, Kubacka I, Xu M, McCullough DA. Leber hereditary optic neuropathy: involvement of the mitochondrial ND1 gene and evidence for an intragenic suppressor mutation. Am J Hum Genet 1991b;48:935–42, Johns DR, Neufeld MJ, Park RD. An ND-6 mitochondrial DNA mutation associated with Leber hereditary optic neuropathy. Biochem Biophys Res Commun 1992;187:1551–7.
[gg]Solano A, Roig M, Vives-Bauza C, Hernandez-Pena J, Garcia-Arumi E, Playan A, Lopez-Perez MJ, Andreu AL, Montoya J. Bilateral striatal necrosis associated with a novel mutation in the mitochondrial ND6 gene. Ann Neurol 2003;54:527–30, Ugalde C, Triepels RH, Coenen MJ, van den Heuvel LP, Smeets R, Uusimaa J, Briones P, Campistol J, Majamaa K, Smeitink JA, Nijtmans LG. Impaired complex I assembly in a Leigh syndrome patient with a novel missense mutation in the ND6 gene. Ann Neurol 2003;54:665–9.
[hh]Chinnery PF, Brown DT, Andrews RM, Singh-Kler R, Riordan-Eva P, Lindley J, Applegarth DA, Turnbull DM, Howell N. The mitochondrial ND6 gene is a hot spot for mutations that cause Leber's hereditary optic neuropathy. Brain 2001;124:209–18.
[ii]Fauser S, Leo-Kottler B, Besch D, Luberichs J. Confirmation of the 14,568 mutation in the mitochondrial ND6 gene as causative in Leber's hereditary optic neuropathy. Ophthalmic Genetics 2002a;23:191–7, Wissinger B, Besch D, Baumann B, Fauser S, Christ-Adler M, Jurklies B, Zrenner E, Leo-Kottler B. Mutation analysis of the ND6 gene in patients with Lebers hereditary optic neuropathy. Biochem Biophys Res Commun 1997;234:511–5.
[jj]Schuelke M, Krude H, Finckh B, Mayatepek E, Janssen A, Schmelz M, Trefz F, Trijbels F, Smeitink J. Septo-optic dysplasia associated with a new mitochondrial cytochrome b mutation. Ann Neurol 2002;51:388–92.
[kk]Emmanuele V, Sotiriou E, Rios PG, Ganesh J, Ichord R, Foley AR, Akman HO, Dimauro S. A novel mutation in the mitochondrial DNA cytochrome b gene (MTCYB) in a patient with mitochondrial encephalomyopathy, lactic acidosis, and strokelike episodes syndrome. J Child Neurol 2013;28:236–42.
[ll]Darin N, Kollberg G, Moslemi AR, Tulinius M, Holme E, Gronlund MA, Andersson S, Oldfors A. Mitochondrial myopathy with exercise intolerance and retinal dystrophy in a sporadic patient with a G583A mutation in the mt tRNA(phe) gene. Neuromusc Disord 2006;16:504–6, Hanna MG, Nelson IP, Morgan-Hughes JA, Wood NW. MELAS: a new disease associated mitochondrial DNA mutation and evidence for further genetic heterogeneity. J Neurol Neurosurg Psychiatry 1998;65:512–7.
[mm]Zhao H, Li R, Wang Q, Yan Q, Deng JH, Han D, Bai Y, Young WY, Guan MX. Maternally inherited aminoglycoside-induced and nonsyndromic deafness is associated with the novel C1494T mutation in the mitochondrial 12S rRNA gene in a large Chinese family. Am J Hum Genet 2004;74:139–152.
[nn]Fischel-Ghodsian N, Prezant TR, Bu X, Oztas S. Mitochondrial ribosomal RNA gene mutation in a patient with sporadic aminoglycoside ototoxicity. Am J Otolaryngol 1993;14:399–403, Hutchin T, Haworth I, Higashi K, Fischel-Ghodsian N, Stoneking M, Saha N, Arnos C, Cortopassi G. A molecular basis for human hypersensitivity to aminoglycoside antibiotics. Nucleic Acids Res 1993;21:4174–9, Prezant TR, Agapian JV, Bohlman MC, Bu X, Oztas S, Qiu WQ, Arnos KS, Cortopassi GA, Jaber L, Rotter JI, Shohat M, Fischel-Ghodsian N. Mitochondrial ribosomal RNA mutation associated with both antibiotic-induced and non-syndromic deafness. Nature Genet 1993;4:289–94.
[oo]Tiranti V, D'Agruma L, Pareyson D, Mora M, Carrara F, Zelante L, Gasparini P, Zeviani M. A novel mutation in the mitochondrial tRNA(Val) gene associated with a complex neurological presentation. Ann Neurol 1998;43:98–101.
[pp]Fraidakis MJ, Jardel C, Allouche S, Nelson I, Aure K, Slama A, Lemiere I, Thenint JP, Hamon JB, Zagnoli F, Heron D, Sedel F, Lombes A. Phenotypic diversity associated with the MT-TV gene m.1644G>A mutation, a matter of quantity. Mitochondrion 2014;15:34–9, Menotti F, Brega A, Diegoli M, Grasso M, Modena MG, Arbustini E. A novel mtDNA point mutation in tRNA(Val) is associated with hypertrophic cardiomyopathy and MELAS. Ital Heart J 2004;5:460–5.

[qq]Goto Y, Nonaka I, Horai S. A mutation in the tRNA$^{Leu(UUR)}$ gene associated with the MELAS subgroup of mitochondrial encephalomyopathies. Nature 1990;348:651–3, Manouvrier S, Rotig A, Hannebique G, Gheerbrandt JD, Royer-Legrain G, Munnich A, Parent M, Grunfeld JP, Largilliere C, Lombes A, Bonnefont JP. Point mutation of the mitochondrial tRNA$^{Leu}$ gene (A 3243 G) in maternally inherited hypertrophic cardiomyopathy, diabetes mellitus, renal failure, and sensorineural deafness. J Med Genet 1995;32:654–6, Massin P, Guillausseau PJ, Vialettes B, Paquis V, Orsini F, Grimaldi AD, Gaudric A. Macular pattern dystrophy associated with a mutation of mitochondrial DNA. Am J Ophthalmol 1995;120:247–8, van den Ouweland JM, Lemkes HHP, Ruitenbeek W, Sandkjujl LA, deVijlder MF, Struyvenberg PAA, van de Kamp JJP, Maassen JA. Mutation in mitochondrial tRNA$^{Leu(UUR)}$ gene in a large pedigree with maternally transmitted type II diabetes mellitus and deafness. Nat Genet 1992;1:368–71.

[rr]Shaag A, Saada A, Steinberg A, Navon P, Elpeleg ON. Mitochondrial encephalomyopathy associated with a novel mutation in the mitochondrial tRNA(leu)(UUR) gene (A3243T). Biochem Biophys Res Commun 1997;233:637–9.

[ss]Moraes CT, Ciacci F, Bonilla E, Jansen C, Hirano M, Rao N, Lovelace RE, Rowland LP, Schon EA. DiMauro S. Two novel pathogenic mitochondrial DNA mutations affecting organelle number and protein synthesis. Is the tRNA$^{Leu(UUR)}$ gene an etiologic hot spot? J Clin Investig 1993;92:2906–15, Sato W, Hayasaka K, Shoji Y, Takahashi T, Takada G, Saito M, Fukawa O, Wachi E. A mitochondrial tRNA(Leu)(UUR) mutation at 3256 associated with mitochondrial myopathy, encephalopathy, lactic acidosis, and stroke-like episodes (MELAS). Biochem Mol Biol Int (Sydney) 1994;33:1055–61.

[tt]Sternberg D, Chatzoglou E, Laforet P, Fayet G, Jardel C, Blondy P, Fardeau M, Amselem S, Eymard B, Lombes A. Mitochondrial DNA transfer RNA gene sequence variations in patients with mitochondrial disorders. Brain 2001;124:984–94.

[uu]Sweeney MG, Brockington M, Weston MJ, Morgan-Hughes JA, Harding AE. Mitochondrial DNA transfer RNA mutation Leu$^{(UUR)}$ A-G 3260: a second family with myopathy and cardiomyopathy. Q J Med 1993;86:435–8, Zeviani M, Gellera C, Antozzi C, Rimoldi M, Morandi L, Villani F, Tiranti V, DiDonato S. Maternally inherited myopathy and cardiomyopathy: association with mutation in mitochondrial DNA tRNA$^{Leu(UUR)}$. Lancet 1991;338:143–7.

[vv]Goto Y, Nonaka I, Horai S. A new mtDNA mutation associated with mitochondrial myopathy, encephalopathy, lactic acidosis and stroke-like episodes (MELAS). Biochim Biophys Acta 1991;1097:238–40, Hayashi J, Ohta S, Takai D, Miyabayashi S, Sakuta R, Goto Y, Nonaka I. Accumulation of mtDNA with a mutation at position 3271 in tRNA$^{Leu(UUR)}$ gene introduced from a MELAS patient to HeLa cells lacking mtDNA results in progressive inhibition of mitochondrial respiratory function. Biochem Biophys Res Commun 1993;197:1049–55, Sakuta R, Goto Y, Horai S, Nonaka I. Mitochondrial DNA mutations at nucleotide positions 3243 and 3271 in mitochondrial myopathy, encephalopathy, lactic acidosis, and stroke-like episodes: a comparative study. J Neurol Sci1993;115:158–60, Tsukuda K, Suzuki Y, Kameoka K, Osawa N, Goto Y, Katagiri H, Asano T, Yazaki Y, Oka Y. Screening of patients with maternally transmitted diabetes for mitochondrial gene mutations in the tRNA[Leu(UUR)] region. Diabet Med 1997;14:1032–7.

[ww]Shoffner JM, Bialer MG, Pavlakis SG, Lott MT, Kaufman A, Dixon J, Teichberg S, Wallace DC. Mitochondrial encephalomyopathy associated with a single nucleotide pair deletion in the mitochondrial tRNA$^{Leu(UUR)}$ gene. Neurology 1995;45:286–92.

[xx]Sternberg D, Chatzoglou E, Laforet P, Fayet G, Jardel C, Blondy P, Fardeau M, Amselem S, Eymard B, Lombes A. Mitochondrial DNA transfer RNA gene sequence variations in patients with mitochondrial disorders. Brain 2001;124:984–94.

[yy]Goto Y. Clinical features of MELAS and mitochondrial DNA mutations. Muscle Nerve 1995;3:S107–12, Goto Y, Tsugane K, Tanabe Y, Nonaka I, Horai S. A new point mutation at nucelotide pair 3291 of the tRNA$^{Leu(UUR)}$ gene in a patient with mitochondrial myopathy, encephalopathy, lactic acidosis, and stroke-like episodes (MELAS). Biochem Biophys Res Commun 1994;202:1624–30.

[zz]Bindoff LA, Howell N, Poulton J, McCullough DA, Morten KJ, Lightowlers RN, Turnbull DM, Weber K. Abnormal RNA processing associated with a novel tRNA mutation in mitochondrial DNA. A potential disease mechanism. J Biol Chem 1993;268:19559–64, Maniura-Weber K, Helm M, Engemann K, Eckertz S, Mollers M, Schauen M, Hayrapetyan A, von Kleist-Retzow JC, Lightowlers RN, Bindoff LA, Wiesner RJ. Molecular dysfunction associated with the human mitochondrial 3302A>G mutation in the MTTL1 (mt-tRNALeu(UUR)) gene. Nucleic Acids Res 2006;34:6404–15, Shoffner JM, Krawiecki N, Cabell, MF, Torroni A, Wallace DC. A novel tRNA$^{Leu(UUR)}$ mutation in childhood mitochondrial myopathy. Poster 949. Am J Hum Genet 1993;53:287.

[aaa]Silvestri G, Santorelli FM, Shanske S, Whitley CB, Schimmenti LA, Smith SA, DiMauro S. A new mtDNA mutation in the tRNA$^{Leu(UUR)}$ gene associated with maternally inherited cardiomyopathy. Human Mutat 1994;3:37–43.

[bbb]Taylor RW, Chinnery PF, Bates MJ, Jackson MJ, Johnson MA, Andrews RM, Turnbull DM. A novel mitochondrial DNA point mutation in the tRNA(Ile) gene: studies in a patient presenting with chronic progressive external ophthalmoplegia and multiple sclerosis. Biochem Biophys Res Commun 1998;243:47–51.

[ccc]Casali C, Santorelli FM, D'Amati G, Bernucci P, DeBiase L, DiMauro S. A novel mtDNA point mutation in maternally inherited cardiomyopathy. Biochem Biophys Res Commun 1995;213:588–93.

[ddd]Schaller A, Desetty R, Hahn D, Jackson CB, Nuoffer JM, Gallati S, Levinger L. Impairment of mitochondrial tRNA(Ile) processing by a novel mutation associated with chronic progressive external ophthalmoplegia. Mitochondrion 2011;11:488–96, Souilem S, Chebel S, Mancuso M, Petrozzi L, Siciliano G, FrihAyed M, Hentati F, Amouri R. A novel mitochondrial tRNA(Ile) point mutation associated with chronic progressive external ophthalmoplegia and hyperCKemia. J Neurol Sci 2011;300:187–90.

[eee]Bataillard M, Chatzoglou E, Rumbach L, Sternberg D, Tournade A, Laforet P, Jardel C, Maisonobe T, Lombes A. Atypical MELAS syndrome associated with a new mitochondrial tRNA glutamine point mutation. Neurology 2001;56:405–7.

[fff]Santorelli FM, Tanji K, Sano M, Shanske S, El-Shahawi M, Kranz-Eble P, DiMauro S, De Vivo DC. Maternally inherited encephalopathy associated with a single-base insertion in the mitochondrial tRNATrp gene. Ann Neurol 1997b;42:256–60.

[ggg]McFarland R, Swalwell H, Blakely EL, He L, Groen EJ, Turnbull DM, Bushby KM, Taylor RW. The m.5650G>A mitochondrial tRNA(Ala) mutation is pathogenic and causes a phenotype of pure myopathy. Neuromusc Disord 2008;18:63–7.

[hhh]Blakely EL, Yarham JW, Alston CL, Craig K, Poulton J, Brierley C, Park S-M, Dean A, Xuereb JH, Anderson KN, Compston A, Allen C, Sharif S, Enevoldson P, Wilson M, Hammans SR, Turnbull DM, McFarland R, Taylor RW. Pathogenic mitochondrial tRNA point mutations: nine novel mutations affirm their importance as a cause of mitochondrial disease. Hum Mutat 2013;34:1260–8.

[iii]Hao H, Moraes CT. A disease-associated G5703A mutation in human mitochondrial DNA causes a conformational change and a marked decrease in steady-state levels of mitochondrial tRNA(Asn). Mol Cell Biol 1997;17:6831–7, Moraes CT, Ciacci F, Bonilla E, Jansen C, Hirano M, Rao N, Lovelace RE, Rowland LP, Schon EA, DiMauro S. Two novel pathogenic mitochondrial DNA mutations affecting organelle number and protein synthesis. Is the tRNALeu(UUR) gene an etiologic hot spot? J Clin Investig 1993;92:2906–15, Vives-Bauza C, Del Toro M, Solano A, Montoya J, Andreu AL, Roig M. Genotype-phenotype correlation in the 5703G>A mutation in the tRNA(ASN) gene of mitochondrial DNA. J Inherit Metab Dis 2003;26:507–8.

[jjj]Jaksch M, Hofmann S, Kleinle S, Liechti-Gallati S, Pongratz DE, Muller-Hocker J, Jedele KB, Meitinger T, Gerbitz KD. A systematic mutation screen of 10 nuclear and 25 mitochondrial candidate genes in 21 patients with cytochrome c oxidase (COX) deficiency shows tRNA(Ser)(UCN) mutations in a subgroup with syndromal encephalopathy. J Med Genet 1998a;35:895–900, Jaksch M, Klopstock T, Kurlemann G, Dorner M, Hofmann S, Kleinle S, Hegemann S, Weissert M, Muller-Hocker J, Pongratz D, Gerbitz KD. Progressive myoclonus epilepsy and mitochondrial myopathy associated with mutations in the tRNA(Ser(UCN)) gene. Ann Neurol 1998b;44:635–40, Schuelke M, Bakker M, Stoltenburg G, Sperner J, von Moers, A. Epilepsia partialis continua associated with a homoplasmic mitochondrial tRNA(Ser(UCN)) mutation. Ann Neurol 1998;44:700–704, Tiranti V, Chariot P, Carella F, Toscano A, Soliveri P, Girlanda P, Carrara F, Fratta GM, Reid FM, Mariotti C, Zeviani M. Maternally inherited hearing loss, ataxia and myoclonus associated with a novel point mutation in mitochondrial $tRNA^{Ser(UCN)}$ gene. Hum Mol Genet 1995;4:1421–7.

[kkk]Grafakou O, Hol FA, Otfried Schwab K, Siers MH, ter Laak H, Trijbels F, Ensenauer R, Boelen C, Smeitink J. Exercise intolerance, muscle pain and lactic acidaemia associated with a 7497G>A mutation in the tRNASer(UCN) gene. J Inherit Metab Dis 2003;26:593–600, Jaksch M, Klopstock T, Kurlemann G, Dorner M, Hofmann S, Kleinle S, Hegemann S, Weissert M, Muller-Hocker J, Pongratz D, Gerbitz KD. Progressive myoclonus epilepsy and mitochondrial myopathy associated with mutations in the tRNA(Ser(UCN)) gene. Ann Neurol 1998b;44:635–40, Mollers M, Maniura-Weber K, Kiseljakovic E, Bust M, Hayrapetyan A, Jaksch M, Helm M, Wiesner RJ, von Kleist-Retzow JC. A new mechanism for mtDNA pathogenesis: impairment of post-transcriptional maturation leads to severe depletion of mitochondrial tRNASer(UCN) caused by T7512C and G7497A point mutations. Nucleic Acids Res 2005;33:5647–58.

[lll]Hutchin TP, Parker MJ, Young ID, Davis AC, Pulleyn LJ, Deeble J, Lench NJ, Markham AF, Mueller RF. A novel mutation in the mitochondrial tRNA(Ser(UCN)) gene in a family with non-syndromic sensorineural hearing impairment. J Med Genet 2000;37:692–4.

[mmm]Sue CM, Tanji K, Hadjigeorgiou G, Andreu AL, Nishino I, Krishna S, Bruno C, Hirano M, Shanske S, Bonilla E, Fischel-Ghodsian N, DiMauro S, Friedman R. Maternally inherited hearing loss in a large kindred with a novel T7511C mutation in the mitochondrial DNA tRNA(Ser(UCN)) gene. Neurology 1999;52:1905–8.

[nnn]Shoffner JM, Lott MT, Lezza AM, Seibel P, Ballinger SW, Wallace DC. Myoclonic epilepsy and ragged-red fiber disease (MERRF) is associated with a mitochondrial DNA $tRNA^{Lys}$ mutation. Cell 1990;61:931–7, Wallace DC, Zheng X, Lott MT, Shoffner JM, Hodge JA, Kelley RI, Epstein CM, Hopkins LC, 1988b. Familial mitochondrial encephalomyopathy (MERRF): Genetic, pathophysiological, and biochemical characterization of a mitochondrial DNA disease. Cell 1990;55:601–10.

[ooo]Masucci JP, Davidson M, Koga Y, Schon EA, King MP. In vitro analysis of mutations causing myoclonus epilepsy with ragged-red fibers in the mitochondrial $tRNA^{Lys}$gene: two genotypes produce similar phenotypes. Mol Cell Biol 1995;15:2872–81, Silvestri G, Moraes CT, Shanske S, Oh SJ, DiMauro S. A new mtDNA mutation in the $tRNA^{Lys}$ gene associated with myoclonic epilepsy and ragged red fibers (MERRF). Am J Hum Genet 1992;51:1213–7, Zeviani M, Muntoni F, Savarese N, Serra G, Tiranti V, Carrara F, Mariotti C, DiDonato S. A MERRF/MELAS overlap syndrome associated with a new point mutation in the mitochondrial DNA $tRNA^{Lys}$ gene. Eur J Hum Genet 1993;1:80–7.

[ppp]Ozawa M, Nishino I, Horai S, Nonaka I, Goto YI. Myoclonus epilepsy associated with ragged-red fibers: a G-to-A mutation at nucleotide pair 8363 in mitochondrial tRNA(Lys) in two families. Muscle Nerve 1997;20:271–8, Santorelli FM, Mak SC, El-Schahawi M, Casali C, Shanske S, Baram TZ, Madrid RE, DiMauro S. Maternally inherited cardiomyopathy and hearing loss associated with a novel mutation in the mitochondrial $tRNA^{Lys}$ gene (G8363A). Am J Hum Genet 1996;58:933–9.

[qqq]Bidooki SK, Johnson MA, Chrzanowska-Lightowlers Z, Bindoff LA, Lightowlers RN. Intracellular mitochondrial triplasmy in a patient with two heteroplasmic base changes. Am J Hum Genet 1997;60:1430–8.

[rrr]Melone MA, Tessa A, Petrini S, Lus G, Sampaolo S, di Fede G, Santorelli FM, Cotrufo R Revelation of a new mitochondrial DNA mutation(G12147A) in a MELAS/MERFF phenotype. Arch Neurol 2004;61:269–72, Taylor RW, Schaefer AM, McDonnell MT, Petty RK, Thomas AM, Blakely EL, Hayes CM, McFarland R, Turnbull DM. Catastrophic presentation of mitochondrial disease due to a mutation in the tRNA(His) gene. Neurology 2004;62:1420–3.

[sss]Cardaioli E, Da Pozzo P, Gallus GN, Malandrini A, Gambelli S, Gaudiano C, Malfatti E, Viscomi C, Zicari E, Berti G, Serni G, Dotti MT, Federico A. A novel heteroplasmic tRNA(Ser(UCN)) mtDNA point mutation associated with progressive external ophthalmoplegia and hearing loss. Neuromusc Disord 2007;17:681–3.

ttt Karadimas CL, Salviati L, Sacconi S, Chronopoulou P, Shanske S, Bonilla E, De Vivo DC, DiMauro S. Mitochondrial myopathy and ophthalmoplegia in a sporadic patient with the G12315A mutation in mitochondrial DNA. Neuromusc Disord 2002;12:865–8.
uuu Cardaioli E, Da Pozzo P, Malfatti E, Gallus GN, Rubegni A, Malandrini A, Gaudiano C, Guidi L, Serni G, Berti G, Dotti MT, Federico A. Chronic progressive external ophthalmoplegia: a new heteroplasmic tRNA(Leu(CUN)) mutation of mitochondrial DNA. J Neurol Sci 2008;272:106–9.
vvv Horvath R, Kemp JP, Tuppen HA, Hudson G, Oldfors A, Marie SK, Moslemi AR, Servidei S, Holme E, Shanske S, Kollberg G, Jayakar P, Pyle A, Marks HM, Holinski-Feder E, Scavina M, Walter MC, Coku J, Gunther-Scholz A, Smith PM, McFarland R, Chrzanowska-Lightowlers ZM, Lightowlers RN, Hirano M, Lochmuller H, Taylor RW, Chinnery PF, Tulinius M, DiMauro S. Molecular basis of infantile reversible cytochrome c oxidase deficiency myopathy. Brain 2009;132:3165–74, Mimaki M, Hatakeyama H, Komaki H, Yokoyama M, Arai H, Kirino Y, Suzuki T, Nishino I, Nonaka I, Goto Y. Reversible infantile respiratory chain deficiency: a clinical and molecular study. Ann Neurol 2010;68:845–54.
www Hanna MG, Nelson I, Sweeney MG, Cooper JM, Watkins PJ, Morgan-Hughes JA, Harding AE. Congenital encephalomyopathy and adult-onset myopathey and diabetes mellitus: different phenotypic associations of a new heteroplasmic mtDNA tRNA glutamic acid mutation. Am J Hum Genet 1995;56:1026–33, Hao H, Bonilla E, Manfredi G, DiMauro S, Moraes CT. Segregation patterns of a novel mutation in the mitochondrial tRNA glutamic acid gene associated with myopathy and diabetes mellitus. Am J Hum Genet 1995;56:1017–25.

present with retinitis pigmentosa at 75% mutant, olivopontocerebellar atrophy at about 85% mutant, and Leigh syndrome at 95%–100% mutant [115–117].

Classic examples of mtDNA protein synthesis mutations are the $tRNA^{Lys}$ nt 8344A>G mutation associated with MERRF syndrome [6,8] and the $tRNA^{Leu(UUR)}$ nt 3243A>G mutation associated with MELAS [118]. Many tRNA mutations changes in the heteroplasmy levels have profound effects on the phenotypic presentation.

For the $tRNA^{Lys}$ 8344A>G mutation, individuals with low heteroplasmy levels may manifest with sensory neural hearing loss and mitochondrial myopathy. However, at high heteroplasmy levels, individuals can present with debilitating myoclonus, cardiomyopathy, and dementia [6,8]. The phenotypic variability of the $tRNA^{Leu(UUR)}$ 3243A>G mutation is even more striking. At about 10%–30% heteroplasmy, the 3243G mutation allele can present as type 1 and type 2 diabetes or autism; at about 50%–80% heteroplasmy, with migraines and neuromuscular diseases including MELAS; and at 90%–100% heteroplasmy, with perinatal lethal disease. These marked differences in cellular and clinical phenotype correlate with changes in mitochondrial and cellular physiology and structure. But the abrupt changes in phenotypic manifestations is the result of mitochondrial signaling to the nucleus, presumably through mitochondrial high-energy intermediates, which results in abrupt changes in nuclear gene transcription profiles. These phase shifts in nuclear gene expression correlate the changes in the clinical manifestations, presumably mediated by alterations in cellular signal transduction and epigenomic signaling [119,120].

Changes in functional elements in the mtDNA control region can also result in maternally transmitted clinical phenotypes. One well-characterized phenotype is cyclic vomiting and migraine headaches [121].

### 10.5.2 Ancient Adaptive mtDNA Variants

Ancient adaptive mtDNA variants can modulate the penetrance of maternally inherited mutations, such as the milder LHON *ND4* 11778G>A (R340H), *ND6* 14484T>C (M64V), and *ND4L* 10663T>C (V65A) [122] mutations. In Europeans, mutations that arise on mtDNA haplogroup J have a significantly increased penetrance [122–125]. This increased penetrance is associated with lower respiration in cybrids in which the 11778G>A (R340H) mutation is on the J background versus on the haplogroup H background [126–128]. mtDNA haplogroups have been associated with increased risk for a wide range of common clinical manifestations [129]. Because these associations increase risk rather than being sufficient in themselves to cause the disease, they will be discussed under the common "complex" disease section.

### 10.5.3 Developmental and Somatic mtDNA Mutations

Finally, developmental and somatic tissue mutations can result in "spontaneous" diseases. De novo mtDNA deletions can arise early in development and result in varying severity phenotypes depending on the distribution and heteroplasmy levels of the deletion. Interestingly, as long as the deletion removes a tRNA gene, the actual size and position of the deletion are less important to the phenotype. Deletions that are at lower heteroplasmy levels present with CPEO [5] but at higher deletion levels with the earlier-onset Kearns-Sayre syndrome (KSS) [130]. In both CPEO and KSS, the deletion is not found in blood presumably because the deletion segregates during mitotic replication of the bone marrow stem cells. Those cells that lose the deletion have a replicative advantage and replace the stem cells that retain the deleted mtDNA [131]. Deletions that are widely distributed throughout the body and at high heteroplasmy levels result in the Pearson marrow-pancreas syndrome. In this syndrome the blood cells have high levels of deletion associated with early childhood transfusion-dependent pancytopenia [132,133]. Presumably in Pearson syndrome the mtDNA deletion level is sufficiently high that few bone marrow stem cells segregate to normal mtDNAs. Occasionally, however, Pearson patients do revert back to more normal bone marrow, relieving the pancytopenia, but these patients progress to KSS.

Most cases of single deletion are spontaneous, but occasionally deletions can be maternally inherited. These maternally inherited rearrangement mutations appear to be mtDNA duplications, which are less lethal then deletions. Within postmitotic tissues, the duplications undergo rearrangement to generate deletions, which then accumulate in the postmitotic tissues to produce the phenotype [134].

Random mtDNA mutations also accumulated in adult stem and postmitotic cells with age. These somatic mtDNA mutations can segregate to higher heteroplasmy progressively eroding mitochondrial function until the bioenergetic capacity of the cell or tissue falls below the expression threshold. Such somatic cell mutations can exacerbate inherited partial bioenergetics defects resulting in the delayed onset and progressive course of diseases and for normal individuals may constitute the aging clock [9,135–137].

### 10.5.4 The Range of mtDNA Disease Phenotypes in MITOMAP and MITOMASTER

A list of the current mtDNA diseases and representative mutations is presented in Table 10.1. Since the late 1980s, diseases associated with mtDNA mutations have been descriptive of the phenotype, so the range of clinical presentations can be deduced from this table. A detailed description of the mtDNA variants associated with these diseases is available in our previous chapter in this series [15], and a complete accounting of human mtDNA variation is available through our information service: MITOMAP and MITOMASTER [16]. MITOMAP currently encompasses 46,092 full-length mtDNA sequences, representing all global populations, with 13,662 nucleotide variants, including those that typify the mtDNA haplogroups. The clinical databases encompass 686 mtDNA mutations, 353 in protein coding regions, and 333 in tRNA and rRNA genes. MITOMASTER provides a comprehensive set of analytical tools for the analysis of mtDNA variation including the capacity to import mtDNA sequences to be interrogated, information on mtDNA variants, and tools to interpret the potential pathophysiological significance of mtDNA variants. In the year ending December 31, 2017, the MITOMAP portal was visited 118,467 times. The distribution of access is provided in Fig. 10.4.

## 10.6 nDNA CODED MITOCHONDRIAL DISEASES

Bioenergetic disease can result from mutations in any one of the hundreds of nDNA genes that code for mitochondrial proteins. Mutations in several hundred nDNA gene loci have already been reported to cause mitochondrial bioenergetic dysfunction [15,87]. Pathogenic mutations have been identified in OXPHOS structural and assembly factors, intermediate metabolism, replication, translation, and regulator genes. When both copies of a chromosomal gene are mutated, severe OXPHOS defects can occur and cause devastating pediatric diseases, the most commonly recognized phenotype being Leigh syndrome [15,87]. However, phenotypes can be highly variable. For example mutations in the complex I *NDUFS2* originally associated with Leigh syndrome can also present with LHON-like hereditary optic neuropathy [138], and mutations in 10 mitochondrial carrier proteins have been associated with a variety of clinical phenotypes [24–26].

Multiple clinical manifestations can result from mutations in nDNA genes. OXPHOS defects can result from mutations in the mitochondrial translation proteins including aminoacyl tRNA synthetases and related mitochondrial translation polypeptides [139,140]. Moreover, protein synthesis defects can result from faulty interactions between mtDNA and nDNA variants.

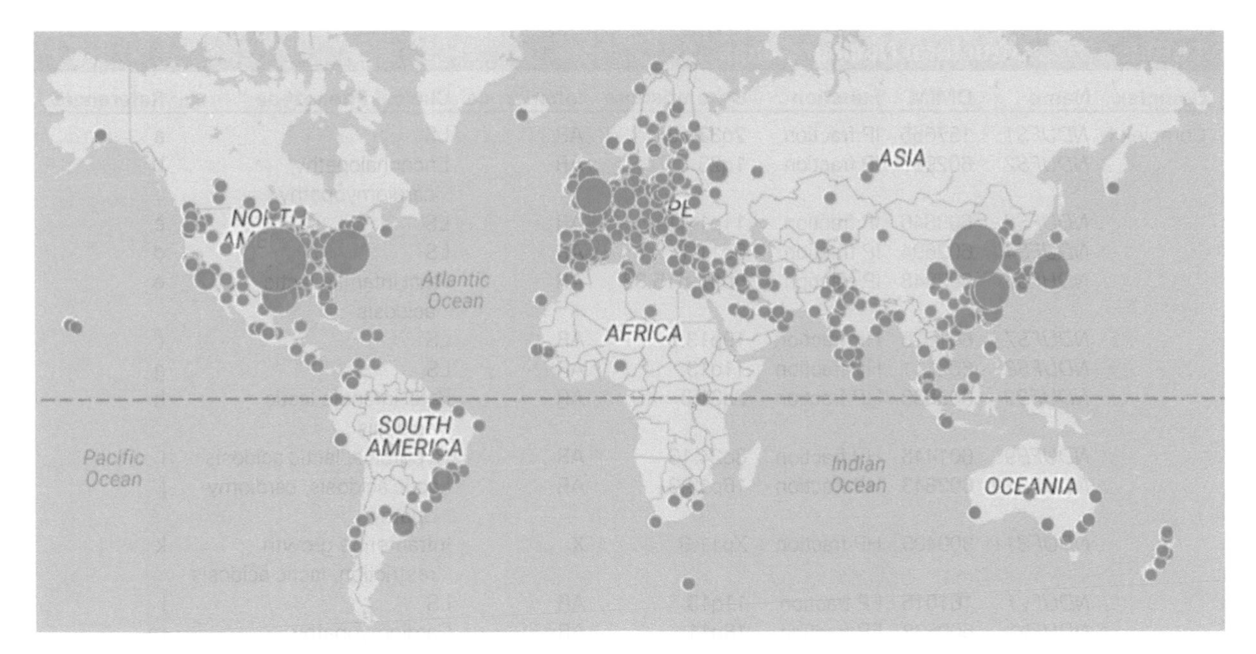


**Figure 10.4** Global access of MITOMAP during 2017.

For example, the mtDNA 12 rRNA 1555A>G variant alone predisposes to aminoglycoside-induced sensorineural hearing loss, but in conjunction with the A10S variant in the TRMU (methyaminomethyl-2-thiouridylate-methyltransferase) modification gene, it can result in inherited deafness [141].

Mutations in the nDNA-coded mtDNA biogenesis genes can cause degenerative diseases by destabilizing mtDNA biogenesis, resulting in multiple deletions and/or mtDNA depletion. Mutations in the mtDNA polymerase γ (*POLG*) are the most common nDNA mutations that cause mtDNA multiple deletions or depletions and can present with a wide variety of phenotypes ranging from CPEO to Alper syndrome [142–144]. Other important mtDNA stability mutations include the Twinkle helicase [145], mitochondrial deoxyguanosine kinase and thymidine kinase 2 [146,147], cytosolic thymidine phosphorylase [148], and the heart-muscle adenine nucleotide (ADP/ATP) translocator (*ANT1*) [43–45], to name a few [15,149].

A representative list of the nDNA mitochondrial genes that have been found to be mutant in OXPHOS-deficient patients is provided in Tables 10.2 and 10.3. Table 10.2 lists the 33 structural OXPHOS genes found to be mutant in patients. This includes 20 complex I genes, all four complex II genes, two complex III genes, four complex IV genes, and three complex V genes. Table 10.3 lists the plethora of additional nDNA mitochondrial genes found to be mutant in patients including 29 OXPHOS complex assembly genes, 37 mitochondrial protein synthesis genes, 16 mitochondrial maintenance genes, 11 iron homeostasis genes, 10 CoQ synthesis genes, and miscellaneous additional genes encoding for proteins involved in other mitochondrial functions. These lists are available online through MITOMAP [16]. The number of clinically relevant nDNA mitochondrial genes is increasing rapidly.

## 10.7 MITOCHONDRIAL ETIOLOGY OF COMPLEX DISEASES

The extraordinary complexity of mtDNA and nDNA mitochondrial genetics and of mitochondrial–nuclear interactions and the commonality between primary mitochondrial disease phenotypes and the phenotypes of the common diseases strongly implicate mitochondrial dysfunction in the etiology of the common "complex" diseases. Mitochondrial dysfunction selectively impairs the organs with the highest bioenergetic demand—the brain, heart, muscle, kidney, endocrine tissues, and liver—and these are the same organs that are affected by the common diseases. Mitochondrial

**TABLE 10.3 Nonstructural Nuclear Genes—cont'd**

| Complex | Name | OMIM | Function | Chromosome | Inheritance | Clinical Phenotype | References |
|---|---|---|---|---|---|---|---|
| | *COX15* | 603646 | Heme A synthesis | 10q24 | AR | Early-onset hypertrophic cardiomyopathy, LS | t |
| | *COX20* | 614698 | Assembly | 1q44 | AR | Ataxia, muscle hypotonia | u |
| | *COA3* | 614775 | Assembly | 17q21.2 | AR | Neuropathy, exercise intolerance | v |
| | *COA5* | 613920 | Assembly | 2q11.2 | AR | Cardioencephalomyopathy | w |
| | *COA6* | 614772 | Assembly | 1q42.2 | AR | Cardioencephalomyopathy | x |
| | *LRPPRC* | 220111 | Assembly | 2p21-p16 | AR | French-Canadian LS | y |
| | *FASTKD2* | 612322 | Role in apoptosis | 2q33.3 | AR | Encephalomyopathy | z |
| | *TACO1* | 612958 | Translational activator of COX1 | 17q22-q24.2 | AR | LS | aa |
| Complex V | *ATPAF2* | 608918 | Assembly | 17p11.2 | AR | Early-onset encephalopathy, lactic acidosis | bb |
| | *TMEM70* | 604273 | Assembly | 8q21.11 | AR | Neonatal encephalopathy, cardiomyopathy | cc |
| MtDNA maintenance | *POLG (PEOA1)* | 174763 | Polymerase gamma mtDNA replication | 15q25 | AD-AR | Alpers syndrome, AD-PEO and AR-PEO, male infertility, SANDO* syndrome, SCAE* | dd |
| | *POLG2 (PEOA4)* | 610131 | Catalytic subunit of DNA polymerase gamma | 17q23-q24 | AD | AD-PEO | ee |
| | *ANT1 (PEOA2)* | 609283 | Adenine nucleotide translocator isoform 1 | 4q35 | AD-AR | AD-PEO, multiple mtDNA deletions | ff |
| | *MPV17* | 137960 | Regulation of mtDNA copy number | 2p23-p21 | AR | Hepatocerebral MDDS | gg |
| | *OPA1* | 165500 | Dynamin-related protein | 3q28-q29 | AD | AD-optic atrophy Multiple deletions | hh |
| | *MFN2* | 609260 | Mitofusin Mitochondrial fusion | 1p36-p35 | AD | Charcot-Marie-Tooth disease-2A2 (CMT2A2) Multiple deletions | ii |

TABLE 10.3 **Nonstructural Nuclear Genes—cont'd**

| Complex | Name | OMIM | Function | Chromosome | Inheritance | Clinical Phenotype | References |
|---|---|---|---|---|---|---|---|
| | *C10ORF2 (PEOA3)* | 609286 | Twinkle helicase | 10q24 | AD | AD-PEO, SANDO syndrome | jj |
| | *TYMP (ECGF1)* | 603041 | Thymidine phosphorylase | 22q13.32-qter | AR | MNGIE, mtDNA depletion | kk |
| | *DGUOK* | 601465 | Deoxyguanosine kinase Mitochondrial dNTP pool maintenance | 2p13 | AR | Hepatocerebral mtDNA depletion syndrome | ll |
| | *RRM2B (PEOA5)* | 604712 | Ribonucleotide reductase M2 B dNTP pool | 8q23.1 | AR | Encephalomyopathic renal tubulopathy MNGIE, AD-PEO | mm |
| | *SUCLA2* | 603921 | Succinate-CoA ligase, ADP-forming, beta Subunit | 13q12.2-q13 | AR | Encephalomyopathy with methylmalonic aciduria | nn |
| | *SUCLG1* | 611224 | Succinate-CoA ligase, alpha subunit | 2p11.2 | AR | Encephalomyopathy with methylmalonic aciduria | oo |
| | *TK2* | 188250 | Thymidine kinase Mitochondrial dNTP pool maintenance | 16q22 | AR | Myopathic mtDNA depletion | pp |
| | *TFAM* | 600438 | mitochondrial transcription factor A | 10q21.1 | AR | Encephalomyopathy mtDNA depletion | qq |
| | *FBXL4* | 605654 | mtDNA maintenance | 6q16.1-q16.2 | AR | Encephalomyopathy and myopathy mtDNA depletion | rr |
| | *MGME1* | 615084 | mtDNA maintenance | 20p11.23 | AR | CPEO and myopathy mtDNA depletion | ss |
| Mitochondrial import | *DDP* | 304700 | Protein import | Xq22 | X-linked | Deafness-dystonia or Mohr-Tranebjaerg syndrome | tt |
| | *DNAJC19* | 608977 | Protein import | 3q26.3 | AR | Cardiomyopathy, ataxia | uu |

Continued

**TABLE 10.3 Nonstructural Nuclear Genes—cont'd**

| Complex | Name | OMIM | Function | Chromosome | Inheritance | Clinical Phenotype | References |
|---|---|---|---|---|---|---|---|
| | *MRPS7* | 611974 | Mitochondrial translation | 17q25.1 | AR | Deafness, hepatic and renal failure | bbbb |
| | *MRPL12* | 602375 | Mitochondrial translation | 17q25.3 | AR | Growth retardation, encephalopathy | cccc |
| | *MRPS16* | 609204 | Mitochondrial translation | 10q22.1 | AR | Neonatal lactic acidosis corpus callosum agenesis | dddd |
| | *MRPS22* | 605810 | Mitochondrial translation | 3q23 | AR | Cardiomyopathy, tubulopathy | eeee |
| | *MRPL44* | 611849 | Mitochondrial translation | 2q36.1 | AR | Cardiomyopathy | ffff |
| Iron homeostasis | *FRDA (FXN)* | 606829 | Frataxin Trinuc.* repeat, | 9q13 | AR | Friedreich ataxia, neuropathy, cardiomyopathy, diabetes | gggg |
| | *ABCB7* | 301310 | Iron transport | Xq13.1-q13.3 | X-linked | X-linked sideroblastic anemia with ataxia | hhhh |
| | *GLRX5* | 205950 | Iron-sulfur cluster biosynthesis | 3p22.1 | AR | Sideroblastic anemia | iiii |
| | *ISCU* | 255125 | Iron-sulfur cluster biosynthesis | 12q23.3 | AR | Myopathy, lactic acidosis, exercise intolerance | jjjj |
| | *BOLA3* | 613183 | Iron-sulfur cluster biosynthesis | 2p13.1 | AR | Encephalomyopathy, cardiomyopathy | kkkk |
| | *NFU1* | 608100 | Iron-sulfur cluster biosynthesis | 2p13.3 | AR | Lactic acidosis multiple respiratory chain deficiency | llll |
| | *ISCA2* | 615317 | Iron-sulfur cluster biosynthesis | 14q24.3 | AR | Leukodystrophy | mmmm |
| | *IBA57* | 615316 | Iron-sulfur cluster biosynthesis | 1q42.13 | AR | Myopathy, encephalopathy | nnnn |
| | *LYRM4* | 613311 | Iron-sulfur cluster biosynthesis | 6p25.1 | AR | Lactic acidosis, failure to thrive | oooo |
| | *LYRM7* | 615831 | Iron-sulfur cluster biosynthesis | 5q23.3-q31.1 | AR | Encephalopathy, lactic acidosis | pppp |
| | *FDXL1* | 614585 | Iron-sulfur cluster biosynthesis | 19p13.2 | AR | Myopathy, lactic acidosis | qqqq |

**TABLE 10.3 Nonstructural Nuclear Genes—cont'd**

| Complex | Name | OMIM | Function | Chromosome | Inheritance | Clinical Phenotype | References |
|---|---|---|---|---|---|---|---|
| Coenzyme Q10 biogenesis | *COQ2* | 609825 | CoQ10 deficiency | 4q21-q22 | AR | Encephalomyopathy, nephropathy | rrrr |
| | *COQ4* | 612898 | CoQ10 deficiency | 9q34.13 | AR | Encephalomyopathy, mental retardation | ssss |
| | *COQ5* | 616359 | CoQ10 deficiency | 12q24.31 | AR | Encephalomyopathy, cerebellar ataxia | tttt |
| | *COQ6* | 614647 | CoQ10 deficiency | 14q24.3 | AR | Nephrotic syndrome, deafness | uuuu |
| | *COQ7* | 601683 | CoQ10 deficiency | 16p12.3 | AR | Hypotonia, cardiac hypertrophy | vvvv |
| | *COQ9* | 612837 | CoQ10 deficiency | 16q13 | AR | Neonatal lactic acidosis seizures, cardiomyopathy | wwww |
| | *APTX* | 606350 | CoQ10 deficiency | 9p13.3 | AR | Cerebellar ataxia Oculomotor apraxia | xxxx |
| | *PDSS1* | 607429 | CoQ10 deficiency | 10p12.1 | AR | Deafness, valvulopathy, mental retardation | yyyy |
| | *PDSS2* | 610564 | CoQ10 deficiency | 6q21 | AR | LS, nephrotic syndrome | zzzz |
| | *CABC1* | 606980 | CoQ10 deficiency | 1q42.2 | AR | Cerebellar ataxia, lactic acidosis | aaaaa |
| Chaperone function | *SPG7* | 607259 | Paraplegin ATPase protease | 16q24.3 | AR | Spastic paraplegia | bbbbb |
| | *HSPD1* | 118190 | Mitochondrial chaperone | 2q33.1 | AR | Spastic paraplegia, leukodystrophy | ccccc |
| Mitochondrial integrity | *DLP1* | 603850 | Mitochondrial and peroxisomal fission | 12p11.21 | AD | Microcephaly, abnormal brain Development, optic atrophy, Lactic acidosis | ddddd |
| | *G4.5 (Tafazzin)* | 302060 | Cardiolipin defect | Xq28 | X-linked | Barth syndrome, X-linked dilated cardiomyopathy | eeeee |
| | *RMRP* | 250250 | RNAse mitochondrial RNA processing | 9p13-p12 | AR | Metaphyseal chondrodysplasia or cartilage-hair hypoplasia | fffff |

Continued

short stature, partial sensorineural deafness, and peripheral neuropathy or with Leigh syndrome. Hum Mutat 2014;35:1285–89. Erratum Hum Mutat 2015:1236–1281.

[ddd]McLaughlin HM, Sakaguchi R, Liu C, Igarashi T, Pehlivan D, Chu K, Iyer R, Cruz P, Cherukuri PF, Hansen NF, Mullikin JC, Program NCS, Biesecker LG, Wilson TE, IonasescuV, Nicholson G, Searby C, Talbot K, Vance JM, Zuchner S, Szigeti K, Lupski JR, Hou YM, Green ED, Antonellis A. Compound heterozygosity for loss-of-function lysyl-tRNA synthetase mutations in a patient with peripheral neuropathy. Am J Hum Genet 2010;87:560–6, Santos-Cortez RL, Lee K, AzeemZ, Antonellis PJ, Pollock LM, Khan S, Irfanullah Andrade-Elizondo PB, Chiu I, Adams MD, Basit S, Smith JD, University of Washington Center for Mendelian G, Nickerson DA, McDermott BM, Jr, Ahmad W, Leal S. Mutations in KARS, encoding lysyl-tRNA synthetase, cause autosomal-recessive nonsyndromic hearing impairment DFNB89. Am J Hum Genet 2013;93:132–40.

[eee]Casey JP, McGettigan P, Lynam-Lennon N, McDermott M, Regan R, Conroy J, Bourke B, O'Sullivan J, Crushell E, Lynch S, Ennis S. Identification of a mutation in LARS as a novel cause of infantile hepatopathy. Mol Genet Metab 2012;106:351–8.

[fff]Pierce SB, Gersak K, Michaelson-Cohen R, Walsh T, Lee MK, Malach D, Klevit RE, King MC, Levy-Lahad E. Mutations in LARS2, encoding mitochondrial leucyl-tRNA synthetase, lead to premature ovarian failure and hearing loss in Perrault syndrome. Am J Hum Genet 2013;92:614–20.

[ggg]Sofou K, Kollberg G, Holmstrom M, Davila M, Darin N, Gustafsson CM, Holme E, Oldfors A, Tulinius M, Asin-Cayuela J. Whole exome sequencing reveals mutations in NARS2 and PARS2, encoding the mitochondrial asparaginyl-tRNA synthetase and prolyl-tRNA synthetase, in patients with Alpers syndrome. Mol Genet Genom Med 2015;3:59–68.

[hhh]see footnote ggg.

[iii]Edvardson S, Shaag A, Kolesnikova O, Gomori JM, Tarassov I, Einbinder T, Saada A, Elpeleg O. Deleterious mutation in the mitochondrial arginyl-transfer RNA synthetase gene is associated with pontocerebellar hypoplasia. Am J Hum Genet 2007;81:857–62.

[jjj]Belostotsky R, Ben-Shalom E, Rinat C, Becker-Cohen R, Feinstein S, Zeligson S, Segel R, Elpeleg O, Nassar S, Frishberg Y. Mutations in the mitochondrial seryl-tRNA synthetase cause hyperuricemia, pulmonary hypertension, renal failure in infancy and alkalosis, HUPRA syndrome. Am J Hum Genet 2011;88:193–200.

[kkk]Diodato D, Melchionda L, Haack TB, Dallabona C, Baruffini E, Donnini C, Granata T, Ragona F, Balestri P, Margollicci M, Lamantea E, Nasca A, Powell CA, Minczuk M, Strom TM, Meitinger T, Prokisch H, Lamperti C, Zeviani M, Ghezzi D. VARS2 and TARS2 mutations in patients with mitochondrial encephalomyopathies. Human Mutat 2014;35:983–9.

[lll]see footnote kkk.

[mmm]Riley LG, Cooper S, Hickey P, Rudinger-Thirion J, McKenzie M, Compton A, Lim SC, Thorburn D, Ryan MT, Giege R, Bahlo M, Christodoulou J. Mutation of the mitochondrial tyrosyl-tRNA synthetase gene, YARS2, causes myopathy, lactic acidosis, and sideroblastic anemia–MLASA syndrome. Am J Hum Genet 2010;87:52–9.

[nnn]Coenen MJ, Antonicka H, Ugalde C, Sasarman F, Rossi R, Heister JG, Newbold RF, Trijbels FJ, van den Heuvel LP, Shoubridge EA, Smeitink JA. Mutant mitochondrial elongation factor G1 and combined oxidative phosphorylation deficiency. N Engl J Med 2004;351:2080–6.

[ooo]Smeitink JA, Elpeleg O, Antonicka H, Diepstra H, Saada A, Smits P, Sasarman F, Vriend G, Jacob-Hirsch J, Shaag A, Rechavi G, Welling B, Horst J, Rodenburg RJ, van den Heuvel B, Shoubridge EA. Distinct clinical phenotypes associated with a mutation in the mitochondrial translation elongation factor EFTs. Am J Hum Genet 2006;79:869–77.

[ppp]Valente L, Tiranti V, Marsano RM, Malfatti E, Fernandez-Vizarra E, Donnini C, Mereghetti P, De Gioia L, Burlina A, Castellan C, Comi GP, Savasta S, Ferrero I, Zeviani M. Infantile encephalopathy and defective mitochondrial DNA translation in patients with mutations of mitochondrial elongation factors EFG1 and EFTu. Am J Hum Genet 2007;80:44–58. Erratum Am J Hum Genet 2007;80:580.

[qqq]Kopajtich R, Nicholls TJ, Rorbach J, Metodiev MD, Freisinger P, Mandel H, Vanlander A, Ghezzi D, Carrozzo R, Taylor RW, Marquard K, Murayama K, Wieland T, Schwarzmayr T, Mayr JA, Pearce SF, Powell CA, Saada A, Ohtake A, Invernizzi F, Lamantea E, Sommerville EW, Pyle A, Chinnery PF, Crushell E, Okazaki Y, Kohda M, Kishita Y, Tokuzawa Y, Assouline Z, Rio M, Feillet F, Mousson de Camaret B, Chretien D, Munnich A, Menten B, Sante T, Smet J, Regal L, Lorber A, Khoury A, Zeviani M, Strom TM, Meitinger T, Bertini ES, Van Coster R, Klopstock T, Rotig A, Haack TB, Minczuk M, Prokisch H. Mutations in GTPBP3 cause a mitochondrial translation defect associated with hypertrophic cardiomyopathy, lactic acidosis, and encephalopathy. Am J Hum Genet 2014;95:708–20.

[rrr]Haack TB, Haberberger B, Frisch EM, Wieland T, Iuso A, Gorza M, Strecker V, Graf E, Mayr JA, Herberg U, Hennermann JB, Klopstock T, Kuhn KA, Ahting U, Sperl W, Wilichowski E, Hoffmann GF, Tesarova M, Hansikova H, Zeman J, Plecko B, Zeviani M, Wittig I, Strom TM, Schuelke M, Freisinger P, Meitinger T, Prokisch H. Molecular diagnosis in mitochondrial complex I deficiency using exome sequencing. J Med Genet 2012;49:277–83.

[sss]Ghezzi D, Baruffini E, Haack TB, Invernizzi F, Melchionda L, Dallabona C, Strom TM, Parini R, Burlina AB, Meitinger T, Prokisch H, Ferrero I, Zeviani M. Mutations of the mitochondrial-tRNA modifier MTO1 cause hypertrophic cardiomyopathy and lactic acidosis. Am J Hum Genet 2012;90:1079–87.

[ttt]Powell CA, Kopajtich R, D'Souza AR, Rorbach J, Kremer LS, Husain RA, Dallabona C, Donnini C, Alston CL, Griffin H, Pyle A, Chinnery PF, Strom TM, Meitinger T, Rodenburg RJ, Schottmann G, Schuelke M, Romain N, Haller RG, Ferrero I, Haack TB, Taylor RW, Prokisch H, Minczuk M. TRMT5 mutations cause a defect in post-transcriptional modification of mitochondrial tRNA associated with multiple respiratory-chain deficiencies. Am J Hum Genet 2015;97:319–28.

[uuu]Metodiev MD, Thompson K, Alston CL, Morris AAM, He L, Assouline Z, Rio M, Bahi-Buisson N, Pyle A, Griffin H, Siira S, Filipovska A, Munnich A, Chinnery PF, McFarland R, Rotig A, Taylor RW. Recessive mutations in TRMT10C cause defects in mitochondrial RNA processing and multiple respiratory chain deficiencies. Am J Hum Genet 2016;98:993–1000. Erratum Am J Hum Genet 2016:1099–246.

[vvv]Zeharia A, Shaag A, Pappo O, Mager-Heckel AM, Saada A, Beinat M, Karicheva O, Mandel H, Ofek N, Segel R, Marom D, Rotig A, Tarassov I, Elpeleg O. Acute infantile liver failure due to mutations in the TRMU gene. Am J Hum Genet 2009;85:401–7.

[www]Valente L, Tiranti V, Marsano RM, Malfatti E, Fernandez-Vizarra E, Donnini C, Mereghetti P, De Gioia L, Burlina A, Castellan C, Comi GP, Savasta S, Ferrero I, Zeviani M. Infantile encephalopathy and defective mitochondrial DNA translation in patients with mutations of mitochondrial elongation factors EFG1 and EFTu. Am J Hum Genet 2007;80:44–58. Erratum: Am. J. Hum. Genet 2007;80:580.

[xxx]Fukumura S, Ohba C, Watanabe T, Minagawa K, Shimura M, Murayama K, Ohtake A, Saitsu H, Matsumoto N, Tsutsumi H. Compound heterozygous GFM2 mutations with Leigh syndrome complicated by arthrogryposis multiplex congenita. J Hum Genet 2015;60:509–13.

[yyy]Antonicka H, Ostergaard E, Sasarman F, Weraarpachai W, Wibrand F, Pedersen AM, Rodenburg RJ, van der Knaap MS, Smeitink JA, Chrzanowska-Lightowlers ZM, Shoubridge EA. Mutations in C12orf65 in patients with encephalomyopathy and a mitochondrial translation defect. Am J Hum Genet 2010;87:115–122, Shimazaki H, Takiyama Y, Ishiura H, Sakai C, Matsushima Y, Hatakeyama H, Honda J, Sakoe K, Naoi T, Namekawa M, Fukuda Y, Takahashi Y, Goto J, Tsuji S, Goto Y, Nakano I, Japan Spastic Paraplegia Research, C. A homozygous mutation of C12orf65 causes spastic paraplegia with optic atrophy and neuropathy (SPG55). J Med Genet 2012;49:777–84.

[zzz]Janer A, Antonicka H, Lalonde E, Nishimura T, Sasarman F, Brown GK, Brown RM, Majewski J, Shoubridge EA. An RMND1 Mutation causes encephalopathy associated with multiple oxidative phosphorylation complex deficiencies and a mitochondrial translation defect. Am J Hum Genet 2012;91:737–43.

[aaaa]Galmiche L, Serre V, Beinat M, Assouline Z, Lebre AS, Chretien D, Nietschke P, Benes V, Boddaert N, Sidi D, Brunelle F, Rio M, Munnich A, Rotig A. Exome sequencing identifies MRPL3 mutation in mitochondrial cardiomyopathy. Hum Mutat 2011;32:1225–31.

[bbbb]Menezes MJ, Guo Y, Zhang J, Riley LG, Cooper ST, Thorburn DR, Li J, Dong D, Li Z, Glessner J, Davis RL, Sue CM, Alexander SI, ArbuckleS, Kirwan P, Keating BJ, Xu X, Hakonarson H, Christodoulou J. Mutation in mitochondrial ribosomal protein S7 (MRPS7) causes congenital sensorineural deafness, progressive hepatic and renal failure and lactic acidemia. Hum Mol Genet 2015;24:2297–307.

[cccc]Serre V, Rozanska A, Beinat M, Chretien D, Boddaert N, MunnichA, Rotig A, Chrzanowska-Lightowlers ZM. Mutations in mitochondrial ribosomal protein MRPL12 leads to growth retardation, neurological deterioration and mitochondrial translation deficiency. Biochim Biophys Acta 2013;1832:1304–12.

[dddd]Miller C, Saada A, Shaul N, Shabtai N, Ben-Shalom E, Shaag A, Hershkovitz E, Elpeleg O. Defective mitochondrial translation caused by a ribosomal protein (MRPS16) mutation. Ann Neurol 2004;56:734–8.

[eeee]Saada A, Shaag A, Arnon S, Dolfin T, Miller C, Fuchs-Telem D, Lombes A, Elpeleg O. Antenatal mitochondrial disease caused by mitochondrial ribosomal protein (MRPS22) mutation. J Med Genet 2007;44:784–6.

[ffff]Carroll CJ, Isohanni P, Poyhonen R, Euro L, Richter U, Brilhante V, Gotz A, Lahtinen T, Paetau A, Pihko H, Battersby BJ, Tyynismaa H, Suomalainen A. Whole-exome sequencing identifies a mutation in the mitochondrial ribosome protein MRPL44 to underlie mitochondrial infantile cardiomyopathy. J Med Genet 2013;50:151–9.

[gggg]Campuzano V, Montermini L, Molto MD, Pianese L, Cossee M, Cavalcanti F, Monros E, Rodius F, Duclos F, Monticelli A, Zara F, Canizares J, Koutnikova H, Bidichandani SI, Gellera C, Brice A, Trouillas P, DeMichele G, Filla A, De Frutos R, Palau F, Patel P, DiDonato S, Mandel J, Cocozza S, Koenig M, Pandolfo M. Friedreich's ataxia: autosomal recessive disease caused by an intronic GAA triplet repeat expansion. Science 1996;271:1423–7, Rotig A, de Lonlay P, Chretien D, Foury F, Koenig M, Sidi D, Munnich A, Rustin P. Aconitase and mitochondrial iron-sulphur protein deficiency in Friedreich ataxia. Nat Genet 1997;17:215–7.

[hhhh]Allikmets R, Raskind WH, Hutchinson A, Schueck ND, Dean M, Koeller DM. Mutation of a putative mitochondrial iron transporter gene (ABC7) in X-linked sideroblastic anemia and ataxia (XLSA/A). Hum Mol Genet 1999;8:743–9.

[iiii]Guernsey DL, Jiang H, Campagna DR, Evans SC, Ferguson M, Kellogg MD, Lachance M, Matsuoka M, Nightingale M, Rideout A, Saint-Amant L, Schmidt PJ, Orr A, Bottomley SS, Fleming MD, Ludman M, Dyack S, Fernandez CV, Samuels ME. Mutations in mitochondrial carrier family gene SLC25A38 cause nonsyndromic autosomal recessive congenital sideroblastic anemia. Nat Genet 2009;41:651–3.

[jjjj]Mochel F, Knight MA, Tong WH, Hernandez D, Ayyad K, Taivassalo T, Andersen PM, Singleton A, Rouault TA, Fischbeck KH, Haller RG. Splice mutation in the iron-sulfur cluster scaffold protein ISCU causes myopathy with exercise intolerance. Am J Hum Genet 2008;82:652–60.

[kkkk]Cameron JM, Janer A, Levandovskiy V, Mackay N, Rouault TA, Tong WH, Ogilvie I, Shoubridge EA, Robinson BH. Mutations in iron-sulfur cluster scaffold genes NFU1 and BOLA3 cause a fatal deficiency of multiple respiratory chain and 2-oxoacid dehydrogenase enzymes. Am J Hum Genet 2011;89:486–95.

[llll]see footnote kkkk.

[mmmm]Al-Hassnan ZN, Al-Dosary M, Alfadhel M, Faqeih EA, Alsagob M, Kenana R, Almass R, Al-Harazi OS, Al-Hindi H, Malibari OI, Almutari FB, Tulbah S, Alhadeq F, Al-Sheddi T, Alamro R, AlAsmari A, Almuntashri M, Alshaalan H, Al-Mohanna FA, Colak D, Kaya N. ISCA2 mutation causes infantile neurodegenerative mitochondrial disorder. J Med Genet 2015;52:186–94.

[nnnn]Ajit Bolar N, Vanlander AV, Wilbrecht C, Van der Aa N, Smet J, De Paepe B, Vandeweyer G, Kooy F, Eyskens F, De Latter E, Delanghe G, Govaert P, Leroy JG, Loeys B, Lill R, Van Laer L, Van Coster R. Mutation of the iron-sulfur cluster assembly gene IBA57 causes severe myopathy and encephalopathy. Hum Mol Genet 2013;22:2590–602, Lossos A, Stumpfig C, Stevanin G, Gaussen M, Zimmerman BE, Mundwiller E, Asulin M, Chamma L, Sheffer R, Misk A, Dotan S, Gomori JM, Ponger P, Brice A, Lerer I, Meiner V, Lill R. Fe/S protein assembly gene IBA57 mutation causes hereditary spastic paraplegia. Neurology 2015;84:659–67.
[oooo]Lim SC, Friemel M, Marum JE, Tucker EJ, Bruno DL, Riley LG, Christodoulou J, Kirk EP, Boneh A, DeGennaro CM, Springer M, Mootha VK, Rouault TA, Leimkuhler S, Thorburn DR, Compton AG. Mutations in LYRM4, encoding iron-sulfur cluster biogenesis factor ISD11, cause deficiency of multiple respiratory chain complexes. Hum Mol Genet 2013;22:4460–73.
[pppp]Invernizzi F, Tigano M, Dallabona C, Donnini C, Ferrero I, Cremonte M, Ghezzi D, Lamperti C, Zeviani MA. homozygous mutation in LYRM7/MZM1L associated with early onset encephalopathy, lactic acidosis, and severe reduction of mitochondrial complex III activity. Hum Mutat 2013;34:1619–22.
[qqqq]Spiegel R, Saada A, Halvardson J, Soiferman D, Shaag A, Edvardson S, Horovitz Y, Khayat M, Shalev SA, Feuk L, Elpeleg O. Deleterious mutation in FDX1L gene is associated with a novel mitochondrial muscle myopathy. Eur J Hum Genet 2014;22:902–6.
[rrrr]Quinzii C, Naini A, Salviati L, Trevisson E, Navas P, Dimauro S, Hirano M. A mutation in *para*-hydroxybenzoate-polyprenyl transferase (COQ2) causes primary coenzyme Q10 deficiency. Am J Hum Genet 2006;78:345–9.
[ssss]Salviati L, Trevisson E, Rodriguez Hernandez MA, Casarin A, Pertegato V, Doimo M, Cassina M, Agosto C, Desbats MA, Sartori G, Sacconi S, Memo L, Zuffardi O, Artuch R, Quinzii C, Dimauro S, Hirano M, Santos-Ocana C, Navas P. Haploinsufficiency of COQ4 causes coenzyme Q10 deficiency. J Med Genet 2012;49:187–91.
[tttt]Malicdan MCV, Vilboux T, Ben-Zeev B, Guo J, Eliyahu A, Pode-Shakked B, Dori A, Kakani S, Chandrasekharappa SC, Ferreira CR, Shelestovich N, Marek-Yagel D, Pri-Chen H, Blatt I, Niederhuber JE, He L, Toro C, Taylor RW, Deeken J, Yardeni T, Wallace DC, Gahl WA, Anikster Y. A novel inborn error of the coenzyme Q10 biosynthesis pathway: cerebellar ataxia and static encephalomyopathy due to COQ5 C-methyltransferase deficiency. Hum Mutat 2018;39:69–79.
[uuuu]Heeringa SF, Chernin G, Chaki M, Zhou W, Sloan AJ, Ji Z, Xie LX, Salviati L, Hurd TW, Vega–WarnerV, Killen PD, Raphael Y, Ashraf S, Ovunc B, Schoeb DS, McLaughlin HM, Airik R, Vlangos CN, Gbadegesin R, Hinkes B, Saisawat P, Trevisson E, Doimo M, Casarin A, Pertegato V, Giorgi G, Prokisch H, Rotig A, Nurnberg G, Becker C, Wang S, Ozaltin F, Topaloglu R, Bakkaloglu A, Bakkaloglu SA, Muller D, Beissert A, Mir S, Berdeli A, Varpizen S, Zenker M, MatejasV, Santos-Ocana C, Navas P, Kusakabe T, Kispert A, Akman S, Soliman NA, Krick S, Mundel P, Reiser J, Nurnberg P, Clarke CF, Wiggins RC, Faul C, Hildebrandt F. COQ6 mutations in human patients produce nephrotic syndrome with sensorineural deafness. J Clin Investig 2011;121:2013–24.
[vvvv]Freyer C, Stranneheim H, Naess K, Mourier A, Felser A, Maffezzini C, Lesko N, Bruhn H, Engvall M, Wibom R, Barbaro M, Hinze Y, Magnusson M, Andeer R, Zetterstrom RH, von Dobeln U, Wredenberg A, Wedell A. Rescue of primary ubiquinone deficiency due to a novel COQ7 defect using 2,4-dihydroxybensoic acid. J Med Genet 2015;52:779–83.
[wwww]Duncan AJ, Bitner-Glindzicz M, Meunier B, Costello H, Hargreaves IP, Lopez LC, Hirano M, Quinzii CM, Sadowski MI, Hardy J, Singleton A, Clayton PT, Rahman S. A nonsense mutation in COQ9 causes autosomal-recessive neonatal-onset primary coenzyme Q10 deficiency: a potentially treatable form of mitochondrial disease. Am J Hum Genet 2009;84:558–66.
[xxxx]Quinzii CM, Kattah AG, Naini A, Akman HO, Mootha VK, DiMauro S, Hirano M. Coenzyme Q deficiency and cerebellar ataxia associated with an aprataxin mutation. Neurology 2005;64:539–41.
[yyyy]Mollet J, Giurgea I, Schlemmer D, Dallner G, Chretien D, Delahodde, A. Bacq D, de Lonlay P, Munnich A, Rotig A. Prenyldiphosphate synthase, subunit 1 (PDSS1) and OH-benzoate polyprenyltransferase (COQ2) mutations in ubiquinone deficiency and oxidative phosphorylation disorders. J Clin Investig 2007;117:765–72.
[zzzz]Lopez LC, Schuelke M, Quinzii CM, Kanki T, Rodenburg RJ, Naini A, Dimauro S, Hirano, M. Leigh syndrome with nephropathy and CoQ10 deficiency due to decaprenyl diphosphate synthase subunit 2 (PDSS2) mutations. Am J Hum Genet 2006;79:1125–9.
[aaaaa]Mollet J, Delahodde A, Serre V, Chretien D, Schlemmer D, Lombes A, Boddaert N, Desguerre I, de Lonlay P, de Baulny HO, Munnich A, Rotig A. CABC1 gene mutations cause ubiquinone deficiency with cerebellar ataxia and seizures. Am J Hum Genet 2008;82:623–30.
[bbbbb]Casari G, De Fusco M, Ciarmatori S, Zeviani M, Mora M, Fernandez P, De Michele G, Filla A, Cocozza S, Marconi R, Durr A, Fontaine B, Ballabio A. Spastic paraplegia and OXPHOS impairment caused by mutations in paraplegin, a nuclear-encoded mitochondrial metalloprotease. Cell 1998;93:973–83.
[ccccc]Magen D, Georgopoulos C, Bross P, Ang D, Segev Y, Goldsher D, Nemirovski A, Shahar E, Ravid S, Luder A, Heno B, Gershoni-Baruch R, Skorecki K, Mandel H. Mitochondrial hsp60 chaperonopathy causes an autosomal-recessive neurodegenerative disorder linked to brain hypomyelination and leukodystrophy. Am J Hum Genet 2008;83:30–42.
[ddddd]Waterham HR, Koster J, van Roermund CW, Mooyer PA, Wanders RJ, Leonard JV. A lethal defect of mitochondrial and peroxisomal fission. N Engl J Med 2007;356:1736–41.
[eeeee]Bione S, D'Adamo P, Maestrini E, Gedeon AK, Bolhuis PA, Toniolo DA novel X-linked gene, G4.5. is responsible for Barth syndrome. Nat Genet 1996;12:385–9, D'Adamo P, Fassone L, Gedeon A, Janssen EA, Bione S, Bolhuis PA, Barth PG, Wilson M, Haan E, Orstavik KH, Patton MA, Green AJ, Zammarchi E, Donati MA, Toniolo D. The X-linked gene G4.5 is responsible for different infantile dilated cardiomyopathies. Am J Hum Genet 1997;61:862–7.
[fffff]Ridanpaa M, Sistonen P, Rockas S, Rimoin DL, Makitie O, Kaitila I. Worldwide mutation spectrum in cartilage-hair hypoplasia: ancient founder origin of the major70A–>G mutation of the untranslated RMRP. Eur J Hum Genet 2002;10:439–47, Ridanpaa M, van Eenennaam H, Pelin K, Chadwick R, Johnson, C, Yuan B, vanVenrooij W, Pruijn G, Salmela R, Rockas S, Makitie O, Kaitila I, de

la Chapelle A. Mutations in the RNA component of RNase MRP cause a pleiotropic human disease, Cartilage-Hair Hypoplasia. Cell 2001;104:195–203.

[ggggg]Matthews PM, Marchington DR, Squier M, Land J, Brown RM, Brown GK. Molecular genetic characterization of an X-linked form of Leigh's syndrome. Ann Neurol 1993;33:652–5.

[hhhhh]Tiranti V, Briem E, Lamantea E, Mineri R, Papaleo E, Degioia L, Forlani F, Rinaldo P, Dickson P, Abu-Libdeh B, Cindro-Heberle L, Owaidha M, Jack RM, Christensen E, Burlina A, Zeviani M. ETHE1 mutations are specific to ethylmalonic encephalopathy. J Med Genet 2006;43:340–6, Tiranti V, D'Adamo P, Briem E, Ferrari G, Mineri R, Lamantea E, Mandel H, Balestri P, Garcia-Silva MT Vollmer B, Rinaldo P, Hahn SH, Leonard J, Rahman S, Dionisi-Vici C, Garavaglia B, Gasparini P, Zeviani M. Ethylmalonic encephalopathy is caused by mutations in ETHE1, a gene encoding a mitochondrial matrix protein. Am J Hum Genet 2004;74:239–52.

[iiiii]Bykhovskaya Y, Casas K, Mengesha E, Inbal A, Fischel-Ghodsian N. Missense mutation in pseudouridine synthase 1 (PUS1) causes mitochondrial myopathy and sideroblastic anemia (MLASA). Am J Hum Genet 2004;74:1303–08.

[jjjjj]Harel T, Yoon WH, Garone C, Gu S, Coban-Akdemir Z, Eldomery MK, Posey JE, Jhangiani SN, Rosenfeld JA, Cho MT, Fox S, Withers M Brooks SM, Chiang T, Duraine L, Erdin S, Yuan B, Shao Y, Moussallem E, Lamperti C, Donati MA, Smith JD, McLaughlin HM, Eng CM, Walkiewicz M, Xia F, Pippucci T, Magini P, Seri M, Zeviani M, Hirano M, Hunter JV, Srour M, Zanigni S, Lewis RA, Muzny DM, Lotze TE, Boerwinkle E, Baylor-Hopkins Center for Mendelian, G, University of Washington Center for Mendelian G, Gibbs RA, Hickey SE, Graham BH, Yang Y, Buhas D, Martin DM, Potocki L, Graziano C, Bellen HJ, Lupski JR Recurrent de novo and biallelic variation of ATAD3A, encoding a mitochondrial membrane protein, results in distinct neurological syndromes. Am J Hum Genet 2016;99:831–45.

OXPHOS is acutely sensitive to toxins linking mitochondrial genetics to environmental challenges. Hence, a mitochondrial etiology of the common diseases explains the organ-specificity, the complex genetics, the age-related onset and progression, and the environmental involvement that contribute to common diseases [9].

A large number of studies have correlated mtDNA haplogroups with predisposition to common "complex diseases" [81]. A meta-analysis of the literature [129] affirmed that European haplogroup H is at increased risk for Alzheimer disease (AD), haplogroup K and haplogroups J-T are protective for Parkinson disease (PD), Asian haplogroups A and F as well as the control region nt 16189C affected increased risk of type 2 diabetes mellitus (T2DM), European haplogroups J and W are associated with increased longevity, and nt 10398G is associated with breast cancer risk. Overall, European haplogroups H, K, and J have significant effects of common disease risk [129].

Considerable evidence has accumulated implicating mitochondrial dysfunction in diabetes and metabolic syndrome [150,151]. Among Europeans, there is the suggestion that haplogroup U [152,153] and haplogroups J/T [154–156] may be at increased risk of T2DM. However, several other studies have not observed associations between mtDNA haplogroups and T2DM in Europeans. This is likely the product of the interaction between mtDNA and nDNA factors, as a carefully controlled study of Israeli Jewish patients with diabetes revealed that subhaplogroup J1 is 2.4-fold underrepresented in T2DM patients whose parents are not diabetic versus patients whose parents are diabetic, suggesting that subhaplogroup J1 increases the risk of T2DM in association with additional nDNA genetic factors [155]. Among Japanese and Koreans, haplogroups D5 and F have been associated with increased risk and N9a with decreased risk of T2DM [157]. These results have been corroborated in gene expression studies in cybrids harboring D5, F, and N9a mtDNAs, which revealed that D5 and F had similar gene expression profiles while N9a cybrids showed an increased expression of OXPHOS genes relative to the D5 and F cybrids [158]. In the Chinese Uyghur, haplogroups H and D4 were at increased risk for T2DM [159], while in the Han Chinese, N9a and M8a [160] and M9 [161] were found to be at increased the risk of diabetes and N9a at increased risk of diabetic nephropathy. In a separate Japanese study, M8a was also associated with diabetes and B4c was associated with obesity [162]. In cybrid studies, haplogroup F1 was associated with significantly lower complex I activity and respiration than B4c and M9 [163]. In relation to single mtDNA nucleotide variants, the 10398G>A variant is associated with increased risk of T2DM in India [164] and Japan [161] while the mtDNA control region variant at nt 16189C in associated with increased T2DM risk when present in multiple mtDNA lineages [161,165].

Haplogroups have also been reported to affect the risk of developing associated diabetes sequelae [153,166]. For example, in Israeli Jews, H is associated with retinopathy, H3 with neuropathy, U3 with nephropathy, and V with renal failure, [167]. In Han Chinese, N9a was associated with increased risk of diabetic nephropathy [160], haplogroup T with increased the risk of coronary artery disease [168], T [169] and B4c [162] with obesity, and K with ischemic stroke [170], while N9a is protective of myocardial infarction in Japanese [171].

Haplogroups J and U are associated with predisposition to age-related macular degeneration [172] while the European haplogroup H mtDNAs correlate with reduced macular degeneration risk [173]. Haplogroups J and T are protective of osteoarthritis in the Spanish [174], and T and U have been associated with reduced sperm motility [175,176]. In athletes, haplogroups J and Uk are enriched in sprinters, haplogroup I is enriched in long distance runners in the Finnish [177], while haplogroup L0 is enriched in Kenyan elite athletes [178]. Haplogroup J has been correlated with increased longevity in Europeans [179–182] and D with increased longevity in Asians [183]. The longevity-associated Asian haplogroup D4b2 harbors a homoplasmic polymorphism in the 12S rRNA-embedded MOTS-c peptide sequence (K14Q) [184], and the normal MOTS-c peptide has been shown to act systemically to suppress inflammation [185].

Differences in mtDNA content have also been correlated with metabolic disease. These include diabetes [186,187], cardiometabolic phenotypes (lipids, glycemic and inflammatory traits, blood pressure) [188], and ischemic stroke [189].

Evidence for a mitochondrial etiology of AD and PD continues to accumulate. Numerous publications have reported systemic mitochondrial complex IV defects in AD [190–193] and complex I defects in PD [194–196]. Direct evidence that mtDNA variants can contribute to AD and PD came with the discovery that an mtDNA $tRNA^{Gln}$ nt 4336A>G mutation is enriched in AD (3.2%), PD (5.3%), and AD/PD (6.8%) patients versus controls (0.4%). This variant arose in Europe about 8500 to 17,000 years ago and is associated with mtDNA haplogroup H5a [197], and its AD/PD association has been confirmed multiple times [151,198,199]. Since the $tRNA^{Gln}$ nt 4336A>G mutation would partially impair mitochondrial protein synthesis, it would preferentially impair the synthesis of the mtDNA complex IV (*COI-III*) and complex I (*ND1-6*) genes, thus predisposing to AD and PD. The mtDNA haplogroup K (Uk) also imparts increased risk of AD [200], and haplogroup Uk containing 143B($TK^-$) cybrids have reduced COX activity [127]. Finally, mitochondrial dysfunction has been verified in cultured cells from AD and Down syndrome with dementia patients [201].

Predisposing mtDNA variants can then be exacerbated by the age-related accumulation of developmental and somatic mutations. Somatic mtDNA mutations have been shown to be increased in AD brains and Down syndrome brains with dementia and to be associated with altered mtDNA transcript levels and mtDNA copy number [72,202–207].

A strong case can also be made for a mitochondrial etiology of PD. A subset of PD patients shows autosomal inheritance, and the cloning of these associated nDNA genes has revealed that most, if not all, are involved in protection of mitochondrial function through mitophagy, mitochondrial inner membrane integrity, mitochondrial redox control, and mitigation of oxidative damage [205,208]. Haplogroup H has been associated with increased risk, and haplogroups J and Uk with decreased risk, for developing PD [209–211], and somatic mtDNA mutations have been found increased in PD brains [205] and in neurons of PD patients [74,75].

Like PD, amyotrophic lateral sclerosis (ALS) is genetically complex. ALS also shows strong indications of underlying mitochondrial dysfunction [212] with ALS fibroblasts showing an overall dysregulated bioenergetics [213]. Several ALS nuclear gene mutations have been found to result in mitochondrial alterations. ALS Cu/Zn superoxide dismutase (*SOD1*) mutations result in multiple alterations in mitochondria function [214,215], and the ALS R15L dominant mutation in the *CHCHD10* gene causes ALS, frontotemporal dementia, and PD and results in defects in complex I assembly, impaired respiration, and impaired mitochondrial bioenergetics–driven proliferation [216]. The protein of the ALS-associated gene, *TDP-43*, is partially localized in the mitochondrion and binds to mitochondrial tRNAs and precursor RNAs encoded by the L-strand, thus regulating the processing of mitochondrial transcripts [217].

Mitochondrial dysfunction has repeatedly been implicated in autism spectrum disorder (ASD). A meta-analysis of publications reporting OXPHOS dysfunction and alterations in mitochondria-related metabolites has concluded that ASD children have a spectrum of mitochondrial alterations of differing severities [218]. Extensive nDNA genomic studies of ASD patients have identified a variety of heterozygous copy number variants (CNVs) [219,220] and loss-of-function (LOF) mutations [221–223]. However, each variant accounts for a only few cases [221,223–226], and the cumulative risk currently accounted for by nDNA mutations is only about 20% [226]. LOF mutations in genes from ASD patients overlap with metabolic and cardiac abnormalities [221] as well as genes implicated in intellectual disability, attention-deficit/hyperactivity disorders, and schizophrenia [226–228]. Hence, ASD is a polygenic disorder [228] with the associated genes being involved in metabolic, neurological, and cardiovascular disorders.

An in-depth study of a well-characterized autism cohort revealed that many of the CNVs delete bioenergetics genes [229] and a number of ASD LOF gene mutations have been found to affect mitochondria-related functions [221,223,230,231]). More recently, both mtDNA haplogroups and heteroplasmic mtDNA mutations have been associated with ASD. Relative to the most common European haplogroup (H-HV), European haplogroups I, J, K, X, T, and U, and Asian-Native American haplogroups A and M are at significantly increased risk of ASD with odds ratios (ORs) ranging from 2.06 to 3.55. Since haplogroups I, J, K, X, T, and U represent about 55% of the European mtDNA lineages, mtDNA haplogroups may make a major contribution to the differential risk of ASD [232]. Moreover, ASD probands are more likely to harbor deleterious heteroplasmic mtDNA mutations than are their unaffected siblings. Nonsynonymous mutations were enriched about 1.5-fold and potentially pathogenic mutations were enriched about 2.2-fold in ASD probands, giving an OR of 2.55 [233].

There is growing evidence that psychiatric disorders also have a bioenergetics etiology [234]. Several meta-analyses and reviews have now reported that mitochondrial dysfunction is associated with neuropsychiatric disease [235–238]. A recent compilation of proteomic data from autopsy brains of neuropsychiatric patients revealed marked changes in mitochondrial and bioenergetic proteins, with 92 differentially expressed proteins in schizophrenia, 95 in bipolar disorder, and 41 in major depressive disorder, five being common to all three [239].

A role for mitochondrial dysfunction is also accumulating for acute illnesses such as infection, inflammation, and trauma. Haplogroup H has been associated with protection against sepsis [240], haplogroup U with increased serum IgE levels [241], and haplogroup Uk with reduced AIDS progression and preservation of CD4 T lymphocyte numbers [242]. Mitochondrial dysfunction is also being linked to inflammation. Mitochondrial dysfunction prompts the release of mitochondrial damage-associated molecular patterns (DAMPs) including mtDNA, *N*-formyl peptides, and cardiolipin [243], and these activate the mitochondrially associated "inflammasome." The inflammasome is composed of the NLRP3, ASC, and pro–caspase-1 polypeptides, and on binding of DAMPs, caspase-1 is activated; it cleaves pro–interleukin (IL)-1β, and the activated IL-1β activates the NF-κB/AP-1 pathway leading to inflammation [244,245]. Mitochondrial dysfunction also alters T-cell biology. T regulatory cells are oxidative while T effector cells are glycolytic, so partial OXPHOS defects preferentially impair T regulatory cells, thus enhancing T effector activity and inflammation [246]. Mitochondrial function is also compromised by trauma as indicated by altered cellular mtDNA copy number [247].

Thus, mitochondrial biology is being implicated in an increasingly broad array of previously intractable clinical problems.

## 10.8 DIAGNOSIS OF MITOCHONDRIAL DISEASES

One of the major challenges of mitochondrial medicine is the accurate diagnosis of mitochondrial disease due to the extreme clinical and genetic heterogeneities. Traditional approaches for diagnosing mitochondrial diseases involved identifying patients with neuromuscular disease, muscle biopsy with histological and ultrastructure analyses, muscle OXPHOS enzyme assays, and $^{31}P$ MRI spectroscopy with exercise stress tests [8]. Several schemes have been proposed for diagnosing primary mitochondrial diseases. A consensus statement was reported [248–251].

With the discovery of mtDNA molecular defects [5–7], molecular diagnostics has become increasingly important. The use of massive parallel sequencing technologies is now permitting the analysis of both mtDNA and nDNA gene sequence variation in a fast and efficient way. This has significantly improved the molecular diagnosis of primary mitochondrial diseases and led to the discovery of many new mitochondrial genes and

8993T>G (L156R) mutation. One implanted ST blastocyst gave rise to an apparently normal boy carrying between 2% and 9% 8993G mutant mtDNA in his various tissues [315,316].

## 10.11 CONCLUSION

It is now 30 years since the first molecular causes of mitochondrial disease were reported. In the interim, an entirely new field of mitochondrial medicine has arisen, which now explains the molecular and biochemical bases of a broad new class of bioenergetic diseases. More importantly, the knowledge gained from the identification and characterization of primary mitochondrial diseases has provided new perspectives on the etiology and genetics on the common metabolic and degenerative diseases and aging. The development of therapeutic modalities for the treatment of primary mitochondrial diseases thus has the potential for application to the common diseases. Hence, the mitochondrial medicine paradigm may revolutionize the conceptual framework of Western medicine in the twenty-first century with broad implications for the health and well-being of patients.

## ACKNOWLEDGMENTS

This work was supported by National Institutes of Health grants MH108592, NS021328, CA182384, and OD010944 and U.S. Department of Defense grant W81XWH-16-1-0400 & 0401 awarded to D.C.W. and supported by grants from: "Fondation pour la Recherche Médicale" DPM20121125554, and AFM Mitoscreen 17122 awarded to V.P.

## REFERENCES

[1] Picard M, Wallace DC, Burelle Y. The rise of mitochondria in medicine. Mitochondrion 2016;30:105–16.
[2] Luft R, Ikkos D, Palmieri G, Ernster L, Afzelius BA. A case of severe hypermetabolism of nonthyroid origin with a defect in the maintenance of mitochondrial respiratory control: a correlated clinical, biochemical, and morphological study. J Clin Investig 1962;41:1776–804.
[3] DiMauro S. Mitochondrial encephalomyopathies. In: Rosenberg, Prusiner SB, DiMauro S, Barchi RL, Kunkel LM, editors. The Molecular and Genetic Basis of Neurological Disease. Stoneham (MA): Butterworth-Heinemann; 1993. p. 665–94.
[4] Morgan-Hughes JA, Hanna MG. Mitochondrial encephalomyopathies: the enigma of genotype versus phenotype. Biochim Biophys Acta 1999;1410:125–45.
[5] Holt IJ, Harding AE, Morgan-Hughes JA. Deletions of muscle mitochondrial DNA in patients with mitochondrial myopathies. Nature 1988;331:717–9.
[6] Shoffner JM, Lott MT, Lezza AM, Seibel P, Ballinger SW, Wallace DC. Myoclonic epilepsy and ragged-red fiber disease (MERRF) is associated with a mitochondrial DNA tRNA$^{Lys}$ mutation. Cell 1990;61:931–7.
[7] Wallace DC, Singh G, Lott MT, Hodge JA, Schurr TG, Lezza AM, Elsas LJ, Nikoskelainen EK. Mitochondrial DNA mutation associated with Leber's hereditary optic neuropathy. Science 1988a;242:1427–30.
[8] Wallace DC, Zheng X, Lott MT, Shoffner JM, Hodge JA, Kelley RI, Epstein CM, Hopkins LC. Familial mitochondrial encephalomyopathy (MERRF): Genetic, pathophysiological, and biochemical characterization of a mitochondrial DNA disease. Cell 1988b;55:601–10.
[9] Wallace DC. Mitochondrial bioenergetic etiology of disease. J Clin Investig 2013;123:1405–12.
[10] Chinnery PF, Johnson MA, Wardell TM, Singh-Kler R, Hayes C, Brown DT, Taylor RW, Bindoff LA, Turnbull DM. The epidemiology of pathogenic mitochondrial DNA mutations. Ann Neurol 2000;48:188–93.
[11] Cree LM, Samuels DC, Chinnery PF. The inheritance of pathogenic mitochondrial DNA mutations. Biochim Biophys Acta 2009;1792:1097–102.
[12] Gorman GS, Schaefer AM, Ng Y, Gomez N, Blakely EL, Alston CL, Feeney C, Horvath R, Yu-Wai-Man P, Chinnery PF, Taylor RW, Turnbull DM, McFarland R. Prevalence of nuclear and mitochondrial DNA mutations related to adult mitochondrial disease. Ann Neurol 2015;77:753–9.
[13] Schaefer AM, McFarland R, Blakely EL, He L, Whittaker RG, Taylor RW, Chinnery PF, Turnbull DM. Prevalence of mitochondrial DNA disease in adults. Ann Neurol 2008;63:35–9.
[14] Elliott HR, Samuels DC, Eden JA, Relton CL, Chinnery PF. Pathogenic mitochondrial DNA mutations are common in the general population. Am J Hum Genet 2008;83:254–60.
[15] Wallace DC, Lott MT, Procaccio V. Mitochondrial Medicine: The Mitochondrial Biology and Genetics of Metabolic and Degenerative Diseases, Cancer, and Aging (Chapter 11). In: Rimoin DL, Pyeritz RE, Korf BR, editors. Emery and Rimoin's Principles and Practice of Medical Genetics. 6th ed. Philadelphia: Churchill Livingstone Elsevier; 2013. p. 1–153.
[16] MITOMAP. A Human Mitochondrial Genome Database. 2018. http://www.mitomap.org.

[17] Pham TD, Pham PQ, Li J, Letai AG, Wallace DC, Burke PJ. Cristae remodeling causes acidification detected by integrated graphene sensor during mitochondrial outer membrane permeabilization. Sci Rep 2016;6:35907.
[18] Mayr JA. Lipid metabolism in mitochondrial membranes. J Inherit Metab Dis 2015;38:137–44.
[19] Bricker DK, Taylor EB, Schell JC, Orsak T, Boutron A, Chen YC, Cox JE, Cardon CM, Van Vranken JG, Dephoure N, Redin C, Boudina S, Gygi SP, Brivet M, Thummel CS, Rutter J. A mitochondrial pyruvate carrier required for pyruvate uptake in yeast, Drosophila, and humans. Science 2012;337:96–100.
[20] Wellen KE, Hatzivassiliou G, Sachdeva UM, Bui TV, Cross JR, Thompson CB. ATP-citrate lyase links cellular metabolism to histone acetylation. Science 2009;324:1076–80.
[21] Gururaja Rao S. Mitochondrial changes in cancer. Handb Exp Pharmacol 2017;240:211–27.
[22] Wallace DC. Mitochondria and cancer. Nat Rev Cancer 2012;12:685–98.
[23] Denton RM. Regulation of mitochondrial dehydrogenases by calcium ions. Biochim Biophys Acta 2009;1787:1309–16.
[24] Palmieri F. The mitochondrial transporter family SLC25: identification, properties and physiopathology. Mol Aspects Med 2013;34:465–84.
[25] Palmieri F. Mitochondrial transporters of the SLC25 family and associated diseases: a review. J Inherit Metab Dis 2014;37:565–75.
[26] Palmieri F, Monne M. Discoveries, metabolic roles and diseases of mitochondrial carriers: A review. Biochim Biophys Acta 2016;1863:2362–78.
[27] Valsecchi F, Ramos-Espiritu LS, Buck J, Levin LR, Manfredi G. cAMP and mitochondria. Physiology (Bethesda) 2013;28:199–209.
[28] Ghafourifar P, Cadenas E. Mitochondrial nitric oxide synthase. Trends Pharm Sci 2005;26:190–5.
[29] Wirth C, Brandt U, Hunte C, Zickermann V. Structure and function of mitochondrial complex I.. Biochim Biophys Acta 2016;1857:902–14.
[30] Zhu J, Vinothkumar KR, Hirst J. Structure of mammalian respiratory complex I. Nature 2016;536:354–8.
[31] Iwata S, Lee JW, Okada K, Lee JK, Iwata M, Rasmussen B, Link TA, Ramaswamy S, Jap BK. Complete structure of the 11-subunit bovine mitochondrial cytochrome bc1 complex. Science 1998;281:64–71.
[32] Solmaz SR, Hunte C. Structure of complex III with bound cytochrome c in reduced state and definition of a minimal core interface for electron transfer. J Biol Chem 2008;283:17542–9.
[33] Saraste M. Oxidative phosphorylation at the fin de siecle. Science 1999;283:1488–93.
[34] Tsukihara T, Aoyama H, Yamashita E, Tomizaki T, Yamaguchi H, Shinzawa-Itoh K, Nakashima R, Yaono R, Yoshikawa S. The whole structure of the 13-subunit oxidized cytochrome c oxidase at 2.8 A. Science 1996;272:1136–44.
[35] Yoshikawa S, Shimada A. Reaction mechanism of cytochrome c oxidase. Chem Rev 2015;115:1936–89.
[36] Letts JA, Sazanov LA. Clarifying the supercomplex: the higher-order organization of the mitochondrial electron transport chain. Nat Struct Mol Biol 2017;24:800–8.
[37] Schagger H. Blue-Native Gels to Isolate Protein Complexes from Mitochondria. In: Pon LA, Schon EA, editors. Mitochondria. San Diego (CA): Academic Press; 2001. p. 231–44.
[38] Schagger H, Pfeiffer K. Supercomplexes in the respiratory chains of yeast and mammalian mitochondria. EMBO J 2000;19:1777–83.
[39] Wu M, Gu J, Guo R, Huang Y, Yang M. Structure of mammalian respiratory supercomplex I1III2IV1. Cell 2016;167:1598–609. e1510.
[40] Walker JE. The ATP synthase: the understood, the uncertain and the unknown. Biochem Soc Trans 2013;41:1–16.
[41] He J, Ford HC, Carroll J, Ding S, Fearnley IM, Walker JE. Persistence of the mitochondrial permeability transition in the absence of subunit c of human ATP synthase. Proc Natl Acad Sci USA 2017;114:3409–14.
[42] Martin JL, Ishmukhametov R, Hornung T, Ahmad Z, Frasch WD. Anatomy of F1-ATPase powered rotation. Proc Natl Acad Sci USA 2014;111:3715–20.
[43] Kaukonen J, Juselius JK, Tiranti V, Kyttala A, Zeviani M, Comi GP, Keranen S, Peltonen L, Suomalainen A. Role of adenine nucleotide translocator 1 in mtDNA maintenance. Science 2000;289:782–5.
[44] Palmieri L, Alberio S, Pisano I, Lodi T, Meznaric-Petrusa M, Zidar J, Santoro A, Scarcia P, Fontanesi F, Lamantea E, Ferrero I, Zeviani M. Complete loss-of-function of the heart/muscle-specific adenine nucleotide translocator is associated with mitochondrial myopathy and cardiomyopathy. Hum Mol Genet 2005;14:3079–88.
[45] Strauss KA, Dubiner L, Simon M, Zaragoza M, Sengupta PP, Li P, Narula N, Dreike S, Platt J, Procaccio V, Ortiz-Gonzalez XR, Puffenberger EG, Kelley RI, Morton DH, Narula J, Wallace DC. Severity of cardiomyopathy associated with adenine nucleotide translocator-1 deficiency correlates with mtDNA haplogroup. Proc Natl Acad Sci USA 2013;110:3253–458.

[46] Baughman JM, Perocchi F, Girgis HS, Plovanich M, Belcher-Timme CA, Sancak Y, Bao XR, Strittmatter L, Goldberger O, Bogorad RL, Koteliansky V, Mootha VK. Integrative genomics identifies MCU as an essential component of the mitochondrial calcium uniporter. Nature 2011;476:341–5.

[47] De Stefani D, Raffaello A, Teardo E, Szabo I, Rizzuto R. A forty-kilodalton protein of the inner membrane is the mitochondrial calcium uniporter. Nature 2011;476:336–40.

[48] Bhola PD, Letai A. Mitochondria - judges and executioners of cell death sentences. Mol Cell 2016;61:695–704.

[49] Sarosiek KA, Ni Chonghaile T, Letai A. Mitochondria: gatekeepers of response to chemotherapy. Trends Cell Biol 2013;23:612–9.

[50] Alavian KN, Beutner G, Lazrove E, Sacchetti S, Park HA, Licznerski P, Li H, Nabili P, Hockensmith K, Graham M, Porter Jr GA, Jonas EA. An uncoupling channel within the c-subunit ring of the F1FO ATP synthase is the mitochondrial permeability transition pore. Proc Natl Acad Sci USA 2014;111:10580–5.

[51] Bernardi P, Rasola A, Forte M, Lippe G. The mitochondrial permeability transition pore: channel formation by F-ATP synthase, integration in signal transduction, and role in pathophysiology. Physiol Rev 2015;95:1111–55.

[52] Giorgio V, von Stockum S, Antoniel M, Fabbro A, Fogolari F, Forte M, Glick GD, Petronilli V, Zoratti M, Szabo I, Lippe G, Bernardi P. Dimers of mitochondrial ATP synthase form the permeability transition pore. Proc Natl Acad Sci USA 2013;110:5887–92.

[53] Jonas EA, Porter Jr GA, Beutner G, Mnatsakanyan N, Alavian KN. Cell death disguised: the mitochondrial permeability transition pore as the c-subunit of the F(1)F(O) ATP synthase. Pharmacol Res 2015;99:382–92.

[54] Teixeira FK, Sanchez CG, Hurd TR, Seifert JR, Czech B, Preall JB, Hannon GJ, Lehmann R. ATP synthase promotes germ cell differentiation independent of oxidative phosphorylation. Nat Cell Biol 2015;17:689–96.

[55] Shanmughapriya S, Rajan S, Hoffman NE, Higgins AM, Tomar D, Nemani N, Hines KJ, Smith DJ, Eguchi A, Vallem S, Shaikh F, Cheung M, Leonard NJ, Stolakis RS, Wolfers MP, Ibetti J, Chuprun JK, Jog NR, Houser SR, Koch WJ, Elrod JW, Madesh M. SPG7 Is an essential and conserved component of the mitochondrial permeability transition pore. Mol Cell 2015;60:47–62.

[56] van der Bliek AM, Shen Q, Kawajiri S. Mechanisms of mitochondrial fission and fusion. Cold Spring Harbor. Perspect Biol 2013;5.

[57] Narendra D, Walker JE, Youle R. Mitochondrial quality control mediated by PINK1 and Parkin: links to parkinsonism. Cold Spring Harbor Perspect Biol 2012;4:a011338.

[58] Wallace DC, Bunn CL, Eisenstadt JM. Cytoplasmic transfer of chloramphenicol resistance in human tissue culture cells. Journal of Cell Biology 1975;67:174–88.

[59] Shuster RC, Rubenstein AJ, Wallace DC. Mitochondrial DNA in anucleate human blood cells. Biochem Biophys Res Commun 1988;155:1360–5.

[60] Wallace DC. Cytoplasmic inheritance of chloramphenicol resistance in mammalian cells. Chapter 12. In: Shay JW, editor. Techniques in Somatic Cell Genetics. New York: Plenum Press; 1982. p. 159–87.

[61] Wallace DC, Bunn CL, Eisenstadt JM. Mitotic segregation of cytoplasmic inherited genes for chloramphenicol resistance in mammalian cells. II: Fusions with human cell lines. Somatic Cell Genet 1977;3:93–119.

[62] Wallace DC, Eisenstadt JM. The expression of cytoplasmically inherited genes for chloramphenicol resistance in interspecific somatic cell hybrids and cybrids. Somatic Cell Genet 1979;5. 573–396.

[63] Wallace DC, Pollack Y, Bunn CL, Eisenstadt JM. Cytoplasmic inheritance in mammalian tissue culture cells. In Vitro 1976;12:758–76.

[64] Case JT, Wallace DC. Maternal inheritance of mitochondrial DNA polymorphisms in cultured human fibroblasts. Somatic Cell Genet 1981;7:103–8.

[65] Giles RE, Blanc H, Cann HM, Wallace DC. Maternal inheritance of human mitochondrial DNA. Proc Natl Acad Sci USA 1980;77:6715–9.

[66] Lee C, Zeng J, Drew BG, Sallam T, Martin-Montalvo A, Wan J, Kim SJ, Mehta H, Hevener AL, de Cabo R, Cohen P. The mitochondrial-derived peptide MOTS-c promotes metabolic homeostasis and reduces obesity and insulin resistance. Cell Metab 2015;21:443–54.

[67] Yen K, Lee C, Mehta H, Cohen P. The emerging role of the mitochondrial-derived peptide humanin in stress resistance. J Mol Endocrinol 2013;50:R11–9.

[68] Brown WM, Prager EM, Wan A, Wilson AC. Mitochondrial DNA sequences in primates: tempo and mode of evolution. J Mol Evol 1982;18:225–39.

[69] Neckelmann N, Li K, Wade RP, Shuster R, Wallace DC. cDNA sequence of a human skeletal muscle ADP/ATP translocator: lack of a leader peptide, divergence from a fibroblast translocator cDNA, and coevolution with mitochondrial DNA genes. Proc Natl Acad Sci USA 1987;84:7580–4.

[70] Wallace DC, Ye JH, Neckelmann SN, Singh G, Webster KA, Greenberg BD. Sequence analysis of cDNAs for the human and bovine ATP synthase b-subunit: mitochondrial DNA genes sustain seventeen times more mutations. Curr Genet 1987;12:81–90.

[71] Corral-Debrinski M, Horton T, Lott MT, Shoffner JM, Beal MF, Wallace DC. Mitochondrial DNA deletions in human brain: regional variability and increase with advanced age. Nat Genet 1992a;2:324–9.

[72] Coskun PE, Wyrembak J, Derbereva O, Melkonian G, Doran E, Lott IT, Head E, Cotman CW, Wallace DC. Systemic mitochondrial dysfunction and the etiology of Alzheimer's disease and down syndrome dementia. J Alzheimers Dis 2010;20(Suppl. 2):S293–310.

[73] Muller-Hocker J, Seibel P, Schneiderbanger K, Kadenbach B. Different in situ hybridization patterns of mitochondrial DNA in cytochrome c oxidase-deficient extraocular muscle fibres in the elderly. Virchows Arch A Pathol Anat Histopathol 1993;422:7–15.

[74] Bender A, Krishnan KJ, Morris CM, Taylor GA, Reeve AK, Perry RH, Jaros E, Hersheson JS, Betts J, Klopstock T, Taylor RW, Turnbull DM. High levels of mitochondrial DNA deletions in substantia nigra neurons in aging and Parkinson disease. Nat Genet 2006;38:515–7.

[75] Kraytsberg Y, Kudryavtseva E, McKee AC, Geula C, Kowall NW, Khrapko K. Mitochondrial DNA deletions are abundant and cause functional impairment in aged human substantia nigra neurons. Nature Genet 2006;38:518–20.

[76] Khrapko K, Bodyak N, Thilly WG, van Orsouw NJ, Zhang X, Coller HA, Perls TT, Upton M, Vijg J, Wei JY. Cell-by-cell scanning of whole mitochondrial genomes in aged human heart reveals a significant fraction of myocytes with clonally expanded deletions. Nucleic Acids Res 1999;27:2434–41.

[77] Wallace DC, Chalkia D. Mitochondrial DNA genetics and the heteroplasmy conundrum in evolution and disease. Cold Spring Harbor. Perspect Biol 2013;5:a021220.

[78] Fan W, Waymire K, Narula N, Li P, Rocher C, Coskun PE, Vannan MA, Narula J, MacGregor GR, Wallace DC. A mouse model of mitochondrial disease reveals germline selection against severe mtDNA mutations. Science 2008;319:958–62.

[79] Sharpley MS, Marciniak C, Eckel-Mahan K, McManus MJ, Crimi M, Waymire K, Lin CS, Masubuchi S, Friend N, Koike M, Chalkia D, MacGregor GR, Sassone-Corsi P, Wallace DC. Heteroplasmy of mouse mtDNA Is genetically unstable and results in altered behavior and cognition. Cell 2012;151:333–43.

[80] Stewart JB, Freyer C, Elson JL, Wredenberg A, Cansu Z, Trifunovic A, Larsson NG. Strong purifying selection in transmission of mammalian mitochondrial DNA. PLoS Biol 2008;6:e10.

[81] Wallace DC. Mitochondrial DNA variation in human radiation and disease. Cell 2015;163:33–8.

[82] Calvo SE, Mootha VK. The mitochondrial proteome and human disease. Annu Rev Genom Hum Genet 2010;11:25–44.

[83] Stojanovski D, Bohnert M, Pfanner N, van der Laan M. Mechanisms of protein sorting in mitochondria. Cold Spring Harbor. Perspect Biol 2012;4:a011320.

[84] Li K, Warner CK, Hodge JA, Minoshima S, Kudoh J, Fukuyama R, Maekawa M, Shimizu Y, Shimizu N, Wallace DC. A human muscle adenine nucleotide translocator gene has four exons, is located on chromosome 4, and is differentially expressed. J Biol Chem 1989;264:13998–4004.

[85] Neckelmann N, Warner CK, Chung A, Kudoh J, Minoshima S, Fukuyama R, Maekawa M, Shimizu Y, Shimizu N, Liu JD, Wallace DC. The human ATP synthase beta subunit gene: sequence analysis, chromosome assignment, and differential expression. Genomics 1989;5:829–43.

[86] Procaccio V, Mousson B, Beugnot R, Duborjal H, Feillet F, Putet G, Pignot-Paintrand I, Lombes A, De Coo R, Smeets H, Lunardi J, Issartel JP. Nuclear DNA origin of mitochondrial complex I deficiency in fatal infantile lactic acidosis evidenced by transnuclear complementation of cultured fibroblasts. J Clin Investig 1999;104:83–92.

[87] Koopman WJ, Willems PH, Smeitink JA. Monogenic mitochondrial disorders. N Engl J Med 2012;366:1132–41.

[88] Potluri P, Davila A, Ruiz-Pesini E, Mishmar D, O'Hearn S, Hancock S, Simon MC, Scheffler I, Wallace DC, Procaccio V. A novel NDUFA1 mutation leads to a progressive mitochondrial complex I-specific neurodegenerative disease. Mol Genet Metab 2009;96:189–95.

[89] Denaro M, Blanc H, Johnson MJ, Chen KH, Wilmsen E, Cavalli Sforza LL, Wallace DC. Ethnic variation in Hpa 1 endonuclease cleavage patterns of human mitochondrial DNA. Proc Natl Acad Sci USA 1981;78:5768–72.

[90] Wallace DC, Brown MD, Lott MT. Mitochondrial DNA variation in human evolution and disease. Gene 1999;238:211–30.

[91] Johnson MJ, Wallace DC, Ferris SD, Rattazzi MC, Cavalli-Sforza LL. Radiation of human mitochondria DNA types analyzed by restriction endonuclease cleavage patterns. J Mol Evol 1983;19:255–71.

[92] Wallace DC. Mitochondrial DNA sequence variation in human evolution and disease. Proc Natl Acad Sci USA 1994;91:8739–46.

[93] Wallace DC. 1994 William Allan Award Address. Mitochondrial DNA variation in human evolution, degenerative disease, and aging. Am J Hum Genet 1995;57:201–23.

[94] Mishmar D, Ruiz-Pesini EE, Golik P, Macaulay V, Clark AG, Hosseini S, Brandon M, Easley K, Chen E, Brown MD, Sukernik RI, Olckers A, Wallace DC. Natural selection shaped regional mtDNA variation in humans. Proc Natl Acad Sci USA 2003;100:171–6.

[95] Ruiz-Pesini E, Mishmar D, Brandon M, Procaccio V, Wallace DC. Effects of purifying and adaptive selection on regional variation in human mtDNA. Science 2004;303:223–6.

[96] Ruiz-Pesini E, Wallace DC. Evidence for adaptive selection acting on the tRNA and rRNA genes of the human mitochondrial DNA. Human Mutat 2006;27:1072–81.

[97] Cann RL, Stoneking M, Wilson AC. Mitochondrial DNA and human evolution. Nature 1987;325:31–6.

[98] Merriwether DA, Clark AG, Ballinger SW, Schurr TG, Soodyall H, Jenkins T, Sherry ST, Wallace DC. The structure of human mitochondrial DNA variation. J Mol Evol 1991;33:543–55.

[99] Chen YS, Torroni A, Excoffier L, Santachiara-Benerecetti AS, Wallace DC. Analysis of mtDNA variation in African populations reveals the most ancient of all human continent-specific haplogroups. Am J Hum Genet 1995;57:133–49.

[100] Schurr TG, Donham BP, Morreale SC, Panter-Brick C, Donham DL, Armelagos GJ, Wallace DC. Genetic diversity in modern African populations and its use for reconstructing ancient and modern population movements. In: Reed DM, editor. Biomolecular Archeology, Genetic Approaches to the Past. Occasional Paper #31. Carbondale: Center for Archaeological Investigations, Southern Illinois University; 2005. p. 169–207.

[101] Torroni A, Huoponen K, Francalacci P, Petrozzi M, Morelli L, Scozzari R, Obinu D, Savontaus ML, Wallace DC. Classification of European mtDNAs from an analysis of three European populations. Genetics 1996;144:1835–50.

[102] Schurr TG, Ballinger SW, Gan YY, Hodge JA, Merriwether DA, Lawrence DN, Knowler WC, Weiss KM, Wallace DC. Amerindian mitochondrial DNAs have rare Asian mutations at high frequencies, suggesting they derived from four primary maternal lineages. Am J Hum Genet 1990;46:613–23.

[103] Brown MD, Hosseini SH, Torroni A, Bandelt HJ, Allen JC, Schurr TG, Scozzari R, Cruciani F, Wallace DC. mtDNA Haplogroup X: an ancient link between Europe/Western Asia and North America? Am J Hum Genet 1998;63:1852–61.

[104] Kazuno AA, Munakata K, Nagai T, Shimozono S, Tanaka M, Yoneda M, Kato N, Miyawaki A, Kato T. Identification of mitochondrial DNA polymorphisms that alter mitochondrial matrix pH and intracellular calcium dynamics. PLoS Genet 2006;2:e128.

[105] Wallace DC, Lott MT. Leber Hereditary Optic Neuropathy: exemplar of an mtDNA disease. Handb Exp Pharmacol 2017;240:339–76.

[106] Holt IJ, Harding AE, Petty RK, Morgan-Hughes JA. A new mitochondrial disease associated with mitochondrial DNA heteroplasmy. Am J Hum Genet 1990;46:428–33.

[107] Sadun AA, La Morgia C, Carelli V. Leber's Hereditary Optic Neuropathy. Curr Treat Options Neurol 2011;13:109–17.

[108] Huoponen K, Vilkki J, Aula P, Nikoskelainen EK, Savontaus ML. A new mtDNA mutation associated with Leber hereditary optic neuroretinopathy. Am J Hum Genet 1991;48:1147–53.

[109] Johns DR, Neufeld MJ, Park RD. An ND-6 mitochondrial DNA mutation associated with Leber hereditary optic neuropathy. Biochem Biophys Res Commun 1992;187:1551–7.

[110] Jun AS, Brown MD, Wallace DC. A mitochondrial DNA mutation at np 14459 of the ND6 gene associated with maternally inherited Leber's hereditary optic neuropathy and dystonia. Proc Natl Acad Sci USA 1994;91:6206–10.

[111] Malfatti E, Bugiani M, Invernizzi F, de Souza CF, Farina L, Carrara F, Lamantea E, Antozzi C, Confalonieri P, Sanseverino MT, Giugliani R, Uziel G, Zeviani M. Novel mutations of ND genes in complex I deficiency associated with mitochondrial encephalopathy. Brain 2007;130:1894–904.

[112] De Vries DD, Van Engelen BG, Gabreels FJ, Ruitenbeek W, Van Oost BA. A second missense mutation in the mitochondrial ATPase 6 gene in Leigh's syndrome. Ann Neurol 1993;34:410–2.

[113] Santorelli FM, Shanske S, Jain KD, Tick D, Schon EA, DiMauro S. A T-C mutation at nt 8993 of mitochondrial DNA in a child with Leigh syndrome. Neurology 1994;44:972–4.

[114] Trounce I, Neill S, Wallace DC. Cytoplasmic transfer of the mtDNA nt 8993 TG (ATP6) point mutation associated with Leigh syndrome into mtDNA-less cells demonstrates cosegregation with a decrease in state III respiration and ADP/O ratio. Proc Natl Acad Sci USA 1994;91:8334–8.

[115] Ortiz RG, Newman NJ, Shoffner JM, Kaufman AE, Koontz DA, Wallace DC. Variable retinal and neurologic manifestations in patients harboring the mitochondrial DNA 8993 mutation. Arch Ophthalmol 1993;111:1525–30.

[116] Santorelli FM, Shanske S, Macaya A, DeVivo DC, DiMauro S. The mutation at nt 8993 of mitochondrial DNA is a common cause of Leigh's syndrome. Ann Neurol 1993;34:827–34.

[117] Tatuch Y, Christodoulou J, Feigenbaum A, Clarke JTR, Wherret J, Smith C, Rudd N, Petrova-Benedict R, Robinson BH. Heteroplasmic mtDNA mutation (T-G) at 8993 can cause Leigh disease when the percentage of abnormal mtDNA is high. Am J Hum Genet 1992;50:852–8.
[118] Goto Y, Nonaka I, Horai S. A mutation in the tRNA$^{Leu(UUR)}$ gene associated with the MELAS subgroup of mitochondrial encephalomyopathies. Nature 1990;348:651–3.
[119] Desquiret-Dumas V, Gueguen N, Barth M, Chevrollier A, Hancock S, Wallace DC, Amati-Bonneau P, Henrion D, Bonneau D, Reynier P, Procaccio V. Metabolically induced heteroplasmy shifting and l-arginine treatment reduce the energetic defect in a neuronal-like model of MELAS. Biochim Biophys Acta 2012;1822:1019–29.
[120] Picard M, Zhang J, Hancock S, Derbeneva O, Golhar R, Golik P, O'Hearn S, Levy S, Potluri P, Lvova M, Davila A, Lin CS, Perin JC, Rappaport EF, Hakonarson H, Trounce IA, Procaccio V, Wallace DC. Progressive increase in mtDNA 3243A>G heteroplasmy causes abrupt transcriptional reprogramming. Proc Natl Acad Sci USA 2014;111:E4033–42.
[121] Wang Q, Ito M, Adams K, Li BU, Klopstock T, Maslim A, Higashimoto T, Herzog J, Boles RG. Mitochondrial DNA control region sequence variation in migraine headache and cyclic vomiting syndrome. Am J Med Genet 2004;131A:50–8.
[122] Brown MD, Starikovskaya E, Derbeneva O, Hosseini S, Allen JC, Mikhailovskaya IE, Sukernik RI, Wallace DC. The role of mtDNA background in disease expression: A new primary LHON mutation associated with Western Eurasian haplogroup J.. Hum Genet 2002;110:130–8.
[123] Brown MD, Sun F, Wallace DC. Clustering of Caucasian Leber hereditary optic neuropathy patients containing the 11778 or 14484 mutations on an mtDNA lineage. Am J Hum Genet 1997;60:381–7.
[124] Brown MD, Torroni A, Reckord CL, Wallace DC. Phylogenetic analysis of Leber's hereditary optic neuropathy mitochondrial DNA's indicates multiple independent occurrences of the common mutations. Hum Mutat 1995;6:311–25.
[125] Torroni A, Petrozzi M, D'Urbano L, Sellitto D, Zeviani M, Carrara F, Carducci C, Leuzzi V, Carelli V, Barboni P, De Negri A, Scozzari R. Haplotype and phylogenetic analyses suggest that one European-specific mtDNA background plays a role in the expression of Leber hereditary optic neuropathy by increasing the penetrance of the primary mutations 11778 and 14484. Am J Hum Genet 1997;60:1107–21.
[126] Carelli V, Achilli A, Valentino ML, Rengo C, Semino O, Pala M, Olivieri A, Mattiazzi M, Pallotti F, Carrara F, Zeviani M, Leuzzi V, Carducci C, Valle G, Simionati B, Mendieta L, Salomao S, Belfort R, Sadun AA, Torroni A. Haplogroup effects and recombination of mitochondrial DNA: novel clues from the analysis of Leber Hereditary Optic Neuropathy pedigrees. Am J Hum Genet 2006;78:564–74.
[127] Gomez-Duran A, Pacheu-Grau D, Lopez-Gallardo E, Diez-Sanchez C, Montoya J, Lopez-Perez MJ, Ruiz-Pesini E. Unmasking the causes of multifactorial disorders: OXPHOS differences between mitochondrial haplogroups. Hum Mol Genet 2010;19:3343–53.
[128] Vergani L, Martinuzzi A, Carelli V, Cortelli P, Montagna P, Schievano G, Carrozzo R, Angelini C, Lugaresi E. MtDNA mutations associated with Leber's hereditary optic neuropathy: studies on cytoplasmic hybrid (cybrid) cells. Biochem Biophys Res Commun 1995;210:880–8.
[129] Marom S, Friger M, Mishmar D. MtDNA meta-analysis reveals both phenotype specificity and allele heterogeneity: a model for differential association. Sci Rep 2017;7:43449.
[130] Zeviani M, Moraes CT, DiMauro S, Nakase H, Bonilla E, Nakase H, Bonilla E, Schon EA, Rowland LP. Deletions of mitochondrial DNA in Kearns-Sayre syndrome. Neurology 1988;38:1339–46.
[131] Shoffner JM, Lott MT, Voljavec AS, Soueidan SA, Costigan DA, Wallace DC. Spontaneous Kearns-Sayre/chronic external ophthalmoplegia plus syndrome associated with a mitochondrial DNA deletion: a slip-replication model and metabolic therapy. Proc Natl Acad Sci USA 1989;86:7952–6.
[132] Rotig A, Colonna M, Blanche S, Fischer A, LeDeist F, Frezal J, Saudubray JM, Munnich A. Deletion of blood mitochondrial DNA in pancytopenia. Lancet 1988;2:567–8.
[133] Rotig A, Colonna M, Bonnefont JP, Blanche S, Fischer A, Saudubray JM, Munnich A. Mitochondrial DNA deletion in Pearson's marrow-pancreas syndrome. Lancet 1989;1:902–3.
[134] Ballinger SW, Shoffner JM, Gebhart S, Koontz DA, Wallace DC. Mitochondrial diabetes revisited. Nat Genet 1994;7:458–9.
[135] Kujoth GC, Hiona A, Pugh TD, Someya S, Panzer K, Wohlgemuth SE, Hofer T, Seo AY, Sullivan R, Jobling WA, Morrow JD, Van Remmen H, Sedivy JM, Yamasoba T, Tanokura M, Weindruch R, Leeuwenburgh C, Prolla TA. Mitochondrial DNA mutations, oxidative stress, and apoptosis in mammalian aging. Science 2005;309:481–4.

[136] Schriner SE, Linford NJ, Martin GM, Treuting P, Ogburn CE, Emond M, Coskun PE, Ladiges W, Wolf N, Van Remmen H, Wallace DC, Rabinovitch PS. Extension of murine life span by overexpression of catalase targeted to mitochondria. Science 2005;308:1909–11.
[137] Trifunovic A, Wredenberg A, Falkenberg M, Spelbrink JN, Rovio AT, Bruder CE, Bohlooly YM, Gidlof S, Oldfors A, Wibom R, Tornell J, Jacobs HT, Larsson NG. Premature ageing in mice expressing defective mitochondrial DNA polymerase. Nature 2004;429:417–23.
[138] Gerber S, Ding MG, Gerard X, Zwicker K, Zanlonghi X, Rio M, Serre V, Hanein S, Munnich A, Rotig A, Bianchi L, Amati-Bonneau P, Elpeleg O, Kaplan J, Brandt U, Rozet JM. Compound heterozygosity for severe and hypomorphic NDUFS2 mutations cause non-syndromic LHON-like optic neuropathy. J Med Genet 2017;54:346–56.
[139] Antonicka H, Choquet K, Lin ZY, Gingras AC, Kleinman CL, Shoubridge EA. A pseudouridine synthase module is essential for mitochondrial protein synthesis and cell viability. EMBO Rep 2017;18:28–38.
[140] Simon M, Richard EM, Wang X, Shahzad M, Huang VH, Qaiser TA, Potluri P, Mahl SE, Davila A, Nazli S, Hancock S, Yu M, Gargus J, Chang R, Al-Sheqaih N, Newman WG, Abdenur J, Starr A, Hegde R, Dorn T, Busch A, Park E, Wu J, Schwenzer H, Flierl A, Florentz C, Sissler M, Khan SN, Li R, Guan MX, Friedman TB, Wu DK, Procaccio V, Riazuddin S, Wallace DC, Ahmed ZM, Huang T, Riazuddin S. Mutations of human NARS2, encoding the mitochondrial asparaginyl-tRNA synthetase, cause nonsyndromic Deafness and Leigh Syndrome. PLoS Genet 2015;11:e1005097.
[141] Meng F, Cang X, Peng Y, Li R, Zhang Z, Li F, Fan Q, Guan AS, Fischel-Ghosian N, Zhao X, Guan MX. Biochemical evidence for a nuclear modifier allele (A10S) in TRMU (Methylaminomethyl-2-thiouridylate-methyltransferase) related to mitochondrial tRNA modification in the phenotypic manifestation of deafness-associated 12S rRNA mutation. J Biol Chem 2017;292:2881–92.
[142] Copeland WC. The mitochondrial DNA polymerase in health and disease. Sub Cell Biochem 2010;50:211–22.
[143] Naviaux RK, Nguyen KV. POLG mutations associated with Alpers' syndrome and mitochondrial DNA depletion. Ann Neurol 2004;55:706–12.
[144] Van Goethem G, Dermaut B, Lofgren A, Martin JJ, Van Broeckhoven C. Mutation of POLG is associated with progressive external ophthalmoplegia characterized by mtDNA deletions. Nature Genet 2001;28:211–2.
[145] Spelbrink JN, Li FY, Tiranti V, Nikali K, Yuan QP, Tariq M, Wanrooij S, Garrido N, Comi G, Morandi L, Santoro L, Toscano A, Fabrizi GM, Somer H, Croxen R, Beeson D, Poulton J, Suomalainen A, Jacobs HT, Zeviani M, Larsson C. Human mitochondrial DNA deletions associated with mutations in the gene encoding Twinkle, a phage T7 gene 4-like protein localized in mitochondria. Nature Genet 2001;28:223–31.
[146] Mandel H, Szargel R, Labay V, Elpeleg O, Saada A, Shalata A, Anbinder Y, Berkowitz D, Hartman C, Barak M, Eriksson S, Cohen N. The deoxyguanosine kinase gene is mutated in individuals with depleted hepatocerebral mitochondrial DNA. Nature Genet 2001;29:337–41.
[147] Saada A, Shaag A, Mandel H, Nevo Y, Eriksson S, Elpeleg O. Mutant mitochondrial thymidine kinase in mitochondrial DNA depletion myopathy. Nature Genet 2001;29:342–4.
[148] Nishino I, Spinazzola A, Hirano M. Thymidine phosphorylase gene mutations in MNGIE, a human mitochondrial disorder. Science 1999;283:689–92.
[149] El-Hattab AW, Craigen WJ, Scaglia F. Mitochondrial DNA maintenance defects. Biochim Biophys Acta 2017;1863:1539–55.
[150] Kwak SH, Park KS, Lee KU, Lee HK. Mitochondrial metabolism and diabetes. J Diabetes Investig 2010;1:161–9.
[151] Wallace DC. A mitochondrial paradigm of metabolic and degenerative diseases, aging, and cancer: a dawn for evolutionary medicine. Annu Rev Genet 2005;39:359–407.
[152] Martikainen MH, Ronnemaa T, Majamaa K. Prevalence of mitochondrial diabetes in southwestern Finland: a molecular epidemiological study. Acta Diabetologica 2013;50:737–41.
[153] Martikainen MH, Ronnemaa T, Majamaa K. Association of mitochondrial DNA haplogroups and vascular complications of diabetes mellitus: A population-based study. Diabetes Vasc Dis Res 2015;12:302–4.
[154] Crispim D, Canani LH, Gross JL, Tschiedel B, Souto KE, Roisenberg I. The European-specific mitochondrial cluster J/T could confer an increased risk of insulin-resistance and type 2 diabetes: an analysis of the m.4216T>C and m.4917A>G variants. Ann Hum Genet 2006;70:488–95.
[155] Feder J, Ovadia O, Blech I, Cohen J, Wainstein J, Harman-Boehm I, Glaser B, Mishmar D. Parental diabetes status reveals association of mitochondrial DNA haplogroup J1 with type 2 diabetes. BMC Med Genet 2009;10:60.
[156] Mohlke KL, Jackson AU, Scott LJ, Peck EC, Suh YD, Chines PS, Watanabe RM, Buchanan TA, Conneely KN, Erdos MR, Narisu N, Enloe S, Valle TT, Tuomilehto J, Bergman RN, Boehnke M, Collins FS. Mitochondrial polymorphisms and susceptibility to type 2 diabetes-related traits in Finns. Hum Genet 2005;118:245–54.

[157] Fuku N, Park KS, Yamada Y, Nishigaki Y, Cho YM, Matsuo H, Segawa T, Watanabe S, Kato K, Yokoi K, Nozawa Y, Lee HK, Tanaka M. Mitochondrial haplogroup N9a confers resistance against type 2 diabetes in Asians. Am J Hum Genet 2007;80:407–15.
[158] Hwang S, Kwak SH, Bhak J, Kang HS, Lee YR, Koo BK, Park KS, Lee HK, Cho YM. Gene expression pattern in transmitochondrial cytoplasmic hybrid cells harboring type 2 diabetes-associated mitochondrial DNA haplogroups. PLoS One 2011;6:e22116.
[159] Jiang W, Li R, Zhang Y, Wang P, Wu T, Lin J, Yu J, Gu M. Mitochondrial DNA mutations associated with type 2 diabetes mellitus in Chinese Uyghur population. Sci Rep 2017;7:16989.
[160] Niu Q, Zhang W, Wang H, Guan X, Lu J, Li W. Effects of mitochondrial haplogroup N9a on type 2 diabetes mellitus and its associated complications. Exp Ther Med 2015;10:1918–24.
[161] Liao WQ, Pang Y, Yu CA, Wen JY, Zhang YG, Li XH. Novel mutations of mitochondrial DNA associated with type 2 diabetes in Chinese Han population. Tohoku J Exp Med 2008;215:377–84.
[162] Guo LJ, Oshida Y, Fuku N, Takeyasu T, Fujita Y, Kurata M, Sato Y, Ito M, Tanaka M. Mitochondrial genome polymorphisms associated with type-2 diabetes or obesity. Mitochondrion 2005;5:15–33.
[163] Ji F, Sharpley MS, Derbeneva O, Alves LS, Qian P, Wang Y, Chalkia D, Lvova M, Xu J, Yao W, Simon M, Platt J, Xu S, Angelin A, Davila A, Huang T, Wang PH, Chuang LM, Moore LG, Qian G, Wallace DC. Mitochondrial DNA variant associated with Leber hereditary optic neuropathy and high-altitude Tibetans. Proc Natl Acad Sci USA 2012;109:7391–6.
[164] Sharma V, Sharma I, Singh VP, Verma S, Pandita A, Singh V, Rai E, Sharma S. mtDNA G10398A variation provides risk to type 2 diabetes in population group from the Jammu region of India. Meta Gene 2014;2:269–73.
[165] Poulton J, Luan J, Macaulay V, Hennings S, Mitchell J, Wareham NJ. Type 2 diabetes is associated with a common mitochondrial variant: evidence from a population-based case-control study. Hum Mol Genet 2002;11:1581–3.
[166] Estopinal CB, Chocron IM, Parks MB, Wade EA, Roberson RM, Burgess LG, Brantley Jr MA, Samuels DC. Mitochondrial haplogroups are associated with severity of diabetic retinopathy. Investig Ophthalmol Vis Sci 2014;55:5589–95.
[167] Achilli A, Olivieri A, Pala M, Hooshiar Kashani B, Carossa V, Perego UA, Gandini F, Santoro A, Battaglia V, Grugni V, Lancioni H, Sirolla C, Bonfigli AR, Cormio A, Boemi M, Testa I, Semino O, Ceriello A, Spazzafumo L, Gadaleta MN, Marra M, Testa R, Franceschi C, Torroni A. Mitochondrial DNA backgrounds might modulate diabetes complications rather than T2DM as a whole. PLoS One 2011;6:e21029.
[168] Kofler B, Mueller EE, Eder W, Stanger O, Maier R, Weger M, Haas A, Winker R, Schmut O, Paulweber B, Iglseder B, Renner W, Wiesbauer M, Aigner I, Santic D, Zimmermann FA, Mayr JA, Sperl W. Mitochondrial DNA haplogroup T is associated with coronary artery disease and diabetic retinopathy: a case control study. BMC Med Genet 2009;10:35.
[169] Nardelli C, Labruna G, Liguori R, Mazzaccara C, Ferrigno M, Capobianco V, Pezzuti M, Castaldo G, Farinaro E, Contaldo F, Buono P, Sacchetti L, Pasanisi F. Haplogroup T is an obesity risk factor: mitochondrial DNA haplotyping in a morbid obese population from southern Italy. BioMed Res Int 2013;2013:631082.
[170] Chinnery PF, Elliott HR, Syed A, Rothwell PM. Mitochondrial DNA haplogroups and risk of transient ischaemic attack and ischaemic stroke: a genetic association study. Lancet Neurol 2010;9:498–503.
[171] Nishigaki Y, Yamada Y, Fuku N, Matsuo H, Segawa T, Watanabe S, Kato K, Yokoi K, Yamaguchi S, Nozawa Y, Tanaka M. Mitochondrial haplogroup N9b is protective against myocardial infarction in Japanese males. Hum Genet 2007;120:827–36.
[172] Udar N, Atilano SR, Memarzadeh M, Boyer DS, Chwa M, Lu S, Maguen B, Langberg J, Coskun P, Wallace DC, Nesburn AB, Khatibi N, Hertzog D, Le K, Hwang D, Kenney MC. Mitochondrial DNA haplogroups associated with age-related macular degeneration. Investig Ophthalmol Vis Sci 2009;50:2966–74.
[173] Jones MM, Manwaring N, Wang JJ, Rochtchina E, Mitchell P, Sue CM. Mitochondrial DNA haplogroups and age-related maculopathy. Arch Ophthalmol 2007;125:1235–40.
[174] Soto-Hermida A, Fernandez-Moreno M, Oreiro N, Fernandez-Lopez C, Rego-Perez I, Blanco FJ. mtDNA haplogroups and osteoarthritis in different geographic populations. Mitochondrion 2014;15:18–23.
[175] Montiel-Sosa F, Ruiz-Pesini E, Enriquez JA, Marcuello A, Diez-Sanchez C, Montoya J, Wallace DC, Lopez-Perez MJ. Differences of sperm motility in mitochondrial DNA haplogroup U sublineages. Gene 2006;368:21–7.
[176] Ruiz-Pesini E, Lapena AC, Diez-Sanchez C, Perez-Martos A, Montoya J, Alvarez E, Diaz M, Urries A, Montoro L, Lopez-Perez MJ, Enriquez JA. Human mtDNA haplogroups associated with high or reduced spermatozoa motility. Am J Hum Genet 2000;67:682–96.
[177] Niemi AK, Majamaa K. Mitochondrial DNA and ACTN3 genotypes in Finnish elite endurance and sprint athletes. Eur J Hum Genet 2005;13:965–9.

NG, Carracedo A, Chahrour MH, Chiocchetti AG, Coon H, Crawford EL, Curran SR, Dawson G, Duketis E, Fernandez BA, Gallagher L, Geller E, Guter SJ, Hill RS, Ionita-Laza J, Jimenz Gonzalez P, Kilpinen H, Klauck SM, Kolevzon A, Lee I, Lei I, Lei J, Lehtimaki T, Lin CF, Ma'ayan A, Marshall CR, McInnes AL, Neale B, Owen MJ, Ozaki N, Parellada M, Parr JR, Purcell S, Puura K, Rajagopalan D, Rehnstrom K, Reichenberg A, Sabo A, Sachse M, Sanders SJ, Schafer C, Schulte-Ruther M, Skuse D, Stevens C, Szatmari P, Tammimies K, Valladares O, Voran A, Li-San W, Weiss LA, Willsey AJ, Yu TW, Yuen RK, Study DDD, Homozygosity Mapping Collaborative for, A., Consortium, U.K., Cook EH, Freitag CM, Gill M, Hultman CM, Lehner T, Palotie A, Schellenberg GD, Sklar P, State MW, Sutcliffe JS, Walsh CA, Scherer SW, Zwick ME, Barett JC, Cutler DJ, Roeder K, Devlin B, Daly MJ, Buxbaum JD. Synaptic, transcriptional and chromatin genes disrupted in autism. Nature 2014;515:209–15.

[222] Kosmicki JA, Samocha KE, Howrigan DP, Sanders SJ, Slowikowski K, Lek M, Karczewski KJ, Cutler DJ, Devlin B, Roeder K, Buxbaum JD, Neale BM, MacArthur DG, Wall DP, Robinson EB, Daly MJ. Refining the role of de novo protein-truncating variants in neurodevelopmental disorders by using population reference samples. Nature Genet 2017;49:504–10.

[223] Krumm N, O'Roak BJ, Shendure J, Eichler EE. A de novo convergence of autism genetics and molecular neuroscience. Trends Neurosci 2014;37:95–105.

[224] Bishop SL, Farmer C, Bal V, Robinson EB, Willsey AJ, Werling DM, Havdahl KA, Sanders SJ, Thurm A. Identification of developmental and behavioral markers associated with genetic abnormalities in autism spectrum disorder. Am J Psychiatry 2017;174:576–85.

[225] Brandler WM, Antaki D, Gujral M, Noor A, Rosanio G, Chapman TR, Barrera DJ, Lin GN, Malhotra D, Watts AC, Wong LC, Estabillo JA, Gadomski TE, Hong O, Fajardo KV, Bhandari A, Owen R, Baughn M, Yuan J, Solomon T, Moyzis AG, Maile MS, Sanders SJ, Reiner GE, Vaux KK, Strom CM, Zhang K, Muotri AR, Akshoomoff N, Leal SM, Pierce K, Courchesne E, Iakoucheva LM, Corsello C, Sebat J. Frequency and complexity of de novo structural mutation in autism. Am J Hum Genet 2016;98:667–79.

[226] Robinson EB, St Pourcain B, Anttila V, Kosmicki JA, Bulik-Sullivan B, Grove J, Maller J, Samocha KE, Sanders SJ, Ripke S, Martin J, Hollegaard MV, Werge T, Hougaard DM, iPsych-SSI Broad Autism Group, Neale BM, Evans DM, Skuse D, Mortensen PB, Borglum AD, Ronald A, Smith GD, Daly MJ. Genetic risk for autism spectrum disorders and neuropsychiatric variation in the general population. Nature Genet 2016;48:552–5.

[227] Autism Spectrum Disorders Working Group of The Psychiatric Genomics Consortium. Meta-analysis of GWAS of over 16,000 individuals with autism spectrum disorder highlights a novel locus at 10q24.32 and a significant overlap with schizophrenia. Mol Autism 2017;8:21.

[228] Weiner DJ, Wigdor EM, Ripke S, Walters RK, Kosmicki JA, Grove J, Samocha KE, Goldstein JI, Okbay A, Bybjerg-Grauholm J, Werge T, Hougaard DM, Taylor J, i, P.-B.A.G., Psychiatric Genomics Consortium Autism, G., Skuse D, Devlin B, Anney R, Sanders SJ, Bishop S, Mortensen PB, Borglum AD, Smith GD, Daly MJ, Robinson EB. Polygenic transmission disequilibrium confirms that common and rare variation act additively to create risk for autism spectrum disorders. Nat Genet 2017;49:978–85.

[229] Smith M, Flodman PL, Gargus JJ, Simon MT, Verrell K, Haas R, Reiner GE, Naviaux R, Osann K, Spence MA, Wallace DC. Mitochondrial and ion channel gene alterations in autism. Biochim Biophys Acta 2012;1817:1796–802.

[230] Sanders SJ, He X, Willsey AJ, Ercan-Sencicek AG, Samocha KE, Cicek AE, Murtha MT, Bal VH, Bishop SL, Dong S, Goldberg AP, Jinlu C, Keaney 3rd JF, Klei L, Mandell JD, Moreno-De-Luca D, Poultney CS, Robinson EB, Smith L, Solli-Nowlan T, Su MY, Teran NA, Walker MF, Werling DM, Beaudet AL, Cantor RM, Fombonne E, Geschwind DH, Grice DE, Lord C, Lowe JK, Mane SM, Martin DM, Morrow EM, Talkowski ME, Sutcliffe JS, Walsh CA, Yu TW, Autism Sequencing Consortium, Ledbetter DH, Martin CL, Cook EH, Buxbaum JD, Daly MJ, Devlin B, Roeder K, State MW. Insights into autism spectrum disorder genomic architecture and biology from 71 risk loci. Neuron 2015;87:1215–33.

[231] Yoon JC, Ng A, Kim BH, Bianco A, Xavier RJ, Elledge SJ. Wnt signaling regulates mitochondrial physiology and insulin sensitivity. Genes Dev 2010;24:1507–18.

[232] Chalkia D, Singh LN, Leipzig J, Lvova M, Derbeneva O, Lakatos A, Hadley D, Hakonarson H, Wallace DC. Association between mitochondrial DNA haplogroup variation and autism spectrum disorders. JAMA Psychiatry 2017;74:1161–8.

[233] Wang Y, Picard M, Gu Z. Genetic evidence for elevated pathogenicity of mitochondrial DNA heteroplasmy in Autism Spectrum Disorder. PLoS Genet 2016;12:e1006391.

[234] Wallace DC. A mitochondrial etiology of neuropsychiatric disorders. JAMA Psychiatry 2017;74:863–4.

[235] Anglin RE, Mazurek MF, Tarnopolsky MA, Rosebush PI. The mitochondrial genome and psychiatric illness. American Journal of Medical Genetics. Part B,. Neuropsychiatr Genet 2012a;159B:749–59.

[236] Anglin RE, Tarnopolsky MA, Mazurek MF, Rosebush PI. The psychiatric presentation of mitochondrial disorders in adults. J Neuropsychiatry Clin Neurosci 2012b;24:394–409.
[237] Manji H, Kato T, Di Prospero NA, Ness S, Beal MF, Krams M, Chen G. Impaired mitochondrial function in psychiatric disorders. Nature Reviews. Neuroscience 2012;13:293–307.
[238] Rosebush PI, Anglin RE, Rasmussen S, Mazurek MF. Mental illness in patients with inherited mitochondrial disorders. Schizophr Res 2017;187:33–7.
[239] Zuccoli GS, Saia-Cereda VM, Nascimento JM, Martins-de-Souza D. The energy metabolism dysfunction in psychiatric disorders postmortem brains: focus on proteomic evidence. Front Neurosci 2017;11:493.
[240] Baudouin SV, Saunders D, Tiangyou W, Elson JL, Poynter J, Pyle A, Keers S, Turnbull DM, Howell N, Chinnery PF. Mitochondrial DNA and survival after sepsis: a prospective study. Lancet 2005;366:2118–21.
[241] Raby BA, Klanderman B, Murphy A, Mazza S, Camargo Jr CA, Silverman EK, Weiss ST. A common mitochondrial haplogroup is associated with elevated total serum IgE levels. J Allergy Clin Immunol 2007;120:351–8.
[242] Hendrickson SL, Hutcheson HB, Ruiz-Pesini E, Poole JC, Lautenberger J, Sezgin E, Kingsley L, Goedert JJ, Vlahov D, Donfield S, Wallace DC, O'Brien SJ. Mitochondrial DNA haplogroups influence AIDS progression. AIDS 2008;22:2429–39.
[243] Dela Cruz CS, Kang MJ. Mitochondrial dysfunction and damage associated molecular patterns (DAMPs) in chronic inflammatory diseases. Mitochondrion 2017. [ePub ahead of print] http://www.sciencedirect.com/science/article/pii/S1567724917301897.
[244] Dan Dunn J, Alvarez LA, Zhang X, Soldati T. Reactive oxygen species and mitochondria: A nexus of cellular homeostasis. Redox Biol 2015;6:472–85.
[245] Nadeau-Vallee M, Obari D, Quiniou C, Lubell WD, Olson DM, Girard S, Chemtob S. A critical role of interleukin-1 in preterm labor. Cytokine Growth Factor Rev 2016;28:37–51.
[246] Angelin A, Gil-de-Gomez L, Dahiya S, Jiao J, Guo L, Levine MH, Wang Z, Quinn 3rd WJ, Kopinski PK, Wang L, Akimova T, Liu Y, Bhatti TR, Han R, Laskin BL, Baur JA, Blair IA, Wallace DC, Hancock WW, Beier UH. Foxp3 reprograms T cell metabolism to function in low-glucose, high-lactate environments. Cell Metab 2017;25:1282–93.
[247] Kilbaugh TJ, Lvova M, Karlsson M, Zhang Z, Leipzig J, Wallace DC, Margulies SS. Peripheral blood mitochondrial DNA as a biomarker of cerebral mitochondrial dysfunction following traumatic brain injury in a porcine model. PLoS One 2015;10:e0130927.
[248] Newman NJ, Yu-Wai-Man P, Sadun AA, Karanjia R, Carelli V. Management of ophthalmologic manifestations of mitochondrial diseases. Genet Med 2017;19.
[249] Parikh S, Goldstein A, Karaa A, Koenig MK, Anselm I, Brunel-Guitton C, Christodoulou J, Cohen BH, Dimmock D, Enns GM, Falk MJ, Feigenbaum A, Frye RE, Ganesh J, Griesemer D, Haas R, Horvath R, Korson M, Kruer MC, Mancuso M, McCormack S, Josee Raboisson M, Reimschisel T, Salvarinova R, Saneto RP, Scaglia F, Shoffner J, Stacpoole PW, Sue CM, Tarnopolsky M, Van Karnebeek C, Wolfe LA, Zolkipli Cunningham Z, Rahman S, Chinnery PF. Response to Newman et al. Genet Med 2017a;19.
[250] Parikh S, Goldstein A, Karaa A, Koenig MK, Anselm I, Brunel-Guitton C, Christodoulou J, Cohen BH, Dimmock D, Enns GM, Falk MJ, Feigenbaum A, Frye RE, Ganesh J, Griesemer D, Haas R, Horvath R, Korson M, Kruer MC, Mancuso M, McCormack S, Raboisson MJ, Reimschisel T, Salvarinova R, Saneto RP, Scaglia F, Shoffner J, Stacpoole PW, Sue CM, Tarnopolsky M, Van Karnebeek C, Wolfe LA, Cunningham ZZ, Rahman S, Chinnery PF. Patient care standards for primary mitochondrial disease: a consensus statement from the Mitochondrial Medicine Society. Genet Med 2017b;19:689.
[251] Parikh S, Goldstein A, Koenig MK, Scaglia F, Enns GM, Saneto R, Anselm I, Cohen BH, Falk MJ, Greene C, Gropman AL, Haas R, Hirano M, Morgan P, Sims K, Tarnopolsky M, Van Hove JLK, Wolfe L, DiMauro S. Diagnosis and management of mitochondrial disease: a consensus statement from the Mitochondrial Medicine Society. Genet Med 2015;17:689–701.
[252] Dinwiddie DL, Smith LD, Miller NA, Atherton AM, Farrow EG, Strenk ME, Soden SE, Saunders CJ, Kingsmore SF. Diagnosis of mitochondrial disorders by concomitant next-generation sequencing of the exome and mitochondrial genome. Genomics 2013;102:148–56.
[253] Tang S, Wang J, Zhang VW, Li FY, Landsverk M, Cui H, Truong CK, Wang G, Chen LC, Graham B, Scaglia F, Schmitt ES, Craigen WJ, Wong LJ. Transition to next generation analysis of the whole mitochondrial genome: a summary of molecular defects. Human Mutat 2013;34:882–93.
[254] Wong LJ. Next generation molecular diagnosis of mitochondrial disorders. Mitochondrion 2013;13:379–87.
[255] Kremer LS, Bader DM, Mertes C, Kopajtich R, Pichler G, Iuso A, Haack TB, Graf E, Schwarzmayr T, Terrile C, Konarikova E, Repp B, Kastenmuller G, Adamski J, Lichtner P, Leonhardt C, Funalot B, Donati A, Tiranti V, Lombes A, Jardel C, Glaser D, Taylor RW, Ghezzi D, Mayr JA, Rotig A, Freisinger P, Distelmaier F, Strom

TM, Meitinger T, Gagneur J, Prokisch H. Genetic diagnosis of Mendelian disorders via RNA sequencing. Nat Commun 2017;8:15824.
[256] Davis R, Liang C, Sue C. A new diagnostic paradigm for mitochondrial disease (Abstract 20). J Clin Neurosci 2014;21:2039.
[257] Morovat A, Weerasinghe G, Nesbitt V, Hofer M, Agnew T, Quaghebeur G, Sergeant K, Fratter C, Guha N, Mirzazadeh M, Poulton J. Use of FGF-21 as a biomarker of mitochondrial disease in clinical practice. J Clin Med 2017;6:80.
[258] Suomalainen A, Elo JM, Pietilainen KH, Hakonen AH, Sevastianova K, Korpela M, Isohanni P, Marjavaara SK, Tyni T, Kiuru-Enari S, Pihko H, Darin N, Ounap K, Kluijtmans LA, Paetau A, Buzkova J, Bindoff LA, Annunen-Rasila J, Uusimaa J, Rissanen A, Yki-Jarvinen H, Hirano M, Tulinius M, Smeitink J, Tyynismaa H. FGF-21 as a biomarker for muscle-manifesting mitochondrial respiratory chain deficiencies: a diagnostic study. Lancet Neurol 2011;10:806–18.
[259] Montero R, Yubero D, Villarroya J, Henares D, Jou C, Rodriguez MA, Ramos F, Nascimento A, Ortez CI, Campistol J, Perez-Duenas B, O'Callaghan M, Pineda M, Garcia-Cazorla A, Oferil JC, Montoya J, Ruiz-Pesini E, Emperador S, Meznaric M, Campderros L, Kalko SG, Villarroya F, Artuch R, Jimenez-Mallebrera C. GDF-15 is elevated in children with mitochondrial diseases and is induced by mitochondrial dysfunction. PLoS One 2016;11:e0148709.
[260] Yatsuga S, Fujita Y, Ishii A, Fukumoto Y, Arahata H, Kakuma T, Kojima T, Ito M, Tanaka M, Saiki R, Koga Y. Growth differentiation factor 15 as a useful biomarker for mitochondrial disorders. Ann Neurol 2015;78:814–23.
[261] Chao de la Barca JM, Simard G, Amati-Bonneau P, Safiedeen Z, Prunier-Mirebeau D, Chupin S, Gadras C, Tessier L, Gueguen N, Chevrollier A, Desquiret-Dumas V, Ferre M, Bris C, Kouassi Nzoughet J, Bocca C, Leruez S, Verny C, Milea D, Bonneau D, Lenaers G, Martinez MC, Procaccio V, Reynier P. The metabolomic signature of Leber's hereditary optic neuropathy reveals endoplasmic reticulum stress. Brain 2016;139:2864–76.
[262] Esterhuizen K, van der Westhuizen FH, Louw R. Metabolomics of mitochondrial disease. Mitochondrion 2017;35:97–110.
[263] Quiros PM, Prado MA, Zamboni N, D'Amico D, Williams RW, Finley D, Gygi SP, Auwerx J. Multi-omics analysis identifies ATF4 as a key regulator of the mitochondrial stress response in mammals. J Cell Biol 2017;216:2027–45.
[264] Vivian CJ, Brinker AE, Graw S, Koestler DC, Legendre C, Gooden GC, Salhia B, Welch DR. Mitochondrial genomic backgrounds affect nuclear DNA methylation and gene expression. Cancer Res 2017;77:6202–14.
[265] Pfeffer G, Majamaa K, Turnbull DM, Thorburn D, Chinnery PF. Treatment for mitochondrial disorders. Cochrane Database System Rev 2012:CD004426.
[266] Murphy MP. Selective targeting of bioactive compounds to mitochondria. Trends Biotechnol 1997;15:326–30.
[267] Smith RA, Kelso GF, James AM, Murphy MP. Targeting coenzyme Q derivatives to mitochondria. Methods Enzymol 2004;382:45–67.
[268] McManus MJ, Murphy MP, Franklin JL. The mitochondria-targeted antioxidant MitoQ prevents loss of spatial memory retention and early neuropathology in a transgenic mouse model of Alzheimer's disease. J Neurosci 2011;31:15703–15.
[269] Pell VR, Chouchani ET, Murphy MP, Brookes PS, Krieg T. Moving forwards by blocking back-flow: the yin and yang of MI therapy. Circ Res 2016;118:898–906.
[270] Carelli V, La Morgia C, Valentino ML, Rizzo G, Carbonelli M, De Negri AM, Sadun F, Carta A, Guerriero S, Simonelli F, Sadun AA, Aggarwal D, Liguori R, Avoni P, Baruzzi A, Zeviani M, Montagna P, Barboni P. Idebenone treatment In Leber's Hereditary Optic Neuropathy. Brain 2011;134:e188.
[271] Mashima Y, Kigasawa K, Wakakura M, Oguchi Y. Do idebenone and vitamin therapy shorten the time to achieve visual recovery in Leber hereditary optic neuropathy? J Neuro Ophthalmol 2000;20:166–70.
[272] Klopstock T, Yu-Wai-Man P, Dimitriadis K, Rouleau J, Heck S, Bailie M, Atawan A, Chattopadhyay S, Schubert M, Garip A, Kernt M, Petraki D, Rummey C, Leinonen M, Metz G, Griffiths PG, Meier T, Chinnery PF. A randomized placebo-controlled trial of idebenone in Leber's hereditary optic neuropathy. Brain 2011;134:2677–86.
[273] Angebault C, Gueguen N, Desquiret-Dumas V, Chevrollier A, Guillet V, Verny C, Cassereau J, Ferre M, Milea D, Amati-Bonneau P, Bonneau D, Procaccio V, Reynier P, Loiseau D. Idebenone increases mitochondrial complex I activity in fibroblasts from LHON patients while producing contradictory effects on respiration. BMC Res Notes 2011;4:557.
[274] Giorgio V, Petronilli V, Ghelli A, Carelli V, Rugolo M, Lenaz G, Bernardi P. The effects of idebenone on mitochondrial bioenergetics. Biochim Biophys Acta 2012;1817:363–9.
[275] Enns GM, Kinsman SL, Perlman SL, Spicer KM, Abdenur JE, Cohen BH, Amagata A, Barnes A, Kheifets V, Shrader WD, Thoolen M, Blankenberg F, Miller G. Initial experience in the treatment of inherited mitochondrial disease with EPI-743. Mol Genet Metab 2012;105:91–102.

[276] Sadun AA, Chicani CF, Ross-Cisneros FN, Barboni P, Thoolen M, Shrader WD, Kubis K, Carelli V, Miller G. Effect of EPI-743 on the clinical course of the mitochondrial disease Leber Hereditary Optic Neuropathy. Arch Neurol 2012;69:331–8.

[277] Pedram A, Razandi M, Wallace DC, Levin ER. Functional estrogen receptors in the mitochondria of breast cancer cells. Mol Biol Cell 2006;17:2125–37.

[278] Pisano A, Preziuso C, Iommarini L, Perli E, Grazioli P, Campese AF, Maresca A, Montopoli M, Masuelli L, Sadun AA, d'Amati G, Carelli V, Ghelli A, Giordano C. Targeting estrogen receptor beta as preventive therapeutic strategy for Leber's hereditary optic neuropathy. Hum Mol Genet 2015;24:6921–31.

[279] Szeto HH. First-in-class cardiolipin-protective compound as a therapeutic agent to restore mitochondrial bioenergetics. Br J Pharmacol 2014;171:2029–50.

[280] Giordano C, Iommarini L, Giordano L, Maresca A, Pisano A, Valentino ML, Caporali L, Liguori R, Deceglie S, Roberti M, Fanelli F, Fracasso F, Ross-Cisneros FN, D'Adamo P, Hudson G, Pyle A, Yu-Wai-Man P, Chinnery PF, Zeviani M, Salomao SR, Berezovsky A, Belfort Jr R, Ventura DF, Moraes M, Moraes Filho M, Barboni P, Sadun F, De Negri A, Sadun AA, Tancredi A, Mancini M, d'Amati G, Loguercio Polosa P, Cantatore P, Carelli V. Efficient mitochondrial biogenesis drives incomplete penetrance in Leber's hereditary optic neuropathy. Brain 2014;137:335–53.

[281] Carelli V, d'Adamo P, Valentino ML, La Morgia C, Ross-Cisneros FN, Caporali L, Maresca A, Loguercio Polosa P, Barboni P, De Negri A, Sadun F, Karanjia R, Salomao SR, Berezovsky A, Chicani F, Moraes M, Moraes Filho M, Belfort Jr R, Sadun AA. Parsing the differences in affected with LHON: genetic versus environmental triggers of disease conversion. Brain 2016;139:e17.

[282] Giordano L, Deceglie S, d'Adamo P, Valentino ML, La Morgia C, Fracasso F, Roberti M, Cappellari M, Petrosillo G, Ciaravolo S, Parente D, Giordano C, Maresca A, Iommarini L, Del Dotto V, Ghelli AM, Salomao SR, Berezovsky A, Belfort Jr R, Sadun AA, Carelli V, Loguercio Polosa P, Cantatore P. Cigarette toxicity triggers Leber's hereditary optic neuropathy by affecting mtDNA copy number, oxidative phosphorylation and ROS detoxification pathways. Cell Death Dis 2015;6:e2021.

[283] Pei L, Wallace DC. Mitochondrial etiology of neuropsychiatric disorders. Biol Psychiatry 2017. [ePub ahead of print] https://doi.org/10.1016/j.biopsych.2017.11.018.

[284] Wallace DC, Fan W. Energetics, epigenetics, mitochondrial genetics. Mitochondrion 2010;10:12–31.

[285] Wallace DC, Fan W, Procaccio V. Mitochondrial energetics and therapeutics. Annu Rev Pathol 2010;5:297–348.

[286] Johri A, Calingasan NY, Hennessey TM, Sharma A, Yang L, Wille E, Chandra A, Beal MF. Pharmacologic activation of mitochondrial biogenesis exerts widespread beneficial effects in a transgenic mouse model of Huntington's disease. Hum Mol Genet 2012;21:1124–37.

[287] Wenz T, Diaz F, Spiegelman BM, Moraes CT. Activation of the PPAR/PGC-1alpha pathway prevents a bioenergetic deficit and effectively improves a mitochondrial myopathy phenotype. Cell Metab 2008;8:249–56.

[288] Wenz T, Diaz F, Spiegelman BM, Moraes CT. Retraction notice to: Activation of the PPAR/PGC-1alpha pathway prevents a bioenergetic deficit and effectively improves a mitochondrial myopathy phenotype. Cell Metab 2016;24:889.

[289] Burte F, Carelli V, Chinnery PF, Yu-Wai-Man P. Disturbed mitochondrial dynamics and neurodegenerative disorders. Nat Rev Neurol 2015;11:11–24.

[290] Georgakopoulos ND, Wells G, Campanella M. The pharmacological regulation of cellular mitophagy. Nat Chem Biol 2017;13:136–46.

[291] Sorrentino V, Menzies KJ, Auwerx J. Repairing mitochondrial dysfunction in disease. Annu Rev Pharmacol Toxicol 2017;58. [ePub ahead of print] https://doi.org/10.1146/annurev-pharmtox-010716-104908.

[292] Flierl A, Chen Y, Coskun PE, Samulski RJ, Wallace DC. Adeno-associated virus-mediated gene transfer of the heart/muscle adenine nucleotide translocator (ANT) in mouse. Gene Ther 2005;12:570–8.

[293] Flierl A, Jackson C, Cottrell B, Murdock D, Seibel P, Wallace DC. Targeted delivery of DNA to the mitochondrial compartment via import sequence-conjugated peptide nucleic acid. Mol Ther 2003;7:550–7.

[294] Yu H, Koilkonda RD, Chou TH, Porciatti V, Ozdemir SS, Chiodo V, Boye SL, Boye SE, Hauswirth WW, Lewin AS, Guy J. Gene delivery to mitochondria by targeting modified adenoassociated virus suppresses Leber's hereditary optic neuropathy in a mouse model. Proc Natl Acad Sci USA 2012;109:E1238–47.

[295] Bonnet C, Kaltimbacher V, Ellouze S, Augustin S, Benit P, Forster V, Rustin P, Sahel JA, Corral-Debrinski M. Allotopic mRNA localization to the mitochondrial surface rescues respiratory chain defects in fibroblasts harboring mitochondrial DNA mutations affecting complex I or v subunits. Rejuvenation Res 2007;10:127–44.

[296] Cwerman-Thibault H, Sahel JA, Corral-Debrinski M. Mitochondrial medicine: to a new era of gene therapy for mitochondrial DNA mutations. J Inherit Metab Dis 2011;34:327–44.

[297] Ellouze S, Augustin S, Bouaita A, Bonnet C, Simonutti M, Forster V, Picaud S, Sahel JA, Corral-Debrinski M. Optimized allotopic expression of the human mitochondrial ND4 prevents blindness in a rat model of mitochondrial dysfunction. Am J Hum Genet 2008;83:373–87.
[298] Guy J, Qi X, Koilkonda RD, Arguello T, Chou TH, Ruggeri M, Porciatti V, Lewin AS, Hauswirth WW. Efficiency and safety of AAV-mediated gene delivery of the human ND4 complex I subunit in the mouse visual system. Investig Ophthalmol Vis Sci 2009;50:4205–14.
[299] Guy J, Qi X, Pallotti F, Schon EA, Manfredi G, Carelli V, Martinuzzi A, Hauswirth WW, Lewin AS. Rescue of a mitochondrial deficiency causing Leber Hereditary Optic Neuropathy. Ann Neurol 2002;52:534–42.
[300] Koilkonda RD, Chou TH, Porciatti V, Hauswirth WW, Guy J. Induction of rapid and highly efficient expression of the human ND4 complex I subunit in the mouse visual system by self-complementary adeno-associated virus. Arch Ophthalmol 2010;128:876–83.
[301] Koilkonda RD, Hauswirth WW, Guy J. Efficient expression of self-complementary AAV in ganglion cells of the ex vivo primate retina. Mol Vis 2009;15:2796–802.
[302] Qi X, Sun L, Lewin AS, Hauswirth WW, Guy J. The mutant human ND4 subunit of complex I induces optic neuropathy in the mouse. Investig Ophthalmol Vis Sci 2007;48:1–10.
[303] Sylvestre J, Margeot A, Jacq C, Dujardin G, Corral-Debrinski M. The role of the 3' untranslated region in mRNA sorting to the vicinity of mitochondria is conserved from yeast to human cells. Mol Biol Cell 2003a;14:3848–56.
[304] Sylvestre J, Vialette S, Corral Debrinski M, Jacq C. Long mRNAs coding for yeast mitochondrial proteins of prokaryotic origin preferentially localize to the vicinity of mitochondria. Genome Biol 2003b;4:R44.
[305] Towheed A, Markantone DM, Crain AT, Celotto AM, Palladino MJ. Small mitochondrial-targeted RNAs modulate endogenous mitochondrial protein expression *in vivo*. Neurobiol Dis 2014;69:15–22.
[306] Wang G, Chen HW, Oktay Y, Zhang J, Allen EL, Smith GM, Fan KC, Hong JS, French SW, McCaffery JM, Lightowlers RN, Morse 3rd HC, Koehler CM, Teitell MA. PNPASE regulates RNA import into mitochondria. Cell 2010;142:456–67.
[307] Wang G, Shimada E, Koehler CM, Teitell MA. PNPASE and RNA trafficking into mitochondria. Biochim Biophys Acta 2012;1819:998–1007.
[308] Craven L, Tuppen HA, Greggains GD, Harbottle SJ, Murphy JL, Cree LM, Murdoch AP, Chinnery PF, Taylor RW, Lightowlers RN, Herbert M, Turnbull DM. Pronuclear transfer in human embryos to prevent transmission of mitochondrial DNA disease. Nature 2010;465:82–5.
[309] Hyslop LA, Blakeley P, Craven L, Richardson J, Fogarty NM, Fragouli E, Lamb M, Wamaitha SE, Prathalingam N, Zhang Q, O'Keefe H, Takeda Y, Arizzi L, Alfarawati S, Tuppen HA, Irving L, Kalleas D, Choudhary M, Wells D, Murdoch AP, Turnbull DM, Niakan KK, Herbert M. Towards clinical application of pronuclear transfer to prevent mitochondrial DNA disease. Nature 2016;534:383–6.
[310] Kang E, Wu J, Gutierrez NM, Koski A, Tippner-Hedges R, Agaronyan K, Platero-Luengo A, Martinez-Redondo P, Ma H, Lee Y, Hayama T, Van Dyken C, Wang X, Luo S, Ahmed R, Li Y, Ji D, Kayali R, Cinnioglu C, Olson S, Jensen J, Battaglia D, Lee D, Wu D, Huang T, Wolf DP, Temiakov D, Belmonte JC, Amato P, Mitalipov S. Mitochondrial replacement in human oocytes carrying pathogenic mitochondrial DNA mutations. Nature 2016;540:270–5.
[311] Paull D, Emmanuele V, Weiss KA, Treff N, Stewart L, Hua H, Zimmer M, Kahler DJ, Goland RS, Noggle SA, Prosser R, Hirano M, Sauer MV, Egli D. Nuclear genome transfer in human oocytes eliminates mitochondrial DNA variants. Nature 2013;493:632–7.
[312] Tachibana M, Amato P, Sparman M, Woodward J, Sanchis DM, Ma H, Gutierrez NM, Tippner-Hedges R, Kang E, Lee HS, Ramsey C, Masterson K, Battaglia D, Lee D, Wu D, Jensen J, Patton P, Gokhale S, Stouffer R, Mitalipov S. Towards germline gene therapy of inherited mitochondrial diseases. Nature 2013;493:627–31.
[313] Tachibana M, Sparman M, Sritanaudomchai H, Ma H, Clepper L, Woodward J, Li Y, Ramsey C, Kolotushkina O, Mitalipov S. Mitochondrial gene replacement in primate offspring and embryonic stem cells. Nature 2009;461:367–72.
[314] Wolf DP, Hayama T, Mitalipov S. Mitochondrial genome inheritance and replacement in the human germline. EMBO J 2017;36:2177–81. Corrigendum: EMBO J. 2017 (Sep) 2136(2117):2659.
[315] Slone J, Zhang J, Huang T. Experience from the first live-birth derived from oocyte nuclear transfer as a treatment strategy for mitochondrial diseases. J Mol Genet Med 2017;11:1000258.
[316] Zhang J, Liu H, Luo S, Lu Z, Chavez-Badiola A, Liu Z, Yang M, Merhi Z, Silber SJ, Munne S, Konstantinidis M, Wells D, Tang JJ, Huang T. Live birth derived from oocyte spindle transfer to prevent mitochondrial disease. Reprod Biomed Online 2017;34:361–8. Corrigendum: Reprod Biomed Online 2017 (Jul) 2035(2011):2049.

11

# Multifactorial Inheritance and Complex Diseases

*Allison Fialkowski, T. Mark Beasley, Hemant K. Tiwari*

Department of Biostatistics, School of Public Health, University of Alabama at Birmingham, Birmingham, AL, United States

## 11.1 INTRODUCTION

If a disease or condition is caused by a single locus of large effect it is called a single-gene or *monogenic* disease, disorder, or, more generically, condition. There are over 10,000 such examples, which include cystic fibrosis, Huntington disease, Duchenne muscular dystrophy, and Marfan syndrome. A single-gene disease may have *locus heterogeneity* if that disease is caused by specific mutations on different genes, but this is more properly considered a special case of an oligogenic disorder. For example, osteogenesis imperfecta is caused by a single mutation in a type I collagen gene on either chromosome 7 or chromosome 17. *Oligogenic* disorders are explained by a few loci with large effects (for examples, see [1]). In contrast to oligogenic traits, *polygenic* inheritance is due to many loci with small effects at each locus. Thus, the term polygenic is generally used to describe multiple factors that are exclusively genetic. Any of these genetic effects, with or without the combination of an environmental effect, can give rise to a *multifactorial* disorder. Therefore, multifactorial diseases are caused by the simultaneous action of multiple genetic and/or environmental factors.

In contrast to *dichotomous* traits (i.e., affected vs. unaffected), *quantitative* traits are measured on a continuous scale, and most of them are thought to be multifactorial (e.g., blood pressure, body mass index). Some quantitative traits may be due to major gene effects with a multifactorial background. Multifactorial inheritance is responsible for the majority of modern deleterious health conditions, such as heart disease and diabetes. Atopic syndrome, diabetes, cancer, spina bifida/anencephaly, pyloric stenosis, cleft palate, congenital hip dysplasia, club foot, and many other diseases and complex phenotypes also result from multifactorial inheritance.

**Definitions and terminology:** The *polygenic model* has its origins in Fisher's seminal work [2], which showed that "many small, equal and additive loci" would result in a Gaussian (or normal) distribution for a phenotype. Similarly, the combined additive effects of many genetic and environmental factors will also produce an approximately Gaussian phenotypic distribution. To illustrate, suppose (naïvely) that a quantitative trait such as percentage body fat is determined by a single gene with two codominant[1] alleles, *A* and *a*, which have equal frequency ($P = .50$). Assume individuals with an *A* allele tend to have a higher value of the trait, while individuals with an *a* allele tend to have a lower value of the trait. If *A* has an additive effect, then there are three distinct phenotypic groups, namely, high (2), intermediate (1), and low (0). If the allele frequencies of *A* and *a* are both

[1] "An allele 'a' is said to be *codominant* with respect to the wild-type allele 'A' if the A/a heterozygote fully expresses both of the phenotypes associated with the a/a and A/A homozygotes" (http://www.informatics.jax.org/glossary/codominant).

Emery and Rimoin's Principles and Practice of Medical Genetics and Genomics: Foundations, https://doi.org/10.1016/B978-0-12-812537-3.00011-1

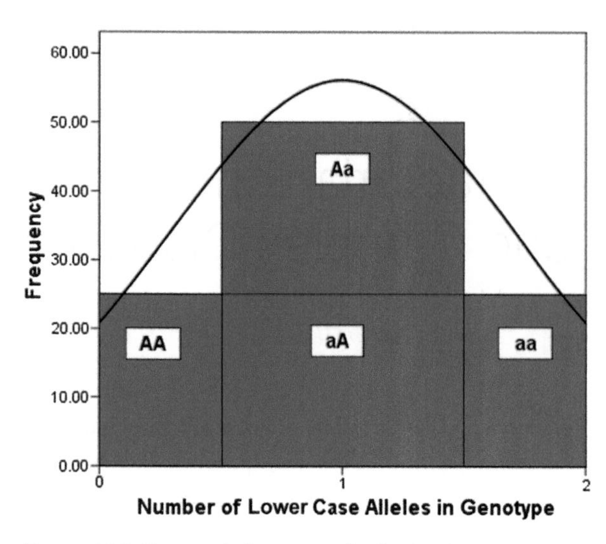


Figure 11.1 Expected phenotype distribution for a trait with a single causal locus with allele frequency of 50% and in Hardy–Weinberg equilibrium.

TABLE 11.1 Frequency Distribution of Genotypic Values for Two Loci With No Linkage Disequilibrium

| | AA | Aa | aa |
|---|---|---|---|
| BB | 0.0625 | 0.1250 | 0.1250 |
| Bb | 0.1250 | 0.2500 | 0.1250 |
| bb | 0.0625 | 0.1250 | 0.0625 |

TABLE 11.2 Genotypic Values of Two-Locus Genotypes

| | AA | Aa | aa |
|---|---|---|---|
| BB | 4 | 3 | 2 |
| Bb | 3 | 2 | 1 |
| bb | 2 | 1 | 0 |

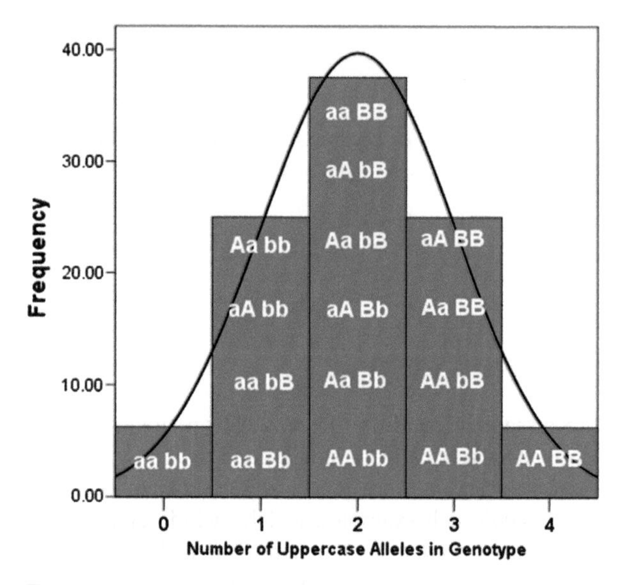


Figure 11.2 Expected phenotype distribution for a trait with two independentlysegregating causal loci of equal effect and allele frequency.

0.50, then 25% of individuals would be expected to be *aa* and of low percentage fat, 50% would be expected to be *Aa* and of moderate percentage fat, and 25% would be expected to be *AA* and of high percentage fat. Fig. 11.1 gives the distribution of the trait in a population.

Now, suppose that the trait is determined by two loci. The second locus also has two codominant alleles, *B* for high and *b* for low expression of the trait, with *B* having an allele frequency of 0.50 and the same effect magnitude as the *A* allele. There are now nine possible genotypes (see Table 11.1).

An individual can possess 0, 1, 2, 3, or 4 "high" trait alleles. Assuming that the combined effects of the two loci are also additivefn22[2], there are five distinct phenotypes with respect to the number of high trait alleles (see Table 11.2).

The trait distribution with respect to genotypic value distribution is shown in Fig. 11.2. As can be seen in Fig. 11.2, even with two loci, the distribution of the phenotype starts to look Gaussian. An example of a three-locus system with equal allele frequencies, no linkage disequilibrium (LD)[3], and equal additive effects, is shown in Fig. 11.3. It can be seen that six diallelic loci are enough to produce population frequencies virtually indistinguishable from a normal curve.

Many traits (or diseases) are treated as dichotomous variables because they appear to be either present or absent (e.g., cancer). By definition, dichotomous variables do not

[2] That is, not *epistatic*, where epistatic refers to an interaction (in the statistical, not necessarily biochemical, sense) between two different loci, such that the effect of genotype at one locus depends on the genotype at another locus.

[3] *Linkage disequilibrium* is defined as the nonrandom association between alleles at *linked* (or adjacent) loci. Two loci are said to be linked if they are sufficiently close on the same chromosome such that they do not segregate independently.

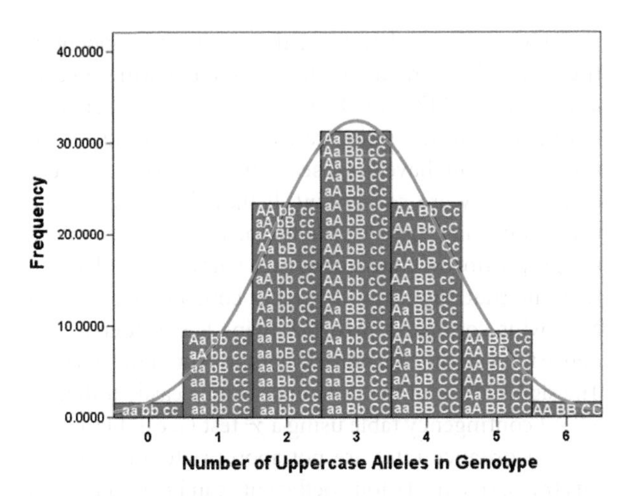


Figure 11.3 Expected phenotype distribution for a trait with three independentlysegregating causal loci of equal effect and allele frequency.

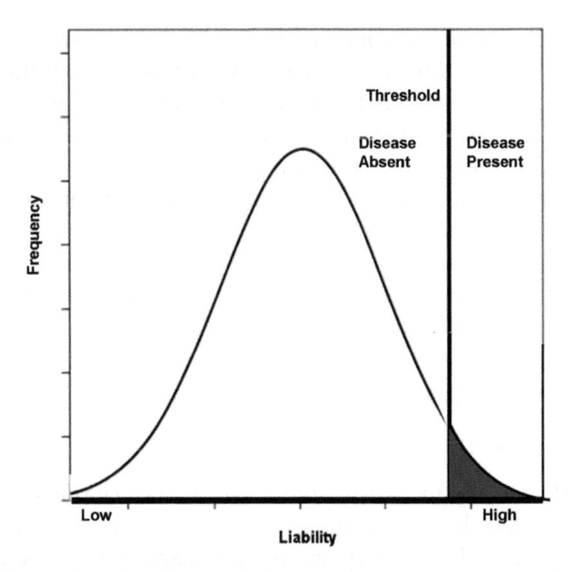


Figure 11.4 Liability distribution for a multifactorial disease. To be affected with the disease, an individual must exceed the threshold.

approximate a Gaussian distribution. These diseases may still be polygenic or multifactorial, because they do not follow the patterns expected of Mendelian (single-gene) diseases. A common explanation is that an underlying *liability* distribution exists for multifactorial diseases [3]. Individuals on the low end of the distribution have little chance of developing the disease because they possess few of the alleles or environmental factors that jointly cause the disease. By contrast, individuals on the high end of the liability distribution have a greater chance of developing the disease because they possess many of the alleles and/or environmental factors that jointly cause the disease. The liability distribution is assumed to be continuous (representing the sum of a large number of independent genetic and environmental factors) and normally distributed within the population. It is also commonplace to assume that all correlations between relatives are due to shared genes but not to shared environment. For multifactorial diseases that are either present or absent, there is a hypothesized *threshold of liability* that must be crossed before the disease is manifest [3].

For example, consider the development of the cleft palate. Early in embryonic development, the palatal arches are in a vertical position. Through embryonic and fetal development, the head grows larger, moving the arches farther apart, and the tongue increases in size, making it more difficult to move. Also, the arches themselves are growing and turning horizontally. There is a critical stage in development by which the two arches must meet and fuse. Head growth, tongue growth, and palatal arch growth are all subject to many genetic and environmental factors. If the two arches start to grow in time, grow at the proper rate, and begin to move soon enough to the horizontal, they will meet and fuse despite head size and tongue growth. The result is no cleft palate. They may fuse well ahead of the critical developmental stage or just barely make it in time; it is impossible to know; however, if they do not meet by the critical stage, a cleft palate results. If they are close together at the critical stage, a small cleft will result, perhaps only a bifurcated uvula. If they are far apart, a more severe cleft will result. That critical difference in liability is called the *threshold*. Beyond the threshold, disease results; below the threshold, normal development is observed. Thus, the underlying liability is distributed as the normal curve shown in Fig. 11.4.

Some diseases may have more than one threshold, and commonly two liability thresholds are present, as defined by factors such as gender, race, age of onset, etc., causing different levels of severity [4]. Examples include pyloric stenosis (sex dimorphism for liability) [5] and orofacial cleft syndrome/cleft lip and palate (two thresholds for fetal mortality and disease) [6,7]. The latter model proposes a lower threshold level of liability resulting in a cleft formation and a higher level causing fetal death (predominantly in males).

It should be emphasized that, like any other loci, the individual loci underlying a polygenic or multifactorial trait are generally assumed to follow the Mendelian principles of *random segregation* and *independent assortment*[4]. The difference is that they act together to influence the trait. Thus, the multifactorial model assumes:

1. several, but not an unlimited number, of loci are involved in the expression of the trait;
2. the loci act in concert in an additive fashion, each adding or subtracting a small amount from the phenotype; and
3. the environment interacts with the genotype to produce the final phenotype.

## 11.2 DETERMINING THE GENETIC COMPONENT OF A TRAIT

Historically, the genetic study of any trait can be divided into four broad categories: familial aggregation, segregation analysis, linkage analysis, and association studies. This paradigm was useful in discovering genes for many monogenic disorders.

### 11.2.1 Familial Aggregation

The first step of any genetic analysis is to establish a genetic component to the disease. Also, one must establish the relative size of the genetic effect in comparison with other sources of variation, such as common household effect and random environmental effect. *Familial aggregation* can be established using family-based or twin/adoption studies. Since family members share genes and environment, familial aggregation of the trait could be due to genetics and environment together. In general, very few traits are influenced only by genes or only by the environment. The detection and estimation of familial aggregation is a first step in the genetic analysis of any multifactorial trait. Twin and adoption studies are traditionally used to determine the genetic component of the trait [8–11]. Because monozygotic (MZ) twins share all their genes, any difference between them for a particular trait should be due solely to environmental effects. If the trait is completely influenced by genes, then MZ twins should have essentially identical expression of the trait; however, this is not true for dizygotic (DZ) twins because, on average, they share only 50% of their genes.

Twin studies to determine the genetic component of the threshold character are based on comparing *concordance rates* of MZ and DZ twins. If both members of a twin pair have the same status of a dichotomous trait (i.e., either both have the disease or both do not have the disease), they are *concordant*. If they do not share the trait status, they are *discordant*. The concordance rate is the proportion of concordant twin pairs among all those with the trait. Significantly higher concordance rates in MZ twins compared with DZ twins is considered evidence for a significant genetic component of the disease. The significance of the difference can be easily tested by a $2 \times 2$ contingency table using a $\chi^2$ test (see Table 11.3).

Concordance rates are not appropriate for continuous traits, so correlation coefficients can be used instead of concordance rates [12].

For continuous traits, the familial aggregation is usually measured by *heritability*, the proportion of variability of the trait explained by genetic variation. Heritability can be defined either using total genetic effects (sum of additive, dominant, and epistatic effects) or using only additive effects. The former quantity is called *heritability in the broad sense* and is given by:

$$h^2 = \mathrm{Var}(G)/\mathrm{Var}(T)$$

where Var(G) and Var(T) are genetic and total variance, respectively. The latter quantity is called *heritability in the narrow sense* and is given by:

$$h^2 = \mathrm{Var}(A)/\mathrm{Var}(T)$$

where Var(A) is the additive genetic variance.

**TABLE 11.3 Using Twin Concordance and Discordance Rates to Test for a Genetic Component of a Disease**

| Twins | Concordant Pair | Discordant Pair | Total Pairs |
|---|---|---|---|
| Monozygotic (MZ) | $n_{11}$ | $n_{12}$ | $n_{MZ}$ |
| Dizygotic (DZ) | $n_{21}$ | $n_{22}$ | $n_{DZ}$ |
| | $n_C$ | $n_D$ | $n$ |

$$\chi_1^2 = \frac{n(n_{11}n_{22} - n_{12}n_{21})^2}{n_C n_D n_{MZ} n_{DZ}}$$

[4] Good descriptions of these principles can be found at https://www.thoughtco.com/mendels-law-373515 and at https://www.thoughtco.com/independent-assortment-373514.

One cannot conclude the number of genes or which genes are involved in the etiology of the trait from a heritability estimate. Although the absence of familial aggregation is generally thought to rule out a genetic contribution to the trait, there are some unlikely, yet plausible, scenarios in which this is not true. These include phenotypic competition within families [13] that counters genetic effects and an extreme form of epistasis referred to by some as emergenesis [14]. It is also important to emphasize that because heritability is a population-specific estimate, it can vary from population to population.

The method to determine the degree of genetic component of a continuous trait is based on a comparison of the variance of the differences between MZ twins and the differences between DZ twins. Since MZ twins share all their genes, the variance of the trait between MZ twins ($V_{MZ}$) must be due to environmental variance ($V_E$), so in this case we have $V_{MZ} = V_E$. The variance of the trait between the DZ twins ($V_{DZ}$) could be due to both environment ($V_E$) and shared genes ($V_G$). So, genetic variance is $V_G = V_{DZ} - V_{MZ}$, and therefore the *heritability*, $h^2$, is defined as:

$$h^2 = \frac{v_{DZ} - v_{MZ}}{V_{MZ}}$$

Heritability ranges between 0 and 1, with 0 meaning a solely environmentallydetermined trait and 1 meaning a completely geneticallydetermined trait.

Adoption studies provide a second familial aggregation strategy for estimating the influence of genes on multifactorial traits. The strategy consists of comparing disease rates among the adopted offspring of affected parents with the rates among adopted offspring of unaffected parents. Certain biases can influence these studies, namely, (1) parental environment could have long-lasting effects on an adopted child, (2) adoption agencies attempt to match the adoptive parents with the natural parents in terms of socioeconomic status, and (3) children might be several years old when adopted, introducing the potential for many environmental confounds. Moreover, these studies are reasonably good at estimating additive genetic effects that are not age-specific, but poor at estimating nonadditive genetic effects or genetic effects that are expressed differently across the age span.

There are many other methods to detect and estimate familial aggregation using family data. For example, the recurrence risk is often used to determine the strength of familial aggregation for a discrete trait. The recurrence risk is the probability that a relative of an affected individual is also affected. The most commonly used measure is the sibling recurrence risk, i.e., the probability that a sibling of an affected individual is also affected. The ratio of the sibling recurrence risk and the overall disease prevalence is called a *sibling relative risk*. It is one of the measures of the magnitude of the genetic contribution to susceptibility for a dichotomous trait (affected vs. unaffected). Examination of relative recurrence risk values for various classes of relatives could suggest that the trait is influenced by multiple loci [15]. For a single ascertainment scheme, the sibling recurrence risk can be calculated from sibling data as follows [16]:

$$K_S = \frac{\sum_{s=1}\sum_{a=1} (a-1)\, n_{s(a)}}{\sum_{s=1}\sum_{a=1} (s-1)\, n_{s(a)}}$$

where $a$ = number of affected sibs in a sibship, $s$ = number of siblings in the sibship, and $n_{s(a)}$ = number of sibships of size $s$ with $a$ affected sibs.

Note that the aforementioned familial aggregation methods use only trait information from the sample. Owing to the availability of genome-wide single-nucleotide polymorphisms (SNP) data, it is now feasible to calculate the heritability using genome-wide SNP markers. One such method was proposed by Visscher et al. [17], who used genome-wide identity of descent (IBD) sharing probability between full sibs using genome-wide SNP data.

### 11.2.2 Segregation Analysis

Once a genetic basis of the trait has been established, the next step has traditionally been to determine the genetic models that explain the segregation of a phenotype (continuous, dichotomous, or ordinal) in a given familial data set via segregation analysis. Segregation analysis requires phenotypic data on related individuals and does not require any molecular data. Segregation analysis is the statistical methodology to determine whether a model with one or more major genes and/or polygenes (i.e., a set of genes, each with a small quantitative effect, that together produce a phenotype) is consistent with the observed pattern of phenotypic inheritance, and to estimate the parameters of the best-fitting genetic model. It entails determining the mode of inheritance (additive, recessive, or dominant), estimating "disease" allele frequency, and

Association is then tested by ascertaining whether a particular marker allele is more frequent among the cases than the controls. A significant result will be observed if the marker is in LD with the disease locus, or from a variety of confounding reasons such as population stratification. Case–control association studies done without controlling for stratification are prone to false positive results with no biological significance. For this reason, association studies were not popular until the mid-1990s, when methods to account for stratification were established using familial data [44]. To control for population stratification, Spielman et al. [44] proposed the *transmission disequilibrium test* (TDT). TDT is a family-based association test in the presence of linkage that controls for population stratification by comparing the allele frequencies among alleles transmitted to an affected offspring with those that are not transmitted to an affected offspring from informative parental matings (i.e., matings with at least one heterozygous parent). This study design requires the collection of family trios that include two parents and an affected offspring. More than 225 extensions and variations of the original TDT have been proposed (see the exhaustive review of TDT procedures in Tiwari et al. [45]). There are a number of software programs available for TDT and/or association analyses using family data, such as FBAT (family-based association test; http://www.biostat.harvard.edu/~fbat/fbat.htm), ASSOC (http://darwin.cwru.edu/sage/), and GASSOC (http://www.mayo.edu/research/labs/statistical-genetics-genetic-epidemiology/software). A complete list of association programs can be found at http://gaow.github.io/genetic-analysis-software/ and http://www.soph.uab.edu/ssg/linkage/associationtdt.

Once the results from the association analyses are deemed adequate, the next step is to screen the candidate genes for DNA sequence variation by direct sequencing. The relevance of the detected mutations is confirmed with additional association studies in the original and other populations, as well as functional assays in vitro (expression studies in different cell lines) and in vivo (transgenic and knockout animal models) [46].

## 11.3 THE INTERNATIONAL HAPMAP PROJECT

In the context presented earlier, studies progress from estimates of heritability, to segregation analysis, to linkage, and then finally to familial association analysis to determine candidate genes for a trait of interest. However, recently this paradigm has changed. With the advent of high-dimensional genotyping technologies using microarrays, the approach for discovering new genetic variants of a disease or trait has changed drastically. In 1996, Lander proposed the "commondisease, commonvariant" (CDCV) hypothesis [47]. The CDCV is based on the idea that the genetic component of common diseases is attributable in part to common allelic variants (i.e., alleles with frequency at least 5%). The HapMap project was initiated to create a dense set of genetic markers to test the CDCV hypothesis. The draft of the complete human genome sequence was completed in 2001 and had a strong effect on advances in genome sequencing technology [48]. The International HapMap Project was an international partnership that was formed in 2002 to help researchers find genes associated with human disease by providing a public database of common genome-wide human variation across populations [49–51]. The first stage of the HapMap project focused on four diverse populations: 30 trios (two parents and one adult child) from the Yoruba people in Ibadan, Nigeria; 30 trios from the Centre d'Etude du PolymorphismeHumain (CEPH) collection of Utah residents of northern and western European ancestry, 45 unrelated individuals from the Han Chinese in Beijing, and 45 unrelated individuals from the Japanese in Tokyo. This project genotyped over 1 million SNPs in phase I and an additional 2.1 million in phase II in the HapMap samples [50,51]. It helped initiate advances in SNP array technologies to make genome-wide association studies (GWAS) feasible and affordable. Affymetrix and Illumina SNP arrays became available to researchers, who initially surveyed approximately 100,000 SNPs, and who now have surveyed 2.5 million SNPs. During the most recent phase, HapMap 3, 1184 individuals representing 11 global populations were genotyped for approximately 1.6 million common SNPs [52](http://www.sanger.ac.uk/resources/downloads/human/hapmap3.html).

As a complement, the 1000 Genomes Project, which ran from 2008 to 2015, was initiated to provide a catalog of low-frequency SNPs and structural and sequence variants in the human genome [53]. The final data set contains information for 2504 individuals from 26 populations (over 88 million variants), using a combination of low-coverage whole-genome sequencing, dense microarray genotyping, and deep exome sequencing.

The results have furthered the understanding of the processes that shape genetic diversity and disease biology, while enabling genotype imputation, array design, and cataloging of variants [54–57]. The International Genome Sample Resource (http://www.internationalgenome.org) provides ongoing support for the 1000 Genomes Project data and incorporates published genomic data, including those from new populations.

## 11.4 GENOME-WIDE ASSOCIATION STUDIES

GWAS are an approach that involves scanning thousands to a few million SNPs across the whole genome on many individuals to find association with a disease or trait. As mentioned earlier, GWAS became a popular choice of genetic studies to detect putative loci associated with a disease or trait because of the availability of high-throughput SNP arrays, decreased cost of genotyping, and methods to correct for population stratification (i.e., systematic differences in allele frequencies between subpopulations in a given population possibly due to different ancestry). Before the HapMap era, investigators were reluctant to conduct association studies using population data because of concerns about population stratification. For example, in case–control studies, we usually test association of a particular SNP by comparing allele frequency between cases and controls. Allele frequencies are known to vary within and between populations depending on genetic ancestry [50,58]. Genetic ancestry becomes a confounding variable leading to spurious associations if allele frequencies are different within or between race/ethnic groups. Methods for correcting population substructure are described later.

### 11.4.1 Study Designs

Any type of data set, such as pedigree data, case–control data, or population data, are all appropriate choices for GWAS. Analysis must adjust for familial correlations in pedigree data and population stratification in population or case–control data sets to control for the confounding due to relatedness or population substructure. Case–control and population data have been commonly used for GWAS because of their availability and convenience of ascertainment; however, there are some issues associated with the case–control design. If the disease is heterogeneous, extra attention should be paid to minimizing heterogeneity in case selection, e.g., selecting the most extreme cases or selecting individuals from a familial disease cohort. There has been controversy regarding the selection of optimal controls. Usually, controls from the same population and residing in the same geographic area are preferred, but can be difficult to ascertain. The Wellcome Trust Case Control Consortium used 3000 UK controls and 2000 cases from each of seven different diseases to show that using common controls was effective, had minimal effect on genotypic distributions, and did not lead to excess false positives [59,60]; however, misclassification error in control selection could affect the power of the association analysis. Specifically, this is true for late-onset diseases because controls have not yet reached the age to develop the disease. This issue can be resolved by increasing the sample size [59]. Population stratification and cryptic relatedness (i.e., relatedness among individuals in the study that is not known to the investigator) can also increase the false positive findings, as previously discussed. The family-based association studies are robust to population stratification, but it is difficult to ascertain all pedigree members, which leads to missing data within families and loss of power compared with case–control designs [61]. There are some other issues with study design selection, and an excellent review is provided by McCarthy et al. [59].

### 11.4.2 Quality Control

The first step of GWAS analysis is the quality control (QC) of the genotypic and phenotypic data. There are a number of procedures needed to ensure the quality of genotype data both at the genotyping laboratory and after calling genotypes using statistical approaches. Here we will assume that the genotyping laboratory has used best practices to remove technical variation, and we present only statistical methods that are used after completion of the genotyping. The QC and association analysis of GWAS data can be performed using the robust, freely available, and open-source software PLINK developed by Purcell et al. [62]. Anderson et al. provide step-by-step PLINK and R commands to implement most of the procedures [63]. In addition, two publications provide excellent reviews of the QC protocol for GWAS data [64,65]. Here, we provide a few important steps of the QC in GWAS using guidelines similar to those in Laurie et al. [64] and Turner et al. [65]. Note that the current genotyping technology is very reliable, but there are still some possibilities of errors when genotyping large number of SNPs.

### 11.4.3 Sex Inconsistency

It is possible that self-reported sex of the individual is incorrect. Sex inconsistency can be checked by comparing the reported sex of each individual with predicted sex by using X-chromosome markers' heterozygosity to determine the sex of the individual empirically.

### 11.4.4 Relatedness and Mendelian Errors

Another kind of error that can occur in genotyping is due to sample mix-up, cryptic relatedness, duplications, and pedigree errors, such as self-reported relationships that are not accurate. To detect sample relatedness, one can calculate three IBD probabilities of sharing 0, 1, and 2 alleles that are IBD for each pair of individuals using software such as PLINK and a kinship coefficient matrix. Individuals sharing zero alleles at every locus are unrelated, individuals sharing one allele IBD at every locus are parent–offspring pairs, individuals sharing two alleles IBD at every locus are MZ twins or a duplicated sample, and on average sibpairs share zero, one, and two alleles IBD with sharing probabilities 0.25, 0.5, and 0.25, respectively. The relationship errors can be corrected by consulting with the self-reported relationships and/or using inferred genetic relationships. Cryptic relatedness can inflate the variance of the test statistic (e.g., if the test statistic is the difference in the overall allele counts between case and control samples in a trend test [66]). The presence of cryptic relatedness in case–control studies increases the false positives in association analysis. Devlin and Roeder provided a method to correct for the variance inflation (see [66] and [67], for details).

### 11.4.5 Batch Effects

For GWAS, samples are processed together for genotyping in a batch. The size and composition of the sample batch depends on the type of the commercial array; for example, an Affymetrix array can genotype up to 96 samples, and an Illumina array can genotype up to 24 samples. To minimize batch effects, samples with different phenotypes, sex, race, and ethnicity should be randomly assigned to plates. The downstream association study can be confounded by the batch effects. There are several methods available to detect any batch effects. The most commonly used method is to compare the average minor allele frequencies and average genotyping call rates across all SNPs for each plate. Most genotyping laboratories perform batch effect detection and usually regenotype the data if there is a batch effect or a large amount of missing data.

### 11.4.6 Marker and Sample Genotyping Efficiency or Call Rate

Marker genotyping efficiency is defined as the proportion of samples with a genotype call for each marker. If large numbers of samples are not called for a particular marker, that is an indication of a poor assay, and the marker should be removed from further analysis. The threshold for removing markers varies from study to study depending on the sample size of the study. Usual recommended call rates are approximately 98%–99%. If the quality of the DNA sample is poor, it leads to a low call rate of genotypes for the individual; i.e., the number of missing genotypes will be large and the sample should be excluded from further analysis. Before performing the association analysis, one should filter out the samples and markers using some threshold for marker and sample call rates.

### 11.4.7 Population Stratification

There are a number of methods proposed to correct for population substructure. Three commonly used methods to correct for the underlying variation in allele frequencies that induces confounding include genomic control [4,66–74], structured association testing [75–77], and principal components (PCs) [78,79]. The genomic control method estimates an inflation factor (ratio of the variance of the test statistic and the variance under the null hypothesis) and adjusts the test statistics for all markers in GWAS downward by the inflation factor. Usually, the inflation factor is calculated using a few hundred loci. Structure association testing [75,76] estimates the ancestry proportions of each individual from the founding population using markers with different allele frequencies in the founder population and then uses these proportions to cluster individuals to create homogeneous groups with similar ancestry profiles for the association analysis. Principal components analysis (PCA) uses thousands of markers to detect population stratification with a program such as Eigenstrat ([78,79], https://data.broadinstitute.org/alkesgroup/EIGENSOFT/). The PCs are entered as covariates into the association model [78,79]. There are two issues with using PCA: how many SNPs to use and how many PCs should be included as covariates in the association analysis.

### 11.4.8 Marker Allele Frequency and Hardy–Weinberg Equilibrium Filter

The Hardy–Weinberg equilibrium (HWE) test compares the observed genotypic proportion at the marker versus the expected proportion. Deviation from HWE at a marker locus can be due to population stratification, inbreeding, selection, nonrandom mating, genotyping error, actual association to the disease or trait under study, or a deletion or duplication polymorphism. However, HWE is typically used to detect genotyping errors. SNPs that do not meet HWE at a certain threshold of significance are usually excluded from further association analysis. It is also important to discard SNPs based on minor allele frequency (MAF). Most GWAS are powered to detect a disease association with common SNPs ($MAF \geq 0.05$). The rare SNPs may lead to spurious results due to the small number of homozygotes for the minor allele, genotyping errors, or population stratification.

## 11.5 IMPUTATION

Genotype imputation is the process of predicting genotypes that are not directly assayed in a sample of individuals by using a reference panel of haplotypes. Imputation methods work by identifying sharing between the reference haplotypes and the underlying haplotypes of the unrelated study subjects using local IBD patterns. Imputation provides finemapping of genomic regions, which increases the ability (power) to find causal SNPs, especially for harder-to-tag rare SNPs. Imputed SNPs that exhibit large associations may become candidates for replication studies. In meta-analyses, when different genotyping chips are used for different cohorts, imputation can equate the set of SNPs across studies. Imputation can also correct genotyping errors and extend to non-SNP variation (i.e., copy number variants, insertions/deletions). Owing to uncertainty in the imputation process, a probability distribution is calculated over all three genotypes, and this should be incorporated into any downstream analysis. The study population, properties of the reference panel and genotyping chip, and allele frequencies will affect genotyping accuracy. Error rates increase as MAF decreases because of the difficulty in tagging rare SNPs. Using a combination of populations can increase accuracy, especially at rare SNPs. Error rates also decrease as reference panel size increases [80,81]. Use of the 1000 Genomes Project reference panel instead of HapMap Project data has been shown to improve imputation accuracy, increase the strength of known associations, and identify novel loci [82,83]; however, not all of the variants present in HapMap data are in 1000 Genomes data [84]. Therefore, the choice of reference panel is an important consideration. Formatted reference panels can be downloaded from software websites.

Li et al. [80] and Marchini and Howie [81] provide excellent reviews of different imputation methods and software. Major programs include IMPUTE2 ([85,86], http://mathgen.stats.ox.ac.uk/impute/impute_v2.html), MaCH ([87], http://csg.sph.umich.edu/abecasis/MaCH/download/), Minimac([88,89], http://genome.sph.umich.edu/wiki/Minimac#Reference), and BEAGLE ([90,91], https://faculty.washington.edu/browning/beagle/beagle.html). The first step of any imputation process should be standard QC of sample data. Next, the data need to be in the correct format (i.e., MERLIN pedigree and data files, one per chromosome) and on the appropriate build (the latest 1000 Genomes reference panel uses NCBI Build 37/HG 19). The University of California at Santa Cruz (UCSC) provides an online liftOver tool for converting genome positions between builds. Markers should be stored by chromosome position, in the forward strand, and encoded as "A", "C", "G", or "T". Minimac imputation uses a two-step approach. First, the samples must be phased into a series of estimated haplotypes (i.e., using MaCH). Second, Minimac performs imputation with these phased haplotypes.

Since the imputation procedure can take weeks to complete, time savings can be achieved by imputing chromosomes in chunks. The ChunkChromosome tool can be used to divide each chromosome into smaller chunks (i.e., 2500-marker chunks, with 500-marker overhang) (http://genome.sph.umich.edu/wiki/ChunkChromosome). After imputation has been completed, imputation quality evaluation should be performed to remove SNPs with poor quality. Different programs provide different-quality statistics, which are not directly comparable. Minimac provides three helpful measurements: (1) looRSQ, the estimated $R^2$ for each SNP; (2) empRSQ, the true $R^2$ comparing imputed and actual genotypes; and (3) empR, the empirical correlation between actual and imputed genotypes (note that negative values indicate the alleles are most likely flipped). The size of the final sample (genotyped + imputed

SNPs) depends on the number of SNPs that pass predetermined thresholds [88]. Tutorials to impute 1000 Genomes SNPs using Minimac or IMPUTE2 can be found at http://genome.sph.umich.edu/wiki/Minimac:_1000_Genomes_Imputation_Cookbook and http://genome.sph.umich.edu/wiki/IMPUTE2:_1000_Genomes_Imputation_Cookbook, respectively.

## 11.6 ASSOCIATION METHODS/ STATISTICAL ANALYSIS

### 11.6.1 Power and Sample Size Calculations

Statistical power is the probability of rejecting the null hypothesis of no association when a true association is present. Power calculation requires a specified effect size and variance, sample size, and significance level (type I error, i.e., $\alpha = 0.05$). A sample size that ensures sufficient power (i.e., 80% power) is critical to the detection of causal variants in complex diseases. GWAS generally require larger sample sizes because association analyses may compare millions of SNPs, increasing false positive rates. In addition, causal SNPs may have small effect sizes that are difficult to find in small samples [92]. Since Mendelian diseases arise from single-gene mutations, the causal mutations have an enormous impact on disease risk, and the large effects can frequently be detected with modest sample sizes. In contrast, the genetic structure underlying common diseases involves multiple risk loci and environmental factors. Doubling the number of subjects generally doubles the number of SNPs passing the $5 \times 10^{-8}$ significance threshold, but the rate of increase and the minimum sample size depend on the disease complexity. In Crohn disease and ulcerative colitis, increasing beyond ~5000 cases was required to see a doubling effect, whereas in type 2 diabetes and breast cancer, the number was ~30,000 [93,94].

The power to detect significant associations at the *P* value threshold is affected by allele frequency, LD, inheritance model, disease prevalence, effect size, and phenotype variation [95–97]. Although most of these are inherent characteristics of the study population, there are factors the investigator can change to maximize power, such as choice of study subjects and methods of phenotype and genotype measurements, data QC, and statistical analyses. In population cohort studies, using a liability threshold model (in which significance is affected by the proportion of variance in liability explained by the locus) can simplify calculations. In case–control studies, it is easier to use an assumed odds ratio and the allele frequency of the putative risk variant. Since many GWAS rely on LD between typed and untyped variants to increase coverage, the extent of LD has a large effect on power. When a proxy SNP that has correlation $R$ with the causal SNP is used, the sample size required to obtain the same amount of power is increased by a factor of $\sim 1/R^2$ [94,95]. Allele frequency is an important consideration because (1) locus allele frequencies determine heterozygosity, which is directly proportional to the phenotypic variance explained by the locus, and (2) the SNPs chosen for commercial SNP arrays represent common variations; rare variants are less likely to have a high correlation ($R^2$) with included SNPs. These factors are relevant when imputation is used because the statistical power for untyped SNPs depends on imputation quality, which is determined by the level of LD between typed and untyped SNPs. Longitudinal studies or multivariate outcomes can help increase power in studies of phenotypes that are not distinctive, rare, stable, and/or highly familial (i.e., hypertension and depression). Assuming the correct effect model is also important since, for example, a test that assumes additive effects has more power than one that assumes dominant effects, if the true effects are indeed additive. Last, appropriate data QC, inclusion of covariates (except perhaps in logistic regression analysis of case–control studies) [98], and correction for confounding factors will increase statistical power [94]. Sham and Purcell provide an excellent review of statistical power and significance testing in large-scale genetic studies, including more detail about Mendelian and complex diseases, tests for rare variants, and exome-sequencing studies [94].

Owing to the complicated nature of LD in the human genome, analytical power calculations are difficult and simulations are frequently used [99]. In situations in which simulations require extensive time and computational resources, Tiwari et al. provided a quick method for power estimation that is applicable to a wide range of genomic studies [100]. This method is based on Elston's Excellent Estimator formula and uses data similar to a user's data, but differs in sample size and/or $\alpha$ level ([100]; download at http://www.ssg.uab.edu/eee-power/).

There are several freely available GWAS power analysis programs. Genetic Power Calculator provides power analysis for common tests, including variance components QTL linkage and association tests, TDT, case–control designs, and sibships [101]. The GAS and CaTS power calculators are for one- or two-stage GWAS [102]. PGA performs power analysis for case–control association studies of candidate genes, fine-mapping studies, and whole-genome scans [103]. R packages include *gap* (genetic analysis package for population- and family-based designs) [104] and *powerpkg* (power analyses for the affected sib pair and the TDT design) ([105], download links for these programs at http://gwa-testdriver.ssg.uab.edu/software.jsp). *GWApower* is another R package that calculates power for GWAS using commercially available genotyping chips and simulation results generated from the HAPGEN program ([99], http://www.well.ox.ac.uk/software). Lee et al. provided derivation of sample size/power for rare variants ([106], https://www.hsph.harvard.edu/skat/download/). GCTA (genome-wide complex trait analysis) provides power calculation for estimating genetic variance or correlation using genome-wide SNPs in GREML analysis ([107,108], http://cnsgenomics.com/shiny/gctaPower/). Finally, QUANTO may be used for geneticepidemiology studies ([109,110], http://biostats.usc.edu/Quanto.html).

## 11.6.2 Discovery Phase of the Genome-Wide Association Study

The choice of the statistical test for association in discovery phase depends on the study design and the phenotype under consideration. In a case–control design, the goal is to compare the allele or genotypic frequencies between cases (affected) and controls (normals). This can be tested with Pearson's $\chi^2$ test, Fisher's exact test, or the Cochran–Armitage test. Pearson's $\chi^2$ tests the null hypothesis of no association between rows and columns of the $2 \times 3$ contingency table consisting of the counts of the three genotypes among cases and controls [111]. Fisher's exact test is similar to Pearson's test, but the deviation from the null hypothesis is calculated exactly from all possible permutations of the data and, thus, does not assume the asymptotic $\chi^2$ distribution [112]. The Cochran–Armitage test for trend is a test of proportions of cases versus controls [113–115] and assumes an additive mode of inheritance that is a linear trend. However, there is a loss of power if the trend is not linear. Freidlin et al. recommended using a maximum of the test statistics obtained from additive, dominant, or recessive effects models [116]. Note that in these statistical procedures, one cannot model covariates such as sex, age, race, age of onset, PCs (from admixture), etc. To accommodate any relevant covariates in the analysis, one can use logistic regression. Logistic regression is more flexible in that it can model covariates, multiple SNPs as main effects, SNP-by-SNP interactions, SNP-by-environment interactions, etc. If the phenotype is continuous, analysis of variance (ANOVA) and general linear model approaches can be employed. One can also use a linear regression framework if extremes of the distribution are used to define case and control status. Huang and Lin have given an efficient association method using extreme phenotypes [117].

The analysis of familial data requires correcting for the dependency of observations. The notable methods include linear mixed model [118–126], *GEMMA* (genome-wide efficient mixed model analysis for association studies [122,123], http://www.xzlab.org/software.html), FBAT (see review by Laird and Lange [124]), ASSOC (a module of the S.A.G.E. software suite [36]), the R packages *GenABEL* (GWA analysis for quantitative, binary, and time-till-event traits [118]) and *ProbABEL* (GWA analysis of imputed genetic data [121], http://www.genabel.org/packages/ProbABEL), lme4 [125], and *GWAF* [126]. After scanning ~2–9 million imputed SNPs (i.e., ~2–9 million statistical tests) to determine significant associations, appropriate multiple testing correction is required to control for false positives and in choosing SNPs for follow-up studies. The guideline for significant association for GWAS is generally a *P*value of $\sim 5 \times 10^{-8}$ [127]. However, it is common practice to use a higher *P* value threshold for follow-up study or replication. Balding [128] provides a comprehensive discussion of the advantages and disadvantages of these methods pertaining to GWAS.

## 11.6.3 Postanalysis Quality Control

After association testing has been done, QQ plots, which plot observed *P* values versus *P* values expected by chance, should be used to determine if there are still issues with the data. For example, there may still be confounding factors (i.e., population stratification, batch effects, or other systematic group differences) that are creating falsepositive results. Traditionally, $-\log_{10}$ (*P*values) are used to spread the points out and facilitate detection of unusual results. The *inflation factor*, $\lambda$, measures the deviation of the observed results from expected results. If $\lambda$ is greater than 1, the QQ plot is inflated, most likely from residual confounding factors. If λ is less than 1, the QQ plot is deflated, most likely from overcorrection in the association analysis (i.e., too many covariates). In either case, test statistics should be adjusted. If most of the *P*values occur on the identity ($y=x$) line, with some falling above, the results indicate that these SNPs may have significant associations with the phenotype. Next, Manhattan plots can be used to display small *P*values by genomic position [129,130]. The interactive program ManhattanPlotter (http://www.biologiaevolutiva.org/~cmorcillo/tools/ManhattanPlotter/ManhattanPlotter.htm) and the R packages *Manhattanly* and *qqman* provide functions to create QQ and Manhattan plots [131,132].

LocusZoom plots enable you to focus on a region and view the strength of association signals, local LD, recombination patterns, and nearby genes (Fig. 11.5). LocusZoom is a program that uses LD information from HapMap phase II or 1000 Genomes and gene information from the UCSC browser ([133], https://statgen.sph.umich.edu/locuszoom/download/). LocusExplorer is another program useful for visualization and exploration of genetic association data ([134], http://www.oncogenetics.icr.ac.uk/LocusExplorer/). Last, cluster plots are used to examine SNPs with significant departure from HWE or a high Mendelian error rate. Well-defined clusters indicate high-quality SNPs, whereas poorlydefined clusters may indicate a poor call rate or other genotyping issue. The Bioconductor package *GWASTools* has functions that execute most of the GWAS QC procedures described here [135,136].

### 11.6.4 Validation and Replication Phase

Some investigators have recommended reanalysis of the original discovery phase GWAS data using a different genotype platform for validation, which has been termed "technical validation" [59]. Technical validation allows the detection of technical errors in genotyping that might give rise to spurious association signals or false positives; however, given the limitations of the resources available to investigators, it may not be feasible. The replication phase or follow-up study is one of the most challenging aspects of the GWAS and is required to control for false positives. Replication in an independent data set with similar genetic background and phenotype is warranted. Usually several hundred or a few thousand SNPs are tested in a replication set depending on the threshold used for significant association *P*value. The statistical methods are the same as in the discovery phase depending on the study design and type of phenotype. SNPs for replication could be selected based on several criteria, including strength of

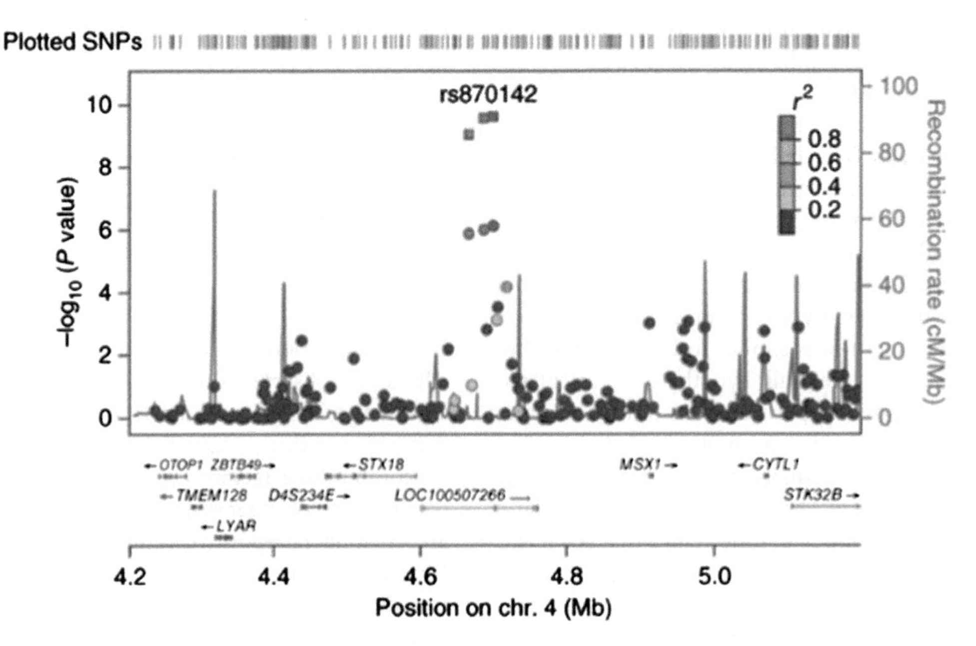


**Figure 11.5** LocusZoom plot displaying the HDL cholesterol–associated region near the *MMAB* gene. (Taken from Kathiresan et al. Nat Genet 2009;41(1):56–65.)

association in GWAS (e.g., $P < 10^{-8}$), size of potential regions and/or candidate genes involved, degree of LD in these regions, presence of coding SNPs of reasonable frequency (>1%), potential biological relevance to the disease/trait and previously identified association with disease/trait with lower threshold (with $P < 10^{-6}$), significant SNPs close to transcription binding sites or close to microRNAs with lower threshold (with $P < 10^{-6}$), and SNPs in disease/trait relevant pathways. Note that the *P* value thresholds as given for selecting potential SNPs for replication are arbitrary and should be decided by an investigator(s) depending on how many SNPs are to be replicated.

## 11.7 ANALYSIS OF RARE VARIANTS USING NEW TECHNOLOGIES

Introduction of the HapMap project and large-scale GWAS studies was driven by the CDCV hypothesis, which was first introduced in the 1990s [47,137]. The CDCV is based on the idea that the genetic components of common diseases are attributable in part to allelic variants that are present in more than 5% of the population. An extension of this hypothesis is that the same variants will be responsible for the disease across multiple populations [138]. The early success of GWAS (i.e., in age-related macular degeneration) seemed to support the theory that a large proportion of the genetic variants underlying complex disease could be explained by the CDCV. It is now becoming apparent that many common variants confer only a small portion of risk individually and explain only a small portion of heritability of common complex diseases [139]. While GWAS have been successful in many ways, identifying hundreds of variants for a large number of traits (http://www.ebi.ac.uk/gwas), there still remains a large proportion of heritability that has yet to be explained.

When the CDCV hypothesis was first introduced, it was not without contention [140]. One of the strongest counterarguments was based on the hypothesis of "common disease, rare variant," which is in essence the antithesis of the CDCV hypothesis [80,141,142]. The rare variant hypothesis proposes that common complex diseases are due to the combined effects of multiple rare variants with moderate to low individual risk. Unlike CDCV, it is generally thought that, due to population history, these rare variants will be population-specific [143]. It is only recently, with the availability of affordable large-scale sequencing technology and advances in analytical methods (discussed later), that scientists have gained the ability to address the role of CDRV in human disease. It is likely that the genetic basis of complex disease is somewhere between the two extremes, with multiple genes interacting together with a variety of common and rare variants and other genetic and environmental factors [144].

As new high-throughput, massively parallel sequencing technologies emerged in 2005, direct sequencing became commonly used to directly interrogate whole genomic sequences for association with disease without prior specification of SNPs currently available on commercial SNP chips [145]. Such technologies overcome some of the shortcomings of GWAS methods, such as ascertainment bias in the set of currently available SNPs and the ability to assay rare or private variants. In addition, greater flexibility exists in the search for variants other than SNPs, such as copy number variants; insertions or deletions, or indels; inversions, etc. Whole-exome sequencing, in which only the sequences of exons are assayed, has been used to discover causal mutations in a number of Mendelian disorders, such as Miller syndrome [146] and hereditary spastic paraparesis [147].

Because of the enormous number of variants introduced from new sequencing technologies, and the small sample sizes typically present, new bioinformatic and statistical methods have been developed to reduce the dimensionality and improve the probability of detection of causal variants. Prior bioinformatic processing may include filtering by IBD methods, if family data are present [148], or filtering based on the expected mode of inheritance [146]. In addition, if only rare variants are desired, then common variants can be filtered out using the SNP database dbSNP [149]. Predicted functional variants (nonsense, missense, splice site variants, indels, frameshift mutations, etc.) can be discerned using tools such as SIFT and PolyPhen [150,151]. Once likely nonfunctional variants have been filtered out, new statistical methods for summarizing the effects of multiple rare variants at a single gene can be applied. Some examples of these methods include the *cohort allelic sums test* method, which compares the number of individuals with mutations within a gene between cases and controls [152]; the *combined multivariate and collapsing* method, which collapses multiple rare variants in conjunction with common variants using multivariate analysis [153]; the *sequence kernel association test*, which tests for association between common or rare variants and a continuous or dichotomous trait while adjusting for covariates [154]; methods that weight the counts of each variant using the estimated standard deviation of the total number of mutations [155,156]; or a method that models these weights in a flexible Bayesian framework [157]. A review of rare variant analysis methods is given in [158]. Whole-exome and whole-genome studies (in which the contribution of noncoding regions to disease can be assayed) are underway as of this writing for complex (multifactorial) diseases, and the next couple of years will show if these technologies can help to fill in the gaps from GWAS, termed "missing heritability"[139], in identifying causal variants underlying multifactorial diseases. In addition, new sequencing technologies offer opportunities for functional characterization studies, such as gene expression profiling using next-generation sequencing [159] and epigenetic profiling [160], and in identifying somatic mutations occurring in cancer [161–163].

## 11.8 STATISTICAL FINE MAPPING OF GENOME-WIDE ASSOCIATION STUDIES DATA SETS TO DETERMINE CAUSAL VARIANTS

Fine mapping to determine the causal variants using GWAS has been investigated by integrating functional annotations with GWAS SNPs in a Bayesian framework [164,165]. Statistical fine mapping involves several steps. (1) Select all SNPs with LD ($>r^2=0.3$), with lead SNPs as potentially causal. (2) Determine the functional annotations using ANNOVAR [166], which can be incorporated into the analysis pipeline. Since the majority of the loci observed in GWAS are located in noncoding regions outside of protein-coding regions [167–169], noncoding annotation is crucial to make sense of GWAS findings. The noncoding annotations can be found using ENCODE [170], the NIH Roadmap Epigenomics Consortium [171], and FANTOM5 [172] projects. (3) Predict coding and noncoding deleteriousness of the variants (strategy is given later). (4) Integrate the functional data probabilistically to prioritize causal variants in statistical fine mapping using a Bayesian framework for both coding and noncoding regions, for example, using a hierarchical model for jointly analyzing GWAS and genomic annotations using a Bayes factor [168]. There are several other approaches using Bayesian frameworksthat have been proposed, namely fGWAS [173–176], PAINTOR [177], PICS [178], CAVIAR [179,180], and GoShifter [181]. There is no gold standard method for fine mapping using functional variant information to determine the causal variants. For example, one could use PAINTOR for fine mapping, which is suitable for transethnic fine-mapping studies, and then validate the findings from PAINTOR with CAVIAR software as well as methodology proposed by van de Bunt using an approximate Bayesian approach [182].

**Coding and noncoding regions functional scores derivation:** Accurate deleteriousness prediction for nonsynonymous variants is important for distinguishing causal mutations from background polymorphisms. Although many deleteriousness prediction methods are developed, their prediction results are sometimes inconsistent with one another and their merits become unclear in practical downstream variant association–based analysis. For example, rs334, a nonsynonymous exonic variant, is well known as a surrogate variant for sickle cell anemia, but the CADD score is 7.277, which is much lower than the recommended score (>10) for conserved or damaging variants. To address this issue, a comprehensive approach of integrating annotation information from different categories of prediction approaches, such as (1) function prediction scores like SIFT [150], PolyPhen2 [151], and MutationTaster (http://www.mutationtaster.org/); (2) conservation scores like GERP++ [183]; and (3) Ensembl scores like CADD [184] and CONDEL (http://bg.upf.edu/fannsdb/help/condel.html), should be used for the annotation of protein consequence, to identify potential nonsynonymous mutations with deleterious consequence on theprotein. In addition, functional scores of noncoding variants (GWAVA [185]) and the recently developed programs EIGEN [186], for prediction of functional impact for coding and noncoding variants, and PINES [187], which is designed specifically for prediction of functional impact for noncoding variants, should be used, since most of the GWAS significant SNPs may be in noncoding regions.

## 11.9 TRANSETHNIC META-ANALYSIS

Transethnic meta-analysis can be performed using MANTRA [188] or PAINTOR [189]. MANTRA is a meta-analysis method developed for transethnic fine mapping, which maximizes homogeneity between closely related populations while allowing for heterogeneity between more diverse groups. Recently, transethnic meta-analysis methods have been used to discover genetic variants associated with a number of complex traits [190–195].

## 11.10 OTHER DATA TYPES AND THEIR ANALYSIS METHODS

### 11.10.1 Gene Expression and RNA-Seq Data

RNA-Seq analysis has become a standard method for global gene expression analysis. Although microarray gene expression statistical methods are well established, there does not exist a gold standard pipeline for analyzing RNA-Seq data. Unlike the microarrays, RNA-Seq does not operate on predetermined selection of cDNA probes, thereby offering a perfect proxy not only for expressed transcript abundance for known genes, but also for the detection and quantification of (1) splice isoforms, (2) novel transcripts, and (3) protein–RNA binding sites. Once RNA abundance is quantified, comparison of gene expression differences and changes between samples representing different treatment/biological conditions has become a well-established application of RNA-Seq. The major steps involved in RNA-Seq data analysis are (1) experimental design consideration, (2) QC, (3) read alignment, (4) quantification of gene and transcript levels, (5) visualization6,(6) differential gene expression, (7) alternative splicing, (8) functional analysis, and (9) gene fusion detection. The experimental design involves randomization and selection of the library type, sequencing depth, and number of replicates. The number of replicates and sequencing depth are crucial to detect significant differences in a transcript or gene between experimental groups. The statistical power to detect differential expression varies with desired effect size, sequencing depth, and number of replicates. A number of power calculation tools are available according to experimental design; for example, the Bioconductor package *RNASeqPower* could be used to calculate power assuming effect size, sequencing depth, and number of replicates ([196], https://bioconductor.org/packages/release/bioc/html/RNASeqPower.html). Also, *RNAseqPS* ([197] with web interface at https://cqs.mc.vanderbilt.edu/shiny/RnaSeqSampleSize/) is a convenient tool to calculate power and sample size, and Scotty (http://scotty.genetics.utah.edu/) calculates the power based on the number of biological replicates, sequencing depth, and cost per replicate and per million reads sequenced. Illumina software generates *FASTQ* files (sequence and quality score). The QC pipeline is crucial in obtaining the highest quality data for all subsequent analyses. QC metrics include sequence quality, GC content, presence of adapters, overrepresentation of *k*-mers and duplicated reads, amount of ribosomal RNA remaining after poly(A) selection, quantification of 3′ end bias, detection of viral RNAs through alignment of sequencing reads to a viral genome database, and percentage of reads that align to the genome and transcriptome. Percentage of mapped reads is a global indicator of overall sequencing accuracy and the presence of contamination. Another important metric is the uniformity of read coverage on exons and the mapped strand. *FASTQC* (http://www.bioinformatics.babraham.ac.uk/projects/fastqc/) is a commonly used tool to perform QC for Illumina RNA-Seq data. The next step involves alignment of reads to a reference genome or mapping to an annotated transcriptome. *TopHat2* [198]/*STAR2* [199]/*Bowtie2* [200] are popular for mapping to gene and transcriptome. The transcript-specific annotation can be done using *GenomicFeatures* ([201], http://www.bioconductor.org/packages/2.12/bioc/html/GenomicFeatures.html). The next step is to estimate gene and transcriptome expression quantification. Programs such as *HTSeq* or *featureCounts* can provide a table of aggregate raw counts of mapped reads [202]. However, the raw read counts are affected by factors such as transcript length, total number of reads, and sequencing biases. Since longer transcripts and deeper sequencing give more reads, the initial solution was proposed to calculate the reads per kilobase per million mapped reads (RPKM) [203]. Subsequent measures were proposed, such as FPKM (fragments per kilobase of exon per million fragments mapped) or TPM (transcripts per kilobase million) [204]. Another issue is that *HTSeq* discards reads mapping to multiple locations, so an alternate method, *RSEM* (RNA-Seq by expectation maximization), that assigns the reads to different locations can be used to quantify the expression from the transcriptome [205]. In addition, the *RSEM* algorithm returns the TPM values for downstream differential gene expression analysis. RPKM, FPKM, and TPM normalize away the most important factor for comparing samples for differential expression analysis. There are a number of normalization methods that have been proposed for adjusting length and total reads, such as *TMM* [206], *DESeq* [207], *UpperQuartile* [208], etc. A good review of comparisons of these approaches is given in Dillies et al. [209]. *NOISeq* is a useful R package that contains a variety of diagnostic plots to identify sources of biases in RNA-Seq data and then applies appropriate normalization procedures in each case [210]. Since RNA-Seq data are read count data, they must be modeled with a Poisson, negative binomial, or zero-inflated negative binomial distribution. One proposed method is a Bayesian hierarchical model that uses a Poisson distribution, conditional on baseline expression and posttreatment expression level fold change [211]. Another method uses a negative binomial model that assumes common dispersion across all genes and executes an exact test for differential expression [212,213]. The negative binomial model was later extended to generalized linear models, making the method applicable to general experiments [214]. Last, the Bioconductor R package *edgeR* [215,216] and *DESeq2* [217] provide several tools for differential gene expression analysis. *TopHat2/Bowtie2* and *STAR2* aligners also offer built-in options to quantify alternate splicing patterns and gene fusion detections. To identify molecular pathways that are differentially expressed between the biological conditions, the top differentially expressed genes are further investigated for their coexpression pattern in several biological, process, and metabolic pathways. As part of downstream functional analysis, tools like *Ingenuity Pathway Analysis* framework [218], *WebGestalt* [219], and *DAVID* [220–223] can be used to conduct pathway-level analysis.

### 11.10.2 Epigenome and Methylation Data

For multifactorial diseases/traits, there often exists the problem of missing heritability [139]. The study of epigenetics may provide vital information in a better understanding of this phenotypic variability among individuals, since epigenetic modifications made to the genome regulate the induction and silencing of gene expression. Methylation of CpG's in CpG islands of promoters, which are typically unmethylated for most genes, is frequently associated with gene silencing. DNA methylation, a type of epigenetic modification, occurs when a methyl group is added to the cytosine base, primarily within the context of CpG dinucleotides. There are two major categories for methylation data creation, namely, microarray-based methods that require hybridization and next-generation sequencing methods. Array hybridization utilizes fluorescence signals to detect methylation levels that can be noisy. The Illumina Infinium methylation arrays have been the most common way to investigate the role of methylation across the human genome for diseases or traits as of this writing. Bisulfite sequencing will become more common as the cost goes down. Here, we will provide a short introduction for both types of data and available software packages starting with array data and then bisulfite sequencing data.

### 11.10.3 Methylation Data From Arrays

Methylation data from arraysin the human genome are measured by Betavalues, which are the ratios of methylated to overall probe intensities, and Mvalues, which are the $\log_2$ ratios of methylated probe to unmethylated probe intensities. The analysis of methylation data by arrays includes probe filtering (detection *P*value, bead count, and SNPs), background correction, adjustment for type II chemistry bias, normalization, cell composition correction, batch effect analysis, batch effect correction, poor performing probes filtering, and differential methylation analysis. There are several R packages available for analyzing methylation data, including *wateRmelon* [224], *methylumi* ([225], https://www.bioconductor.org/packages/release/bioc/html/methylumi.html), *minify* [226,227], *ChAMP* [228], and *RnBeads* [229], to name a few. All of these packages allow the user to import raw IDAT files or tabular methylation values; however, *methylumi* and *wateRmelon* do not provide a complete pipeline from raw data to identification of differentially methylated regions (DMRs). Some packages, such as *minifi* and *RnBeads*, provide a complete pipeline for QC to statistical testing for DMRs. The latest version of the *ChAMP* package offers additional features like detection of differentially methylated blocks, quantification of copynumber events from 450k/EPIC arrays, and *gene enrichment set analysis*modules.

Sequencing approaches for DNA methylation fall into two main categories, capture-based or enrichment-based sequencing and bisulfiteconversion–based sequencing (BS-Seq). Capture-based sequencing uses methyl-binding proteins or antibodies to capture methylated DNA followed by sequencing. Capture-based sequencing falls into three categories: methods based on methylated DNA immunoprecipitation (such as *MeDIP-Seq* [230]), restriction enzymes sensitive to DNA methylation (such as *MRE-Seq* [231]), and methyl-binding domains to enrich for methylated DNA (such as *methylCap-Seq* [231], *MBD-Seq* [232], and *MIRA-Seq* [233–235]). The drawback of capture-based sequencing is that it has low resolution for methylation detection over 100- to 1000-bp regions. In BS-Seq, bisulfite treatment converts unmethylated cytosines (C's) to thymine (T's) (via uracil), while methylated C's remain unchanged. In BS-Seq, there is no enrichment for the methylated DNA since the whole genome is treated with bisulfite, captured, and then sequenced, providing single-base-pair resolution and methylation status of each CpG site. Therefore, BS-Seq follows fewer steps: treat fragmented DNA with bisulfite (unmethylated C will be converted to U and amplified as T, and methylated C will be protected and remain C, and no change to other bases), amplify the treated DNA, sequence the DNA sequence, align sequence reads to the genome, and then analyze the data. Another affordable alternative to genome-wide methylation sequencing is reduced representation bisulfite sequencing (RRBS) [236,237]. While RRBS is enriched for CpG-rich regions of the genome and hence its high-resolution coverage offerings, its counterpart, whole-genome bisulfite sequencing (WGBS), is a better option for comprehensive cytosine modification profiling. Targeted bisulfite sequencing uses the best of bisulfite sequencing with high-throughput sequencing, but works on a predetermined set of genomic regions of interest that revolve around regions of differentially methylated gene-regulatory elements. Understanding the intricate interactions between gene transcription factor binding sites (TBFS) and DNA methylation can also be well studied using both targeted and WGBS approaches. Packages like *MeDReaders* [238] offer a platform for exploring already cataloged interactions between TBFS and DNA methylation as well as overlapping methylation findings from one's own research. There is no gold standard for analysis of the BS-Seq data. There are several tools available for data preprocessing (for experimental design, read QC, bisulfite conversion rate estimation, base calling, and adapter rimming), data processing (read alignment, methylation scoring, QC, quantitative score assessment), data analysis (DMR detection, transcript binding prediction, interpretation, etc.), and data visualization (https://omictools.com/bs-seq-category). The two most common software systems used for alignment of BS-Seq data are *BISMARK* [239] and *BSMAP* [239]. The goal of the experiments is to identify methylated regions or loci, so the approaches that borrow information across sites are preferred. There are two common approaches used, namely, the smoothing approach and the Bayesian hierarchical models. A useful software package, *bsseq*, includes smoothing, smoothed *t*-test, Fisher's exact test, DMR identification, and a tool for visualization of results [240]. In contrast, the Bayesian hierarchical technique models biological and technical variation separately using a Betabinomial hierarchical model [241]. Robinson et al. have provided a very nice review of methylation methods for both array-based and sequencing-based methods [242]. Betabinomial distributions have been used to model methylated reads in several software packages, such as *DSS* [241], *BiSeq* [243,244], *MOABS* [245], *RADMeth* [246], and *MethylSig* [247], to identify differentially methylated loci or DMRs. A broader outlook on the involved methylation technologies and their analysis tools is given in Kurdyukov and Bullock [248]. Understanding the coverage of methylation loci, regions, and blocks on a genomic landscape to infer their distribution across gene, promoter, regulatory, intronic, and exonic regions involves the integration of genomic annotation datasets. Software packages like Genomation [249] and ChIPpeakAnno [250] offer annotation datasets that both RRBS and WGBS analysis modules can benefit from in their cytosine modification findings.

### 11.10.4 Proteome and Protein Data

Proteomics includes "the systematic identification and quantification of the complete complement of proteins (the proteome) of a biological system (cell, tissue, organ, biological fluid, or organism) at a specific point in time."[6] Processing and analysis of proteome data is a complex, multistep process that involves liquid chromatography (LC) coupled to mass spectrometry (MS). The two most common approaches are shotgun MS, which achieves a deep coverage of the proteome, or targeted MS, which uses a defined set of peptides [251]. The Seattle Proteome Center provides the Trans-Proteomic Pipeline, a group of tools for MS-based proteomics, including statistical validation, quantitation, visualization, and

[6] https://www.nature.com/subjects/proteomic-analysis.

data conversion. First, MS spectra are assigned to a database search (i.e., X!Tandem, SpectraST, SEQUEST, or Mascot). *PeptideProphet* then validates the peptide assignments and computes a probability that each is correct [252]. Further peptide-level validation may be performed with *iProphet* [253]. In the case of quantitation experiments, *ASAPRatio* measures the relative expression levels of peptides and proteins from isotopicallylabeled samples [254]. *ProteinProphet* then groups peptides by their corresponding protein(s) to calculate probabilities that those proteins were present in the original sample [255]. *Pep3D* is a useful tool for visualizing LC–MS data and results from *PeptideProphet* [256]. Finally, *reSpect* may identify more peptides from existing spectra without collecting further data [257]. Information on the use and download of these tools can be found at [258] http://tools.proteomecenter.org/wiki/index.php?title=TPP_Tutorial. A bioinformatic analysis of the proteomic data pipeline can be accessed in Schmidt et al. [254].

Functional analysis of a large protein list begins by connecting the protein name to a unique identifier (i.e., Entrez Gene, Unigene, UniProt, etc.) and its associated Gene Ontology (GO) term (part of a biological process, molecular function, or cellular component) ([259], http://www.geneontology.org). *GO-term enrichment analysis* (DAVID [220–223]), *Babelomics* 5 ([260], http://babelomics.bioinfo.cipf.es/), and *GSEA* ([261,262], http://software.broadinstitute.org/gsea/index.jsp) then compare the proportion of specific GOterms in the sample with the natural abundance in the reference dataset [263]. This is followed by pathway analysis, protein–protein interaction analysis, and protein domain and motif analysis [251]. A comprehensive list of biological pathway and molecular interaction resources can be found at *Pathguide* ([264], http://www.pathguide.org/) For proteins of unknown function, a *BLAST* search against the database of known protein sequences can determine if proteins with similar amino acid sequences have been described in other organisms ([265], https://blast.ncbi.nlm.nih.gov/Blast.cgi). The Proteome Exchange project (http://www.proteomeexchange.org) operates the databases PRIDE, Proteome Commons, and Peptide Atlas, which provide stored proteomic datasets [266–268]. An extensive list of gene and protein analysis programs and databases can also be found at http://www.humgen.nl/programs.html.

### 11.10.5 Metabolome and Metabolite Data

Metabolomics is the study of the metabolite composition of a cell type, tissue, or biological fluid. Metabolomics is the fastest growing area of research. Metabolites represent intermediate phenotypes that lead to clinical phenotypes. Metabolomic analyses consist of several steps, namely designing the experiments, sample preparation (grinding, freeze-drying, dilution, etc.), metabolite extraction (targeted or untargeted), derivatization (used only for gas chromatography [GC]), metabolite separation (using LC, GC, or capillary electrophoresis), detection (mass spectroscopy, near-infrared spectroscopy, or nuclear magnetic resonance [NMR]), and data processing and statistical analysis. The most common techniques used in data acquisition are mass spectroscopy and NMR. The aim of the data processing is to extract biologically relevant metabolite features from the data. A good understanding of extraction methods is crucial in minimizing the risk of false positive results in downstream statistical analysis. Thus, processing consists of noise filtering, peak detection, peak alignment, normalization, and deconvolution. A feature is typically a peak or signal that represents a metabolite compound. Once the metabolite features are quantified, then the downstream statistical analysis could follow with univariate or multivariate statistical analyses. Commonly used univariate methods include Student's *t*test and ANOVA to assess differences between two or more groups (case–control or treatments assigned to different groups). These methods assume that metabolite features are normally distributed and there is no heteroscedasticity (i.e., variability among different groups is unequal), so they may not be appropriate if the statistical assumptions are not met. We can test for these assumptions using the Kolmogorov–Smirnov normality test or Bartlett's homogeneity of variances test. Another option is to use nonparametric tests, such as Mann–Whitney *U* test or Kruskal–Wallis one-way ANOVA, that do not depend on the normality or homogeneity of variance. If the outcome of interest is continuous, linear regression models can be used. Univariate analysis tests each feature separately, leading to the multiple testing problem. Several well-known approaches have been used to adjust for multiple testing, namely Bonferroni correction or false discovery rate (FDR) [269–272]. The Bonferroni correction assumes the independence of each hypothesis test, which is not true in general for untargeted metabolomics studies.

Also, Bonferroni correction is the most conservative. Thus, FDR has been favored over Bonferroni correction for metabolomics studies. In contrast to univariate methods, multivariate methods do not assume independence of metabolites and model a group of metabolites clustered together, resulting in reduction of multidimensionality. Multivariate methods are classified into supervised and unsupervised methods. Supervised methods, which use both metabolite and trait information, include support vector machines, neural networks, linear discriminant analysis, regression analysis, logistic regression analysis, regression trees, naïve Bayes, inductive logic programming, etc. Partial least squares (PLS) [273], a supervised method, is most commonly used in identifying the metabolomics patterns associated with the variable of interest. The advantage of PLS is that it can be used as a regression analysis for continuous outcomes or discriminant analysis [274], for binary outcomes. The drawback of the PLS is that some metabolites that are not associated with the outcome variable can influence the results. To alleviate the problem, orthogonal PLS can be used; since data variance is factorized into two orthogonal components, one is correlated with the outcome of interest and the other is uncorrelated [275]. The most commonly used unsupervised methodin metabolomics is PCA. PCA creates new variables (PCs) by linear combinations of metabolites while maximizing the variance and minimizing the covariance between components as well as reducing the data dimension. This new variable can be correlated with the outcome of interest. Other unsupervised methods, such as *hierarchical clustering analysis*, have been used to analyze metabolites. Software to perform the processing of the metabolite data sets is provided by individual instrument manufactures. There are some R packages that can be used to analyze metabolomics data, such as *XCMS* (available online as well as command line level using R), a very powerful software for the analysis of metabolomics data sets written in R ([276], https://xcmsonline.scripps.edu), and *MetaboAnalyst* ([277], www.metaboanalyst.ca).

## 11.11 FUTURE DIRECTIONS/INTEGRATION

Integrating information from the genome, epigenome, transcriptome, proteome, metabolome, phenome, etc., to understand the biological mechanisms of multifactorial traits is challenging. Integration of different omic data sets faces issues such as computational burden, high dimensionality of data, connectivity, data heterogeneity and comparability, visualization, multiple testing, and downstream analysis with different pathways and networks. For example, in expression quantitative trait (eQTL) analysis, associations between genotypes from GWAS and transcriptome profiles are sought to unravel the genetic basis of multifactorial diseases. Similarly, association of genotypes can be tested with methylation (meQTL), protein (pQTL), and metabolites (mQTL) (see Fig. 11.6).

*MatrixEQTL* is a computationally efficient software and can be used for eQTL, meQTL, pQTL, and mQTL studies [278]. Several tools have been developed to help integrate multiple omic data sets based on pathway enrichment analysis from gene or protein expression and metabolomics data (*IMPALA* [279]); transcriptome, proteome, and metabolome data sets (*iPEAP* [280]); and transcriptome and metabolomic data (*MetaboAnalyst* [277]). Pathway-based approaches rely on known pathways, which may not represent the complexity of biological systems. Another set of tools has been developed based on biological network analysis; for example, *SAMNetWeb* [281] and *pwOmics* [282] model transcriptomics and proteomics data sets, *MetScape* [283] models gene expression and metabolite data, and *Grinn* [284] uses genomic, proteomic, and metabolomic data sets.

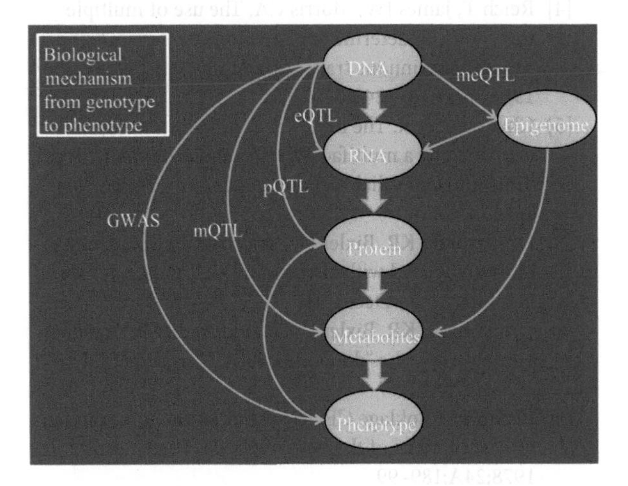


**Figure 11.6** Relationships between various types of genetic data and the molecular biology pathway.*eQTL*, expression quantitative trait loci; *GWAS*, genome-wide association studies; *meQTL*, methylation QTL; *mQTL*, metabolites QTL; *pQTL*, protein QTL.

Another notable software to analyze multivariate mQTL data in a GWAS framework, *xMWAS*, integrates metabolomics, transcriptomic data, and network analysis using multilevel sparse PLS regression [285]. Integration of omics is a rapidly evolving field of research and a holy grail of systems biology.

## 11.12 CONCLUSIONS

Genetic modeling is a challenging art and science. Advances in molecular technology and statistical methodology[7] and increasing availability of large samples allow many new investigations to be undertaken on unprecedented scales. Interpretation of the resulting findings remains both difficult and one of the more exciting challenges facing today's biomedical researchers.

## REFERENCES


[1] Badano JL, Katsanis N. Beyond Mendel: an evolving view of human genetic disease transmission. Nat Rev Genet October 2002;3(10):779–89.

[2] Fisher RA. The correlation between relatives on the supposition of Mendelian inheritance. Trans R Soc Edinburgh 1918;52:399–433.

[3] Falconer DS. The inheritance of liability to certain diseases, estimated from the incidence among relatives. Ann Hum Genet August 1965;29:51–76.

[4] Reich T, James JW, Morris CA. The use of multiple thresholds in determining the mode of transmission of semi-continuous traits. Ann Hum Genet November 1972;36(2):163–84.

[5] Chakraborty R. The inheritance of pyloric stenosis explained by a multifactorial threshold model with sex dimorphism for liability. Genet Epidemiol 1986;3(1): 1–15.

[6] Dronamraju KR, Bixler D, Majumder PP. Fetal mortality associated with cleft lip and cleft palate. Johns Hopkins Med J December 1982;151(6):287–9.

[7] Dronamraju KR, Bixler D. Fetal mortality in oral cleft families (IV): the "doubling effect". Clin Genet July 1983;24(1):22–5.

[8] Elston RC, Boklage CE. An examination of fundamental assumptions of the twin method. Prog Clin Biol Res 1978;24A:189–99.


[7] A steady stream of videos offering tutelage on these advances can be freely seen at http://www.soph.uab.edu/ssg/courses/.


[9] Hopper JL. Twin concordance. In: Encyclopedia of biostatistics, vol. 6. New York: John Wiley; 1998. p. 4626–9.

[10] Karlin S, Cameron EC, Williams PT. Sibling and parent–offspring correlation estimation with variable family size. Proc Natl Acad Sci U S A May 1981;78(5):2664–8.

[11] Neale MC. Adoption studies. In: Encyclopedia of biostatistics, vol. 1. New York: John Wiley; 1998. p. 77–81.

[12] Neale MC, Cardon LR. Methodology for genetic studies of twins and families. London: Kluwer; 1992.

[13] Carey G. Sibling imitation and contrast effects. Behav Genet May 1986;16(3):319–41.

[14] Lykken DT, McGue M, Tellegen A, Bouchard Jr TJ. Emergenesis. Genetic traits that may not run in families. Am Psychol December 1992;47(12):1565–77.

[15] Risch N. Linkage strategies for genetically complex traits. I. Multilocus models. Am J Hum Genet February 1990;46(2):222–8.

[16] Olson JM, Cordell HJ. Ascertainment bias in the estimation of sibling genetic risk parameters. Genet Epidemiol March 2000;18(3):217–35.

[17] Visscher PM, Medland SE, Ferreira MA, Morley KI, Zhu G, Cornes BK, Montgomery GW, Martin NG. Assumption-free estimation of heritability from genome-wide identity-by-descent sharing between full siblings. PLoS Genet March 2006;2(3):e41. [Epub 2006 Mar 24].

[18] Morton NE. Sequential tests for the detection of linkage. Am J Hum Genet September 1955;7(3):277–318.

[19] Elston RC, Stewart J. A general model for the genetic analysis of pedigree data. Hum Hered 1971;21(6):523–42.

[20] Elston RC, Rao DC. Statistical modeling and analysis in human genetics. Annu Rev Biophys Bioeng 1978;7:253–86. [Review].

[21] Lander ES, Green P. Construction of multilocus genetic linkage maps in humans. Proc Natl Acad Sci U S A April 1987;84(8):2363–7.

[22] Haseman JK, Elston RC. The investigation of linkage between a quantitative trait and a marker locus. Behav Genet March 1972;2(1):3–19.

[23] Penrose LS. The detection of autosomal linkage in data which consist of pairs of brothers and sisters of unspecified parentage. Ann Eugen (London) 1935;6:133–8.

[24] Amos CI, Elston RC, Wilson AF, Bailey-Wilson JE. A more powerful robust sib-pair test of linkage for quantitative traits. Genet Epidemiol 1989;6(3):435–49.

[25] Olson JM, Wijsman EM. Linkage between quantitative trait and marker loci: methods using all relative pairs. Genet Epidemiol 1993;10(2):87–102.

[26] Drigalenko E. How sib pairs reveal linkage. Am J Hum Genet October 1998;63(4):1242–5.
[27] Forrest WF. Weighting improves the "new Haseman-Elston" method. Hum Hered 2001;52(1):47–54.
[28] Gerhard D, Hothorn LA. Rank transformation in Haseman-Elston regression using scores for location-scale alternatives. Hum Hered 2010;69(3):143–51.
[29] Sham PC, Purcell S. Equivalence between Haseman-Elston and variance-components linkage analyses for sib pairs. Am J Hum Genet June 2001;68(6):1527–32.
[30] Sham PC, Purcell S, Cherny SS, Abecasis GR. Powerful regression-based quantitative-trait linkage analysis of general pedigrees. Am J Hum Genet August 2002;71(2):238–53.
[31] Shete S, Jacobs KB, Elston RC. Adding further power to the Haseman and Elston method for detecting linkage in larger sibships: weighting sums and differences. Hum Hered 2003;55(2–3):79–85.
[32] Visscher PM, Hopper JL. Power of regression and maximum likelihood methods to map QTL from sib-pair and DZ twin data. Ann Hum Genet November 2001;65(Pt 6):583–601.
[33] Wang T, Elston RC. A modified revisited Haseman-Elston method to further improve power. Hum Hered 2004;57(2):109–16.
[34] Wright FA. The phenotypic difference discards sib-pair QTL linkage information. Am J Hum Genet March 1997;60(3):740–2.
[35] Xu X, Weiss S, Xu X, Wei LJ. A unified Haseman-Elston method for testing linkage with quantitative traits. Am J Hum Genet October 2000;67(4):1025–8.
[36] S.A.G.E. 6.x. Statistical analysis for genetic epidemiology. 2010. http://darwin.cwru.edu/sage/.
[37] Kruglyak L, Daly MJ, Reeve-Daly MP, Lander ES. Parametric and nonparametric linkage analysis: a unified multipoint approach. Am J Hum Genet March 1996;58:1347–63.
[38] Abecasis GR, Cherny SS, Cookson WO, Cardon LR. Merlin–rapid analysis of dense genetic maps using sparse gene flow trees. Nat Genet January 2002;30(1):97–101.
[39] Almasy L, Blangero J. Multipoint quantitative-trait linkage analysis in general pedigrees. Am J Hum Genet May 1998;62(5):1198–211.
[40] Amos CI. Robust variance-components approach for assessing genetic linkage in pedigrees. Am J Hum Genet March 1994;54(3):535–43.
[41] Amos CI, Zhu DK, Boerwinkle E. Assessing genetic linkage and association with robust components of variance approaches. Ann Hum Genet March 1996;60(Pt 2):143–60.
[42] Goldgar DE. Multipoint analysis of human quantitative genetic variation. Am J Hum Genet December 1990;47(6):957–67.
[43] Schork NJ. Extended multipoint identity-by-descent analysis of human quantitative traits: efficiency, power, and modeling considerations. Am J Hum Genet December 1993;53(6):1306–19.
[44] Spielman RS, McGinnis RE, Ewens WJ. Transmission test for linkage disequilibrium: the insulin gene region and insulin-dependent diabetes mellitus (IDDM). Am J Hum Genet March 1993;52(3):506–16.
[45] Tiwari HK, Barnholtz-Sloan J, Wineinger N, Padilla MA, Vaughan LK, Allison DB. Review and evaluation of methods correcting for population stratification with a focus on underlying statistical principles. Hum Hered 2008;66(2):67–86.
[46] Page GP, George V, Go RC, Page PZ, Allison DB. "Are we there yet?": Deciding when one has demonstrated specific genetic causation in complex diseases and quantitative traits. Am J Hum Genet October 2003;73(4):711–9. [Review].
[47] Lander ES. The new genomics: global views of biology. Science October 25, 1996;274(5287):536–9.
[48] Lander ES, Linton LM, Birren B, Nusbaum C, Zody MC, Baldwin J, Devon K, Dewar K, Doyle M, FitzHugh W, Funke R, Gage D, Harris K, Heaford A, Howland J, Kann L, Lehoczky J, LeVine R, McEwan P, McKernan K, Meldrim J, Mesirov JP, Miranda C, Morris W, Naylor J, Raymond C, Rosetti M, Santos R, Sheridan A, Sougnez C, Stange-Thomann N, Stojanovic N, Subramanian A, Wyman D, Rogers J, Sulston J, Ainscough R, Beck S, Bentley D, Burton J, Clee C, Carter N, Coulson A, Deadman R, Deloukas P, Dunham A, Dunham I, Durbin R, French L, Grafham D, Gregory S, Hubbard T, Humphray S, Hunt A, Jones M, Lloyd C, McMurray A, Matthews L, Mercer S, Milne S, Mullikin JC, Mungall A, Plumb R, Ross M, Shownkeen R, Sims S, Waterston RH, Wilson RK, Hillier LW, McPherson JD, Marra MA, Mardis ER, Fulton LA, Chinwalla AT, Pepin KH, Gish WR, Chissoe SL, Wendl MC, Delehaunty KD, Miner TL, Delehaunty A, Kramer JB, Cook LL, Fulton RS, Johnson DL, Minx PJ, Clifton SW, Hawkins T, Branscomb E, Predki P, Richardson P, Wenning S, Slezak T, Doggett N, Cheng JF, Olsen A, Lucas S, Elkin C, Uberbacher E, Frazier M, Gibbs RA, Muzny DM, Scherer SE, Bouck JB, Sodergren EJ, Worley KC, Rives CM, Gorrell JH, Metzker ML, Naylor SL, Kucherlapati RS, Nelson DL, Weinstock GM, Sakaki Y, Fujiyama A, Hattori M, Yada T, Toyoda A, Itoh T, Kawagoe C, Watanabe H, Totoki Y, Taylor T, Weissenbach J, Heilig R, Saurin W, Artiguenave F, Brottier

P, Bruls T, Pelletier E, Robert C, Wincker P, Smith DR, Doucette-Stamm L, Rubenfield M, Weinstock K, Lee HM, Dubois J, Rosenthal A, Platzer M, Nyakatura G, Taudien S, Rump A, Yang H, Yu J, Wang J, Huang G, Gu J, Hood L, Rowen L, Madan A, Qin S, Davis RW, Federspiel NA, Abola AP, Proctor MJ, Myers RM, Schmutz J, Dickson M, Grimwood J, Cox DR, Olson MV, Kaul R, Raymond C, Shimizu N, Kawasaki K, Minoshima S, Evans GA, Athanasiou M, Schultz R, Roe BA, Chen F, Pan H, Ramser J, Lehrach H, Reinhardt R, McCombie WR, de la Bastide M, Dedhia N, Blöcker H, Hornischer K, Nordsiek G, Agarwala R, Aravind L, Bailey JA, Bateman A, Batzoglou S, Birney E, Bork P, Brown DG, Burge CB, Cerutti L, Chen HC, Church D, Clamp M, Copley RR, Doerks T, Eddy SR, Eichler EE, Furey TS, Galagan J, Gilbert JG, Harmon C, Hayashizaki Y, Haussler D, Hermjakob H, Hokamp K, Jang W, Johnson LS, Jones TA, Kasif S, Kaspryzk A, Kennedy S, Kent WJ, Kitts P, Koonin EV, Korf I, Kulp D, Lancet D, Lowe TM, McLysaght A, Mikkelsen T, Moran JV, Mulder N, Pollara VJ, Ponting CP, Schuler G, Schultz J, Slater G, Smit AF, Stupka E, Szustakowski J, Thierry-Mieg D, Thierry-Mieg J, Wagner L, Wallis J, Wheeler R, Williams A, Wolf YI, Wolfe KH, Yang SP, Yeh RF, Collins F, Guyer MS, Peterson J, Felsenfeld A, Wetterstrand KA, Patrinos A, Morgan MJ, de Jong P, Catanese JJ, Osoegawa K, Shizuya H, Choi S, Chen YJ. International human genome sequencing Consortium. Initial sequencing and analysis of the human genome. Nature February 15, 2001;409(6822):860–921.

[49] International HapMap Consortium. The International HapMap project. Nature December 18, 2003;426(6968):789–96.

[50] International HapMap Consortium. A haplotype map of the human genome. Nature October 27, 2005;437(7063):1299–320.

[51] International HapMap Consortium, Frazer KA, Ballinger DG, Cox DR, Hinds DA, Stuve LL, Gibbs RA, Belmont JW, Boudreau A, Hardenbol P, Leal SM, Pasternak S, Wheeler DA, Willis TD, Yu F, Yang H, Zeng C, Gao Y, Hu H, Hu W, Li C, Lin W, Liu S, Pan H, Tang X, Wang J, Wang W, Yu J, Zhang B, Zhang Q, Zhao H, Zhao H, Zhou J, Gabriel SB, Barry R, Blumenstiel B, Camargo A, Defelice M, Faggart M, Goyette M, Gupta S, Moore J, Nguyen H, Onofrio RC, Parkin M, Roy J, Stahl E, Winchester E, Ziaugra L, Altshuler D, Shen Y, Yao Z, Huang W, Chu X, He Y, Jin L, Liu Y, Shen Y, Sun W, Wang H, Wang Y, Wang Y, Xiong X, Xu L, Waye MM, Tsui SK, Xue H, Wong JT, Galver LM, Fan JB, Gunderson K, Murray SS, Oliphant AR, Chee MS, Montpetit A, Chagnon F, Ferretti V, Leboeuf M, Olivier JF, Phillips MS, Roumy S, Sallée C, Verner A, Hudson TJ, Kwok PY, Cai D, Koboldt DC, Miller RD, Pawlikowska L, Taillon-Miller P, Xiao M, Tsui LC, Mak W, Song YQ, Tam PK, Nakamura Y, Kawaguchi T, Kitamoto T, Morizono T, Nagashima A, Ohnishi Y, Sekine A, Tanaka T, Tsunoda T, Deloukas P, Bird CP, Delgado M, Dermitzakis ET, Gwilliam R, Hunt S, Morrison J, Powell D, Stranger BE, Whittaker P, Bentley DR, Daly MJ, de Bakker PI, Barrett J, Chretien YR, Maller J, McCarroll S, Patterson N, Pe'er I, Price A, Purcell S, Richter DJ, Sabeti P, Saxena R, Schaffner SF, Sham PC, Varilly P, Altshuler D, Stein LD, Krishnan L, Smith AV, Tello-Ruiz MK, Thorisson GA, Chakravarti A, Chen PE, Cutler DJ, Kashuk CS, Lin S, Abecasis GR, Guan W, Li Y, Munro HM, Qin ZS, Thomas DJ, McVean G, Auton A, Bottolo L, Cardin N, Eyheramendy S, Freeman C, Marchini J, Myers S, Spencer C, Stephens M, Donnelly P, Cardon LR, Clarke G, Evans DM, Morris AP, Weir BS, Tsunoda T, Mullikin JC, Sherry ST, Feolo M, Skol A, Zhang H, Zeng C, Zhao H, Matsuda I, Fukushima Y, Macer DR, Suda E, Rotimi CN, Adebamowo CA, Ajayi I, Aniagwu T, Marshall PA, Nkwodimmah C, Royal CD, Leppert MF, Dixon M, Peiffer A, Qiu R, Kent A, Kato K, Niikawa N, Adewole IF, Knoppers BM, Foster MW, Clayton EW, Watkin J, Gibbs RA, Belmont JW, Muzny D, Nazareth L, Sodergren E, Weinstock GM, Wheeler DA, Yakub I, Gabriel SB, Onofrio RC, Richter DJ, Ziaugra L, Birren BW, Daly MJ, Altshuler D, Wilson RK, Fulton LL, Rogers J, Burton J, Carter NP, Clee CM, Griffiths M, Jones MC, McLay K, Plumb RW, Ross MT, Sims SK, Willey DL, Chen Z, Han H, Kang L, Godbout M, Wallenburg JC, L'Archevêque P, Bellemare G, Saeki K, Wang H, An D, Fu H, Li Q, Wang Z, Wang R, Holden AL, Brooks LD, McEwen JE, Guyer MS, Wangc VO, Peterson JL, Shi M, Spiegel J, Sung LM, Zacharia LF, Collins FS, Kennedy K, Jamieson R, Stewart J. A second generation human haplotype map of over 3.1 million SNPs. Nature October 18, 2007;449(7164):851–61.

[52] International HapMap 3 Consortium, Altshuler DM, Gibbs RA, Peltonen L, Altshuler DM, Gibbs RA, Peltonen L, Dermitzakis E, Schaffner SF, Yu F, Peltonen L, Dermitzakis E, Bonnen PE, Altshuler DM, Gibbs RA, de Bakker PI, Deloukas P, Gabriel SB, Gwilliam R, Hunt S, Inouye M, Jia X, Palotie A, Parkin M, Whittaker P, Yu F, Chang K, Hawes A, Lewis LR, Ren Y, Wheeler D, Gibbs RA, Muzny DM, Barnes C, Darvishi K, Hurles M, Korn JM, Kristiansson K, Lee C, McCarrol SA, Nemesh J, Dermitzakis E, Keinan A, Montgomery SB, Pollack S, Price AL, Soranzo N, Bonnen PE, Gibbs RA, Gonzaga-Jauregui C, Keinan A, Price AL, Yu F, Anttila V, Brodeur W, Daly MJ, Leslie S, McVean G, Moutsianas L, Nguyen H, Schaffner SF,

Zhang Q, Ghori MJ, McGinnis R, McLaren W, Pollack S, Price AL, Schaffner SF, Takeuchi F, Grossman SR, Shlyakhter I, Hostetter EB, Sabeti PC, Adebamowo CA, Foster MW, Gordon DR, Licinio J, Manca MC, Marshall PA, Matsuda I, Ngare D, Wang VO, Reddy D, Rotimi CN, Royal CD, Sharp RR, Zeng C, Brooks LD, McEwen JE. Integrating common and rare genetic variation in diverse human populations. Nature September 2, 2010;467(7311):52–8.

[53] Genomes Project Consortium. A map of human genome variation from population-scale sequencing. Nature October 28, 2010;467(7319):1061–73. [Erratum in: Nature May 26, 2011;473(7348):544].

[54] The 1000 Genomes Project Consortium. An integrated map of genetic variation from 1,092 human genomes. Nature November 2012;491:56–65.

[55] The 1000 Genomes Project Consortium. A global reference for human genetic variation. Nature October 2015;526:68–74.

[56] Sudmant PH, Rausch T, Gardner EJ, Handsaker RE, Abyzov A, Huddleston J, Korbel JO. An integrated map of structural variation in 2,504 human genomes. Nature October 2015;526(7571):75–81.

[57] Birney E, Soranzo N. Human genomics: the end of the start for population sequencing. Nature October 2015;526:52–3.

[58] Stephens JC, Schneider JA, Tanguay DA, Choi J, Acharya T, Stanley SE, Jiang R, Messer CJ, Chew A, Han JH, Duan J, Carr JL, Lee MS, Koshy B, Kumar AM, Zhang G, Newell WR, Windemuth A, Xu C, Kalbfleisch TS, Shaner SL, Arnold K, Schulz V, Drysdale CM, Nandabalan K, Judson RS, Ruano G, Vovis GF. Haplotype variation and linkage disequilibrium in 313 human genes. Science July 20, 2001;293(5529):489–93.

[59] McCarthy MI, Abecasis GR, Cardon LR, Goldstein DB, Little J, Ioannidis JP, Hirschhorn JN. Genome-wide association studies for complex traits: consensus, uncertainty and challenges. Nat Rev Genet May 2008;9(5):356–69. [Review].

[60] Wellcome Trust Case Control Consortium. Genome-wide association study of 14,000 cases of seven common diseases and 3,000 shared controls. Nature June 7, 2007;447(7145):661–78.

[61] Risch N, Merikangas K. The future of genetic studies of complex human diseases. Science September 13, 1996;273(5281):1516–7.

[62] Purcell S, Neale B, Todd-Brown K, Thomas L, Ferreira MA, Bender D, Maller J, Sklar P, de Bakker PI, Daly MJ, Sham PC. PLINK: a tool set for whole-genome association and population-based linkage analyses. Am J Hum Genet September 2007;81(3):559–75.

[63] Anderson CA, Pettersson FH, Clarke GM, Cardon LR, Morris AP, Zondervan KT. Data quality control in genetic case-control association studies. Nat Protoc September 2010;5(9):1564–73. https://doi.org/10.1038/nprot.2010.116. [Epub 2010 Aug 26].

[64] Laurie CC, Doheny KF, Mirel DB, Pugh EW, Bierut LJ, Bhangale T, Boehm F, Caporaso NE, Cornelis MC, Edenberg HJ, Gabriel SB, Harris EL, Hu FB, Jacobs KB, Kraft P, Landi MT, Lumley T, Manolio TA, McHugh C, Painter I, Paschall J, Rice JP, Rice KM, Zheng X, Weir BS. GENEVA Investigators. Quality control and quality assurance in genotypic data for genome-wide association studies. Genet Epidemiol September 2010;34(6):591–602.

[65] Turner S, Armstrong LL, Bradford Y, Carlson CS, Crawford DC, Crenshaw AT, de Andrade M, Doheny KF, Haines JL, Hayes G, Jarvik G, Jiang L, Kullo IJ, Li R, Ling H, Manolio TA, Matsumoto M, McCarty CA, McDavid AN, Mirel DB, Paschall JE, Pugh EW, Rasmussen LV, Wilke RA, Zuvich RL, Ritchie MD. Quality control procedures for genome-wide association studies. Curr Protoc Hum Genet 2011 Jan; Chapter 1:Unit1.19;68:1.19.1–1.19.18. https://doi.org/10.1002/0471142905.hg0119s68.

[66] Devlin B, Roeder K. Genomic control for association studies. Biometrics December 1999;55(4):997–1004.

[67] Voight BF, Pritchard JK. Confounding from cryptic relatedness in case-control association studies. PLoS Genet September 2005;1(3):e32.

[68] Devlin B, Roeder K, Wasserman L. Genomic control, a new approach to genetic-based association studies. Theor Popul Biol November 2001;60(3):155–66. [Review].

[69] Bacanu SA, Devlin B, Roeder K. The power of genomic control. Am J Hum Genet June 2000;66(6):1933–44.

[70] Dadd T, Weale ME, Lewis CM. A critical evaluation of genomic control methods for genetic association studies. Genet Epidemiol May 2009;33(4):290–8. [Review].

[71] Devlin B, Bacanu SA, Roeder K. Genomic control to the extreme. Nat Genet November 2004;36(11):1129–30.

[72] Reich DE, Goldstein DB. Detecting association in a case-control study while correcting for population stratification. Genet Epidemiol January 2001;20(1):4–16. [Review].

[73] Zheng G, Freidlin B, Li Z, Gastwirth JL. Genomic control for association studies under various genetic models. Biometrics March 2005;61(1):186–92.

[74] Zheng G, Freidlin B, Gastwirth JL. Robust genomic control for association studies. Am J Hum Genet February 2006;78(2):350–6.

[75] Pritchard JK, Rosenberg NA. Use of unlinked genetic markers to detect population stratification in association studies. Am J Hum Genet July 1999;65(1):220–8.
[76] Pritchard JK, Stephens M, Rosenberg NA, Donnelly P. Association mapping in structured populations. Am J Hum Genet July 2000;67(1):170–81.
[77] Redden D, Divers J, Vaughan L, Tiwari H, Beasley T, Fernandez J, Kimberly R, Feng R, Padilla M, Lui N, Miller M, Allison D. Regional admixture mapping and structured association testing: conceptual unification and an extensible general linear model. PLoS Genet August 25, 2006;2(8):e137.
[78] Patterson N, Price AL, Reich D. Population structure and eigenanalysis. PLoS Genet December 2006;2(12):e190.
[79] Price AL, Patterson NJ, Plenge RM, Weinblatt ME, Shadick NA, Reich D. Principal components analysis corrects for stratification in genome-wide association studies. Nat Genet August 2006;38(8):904–9.
[80] Li Y, Willer C, Sanna S, Abecasis G. Genotype imputation. Annu Rev Genom Hum Genet 2009;10:387–406.
[81] Marchini J, Howie B. Genotype imputation for genome-wide association studies. Nat Rev Genet July 2010;11:499–511.
[82] Zheng H-F, Rong J-J, Liu M, Han F, Zhang XW, Richards JB, Wang L. Performance of genotype imputation for low frequency and rare variants from the 1000 genomes. PLoS One January 2015;10(1):e0116487.
[83] Wood AR, Perry JRB, Tanaka T, Hernandez DG, Zheng HF, Melzer D, Frayling TM. Imputation of variants from the 1000 genomes project modestly improves known associations and can identify low-frequency variant - phenotype associations undetected by HapMap based imputation. PLoS One May 2013;8(5):e64343.
[84] Buchanan CC, Torstenson ES, Bush WS, Ritchie MD. A comparison of cataloged variation between international HapMap Consortium and 1000 genomes project data. J Am Med Inf Assoc 2012 Mar-Apr;19(2):28994.
[85] Marchini J, Howie B, Myers S, McVean G, Donnelly P. A new multipoint method for genome-wide association studies by imputation of genotypes. Nat Genet 2007;39:906–13.
[86] Howie BN, Donnelly P, Marchini J. A flexible and accurate genotype imputation method for the next generation of genome-wide association studies. PLoS Genet 2009;5:e1000529.
[87] Li Y, Willer CJ, Ding J, Scheet P, Abecasis GR. MaCH: using sequence and genotype data to estimate haplotypes and unobserved genotypes. Genet Epidemiol December 2010;34(8):816–34.
[88] Howie B, Fuchsberger C, Stephens M, Marchini J, Abecasis GR. Fast and accurate genotype imputation in genome-wide association studies through pre-phasing. Nat Genet July 2012;44(8):955–9.
[89] Fuchsberger C, Abecasis GR, Hinds DA. minimac2: faster genotype imputation. Bioinformatics March 2015;31(5):782–4.
[90] Browning SR, Browning BL. Rapid and accurate haplotype phasing and missing data inference for whole genome association studies by use of localized haplotype clustering. Am J Hum Genet 2007;81:1084–97.
[91] Browning SR, Browning BL. Genotype imputation with millions of reference samples. Am J Hum Genet 2016;98:116–26.
[92] Hong EP, Park JW. Sample size and statistical power calculation in genetic association studies. Genom Inform 2012;10(2):117–22.
[93] Visscher PM, Brown MA, McCarthy MI, Yang J. Five years of GWAS discovery. Am J Hum Genet January 2012;90:7–24.
[94] Sham PC, Purcell SM. Statistical power and significance testing in large-scale genetic studies. Nat Rev Genet April 2014;15:335–46.
[95] Scherag A, Müller HH, Dempfle A, Hebebrand J, Schäfer H. Data adaptive interim modification of sample sizes for candidate-gene association studies. Hum Hered 2003;56:56–62.
[96] Gordon D, Levenstien MA, Finch SJ, Ott J. Errors and linkage disequilibrium interact multiplicatively when computing sample sizes for genetic case-control association studies. Pac Symp Biocomput 2003:490–501.
[97] Pfeiffer RM, Gail MH. Sample size calculations for population and family-based case-control association studies on marker genotypes. Genet Epidemiol 2003;25:136–48.
[98] Pirinen M, Donnelly P, Spencer CCA. Including known covariates can reduce power to detect genetic effects in case-control studies. Nat Genet July 2012;44:848–51.
[99] Spencer C, Su Z, Donnelly P, Marchini J. Designing Genome-Wide Association Studies: sample size, power, and the choice of genotyping chip. PLoS Genet May 2009;5(5):e1000477.
[100] Tiwari HK, Birkner T, Moondan A, Zhang S, Page GP, Patki A. Accurate and flexible power calculations on the spot: applications to genomic research. Stat Interface 2011;4(3):353–8.
[101] Purcell S, Cherny SS, Sham PC. Genetic power calculator: design of linkage and association genetic mapping studies of complex traits. Bioinformatics 2003;19(1):149–50.

[102] Skol AD, Scott LJ, Abecasis GR, Boehnke M. Joint analysis is more efficient than replication-based analysis for two-stage genome-wide association studies. Nat Genet 2006;38:209–13.
[103] Menashe I, Rosenberg PS, Chen BE. PGA: power calculator for case-control genetic association analyses. BMC Genet May 2008;9:36.
[104] Zhao JH. Gap: genetic analysis package. R package version 1. 2017. p. 1–17. https://CRAN.R-project.org/package=gap.
[105] Weeks DE. powerpkg: power analyses for the affected sib pair and the TDT design. R package version 1. 2012. p. 5. https://CRAN.R-project.org/package=powerpkg.
[106] Lee S, Wu MC, Lin X. Optimal tests for rare variant effects in sequencing association studies. Biostatistics September 2012;13:762–75.
[107] Yang J, Lee SH, Goddard ME, Visscher PM. GCTA: a tool for genome-wide complex trait analysis. Am J Hum Genet January 2011;88(1):76–82.
[108] Visscher PM, Hemani G, Vinkhuyzen AAE, Chen GB, Lee SH, Wray NR, Goddard ME, Yang J. Statistical power to detect genetic (Co)Variance of complex traits using SNP data in unrelated samples. PLoS Genet 2014;10(4):e1004269.
[109] Gauderman WJ. Sample size requirements for matched case–control studies of gene–environment interaction. Stat Med January 2002;21(1):35–50.
[110] Gauderman WJ. Sample size requirements for association studies of gene-gene interaction. Am J Epidemiol March 2002;155(5):478–84.
[111] Pearson K. On the criterion that a given system of deviations from the probable in the case of a correlated system of variables is such that it can be reasonably supposed to have arisen from random sampling. Phil Mag 1900;50(302):157–75. [Series 5].
[112] Fisher RA. On the interpretation of $\chi^2$ from contingency tables, and the calculation of P. J Roy Stat Soc 1922;85(1):87–94.
[113] Armitage P. Tests for linear trends in proportions and frequencies. Biometrics 1955;11(3):375–86.
[114] Cochran WG. Some methods for strengthening the common chi-square tests. Biometrics 1954;10(4):417–51.
[115] Sasieni P. From genotypes to genes: doubling the sample size. Biometrics December 1997;53(4):1253–61.
[116] Freidlin B, Zheng G, Li Z, Gastwirth JL. Trend tests for case-control studies of genetic markers: power, sample size and robustness. Hum Hered 2002;53(3):146–52.
[117] Huang BE, Lin DY. Efficient association mapping of quantitative trait loci with selective genotyping. Am J Hum Genet 2007;80:567–76.
[118] Aulchenko YS, de Koning DJ, Haley C. Genomewide rapid association using mixed model and regression: a fast and simple method for genomewide pedigree-based quantitative trait loci association analysis. Genetics September 2007;177(1):577–85.
[119] Kang HM, Zaitlen NA, Wade CM, Kirby A, Heckerman D, Daly MJ, Eskin E. Efficient control of population structure in model organism association mapping. Genetics March 2008;178(3):1709–23.
[120] Zhang Z, Ersoz E, Lai CQ, Todhunter RJ, Tiwari HK, Gore MA, Bradbury PJ, Yu J, Arnett DK, Ordovas JM, Buckler ES. Mixed linear model approach adapted for genome-wide association studies. Nat Genet April 2010;42(4):355–60.
[121] Aulchenko YS, Struchalin MV, van Duijn CM. ProbABEL package for genome-wide association analysis of imputed data. BMC Bioinf 2010;11:1345.
[122] Zhou X, Stephens M. Genome-wide efficient mixed-model analysis for association studies. Nat Genet June 17, 2012;44(7):821–4. https://doi.org/10.1038/ng.2310.
[123] Zhou X, Stephens M. Efficient multivariate linear mixed model algorithms for genome-wide association studies. Nat Methods April 2014;11(4):407–9. https://doi.org/10.1038/nmeth. 2848. [Epub 2014 Feb 16].
[124] Laird NM, Lange C. Family-based designs in the age of large-scale gene-association studies. Nat Rev Genet May 2006;7(5):385–94. [Review].
[125] Bates D, Maechler M, Bolker B, Walker S. Fitting linear mixed-effects models using lme4. J Stat Software 2015;67(1):1–48. https://doi.org/10.18637/jss.v067.i01.
[126] Chen MH, Yang Q. GWAF: an R package for genome-wide association analyses with family data. Bioinformatics February 15, 2010;26(4):580–1. https://doi.org/10.1093/bioinformatics/btp710. [Epub 2009 Dec 29].
[127] Hoggart CJ, Clark TG, De Iorio M, Whittaker JC, Balding DJ. Genome-wide significance for dense SNP and resequencing data. Genet Epidemiol February 2008;32(2):179–85.
[128] Balding DJ. A tutorial on statistical methods for population association studies. Nat Rev Genet October 2006;7(10):781–91. [Review].
[129] Corvin A, Craddock N, Sullivan PF. Genome-wide association studes: aprimer. Psychol Med July 2010;40(7):1063–77.
[130] Hinrichs AL, Larkin EK, Suarez BK. Population stratification and patterns of linkage disequilibrium. Genet Epidemiol 2009;33(Suppl. 1):S88–92.
[131] Bhatnagar S. Interactive Q-Q and manhattan plots using Plotly.js. 2016 Nov. http://sahirbhatnagar.com/manhattanly/.

[132] Turner SD. qqman: an R package for visualizing GWAS results using Q-Q and manhattan plots. 2014. https://doi.org/10.1101/005165. biorXiv.
[133] Pruim RJ, Welch RP, Sanna S, et al. LocusZoom: regional visualization of genome-wide association scan results. Bioinformatics September 2010;26(18):2336–7.
[134] Dadev T, Leongamornlert DA, Saunders EJ, Eeles R, Kote-Jarai Z. LocusExplorer: a user-friendly tool for integrated visualization of human genetic association data and biological annotations. Bioinformatics 2016;32(6):949–51.
[135] Schillert A, Schwarz DF, Vens M, Szymczak S, König IR, Ziegler A. ACPA: automated cluster plot analysis of genotype data. BMC Proc 2009;3(Suppl. 7):S58.
[136] Gogarten SM, Bhangale T, Conomos MP, Laurie CA, McHugh CP, Painter I, Zheng X, Crosslin DR, Levine D, Lumley T, Nelson SC, Rice K, Shen J, Swarnkar R, Weir BS, Laurie CC. GWASTools: an R/Bioconductor package for quality control and analysis of genome-wide association studies. Bioinformatics December 2012;28(24):3329–31.
[137] Chakravarti A. Population genetics–making sense out of sequence. Nat Genet January 1999;21(1 Suppl.):56–60. [Review].
[138] Lohmueller KE, Mauney MM, Reich D, Braverman JM. Variants associated with common disease are not unusually differentiated in frequency across populations. Am J Hum Genet January 2006;78(1):130–6.
[139] Manolio TA, Collins FS, Cox NJ, Goldstein DB, Hindorff LA, Hunter DJ, McCarthy MI, Ramos EM, Cardon LR, Chakravarti A, Cho JH, Guttmacher AE, Kong A, Kruglyak L, Mardis E, Rotimi CN, Slatkin M, Valle D, Whittemore AS, Boehnke M, Clark AG, Eichler EE, Gibson G, Haines JL, Mackay TF, McCarroll SA, Visscher PM. Finding the missing heritability of complex diseases. Nature October 8, 2009;461(7265):747–53. [Review].
[140] Terwilliger JD, Hiekkalinna T. An utter refutation of the "fundamental theorem of the HapMap". Eur J Hum Genet April 2006;14(4):426–37.
[141] Terwilliger JD, Göring HH. Update to Terwilliger and Göring's "Gene mapping in the 20th and 21st centuries" (2000): gene mapping when rare variants are common and common variants are rare. Hum Biol December 2009;81(5–6):729–33.
[142] Pritchard JK, Cox NJ. The allelic architecture of human disease genes: common disease-common variant… or not? Hum Mol Genet October 1, 2002;11(20):2417–23.
[143] Bodmer W, Bonilla C. Common and rare variants in multifactorial susceptibility to common diseases. Nat Genet June 2008;40(6):695–701.
[144] Zondervan KT, Cardon LR. The complex interplay among factors that influence allelic association. Nat Rev Genet February 2004;5(2):89–100.
[145] Mardis ER. A decade's perspective on DNA sequencing technology. Nature February 10, 2011;470(7333):198–203.
[146] Ng SB, Buckingham KJ, Lee C, Bigham AW, Tabor HK, Dent KM, Huff CD, Shannon PT, Jabs EW, Nickerson DA, Shendure J, Bamshad MJ. Exome sequencing identifies the cause of a mendelian disorder. Nat Genet January 2010;42(1):30–5.
[147] Erlich Y, Edvardson S, Hodges E, Zenvirt S, Thekkat P, Shaag A, Dor T, Hannon GJ, Elpeleg O. Exome sequencing and disease-network analysis of a single family implicate a mutation in KIF1A in hereditary spastic paraparesis. Genome Res May 2011;21(5):658–64.
[148] Rödelsperger C, Krawitz P, Bauer S, Hecht J, Bigham AW, Bamshad M, de Condor BJ, Schweiger MR, Robinson PN. Identity-by-descent filtering of exome sequence data for disease-gene identification in autosomal recessive disorders. Bioinformatics March 15, 2011;27(6):829–36.
[149] Ng SB, Turner EH, Robertson PD, Flygare SD, Bigham AW, Lee C, Shaffer T, Wong M, Bhattacharjee A, Eichler EE, Bamshad M, Nickerson DA, Shendure J. Targeted capture and massively parallel sequencing of 12 human exomes. Nature September 10, 2009;461(7261):272–6.
[150] Vaser R, Adusumalli S, Leng SN, Sikic M, Ng PC. SIFT missense predictions for genomes. Nat Protoc January 2016;11(1):1–9. https://doi.org/10.1038/mprof.2015.123. [Epub 2015 Dec 3].
[151] Sunyaev S, Ramensky V, Koch I, Lathe 3rd W, Kondrashov AS, Bork P. Prediction of deleterious human alleles. Hum Mol Genet March 15, 2001;10(6):591–7.
[152] Morgenthaler S, Thilly WG. A strategy to discover genes that carry multi-allelic or mono-allelic risk for common diseases: a cohort allelic sums test (CAST). Mutat Res February 3, 2007;615(1–2):28–56.
[153] Li B, Leal SM. Methods for detecting associations with rare variants for common diseases: application to analysis of sequence data. Am J Hum Genet September 2008;83(3):311–21.
[154] Wu MC, Lee S, Cai T, Li Y, Boehnke M, Lin X. Rare-variant association testing for sequencing data with the sequence kernel association test. Am J Hum Genet July 2011;89(1):82–93.
[155] Madsen BE, Browning SR. A groupwise association test for rare mutations using a weighted sum statistic. PLoS Genet February 2009;5(2):e1000384.
[156] Price AL, Kryukov GV, de Bakker PI, Purcell SM, Staples J, Wei LJ, Sunyaev SR. Pooled association tests for rare variants in exon-resequencing studies. Am J Hum Genet June 11, 2010;86(6):832–8.
[157] Yi N, Zhi D. Bayesian analysis of rare variants in genetic association studies. Genet Epidemiol January 2011;35(1):57–69.

[158] Bansal V, Libiger O, Torkamani A, Schork NJ. Statistical analysis strategies for association studies involving rare variants. Nat Rev Genet November 2010;11(11):773–85.

[159] Ansorge WJ. Next-generation DNA sequencing techniques. Nat Biotechnol April 2009;25(4):195–203.

[160] Hirst M, Marra MA. Next generation sequencing based approaches to epigenomics. Brief Funct Genom December 2010;9(5–6):455–65. [Review].

[161] Meyerson M, Gabriel S, Getz G. Advances in understanding cancer genomes through second-generation sequencing. Nat Rev Genet October 2010;11(10):685–96. [Review].

[162] Timmermann B, Kerick M, Roehr C, Fischer A, Isau M, Boerno ST, Wunderlich A, Barmeyer C, Seemann P, Koenig J, Lappe M, Kuss AW, Garshasbi M, Bertram L, Trappe K, Werber M, Herrmann BG, Zatloukal K, Lehrach H, Schweiger MR. Somatic mutation profiles of MSI and MSS colorectal cancer identified by whole exome next generation sequencing and bioinformatics analysis. PLoS One December 22, 2010;5(12):e15661.

[163] Wei X, Walia V, Lin JC, Teer JK, Prickett TD, Gartner J, Davis S, NISC Comparative Sequencing Program, Stemke-Hale K, Davies MA, Gershenwald JE, Robinson W, Robinson S, Rosenberg SA, Samuels Y. Exome sequencing identifies GRIN2A as frequently mutated in melanoma. Nat Genet May 2011;43(5):442–6. [Epub 2011 Apr 15].

[164] Wellcome Trust Case Control Consortium, Maller JB, McVean G, Byrnes J, Vukcevic D, Palin K, Su Z, Howson JM, Auton A, Myers S, Morris A, Pirinen M, Brown MA, Burton PR, Caulfield MJ, Compston A, Farrall M, Hall AS, Hattersley AT, Hill AV, Mathew CG, Pembrey M, Satsangi J, Stratton MR, Worthington J, Craddock N, Hurles M, Ouwehand W, Parkes M, Rahman N, Duncanson A, Todd JA, Kwiatkowski DP, Samani NJ, Gough SC, McCarthy MI, Deloukas P, Donnelly P. Bayesian refinement of association signals for 14 loci in 3 common diseases. Nat Genet December 2012;44(12):1294–301. https://doi.org/10.1038/ng. 2435. [Epub 2012 Oct 28].

[165] Gaulton KJ, Ferreira T, Lee Y, Raimondo A, Mägi R, Reschen ME, Mahajan A, Locke A, Rayner NW, Robertson N, Scott RA, Prokopenko I, Scott LJ, Green T, Sparso T, Thuillier D, Yengo L, Grallert H, Wahl S, Frånberg M, Strawbridge RJ, Kestler H, Chheda H, Eisele L, Gustafsson S, Steinthorsdottir V, Thorleifsson G, Qi L, Karssen LC, van Leeuwen EM, Willems SM, Li M, Chen H, Fuchsberger C, Kwan P, Ma C, Linderman M, Lu Y, Thomsen SK, Rundle JK, Beer NL, van de Bunt M, Chalisey A, Kang HM, Voight BF, Abecasis GR, Almgren P, Baldassarre D, Balkau B, Benediktsson R, Blüher M, Boeing H, Bonnycastle LL, Bottinger EP, Burtt NP, Carey J, Charpentier G, Chines PS, Cornelis MC, Couper DJ, Crenshaw AT, van Dam RM, Doney AS, Dorkhan M, Edkins S, Eriksson JG, Esko T, Eury E, Fadista J, Flannick J, Fontanillas P, Fox C, Franks PW, Gertow K, Gieger C, Gigante B, Gottesman O, Grant GB, Grarup N, Groves CJ, Hassinen M, Have CT, Herder C, Holmen OL, Hreidarsson AB, Humphries SE, Hunter DJ, Jackson AU, Jonsson A, Jørgensen ME, Jørgensen T, Kao WH, Kerrison ND, Kinnunen L, Klopp N, Kong A, Kovacs P, Kraft P, Kravic J, Langford C, Leander K, Liang L, Lichtner P, Lindgren CM, Lindholm E, Linneberg A, Liu CT, Lobbens S, Luan J, Lyssenko V, Männistö S, McLeod O, Meyer J, Mihailov E, Mirza G, Mühleisen TW, Müller-Nurasyid M, Navarro C, Nöthen MM, Oskolkov NN, Owen KR, Palli D, Pechlivanis S, Peltonen L, Perry JR, Platou CG, Roden M, Ruderfer D, Rybin D, van der Schouw YT, Sennblad B, Sigurðsson G, Stančáková A, Steinbach G, Storm P, Strauch K, Stringham HM, Sun Q, Thorand B, Tikkanen E, Tonjes A, Trakalo J, Tremoli E, Tuomi T, Wennauer R, Wiltshire S, Wood AR, Zeggini E, Dunham I, Birney E, Pasquali L, Ferrer J, Loos RJ, Dupuis J, Florez JC, Boerwinkle E, Pankow JS, van Duijn C, Sijbrands E, Meigs JB, Hu FB, Thorsteinsdottir U, Stefansson K, Lakka TA, Rauramaa R, Stumvoll M, Pedersen NL, Lind L, Keinanen-Kiukaanniemi SM, Korpi-Hyövälti E, Saaristo TE, Saltevo J, Kuusisto J, Laakso M, Metspalu A, Erbel R, Jöcke KH, Moebus S, Ripatti S, Salomaa V, Ingelsson E, Boehm BO, Bergman RN, Collins FS, Mohlke KL, Koistinen H, Tuomilehto J, Hveem K, Njølstad I, Deloukas P, Donnelly PJ, Frayling TM, Hattersley AT, de Faire U, Hamsten A, Illig T, Peters A, Cauchi S, Sladek R, Froguel P, Hansen T, Pedersen O, Morris AD, Palmer CN, Kathiresan S, Melander O, Nilsson PM, Groop LC, Barroso I, Langenberg C, Wareham NJ, O'Callaghan CA, Gloyn AL, Altshuler D, Boehnke M, Teslovich TM, McCarthy MI, Morris AP, DIAbetes Genetics Replication And Meta-analysis (DIAGRAM) Consortium. Genetic fine mapping and genomic annotation defines causal mechanisms at type 2 diabetes susceptibility loci. Nat Genet December 2015;47(12):1415–25. https://doi.org/10.1038/ng. 3437. [Epub 2015 Nov 9].

[166] Wang K, Li M, Hakonarson H. ANNOVAR: functional annotation of genetic variants from high-throughput sequencing data. Nucleic Acids Res September 2010;38(16):e164. https://doi.org/10.1093/nar/gkq603. [Epub 2010 Jul 3].

[167] Hindorff LA, Sethupathy P, Junkins HA, Ramos EM, Mehta JP, Collins FS, Manolio TA. Potential etiologic and functional implications of genome-wide association loci for human diseases and traits. Proc Natl Acad Sci U S A June 9, 2009;106(23):9362–7. https://doi.org/10.1073/pnas.0903103106. [Epub 2009 May 27].

[168] Pickrell JK. Joint analysis of functional genomic data and genome-wide association studies of 18 human traits. Am J Hum Genet April 3, 2014;94(4):559–73. https://doi.org/10.1016/j.ajhg.2014.03.004. [Erratum in: Am J Hum Genet July 3, 2014;95(1):126].

[169] Gusev A, Lee SH, Trynka G, Finucane H, Vilhjálmsson BJ, Xu H, Zang C, Ripke S, Bulik-Sullivan B, Stahl E, Schizophrenia Working Group of the Psychiatric Genomics Consortium, SWE-SCZ Consortium, Kähler AK, Hultman CM, Purcell SM, McCarroll SA, Daly M, Pasaniuc B, Sullivan PF, Neale BM, Wray NR, Raychaudhuri S, Price AL, Schizophrenia Working Group of the Psychiatric Genomics Consortium, SWE-SCZ Consortium. Partitioning heritability of regulatory and cell-type-specific variants across 11 common diseases. Am J Hum Genet November 6, 2014;95(5):535–52. https://doi.org/10.1016/j.ajhg.2014.10.004. [Epub 2014 Nov 6].

[170] ENCODE Project Consortium. An integrated encyclopedia of DNA elements in the human genome. Nature September 6, 2012;489(7414):57–74. https://doi.org/10.1038/nature11247.

[171] Kawai J, Shinagawa A, Shibata K, Yoshino M, Itoh M, Ishii Y, Arakawa T, Hara A, Fukunishi Y, Konno H, Adachi J, Fukuda S, Aizawa K, Izawa M, Nishi K, Kiyosawa H, Kondo S, Yamanaka I, Saito T, Okazaki Y, Gojobori T, Bono H, Kasukawa T, Saito R, Kadota K, Matsuda H, Ashburner M, Batalov S, Casavant T, Fleischmann W, Gaasterland T, Gissi C, King B, Kochiwa H, Kuehl P, Lewis S, Matsuo Y, Nikaido I, Pesole G, Quackenbush J, Schriml LM, Staubli F, Suzuki R, Tomita M, Wagner L, Washio T, Sakai K, Okido T, Furuno M, Aono H, Baldarelli R, Barsh G, Blake J, Boffelli D, Bojunga N, Carninci P, de Bonaldo MF, Brownstein MJ, Bult C, Fletcher C, Fujita M, Gariboldi M, Gustincich S, Hill D, Hofmann M, Hume DA, Kamiya M, Lee NH, Lyons P, Marchionni L, Mashima J, Mazzarelli J, Mombaerts P, Nordone P, Ring B, Ringwald M, Rodriguez I, Sakamoto N, Sasaki H, Sato K, Schönbach C, Seya T, Shibata Y, Storch KF, Suzuki H, Toyo-oka K, Wang KH, Weitz C, Whittaker C, Wilming L, Wynshaw-Boris A, Yoshida K, Hasegawa Y, Kawaji H, Kohtsuki S, Kohtsuki S, Hayashizaki Y, RIKEN Genome Exploration Research Group Phase II Team, the FANTOM Consortium. Functional annotation of a full-length mouse cDNA collection. Nature February 8, 2001;409(6821):685–90.

[172] Romanoski CE, Glass CK, Stunnenberg HG, Wilson L, Almouzni G. Epigenomics: roadmap for regulation. Nature February 19, 2015;518(7539):314–6. https://doi.org/10.1038/518314a. [No abstract available].

[173] Jiangtao L, Arthur B, Kwangi A, Kiranmoy D, Jiahan L, Zhong W, Yao L, Rongling W. Functional genome-wide association studies of longitudinal traits. In: Chow SC, editor. Handbook of Adaptive Designs in Pharmaceutical and Clinical Development. London, UK: Wiley; 2010.

[174] Li J, Das K, Fu G, Li R, Wu R. The Bayesian lasso for genome-wide association studies. Bioinformatics February 15, 2011;27(4):516–23. https://doi.org/10.1093/bioinformatics/btq688. [Epub 2010 Dec 14].

[175] Das K, Li J, Wang Z, Tong C, Fu G, Li Y, Xu M, Ahn K, Mauger D, Li R, Wu R. A dynamic model for genome-wide association studies. Hum Genet June 2011;129(6):629–39. https://doi.org/10.1007/s00439-011-0960-6. [Epub 2011 Feb 4].

[176] Li J, Wang Z, Li R, Wu R. Bayesian group lasso for nonparametric varying-coefficient models with application to functional genome-wide association studies. Ann Appl Stat June 2015;9(2):640–64.

[177] Kichaev G, Yang WY, Lindstrom S, Hormozdiari F, Eskin E, Price AL, Kraft P, Pasaniuc B. Integrating functional data to prioritize causal variants in statistical fine-mapping studies. PLoS Genet October 30, 2014;10(10):e1004722. https://doi.org/10.1371/journal.pgen.1004722. [eCollection 2014 Oct].

[178] Farh KK, Marson A, Zhu J, Kleinewietfeld M, Housley WJ, Beik S, Shoresh N, Whitton H, Ryan RJ, Shishkin AA, Hatan M, Carrasco-Alfonso MJ, Mayer D, Luckey CJ, Patsopoulos NA, De Jager PL, Kuchroo VK, Epstein CB, Daly MJ, Hafler DA, Bernstein BE. Genetic and epigenetic fine mapping of causal autoimmune disease variants. Nature February 19, 2015;518(7539):337–43. https://doi.org/10.1038/nature13835. [Epub 2014 Oct 29].

[179] Hormozdiari F, Kostem E, Kang EY, Pasaniuc B, Eskin E. Identifying causal variants at loci with multiple signals of association. Genetics October 2014;198(2):497–508. https://doi.org/10.1534/genetics.114.167908. [Epub 2014 Aug 7].

[180] Hormozdiari F, Kichaev G, Yang WY, Pasaniuc B, Eskin E. Identification of causal genes for complex traits. Bioinformatics June 15, 2015;31(12):i206–13. https://doi.org/10.1093/bioinformatics/btv240.

[181] Trynka G, Westra HJ, Slowikowski K, Hu X, Xu H, Stranger BE, Klein RJ, Han B, Raychaudhuri S. Disentangling the effects of colocalizing genomic annotations to functionally prioritize non-coding variants within complex-trait loci. Am J Hum Genet July 2, 2015;97(1):139–52. https://doi.org/10.1016/j.ajhg.2015.05.016.

[182] van de Bunt M, Cortes A, IGAS Consortium, Brown MA, Morris AP, McCarthy MI. Evaluating the performance of fine-mapping strategies at common variant GWAS loci. PLoS Genet September 25,

2015;11(9):e1005535. https://doi.org/10.1371/journal.pgen.1005535. [eCollection 2015].

[183] Davydov EV, Goode DL, Sirota M, Cooper GM, Sidow A, Batzoglou S. Identifying a high fraction of the human genome to be under selective constraint using GERP++. PLoS Comput Biol December 2, 2010;6(12):e1001025. https://doi.org/10.1371/journal.pcbi.10010125.

[184] Kircher M, Witten DM, Jain P, O'Roak BJ, Cooper GM, Shendure J. A general framework for estimating the relative pathogenicity of human genetic variants. Nat Genet February 2, 2014. https://doi.org/10.1038/ng.2892.

[185] Ritchie GR, Dunham I, Zeggini E, Flicek P. Functional annotation of noncoding sequence variants. Nat Methods March 2014;11(3):294–6. https://doi.org/10.1038/nmeth. 2832. [Epub 2014 Feb 2].

[186] Ionita-Laza I, McCallum K, Xu B, Buxbaum JD. A spectral approach integrating functional genomic annotations for coding and noncoding variants. Nat Genet February 2016;48(2):214–20. https://doi.org/10.1038/ng.3477.

[187] Bodea CA, Mitchell AA, Runz H, Sunyaev SR. Phenotype-specific information improves prediction of functional impact for noncoding variants. bioRxiv. https://doi.org/10.1101/083642.

[188] Morris AP. Transethnic meta-analysis of genomewide association studies. Genet Epidemiol December 2011;35(8):809–22. https://doi.org/10.1002/gepi.20630.

[189] Kichaev G, Pasaniuc B. Leveraging functional-annotation data in trans-ethnic fine-mapping studies. Am J Hum Genet August 6, 2015;97(2):260–71. https://doi.org/10.1016/j.ajhg.2015.06.007. [Epub 2015 Jul 16. Erratum in: Am J Hum Genet Aug 6, 2015;97(2):353].

[190] Keller MF, Reiner AP, Okada Y, van Rooij FJ, Johnson AD, Chen MH, Smith AV, Morris AP, Tanaka T, Ferrucci L, Zonderman AB, Lettre G, Harris T, Garcia M, Bandinelli S, Qayyum R, Yanek LR, Becker DM, Becker LC, Kooperberg C, Keating B, Reis J, Tang H, Boerwinkle E, Kamatani Y, Matsuda K, Kamatani N, Nakamura Y, Kubo M, Liu S, Dehghan A, Felix JF, Hofman A, Uitterlinden AG, van Duijn CM, Franco OH, Longo DL, Singleton AB, Psaty BM, Evans MK, Cupples LA, Rotter JI, O'Donnell CJ, Takahashi A, Wilson JG, Ganesh SK, Nalls MA. Trans-ethnic meta-analysis of white blood cell phenotypes. Hum Mol Genet December 20, 2014;23(25):6944–60.

[191] Ng MC, Shriner D, Chen BH, Li J, Chen WM, Guo X, Liu J, Bielinski SJ, Yanek LR, Nalls MA, Comeau ME, Rasmussen-Torvik LJ, Jensen RA, Evans DS, Sun YV, An P, Patel SR, Lu Y, Long J, Armstrong LL, Wagenknecht L, Yang L, Snively BM, Palmer ND, Mudgal P, Langefeld CD, Keene KL, Freedman BI, Mychaleckyj JC, Nayak U, Raffel LJ, Goodarzi MO, Chen YD, Taylor Jr HA, Correa A, Sims M, Couper D, Pankow JS, Boerwinkle E, Adeyemo A, Doumatey A, Chen G, Mathias RA, Vaidya D, Singleton AB, Zonderman AB, Igo Jr RP, Sedor JR, Kabagambe EK, Siscovick DS, McKnight B, Rice K, Liu Y, Hsueh WC, Zhao W, Bielak LF, Kraja A, Province MA, Bottinger EP, Gottesman O, Cai Q, Zheng W, Blot WJ, Lowe WL, Pacheco JA, Crawford DC, Grundberg E, Rich SS, Hayes MG, Shu XO, Loos RJ, Borecki IB, Peyser PA, Cummings SR, Psaty BM, Fornage M, Iyengar SK, Evans MK, Becker DM, Kao WH, Wilson JG, Rotter JI, Sale MM, Liu S, Rotimi CN, Bowden DW. Meta-analysis of genome-wide association studies in African Americans provides insights into the genetic architecture of type 2 diabetes. PLoS Genet August 2014;10(8):e1004517.

[192] Cornelis MC, Byrne EM, Esko T, Nalls MA, Ganna A, Paynter N, Monda KL, Amin N, Fischer K, Renstrom F, Ngwa JS, Huikari V, Cavadino A, Nolte IM, Teumer A, Yu K, Marques-Vidal P, Rawal R, Manichaikul A, Wojczynski MK, Vink JM, Zhao JH, Burlutsky G, Lahti J, Mikkila V, Lemaitre RN, Eriksson J, Musani SK, Tanaka T, Geller F, Luan J, Hui J, Magi R, Dimitriou M, Garcia ME, Ho WK, Wright MJ, Rose LM, Magnusson PK, Pedersen NL, Couper D, Oostra BA, Hofman A, Ikram MA, Tiemeier HW, Uitterlinden AG, van Rooij FJ, Barroso I, Johansson I, Xue L, Kaakinen M, Milani L, Power C, Snieder H, Stolk RP, Baumeister SE, Biffar R, Gu F, Bastardot F, Kutalik Z, Jacobs Jr DR, Forouhi NG, Mihailov E, Lind L, Lindgren C, Michaelsson K, Morris A, Jensen M, Khaw KT, Luben RN, Wang JJ, Mannisto S, Perala MM, Kahonen M, Lehtimaki T, Viikari J, Mozaffarian D, Mukamal K, Psaty BM, Doring A, Heath AC, Montgomery GW, Dahmen N, Carithers T, Tucker KL, Ferrucci L, Boyd HA, Melbye M, Treur JL, Mellstrom D, Hottenga JJ, Prokopenko I, Tonjes A, Deloukas P, Kanoni S, Lorentzon M, Houston DK, Liu Y, Danesh J, Rasheed A, Mason MA, Zonderman AB, Franke L, Kristal BS, Karjalainen J, Reed DR, Westra HJ, Evans MK, Saleheen D, Harris TB, Dedoussis G, Curhan G, Stumvoll M, Beilby J, Pasquale LR, Feenstra B, Bandinelli S, Ordovas JM, Chan AT, Peters U, Ohlsson C, Gieger C, Martin NG, Waldenberger M, Siscovick DS, Raitakari O, Eriksson JG, Mitchell P, Hunter DJ, Kraft P, Rimm EB, Boomsma DI, Borecki IB, Loos RJ, Wareham NJ, Vollenweider P, Caporaso N, Grabe HJ, Neuhouser ML, Wolffenbuttel BH, Hu FB, Hypponen E, Jarvelin MR, Cupples LA, Franks PW, Ridker PM, van Duijn CM, Heiss G, Metspalu A, North KE, Ingelsson E, Nettleton JA, van Dam RM, Chasman DI. Genome-wide

[226] Aryee MJ, Jaffe AE, Corrada-Bravo H, Ladd-Acosta C, Feinberg AP, Hansen KD, Irizarry RA. Minfi: a flexible and comprehensive Bioconductor package for the analysis of Infinium DNA methylation microarrays. Bioinformatics May 15, 2014;30(10):1363–9. https://doi.org/10.1093/bioinformatics/btu049. [Epub 2014 Jan 28].
[227] Fortin JP, Triche Jr TJ, Hansen KD. Preprocessing, normalization and integration of the Illumina HumanMethylationEPIC array with minfi. Bioinformatics February 15, 2017;33(4):558–60. https://doi.org/10.1093/bioinformatics/btw691.
[228] Morris TJ, Butcher LM, Feber A, Teschendorff AE, Chakravarthy AR, Wojdacz TK, Beck S. ChAMP: 450k chip analysis methylation pipeline. Bioinformatics February 1, 2014;30(3):428–30. https://doi.org/10.1093/bioinformatics/btt684. [Epub 2013 Dec 12].
[229] Assenov Y, Müller F, Lutsik P, Walter J, Lengauer T, Bock C. Comprehensive analysis of DNA methylation data with RnBeads. Nat Methods November 2014;11(11):1138–40. https://doi.org/10.1038/nmeth.3115. [Epub 2014 Sep. 28].
[230] Weber M, Davies JJ, Wittig D, Oakeley EJ, Haase M, Lam WL, Schübeler D. Chromosome-wide and promoter-specific analyses identify sites of differential DNA methylation in normal and transformed human cells. Nat Genet August 2005;37(8):853–62. [Epub 2005 Jul 10].
[231] Maunakea AK, Nagarajan RP, Bilenky M, Ballinger TJ, D'Souza C, Fouse SD, Johnson BE, Hong C, Nielsen C, Zhao Y, Turecki G, Delaney A, Varhol R, Thiessen N, Shchors K, Heine VM, Rowitch DH, Xing X, Fiore C, Schillebeeckx M, Jones SJ, Haussler D, Marra MA, Hirst M, Wang T, Costello JF. Conserved role of intragenic DNA methylation in regulating alternative promoters. Nature July 8, 2010;466(7303):253–7. https://doi.org/10.1038/nature09165.
[232] Serre D, Lee BH, Ting AH. MBD-isolated Genome Sequencing provides a high-throughput and comprehensive survey of DNA methylation in the human genome. Nucleic Acids Res January 2010;38(2):391–9. https://doi.org/10.1093/nar/gkp992. [Epub 2009 Nov 11].
[233] Rauch TA, Pfeifer GP. The MIRA method for DNA methylation analysis. Meth Mol Biol 2009;507:65–75.
[234] Rauch T, Pfeifer GP. Methods for assessing genome-wide DNA methylation. In: Tollefsbol T, editor. Handbook of Epigenetics. Amsterdam, The Netherlands: Elsevier; 2010. p. 135–47.
[235] Jung M, Kadam S, Xiong W, Rauch TA, Jin SG, Pfeifer GP. MIRA-seq for DNA methylation analysis of CpG islands. Epigenomics August 2015;7(5):695–706. https://doi.org/10.2217/epi.15.33. [Epub 2015 Apr 17].
[236] Meissner A, Gnirke A, Bell GW, Ramsahoye B, Lander ES, Jaenisch R. Reduced representation bisulfite sequencing for comparative high-resolution DNA methylation analysis. Nucleic Acids Res October 13, 2005;33(18):5868–77.
[237] Gu H, Smith ZD, Bock C, Boyle P, Gnirke A, Meissner A. Preparation of reduced representation bisulfite sequencing libraries for genome-scale DNA methylation profiling. Nat Protoc April 2011;6(4):468–81. https://doi.org/10.1038/nprot.2010.190. [Epub 2011 Mar 18].
[238] Wang G, Luo X, Wang J, Wan J, Xia S, Zhu H, Qian J, Wang Y. MeDReaders: a database for transcription factors that bind to methylated DNA. Nucleic Acids Res January 4, 2018;46(D1):D146–51. https://doi.org/10.1093/nar/gkx1096.
[239] Krueger F, Andrews SR. Bismark: a flexible aligner and methylation caller for Bisulfite-Seq applications. Bioinformatics June 1, 2011;27(11):1571–2. https://doi.org/10.1093/bioinformatics/btr167. [Epub 2011 Apr 14].
[240] Hansen KD, Langmead B, Irizarry RA. BSmooth: from whole genome bisulfite sequencing reads to differentially methylated regions. Genome Biol October 3, 2012;13(10):R83. https://doi.org/10.1186/gb-2012-13-10-r83.
[241] Feng H, Conneely KN, Wu H. A Bayesian hierarchical model to detect differentially methylated loci from single nucleotide resolution sequencing data. Nucleic Acids Res April 2014;42(8):e69. https://doi.org/10.1093/nar/gku154. [Epub 2014 Feb 22].
[242] Robinson MD, Kahraman A, Law CW, Lindsay H, Nowicka M, Weber LM, Zhou X. Statistical methods for detecting differentially methylated loci and regions. Front Genet September 16, 2014;5:324. https://doi.org/10.3389/fgene.2014.00324. [eCollection 2014. Review].
[243] Hebestreit K, Dugas M, Klein HU. Detection of significantly differentially methylated regions in targeted bisulfite sequencing data. Bioinformatics July 1, 2013;29(13):1647–53. https://doi.org/10.1093/bioinformatics/btt263. [Epub 2013 May 8].
[244] Hebestreit K, Klein H. BiSeq: processing and analyzing bisulfite sequencing data. 2015. R package version 1.18.0.
[245] Sun D, Xi Y, Rodriguez B, Park HJ, Tong P, Meong M, Goodell MA, Li W. MOABS: model based analysis of bisulfite sequencing data. Genome Biol February 24, 2014;15(2):R38. https://doi.org/10.1186/gb-2014-15-2-r38.
[246] Dolzhenko E, Smith AD. Using beta-binomial regression for high-precision differential methylation analysis in multifactor whole-genome bisulfite sequencing

experiments. BMC Bioinf June 24, 2014;15:215. https://doi.org/10.1186/1471-2105-15-215.

[247] Park Y, Figueroa ME, Rozek LS, Sartor MA. MethylSig: a whole genome DNA methylation analysis pipeline. Bioinformatics September 1, 2014;30(17):2414–22. https://doi.org/10.1093/bioinformatics/btu339. [Epub 2014 May 16].

[248] Kurdyukov S, Bullock M. DNA methylation analysis: choosing the right method. Biology January 6, 2016;5(1). https://doi.org/10.3390/biology5010003. pii: E3 [Review].

[249] Akalin A, Franke V, Vlahoviček K, Mason CE, Schübeler D. Genomation: a toolkit to summarize, annotate and visualize genomic intervals. Bioinformatics April 1, 2015;31(7):1127–9. https://doi.org/10.1093/bioinformatics/btu775. [Epub 2014 Nov 21].

[250] Zhu LJ, Gazin C, Lawson ND, Pagès H, Lin SM, Lapointe DS, Green MR. ChIPpeakAnno: a bioconductor package to annotate ChIP-seq and ChIP-chip data. BMC Bioinf May 11, 2010;11:237. https://doi.org/10.1186/1471-2105-11-237.

[251] Schmidt A, Forne I, Imhof A. Bioinformatic analysis of proteome data. BMC Syst Biol 2014;8(Suppl. 2):S3.

[252] Keller A, Nesvizhskii AI, Kolker E, Aebersold R. Empirical statistical model to estimate the accuracy of peptide identifications made by MS/MS and database search. Anal Chem October 2002;74(20):5383–92.

[253] Shteynberg D, Deutsch EW, Lam H, Eng JK, Sun Z, Tasman N, Mendoza L, Moritz RL, Aebersold R, Nesvizhskii AI. iProphet: multi-level integrative analysis of shotgun proteomic data improves peptide and protein identification rates and error estimates. Mol Cell Proteomics December 2011;10(12):M111. 007690.

[254] Li XJ, Zhang H, Ranish JA, Aebersold R. Automated statistical analysis of protein abundance ratios from data generated by stable-isotope dilution and tandem mass spectrometry. Anal Chem December 2003;75(23):6648–57.

[255] Nesvizhskii AI, Keller A, Kolker E, Aebersold R. A statistical model for identifying proteins by tandem mass spectrometry. Anal Chem September 2003;75(17):4646–58.

[256] Li X-J, Pedrioli PGA, Eng J, Martin D, Yi EC, Lee H, Aebersold R. A tool to visualize and evaluate data obtained by liquid chromatography/electrospray ionization/mass spectrometry. Anal Chem July 2004;76(13):3856–60.

[257] Shteynberg D, Mendoza L, Hoopmann MR, Sun Z, Schmidt F, Deutsch EW, Moritz RL. reSpect: software for identification of high and low abundance ion species in chimeric tandem mass spectra. J Am Soc Mass Spectrom November 2015;26(11):1837–47.

[258] Deutsch EW, Mendoza L, Shteynberg D, Farrah T, Lam H, Tasman N, Sun Z, Nilsson E, Pratt B, Prazen B, Eng JK, Martin DB, Nesvizhskii A, Aebersold R. A guided tour of the trans-proteomic pipeline. Proteomics March 2010;10(6):1150–9.

[259] Ashburner M, Ball CA, Blake JA, Botstein D, Butler H, Cherry JM, Davis AP, Dolinski K, Dwight SS, Eppig JT, et al. Gene ontology: tool for the unification of biology. The Gene Ontology Consortium. Nat Genet 2000;25(1):25–9.

[260] Alonso R, Salavert F, Garcia-Garcia F, Carbonell-Caballero J, Bleda M, Garcia-Alonso L, Sanchis-Juan A, Perez-Gil D, Marin-Garcia P, Sanchez R, Cubuk C, Hidalgo MR, Amadoz A, Hernansaiz-Ballesteros RD, Alemán A, Tarraga J, Montaner D, Medina I, Dopazo J. Babelomics 5.0: functional interpretation for new generations of genomic data. Nucleic Acids Res 2015;43(W1):W117–21.

[261] Subramanian A, Kuehn H, Gould J, Tamayo P, Mesirov JP. GSEA-P: a desktop application for gene set enrichment analysis. Bioinformatics December 1, 2007;23(23):3251–3.

[262] Subramanian A, Tamayo P, Mootha VK, Mukherjee S, Ebert BL, Gillette MA, Paulovich A, Pomeroy SL, Golub TR, Lander ES, Mesirov JP. Gene set enrichment analysis: a knowledge-based approach for interpreting genome-wide expression profiles. Proc Natl Acad Sci U S A October 2005;102(43):15545–50.

[263] Malik R, Dulla K, Nigg E, Körner R. From proteome lists to biological impact - tools and strategies for the analysis of large MS data sets. Proteomics 2010;10:1270–83.

[264] Bader G, Cary M, Sander C. Pathguide: a pathway resource list. Nucleic Acids Res 2006;34(Database):D504–6.

[265] Altschul SF, Gish W, Miller W, Myers EW, Lipman DJ. Basic local alignment search tool. J Mol Biol 1990;215(3):403–10.

[266] Desiere F, Deutsch EW, Nesvizhskii AI, Mallik P, King NL, Eng JK, Aderem A, Boyle R, Brunner E, Donohoe S, et al. Integration with the human genome of peptide sequences obtained by high-throughput mass spectrometry. Genome Biol 2004:6.

[267] Falkner JA, Ulintz PJ, Andrews PC. A code and data archival and dissemination tool for the proteomics community. Am Biotechnol Lab 2006.

[268] Vizcaíno AJ, Côté RG, Csordas A, Dianes JA, Fabregat A, Foster JM, Griss J, Alpi E, Birim M, Contell J, et al. The Proteomics Identifications (PRIDE) database and associated tools: status in 2013. Nucleic Acids Res 2012;41(D1):D1063–9.

[269] Benjamini Y, Hochberg Y. Controlling the false discovery rate: apractical and powerful approach to multiple testing. J R Stat Soc 1995;57(No. 1):289–300.

[270] Benjamini Y, Daniel Yekutieli D. The control of the false discovery rate in multiple testing under dependency. Ann Stat Aug., 2001;29(No. 4):1165–88.

[271] Storey JD. The positive false discovery rate: a Bayesian interpretation and the $q$-value. Ann Stat 2003;31(Number 6):2013–35.

[272] Storey JD. False discovery rate. In: Lovric M, editor. International Encyclopedia of Statistical Science. Berlin, Heidelberg: Springer; 2011.

[273] Fonville JM, Richards SE, Barton RH, Boulange CL, Ebbels TMD, Nicholson JK, Holms E, Dumas ME. The evolution of partial least squares models and related chemometric approaches in metabonomics and metabolic phenotyping. J Chemometr 2010;24(11–12):636–49. https://doi.org/10.1002/cem.1359.

[274] Barker M, Rayens W. Partial least squares for discrimination. J Chemometr 2003;17:166–73. https://doi.org/10.1002/chem.785.

[275] Trygg J, Wold S. Orthogonal projections to latent structures (O-PLS). J Chemom 2002;16:119–28. https://doi.org/10.1002/cem.695.

[276] Smith C, Want EJ, O'Maille G, Abagyan R, Siuzdak GE. XCMS: processing mass spectrometry data for metabolite profiling using nonlinear peak alignment, matching, and identification. Anal Chem 2006;78:779.

[277] Xia J, Sinelnikov IV, Han B, Wishart DS. MetaboAnalyst 3.0-making metabolomics more meaningful. Nucleic Acids Res 2015;43:W251.

[278] Shabalin AA. Matrix eQTL: ultra fast eQTL analysis via large matrix operations. Bioinformatics May 15, 2012;28(10):1353–8. https://doi.org/10.1093/bioinformatics/bts163. [Epub 2012 Apr 6].

[279] Kamburov A, Cavill R, Ebbels TM, Herwig R, Keun HC. Integrated pathway-level analysis of transcriptomics and metabolomics data with IMPaLA. Bioinformatics October 15, 2011;27(20):2917–8. https://doi.org/10.1093/bioinformatics/btr499. [Epub 2011 Sep. 4].

[280] Sun H, Wang H, Zhu R, Tang K, Gong Q, Cui J, Cao Z, Liu Q. iPEAP: integrating multiple omics and genetic data for pathway enrichment analysis. Bioinformatics March 1, 2014;30(5):737–9. https://doi.org/10.1093/bioinformatics/btt576. [Epub 2013 Oct 3].

[281] Gosline SJ, Oh C, Fraenkel E. SAMNetWeb: identifying condition-specific networks linking signaling and transcription. Bioinformatics April 1, 2015;31(7):1124–6. https://doi.org/10.1093/bioinformatics/btu748. [Epub 2014 Nov 19].

[282] Wachter A, Beißbarth T. pwOmics: an R package for pathway-based integration of time-series omics data using public database knowledge. Bioinformatics September 15, 2015;31(18):3072–4. https://doi.org/10.1093/bioinformatics/btv323. [Epub 2015 May 21].

[283] Karnovsky A, Weymouth T, Hull T, Tarcea VG, Scardoni G, Laudanna C, Sartor MA, Stringer KA, Jagadish HV, Burant C, Athey B, Omenn GS. Metscape 2 bioinformatics tool for the analysis and visualization of metabolomics and gene expression data. Bioinformatics February 1, 2012;28(3):373–80. https://doi.org/10.1093/bioinformatics/btr661. [Epub 2011 Nov 30].

[284] Wanichthanarak K, Fahrmann JF, Grapov D. Genomic, proteomic, and metabolomic data integration strategies. Biomark Insights 2015;10:1–6. https://doi.org/10.4137/BMI.S29511.

[285] Uppal K, Go Y-M, Jones DP. xMWAS: an R package for data-driven integration and differential network analysis. Mar. 30, 2017. https://doi.org/10.1101/122432. bioRxiv.

## FURTHER READING

Brinkman AB, Simmer F, Ma K, Kaan A, Zhu J, Stunnenberg HG. Whole-genome DNA methylation profiling using MethylCap-seq. Methods November 2010;52(3):232–6. https://doi.org/10.1016/j.ymeth.2010.06.012. [Epub 2010 Jun 11].

Sham PC, Cherny SS, Purcell S, Hewitt JK. Power of linkage versus association analysis of quantitative traits, by use of variance-components models, for sibship data. Am J Hum Genet May 2000;66:1616–30.

Xi Y, Li W. BSMAP: whole genome bisulfite sequence MAPping program. BMC Bioinf July 27, 2009;10:232. https://doi.org/10.1186/1471-2105-10-232.

# Population Genetics

*H. Richard Johnston[1], Bronya J.B. Keats[2], Stephanie L. Sherman[1]*

[1]Department of Human Genetics, Emory University School of Medicine, Atlanta, GA, United States
[2]Department of Genetics (Emeritus), Louisiana State University Health Sciences Center, New Orleans, LA, United States

Population genetics is the study of genetic variation within and among populations and the evolutionary factors that explain this variation. Its foundation is the Hardy–Weinberg law, which is maintained as long as the population size is large, mating is at random, and mutation, selection, and migration are negligible. If these assumptions are violated, allele frequencies and genotype frequencies may change from one generation to the next. Ethnic variation in allele frequencies is found throughout the genome, and by examining this genetic diversity, evolutionary patterns can be inferred, and variants contributing to common diseases can be identified. As a result of major international initiatives, extensive databases containing millions of genetic variants are available. Together with automated technology for genotyping, sequencing, and bioinformatic analysis, these datasets provide the population geneticist with a huge set of densely mapped polymorphisms for reconciling genome variation with population histories of bottlenecks, admixture, and migration, for revealing evidence of natural selection, and for advancing the understanding of many diseases.

## 12.1 INTRODUCTION

With the monumental scientific advances that have resulted from the Human Genome Project, the genetic composition of populations can now be examined in detail. Thousands of rare alleles are known to be disease-causing variants and for most of these diseases [e.g., cystic fibrosis, Tay–Sachs, phenylketonuria (PKU), hemophilia A, and familial hypercholesterolemia], many different pathogenic variants are found within the same gene. In general, the frequencies of these rare disease alleles differ among populations, as do the frequencies of common alleles (>1%), many of which are associated with common diseases such as Crohn's disease, diabetes, coronary disease, celiac disease, multiple sclerosis, and macular degeneration. The principles of population genetics attempt to explain the genetic diversity in present populations and the changes in allele and genotype frequencies over time. Population genetic studies facilitate the identification of alleles associated with disease risk and provide insight into the effect of medical intervention on the population frequency of a disease. Allele and genotype frequencies depend on factors such as mating patterns, population size and distribution, mutation, migration, and selection. By making specific assumptions about these factors, the Hardy–Weinberg law, a fundamental principle of population genetics, provides a model for calculating genotype frequencies from allele frequencies for a random mating population in equilibrium.

## 12.2 HARDY–WEINBERG LAW

The allele frequencies at a locus can always be calculated from the genotype frequencies, but the converse is not necessarily true. The Hardy–Weinberg law states that for a single autosomal locus in a large population in which

Emery and Rimoin's Principles and Practice of Medical Genetics and Genomics: Foundations, https://doi.org/10.1016/B978-0-12-812537-3.00012-3

(1) mating takes place at random with respect to genotype, (2) allele frequencies are the same in males and females, and (3) mutation, selection, and migration are negligible, genotype frequencies can be calculated from allele frequencies after one generation, regardless of the allele and genotype frequencies in the initial population. This is not true for a single X-linked locus or for any set of loci considered jointly; for these loci, the establishment of this relationship between allele and genotype frequencies takes more than one generation.

### 12.2.1 Autosomal Locus

Consider a locus with two alleles, $A_1$ and $A_2$, and suppose the population frequencies of the three genotypes $A_1A_1$, $A_1A_2$, $A_2A_2$ are $p_{11}$, $p_{12}$, $p_{22}$, respectively, where $p_{11}+p_{12}+p_{22}=1$, then, in this initial population, the frequency of $A_1$ is $p_{11}+½p_{12}$ and the frequency of $A_2$ is $p_{22}+½p_{12}$. Random mating is approximately equivalent to random union of gametes. Thus, random mating within this initial population results in the following genotype frequencies in the next generation:

$$\text{Frequency of } A_1A_1 = (p_{11} + ½\, p_{12})^2$$

$$\text{Frequency of } A_1A_2 = 2\,(p_{11} + ½\, p_{12})\,(p_{22} + ½\, p_{12})$$

$$\text{Frequency of } A_2A_2 = (p_{22} + ½\, p_{12})^2$$

The genotype frequencies in this second generation may be different from those in the first generation. However, calculation of the allele frequencies from the genotype frequencies in the second generation gives

$$\text{Frequency of } A_1 = (p_{11} + ½\, p_{12})^2 + (p_{11} + ½\, p_{12})(p_{22} + ½\, p_{12}) = p_{11}\,(p_{11} + p_{12} + p_{22}) + ½\, p_{12}(p_{11} + p_{12} + p_{22}) = p_{11} + ½\, p_{12}$$

Similarly, the frequency of $A_2$ is $p_{22}+½p_{12}$, which is equal to $1-(p_{11}+½p_{12})$. These allele frequencies are identical to those in the first generation. In other words, if the allele frequencies are $p=p_{11}+½p_{12}$ and $q=1-p=p_{22}+½p_{12}$, then after one generation of random mating, the genotype frequencies are $p^2$, $2pq$, and $q^2$. These frequencies are the Hardy–Weinberg proportions, and the population is said to be in Hardy–Weinberg equilibrium.

**TABLE 12.1 Table Establishment of Equilibrium in One Generation for an Autosomal Locus**

| | Mating Frequency | OFFSPRING GENOTYPE FREQUENCIES | | |
|---|---|---|---|---|
| | | $A_1A_1$ | $A_1A_2$ | $A_2A_2$ |
| $A_1A_1 \times A_1A_1$ | $(0.1)^2$ | $(0.1)^2$ | 0 | 0 |
| $A_1A_1 \times A_1A_2$ | 2(0.1)(0.2) | (0.1)(0.2) | (0.1)(0.2) | 0 |
| $A_1A_1 \times A_2A_2$ | 2(0.1)(0.7) | 0 | 2(0.1)(0.7) | 0 |
| $A_1A_2 \times A_1A_2$ | $(0.2)^2$ | $1/4(0.2)^2$ | $1/2(0.2)^2$ | $1/4(0.2)^2$ |
| $A_1A_2 \times A_2A_2$ | 2(0.2)(0.7) | 0 | (0.2)(0.7) | (0.2)(0.7) |
| $A_2A_2 \times A_2A_2$ | $(0.7)^2$ | 0 | 0 | $(0.7)^2$ |
| Total | 1 | 0.04 | 0.32 | 0.64 |

Table 12.1 presents a numerical example, in which the initial population comprises 20, 40, and 140 individuals with genotypes $A_1A_1$, $A_1A_2$, and $A_2A_2$, respectively. The genotype frequencies are

$$\text{Frequency } A_1A_1 = p_{11} = 20/200 = 0.10$$

$$\text{Frequency } A_1A_2 = p_{12} = 40/200 = 0.20$$

$$\text{Frequency } A_2A_2 = p_{22} = 140/200 = 0.70$$

and, therefore, the allele frequencies are

$$\text{Frequency } A_1 = 0.10 + ½\,(0.20) = 0.2$$

$$\text{Frequency } A_2 = 0.70 + ½\,(0.20) = 0.8$$

Random union of gametes results in the following genotype frequencies in the next generation:

$$\text{Frequency } A_1A_1 = 0.2^2 = 0.04$$

$$\text{Frequency } A_1A_2 = 2\,(0.2)\,(0.8) = 0.32$$

$$\text{Frequency } A_2A_2 = 0.8^2 = 0.64$$

Note that these genotype frequencies are different from those in the initial population. To confirm that these results are correct, Table 12.1 shows the genotype frequencies in the offspring that result from each of the six possible mating types. For example, all the offspring of the mating $A_1A_1 \times A_1A_1$ must be $A_1A_1$, while for the mating type $A_1A_1 \times A_1A_2$, each of the two offspring genotypes, $A_1A_1$ and $A_1A_2$, has a probability of one half (Table 12.1). Summing-up the columns in Table 12.1 gives the frequencies of each of the genotypes in the second generation. These frequencies are the same as those obtained by random union of gametes, and the allele frequencies calculated from these genotype frequencies are:

$$\text{Frequency } A_1 = 0.04 + ½\ (0.32) = 0.2$$

$$\text{Frequency } A_2 = 0.64 + ½\ (0.32) = 0.8$$

Repeating these steps will give identical genotype and allele frequencies in the third generation to those in the second generation. Note that only the genotype frequencies change in the establishment of equilibrium; the allele frequencies in the initial population remain the same in subsequent generations.

The chi-square goodness-of-fit test may be used to determine whether the observed numbers of each genotype are significantly different from those expected under Hardy–Weinberg equilibrium. The total number of individuals is 200, so the expected numbers for the three genotypes are $(0.2)^2 200 = 8$, $2(0.2)(0.8)200 = 64$, and $(0.8)^2 200 = 128$, compared with the observed numbers of 20, 40, and 140. The test value is $(20-8)^2/8 + (40-64)^2/64 + (140-128)^2/128 = 28$. This value is compared to the chi-square distribution with one degree of freedom. (There are three classes, but the total number of individuals is known and also the allele frequencies are known. Therefore, there is only one independent class and one degree of freedom.) In general, the number of degrees of freedom is equal to the number of genotypes minus the number of alleles. The 99.9th percentile of the chi-square distribution with one degree of freedom is 10.83.

Thus, the observed numbers of each genotype in the initial population are significantly different at the 1% level from those expected under Hardy–Weinberg equilibrium. However, after one generation of random mating, the observed and expected numbers are the same.

Calculation of allele frequencies from genotype frequencies is straightforward when all three genotypes are observable, but, in the case of recessive diseases, such as cystic fibrosis, only two phenotype classes are observed. If, however, equilibrium is assumed, the frequency of affected individuals is $q^2$; thus, the square root of this frequency is the frequency of the disease allele. The frequency of heterozygotes (carriers) is $2(1-q)q$, and the proportion of carriers among unaffected individuals in the population is

$$[2(1-q)\,q]\,/\,(1-q^2) = 2q/(1+q)$$

For example, in populations of European ancestry, the frequency of cystic fibrosis is estimated to be 1/2000; thus, the frequency of the abnormal allele is 0.022 and the normal allele is 0.978. The frequency of heterozygotes is therefore $2 \times 0.022 \times 0.978$, which is about 1/23. That is, approximately 4% of the populations are carriers, but less than 0.1% are affected. Several different pathogenic variants have been described in the cystic fibrosis gene. Each one of these is a disease allele; thus, the frequency of 0.022 is actually the sum of the frequencies of all the disease alleles in the cystic fibrosis gene.

The Hardy–Weinberg principle may be extended to more than two alleles. In general, for $n$ alleles, $A_1, A_2, \ldots, A_n$, with frequencies $p_1, p_2, \ldots, p_n$, the genotype frequencies are $p_i^2$ for homozygotes $A_iA_i$ and $2p_ip_j$ for heterozygotes $A_iA_j$. The heterozygosity value ($H$) for a locus is the total frequency of heterozygotes, and it may be written as

$$H = \sum 2p_ip_j = 1 - \sum p_i^2$$

For two alleles, the maximum heterozygosity is 0.5, for five alleles it is 0.8, and for 10 alleles it is 0.9. In other words, for a locus to have a heterozygosity of 80%, it must have at least five alleles. (The maximum heterozygosity is reached when the alleles have equal frequencies.)

**Example**

Suppose a locus has five alleles (designated 1, 2, 3, 4, 5) with frequencies 0.5, 0.3, 0.1, 0.08, 0.02. What are the genotype frequencies when Hardy–Weinberg equilibrium is established? What is the heterozygosity value (H) at this locus?

With n alleles, there are $n(n+1)/2$ genotypes. Thus, for five alleles there are 15 genotypes. The frequencies of the five homozygotes, 1–1, 2–2, 3–3, 4–4, 5–5, are 0.25, 0.09, 0.01, 0.0064, 0.0004, respectively. The frequencies of the 10 heterozygotes, 1–2, 1–3, 1–4, 1–5, 2–3, 2–4, 2–5, 3–4, 3–5, 4–5, are 0.3, 0.1, 0.08, 0.02, 0.06, 0.048, 0.012, 0.016, 0.004, 0.0032, respectively.

$$\text{Heterozygosity } (H) = 1 - (0.25 + 0.09 + 0.01 + 0.0064 + 0.0004) = 0.6432$$

**TABLE 12.2 Approach to Equilibrium for a Locus on the X Chromosome**

| Generation | $p_m$ | $p_f$ |
|---|---|---|
| 0 | 0.33 | 0.57 |
| 1 | 0.57 | 0.45 |
| 2 | 0.45 | 0.51 |
| 3 | 0.51 | 0.48 |
| 4 | 0.48 | 0.495 |
| 5 | 0.495 | 0.4875 |
| 6 | 0.4875 | 0.49125 |
| 7 | 0.49125 | 0.489375 |
| 8 | 0.489375 | 0.4903125 |
| 9 | 0.4903125 | 0.48984375 |
| 10 | 0.48984375 | 0.490078125 |
| 11 | 0.490078125 | 0.4899609375 |
| 12 | 0.4899609375 | 0.49001953125 |
| – | | |
| – | | |
| – | | |
| Equilibrium | 0.49 | 0.49 |

## 12.2.2 X-Linked Locus

The genotype frequencies at a locus on the X chromosome differ in the two sexes because males have only one X chromosome, whereas females have two X chromosomes. Thus, in males, the genotype frequency is equal to the allele frequency. For Hardy–Weinberg equilibrium, the allele frequencies in males must be equal to those in females. Suppose the frequency of the $A_1$ allele is $p_m$ in males and $p_f$ in females in the first generation. By the principles of X-linked inheritance, the frequency of this allele in males in the second generation must be $p_f$ because males get their X chromosomes from their mothers. By contrast, for females in the second generation the frequency of the $A_1$ allele is $\frac{1}{2}(p_m+p_f)$ because females get one X chromosome from each parent. The difference between the male and female frequencies in this generation is $p_f-\frac{1}{2}(p_m+p_f)=\frac{1}{2}(p_f-p_m)$, which is one half of the difference in the first generation. Similarly, in the third generation, the male allele frequency is $\frac{1}{2}(p_m+p_f)$, while the female frequency is $\frac{1}{4}p_m+\frac{3}{4}p_f$, and the difference is $\frac{1}{4}(p_f-p_m)$. With each generation, the difference between the male and female frequencies becomes smaller, and equilibrium is reached when they are the same. The equilibrium allele frequency of $A_1$ is equal to $\frac{2}{3}p_f+\frac{1}{3}p_m$ in both sexes. Table 12.2 shows the approach to equilibrium for an X-linked locus when the initial allele frequencies are 0.33 in males and 0.57 in females. With each generation, the difference between the frequencies in males and females is reduced, and they approach the equilibrium frequency of $\frac{1}{3}(0.33)+\frac{2}{3}(0.57)=0.49$. If the frequencies of the two alleles at the locus ($A_1$ and $A_2$) are $p=\frac{1}{3}p_m+\frac{2}{3}p_f$ and $q=1-p$, the equilibrium genotype frequencies are $p$ and $q$ in males and $p^2$, $2pq$, and $q^2$ in females. Table 12.3 gives the frequency of each possible mating type and the expected offspring genotype frequencies for males and females. Summing-up these genotype frequencies shows that the equilibrium frequencies are maintained in the next generation (Table 12.3).

**TABLE 12.3 Genotype Equilibrium Frequencies for an X-Linked Locus**

| | | OFFSPRING GENOTYPE FREQUENCIES | | | | |
|---|---|---|---|---|---|---|
| | | MALE | | FEMALE | | |
| | Mating Frequency | $A_1$ | $A_2$ | $A_1A_1$ | $A_1A_2$ | $A_2A_2$ |
| $A_1 \times A_1A_1$ | $p^3$ | $p^3$ | 0 | $p^3$ | 0 | 0 |
| $A_1 \times A_1A_2$ | $2p^2q$ | $p^2q$ | $p^2q$ | $p^2q$ | $p^2q$ | 0 |
| $A_1 \times A_2A_2$ | $pq^2$ | 0 | $pq^2$ | 0 | $pq^2$ | 0 |
| $A_2 \times A_1A_1$ | $p^2q$ | $p^2q$ | 0 | 0 | $p^2q$ | 0 |
| $A_2 \times A_1A_2$ | $2pq^2$ | $pq^2$ | 0 | 0 | $pq^2$ | $pq^2$ |
| $A_2 \times A_2A_2$ | $q^3$ | 0 | $q^3$ | 0 | 0 | $q^3$ |
| Total | | $p$ | $q$ | $p^2$ | $2pq$ | $q^2$ |

**Example**

Suppose the frequency of an allele at an X-linked locus is 0.03 in males and 0.06 in females. What is the equilibrium allele frequency? Furthermore, suppose that this allele is responsible for a recessive trait. What are the equilibrium frequencies of this trait in males and females, and what is the frequency of heterozygous (carrier) females?

The equilibrium frequency of the allele is $\frac{1}{3}(0.03)+\frac{2}{3}(0.06)=0.05$. Thus, the frequency of males with the trait is 0.05 and the frequency of females with the trait is $(0.05)^2=0.0025$. The frequency of carrier females is $2(0.05)(0.95)=0.095$.

## 12.2.3 Two Loci

Equilibrium is reached after one generation of random mating for a single autosomal locus and over several generations for an X-linked locus. However, the approach to equilibrium may be much longer for two loci considered

**TABLE 12.4 Joint Genotype Frequencies for Two Loci**

| Genotype | Frequency | Equilibrium Frequency |
|---|---|---|
| $A_1A_1B_1B_1$ | $g_{11}^2$ | $p_1^2q_1^2$ |
| $A_1A_1B_1B_2$ | $2g_{11}g_{12}$ | $2p_1^2q_1q_2$ |
| $A_1A_1B_2B_2$ | $g_{12}^2$ | $p_1^2q_2^2$ |
| $A_1A_2B_1B_1$ | $2g_{11}g_{21}$ | $2p_1p_2q_1^2$ |
| $A_1A_2B_1B_2$ | $2g_{11}g_{22} + 2g_{12}g_{21}$ | $4p_1p_2q_1q_2$ |
| $A_1A_2B_2B_2$ | $2g_{12}g_{22}$ | $2p_1p_2q_2^2$ |
| $A_2A_2B_1B_1$ | $g_{21}^2$ | $p_2^2q_1^2$ |
| $A_2A_2B_1B_2$ | $2g_{21}g_{22}$ | $2p_2^2q_1q_2$ |
| $A_2A_2B_2B_2$ | $g_{22}^2$ | $p_2^2q_2^2$ |

jointly, and the number of generations depends on the recombination fraction. Suppose the first locus has alleles $A_1$ and $A_2$, with frequencies $p_1$ and $p_2$, and the second locus has alleles $B_1$ and $B_2$, with frequencies $q_1$ and $q_2$, respectively. The four possible gametes are $A_1B_1$, $A_1B_2$, $A_2B_1$, $A_2B_2$; let their frequencies in the population be $g_{11}$, $g_{12}$, $g_{21}$, $g_{22}$, where $p_1 = g_{11} + g_{12}$, $p_2 = g_{21} + g_{22}$, $q_1 = g_{11} + g_{21}$, and $q_2 = g_{12} + g_{22}$. Allowing these gametes to unite at random gives the genotype frequencies in the next generation (Table 12.4). Now consider the gametic output of this population. In doing so, we must take into account the fact that the frequency of gametes produced by the double heterozygote ($A_1A_2B_1B_2$) depends on the recombination fraction, $\theta$ (Table 12.4). If the phase is $A_1B_1/A_2B_2$, then $A_1B_1$ and $A_2B_2$ are nonrecombinants, and $A_1B_2$ and $A_2B_1$ are recombinants. Conversely, if phase is $A_1B_2/A_2B_1$, then $A_1B_2$ and $A_2B_1$ are nonrecombinants, and $A_1B_1$ and $A_2B_2$ are recombinants. Therefore, the frequency of $A_1B_1$ gametes from double heterozygotes is $g_{11}g_{22}(1-\theta) + g_{12}g_{21}\theta$. In addition, all the gametes produced by individuals with the genotype $A_1A_1B_1B_1$, and one half of those produced by individuals with the genotypes $A_1A_1B_1B_2$ and $A_1A_2B_1B_1$ will be $A_1B_1$. Thus, the total frequency of $A_1B_1$ gametes in this generation is $g_{11}^2 + g_{11}g_{12} + g_{11}g_{21} + g_{11}g_{22}(1-\theta) + g_{12}g_{21}\theta$, which may be written as $g_{11} - \theta D$, where $D = g_{11}g_{22} - g_{12}g_{21}$. $D$ is called the coefficient of linkage disequilibrium (LD) and is a measure of allelic association. Similar calculations may be done for each of the gametic types, and the frequencies obtained are $g_{12} + \theta D$, $g_{21} + \theta D$, and $g_{22} - \theta D$ for $A_1B_2$, $A_2B_1$, and $A_2B_2$, respectively.

If the loci are unlinked, $\theta = ½$, and the change in gametic frequency from one generation to the next is $½D$. For linked loci the change is $\theta D$. Thus, the more closely two loci are linked, the slower is the approach to equilibrium. The coefficient of LD after $t$ generations may be written as

$$D_t = (1-\theta)\,D_{t-1} = (1-\theta)^t D_0$$

which approaches zero as $t$ trends to infinity.

At equilibrium, $D$ is equal to zero and the genotype and gametic frequencies are products of the allele frequencies (see Table 12.4). The gametic frequencies may be written as $g_{11} = p_1q_1 + D$, $g_{12} = p_1q_2 - D$, $g_{21} = p_2q_1 - D$, and $g_{22} = p_2q_2 + D$. Each of these gametic frequencies must be greater than or equal to zero. Thus, $D$ must be greater than or equal to both $-p_1q_1$ and $-p_2q_2$, and $D$ must be less than or equal to both $p_1q_2$ and $p_2q_1$. These results may be written as

$$D_{\min} = \max(-p_1q_1, -p_2q_2)$$

$$D_{\max} = \min(p_1q_2, p_2q_1)$$

For two loci each with two alleles, $D$ must lie between $-0.25$ and $0.25$, and it can reach these extreme values only if the frequencies of the four alleles are 0.5. Thus, the value of $D$ is dependent on allele frequencies, meaning that $D$ values for different pairs of loci are not comparable. The value of the standardized measure, $D' = D/D_{extreme}$, where $D_{extreme} = -D_{min}$ if $D<0$ and $D_{max}$ if $D>0$, is less dependent on the allele frequencies and lies between $-1$ and 1.

The statistic, $\delta$, is another measure of LD that is useful for estimating the location of a disease locus if a single mutation is likely. The formula is

$$\delta = (p_D - p_N)/(1 - p_N)$$

where $p_D$ is the frequency of the associated allele on disease chromosomes and $p_N$ is the frequency of this allele on normal chromosomes. This value represents an estimate of the proportion of disease chromosomes bearing the original associated allele. If there is a single mutation, the proportion of chromosomes carrying this mutation is the same for all marker loci, so differences in $\delta$ across loci should largely represent effects of recombination. Thus, $\delta$ can be used to determine the most likely location of the disease locus among a set of tightly linked marker loci.

The effectiveness of LD in locating disease mutations was demonstrated with the identification of the cystic fibrosis F508del mutation in 1989. Over the next decade, it was used successfully in the identification of ethnic-specific disease mutations in the Ashkenazi Jewish, Finnish, Acadian, Roma, and other isolated and endogamous populations.

Note that LD is one possible explanation for the association between a phenotype and a marker allele in a population. In this case, the disease locus is tightly linked to the marker locus. However, population association does not necessarily mean tight linkage, and vice versa. Other possible reasons for association are pleiotropy (multiple effects of the same gene), such as the association between stomach cancer and the A allele of the ABO blood group, and departures from random mating due to events such as racial admixture, stratification, inbreeding, and assortative mating.

**Examples**

1. Suppose the frequencies of the gametes $A_1B_1$, $A_1B_2$, $A_2B_1$, $A_2B_2$ are 0.5, 0.1, 0.3, 0.1, respectively. What is the value of D after one generation of random mating if (1) the two loci are unlinked and (2) the recombination fraction between the two loci is 0.01?
   The value of D in the original population is $(0.5)(0.1)-(0.1)(0.3)=0.02$. After one generation, $D=(1-0.5)(0.02)=0.01$ if the two loci are unlinked, and $D=(1-0.01)(0.02)=0.0198$ if the recombination fraction is 0.01.
2. How many generations are required for the value of D to be one-half its initial value?
   $D_t/D_0=(1-\theta)^t=½$; therefore, $t=\log(½)/\log(1-\theta)$. Thus, for $\theta$ equal to 0.3, 0.1, 0.01, and 0.001, the numbers of generations required are approximately 2, 7, 69, and 693, respectively. Note that for unlinked loci, D is halved in one generation, as seen in Example 1.

## 12.3 FACTORS THAT AFFECT HARDY–WEINBERG EQUILIBRIUM

The assumption of a large, random mating population is fundamental to Hardy–Weinberg equilibrium. If mating is not at random, the allele frequencies at a locus (say, $p$ and $q$) in the population do not change from one generation to the next, but the genotype frequencies are not $p^2$, $2pq$, and $q^2$. Evolutionary forces such as random genetic drift, mutation, selection, and migration, however, will change allele frequencies (and consequently genotype frequencies) from one generation to the next.

### 12.3.1 Factors That Affect Genotype Frequencies but Not Allele Frequencies

Random mating has been assumed so far in all the derivations. If gametes do not unite at random, the genotype frequencies are not in Hardy–Weinberg proportions and cannot be derived simply from allele frequencies. Consanguinity (inbreeding), assortative mating, and stratification (e.g., ethnic subgroups within a population) are examples of nonrandom mating. In these situations, the frequency of homozygotes is increased at the expense of heterozygotes, and the genotype frequencies may be significantly different from Hardy–Weinberg expectations. Allele frequencies, however, do not change.

#### 12.3.1.1 Consanguinity and Inbreeding

Individuals who are related genetically are termed consanguineous, and the offspring of mating between such individuals are said to be inbred. Inbreeding increases the frequency of homozygous genotypes and decreases the frequency of heterozygous genotypes in the population. The offspring of consanguineous marriages have an increased risk of having recessive disorders over that of the general population. The increase in risk depends on the population frequency of the disease allele and the degree of relationship between the parents. In cultures in which uncle–niece and first- and second-cousin marriages are encouraged, recessive disorders that are rare in most randomly mating populations may be relatively common. The coefficient of inbreeding ($F$) for a child of a consanguineous marriage is the probability that the child receives two alleles at a given locus that are both from the same ancestor and are, thus, identical by descent (autozygous). For example, half first cousins share a grandparent in common. The probability that a child of half first cousins is homozygous by descent at a locus is $F=(½)^5=1/32$. In general, for autosomal loci, the inbreeding coefficient for an individual is $F=(½)^{(n1+n2+1)}$, where $n_1$ and $n_2$ are the numbers of generations separating the individuals in the consanguineous mating from their common ancestor. (This formula assumes that the common ancestor is not inbred.) Half first cousins are separated from their common grandparent by two generations. Thus, the exponent is $2+2+1=5$. Table 12.5 gives the estimated proportion of alleles shared by consanguineous individuals that are identical by descent as well as the coefficient of inbreeding for the offspring of these consanguineous matings

**TABLE 12.5 Proportion of Alleles Shared by Related Individuals That Are Identical by Descent and the Inbreeding Coefficient (*F*) in the Offspring of Various Types of Consanguineous Mating**

| Type of Mating | Proportion of Shared Alleles | *F* |
|---|---|---|
| Parent–offspring | 1/2 | 1/4 |
| Brother–sister | 1/2 | 1/4 |
| Half sibs | 1/4 | 1/8 |
| Uncle–niece, aunt–nephew | 1/4 | 1/8 |
| First cousins | 1/8 | 1/16 |
| Double first cousins | 1/4 | 1/8 |
| Half first cousins | 1/16 | 1/32 |
| First cousins once removed | 1/16 | 1/32 |
| Second cousins | 1/32 | 1/64 |
| Second cousins once removed | 1/64 | 1/128 |
| Third cousins | 1/128 | 1/256 |

(Table 12.5). If a child is inbred through more than one line of descent, the total coefficient of inbreeding is the sum of each of the separate coefficients. For example, first cousins are related through two grandparents. Thus, the inbreeding coefficient for the offspring of first cousins is $F = (½)^5 + (½)^5 = (½)^4 = 1/16$. The coefficient of inbreeding is also an estimate of the proportion of loci at which an individual is autozygous.

The coefficient of inbreeding for X-linked loci depends on the number of males in the lines of descent and is always zero for male offspring, because they have only one X chromosome. In order to calculate the inbreeding coefficient for daughters of first cousins, four possibilities need to be considered for the first cousins: their fathers are brothers, their mothers are sisters, the father of the male cousin and the mother of the female cousin are siblings, or vice versa. If the fathers are brothers, the first cousins cannot share any X-linked alleles in common because the male first cousin did not inherit an X chromosome from his father. Thus, female offspring of this type of first-cousin mating are not inbred for X-linked loci and have an inbreeding coefficient of zero. Similarly, if the first cousins are offspring of a brother and a sister with the father being the son of the brother and the mother being the daughter of the sister, the inbreeding coefficient for their daughters is zero because the first cousins cannot share any X-linked alleles in common. On the other hand, if the mothers of the first cousins are sisters, then the inbreeding coefficient for X-linked loci in their daughters is greater than that for autosomal loci because a male transmits the X chromosome he received from his mother to all his daughters. Thus, the inbreeding coefficient in this situation is $(½)^3 + (½)^4 = 3/16$. The fourth possibility is that the first cousins are offspring of a brother and sister, with the sister being the mother of the male and the brother being the father of the female. In this case, the inbreeding coefficient for X-linked loci in female offspring is $(½)^3 = 1/8$.

Genotype frequencies in inbred populations cannot be calculated from the allele frequencies alone, but they can be obtained if the average inbreeding coefficient in the population is known. The amount of inbreeding in the population may be measured in terms of the decrease in heterozygosity relative to a random mating population. If the allele frequencies at a locus are $p$ and $q$, then under random mating the frequency of heterozygotes is $2pq$. Suppose the frequency of heterozygotes in the inbred population is $H$. Then the inbreeding coefficient for the population is $F = (2pq - H)/2pq$. Therefore, $H = 2pq - 2pqF$. The frequencies of the two types of homozygotes in the inbred population can then be calculated to be $p^2 + pqF$ and $q^2 + pqF$. If the inbreeding coefficient is zero (i.e., random mating), the genotype frequencies are those expected for Hardy–Weinberg equilibrium. On the other hand, if there is complete inbreeding ($F = 1$), the frequency of heterozygotes is zero, and the population consists only of homozygotes with frequencies of $p$ and $q$. However, note that the allele frequencies will not change from one generation to the next, regardless of the value of the inbreeding coefficient in the population.

**Example**

Suppose the frequency of an autosomal recessive disease is 1/40,000 in the general population. What is the expected frequency of the disease among the offspring of first cousins?

The frequency of the deleterious allele is 1/200, the square root of the frequency of the disease. The inbreeding coefficient for offspring of first-cousin marriages is 1/16. Thus, the frequency of the disease among the offspring of first cousins is $1/40{,}000 + (199/200)(1/200)(1/16) = 1/2977$.

#### 12.3.1.2 Assortative Mating

Assortative mating is the tendency for people to choose mates who are more similar (positive) or dissimilar

**TABLE 12.6 Selection Against the $A_2A_2$ Genotype at an Autosomal Locus**

| Genotype | $A_1A_1$ | $A_1A_2$ | $A_2A_2$ |
|---|---|---|---|
| Frequency before selection | $p^2$ | $2pq$ | $q^2$ |
| Relative fitness | 1 | 1 | $1-s$ |
| Frequency after selection | $p^2$ | $2pq$ | $q^2(1-s)$ |

After one generation of selection

Frequency $A_1=(p^2+pq)/[p^2+2pq+q^2(1-s)]=p/(1-sq^2)$

Frequency $A_2=[pq+(1-s)q^2]/(1-sq^2)$

If $s=1$ (i.e., complete selection against the $A_2A_2$ genotype), then

After one generation of selection

Frequency $A_1=(p^2+pq)/(p^2+2pq)=1/(1+q)$

Frequency $A_2=pq/(p^2+2pq)=q/(1+q)$

After $t$ generations of selection

Frequency $A_1=[1+(t-1)q]/(1+tq)$

Frequency $A_2=q/(1+tq)$

generations if the mutation rate is $10^{-4}$ and 69,315 generations if the mutation rate is $10^{-5}$.

2. Suppose the mutation rate from $A_1$ to $A_2$ is $10^{-4}$ and the reverse rate is $10^{-5}$. What is the equilibrium frequency of $A_1$?

   The equilibrium frequency of $A_1$ is $10^{-5}/(10^{-4}+10^{-5})=0.091$. However, to reach this equilibrium frequency may take tens of thousands of generations, depending on the initial allele frequencies.

#### 12.3.2.3 Selection

The fitness of an individual is defined as an ability to survive and reproduce. The process by which the frequencies of genotypes in individuals with greater fitness increase in the population is natural selection. It acts to decrease the frequencies of the less-fit genotypes. The relative fitness is defined as $1-s$, where $s$ is the selection coefficient against the deleterious genotype. Thus, the most-fit genotype has a relative fitness of 1 (and a selection coefficient of 0).

Consider the situation where there are three genotypes, $A_1A_1$, $A_1A_2$, $A_2A_2$, at a locus with relative fitnesses of 1, 1, $1-s$, respectively. That is, there is selection against the $A_2A_2$ homozygote. (If $s=1$, the selection is complete, meaning that individuals with the $A_2A_2$ genotype do not reproduce.) Table 12.6 shows the change in allele frequencies from one generation to the next. In the case in which $s=1$, the frequencies after $t$ generations can be written in terms of the initial allele frequencies (Table 12.6). Substituting in the formula given in Table 12.6 shows that when the $A_2A_2$ homozygote does not reproduce, the number of generations required to reduce the frequency of $A_2$ to one half its initial value is equal to the reciprocal of its initial value. Thus, if the frequency of $A_2$ is 0.01, it will take 100 generations of complete selection against the $A_2A_2$ homozygote to reduce the frequency of $A_2$ to 0.005. In other words, lack of reproduction of individuals with a rare recessive disease does not lead to a rapid reduction in the frequency of the deleterious allele from one generation to the next.

Now consider the situation in which there is partial selection against the $A_2A_2$ genotype. The allele frequencies after $t$ generations cannot be written in terms of the initial frequencies, but the decrease in the frequency of the $A_2$ allele from one generation to the next can be calculated. This decrease is equal to $sq^2(1-q)/(1-sq^2)$, and the number of generations required to change the frequency of $A_2$ from its initial value to a new value can be approximated. For example, if $s=0.001$ and the initial frequency of $A_2$ is 0.01, more than 100,000 generations will be required to reduce the frequency to 0.005. This example makes the point that even if the selective disadvantage of a genotype is very small, the allele frequencies in the population will gradually change. For the same selection coefficient ($s=0.001$), 11,665 generations are required to reduce the frequency of $A_2$ from 0.7 to 0.1. If there is selection against the heterozygous genotype ($A_1A_2$) as well as the $A_2A_2$ genotype, with $s=0.001$ for $A_2A_2$ and $s=0.0005$ for $A_1A_2$, then 6156 generations are required to reduce the frequency of $A_2$ from 0.7 to 0.1.

In the case in which there is selection favoring the heterozygote over both homozygotes, an equilibrium state is reached for the allele frequencies. Table 12.7 shows the change in allele frequencies from one generation to the next. At equilibrium, $s_1p=s_2q$, so that $p=s_2/(s_1+s_2)$ and $q=s_1/(s_1+s_2)$ (Table 12.7).

This equilibrium is stable and is called a balanced polymorphism. This type of selection is known as overdominance. If, on the other hand, selection is against the heterozygote, the equilibrium is unstable, and the selection is known as underdominance. The equilibrium frequencies are the same, but if a disturbance occurs such that $q>s_1/(s_1+s_2)$, $q$ will increase further rather than returning to its equilibrium value. The reverse is also true, so eventually one allele or the other will be eliminated.

**TABLE 12.7 Selection Favoring the Heterozygous Genotype at an Autosomal Locus**

| Genotype | $A_1A_1$ | $A_1A_2$ | $A_2A_2$ |
|---|---|---|---|
| Frequency before selection | $p^2$ | $2pq$ | $q^2$ |
| Relative fitness | $1-s_1$ | 1 | $1-s_2$ |
| Frequency after selection | $p^2(1-s_1)$ | $2pq$ | $q^2(1-s^2)$ |

After one generation of selection

Frequency $A_1 = (p - s_1p^2)/(1 - s_1p^2 - s_2q^2)$

Frequency $A_2 = (q - s_2q^2)/(1 - s_1p^2 - s_2q^2)$

The change in the frequency of $A_2$ from one generation to the next is $pq(s_1p - s_2q)/(1 - s_1p^2 - s_2q^2)$. Equating this quantity to zero gives the equilibrium allele frequencies, which are

Frequency $A_1 = s_2/(s_1 + s_2)$

Frequency $A_2 = s_1/(s_1 + s_2)$

Let us now consider a balance between mutation and selection. Suppose the mutation rate from $A_1$ to $A_2$ is $\mu$, and the relative fitnesses of the genotypes $A_1A_1$, $A_1A_2$, $A_2A_2$, are 1, 1, $1-s$, respectively. As shown in Table 12.6, the frequency of $A_1$ after selection is $p/(1-sq^2)$. Thus, the increase in frequency of $A_2$ due to mutation from $A_1$ to $A_2$ is $\mu p/(1-sq^2)$, while the decrease due to selection is $sq^2(1-q)/(1-sq^2)$. At equilibrium, $\mu p/(1-sq^2) = sq^2(1-q)/(1-sq^2)$, which simplifies to $q = \sqrt{(\mu/s)}$. This equilibrium is stable and $q = \sqrt{\mu}$ when $s = 1$. Thus, for a lethal recessive disease and a mutation rate of $10^{-6}$, the equilibrium frequency of the deleterious allele is 1/1000.

In the case of a deleterious dominant phenotype, the fitness of both the homozygote and the heterozygote is reduced. With selection coefficients of $1-s$, $1-s$, 1, the increase in the frequency of $A_1$ due to mutation is equal to the decrease due to selection when $q = \mu/s$, which reduces to $q = \mu$ for $s = 1$. If individuals with a dominant disease do not reproduce, the frequency of the deleterious allele in the next generation is equal to the mutation rate. Examples of such disorders are atelosteogenesis and thanatophoric dysplasia, which are both lethal forms of short-limbed dwarfism. In the case of achondroplasia, fitness is not zero, but it is considerably lower than one, and is estimated to be about 0.2. Thus, the equilibrium frequency of the deleterious allele is $10^{-5}/0.8 = 1.25 \times 10^{-5}$, or slightly higher than the mutation rate.

**TABLE 12.8 X-Linked Locus: Selection Against the $A_2A_2$ Genotype in Females and the $A_2$ Genotype in Males**

**Females**

| Genotype | $A_1A_1$ | $A_1A_2$ | $A_2A_2$ |
|---|---|---|---|
| Frequency before selection | $p^2$ | $2pq$ | $q^2$ |
| Relative fitness | 1 | 1 | $1-s$ |
| Frequency after selection | $p^2$ | $2pq$ | $q^2(1-s)$ |

**Males**

| Genotype | $A_1$ | $A_2$ |
|---|---|---|
| Frequency before selection | $p$ | $q$ |
| Relative fitness | 1 | $1-s$ |
| Frequency after selection | $p$ | $q(1-s)$ |

After one generation of selection in males

Frequency $A_1 = p/(1 - sq)$

Frequency $A_2 = (q - sq)/(1 - sq)$

Selection against genotypes at loci on the X chromosome needs to be tabulated separately for males and females because males have only one allele at an X-linked locus. Table 12.8 shows the case in which the $A_2A_2$ genotype and the $A_2$ genotype are selected against in females and males, respectively. The decrease in frequency of $A_2$ due to reduced fitness in females is extremely small compared with the decrease due to reduced fitness in males. Thus, we will only consider males (Table 12.8). The loss of $A_2$ alleles is equal to $sq(1-q)/(1-sq)$, which is $q$ if $s = 1$. In other words, if selection is complete, all male $A_2$ alleles are lost in each generation. Because males have only one allele and females have two, this loss represents one third of the $A_2$ alleles in the population. If the mutation rate from $A_1$ to $A_2$ is $\mu$, the increase in frequency of $A_2$ due to mutation in males is $\mu p/(1-sq)$. But mutation in males represents only one third of the mutations that are occurring in the population. Thus, an increase in frequency due to mutation balances a decrease due to selection when $3\mu p/(1-sq) = sq(1-q)/(1-sq)$, which reduces to $q = 3\mu/s$. For an X-linked recessive lethal, $s = 1$, and $\mu = q/3$. In other words, one third of the deleterious alleles in the population, and, thus, one third of cases of diseases, such as Duchenne muscular dystrophy, are new mutations. In less severe X-linked disorders, the proportion of cases that are new mutations is not as high; for example, the relative fitness of individuals with hemophilia A is about 70%. Therefore, the proportion of new mutations is $0.3q/3$, meaning phenomenon have been helpful in defining the precise location of disease genes. The complexity of patterns of LD and the extent and explanation of variability among populations are critical factors that are only now becoming understood [10].

A large number of genome-wide association studies have been performed using the millions of SNPs that are now available. These studies have identified genomic regions containing genetic risk factors for many complex diseases, for example, Parkinson disease [11]. However, for the majority of such diseases that have been studied so far, much of the genetic influence on them remains to be explained [12].

Current work in the field of population genetics involves studying the differences in disease-causing variants between major population groups. As many GWAS studies have only been conducted in European populations, there are open questions as to the utility of results from these studies in other populations. The more we learn, the more it appears that studying diseases in multiple populations is going to be critical to discerning the true nature of causal variants for each ancestral group [13,14].

The availability of massive databases of genetic variation and automated technology for genotyping, sequencing, and bioinformatic analysis [15,16] is significantly enhancing collaborative efforts between population geneticists and molecular geneticists and advancing understanding of many diseases. As we would anticipate, thousands of interesting new questions are being raised and the tools to answer many of them are now at our fingertips.

Improved sequencing has led to significantly lower costs per genome as well as improved sequencing quality, allowing researchers to choose whole-genome sequencing as a study design more often. There is now a massively ballooning set of fully sequenced genomes for study. The proliferation of these modern, large-scale genome-wide sequencing studies is generating variant information on an incredible scale, enabling population geneticists to ask and answer questions that have previously only been theoretical in nature.

## REFERENCES

[1] Cox MP, Hammer MF. A question of scale: human migrations writ large and small. BMC Biol 2010;8(98).

[2] Green RE, Krause J, Briggs AW, et al. A draft sequence of the Neandertal genome. Science 2010;328:710–22.

[3] Tishkoff SA, Reed FA, Friedlaender FR, et al. The genetic structure and history of Africans and African Americans. Science 2009;324:1035–44.

[4] Zakharia F, Basu A, Absher D, et al. Characterizing the admixed African ancestry of African Americans. Genome Biol 2009;10:R141.

[5] The International HapMap 3 Consortium. Integrating common and rare genetic variation in diverse human populations. Nature 2010;467:52–8.

[6] Sudman PH, Kitzman JO, Antonacci F, et al. Diversity of human copy number variation and multicopy genes. Science 2010;330:641–6.

[7] The 1000 Genomes Consortium. A map of human genome variation from population-scale sequencing. Nature 2010;467:1061–73.

[8] The 1000 Genomes Project Consortium. A global reference for human genetic variation. Nature 2015;526:68–74.

[9] Lomelin D, Jorgenson E, Risch N. Human genetic variation recognizes functional elements in noncoding sequence. Genome 2009;20:311–9.

[10] Altshuler D, Daly MJ, Lander ES. Genetic mapping in human disease. Science 2008;322:881–8.

[11] The UK Parkinson's Disease Consortium, The Wellcome Trust Case Control Consortium 2. Dissection of the genetics of Parkinson's disease identifies an additional association 5′ of SNCA and multiple associated haplotypes at 17q21. Hum Mol Genet 2011;20:345–53.

[12] Manolio TA, Collins FS, Cox NJ, et al. Finding the missing heritability of complex diseases. Nature 2009;461:747–53.

[13] Mathias RA, Taub MA, Gignoux CR, et al. A continuum or admixture in the Western Hemisphere revealed by the African Diaspora genome. Nat Commun 2016;7:12522.

[14] Martin AR, Gignoux CR, Walters RK, et al. Human demographic history impacts genetic risk prediction across diverse populations. Am J Hum Genet 2017;03:635–49.

[15] Johnston HR, Chopra P, Wingo TS, et al. PEMapper and PECaller provide a simplified approach to whole-genome sequencing. Proc Natl Acad Sci Unit States Am 2017;114(10):E1923–32.

[16] McKenna A, Hanna M, Banks E, et al. The Genome Analysis Toolkit; a MapReduce framework for analyzing next-generation DNA sequencing data. Genome Res 2010;20:1297–303.

**Relevant Websites**

International HapMap project. http://hapmap.ncbi.nlm.nih.gov/.

A catalog of published genome-wide association studies. http://www.ebi.ac.uk/gwas/.

## FURTHER READING

Hindorff LA, Sethupathy P, Junkins HA, et al. Potential etiologic and functional implications of genome-wide association loci for human diseases and traits. Proc Natl Acad Sci USA May 27, 2009.

Maniolo T, Brooks LD, Collins FS. A HapMap harvest of insights into the genetics of common disease. J Clin Invest 2008;118:1590–605.

Psychiatric GWAS Consortium Coordinating Committee. Genomewide association studies: history, rationale, and prospects for psychiatric disorders. Am J Psychiatr 2009;166:540–56.

13

# Pathogenetics of Disease

*Reed E. Pyeritz*

Perelman School of Medicine at the University of Pennsylvania, Philadelphia, PA, United States

*To wrest from nature the secrets which have perplexed philosophers in all ages, to track to their sources the causes of disease, to correlate the vast stores of knowledge, that they may be quickly available for the prevention and cure of disease—these are our ambitions.*

**Osler W. Chauvinism in Medicine. *Montreal Med J* 1902;31:684–99.**

*The great ease with which molecular information can be collected on the genomes of higher organisms will tempt many. We can inevitably expect vast compendia of sequences but, without functional reference, these compendia will be uninterpretable, like an undeciphered ancient language. Many people and many computers will play games with these sequences, but we will have to find out by experiment what the sequences do and how the products they make participate in the physiology and development of the organism. Thus, although the analysis of the genotype has been taken care of, we still need better ways of analyzing phenotypes. Many of us are ultimately interested in the causal analysis of development and the reduction of the complex phenotypes of higher organisms to the level of gene products. This is still the major problem of biology. We must understand what cells can do because all of what we are is generated by cells growing, moving and differentiating.*

**Brenner S. Human Genetics: Possibilities and Relatives. *Ciba Found Symp Excerpta Medica* 1973;66:1–3.**

*The effort spent on the identification of genes is likely to prove only a small fraction of that required to work out their normal function in the tissues in which they are expressed. Yet that is where clues to the treatment and prophylaxis of disease are most likely to arise.*

**Maddox J. Genes and Patent Law. *Nature* 1994;37:1270.**

## 13.1 INTRODUCTION

The foregoing quotations emphasize that the general theme of this chapter—all that occurs between the gene and the bedside—is not a new one. The true promise of the Human Genome Project only began to be realized when our genome was sequenced, and the really hard work persists [1,2].

At the outset, the definitions of four terms are fundamental to everything that follows. All deal with the causation of phenotypes, and they are distinguished as to method and scope of enquiry.

*Etiology* is the study of the causes of a phenomenon and, in the medical context, of disease. Its method is to discover the association between factors that are thought to be causes and certain features that we wish to explain. The goal and method of the discipline are strictly empirical, with at best minor interest in discerning the actual mechanisms involved.

*Genetic etiology* is a more specialized topic that deals with the properties of the genetic causal factors of disease and how they behave. Mendel's laws, which were

Emery and Rimoin's Principles and Practice of Medical Genetics and Genomics: Foundations, https://doi.org/10.1016/B978-0-12-812537-3.00013-5

question of which information and which kinds of mental process shall be assigned to the left side and which to the right. Wholly indifferent, random assignment is theoretically possible [18]. However, although the facts have been distorted in the past by social prejudice, there still seems to be a higher rate of "left brain dominance," and although the patterns are not altogether clear, there is evidence that genetic factors are at work. The genetics of handedness, as a surrogate for cerebral dominance, has been studied for decades. According to one hypothesis, mothers tend to support infants using their left arm, perhaps so to sooth them with the sound of the maternal heart. Thus, mothers who were by nature right-handed (dextral) would find some evolutionary favor. Another hypothesis suggests that left-handed warriors were more successful. What is clear, however, is that left-handedness is less common in all human populations, despite some geographic variation. Thus, whatever selective pressures favor left-handedness must be balanced by some negative ones for the trait to remain less common. The prevalence of left-handedness decreases with age [19]. Whether sinistrality itself reduces life span, or serves as a marker for underlying neuropathology (perhaps related to cerebral dominance), remains unclear [20].

### 13.3.1 Elaborateness of Repair

Ontogenic robustness and recuperative power are reciprocally related. The brain is the most highly organized structure but, while having some functional capacity to recuperate from damage by the fluidity in allocating space to functions, only recently has it been shown to have any regenerative power through stem cells. The structure of the kidney may be less critical than that of the heart but, like the heart, it has little capacity to repair damage to its architecture. Anatomically at least, the liver is both less critically structured and more robust, but has much greater forces of recuperation. Tissues such as skin, bone marrow, spermatogonia, and intestinal endothelium are still less elaborate and are therefore so ready for regeneration as to be notorious sites of sensitivity to mutagens.

### 13.3.2 Life History

A natural feature of the impact of a disorder is how it affects well-being, fertility, and length of life. These three, although distinct, are obviously connected; yet the traditional methods of analysis pay little attention to this fact. Discussions on fitness make much of the fact that clinical, genetic, evolutionary, athletic, and moral fitness are so different that they must be discussed separately. Indeed, excellence in one may go with mediocrity, even total incompetence in another. The super athlete may be sterile and morally bankrupt. The puny may live a long life free of disability. The fertile may be negligent in the care of their progeny. But it is absurd to suppose the several types of fitness to be totally unconnected. The issues are too large to deal with here, and we consider only the relationship between clinical fitness and length of life.

It is useful to distinguish between the unfolding of a disease and the impact of its complications. The wholly static disorders are mostly trivial: for example, red–green color blindness, tone deafness, pentosuria, synophrys, and the like. On the other hand, the usual patterns of deterioration may vary greatly. In severe cardiac malformations, the disability is evident at birth or soon after. Even so, major complications such as pulmonary hypertension and reversals of flow through a patent ductus arteriosus or an atrial septal defect may occur late. In hereditary polyposis coli, it is the course of the disease itself—the long latent phase, progressive polyposis, and malignancy—that is the chief concern, while other complications (except those due to therapy) are minor. By contrast, Alzheimer disease (OMIM*104300) often appears late enough that it is frequently obscured by intercurrent and competing diseases. Indeed, for centuries "senile dementia" was viewed by some as a concomitant of aging, and not as a disease. But while the duration of Alzheimer disease is typically less than a decade depending on diagnostic finesse and management, it is surprisingly difficult to find evidence that the patients ever have neurologic deaths. They seem to die of the complications (e.g., trauma, intercurrent infections, malnutrition) that occur at much increased risk [21].

The notion that well-being is eroded by overt catastrophes (e.g., strokes) or those imperceptible insults (e.g., chronic pyelonephritis) that we identify as "wear and tear" is so appealing that we are led, almost unconsciously, to accept multistage models. Where the genetic disorder becomes manifest early (as in Duchenne muscular dystrophy), the competing risks are small, and the pattern of deterioration is dominated by a single class of insults, with the typical survivorship curve positively skewed (i.e., with a long tail to the right), to an important extent because of interindividual genetic heterogeneity.

When the onset of disease is late, the patient shows the characteristic multiplex pathology so familiar to the geriatrician: damage in several body systems, so that it is often difficult to say which is the final cause of death. The survivorship is then often negatively skewed. That class of survivorship in which death is due to whichever of several partially damaged systems fails first is a "bingo" model, which is common but particularly difficult to examine either practically or quantitatively.

## 13.4 PATHOGENETICS OF REFINED TRAITS

A crucial process in molecular biology is generating the primary sequence of the polypeptide in the proper cell at the right time. This is attained by the elaborate apparatus of genetic coding, transcription and its regulation, and translation, which is highly conserved in evolutionary time, but the high organization largely ends at that stage. Once formed, the polypeptide assumes its secondary and higher structure by processes that are little understood; aside from posttranslational modifications catalyzed by enzymes, there seems to be little need to direct these processes. The polypeptide quickly assumes a stable low-energy state. Whether it ever becomes completely fixed is not readily established. But in or near that state, it functions most efficiently. The subsequent fate of the polypeptide may be largely random. For example, the theory of red cell survival suggests that the cell is eventually destroyed by random wear and tear, and the hemoglobin with it. However, survival of the whole is still shortened by some mutant forms of the primary structure of hemoglobin or of components of the erythrocyte wall.

The speed at which the polypeptide is made is certainly important. For instance, sickle hemoglobin is manufactured more slowly than the wild-type, such as to lead to a representation in the heterozygote in a ratio of 2:1 to 3:1. Furthermore, in heterozygotes, A and S hemoglobins tend to be concentrated in particular cells. However, one does not ordinarily regard translation (as opposed to transcription) primarily as a timed or quantitative process. Posttranslational modification is also sensitive to time. A mutation that results in substitution of a glycine in the triple-helical domain of type I procollagen results in slower winding of the helix. This in turn exposes for a longer time critical amino acids to the enzymes that catalyze modifications, such as glycosylation. The net result is a much more "damaged" molecule than a simple amino acid substitution might predict. This, in turn, is reflected in the degrees of severity of osteogenesis imperfect.

But there is even a higher-order effect possible when a protein is malfolded or otherwise damaged as it traverses the cellular machinery. When the endoplasmic reticulum encounters a malfolded protein, processing slows; if severe, a situation of "ER stress" ensues, which can lead to marked cellular dysfunction, even cell death [22–24]. Interestingly, the cellular phenotype may be the same for different mutations that affect entirely separate proteins. Understanding the importance of ER stress to the overall phenotype may afford a generic approach to therapy, whereby refolding of the mutant protein is facilitated.

Where the components are interchangeable (e.g., $\beta^A$- and $\beta^S$-globins), systems are appropriately described by their corporate properties. Where the numbers are large (e.g., numbers of erythrocytes), the usual device is the probabilistic model; and where the numbers are even larger (e.g., molecules), deterministic methods greatly simplify the analysis with negligible loss of accuracy. However, whatever the value of deterministic models in microbial populations, they have little place in studies of human beings; even in molecular studies, they must be handled with circumspection. This is a major difference between classic population genetics and the highly individualized character of medical genetics.

## 13.5 PATHWAYS AND MULTIPLE-STAGE PROCESSES

Two highly refined approaches have much in common and may, in certain circumstances, be united by a single theory.

### 13.5.1 Simple Pathways

The simplest possible process involves synthesis of B from A by enzyme ab (Fig. 13.1). There are three potential deleterious consequences:

- Precursor toxicity: Because ab fails, A accumulates and proves harmful. Alkaptonuria (OMIM*203500)

$$A \xrightarrow{ab} B$$

**Figure 13.1** An enzyme, ab, catalyzes conversion of substrate A to product B.

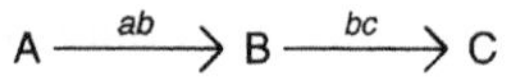

$$A \xrightarrow{ab} B \xrightarrow{bc} C$$

**Figure 13.2** An enzyme, ab, catalyzes conversion of substrate A to intermediate B, which is converted to final product C by enzyme bc. A defect in ab impairs production of C. The gene specifying ab is epistatic to that encoding bc.

is such a disorder, as are most enzymopathies in catabolic pathways, such as lysosomal storage disorders.

- Product deficit: Because ab fails, B is reduced or absent. Examples include the various forms of albinism due to failure to produce pigment (e.g., OMIM*203100) and most enzymopathies involving posttranslational processing of proteins. In a few mammalian species, the inability due to deficiency of one enzyme, L-gulonolactone oxidase, to synthesize ascorbic acid is another example. Does deficiency of this enzyme in all humans exclude it from the category of "disease"? At a minimum, deficiency of this enzyme creates a risk factor for scurvy in all of us.
- Combined product deficit and precursor excess: The glycogen storage disorders are examples (e.g., OMIM*232200). The glycogen that accumulates disrupts cellular and tissue processes, while failure to release glucose from glycogen leads to hypoglycemia. Phenylketonuria (OMIM*261600) is another such example; phenylalanine is toxic in excess, and synthesis of tyrosine is impaired, resulting in the pleiotropic manifestations of phenylalanine hydroxylase deficiency.

This elemental pattern extends to the three-step process: A → B → C (Fig. 13.2). If A is absent, then B is lacking, and C cannot be synthesized. This suppression is epistasis; the gene governing the first step is epistatic to that governing the second. The classic example in humans is the rather trivial Bombay blood group phenomenon (OMIM*211100) in which the failure to generate H substance destroys all means of expressing the ABO blood group phenotype.

Consider a typical multistage metabolic pathway, such as the synthesis of cholesterol or thyroid hormone. Each step is under the control of an enzyme. It will be at once evident that total failure at one or more steps means total blockade of later stages and that substrate accumulates before the first failed step. The gene for the enzyme at any step is therefore epistatic to all subsequent steps. The combined effect of defects in all genes will be the same as that of any subset of defective genes.

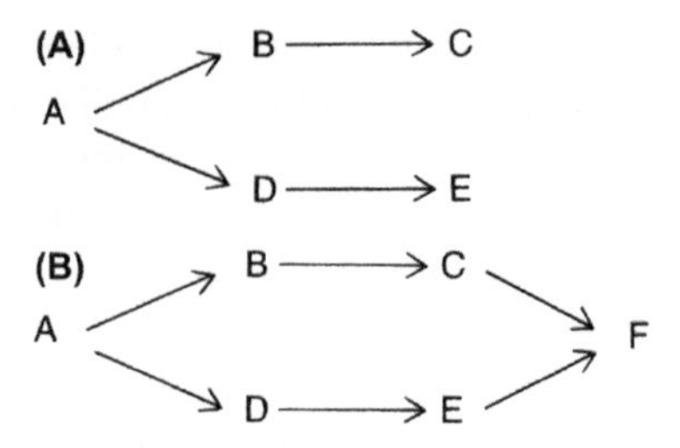


**Figure 13.3** Metabolic pathways with branches. (A) Open branched pathway. (B) Closed branched pathway.

In this it is quite different from the usual additivity of traits in Galton–Fisher theory.

## 13.5.2 Branching Pathways

Two kinds of branching pathways, the open and the closed, can be distinguished. In the open type (Fig. 13.3A), the branches do not rejoin and pool their products; thus, they compete for substrate, and the flow through each is correspondingly decremented. In the closed type (Fig. 13.3B), the paths rejoin, and the result is a parastasis; two or more pathways run in parallel, which accelerates the entire process and acts as a fail-safe device should any of them fail. This scheme can be used as an advantage in treatment, such as by promoting remethylation of homocysteine to methionine through an alternative pathway dependent on the cofactor betaine. Those classic inborn errors of metabolism that lack adequate alternate pathways are the most severe clinically. Their rarity argues that metabolic processes without auxiliary paths are the exception, and the selective disadvantages may explain why.

On the other hand, a defect in one branch of a pathway may generate all or most of its pathology by leading to overflow through the alternative branch. For example, a defect in the enzyme hypoxanthine-guanine phosphoribosyltransferase (OMIM*308000) leads to overproduction of phosphoribosylpyrophosphate. This in turn drives overproduction of purines, which leads to hyperuricemia, hyperuricosuria, and gout.

## 13.5.3 Pathways with Feedback

Metabolic pathways may be actively regulated in some cases by demands downstream. Negative feedback, positive feedback, or both can achieve a desired rate of processing or level of synthesis. This represents a form of physiologic homeostasis. Production of most hormones involves feedback at multiple levels. For example, estrogen is secreted by ovarian follicular cells in response to the anterior pituitary hormone and follicle-stimulating hormone (FSH); estrogen in turn feeds back on both the hypothalamus, to inhibit production of gonadotropin-releasing hormone, and on the anterior pituitary, to inhibit release of FSH, thereby modulating estrogen production and preparing the endometrium for implantation. Once an embryo implants in the endometrium, synthesis of chorionic gonadotropin signals the ovary to continue production of progesterone, which maintains the endometrium as a nurturing environment for continuing the pregnancy.

### 13.5.4 Multiple-Hit Processes

A metabolic process involving several steps may, where necessary, be viewed in more quantitative terms. In the synthesis of insulin in response to a carbohydrate load, multiple steps are involved, but physiologic impact occurs only in the final step that results in a physiologically active form. The lag time for the response, then, is that for the transit through the entire system, and the characteristics of inert precursors make little difference. In this sense, what matters is only the process as a whole; permutations of the components do not matter.

### 13.5.5 Multiple-Compartment Models

To deal with chemical processes in the foregoing fashion has certain uses, notably, in understanding steady-state processes. A more quantitative approach is necessary where changes need to be more rapid. Many of the processes of converting one class of compound into another are of so-called zero-order kinetics, that is, other things being equal, the rate of transfer equals the concentration of the substrate multiplied by the Michaelis constant, $m$, of the enzyme. In a system without replenishment of substrate, these conditions define the negative exponential process, and the mean time for conversion, the waiting time, is precisely $1/m$. This pattern may be viewed from two perspectives: first, as a chemist sees it, deterministically conforming to the law of mass action; and second, as a probabilist sees it, a random exponential (one-hit) process in which every eligible molecule is at a fixed instantaneous hazard of change. As long as the number of molecules remains large, the distinction hardly matters. But the probabilistic model is more appealing because of its wider relevance in biology (e.g., when the number of decaying items—molecules, body cells, recurrent bodily insults—is small, and random uncertainty may no longer be safely ignored). Assuming that each step operates independently, the transit time through the metabolic chain is then the sum of the waiting times of each step. The whole is termed a multiple-hit process. Its mean value is the sum of the waiting times, that is, the sum of the reciprocals of the Michaelis constants.

The smaller any particular Michaelis constant compared to constants for steps elsewhere in the chain, the larger is its reciprocal and the greater its impact on the whole transit time. Moreover, other things being equal, for any given variation in $m$, the smaller the mean, the larger the variation in its reciprocal, and the more sensitively the impact of variations in it will be detected. At some, quite arbitrarily small, value of $m$ and large value for its reciprocal, this dominating step is termed the rate-limiting step; viewed genetically, if this step is itself Mendelian, the whole will be termed Mendelian. It will be evident that the conventional distinction between Mendelian and Galtonian ("multifactorial") traits is both vague and arbitrary.

## 13.6 MOLECULAR PATHOGENETICS

Describing the defect in a genetic disorder at the level of the mutation specifies its etiology. Anything more remote from the mutation represents phenotype and at least the first layer of complexity in the pathogenetics. At the most remote layer is the ultimate phenotype. Any intermediate phenotype reflects the pathogenetics. Thus, the resolution or sensitivity of the methods being brought to bear on the investigation of how the disorder arises determines how closely the mutation can be approached. The clinician has long had to deal with crude tools—stethoscope, tape measure, electrocardiogram, radiograph, urinalysis—to define the phenotype; what generally results is a perception of pathogenesis that is shallow, often complex, even confused. At the bedside, and even in the clinical laboratory, one usually sees only the leaves on the pathogenetic tree. Advances in clinical chemistry, biochemical genetics, cytogenetics, immunology, noninvasive and invasive imaging of many types, and pathology have all led to a more radical, hence sensitive, discernment of what is wrong with the patient and elucidation of pathogenesis. All these advances have facilitated in brushing aside the leaves and clambering part way down branches toward the trunk of the pathogenetic tree (Fig. 13.4).

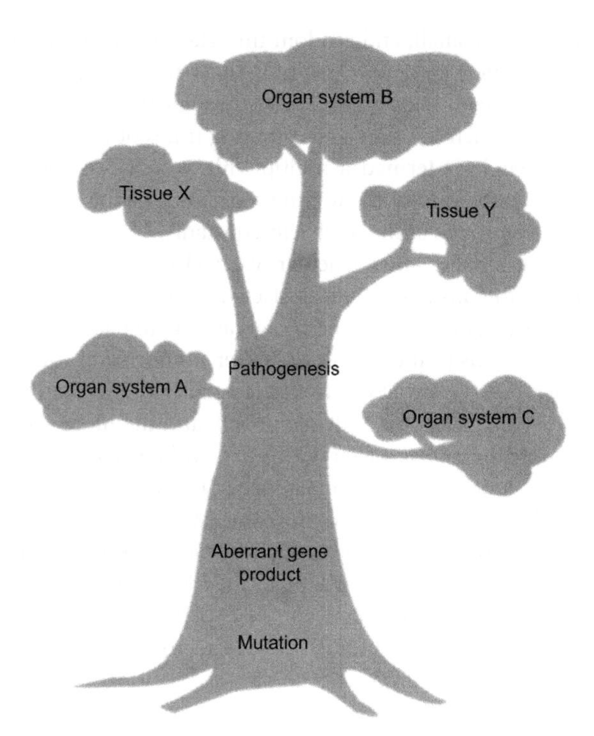


**Figure 13.4** Pathogenetic tree for a Mendelian condition. Leaves correspond to the phenotypic features, detectable by bedside investigation. Branches represent the pathogenetic pathways leading to organ- and tissue-specific pathology. The trunk corresponds to the gene product. Roots indicate the cause, in this case, the mutation.

Fig. 13.4 also illustrates a fundamental characteristic of many human phenotypes termed *pleiotropy*. This word encompasses several concepts in biology; here it refers to multiple, even seemingly unrelated, aspects of the same syndrome. Indeed, *syndrome* embodies this notion of several clinical properties "running together." Each of the leaves on the pathogenetic tree represents an aspect of the phenotype, connected through the limbs of pathogenesis. The analogy breaks down in that, while all leaves appear similar, the clinical details of the phenotype may be quite diverse. For example, dislocated lens, elongated digits, dural ectasia, and aortic root aneurysm are cardinal manifestations of Marfan syndrome (OMIM*154700), but outwardly bear no connection to each other. These features are all rooted in mutations in the gene (*FBN1*) encoding a large structural protein, fibrillin-1, and at the first level of pathogenesis, share defects in an extracellular structure, the microfibril (corresponding to the trunk in the figure). However, microfibrils have distinct functions, including regulating TGF-β signaling pathways and providing structural integrity to certain structures, so the trunk branches in two. Further, since these two distinct functions vary in different organs and tissues, each of the limbs of the tree heads in its own direction. The molecular bases of pleiotropy are as diverse as the features of some syndromes [25]. For example, the complexity of some Mendelian inborn errors of metabolism is due to the mutant enzyme having functional roles beyond that in the specific metabolic pathway at issue. The extent of such "moonlighting enzymes" in human biology is uncertain [26].

The phenotype can be explored systematically, beginning with the first product of the mutant gene, mRNA. Various defects in the structure and amount of a given mRNA can be described, albeit only with considerable effort and sophistication in techniques. The thalassemias constitute one group of diseases that beautifully illustrate the molecular pathology of mRNA.

It is more feasible and instructive to focus on the stable product of most genes, the protein, and describe the types of molecular pathology that arise from variation. In the most fundamental terms, a mutation can affect the quantity of a protein, the quality of a protein, and occasionally both aspects. The quantity of a protein synthesized by a gene is regulated at the level of transcription, by promoters, enhancers, and other locus control elements, and at the level of translation. DNA variants in any of a number of sites, *cis* and *trans*, to the gene of interest can affect the amount of protein produced. Usually, but not always, production from the variant alleles associated with a disease (often termed mutations) is decreased. One class of variation that has garnered considerable attention is the expansion of a trinucleotide repeat within or, more commonly, outside the actual coding sequence. The number of repeats may be inversely proportional to transcription of the gene. Furthermore, the more repeats, often the more severe the phenotypic change.

A change in the primary structure, the amino acid sequence, of a protein may alter its function (i.e., the quality of the protein). The study of diverse variants of the same protein has greatly advanced the understanding of molecular pathogenetics and investigating authentic relationships between genotype and phenotype. This inquiry calls for a new commitment to meaning and the authenticity of the descriptors, matters that are much more sophisticated than the correlations and coefficients that have dominated classic Galton–Fisher theory. The latter may lead to such paradoxical results as that, a nearly perfect correspondence between genotype and phenotype, may nevertheless yield a zero correlation and hence zero heritability. How the quality of a protein can be affected depends, in the first instance, on its normal function.

A DNA variant can change both the quality and quantity of a protein. For example, a change in primary sequence might affect the stability of the protein and lead to enhanced (or retarded) degradation. In some situations, the amount of the altered protein is crucial to the severity of the phenotype, especially in a dominant-negative scenario.

Proteins can be divided into three classes based on function: those whose essential functions involve interactions with small molecules, such as enzymes, receptors, and transporters; those that perform regulatory roles, such as transcription factors and hormones; and those that function in complex systems, often in a structural role, and often in association with other proteins.

Most proteins have one or more domains associated with specific functions. Not surprisingly, proteins in the same class often have domains in common, and there is remarkable conservation of sequences among domains. For example, transcription factors all have one or more amino acid sequence motifs (such as leucine zipper or zinc finger) that facilitate binding of the protein to DNA sequences. Cellular receptor molecules have domains that enable interaction with the lipid bilayer of the cell membrane, an extracellular domain that binds a ligand, and often a domain that resides in the cytoplasm, perhaps exhibiting kinase activity. Some molecules have many domains, some of which are composed of dozens of repeated motifs—witness fibrillin (OMIM*134797), lipoprotein(a) (OMIM*152200), and plasminogen (OMIM*173350). The conservation of domains has facilitated discovery of the cause of numerous diseases through positional cloning. Thus, when a newly identified open reading frame is sequenced and found to be a strong candidate for the cause of a disorder, generally through identification of a variant, the logical question is what the function of the protein encoded by the gene might be. If the coding sequence specifies an amino acid motif typical of a zinc finger, the protein is likely to be a transcription factor. This process is aided considerably by large databases that incorporate knowledge of genetic sequences and protein structure and function from both humans and all other organisms.

Proteins that interact with themselves (to form multimers) or with other proteins are subject to enhancement of a pathologic effect when one copy of their gene is mutant. Even though the patient is heterozygous for the mutation, the defective protein, by interacting with the product of its normal allele, or the products of other nonmutant genes, consumes these normal proteins; the result is a much more severe phenotype than would be expected from having half-normal levels of the normal protein. This is termed the dominant-negative effect, and is rather common. The irony is that a "more severe" variant, such as one that eliminates transcription from the variant allele altogether (i.e., a null allele), has less effect on phenotype than does a missense variant that leads to normal transcription and translation of a mutant protein.

Within each of the three classes of proteins, a mutation can have one of four consequences: quantitative increase or decrease in function and qualitative gain or loss of function. Each of these consequences can have a number of molecular explanations.

A quantitative increase in function can be due to a regulatory mutation. An example is loss of sensitivity to inhibition, such as by a repressor molecule. A variant could also affect the active site of an enzyme, such that its $V_{max}$ was increased, or the binding site of a hormone, such that the $K_M$ was lowered.

A quantitative decrease in function could operate by the converse of any of these mechanisms. The extreme of the spectrum of decreased function is loss of function, perhaps the easiest to conceptualize, and certainly the most prevalent consequence. For example, most inborn errors of metabolic pathways result from an enzymatic failure. The enzymopathy can be due to a variant in or around the locus encoding that enzyme, resulting in a qualitative or quantitative defect as described earlier; to abnormal posttranslational processing of the nascent enzyme; to abnormal subcellular localization or extracellular trafficking; to altered affinities for substrates or cofactors; or to altered responsiveness to allosteric regulators of activity. Other examples of loss-of-function phenotypes include familial hypercholesterolemia due to many of the defects in the low-density lipoprotein receptor, and neoplasia due to defects in tumor suppressor genes such as the retinoblastoma or neurofibromatosis type 1 genes. Strains of mice bearing gene "knockouts" represent specified loss-of-function mutations; these are especially popular tools for studying development and neoplasia.

Quantitative and qualitative loss of function clearly overlap. A variant that reduces the ability of an enzyme to bind substrate also might lead to enhanced degradation and a reduced steady-state amount of the protein molecule.

Variants that cause a gain in function, that is, a function not intrinsic to the wild-type protein, are less common. The diverse familial amyloidoses are examples, in which a change in amino acid sequence of one or another protein (e.g., transthyretin) results in enhanced stability of the protein and abnormal tissue deposition (OMIM*176300).

The least commonly recognized molecular phenotype, also qualitative, is a change in function. One example is the product of the fusion of *BCR* and *ABL* in chronic myelogenous leukemia. Another example is the *p53* protein, which when mutated in some ways, assumes regulatory capabilities foreign to the normal product.

As useful as these protein phenotypes are for classification (and education), there are limitations in making the intellectual leap to the next level of pathogenetic complexity. For example, gene knockout mutations are relatively easy to generate in mice, and increasingly in other species. Many investigators see this technique as a facile way to isolate the physiologic role of a particular gene product, to generate an animal model of a given disease, or to serve as the background strain into which a defined mutation is introduced. There is no question that the approach has been brilliantly successful in a number of instances; however, the pitfalls have been underemphasized. For example, some mice homozygous for the absence of transforming growth factor-$\beta_1$ (TGF-$\beta_1$) are born normal in appearance and survived, both unexpected results given the prominent role of this cytokine in many aspects of development. The reason is "rescue" of the embryo by maternal TGF-$\beta_1$ that presumably crosses the placenta in sufficient quantity to replace the deficient fetal sources. But at a more fundamental level, the "null" mutant animal cannot be viewed as an artificial isolated system focused on that deficient gene product. Rather, the mutant strain is a complex homeostatic system capable of responding to loss of a specific protein, even compensating for it. Thus, if the null strain shows no phenotype, it would be inappropriate to conclude that the missing protein is not important to the physiology of a given system (physiologic or developmental).

The actual effect of the variant may be loss of function at the protein level, but gain of function at the cellular level. For example, Rett syndrome (OMIM*312750) is a pleiotropic, severe neurologic disorder that primarily affects girls. The cause is mutation of the *MECP2* locus, which encodes a protein that represses transcription of other genes. Mutations in *MECP2* (OMIM*300005) that inactivate the protein result in enhanced or inappropriate production of proteins in various tissues, most obviously the brain.

## 13.7 CONCLUSIONS

All of the steps that occur between causation and the bedside constitute pathogenesis. When considering the genetic causal factors and variations in pathogenesis, the term "pathogenetics" applies. The phenotype can be studied at a coarse level, such as the clinical features, and this is the object of evolution. Alternatively, the phenotype can be studied at more and more refined levels, for example, biochemical pathways, to specific enzymatic defects, and to aberrations of messenger RNA processing. These intermediate steps, which require traditional disciplines of pathology, physiology, and biochemistry, help to elucidate pathogenetics. In the vast majority of instances, only through understanding pathogenetics will novel and effective therapies emerge.

The prognosis of a disease is largely a matter of pathogenesis. For instance, its age of onset, the rapidity of its course, and the vulnerable points at which disease and complications may occur all depend on details that in principle, as much as in fact, may be difficult to infer even from the most detailed knowledge of the basic defect. Some knowledge of the prognosis may come from "black box" empirical inquiries—the natural history of myotonic dystrophy, for instance—but this course calls for extensive data, and there may be disturbing discrepancies between one study and another that are not readily reconciled. If the pathogenesis is understood, even partially, more incisive methods may be available, including direct measurements of the progress of components of the disease. For instance, the pathogenesis of familial polyposis coli is not clearly established, but currently the course of this disease and its response to treatment are easier to study than Alzheimer dementia. Refined studies at the molecular level make for very precise statements about etiology. It is tempting, but rather treacherous, to view pathogenesis in the same way. But where the concern lies in either the assessment of morbidity or the study of the population and eugenic behavior of the mutant, to attach too much weight to refined biochemistry may push the precision of the statement at the expense of its significance. For the overt clinical pattern and the target of selection are very coarse matters; the many modifying factors, which to the basic scientist are largely a nuisance, may have important attenuating effects on the course of the disorder.

Many advances in therapeutics have resulted from largely empirical reasoning as to choosing an approach and from an understanding of natural history in judging whether the therapy was successful. A more rational approach to targeting therapy is based on an understanding of pathogenesis. Some fondly held the hope of circumventing "indirect" therapies for genetic disorders by simply replacing the defective gene. But considerable experience has amply shown the general fallacy in this approach. Until the molecular pathogenesis of a disorder is elucidated, the effects of simply adding back, or even replacing, a gene that should have been functioning perhaps from conception will be as empirical as anything physicians had available in the 18th century.

## ACKNOWLEDGMENTS

The original chapter on which this revision is based owes much to the research and insights of the late Edmond A. Murphy, MD, ScD.

## REFERENCES

[1] Green ED, Guyer MS. Charting a course for genomic medicine from base pairs to bedside. Nature 2011;470:204–13.
[2] Lander ES. Initial impact of the sequencing of the human genome. Nature 2011;470:187–97.
[3] Barondes RD. Extrahuman sources of polio virus; new concept on the pathogenesis of the viruses. Mil Surg 1949;105:400–8.
[4] Barondes RD. Duodenal ulcer; pathogenetics, and the re-evaluation of therapeusis. Mil Surg 1951;109:720–31.
[5] Wolf U. The genetic contribution to the phenotype. Hum Genet 1995;595:127.
[6] Fisher RA. The correlation between relatives on the supposition of Mendelian inheritance. Trans R Soc Edinburgh 1918;52:399–433.
[7] Bernard C. De la Physiologie Gènèrale. Paris: Hachette; 1872.
[8] Cannon WB. The wisdom of the body. New York: Norton; 1932.
[9] Wiener N. Cybernetics or control and communication in the animal and the machine. New York: John Wiley & Sons; 1948.
[10] Murphy EA, Pyeritz RE. Homeostasis. VII. A conspectus. Am J Med Genet 1986;24:735–51.
[11] Hebebrand J, Remschmidt H. Anorexia nervosa viewed as an extreme weight condition: genetic implications. Hum Genet 1995;595:1–11.
[12] Waddington CH. The strategy of the genes. London: Allen & Unwin; 1957.
[13] Murphy EA, Berger KR, Trojak JE, Sagawa Y. Angular homeostasis. V. Some issues in genetics, ontogeny and evolution. Am J Med Genet 1988;31:963–79.
[14] Murphy EA, Berger KR, Pyeritz RE, Sagawa Y. Angular homeostasis. VI. Threshold processes with bivariate liabilities. Am J Med Genet 1990;36:115–21.
[15] Mallo M, Wellik DM, Deschamps J. Hoxgenes and regional patterning of the vertebrate body plan. Dev Biol 2010;344:7–15.
[16] Pauli A, Rinn JL, Schier AF. Non-coding RNAs as regulators of embryogenesis. Nat Rev Genet 2011;12: 136–49.
[17] Sampath K, Ephrussi A. Development 2016;143:1234–41.
[18] Laland KN, Kumm J, Van Horn JD, Feldman MW. A gene-culture model of human handedness. Behav Genet 1995;25:433–45.
[19] Llaurens V, Raymond M, Faurie C. Why are some people left-handed? An evolutionary perspective. Philos Trans R Soc B 2009;364:881–94.
[20] Halpern DF, Coren S. Handedness and life span. N Engl J Med 1991;324:998.
[21] Spalletta G, Long JD, Robinson RG, et al. Longitudinal neuropsychiatric predictors of death in Alzheimer's disease. J Alzheimers Dis 2015;48:627–36.
[22] Macario AJL, de Macario EC. Sick chaperones, cellular stress, and disease. N Engl J Med 2005;353:1489–501.
[23] Tabas I, Ron D. Integrating the mechanisms of apoptosis induced by endoplasmic reticulum stress. Nat Cell Biol 2011;13:184–90.
[24] Xu C, Bailly-Maitre B, Reed JC. Endoplasmic reticulum stress: cell life and death decisions. J Clin Invest 2005;115:2656–64.
[25] Pyeritz RE. Pleiotrophy revisited. J Med Genet 1989;34:124–34.
[26] Siriam G, Martinez JA, McCabe ER, et al. Single-gene disorders: what role could moonlighting enzymes play? Am J Hum Genet 2005;76:911–94.
[27] Brenner S. Human genetics: possibilities and relatives. Ciba Found Symp Excerpta Medica 1973;66:1–3.
[28] Maddox J. Genes and patent law. Nature 1994;37:1270.
[29] Osler W. Chauvinism in medicine. Montreal Med J 1902;31:684–99.

# 14

# Twins and Twinning

*Mark P. Umstad*[1,2], *Lucas Calais-Ferreira*[3,4], *Katrina J. Scurrah*[3], *Judith G. Hall*[5], *Jeffrey M. Craig*[6,7]

[1]Department of Maternal-Fetal Medicine, The Royal Women's Hospital, Melbourne, VIC, Australia
[2]University Department of Obstetrics and Gynaecology, University of Melbourne, Melbourne, VIC, Australia
[3]Centre for Epidemiology and Biostatistics, Melbourne School of Population and Global Health, University of Melbourne, Melbourne, VIC, Australia
[4]CAPES Foundation, Ministry of Education, Brasilia, Brazil
[5]Departments of Medical Genetics and Pediatrics, British Columbia's Children's Hospital, Vancouver, BC, Canada
[6]Deakin University School of Medicine, Geelong, VIC, Australia
[7]Murdoch Children's Research Institute, Royal Children's Hospital, Parkville, VIC, Australia

## 14.1 INTRODUCTION

Twins have sparked curiosity in our society since ancient history, such as the Biblical accounts of Esau and Jacob, a pair of twins who were physically similar in spite of striking differences in their personalities. In science, Galton [1] is recognized as the pioneer in involving twins as participants in what we know today as twin studies. However, it is arguable whether he had a correct understanding of zygosity and its implications in relation to Mendel's laws of inheritance which were, back then, still largely undiscovered [2].

The existence of two types of twins was first noted in the 19th century [3], but the realization that monozygotic (MZ or "identical") twin pairs share close to 100% of their genetic material, while dizygotic (DZ or "fraternal") pairs share on average 50% of genetic variation, as any other sibling, was only conceptualized in the early 1900s [4]. Fisher [5] then developed quantitative genetic theory, which allowed for the quantification of similarity in human traits within groups of MZ and DZ twin pairs, which can be seen as an important increment in the "collective" creation of the classical twin design and its early applications in the 1920s. The first twin registries began to be established about 30 years later and facilitated the collection of valuable prospective and retrospective longitudinal data.

Since then, the fast-paced development of technology and analytical techniques has substantially advanced the core understanding of the twinning process and consequently the applicability of twin designs in genetic epidemiology. The further understanding of atypical twinning and the importance of differences in placentation and chorionicity have played a role in shaping the perception that traditional models of twinning may be no longer sufficient to explain the phenomena [6] (see Section 14.2). This perception has an obvious effect not only on medical practice related to twins, but also on some of the assumptions under which the twin study designs operate. In spite of such challenges, twin studies continue to be recognized as an important tool in any health researcher's repertoire, especially in the current era of omics and molecular studies [7].

Emery and Rimoin's Principles and Practice of Medical Genetics and Genomics: Foundations, https://doi.org/10.1016/B978-0-12-812537-3.00014-7

This chapter summarizes what is known about the different types of twins, including frequency within the population; the mechanisms of twinning, including genetic contributions; and practical advice for working with twins as patients. It also discusses the value of twin research, including landmark studies and modern applications, and showcases how global collaborative networks of twin registries and researchers are reshaping the way in which twin studies can be relevant to the understanding of human traits and diseases toward prevention, prognosis, and treatment.

## 14.2 TYPICAL TWINNING IN HUMANS

The traditional model of twinning proposes that DZ twins result from fertilization of two distinct ova by two separate spermatozoa and MZ twins are the product of a single ovum and sperm fertilization that subsequently divides to form two embryos. The most widely accepted model of MZ twinning is based on the unproven hypothesis of postzygotic division of the conceptus [6]. In this model the numbers of fetuses, chorions, and amnions are determined by the timing of embryo splitting (Table 14.1, Figs. 14.1 and 14.2).

Herranz [8] argued that the postzygotic splitting model lacked scientific evidence and that factors initiating cleavage have not been specified. He also noted that the rate of postzygotic splitting becomes more unlikely with the passage of time and that splitting has never been observed in vitro. He proposed an alternative theory of MZ twinning based on two principles:

1. MZ twinning occurs at the first cleavage division of the zygote; and

**TABLE 14.1 Chorionicity and Amnionicity by Time of Zygote Splitting**

| Zygosity | Twins | Time of Split | Chorions | Amnions | Fetal Mass |
|---|---|---|---|---|---|
| Dizygotic | DC DA | No split | 2 | 2 | 2 |
| Monozygotic | DC DA | Days 1–3 | 2 | 2 | 2 |
| Monozygotic | MC DA | Days 3–8 | 1 | 2 | 2 |
| Monozygotic | MC MA | Days 8–13 | 1 | 1 | 2 |
| Monozygotic | Conjoined | >Day 13 | 1 | 1 | 1 |

*DA*, diamniotic; *DC*, dichorionic; *MA*, monoamniotic; *MC*, monochorionic.

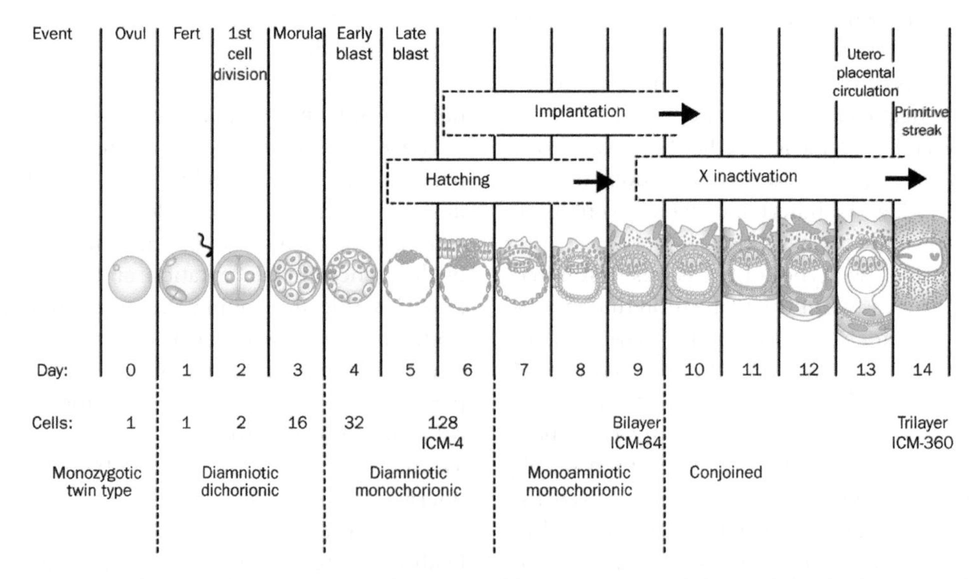


**Figure 14.1 Process of Monozygotic Twinning During Postfertilization.** Ovul, ovulation; fert, fertilization; divis; division; blast; blastulation. *Reproduced from Hall, J. (2003). Twinning. The Lancet, 362(9385), 735-743. doi: 10.1016/s0140-6736(03)14237-7, with permission from Elsevier.*

2. subsequent chorionicity and amnionicity is determined by the degree of fusion of embryonic membranes within the zona pellucida [8].

Both the traditional "fission model" and the "fusion model" of Herranz are unsubstantiated [9].

## 14.3 ATYPICAL TWINNING

Insights and challenges to the traditional models of twinning are seen in the variety of atypical or unusual twins. These variations from the usual ideas of twinning are discussed in this section.

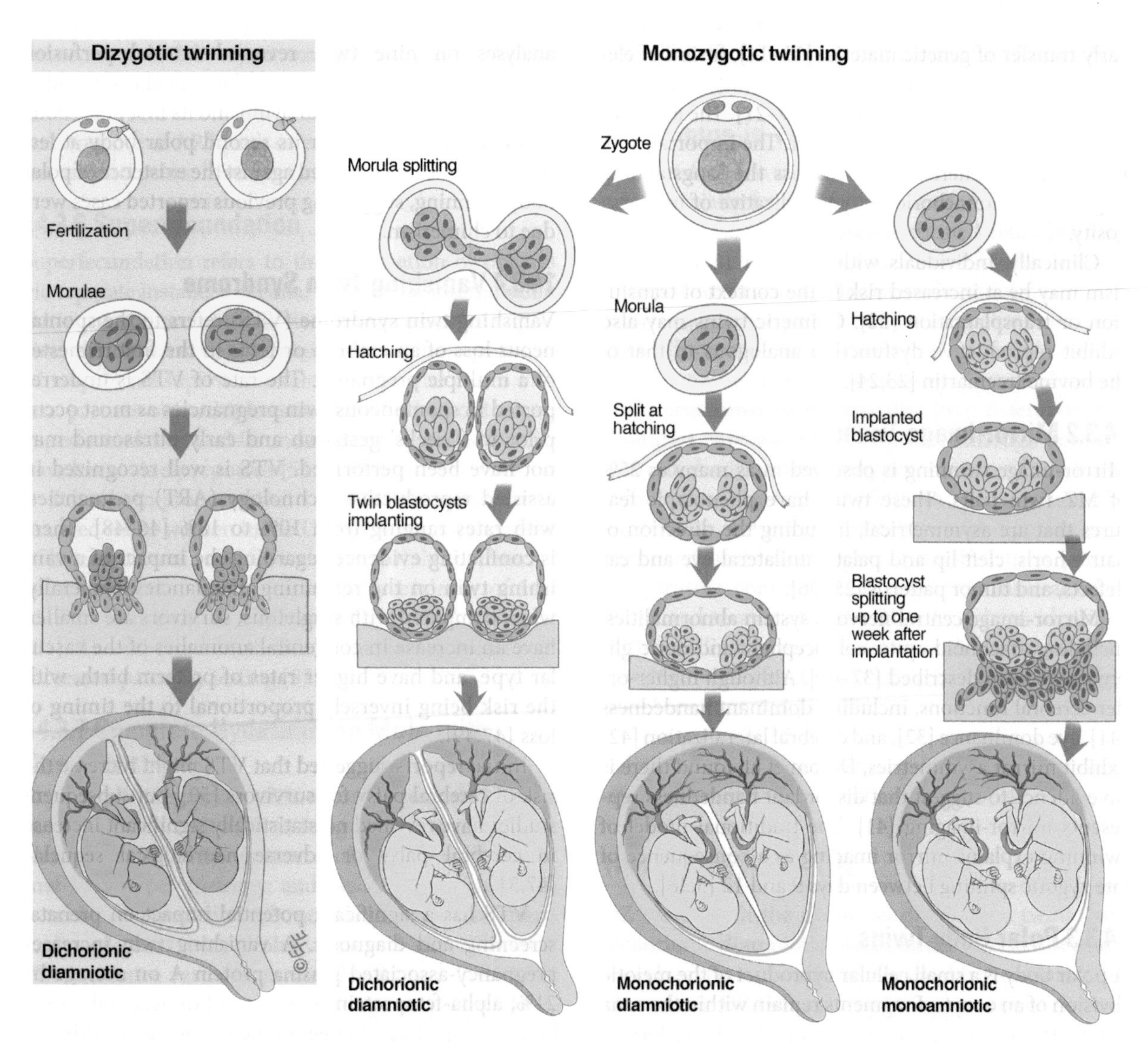


**Figure 14.2 The Formation of the Main Types of Twins.** DZ twins result from two separate fertilization events and are dichorionic and diamniotic. MZ twins result from the splitting of a single embryo early in gestation. Approximately one-third of MZ twins split early and are dichorionic and diamniotic. Approximately two-thirds of MZ twins split later and are monochorionic and diamniotic. Approximately 1% of twins split even later and share a single chorion and amnion. *Reproduced from McNamara, H.C., Kane, S.C., Craig, J.M., Short, R.V., Umstad, M.P., 2016. A review of the mechanisms and evidence for typical and atypical twinning. Am J Obstet Gynecol 214, 172–191.*

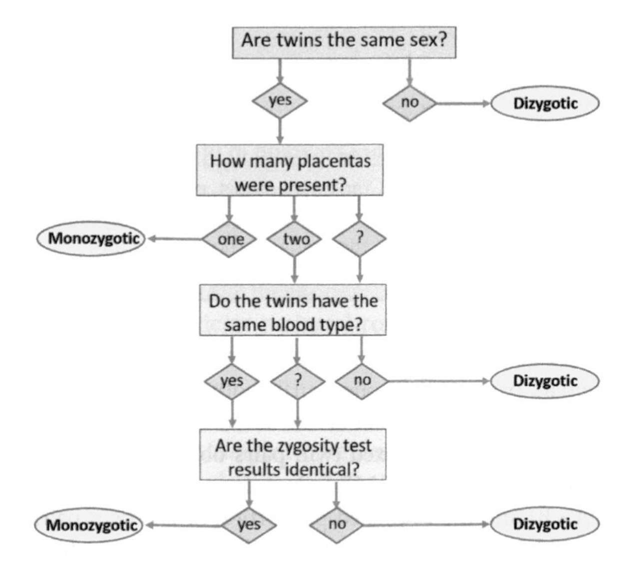


**Figure 14.3** Decision Tree for Determining Zygosity in Twins. *Number of placentas should be determined by ultrasound prior to 12 weeks' gestation and confirmed at birth by physical examination of membranes.

genetic zygosity testing (Fig. 14.3) [3,109,110]. Blood type is also sometimes useful. Chorionicity is most accurately determined by the thickness of the membrane between the twins as determined by ultrasound examination, most accurately between weeks 6 and 9 of gestation, but it may be determined up to week 14 [111]. In dichorionic twins a thick membrane forms a lambda shape and separates the twins. In monochorionic twins this membrane is much thinner and joins the placenta to form a "T" shape. Ultrasounds taken later in gestation are less reliable due to the increased crowding of twins in the uterus. Sonography may provide additional clarification if needed. Physical examination of the intertwin membranes at birth should also be used to determine chorionicity. This will provide confirmation of early ultrasound data and determination of chorionicity in twins without early ultrasound information. Again, dichorionic membranes are thick, opaque, and can be pulled apart, whereas monochorionic membranes are thin, semitransparent, and inseparable. When sex and chorionicity are known, we can assign different-sex twins as DZ and same-sex monochorionic twins as MZ (with rare exceptions in cases of chromosomal abnormalities). The steps in Fig. 14.3 can be used to accurately determine twin zygosity in almost all pairs irrespective of age. When blood type is known, we can assign twins with different blood types as DZ. In same-sex twins, if chorionicity and blood type are unknown then a zygosity test is required to accurately determine zygosity. Same-sex dichorionic twin pairs with the same blood type also require a zygosity test to determine their zygosity.

## 14.8.1 Incorrect Assumptions about Zygosity

Our research [112] and that of others [113,114] have found that a substantial proportion of parents and twins are misinformed about their zygosity status, and that this misinformation may come from parents or medical professionals.

Incorrect assumptions lead to misclassification of twins by healthcare professionals and the twins and their families. The main two incorrect assumptions being that (1) all dichorionic twin pairs are DZ and (2) genetically identical MZ twins must have identical phenotypes. The false belief that all dichorionic twins are DZ can lead to one-third of MZ twins being incorrectly categorized as DZ. In addition to this, the number of chorions can be difficult to determine from ultrasounds. Dichorionic twin pairs may have fused placentas, which can be mistaken for a single placenta without careful examination at birth. This can lead to misclassification of DZ pairs as MZ. MZ twins usually look and behave more similarly than DZ twins due to their greater genetic similarity. However, MZ twins are often not physically and behaviorally identical due to differences in the environments they encounter from conception onwards (see Section 14.11.2). This can lead to the misclassification of phenotypically different MZ pairs as DZ.

## 14.8.2 The Importance of Zygosity Knowledge

Accurate knowledge of zygosity is important to twins and people with whom they come into contact. Below we summarize the range of medical, personal, financial, scientific, legal, and ethical reasons supporting the testing and reporting of twin zygosity [106,115,116].

MZ pairs are perfectly compatible organ donors for one another, requiring much less post-transplant immunosuppression than DZ twins and have better chances of long-term survival. As nearly all diseases have at least some genetic component to their origin, the diagnosis of a disease in one twin means the co-twin is at increased risk of that disease, more so for MZ pairs than DZ pairs. Genetic sequence data will almost always be the same for MZ co-twins and the implications of this for the co-twin should be considered if testing is undertaken. In the event that one or both twins die before or soon after birth, it is vital that parents and/or the surviving twin have this fundamental information about twin zygosity as it bears upon the immediate bereavement response, the long-term identity of the surviving twin, and future family planning. Accurate determination of zygosity is also important postnatally for estimation of the likelihood of the mother or close female relatives giving birth to further sets of twins, because only DZ twins can run in families. Zygosity knowledge is also important for understanding the physical and behavioral differences and similarities between twins.

Increased understanding of zygosity helps define social relationships and helps define twins as individuals. It helps avoid embarrassment over uncertainty when asked about by zygosity by family, friends, teachers, and strangers. It can provide peace of mind and positive emotional responses for twins and their families. Some twins experience significant emotional stress if they discover later in life that their belief about their zygosity is incorrect. Knowledge of zygosity is also a prerequisite for twin research, and many twins feel they are unable to participate because of this [112].

Knowledge of a genetic disorder manifested in only one of a pair of identical twins is likely to lead to early detection in the second twin, thus leading to improved health outcomes and potentially savings in costs for treatment and management compared to detection at a later stage of disease. Costs incurred for zygosity testing would be outweighed, in the cost of suspected genetic disorder, by the savings from genetic testing of the second twin after a genetic diagnosis in the first (this will only happen with MZ twins).

Accurate knowledge of zygosity will affect the results and findings of medical research involving twins and saves both time and expense for researchers and participants as additional testing would not be required.

The International Council for Multiple Birth Organisations (ICOMBO) and the International Society of Twin Studies (ISTS), in their *Declaration of Rights* state that "Parents have a right to expect accurate recording of placentation, determination of chorionicity and amnionicity via ultrasound, and the diagnosis of zygosity of same sex multiples at birth" and that "older, same sex multiples of undetermined zygosity have a right to testing to ascertain their zygosity" and "Zygosity should be respected as any other human trait and deserves the same privacy rules." (http://icombo.org/wp-content/uploads/2010/11/Declaration-of-Rights-2014.pdf). Respect for the individual recognizes the importance of the concept of identity for a person, which is important for wellbeing, and for avoiding the harm of misinformation. Accurate knowledge of zygosity at birth also avoids any erroneous assumptions from the outset, including those that would result in damaging psychological or emotional impact if zygosity assumptions are proven incorrect, for example if twins had assumed incorrectly that they were identical.

In summary, research has shown that accurate knowledge of twin zygosity can be very reassuring for both twins and their families [112,113].

## 14.9 THE ETIOLOGY OF TWINNING

### 14.9.1 Genetic Causes of MZ Twinning

There have been a number of reports of families in which MZ twinning occurs more frequently than expected [117–121], which has been termed "familial MZ twinning." Interestingly, there does not seem to be an increase in congenital anomalies among the MZ twins in these families. Because familial MZ twinning has been reported on both the maternal and paternal sides of the family, it has also been suggested that it may be caused by a single gene effect that is unaffected by the sex of the parent transmitting the gene [120,122]. However, data from Lichtenstein and colleagues [123] have suggested that there is no paternal effect on familial MZ twinning.

Recently, a gene has been characterized that is likely to play a role in MZ twinning. *PITX2*, a transcription factor, was shown to be involved in the formation of embryonic axis formation in a model of cleavage-induced "experimental" twinning in chickens [124]. The authors concluded that this meant that the pathway associated with this gene "guarantees" that the two products of an early embryonic split are each "guaranteed" to form separate embryos, and that the opposing axes may even explain a proportion of mirror twins. Genetic studies of rare familial MZ twinning may shed light on whether variants of this and other genes influence the likelihood of MZ twinning in such cases.

### 14.9.2 Genetic Causes of DZ Twinning

There are many reports of familial DZ twinning [125]. The female members of these families are thought to have an inherited predisposition to multiple ovulation and in turn have a higher number of DZ twin pairs when compared to the general population [126]. The risk of having twins is up to 2.5 times higher for a woman with a sister with DZ twins than it is for the general population [127,128]. An established association between higher gonadotrophin levels and higher incidence of DZ twins in certain families is thought to be the basis for familial DZ twinning. While there appears to be some controversy whether this is an autosomal maternal or paternal effect [123,127], in reality a twinning gene could be inherited through either the maternal or paternal side, although it will only be expressed in females. It is possible that some genetic disorders may also predispose to DZ twinning [129].

Studies of species other than humans have revealed a number of genes that contribute to DZ twinning. The study of sheep has proved particularly informative, as sheep typically give birth to a single offspring at a time, but some strains have high incidences of multiple births [130]. To date, three genes have been confirmed to influence DZ twinning rates in sheep: the growth differentiation factor 9 (*GDF9*) and bone morphogenetic protein 15 (*BMP15*) genes, both of which are expressed in the oocyte and are essential for follicle development, and the bone morphogenetic protein receptor 1B (*BMPR1B*), the receptor for *BMP15* and expressed in multiple cell types in the ovary. Interestingly, while mutations in *GDF9* and *BMP15* can increase twinning rates when one copy is present (i.e., when the mother is heterozygous for the mutation), they can also cause female infertility when two copies are present in one individual (i.e., in homozygous form) [131,132].

At present, only mutations in *GDF9* appear to influence DZ twinning in humans, although such mutations are rare. Screening these genes in large numbers of DZ twinning families has revealed a loss-of-function mutation and a two-base deletion in *GDF9* in heterozygous form in three families [133,134], and it appears that overall genetic variation in *GDF9* is more common in mothers of DZ twins than it is in controls [134]. No such effect has been found for *BMP15* [135] or *BMPR1B* [136]. Interestingly, both *GDF9* and *BMP15* have been implicated in premature ovarian failure [137].

There are a number of additional genes known to have roles in ovulation and DZ twinning in humans [79]. Genetic variants that result in changes in amino acids in the follicle-stimulating hormone receptor (FSHR) protein were suggested to contribute to DZ twinning [138] and a variant with a known functional effect located in the promoter of the *FSHR* gene was found to segregate with the DZ twinning phenotype in one large family [139], but further studies found no evidence for the involvement of this gene [140]. A variant of the *FSHB* gene, which codes for the beta subunit of FSH, and *SMAD3*, the product of which is involved in gonadal responsiveness of FSH, were recently found to be associated with a higher rate of spontaneous DZ twinning [141]. Both variants are also associated with other aspects of female fertility, such as earlier age at first and last child, confirming the link between fertility and DZ twinning.

Evidence for the involvement of serine proteinase inhibitor clade A member 1 (*SERPINA1*, commonly known as alpha-1-antitrypsin [142,143]), peroxisome proliferator-activated receptor gamma (*PPARG* [144]) and the fragile X (FRAXA) "premutation" [145,146]: has not been borne out in later studies [79]. Family-based "linkage" studies have found no evidence for increased levels of genetic sharing among family members over chromosomal regions in which such candidate genes are located [139,147,148]. Linkage studies have, however, indicated chromosomal locations that may harbor new candidate genes for DZ twinning [139,147,149]. A study including 525 DZ twinning families suggests the presence of such genes on a number of chromosomes, most notably, chromosomes 6, 12, and 20, in Australian and Dutch DZ twinning families and confirmed that DZ twinning is a complex trait likely to be influenced by multiple genes [139]. Much work remains to be done to find the genes underlying the tendency to human DZ twinning.

### 14.9.3 Other Causes of Twinning

Several nongenetic mechanisms of MZ twinning have been proposed. The finding that mammalian female embryos are somewhat behind male embryos in the number of cells present at a certain stage during early stages of embryonic development [85] and the fact that there is a slightly higher incidence of female MZ twins, particularly among conjoined twins, in which the twinning process is assumed to occur relatively late in the very early embryonic developmental process, support the suggestion that MZ twinning is somehow related to delayed implantation and that the timing of different developmental clocks plays a critical role in MZ twinning.

Several authors have suggested that skewed X-chromosomal inactivation may play a role in female MZ twinning if, during embryogenesis, two different foci were to arise: one expressing the maternal X and the other expressing the paternal X. A number of female MZ twins have been discordant for a variety of X-linked recessive diseases [150], suggesting, and often demonstrating, nonrandom X-inactivation. Goodship and colleagues [151] have tested the hypothesis that skewed X-inactivation can trigger MZ twinning in females by studying umbilical cord tissue in female MZ twins. They observed random X-inactivation in most pairs of female MZ twins, but some showed marked skewing. Thus, it would appear that skewed X-inactivation does not explain all female MZ twinning but could be responsible for the excess of MZ female twins. Interestingly, Tan and colleagues [152] have shown that X-inactivation occurs at different times in different tissues postimplantation in the mouse embryo. These findings suggest that since X-inactivation occurs at the time of tissue differentiation, X-inactivation in blood and skin may not be representative of the rest of the tissue in an organism. To properly determine the exact role of X-inactivation in female MZ twinning, it would be necessary to study many different tissues. Bamforth and colleagues [153] have studied the parent-of-origin of X-inactivation in placental membranes and umbilical cords in twins and triplets. The chorion did show asymmetric X-inactivation in MZ dichorionic twins. Of course, the chorion is not representative of the whole embryo, but may represent processes that occurred early in development. The study also suggested that monochorionic MZ twins may react differently from dichorionic ones, reflecting the importance of timing in the MZ twinning process.

Observations of discordance in the expression of genetic material in MZ twins have suggested the intriguing possibility that some cases of MZ twinning may occur because of epigenetic events [154,155] (explained in Section 14.10.2). Weksberg and colleagues [156] found differential imprinting in female MZ twins discordant for Beckwith–Wiedemann syndrome and proposed that in such cases either unequal splitting of the inner cell mass or, alternatively, a lack of maintenance of DNA methylation, leads to a loss of imprinting that predisposes to MZ twinning. Such discordance would be expected to arise early in development among cells from a single zygote. A discordance of expression of genetic information could then lead to division of the zygote into two separate embryos during a specific period, early in development, perhaps from the stage of eight cells to approximately 360 cells in the inner cell mass, when differentiation and primitive streak formation begin. After birth, this discordance of genetic information could be mosaic in each twin, but it could be present to different degrees in the two different twins, sufficient to cause observable phenotypic discordance. In other words, once measurement error and environmental influences are accounted for, genetic discordance or differences in the expression of genetic information should be suspected in cases of discordant MZ twins. This topic is covered in more detail in Section 14.11.2.

Although temperature, delay from the time of ovulation until fertilization or implantation, oxygen supply, and various teratogenic agents have been shown to affect MZ twinning rates in other animals [157], no such factor has been associated with MZ twinning rates in humans. A recent association between an increase of twin births (both MZ and DZ) and periconceptual vitamin supplementation, specifically folic acid, has been reported by Czeizel and colleagues [158], suggesting that adequate maternal nutrition is important for survival to birth for human twins (e.g., loss or conversion to a singleton may occur with inadequate nutrition). The population rate of spontaneous MZ twins seems to be increasing [75], and a 3–5 times increase in MZ twin births has been seen with ART [159,160], perhaps related to ovarian stimulation, disturbance of the zona pellucida, or culturing conditions or handling (e.g., blastocyst transfer) during ART procedures [159]. MZ twins and MZ triplets are frequent among spontaneous triplets [161].

Stockard [162] suggested that MZ twinning may be due to a lack of oxygen prior to implantation, which causes developmental arrest and splitting in the zygote. His work was supported by the finding that the implantation of the ovum is delayed in the armadillo, which results in MZ quadruplets or octuplets [163], and by studies in rabbit and roe deer showing that twinning in these animals is also associated with delayed implantation [164]. These findings suggested that MZ twinning is associated with disturbance of development clocks or thresholds and that delayed fertilization or delayed implantation may play a role in MZ twinning.

On the basis of observations of a higher-than-expected incidence of MZ twins after ART [99,165], Edwards and colleagues [166] suggested that abnormalities or rupture of the zona pellucida may lead to herniation of the blastocyst and predispose to MZ twinning. Boklage [167–169] estimated that differentiation of the chorion occurs at approximately the fourth day after fertilization and that, in monochorionic MZ twins, the physical separation of two embryos is unlikely if the zona is still intact when the chorion begins to develop. Boklage suggested that if the zona is intact, rather than a physical separation of the MZ twins, there may be "developmental" separation, rendering two groups of cells within a morula that organize themselves separately and continue with embryogenesis separately. A recent study investigating the elevated frequency of MZ twinning resulting from ART [170] found that extended culture (or embryo stage of transfer) was a major risk factor. In one report [171], extended culture of spare 2–10-cell ART embryos resulted in two cases of ectopic adhesion of cells from a blastocyst's inner cell mass to the opposing inner trophectoderm wall, which was followed in one by blastocyst splitting and both products hatching. It was proposed that blastocyst collapse and re-expansion observed in mouse blastocysts [172] could occasionally result in adhesion of the inner cell mass to the opposite wall, to which it would transfer a portion of the inner cell mass, triggering MZ splitting.

Other investigators have suggested that twinning itself may be a type of congenital anomaly or an abnormality of development, with the "twinning" fertilized egg (i.e., a fertilized egg resulting in twins) developing at a different rate and in a different way, as compared to a "normal" fertilized egg (i.e., an egg resulting in a singleton). There must be a relatively narrow window during which MZ twinning can occur (normally only up until 11–13 days postfertilization, when the primitive streak forms), and there are a number of different events taking place during post fertilization, including hatching, implantation, genomic imprinting, and X-inactivation (see Fig. 14.1). If the twinning zygote is maturing at a different rate than the normal zygote, the timing for all these events may be shifted and may even occur in an order different from the predicted normal timing for singletons.

## 14.10 GENETIC AND EPIGENETIC DIFFERENCES WITHIN PAIRS OF MZ TWINS

MZ twins have been described as natural clones; however, this is incorrect. In the course of the development of every large multicellular organism, somatic mutations as well as epigenetic and stochastic processes lead to distinct genetic differences between MZ twins [109,173–175].

### 14.10.1 Genetic Differences Within Pairs of MZ Twins

Genetic differences within MZ twin pairs include differences in chromosomal aneuploidy [176–178], uniparental disomy [179], chromosomal rearrangement [180], triplet expansion [181], or nuclear [182,183] or mitochondrial [184,185] point mutations that have occurred postzygotically [175]. Within-pair differences in telomere crossover [186] and length [187] have also been observed.

A small number of case studies using a candidate gene approach have been particularly informative clinically. For example, Vadlamudi and colleagues identified a de novo mutation in the sodium channel alpha 1 subunit gene *SCN1A* as the likely cause of the epileptic disorder Dravet syndrome in the affected twin of a discordant MZ pair [188]. The extent to which DNA sequence has been shown to differ within any given MZ twin pair, whether phenotypically similar or discordant, has depended on the genomic platform used. Medium-to-large-scale studies of single nucleotide polymorphism (SNP) arrays, typically measuring 500,000–1,000,000 SNPs, have shown that fewer than 1% of phenotypically normal MZ twins show genetic discordance at single nucleotides and copy number variants [189–191]. The rate of discordance in disease-discordant pairs detected using SNP arrays may be higher [192–194], although numbers of twin pairs in such studies have been low (typically less than 10 pairs per study) and validation of sequence differences using locus-specific techniques has not been performed for all putative variants. Studies of disease-discordant MZ twin pairs on a similar scale have been performed on coding regions using exome sequencing; however, these have either failed to validate putative within-pair differences [195–197] or detected only mosaic variants [198]. More recently, whole-genome sequencing has enabled sequence comparisons across coding and noncoding regions within pairs of MZ twins. Although such studies are still finding null results with small numbers of twin pairs [199,200], two studies are worth highlighting. Francioli and colleagues derived whole-genome sequences from 11 pairs of MZ twins and their parents, eight pairs of DZ twins and their parents, and 231 child–parents trios [201].

Validated variants in MZ twins enabled them to estimate that germline and postzygotic somatic mutations occur at a ratio of approximately 33:1. They also found that only 7% of such variants detected in single twins were specific to that twin only and were more likely than expected to be located in coding regions. Weber-Lehman and colleagues performed whole-genome sequencing on a pair of father and son and the father's twin brother [202]. The authors found five validated SNPs in sperm DNA from the father and in blood DNA from his son, all of which were not present in the father's co-twin. Although none were located in coding regions or in SNP databases, three of the five were located in putative regulatory regions. Only one of the five variant SNPs were found in the father's blood and four were found in his buccal cells.

In summary, we know that there are likely to be genetic differences within many pairs of MZ twins, but until larger studies that compare multiple tissues at different sequencing read depths, we will not know the true frequency of such events or their contribution, if any, to phenotype.

## 14.10.2 Epigenetic Differences Within MZ Twin Pairs

Epigenetics is the study of changes to gene activity, perpetuated through cell division, that are not accompanied by changes to the DNA sequence. Epigenetic change, which is associated with both early development and aging, involves the addition and removal of small molecules onto DNA and histones—the DNA-packaging proteins. The most understood epigenetic change is DNA methylation—the addition of a methyl group ($CH_3$) to the cytosine nucleotide of a cytosine–guanine (CpG) dinucleotide. Studies involving genome-wide analysis of DNA methylation have started to show promise in identifying causal mechanisms, and diagnostic biomarkers for complex human diseases [203].

Twin studies have shown that throughout mammalian genomes, epigenetic state is influenced mostly by variation in nonshared environment (75–85% of the variance explained), to a lesser extent by genetic variation (10–20% of the variance explained), with variation in shared environment having the smallest influence [204–208]. Like other traits, the epigenetic state of individual locations throughout the epigenome (the sum total of epigenetic marks in any given tissue) can vary greatly in their components of variance.

Twin studies have demonstrated within-pair epigenetic differences are present at birth, whether at individual loci [209] or throughout the epigenome [210,211]. These results also showed that some MZ twin pairs can be more epigenetically discordant than DZ twins and some unrelated individuals, emphasizing the formative nature of the nonshared intrauterine environment. Also of note, the same studies found that genes with the highest levels of within-pair epigenetic discordance were enriched in those associated with development and morphogenesis. In agreement with this, in an MZ twin pair discordant for the developmental anomaly of caudal duplication, the twins differed in DNA methylation state in a gene, *AXIN1*, whose genetic state had previously been associated with the disorder [212].

As tissues age, their epigenetic state changes as a function of age and environment, and this has been termed epigenetic drift [213]. A cross-sectional study of MZ twins of different ages found that epigenetic discordance on a genome-wide scale increases with age [214]; however, these findings were not replicated in another cross-sectional study [210] or longitudinal studies of infants [215] and adults [216]. Although we do know that the epigenetic state of a small subset of the epigenome does correlate highly with age [217,218], we also know that factors such as tissue and genomic location also influence epigenetic change over time [219,220]. More longitudinal epigenome-wide studies are needed to resolve this issue.

It is also worth noting that nonshared environment includes twin-specific environmental factors, measurement error, and stochastic factors, the latter being an intrinsic part of eukaryotic development [221–223]. This implies that we need to improve the accuracy of measuring phenotypes in twin studies and need to test hypotheses that specific (intrauterine) nonshared environmental factors are associated with the epigenetic state. To date, MZ twin discordance for birth weight, an easily measurable proxy for intrauterine growth and viability, has been associated with epigenetic discordance in genes involved in metabolism when measured at birth [210] or in adulthood [224], and in the growth-associated gene *IGF1R* when measured in adulthood [225]; however, two other epigenome-wide studies found no evidence for differential DNA methylation levels in adult twins discordant for birth weight. Again, more studies are needed to resolve this issue.

### 14.10.3 Nonshared Environment and Chronic Disease

Twins studies have made a tremendous contribution to the understanding of the causes of chronic disorders, from cancers to neurodevelopmental conditions such as schizophrenia, allergies, and cardiometabolic disorders such as cardiovascular disease [7,173,204,226,227]. Such disorders originate in very early life [228–230] and like the epigenetic state itself, are mainly influenced by the nonshared environment [231]. This finding was initially surprising because maternal diet and lifestyle had been assumed to be influential on the developing fetus; however, multiple nonshared factors have now been shown to influence risk for chronic disease [116,173,232–234]. Most of these factors are associated with the "fetoplacental unit" [235]—the placenta, cords, and fetus. Such factors include uterine implantation site, the placental location at which the umbilical cord is inserted, and the physical characteristics of the umbilical cord such as length, width, and torsion. All these factors have the potential to affect the growth rate of individual twins via the transplacental transport of oxygen, nutrients, and teratogens.

MZ twins can be discordant for inflammatory state [236–240], which is a major contributor to chronic disease risk [241,242]. Twin studies have shown that the nonshared environment dominates as the largest component of variance of immune factors in adults [243]. Prenatal inflammation can occur in the umbilical cord or placenta of a single MZ twin [237,244] and soluble inflammatory factors can pass to the associated fetus [245]. More longitudinal twin birth cohorts are required to improve our understanding of the way the nonshared environment influences health and disease and its epigenetic mediators.

## 14.11 TWIN RESEARCH: DESIGNS AND ANALYTIC APPROACHES

Studying twins can provide insights about the health of both twins and nontwin individuals—and whole populations. A number of different study designs involving twins are possible and these have different aims, statistical analysis techniques, model assumptions, advantages, and limitations.

### 14.11.1 The "Classic Twin Design"

One of the most common twin designs, and one that researchers first think of when considering a twin study, is the "classic twin design." Statistically, familial resemblance can be quantified by estimating the correlation, either for all types of twins together or separately for MZ and DZ twins. Correlations are always between −1 and 1, and correlations greater than 0 for pairs of twins suggest that familial resemblance exists for the particular trait. The "classic twin model" initially estimates correlations separately for MZ and DZ pairs and compares these. If the correlation is greater for MZ twins than for DZ twins, this is consistent with (but does not prove) the existence of genetic effects influencing variation. Even when results are consistent with genetic effects, the specific genes responsible usually cannot be identified simply by studying disease traits. In addition, this conclusion relies on several strong assumptions: (1) MZ and DZ twins share their environment to the same extent (the "equal environments" assumption [246]) and (2) the only difference between MZ and DZ twins is the proportion of genes they share. The first assumption does not mean that all pairs of twins are assumed to share their environments equally, just that the extent of sharing does not depend on zygosity, and that on average MZ pairs do not share their environment more (or less) than DZ pairs.

The trait often depends on measured variables such as age and sex, and in such cases should be adjusted for these before correlations are estimated. This usually requires fitting statistical models which estimate the effects of the measured variables on the trait of interest and the correlation of the residuals in MZ and DZ pairs simultaneously.

If correlations for MZ and DZ pairs are statistically significantly different, additional models which divide the residual variation into components due to shared genetic effects, shared environmental effects, and unshared effects can be fitted. These "variance components models" can be fitted using maximum likelihood estimation [247] or via a structural equations approach [248], and should also incorporate adjustments for measured variables. The residual variance (which includes variation due to measurement error and stochastic factors as well as factors unique to individual twins), and the amounts of variation attributable to shared genes and shared environmental effects can also provide information, as can opposite-sex fraternal pairs, for example, about whether different genes or environmental effects are influencing variation of a trait in males and females. Often, the components of variance are reported as proportions of the total variance, and in this case the proportion of shared genetic effects is referred to as the *heritability*. Although it is commonly reported, focusing on the heritability loses a lot of information and is not recommended [249,250].

Correlations are usually estimated for continuous traits such as height, but it is also often of interest to estimate how similar twins are for a categorical outcome, such as a diagnosis (or not) of schizophrenia or of metastatic prostate cancer, or severity of baldness (none, mild, moderate, or severe). This similarity is usually quantified using the *concordance* (either casewise or pairwise [251]). In the case of schizophrenia, the pairwise concordance for MZ twins is 0.48, meaning that if one member of a pair receives a diagnosis there is close to a 50% chance that the other one will too [252]. For DZ twins the concordance is 0.17. If we assume again that the two twin types share their environments equally and only differ in the proportion of genes they share, this is again consistent with genetic influences on liability of schizophrenia, but factors in the environment matter too; otherwise the concordance for MZ twins would be 1.00. Variance components models can also be fitted for binary traits, and these estimate the proportions of variation on the liability scale, which makes interpretation of results more challenging. These models for binary outcomes also have lower power. Alternatives exist, including estimation of "intrinsic correlation" after adjustment for measured variables such as age, which can affect concordances [253,254], and generalized linear mixed models (GLMMs) fitted using Markov chain Monte Carlo (MCMC) methods [255]. Methods for outcomes which are best represented as ordinal, categorical [256,257], or as censored survival times [258,259] also exist.

Advantages of the classic twin design include the ability to include all types of twins regardless of their measured outcomes or exposures, the ability to estimate variation and covariation as well as correlations and heritability, and the capacity to assess effects of measured covariates, such as age and sex, on both the trait and the variances/covariances. This approach also has limitations, perhaps most importantly, regarding the equal environments assumption, which is crucial to the model and yet difficult to test. The model also has low power to detect shared environmental effects (see, e.g., [246]) and assumes that any differences in correlations are due solely to differences in genetics.

### 14.11.2 Other Types of Twin Designs

Studies with twins extend beyond the classic twin design and are very useful in epidemiology generally, not just in genetic epidemiology. Some examples of other designs and of the types of questions that twin research can address are included in Table 14.3. Most designs require statistical models which allow for correlated observations to be fitted, such as GLMMs, also known as mixed effects models, or models fitted using generalized estimating equations (GEEs). These models also have advantages and disadvantages. For example, the GEE approach is appropriate when interest focuses on the association of measured covariates with the outcome, while GLMMs are required if estimation of variation and covariation is also of interest, but both these models are appropriate for continuous and binary outcomes and both can be fitted in standard statistical software packages such as Stata [274]. The most appropriate approach and model will depend on the specific research question and available data.

Each of these study types also has advantages and disadvantages. In outcome-discordant twin studies, the case-control pairs are matched for both measured and unmeasured factors—for age, genetic factors (perfectly for MZ pairs; 50% for DZ), nongenetic familial factors (not necessarily to the same degree for MZ and DZ pairs), mother, father, uterus and, perhaps, placenta, sex (if only same-sex pairs are included), calendar year, and season of birth. Although this type of study can be less costly and time-consuming compared with cohort studies, it has similar limitations to standard case-control studies (potential recall bias, inefficient for rare exposures). In contrast, exposure-discordant twin studies also match for both unmeasured and measured factors (other than the exposure of interest), and have the potential to allow causal inferences, but may not be representative of the general twin population. This study type has similar advantages and limitations to standard matched cohort studies, namely that it is advantageous for rare exposures but that it can be difficult to find and recruit exposure-discordant twin pairs. For example, over 2000 pairs of female twins were screened to find 20 pairs discordant by 20 or more pack-years of smoking [266]. If twins are included in an intervention study, they will be matched for genes, their participation may be enhanced by the pairing, and they may be a highly motivated group, but the close bond between twins may mean a potential failure to adhere to study protocol (e.g., by discussing or swapping treatments).

**TABLE 14.3 Designs and Statistical Approaches in Twin Research**

| Design | Examples/Applications | Statistical Model/Analytic Approach |
|---|---|---|
| **Co-twin control study—Disease discordant**: matched case-control design, generally with identical twins discordant for a trait | Earlier onset of puberty is not associated with breast cancer risk in disease-discordant twins [260] | Conditional logistic regression or binary GLMMs/GEEs |
| **Co-twin control study—Exposure-discordant**: identical twins discordant for a health-related characteristic | Military service in Vietnam is associated with higher risk of post-traumatic stress symptoms than military service elsewhere [261] | Linear or logistic regression with adjustment for within-pair correlation (e.g., GEEs or mixed effects models) |
| **Randomized controlled trial**: identical twins, naturally matched for age, sex, and genes, given the same or different interventions (or possibly both, if a crossover design is used) | Calcium supplementation in adolescence has little effect on bone density [262] | As for exposure-discordant studies |
| **Epigenetics**: MZ twins share the same DNA but the way in which this DNA operates can differ | Birth weight is associated with epigenetic differences in growth and metabolism genes [210] | As for exposure-discordant studies |
| **Causal models** in twins | The observed association between BMI and mortality is unlikely to be causal and appears largely due to shared confounding [263,264] | As for exposure-discordant studies |
| **Within-between study** (related to causal models) | The association between birth weight and cord blood erythropoietin appears due to individual rather than pair-specific factors [265] | Linear or logistic regression with adjustment for within-pair correlation (e.g., GEEs or mixed effects models) |
| **Differences study**: Both exposure and outcome are continuous | Differences in smoking consumption predict differences in bone density [266] | Linear regression |
| **Issues specific to twins**: Raising twins can be challenging as well as rewarding, and research can help parents with day-to-day decisions | For most twin pairs, staying together in the first year of school is a critical social support [267] | Regression-based approach; depends on research question and outcome type |
| **Longitudinal designs**: Following twins over growth and development can identify changing patterns of genetic and environmental influence | Higher cumulative exposure to solvents over the lifetime is associated with increased risk of Parkinson's disease [268]. Many twin registries have longitudinal data on twins [269–271] | Regression-based approach accounting for correlation (e.g., mixed effects models) |
| **Multivariate designs**: Studying two or more traits simultaneously can identify shared and separate genetic and environmental influences | Genes that influence children's reading abilities are in part shared with genes that influence mathematics abilities [272] | Multivariate variance components models |
| **Extended twin family study**: Inclusion of other family members, such as parents and siblings, extends the range of genetic and environmental factors that can be understood | Different patterns of covariation are apparent for height, BMI, and blood pressure [273] | Variance components models |

### 14.11.3 General Statistical Issues

General statistical principles apply when analyzing data from twins and families. Good statistical practice in this context includes thorough exploration of data prior to model fitting, being aware of and testing model assumptions, reporting estimates, 95% confidence intervals and *P*-values, starting with simple analyses and models, and building on these, adjusting for measured variables before considering unmeasured effects, and remembering that analyses of continuous outcomes are usually more powerful than those of binary outcomes.

### 14.11.4 Summary

While the classic twin design has been applied for around 100 years as a powerful tool for disentangling the etiology of complex traits and diseases, researchers now increasingly recognize the value of including twins in many other types of studies, particularly in those related to health and medical research. Technological advances such as epigenetic, microbiome, and other "omics" platforms have resulted in new opportunities and possibilities for studies and study designs [7], and new statistical models and approaches (such as causal inference utilizing data from twins) being developed, suggesting that twin studies will continue to remain highly relevant in medical research.

## 14.12 TWIN REGISTRIES AND INTERNATIONAL COLLABORATION

The first twin registries were set up in Scandinavia around the middle of the 20th century, aimed at collecting and maintaining contact information from twins who were identified through national birth registers, and accessing their longitudinal health data through links with medical data sources to conduct twin studies. The value of such resources has been since recognized based on a number of other twin registries established in developed and developing countries throughout the years [275]. Their relevance expands beyond the collection and analysis of twin data and biospecimens toward further development of the twin methodology and novel applications in health and social sciences.

While some twin registries have relied upon population-based recruiting strategies such as identifying twins from national birth registers [276,277], drivers' licensing departments [278], regional health systems [279], and war veterans' databases [269], others have opted for a volunteer-based approach, using traditional and social media channels [280] and twin festivals to improve recruitment and engagement with members [270]. In all twin registries, zygosity determination strategies have also varied from questionnaire self-reports to DNA testing in twin registries where there is a large amount of genotypes collected [281].

Global collaboration in twin research has also been a task facilitated by twin registries, which have formed the International Network of Twin Registries [282] with that purpose. The network aims to build a worldwide catalogue of available twin data and biospecimens to be used in ethically approved future studies, and it is formally linked to the International Society of Twin Studies (ISTS), the main professional society in twin research, which is also responsible for the *Twin Research and Human Genetics* journal. A good example of what can be achieved in multicenter twin studies is the CODATwins project, a consortium with mostly anthropometrical data on 434,723 twins in 22 different countries [283]. Such collaborations will be especially relevant for achieving the necessary sample sizes to find discordant twin pairs for specific traits and conditions and to study rare diseases, as they also provide means for expert knowledge generation and exchange.

## 14.13 CONCLUSIONS

In the field of twins and twinning, there is still a great deal to be learned. The development of new DNA molecular and cytogenetic techniques, the use of prenatal diagnosis such as chorionic villus sampling, amniocentesis, and ultrasound examination in humans, as well as embryo pathology, histology, and genetic advances all give clues to the increased understanding of MZ and DZ twins and the twinning process itself.

## REFERENCES

[1] Galton F. The history of twins, as a criterion of the relative powers of nature and nurture. Trubner and Co.; 1876. p. 391.

[2] Fisher RA. Has Mendel's work been rediscovered? Ann Sci 1936;1:115.

[3] Hall J. Twinning. Lancet 2003a;362:735–43.

[4] Mayo O. Early research on human genetics using the twin method: who really invented the method? Twin Res Hum Genet 2009;12:237–45.

[5] Fisher RA. The correlation between relatives on the supposition of Mendelian inheritance. Trans R Soc Edinburgh 1918;52:399–433.
[6] McNamara HC, Kane SC, Craig JM, Short RV, Umstad MP. A review of the mechanisms and evidence for typical and atypical twinning. Am J Obstet Gynecol 2016;214:172–91.
[7] van Dongen J, Slagboom PE, Draisma HH, Martin NG, Boomsma DI. The continuing value of twin studies in the omics era. Nat Rev Genet 2012;13:640–53.
[8] Herranz G. The timing of monozygotic twinning: a criticism of the common model. Zygote 2013:1–14.
[9] Denker HW. Comment on G. Herranz: the timing of monozygotic twinning: a criticism of the common model. Zygote (2013). Zygote 2013:1–3.
[10] Ginsberg NA, Ginsberg S, Rechitsky S, Verlinsky Y. Fusion as the etiology of chimerism in monochorionic dizygotic twins. Fetal Diagn Ther 2005;20:20–2.
[11] Assaf SA, Randolph LM, Benirschke K, Wu S, Samadi R, Chmait RH. Discordant blood chimerism in dizygotic monochorionic laser-treated twin–twin transfusion syndrome. Obstet Gynecol 2010;116:483–5.
[12] Chen K, Chmait RH, Vanderbilt D, Wu S, Randolph L. Chimerism in monochorionic dizygotic twins: case study and review. Am J Med Genet 2013;161A:1817–24.
[13] Fumoto S, Hosoi K, Ohnishi H, Hoshina H, Yan K, Saji H, Oka A. Chimerism of buccal membrane cells in a monochorionic dizygotic twin. Pediatrics 2014;133:e1097–1100.
[14] Umstad MP, Short RV, Wilson M, Craig JM. Chimaeric twins: why monochorionicity does not guarantee monozygosity. Aust N Z J Obstet Gynaecol 2012;52:305–7.
[15] Miura K, Niikawa N. Do monochorionic dizygotic twins increase after pregnancy by assisted reproductive technology? J Hum Genet 2005;50:1–6.
[16] Nylander PP, Osunkoya BO. Unusual monochorionic placentation with heterosexual twins. Obstet Gynecol 1970;36:621–5.
[17] Tarkowski AK, Wojewodzka M. A method for obtaining chimaeric mouse blastocysts with two separate inner cell masses: a preliminary report. J Embryol Exp Morphol 1982;71:215–21.
[18] Williams CA, Wallace MR, Drury KC, Kipersztok S, Edwards RK, Williams RS, Haller MJ, Schatz DA, Silverstein JH, Gray BA, Zori RT. Blood lymphocyte chimerism associated with IVF and monochorionic dizygous twinning: case report. Hum Reprod 2004;19:2816–21.
[19] Safran A, Reubinoff BE, Porat Katz A, Werner M, Friedler S, Lewin A. Intracytoplasmic sperm injection allows fertilization and development of a chromosomally balanced embryo from a binovular zona pellucida. Hum Reprod 1998;13:2575–8.
[20] Van de Leur SJ, Zeilmaker GH. Double fertilization in vitro and the origin of human chimerism. Fertil Steril 1990;54:539–40.
[21] Vicdan K, Işik AZ, Dagli HG, Kaba A, Kişnişçi H. Fertilization and development of a blastocyst-stage embryo after selective intracytoplasmic sperm injection of a mature oocyte from a binovular zona pellucida: a case report. J Assist Reprod Genet 1999;16:355–7.
[22] Walker SP, Meagher S, White SM. Confined blood chimerism in monochorionic dizygous (MCDZ) twins. Prenat Diagn 2007;27:369–72.
[23] Choi DH, Kwon H, Lee SD, Moon MJ, Yoo EG, Lee KH, Hong YK, Kim G. Testicular hypoplasia in monochorionic dizygous twin with confined blood chimerism. J Assist Reprod Genet 2013;30:1487–91.
[24] Short RV. The bovine freemartin: a new look at an old problem. Philos Trans R Soc Lond B Biol Sci 1970;259:141–7.
[25] Dirani M, Chamberlain M, Garoufalis P, Chen CY, Guymer RH, Baird PN. Mirror-image congenital esotropia in monozygotic twins. J Pediatr Ophthalmol Strabismus 2006;43:170–1.
[26] Goto T, Nemoto T, Okuma T, Kobayashi H, Funata N. Mirror-image solitary bone cyst of the humerus in a pair of mirror-image monozygotic twins. Arch Orthop Trauma Surg 2008;128:1403–6.
[27] Hu JT, Liu T, Qian J, Zhang YB, Zhou X, Zhang QG. Occurrence of different external ear deformities in monozygotic twins: report of 2 cases. Plast Reconstr Surg Glob Open 2014;2:e206.
[28] Karaca C, Yilmaz M, Karatas O, Menderes A, Karademir S. Mirror imaging cleft lip in monozygotic twins. Eur J Plast Surg 1995;18:260–1.
[29] Morison D, Reyes CV, Skorodin MS. Mirror-image tumors in mirror-image twins. Chest 1994;106:608–10.
[30] Novak RW. Laryngotracheoesophageal cleft and unilateral pulmonary hypoplasia in twins. Pediatrics 1981;67:732–4.
[31] Riess A, Dufke A, Riess O, Beck Woedl S, Fode B, Skladny H, Klaes R, Tzschach A. Mirror-image asymmetry in monozygotic twins with kabuki syndrome. Mol Syndromol 2012;3:94–7.
[32] Rife DC. Genetic studies of monozygotic twins: III. Mirror-imaging. J Hered 1933;24:443–6.
[33] Satoh K, Shibata Y, Tokushige H, Onizuka T. A mirror image of the first and second branchial arch syndrome associated with cleft lip and palate in monozygotic twins. Br J Plast Surg 1995;48:601–5.

[34] Sperber GH, Machin GA, Bamforth FJ. Mirror-image dental fusion and discordance in monozygotic twins. Am J Med Genet 1994;51:41–5.

[35] Springer SP, Searleman A. Laterality in twins: the relationship between handedness and hemispheric asymmetry for speech. Behav Genet 1978;8:349–57.

[36] Wang ED, Xu X, Dagum AB. Mirror-image trigger thumb in dichorionic identical twins. Orthopedics 2012;35:e981–3.

[37] Helland CA, Wester K. Monozygotic twins with mirror image cysts: indication of a genetic mechanism in arachnoid cysts? Neurology 2007;69:110–1.

[38] Nigro MA, Wishnow R, Maher L. Colpocephaly in identical twins. Brain Dev 1991;13:187–9.

[39] Pascual-Castroviejo I, Verdú A, Román M, De la Cruz-Medina M, Villarejo F. Optic glioma with progressive occlusion of the aqueduct of sylvius in monozygotic twins with neurofibromatosis. Brain Dev 1988;10:24–9.

[40] Zhou JY, Pu JL, Chen S, Hong Y, Ling CH, Zhang JM. Mirror-image arachnoid cysts in a pair of monozygotic twins: a case report and review of the literature. Int J Med Sci 2011;8:402–5.

[41] Derom C, Thiery E, Vlietinck R, Loos R, Derom R. Handedness in twins according to zygosity and chorion type: a preliminary report. Behav Genet 1996;26:407–8.

[42] Sommer IE, Ramsey NF, Bouma A, Kahn RS. Cerebral mirror-imaging in a monozygotic twin. Lancet 1999;354:1445–6.

[43] Schmerler S, Wessel GM. Polar bodies-more a lack of understanding than a lack of respect. Mol Reprod Dev 2011;78:3–8.

[44] Bieber FR, Nance WE, Morton CC, Brown JA, Redwine FO, Jordan RL, Mohanakumar T. Genetic studies of an acardiac monster: evidence of polar body twinning in man. Science 1981;213:775–7.

[45] Fisk NM, Ware M, Stanier P, Moore G, Bennett P. Molecular genetic etiology of twin reversed arterial perfusion sequence. Am J Obstet Gynecol 1996;174:891–4.

[46] La Sala GB, Villani MT, Nicoli A, Gallinelli A, Nucera G, Blickstein I. Effect of the mode of assisted reproductive technology conception on obstetric outcomes for survivors of the vanishing twin syndrome. Fertil Steril 2006;86:247–9.

[47] Pinborg A, Lidegaard O, la Cour Freiesleben NI, Andersen AN. Consequences of vanishing twins in IVF/ICSI pregnancies. Hum Reprod 2005;20:2821–9.

[48] Rodríguez-González M, Serra V, Garcia-Velasco JA, Pellicer A, Remohí J. The 'vanishing embryo' phenomenon in an oocyte donation programme. Hum Reprod 2002;17:798–802.

[49] Pinborg A, Lidegaard O, la Cour Freiesleben NI, Andersen AN. Vanishing twins: a predictor of small-for-gestational age in IVF singletons. Hum Reprod 2007;22:2707–14.

[50] Pharoah PO, Cooke RW. A hypothesis for the aetiology of spastic cerebral palsy - the vanishing twin. Dev Med Child Neurol 1997;39:292–6.

[51] Newton R, Casabonne D, Johnson A, Pharoah P. A case-control study of vanishing twin as a risk factor for cerebral palsy. Twin Res 2003;6:83–4.

[52] Huang T, Boucher K, Aul R, Rashid S, Meschino WS. First and second trimester maternal serum markers in pregnancies with a vanishing twin. Prenat Diagn 2015;35:90–6.

[53] Blickstein I. Superfecundation and superfetation: lessons from the past on early human development. J Matern Fetal Neonatal Med 2003;14:217–9.

[54] Czyz W, Morahan JM, Ebers GC, Ramagopalan SV. Genetic, environmental and stochastic factors in monozygotic twin discordance with a focus on epigenetic differences. BMC Med 2012;10:93.

[55] Amsalem H, Tsvieli R, Zentner BS, Yagel S, Mitrani-Rosenbaum S, Hurwitz A. Monopaternal superfecundation of quintuplets after transfer of two embryos in an in vitro fertilization cycle. Fertil Steril 2001;76:621–3.

[56] Peigné M, Andrieux J, Deruelle P, Vuillaume I, Leroy M. Quintuplets after a transfer of two embryos following in vitro fertilization: a proved superfecundation. Fertil Steril 2011;95(2124):e2113–24. e2116.

[57] Girela E, Lorente JA, Alvarez JC, Rodrigo MD, Lorente M, Villanueva E. Indisputable double paternity in dizygous twins. Fertil Steril 1997;67:1159–61.

[58] Harris DW. Superfecundation. J Reprod Med 1982;27:39.

[59] Terasaki PI, Gjertson D, Bernoco D, Perdue S, Mickey MR, Bond J. Twins with two different fathers identified by HLA. N Engl J Med 1978;299:590–2.

[60] Wenk RE, Houtz T, Chiafari F, Brooks M. Superfecundation identified by HLA, protein, and VNTR DNA polymorphisms. Transfus Med 1991;1:253–5.

[61] James WH. The incidence of superfecundation and of double paternity in the general population. Acta Genet Med Gemellol 1993;42:257–62.

[62] Bristow RE, Shumway JB, Khouzami AN, Witter FR. Complete hydatidiform mole and surviving coexistent twin. Obstet Gynecol Surv 1996;51:705–9.

[63] Fishman DA, Padilla LA, Keh P, Cohen L, Frederiksen M, Lurain JR. Management of twin pregnancies consisting of a complete hydatidiform mole and normal fetus. Obstet Gynecol 1998;91:546–50.

[64] Massardier J, Golfier F, Journet D, Frappart L, Zalaquett M, Schott A, Lenoir V, Dupuis O, Hajri T, Raudrant D. Twin pregnancy with complete hydatidiform mole and coexistent fetus: obstetrical and oncological outcomes in a series of 14 cases. Eur J Obstet Gynecol Reprod Biol 2009;143:84–7.
[65] Sebire NJ, Foskett M, Paradinas FJ, Fisher RA, Francis RJ, Short D, Newlands ES, Seckl MJ. Outcome of twin pregnancies with complete hydatidiform mole and healthy co-twin. Lancet 2002;359:2165–6.
[66] Niemann I, Sunde L, Petersen LK. Evaluation of the risk of persistent trophoblastic disease after twin pregnancy with diploid hydatidiform mole and coexisting normal fetus. Am J Obstet Gynecol 2007;197:45.e41–5.
[67] George V, Khanna M, Dutta T. Fetus in fetu. J Pediatr Surg 1983;18:288–9.
[68] Spencer R. Parasitic conjoined twins: external, internal (fetuses in fetu and teratomas), and detached (acardiacs). Clin Anat 2001;14:428–44.
[69] Brand A, Alves MC, Saraiva C, Loio P, Goulão J, Malta J, Palminha JM, Martins M. Fetus in fetu–diagnostic criteria and differential diagnosis–a case report and literature review. J Pediatr Surg 2004;39:616–8.
[70] Gerber RE, Kamaya A, Miller SS, Madan A, Cronin DM, Dwyer B, Chueh J, Conner KE, Barth RA. Fetus in fetu: 11 fetoid forms in a single fetus: review of the literature and imaging. J Ultrasound Med 2008;27: 1381–7.
[71] Hoeffel CC, Nguyen KQ, Phan HT, Truong NH, Nguyen TS, Tran TT, Fornes P. Fetus in fetu: a case report and literature review. Pediatrics 2000;105:1335–44.
[72] Huddle LN, Fuller C, Powell T, Hiemenga JA, Yan J, Deuell B, Lyders EM, Bodurtha JN, Papenhausen PR, Jackson-Cook CK, Pandya A, Jaworski M, Tye GW, Ritter AM. Intraventricular twin fetuses in fetu. J Neurosurg Pediatr 2012;9:17–23.
[73] Escobar MA, Rossman JE, Caty MG. Fetus-in-fetu: report of a case and a review of the literature. J Pediatr Surg 2008;43:943–6.
[74] Hopkins KL, Dickson PK, Ball TI, Ricketts RR, O'Shea PA, Abramowsky CR. Fetus-in-fetu with malignant recurrence. J Pediatr Surg 1997;32:1476–9.
[75] Bressers WM, Eriksson AW, Kostense PJ, Parisi P. Increasing trend in the monozygotic twinning rate. Acta Genet Med Gemellol 1987;36:397–408.
[76] Nylander PP. The twinning incidence of Nigeria. Acta Genet Med Gemellol 1979;28:261–3.
[77] Imaizumi Y. Triplets and higher order multiple births in Japan. Acta Genet Med Gemellol 1990;39:295–306.
[78] Campbell DM, Campbell AJ, MacGillivray I. Maternal characteristics of women having twin pregnancies. J Biosoc Sci 1974;6:463–70.
[79] Hoekstra C, Zhao ZZ, Lambalk CB, Willemsen G, Martin NG, Boomsma DI, Montgomery GW. Dizygotic twinning. Hum Reprod Update 2008;14:37–47.
[80] Nylander PP. Biosocial aspects of multiple births. J Biosoc Sci Suppl 1971:29–38.
[81] Umstad MP, Hale L, Wang YA, Sullivan EA. Multiple deliveries: the reduced impact of in vitro fertilisation in Australia. Aust N Z J Obstet Gynaecol 2013;53:158–64.
[82] James WH. Sex ratio in twin births. Ann Hum Biol 1975;2:365–78.
[83] James WH. Sex ratio and placentation in twins. Ann Hum Biol 1980b;7:273–6.
[84] James WH. Gestational age in twins. Arch Dis Child 1980a;55:281–4.
[85] Tsunoda Y, Tokunaga T, Sugie T, Katsumata M. Production of monozygotic twins following the transfer of bisected embryos in the goats. Theriogenology 1985;24:337–43.
[86] Bryan E, Little J, Burn J. Congenital anomalies in twins. Baillieres Clin Obstet Gynaecol 1987;1:697–721.
[87] Mastroiacovo P, Castilla EE, Arpino C, Botting B, Cocchi G, Goujard J, Marinacci C, Merlob P, Metneki J, Mutchinick O, Ritvanen A, Rosano A. Congenital malformations in twins: an international study. Am J Med Genet 1999;83:117–24.
[88] Doyle PE, Beral V, Botting B, Wale CJ. Congenital malformations in twins in England and Wales. J Epidemiol Community Health 1991;45:43–8.
[89] Kallen B. Congenital malformations in twins: a population study. Acta Genet Med Gemellol 1986;35:167–78.
[90] Schinzel AA, Smith DW, Miller JR. Monozygotic twinning and structural defects. J Pediatr 1979;95:921–30.
[91] Derom C, Derom R, Loos RJ, Jacobs N, Vlietinck R. Retrospective determination of chorion type in twins using a simple questionnaire. Twin Res 2003;6:19–21.
[92] Burn J, Corney G. Zygosity determination and the types of twinning. In: MacGillivray I, Campbell DM, Thompson B, editors. Twinning and twins. Chichester: John Wiley & Sons; 1988.
[93] Forget-Dubois N, Perusse D, Turecki G, Girard A, Billette JM, Rouleau G, Boivin M, Malo J, Tremblay RE. Diagnosing zygosity in infant twins: physical similarity, genotyping, and chorionicity. Twin Res 2003;6:479–85.
[94] Weinberg W. Contribution on the physiology and pathology of multiple birth in man. Pflugers Arch Physiol 1901;88:346.
[95] Dallapiccola B, Stomeo C, Ferranti G, Di Lecce A, Purpura M. Discordant sex in one of three monozygotic triplets. J Med Genet 1985;22:6–11.
[96] Edwards JH, Dent T, Kahn J. Monozygotic twins of different sex. J Med Genet 1966;3:117–23.

[97] Kurosawa K, Kuromaru R, Imaizumi K, Nakamura Y, Ishikawa F, Ueda K, Kuroki Y. Monozygotic twins with discordant sex. Acta Genet Med Gemellol 1992;41:301–10.
[98] Akane A, Matsubara K, Shiono H, Yamada M, Nakagome Y. Diagnosis of twin zygosity by hypervariable RFLP markers. Am J Med Genet 1991;41:96–8.
[99] Derom C, Bakker E, Vlietinck R, Derom R, Van den Berghe H, Thiery M, Pearson P. Zygosity determination in newborn twins using DNA variants. J Med Genet 1985;22:279–82.
[100] Hill AV, Jeffreys AJ. Use of minisatellite DNA probes for determination of twin zygosity at birth. Lancet 1985;2:1394–5.
[101] Becker A, Busjahn A, Faulhaber HD, Bahring S, Robertson J, Schuster H, Luft FC. Twin zygosity. Automated determination with microsatellites. J Reprod Med 1997;42:260–6.
[102] Hannelius U, Gherman L, Makela VV, Lindstedt A, Zucchelli M, Lagerberg C, Tybring G, Kere J, Lindgren CM. Large-scale zygosity testing using single nucleotide polymorphisms. Twin Res Hum Genet 2007;10:604–25.
[103] Bianchi DW, Fisk NM. Fetomaternal cell trafficking and the stem cell debate: gender matters. J Am Med Assoc 2007;297:1489–91.
[104] Erlich Y. Blood ties: chimerism can mask twin discordance in high-throughput sequencing. Twin Res Hum Genet 2011;14:137–43.
[105] Bajoria R, Kingdom J. The case for routine determination of chorionicity and zygosity in multiple pregnancy. Prenat Diagn 1997;17:1207–25.
[106] Craig JM, Segal NL, Umstad MP, Cutler TL, Keogh LA, Hopper JL, Rankin M, Denton J, Derom CA, Sumathipala A, Harris JR, International Society for Twin S, International Council of Multiple Birth O. Zygosity testing should be encouraged for all same-sex twins: FOR: a genetic test is essential to determine zygosity. BJOG 2015;122:1641.
[107] Derom R, Vlietinck RF, Derom C, Keith LG, Van Den Berghe H. Zygosity determination at birth: a plea to the obstetrician. J Perinat Med 1991;19(Suppl 1):234–40.
[108] Machin GA. Why is it important to diagnose chorionicity and how do we do it? Best Pract Res Clin Obstet Gynaecol 2004;18:515–30.
[109] Machin G. Non-identical monozygotic twins, intermediate twin types, zygosity testing, and the non-random nature of monozygotic twinning: a review. Am J Med Genet C Semin Med Genet 2009b;151C:110–27.
[110] Segal NL. Zygosity testing: laboratory and the investigator's judgment. Acta Genet Med Gemellol 1984;33:515–21.
[111] Maruotti GM, Saccone G, Morlando M, Martinelli P. First-trimester ultrasound determination of chorionicity in twin gestations using the lambda sign: a systematic review and meta-analysis. Eur J Obstet Gynecol Reprod Biol 2016;202:66–70.
[112] Cutler TL, Murphy K, Hopper JL, Keogh LA, Dai Y, Craig JM. Why accurate knowledge of zygosity is important to twins. Twin Res Hum Genet 2015;18:298–305.
[113] Bamforth F, Machin G. Why zygosity of multiple births is not always obvious: an examination of zygosity testing requests from twins or their parents. Twin Res 2004;7:406–11.
[114] van Jaarsveld CH, Llewellyn CH, Fildes A, Fisher A, Wardle J. Are my twins identical: parents may be misinformed by prenatal scan observations. BJOG 2012;119:517–8.
[115] Craig JM, All A. Re: zygosity testing should be encouraged for all same-sex twins. AGAINST: the benefit of this knowledge should be weighed against the potential pitfalls. BJOG 2016;123:1560–1.
[116] Keith L, Machin G. Zygosity testing. Current status and evolving issues. J Reprod Med 1997;42:699–707.
[117] Cyranoski D. Developmental biology: two by two. Nature 2009;458:826–9.
[118] Harvey MA, Huntley RM, Smith DW. Familial monozygotic twinning. J Pediatr 1977;90:246–7.
[119] Machin G. Familial monozygotic twinning: a report of seven pedigrees. Am J Med Genet C Semin Med Genet 2009a;151C:152–4.
[120] Shapiro LR, Zemek L, Shulman MJ. Genetic etiology for monozygotic twinning. Birth Defects Orig Artic Ser 1978;14:219–22.
[121] St Clair JB, Golubovsky MD. Paternally derived twinning: a two century examination of records of one Scottish name. Twin Res 2002;5:294–307.
[122] Michels VV, Riccardi VM. Twin recurrence and amniocentesis: male and MZ heritability factors. Birth Defects Orig Artic Ser 1978;14:201–11.
[123] Lichtenstein P, Kallen B, Koster M. No paternal effect on monozygotic twinning in the Swedish Twin Registry. Twin Res 1998;1:212–5.
[124] Torlopp A, Khan MA, Oliveira NM, Lekk I, Soto-Jimenez LM, Sosinsky A, Stern CD. The transcription factor Pitx2 positions the embryonic axis and regulates twinning. Elife 2014;3:e03743.
[125] Parisi P, Gatti M, Prinzi G, Caperna G. Familial incidence of twinning. Nature 1983;304:626–8.
[126] Meulemans WJ, Lewis CM, Boomsma DI, Derom CA, Van den Berghe H, Orlebeke JF, Vlietinck RF, Derom RM. Genetic modelling of dizygotic twinning in pedigrees of spontaneous dizygotic twins. Am J Med Genet 1996;61:258–63.

[127] Lewis CM, Healey SC, Martin NG. Genetic contribution to DZ twinning. Am J Med Genet 1996;61:237–46.
[128] Samra JS, Hampton N, Fitzgibbon MN, Obhrai MS. The second twin. Lancet 1990;336:883.
[129] Healey SC, Duffy DL, Martin NG, Turner G. Is fragile X syndrome a risk factor for dizygotic twinning? Am J Med Genet 1997;72:245–6.
[130] Montgomery GW, McNatty KP, Davis GH. Physiology and molecular genetics of mutations that increase ovulation rate in sheep. Endocr Rev 1992;13:309–28.
[131] Galloway SM, McNatty KP, Cambridge LM, Laitinen MP, Juengel JL, Jokiranta TS, McLaren RJ, Luiro K, Dodds KG, Montgomery GW, Beattie AE, Davis GH, Ritvos O. Mutations in an oocyte-derived growth factor gene (BMP15) cause increased ovulation rate and infertility in a dosage-sensitive manner. Nat Genet 2000;25:279–83.
[132] Hanrahan JP, Gregan SM, Mulsant P, Mullen M, Davis GH, Powell R, Galloway SM. Mutations in the genes for oocyte-derived growth factors GDF9 and BMP15 are associated with both increased ovulation rate and sterility in Cambridge and Belclare sheep (*Ovis aries*). Biol Reprod 2004;70:900–9.
[133] Montgomery GW, Zhao ZZ, Marsh AJ, Mayne R, Treloar SA, James M, Martin NG, Boomsma DI, Duffy DL. A deletion mutation in GDF9 in sisters with spontaneous DZ twins. Twin Res 2004;7:548–55.
[134] Palmer JS, Zhao ZZ, Hoekstra C, Hayward NK, Webb PM, Whiteman DC, Martin NG, Boomsma DI, Duffy DL, Montgomery GW. Novel variants in growth differentiation factor 9 in mothers of dizygotic twins. J Clin Endocrinol Metab 2006;91:4713–6.
[135] Zhao ZZ, Painter JN, Palmer JS, Webb PM, Hayward NK, Whiteman DC, Boomsma DI, Martin NG, Duffy DL, Montgomery GW. Variation in bone morphogenetic protein 15 is not associated with spontaneous human dizygotic twinning. Hum Reprod 2008;23:2372–9.
[136] Luong HT, Chaplin J, McRae AF, Medland SE, Willemsen G, Nyholt DR, Henders AK, Hoekstra C, Duffy DL, Martin NG, Boomsma DI, Montgomery GW, Painter JN. Variation in BMPR1B, TGFRB1 and BMPR2 and control of dizygotic twinning. Twin Res Hum Genet 2011;14:408–16.
[137] Dixit H, Rao L, Padmalatha V, Raseswari T, Kapu AK, Panda B, Murthy K, Tosh D, Nallari P, Deenadayal M, Gupta N, Chakrabarthy B, Singh L. Genes governing premature ovarian failure. Reprod Biomed Online 2010;20:724–40.
[138] Al-Hendy A, Moshynska O, Saxena A, Feyles V. Association between mutations of the follicle-stimulating-hormone receptor and repeated twinning. Lancet 2000;356:914.
[139] Painter JN, Willemsen G, Nyholt D, Hoekstra C, Duffy DL, Henders AK, Wallace L, Healey S, Cannon-Albright LA, Skolnick M, Martin NG, Boomsma DI, Montgomery GW. A genome wide linkage scan for dizygotic twinning in 525 families of mothers of dizygotic twins. Hum Reprod 2010;25:1569–80.
[140] Montgomery GW, Duffy DL, Hall J, Kudo M, Martin NG, Hsueh AJ. Mutations in the follicle-stimulating hormone receptor and familial dizygotic twinning. Lancet 2001;357:773–4.
[141] Mbarek H, Steinberg S, Nyholt DR, Gordon SD, Miller MB, McRae AF, Hottenga JJ, Day FR, Willemsen G, de Geus EJ, Davies GE, Martin HC, Penninx BW, Jansen R, McAloney K, Vink JM, Kaprio J, Plomin R, Spector TD, Magnusson PK, Reversade B, Harris RA, Aagaard K, Kristjansson RP, Olafsson I, Eyjolfsson GI, Sigurdardottir O, Iacono WG, Lambalk CB, Montgomery GW, McGue M, Ong KK, Perry JR, Martin NG, Stefansson H, Stefansson K, Boomsma DI. Identification of common genetic variants influencing spontaneous dizygotic twinning and female fertility. Am J Hum Genet 2016;98:898–908.
[142] Boomsma DI, Frants RR, Bank RA, Martin NG. Protease inhibitor (Pi) locus, fertility and twinning. Hum Genet 1992;89:329–32.
[143] Lieberman J, Borhani NO, Feinleib M. Twinning as a heterozygous advantage for alpha1-antitrypsin deficiency. Prog Clin Biol Res 1978;24 Pt B:45–54.
[144] Jauniaux E, Elkazen N, Leroy F, Wilkin P, Rodesch F, Hustin J. Clinical and morphologic aspects of the vanishing twin phenomenon. Obstet Gynecol 1988;72:577–81.
[145] Fryns JP. The female and the fragile X. A study of 144 obligate female carriers. Am J Med Genet 1986;23:157–69.
[146] Kenneson A, Warren ST. The female and the fragile X reviewed. Semin Reprod Med 2001;19:159–65.
[147] Derom C, Jawaheer D, Chen WV, McBride KL, Xiao X, Amos C, Gregersen PK, Vlietinck R. Genome-wide linkage scan for spontaneous DZ twinning. Eur J Hum Genet 2006;14:117–22.
[148] Duffy D, Montgomery G, Treloar S, Birley A, Kirk K, Boomsma D, Beem L, de Geus E, Slagboom E, Knighton J, Reed P, Martin N. IBD sharing around the PPARG locus is not increased in dizygotic twins or their mothers. Nat Genet 2001;28:315.
[149] Busjahn A, Knoblauch H, Faulhaber HD, Aydin A, Uhlmann R, Tuomilehto J, Kaprio J, Jedrusik P, Januszewicz A, Strelau J, Schuster H, Luft FC, Muller-Myhsok B. A region on chromosome 3 is linked to dizygotic twinning. Nat Genet 2000;26:398–9.

[150] Valleix S, Vinciguerra C, Lavergne JM, Leuer M, Delpech M, Negrier C. Skewed X-chromosome inactivation in monochorionic diamniotic twin sisters results in severe and mild hemophilia A. Blood 2002;100:3034–6.
[151] Goodship J, Carter J, Burn J. X-inactivation patterns in monozygotic and dizygotic female twins. Am J Med Genet 1996;61:205–8.
[152] Tan SS, Williams EA, Tam PP. X-chromosome inactivation occurs at different times in different tissues of the post-implantation mouse embryo. Nat Genet 1993;3:170–4.
[153] Bamforth F, Machin G, Innes M. X-chromosome inactivation is mostly random in placental tissues of female monozygotic twins and triplets. Am J Med Genet 1996;61:209–15.
[154] Hall JG. Twinning: mechanisms and genetic implications. Curr Opin Genet Dev 1996;6:343–7.
[155] Shur N. The genetics of twinning: from splitting eggs to breaking paradigms. Am J Med Genet C Semin Med Genet 2009;151C:105–9.
[156] Weksberg R, Shuman C, Caluseriu O, Smith AC, Fei YL, Nishikawa J, Stockley TL, Best L, Chitayat D, Olney A, Ives E, Schneider A, Bestor TH, Li M, Sadowski P, Squire J. Discordant KCNQ1OT1 imprinting in sets of monozygotic twins discordant for Beckwith-Wiedemann syndrome. Hum Mol Genet 2002;11:1317–25.
[157] Kaufman MH, O'Shea KS. Induction of monozygotic twinning in the mouse. Nature 1978;276:707–8.
[158] Czeizel AE, Metneki J, Dudas I. Higher rate of multiple births after periconceptional vitamin supplementation. N Engl J Med 1994;330:1687–8.
[159] Aston KI, Peterson CM, Carrell DT. Monozygotic twinning associated with assisted reproductive technologies: a review. Reproduction 2008;136:377–86.
[160] Sills ES, Tucker MJ, Palermo GD. Assisted reproductive technologies and monozygous twins: implications for future study and clinical practice. Twin Res 2000;3:217–23.
[161] Elizur SE, Levron J, Shrim A, Sivan E, Dor J, Shulman A. Monozygotic twinning is not associated with zona pellucida micromanipulation procedures but increases with high-order multiple pregnancies. Fertil Steril 2004;82:500–1.
[162] Stockard CR. Developmental rate and structural expression: an experimental study of twins, 'double monsters' and single deformities, and the interaction among embryonic organs during their origin and development. Am J Anat 1921;28:115–277.
[163] Storrs EE, Williams RJ. A study of monozygous quadruplet armadillos in relation to mammalian inheritance. Proc Natl Acad Sci U S A 1968;60:910–4.
[164] Bulmer M. The biology of twinning in man. Oxford: Clarendon Press; 1970.
[165] Yovich JL, Stanger JD, Grauaug A, Barter RA, Lunay G, Dawkins RL, Mulcahy MT. Monozygotic twins from in vitro fertilization. Fertil Steril 1984;41:833–7.
[166] Edwards RG, Mettler L, Walters DE. Identical twins and in vitro fertilization. J In Vitro Fert Embryo Transf 1986;3:114–7.
[167] Boklage CE. On the timing of monozygotic twinning events. In: Gedda L, Parisi P, Nance WE, editors. Twin research. New York: Alan R. Liss; 1981. p. 155–65.
[168] Boklage CE. The organization of the oocyte and embryogenesis in twinning and fusion malformations. Acta Genet Med Gemellol 1987a;36:421–31.
[169] Boklage CE. Twinning, nonrighthandedness, and fusion malformations: evidence for heritable causal elements held in common. Am J Med Genet 1987b;28:67–84.
[170] Knopman JM, Krey LC, Oh C, Lee J, McCaffrey C, Noyes N. What makes them split? Identifying risk factors that lead to monozygotic twins after in vitro fertilization. Fertil Steril 2014;102:82–9.
[171] Payne D, Okuda A, Wakatsuki Y, Takeshita C, Iwata K, Shimura T, Yumoto K, Ueno Y, Flaherty S, Mio Y. Time-lapse recording identifies human blastocysts at risk of producing monozygotic twins. Hum Reprod 2007;22:i9–10.
[172] Niimura S. Time-lapse videomicrographic analyses of contractions in mouse blastocysts. J Reprod Dev 2003;49:413–23.
[173] Martin N, Boomsma D, Machin G. A twin-pronged attack on complex traits. Nat Genet 1997;17:387–92.
[174] Silva S, Martins Y, Matias A, Blickstein I. Why are monozygotic twins different? J Perinat Med 2011;39:195–202.
[175] Zwijnenburg PJ, Meijers-Heijboer H, Boomsma DI. Identical but not the same: the value of discordant monozygotic twins in genetic research. Am J Med Genet B Neuropsychiatr Genet 2010;153B:1134–49.
[176] Gilgenkrantz S, Janot C. Monozygotic twins discordant for trisomy 21 or chimeric dizygotic twins? Am J Med Genet 1983;15:159–60.
[177] Marcus-Soekarman D, Hamers G, Velzeboer S, Nijhuis J, Loneus WH, Herbergs J, de Die-Smulders C, Schrander-Stumpel C, Engelen J. Mosaic trisomy 11p in monozygotic twins with discordant clinical phenotypes. Am J Med Genet 2004;124A:288–91.
[178] Rogers JG, Voullaire L, Gold H. Monozygotic twins discordant for trisomy 21. Am J Med Genet 1982;11:143–6.
[179] West PM, Love DR, Stapleton PM, Winship IM. Paternal uniparental disomy in monozygotic twins discordant for hemihypertrophy. J Med Genet 2003;40:223–6.

[180] Wakita Y, Narahara K, Tsuji K, Yokoyama Y, Ninomiya S, Murakami R, Kikkawa K, Seino Y. De novo complex chromosome rearrangement in identical twins with multiple congenital anomalies. Hum Genet 1992;88:596–8.

[181] Kruyer H, Mila M, Glover G, Carbonell P, Ballesta F, Estivill X. Fragile X syndrome and the (CGG)n mutation: two families with discordant MZ twins. Am J Hum Genet 1994;54:437–42.

[182] Kondo S, Schutte BC, Richardson RJ, Bjork BC, Knight AS, Watanabe Y, Howard E, de Lima RL, Daack-Hirsch S, Sander A, McDonald-McGinn DM, Zackai EH, Lammer EJ, Aylsworth AS, Ardinger HH, Lidral AC, Pober BR, Moreno L, Arcos-Burgos M, Valencia C, Houdayer C, Bahuau M, Moretti-Ferreira D, Richieri-Costa A, Dixon MJ, Murray JC. Mutations in IRF6 cause Van der Woude and popliteal pterygium syndromes. Nat Genet 2002;32:285–9.

[183] Robertson SP, Thompson S, Morgan T, Holder-Espinasse M, Martinot-Duquenoy V, Wilkie AO, Manouvrier-Hanu S. Postzygotic mutation and germline mosaicism in the otopalatodigital syndrome spectrum disorders. Eur J Hum Genet 2006;14:549–54.

[184] Biousse V, Brown MD, Newman NJ, Allen JC, Rosenfeld J, Meola G, Wallace DC. De novo 14484 mitochondrial DNA mutation in monozygotic twins discordant for Leber's hereditary optic neuropathy. Neurology 1997;49:1136–8.

[185] Blakely EL, He L, Taylor RW, Chinnery PF, Lightowlers RN, Schaefer AM, Turnbull DM. Mitochondrial DNA deletion in "identical" twin brothers. J Med Genet 2004;41:e19.

[186] Shaffer LG, Kashork CD, Bacino CA, Benke PJ. Caution: telomere crossing. Am J Med Genet 1999;87:278–80.

[187] Graakjaer J, Bischoff C, Korsholm L, Holstebroe S, Vach W, Bohr VA, Christensen K, Kolvraa S. The pattern of chromosome-specific variations in telomere length in humans is determined by inherited, telomere-near factors and is maintained throughout life. Mech Ageing Dev 2003;124:629–40.

[188] Vadlamudi L, Dibbens LM, Lawrence KM, Iona X, McMahon JM, Murrell W, Mackay-Sim A, Scheffer IE, Berkovic SF. Timing of de novo mutagenesis–a twin study of sodium-channel mutations. N Engl J Med 2010;363:1335–40.

[189] Abdellaoui A, Ehli EA, Hottenga JJ, Weber Z, Mbarek H, Willemsen G, van Beijsterveldt T, Brooks A, Hudziak JJ, Sullivan PF, de Geus EJ, Davies GE, Boomsma DI. CNV concordance in 1,097 MZ twin pairs. Twin Res Hum Genet 2015;18:1–12.

[190] Li R, Montpetit A, Rousseau M, Wu SY, Greenwood CM, Spector TD, Pollak M, Polychronakos C, Richards JB. Somatic point mutations occurring early in development: a monozygotic twin study. J Med Genet 2014;51:28–34.

[191] McRae AF, Visscher PM, Montgomery GW, Martin NG. Large autosomal copy-number differences within unselected monozygotic twin pairs are rare. Twin Res Hum Genet 2015;18:13–8.

[192] Breckpot J, Thienpont B, Gewillig M, Allegaert K, Vermeesch JR, Devriendt K. Differences in copy number variation between discordant monozygotic twins as a model for exploring chromosomal mosaicism in congenital heart defects. Mol Syndromol 2012;2:81–7.

[193] Castellani CA, Awamleh Z, Melka MG, O'Reilly RL, Singh SM. Copy number variation distribution in six monozygotic twin pairs discordant for schizophrenia. Twin Res Hum Genet 2014;17:108–20.

[194] Ehli EA, Abdellaoui A, Hu Y, Hottenga JJ, Kattenberg M, van Beijsterveldt T, Bartels M, Althoff RR, Xiao X, Scheet P, de Geus EJ, Hudziak JJ, Boomsma DI, Davies GE. De novo and inherited CNVs in MZ twin pairs selected for discordance and concordance on Attention Problems. Eur J Hum Genet 2012;20:1037–43.

[195] Lu P, Wang P, Li L, Xu C, Liu JC, Guo X, He D, Huang H, Cheng Z. Exomic and epigenomic analyses in a pair of monozygotic twins discordant for cryptorchidism. Twin Res Hum Genet 2017;20:349–54.

[196] Petersen BS, Spehlmann ME, Raedler A, Stade B, Thomsen I, Rabionet R, Rosenstiel P, Schreiber S, Franke A. Whole genome and exome sequencing of monozygotic twins discordant for Crohn's disease. BMC Genom 2014;15:564.

[197] Zhang R, Thiele H, Bartmann P, Hilger AC, Berg C, Herberg U, Klingmuller D, Nurnberg P, Ludwig M, Reutter H. Whole-exome sequencing in nine monozygotic discordant twins. Twin Res Hum Genet 2016;19:60–5.

[198] Morimoto Y, Ono S, Imamura A, Okazaki Y, Kinoshita A, Mishima H, Nakane H, Ozawa H, Yoshiura KI, Kurotaki N. Deep sequencing reveals variations in somatic cell mosaic mutations between monozygotic twins with discordant psychiatric disease. Hum Genome Var 2017;4:17032.

[199] Meltz Steinberg K, Nicholas TJ, Koboldt DC, Yu B, Mardis E, Pamphlett R. Whole genome analyses reveal no pathogenetic single nucleotide or structural differences between monozygotic twins discordant for amyotrophic lateral sclerosis. Amyotroph Lateral Scler Frontotemporal Degener 2015;16:385–92.

[200] Tang J, Fan Y, Li H, Xiang Q, Zhang DF, Li Z, He Y, Liao Y, Wang Y, He F, Zhang F, Shugart YY, Liu C, Tang Y, Chan RCK, Wang CY, Yao YG, Chen X. Whole-genome sequencing of monozygotic twins discordant for schizophrenia indicates multiple genetic risk factors for schizophrenia. J Genet Genomics 2017;44:295–306.

[201] Francioli LC, Polak PP, Koren A, Menelaou A, Chun S, Renkens I, Genome of the Netherlands C, van Duijn CM, Swertz M, Wijmenga C, van Ommen G, Slagboom PE, Boomsma DI, Ye K, Guryev V, Arndt PF, Kloosterman WP, de Bakker PI, Sunyaev SR. Genome-wide patterns and properties of de novo mutations in humans. Nat Genet 2015;47:822–6.

[202] Weber-Lehmann J, Schilling E, Gradl G, Richter DC, Wiehler J, Rolf B. Finding the needle in the haystack: differentiating "identical" twins in paternity testing and forensics by ultra-deep next generation sequencing. Forensic Sci Int Genet 2014;9:42–6.

[203] Mikeska T, Craig JM. DNA methylation biomarkers: cancer and beyond. Genes 2014;5:821–64.

[204] Bell JT, Saffery R. The value of twins in epigenetic epidemiology. Int J Epidemiol 2012;41:140–50.

[205] Busche S, Shao X, Caron M, Kwan T, Allum F, Cheung WA, Ge B, Westfall S, Simon MM, Multiple Tissue Human Expression R, Barrett A, Bell JT, McCarthy MI, Deloukas P, Blanchette M, Bourque G, Spector TD, Lathrop M, Pastinen T, Grundberg E. Population whole-genome bisulfite sequencing across two tissues highlights the environment as the principal source of human methylome variation. Genome Biol 2015;16:290.

[206] McRae AF, Powell JE, Henders AK, Bowdler L, Hemani G, Shah S, Painter JN, Martin NG, Visscher PM, Montgomery GW. Contribution of genetic variation to transgenerational inheritance of DNA methylation. Genome Biol 2014;15:R73.

[207] van Dongen J, Nivard MG, Willemsen G, Hottenga JJ, Helmer Q, Dolan CV, Ehli EA, Davies GE, van Iterson M, Breeze CE, Beck S, Consortium B, Suchiman HE, Jansen R, van Meurs JB, Heijmans BT, Slagboom PE, Boomsma DI. Genetic and environmental influences interact with age and sex in shaping the human methylome. Nat Commun 2016;7:11115.

[208] Yet I, Tsai PC, Castillo-Fernandez JE, Carnero-Montoro E, Bell JT. Genetic and environmental impacts on DNA methylation levels in twins. Epigenomics 2016;8:105–17.

[209] Ollikainen M, Smith KR, Joo EJ, Ng HK, Andronikos R, Novakovic B, Abdul Aziz NK, Carlin JB, Morley R, Saffery R, Craig JM. DNA methylation analysis of multiple tissues from newborn twins reveals both genetic and intrauterine components to variation in the human neonatal epigenome. Hum Mol Genet 2010;19:4176–88.

[210] Gordon L, Joo JE, Powell JE, Ollikainen M, Novakovic B, Li X, Andronikos R, Cruickshank MN, Conneely KN, Smith AK, Alisch RS, Morley R, Visscher PM, Craig JM, Saffery R. Neonatal DNA methylation profile in human twins is specified by a complex interplay between intrauterine environmental and genetic factors, subject to tissue-specific influence. Genome Res 2012;22:1395–406.

[211] Gordon L, Joo JH, Andronikos R, Ollikainen M, Wallace EM, Umstad MP, Permezel M, Oshlack A, Morley R, Carlin JB, Saffery R, Smyth GK, Craig JM. Expression discordance of monozygotic twins at birth: effect of intrauterine environment and a possible mechanism for fetal programming. Epigenetics 2011;6:579–92.

[212] Oates NA, van Vliet J, Duffy DL, Kroes HY, Martin NG, Boomsma DI, Campbell M, Coulthard MG, Whitelaw E, Chong S. Increased DNA methylation at the AXIN1 gene in a monozygotic twin from a pair discordant for a caudal duplication anomaly. Am J Hum Genet 2006;79:155–62.

[213] Martin GM. Epigenetic drift in aging identical twins. Proc Natl Acad Sci U S A 2005;102:10413–4.

[214] Fraga MF, Ballestar E, Paz MF, Ropero S, Setien F, Ballestar ML, Heine-Suner D, Cigudosa JC, Urioste M, Benitez J, Boix-Chornet M, Sanchez-Aguilera A, Ling C, Carlsson E, Poulsen P, Vaag A, Stephan Z, Spector TD, Wu YZ, Plass C, Esteller M. Epigenetic differences arise during the lifetime of monozygotic twins. Proc Natl Acad Sci U S A 2005;102:10604–9.

[215] Martino D, Loke YJ, Gordon L, Ollikainen M, Cruickshank MN, Saffery R, Craig JM. Longitudinal, genome-scale analysis of DNA methylation in twins from birth to 18 months of age reveals rapid epigenetic change in early life and pair-specific effects of discordance. Genome Biol 2013;14:R42.

[216] Zhang N, Zhao S, Zhang SH, Chen J, Lu D, Shen M, Li C. Intra-monozygotic twin pair discordance and longitudinal variation of whole-genome scale DNA methylation in adults. PLoS One 2015;10:e0135022.

[217] Jones MJ, Goodman SJ, Kobor MS. DNA methylation and healthy human aging. Aging Cell 2015;14:924–32.

[218] Jylhava J, Pedersen NL, Hagg S. Biological age predictors. EBioMedicine 2017;21:29–36.

[219] Day K, Waite LL, Thalacker-Mercer A, West A, Bamman MM, Brooks JD, Myers RM, Absher D. Differential DNA methylation with age displays both common and dynamic features across human tissues that are influenced by CpG landscape. Genome Biol 2013;14:R102.

[220] Tan Q, Heijmans BT, Hjelmborg JV, Soerensen M, Christensen K, Christiansen L. Epigenetic drift in the aging genome: a ten-year follow-up in an elderly twin cohort. Int J Epidemiol 2016;45:1146–1158.
[221] Gartner K. A third component causing random variability beside environment and genotype. A reason for the limited success of a 30 year long effort to standardize laboratory animals? Int J Epidemiol 2012;41:335–41.
[222] Pujadas E, Feinberg AP. Regulated noise in the epigenetic landscape of development and disease. Cell 2012;148:1123–31.
[223] Whitelaw NC, Chong S, Whitelaw E. Tuning in to noise: epigenetics and intangible variation. Dev Cell 2010;19:649–50.
[224] Chen M, Baumbach J, Vandin F, Rottger R, Barbosa E, Dong M, Frost M, Christiansen L, Tan Q. Differentially methylated genomic regions in birth-weight discordant twin pairs. Ann Hum Genet 2016;80:81–7.
[225] Tsai PC, Van Dongen J, Tan Q, Willemsen G, Christiansen L, Boomsma DI, Spector TD, Valdes AM, Bell JT. DNA methylation changes in the IGF1R gene in birth weight discordant adult monozygotic twins. Twin Res Hum Genet 2015;18:635–46.
[226] Chiarella J, Tremblay RE, Szyf M, Provencal N, Booij L. Impact of early environment on children's mental health: lessons from DNA methylation studies with monozygotic twins. Twin Res Hum Genet 2015:1–12.
[227] Craig JM. Epigenetics in twin studies. Med Epigenetics 2013;1:70–7.
[228] Barker DJ, Osmond C. Low birth weight and hypertension. BMJ 1988;297:134–5.
[229] Gluckman PD, Hanson MA, Buklijas T. A conceptual framework for the developmental origins of health and disease. J Dev Orig Health Dis 2010;1:6–18.
[230] Woo Baidal JA, Locks LM, Cheng ER, Blake-Lamb TL, Perkins ME, Taveras EM. Risk factors for childhood obesity in the first 1,000 Days: a systematic review. Am J Prev Med 2016;50:761–79.
[231] Rappaport SM. Genetic factors are not the major causes of chronic diseases. PLoS One 2016;11:e0154387.
[232] Machin GA. Some causes of genotypic and phenotypic discordance in monozygotic twin pairs. Am J Med Genet 1996;61:216–28.
[233] Plomin R. Commentary: why are children in the same family so different? Non-shared environment three decades later. Int J Epidemiol 2011;40:582–92.
[234] Stromswold K. Why aren't identical twins linguistically identical? Genetic, prenatal and postnatal factors. Cognition 2006;101:333–84.
[235] Dwyer T, Blizzard L, Morley R, Ponsonby AL. Within pair association between birth weight and blood pressure at age 8 in twins from a cohort study. Br Med J 1999;319:1325–9.
[236] Bekhit MT, Greenwood PA, Warren R, Aarons E, Jauniaux E. In utero treatment of severe fetal anaemia due to parvovirus B19 in one fetus in a twin pregnancy–a case report and literature review. Fetal Diagn Ther 2009;25:153–7.
[237] Dickinson JE, Keil AD, Charles AK. Discordant fetal infection for parvovirus B19 in a dichorionic twin pregnancy. Twin Res Hum Genet 2006;9:456–9.
[238] Jamieson DJ, Read JS, Kourtis AP, Durant TM, Lampe MA, Dominguez KL. Cesarean delivery for HIV-infected women: recommendations and controversies. Am J Obstet Gynecol 2007;197:S96–100.
[239] Pimentel JD, Szymanski LJ, Samuel LP, Meier FA. Discordant Streptococcus agalactiae (Group B streptococcus) gestational infection in monochorionic/diamniotic and dichorionic/diamniotic twins. Fetal Pediatr Pathol 2012;31:176–83.
[240] Schiesser M, Sergi C, Enders M, Maul H, Schnitzler P. Discordant outcomes in a case of parvovirus b19 transmission into both dichorionic twins. Twin Res Hum Genet 2009;12:175–9.
[241] Pawelec G, Goldeck D, Derhovanessian E. Inflammation, ageing and chronic disease. Curr Opin Immunol 2014;29:23–8.
[242] Tabas I, Glass CK. Anti-inflammatory therapy in chronic disease: challenges and opportunities. Science 2013;339:166–72.
[243] Brodin P, Jojic V, Gao T, Bhattacharya S, Angel CJ, Furman D, Shen-Orr S, Dekker CL, Swan GE, Butte AJ, Maecker HT, Davis MM. Variation in the human immune system is largely driven by non-heritable influences. Cell 2015;160:37–47.
[244] Jacques SM, Qureshi F. Chronic villitis of unknown etiology in twin gestations. Pediatr Pathol 1994;14:575–84.
[245] Phung DT, Blickstein I, Goldman RD, Machin GA, LoSasso RD, Keith LG. The Northwestern Twin Chorionicity Study: I. Discordant inflammatory findings that are related to chorionicity in presenting versus nonpresenting twins. Am J Obstet Gynecol 2002;186:1041–5.
[246] Hopper JL. Why 'common environmental effects' are so uncommon in the literature. In: Spector TD, Snieder H, MacGregor AJ, editors. Advances in twin and sib-pair analysis. London: Oxford University Press; 2000. p. 151–65.
[247] Lange K, Weeks D, Boehnke M. Programs for pedigree analysis: MENDEL, Fisher and dGene. Genet Epidemiol 1988;5:471–2.
[248] Rijsdijk FV, Sham PC. Analytic approaches to twin data using structural equation models. Brief Bioinform 2002;3:119–33.

[249] Fisher RA. Limits to intensive production in animals. Br Agric Bull 1951;4:217–8.
[250] Hopper JL. Heritability, Encyclopedia of Biostatistics. In: Amitage P, Colton T, editors. John Wiley & Sons, Ltd; 2005.
[251] Witte JS, Carlin JB, Hopper JL. Likelihood-based approach to estimating twin concordance for dichotomous traits. Genet Epidemiol 1999;16:290–304.
[252] Plomin R, DeFries JC, Knopik VS, Neiderhiser JM. Behavioral Genetics (6th Edition) Worth Publishers. New York, 2013, ISBN-10: 1-4292-4215-9.
[253] Hannah MC, Hopper JL, Mathews JD. Twin concordance for a binary trait. I. Statistical models illustrated with data on drinking status. Acta Genet Med Gemellol 1983;32:127–37.
[254] Munteanu SE, Menz HB, Wark JD, Christie JJ, Scurrah KJ, Bui M, Erbas B, Hopper JL, Wluka AE. Hallux valgus, by nature or nurture? A twin study. Arthritis Care Res 2016.
[255] Burton P, Tiller K, Gurrin L, Cookson W, Musk A, Palmer L. Genetic variance components analysis for binary phenotypes using generalized linear mixed models (GLMMs) and Gibbs sampling. Genet Epidemiol 1999;17(118):140.
[256] Nyholt DR, Gillespie NA, Heath AC, Martin NG. Genetic basis of male pattern baldness. J Invest Dermatol 2003;121:1561–4.
[257] Zaloumis SG, Scurrah KJ, Harrap SB, Ellis JA, Gurrin LC. Non-proportional odds multivariate logistic regression of ordinal family data. Biom J 2015;57:286–303.
[258] Scurrah KJ, Palmer LJ, Burton PR. Variance components analysis for pedigree-based censored survival data using generalized linear mixed models (GLMMs) and Gibbs sampling in BUGS. Genet Epidemiol 2000;19:127–48.
[259] Yashin I, Vaupel J, Iachine I. Correlated individual frailty: an advantageous approach to survival analysis of bivariate data. Math Popul Stud 1995;5:145–59.
[260] Hamilton AS, Mack TM. Puberty and genetic susceptibility to breast cancer in a case-control study in twins. N Engl J Med 2003;348:2313–22.
[261] Goldberg J, Fischer M. Co-twin control methods, encyclopedia of statistics in behavioral science. John Wiley & Sons, Ltd; 2005.
[262] Nowson CA, Green RM, Hopper JL, Sherwin AJ, Young D, Kaymakci B, Guest CS, Smid M, Larkins RG, Wark JD. A co-twin study of the effect of calcium supplementation on bone density during adolescence. Osteoporos Int 1997;7:219–25.
[263] Sjölander A, Frisell T, Öberg S. Causal interpretation of between-within models for twin research, epidemiologic methods. 2012. p. 217.
[264] Sjolander A, Lichtenstein P, Larsson H, Pawitan Y. Between-within models for survival analysis. Stat Med 2013;32:3067–76.
[265] Carlin JB, Gurrin LC, Sterne JA, Morley R, Dwyer T. Regression models for twin studies: a critical review. Int J Epidemiol 2005;34:1089–99.
[266] Hopper JL, Seeman E. The bone density of female twins discordant for tobacco use. N Engl J Med 1994;330:387–92.
[267] Staton S, Thorpe K, Thompson C, Danby S. To separate or not to separate? Parental decision-making regarding the separation of twins in the early years of schooling. J Early Child Res 2012;10:196–208.
[268] Goldman SM, Quinlan PJ, Ross GW, Marras C, Meng C, Bhudhikanok GS, Comyns K, Korell M, Chade AR, Kasten M, Priestley B, Chou KL, Fernandez HH, Cambi F, Langston JW, Tanner CM. Solvent exposures and Parkinson's disease risk in twins. Ann Neurol 2012;71:776–84.
[269] Gatz M, Harris JR, Kaprio J, McGue M, Smith NL, Snieder H, Spiro 3rd A, Butler DA, Institute of Medicine Committee on Twins S. Cohort profile: the national Academy of sciences-national research Council twin registry (NAS-NRC twin registry). Int J Epidemiol 2015;44:819–25.
[270] Hopper JL, Foley DL, White PA, Pollaers V. Australian twin registry: 30 years of progress. Twin Res Hum Genet 2013;16:34–42.
[271] Moayyeri A, Hammond CJ, Hart DJ, Spector TD. The UK adult twin registry (TwinsUK resource). Twin Res Hum Genet 2013;16:144–9.
[272] Davis OSP, Band G, Pirinen M, Haworth CMA, Meaburn EL, Kovas Y, Harlaar N, Docherty SJ, Hanscombe KB, Trzaskowski M, Curtis CJC, Strange A, Freeman C, Bellenguez C, Su Z, Pearson R, Vukcevic D, Langford C, Deloukas P, Hunt S, Gray E, Dronov S, Potter SC, Tashakkori-Ghanbaria A, Edkins S, Bumpstead SJ, Blackwell JM, Bramon E, Brown MA, Casas JP, Corvin A, Duncanson A, Jankowski JAZ, Markus HS, Mathew CG, Palmer CNA, Rautanen A, Sawcer SJ, Trembath RC, Viswanathan AC, Wood NW, Barroso I, Peltonen L, Dale PS, Petrill SA, Schalkwyk LS, Craig IW, Lewis CM, Price TS, Donnelly P, Plomin R, Spencer CCA. The correlation between reading and mathematics ability at age twelve has a substantial genetic component. Nat Commun 2014;5:4204.
[273] Harrap SB, Stebbing M, Hopper JL, Hoang HN, Giles GG. Familial patterns of covariation for cardiovascular risk factors in adults: the Victorian Family Heart Study. Am J Epidemiol 2000;152:704–15.
[274] StataCorp. Stata statistical software: release 14. StataCorp LLC College Station, T; 2015.

[275] Hur YM, Craig JM. Twin registries worldwide: an important resource for scientific research. Twin Res Hum Genet 2013;16:1–12.

[276] Kaprio J. The Finnish twin cohort study: an update. Twin Res Hum Genet 2013;16:157–62.

[277] Skytthe A, Christiansen L, Kyvik KO, Bodker FL, Hvidberg L, Petersen I, Nielsen MM, Bingley P, Hjelmborg J, Tan Q, Holm NV, Vaupel JW, McGue M, Christensen K. The Danish Twin Registry: linking surveys, national registers, and biological information. Twin Res Hum Genet 2013;16:104–11.

[278] Strachan E, Hunt C, Afari N, Duncan G, Noonan C, Schur E, Watson N, Goldberg J, Buchwald D. University of Washington Twin Registry: poised for the next generation of twin research. Twin Res Hum Genet 2013;16:455–62.

[279] Ordonana JR, Rebollo-Mesa I, Carrillo E, Colodro-Conde L, Garcia-Palomo FJ, Gonzalez-Javier F, Sanchez-Romera JF, Aznar Oviedo JM, de Pancorbo MM, Perez-Riquelme F. The Murcia Twin Registry: a population-based registry of adult multiples in Spain. Twin Res Hum Genet 2013;16:302–6.

[280] Ferreira PH, Oliveira VC, Junqueira DR, Cisneros LC, Ferreira LC, Murphy K, Ordoñana JR, Hopper JL, Teixeira-Salmela LF. The Brazilian twin registry. Twin Res Hum Genet 2016;19:687–91.

[281] Moayyeri A, Hammond CJ, Valdes AM, Spector TD. Cohort Profile: TwinsUK and healthy ageing twin study. Int J Epidemiol 2013;42:76–85.

[282] Buchwald D, Kaprio J, Hopper JL, Sung J, Goldberg J, Fortier I, Busjhan A, Sumathipala A, Cozen W, Mack T, Craig JM, Harris JR. International network of twin registries (INTR): building a platform for international collaboration. Twin Res Hum Genet 2014;17:574–7.

[283] Silventoinen K, Jelenkovic A, Sund R, Honda C, Aaltonen S, Yokoyama Y, Kaprio J. The CODATwins Project: The Cohort Description of Collaborative Project of Development of Anthropometrical Measures in Twins to Study Macro-Environmental Variation in Genetic and Environmental Effects on Anthropometric Traits. Twin Res Hum Gen 2015;18:348–60.

15

# The Biological Basis of Aging: Implications for Medical Genetics

*Junko Oshima[1], Fuki M. Hisama[2], George M. Martin[1]*

[1]Department of Pathology, University of Washington, Seattle, WA, United States
[2]Division of Medical Genetics, Department of Medicine, University of Washington, Seattle, WA, United States

## 15.1 INTRODUCTION

Evolutionary biology has provided a robust theory to explain why we age, but we have much less confidence that we understand how we age, by which we mean the proximal molecular mechanisms of aging. Medical geneticists are in a good position to advance our knowledge of such mechanisms. The first strategy is the time-honored approach of mapping, identifying and characterizing the relevant gene actions underlying important late-onset disorders of aging, such as dementias of the Alzheimer type, atherosclerosis, ocular cataracts, type II diabetes mellitus, osteoporosis, osteoarthritis, and cancer. Significant progress has been made using this strategy, including major advances in our understanding of segmental progeroid syndromes such as the Werner and Hutchinson–Gilford syndromes. The second strategy is to investigate the genetic basis for unusually well-preserved structure and function during the latter half of the usual life span. Unfortunately, medical geneticists have been too preoccupied with disease and, with some notable exceptions (e.g., studies of centenarians), physicians and geneticists have not shown a strong inclination to investigate exceptional well-preserved late-life phenotypes. The new statistical and molecular tools at our disposal are impressive, but they have not been matched by the application of sensitive functional assays [2].

## 15.2 WHAT IS AGING?

Some gerontologists, particularly those interested in the aging of plants, sharply differentiate between the terms aging and senescing. Senescent changes, they would argue, are those structural and functional changes that occur near the end of the life cycle of a cell, tissue, organ, or organism, and are associated with the impending death of the tissue or organism. By contrast, the term aging would be used for any change in structure or function throughout the life cycle. In other words, some would argue that "aging begins at birth." Most gerontologists who work with mammals and human subjects, however, use the two terms more or less interchangeably. While different scholars define human aging in various ways, most include exponential increases of age-specific mortality rate and declines of the physiological functions as general characteristics of human aging processes. These parameters were primarily derived from cross-sectional population studies. Researchers of basic biology of human aging at the organismal level have a major difficulty that stems from the fact that, unlike model animals, humans are genetically heterogeneous and undergo behavioral and environmental changes throughout their lifetimes. Moreover, due to the relatively long life span of humans, the cohort studies of longevity, for example, take many years to complete. One approach to testing the validity of the population studies of human longevity and aging is to compare the findings

Emery and Rimoin's Principles and Practice of Medical Genetics and Genomics: Foundations, https://doi.org/10.1016/B978-0-12-812537-3.00015-9

of other model organisms such as yeast (*Saccharomyces cerevisiae*), fruit fly (*Drosophila melanogaster*), worm (*Caenorhabditis elegans*), and mice, with those from *Homo sapiens* [3]. They continue to create a large segment of the foundations for the progress we have seen in human biology, an example of which is the recent comparative studies on proteostasis under conditions of stress, research that was initiated by the discovery of the remarkable longevities of species of bivalve mollusks [4]. Moreover, their contributions have typically been more incisive, in part because of the experimental tractability of their materials.

None would deny the importance of development in determining the subsequent life history of an organism. A minority of gerontologists in fact embrace the idea that aging is "programmed," despite cogent arguments that defend what is sometimes referred to as the "classical" evolutionary biological theory of aging, a theory (discussed below) that is based upon the nonadaptive nature of biological aging [5]. Most gerontologists are concerned with declines in structure and function that gradually and insidiously unfold after the organism has achieved the young, mature adult phenotype. At the level of populations, these functional declines translate into an exponential increase in the force of mortality over unit time—the hazard function or instantaneous mortality rate [6,7]. This is the famous Gompertz relationship [8]. This was modified by Makeham [9], who included a constant, $A$, to account for kinetic departures presumed to have resulted from causes of death during the early life history that were age-independent. The Gompertz–Makeham equation can thus be given as the sum of two types of mortalities, age-independent and age-dependent, the latter exhibiting exponential kinetics over the adult life span:

$$\mu_x = A + Re^{\alpha x}$$

where $\mu_x$ is the force of mortality at a given age, $x$; $A$ is the Makeham constant; $R$ is the hypothetical value for the force of mortality at birth, the lowest force of mortality, or the $Y$ intercept in a graphic plot of age ($X$-axis) versus force of mortality ($Y$-axis) (Fig. 15.1); $e$ is an exponent; and $\mu$ is a constant representing the slope of the graphical plot (Fig. 15.1). Fig. 15.1 illustrates differing rates of exponential increases in the force of mortality for two noninbred wildtype murine species despite comparable values for $R$. These two species, *Peromyscus leucopus* and *Mus musculus*, are of approximately the same size and were

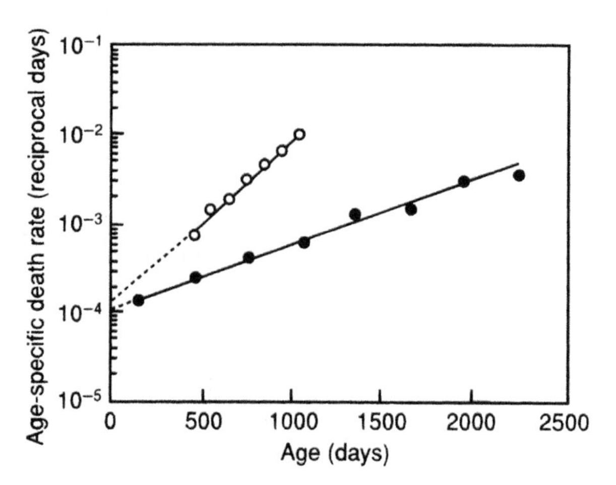


**Figure 15.1** Gompertz function plot of the age-specific mortality rates for combined sexes of two different murine species of contrasting maximum life-span potentials but of comparable size. Both species were wild-type and randomly bred from small cohorts captured near the Argonne National Laboratories (Argonne, IL) by the late George A. Sacher. They were housed under essentially identical conditions (caging, bedding, humidity, temperature, diet) in adjacent animal rooms with no special efforts to establish specific pathogen-free conditions (G. A. Sacher, personal communication to G. M. Martin). The longer-lived species (●), *P. leucopus*, was found to have a maximum life span of about 8 years, approximately twice that of *M. musculus* (m). (From Sacher G.A. Evolution of longevity and survival characteristics in mammals. In: Schneider E.L, editor. The genetics of aging. New York: Plenum Press; 1978.)

housed and fed throughout their lifetimes under identical conditions [10]. The maximum life span of *Peromyscus* sp. was found to be about 8 years, about twice that of *Mus* sp. These data illustrate the importance of genetic factors in the determination of approximate life potential. Given the considerable evolutionary distance between these two species (at least 15 million years), this is not a surprising result. Mortality rates of different human populations in the 20th century also followed the Gompertz–Makeham relationship until they reach very old age [6,7,11].

Experiments employing very large populations of aging cohorts of fruit flies and medflies have been reported as showing dramatic departures from Gompertz kinetics within the oldest cohorts, with apparent decreases in the force of mortality at very advanced ages [12]. Very aged flies, however, may become virtually immobilized and may therefore be protected from environmental hazards related, for example, to attempts at flight. Declines in age-specific mortality rates have also been seen for very aged individuals in human populations [13]. Perhaps this observation might also be explained by behavioral changes (e.g., a more protective environment) in extreme old age. It will nevertheless be prudent to explore various non-Gompertzian models of mortality in human populations that adequately describe mortality at extreme ages. The existence and estimate of the upper limit of the human longevity or maximum life span of humans has been a focus of debates, although it is generally believed to be around 125 years based on the verified longest-lived human [14].

Functional declines can be documented in virtually every organ system starting shortly after sexual maturation. Most physiologic declines, at least in cross-sectional studies, exhibit linear declines, the slopes of which are variable [15]. Declines in the various physiological processes (and underlying molecular and biochemical processes) that maintain optimum functions are likely to "set the stage" for the plethora of late-life disorders and diseases, some 87 of which have recently been tabulated, all of which are subject to both genetic and environmental modulations [3]. Observations of exponential increases in the force of mortality within populations should not lead one to conclude that underlying processes of aging or incidences of geriatric diseases necessarily exhibit exponential kinetics. Consider, for example, the world records of marathon runners, which select for the most robust, physically fit members of our population. This is an attractive assay, as it tests for fitness of multiple organ systems and one's ability to maintain metabolic homeostasis. Declines are observable during the fourth decade, later than what is the case for sprinters. This probably occurs, in part, because it takes considerably more training and experience in perfecting one's optimal pacing for a marathon. It also takes years for the gradual development of such compensatory processes as cardiac muscle hypertrophy. For a remarkably wide range of sports, peak activity occurs during the third decade [16].

Many major diseases of late life, however, show exponential increases in age-specific incidence and prevalence, although there may be slight declines in age-specific incidence at very advanced ages, raising the question of selection for genotypic resistance against specific late-life disorders. Alzheimer disease serves as a good example [17,18]. Fig. 15.2 summarizes the results of several community-based studies of the age-specific incidence of late-life dementias, most of which are due to dementias of the Alzheimer type [19].

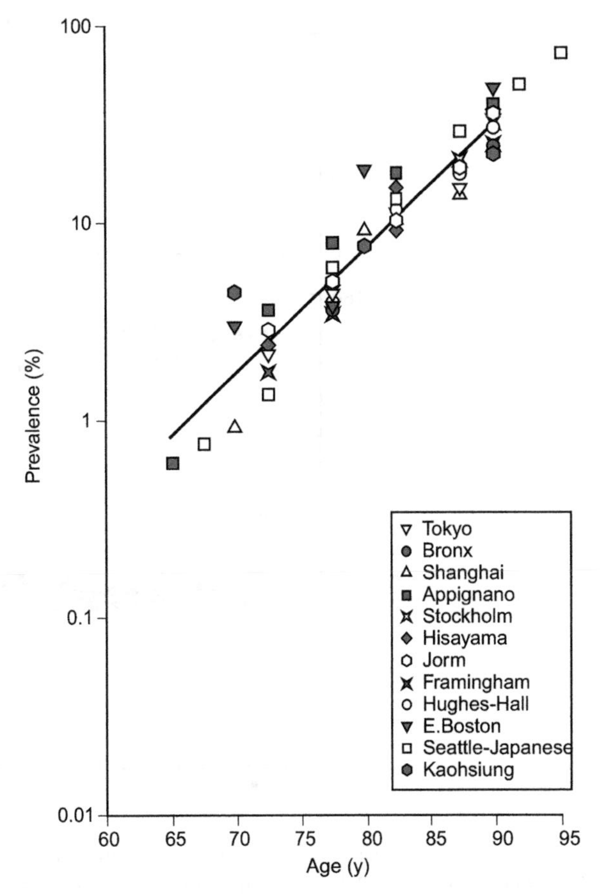


**Figure 15.2** Evidence for exponential increases in the age-specific incidence of probable Alzheimer disease in six different community-based studies. (From Breteler MMB, et al. Epidemiology of Alzheimer disease. Epidemiol Rev 1992;14:59–82.)

Longitudinal studies of physiologic parameters may exhibit striking variation among individuals [20]. Fig. 15.3 illustrates the case of a measure of renal function. By this measure, some individuals show no evidence of a decline in renal function; some may have superior compensations for structural alterations [21]. Are any of these varied patterns of functional decline (or lack of decline) in apparently normal aging human subjects determined, in part, by constitutional differences in the genotype? Essentially no research has been carried out to address this important question. Medical geneticists have an obvious bias in favor of the discovery of deleterious allelic variants. There is a great need to define allelic variants that, in ordinary environments, are associated with

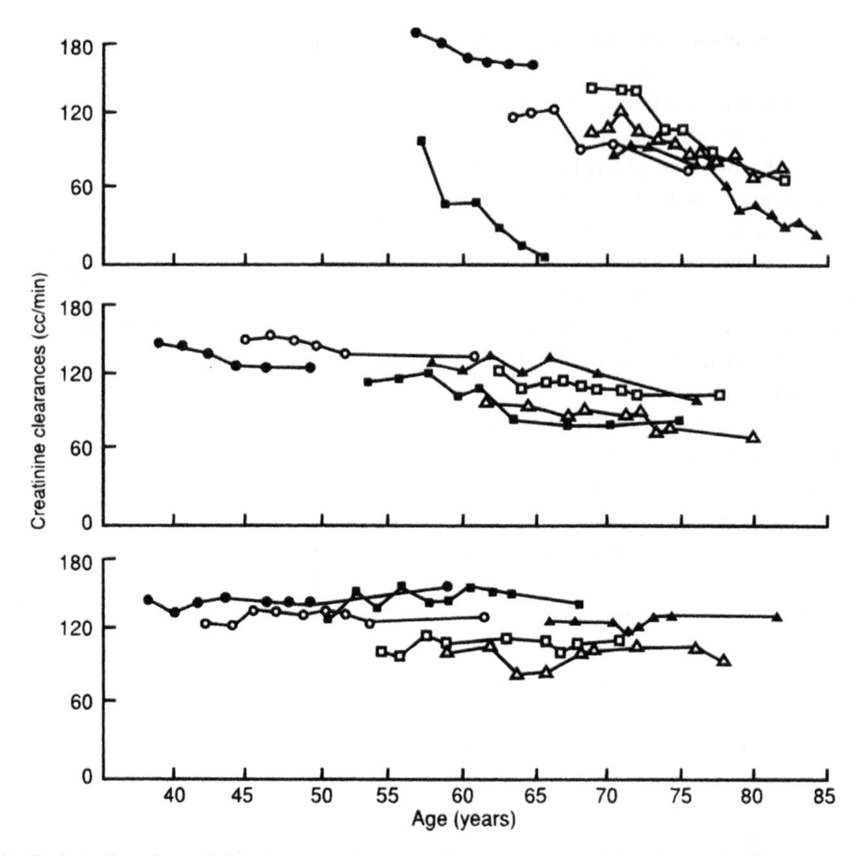


**Figure 15.3** Longitudinal studies of creatinine clearance (an approximate measure of the glomerular filtration rate) for a representative sample of a subset of 446 clinically normal male volunteers in the Baltimore Longitudinal Study of Aging of the National Institute on Aging followed between 1958 and 1981. The results could be classified in one of three major patterns. The top panel illustrates substantial rates of decline in this measure of renal function for six representative subjects who were followed for 8–14 years. The middle panel illustrates a pattern of slight, but significant, decline for six representative subjects followed for 11–22 years. For the six representative subjects in the bottom panel, who were followed for periods of 15–21 years, there were no apparent declines in this measure of renal function. (From Lindeman RD, et al. Longitudinal studies on the rate of decline in renal function with age. J Am Geriatr Soc 1985;33:278–85.)

the maintenance of enhanced structure and function during aging. One such example may be the ApoE2 allele, the prevalence of which is significantly increased in centenarians [22]. The APOE2 results are understandable in that there is evidence that carriers are provided with some protection against Alzheimer disease [23].

There is very strong evidence indicating a major role for the constitutional genotype in the susceptibility to various familial and "sporadic" forms of Alzheimer disease. In addition to the apparent protection by the E2 allele of ApoE noted above, individuals carrying the E4 allele, particularly homozygotes, are at elevated risk to develop the disease [24]. ApoE4 may act as an age-of-onset modifier for the common relatively late-onset forms of the disorder. Although deserving of additional research, there is evidence for multiple mechanisms underlying the effects of the E4 allele [25]. This important subject is considered in much more detail in a later volume. The lay perception that aging is accompanied by a global loss of cognitive function is certainly incorrect. Only selected regions of the nervous system appear to be particularly susceptible.

The other type of terminally differentiated cell receiving special attention from gerontologists is the multinucleated skeletal muscle cell. Structural and functional declines in skeletal muscle vary from muscle to muscle, with weight-bearing muscles being more susceptible; the rates of these declines accelerate after about age 70 [26]. At least some proportion of the pathology is likely to be related to denervation atrophy [27]. Disuse atrophy is also an important component [28].

Many postreplicative aging cell types gradually accumulate a mixture of complex fluorescent pigments called lipofuscins. These are likely to vary in composition from tissue to tissue. Most investigators believe that all lipofuscins are the products of lipid peroxidation reactions. They could therefore be regarded as evidence in support of theories of aging that invoke oxidative alterations of macromolecules. That lipofuscins are markers of some underlying aging process is a theory supported by three lines of evidence. First, they appear to be almost invariable features of aging in an amazing variety of organisms, including certain strains of fungi under certain growth conditions (e.g., *Podospora anserina* and *Neurospora crassa*), paramecia, nematodes, snails, fruit flies, houseflies, frogs, parrots, house mice, rats, guinea pigs, cats, dogs, pigs, monkeys, and humans [29,30]. Second, quantitative studies of lipofuscin rates of accumulation in the hearts of dogs and humans indicate appropriate correlations with the lifespan potentials of those species [31]. (No such correlations have been observed, however, among cardiac tissues from a group of primates of contrasting life spans [32].) Third, age-related increases in concentrations of at least some classes of lipofuscins are blunted by caloric restriction, an intervention known to increase life span in mammals [33,34].

The importance of extracellular aging has been emphasized by the late Robert R. Kohn [35]. Long-lived proteins, such as lens crystallines and collagens, are particularly susceptible to a variety of posttranslational alterations; these can result in changes of amino acids [36]. Diabetics, who have many progeroid features, are particularly susceptible to glycation of proteins [37]. Advanced glycation end products may also play an important role in the genesis of osteoarthritis [38]. Modified matrix components could perturb cell–matrix interactions and hence change cell function. Such a scenario has been suggested to play a role in the genesis of atherosclerosis [39,40].

## 15.3 WHY DO WE AGE?

Evolutionary biologists believe that they have an answer to the ultimate cause of aging in age-structured populations (i.e., populations that consist, at any given time, of cohorts of varying chronological ages) [41]. Simply put, we age because senescent phenotypes escape the force of natural selection [42]. This theory was developed for the case of species with age-structured populations, a situation that occurs when there are serial episodes of reproduction in an individual's lifetime, as opposed to one massive "big-bang" production of progeny in short-lived animals [43]. For human populations, the late William Hamilton showed that the force of natural selection for or against alleles that do not reach phenotypic expression until about the age around 45 years is essentially nil [44,45]. More recently, a mathematically rigorous challenge to Hamilton's theory has been published [46]. It describes a scenario whereby the force of natural selection can actually increase during aging. There are in fact some species of fish that continue to grow and, as such, become more like predators than prey. Under those circumstances, it is easy to imagine declines in the force of natural selection with age.

August Weissmann, one of the giants of 19th century biology, postulated a limited life span of somatic cells, from which he proposed programmed death theory with the idea that aging is good for the species in that it results in enhanced resources for the young. While a finite replicative capacity of somatic cells was later confirmed experimentally [47], there has been no good evidence that aging evolved because it was adaptive for the species or the individual. Essentially all population geneticists who have considered this issue have concluded that aging is nonadaptive. A striking demonstration that single gene mutations can extend the life spans of nematodes, fruit flies, and mice—sometimes dramatically [1,48]—can be interpreted as providing evidence against the classical evolutionary biologic theory of aging, which would predict a highly polygenic modulation. We have since learned, however, that a remarkable number of mutations at single loci can enhance the life spans of model organisms, notably of *C. elegans*, where numerous single gene mutations and genes that have been downregulated by RNAi have been shown to provide substantial increases in life span [49]. A caveat in the interpretation of such studies, however, is that these increases in life span have typically been only examined under conventional laboratory conditions. When challenged by competition experiments with wildtype organisms under other conditions, such as "feast or famine" conditions of the availability of their bacterial

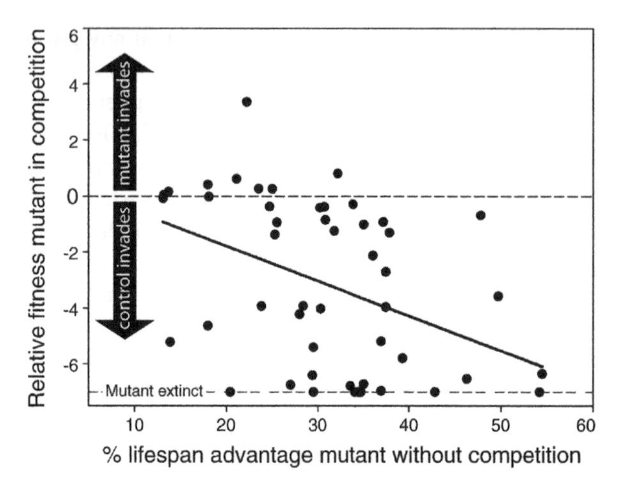


**Figure 15.4** Genetic variants of model organisms (in this case, yeast) that exhibit enhanced life spans typically exhibit reductions in relative fitness in competition experiments. (From Briga M, Verhulst S. What can long-lived mutants tell us about mechanisms causing aging and lifespan variation in natural environments? Exp Gerontol 2015;71:21–6.)

diets, that advantage was shown to have disappeared in *age1*, the first mutation shown to enhance life span in *C. elegans* [50]. In a systematic competition study involving long-lived yeast mutants versus the parental wildtype strain, the wildtype typically exhibited greater fitness [51] (Fig. 15.4). There is a lesson here for clinical investigators interested in translational research towards the enhancements of human health spans and life spans based upon laboratory experiments in model organisms. Human subjects not only show enormous differences in their genomes and epigenomes, they may often share strikingly different environmental exposures and may lose fitness even within comparable environments when competed with nonmutant individuals.

We have known for many decades that a single environmental manipulation, caloric restriction (or, more conservatively, dietary restriction, since restriction of a single amino acid, methionine, can give comparable results) [52], can substantially enhance life span in a remarkably wide range of species [53]. Once aging, some would argue that these observations support a "programmed" mechanism of aging—that aging involves sequential, determinative changes in gene expression that actively produce aging. One interpretation of both the single gene mutation and caloric restriction experiments, however, is that all or many of them are examples of *diapauses*—time-outs from the business of reproduction during "bad times"—be they nutritional, climatic, or other environmental challenges.

In terms of genetic mechanisms that form the basis of the classical evolutionary theory of aging, two ideas currently dominate the field. The first, championed by the late Peter Medawar, is generally referred to as "mutation accumulation" [54]. This is an unfortunate name, as the mutations in question are not somatic mutations developing during the life span, but germline mutations that do not reach phenotypic expression until late in the life course, when the force of natural selection would be attenuated. Huntington disease is the prototypical example. Haldane was puzzled by the surprisingly high prevalence of this disorder, which exceeds 15 per 100,000 in some western European populations [55], while germline mutations typically have frequencies of about one in a million. Haldane suggested that the reason the mutation survived in the population was because of its delayed manifestations, thus escaping the force of natural selection [56]. If that were the case, there would be selection for "suppressor alleles" that progressively delayed the age of phenotypic expression. Medawar concluded that many such suppressors might only delay these deleterious effects [54]; eventually, however, the delayed age of expression would be such that there could be little or no influence of natural selection. This scenario, especially when coupled with the other mechanisms discussed below, would result in an enormous degree of heterogeneity in patterns of aging among individuals in out-breeding populations. Thus, each of us may be essentially unique in precisely how we age.

The second dominant idea was first clearly enunciated by George C. Williams [57], and has been referred to as antagonistic pleiotropy; it has been elaborated by Michael R. Rose [42]. By this view, some varieties of genes might have been selected because of good effects early in the life span, but may also have deleterious effects late in the life span, thus contributing to aging phenotypes. As one potential example, Williams speculated that alleles selected because of enhanced incorporation of calcium into bones might be responsible for forms of calcific arteriosclerosis, when acting over long periods. There have been many suggested examples in the literature, but they have been hard to definitively establish. Potential examples include atherosclerosis [58], the role of the apolipoprotein E4 allele in Alzheimer disease [59], common late-life cancers [60], and immunosenescence [61]. Surprisingly, a single example of what has been called "reverse" antagonistic pleiotropy has now been documented in mice, at least under standard laboratory conditions [62]. Medical geneticists are in a good position to suggest a number of other examples, and perhaps to provide supportive evidence. Such research has the potential to illuminate the most basic aspects of the aging problem. It may also have important translational implications.

Another conceptual formulation, one that overlaps with what has been discussed above, is that there is, inevitably, a trade-off of energetic resources expended by an organism for purposes of reproduction and resources devoted to the maintenance of the macromolecular integrity of the organism. Examples include repair of DNA, scavenging of abnormal proteins, and replacement of effete somatic cells; this is the disposable soma theory of Tom Kirkwood [63]. These ideas can be generalized as life history "optimization" theories of aging [64]. Experimental evidence in *Drosophila* sp. supports both optimization theories and the mutation accumulation theory of aging [64,65]. The relative quantitative contributions of each theory, however, particularly in *H. sapiens*, are completely unknown. As pointed out by Partridge and Barton [64], the resolution of this issue has potentially profound implications for the future life history of our species. If optimization mechanisms predominate, any lifespan extensions may be offset by the trade-off of lower early fertility, delayed maturation, and potential increases in early life history morbidity and mortality. If mutation accumulation mechanisms prevail, enhanced life span attributable to the elimination of such constitutional mutations would presumably have few effects on early lifespan structure and function. Given a major role for optimization theories, a continuation of the present secular trends of elective delays in the ages of reproduction in the developed societies would predict the emergence, by indirect selection, of increased life spans and related declines in early fertility after several centuries of continued evolution of our species. It has been well documented that advanced parental age is associated with increases in germline mutations. Of particular relevance to our interest in late-life disorders is the evidence that there is a large paternal age effect for point mutations [66]. Such secular trends could therefore be associated with increases in germline mutations, with potential deleterious effects in subsequent generations. It has been difficult, however, to confirm a relationship of paternal age and the occurrence of nonfamilial varieties of such common polygenic late-life disorders such as Alzheimer disease or prostate cancer [67].

At a more fundamental level, one can ask why one observes such striking variations in the life spans of various mammalian species. While such variation is obviously related to the constitutional genotype, it does not necessarily follow that aging is "programmed"—at least in the sense of concerted, determinative, sequential gene action comparable to what one observes in development. The most satisfying idea invokes differential impacts of environmental hazards (e.g., accidents, predation, drought, starvation, infectious diseases) during the emergence and maintenance of various species. This is nicely articulated in a popular book on aging [68]. Species with comparatively high hazard functions would be expected to evolve life history strategies that emphasize rapid maturation, high fecundity, early fecundity, and short life spans. An attenuation of those hazards could set the stage for the emergence of sibling species with a more leisurely rate of maturation, lower early fecundity, and longer life spans. One of the few field biology studies to examine this idea has in fact provided strong support for that hypothesis [69]. Contrary findings have been reported, however, for different species in different ecologies [70].

The evolutionary formulations of the nature of aging have a number of interesting and important implications, in addition to those noted above. Let us summarize some of these propositions:

1. Stochastic processes: These are likely to play a major role in senescence. This follows from the conclusion that one is not dealing with a determinative sequence of concerted gene action but rather with an epiphenomenon of selection for gene action designed for reproductive fitness. Consider the analogy with a spacecraft engineered to function for a given period of time in order to complete a specific mission. Engineering specifications for indefinite maintenance of the craft would be prohibitively expensive or impossible. One would therefore anticipate an element of chance as to which components will initially exhibit structural and functional failures and when such failures will be detectible with the available diagnostic facilities. Many major geriatric diseases of humans (e.g., cancer, strokes, coronary thrombosis) are surely based on stochastic events. In the case of malignant neoplasms, selection for a series of random somatic

mutations is the key to the understanding of the pathogenesis, although arguments can be made that the first step in neoplasia may be age-related hyperplasia resulting from variegated gene expression involving cell cycle regulatory loci [71]. Overall longevity also is subject, in part, to stochastic laws. There are numerous examples in which investigators have rigorously controlled both environment and genotype, yet have observed marked variations in longevity. The most convincing example comes from studies of *C. elegans*, which can be grown (except for yeast extracts) in chemically defined media in suspension cultures, thus ensuring rigorous control of the environment [72].

2. Polygenic basis: There is a polygenic basis for aging. There is no reason to believe that the optimization theories or mutation accumulation theories involve only a few genes. Indeed, for the case of the successful experiments involving indirect selection for increased life spans in genetically heterogeneous wildtype stocks of *D. melanogaster*, genetic analysis indicated genes on all of the major chromosomes [73]. Martin [74] estimated (as an upper limit) that allelic variation or mutation at close to 7% of loci of the human genome has the potential to modulate varying aspects of the senescent phenotype. A different and more conservative estimate—the number of genes likely to have evolved in the hominoid lineage leading to humans (and thus could be associated with the increased life span of *H. sapiens*)—gave a figure of 0.6% of functional genes [75]. Neither estimate can be characterized as oligogenic.
3. Multiple mechanisms: There are likely to be multiple mechanisms of aging, although there would be selective pressure toward some degree of synchronization of the ages of expression of phenotypic effects resulting from independent mechanisms. This proposition follows from the randomness of the accumulated constitutional mutations and from the great variety of types of gene action that could be involved in "trade-off" types of gene action. Against this proposition, however, is the fact that a single environmental manipulation—dietary or caloric restriction—regularly leads to lifespan extension in rodents, or at least those that have been selected for the easy life of the laboratory setting [53]. We have little information on the effects of caloric restriction in the wild, however. The life course histories of organisms, whose evolutionary history reflected exceptionally high environmental hazard functions, such as the mice and rats used for most of the calorie-restriction experiments, are quite distinct from those of the higher primates [68]. It is therefore not at all clear that dietary restriction would make a significant impact on the lifespan potentials of human subjects. Evidence is pointing toward the conclusion that dietary restriction may delay the onset of age-associated pathologies and reduce the incidence of common age-related disorders and age-related deaths [76]. We shall have to await the final outcome of current research in rhesus and squirrel monkeys (reviewed by Roth et al. [77]) to know how likely such an effect will obtain for our species. The most recent analysis of the data from two major studies supports the conclusion that this intervention enhances both life span and health span in Rhesus monkeys [78]. Meanwhile, common sense tells us that we should avoid gluttony!
4. Species specificity: There is likely to be a degree of species specificity in relevant gene actions. We have already developed certain of these arguments, but let us consider an extended argument. If aging is in fact an epiphenomenon—a byproduct of selection for alleles ensuring an optimal degree of reproductive fitness in a given environment—there is no a priori reason to expect identical scenarios of gene action among very different species. Consider the striking differences in the behavioral patterns among different species that lead to successful matings. There is surely a wide variety of different loci involved, and those that are operative in fruit flies must surely differ from those that are relevant for man! Nevertheless, it is quite possible that there are a number of common mechanisms among groups of related species (including all mammals), and it is even conceivable that such global mechanisms as oxidative damage to macromolecules underlie the aging of all or most organisms. This is the rationale for carrying out comparative gerontological research. The first "public" mechanisms of aging were documented in experiments in *C. elegans*, *D. melanogaster*, and *M. musculus domesticus* (Fig. 15.5) [48,65]. Remarkably, "leaky" mutations in comparable neuroendocrine signal transduction pathways involving insulin-like growth factor (IGF1) receptors and the nuclear translocation of a transcription factor can

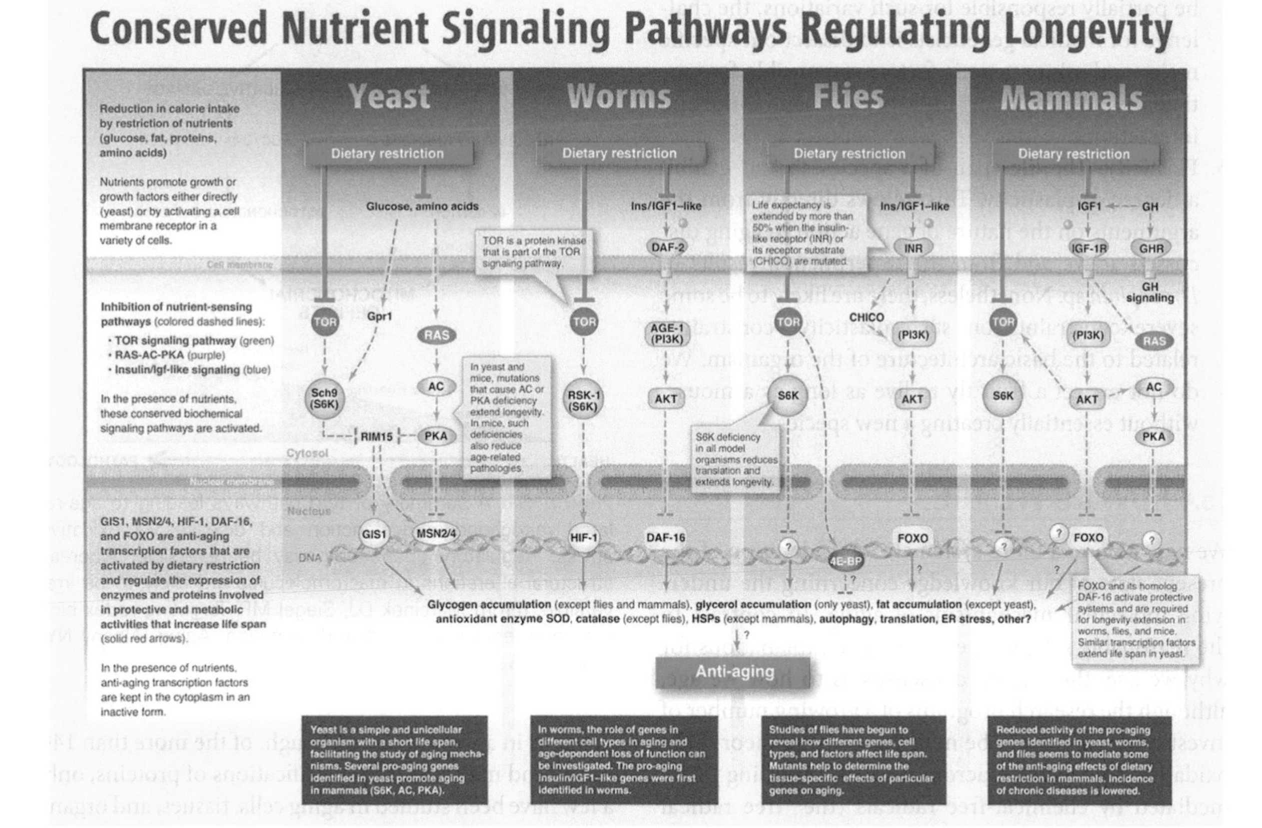


**Figure 15.5** Evidence for a partially conserved "public" mechanism for the modulation of aging in yeast, worms, fruit flies, laboratory mice, and, possibly humans. These signal transduction pathways evolved to modulate metabolism under transient adverse environmental conditions, during which gene actions that enhance the protection of somatic cells are up regulated while further growth, development, or reproduction is postponed. Certain mutations in these pathways were found to substantially enhance the life spans of laboratory strains of model organisms amenable to genetic analysis. Mutations so far studied in this general pathway in humans result in pathology. The possible effects of a wider spectrum of such mutations upon the longevity of humans are not yet understood. (From Fontana L, Partridge L, Longo VD. Extending healthy life span from yeast to humans. Science 2010;328:321–6.)

lead to substantial extensions of life spans [1]. At first blush, these observations would appear to contradict the conclusion, discussed above, that aging is under polygenic controls and that multiple mechanisms are at work. The authors' interpretation of these important discoveries, however, is that they are examples of diapauses—nature's "time-outs" from the business of reproduction when faced with conditions of nutritional, climatic, or other environmental challenges [3]. These can best be considered as "reprieves"; they will be eventually trumped by other gene actions that, unlike diapauses, have escaped the force of natural selection.

5. Intraspecific variations: There are likely to be significant intraspecific variations in phenotypic patterns of aging, particularly in humans. This also follows from many of the arguments discussed above. Given the polygenic nature of aging, the likelihood of a variety of mechanisms, a strong stochastic component, the realization that one is dealing with alterations in all body systems, and the enormous genetic and environmental heterogeneity in our species, one would certainly predict substantial differences in the way it plays out in individual subjects. Every clinician has witnessed this phenomenon first-hand. While differential impacts of the environment are likely to

be partially responsible for such variations, the challenge for medical geneticists is to dissect out specific major and minor genetic factors responsible for particularly favorable or unfavorable nature–nurture interactions.

6. Plasticity: The life span of a species should exhibit a degree of plasticity. This follows directly from the arguments on the nature of gene action in aging discussed above and from the experimental results in *Drosophila* sp. Nonetheless, there are likely to be some severe constraints on such plasticity—constraints related to the basic architecture of the organism. We do not expect a fruit fly to live as long as a mouse, without essentially creating a new species.

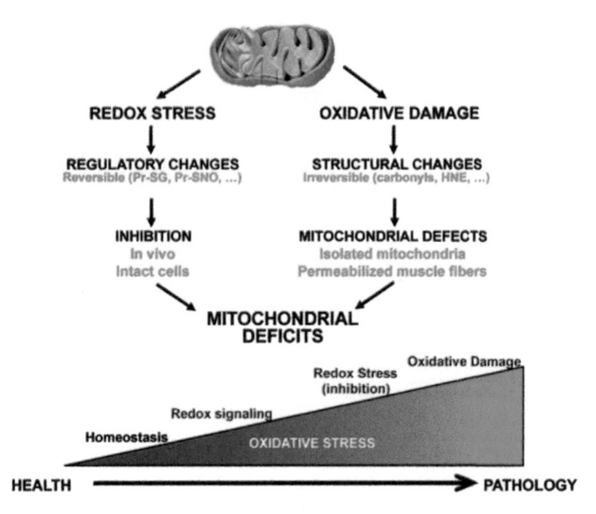


**Figure 15.6** A summary of two pathways leading to age-related mitochondrial dysfunction and disease via oxidative stress: Regulatory aberrations may be reversible, whereas structural alterations in macromolecules are likely to be irreversible. (From Marcinek DJ, Siegel MP. Targeting redox biology to reverse mitochondrial dysfunction. Aging (Albany NY) 2013;5:588–9.)

## 15.4 HOW DO WE AGE?

We now turn to a more systematic consideration of the present state of our knowledge concerning the underlying molecular mechanisms of aging. In contrast to the reasonably satisfying evolutionary explanations for why we age, there is no consensus as to how we age, although the research programs of a growing number of investigators appear to be motivated by the theory that oxidative damage to macromolecules, including those mediated by chemical-free radicals (the "free radical theory of aging") (reviewed by Muller et al. [79]), are of paramount importance. Even that canonical theory has likely been oversimplified, however, as alterations in redox signaling in mitochondria are considered to be of importance to mitochondrial dysfunction [80,81] (Fig. 15.6).

### 15.4.1 Alterations in Proteins

In 1963, Orgel introduced the protein synthesis error catastrophe theory of aging [82,83]. It was proposed that transcriptional and/or translational errors in the synthesis of proteins that were themselves used for the synthesis of proteins (e.g., DNA-dependent RNA polymerases, ribosomal proteins, etc.) could result in an exponential cascade of errors involving essentially all proteins, leading to cell and organismal death. Biosynthetic errors in protein synthesis appear to be rare, however, even in old organisms [84]. Although most gerontologists have abandoned this theory, very few tests of the theory have been carried out with postreplicative cells in vivo [85]. By contrast, there is a growing body of evidence indicating the prevalence of posttranslational modifications in proteins in aging tissues; although, of the more than 140 major and minor known modifications of proteins, only a few have been studied in aging cells, tissues, and organisms [86,87]. Beginning with a classic paper on senescent nematodes by Gershon and Gershon [88], many studies have demonstrated an accumulation of immunologically detectable, but enzymatically inactive, enzyme molecules in various mammalian tissues. These may result from a variety of posttranslational modifications, including subtle conformational changes [84]. There is currently a great deal of interest in oxidative alterations [89]. Metal-catalyzed oxidation systems have the potential to inactivate enzymes oxidatively via attacks on the side chains of certain amino acids, with the formation of carbonyl derivatives. The side chains of histidine, arginine, lysine, and proline are particularly susceptible. The sulfhydryl groups of methionine are also susceptible to oxidation. Other posttranslational changes that can be observed in aging cells include racemization, deamidation, isomerization, phosphorylation, and glycation.

Many gerontologists believe that glycation, the spontaneous nonenzymatic reaction of glucose with proteins and nucleic acids, may be a major factor in the development of certain age decrements, as well as complications of diabetes mellitus. Glycation is the slow, spontaneous reaction of the aldehydic form of glucose with free amino groups to form a Schiff base, which subsequently rearranges to form a stable Amadori product. Subsequent reactions, possibly involving oxygen radicals, generate more complex products referred to as advanced glycosylation end (AGE) products. Some of these compounds, including pentosidine, have been characterized. Antibodies to the AGE products have been generated and used to map their distribution to neuritic plaques and tangles and to other sites [37]. Because the levels of AGE products increase with age and with elevated blood glucose, crosslink proteins, and change their physical and biologic properties, they are thought to underlie the development of atherosclerosis, cataracts, and peripheral neuropathies. In addition, macrophage receptors bind to the AGE products and initiate the secretion of inflammatory cytokines such as the tumor necrosis factor [90]. Thus, glycation represents a progressive age change linked to age-associated disabilities. Support for these ideas has come from experiments in aging dogs, in which it was possible to reverse myocardial stiffness and improve cardiac function by the administration of an experimental compound known to break the crosslinks associated with the formation of advanced glycation end products [91].

Calorically restricted rodents, which have substantially increased life spans, exhibit evidence of both enhanced defenses against reactive oxygen species and reduced levels of protein glycation (associated with decreased levels of plasma glucose). Such results suggest that both the free radical theory of aging and the glycation theory of aging may be operative and potentially synergistic [92]. A number of different types of amyloids accumulate in mammalian tissues during aging [93]. In their advanced states, they are detected extracellularly as protein aggregates associated with proteoglycans and other proteins. Each type is derived from a different precursor protein. These include the beta-amyloid protein of Alzheimer disease and the aging brain; a transthyretin-derived amyloid in peripheral nerve tissues, autonomic nervous system, choroid plexus, cardiovascular system, and kidneys; atrial amyloid derived from the atrial natriuretic peptide; the amylin-derived amyloid in the pancreatic islets of Langerhans [94]; systemic amyloid AA derived from apolipoprotein A-II [95]; and possibly unique types of amyloid in the anterior pituitary gland, intervertebral discs, the aortic intima and media, aortic heart valves, and the adrenal cortex. In certain of these conditions, variants in the precursor protein greatly accelerate the rates of deposition of the derivative amyloids. This has been particularly well demonstrated for the case of beta amyloid [96].

It is a challenge for the future to discover common denominators underlying this remarkable propensity of mammalian tissues to accumulate these different types of abnormal proteins. Obvious approaches would include more detailed studies of alterations in protein turnover with age (including the turnover of amyloid deposits) and how such turnover might be modulated by endocrine and neuroendocrine factors. Another promising and relatively new area of research seeks to define gene products that function in the repair of altered proteins. An example is the catalysis of the transfer of a methyl group from *S*-adenosylmethionine to L-aspartyl and D-aspartyl residues by protein carboxyl methyltransferases (ED 2.1.1.77). These enzymes have the potential to repair abnormal proteins via the conversion of L-isoaspartyl residues to L-aspartyl residues [97]. This enzyme is polymorphic in humans, raising the question of the differential repair of such classes of altered proteins during aging in human populations [98].

Research on the maintenance of the integrity of proteins and protein complexes (protein homeostasis or "proteostasis") is among the fastest-growing fields in geroscience. It is now being pursued in terms of networks of gene actions that modulate protein synthesis, folding, transport, heteromeric protein complex formations, and degradation [99].

### 15.4.2 Alterations in DNA

#### 15.4.2.1 Nuclear DNA—Epigenetic Events

Given the fact that, for most genetic loci, only two alleles are present, nuclear DNA would appear to be a particularly vulnerable target for damage during aging. Historically, the first specific type of somatic "mutational" theories of aging was proposed by a physicist, Leo Szilard [100]. He envisioned random "hits" that would inactivate entire chromosomes or chromosome arms. In modern terms, such inactivations could conceivably be associated with epigenetic events, as for the case of the random inactivation of one of the two X chromosomes of the human female during embryonic development and the processes of parental genomic imprinting. There is no good evidence of widespread heterochromatizations or inactivation of large chromatin domains during aging. In fact, at least for the case of mice, there is evidence of a reactivation of certain gene loci on a previously inactive X chromosome during aging [101–103]. No such reactivation could be demonstrated for the case of the *HPRT* locus of heterozygous human females [104]. Reactivation has also been demonstrated for a genomically imprinted autosomal locus in mice [101]. Global losses of 5-methyl cytosine have been demonstrated in aging fibroblast cultures [105] and in tissues of two species of aging rodents [106], but there have been few studies of altered methylation in specific domains of specific genes during aging. In one such study, hypermethylation was mapped to the proximal 5′ spacer domain of ribosomal DNA genes of aging mice; silver stains of cytogenetic preparations revealed that the ribosomal gene cluster on chromosome 16 was preferentially inactivated [106]. It remains to be seen, however, whether this remarkable result reflects some developmental, adaptive process in laboratory mice or in the particular strain of mice investigated, as the biochemical changes were observed as early as 6 months. A form of gene-specific methylation of CpG islands has clearly been established to progress steadily into old age in human subjects. It is associated with the silencing of the estrogen receptor gene of a subset of cells of the colonic mucosa [107]. A striking finding in that study was that, of a set of 45 colorectal human tumors examined, including those in very early stages of oncogenesis, estrogen receptor expression was either diminished or absent. Moreover, the introduction and expression of an estrogen receptor gene in a line of colon carcinoma cells resulted in marked growth suppression. This important paper therefore demonstrates a link between a presumably epigenetically based progressive repression of a specific gene during aging and the susceptibility to the development of a common type of cancer of aging.

Using the yeast model of replicative aging, Lenny Guarente and his colleagues highlighted a key role of NAD-dependent histone deacetylation in the regulation of energy metabolism, genomic silencing, and aging. There continues to be a great deal of current research in various organisms on homologs (sirtuins) of the yeast Sir2 gene responsible for histone deacetylations [108,109]. A variety of other changes in gene expression occur throughout the life span, notably changes in the methylome, but it remains to be seen which of these alterations are of primary significance to one or more aging processes and which are merely epiphenomena. One such approach is to explore the effects of caloric restriction [110]. A marked transcriptional stress response, with lowered expression of metabolic and biosynthetic genes, is found in aging mouse tissues. These alterations are ameliorated in calorically restricted mice.

Age-related changes of methylation at CpG regions of the genome have received a great deal of attention because of the pioneering research by Steve Horvath and colleagues on the concept of the "Epigenetic Clock" [111]. Patterns of genomic methylation (the "methylome") predicted longevity within several ethnically distinct populations independent of chronological ages. These results were even more robust when incorporating data on age-related shifts in peripheral blood cell compositions [111].

#### 15.4.2.2 Nuclear DNA—Mutational Events

In 1961, a now classic paper appeared, casting doubt on the validity of somatic mutational theories of aging [112]. Taking advantage of the occurrence of a species of wasp in nature—the males of which exist as either haploid or diploid organisms—it was found that there was no difference in the life spans of these organisms. As expected, however, the haploid wasps were much more susceptible to the effects of ionizing radiation. These results were strong evidence against a role for recessive mutations in insects. They did not rule out, however, some role for a combination of dominant and recessive mutations in the aging of such organisms. Moreover, the interpretations are complicated by the occurrence of polyploid cell types. Finally, those experiments told nothing about the role of somatic mutations of replicating populations of cells in the limitation of life span, since wasps and other insects, with the exception of gonadal tissues and, in *Drosophila*, certain intestinal cells, consist of postreplicative cells. For the case of such replicating populations of cells, there is now compelling evidence that somatic mutations constitute a link between the biology of aging and the biology of cancer. Thus, while more data are required, there are reasonable correlations of species-specific life spans and rates of development of various neoplasms (e.g., [112]). Among those genes that evolved in association with the relatively long life spans of *H. sapiens* (the longest lived of all mammalian species), there must be loci conferring enhanced genomic stability in comparison, for example, with those of *M. musculus domesticus*. Moreover, there may be considerable species differences in the patterns of somatic mutation. For mice, for example, there is evidence of a marked susceptibility to cytologically detectible chromosomal mutations during aging [113], while in comparable cell types (renal tubular epithelial cells), there is little evidence for the accumulation of mutations (presumably intragenic) at the *HPRT* locus, a target chosen because of the lack of evidence for selection against such mutations in renal tissue [114]. Mutations have been shown to accumulate at the *HPRT* locus of T lymphocytes and in the renal tubular epithelial cells of aging human subjects, however, with higher frequencies of mutation being observed in the epithelial cell type [115,116]. The lymphocyte study showed that deletions were relatively common [117]. Such accumulations could be attributable to chronology rather than to intrinsic biologic aging. We will require additional research in mammals of contrasting lifespan potentials to address this question; an approach using comparable transgenic reporter constructs may be promising, if these could be comparably buffered from position effects [118].

#### 15.4.2.3 Nuclear DNA—Molecular Misreading

What originally appeared to be the accumulation of frameshift mutations in DNA in aging mammals now appears to be the result of transcriptional errors at particularly vulnerable sites, especially those with runs of GAGAG. van Leeuwen has named this phenomenon "molecular misreading." The process can impact upon the fidelity of transcription of such important loci as the beta-amyloid precursor protein and ubiquitin B [119].

#### 15.4.2.4 Telomeric DNA

Perhaps the most robust age changes noted in the nuclear genome of normal somatic cells are alterations in telomere length, leading to replicative senescence (reviewed by de Lange [120]). Telomeres have a highly repetitive structure (TTAGGG in humans and mice) that extends for many thousands of nucleotides at the ends of chromosomes. The telomeres stabilize chromosomal structure and their loss leads to various cytologic aberrations and the arrest of cell division. Current concepts suppose that telomeres in cells of the germline and in many neoplastic cells are added to the ends of chromosomes de novo by a unique enzyme referred to as telomerase, which uses an associated RNA to code for the hexanucleotide repeats. It appears that telomerase is lost or its concentration greatly diminished in the progeny of somatic stem cells (but not in cells of the germline). Somatic cell telomeres are then duplicated during cell division by DNA polymerases without the assistance of telomerase. It is characteristic of DNA polymerases that they fail to copy some 50–200 terminal bases of the trailing strand and the telomeres are shortened by this amount with every cell division. The shortening of telomeres is strikingly apparent when one examines the telomeric/subtelomeric DNA isolated by appropriate restriction enzyme cleavage from normal human fibroblasts that have undergone large numbers of cell divisions in culture. Exit from the cell cycle may occur as when only a few chromosome arms reach a critical level of shortening, thus activating cell cycle checkpoints [121].

The development of a PCR-based method for the assay of telomere lengths has led to a remarkable association of short telomeres in DNA from peripheral blood of elderly human subjects with mortality; the results were largely attributable to earlier deaths from cardiovascular and infectious diseases [122]. Shorter telomere lengths have also been reported in mothers of chronically ill children; the authors surmised that life stress could shorten life span [123]. An alternative interpretation, however, is that the mothers of many chronically ill children are more frequently exposed to infectious agents, thus driving proliferation of lymphocytes, resulting in shorter telomeres. The relationships between various forms of stress and telomere lengths are the subject of continuing research [124].

#### 15.4.2.5 Mitochondrial DNA

The venerable "Free Radical Theory of Aging" had its origins with the work of Rachel Gershman and Daniel Gilbert on "oxygen toxicity" in 1954 [125,126] and subsequent papers. Harman proposed that aging resulted from the cumulative damaging effects of the byproducts of aerobic metabolism, namely reactive oxygen species (ROS), on mitochondrial DNA itself as well as proteins, leading to cellular deterioration and organ dysfunction. Over the years, advances in understanding mitochondrial genetics and biology has led to a more multifaceted and complex picture of mitochondrial dysfunction and aging, including the role of oxygen-derived free radicals in redox signaling as discussed above and summarized in Fig. 15.6.

mtDNA is closed circular DNA of some 16,569 nucleotides that codes for some of the mitochondrial proteins plus the tRNAs and rRNAs used for mitochondrial protein synthesis. Other components of the mitochondria are coded for by nuclear genes and are transported to the mitochondria. Essentially, all mtDNA molecules are maternal in origin; thus, mtDNA genetic diseases are maternally transmitted.

It is well known that mtDNA rearrangements (single deletions or multiple deletions) cause numerous mitochondrial diseases, for example, Leber hereditary optic neuropathy (LHON), Kearns–Sayre syndrome (KSS), or progressive external ophthalmoplegia (PEO). In addition, mitochondrial rearrangements are found at low levels in healthy tissues with no signs of mitochondrial disease, and accumulate with aging in postmitotic tissues, where their relationship to aging remains an area of debate. Based on the frequency and consistent location of common specific age-related deletions, one can postulate that the sequences between the direct repeats are looped out following damage to the DNA by a slip replication mechanism. The damage to mtDNA molecules may be initiated by oxygen radicals generated as a byproduct of the oxidative phosphorylation reactions carried out by the mitochondria. The proximity of mtDNA to the sources of oxygen radicals, plus the lack of associated histones, would make mtDNA more vulnerable than nuclear DNA. Thus, the age-related changes observed could be due to increased damage and/or reduced repair. One important mechanism for repair is the proofreading domain of DNA polymerase gamma, the enzyme that replicates mtDNA. Support for the importance of this function has come from the synthesis of mice with knock-in mutations in that domain; these transgenic mice exhibited progeroid features [127].

More recently, a compelling case for a relationship between mitochondrial function and neurodegeneration has been highlighted by the discovery of a number of genes, in which pathogenic variants result in hereditary forms of Parkinson disease. Parkinson disease (PD) is the second most common neurodegenerative disorder after Alzheimer disease, and over 90% of cases are sporadic. Pathogenic variants were first identified in parkin in Japanese families with juvenile PD, and are the most common cause of autosomal recessive PD. Parkin encodes an E3 ubiquitin ligase which is recruited to regions of mitochondrial damage where it ubiquitinates outer mitochondrial membrane proteins. The discovery of *PINK1*, a serine/threonine kinase which phosphorylates a key protein in mitochondrial trafficking, and regulates mitochondrial quality control, added further evidence to the link between mitochondrial dysfunction and Parkinson disease. Specifically, PINK1 and parkin function to target and degrade damaged mitochondria through a specific form of autophagy termed mitophagy. A more detailed picture of mitochondrial function in PD pathogenesis has emerged with the discovery of additional hereditary forms of PD which are caused by pathogenic variants in *DJ-1*, *ATP13A2*, *FBX07*, *DNAJC6*, *SYNJ1*, and *PLA2G6* [128]. Collectively, these have implicated mitochondrial transport, fission/fusion, biogenesis, quality control, and mitophagy as underlying causes of neurodegeneration, certainly in hereditary forms, and potentially in sporadic forms of PD. Strategies which enhance parkin expression or increase PINK1 activity are emerging as promising targets for new therapeutics.

In summary, the link between mitochondrial biology and aging continues to be an important topic, and has expanded beyond the role of reactive oxygen species; currently, multiple aspects of mitochondrial physiology are under investigation including apoptosis, senescence, calcium-dependent signaling, and mitochondrial trafficking and turnover [129]. Age-related alterations in the proteostasis of the heteromeric protein complexes are also worthy of research, particularly given their complex origins from both nuclear and mitochondrial gene products.

#### 15.4.2.6 Germline Mutations

Medical geneticists are well aware of the increased risk to the conceptus of chromosomal types of mutations (mainly aneuploidies) as functions of maternal age. This is, of course, the basis for the clinical practice of counseling women of the availability of prenatal diagnosis. The relationship of paternal age to the increased risk of certain types of mutations has also been well documented and considered to be driven by a variety of mechanisms [130]. These important subjects have in fact received substantially less attention by the gerontological community than the question of somatic mutation and aging.

### 15.4.3 Alterations in Lipids

Given the seminal importance of membranes in cell biology, alterations in the structure of membrane lipids could constitute a primary mechanism of age-related cellular dysfunction and cell death. Most research in this field has addressed the issue of lipid peroxidation, an integral component of the free radical theory of aging. Aspects of this idea have been discussed above, including the ubiquitous nature of lipofuscin pigments as a biologic marker of aging. A second line of research

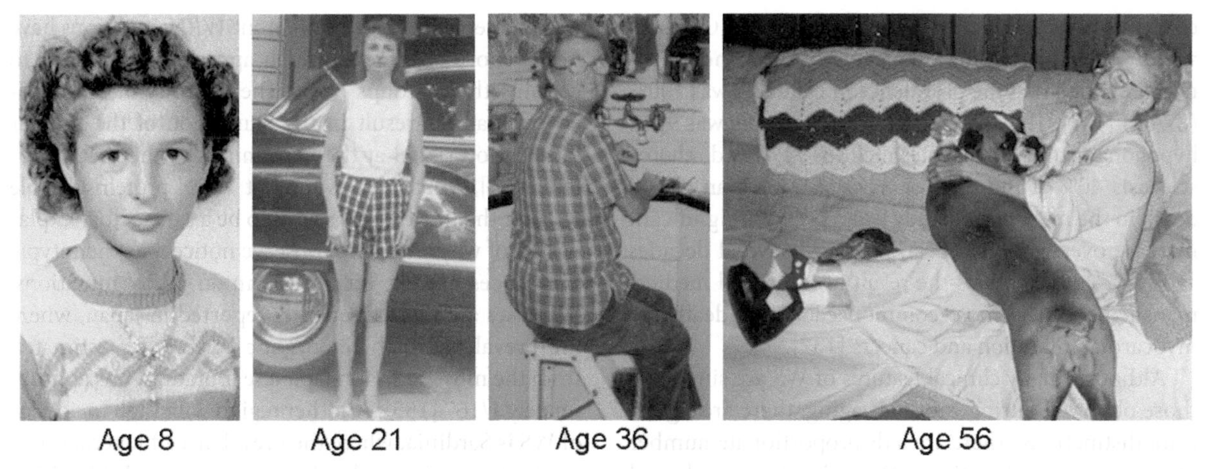


**Figure 15.7** Werner syndrome patients with homozygous *WRN* mutations. (From Hisama FM, Bohr VA, Oshima J. WRN's tenth anniversary. Sci Aging Knowledge Environ 2006;28:pe18.)

in this field has emphasized age-related increases, in various cell types, of the cholesterol-to-phospholipid ratios of plasma cell membranes, with a consequent decrease in membrane fluidity. At least in some cell types, such as neurons of the dorsal root ganglia, the decline in membrane fluidity, as measured by lateral diffusion coefficients, is related to development rather than to postmaturational aging [131]. Studies of replicative senescence in yeast have emphasized the importance of alterations in the lipid membranes of mitochondria [132], thus integrating this topic with the preceding topic on mitochondrial alterations in aging. Such overlaps of various mechanisms of aging are likely to be the rule rather than the exception.

## 15.5 PROGEROID SYNDROMES OF HUMANS

Having reviewed the state of our knowledge of the biology and pathobiology of aging, we can now consider spontaneous mutations in humans that may modulate the aging phenotype. As we have seen, however, evolutionary theory would argue that no single mutation or polymorphism is likely to modulate all aspects of the senescent phenotype. In a systematic survey of several editions of McKusick's catalog of the Mendelian inheritance of humans, one of us (GMM) indeed concluded that no single mutation has yet been identified that could be characterized as a global progeria [74]. A number of mutations, however, could be characterized as "segmental progeroid mutations," in that multiple segments of the complex senescent phenotype appear to have been affected, whereas unimodal syndromes predominantly affect a single organ (e.g., dementias of the Alzheimer type) [74,133]. The responsible mutations include those that impact genomic stability, nuclear structure, numbers of triplet-repeats, and alterations in lipid and carbohydrate metabolism. Some chromosomal aneuploidies (e.g., trisomy 21) also exhibit segmental progeroid features [74]. The two best-known examples of segmental progeroid syndromes are Werner syndrome (WS) and Hutchinson–Gilford progeria syndrome (HGPS), which are discussed in more detail in the following.

### 15.5.1 Werner Syndrome

The clinical phenotype of WS (OMIM# 277700) has been succinctly summarized as a "caricature of aging" (Fig. 15.7) [134,135]. WS patients usually develop normally until they reach the second decade of life. The first clinical sign is a lack of the pubertal growth spurt during the teen years. In their 20s and 30s, patients begin to exhibit a general appearance of accelerated aging with skin atrophy, loss of subcutaneous fat, and loss and graying of hair. They also develop common age-related disorders including type II diabetes mellitus, bilateral ocular cataracts (requiring surgery at a median age of 30), osteoporosis; gonadal atrophy (with early loss of fertility), premature and severe forms of arteriosclerosis (including atherosclerosis, arteriolosclerosis, and medial calcinosis); and peripheral neuropathy. Multiple cancers can be observed by middle age [136]. Our survey of WS patients with a molecularly confirmed diagnosis revealed that the prevalence of cataracts was 100% (87/87) [137]. The prevalence of osteoporosis was 91%, hypogonadism 80%, diabetes mellitus 71%, and atherosclerosis 40% at the time of diagnosis. Median age of death in the most recent study was 54 years, a significant increase over what had been observed several decades ago [137–140], perhaps the result of improved medical management. The most common causes of death are myocardial infarction and cancers [137].

Although many clinical features of WS are similar to those observed during "normal" aging, there are significant distinctions. There are a disproportionate number of sarcomas in WS patients: the ratio of mesenchymal cancers to epithelial cancers in WS is approximately 1:1 as compared to 1:10 in the general population [136]. Alzheimer-type dementia is not common in WS [141]. The long bones of the limbs, especially of the lower limbs, are particularly vulnerable to osteoporosis, whereas in ordinary aging the vertebral column is particularly vulnerable, especially in females [142]. There is also a peculiar osteosclerosis of the distal phalanges that is not seen during ordinary aging [143]. Necrotic skin ulcers and necrosis around ankles and occasionally around elbows, which eventually may require amputation, are characteristic to WS, but rarely seen during usual aging.

Classical WS is caused by mutations of the *WRN* gene on chromosome 8. The locus spans approximately 250 kb and consists of 35 exons, 34 of which are protein coding [144]. *WRN* encodes a 180-kDa multifunctional nuclear protein that belongs to the RecQ family of helicases [145]. A structural study revealed a unique interaction between the RecQC-terminal domain of WRN protein and the DNA substrates during base separation [146]. In contrast to other members of the RecQ family, WRN protein includes an N-terminal domain that codes for exonuclease activity [147]. A single-strand DNA annealing activity in the C-terminal region has also been reported [148]. Its preferred substrates resemble various DNA metabolic intermediates, substrates for which its helicase and exonuclease activities function in a coordinated manner, suggestive of roles in DNA repair, recombination, and replication [149,150]. The WRN protein is also involved in telomere maintenance [151], which explains the accelerated telomere shortening of fibroblasts derived from WS patients [152].

To date, more than 80 different *WRN* mutations have been reported, some of which appear to be specific to certain ethnic groups [153]. The majority of these disease mutations result in the truncation of the nuclear localization signal at the C-terminus of the WRN protein [154], which makes mutant WRN proteins unable to enter the nuclei. This seems to be a satisfactory explanation of why we do not observe noticeable phenotypic differences among various common *WRN* mutations. WS cases are most frequently reported in Japan, where the prevalence of heterozygotic carriers, as estimated from the most common Japanese mutation, was approximately 1/167 [155]. Another region with high incidence of WS is Sardinia, where the prevalence of heterozygous carriers was estimated to be of the order of 1/120 [156]. Frequencies of *WRN* mutations in other populations are unknown, as WS may often escape diagnosis.

Some evolutionary biologists would argue that WS is a poor model of aging, in that it is clear that it would not fit the definition of a set of phenotypes that have escaped the force of natural selection [42].

### 15.5.2 Hutchinson–Gilford Progeria Syndrome

HGPS (OMIM# 176670) is a childhood-onset progeria (Fig. 15.8). It was first described by Jonathan Hutchinson [157] in a boy with baldness and atrophic skin. Hastings Gilford [158] then described a patient with accelerated aging and ateleiosis who died with symptoms of angina pectoris at age 18. HGPS patients are typically normal at birth, but their growth soon falls below the normal range within the first 3–6 months of life. In addition, the children develop accelerated, degenerative changes of the cutaneous, musculoskeletal, and cardiovascular systems [159–161]. The pathognomonic appearance develops usually by age 2 years, and includes: baldness with loss of eyelashes and eyebrows, prominent eyes, convex nasal bridge, small jaw, and generalized loss of subcutaneous fat, resulting in an overall prematurely aged appearance. Historically, most patients survived only to the early teens, with a median age of death in HGPS patients of 16.4 years. Virtually all the patients succumb to myocardial infarction, strokes, or congestive cardiac failure [162]. This is in contrast to patients with the adult-onset progeroid syndrome, WS, whose onset is after puberty and who live until the sixth decade or beyond. Malignancies, ocular cataracts, and Alzheimer-type dementia are also not commonly seen in HGPS, perhaps because they die at such early ages.

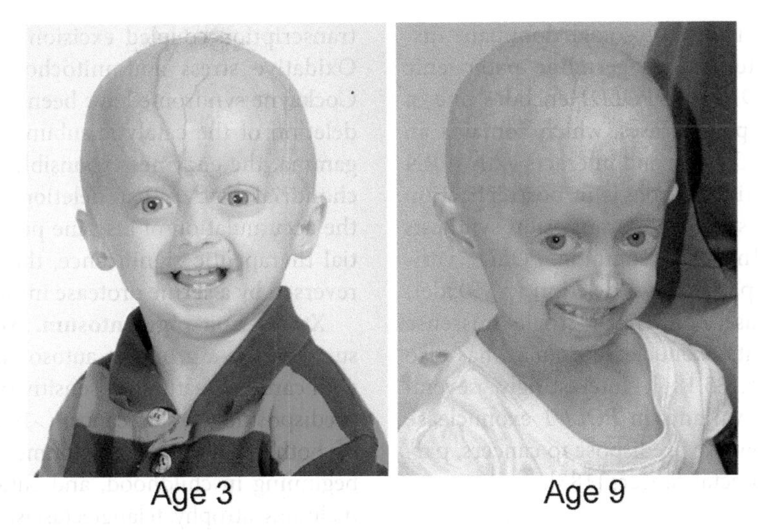


**Figure 15.8** Hutchinson–Gilford progeria syndrome. (Courtesy of the Progeria Research Foundation.)

HGPS is caused in nearly all cases by de novo heterozygous mutations in *LMNA*, which encodes nuclear intermediate filaments, lamin A and C [163,164]. Lamin A and C, generated by alternative splicing of *LMNA*, undergo dimerization and head-to-tail assembly to form nuclear lamina that lies on the inner surface of the inner nuclear membrane [165]. Point mutations within *LMNA* exon 11 found in HGPS create a cryptic splicing site and generate a 50-amino-acid in-frame deletion that includes the proteolytic site required for the maturation of prelamin A to lamin A [163,164]. Unlike wild-type lamin A, this in-frame deletion mutant, termed progerin, retains the farnesyl moiety at its C-terminus. The resulting accumulation of progerin is thought to be responsible for the phenotypic presentation of HGPS [166]. At a cellular level, the presence of progerin is shown to cause structural abnormalities and/or fragility of nuclei [167], aberrant reorganization of the heterochromatin and epigenetic changes [168], genomic instability [169], and impaired telomere maintenance [170]. Age-associated accumulation of small amounts of progerin has been demonstrated in human fibroblasts and coronary arteries, suggesting the possibility that progerin may be, in part, involved in development of the age-related pathologies in normal individuals [162,171].

*LMNA* mutations are also responsible for a group of disorders, termed laminopathies, including Emery–Dreifuss muscular dystrophy, dilated cardiomyopathy type 1A (DCM1A) with or without atrioventricular conduction disturbance, limb-girdle muscular dystrophy type 1B (LGMD1B), Charcot–Marie–Tooth disease type 2 (CMT2), Dunnigan-type familial partial lipodystrophy, mandibuloacral dysplasia, restrictive dermopathy (RD), and atypical forms of Werner syndrome [172]. The atypical Werner syndrome patients have tested negative for biallelic pathogenic WRN variants, have short stature, and adult onset of progeroid features, but with accelerated cardiovascular disease in adulthood. Two such patients were investigated and found to express progerin, albeit at much lower levels than found in classic HGPS patients [173].

Farnesyltransferase inhibitors have been shown to ameliorate HGPS phenotypes in cell cultures and in mouse models [174–177]. Results of a clinical trial in humans were published in 2018 [178,179]. The rationale for this approach has been challenged, however, by the finding that nonfarnesylated progerin can elicit HGPS-like phenotypes in mice [180].

### 15.5.3 MDPL Syndrome

MDPL syndrome (mandibular hypoplasia, deafness, progeroid features, lipodystrophy) usually presents in the first or second decades of life [181]. MDPL patients begin to develop prominent loss of subcutaneous fat, a characteristic facial appearance, metabolic abnormalities, including diabetes mellitus, and progeroid features. Sensorineural deafness is seen in most cases. Undescended testes and hypogonadism have been reported in males but females may be fertile.

MDPL syndrome is an autosomal dominant disorder caused by heterozygous germline pathogenic variants in the *POLD1* gene. *POLD1* encodes one of the main replicative polymerases, which contains an intrinsic exonuclease domain and interacts with WRN protein [182]. Its additional role is in postreplication repair of the lagging strand as a translesion synthesis (TLS) polymerase. The most common *POLD1* variant found in MDPL patients is a deletion (p.S605del) within the polymerase domain. A single missense mutation located in its exonuclease domain has also been identified [181,183,184]. Interestingly, several germline pathogenic variants in *POLD1* exonuclease domains are also known to predispose to cancers, particularly familial colorectal cancers [185].

### 15.5.4 Rare Genomic Instability Disorders Resulting in Segmental Progeroid Phenotypes

The role of genomic instability in producing a wide variety of segmental progeroid phenotypes has been revealed by additional rare genetic syndromes, a few illustrative examples of which are given here.

**Ataxia-telangiectasia (AT).** The progeroid features caused by biallelic pathogenic ATM gene variants in AT individuals include: an increased cancer risk, graying hair, immunodeficiencies, reduced fertility, and neurological signs, such as ataxia and oculomotor apraxia. Women who are heterozygous for ATM gene pathogenic variants are at usually moderately increased risk of breast cancer. Cultured cells from AT patients exhibit an increased frequency of spontaneous chromosomal anomalies including double-stranded breaks and telomeric abnormalities, and demonstrate a characteristic and marked sensitivity to the cytotoxic effects of ionizing radiation.

**Cockayne syndrome (CS).** Loss of subcutaneous adipose tissue, hypertension, atherosclerosis and arteriolosclerosis, age-related renal pathology, and cognitive decline are among the clinical features of Cockayne syndrome, which contribute to its inclusion as a segmental progeroid syndrome [186]. Mutations in at least five loci have been associated with CS—*CSA*, *CSB*, *XPB*, *XPD*, and *XPG*—thus documenting pathogenetic overlaps with xeroderma pigmentosa [187]. About two-thirds of Cockayne patients have mutations at *CSB* and about one-third at the *CSA* locus [186]. The underlying pathogenesis involves defects in transcription-coupled excision repair of DNA [188]. Oxidative stress and mitochondrial dysfunction in Cockayne syndrome have been shown to be related to deletion of the catalytic subunit of DNA polymerase gamma, the enzyme responsible for replicating mitochondrial DNA. That deletion was associated with the accumulation of a serine protease; of great potential therapeutic significance, the phenotype could be reversed by a serine protease inhibitor [189].

**Xeroderma pigmentosum.** Xeroderma pigmentosum (XP) is a group of autosomal recessive disorders with cardinal features of sensitivity to sunlight, marked predisposition to skin cancer (>1000-fold increased risk of both melanoma and nonmelanoma-type cancers) beginning in childhood, and cutaneous abnormalities including atrophy, telangiectasias, actinic keratosis, and pigmentary changes. Ocular features include: corneal abnormalities, visual impairment, and tumors. Neurological abnormalities are seen in a small subset of patients, and features include microcephaly, intellectual disability, hearing loss, and impaired motor function.

XP results in most cases from a defect in nucleotide excision repair (NER), the major DNA repair mechanism to remove helix distorting lesions such as UV-induced pyrimidine dimers, and bulky adducts induced by certain chemicals. NER requires the coordination of more than 30 polypeptides acting in two NER subpathways: global genome repair, and transcription-coupled repair.

Xeroderma pigmentosum was defined classically by eight complementation groups (XPA–XPG, and XPV), however, the discovery of the underlying molecular genetic causes of XP has revealed significant phenotypic overlap with other defined disorders of DNA repair such as Cockayne syndrome, Fanconi anemia, trichothiodystrophy, and cerebro-oculo-facio-skeletal syndrome (COFS).

**XPF/progeroid syndrome.** A single patient has been reported who was normal at birth except for marked sun sensitivity, but then had learning disability, hearing loss, optic atrophy by age 6 years, a progeroid appearance by age 10 years [190]. As a teen, he was found to have microcephaly, renal insufficiency, hypertension, and dry atrophic, irregularly pigmented skin with sunburn, but without skin cancer. Skin fibroblasts showed a severe reduction in UV-induced DNA damage repair, consistent with a diagnosis of xeroderma pigmentosum, but the mild skin findings and progeroid features with multiorgan involvement were not characteristic. The parents were consanguineous, and the patient was found to be homozygous for p.R153P in the ERCC4 gene. It is thought that the p.R153P variant results in cell death, which allows progeroid features, but milder variants result in accumulation of somatic mutations, and the development of cancer, rather than cell death.

**COATS plus disease.** COATS plus disease, also called cerebroretinal microangiopathy with calcifications and cysts-1 (CRMCC1), is a developmental disorder characterized by intracranial calcifications, leukoencephalopathy, and retinal telangiectasia (COATS disease). It is caused by biallelic mutations of CTC1 gene that encodes the conserved telomere maintenance component 1 [191]. While null or truncated CTC1 mutations are generally associated with lethality, patients with missense mutations may present with a combination of milder phenotypes including progeroid features and recurrent fractures [192].

### 15.5.5 Disorders of Lipid and Carbohydrate Metabolism Resulting in Segmental Progeroid Phenotypes

Generalized lipodystrophies can be genetic or acquired, and are associated with profound metabolic disturbances that increase in prevalence with age in the normal population including: insulin resistance, fatty liver, hypertriglyceridemia, and type II diabetes mellitus. Lipodystrophies therefore warrant consideration as segmental progeroid syndromes.

A valuable review of the pathophysiologies of a wide range of both genetic and acquired lipodystrophies has recently been published, including several genetic variants of the Seip syndrome [193]. All currently recognized forms of the latter are autosomal recessive in nature. The type 1 disorder is due to mutations at *AGPAT2*, which codes for 1-acylglycerol-3-phosphate O-acyltransferase-2, an enzyme involved in de novo phospholipid biosynthesis. Type 2 is caused by mutations at *BSCL2* (Berardinelli–Seip congenital lipodystrophy 2, also known as Seipin), which codes for a transmembrane protein that, like *AGPAT2*, is localized to the endoplasmic reticulum and participates in the control of lipid droplet formation and adipocyte differentiation. The type 3 disorder involves mutations at *CAV1*, or caveolin 1, a plasma membrane scaffolding protein and oncogene. Finally, the type 4 disorder is associated with mutations at *PTRF*, coding for the polymerase I and transcript release factor, which is required for dissociation of a transcription complex and is also involved in the organization of the caveolae of plasma membranes. These mutations result in variable expressions of striking losses of normal adipose tissue and abnormal accumulations of lipids in various viscera, including skeletal muscle, liver, and heart. In addition to the regional atrophy of subcutaneous tissues that is so common in normative aging, one also observes type II diabetes mellitus [194], cardiovascular lesions [195] often associated with lipid abnormalities, sometimes with multiple xanthomas [196], psychomotor abnormalities [197] and what some regard as secondary abnormalities of mitochondrial oxidative phosphorylation [198]. Gastrointestinal polyps, a common benign feature of normative aging, have also been observed [199].

Like all segmental progeroid syndromes, there are of course discordances with what one observes in normative aging, the most dramatic of which is the striking muscular hypertrophy associated with Berardinelli–Seip congenital lipodystrophy.

### 15.5.6 Miscellaneous Disorders Resulting in Segmental Progeroid Phenotypes

A number of additional genetic disorders result in a segmental progeroid phenotype, including myotonic dystrophy (cataracts and muscle atrophy), and trisomy 21 (premature Alzheimer disease and prematurely aged appearance). Type II diabetes is a common, complex disorder with a dramatic increase in prevalence in the past two generations in industrialized countries, and in recent years, is becoming an increasingly common diagnosis in the pediatric population. The rapid change in the prevalence cannot be attributed to underlying genetic alterations, but rather largely to lifestyle and environmental factors, including sugar-laden foods and beverages, and sedentary lifestyle. The reason it is worth bringing to attention here is that type II diabetes has premature aging effects on many organs including: the cardiovascular system, the renal system, the nervous system, and causes retinopathy. It is entirely possible that the cumulative effects of a large population of children and young adults with many more years of type II diabetes than previously could result in shortened life spans and health spans in the next few generations.

## 15.6 PRO-LONGEVITY LOCI AND "ANTIGEROID" SYNDROMES

Similar to the absence of a global "progeroid syndrome", i.e., one that recapitulates *all* of the features of usual aging, there is also no global "antigeroid" syndrome yet discovered in humans. Given the discussion above on the polygenic mechanisms of aging, it would seem unlikely that allelic variants at a single locus would lead to such a syndrome.

It is the case, however, that there are many human subjects who remain healthy, cognitively and physically active, well into their 80s, 90s, and beyond. These have included research subjects recruited by the New England Centenarian study and the Institute of Aging Research Longevity Genes Project of the Albert Einstein College of Medicine.

Unlike the case for model organisms such as *C. elegans*, in which single gene variants can lead to a doubling of life span, no such rare variants of large effect have been discovered to date among populations of human centenarians. Nonetheless, there have been some tantalizing genetic associations. For example, a deletion of exon 3 (d3) in the human growth hormone receptor is a polymorphism found in approximately 25% of the population. The frequency of homozygosity for this allele is tripled from 4% in controls to 12% in male centenarians. Multivariate regression analysis indicated that the d3/d3 genotype increased life span by 10 years. Given the role of growth hormone and insulin-like growth factor signaling in the regulation of life spans of a number of species, this finding supports a similar role in our own species. Surprisingly, however, no enrichment of this allele was found among female centenarians [200].

### 15.6.1 Dementias of the Alzheimer Type (DAT)

The presence of one or two *APOE4* alleles has been shown to be by far the major genetic risk factor for sporadic, late-onset forms of DAT. The complementary observation that the *APOE2* allele is protective, however [201], has received comparatively much less attention, as evidenced by searches of PubMed for "APOE4 and Alzheimer's disease" versus "APOE2 and Alzheimer's disease." As of this writing, those numbers are, respectively, 3201 versus 432. This, plus the fact that it is so difficult to find other well-established examples of unimodal antigeroid alleles, supports the need for much more research on this topic. For the present example, the importance of such research is not only relevant to the disease entities in question (DAT), it is also relevant to the broader issue of gene actions related to the heritability of longevity. The *APOE2* allele is among those which contribute to this heritability, which has been estimated to be of the order of 25–33% [202].

A number of suggested mechanisms of gene action have been reviewed for the *APOE2* allele [202b]. Of particular interest are gene actions that impact upon the structure and function of dendrites and possible antioxidant functions.

### 15.6.2 Atherosclerosis

Although the *APOE2* allele has the potential to enhance longevity (Suri et al., 2013), it is perhaps surprising that homozygosity for that allele has been associated with dysbetalipoproteinemia (type III hyperlipoproteinemia), a disorder that accelerates atherogenesis. The common E3 allele might therefore be considered to protect from atherogenesis.

There are many potentially fruitful areas of research regarding gene actions that protect human subjects from atherosclerosis, a major contributor to death from cardiovascular diseases. This has been well established, most notably at the PCSK9 locus. Pathogenic missense variants in *PCSK9*, which encodes proprotein convertase subtilisin kexin type 9, a serine protease, were reported to result in a rare familial form of hypercholesterolemia in 2003 [203]. The mechanism was thought to be a gain of function which promotes degradation of the hepatic low-density lipoprotein (LDL) receptors, which normally act to clear circulating LDL. This observation suggested the possibility of loss-of-function variants in PCSK9 which would be predicted to increase total cell surface LDL receptors and result in decreased cholesterol. In fact, two premature truncating variants were discovered in ~2% of African-Americans with low plasma LDL, and associated with 30–40% reductions in LDL and an 80% reduction in coronary artery disease [204]. In a relatively short period of time, this discovery led to the development of a new class of effective (albeit expensive) cholesterol-lowering drugs: monoclonal antibodies against PCSK9.

This success story of the role of human genetics in leading to development of a novel drug focused attention on another potential target: *ANGPTL3*, which encodes angiopoietin-like protein 3. Investigation of a family with loss-of-function variants in *ANGPTL3* found affected members had combined hypolipidemia with reduced triglycerides, as well as low LDL and HDL cholesterol, and resistance to atherosclerosis [205]. Subsequently, in clinical trials with monoclonal antibodies against ANGPTL3 or antisense oligonucleotides against ANGPTL3 mRNA, both approaches resulted in dramatic reductions in triglycerides, and significant reductions in cholesterol [206,207].

Finally, systematic studies of "human knockouts" in consanguineous populations have identified individuals homozygous for loss-of-function (LOF) variants in APOC3 encoding apolipoprotein C3. Deep phenotyping of one family in which both parents and all of their children were homozygous for LOF APOC3 showed absent plasma apoC3 protein, lower triglycerides, higher HDL cholesterol and similar LDL cholesterol, and blunting of the postprandial rise in triglycerides after a fatty meal [208].

### 15.6.3 Genetic Resistance to Environmental Carcinogens

There is no doubt that one of the secrets to the avoidance of cancer (especially lung cancer) and to increasing one's chances of living to the 10th and 11th decades of life is to avoid cigarette smoke [209]. But why do some heavy cigarette smokers live well into their 10th and 11th decades free of lung cancer [210]? There is a very large literature on candidate polymorphic variants that can provide protection against some of the large numbers of carcinogenic compounds in cigarette smoke, a review of which is beyond the scope of this chapter. Polygenic models [211,212] are likely to be the most satisfactory approaches to uncovering various patterns of resistance and susceptibility as we approach the era of whole-genome sequencing and precision medicine [213].

### 15.6.4 Human Allelic Variants Homologous to Pro-Longevity Genes in Model Organisms

There has been a surge of interest in testing the hypothesis that the ability to achieve remarkable longevities in centenarians is due to the inheritance of alleles at a few loci of major relevance. A priori, one would predict that such research would be quite risky, given the arguments made earlier in this chapter that life span is under highly polygenic modulations and that it is also determined, in part, by stochastic events. It is the case, however, also noted above, that atherosclerosis (and the associated heart attacks and strokes) is a major limitation of human life span in developed societies. Therefore, it is perhaps not surprising that an association of unusual longevity with variant alleles for lipoprotein metabolism has been observed [214]. More recent studies demonstrated the association of polymorphisms in the forkhead box class O (FOXO) family of transcription factors among several independent centenarian populations [215]. The FOXO genes are key regulators of the insulin-IGF1 signaling pathway (Fig. 15.5).

There has also been considerable interest lately in Laron dwarfism because of their mutations in the growth hormone pathway [216,217]. People with Laron dwarfism in Ecuadorian villages are resistant to cancer and diabetes and are somewhat protected against aging. This is consistent with findings in mice with a defective growth hormone receptor gene, suggesting this "public mechanism" of aging may apply to our species [218].

## 15.7 CONCLUSIONS AND FUTURE DIRECTIONS

The careful phenotypic characterization of both segmental progeroid syndromes [74] and unimodal progeroid syndromes [219] by medical geneticists and others have greatly contributed to the development of various hypotheses of gene actions involved in biological aging and geriatric disorders. Regarding fundamental mechanisms of aging, the segmental progeroid syndromes have provided particularly strong support for the role of genomic instability [220]. This mechanism of aging may have deep evolutionary roots [221]. Regarding specific geriatric disorders, very specific pathogenetic pathways have been discovered, a cogent example of which is the elucidation of the role of beta amyloid in the pathogenesis of all forms of dementias of the Alzheimer type [222]. It is important to point out, however, that the origins of that hypothesis did not come from the study of the common sporadic, late-onset forms of the disorder, but were the result of studies by medical geneticists of rare pedigrees with autosomal dominant mutations in that pathway leading to early-onset forms of the disorder [96]. There is an important lesson here for investigators interested in other common geriatric disorders.

Regarding future directions by medical geneticists interested in the pathobiology of aging and age-directed disorders, we wish to emphasize what has been a comparatively neglected approach—a genetic analysis of individuals exhibiting unusual resistance to segmental or unimodal patterns of aging and age-related disorders—i.e., the search for antigeroid allelic variants. Given the classical evolutionary biological theories of aging [54,57], it would seem prudent to carry longitudinal studies of phenotypes that begin to emerge after the steep decline in the force of natural selection—i.e., beginning in early middle age [2].

Medical geneticists have great opportunities to capitalize upon the increasing pace of our understanding of the epigenome, including the development of tools for the epigenetic analysis of single cells [223]. These have the potential to elucidate stochastic variations in gene expression during aging [71].

Finally, although we have been focusing upon phenotypes in middle and old age, it will be important to keep in mind that fact that how well one builds an organism makes a great deal of difference in how well that organism functions and how long it lasts. Geroscientists, including geneticists interested in variations in rates and patterns of aging, should certainly not neglect developmental biology. This research should enthusiastically embrace epigenetic research on intergenerational and transgenerational inheritance [224]; it would be hard to overemphasize the public health significance of such research.

## REFERENCES


[1] Mazucanti CH, Cabral-Costa JV, Vasconcelos AR, Andreotti DZ, Scavone C, Kawamoto EM. Longevity pathways (mTOR, SIRT, insulin/IGF-1) as key modulatory targets on aging and neurodegeneration. Curr Top Med Chem 2015;15:2116–38.

[2] Martin GM. Help wanted: physiologists for research on aging. Sci Aging Knowledge Environ 2002:vp2.

[3] Martin GM. Modalities of gene action predicted by the classical evolutionary biological theory of aging. Ann N Y Acad Sci 2007;1100:14–20.

[4] Treaster SB, Chaudhuri AR, Austad SN. Longevity and GAPDH stability in bivalves and mammals: a convenient marker for comparative gerontology and proteostasis. PLoS One 2015;10:e0143680.

[5] Kowald A, Kirkwood TB. Can aging be programmed? A critical literature review. Aging Cell 2016.

[6] Gavrilov LA, Gavrilova NS. The biology of life span: a quantitative approach. New York: Harwood Academic; 1991.

[7] Gavrilov LA, Gavrilova NS. The quest for a general theory of aging and longevity. Sci Aging Knowl Environ 2003;2003:RE5.

[8] Gompertz B. On the nature of the function expressive of the law of human mortality and on a new mode of determining life contingencies. Philos Trans R Soc Lond Biol Ser A 1825;115:513–85.

[9] Makeham WM. On the law of mortality and the construction of annuity tables. J Inst Actuar 1860;8: 301–10.

[10] Sacher GA. Evolution of longevity and survival characteristics in mammals. In: Schneider EL, editor. The genetics of aging. New York: Plenum Press; 1978.

[11] Milne EM. Dynamics of human mortality. Exp Gerontol 2010;45:180–7.

[12] Carey JR, Liedo P, Muller HG, Wang JL, Vaupel JW. Dual modes of aging in Mediterranean fruit fly females. Science 1998;281:996–8.

[13] Vaupel JW, Carey JR, Christensen K, Johnson TE, Yashin AI, Holm NV, Iachine IA, Kannisto V, Khazaeli AA, Liedo P, Longo VD, Zeng Y, Manton KG, Curtsinger JW. Biodemographic trajectories of longevity. Science 1998;280:855–60.

[14] Robine JM, Allard M. The oldest human. Science 1998;279:1834–5.

[15] Schock NW. Systems integration. In: Finch CE, Hayflick L, editors. Handbook of the biology of aging. New York: Van Nostrand-Reinhold; 1977.

[16] Schulz R, Curnow C. Peak performance and age among superathletes: track and field, swimming, baseball, tennis, and golf. J Gerontol 1988;43:P113–20.

[17] Breteler MM, Claus JJ, van Duijn CM, Launer LJ, Hofman A. Epidemiology of Alzheimer's disease. Epidemiol Rev 1992;14:59–82.

[18] Ritchie K, Kildea D. Is senile dementia "age-related" or "ageing-related"?-evidence from meta-analysis of dementia prevalence in the oldest old. Lancet 1995;346:931–4.

[19] Katzman R, Kawas C. Risk factors for Alzheimer's disease. Neurosci News 1998;1:27–34.

[20] Nelson EA, Dannefer D. Aged heterogeneity: fact or fiction? The fate of diversity in gerontological research. Gerontol 1992;32:17–23.

[21] Lindeman RD. Is the decline in renal function with normal aging inevitable? Geriatr Nephrol Urol 1998;8:7–9.

[22] Schachter F, Faure-Delanef L, Guenot F, Rouger H, Froguel P, Lesueur-Ginot L, Cohen D. Genetic associations with human longevity at the APOE and ACE loci. Nat Genet 1994;6:29–32.

[23] Higgins GA, Large CH, Rupniak HT, Barnes JC. Apolipoprotein E and Alzheimer's disease: a review of recent studies. Pharmacol Biochem Behav 1997;56: 675–85.

[24] Corder EH, Saunders AM, Strittmatter WJ, Schmechel DE, Gaskell PC, Small GW, Roses AD, Haines JL, Pericak-Vance MA. Gene dose of apolipoprotein E type 4 allele and the risk of Alzheimer's disease in late onset families. Science 1993;261:921–3.
[25] Yu JT, Tan L, Hardy J. Apolipoprotein E in Alzheimer's disease: an update. Annu Rev Neurosci 2014;37:79–100.
[26] Carmeli E, Reznick AZ. The physiology and biochemistry of skeletal muscle atrophy as a function of age. Proc Soc Exp Biol Med 1994;206:103–13.
[27] Gonzalez-Freire M, de Cabo R, Studenski SA, Ferrucci L. The neuromuscular junction: aging at the crossroad between nerves and muscle. Front Aging Neurosci 2014;6:208.
[28] Wall BT, Dirks ML, van Loon LJ. Skeletal muscle atrophy during short-term disuse: implications for age-related sarcopenia. Ageing Res Rev 2013;12:898–906.
[29] Lopez-Torres M, Perez-Campo R, Fernandez A, Barba C, Barja de Quiroga G. Brain glutathione reductase induction increases early survival and decreases lipofuscin accumulation in aging frogs. J Neurosci Res 1993;34:233–42.
[30] Martin GM. Interactions of aging and environmental agents: the gerontological perspective. Prog Clin Biol Res 1987;228:25–80.
[31] Martin GM. Cellular aging–postreplicative cells. A review (Part II). Am J Pathol 1977;89:513–30.
[32] Nakano M, Mizuno T, Gotoh S. Accumulation of cardiac lipofuscin in crab-eating monkeys (*Macaca fasicularis*): the same rate of lipofuscin accumulation in several species of primates. Mech Ageing Dev 1993;66:243–8.
[33] Katz ML, White HA, Gao CL, Roth GS, Knapka JJ, Ingram DK. Dietary restriction slows age pigment accumulation in the retinal pigment epithelium. Invest Ophthalmol Vis Sci 1993;34:3297–302.
[34] Rao G, Xia E, Nadakavukaren MJ, Richardson A. Effect of dietary restriction on the age-dependent changes in the expression of antioxidant enzymes in rat liver. J Nutr 1990;120:602–9.
[35] Kohn RR. Extracellular aging. In: Kohn RR, editor. Principles of mammalian aging. Englewood Cliffs, NJ: Prentice-Hall; 1978.
[36] Sell DR, Monnier VM. Conversion of arginine into ornithine by advanced glycation in senescent human collagen and lens crystallins. J Biol Chem 2004;279:54173–84.
[37] Yan SF, Ramasamy R, Naka Y, Schmidt AM. Glycation, inflammation, and RAGE: a scaffold for the macrovascular complications of diabetes and beyond. Circ Res 2003;93:1159–69.
[38] Saudek DM, Kay J. Advanced glycation endproducts and osteoarthritis. Curr Rheumatol Rep 2003;5:33–40.
[39] Barnes 2nd RH, Akama T, Ohman MK, Woo MS, Bahr J, Weiss SJ, Eitzman DT, Chun TH. Membrane-tethered metalloproteinase expressed by vascular smooth muscle cells limits the progression of proliferative atherosclerotic lesions. J Am Heart Assoc 2017;6.
[40] Bilato C, Crow MT. Atherosclerosis and the vascular biology of aging. Aging 1996;8:221–34.
[41] Charlesworth B. Evolution in age-structured populations. Cambridge: Cambridge University Press; 1980.
[42] Rose MR. Evolutionary biology of aging. New York: Oxford University Press; 1991.
[43] Diamond JM. Big-bang reproduction and ageing in male marsupial mice. Nature 1982;298:115–6.
[44] Hamilton WD. The moulding of senescence by natural selection. J Theor Biol 1966;12:12–45.
[45] Martin GM, Austad SN, Johnson TE. Genetic analysis of ageing: role of oxidative damage and environmental stresses. Nat Genet 1996;13:25–34.
[46] Baudisch A. Hamilton's indicators of the force of selection. Proc Natl Acad Sci U S A 2005;102:8263–8.
[47] Hayflick L, Moorhead PS. The serial cultivation of human diploid cell strains. Exp Cell Res 1961;25:585–621.
[48] Fontana L, Partridge L, Longo VD. Extending healthy life span–from yeast to humans. Science 2010;328:321–6.
[49] Uno M, Nishida E. Lifespan-regulating genes in *C. elegans*. NPJ Aging Mech Dis 2016;2:16010.
[50] Walker DW, McColl G, Jenkins NL, Harris J, Lithgow GJ. Evolution of lifespan in *C. elegans*. Nature 2000;405:296–7.
[51] Briga M, Verhulst S. What can long-lived mutants tell us about mechanisms causing aging and lifespan variation in natural environments? Exp Gerontol 2015;71:21–6.
[52] Johnson JE, Johnson FB. Methionine restriction activates the retrograde response and confers both stress tolerance and lifespan extension to yeast, mouse and human cells. PLoS One 2014;9:e97729.
[53] Masoro EJ. Dietary restriction-induced life extension: a broadly based biological phenomenon. Biogerontology 2006;7:153–5.
[54] Medawar PB. An unsolved problem of biology. London: HK Lewis; 1952.
[55] Warby SC, Graham RK, Hayden MR. Huntington disease, GeneReviews. Seattle: University of Washington; 2010.
[56] Haldane JBS. New paths in genetics. New York: Harper and Brothers; 1942.
[57] Williams GC. Pleiotropy, natural selection, and the evolution of senescence. Evolution 1957;11:398–411.
[58] Martin GM. Atherosclerosis is the leading cause of death in the developed societies. Am J Pathol 1998;153:1319–20.

[59] Martin GM. APOE alleles and lipophilic pathogens. Neurobiol Aging 1999;20:441–3.

[60] Campisi J. Aging, tumor suppression and cancer: high wire-act!. Mech Ageing Dev 2005;126:51–8.

[61] Cicin-Sain L, Messaoudi I, Park B, Currier N, Planer S, Fischer M, Tackitt S, Nikolich-Zugich D, Legasse A, Axthelm MK, Picker LJ, Mori M, Nikolich-Zugich J. Dramatic increase in naive T cell turnover is linked to loss of naive T cells from old primates. Proc Natl Acad Sci U S A 2007;104:19960–5.

[62] Basisty N, Dai DF, Gagnidze A, Gitari L, Fredrickson J, Maina Y, Beyer RP, Emond MJ, Hsieh EJ, MacCoss MJ, Martin GM, Rabinovitch PS. Mitochondrial-targeted catalase is good for the old mouse proteome, but not for the young: 'reverse' antagonistic pleiotropy? Aging Cell 2016;15:634–45.

[63] Kirkwood TB, Rose MR. Evolution of senescence: late survival sacrificed for reproduction. Philos Trans R Soc Lond B Biol Sci 1991;332:15–24.

[64] Partridge L, Barton NH. Optimality, mutation and the evolution of ageing. Nature 1993;362:305–11.

[65] Partridge L, Gems D. Mechanisms of ageing: public or private? Nat Rev Genet 2002;3:165–75.

[66] Crow JF. Spontaneous mutation in man. Mutat Res 1999;437:5–9.

[67] Jung A, Schuppe HC, Schill WB. Are children of older fathers at risk for genetic disorders? Andrologia 2003;35:191–9.

[68] Austad SN. Why we age. New York: John Wiley and Sons; 1997.

[69] Austad SN. Retarded senescence in an insular population of Virginia opossums (*Didelphis virginiana*). J Zool 1993;229:695–708.

[70] Reznick DN, Bryant MJ, Roff D, Ghalambor CK, Ghalambor DE. Effect of extrinsic mortality on the evolution of senescence in guppies. Nature 2004;431:1095–9.

[71] Martin GM. Stochastic modulations of the pace and patterns of ageing: impacts on quasi-stochastic distributions of multiple geriatric pathologies. Mech Ageing Dev 2012;133:107–11.

[72] Vanfleteren JR, De Vreese A, Braeckman BP. Two-parameter logistic and Weibull equations provide better fits to survival data from isogenic populations of *Caenorhabditis elegans* in axenic culture than does the Gompertz model. J Gerontol A Biol Sci Med Sci 1998;53:B393–403. discussion B404–398.

[73] Luckinbill LS, Graves JL, Reed AH, Koetsawang S. Localizing genes that defer senescence in *Drosophila melanogaster*. Heredity 1988;60(Pt 3):367–74.

[74] Martin GM. Genetic syndromes in man with potential relevance to the pathobiology of aging. Birth Defects Orig Artic Ser 1978;14:5–39.

[75] Cutler RG. Evolution of human longevity and the genetic complexity governing aging rate. Proc Natl Acad Sci U S A 1975;72:4664–8.

[76] Colman RJ, Anderson RM, Johnson SC, Kastman EK, Kosmatka KJ, Beasley TM, Allison DB, Cruzen C, Simmons HA, Kemnitz JW, Weindruch R. Caloric restriction delays disease onset and mortality in rhesus monkeys. Science 2009;325:201–4.

[77] Roth GS, Mattison JA, Ottinger MA, Chachich ME, Lane MA, Ingram DK. Aging in rhesus monkeys: relevance to human health interventions. Science 2004;305:1423–6.

[78] Mattison JA, Colman RJ, Beasley TM, Allison DB, Kemnitz JW, Roth GS, Ingram DK, Weindruch R, de Cabo R, Anderson RM. Caloric restriction improves health and survival of rhesus monkeys. Nat Commun 2017;8:14063.

[79] Muller FL, Lustgarten MS, Jang Y, Richardson A, Van Remmen H. Trends in oxidative aging theories. Free Radic Biol Med 2007;43:477–503.

[80] Brand MD. Mitochondrial generation of superoxide and hydrogen peroxide as the source of mitochondrial redox signaling. Free Radic Biol Med 2016;100:14–31.

[81] Marcinek DJ, Siegel MP. Targeting redox biology to reverse mitochondrial dysfunction. Aging 2013;5:588–9.

[82] Orgel LE. The maintenance of the accuracy of protein synthesis and its relevance to ageing. Proc Natl Acad Sci U S A 1963;49:517–21.

[83] Orgel LE. The maintenance of the accuracy of protein synthesis and its relevance to ageing: a correction. Proc Natl Acad Sci U S A 1970;67:1476.

[84] Rothstein M. An overview of age-related changes in proteins. Prog Clin Biol Res 1989;287:259–67.

[85] Martin GM, Bressler SL. Transcriptional infidelity in aging cells and its relevance for the Orgel hypothesis. Neurobiol Aging 2000;21:897–900. discussion 903–894.

[86] Rattan SI. Synthesis, modification and turnover of proteins during aging. Adv Exp Med Biol 2010;694:1–13.

[87] Rattan SI, Derventzi A, Clark BF. Protein synthesis, posttranslational modifications, and aging. Ann N Y Acad Sci 1992;663:48–62.

[88] Gershon H, Gershon D. Detection of inactive enzyme molecules in ageing organisms. Nature 1970;227: 1214–7.

[89] Stadtman ER. Protein oxidation and aging. Free Radic Res 2006;40:1250–8.

[90] Kirstein M, Aston C, Hintz R, Vlassara H. Receptor-specific induction of insulin-like growth factor I in human monocytes by advanced glycosylation end product-modified proteins. J Clin Invest 1992;90:439–46.

[91] Asif M, Egan J, Vasan S, Jyothirmayi GN, Masurekar MR, Lopez S, Williams C, Torres RL, Wagle D, Ulrich P, Cerami A, Brines M, Regan TJ. An advanced glycation endproduct cross-link breaker can reverse age-related increases in myocardial stiffness. Proc Natl Acad Sci U S A 2000;97:2809–13.
[92] Kristal BS, Yu BP. An emerging hypothesis: synergistic induction of aging by free radicals and Maillard reactions. J Gerontol 1992;47:B107–14.
[93] Buxbaum JN. The systemic amyloidoses. Curr Opin Rheumatol 2004;16:67–75.
[94] Edwards BJ, Morley JE. Amylin. Life Sci 1992;51: 1899–912.
[95] Higuchi K, Naiki H, Kitagawa K, Hosokawa M, Takeda T. Mouse senile amyloidosis. ASSAM amyloidosis in mice presents universally as a systemic age-associated amyloidosis. Virchows Arch B Cell Pathol Incl Mol Pathol 1991;60:231–8.
[96] Tcw J, Goate AM. Genetics of beta-amyloid precursor protein in Alzheimer's disease. Cold Spring Harb Perspect Med 2017;7.
[97] Clarke S. Aging as war between chemical and biochemical processes: protein methylation and the recognition of age-damaged proteins for repair. Ageing Res Rev 2003;2:263–85.
[98] DeVry CG, Clarke S. Polymorphic forms of the protein L-isoaspartate (D-aspartate) O-methyltransferase involved in the repair of age-damaged proteins. J Hum Genet 1999;44:275–88.
[99] Sala AJ, Bott LC, Morimoto RI. Shaping proteostasis at the cellular, tissue, and organismal level. J Cell Biol 2017;216:1231–41.
[100] Szilard L. On the nature of the aging process. Proc Natl Acad Sci U S A 1959;45:30–45.
[101] Bennett-Baker PE, Wilkowski J, Burke DT. Age-associated activation of epigenetically repressed genes in the mouse. Genetics 2003;165:2055–62.
[102] Cattanach BM. Position effect variegation in the mouse. Genet Res 1974;23:291–306.
[103] Wareham KA, Lyon MF, Glenister PH, Williams ED. Age related reactivation of an X-linked gene. Nature 1987;327:725–7.
[104] Migeon BR, Axelman J, Beggs AH. Effect of ageing on reactivation of the human X-linked HPRT locus. Nature 1988;335:93–6.
[105] Wilson VL, Jones PA. DNA methylation decreases in aging but not in immortal cells. Science 1983;220:1055–7.
[106] Wilson VL, Smith RA, Ma S, Cutler RG. Genomic 5-methyldeoxycytidine decreases with age. J Biol Chem 1987;262:9948–51.
[107] Issa JP, Ottaviano YL, Celano P, Hamilton SR, Davidson NE, Baylin SB. Methylation of the oestrogen receptor CpG island links ageing and neoplasia in human colon. Nat Genet 1994;7:536–40.
[108] Imai SI, Guarente L. It takes two to tango: NAD+ and sirtuins in aging/longevity control. NPJ Aging Mech Dis 2016;2:16017.
[109] Watroba M, Dudek I, Skoda M, Stangret A, Rzodkiewicz P, Szukiewicz D. Sirtuins, epigenetics and longevity. Ageing Res Rev 2017;40:11–9.
[110] Linford NJ, Beyer RP, Gollahon K, Krajcik RA, Malloy VL, Demas V, Burmer GC, Rabinovitch PS. Transcriptional response to aging and caloric restriction in heart and adipose tissue. Aging Cell 2007;6(5):673–88.
[111] Chen BH, Marioni RE, Colicino E, Peters MJ, Ward-Caviness CK, Tsai PC, Roetker NS, Just AC, Demerath EW, Guan W, Bressler J, Fornage M, Studenski S, Vandiver AR, Moore AZ, Tanaka T, Kiel DP, Liang L, Vokonas P, Schwartz J, Lunetta KL, Murabito JM, Bandinelli S, Hernandez DG, Melzer D, Nalls M, Pilling LC, Price TR, Singleton AB, Gieger C, Holle R, Kretschmer A, Kronenberg F, Kunze S, Linseisen J, Meisinger C, Rathmann W, Waldenberger M, Visscher PM, Shah S, Wray NR, McRae AF, Franco OH, Hofman A, Uitterlinden AG, Absher D, Assimes T, Levine ME, Lu AT, Tsao PS, Hou L, Manson JE, Carty CL, LaCroix AZ, Reiner AP, Spector TD, Feinberg AP, Levy D, Baccarelli A, van Meurs J, Bell JT, Peters A, Deary IJ, Pankow JS, Ferrucci L, Horvath S. DNA methylation-based measures of biological age: meta-analysis predicting time to death. Aging 2016;8:1844–65.
[112] Clark AM, Rubin MA. The modification by x-irradiation of the life span of haploids and diploids of the wasp, Habrobracon sp. Radiat Res 1961;15:244–53.
[113] Martin GM, Smith AC, Ketterer DJ, Ogburn CE, Disteche CM. Increased chromosomal aberrations in first metaphases of cells isolated from the kidneys of aged mice. Isr J Med Sci 1985;21:296–301.
[114] Horn PL, Turker MS, Ogburn CE, Disteche CM, Martin GM. A cloning assay for 6-thioguanine resistance provides evidence against certain somatic mutational theories of aging. J Cell Physiol 1984;121:309–15.
[115] Martin GM, Ogburn CE, Colgin LM, Gown AM, Edland SD, Monnat Jr RJ. Somatic mutations are frequent and increase with age in human kidney epithelial cells. Hum Mol Genet 1996;5:215–21.
[116] Trainor KJ, Wigmore DJ, Chrysostomou A, Dempsey JL, Seshadri R, Morley AA. Mutation frequency in human lymphocytes increases with age. Mech Ageing Dev 1984;27:83–6.
[117] Turner DR, Morley AA, Haliandros M, Kutlaca R, Sanderson BJ. In vivo somatic mutations in human lymphocytes frequently result from major gene alterations. Nature 1985;315:343–5.

[118] Dolle ME, Snyder WK, Dunson DB, Vijg J. Mutational fingerprints of aging. Nucleic Acids Res 2002;30:545–9.
[119] Gerez L, de Haan A, Hol EM, Fischer DF, van Leeuwen FW, van Steeg H, Benne R. Molecular misreading: the frequency of dinucleotide deletions in neuronal mRNAs for beta-amyloid precursor protein and ubiquitin B.. Neurobiol Aging 2005;26:145–55.
[120] de Lange T. How telomeres solve the end-protection problem. Science 2009;326:948–52.
[121] Shay JW, Wright WE. Hallmarks of telomeres in ageing research. J Pathol 2007;211:114–23.
[122] Cawthon RM, Smith KR, O'Brien E, Sivatchenko A, Kerber RA. Association between telomere length in blood and mortality in people aged 60 years or older. Lancet 2003;361:393–5.
[123] Epel ES, Blackburn EH, Lin J, Dhabhar FS, Adler NE, Morrow JD, Cawthon RM. Accelerated telomere shortening in response to life stress. Proc Natl Acad Sci U S A 2004;101:17312–5.
[124] Fair B, Mellon SH, Epel ES, Lin J, Revesz D, Verhoeven JE, Penninx BW, Reus VI, Rosser R, Hough CM, Mahan L, Burke HM, Blackburn EH, Wolkowitz OM. Telomere length is inversely correlated with urinary stress hormone levels in healthy controls but not in un-medicated depressed individuals-preliminary findings. J Psychosom Res 2017;99:177–80.
[125] Gerschman R, Nye SW, Gilbert DL, Dwyer P, Fenn WO. Studies on oxygen poisoning: protective effect of beta-mercaptoethylamine. Proc Soc Exp Biol Med 1954;85:75–7.
[126] Harman D. Aging: a theory based on free radical and radiation chemistry. J Gerontol 1956;11:298–300.
[127] Trifunovic A, Wredenberg A, Falkenberg M, Spelbrink JN, Rovio AT, Bruder CE, Bohlooly YM, Gidlof S, Oldfors A, Wibom R, Tornell J, Jacobs HT, Larsson NG. Premature ageing in mice expressing defective mitochondrial DNA polymerase. Nature 2004;429:417–23.
[128] Scott L, Dawson VL, Dawson TM. Trumping neurodegeneration: targeting common pathways regulated by autosomal recessive Parkinson's disease genes. Exp Neurol 2017.
[129] Gonzalez-Freire M, de Cabo R, Bernier M, Sollott SJ, Fabbri E, Navas P, Ferrucci L. Reconsidering the role of mitochondria in aging. J Gerontol A Biol Sci Med Sci 2015;70:1334–42.
[130] Herati AS, Zhelyazkova BH, Butler PR, Lamb DJ. Age-related alterations in the genetics and genomics of the male germ line. Fertil Steril 2017;107:319–23.
[131] Horie H, Kawasaki Y, Takenaka T. Lateral diffusion of membrane lipids changes with aging in C57BL mouse dorsal root ganglion neurons from a fetal stage to an aged stage. Brain Res 1986;377(2):246–50.
[132] Medkour Y, Dakik P, McAuley M, Mohammad K, Mitrofanova D, Titorenko VI. Mechanisms underlying the essential role of mitochondrial membrane lipids in yeast chronological aging. Oxid Med Cell Longev 2017;2017:2916985.
[133] Martin GM, Oshima J. Lessons from human progeroid syndromes. Nature 2000;408:263–6.
[134] Oshima J, Martin GM, Hisama FM. Werner syndrome. In: Pagon RA, Adam MP, Ardinger HH, Wallace SE, Amemiya A, Bean LJH, Bird TD, Fong CT, Mefford HC, Smith RJH, Stephens K, editors. GeneReviews(R), Seattle (WA). 2014.
[135] Takemoto M, Mori S, Kuzuya M, Yoshimoto S, Shimamoto A, Igarashi M, Tanaka Y, Miki T, Yokote K. Diagnostic criteria for Werner syndrome based on Japanese nationwide epidemiological survey. Geriatr Gerontol Int 2013;13:475–81.
[136] Goto M, Miller RW, Ishikawa Y, Sugano H. Excess of rare cancers in Werner syndrome (adult progeria). Cancer Epidemiol Biomark Prev 1996;5:239–46.
[137] Huang S, Lee L, Hanson NB, Lenaerts C, Hoehn H, Poot M, Rubin CD, Chen DF, Yang CC, Juch H, Dorn T, Spiegel R, Oral EA, Abid M, Battisti C, Lucci-Cordisco E, Neri G, Steed EH, Kidd A, Isley W, Showalter D, Vittone JL, Konstantinow A, Ring J, Meyer P, Wenger SL, von Herbay A, Wollina U, Schuelke M, Huizenga CR, Leistritz DF, Martin GM, Mian IS, Oshima J. The spectrum of WRN mutations in Werner syndrome patients. Hum Mutat 2006;27:558–67.
[138] Epstein CJ, Martin GM, Schultz AL, Motulsky AG. Werner's syndrome a review of its symptomatology, natural history, pathologic features, genetics and relationship to the natural aging process. Medicine (Baltim) 1966;45:177–221.
[139] Goto M. Hierarchical deterioration of body systems in Werner's syndrome: implications for normal ageing. Mech Ageing Dev 1997;98:239–54.
[140] Goto M, Ishikawa Y, Sugimoto M, Furuichi Y. Werner syndrome: a changing pattern of clinical manifestations in Japan (1917~2008). Biosci Trends 2013;7:13–22.
[141] Postiglione A, Soricelli A, Covelli EM, Iazzetta N, Ruocco A, Milan G, Santoro L, Alfano B, Brunetti A. Premature aging in Werner's syndrome spares the central nervous system. Neurobiol Aging 1996;17:325–30.
[142] Mori S, Zhou H, Yamaga M, Takemoto M, Yokote K. Femoral osteoporosis is more common than lumbar osteoporosis in patients with Werner syndrome. Geriatr Gerontol Int 2017;17:854–6.
[143] Goto M, Kindynis P, Resnick D, Sartoris DJ. Osteosclerosis of the phalanges in Werner syndrome. Radiology 1989;172:841–3.

[144] Yu CE, Oshima J, Fu YH, Wijsman EM, Hisama F, Alisch R, Matthews S, Nakura J, Miki T, Ouais S, Martin GM, Mulligan J, Schellenberg GD. Positional cloning of the Werner's syndrome gene. Science 1996;272:258–62.
[145] Gray MD, Shen JC, Kamath-Loeb AS, Blank A, Sopher BL, Martin GM, Oshima J, Loeb LA. The Werner syndrome protein is a DNA helicase. Nat Genet 1997;17:100–3.
[146] Kitano K, Kim SY, Hakoshima T. Structural basis for DNA strand separation by the unconventional winged-helix domain of RecQ helicase WRN. Structure 2010;18:177–87.
[147] Huang S, Li B, Gray MD, Oshima J, Mian IS, Campisi J. The premature ageing syndrome protein, WRN, is a 3′→5′ exonuclease. Nat Genet 1998;20:114–6.
[148] Muftuoglu M, Kulikowicz T, Beck G, Lee JW, Piotrowski J, Bohr VA. Intrinsic ssDNA annealing activity in the C-terminal region of WRN. Biochemistry 2008;47:10247–54.
[149] Brosh Jr RM, Opresko PL, Bohr VA. Enzymatic mechanism of the WRN helicase/nuclease. Methods Enzymol 2006;409:52–85.
[150] Croteau DL, Popuri V, Opresko PL, Bohr VA. Human RecQ helicases in DNA repair, recombination, and replication. Annu Rev Biochem 2014;83:519–52.
[151] Crabbe L, Jauch A, Naeger CM, Holtgreve-Grez H, Karlseder J. Telomere dysfunction as a cause of genomic instability in Werner syndrome. Proc Natl Acad Sci U S A 2007;104:2205–10.
[152] Opresko PL. Telomere ResQue and preservation–roles for the Werner syndrome protein and other RecQ helicases. Mech Ageing Dev 2008;129:79–90.
[153] Yokote K, Chanprasert S, Lee L, Eirich K, Takemoto M, Watanabe A, Koizumi N, Lessel D, Mori T, Hisama FM, Ladd PD, Angle B, Baris H, Cefle K, Palanduz S, Ozturk S, Chateau A, Deguchi K, Easwar TK, Federico A, Fox A, Grebe TA, Hay B, Nampoothiri S, Seiter K, Streeten E, Pina-Aguilar RE, Poke G, Poot M, Posmyk R, Martin GM, Kubisch C, Schindler D, Oshima J. WRN mutation update: mutation spectrum, patient registries, and translational prospects. Hum Mutat 2017;38:7–15.
[154] Suzuki T, Shiratori M, Furuichi Y, Matsumoto T. Diverged nuclear localization of Werner helicase in human and mouse cells. Oncogene 2001;20:2551–8.
[155] Satoh M, Imai M, Sugimoto M, Goto M, Furuichi Y. Prevalence of Werner's syndrome heterozygotes in Japan. Lancet 1999;353:1766.
[156] Masala MV, Scapaticci S, Olivieri C, Pirodda C, Montesu MA, Cuccuru MA, Pruneddu S, Danesino C, Cerimele D. Epidemiology and clinical aspects of Werner's syndrome in North Sardinia: description of a cluster. Eur J Dermatol 2007;17:213–6.
[157] Hutchinson J. Congenital absence of hair and mammary glands with atrophic condition of the skin and its appendages, in a boy whose mother had been almost wholly bald from alopecia areata from the age of six. Med Chir Trans 1886;69:473–7.
[158] Gilford H. Ateleiosis and progeria: continuous youth and premature old age. Br Med J 1904;2:914–8.
[159] Brown WT, Kieras FJ, Houck Jr GE, Dutkowski R, Jenkins EC. A comparison of adult and childhood progerias: Werner syndrome and Hutchinson-Gilford progeria syndrome. Adv Exp Med Biol 1985;190: 229–44.
[160] Gordon LB, Brown WT, Collins FS. Hutchinson-gilford progeria syndrome. In: Pagon RA, Adam MP, Ardinger HH, Wallace SE, Amemiya A, Bean LJH, Bird TD, Fong CT, Mefford HC, Smith RJH, Stephens K, editors. GeneReviews(R), 2015/01/08 ed, Seattle (WA). 2015.
[161] Merideth MA, Gordon LB, Clauss S, Sachdev V, Smith AC, Perry MB, Brewer CC, Zalewski C, Kim HJ, Solomon B, Brooks BP, Gerber LH, Turner ML, Domingo DL, Hart TC, Graf J, Reynolds JC, Gropman A, Yanovski JA, Gerhard-Herman M, Collins FS, Nabel EG, Cannon 3rd RO, Gahl WA, Introne WJ. Phenotype and course of Hutchinson-Gilford progeria syndrome. N Engl J Med 2008;358:592–604.
[162] Olive M, Harten I, Mitchell R, Beers JK, Djabali K, Cao K, Erdos MR, Blair C, Funke B, Smoot L, Gerhard-Herman M, Machan JT, Kutys R, Virmani R, Collins FS, Wight TN, Nabel EG, Gordon LB. Cardiovascular pathology in Hutchinson-Gilford progeria: correlation with the vascular pathology of aging. Arterioscler Thromb Vasc Biol 2010;30:2301–9.
[163] De Sandre-Giovannoli A, Bernard R, Cau P, Navarro C, Amiel J, Boccaccio I, Lyonnet S, Stewart CL, Munnich A, Le Merrer M, Levy N. Lamin a truncation in Hutchinson-Gilford progeria. Science 2003;300:2055.
[164] Eriksson M, Brown WT, Gordon LB, Glynn MW, Singer J, Scott L, Erdos MR, Robbins CM, Moses TY, Berglund P, Dutra A, Pak E, Durkin S, Csoka AB, Boehnke M, Glover TW, Collins FS. Recurrent de novo point mutations in lamin A cause Hutchinson-Gilford progeria syndrome. Nature 2003;423:293–8.
[165] Broers JL, Ramaekers FC, Bonne G, Yaou RB, Hutchison CJ. Nuclear lamins: laminopathies and their role in premature ageing. Physiol Rev 2006;86:967–1008.
[166] Goldman RD, Shumaker DK, Erdos MR, Eriksson M, Goldman AE, Gordon LB, Gruenbaum Y, Khuon S, Mendez M, Varga R, Collins FS. Accumulation of mutant lamin A causes progressive changes in nuclear architecture in Hutchinson-Gilford progeria syndrome. Proc Natl Acad Sci U S A 2004;101: 8963–8.

[167] Dahl KN, Scaffidi P, Islam MF, Yodh AG, Wilson KL, Misteli T. Distinct structural and mechanical properties of the nuclear lamina in Hutchinson-Gilford progeria syndrome. Proc Natl Acad Sci U S A 2006;103:10271–6.
[168] Shimi T, Pfleghaar K, Kojima S, Pack CG, Solovei I, Goldman AE, Adam SA, Shumaker DK, Kinjo M, Cremer T, Goldman RD. The A- and B-type nuclear lamin networks: microdomains involved in chromatin organization and transcription. Genes Dev 2008;22:3409–21.
[169] Liu Y, Rusinol A, Sinensky M, Wang Y, Zou Y. DNA damage responses in progeroid syndromes arise from defective maturation of prelamin A. J Cell Sci 2006;119:4644–9.
[170] Benson EK, Lee SW, Aaronson SA. Role of progerin-induced telomere dysfunction in HGPS premature cellular senescence. J Cell Sci 2010;123:2605–12.
[171] Scaffidi P, Misteli T. Lamin A-dependent nuclear defects in human aging. Science 2006;312:1059–63.
[172] Worman HJ, Bonne G. "Laminopathies": a wide spectrum of human diseases. Exp Cell Res 2007;313:2121–33.
[173] Hisama FM, Lessel D, Leistritz D, Friedrich K, McBride KL, Pastore MT, Gottesman GS, Saha B, Martin GM, Kubisch C, Oshima J. Coronary artery disease in a Werner syndrome-like form of progeria characterized by low levels of progerin, a splice variant of lamin A. Am J Med Genet 2011;155A:3002–6.
[174] Capell BC, Olive M, Erdos MR, Cao K, Faddah DA, Tavarez UL, Conneely KN, Qu X, San H, Ganesh SK, Chen X, Avallone H, Kolodgie FD, Virmani R, Nabel EG, Collins FS. A farnesyltransferase inhibitor prevents both the onset and late progression of cardiovascular disease in a progeria mouse model. Proc Natl Acad Sci U S A 2008;105:15902–7.
[175] Gabriel D, Gordon LB, Djabali K. Temsirolimus partially rescues the Hutchinson-Gilford progeria cellular phenotype. PLoS One 2016;11:e0168988.
[176] Yang SH, Chang SY, Andres DA, Spielmann HP, Young SG, Fong LG. Assessing the efficacy of protein farnesyltransferase inhibitors in mouse models of progeria. J Lipid Res 2010;51:400–5.
[177] Yang SH, Meta M, Qiao X, Frost D, Bauch J, Coffinier C, Majumdar S, Bergo MO, Young SG, Fong LG. A farnesyltransferase inhibitor improves disease phenotypes in mice with a Hutchinson-Gilford progeria syndrome mutation. J Clin Invest 2006;116:2115–21.
[178] Gordon LB, Kleinman ME, Miller DT, Neuberg DS, Giobbie-Hurder A, Gerhard-Herman M, Smoot LB, Gordon CM, Cleveland R, Snyder BD, Fligor B, Bishop WR, Statkevich P, Regen A, Sonis A, Riley S, Ploski C, Correia A, Quinn N, Ullrich NJ, Nazarian A, Liang MG, Huh SY, Schwartzman A, Kieran MW. Clinical trial of a farnesyltransferase inhibitor in children with Hutchinson-Gilford progeria syndrome. Proc Natl Acad Sci U S A 2012;109:16666–71.
[179] Gordon LB, Shappell H, Massaro J, D'Agostino RB, Brazier Sr., J, Campbell SE, Kleinman ME, Kieran MW. Association of lonafarnib treatment vs no treatment with mortality rate in patients with Hutchinson-Gilford progeria syndrome. JAMA 2018;319(16):1687–95.
[180] Yang SH, Chang SY, Ren S, Wang Y, Andres DA, Spielmann HP, Fong LG, Young SG. Absence of progeria-like disease phenotypes in knock-in mice expressing a non-farnesylated version of progerin. Hum Mol Genet 2011;20:436–44.
[181] Weedon MN, Ellard S, Prindle MJ, Caswell R, Lango Allen H, Oram R, Godbole K, Yajnik CS, Sbraccia P, Novelli G, Turnpenny P, McCann E, Goh KJ, Wang Y, Fulford J, McCulloch LJ, Savage DB, O'Rahilly S, Kos K, Loeb LA, Semple RK, Hattersley AT. An in-frame deletion at the polymerase active site of POLD1 causes a multisystem disorder with lipodystrophy. Nat Genet 2013;45:947–50.
[182] Kamath-Loeb AS, Shen JC, Schmitt MW, Loeb LA. The Werner syndrome exonuclease facilitates DNA degradation and high fidelity DNA polymerization by human DNA polymerase delta. J Biol Chem 2012;287:12480–90.
[183] Lessel D, Hisama FM, Szakszon K, Saha B, Sanjuanelo AB, Salbert BA, Steele PD, Baldwin J, Brown WT, Piussan C, Plauchu H, Szilvassy J, Horkay E, Hogel J, Martin GM, Herr AJ, Oshima J, Kubisch C. POLD1 germline mutations in patients initially diagnosed with Werner syndrome. Hum Mutat 2015;36:1070–9.
[184] Pelosini C, Martinelli S, Ceccarini G, Magno S, Barone I, Basolo A, Fierabracci P, Vitti P, Maffei M, Santini F. Identification of a novel mutation in the polymerase delta 1 (POLD1) gene in a lipodystrophic patient affected by mandibular hypoplasia, deafness, progeroid features (MDPL) syndrome. Metabolism 2014;63:1385–9.
[185] Palles C, Cazier JB, Howarth KM, Domingo E, Jones AM, Broderick P, Kemp Z, Spain SL, Guarino E, Salguero I, Sherborne A, Chubb D, Carvajal-Carmona LG, Ma Y, Kaur K, Dobbins S, Barclay E, Gorman M, Martin L, Kovac MB, Humphray S, Consortium C, Consortium WGS, Lucassen A, Holmes CC, Bentley D, Donnelly P, Taylor J, Petridis C, Roylance R, Sawyer EJ, Kerr DJ, Clark S, Grimes J, Kearsey SE, Thomas HJ, McVean G, Houlston RS, Tomlinson I. Germline mutations affecting the proofreading domains of POLE and POLD1 predispose to colorectal adenomas and carcinomas. Nat Genet 2013;45:136–44.

[186] Laugel V. Cockayne syndrome: the expanding clinical and mutational spectrum. Mech Ageing Dev 2013;134:161–70.

[187] Jaarsma D, van der Pluijm I, van der Horst GT, Hoeijmakers JH. Cockayne syndrome pathogenesis: lessons from mouse models. Mech Ageing Dev 2013;134:180–95.

[188] Marteijn JA, Lans H, Vermeulen W, Hoeijmakers JH. Understanding nucleotide excision repair and its roles in cancer and ageing. Nat Rev Mol Cell Biol 2014;15:465–81.

[189] Chatre L, Biard DS, Sarasin A, Ricchetti M. Reversal of mitochondrial defects with CSB-dependent serine protease inhibitors in patient cells of the progeroid Cockayne syndrome. Proc Natl Acad Sci U S A 2015;112(22):E2910–9.

[190] Niedernhofer LJ, Garinis GA, Raams A, Lalai AS, Robinson AR, Appeldoorn E, Odijk H, Oostendorp R, Ahmad A, van Leeuwen W, Theil AF, Vermeulen W, van der Horst GT, Meinecke P, Kleijer WJ, Vijg J, Jaspers NG, Hoeijmakers JH. A new progeroid syndrome reveals that genotoxic stress suppresses the somatotroph axis. Nature 2006;444:1038–43.

[191] Anderson BH, Kasher PR, Mayer J, Szynkiewicz M, Jenkinson EM, Bhaskar SS, Urquhart JE, Daly SB, Dickerson JE, O'Sullivan J, Leibundgut EO, Muter J, Abdel-Salem GM, Babul-Hirji R, Baxter P, Berger A, Bonafe L, Brunstom-Hernandez JE, Buckard JA, Chitayat D, Chong WK, Cordelli DM, Ferreira P, Fluss J, Forrest EH, Franzoni E, Garone C, Hammans SR, Houge G, Hughes I, Jacquemont S, Jeannet PY, Jefferson RJ, Kumar R, Kutschke G, Lundberg S, Lourenco CM, Mehta R, Naidu S, Nischal KK, Nunes L, Ounap K, Philippart M, Prabhakar P, Risen SR, Schiffmann R, Soh C, Stephenson JB, Stewart H, Stone J, Tolmie JL, van der Knaap MS, Vieira JP, Vilain CN, Wakeling EL, Wermenbol V, Whitney A, Lovell SC, Meyer S, Livingston JH, Baerlocher GM, Black GC, Rice GI, Crow YJ. Mutations in CTC1, encoding conserved telomere maintenance component 1, cause Coats plus. Nat Genet 2012;44:338–42.

[192] Gu P, Chang S. Functional characterization of human CTC1 mutations reveals novel mechanisms responsible for the pathogenesis of the telomere disease Coats plus. Aging Cell 2013;12:1100–9.

[193] Nolis T. Exploring the pathophysiology behind the more common genetic and acquired lipodystrophies. J Hum Genet 2014;59:16–23.

[194] Lawson MA. Lipoatrophic diabetes: a case report with a brief review of the literature. J Adolesc Health 2009;44:94–5.

[195] Nelson MD, Victor RG, Szczepaniak EW, Simha V, Garg A, Szczepaniak LS. Cardiac steatosis and left ventricular hypertrophy in patients with generalized lipodystrophy as determined by magnetic resonance spectroscopy and imaging. Am J Cardiol 2013;112:1019–24.

[196] Machado PV, Daxbacher EL, Obadia DL, Cunha EF, Alves Mde F, Mann D. Do you know this syndrome? Berardinelli-Seip syndrome. An Bras Dermatol 2013;88:1011–3.

[197] Wei S, Soh SL, Qiu W, Yang W, Seah CJ, Guo J, Ong WY, Pang ZP, Han W. Seipin regulates excitatory synaptic transmission in cortical neurons. J Neurochem 2013;124:478–89.

[198] Jeninga EH, de Vroede M, Hamers N, Breur JM, Verhoeven-Duif NM, Berger R, Kalkhoven E. A patient with congenital generalized lipodystrophy due to a novel mutation in BSCL2: indications for secondary mitochondrial dysfunction. JIMD Rep 2012;4:47–54.

[199] Agrawala RK, Choudhury AK, Mohanty BK, Baliarsinha AK. Berardinelli-Seip congenital lipodystrophy: an autosomal recessive disorder with rare association of duodenocolonic polyps. J Pediatr Endocrinol Metab 2014;27:989–91.

[200] Ben-Avraham D, Govindaraju DR, Budagov T, Fradin D, Durda P, Liu B, Ott S, Gutman D, Sharvit L, Kaplan R, Bougneres P, Reiner A, Shuldiner AR, Cohen P, Barzilai N, Atzmon G. The GH receptor exon 3 deletion is a marker of male-specific exceptional longevity associated with increased GH sensitivity and taller stature. Sci Adv 2017;3:e1602025.

[201] Corder EH, Saunders AM, Risch NJ, Strittmatter WJ, Schmechel DE, Gaskell Jr PC, Rimmler JB, Locke PA, Conneally PM, Schmader KE, et al. Protective effect of apolipoprotein E type 2 allele for late onset Alzheimer disease. Nat Genet 1994;7:180–4.

[202] Drenos F, Kirkwood TB. Selection on alleles affecting human longevity and late-life disease: the example of apolipoprotein E. PLoS One 2010;5:e10022.

[202b] Suri S, Heise V, Trachtenberg AJ, Mackay CE. The forgotten APOE allele: a review of the evidence and suggested mechanisms for the protective effect of APOE ε2. Neurosci Biobehav Rev 2013 Dec;37(10 Pt 2):2878–86. https://doi.org/10.1016/j.neubiorev.2013.10.010. Epub 2013 Oct 29. Review. PMID: 24183852.

[203] Abifadel M, Varret M, Rabes JP, Allard D, Ouguerram K, Devillers M, Cruaud C, Benjannet S, Wickham L, Erlich D, Derre A, Villeger L, Farnier M, Beucler I, Bruckert E, Chambaz J, Chanu B, Lecerf JM, Luc G, Moulin P, Weissenbach J, Prat A, Krempf M, Junien C, Seidah NG, Boileau C. Mutations in PCSK9 cause autosomal dominant hypercholesterolemia. Nat Genet 2003;34:154–6.

[204] Horton JD, Cohen JC, Hobbs HH. PCSK9: a convertase that coordinates LDL catabolism. J Lipid Res 2009;50(Suppl.):S172–7.

[205] Musunuru K, Pirruccello JP, Do R, Peloso GM, Guiducci C, Sougnez C, Garimella KV, Fisher S, Abreu J, Barry AJ, Fennell T, Banks E, Ambrogio L, Cibulskis K, Kernytsky A, Gonzalez E, Rudzicz N, Engert JC, DePristo MA, Daly MJ, Cohen JC, Hobbs HH, Altshuler D, Schonfeld G, Gabriel SB, Yue P, Kathiresan S. Exome sequencing, ANGPTL3 mutations, and familial combined hypolipidemia. N Engl J Med 2010;363:2220–7.

[206] Dewey FE, Gusarova V, Dunbar RL, O'Dushlaine C, Schurmann C, Gottesman O, McCarthy S, Van Hout CV, Bruse S, Dansky HM, Leader JB, Murray MF, Ritchie MD, Kirchner HL, Habegger L, Lopez A, Penn J, Zhao A, Shao W, Stahl N, Murphy AJ, Hamon S, Bouzelmat A, Zhang R, Shumel B, Pordy R, Gipe D, Herman GA, Sheu WHH, Lee IT, Liang KW, Guo X, Rotter JI, Chen YI, Kraus WE, Shah SH, Damrauer S, Small A, Rader DJ, Wulff AB, Nordestgaard BG, Tybjaerg-Hansen A, van den Hoek AM, Princen HMG, Ledbetter DH, Carey DJ, Overton JD, Reid JG, Sasiela WJ, Banerjee P, Shuldiner AR, Borecki IB, Teslovich TM, Yancopoulos GD, Mellis SJ, Gromada J, Baras A. Genetic and pharmacologic inactivation of ANGPTL3 and cardiovascular disease. N Engl J Med 2017;377: 211–21.

[207] Graham MJ, Lee RG, Brandt TA, Tai LJ, Fu W, Peralta R, Yu R, Hurh E, Paz E, McEvoy BW, Baker BF, Pham NC, Digenio A, Hughes SG, Geary RS, Witztum JL, Crooke RM, Tsimikas S. Cardiovascular and metabolic effects of ANGPTL3 antisense oligonucleotides. N Engl J Med 2017;377:222–32.

[208] Saleheen D, Natarajan P, Armean IM, Zhao W, Rasheed A, Khetarpal SA, Won HH, Karczewski KJ, O'Donnell-Luria AH, Samocha KE, Weisburd B, Gupta N, Zaidi M, Samuel M, Imran A, Abbas S, Majeed F, Ishaq M, Akhtar S, Trindade K, Mucksavage M, Qamar N, Zaman KS, Yaqoob Z, Saghir T, Rizvi SNH, Memon A, Hayyat Mallick N, Ishaq M, Rasheed SZ, Memon FU, Mahmood K, Ahmed N, Do R, Krauss RM, MacArthur DG, Gabriel S, Lander ES, Daly MJ, Frossard P, Danesh J, Rader DJ, Kathiresan S. Human knockouts and phenotypic analysis in a cohort with a high rate of consanguinity. Nature 2017;544:235–9.

[209] Wilhelmsen L, Dellborg M, Welin L, Svardsudd K. Men born in 1913 followed to age 100 years. Scand Cardiovasc J 2015;49:45–8.

[210] Rajpathak SN, Liu Y, Ben-David O, Reddy S, Atzmon G, Crandall J, Barzilai N. Lifestyle factors of people with exceptional longevity. J Am Geriatr Soc 2011;59: 1509–12.

[211] Dragani TA, Canzian F, Pierotti MA. A polygenic model of inherited predisposition to cancer. FASEB J 1996;10:865–70.

[212] Galvan A, Falvella FS, Spinola M, Frullanti E, Leoni VP, Noci S, Alonso MR, Zolin A, Spada E, Milani S, Pastorino U, Incarbone M, Santambrogio L, Gonzalez Neira A, Dragani TA. A polygenic model with common variants may predict lung adenocarcinoma risk in humans. Int J Cancer 2008;123:2327–30.

[213] Esplin ED, Oei L, Snyder MP. Personalized sequencing and the future of medicine: discovery, diagnosis and defeat of disease. Pharmacogenomics 2014;15:1771–90.

[214] Barzilai N, Atzmon G, Schechter C, Schaefer EJ, Cupples AL, Lipton R, Cheng S, Shuldiner AR. Unique lipoprotein phenotype and genotype associated with exceptional longevity. J Am Med Assoc 2003;290:2030–40.

[215] Kleindorp R, Flachsbart F, Puca AA, Malovini A, Schreiber S, Nebel A. Candidate gene study of FOXO1, FOXO4 and FOXO6 reveals no association with human longevity in Germans. Aging Cell 2011.

[216] Guevara-Aguirre J, Balasubramanian P, Guevara-Aguirre M, Wei M, Madia F, Cheng CW, Hwang D, Martin-Montalvo A, Saavedra J, Ingles S, de Cabo R, Cohen P, Longo VD. Growth hormone receptor deficiency is associated with a major reduction in pro-aging signaling, cancer, and diabetes in humans. Sci Transl Med 2011;3:70ra13.

[217] Krzisnik C, Grguric S, Cvijovic K, Laron Z. Longevity of the hypopituitary patients from the island Krk: a follow-up study. Pediatr Endocrinol Rev 2010;7:357–62.

[218] Bartke A. Single-gene mutations and healthy ageing in mammals. Philos Trans R Soc Lond B Biol Sci 2011;366:28–34.

[219] Martin GM. Syndromes of accelerated aging. Natl Canc Inst Monogr 1982;60:241–7.

[220] Martin GM. Genetic modulation of senescent phenotypes in *Homo sapiens*. Cell 2005;120:523–32.

[221] Xie Z, Jay KA, Smith DL, Zhang Y, Liu Z, Zheng J, Tian R, Li H, Blackburn EH. Early telomerase inactivation accelerates aging independently of telomere length. Cell 2015;160:928–39.

[222] Nhan HS, Chiang K, Koo EH. The multifaceted nature of amyloid precursor protein and its proteolytic fragments: friends and foes. Acta Neuropathol 2015;129:1–19.

[223] Schwartzman O, Tanay A. Single-cell epigenomics: techniques and emerging applications. Nat Rev Genet 2015;16:716–26.

[224] Ambeskovic M, Roseboom TJ, Metz GAS. Transgenerational effects of early environmental insults on aging and disease incidence. Neurosci Biobehav Rev 2017.

16

# Pharmacogenomics

*Daniel W. Nebert*[1,2,3], *Ge Zhang*[2,3]

[1]Department of Environmental Health and Center for Environmental Genetics, University of Cincinnati School of Medicine, Cincinnati, OH, United States
[2]Department of Pediatrics, University of Cincinnati School of Medicine, Cincinnati, OH, United States
[3]Division of Human Genetics, Cincinnati Children's Hospital Medical Center, Cincinnati, OH, United States

## 16.1 INTRODUCTION

*Individualized drug therapy* represents a major portion of *personalized medicine*. The visionary human geneticist Arno G. Motulsky is credited with being first to propose in a publication that "*drug responses* will depend upon each patient's genetic make-up" [146]. Another human geneticist, Friedrich O. Vogel, coined the term *pharmacogenetics* and defined it as "the study of heritable variability in drug response" [229], or, simply, "gene–drug interactions."

In the 1990s, the term *pharmacogenomics* was introduced; this came about as a direct offshoot of the Human Genome Project. Pharmacogenomics is defined as "the study of how drugs interact with the *total genome*, to influence biological pathways and processes" [148], or, simply, "drug–genome interactions." This field should help identify new drug targets and thus be instrumental in designing new drugs. Today, the terms pharmacogenetics and pharmacogenomics are used interchangeably; hence, in this chapter, we use the abbreviation "PGx" to incorporate them both.

In large part, our genetic make-up determines our *individual drug response*. However, each individual's drug response is *holistic* in that it actually encompasses five contributing influences: (1) *genotype* (DNA single-nucleotide variants, insertions, deletions, duplications, and inversions); (2) *epigenetic effects* (DNA methylation, RNA interference, histone modifications, and chromatin remodeling); (3) *endogenous influences* (age, gender, ethnicity, exercise, various disease states, and functional status of kidneys and other organs); (4) *environmental factors* (diet, cigarette smoking, lifestyle, drug–drug interactions, and significant exposure to occupational chemicals and other environmental pollutants); and (5) *microbiome differences* specific to each person.

Thus, each patient has his own unique "*PGx profile*"—just as each of us has his own distinct pattern of DNA microsatellite differences, profile of *single-nucleotide polymorphisms*, or *variants* (SNPs; SNVs), finger prints, and even the biometric properties of the eye's iris. Besides the genetic component, however, the other four categories listed above are environment- and time-dependent, i.e., a constantly moving target. A patient's response to a drug today might differ from that same patient's response tomorrow, next month, or next year. Thus, it is best to keep in mind that any individual's drug response is never entirely static.

### 16.1.1 Types of Drug Responses

*Interindividual variability in drug response* is defined as an "effect of varying intensity occurring in different individuals receiving a specified drug dose," or "requirement of a range of doses (concentrations) in order to produce an effect of specified intensity in each patient" [11]. Classifications of drug response include: (1) the desired

Emery and Rimoin's Principles and Practice of Medical Genetics and Genomics: Foundations, https://doi.org/10.1016/B978-0-12-812537-3.00016-0

beneficial effect (*efficacy*); (2) *adverse effect*; (3) no effect (*therapeutic failure*); and (4) *toxic effect*. The latter two effects are particularly dependent on drug dosage.

One goal of PGx intends to develop rational approaches to optimize drug therapy with respect to each patient's genotype. Another goal of PGx is to ensure maximum efficacy, combined with minimal adverse effects in each individual. If one considers, in addition, the above-mentioned issues of epigenetics, endogenous influences, environmental effects, and each patient's changing microbiome—it is easy to understand how complicated *genetic risk prediction of drug response* can be for each individual patient.

### 16.1.2 Adverse Drug Reactions

Two decades ago, it was reported that *adverse drug reactions* (ADRs) in the US rank as approximately the fifth leading cause of death [111]. ADRs have been divided into categories (Table 16.1): (1) dose-dependent; (2) dose-independent; (3) dose- and time-dependent (cumulative); and (4) time-related withdrawal reactions [39]. Dose-independent ADRs comprise idiosyncratic drug reactions and allergic reactions; the latter is not addressed in this chapter. A better understanding of PGx should help reduce morbidity and mortality caused by ADRs, but should especially help prevent ADRs caused by *idiosyncratic dose-**in**dependent drug reactions*.

In this chapter, we first introduce the topic of clinical pharmacology, followed by the history of genetics and its impact on PGx. Then we present several phenotype examples of early PGx studies, followed by more complex examples. Finally, given the complexity of the genome and interindividual differences, we attempt to clarify some of the challenges and show how difficult it will be—anytime in the near future—to be able to predict each patient's individual drug response, given each person's unique genome.

| TABLE 16.1 **A Possible Classification of Adverse Drug Reactions (ADRs)** |
|---|
| Dose-dependent |
| Dose-independent |
| Idiosyncratic drug reactions |
| Allergic reactions |
| Dose- and time-dependent (cumulative) |
| Time-related withdrawal reactions |

## 16.2 FUNDAMENTAL ASPECTS OF CLINICAL PHARMACOLOGY

The objective of clinical pharmacologists, in treating a patient with a drug, is to maintain optimal plasma levels of the *active principle* (drug) in the *therapeutic range* (Fig. 16.1). If the dose of drug is too small, the interval of administration inadequate, bioavailability of the active drug too low, or the active drug too rapidly metabolized and thus quickly cleared—then the active drug level in the blood might never reach its effective concentration. This would lead to absence of the expected response (*therapeutic failure*). On the contrary, if the dose of drug is too large, intervals of administration too short, or the active drug poorly metabolized and thus too slowly excreted, this accumulation can lead to *toxic concentrations*. Levels that cause toxicity will lead to ADRs; such a drug response might occur because of drug accumulation at toxic levels in blood, and accumulation might occur in one or more critical target organs, or in both blood and target organs. Dose-***in***dependent ADRs can happen at any dosage and are generally unpredictable.

The above scenarios exist if the *parent drug* is the component responsible for efficacy as well as toxicity. If a *metabolite* is the *active principle* (i.e., that which causes the efficacy) and it is also the toxicant, then one can replace the word "drug" (above) with

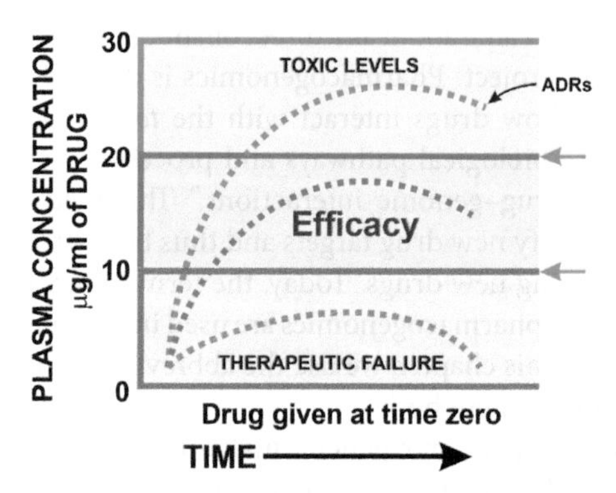


**Figure 16.1** Theoretical plasma concentration curves for any drug, as a function of time after administration of the dose. In this hypothetical case, the horizontal line (at 10 µg/mL) is the minimum *effective concentration*; the horizontal line (at 20 µg/mL) is the point at which *toxicity* occurs. (From previous Emery and Rimoin edition; Nebert DW, Vesell ES. Chapter 19-"Pharmacogenetics and pharmacogenomics". In: Emery and Rimoin's principles and practice of medical genetics. 6th ed. Oxford: Academic Press; 2013. pp. 1–27.)

"metabolite" and describe the same events leading to *efficacy* versus *therapeutic failure* versus *toxic levels* [158]. A classic example in which the metabolite is more biologically active than the parent drug is the parent drug codeine, which becomes activated by the CYP2D6 enzyme into the more biologically active metabolite, morphine [54].

## 16.2.1 Pharmacokinetics (PK) and Pharmacodynamics (PD)

Pharmacokinetics (PK) represents "what the body does to the drug," whereas pharmacodynamics (PD) can be defined as "what the drug does to the body," or, more specifically, to target cell-types, tissues, or organs (Fig. 16.2). The end result of all collective PK and PD processes will lead to drug efficacy, therapeutic failure, or toxicity. Sometimes pharmacologists combine therapeutic failure and toxicity into the common category of "ADRs."

Variations in the drug's PK phase incorporate the processes of ***a***bsorption (uptake), ***d***istribution, ***m***etabolism, and ***e***xcretion. Pharmaceutical companies describe these four processes as "ADME"; these companies often have consolidated specific laboratories into "ADME sections, or divisions, of research."

PK differences are generally detected by determinations of the active drug (or metabolite) in blood and less often drug levels in urine (or saliva, sweat, feces, milk, semen). For most systemic drugs, concentration of the (unbound, or free) active principle in blood is almost always proportional to the concentration of *active principle* in the target tissue. An exception would include certain chemotherapeutic agents, designed to become concentrated in cancer cells while avoiding nearby noncancer cells.

Traditionally, most PGx differences that were initially identified—from the 1940s into the 1990s—involved genes participating in drug metabolism ("PK genes"). Since the mid-1980s, PGx differences began to be described also in genes involved in distribution (binding, transport), absorption, and excretion. It seems reasonable to include cell-surface (both influx and outflux) transporter genes as PK genes, because cellular uptake and excretion are usually essential to the process of absorption in the ADME equation.

Variations in the PD phase occur "downstream" of the PK phase (Fig. 16.2); PD genes would include those that influence drug-action—such as binding to target receptors. There are examples of "recycling" of the active principle from the PD phase back to the PK phase. Thus, PD gene differences might be seen in: transporters and channels across the different intracellular membranes between organelles and in the cytosol and nucleus; tissue- and cell type-specific transcription

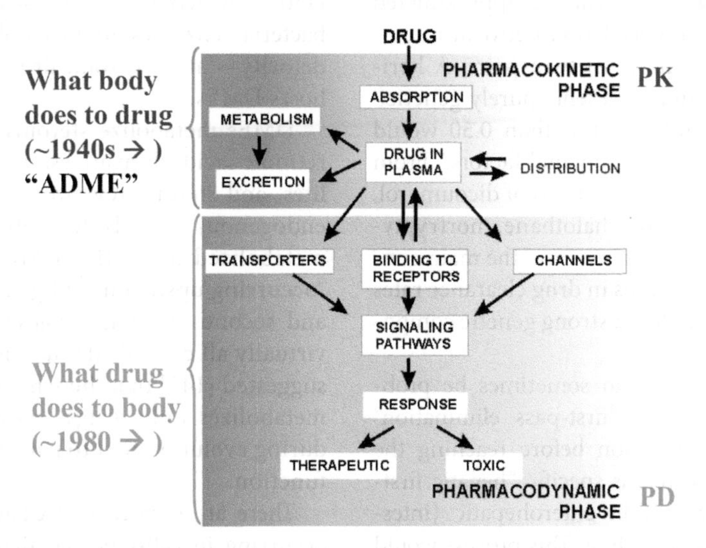


**Figure 16.2** Fundamentals of clinical pharmacology. Processes involve the *pharmacokinetic* (PK) phase, and the *pharmacodynamic* (PD) phase—of any drug or over-the-counter preparation. (From previous Emery and Rimoin edition; Nebert DW, Vesell ES. Chapter 19-"Pharmacogenetics and pharmacogenomics". In: Emery and Rimoin's principles and practice of medical genetics. 6th ed. Oxford: Academic Press; 2013. pp. 1–27.)

factors; components of signal transduction pathways; nucleic acid and protein repair processes; molecular "chaperones"; and cell infrastructure (e.g., nuclear matrix, membranes, and subcellular organelles such as endoplasmic reticulum, Golgi bodies, melanosomes, or peroxisomes). Until the 1980s, what happened in the PD phase was largely a "black box." Due to the rapid advances in molecular biology since the 1980s, however, downstream drug-signaling pathways have now become better understood, and the field of PD has exploded (reviewed in [158]).

Historically, drugs or chemicals were given to patients, human healthy volunteers, or laboratory animals—and then differences in urinary metabolite profiles were determined (e.g., [10,233]). Subsequently, PK variations in laboratory animals [226] and ultimately in humans [228]—reflected as differences in plasma drug levels or rate of clearance—led to the appreciation of *genetic variability*, presumed to be in drug-metabolizing enzymes (DMEs). These types of studies were carried out in human volunteers or patients, because blood samples are easier to obtain than tissue biopsies.

### 16.2.2 Plasma Clearance of a Drug

In the late 1960s, comparisons of drug clearances in monozygotic versus dizygotic twin-pairs were performed. The *heritability index* can be approximated as twice the difference in correlation between monozygotic (MZ) and dizygotic (DZ) twin-pairs. A heritability index of 1.0 would represent "purely genetic," whereas a heritability index of less than 0.50 would indicate "predominantly environmental factors." From twin studies—of plasma concentrations of dicoumarol, phenylbutazone, desipramine, halothane, nortriptyline, oxyphenbutazone, and antipyrine—the main conclusion was that large variations in drug clearance rates among healthy subjects reflect a strong genetic component [227].

Plasma clearance studies can sometimes be problematic if the drug exhibits "first-pass elimination" (i.e., metabolism or degradation before reaching the systemic circulation), as well as specific "hepatic first-pass kinetics," i.e., involved in enterohepatic (intestine-to-liver-to-intestine) recycling. This process would be visualized in a graph as a series of rises and falls in the drug's concentration in blood (plotted on the Y-axis) as a function of time (on the X-axis).

### 16.2.3 Extrahepatic PGx Differences and Endogenous Functions

Early on, a common misconception was that DMEs exist almost exclusively, or entirely, in liver. Another fallacy was that only drugs, and not endogenous compounds, are substrates for DMEs. Both of these myths are now realized to be wrong (reviewed in [147,157]). For example, CYP3A4, the most abundant cytochrome P450 monooxygenase in liver, is also present in large concentrations in the gastrointestinal (GI) tract. Numerous DMEs are located in lung and kidney. DMEs are even found in the ciliary body [191] and cornea [186] of the eye.

Many DMEs exist in blood cells, vascular endothelial cells, and virtually all cell-types in the body, participating also in the activation and deactivation of *lipid mediators* (LMs) of the *arachidonic acid*, *docosahexaenoic acid*, and *eicosapentaenoic acid* cascades. The end-result of these LM cascades leads to regulation of cell division, cell adhesion and migration, pre- and postinflammatory responses, cell migration, bronchoconstriction, vasodilation, and numerous other developmental and homeostatic mechanisms (reviewed in [154,157]). Because DMEs exist in essentially all cell-types of the brain, they participate in key roles in neuroendocrine functions. Emergence of the "brain–gut–microbiome" (reviewed in [150]) has become a very recent consideration in PGx studies, because of the capacity of gut bacterial enzymes to metabolize—both activate and detoxify—(at least) some drugs, similarly to that of the host's DMEs.

DMEs metabolize steroids, fatty acids, bile acids, retinoic acid derivatives, vitamin D and metabolites, and sterols (reviewed in [157]). Because many endogenous ligands for intracellular receptors are metabolized by DMEs, DMEs can be considered as "occurring upstream" of ligand–receptor interactions and second-messenger pathways that are pivotal to virtually all critical-life functions. In fact, it has been suggested [147] that there might not be any DME that metabolizes only drugs—without having originated during evolution first to participate in an endogenous function.

There are hundreds of examples of DME activities occurring in cell-types at higher levels than those in liver, or of DME metabolism occurring exclusively in a tissue other than liver. For example, phenytoin hydroxylation is ~50-fold greater in human oral mucosa than

**"Therapeutic Index"**

$$\frac{\text{Toxic dose } (TD_{50})}{\text{Effective dose } (ED_{50})} = 20 \text{ [large window] A}$$

$$\frac{\text{Toxic dose } (TD_{50})}{\text{Effective dose } (ED_{50})} = 2 \text{ [narrow window] B}$$

**If genetic differences in drug response are 10-fold, then drug A → no problem; drug B → ADRs**

**Figure 16.3** Simple equations illustrating a large versus small "therapeutic window" or "therapeutic index." (From previous Emery and Rimoin edition; Nebert DW, Vesell ES. Chapter 19-"Pharmacogenetics and pharmacogenomics". In: Emery and Rimoin's principles and practice of medical genetics. 6th ed. Oxford: Academic Press; 2013. pp. 1–27.)

in liver [253]. Some DMEs exist at high concentrations in nasal mucosa [120]. In conclusion of this section, it should be appreciated that variability in drug response can occur in any tissue, as well as any cell-type—and not solely in liver. Furthermore, some specific DME activities can be much higher in a particular tissue than in liver, or exclusively in one or another cell-type instead of liver. Moreover, all that has been described herein for PK genes also holds true for PD genes.

### 16.2.4 Therapeutic Index (or "Window")

Most early PGx studies owe their success to pharmacogeneticists' selection of a drug having a narrow *therapeutic index*. If a drug shows a wide therapeutic window, it is unlikely to cause toxicity in a significant subset of any human population; therefore, this would be of little concern to public health, and the need to prevent ADRs would be small. For example, if the dose causing toxicity is 20 times greater than the dose needed to be effective (Fig. 16.3), and genetic differences in handling this drug are never greater than 10-fold across all human populations, this drug would be of little concern to pharmacogeneticists. On the other hand, if the dose causing toxicity is only twice the dose for efficacy—and PGx differences in handling this drug are 10-fold—then this drug would be an important candidate to study, in order to understand ADRs that might lead to morbidity and mortality. Therefore, the therapeutic index can be altered by both dosage and genotype.

### 16.2.5 Genetics of Drug Response

All the above-mentioned drug responses (*efficacy*, *therapeutic failure*, *adverse effect*, and *toxic effect*) can be regarded in genetic terms as *phenotypes* (or *traits*), which in this chapter collectively we call "*PGx traits*." Any amount of success—in predicting outcome of a drug before treating the patient—will depend largely on the "genetic basis (*genotype*) of the PGx trait," which will be influenced by the number of genes and genetic variants contributing to that *phenotype*, the allele frequency, the *effect-size* of each contributing genetic variant [166], and interactions between these genetic factors with the other environmental factors listed above.

The underlying genetic contribution to any phenotype represents the *genetic architecture*. One's "genetic architecture" encompasses: the gene(s) and their *cis*-regulatory regions (introns, plus 5′- and 3′-flanking DNA segments near the gene), and *trans*-regulatory regions (DNA segments hundreds of kilobases away from the gene, or on other chromosomes); the number of alleles in any human population studied; distribution of allelic and mutational effects; and patterns of *pleiotropy*, *dominance*, and *epistasis* [70]. This definition of *genetic architecture* does not include *epigenetics* per se, but epigenetics can obviously influence the "mapping," or association, of the genotype to a phenotype.

Therefore, epigenetics is not explicitly excluded from *genetic architecture*. The major feature should be whether the "epigenetic effect" is *transgenerational* (examples are given later in the chapter) or not. If an epigenetic modification is not transgenerational, then its effect (e.g., developmental) on a phenotype pertains only to the patient himself, and it should not influence the *evolutionary trajectory* (i.e., heritable properties) of the trait. On the other hand, if the epigenetic modification has a transgenerational effect (e.g., imprinting, changes in risk of obesity, changes in risk of type II diabetes due to parent or grandparent influence), then it should be included as part of the genetic architecture. In summary, the overall outcome of a drug response will depend on the *genetic architecture* of the PGx trait—plus the *epigenetic, endogenous, environmental,* and *microbiome* factors mentioned earlier.

Recent genome-wide PGx studies have suggested (reviewed in [248]) that the genetic basis of variability in drug response can be grouped into three categories: (1) *monogenic* (Mendelian) *traits* that include many of the early examples of inherited disorders, as well as some severe idiosyncratic ADRs typically influenced by one or a few rare large-effect variants; (2) *predominantly oligogenic traits* that represent variability mainly elicited by a small number of major (PK or PD) genes; and (3) *complex PGx traits*—produced mostly by innumerable small-effect variants, together with epigenetic, endogenous, environmental, and microbiome influences. These three categories should not necessarily be considered as "distinct" from one another, but rather as an overlapping gradient. These categories will be detailed throughout the remainder of this chapter.

It is now realized that the vast majority of drug responses represent "group (3)," *PGx multifactorial traits,* similar in many ways to quantitative traits—such as, for example, height, weight, body mass index, blood pressure, and serum cholesterol levels. Multifactorial drug responses are also comparable to numerous complex diseases (e.g., type II diabetes, metabolic syndrome, coronary artery disease, bipolar disorder, schizophrenia, asthma, and cancer). Obviously, a major difference between PGx studies and complex-disease studies is that any patient who has never been challenged with a particular drug will not know his phenotype for that drug.

## 16.3 HISTORY OF GENETICS RELEVANT TO PGx

The complexity of understanding PGx traits, over decades of time, has paralleled our progressively better understanding of the complexity of human genetics. Hence, most traits were regarded in Mendelian terms between 1860 and 1920, then predominantly oligogenic traits became more appreciated between 1920 and the late 1980s, and finally, consideration of the extreme complexity of multifactorial traits came to the forefront after 1990. This progression of conceptual thinking will be emphasized throughout the rest of this chapter.

At the laboratory bench, from the 1940s onward, enzyme assays of animal tissues and cell fractions became popular. In the 1940s and 1950s, tissue fractions that could be studied were simply tissue homogenates. After the invention and availability of high-speed centrifuges by the late 1950s, DME activities in microsomal versus mitochondrial versus cytosolic versus nuclear fractions became increasingly easy to study.

With advances in molecular-biology methodologies, DME genes began to be characterized in the 1980s. With the explosion in knowledge derived from the Human Genome Project (1990), elucidation of genetic differences in measurements other than DME gene studies became more commonplace. For example, receptor assays began in the late 1970s and rapidly advanced during the 1980s; transporter and ion-channel assays were initiated in the 1980s and became popular in the 1990s; analyses of signal transduction pathways, post-translational modifications, and many other subcellular processes have greatly expanded especially during the past three decades. A technical consideration in molecular biology studies during the 1980s and 1990s was that DME genes are in general smaller in length, spanning ~5–20 kb, whereas receptor and transporter genes are usually larger (~50 to >100 kb in length); cloning, sequencing, and characterization of DME genes were therefore technically easier than for the much larger genes.

### 16.3.1 Monogenic Traits

In the 1860s, Gregor Mendel introduced "dominant-versus-recessive" classical genetics—studying garden peas (e.g., red flower color *dominant*, white *recessive*). Whereas the $F_1$ cross yields all red flowers, the $F_2$ generation yields three red (one *R/R* homozygote and two *R/r* heterozygotes) and one white flower color (*r/r*). The $F_2$ population distribution thus illustrates the classical "Mendelian pattern of inheritance" (Fig. 16.4A).

In the early 1900s, Sir Archibald Garrod described four clinical "inborn-errors-of-metabolism":

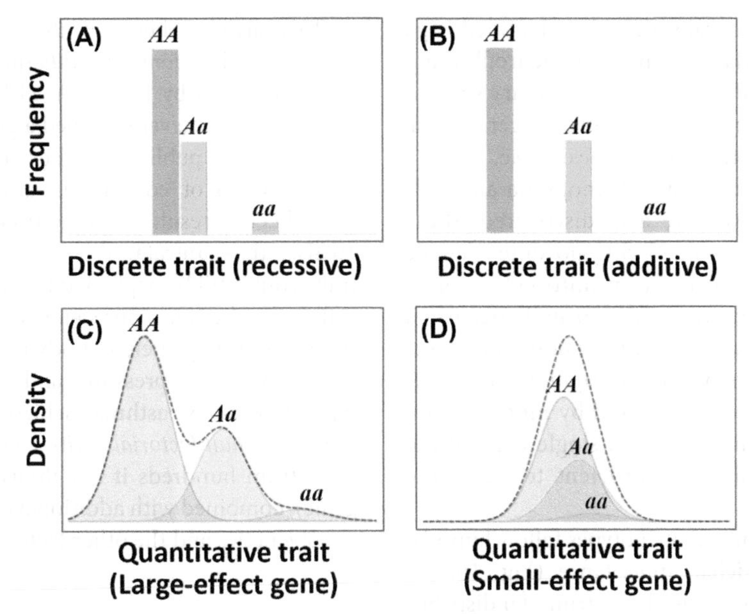


**Figure 16.4** Phenotypic distribution of different traits. (A) Recessive Mendelian trait with two discrete phenotypes. (B) Distinct codominant Mendelian trait with three discrete phenotypes. (C) Quantitative trait—controlled predominantly by one large-effect gene, and unquestionably additional modifiers showing a continuous distribution with three distinct modes. (D) Quantitative trait influenced by genetic (innumerable small-effect genes)—plus epigenetic, endogenous, environmental, and perhaps microbiome effects; in other words, a polygenic trait that follows a normal distribution. (Modified from Zhang G, Nebert DW. Personalized medicine: genetic risk prediction of drug response. Pharmacol Ther 2017;175:75–90.)

albinism, alkaptonuria, cystinuria, and pentosuria. Each of these conditions was found to be inherited as an *autosomal recessive* trait, showing a pattern of inheritance similar to that of the white flower color of Mendel's garden pea. Garrod is recognized as spearheading the era of human genetics; the underlying tenet was "one gene, one disease," or "one wild-type (healthy) allele, and one disease (mutant) allele." For each pregnancy, two asymptomatic parents—who are both "carriers," heterozygous for a disease allele—exhibit a one-in-four chance of producing a child having both disease alleles and, hence, inheriting the unwanted genetic disorder.

Each gene has two alleles in a chromosome pair—one from either parent. The Hardy–Weinberg equilibrium (HWE; $p^2+2pq+q^2=1$) was originally established to describe the expected genotype frequencies under random mating. Usually $p$ was used to refer to the frequency of the "wild-type" allele and $q$ for the frequency of the "mutant," or variant, allele. For example, if $q=0.10$, this means that the percentage of individuals homozygous for that recessive trait will be $q^2=0.01$, i.e., one in 100. On the other hand, if $q=0.01$, then the percentage of individuals who are homozygous for that recessive trait will be $q^2=0.0001$, i.e., 1 in 10,000.

Thus, due to early technical difficulties in recruiting and studying populations of 10,000 or greater, for most clinical studies the lowest frequency of variant alleles usually studied used to be 0.05, i.e., *common variants.* After it was realized that multiple alleles exist for every gene, $q$ was then defined as the sum of all variant alleles. The term $q$ has now been replaced by MAF (*minor allele frequency*), referring to "the frequency at which any particular variant allele—other than the major allele—occurs in a given population."

Garrod's four disorders, described above, which normally skip a generation, were among the first *large-effect* single-gene mutations causing severe clinical disorders. During the next several decades, many additional examples of large-effect *autosomal recessive* clinical diseases were described (e.g., maple syrup urine disease, sickle-cell disease, Gaucher disease, phenylketonuria, cystic fibrosis, and congenital adrenal hyperplasia).

*Autosomal dominant* traits were also identified (e.g., Huntington disease, Marfan syndrome, neurofibromatosis, achondroplastic dwarfism, and hereditary spherocytosis); these traits typically appear in each generation, because the heterozygote manifests the disease.

*X-linked recessive* traits (e.g., hemophilia and red-green color blindness) were also discovered; these represent mutations on the X chromosome. Female carriers show a 50% chance of transmitting the defective allele to offspring; the disorder usually occurs in males but not females. Also, *X-linked dominant* traits were identified (e.g., incontinentia pigmenti and Coffin–Lowry syndrome)—also caused by mutations on the X chromosome; in this case, a single copy of the defective dominant allele is sufficient to cause the syndrome.

All the above-mentioned phenotypes follow bimodal distributions of Mendelian inheritance (Fig. 16.4A). Large-effect alleles can also lead to a trimodal distribution (Fig. 16.4B), in which additive traits from both parents result in an intermediate phenotype. An additional complexity to virtually all Mendelian diseases includes "modifier genes." In other words, any number of additional genes can affect age-of-onset and/or degree-of-severity of the disorder, thereby causing overlapping of phenotypes when plotting a quantitative trait, as illustrated in Fig. 16.4C.

### 16.3.2 Resolution of Multifactorial Traits with Mendelian Inheritance

In addition to the relatively simple Mendelian traits having distinct patterns of inheritance, described to this point, many traits (e.g., height, body mass index, intellectual ability, and serum uric acid levels) exhibit a gradient variation in any population; moreover, one sees stronger similarity within any family than between families. The inheritance pattern of these continuous traits represented a dilemma, not readily explained by any simple Mendelian distribution.

Between 1880 and 1920, a group of biometric statisticians argued against Mendelian inheritance, because they saw that most phenotypic variation was continuous rather than bimodal or trimodal. Hence, because any phenotype of an offspring is approximately the average of that seen in his two parents—a "blending model" seemed more appropriate to explain the inheritance pattern of continuous traits.

This division between the *Mendelian inheritance school* and the *biometrics inheritance school* was most clearly resolved by Robert A. Fisher [47], a mathematician who had never obtained a graduate degree. In his breakthrough publication, Fisher presented evidence that a gradient of "continuous variation" could represent the collective result of many discrete genetic loci (Fig. 16.4D); thus, intrafamily resemblance of continuous traits could still be explained by Mendelian inheritance. It therefore became appreciated that most *human quantitative traits* (e.g., height, body mass index, serum lipid levels, and blood pressure) and *complex diseases* (e.g., type II diabetes, asthma, schizophrenia, and cancer) represent *multifactorial traits*—which reflect contributions from hundreds if not thousands of genes (polygenic), combined with additional modifying effects such as epigenetics and the other factors listed earlier.

### 16.3.3 Beginning of the Genomics Era

Following the advances in molecular biology, recombinant-DNA cloning, and DNA sequencing that began in the 1970s—the field of genomics quickly helped geneticists appreciate that "monogenic diseases" virtually always represent numerous "disease alleles." Among the earliest breakthroughs was *phenylketonuria* (PKU), described as an *autosomal recessive* disorder caused by phenylalanine hydroxylase (PAH) deficiency. After cloning the *PAH* gene from one chromosome of a "carrier" parent of a PKU child [239], the Savio Woo lab reported that the *PAH* disease mutation was located at the 5′ splice-donor site of intron 12 [135]. This discovery was initially described as "***the*** disease allele" for PKU. However, within months, a second mutation (this one changing an amino acid; i.e., *nonsynonymous*, or *missense*, mutation) was reported. Three years later [22], 18 distinct mutations had been identified. Soon the concept of *allelic heterogeneity* was widely accepted as "the norm" for virtually all genes in which single-nucleotide alterations, as well as insertions and deletions (indels), or duplications or inversions of DNA segments—can cause serious disease.

As of October 2017, distinct mutations (in and near the *PAH* gene) that cause variable symptoms of PKU, reported worldwide, total at least 1040 (http://www.biopku.org/home/pah.asp), with many reports of *ethnic differences* in allelic frequencies. Similar findings have been described for most other Mendelian disorders.

### 16.3.4 Single Nucleotide Polymorphisms/ Variants (SNPs, SNVs)

Following initiation of the Human Genome Project in 1990, the field of genomics advanced exponentially. Since the early 1980s, yeast, fly, and worm geneticists had used the term "nucleotide substitution" for a mutation. However, in the mid-1990s several human genetics laboratories coined the term "s*ingle nucleotide polymorphism*" (SNP), and "SNiPping through the DNA" sounded exciting. Consequently, "SNP fever" was launched. In retrospect, a better name for "SNP" would have been "single nucleotide variant," and, in fact, in recent years, the use of "SNV" has increased in popularity.

In the mid-1990s, dozens of publications began demonstrating "statistically significant" associations (with $P$ values < 0.05) between one or several SNPs and a complex disease—such as Alzheimer disease, type II diabetes, asthma, or autism spectrum disorder. Very soon thereafter, studies from other laboratories reported they were unable to corroborate those initial findings. It was quickly realized that the genetics of complex diseases would not be nearly as simple as that of Mendelian diseases.

Methods such as linkage studies, which had been successful in identifying major genes, were found to have limited power in detecting genes of modest effect or lower penetrance. Subsequently, a new method of "genotype-phenotype association studies" arose. Searching concurrently for all candidate genes associated with a trait would have greater power, even if this meant testing every gene in the genome. The landmark publication by Risch and Merikangas [179] described such a genome-wide approach—including the proposal to use permutation analysis and multilocus testing, with a $P$ value of $<5.0 \times 10^{-8}$ (<5.0e–08) as the "statistically significant cut-off" for any genetic variants in the 3-billion-base-pair human haploid genome.

The vast majority of drug responses (efficacy, therapeutic failure, adverse effect, and toxic effect) involves not "Mendelian" or "predominantly oligogenic"—but rather *polygenic, multifactorial phenotypes*. What follows is therefore a brief overview of genome-wide association studies (GWAS), missing heritability, and rare versus common variants. It will be important to appreciate these concepts, in order to acquaint the reader later with the particularly problematic properties of multifactorial traits.

### 16.3.5 Genome-Wide Association Studies (GWAS)

Associations between five SNPs in the lymphotoxin-α gene (*LTA*) and myocardial infarction were reported, by using ~93,000 gene-based SNP markers [164]; this is purportedly the earliest published GWAS. Another early GWAS included >116,000 SNPs and demonstrated an association between the complement factor H gene (*CFH*) and age-related macular degeneration [103]. Currently, DNA-chip platforms containing 1 million to 5 million SNPs are available, easy to use, and relatively inexpensive. At the last count (https://www.ebi.ac.uk/gwas/), >61,000 SNP–trait associations have been reported in >3300 studies. These GWAS—with $P$ values ranging from $<10^{-8}$ to $<10^{-600}$—underscore the value of using stringent statistical significance levels when one is testing >1 million SNPs genome-wide of large cohorts comprising thousands, or even hundreds of thousands, of subjects.

GWAS quickly became much more reliable for genotype–phenotype association tests than commonly published studies involving one or several SNPs in small cohorts. These latter publications, using several dozen or even several hundred individuals, are highly prone to the statistical artifacts of *type I errors* ("false-positive"; erroneous rejection of a true null hypothesis) and *type II errors* ("false-negative"; incorrect acceptance of a false null hypothesis). Unfortunately, such published useless data continue to flood the literature, and have been variously called "the incidentalome" [105] and "$P$<.05 false-positive/false-negative studies" [158].

Several parameters (e.g., effect-size, allele frequency, significance level, and sample size) will affect the statistical power for any genotype–phenotype association study. Clearly, the larger the numbers of cases and controls, the greater the statistical power. Moreover, as any MAF increases in frequency, fewer subjects are usually necessary in the study group, and the level of detectable contribution by an SNP to the trait will be lower. As any MAF decreases, greater numbers per group will be needed, and the level of detectable contribution by an SNP to the trait will need to be higher. GWAS will almost never have sufficient statistical power to detect epistasis (gene × gene interactions; G × G) [7,183] or gene–environment (G × E) interactions [214]. GWAS studies of PGx traits are covered later in this chapter.

### 16.3.6 Variance Explained Versus "Missing Heritability"

Findings from GWAS eventually became unsatisfying to some investigators—who preferred clear-cut data that unequivocally quantified the total number of genes contributing to a multifactorial quantitative trait such as height or body mass index, or to explain the cause of a complex disease [62]. For most complex diseases or PGx traits, even DNA variants found together in a GWAS (e.g., using polygenic risk scores) typically explain only a small proportion of phenotypic variance ($R^2$) and therefore have limited clinical predictive value [132,248]. The absent proportion became known as "missing heritability" [110,133]. To make matters more unsatisfactory, although the "revealed heritability" continued to grow as the sizes of GWAS cohorts became increasingly larger, the *variance explained* rarely reached more than 20%–25% for various diseases as well as quantitative multifactorial traits [110].

Three overlapping theories to explain "missing heritability" were then proposed [58]: the "infinitesimal model" (large number of variants across the entire allele frequency spectrum of small-effects); the "rare variant model" (multiple large-effect rare variants that are poorly tagged by genotyping arrays); and the "broad-sense heritability model" (contributions from G×G, G×E, and/or epigenetic interactions). It is now clear that—as the GWAS cohort sizes continue to get larger—more and more *small-effect DNA variants*, in addition to credible candidate genes contributing to any multifactorial trait, will become statistically significant. However, even if the entire population on our planet could be studied, it now appears likely that *variance explained* will still not reach 100% for many of these complex diseases and quantitative multifactorial traits.

Yet, it is important to emphasize that some GWAS data have identified potential novel therapeutic targets for treating a complex disease. Similarly, some PGx GWAS data might uncover potential drug targets for improving efficacy or treating an ADR, by learning something about its mode-of-action, without necessarily understanding any precise mechanism-of-action.

## 16.4 EARLY PGX EXAMPLES

Due to space limitations, fewer than a dozen cases are described in this section. For additional examples, please see Table 16.2 and references therein. Histories of how some of these PGx traits came to be discovered make for entertaining stories at cocktail hours and parties.

### 16.4.1 *N*-Acetylation Polymorphism (*NAT2* Gene)

Originally called the "*isoniazid acetylation polymorphism*," this PGx disorder was first noticed clinically in the 1940s when patients, who had converted from a negative to a positive tuberculin test, were routinely prescribed isoniazid. A high incidence of peripheral neuropathy was noted. This is an example of the *active principle* (parent drug) reaching toxic levels (Fig. 16.1)—when the major detoxification enzyme in the isoniazid metabolic pathway is defective.

Isoniazid was administered to volunteers, and their plasma isoniazid levels measured 6 h later (Fig. 16.5); the *bimodal distribution* found by the Victor McKusick lab [174] is most similar to that illustrated in Fig. 16.4A. The phenotypes were termed "slow acetylators" and "rapid acetylators" (i.e., slow vs. rapid plasma clearance of isoniazid). The true biological parents of slow-acetylator children were always slow acetylators—indicating that slow acetylators are homozygous for the "slow-acetylator" allele (*r*), whereas rapid acetylators are either heterozygous or homozygous for the "rapid" (*R*) allele. Hence, the *slow phenotype* is inherited as an *autosomal recessive* trait. The frequency of the *r* allele was ~0.72 in the US population that was studied [174]; i.e., if $q = 0.72$, then $q^2 = 0.52$. This means that about one in every two individuals (in this population) is homozygous for *r*/*r*, manifesting the slow-acetylator trait. This study also reflects the thinking at the time (Section 16.3.1): "one wild-type (healthy) allele, and one disease allele."

Isoniazid *N*-acetyltransferase *variability* represents an example of a PK gene polymorphism. Three decades later, it was determined that there are two *N*-acetyltransferase genes (*NAT1, NAT2*), located in tandem on human chromosome (Chr) 8p22. The *NAT2* gene was responsible for the rapid- versus slow-acetylator phenotypes; when isoniazid and other arylamine substrates were studied, the NAT2 enzyme was found to exhibit a 10-fold lower $K_m$ than NAT1. Several *NAT2* slow-acetylator variant alleles were found to encode a stable protein having little or no enzymic activity [9].

A systematic allele nomenclature system for many human PK genes was initiated [30] and can now be found online (https://www.pharmvar.org/). A consensus nomenclature system for the *NAT1* and *NAT2* alleles

TABLE 16.2 **History: Early Mendelian PGx Disorders**[a]

| Disorder or Trait | Major Gene Known/Identified | Breakthrough Reference(s) |
|---|---|---|
| Phenylthiourea–nontaster | *TAS2R1* | [101,196] |
| Hypocatalasemia | *CAT* | [207] |
| Atypical serum cholinesterase | *BCHE* | [95] |
| Glucose-6-phosphate dehydrogenase deficiency | *G6PD* | [134] |
| Isoniazid slow *N*-acetylation | *NAT2* | [9,174] |
| Fish-odor syndrome trimethylaminuria | *FMO3* | [74,77] |
| Debrisoquine/sparteine oxidation poor metabolizer | *CYP2D6* | [40,67,128] |
| Serum paraoxonase low activity | *PON1* | [56,78] |
| Thiopurine methyltransferase deficiency | *TPMT* | [235] |
| Sensitivity to alcohol | *ALDH2* | [213] |
| *S*-mephenytoin oxidation deficiency | *CYP2C19* | [34,107] |
| Sulfotransferase deficiency | *SULT1A1, SULT1A2* | [234] |
| Nicotine oxidase deficiency | *CYP2B2* | [245] |
| P-glycoprotein transporter defect | *ABCB1* | [102] |
| Malignant hyperthermia | *RYR1* | [127] |
| Quinone oxidoreductase defect | *NQO1* | [218] |
| Peptide transporter defect | *TAP2* | [173] |
| Phenytoin, warfarin oxidation defect | *CYP2C9* | [33] |
| Debrisoquine ultrametabolizer | *CYP2D6*1XN* | [89] |
| Warfarin metabolism | *CYP2C9* | [63] |
| Epoxide hydrolase deficiency | *EPHX1* | [71] |
| Glutathione *S*-transferase null alleles | *GSTM1*0, GSTT1*0* | [98,236] |
| Long-QT syndrome | *KCNH2* | [23] |
| Dihydropyrimidine dehydrogenase deficiency | *DPYD* | [140] |
| Chlorzoxazone hydroxylation defect | *CYP2E1* | [76] |
| Peptide transporter defect | *TAP1* | [175] |
| Sulfonylurea receptor defect | *ABCC8* | [69] |
| Calcium channel defect | *CACNA1A* | [247] |
| Androstane glucuronosyl conjugation | *UGT2B4* | [116] |
| Congenital long-QT syndrome | *SCN5A* | [232] |
| *S*-oxazepam glucuronosyl conjugation | *UGT2B7* | [202] |
| Paclitaxel hydroxylase deficiency | *CYP2C8* | [26] |
| Chlorpyrifos oxidation deficiency | *CYP3A4* | [25] |
| Nicotine metabolism alterations | *CYP2A6* | [242a] |
| Acrodermatitis enteropathica | *SLC39A4* | [231] |
| Nifedipine oxidation deficiency | *CYP3A5* | [114] |
| Cyclophosphamide metabolism deficiency | *CYP2B6* | [109] |
| Hyperinsulinemic hypoglycemia | *SLC16A1* | [162] |
| Warfarin resistance | *VKORC1* | [181] |
| Hereditary folate malabsorption | *SLC46A1* | [250] |
| Warfarin metabolism | *CYP4F2* | [12] |

[a]This list (not intended to be all-inclusive) in each case compares the consensus allele with one or more variant alleles that lead to a defective gene product. Other variants in PK and PD genes can be found at https://www.pharmvar.org/. The result is decreased metabolism or transporter (PK gene), or receptor or channel function (PD gene). The clinical consequence in most homozygous affected subjects is *toxicity*, due to drug accumulation with enhanced drug activity. Occasionally, decreased drug activity (*therapeutic failure*) ensues if the variant reflects ultrarapid drug metabolism or if, for activity, the drug requires metabolic conversion to an active form and this conversion is decreased in the variant. Some of the traits listed here might concern primarily environmental toxicants (e.g., *TAS2R1, CAT, FMO3,* and *PON1*) more so than prescribed drugs.

Modified from Nebert DW, Vesell ES. Chapter 19-"Pharmacogenetics and pharmacogenomics". In: Emery and Rimoin's principles and practice of medical genetics. 6th ed. Oxford: Academic Press; 2013. pp. 1–27.

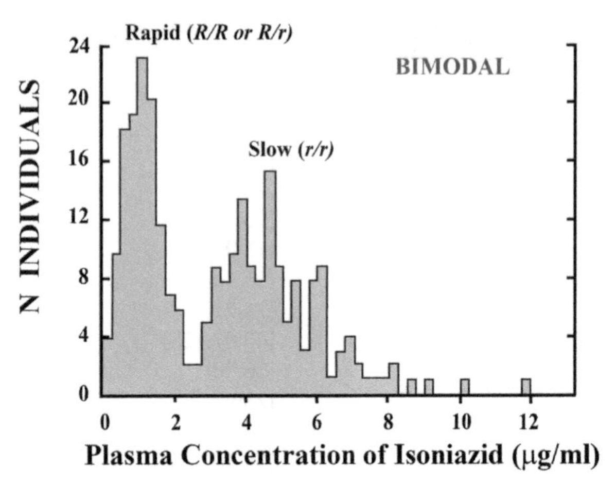


**Figure 16.5** Plasma isoniazid concentrations 6h after the drug was given. Results were obtained in 267 members of 53 complete family units. All subjects received 9.8mg isoniazid per kg body weight [174]. (From previous Emery and Rimoin edition; Nebert DW, Vesell ES. Chapter 19-"Pharmacogenetics and pharmacogenomics". In: Emery and Rimoin's principles and practice of medical genetics. 6th ed. Oxford: Academic Press; 2013. pp. 1–27.)

was developed in 1995 (http://nat.mbg.duth.gr/Human%20NAT2%20alleles_2013.htm); as of the time of this writing, it was last updated in April 2016, with more than 170 named alleles. The (rapid-acetylator) consensus allele is *NAT2*4*, with allele numbers designated (e.g., *NAT2*6A*, *NAT2*7D*, *NAT2*14J*, *NAT2*14K*, etc., with the highest allele number to date as *NAT2*27*).

## 16.4.2 Debrisoquine/Sparteine Oxidation Polymorphism (*CYP2D6* Gene)

In the 1970s, the debrisoquine/sparteine polymorphism (Fig. 16.6) was independently discovered by two groups. The Robert L. Smith laboratory in England [4] studied oxidative metabolism of the antihypertensive agent, debrisoquine. Soon after the drug was available in the UK, Smith noticed that debrisoquine caused a remarkably high incidence of ADRs; he correctly surmised that the combination of a narrow therapeutic index (described above; Fig. 16.3), with an underlying genetic variation in metabolism, might be responsible. Smith and three laboratory colleagues took the "recommended prescribed" dose of debrisoquine; Smith himself became hypotensive, and his urinary 4-hydroxy metabolite was ~20-fold lower than that of his three colleagues who appeared unaffected by that

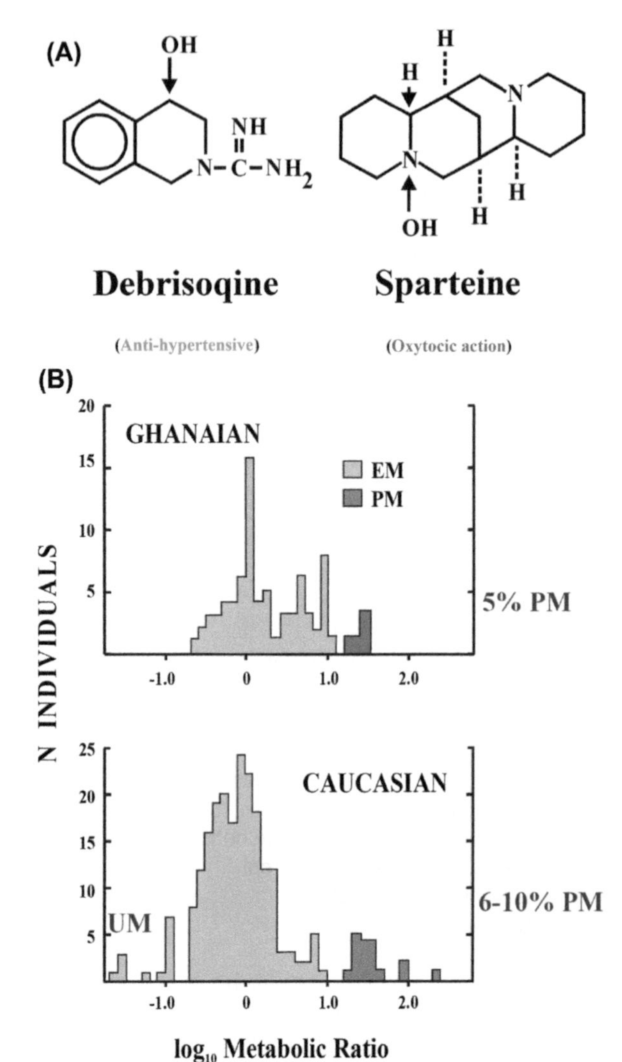


**Figure 16.6** Illustration of the *CYP2D6* polymorphism. (A) Chemical structures and major metabolites of debrisoquine and sparteine, two substrates of CYP2D6. (B) Frequency of the efficient-metabolizer (EM) and poor-metabolizer (PM) phenotypes in a population from Ghana (top), and frequency of the EM, PM, and ultrametabolizer (UM) phenotypes in a population of Caucasians from the United Kingdom (bottom). Urinary "metabolic ratio" (MR) is defined as the "parent drug debrisoquine divided by hydroxylated debrisoquine metabolites." Because PM individuals show less metabolism than EM and especially UM subjects, this higher ratio places PM subjects to the far right [240]. (From previous Emery and Rimoin edition; Nebert DW, Vesell ES. Chapter 19-"Pharmacogenetics and pharmacogenomics". In: Emery and Rimoin's principles and practice of medical genetics. 6th ed. Oxford: Academic Press; 2013. pp. 1–27.)

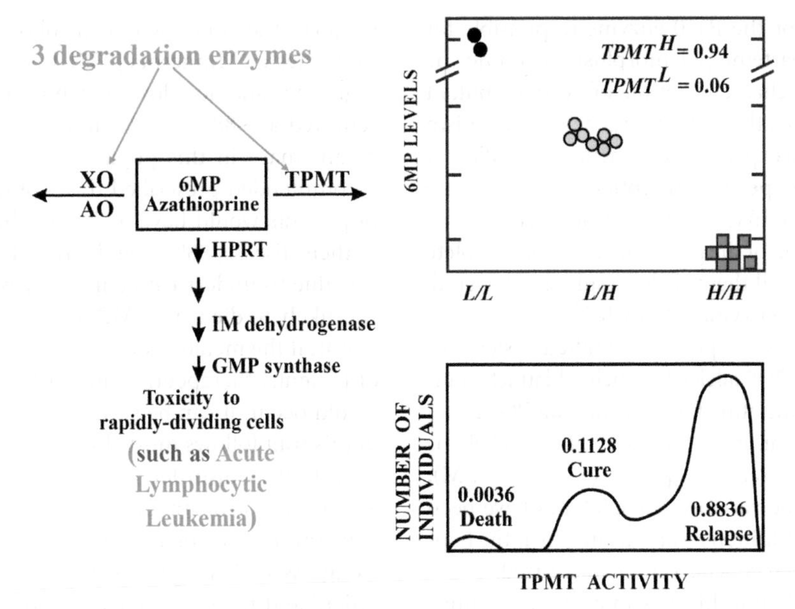


**Figure 16.7** Diagram of the 6-mercaptopurine (6MP), azathioprine, or 6-thioguanine *drug response* phenotype. *Toxicity* by these chemotherapeutic agents occurs in all cells, but especially in rapidly dividing cells such as acute lymphocytic leukemia (ALL) white cells—due to disruption of purine biosynthesis. Given the "recommended" prescribed dose of 6MP, the patients' *drug response* is plotted as a function of the three proposed genotypes (upper right); and the patients' *drug response* is plotted as the number of individuals exhibiting *toxicity* as mortality, *efficacy* as being cured, and *therapeutic failure* as having a relapse of the disease (lower right). XO, xanthine dehydrogenase encoded by the *XDH* gene. AO, adenine oxidases-1 and -2 encoded by the *DUOX1* and *DUOX2* genes. TPMT, thiopurine methyltransferase. All three of these enzymes participate in detoxification of these purine analogs. About three in 1000 Caucasians are found to be homozygous for the *L/L* phenotype, 13% are *L/H* heterozygotes, and 88% are homozygous for the *H/H* phenotype [235]. (From previous Emery and Rimoin edition; Nebert DW, Vesell ES. Chapter 19-"Pharmacogenetics and pharmacogenomics". In: Emery and Rimoin's principles and practice of medical genetics. 6th ed. Oxford: Academic Press; 2013. pp. 1–27.)

dose. This is another example of a pharmacologically active drug reaching toxic levels due to insufficient detoxification to an inactive metabolite (Fig. 16.1). A larger population was screened and—similar to the isoniazid polymorphism—showed a *bimodal distribution* (much like Fig. 16.4A), separating "poor-metabolizer" (PM) from "extensive-metabolizer" (EM) subjects.

Michel Eichelbaum, for his 1975 thesis in the Hans J. Dengler lab in Germany, studied human metabolism of the oxytocic drug, sparteine. This drug was known to cause erratic and excessive uterine contractions in some, but not most, women; the urinary ratio of sparteine to the dehydrosparteines showed a *bimodal distribution* [41]. This variability in debrisoquine/sparteine metabolism is another example of a PK enzyme polymorphism.

Compared with EM-phenotype subjects that metabolize the drug 10–50 times more effectively, the PM phenotype for debrisoquine [80] occurs in 6%–10% of people of European descent (Fig. 16.6B). The incidence of the PM phenotype was found to be ~5% in an African population (Fig. 16.6B) and <1% in Asians. Subsequently, an "ultrarapid metabolizer" (UM) phenotype was also described (which actually would account for those few samples seen at the *far left* in Fig. 16.6B). This phenotype was found to be caused by multiple copies of the *CYP2D6* gene—from two, to as many as 13 copies [142]. The incidence of the UM phenotype is ~0.8% in northern Europeans, but 21% in Saudi Arabians, and 29% in Ethiopians [82]; the reason for this very high UM phenotype frequency in Saudi Arabia and Ethiopia is not known, but is most likely the result of either a genetic bottleneck or a selective environmental pressure such as diet.

*CYP2D6* codes for the P450 enzyme responsible for the debrisoquine/sparteine polymorphism. Cloning the *CYP2D6* gene and characterization of several mutant alleles [67] represented the first time a genetic mechanism was demonstrated to explain a PGx phenotype. Different PM alleles—due to specific nucleotide changes—were shown to code for: an inactive enzyme, an unstable protein, incorrect splicing of the gene transcript, or complete deletion of the gene. All these alleles resulted in lowered, or completely absent, enzyme activity [67].

As mentioned above, a proposed unified system for naming human *CYP2D6* alleles [30] helped launch standardized allele nomenclature for many human PGx genes (https://www.pharmvar.org/). The *CYP2D6*1* allele is the consensus, or reference, sequence (wild-type, EM); currently, > 200 allelic variants or haplotypes have been reported—plus ~30 additional variants in which the haplotype has not yet been conclusively characterized.

The *CYP2D6* polymorphism is important in elimination of >20% of commonly prescribed drugs, as well as many over-the-counter drugs; St. John's wort and grapefruit juice are the two most popular examples. The debrisoquine "panel" now comprises >120 drugs, such as: tricyclics and other antidepressants including serotonin-reuptake inhibitors and monoamine-oxidase inhibitors; neuroleptics; antiarrhythmics and antihypertensives (including beta-blockers); the antiestrogen tamoxifen; and opiates (cf. the website designed by David Flockhart, http://medicine.iupui.edu/clinpharm/ddis/main-table/).

The analgesics codeine, hydrocodone, and oxycodone are cleared via CYP2D6-mediated *O*-demethylation; the rate of clearance can occur ~200-fold more rapidly in EM than in PM patients. Formation of morphine, by way of CYP2D6-mediated *O*-demethylation of codeine, is central to codeine's analgesic PD effects. Patients who lack CYP2D6, or whose CYP2D6 is inhibited, would not be expected to benefit from codeine, whereas CYP2D6 UM-phenotype patients would be at greater risk of serious toxicity [54]. Thus, phenotyping for CYP2D6, and avoidance of CYP2D6 inhibitors, has been proposed for chronic pain patients [13].

### 16.4.3 Thiopurine Methyltransferase Polymorphism (*TPMT* Gene)

TPMT plays a pivotal role in detoxification of 6-mercaptopurine (6MP), commonly used in chemotherapy for childhood acute lymphocytic leukemia (ALL). In the original study of a Caucasian cohort of ~500 volunteers in Rochester, Minnesota [235], frequencies of high/high, high/low, and low/low metabolism phenotypes were reported as ~88%, ~11%, and ~0.4%, respectively. This meant that—in this population when the "commonly recommended prescribed dose" of 6MP is given—11% of patients would have high probability of being cured of their disease, 88% would have relapses in their leukemia due to undertreatment, and 1 out of ~300 patients was likely to die from 6MP toxicity (Fig. 16.7). In other words, if the metabolism of 6MP is too extensive in 88% of patients, therapeutic failure (illustrated in Fig. 16.1) would occur. It can be seen (Fig. 16.7) that distribution of this trait follows most closely that of Fig. 16.4B.

This PGx disorder is very dramatic because it can lead to life-or-death clinical situations. Accordingly, in 1994 the *TPMT* polymorphism was presented to the US Congress as "the quintessential pharmacogenetic disorder," and increased federal funding for PGx research was requested. Because the TPMT defect can lead to dire consequences, ALL patients are now routinely phenotyped for red-cell TPMT activity prior to initiation of 6MP chemotherapy. $TPMT^{H/H}$ individuals generally show a favorable response with a four-times-larger dose, and $TPMT^{L/L}$ patients with a 10- to 15-times-smaller dose. This regimen has resulted in substantially higher cure rates and longer survival rates for childhood ALL.

It should be emphasized that "red-cell TPMT activity" is a phenotyping test—not a genotyping test. The *TPMT* gene spans 26.8 kb. At least 50 allelic variants have now been identified, more than half of which alter TPMT activity (https://databases.lovd.nl/shared/genes/TPMT), resulting in very low or negligible catalytic activity. This is an example in which the phenotyping test is superior to any genotyping test, due to the ever-present possibility that a disease-causing variant lies beyond those variants that have been discovered (or any nongenetic factor, such as blood transfusion, affects the phenotype).

Azathioprine and 6-thioguanine are other TPMT substrates (Fig. 16.7). Azathioprine is widely used as an immunosuppressant in conditions as diverse as systemic lupus erythematosus and organ transplantation. Thioguanine is one of the agents used in treating chronic myelocytic leukemia. As with 6MP, azathioprine and 6-thioguanine can be lethal to the 1-in-300 homozygous $TPMT^{L/L}$ patient—if that individual receives the "commonly recommended prescribed" dose.

Although >80% long-term survival rates in ALL have resulted from the TPMT-phenotyping test, morbidity due to drug-related myelotoxicity has continued to be problematic. Two additional relevant genes encode enzymes in the purine biosynthesis pathway: inosine triphosphatase (*ITPA*) and nudix hydrolase-15 (*NUDT15*), and it has been discovered that variants in both genes can lower the metabolism of purine analogs. One *NUDT15* variant was recently identified as a novel polymorphism linked to 6MP-induced leukopenia in inflammatory bowel disease and ALL patients [193]. There are 22 nudix hydrolase genes (*NUDT*) in the human genome. Genetic variants of *TPMT*, *ITPA*, and *NUDT15* have now been shown to affect 6MP (also, azathioprine and 6-thioguanine) metabolism, and ethnic differences of course exist in all three genes. What therefore began as a simplistic scenario—large-effect *TPMT* alleles altering an enzyme in the purine biosynthesis pathway—has now evolved into a more complex picture of PGx; this has led to further issues to prevent *drug toxicity* and the challenges of personalized medicine.

### 16.4.4 *S*-Mephenytoin Polymorphism (*CYP2C19* Gene)

CYP2C19 participates in metabolism of at least four dozen commonly prescribed drugs—including antiepileptics such as *S*-mephenytoin and diazepam, proton-pump inhibitors such as omeprazole, and other drugs such as amitryptyline, citalopram, and propranolol (http://medicine.iupui.edu/clinpharm/ddis/). The CYP2C19 story is similar to the three above-described examples: the EM phenotype appears to be dominant over the PM phenotype, and almost always the parent drug in PM patients exhibits ADRs due to overdose (toxic levels; illustrated in Fig. 16.1).

At least 44 *CYP2C19* mutant alleles have been described, and another 40 haplotypes are not yet completely characterized (https://www.pharmvar.org/). The consensus, or wild-type, *CYP2C19*1* allele results in normal enzyme activity. Most of the mutant alleles (e.g., *CYP2C19*2* and **3* alleles) encode a protein having little or no activity. The *CYP2C19*17* allele is responsible for a CYP2C19 that exhibits increased levels of enzymatic activity.

Voriconazole has become important for treatment of invasive fungal infections (e.g., aspergillosis and candidiasis). *CYP2C19* polymorphisms appear to account for the largest portion of variability in response to voriconazole. A role for *CYP2C19* genotyping to guide the initial voriconazole dosing, followed by therapeutic-drug monitoring, has been proposed as a means of increasing the likelihood of achieving efficacy while avoiding toxicity [163].

CYP2C19 also catalyzes bioactivation of clopidogrel, an antiplatelet prodrug. Loss-of-function alleles (e.g., *CYP2C19*2*) thus impair formation of the *active principle* (i.e., metabolite), resulting in decreased platelet inhibition and increased risk for adverse cardiovascular events, especially in PM patients undergoing percutaneous coronary intervention [91]. Therefore, alternative antiplatelet therapy (e.g., prasugrel, ticagrelor) is now recommended for patients who are CYP2C19 PMs or intermediate metabolizers (IMs)—if there are no contraindications [188].

### 16.4.5 Glutathione *S*-Transferase Polymorphisms (*GST* Genes)

Encoded by PK genes, the glutathione *S*-transferases (GSTs) are conjugation enzymes that add glutathione to many drugs and chemicals. High GST activity can lead to rapid detoxification rates of antibiotics and chemotherapeutic agents [217]. Usually considered to be detoxification enzymes, it should be noted that GSTs can also be involved in bioactivation [144]. The *GST* gene family (https://www.genenames.org) comprises 17 genes in six subfamilies: *GSTA*, *GSTM*, *GSTO*, *GSTP*, *GSTT*, and *GSTZ* [155]. Human populations exhibit high frequencies for total deletion of the *GSTM1* or *GSTT1* genes (so-called "null alleles" *GSTM1*0*, *GSTT1*0*); the incidence of GST-null individuals ranges between 20% and 50% in East Asian populations, and varies among different ethnic populations.

During the past several decades, many dozens of genotype–phenotype associations have been reported—between cancer or toxicity and SNPs in the *NAT2* or *NAT1* genes, *CYP2D6* gene, *CYP2C19*, or in the *GSTM1*0* or *GSTT1*0* null alleles. Clearly, perhaps especially without glutathione conjugation, it seems reasonable to expect that genes encoding enzymes that detoxify drugs or environmental toxicants might be identifiable in genotype–phenotype association studies involving toxicity or cancer. However, as emphasized repeatedly in this chapter, virtually all studies involving one or a few SNPs associated with multifactorial traits such as cancer or drug toxicity in relatively small cohorts represent statistically underpowered false-positive data [105,152,158].

### 16.4.6 UDP Glucuronosyltransferase-1A1 Polymorphism (*UGT1A1* Gene)

*UGT1A1* codes for the enzyme UGT1A1 that metabolizes irinotecan (commonly used for metastatic colorectal cancer), as well as many other drugs. Homozygotes having the poor-metabolizer *UGT1A1*28* allele were shown to be at high risk for irinotecan-induced neutropenia. In fact, on the irinotecan product label, it is recommended for *UGT1A1*28* homozygotes to decrease the starting dose of this drug. However, due to lack of sufficient prospective data, it remains uncertain whether this recommended dose-reduction will result in decreased toxicity. Combined toxicity analysis has indicated that most patients who experience grade 3 or 4 diarrhea and/or neutropenia are not homozygous for the *UGT1A1*28* allele [108].

### 16.4.7 Dihydropyrimidine Dehydrogenase Polymorphism (*DPYD* Gene)

The fluoropyrimidines are frequently prescribed anticancer drugs, and they are inactivated by hepatic dihydropyrimidine dehydrogenase (DPYD). As much as 5% of cancer populations exhibit DPYD deficiency. This information is considered to have practical value—and might even be cost-effective—for patients receiving these anticancer fluoropyrimidine substrates; the *DPYD*2A* low-activity allele is highly associated with 5-FU-induced severe and life-threatening toxicity [35]. There is convincing evidence to implement prospective *DPYD* genotyping with an up-front dose adjustment in DPYD-deficient patients [124].

In each of these examples in which the active parent drug causes toxicity if not adequately metabolized, or if the PGx assay reveals diminished enzymatic activity or deficiency, then lower initial doses, or alternate drugs, are usually recommended.

### 16.4.8 Abacavir-Induced Hypersensitivity (*HLA* Loci)

Abacavir is an HIV-1 nucleoside-analog reverse-transcriptase inhibitor used to treat human immunodeficiency virus (HIV) infections. In an early example of identification of a PGx disorder by a "candidate-gene-region" study, 18 abacavir-treated patients exhibited a life-threatening hypersensitivity syndrome, out of 185 patients receiving abacavir [130]. Compared with 167 abacavir-resistant controls, SNP-typing of loci in the major histocompatibility complex (MHC) region revealed a very strong association [odds ratio (OR) = 117] with the *HLA-B*57:01* allele, and also in combination with the *HLA-DR7* and *HLA-DQ3* loci (OR = 73).

These *HLA* loci represent immune-response genes. They encode specific cell-surface molecules responsible for presentation of endogenous peptides to cells of the immune system. This study of a highly significant association of abacavir-induced hypersensitivity with *HLA-B*57:01* [130] was later confirmed in a much larger double-blind prospective randomized study that involved ~2000 patients from 19 countries [131]. However, even though the OR for the abacavir-hypersensitivity syndrome is very large, penetration of this phenotype is very weak—and therefore PGx testing would never be economically feasible.

### 16.4.9 Warfarin Polymorphisms (*CYP2C9*, *VKORC1*, and *CYP4F2* Genes)

Up to this point, our examples have been predominantly single-gene large-effect responses of a specific subpopulation to a drug, or to different types of drugs that are substrates of the encoded enzyme. Optimizing warfarin, coumarin, or acenocoumarol dosage for anticoagulation therapy is clinically of extreme importance, because of the dangers of either too little drug (causing clotting) or too much drug (causing unwanted hemorrhaging). Warfarin metabolism is an early oligogenic example in which a substantial contribution of several genes was found—resulting in rapid versus slow PK and/or PD of coumarins. *CYP2C9* [63] and *VKORC1* [24] polymorphisms were independently discovered in candidate-gene studies. These two genes were subsequently confirmed in GWAS [20,208], along with the additional discovery of *CYP4F2* [12].

The *CYP2C9* and *CYP4F2* polymorphisms represent (relatively) large-effect PK genes coding for enzymes involved in metabolism. *VKORC1* is considered to be a large-effect PD gene. The *VKORC1* gene codes for vitamin K-epoxide reductase complex subunit-1, which is targeted directly by coumarins. Coumarins are considered as vitamin K antagonists [84], and, as such, are potent inhibitors of this reductase complex; inhibition of the complex results in depletion of reduced vitamin K, which is essential for normal coagulation. The epoxide reductase has therefore been considered as a drug target for coumarins—especially because, except for the vitamin K substrate, the encoded epoxide reductase appears not to metabolize any other drug.

When variants of all three genes are combined, the combination provides ~45% of *variance explained*, i.e., the patient's total variability in drug response that can be accounted for [27]. Thus, the remaining ~55% of variability in coumarin response must originate from contributions by other genes and/or environmental factors. This oligogenic example, over a decade ago, was an excellent illustration of the growing complexity of the goal of predicting PGx phenotypes such as *drug efficacy* or *toxicity*—which has become increasingly appreciated during this past decade.

## 16.4.10 Ethnic Differences in Drug Metabolism

Numerous examples of *ethnic differences* in drug response are known [94] and several have been mentioned above. In several of these cases, the interethnic variability is sufficiently striking that PGx assays for a drug, or family of drugs, are recommended for one ethnic group, while being of considerably less importance to another ethnic group. Table 16.3 lists some examples of ethnic differences.

Among the earliest important ethnic differences discovered was the rapid-acetylator versus slow-acetylator phenotype (Table 16.3). Frequencies of the slow-acetylator allele range worldwide from less than 10% in Japanese populations to more than 90% in some Mediterranean peoples.

Mitochondrial aldehyde dehydrogenase-2 (ALDH2) deficiency is an interesting early example (Table 16.4). The incidence of the *ALDH2* Glu504Lys mutation ranges between 25% and 45% in most East Asian populations, yet is virtually never seen among Africans or Caucasians. A purportedly different *ALDH2* allele, also resulting in a lack of ALDH activity, was found in South American Amerindians [94], probably due to a founder effect or genetic bottleneck. These data led to speculation that ALDH2 deficiency might have arisen only in populations that traditionally have not been commonly exposed to ethanol. In other words, East Asians have boiled water for at least the past 30 centuries, compared with African and Caucasian populations that had used alcohol for enjoyment and preservation of foods for many earlier centuries.

For more than 100 years, nitroglycerin has been clinically used to treat angina and heart failure; it was recently

**TABLE 16.3 Frequency of *N*-Acetylator *NAT2* PM Phenotypes in Different Ethnic Populations**

| Ethnic Population | No. of Studies | Frequency of PM Phenotypes |
|---|---|---|
| Japanese | 7 | 0.09 |
| Eskimo | 4 | 0.23 |
| South Pacific Islands | 5 | 0.35 |
| Korean/Chinese | 14 | 0.37 |
| North and South Amerindian | 10 | 0.50 |
| African[a] | 19 | 0.71 |
| Central and West Asian | 22 | 0.74 |
| European | 50 | 0.75 |
| Egyptian | 2 | 0.96 |

[a]Excluding the !Kung Bushmen of Southern Africa, in which the PM frequency is 0.18.

Data modified and condensed from Kalow W, Bertilsson L. Interethnic factors affecting drug response. Adv Drug Res 1994;25:1–53.

**TABLE 16.4 Distribution of the ALDH2 Deficiency Phenotype in Different Ethnic Populations**

| Ethnic Population | Percent Having ALDH2 Deficiency[a] |
|---|---|
| Japanese | 44 |
| Central, East, and Southeast Asian | 25–50 |
| South Amerindian | 40–45[b] |
| North Amerindian | 2–5 |
| European, Mideast, and African | <0.1 |

[a]The mutation in Asians and North American Amerindians appears to be solely Glu504Lys, which causes a complete loss of ALDH2 activity in that subunit. Interestingly, the Lys504 allele contributes in large part to the lack of a clinically efficacious response to sublingual nitroglycerin [117]; this is particularly important among East Asian populations, 30%–50% of whom carry the *ALDH2*2* mutant allele. ALDH comprises four subunits; if one or more of the subunits are encoded by the *ALDH2*2* allele, then the entire tetramer is inactive. Thus, the *ALDH2*1/*2* heterozygote exhibits $(1/2)^5 = 1/16$, or 6.25%, of activity of the *ALDH2*1/*1* homozygous individual.

[b]Mutation in Amerindians from South America is purportedly different from that in Asians; to our knowledge, however, the DNA sequence of the Amerindian *ALDH2* allele from South America has never been published.

Data modified and condensed from Goedde HW, Agarwal DP. Aldehyde oxidation: ethnic variations in metabolism and response. Prog Clin Biol Res 1986;214:113–38.

discovered that mitochondrial ALDH2 is responsible for formation of nitric oxide [126], the metabolite required for nitroglycerin efficacy. Subsequently, it was shown that the catalytic efficiency of nitroglycerin metabolism of the consensus allele *ALDH2*1* (encoding the Glu504 protein) is ~10-fold higher than that of the mutant Lys504 enzyme. It has thus been recommended that *ALDH2*-genotyping be considered when administering nitroglycerin to patients—especially Asians, in whom 30%–50% carry the inactive *ALDH2*2* mutant allele [117].

Ethnic differences in *CYP2D6* alleles (https://www.pharmvar.org/) and many other P450 genes exist. CYP2C19 is one of the most important PGx examples of ethnic variability. CYP2D6 and CYP2C19 were discussed previously, and their wide range of drug substrates is what makes any substantial ethnic differences more relevant to clinical pharmacology. Ethnic differences for the CYP2C19 enzyme encoded by this PK gene are very striking; e.g., the East Asian PM subset is ~33% and the Oceanian PM subset >50%, whereas the Caucasian PM subset represents <6% [143].

The incidence of the *CYP2C19*2A* and *2B alleles (responsible for splicing defects) is 2%–5% in Caucasians, yet 20%–30% in Asians. This is an excellent example in which the physician must be more careful in prescribing "any drug in the CYP2C19 repertoire" for patients of Asian ancestry than for those having primarily Caucasian ancestry. Similarly, East Asian drug companies have recognized the importance of the *CYP2C19* polymorphism in their populations more so than companies in predominantly non-Asian countries.

### 16.4.11 Why Might Ethnic Differences in Drug Metabolism Exist?

It had been proposed [66] that PGx genes in animals might have originated because animals eat plants, and plant metabolites are similar in their intramolecular composition and molecular size to drugs. The Great Human Diaspora "Out of Africa" occurred over many tens of thousands of years of evolution. Ethnic groups originated as populations living in geographic isolation for >10,000 years and subsisting on distinct foods and diets relevant to that geographic region and their culture (lifestyle). It seems feasible that the Great Human Diaspora would explain today's observations of interethnic differences in drug metabolism and drug response [153].

Examination of *NAT2* variants in ~15,000 subjects from 128 populations [182], revealed a higher prevalence of the slow-acetylator NAT2 phenotype in populations practicing agriculture-and-herding, when compared with those relying mostly on hunting-and-gathering. Perpetuation of mutant alleles, resulting in monogenic disorders and in some PGx genes, can also result from enhanced resistance in the heterozygote to certain infections [153,238].

How long might it take for a genome to adapt to dietary selective pressures? By means of mutations (nucleotide substitutions, indels, inversions, duplications, crossing-over events, etc.), genetic drift, and natural selection, new gene alleles will become fixed and passed on to the next generation—if the new allele confers reproductive and ecological advantages, or is neutral (i.e., no detrimental effect), to the species. The response of a genome to environmental pressures, over a minimal number of generations, has been variously described as *molecular drive* [38], *meiotic drive* [211], cryptic genetic variation [59], and *decanalization* [57]. This gene–environment response most likely involves both genetics and epigenetics and probably also plays a role in the processes of drug- and/or plant-induced efficacy and toxicity.

Whereas mutational changes happen slowly over many dozens of generations, epigenetic changes (i.e., no alterations in DNA sequence) occur rapidly—in response to severe environmental challenges [104,178]. If the population of a species decreases dramatically, allelic frequencies in that population will change even more radically. Within a population, the emergence of individuals resistant to environmental changes has been shown experimentally, in various organisms from prokaryotes to insects and vertebrates, but why this happens remains basically obscure.

For example, Atlantic tomcod fish (*Microgadus tomcod*), living in the polluted Hudson River for 50–100 years (50–100 generations), have developed resistance to polychlorinated biphenyls (PCBs); a 6-nucleotide deletion in the *AHR2* gene—encoding an aryl hydrocarbon receptor-2 protein having poor-affinity for planar PCBs—was the basis for PCB resistance [237]. Depending on the organism studied, between nine and 45 generations appear to be required [151]. For humans, nine to 45 generations would extrapolate to times between ~200 and ~1000 years.

The five major *Homo sapiens* geographically isolated subgroups [251] are estimated to have diverged from one another between 20,000 and 45,000 years ago.

Based on studies in various animal species, 10,000 years of geographic isolation are believed to be sufficient for striking differences in PGx genes to have arisen from selective pressures—such as tribal differences in diet, or exposure to other environmental signals (e.g., altered climate and altitude). For example, it would seem likely that a tribe subsisting for ~10,000 years on a diet principally of goat meat and milk products on a high desert, might have different selective pressures than that of a tribe eating tropical fruit and fish at the seashore. Among the best examples would be the dramatic ethnic differences seen in lactase deficiency and response to milk products (reviewed in [86]). During a time period of ~10,000 years, it therefore seems feasible that striking differences in allelic frequencies of PGx genes would likely have arisen.

## 16.5 PHARMACOGENOMICS

As stated early in this chapter, *pharmacogenomics* began as a field distinct from pharmacogenetics and was defined as "the study of how drugs interact with the *total genome*, to influence biological pathways and processes" [148]; in other words, "drug–genome interactions." This field is expected to help identify new druggable targets and thus be instrumental in designing new drugs.

Advances in genome technologies, along with the development of statistical packages for large-scale data analysis, have enabled huge collaborative groups of investigators to carry out GWAS involving drug responses. Since 2007, dozens of PGx GWAS—having sufficiently large numbers in their cohorts—have led to the identification of numerous genes having SNPs associated with various responses of drug efficacy, ADRs, and toxicity. GWAS are similar to "fishing expeditions," i.e., they are hypothesis-free and thus do not require a priori assumptions about chromosomal locations of functional variants [137,201]. Accordingly, GWAS have provided an unbiased, powerful tool for the systematic discovery of genetic variants associated with a growing number of PGx traits [145].

### 16.5.1 ADRs can Be Indistinguishable From Complex Diseases

As we introduced early in this chapter, recent genome-wide PGx studies have refined our thinking [248]. We believe it is reasonable to classify the genetic basis of variability in drug response—together with epigenetic, endogenous, environmental, and microbiome effects—into three groups. Moreover, these groups should not be considered separate from one another, but rather a gradient. (1) We first presented more than a half-dozen *monogenic* (Mendelian) *traits*, and described early classical examples of severe toxicity or idiosyncratic dose-independent ADRs, typically reflecting one or a few rare coding large-effect variants, usually in PK genes or immune-response genes. (2) Among the *predominantly oligogenic traits*, we presented warfarin metabolism as an excellent example of variability caused in moderate amounts mostly by a relatively small number of large-effect (PK or PD) genes. (3) *Complex PGx traits*, which comprise the remainder of this chapter, represent typically the contribution of large numbers of small-effect variants; discovery of such genes contributing to complex PGx traits will almost always require GWAS to enable identification.

The clinician should be aware that idiosyncratic dose-independent ADRs and complex diseases are traits that can often be difficult to distinguish from one another (Fig. 16.8). In the case of a complex disease, after a stimulus (or stimuli) that triggers the disorder, there is a cascade of downstream effects leading to the phenotype (the complex disease), which can be rapid, but also might develop slowly over many years. In the case of an ADR, the drug elicits the stimulus (or stimuli) that sets into motion the downstream cascade of effects causing the phenotype (the ADR); this usually occurs in a matter of a few hours, days, or weeks. Table 16.5 is a partial list of common ADRs known to occur in subpopulations of patients—after they have received the recommended dosage of a commonly prescribed drug.

How does this happen? This is among the most intriguing mysteries in clinical pharmacology. How does a small-molecular-weight drug—given to some patients, but not the majority of patients in any cohort—cause an ADR that is often indistinguishable from the appearance of a complex disease? For example, sitagliptin, a reversible inhibitor of dipeptidyl-peptidase-4, is approved by the FDA to treat type II diabetes; yet, a small subset taking the recommended prescribed dose develops acute pancreatitis (reviewed in [187]). In a meta-analysis study of >50,000 patients with type II diabetes and >270,000 controls [122], it was shown that treatment with LDL-cholesterol-lowering drugs was correlated with higher risk of type II diabetes—and this trait was significantly associated with SNPs in or near the *NPC1L1* gene and at least four other genes.

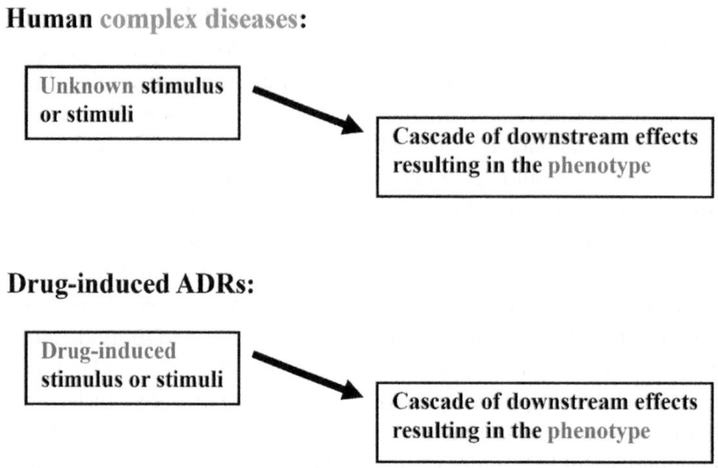

**Human complex diseases:**

Unknown stimulus or stimuli

Cascade of downstream effects resulting in the phenotype

**Drug-induced ADRs:**

Drug-induced stimulus or stimuli

Cascade of downstream effects resulting in the phenotype

**Figure 16.8** Similarities between human complex diseases and drug-induced ADRs, both in their genetic origins and their phenotypic manifestations. (From previous Emery and Rimoin edition; Nebert DW, Vesell ES. Chapter 19-"Pharmacogenetics and pharmacogenomics". In: Emery and Rimoin's principles and practice of medical genetics. 6th ed. Oxford: Academic Press; 2013. pp. 1–27.)

Treatment of malaria, lupus erythematosus, or rheumatoid arthritis with hydroxychloroquine can lead to acute pancreatitis. In a subpopulation of patients receiving many psychotropic drugs (e.g., valproic acid), undesirable weight gain occurs as a dose-independent ADR; in a smaller subset, hepatic steatosis as a dose-independent ADR has also been found (reviewed in [2]). In a small subpopulation of patients taking bisphosphonates for osteoporosis, increased risk of esophageal and gastric cancer has been repeatedly reported; however, a thorough review and meta-analysis of this association [242] has not found any significantly increased risk.

Complex PGx traits include not only dose-independent idiosyncratic ADRs and many instances of drug toxicity, but also drug efficacy. Hence, it seems reasonable to conclude that attempts to dissect differences in drug response (*efficacy, therapeutic failure, ADRs, toxicity*) will be very similar to attempts to dissect differences in complex diseases; in other words, both represent polygenic multifactorial traits.

## 16.5.2 Genome-Wide Association Studies of ADRs

It goes without saying that any recruitment of a sufficiently large cohort to study a PGx trait (i.e., patients receiving the same drug and preferably a similar dose of that drug) will be far more difficult than recruiting sufficient numbers of patients or volunteers to study a complex disease, such as type II diabetes—or a quantitative trait such as height or body mass index. Nonetheless, during the past decade, a growing number of GWAS (Table 16.6) have identified novel genetic loci associated with severe ADRs [28].

### 16.5.2.1 Statin-Induced Myopathy

Among the earliest PGx GWAS was simvastatin-treated patients that had developed myopathy [121]. Initially, the authors had screened ~300,000 markers in 85 cases and 90 controls selected from a clinical trial of ~12,000 participants; their findings were then replicated in a second cohort of ~20,000 subjects. (For a growing number of journals, "replication in a second cohort" has become a standard requirement for acceptance of a GWAS publication.) Simvastatin-induced myopathy was found to be associated with a mutation that changed an amino acid (i.e., a *nonsynonymous variant*) in the *SLCO1B1* gene. Being homozygous for this variant was found to confer an odds ratio (OR) of 16.9 for patients that developed simvastatin-caused myopathy, compared with patients not having this variant and receiving simvastatin without developing myopathy.

The *SLCOB1* gene encodes an organic-anion SLC transporter, presumed to function in cellular uptake of statins [106], and therefore it appears

**TABLE 16.5 Archetypal ADRs That can Occur in Patients Receiving Commonly Prescribed Drugs**

| Organ or System | Possible ADRs (Multifactorial Traits) |
|---|---|
| Central nervous system | Headache; fainting; hallucinations; stroke; mental or mood changes (e.g., new or worsening anxiety; nervousness; agitation; suicidal thoughts; confusion depression; restlessness; sleeplessness; inability to concentrate); memory loss; new or worsening nightmares; tremor; seizures; irreversible brain damage; transient psychotic episodes; toxic psychosis; ataxia; cogwheel rigidity; speech disorder (dysphasia); irritability; panic attacks; blacking out due to hypotension; progressive multifocal leukoencephalopathy |
| Eye | Acute-angle closure glaucoma; changes in vision; blurred vision; loss of vision; photosensitivity; phototoxicity; dry eye; periorbital edema; increased tearing; corneal keratitis; cataracts |
| Gastrointestinal tract | Constipation; severe or persistent diarrhea; gas; nausea; vomiting; severe or persistent stomach pain/cramps; difficulty in swallowing; abdominal cramps; bloody or tarry stools; heartburn (dyspepsia); indigestion; ulcers of mouth, esophagus or colon; gingival overgrowth |
| Heart | Hypotension; hypertension; chest pain; shortness of breath; heart attack; angina; heart failure; edema; swelling below the knees; heart block; cardiac arrhythmias; sinus tachycardia; palpitations; atrial fibrillation; ventricular fibrillation; postural hypotension; cardiac tamponade |
| Hematological system | Anemia; unusual bruising or bleeding; excessive bleeding (sometimes can be fatal); clotting disorders; bone marrow suppression |
| Immune system | Severe allergic reactions (e.g., rash; hives; pruritis; difficulty breathing; tightness in chest; swelling of the mouth, face, lips, or tongue); unusual hoarseness; immunosuppression; autoimmune diseases |
| Inner ear | Tinnitus; dizziness; light-headedness; hearing loss |
| Kidney | Decreased or painful urination; changes in glomerular filtration rate; changes in creatinine levels; renal insufficiency or failure; hypertension; hypotension; hyponatremia; hyperkalemia; nephrogenic systemic fibrosis |
| Liver | Symptoms of liver problems (e.g., dark urine; loss of appetite; pale stools; jaundice); chemical hepatitis; liver failure |
| Musculoskeletal system | Pain, soreness, redness, swelling, weakness, or bruising of a tendon or joint area; muscle pain or weakness; osteonecrosis of the jaw; ankle swelling; inability to move or bear weight on a joint or tendon area; irreversible tendon damage; spontaneous tendon ruptures; tremor; ankylosing spondylitis; rhabdomyolysis; osteopenia; osteoporosis; leg cramps |
| Pancreas | Symptoms similar to diabetes (e.g., high blood sugar; dizziness; fainting; rapid breathing; flushing; increased thirst, hunger, or urination; increased sweating; vision changes); hypoglycemia; acute pancreatitis |
| Peripheral nervous system | Symptoms of nerve problems (e.g., changes in perception of heat or cold; decreased sensation of touch; unusual burning, numbness, tingling, pain, or weakness of the arms, hands, legs, or feet); tremors; irreversible peripheral neuropathy |
| Reproductive system | Vaginal discharge, irritation, or odor; hypomenorrhea; hypermenorrhea; dysmenorrhea; erectile dysfunction; loss of libido; increased libido (males or females); priapism |
| Respiratory tract, lower | Shortness of breath; asthmatic-like wheezing; pulmonary edema; pulmonary thrombosis; lower respiratory infection; cough |
| Respiratory tract, upper | Fever, chills, sore throat, or unusual cough; runny nose; dry nose and throat; stuffy nose and congestion; upper respiratory infection; malignant hyperpyrexia |
| Skin | Moderate or severe sunburn; red, swollen, blistered, or peeling skin; irreversible skin damage; Stephen–Johnson syndrome; toxic epidermal necrolysis; hives; eczema; edema |
| "Systemic" (?) | Esophageal cancer, urinary bladder cancer, and other types of cancer; signs and symptoms of cardiac failure; facial flushing; obesity; anorexia; weight loss; weight gain; bigorexia; drug–drug interactions; food–drug interactions; death |
| DNA changes | DNA damage; mutations; increased risk of cancer |

Modified from Nebert DW, Vesell ES. Chapter 19-"Pharmacogenetics and pharmacogenomics". In: Emery and Rimoin's principles and Practice of medical genetics. 6th ed. Oxford: Academic Press; 2013. pp. 1–27.

TABLE 16.6 **Selected PGx GWAS With Genome-Wide Statistically Significant Findings of $P<5.0\times10^{-8}$ ($P<$5e−8)[a]**

| Drug | Response | Gene(s) | Effect | *P*-Value | Year | Reference |
|---|---|---|---|---|---|---|
| **ADR (Resembles Mendelian Trait With Incomplete Penetrance)** | | | | | | |
| Statin | Myopathy | *SLCO1B1* | OR = 4.5 | 4.00e−09 | 2008 | [121] |
| Flucloxacillin | DILI[b] | *HLA-B*57:01* | OR = 80.6 | 8.70e−33 | 2009 | [31] |
| Lumiracoxib | DILI | *HLA-DRB1*15:01* | OR = 5.0 | 6.80e−25 | 2010 | [192] |
| Carbamazepine | Hypersensitivity | *HLA-DRB*15:02* | OR = 12.4 | 3.50e−08 | 2011 | [138] |
| Heparin | Thrombocytopenia | *TDAG8* | OR = 18.6[c] | 3.18e−09 | 2015 | [97] |
| Anthracycline | Cardiotoxicity | *RARG* | OR = 4.7 | 5.90e−08 | 2015 | [3] |
| Lapatinib | DILI | *HLA-DRB1*07:01* | OR = 14[d] | 7.8e−11 | 2016 | [165] |
| Asparaginase | Hypersensitivity | *NFATC2* | OR = 3.11 | 4.10e−08 | 2015 | [46] |
| Cisplatin | Hearing loss | *ACYP2* | HR = 4.5 | 3.90e−08 | 2015 | [243] |
| **Dosage/Efficacy (Having Major Gene Effect)** | | | | | | |
| Warfarin | Dosage | *VKORC1* and *CYP2C9* | 34% | 6.20e−13 | 2008 | [20] |
| Warfarin | Dosage | *VKORC1*, *CYP2C9* and *CYP4F2* | 43.1% | <1e−78 | 2009 | [208] |
| Acenocoumarol | Dosage | *VKORC1*, *CYP2C9*, *CYP4F2* and *CYP2C18* | 48.80% | <5e−8 | 2009 | [212] |
| PegIFN-alpha | Viral clearance | *IL28B* | OR = 7.3 | 1.00e−38 | 2009 | [55] |
| | | *IL28B* | OR = 1.98 | 9.30e−09 | 2009 | [204] |
| | | *IL28B* | OR = 27.4 | 2.70e−32 | 2009 | [210] |
| Citalopram and escitalopram | Concentration | *CYP2C19* and *CYP2D6* | NA | <5e−8 | 2014 | [88] |
| | TPMT activity | *TPMT* | | 1.2e−72 | 2017 | [209] |
| Clopidogrel | Platelet reactivity | *CYP2C19* | | 5.1e−40 | 2018 | [6] |
| Clopidogrel | Clopidogrel active metabolite levels | *CYP2C19* and two other loci | | 9.5e−15, 3.3e−11 1.3e−8 | 2017 | [5] |
| **Complex PGx Traits (Influenced by Multiple Genetic Variants Each With Small Effect)** | | | | | | |
| Statins | LDL response | *SORT1/CELSR2/PSRC1*, *SLCO1B1*, *APOE*, *LPA* | 1.3%–5.2%[d] | <5e−8 | 2014 | [172] |
| Metformin | HbA1c <7% | *ATM* | OR = 1.35 | 2.90e−09 | 2011 | [60] |
| | | *SLC2A2* | 0.17% greater | 6.60e−14 | 2016 | [252] |
| Antidepressant | Questionnaire physician's opinion | *HPRTP4* | OR = 1.36 | 5.03 e−08 | 2015 | [8] |
| Methadone | Dosage | *OPRM1* | | 2.8 e−8 | 2017 | [195] |
| Corticosteroids | Adrenal suppression | *PDGFD* | OR = 4.05 | 3.5 e−10 | 2018 | [73] |

[a]Condensed from [248] and updated.
[b]DILI, drug-induced liver injury.
[c]OR of homozygous.
[d]Percentage of extra LDL-C lowering in carriers versus noncarriers of the SNP.

to be a credible candidate gene. Association of this *SLCO1B1* variant with simvastatin-triggered myopathy (Table 16.6) was subsequently reproduced in further clinical studies [230]. Consequently, a simvastatin-dosing regimen, based on this *SLCO1B1* genotype, has been recommended [176] by the Clinical Pharmacogenomics Implementation Consortium (CPIC). Interestingly, for whatever reason, association of this *SLCO1B1* variant with myopathy has been less conclusive for other statins [161]. Later PGx GWAS also showed that the same gene variant is a primary determinant of methotrexate pharmacokinetics and clinical effects [177,219].

#### 16.5.2.2 Drug-Induced Liver Injury (DILI) and/or Hypersensitivity

Another early PGx GWAS identified the *HLA-B*57:01* allele as a major risk factor for flucloxacillin-induced liver injury [31]; this ADR is quite rare (~8.5 in every 100,000). The authors found a strong correlation with a variant SNP in complete *linkage disequilibrium* (LD) with the *HLA-B*57:01* allele, having an OR of 80.6 (Table 16.6). Despite the strong association, however, only 1 in every 500–1000 flucloxacillin-treated patients having this allele will develop liver injury; this infrequent occurrence of DILI limits the clinical utility, because the test will have a very high rate of false positives. This genotype–phenotype association study is an excellent example in which *incomplete penetrance* makes it difficult for the physician who must decide, on a patient-by-patient basis, who should receive this drug.

Other DILI traits strongly correlated with different *HLA* haplotypes (Table 16.6) were reported for patients taking lumiracoxib [192] and lapatinib [165]. Another dose-independent ADR phenotype associated with an *HLA* locus was carbamazepine-induced hypersensitivity [138]. Yet, the clinically predictive values, for typing each of these *HLA* alleles associated with these ADRs, are regrettably very low (reviewed in [29]).

In addition to relationships with the *HLA* region, PGx GWAS have identified a number of additional associations with various ADRs [46,97,243]; remarkably, none of these loci (Table 16.6) would have been suspected—based on established data about known functions of each drug's PK and/or PD. Although these GWAS probably do not have immediate clinical utility, they do identify new biological pathways that might underlie the mode-of-action or mechanism of toxicity for each of these drugs.

#### 16.5.2.3 Drug-Induced QT-Interval Prolongation and Osteonecrosis of the Jaw

It should be emphasized that not all PGx GWAS have produced robust associations. For example, a number of GWAS of drug-induced QT-interval prolongation did not produce consistent results (reviewed in [160]). On the other hand, bisphosphonate-induced osteonecrosis of the jaw was reported to be associated with *CYP2C8* variants [184]; CYP2C8 is a plausible candidate because it metabolizes several anticancer drugs. However, additional studies [42,99,203] have not been able to reproduce those data. Osteonecrosis of the jaw is well known to display an extremely heterogeneous range of phenotypes—a phenomenon seen in numerous other complex diseases; thus, unless one has a very large sample size, as well as an unequivocal diagnosis of the phenotype being studied [225], identification of statistically significant loci that can be replicated will continue to be problematic.

### 16.5.3 Genome-Wide Association Studies of Drug Efficacy

#### 16.5.3.1 Efficacy of Anticoagulants

As discussed earlier, the coumarins including warfarin are widely used anticoagulants; however, their regimens are complicated—because of the wide-ranging interindividual variability in dosage and narrow therapeutic "window" (as illustrated in Fig. 16.3). Trying to perfect the dosage of coumarins including warfarin is therefore a very active area of PGx research. Before the GWAS era, candidate-gene associations and molecular studies had already identified the *CYP2C9* and *VKORC1* genes as being important, as detailed in Section 16.4.9.

Several GWAS (Table 16.6) have now validated those associations of *CYP2C9* and *VKORC1* with warfarin efficacy. Moreover, these GWAS have provided more accurate estimates of the amounts of contribution of these two genes in defining the *variance explained*: the contribution of *VKORC1* is ~30% and that of *CYP2C9* is ~12% [20,208]. In addition, the latter GWAS, having a larger sample-size [208], provided sufficient statistical power to uncover a third locus with smaller effect—i.e., the contribution of 1.1% by the *CYP4F2* gene.

Nonetheless, due to its relatively small contribution, the *CYP4F2* polymorphism has not been included in most dosing-recommendation algorithms [90,171]. For other coumarin anticoagulants, GWAS have also identified similar genetic determinants [212]. Furthermore,

GWAS of anticoagulants have been reported to be different in several different ethnic groups [14,167,169].

#### 16.5.3.2 Efficacy of Hepatitis C Virus Infection

The study of variability in response to interferon-α treatment of hepatitis C virus (HCV) infection is another GWAS example of drug efficacy. Three independent studies [55,204,210] described DNA variants near the interleukin-28B gene (*IL28B*) associated with response to interferon-α treatment. The degree of efficacious response was measured by the *sustained virological response* (SVR), i.e., the absence of detectable virus at the end of clinical follow-up. The same variant was also reported to be associated with spontaneous clearance of HCV [215]; these data suggest that the variant allele might directly influence the host immune response—with or without drug intervention.

The advantageous *IL28B* allele (a cytidine at nucleotide position rs12979860) is more frequent in Caucasians and Asians than in Africans; this finding would explain an earlier report [17] that, in trials of interferon-α treatment, HCV-infected African-Americans show relatively worse outcomes than Caucasians or Asians. Despite an estimated large effect size (OR=7.3), the imperfect predictive power—combined with the lack of alternative treatment regimens—has resulted in limited immediate clinical utility for determining the *IL28B* genotype, with regard to decisions involving personalized treatment [79].

An additional caveat is that the *direct antiviral agents* (DAAs), in combination with PEGylated interferon/ribavirin, represent the new standard of HCV treatment; this makes it important to re-evaluate the clinical utility of the *IL28B* genotype [216]. Whereas *IL28B* genotyping will continue to provide useful clinical information with regard to the predicted treatment response, it is likely that the *IL28B* genotype assay will eventually lose its usefulness—primarily because HCV therapy has begun to distance itself from interferon-based regimens [87,119].

#### 16.5.3.3 Mutational Landscape of G-Protein-Coupled Receptor (GPCR) Drug Targets

In an avante gard data-mining in silico approach—rather than GWAS—the first study of its kind searched for DNA variants in or near each of the 108 G-protein-coupled receptor genes (*GPCRs*); these 108 genes are known targets of 475 Food and Drug Administration (FDA)-approved drugs, which account for a global sales volume of > US$180 billion annually [72]. Each of the genomes of almost 68,500 individuals was then separately investigated for missense variants in and near each of the *GPCR* genes and the clinical associations with altered drug response were gleaned from the literature. To estimate the de novo missense mutation rate within these *GPCR* genes, authors also identified de novo mutations from >1700 control trios (having no reported pathological conditions), which were compiled from 10 different studies registered in the "de novo-database," a collection of germline de novo variants (http://denovo-db.gs.washington.edu/denovo-db/).

In proof-of-principle, the authors then experimentally showed that certain variants of the mu-opioid and cholecystokinin receptors resulted in altered drug responses and/or idiosyncratic dose-independent ADRs. These data—on just two of the 108 *GPCR* genes—underscore the need to characterize SNPs among all 108 of the *GPCR* genes. The authors suggest that the ultimate results of this kind of in silico study "might enhance prescription precision, improve patients' quality-of-life, and remove some of the economic and societal burden caused by variability in drug response." It is anticipated that such "dry-lab" data-mining studies, such as this landmark publication [72], are likely to become a new major approach to PGx research in the near future.

### 16.5.4 Genome-Wide Association Studies of Complex PGx Traits

PGx GWAS have been attempted in studies of the treatment of common complex diseases such as type II diabetes, dyslipidemia, hypertension, and psychiatric disorders. Although some statistically significant findings have been reported, none of the identified PGx associations has clinical predictive value.

#### 16.5.4.1 GWAS of Type II Diabetes Treatment

In GWAS examining the *glycemic-index response* to metformin in >1000 individuals with type II diabetes, for example, the GoDarts Group [60] identified several variants in the ataxia-telangiectasia-mutated gene (*ATM*) that are associated with treatment success; in an independent cohort with almost 1800 samples, they were able to replicate this association. For each minor allele reflecting the most strongly associated variant, the combined effect was reported to be 1.35 times higher than that of the major allele—for reaching the treatment goal of lowering glucose-bound hemoglobin (HbA1c) levels by at least 7%.

It was exciting to see the correlation of metformin efficacy with this *ATM* variant replicated in multiple additional cohorts [224]. However, the variant of the *ATM* gene appears to have an insufficient impact on the predictive value, with regard to metformin action on the glycemic index control [48].

Nevertheless—for reasons not clear—several additional research findings do indicate a substantial role of *ATM* variants in glucose homeostasis. For example, ataxia telangiectasia patients sometimes manifest a severe form of diabetes. Moreover, *Atm(−/−)* knockout mice exhibit insulin resistance and abnormal adipose distribution [206]. In a GWAS comprising ~13,000 participants [252], the Metformin Genetics (MetGen) Consortium identified an additional association between a variant in a facilitated glucose-transporter gene (*SLC2A2*) and the response by type II diabetes patients to metformin treatment (Table 16.6); the SLC2A2 transporter is considered a credible candidate.

#### 16.5.4.2 GWAS of Cardiovascular Disease Treatment

The cardiovascular response to efficacy of statins has been extensively tested by multiple GWAS [15,37,75,172] (reviewed in [115]). The GWAS sample sizes ranged between ~2000 and >40,000 subjects. These studies have identified common variants in the *LPA*, *APOE*, *SLCO1B1*, *SORT1*, and *ABCG2* genes most robustly associated with a favorable statin-induced response to low-density-lipoprotein-cholesterol (LDL-C) levels. However, none of the variants was statistically significantly correlated with modifications in risk reduction for cardiovascular events. These data suggest "limited clinical utility."

### 16.5.5 Genome-Wide Association Studies of Unsuccessful PGx Examples

#### 16.5.5.1 GWAS Dilemma of Hypertension Treatment

Probably the most prevalent modifiable risk factor for worldwide disease burden is elevated blood pressure. Accordingly, many PGx GWAS have attempted to identify genetic variants that might cause the largely inconsistent responses to various antihypertensive medications. Despite extreme efforts, PGx GWAS attempting to find links between genotypes and efficacy of antihypertensive agents have not generated robust independently replicable findings [125,141]; such failures likely reflect the extreme heterogeneity and complexity of hypertensive disease.

Elevated blood pressure has multiple etiologies and involves numerous physiological changes throughout one's lifetime. Such confounders, of course, create limitations in any genotype–phenotype association study. Additional caveats include the multiple classes of drugs that are used to treat hypertension (diuretics, ACE inhibitors, angiotensin II receptor blockers, calcium channel blockers, beta-blockers)—alone, and in combinations. These limitations further complicate "clean" PGx studies—not to mention individual variability caused by fluctuations of blood pressure (e.g., changes in diet, stress, exercise, time-of-day, combinations of prescribed drugs, usage of over-the-counter drugs, amount of cigarette-smoking, alcohol intake, etc.).

Several PGx GWAS have reported variants in plausible genes associated with thiazide diuretics [16,220,222], angiotensin II receptor blockers [51,221], and beta-blockers [64,65]. More recently, the PEAR (Pharmacogenomic Evaluation of Antihypertensive Responses) group reported genetic variants associated with chlorthalidone-induced increases in blood glucose levels [194] and with heart rate response to beta-blockers [189]. However, most of these GWAS involve relatively small sample sizes, and thus are underpowered to establish robust associations having genome-wide significance; furthermore, the findings will need to be replicated in future independent studies.

Huge international consortia, aimed at increasing opportunities for discovery and replication by assembling large samples, have been established [19]. Even though large consortium studies might be able to identify genotype–phenotype associations for hypertensive drug efficacy in the future, however, it is difficult to envision that—even in combination—any of these newly identified small-effect variants will ever achieve important clinical predictive power. (However, new drug targets might be identified.)

#### 16.5.5.2 GWAS Dilemma of Psychotropic Drugs

Just as is true for hypertensive drug treatment, GWAS of antidepressant treatment response have shown little promise of success [8,43,53,83,223]. Besides patient *compliance*, one specific challenge in PGx studies of psychotropic medications is the *evaluation* of disease conditions and *quantitative measurements* of treatment responses—efficacies of antidepressant drugs are all based on multidimensional and (the physician's and the patient's) subjective psychiatric diagnostic criteria.

For many years, these "soft" measurements have been criticized for their lack of reliability. One must trust physicians' consensus of the diagnosis and treatment response, determined by different clinicians, as well as by how honestly the patient answers the questionnaire (reviewed in [1]). This problem—i.e., dealing with an "equivocal phenotype" instead of an "unequivocal phenotype"—has previously been emphasized [158].

An excellent example of questionable diagnostic reliability exists with the major depressive disorder (MDD) and how to quantify treatment efficacy [52]. Clinically based psychiatric criteria clearly represent a gradient, rather than any biologically homogeneous condition; one solution offered to help in classifying an unequivocal trait is the "extreme discordant phenotype" (EDP) method of analysis [149,249].

Consequently, difficulties in phenotypic definitions of psychiatric conditions are highly heterogeneous, and measurements of treatment responses are decidedly inaccurate. A recent large GWAS meta-analysis [241] identified 44 loci associated with MDD; the revealed *genetic architecture* suggests MDD is not a distinct pathological condition, but rather an anthropocentric clinical construct associated with a wide range of diverse outcomes—the end result of which is a complex process of intertwined genetic and environmental effects.

On the other hand, a GWAS evaluating *plasma concentrations* of the parent drugs citalopram and escitalopram, and their *metabolites*, was successful; in 435 MDD patients, investigators succeeded in identifying highly significant associations with *CYP2C19* and the *CYP2D6* gene variants [88]. This study underscores an important distinction between *subjective* psychiatric clinical phenotypes and *quantitative measurements* of blood or urine drug- or metabolite-level phenotypes; as mentioned earlier, this concept of an equivocal phenotype versus unequivocal phenotype has previously been emphasized [158]. One therefore can recognize the difficulties in attempting to use PGx GWAS—as well as pharmacometabolomic association studies, such as those described for MDD [88]—to predict drug efficacy, or risk of ADRs, or genetic risk of many complex clinical responses.

## 16.6 RESPONSE TO DRUGS OTHER THAN GENOTYPE OF THE PATIENT

Early in this chapter, we described four additional factors other than the *genetic architecture* of each patient that can influence drug response. These include *epigenetic effects, endogenous influences, environmental factors,* and *interindividual differences in the microbiome.* These are briefly covered below.

### 16.6.1 Epigenetics

Epigenetic effects on drug response are likely to be important, but, currently, data showing the extent to which this happens are very much limited. Examples of "epigenetic effects" include: DNA-methylation patterns, RNA-interference regulatory processes, histone modifications, and chromatin remodeling. Ready-to-use kits are now available—and becoming increasingly less expensive—to study genome-wide DNA methylation, as well as microRNA assays. Modifications of histones and remodeling of chromatin, on the other hand, are two fields of active research that continue to be advanced at the present time; currently, there are not yet any simple screening procedures. Epigenetics might provide a new framework in our search for etiological factors contributing to complex diseases, drug efficacy, and ADRs.

Genetic differences in epigenetic effects are expected to be found. For example, a comparison of DNA-methylation in buccal mucosal cells [96] found highly significantly ($P = 1.2 \times 10^{-294}$) less epigenetic variability between monozygotic twin pairs than between dizygotic twin pairs.

Some epigenetic variants can be inherited by offspring (and the offspring's offspring)—which apparently represents a *transgenerational* mechanism [32] for "biological heredity apparently not based on DNA sequence." For example, during the Dutch Famine of 1945, risk of neurodevelopmental disorders was increased in grandchildren whose grandparents were exposed prenatally [205]. Famine in a small Swedish village—at the time that males were entering puberty—appears to send an unknown message to their grandsons, curiously resulting in less risk of type II diabetes; famine in the same population, at the time that females' oocytes are forming during the third trimester in utero, increases risk of obesity and type II diabetes in those babies' granddaughters [168].

Grandmothers who smoke cigarettes during pregnancy are associated with a fourfold increased risk of asthma in their granddaughters [118]. Germ cells can also carry epigenetic effects from the grandmother's diet [18]. Individuals whose grandparents suffered malnutrition in utero during the 1959–61 Chinese Famine were reported to have increased risk of schizophrenia [244]; however, another analysis of these same data suggests this result might have reflected an epidemiological artifact called *population stratification* [198]. Chronic stress in pregnant mothers [92], as well as dietary effects in fathers' sperm [159], were reported to be sufficient to induce changes in the *epigenetic landscape* of the developing embryo and fetus in utero.

Development of the *chromosome-conformation capture* method [36] has demonstrated that chromatin is partitioned into active and inactive compartments (reviewed in [200]). This type of chromatin remodeling would fall under the category of "epigenetic effects on phenotype." Understanding the impact of "three-dimensional genomics" on multifactorial traits—such as complex diseases (e.g., cancer, innate immunity), quantitative traits (e.g., height, body mass index), and drug efficacy, ADRs, and toxicity—is predicted to become relevant to PGx in the future.

Epigenetic changes are well known to be involved in various complex diseases, including cancer and asthma [45]. However, the extent to which epigenetic factors—both developmental and transgenerational—might affect PGx phenotypes (efficacy, therapeutic failure, ADRs, and/or toxicity) is unknown.

### 16.6.2 Endogenous Influences

Any drug's PK and PD phenotypes can be expected to change as a function of *age*. Changes in the neonate and infant during the first few months of life occur much more rapidly than after 1 year of age. Age-related effects are then less striking throughout childhood, adolescence, and most of adulthood. In *geriatric populations*, physicians need to pay more attention to increased inter-individual variability; this is in part due to decreased *renal excretion* (as much as 50%) in about two-thirds of elderly patients over the age of 75, but also due to confounding factors such as *coronary heart disease* and *hypertension* (reviewed in [139,190]).

There are *hormonal differences* in PK and PD traits in men, women, and pregnant women [197]. *Ethnic differences* in drug response have been repeatedly mentioned throughout this chapter, as well as especially addressed in Section 16.4.10. Lastly, vigorous *exercise* diminishes blood flow in the liver—which will mostly affect the degradation of drugs having *hepatic high-clearance rates* (e.g., lidocaine, nitrates); on the other hand, metabolic breakdown of hepatic low-clearance drugs (e.g., amobarbital, antipyrine, and diazepam) is not significantly affected by exercise [246]. Strenuous exercise can also lower renal plasma flow, urinary excretion rates, and urine pH—which would explain why serum levels of drugs that are eliminated through the kidneys increase during physical stress.

### 16.6.3 Environmental Factors

Environmental agents, when present in substantial amounts, can affect drug response (*efficacy*, *therapeutic failure*, *adverse effect*, and *toxic effect*); clearly, these effects might influence, or be "confounders" in, genetic association studies of PGx traits. Examples include over-the-counter medications, dietary factors, drug-drug interactions, cigarette smoking, heavy alcohol usage, occupation-related chemicals in the workplace, and living in regions where a hazardous substance, or mixture of toxic substances, is/are present in considerable amounts. For over-the-counter medications and dietary factors, St. John's wort and other botanical preparations [199], as well as grapefruit juice [113], are among the most-often mentioned examples. Miniscule amounts of environmental *bis*-phenol A, manganese, and polyfluoroalkyl pollution have been extensively studied, but any significant effects on PGx variability would seem highly unlikely.

In the case of drug–drug interactions [100], obviously if a drug—not being analyzed in the GWAS cohort—is being taken by the patient, and is a substrate for a (PK or PD) gene target that competes with the drug being studied, this would be a caveat. This other drug will interfere with the "purity" of the GWAS. In other words, the phenotype will be less certain, which, in turn, could impact the power of the overall genotype–phenotype association study.

Cigarette smoking or hazardous occupational chemicals might sometimes affect drug response. For cigarette smoking, an accurate "smoking history" (i.e., cigarette-pack-years) is usually a reasonably quantifiable effect on drug response that can be studied. For occupation-related chemicals in the workplace, precise quantification (i.e., the amount of exposure to any individual work as a function of time) is more problematic than cigarette-smoking history. For patients living in regions where considerable amounts of a hazardous substance are present (in the air, ground, or water), any influence of drug response becomes even less quantifiable, i.e., more uncertain.

In PGx studies, the dose of a drug, and how long it has been taken by the patient (presuming that the patient has been *compliant*), is more quantitative than exposure to any of these above-mentioned groups of environmental agents. Nonetheless, it is worth mentioning that there are GWAS of environmental stimuli that have identified specific genetic loci. For example, a meta-analysis representing five population-based GWAS of ~47,000 "habitual coffee drinkers" of European descent—adjusted for age, sex, and smoking history—found two statistically significant genes [21]: *AHR* ($P = 2.4 \times 10^{-19}$) and the *CYP1A2_1A1* locus ($P = 5.2 \times 10^{-14}$). Both of these loci are credible biological candidates because CYP1A2 metabolizes caffeine, and AHR regulates cigarette-smoke-inducible CYP1A1 and CYP1A2 enzymatic activities that participate in the metabolism of chemicals in cigarette smoke. A second example is a GWAS of ~1300 arsenic-exposed Bangladeshi individuals [170], in which five highly significant SNPs in and near the arsenite methyltransferase gene (*AS3MT*) showed independent associations with arsenic-caused toxic effects.

### 16.6.4 Microbiome Differences

Benign bacteria (and a few viruses and fungi)—which normally inhabit our bodies synergistically—have been termed "the microbiome" [112]. Although the microbiome has largely been overlooked (except for attempts to suppress or eradicate microorganisms), these microorganisms constitute ~90% of the total number of cells associated with our bodies, i.e., human cells comprise merely the remaining 10% [185]. Because of recent advances in genomics technology, we have only begun to appreciate contributions of the microbiome to modifications of health and disease; this includes conditions once believed to be genetically encoded purely by the host's chromosomes [68].

Although many drugs have GI side-effects and the gut microbiome itself is pivotal for human health [93], the role of the microbiome in these processes has rarely been considered. Recent studies have demonstrated that drugs designed to target human cells, rather than microbes—such as the antidiabetic metformin [50], proton-pump inhibitors [81,85], nonsteroidal antiinflammatory drugs [180], and atypical antipsychotics [49]—have been associated with alterations in composition of the microbiome. A larger cohort study [44] suggests that medications can also substantially alter the GI microbiome composition in more general ways.

Recently, in a screen of >1000 marketed drugs against 40 representative gut bacterial strains, 24% of drugs designed for targets in the human host, including members of all therapeutic classes mentioned above, were found to inhibit in vitro the growth of at least one bacterial strain [129]. These recent advances indicate a growing trend of appreciation for future research on *drug–microbiome interactions*.

Low-molecular-weight chemicals, produced and enzymatically altered in the gut flora, are chemically similar in structure to drugs, as well as to ligands that activate the host's endogenous receptors, and lipid mediators (LMs) that participate in second-messenger pathways mediated by the arachidonic acid, eicosapentaenoic acid, and docosahexaenoic acid cascades (reviewed in [150]). Thus, the degree to which various drugs and drug metabolites administered to the host might be affected by gut flora chemicals is largely unappreciated at the present time. It therefore is reasonable to expect that the microbiome will influence drug response (*efficacy*, *therapeutic failure*, *adverse effect*, and *toxic effect*); in turn, such effects are clearly likely to have an effect on GWAS of PGx.

A specialized PGx aspect of *neuropharmacogenomics* involves the recent realization of the importance of the "brain–gut–microbiome" (reviewed in [123,136]). It is now clear that there is bidirectional communication between the GI tract and the central nervous system, and that the CNS provides communication pathways between intestinal microbiota and the patient's neural circuitry. Variations in GI tract and CNS function via the microbiome are now proposed to include such ill-defined traits as "mood," "behavior," "suicidal tendencies," "obsessive-compulsive disorder," cognitive functions, appetite, and autism spectrum disorder. Production of bioactive compounds by microbiota, and their potential probiotic activities includes neuroactive molecules such as histamine, serotonin, catecholamines, and trace amines [123,136].

In the near future, we therefore expect specific studies on effects of the brain–gut–microbiome on drug-response phenotypes. Perhaps more importantly, the outcome of response to psychotropic drugs might be particularly affected. Hence, problems of GWAS involving psychotropic drugs (as detailed in Section 16.5.5.2) might be due, in part, to the impact of the brain–gut–microbiome. We predict this topic will become a major thrust of PGx research in the near future.

To summarize Section 16.6, these nongenetic factors—*epigenetic effects*, *endogenous influences*, *environmental factors*, and *interindividual differences in the microbiome*—contribute to the overall phenotype in PGx studies. The extent to which each of these contributes, to each individual volunteer or patient in the cohort, remains to be determined but will become an important area of research in the future.

## 16.7 FDA RECOMMENDATIONS FOR PGx GENOTYPING

The US Food and Drug Administration (FDA) recommends that, before prescribing, physicians should genotype their patients for specific biomarkers. Currently, there is FDA-approved information on the labels of more than 260 drugs and various medications https://www.fda.gov/Drugs/ScienceResearch/ResearchAreas/Pharmacogenetics/ucm083378.htm (last updated February 8, 2018). The FDA suggests that these genetic biomarkers might help physicians identify patients in whom commonly prescribed drugs might be "less efficacious, insufficiently metabolized, or more likely to be toxic." In recent surveys, most physicians agree that genetic profiles of patients can affect drug therapy. However, only about one in 10 physicians feel they have been adequately educated about using such genetic biomarkers in patients; therefore, few physicians actually order these tests.

Furthermore, the usual range, on average (~1.8–2.0-fold), of increased risk associated with positive results of these tests—might not seem large enough to cause sufficient concern. There are a number of reasons why physicians avoid PGx genomics-testing usage and implementation, and these include: the need to demonstrate clinical utility unequivocally and more clearly; the continuous introduction of new FDA-approved drugs on the market almost monthly; the almost daily flood of bewildering, indigestible volumes of new genomics information to which physicians are exposed; and the fact that not many drugs exhibit a therapeutic index (Fig. 16.3) that is substantially higher than ~1.8–2-fold (e.g., compared to the safety of aspirin). Moreover, genomics-testing results are often irreproducible, or they vary between different ethnic populations; this often makes insurance reimbursements difficult to process. Hence, "individualized drug therapy" and "personalized medicine" based entirely on DNA-sequence testing still seem far from becoming a clinical reality.

## 16.8 CONCLUSIONS

1. A major portion of *personalized medicine* is *individualized drug therapy*; however, because we continue to see how incredibly complex the human genome is—and how unique each individual is—confident prediction of most drug responses is not possible in the foreseeable future.
2. *Pharmacogenetics* was originally defined as "gene–drug interactions" and *pharmacogenomics* defined as "drug–genome interactions," but now both fields have become combined into the latter term, pharmacogenomics (PGx).
3. Each individual's drug response is *holistic*, encompassing five contributing factors: *genotype*, *epigenetic effects*, *endogenous influences*, *environmental factors*, and *microbiome differences*. Whereas genotype is virtually always constant, the remaining four factors are dynamic and nongenetic.
4. Drug responses can include: *efficacy*, *adverse reaction*, *therapeutic failure*, and *toxic effect*. Any drug response can be regarded as a *phenotypic trait*.
5. Adverse drug reactions (ADRs) in the US have been reported to rank as high as the fifth leading cause of death.
6. Drug metabolism and drug target responses exist in every cell type of the body, and are often strikingly different from one cell type to another.
7. Variability in interindividual drug response, while it can be seen as a gradient, can be classified as: *monogenic* (Mendelian) *traits*, typically influenced by one or a few rare coding variants; *predominantly oligogenic traits* that usually represent variability largely elicited by a small number of major pharmacogenes; and *complex PGx traits*—produced mostly by innumerable small-effect variants. This last category is by far the most common.
8. Since the mid-2000s, PGx genome-wide association studies (GWAS) involving large cohorts were found to be able to detect genes not only in the *monogenic* and *oligogenic* categories, but also some genetic variants in the *complex PGx* category. The larger the cohort, the more genetic variants are revealed; however, each of these variants usually contributes a small effect to the trait and has no clinical predictive value—even when combined.
9. Recruiting a large number of individuals to study a quantitative trait (e.g., height, blood pressure, or body mass index) or a complex disease (e.g., type II diabetes, schizophrenia, or ulcerative colitis) is less difficult than identifying large numbers of patients being treated with a specific drug.

10. One major difference between pharmacogenomics and complex-disease GWAS is that any patient who has not been challenged with a particular drug will never know his phenotype for that drug.
11. Statistical predictions are best applied at the population level. For the individual, we need to focus on gene–gene, gene–environment, and drug–drug interactions in that particular patient. Thus, overall effects from DNA variants that show statistical significance in a large cohort are generally too small to predict in the individual patient.
12. Epigenetics includes DNA-methylation effects, RNA-interference processes, histone modifications, and chromatin remodeling. There are assays now available for the first two categories; the latter two categories are more complicated and will require many more years of study before assays become readily available.
13. Each patient's microbiota, as well as the brain–gut–microbiome specifically, might contribute more to drug-response phenotypes than is currently appreciated.
14. As detailed in this chapter, given all the complexity of each individual's *genetic architecture*—as well as *epigenetic effects, endogenous influences, environmental factors,* and *differences in the microbiome*—it can be seen how difficult *individualized drug therapy* will be in the foreseeable future.

## ACKNOWLEDGMENTS

We thank our colleagues for valuable discussions and a careful reading of this manuscript. This work was funded, in part, by NIH Grant P30 ES006096.

## REFERENCES

[1] Aboraya A, Rankin E, France C, El-Missiry A, John C. Reliability of psychiatric diagnosis revisited: the clinician's guide to improve reliability of psychiatric diagnosis. Psychiatry 2006;3:41–50.

[2] Amacher DE, Chalasani N. Drug-induced hepatic steatosis. Semin Liver Dis 2014;34:205–14.

[3] Aminkeng F, Bhavsar AP, Visscher H, Rassekh SR, Li Y, Lee JW, Brunham LR, Caron HN, van Dalen EC, Kremer LC, van der Pal HJ, Amstutz U, Rieder MJ, Bernstein D, Carleton BC, Hayden MR, Ross CJ. A coding variant in *RARG* confers susceptibility to anthracycline-induced cardiotoxicity in childhood cancer. Nat Genet 2015;47:1079–84.

[4] Angelo M, Dring LG, Lancaster R, Latham A, Smith RL. Proceedings: a correlation between the response to debrisoquine and the amount of unchanged drug excreted in the urine. Br J Pharmacol 1975;55:264P.

[5] Backman JD, O'Connell JR, Tanner K, Peer CJ, Figg WD, Spencer SD, Mitchell BD, Shuldiner AR, Yerges-Armstrong LM, Horenstein RB, Lewis JP. Genome-wide analysis of clopidogrel active metabolite levels identifies novel variants that influence antiplatelet response. Pharmacogenet Genom 2017;27:159–63.

[6] Bergmeijer TO, Reny JL, Pakyz RE, Gong L, Lewis JP, Kim EY, Aradi D, Fernandez-Cadenas I, Horenstein RB, Lee MTM, Whaley RM, Montaner J, Gensini GF, Cleator JH, Chang K, Holmvang L, Hochholzer W, Roden DM, Winter S, Altman RB, Alexopoulos D, Kim HS, Dery JP, Gawaz M, Bliden K, Valgimigli M, Marcucci R, Campo G, Schaeffeler E, Dridi NP, Wen MS, Shin JG, Simon T, Fontana P, Giusti B, Geisler T, Kubo M, Trenk D, Siller-Matula JM, Ten Berg JM, Gurbel PA, Hulot JS, Mitchell BD, Schwab M, Ritchie MD, Klein TE, Shuldiner AR, Investigators I. Genome-wide and candidate gene approaches of clopidogrel efficacy using pharmacodynamic and clinical end points — rationale and design of the International Clopidogrel Pharmacogenomics Consortium (ICPC). Am Heart J 2018;198:152–9.

[7] Bhattacharjee S, Wang Z, Ciampa J, Kraft P, Chanock S, Yu K, Chatterjee N. Using principal components of genetic variation for robust and powerful detection of gene-gene interactions in case-control and case-only studies. Am J Hum Genet 2010;86:331–42.

[8] Biernacka JM, Sangkuhl K, Jenkins G, Whaley RM, Barman P, Batzler A, Altman RB, Arolt V, Brockmoller J, Chen CH, Domschke K, Hall-Flavin DK, Hong CJ, Illi A, Ji Y, Kampman O, Kinoshita T, Leinonen E, Liou YJ, Mushiroda T, Nonen S, Skime MK, Wang L, Baune BT, Kato M, Liu YL, Praphanphoj V, Stingl JC, Tsai SJ, Kubo M, Klein TE, Weinshilboum R. The International SSRI Pharmacogenomics Consortium (ISPC): a genome-wide association study of antidepressant treatment response. Transl Psychiatry 2015;5:e553.

[9] Blum M, Grant DM, McBride W, Heim M, Meyer UA. Human arylamine *N*-acetyltransferase genes: isolation, chromosomal localization, and functional expression. DNA Cell Biol 1990;9:193–203.

[10] Brodie BB, Aronow L, Axelrod J. The fate of benzazoline (priscoline) in dog and man and a method for its estimation in biological material. J Pharmacol Exp Therapeut 1952;106:200–7.

[11] Brunton L, Chabner B, Knollman B. Goodman & Gilman's the pharmacological basis of therapeutics. McGraw-Hill Companies, Inc; 2011. [Printed in China].

[12] Caldwell MD, Awad T, Johnson JA, Gage BF, Falkowski M, Gardina P, Hubbard J, Turpaz Y, Langaee TY, Eby C, King CR, Brower A, Schmelzer JR, Glurich I, Vidaillet HJ, Yale SH, Qi ZK, Berg RL, Burmester JK. *CYP4F2* genetic variant alters required warfarin dose. Blood 2008;111:4106–12.

[13] Caraco Y, Sheller J, Wood AJ. Pharmacogenetic determination of the effects of codeine and prediction of drug interactions. J Pharmacol Exp Therapeut 1996;278:1165–74.

[14] Cha PC, Mushiroda T, Takahashi A, Kubo M, Minami S, Kamatani N, Nakamura Y. Genome-wide association study identifies genetic determinants of warfarin responsiveness for Japanese. Hum Mol Genet 2010;19:4735–44.

[15] Chasman DI, Giulianini F, MacFadyen J, Barratt BJ, Nyberg F, Ridker PM. Genetic determinants of statin-induced low-density lipoprotein cholesterol reduction. Justification for Use of statins in Prevention: an Intervention Trial Evaluating Rosuvastatin (JUPITER) trial. Circ Cardiovasc Genet 2012;5:257–64.

[16] Chittani M, Zaninello R, Lanzani C, Frau F, Ortu MF, Salvi E, Fresu G, Citterio L, Braga D, Piras DA, Carpini SD, Velayutham D, Simonini M, Argiolas G, Pozzoli S, Troffa C, Glorioso V, Kontula KK, Hiltunen TP, Donner KM, Turner ST, Boerwinkle E, Chapman AB, Padmanabhan S, Dominiczak AF, Melander O, Johnson JA, Cooper-DeHoff RM, Gong Y, Rivera NV, Condorelli G, Trimarco B, Manunta P, Cusi D, Glorioso N, Barlassina C. *TET2* and *CSMD1* genes affect SBP response to hydrochlorothiazide in never-treated essential hypertensives. J Hypertens 2015;33:1301–9.

[17] Conjeevaram HS, Fried MW, Jeffers LJ, Terrault NA, Wiley-Lucas TE, Afdhal N, Brown RS, Belle SH, Hoofnagle JH, Kleiner DE, Howell CD. Peginterferon and ribavirin treatment in African-American and Caucasian-American patients with hepatitis-C genotype 1. Gastroenterology 2006;131:470–7.

[18] Cooney CA. Germ cells carry the epigenetic benefits of grandmother's diet. Proc Natl Acad Sci U S A 2006;103:17071–2.

[19] Cooper-DeHoff RM, Johnson JA. Hypertension pharmacogenomics: in search of personalized treatment approaches. Nat Rev Nephrol 2016;12:110–22.

[20] Cooper GM, Johnson JA, Langaee TY, Feng H, Stanaway IB, Schwarz UI, Ritchie MD, Stein CM, Roden DM, Smith JD, Veenstra DL, Rettie AE, Rieder MJ. Genome-wide scan for common genetic variants with a large influence on warfarin maintenance dose. Blood 2008;112:1022–7.

[21] Cornelis MC, Monda KL, Yu K, Paynter N, Azzato EM, Bennett SN, Berndt SI, Boerwinkle E, Chanock S, Chatterjee N, Couper D, Curhan G, Heiss G, Hu FB, Hunter DJ, Jacobs K, Jensen MK, Kraft P, Landi MT, Nettleton JA, Purdue MP, Rajaraman P, Rimm EB, Rose LM, Rothman N, Silverman D, Stolzenberg-Solomon R, Subar A, Yeager M, Chasman DI, van Dam RM, Caporaso NE. Genome-wide meta-analysis identifies regions on 7p21 (*AHR*) and 15q24 (*CYP1A2*) as determinants of habitual caffeine consumption. PLoS Genet 2011;7:e1002033.

[22] Cotton RG. Heterogeneity of phenylketonuria at the clinical, protein and DNA levels. J Inherit Metab Dis 1990;13:739–50.

[23] Curran ME, Splawski I, Timothy KW, Vincent GM, Green ED, Keating MT. A molecular basis for cardiac arrhythmia: HERG mutations cause long-QT syndrome. Cell 1995;80:795–803.

[24] D'Andrea G, D'Ambrosio RL, Di PP, Chetta M, Santacroce R, Brancaccio V, Grandone E, Margaglione M. Polymorphism in the *VKORC1* gene associated with an interindividual variability in the dose-anticoagulant effect of warfarin. Blood 2005;105:645–9.

[25] Dai D, Tang J, Rose R, Hodgson E, Bienstock RJ, Mohrenweiser HW, Goldstein JA. Identification of variants of *CYP3A4* and characterization of their abilities to metabolize testosterone and chlorpyrifos. J Pharmacol Exp Therapeut 2001;299:825–31.

[26] Dai D, Zeldin DC, Blaisdell JA, Chanas B, Coulter SJ, Ghanayem BI, Goldstein JA. Polymorphisms in human *CYP2C8* decrease metabolism of the anticancer drug paclitaxel and arachidonic acid. Pharmacogenetics 2001;11:597–607.

[27] Daly AK. Pharmacogenomics of anticoagulants: steps toward personal dosage. Genome Med 2009;1:10.

[28] Daly AK. Using genome-wide association studies to identify genes important in serious adverse drug reactions. Annu Rev Pharmacol Toxicol 2012;52:21–35.

[29] Daly AK. Human leukocyte antigen (*HLA*) pharmacogenomic tests: potential and pitfalls. Curr Drug Metabol 2014;15:196–201.

[30] Daly AK, Brockmoller J, Broly F, Eichelbaum M, Evans WE, Gonzalez FJ, Huang JD, Idle JR, Ingelman-Sundberg M, Ishizaki T, Jacqz-Aigrain E, Meyer UA, Nebert DW, Steen VM, Wolf CR, Zanger UM. Nomenclature for human *CYP2D6* alleles. Pharmacogenetics 1996;6:193–201.

[31] Daly AK, Donaldson PT, Bhatnagar P, Shen Y, Pe'er I, Floratos A, Daly MJ, Goldstein DB, John S, Nelson MR, Graham J, Park BK, Dillon JF, Bernal W, Cordell HJ, Pirmohamed M, Aithal GP, Day CP. *HLA-B*5701* genotype is a major determinant of drug-induced liver injury due to flucloxacillin. Nat Genet 2009;41:816–9.

[32] Daxinger L, Whitelaw E. Transgenerational epigenetic inheritance: more questions than answers. Genome Res 2010;20:1623–8.

[33] de Morais SM, Schweikl H, Blaisdell J, Goldstein JA. Gene structure and upstream regulatory regions of human *CYP2C9* and *CYP2C18*. Biochem Biophys Res Commun 1993;194:194–201.

[34] de Morais SM, Wilkinson GR, Blaisdell J, Nakamura K, Meyer UA, Goldstein JA. The major genetic defect responsible for the polymorphism of *S*-mephenytoin metabolism in humans. J Biol Chem 1994;269: 15419–22.

[35] Deenen MJ, Meulendijks D, Cats A, Sechterberger MK, Severens JL, Boot H, Smits PH, Rosing H, Mandigers CM, Soesan M, Beijnen JH, Schellens JH. Upfront genotyping of *DPYD*2A* to individualize fluoropyrimidine therapy: a safety and cost analysis. J Clin Oncol 2016;34:227–34.

[36] Denker A, de Laat W. The second decade of 3C technologies: detailed insights into nuclear organization. Genes Dev 2016;30:1357–82.

[37] Deshmukh HA, Colhoun HM, Johnson T, McKeigue PM, Betteridge DJ, Durrington PN, Fuller JH, Livingstone S, Charlton-Menys V, Neil A, Poulter N, Sever P, Shields DC, Stanton AV, Chatterjee A, Hyde C, Calle RA, Demicco DA, Trompet S, Postmus I, Ford I, Jukema JW, Caulfield M, Hitman GA. Genome-wide association study of genetic determinants of LDL-Chol response to atorvastatin therapy: importance of *LPA*. J Lipid Res 2012;53:1000–11.

[38] Dover GA. Molecular drive in multigene families: how biological novelties arise, spread, and are assimilated. Trends Genet 1986;2:159–65.

[39] Edwards IR, Aronson JK. Adverse drug reactions: definitions, diagnosis, and management. Lancet 2000;356:1255–9.

[40] Eichelbaum M. Ein neuendeckte defect im ArzneiMittelstoffwechsel des Menschen: die fahlende *N*-Oxydation des Spartein [thesis]. University of Bonn; 1975.

[41] Eichelbaum M, Spannbrucker N, Steincke B, Dengler HJ. Defective *N*-oxidation of sparteine in man: a new pharmacogenetic defect. Eur J Clin Pharmacol 1979;16:183–7.

[42] English BC, Baum CE, Adelberg DE, Sissung TM, Kluetz PG, Dahut WL, Price DK, Figg WD. A SNP in *CYP2C8* is not associated with development of bisphosphonate-related osteonecrosis of the jaw in men with castrate-resistant prostate cancer. Therapeut Clin Risk Manag 2010;6:579–83.

[43] Fabbri C, Corponi F, Souery D, Kasper S, Montgomery S, Zohar J, Rujescu D, Mendlewicz J, Serretti A. The genetics of treatment-resistant depression: a critical review and future perspectives. Int J Neuropsychopharmacol 2018. [Epub ahead of print].

[44] Falony G, Joossens M, Vieira-Silva S, Wang J, Darzi Y, Faust K, Kurilshikov A, Bonder MJ, Valles-Colomer M, Vandeputte D, Tito RY, Chaffron S, Rymenans L, Verspecht C, De Sutter L, Lima-Mendez G, D'Hoe K, Jonckheere K, Homola D, Garcia R, Tigchelaar EF, Eeckhaudt L, Fu J, Henckaerts L, Zhernakova A, Wijmenga C, Raes J. Population-level analysis of gut microbiome variation. Science 2016;352:560–4.

[45] Feinberg AP. Phenotypic plasticity and the epigenetics of human disease. Nature 2007;447:433–40.

[46] Fernandez CA, Smith C, Yang W, Mullighan CG, Qu C, Larsen E, Bowman WP, Liu C, Ramsey LB, Chang T, Karol SE, Loh ML, Raetz EA, Winick NJ, Hunger SP, Carroll WL, Jeha S, Pui CH, Evans WE, Devidas M, Relling MV. Genome-wide analysis links *NFATC2* with asparaginase hypersensitivity. Blood 2015;126:69–75.

[47] Fisher RA. The correlation between relatives on the supposition of Mendelian inheritance. Trans Roy Soc Edinb 1919;52:399–433.

[48] Florez JC, Jablonski KA, Taylor A, Mather K, Horton E, White NH, Barrett-Connor E, Knowler WC, Shuldiner AR, Pollin TI, Diabetes Prevention Program Research G. The C allele of *ATM* rs11212617 does not associate with metformin response in the Diabetes Prevention Program. Diabetes Care 2012;35:1864–7.

[49] Flowers SA, Evans SJ, Ward KM, McInnis MG, Ellingrod VL. Interaction between atypical antipsychotics and the gut microbiome in a bipolar disease cohort. Pharmacotherapy 2017;37:261–7.

[50] Forslund K, Hildebrand F, Nielsen T, Falony G, Le Chatelier E, Sunagawa S, Prifti E, Vieira-Silva S, Gudmundsdottir V, Pedersen HK, Arumugam M, Kristiansen K, Voigt AY, Vestergaard H, Hercog R, Costea PI, Kultima JR, Li J, Jorgensen T, Levenez F, Dore J, Meta HITc, Nielsen HB, Brunak S, Raes J, Hansen T, Wang J, Ehrlich SD, Bork P, Pedersen O. Disentangling type-2 diabetes and metformin treatment signatures in the human gut microbiota. Nature 2015;528:262–6.

[51] Frau F, Zaninello R, Salvi E, Ortu MF, Braga D, Velayutham D, Argiolas G, Fresu G, Troffa C, Bulla E, Bulla P, Pitzoi S, Piras DA, Glorioso V, Chittani M, Bernini G, Bardini M, Fallo F, Malatino L, Stancanelli B, Regolisti G, Ferri C, Desideri G, Scioli GA, Galletti F, Sciacqua A, Perticone F, Degli EE, Sturani A, Semplicini A, Veglio F, Mulatero P, Williams TA, Lanzani C, Hiltunen TP, Kontula K, Boerwinkle E, Turner ST, Manunta P, Barlassina C, Cusi D, Glorioso N. Genome-wide association study identifies *CAMKID* variants involved in blood pressure response to losartan: the SOPHIA study. Pharmacogenomics 2014;15:1643–52.

[52] Freedman R, Lewis DA, Michels R, Pine DS, Schultz SK, Tamminga CA, Gabbard GO, Gau SS, Javitt DC, Oquendo MA, Shrout PE, Vieta E, Yager J. The initial field trials of DSM-5: new blooms and old thorns. Am J Psychiatr 2013;170:1–5.
[53] Garriock HA, Kraft JB, Shyn SI, Peters EJ, Yokoyama JS, Jenkins GD, Reinalda MS, Slager SL, McGrath PJ, Hamilton SP. A genome-wide association study of citalopram response in major depressive disorder. Biol Psychiatr 2010;67:133–8.
[54] Gasche Y, Daali Y, Fathi M, Chiappe A, Cottini S, Dayer P, Desmeules J. Codeine intoxication associated with ultrarapid CYP2D6 metabolism. N Engl J Med 2004;351:2827–31.
[55] Ge D, Fellay J, Thompson AJ, Simon JS, Shianna KV, Urban TJ, Heinzen EL, Qiu P, Bertelsen AH, Muir AJ, Sulkowski M, McHutchison JG, Goldstein DB. Genetic variation in *IL28B* predicts hepatitis C treatment-induced viral clearance. Nature 2009;461:399–401.
[56] Geldmacher-v Mallinckrodt M, Hommel G, Dumbach J. On the genetics of the human serum paraoxonase (EC 3.1.1.2). Hum Genet 1979;50:313–26.
[57] Gibson G. Decanalization and the origin of complex disease. Nat Rev Genet 2009;10:134–40.
[58] Gibson G. Rare and common variants: twenty arguments. Nat Rev Genet 2011;13:135–45.
[59] Gibson G, Dworkin I. Uncovering cryptic genetic variation. Nat Rev Genet 2004;5:681–90.
[60] GoDarts, Group UDPS, Wellcome Trust Case Control C, Zhou K, Bellenguez C, Spencer CC, Bennett AJ, Coleman RL, Tavendale R, Hawley SA, Donnelly LA, Schofield C, Groves CJ, Burch L, Carr F, Strange A, Freeman C, Blackwell JM, Bramon E, Brown MA, Casas JP, Corvin A, Craddock N, Deloukas P, Dronov S, Duncanson A, Edkins S, Gray E, Hunt S, Jankowski J, Langford C, Markus HS, Mathew CG, Plomin R, Rautanen A, Sawcer SJ, Samani NJ, Trembath R, Viswanathan AC, Wood NW, investigators M, Harries LW, Hattersley AT, Doney AS, Colhoun H, Morris AD, Sutherland C, Hardie DG, Peltonen L, McCarthy MI, Holman RR, Palmer CN, Donnelly P, Pearson ER. Common variants near *ATM* are associated with glycemic response to metformin in type-2 diabetes. Nat Genet 2011;43:117–20.
[61] Goedde HW, Agarwal DP. Aldehyde oxidation: ethnic variations in metabolism and response. Prog Clin Biol Res 1986;214:113–38.
[62] Goldstein DB. Common genetic variation and human traits. N Engl J Med 2009;360:1696–8.
[63] Goldstein JA, de Morais SM. Biochemistry and molecular biology of the human *CYP2C* subfamily. Pharmacogenetics 1994;4:285–99.
[64] Gong Y, McDonough CW, Beitelshees AL, El RN, Hiltunen TP, O'Connell JR, Padmanabhan S, Langaee TY, Hall K, Schmidt SO, Curry Jr RW, Gums JG, Donner KM, Kontula KK, Bailey KR, Boerwinkle E, Takahashi A, Tanaka T, Kubo M, Chapman AB, Turner ST, Pepine CJ, Cooper-DeHoff RM, Johnson JA. *PTPRD* gene associated with blood pressure response to atenolol and resistant hypertension. J Hypertens 2015;33:2278–85.
[65] Gong Y, Wang Z, Beitelshees AL, McDonough CW, Langaee TY, Hall K, Schmidt SO, Curry Jr RW, Gums JG, Bailey KR, Boerwinkle E, Chapman AB, Turner ST, Cooper-DeHoff RM, Johnson JA. Pharmacogenomic genome-wide meta-analysis of blood pressure response to β-blockers in hypertensive African Americans. Hypertension 2016;67:556–63.
[66] Gonzalez FJ, Nebert DW. Evolution of the P450 gene superfamily: animal-plant 'warfare', molecular drive, and human genetic differences in drug oxidation. Trends Genet 1990;6:182–6.
[67] Gonzalez FJ, Skoda RC, Kimura S, Umeno M, Zanger UM, Nebert DW, Gelboin HV, Hardwick JP, Meyer UA. Characterization of the common genetic defect in humans deficient in debrisoquine metabolism. Nature 1988;331:442–6.
[68] Grice EA, Segre JA. The human microbiome: our second genome. Annu Rev Genom Hum Genet 2012;13:151–70.
[69] Hansen T, Echwald SM, Hansen L, Moller AM, Almind K, Clausen JO, Urhammer SA, Inoue H, Ferrer J, Bryan J, Aguilar-Bryan L, Permutt MA, Pedersen O. Decreased tolbutamide-stimulated insulin secretion in healthy subjects with sequence variants in the high-affinity sulfonylurea receptor gene. Diabetes 1998;47:598–605.
[70] Hansen TF. The evolution of genetic architecture. Annu Rev Ecol Evol Systemat 2006;37:123–57.
[71] Hassett C, Aicher L, Sidhu JS, Omiecinski CJ. Human microsomal epoxide hydrolase: genetic polymorphism and functional expression in vitro of amino acid variants. Hum Mol Genet 1994;3:421–8.
[72] Hauser AS, Chavali S, Masuho I, Jahn LJ, Martemyanov KA, Gloriam DE, Babu MM. Pharmacogenomics of GPCR drug targets. Cell 2018;172:41–54. e19.
[73] Hawcutt DB, Francis B, Carr DF, Jorgensen AL, Yin P, Wallin N, O'Hara N, Zhang EJ, Bloch KM, Ganguli A, Thompson B, McEvoy L, Peak M, Crawford AA, Walker BR, Blair JC, Couriel J, Smyth RL, Pirmohamed M. Susceptibility to corticosteroid-induced adrenal suppression: a genome-wide association study. Lancet Respir Med 2018;6:442–50.
[74] Hernandez D, Addou S, Lee D, Orengo C, Shephard EA, Phillips IR. Trimethylaminuria and a human *FMO3* mutation database. Hum Mutat 2003;22:209–13.

[75] Hopewell JC, Parish S, Offer A, Link E, Clarke R, Lathrop M, Armitage J, Collins R. Impact of common genetic variation in response to simvastatin therapy among 18,705 participants in the Heart Protection Study. Eur Heart J 2013;34:982–92.
[76] Hu Y, Oscarson M, Johansson I, Yue QY, Dahl ML, Tabone M, Arinco S, Albano E, Ingelman-Sundberg M. Genetic polymorphism of human *CYP2E1*: characterization of two variant alleles. Mol Pharmacol 1997;51:370–6.
[77] Humbert JA, Hammond KB, Hathaway WE. Trimethylaminuria: the fish-odour syndrome. Lancet 1970;2:770–1.
[78] Humbert R, Adler DA, Disteche CM, Hassett C, Omiecinski CJ, Furlong CE. The molecular basis of the human serum paraoxonase activity polymorphism. Nat Genet 1993;3:73–6.
[79] Iadonato SP, Katze MG. Genomics: hepatitis C virus gets personal. Nature 2009;461:357–8.
[80] Idle JR, Smith RL. Polymorphisms of oxidation at carbon centers of drugs and their clinical significance. Drug Metabol Rev 1979;9:301–17.
[81] Imhann F, Bonder MJ, Vich Vila A, Fu J, Mujagic Z, Vork L, Tigchelaar EF, Jankipersadsing SA, Cenit MC, Harmsen HJ, Dijkstra G, Franke L, Xavier RJ, Jonkers D, Wijmenga C, Weersma RK, Zhernakova A. Proton-pump inhibitors affect the gut microbiome. Gut 2016;65:740–8.
[82] Ingelman-Sundberg M, Oscarson M, McLellan RA. Polymorphic human cytochrome P450 enzymes: an opportunity for individualized drug treatment. Trends Pharmacol Sci 1999;20:342–9.
[83] Ising M, Lucae S, Binder EB, Bettecken T, Uhr M, Ripke S, Kohli MA, Hennings JM, Horstmann S, Kloiber S, Menke A, Bondy B, Rupprecht R, Domschke K, Baune BT, Arolt V, Rush AJ, Holsboer F, Muller-Myhsok B. A genomewide association study points to multiple loci that predict anti-depressant drug treatment outcome in depression. Arch Gen Psychiatr 2009;66:966–75.
[84] Jackson CM, Suttie JW. Recent developments in understanding the mechanism of vitamin K and vitamin K-antagonist drug action and consequences of vitamin K action in blood coagulation. Prog Hematol 1977;10:333–59.
[85] Jackson MA, Goodrich JK, Maxan ME, Freedberg DE, Abrams JA, Poole AC, Sutter JL, Welter D, Ley RE, Bell JT, Spector TD, Steves CJ. Proton pump inhibitors alter the composition of the gut microbiota. Gut 2016;65:749–56.
[86] Jarvela I, Torniainen S, Kolho KL. Molecular genetics of human lactase deficiencies. Ann Med 2009;41:568–75.
[87] Jensen DM, Pol S. *IL28B* genetic polymorphism testing in the era of direct acting antivirals therapy for chronic hepatitis C: ten years too late? Liver Int 2012;32(Suppl. 1):74–8.
[88] Ji Y, Schaid DJ, Desta Z, Kubo M, Batzler AJ, Snyder K, Mushiroda T, Kamatani N, Ogburn E, Hall-Flavin D, Flockhart D, Nakamura Y, Mrazek DA, Weinshilboum RM. Citalopram and escitalopram plasma drug and metabolite concentrations: genome-wide associations. Br J Clin Pharmacol 2014;78:373–83.
[89] Johansson I, Lundqvist E, Bertilsson L, Dahl ML, Sjoqvist F, Ingelman-Sundberg M. Inherited amplification of an active gene in the cytochrome P450 *CYP2D* locus as a cause of ultrarapid metabolism of debrisoquine. Proc Natl Acad Sci U S A 1993;90:11825–9.
[90] Johnson JA, Gong L, Whirl-Carrillo M, Gage BF, Scott SA, Stein CM, Anderson JL, Kimmel SE, Lee MT, Pirmohamed M, Wadelius M, Klein TE, Altman RB, Clinical Pharmacogenetics Implementation. Clinical Pharmacogenetics Implementation Consortium (CPIC) guidelines for CYP2C9 and VKORC1 genotypes and warfarin dosing. Clin Pharmacol Ther 2011;90:625–9.
[91] Johnson JA, Roden DM, Lesko LJ, Ashley E, Klein TE, Shuldiner AR. Clopidogrel: a case for indication-specific pharmacogenetics. Clin Pharmacol Ther 2012;91:774–6.
[92] Johnstone SE, Baylin SB. Stress and the epigenetic landscape: a link to the pathobiology of human diseases? Nat Rev Genet 2010;11:806–12.
[93] Kahrstrom CT, Pariente N, Weiss U. Intestinal microbiota in health and disease. Nature 2016;535:47.
[94] Kalow W, Bertilsson L. Interethnic factors affecting drug response. Adv Drug Res 1994;25:1–53.
[95] Kalow W, Genest K. A method for the detection of atypical forms of human serum cholinesterase: determination of dibucaine numbers. Can J Biochem Physiol 1957;35:339–46.
[96] Kaminsky ZA, Tang T, Wang SC, Ptak C, Oh GH, Wong AH, Feldcamp LA, Virtanen C, Halfvarson J, Tysk C, McRae AF, Visscher PM, Montgomery GW, Gottesman II, Martin NG, Petronis A. DNA-methylation profiles in monozygotic and dizygotic twins. Nat Genet 2009;41:240–5.
[97] Karnes JH, Cronin RM, Rollin J, Teumer A, Pouplard C, Shaffer CM, Blanquicett C, Bowton EA, Cowan JD, Mosley JD, Van Driest SL, Weeke PE, Wells QS, Bakchoul T, Denny JC, Greinacher A, Gruel Y, Roden DM. A genome-wide association study of heparin-induced thrombocytopenia using an electronic medical record. Thromb Haemostasis 2015;113:772–81.

[98] Katoh T. The frequency of glutathione-*S*-transferase M1 (*GSTM1*) gene deletion in patients with lung and oral cancer. Sangyo Igaku 1994;36:435–9.

[99] Katz J, Gong Y, Salmasinia D, Hou W, Burkley B, Ferreira P, Casanova O, Langaee TY, Moreb JS. Genetic polymorphisms and other risk factors associated with bisphosphonate-induced osteonecrosis of the jaw. Int J Oral Maxillofac Surg 2011;40:605–11.

[100] Kim BY, Sharafoddini A, Tran N, Wen EY, Lee J. Consumer mobile apps for potential drug-drug interaction check: systematic review and content analysis using the Mobile App Rating Scale (MARS). JMIR Mhealth Uhealth 2018;6:e74.

[101] Kim UK, Jorgenson E, Coon H, Leppert M, Risch N, Drayna D. Positional cloning of the human quantitative trait locus underlying taste sensitivity to phenylthiocarbamide. Science 2003;299:1221–5.

[102] Kioka N, Tsubota J, Kakehi Y, Komano T, Gottesman MM, Pastan I, Ueda K. P-glycoprotein gene (MDR1) cDNA from human adrenal: normal P-glycoprotein carries Gly185 with an altered pattern of multidrug resistance. Biochem Biophys Res Commun 1989;162:224–31.

[103] Klein RJ, Zeiss C, Chew EY, Tsai JY, Sackler RS, Haynes C, Henning AK, SanGiovanni JP, Mane SM, Mayne ST, Bracken MB, Ferris FL, Ott J, Barnstable C, Hoh J. Complement factor H polymorphism in age-related macular degeneration. Science 2005;308:385–9.

[104] Klironomos FD, Berg J, Collins S. How epigenetic mutations can affect genetic evolution: model and mechanism. Bioessays 2013;35:571–8.

[105] Kohane IS, Masys DR, Altman RB. The incidentalome: a threat to genomic medicine. J Am Med Ass 2006;296:212–5.

[106] König J, Seithel A, Gradhand U, Fromm MF, ö. Pharmacogenomics of human (*SLCO* gene) OATP transporters. Naunyn-Schmiedeberg's Arch Pharmacol 2006;372:432–43.

[107] Kupfer A, Preisig R. Pharmacogenetics of mephenytoin: a new drug hydroxylation polymorphism in man. Eur J Clin Pharmacol 1984;26:753–9.

[108] Kweekel D, Guchelaar HJ, Gelderblom H. Clinical and pharmacogenetic factors associated with irinotecan toxicity. Canc Treat Rev 2008;34:656–69.

[109] Lamba V, Lamba J, Yasuda K, Strom S, Davila J, Hancock ML, Fackenthal JD, Rogan PK, Ring B, Wrighton SA, Schuetz EG. Hepatic CYP2B6 expression: gender and ethnic differences and relationship to *CYP2B6* genotype and CAR (constitutive androstane receptor) expression. J Pharmacol Exp Therapeut 2003;307:906–22.

[110] Lander ES. Initial impact of the sequencing of the human genome. Nature 2011;470:187–97.

[111] Lazarou J, Pomeranz BH, Corey PN. Incidence of adverse drug reactions in hospitalized patients: a meta-analysis of prospective studies. J Am Med Ass 1998;279:1200–5.

[112] Lederberg J, McCray AT. 'Ome Sweet 'Omics — a genealogical treasury of words. Scientist 2001;15:8.

[113] Lee JW, Morris JK, Wald NJ. Grapefruit juice and statins. Am J Med 2016;129:26–9.

[114] Lee SJ, Usmani KA, Chanas B, Ghanayem B, Xi T, Hodgson E, Mohrenweiser HW, Goldstein JA. Genetic findings and functional studies of human *CYP3A5* single-nucleotide polymorphisms in different ethnic groups. Pharmacogenetics 2003;13:461–72.

[115] Leusink M, Onland-Moret NC, de Bakker PI, de Boer A, Maitland-van der Zee AH. Seventeen years of statin pharmacogenetics: a systematic review. Pharmacogenomics 2016;17:163–80.

[116] Levesque E, Beaulieu M, Hum DW, Belanger A. Characterization and substrate specificity of UGT2B4 (E458): a UDP-glucuronosyltransferase encoded by a polymorphic gene. Pharmacogenetics 1999;9:207–16.

[117] Li Y, Zhang D, Jin W, Shao C, Yan P, Xu C, Sheng H, Liu Y, Yu J, Xie Y, Zhao Y, Lu D, Nebert DW, Harrison DC, Huang W, Jin L. Mitochondrial aldehyde dehydrogenase-2 (ALDH2) Glu504Lys polymorphism contributes to the variation in efficacy of sublingual nitroglycerin. J Clin Invest 2006;116:506–11.

[118] Li YF, Langholz B, Salam MT, Gilliland FD. Maternal and grandmaternal smoking patterns are associated with early childhood asthma. Chest 2005;127:1232–41.

[119] Liapakis A, Jesudian AB. Is there clinical utility to *IL28B* genotype testing in the treatment of chronic hepatitis C virus infection? Pharmacogenomics 2012;13:1317–9.

[120] Ling G, Gu J, Genter MB, Zhuo X, Ding X. Regulation of cytochrome P450 gene expression in the olfactory mucosa. Chem Biol Interact 2004;147:247–58.

[121] Link E, Parish S, Armitage J, Bowman L, Heath S, Matsuda F, Gut I, Lathrop M, Collins R. *SLCO1B1* variants and statin-induced myopathy — a genomewide study. N Engl J Med 2008;359:789–99.

[122] Lotta LA, Sharp SJ, Burgess S, Perry JRB, Stewart ID, Willems SM, Luan J, Ardanaz E, Arriola L, Balkau B, Boeing H, Deloukas P, Forouhi NG, Franks PW, Grioni S, Kaaks R, Key TJ, Navarro C, Nilsson PM, Overvad K, Palli D, Panico S, Quiros JR, Riboli E, Rolandsson O, Sacerdote C, Salamanca EC, Slimani N, Spijkerman AM, Tjonneland A, Tumino R, van der AD, van der Schouw YT, McCarthy MI, Barroso I, O'Rahilly S, Savage DB, Sattar N, Langenberg C, Scott RA, Wareham NJ. Association between low-density

lipoprotein cholesterol-lowering genetic variants and risk of type-2 diabetes: a meta-analysis. J Am Med Ass 2016;316:1383–91.

[123] Luna RA, Savidge TC, Williams KC. The brain-gut-microbiome axis: what role does it play in autism spectrum disorder? Curr Dev Disord Rep 2016;3:75–81.

[124] Lunenburg CA, Henricks LM, Guchelaar HJ, Swen JJ, Deenen MJ, Schellens JH, Gelderblom H. Prospective *DPYD* genotyping to reduce the risk of fluoropyrimidine-induced severe toxicity: ready for prime-time. Eur J Canc 2016;54:40–8.

[125] Lupoli S, Salvi E, Barcella M, Barlassina C. Pharmacogenomics considerations in the control of hypertension. Pharmacogenomics 2015;16:1951–64.

[126] Mackenzie IS, Maki-Petaja KM, McEniery CM, Bao YP, Wallace SM, Cheriyan J, Monteith S, Brown MJ, Wilkinson IB. Aldehyde dehydrogenase-2 plays a role in the bioactivation of nitroglycerin in humans. Arterioscler Thromb Vasc Biol 2005;25:1891–5.

[127] MacLennan DH, Duff C, Zorzato F, Fujii J, Phillips M, Korneluk RG, Frodis W, Britt BA, Worton RG. Ryanodine receptor gene is a candidate for predisposition to malignant hyperthermia. Nature 1990;343:559–61.

[128] Mahgoub A, Idle JR, Dring LG, Lancaster R, Smith RL. Polymorphic hydroxylation of debrisoquine in man. Lancet 1977;2:584–6.

[129] Maier L, Pruteanu M, Kuhn M, Zeller G, Telzerow A, Anderson EE, Brochado AR, Fernandez KC, Dose H, Mori H, Patil KR, Bork P, Typas A. Extensive impact of non-antibiotic drugs on human gut bacteria. Nature 2018;555:623–8.

[130] Mallal S, Nolan D, Witt C, Masel G, Martin AM, Moore C, Sayer D, Castley A, Mamotte C, Maxwell D, James I, Christiansen FT. Association between presence of *HLA-B*5701, HLA-DR7*, and *HLA-DQ3* and hypersensitivity to HIV-1 reverse-transcriptase inhibitor abacavir. N Engl J Med 2008;358:568–79.

[131] Mallal S, Phillips E, Carosi G, Molina J-M, Workman C., Tomažiè J, Jägel-Guedes E, Rugina S, Kozyrev O, Cid JF, Hay P, Nolan D, Hughes S, Hughes A, Ryan S, Fitch N, Thorborn D, Benbow A, Team P-S. *HLA-B*5701* screening for hypersensitivity to abacavir. Virol J 2008;5:88.

[132] Manolio TA. Bringing genome-wide association findings into clinical use. Nat Rev Genet 2013;14:549–58.

[133] Manolio TA, Collins FS, Cox NJ, Goldstein DB, Hindorff LA, Hunter DJ, McCarthy MI, Ramos EM, Cardon LR, Chakravarti A, Cho JH, Guttmacher AE, Kong A, Kruglyak L, Mardis E, Rotimi CN, Slatkin M, Valle D, Whittemore AS, Boehnke M, Clark AG, Eichler EE, Gibson G, Haines JL, Mackay TF, McCarroll SA, Visscher PM. Finding the missing heritability of complex diseases. Nature 2009;461:747–53.

[134] Marks PA, Gross RT. Erythrocyte glucose-6-phosphate dehydrogenase deficiency: evidence of differences between Negroes and Caucasians with respect to this genetically-determined trait. J Clin Invest 1959;38: 2253–62.

[135] Marvit J, DiLella AG, Brayton K, Ledley FD, Robson KJ, Woo SL. GT to AT transition at a splice-donor site causes skipping of the preceding exon in phenylketonuria. Nucleic Acids Res 1987;15:5613–28.

[136] Mazzoli R, Pessione E. The neuro-endocrinological role of microbial glutamate and GABA signaling. Front Microbiol 2016;7:1934.

[137] McCarthy MI, Abecasis GR, Cardon LR, Goldstein DB, Little J, Ioannidis JP, Hirschhorn JN. Genome-wide association studies for complex traits: consensus, uncertainty, and challenges. Nat Rev Genet 2008;9:356–69.

[138] McCormack M, Alfirevic A, Bourgeois S, Farrell JJ, Kasperaviciute D, Carrington M, Sills GJ, Marson T, Jia X, de Bakker PI, Chinthapalli K, Molokhia M, Johnson MR, O'Connor GD, Chaila E, Alhusaini S, Shianna KV, Radtke RA, Heinzen EL, Walley N, Pandolfo M, Pichler W, Park BK, Depondt C, Sisodiya SM, Goldstein DB, Deloukas P, Delanty N, Cavalleri GL, Pirmohamed M. *HLA-A*3101* and carbamazepine-induced hypersensitivity reactions in Europeans. N Engl J Med 2011;364:1134–43.

[139] McLachlan AJ, Pont LG. Drug metabolism in older people — a key consideration in achieving optimal outcomes with medicines. J Gerontol A Biol Sci Med Sci 2012;67:175–80.

[140] Meinsma R, Fernandez-Salguero P, Van Kuilenburg AB, Van Gennip AH, Gonzalez FJ. Human polymorphism in drug metabolism: mutation in the dihydropyrimidine dehydrogenase gene results in exon skipping and thymine uracilurea. DNA Cell Biol 1995;14:1–6.

[141] Menni C. Blood pressure pharmacogenomics: gazing into a misty crystal ball. J Hypertens 2015;33:1142–3.

[142] Meyer UA. Pharmacogenetics: the slow, the rapid, and the ultrarapid. Proc Natl Acad Sci U S A 1994;91: 1983–4.

[143] Mizutani T. PM frequencies of major CYPs in Asians and Caucasians. Drug Metab Rev 2003;35:99–106.

[144] Monks TJ, Anders MW, Dekant W, Stevens JL, Lau SS, van Bladeren PJ. Glutathione conjugate mediated toxicities. Toxicol Appl Pharmacol 1990;106:1–19.

[145] Motsinger-Reif AA, Jorgenson E, Relling MV, Kroetz DL, Weinshilboum R, Cox NJ, Roden DM. Genome-wide association studies in pharmacogenomics: successes and lessons. Pharmacogenet Genom 2013;23:383–94.

[146] Motulsky AG. Drug reactions, enzymes, and biochemical genetics. J Am Med Ass 1957;165:835–7.

[147] Nebert DW. Proposed role of drug-metabolizing enzymes: regulation of steady state levels of the ligands that effect growth, homeostasis, differentiation, and neuroendocrine functions. Mol Endocrinol 1991;5:1203–14.

[148] Nebert DW. Pharmacogenetics and pharmacogenomics: why is this relevant to the clinical geneticist? Clin Genet 1999;56:247–58.

[149] Nebert DW. Extreme discordant phenotype methodology: an intuitive approach to clinical pharmacogenetics. Eur J Pharmacol 2000;410:107–20.

[150] Nebert DW. Aryl hydrocarbon receptor (AHR): "pioneer member" of the basic-helix/loop/helix per-Arnt-sim (bHLH/PAS) family of "sensors" of foreign and endogenous signals. Prog Lipid Res 2017;67:38–57.

[151] Nebert DW, Carvan 3rd MJ. Ecogenetics: from ecology to health. Toxicol Ind Health 1997;13:163–92.

[152] Nebert DW, Dalton TP. The role of cytochrome P450 enzymes in endogenous signalling pathways and environmental carcinogenesis. Nat Rev Canc 2006;6:947–60.

[153] Nebert DW, Dieter MZ. The evolution of drug metabolism. Pharmacology 2000;61:124–35.

[154] Nebert DW, Karp CL. Endogenous functions of aryl hydrocarbon receptor (AHR): intersection of cytochrome P450 1 (CYP1)-metabolized eicosanoids and AHR biology. J Biol Chem 2008;283:36061–5.

[155] Nebert DW, Vasiliou V. Analysis of the glutathione S-transferase (*GST*) gene family. Hum Genom 2004;1: 460–4.

[156] Nebert DW, Vesell ES. Chapter 19-"Pharmacogenetics and pharmacogenomics". In: Emery and Rimoin's principles and practice of medical genetics. Oxford: Academic Press; 2013. p. 1–27.

[157] Nebert DW, Wikvall K, Miller WL. Human cytochromes P450 in health and disease. Philos Trans R Soc Lond B Biol Sci 2013;368:20120431.

[158] Nebert DW, Zhang G, Vesell ES. From human genetics and genomics to pharmacogenetics and pharmacogenomics: past lessons, future directions. Drug Metab Rev 2008;40:187–224.

[159] Ng SF, Lin RC, Laybutt DR, Barres R, Owens JA, Morris MJ. Chronic high-fat diet in fathers programs β-cell dysfunction in female rat offspring. Nature 2010;467:963–6.

[160] Niemeijer MN, van den Berg ME, Eijgelsheim M, Rijnbeek PR, Stricker BH. Pharmacogenetics of drug-induced QT-interval prolongation: an update. Drug Saf 2015;38:855–67.

[161] Niemi M. Transporter pharmacogenetics and statin toxicity. Clin Pharmacol Ther 2010;87:130–3.

[162] Otonkoski T, Kaminen N, Ustinov J, Lapatto R, Meissner T, Mayatepek E, Kere J, Sipila I. Physical exercise-induced hyperinsulinemic hypoglycemia is an autosomal-dominant trait characterized by abnormal pyruvate-induced insulin release. Diabetes 2003;52:199–204.

[163] Owusu Obeng A, Egelund EF, Alsultan A, Peloquin CA, Johnson JA. *CYP2C19* polymorphisms and therapeutic drug monitoring of voriconazole: are we ready for clinical implementation of pharmacogenomics? Pharmacotherapy 2014;34:703–18.

[164] Ozaki K, Ohnishi Y, Iida A, Sekine A, Yamada R, Tsunoda T, Sato H, Sato H, Hori M, Nakamura Y, Tanaka T. Functional SNPs in the lymphotoxin-α gene (*LTA*) that are associated with susceptibility to myocardial infarction. Nat Genet 2002;32:650–4.

[165] Parham LR, Briley LP, Li L, Shen J, Newcombe PJ, King KS, Slater AJ, Dilthey A, Iqbal Z, McVean G, Cox CJ, Nelson MR, Spraggs CF. Comprehensive genome-wide evaluation of lapatinib-induced liver injury yields single genetic signal centered on known risk allele *HLA-DRB1*07:01*. Pharmacogenomics J 2016;16:180–5.

[166] Park JH, Gail MH, Weinberg CR, Carroll RJ, Chung CC, Wang Z, Chanock SJ, Fraumeni Jr JF, Chatterjee N. Distribution of allele frequencies and effect-sizes and their interrelationships for common genetic susceptibility variants. Proc Natl Acad Sci U S A 2011;108:18026–31.

[167] Parra EJ, Botton MR, Perini JA, Krithika S, Bourgeois S, Johnson TA, Tsunoda T, Pirmohamed M, Wadelius M, Limdi NA, Cavallari LH, Burmester JK, Rettie AE, Klein TE, Johnson JA, Hutz MH, Suarez-Kurtz G. Genome-wide association study of warfarin maintenance dose in a Brazilian sample. Pharmacogenomics 2015;16:1253–63.

[168] Pembrey ME. Time to take epigenetic inheritance seriously. Eur J Hum Genet 2002;10:669–71.

[169] Perera MA, Cavallari LH, Limdi NA, Gamazon ER, Konkashbaev A, Daneshjou R, Pluzhnikov A, Crawford DC, Wang J, Liu N, Tatonetti N, Bourgeois S, Takahashi H, Bradford Y, Burkley BM, Desnick RJ, Halperin JL, Khalifa SI, Langaee TY, Lubitz SA, Nutescu EA, Oetjens M, Shahin MH, Patel SR, Sagreiya H, Tector M, Weck KE, Rieder MJ, Scott SA, Wu AH, Burmester JK, Wadelius M, Deloukas P, Wagner MJ, Mushiroda T, Kubo M, Roden DM, Cox NJ, Altman RB, Klein TE, Nakamura Y, Johnson JA. Genetic variants associated with warfarin dose in African-American individuals: a genome-wide association study. Lancet 2013;382: 790–6.

[170] Pierce BL, Kibriya MG, Tong L, Jasmine F, Argos M, Roy S, Paul-Brutus R, Rahaman R, Rakibuz-Zaman M, Parvez F, Ahmed A, Quasem I, Hore SK, Alam S, Islam T, Slavkovich V, Gamble MV, Yunus M, Rahman M, Baron JA, Graziano JH, Ahsan H. Genome-wide

association study identifies chromosome 10q24.32 variants associated with arsenic metabolism and toxicity phenotypes in Bangladesh. PLoS Genet 2012;8:e1002522.
[171] Pirmohamed M, Kamali F, Daly AK, Wadelius M. Oral anticoagulation: a critique of recent advances and controversies. Trends Pharmacol Sci 2015;36:153–63.
[172] Postmus I, Trompet S, Deshmukh HA, Barnes MR, Li X, Warren HR, Chasman DI, Zhou K, Arsenault BJ, Donnelly LA, Wiggins KL, Avery CL, Griffin P, Feng Q, Taylor KD, Li G, Evans DS, Smith AV, de Keyser CE, Johnson AD, de Craen AJ, Stott DJ, Buckley BM, Ford I, Westendorp RG, Slagboom PE, Sattar N, Munroe PB, Sever P, Poulter N, Stanton A, Shields DC, O'Brien E, Shaw-Hawkins S, Chen YD, Nickerson DA, Smith JD, Dube MP, Boekholdt SM, Hovingh GK, Kastelein JJ, McKeigue PM, Betteridge J, Neil A, Durrington PN, Doney A, Carr F, Morris A, McCarthy MI, Groop L, Ahlqvist E, Bis JC, Rice K, Smith NL, Lumley T, Whitsel EA, Sturmer T, Boerwinkle E, Ngwa JS, O'Donnell CJ, Vasan RS, Wei WQ, Wilke RA, Liu CT, Sun F, Guo X, Heckbert SR, Post W, Sotoodehnia N, Arnold AM, Stafford JM, Ding J, Herrington DM, Kritchevsky SB, Eiriksdottir G, Launer LJ, Harris TB, Chu AY, Giulianini F, MacFadyen JG, Barratt BJ, Nyberg F, Stricker BH, Uitterlinden AG, Hofman A, Rivadeneira F, Emilsson V, Franco OH, Ridker PM, Gudnason V, Liu Y, Denny JC, Ballantyne CM, Rotter JI, Adrienne CL, Psaty BM, Palmer CN, Tardif JC, Colhoun HM, Hitman G, Krauss RM, Wouter JJ, Caulfield MJ. Pharmacogenetic meta-analysis of genome-wide association studies of LDL-cholesterol response to statins. Nat Commun 2014;5:5068.
[173] Powis SH, Mockridge I, Kelly A, Kerr LA, Glynne R, Gileadi U, Beck S, Trowsdale J. Polymorphism in a second *ABC* transporter gene located within the class II region of the human major histocompatibility complex. Proc Natl Acad Sci U S A 1992;89:1463–7.
[174] Price Evans DA, Manley KA, McKusick VA. Genetic control of isoniazid metabolism in man. Br Med J 1960;2:485–91.
[175] Quadri SA, Singal DP. Peptide transport in human lymphoblastoid and tumor cells: effect of transporter associated with antigen presentation (TAP) polymorphism. Immunol Lett 1998;61:25–31.
[176] Ramsey LB, Johnson SG, Caudle KE, Haidar CE, Voora D, Wilke RA, Maxwell WD, McLeod HL, Krauss RM, Roden DM, Feng Q, Cooper-DeHoff RM, Gong L, Klein TE, Wadelius M, Niemi M. Clinical pharmacogenetics implementation consortium guideline for SLCO1B1 and simvastatin-induced myopathy: 2014 update. Clin Pharmacol Ther 2014;96:423–8.
[177] Ramsey LB, Panetta JC, Smith C, Yang W, Fan Y, Winick NJ, Martin PL, Cheng C, Devidas M, Pui CH, Evans WE, Hunger SP, Loh M, Relling MV. Genome-wide study of methotrexate clearance replicates *SLCO1B1*. Blood 2013;121:898–904.
[178] Richards EJ. Inherited epigenetic variation -- revisiting soft inheritance. Nat Rev Genet 2006;7:395–401.
[179] Risch N, Merikangas K. The future of genetic studies of complex human diseases. Science 1996;273:1516–7.
[180] Rogers MAM, Aronoff DM. The influence of non-steroidal anti-inflammatory drugs on the gut microbiome. Clin Microbiol Infect 2016;22:178. e171-178 e179.
[181] Rost S, Fregin A, Ivaskevicius V, Conzelmann E, Hortnagel K, Pelz HJ, Lappegard K, Seifried E, Scharrer I, Tuddenham EG, Muller CR, Strom TM, Oldenburg J. Mutations in *VKORC1* cause warfarin resistance and multiple coagulation factor deficiency type-2. Nature 2004;427:537–41.
[182] Sabbagh A, Darlu P, Crouau-Roy B, Poloni ES. Arylamine *N*-acetyltransferase 2 (*NAT2*) genetic diversity and traditional subsistence: a worldwide population survey. PLoS One 2011;6:e18507.
[183] Sackton TB, Hartl DL. Genotypic context and epistasis in individuals and populations. Cell 2016;166:279–87.
[184] Sarasquete ME, Garcia-Sanz R, Marin L, Alcoceba M, Chillon MC, Balanzategui A, Santamaria C, Rosinol L, de la Rubia J, Hernandez MT, Garcia-Navarro I, Lahuerta JJ, Gonzalez M, San Miguel JF. Bisphosphonate-related osteonecrosis of the jaw is associated with polymorphisms of cytochrome P450 *CYP2C8* in multiple myeloma: a genome-wide single-nucleotide polymorphism analysis. Blood 2008;112:2709–12.
[185] Savage DC. Microbial ecology of the gastrointestinal tract. Annu Rev Microbiol 1977;31:107–33.
[186] Schwartzman ML, Davis KL, Nishimura M, Abraham NG, Murphy RC. The cytochrome P450 metabolic pathway of arachidonic acid in the cornea. Adv Prostag Thromb Leukot Res 1991;21A:185–92.
[187] Scott LJ. Sitagliptin: a review in type-2 diabetes. Drugs 2017;77:209–24.
[188] Scott SA, Sangkuhl K, Stein CM, Hulot JS, Mega JL, Roden DM, Klein TE, Sabatine MS, Johnson JA, Shuldiner AR, Clinical Pharmacogenetics Implementation. Clinical Pharmacogenetics Implementation Consortium guidelines for CYP2C19 genotype and clopidogrel therapy: 2013 update. Clin Pharmacol Ther 2013;94:317–23.
[189] Shahin MH, Conrado DJ, Gonzalez D, Gong Y, Lobmeyer MT, Beitelshees AL, Boerwinkle E, Gums JG, Chapman A, Turner ST, Cooper-DeHoff RM, Johnson JA. Genome-wide association approach identified novel genetic predictors of heart rate response to β-blockers. J Am Heart Ass 2018;7:e006463.
[190] Shi S, Klotz U. Age-related changes in pharmacokinetics. Curr Drug Metabol 2011;12:601–10.

[191] Shichi H, Nebert DW. Genetic differences in drug metabolism associated with ocular toxicity. Environ Health Perspect 1982;44:107–17.

[192] Singer JB, Lewitzky S, Leroy E, Yang F, Zhao X, Klickstein L, Wright TM, Meyer J, Paulding CA. A genome-wide study identifies *HLA* alleles associated with lumiracoxib-related liver injury. Nat Genet 2010;42:711–4.

[193] Singh M, Bhatia P, Khera S, Trehan A. Emerging role of *NUDT15* polymorphisms in 6-mercaptopurine metabolism and dose related toxicity in acute lymphoblastic leukaemia. Leuk Res 2017;62:17–22.

[194] Singh S, McDonough CW, Gong Y, Alghamdi WA, Arwood MJ, Bargal SA, Dumeny L, Li WY, Mehanna M, Stockard B, Yang G, de Oliveira FA, Fredette NC, Shahin MH, Bailey KR, Beitelshees AL, Boerwinkle E, Chapman AB, Gums JG, Turner ST, Cooper-DeHoff RM, Johnson JA. Genome-wide association study identifies the *HMGCS2* locus to be associated with chlorthalidone-induced glucose increase in hypertensive patients. J Am Heart Ass 2018;7:e007339.

[195] Smith AH, Jensen KP, Li J, Nunez Y, Farrer LA, Hakonarson H, Cook-Sather SD, Kranzler HR, Gelernter J. Genome-wide association study of therapeutic opioid dosing identifies a novel locus upstream of *OPRM1*. Mol Psychiatr 2017;22:346–52.

[196] Snyder LH. Studies in human inheritance. IX, the inheritance of taste deficiency in man. Ohio J Sci 1932;32:436–68.

[197] Soldin OP, Mattison DR. Sex differences in pharmacokinetics and pharmacodynamics. Clin Pharmacokinet 2009;48:143–57.

[198] Song S, Wang W, Hu P. Famine, death, and madness: schizophrenia in early adulthood after prenatal exposure to the Chinese Great Leap Forward Famine. Soc Sci Med 2009;68:1315–21.

[199] Sprouse AA, van Breemen RB. Pharmacokinetic interactions between drugs and botanical dietary supplements. Drug Metab Dispos 2016;44:162–71.

[200] Stadhouders R. Expanding the toolbox for 3D genomics. Nat Genet 2018;50:634–5.

[201] Stranger BE, Stahl EA, Raj T. Progress and promise of genome-wide association studies for human complex trait genetics. Genetics 2011;187:367–83.

[202] Strassburg CP, Kneip S, Topp J, Obermayer-Straub P, Barut A, Tukey RH, Manns MP. Polymorphic gene regulation and interindividual variation of UDP-glucuronosyltransferase activity in human small intestine. J Biol Chem 2000;275:36164–71.

[203] Such E, Cervera J, Terpos E, Bagan JV, Avaria A, Gomez I, Margaix M, Ibanez M, Luna I, Cordon L, Roig M, Sanz MA, Dimopoulos MA, de la Rubia J. *CYP2C8* gene polymorphism and bisphosphonate-related osteonecrosis of the jaw in patients with multiple myeloma. Haematologica 2011;96:1557–9.

[204] Suppiah V, Moldovan M, Ahlenstiel G, Berg T, Weltman M, Abate ML, Bassendine M, Spengler U, Dore GJ, Powell E, Riordan S, Sheridan D, Smedile A, Fragomeli V, Muller T, Bahlo M, Stewart GJ, Booth DR, George J. *IL28B* is associated with response to chronic hepatitis-C interferon-α and ribavirin therapy. Nat Genet 2009;41:1100–4.

[205] Susser E, Hoek HW, Brown A. Neurodevelopmental disorders after prenatal famine: the story of the Dutch Famine Study. Am J Epidemiol 1998;147:213–6.

[206] Takagi M, Uno H, Nishi R, Sugimoto M, Hasegawa S, Piao J, Ihara N, Kanai S, Kakei S, Tamura Y, Suganami T, Kamei Y, Shimizu T, Yasuda A, Ogawa Y, Mizutani S. *ATM* regulates adipocyte differentiation and contributes to glucose homeostasis. Cell Rep 2015;10:957–67.

[207] Takahara S. Progressive oral gangrene probably due to lack of catalase in the blood (acatalasaemia): report of nine cases. Lancet 1952;2:1101–4.

[208] Takeuchi F, McGinnis R, Bourgeois S, Barnes C, Eriksson N, Soranzo N, Whittaker P, Ranganath V, Kumanduri V, McLaren W, Holm L, Lindh J, Rane A, Wadelius M, Deloukas P. Genome-wide association study confirms *VKORC1, CYP2C9*, and *CYP4F2* as principal genetic determinants of warfarin dose. PLoS Genet 2009;5:e1000433.

[209] Tamm R, Magi R, Tremmel R, Winter S, Mihailov E, Smid A, Moricke A, Klein K, Schrappe M, Stanulla M, Houlston R, Weinshilboum R, Mlinaric Rascan I, Metspalu A, Milani L, Schwab M, Schaeffeler E. Polymorphic variation in *TPMT* is the principal determinant of TPMT phenotype: a meta-analysis of three genome-wide association studies. Clin Pharmacol Ther 2017;101:684–95.

[210] Tanaka Y, Nishida N, Sugiyama M, Kurosaki M, Matsuura K, Sakamoto N, Nakagawa M, Korenaga M, Hino K, Hige S, Ito Y, Mita E, Tanaka E, Mochida S, Murawaki Y, Honda M, Sakai A, Hiasa Y, Nishiguchi S, Koike A, Sakaida I, Imamura M, Ito K, Yano K, Masaki N, Sugauchi F, Izumi N, Tokunaga K, Mizokami M. Genome-wide association of *IL28B* with response to pegylated interferon-α and ribavirin therapy for chronic hepatitis-C. Nat Genet 2009;41:1105–9.

[211] Taylor DR, Ingvarsson PK. Common features of segregation distortion in plants and animals. Genetica 2003;117:27–35.

[212] Teichert M, Eijgelsheim M, Rivadeneira F, Uitterlinden AG, van Schaik RH, Hofman A, De Smet PA, van GT, Visser LE, Stricker BH. A genome-wide association study of acenocoumarol maintenance dosage. Hum Mol Genet 2009;18:3758–68.

[213] Teng YS. Human liver aldehyde dehydrogenase in Chinese and Asiatic Indians: gene deletion and its possible implications in alcohol metabolism. Biochem Genet 1981;19:107–14.

[214] Thomas D. Gene-environment-wide association studies: emerging approaches. Nat Rev Genet 2010;11: 259–72.

[215] Thomas DL, Thio CL, Martin MP, Qi Y, Ge D, O'Huigin C, Kidd J, Kidd K, Khakoo SI, Alexander G, Goedert JJ, Kirk GD, Donfield SM, Rosen HR, Tobler LH, Busch MP, McHutchison JG, Goldstein DB, Carrington M. Genetic variation in *IL28B* and spontaneous clearance of hepatitis C virus. Nature 2009;461:798–801.

[216] Thompson AJ, McHutchison JG. Will *IL28B* polymorphism remain relevant in the era of direct-acting antiviral agents for hepatitis C virus? Hepatology 2012;56:373–81.

[217] Townsend DM, Tew KD. The role of glutathione-*S*-transferase in anti-cancer drug resistance. Oncogene 2003;22:7369–75.

[218] Traver RD, Horikoshi T, Danenberg KD, Stadlbauer TH, Danenberg PV, Ross D, Gibson NW. NAD(P) H:quinone oxidoreductase gene expression in human colon carcinoma cells: characterization of a mutation which modulates DT-diaphorase activity and mitomycin sensitivity. Canc Res 1992;52:797–802.

[219] Trevino LR, Shimasaki N, Yang W, Panetta JC, Cheng C, Pei D, Chan D, Sparreboom A, Giacomini KM, Pui CH, Evans WE, Relling MV. Germline genetic variation in an organic anion transporter polypeptide associated with methotrexate pharmacokinetics and clinical effects. J Clin Oncol 2009;27:5972–8.

[220] Turner ST, Bailey KR, Fridley BL, Chapman AB, Schwartz GL, Chai HS, Sicotte H, Kocher JP, Rodin AS, Boerwinkle E. Genomic association analysis suggests chromosome 12 locus influencing anti-hypertensive response to thiazide diuretic. Hypertension 2008;52:359–65.

[221] Turner ST, Bailey KR, Schwartz GL, Chapman AB, Chai HS, Boerwinkle E. Genomic association analysis identifies multiple loci influencing anti-hypertensive response to an angiotensin II receptor blocker. Hypertension 2012;59:1204–11.

[222] Turner ST, Boerwinkle E, O'Connell JR, Bailey KR, Gong Y, Chapman AB, McDonough CW, Beitelshees AL, Schwartz GL, Gums JG, Padmanabhan S, Hiltunen TP, Citterio L, Donner KM, Hedner T, Lanzani C, Melander O, Saarela J, Ripatti S, Wahlstrand B, Manunta P, Kontula K, Dominiczak AF, Cooper-DeHoff RM, Johnson JA. Genomic association analysis of common variants influencing anti-hypertensive response to hydrochlorothiazide. Hypertension 2013;62:391–7.

[223] Uher R, Perroud N, Ng MY, Hauser J, Henigsberg N, Maier W, Mors O, Placentino A, Rietschel M, Souery D, Zagar T, Czerski PM, Jerman B, Larsen ER, Schulze TG, Zobel A, Cohen-Woods S, Pirlo K, Butler AW, Muglia P, Barnes MR, Lathrop M, Farmer A, Breen G, Aitchison KJ, Craig I, Lewis CM, McGuffin P. Genome-wide pharmacogenetics of anti-depressant response in the GENDEP project. Am J Psychiatr 2010;167:555–64.

[224] van Leeuwen N, Nijpels G, Becker ML, Deshmukh H, Zhou K, Stricker BH, Uitterlinden AG, Hofman A, van 't RE, Palmer CN, Guigas B, Slagboom PE, Durrington P, Calle RA, Neil A, Hitman G, Livingstone SJ, Colhoun H, Holman RR, McCarthy MI, Dekker JM, 't Hart LM, Pearson ER. A gene variant near ATM is significantly associated with metformin treatment response in type-2 diabetes: replication and meta-analysis of five cohorts. Diabetologia 2012;55:1971–7.

[225] Vasilevsky NA, Foster ED, Engelstad ME, Carmody L, Might M, Chambers C, Dawkins HJS, Lewis J, Della Rocca MG, Snyder M, Boerkoel CF, Rath A, Terry SF, Kent A, Searle B, Baynam G, Jones E, Gavin P, Bamshad M, Chong J, Groza T, Adams D, Resnick AC, Heath AP, Mungall C, Holm IA, Rageth K, Brownstein CA, Shefchek K, McMurry JA, Robinson PN, Kohler S, Haendel MA. Plain-language medical vocabulary for precision diagnosis. Nat Genet 2018;50:474–6.

[226] Vesell ES. Factors altering the responsiveness of mice to hexobarbital. Pharmacology 1968;1:81–97.

[227] Vesell ES. Pharmacogenetics. N Engl J Med 1972;287:904–9.

[228] Vesell ES, Page JG. Genetic control of drug levels in man: phenylbutazone. Science 1968;159:1479–80.

[229] Vogel F. Moderne probleme der Humangenetik. In: Heilmeyer L, Schoen R, de Rudder B, editors. Ergebnisse der Inneren Medizin und Kinderheilkunde. Berlin, Heidelberg: Springer Berlin Heidelberg; 1959. p. 52–125.

[230] Voora D, Shah SH, Spasojevic I, Ali S, Reed CR, Salisbury BA, Ginsburg GS. The *SLCO1B1**5* genetic variant is associated with statin-induced side effects. J Am Coll Cardiol 2009;54:1609–16.

[231] Wang K, Zhou B, Kuo YM, Zemansky J, Gitschier J. A novel member of a zinc transporter family is defective in acrodermatitis enteropathica. Am J Hum Genet 2002;71:66–73.

[232] Wei J, Wang DW, Alings M, Fish F, Wathen M, Roden DM, George Jr AL. Congenital long-QT syndrome caused by a novel mutation in a conserved acidic domain of the cardiac $Na^+$ channel. Circulation 1999;99:3165–71.

[233] Weiner M, Shapiro S, Axelrod J, Cooper JR, Brodie BB. The physiological disposition of dicumarol in man. J Pharmacol Exp Therapeut 1950;99:409–20.

[234] Weinshilboum R. Phenol sulfotransferase inheritance. Cell Mol Neurobiol 1988;8:27–34.

[235] Weinshilboum RM, Sladek SL. Mercaptopurine pharmacogenetics: monogenic inheritance of erythrocyte thiopurine methyltransferase activity. Am J Hum Genet 1980;32:651–62.

[236] Wiencke JK, Pemble S, Ketterer B, Kelsey KT. Gene deletion of glutathione S-transferase theta: correlation with induced genetic damage and potential role in endogenous mutagenesis. Cancer Epidemiol Biomark Prev 1995;4:253–9.

[237] Wirgin I, Roy NK, Loftus M, Chambers RC, Franks DG, Hahn ME. Mechanistic basis of resistance to PCBs in Atlantic tomcod from the Hudson River. Science 2011;331:1322–5.

[238] Wolfe ND, Dunavan CP, Diamond J. Origins of major human infectious diseases. Nature 2007;447:279–83.

[239] Woo SL, Lidsky AS, Guttler F, Chandra T, Robson KJ. Cloned human phenylalanine hydroxylase gene allows prenatal diagnosis and carrier detection of classical phenylketonuria. Nature 1983;306:151–5.

[240] Woolhouse NM, Andoh B, Mahgoub A, Sloan TP, Idle JR, Smith RL. Debrisoquine hydroxylation polymorphism among Ghanaians and Caucasians. Clin Pharmacol Ther 1979;26:584–91.

[241] Wray NR, Ripke S, Mattheisen M, Trzaskowski M, Byrne EM, Abdellaoui A, Adams MJ, Agerbo E, Air TM, Andlauer TMF, Bacanu SA, Baekvad-Hansen M, Beekman AFT, Bigdeli TB, Binder EB, Blackwood DRH, Bryois J, Buttenschon HN, Bybjerg-Grauholm J, Cai N, Castelao E, Christensen JH, Clarke TK, Coleman JIR, Colodro-Conde L, Couvy-Duchesne B, Craddock N, Crawford GE, Crowley CA, Dashti HS, Davies G, Deary IJ, Degenhardt F, Derks EM, Direk N, Dolan CV, Dunn EC, Eley TC, Eriksson N, Escott-Price V, Kiadeh FHF, Finucane HK, Forstner AJ, Frank J, Gaspar HA, Gill M, Giusti-Rodriguez P, Goes FS, Gordon SD, Grove J, Hall LS, Hannon E, Hansen CS, Hansen TF, Herms S, Hickie IB, Hoffmann P, Homuth G, Horn C, Hottenga JJ, Hougaard DM, Hu M, Hyde CL, Ising M, Jansen R, Jin F, Jorgenson E, Knowles JA, Kohane IS, Kraft J, Kretzschmar WW, Krogh J, Kutalik Z, Lane JM, Li Y, Li Y, Lind PA, Liu X, Lu L, MacIntyre DJ, MacKinnon DF, Maier RM, Maier W, Marchini J, Mbarek H, McGrath P, McGuffin P, Medland SE, Mehta D, Middeldorp CM, Mihailov E, Milaneschi Y, Milani L, Mill J, Mondimore FM, Montgomery GW, Mostafavi S, Mullins N, Nauck M, Ng B, Nivard MG, Nyholt DR, O'Reilly PF, Oskarsson H, Owen MJ, Painter JN, Pedersen CB, Pedersen MG, Peterson RE, Pettersson E, Peyrot WJ, Pistis G, Posthuma D, Purcell SM, Quiroz JA, Qvist P, Rice JP, Riley BP, Rivera M, Saeed Mirza S, Saxena R, Schoevers R, Schulte EC, Shen L, Shi J, Shyn SI, Sigurdsson E, Sinnamon GBC, Smit JH, Smith DJ, Stefansson H, Steinberg S, Stockmeier CA, Streit F, Strohmaier J, Tansey KE, Teismann H, Teumer A, Thompson W, Thomson PA, Thorgeirsson TE, Tian C, Traylor M, Treutlein J, Trubetskoy V, Uitterlinden AG, Umbricht D, Van der Auwera S, van Hemert AM, Viktorin A, Visscher PM, Wang Y, Webb BT, Weinsheimer SM, Wellmann J, Willemsen G, Witt SH, Wu Y, Xi HS, Yang J, Zhang F, eQtlgen, andMe, Arolt V, Baune BT, Berger K, Boomsma DI, Cichon S, Dannlowski U, de Geus ECJ, DePaulo JR, Domenici E, Domschke K, Esko T, Grabe HJ, Hamilton SP, Hayward C, Heath AC, Hinds DA, Kendler KS, Kloiber S, Lewis G, Li QS, Lucae S, Madden PFA, Magnusson PK, Martin NG, McIntosh AM, Metspalu A, Mors O, Mortensen PB, Muller-Myhsok B, Nordentoft M, Nothen MM, O'Donovan MC, Paciga SA, Pedersen NL, Penninx B, Perlis RH, Porteous DJ, Potash JB, Preisig M, Rietschel M, Schaefer C, Schulze TG, Smoller JW, Stefansson K, Tiemeier H, Uher R, Volzke H, Weissman MM, Werge T, Winslow AR, Lewis CM, Levinson DF, Breen G, Borglum AD, Sullivan PF, Major Depressive Disorder Working Group of the Psychiatric Genomics C. Genome-wide association analyses identify 44 risk variants and refine the genetic architecture of major depression. Nat Genet 2018;50:668–81.

[242] Wright E, Schofield PT, Molokhia M. Bisphosphonates and evidence for association with esophageal and gastric cancer: a systematic review and meta-analysis. BMJ Open 2015;5:e007133.

[242a] Xu C, Goodz S, Sellers EM, Tyndale RF. *CYP2A6* genetic variation and potential consequences. Advanc Drug Deliv Rev 2002;54:1245–1256.

[243] Xu H, Robinson GW, Huang J, Lim JY, Zhang H, Bass JK, Broniscer A, Chintagumpala M, Bartels U, Gururangan S, Hassall T, Fisher M, Cohn R, Yamashita T, Teitz T, Zuo J, Onar-Thomas A, Gajjar A, Stewart CF, Yang JJ. Common variants in *ACYP2* influence susceptibility to cisplatin-induced hearing loss. Nat Genet 2015;47:263–6.

[244] Xu MQ, Sun WS, Liu BX, Feng GY, Yu L, Yang L, He G, Sham P, Susser E, St Clair D, He L. Prenatal malnutrition and adult schizophrenia: further evidence from the 1959-1961 Chinese famine. Schizophr Bull 2009;35:568–76.

[245] Yamano S, Nhamburo PT, Aoyama T, Meyer UA, Inaba T, Kalow W, Gelboin HV, McBride OW, Gonzalez FJ. cDNA cloning and sequence and cDNA-directed expression of human P450 IIB1: identification of a normal and two variant cDNAs derived from the *CYP2B* locus on chromosome 19 and differential expression of the IIB mRNAs in human liver. Biochemistry 1989;28:7340–8.

[246] Ylitalo P. Effect of exercise on pharmacokinetics. Ann Med 1991;23:289–94.

[247] Yue Q, Jen JC, Thwe MM, Nelson SF, Baloh RW. De novo mutation in *CACNA1A* causes acetazolamide-responsive episodic ataxia. Am J Med Genet 1998;77:298–301.

[248] Zhang G, Nebert DW. Personalized medicine: genetic risk prediction of drug response. Pharmacol Ther 2017;175:75–90.

[249] Zhang G, Nebert DW, Chakraborty R, Jin L. Statistical power of association using the extreme discordant phenotype design. Pharmacogenet Genom 2006;16:401–13.

[250] Zhao R, Min SH, Qiu A, Sakaris A, Goldberg GL, Sandoval C, Malatack JJ, Rosenblatt DS, Goldman ID. The spectrum of mutations in the *PCFT* gene, coding for an intestinal folate transporter, the basis for hereditary folate malabsorption. Blood 2007;110:1147–52.

[251] Zhivotovsky LA, Rosenberg NA, Feldman MW. Features of evolution and expansion of modern humans, inferred from genomewide microsatellite markers. Am J Hum Genet 2003;72:1171–86.

[252] Zhou K, Yee SW, Seiser EL, van LN, Tavendale R, Bennett AJ, Groves CJ, Coleman RL, van der Heijden AA, Beulens JW, de Keyser CE, Zaharenko L, Rotroff DM, Out M, Jablonski KA, Chen L, Javorsky M, Zidzik J, Levin AM, Williams LK, Dujic T, Semiz S, Kubo M, Chien HC, Maeda S, Witte JS, Wu L, Tkac I, Kooy A, van Schaik RH, Stehouwer CD, Logie L, Sutherland C, Klovins J, Pirags V, Hofman A, Stricker BH, Motsinger-Reif AA, Wagner MJ, Innocenti F, Hart LM, Holman RR, McCarthy MI, Hedderson MM, Palmer CN, Florez JC, Giacomini KM, Pearson ER. Variation in the glucose transporter gene *SLC2A2* is associated with glycemic response to metformin. Nat Genet 2016;48:1055–9.

[253] Zhou LX, Pihlstrom B, Hardwick JP, Park SS, Wrighton SA, Holtzman JL. Metabolism of phenytoin by the gingiva of normal humans: possible role of reactive metabolites of phenytoin in the initiation of gingival hyperplasia. Clin Pharmacol Ther 1996;60:191–8.

## FURTHER READING

Ahern TP. Pharmacoepidemiology in pharmacogenetics. Adv Pharmacol 2018;83:109–30.

Boyle EA, Li YI, Pritchard JK. An expanded view of complex traits: from polygenic to omnigenic. Cell 2017;169: 1177–86.

Daly AK. Using genome-wide association studies to identify genes important in serious adverse drug reactions. Annu Rev Pharmacol Toxicol 2012;52:21–35.

Doestzada M, Vila AV, Zhernakova A, Koonen DPY, Weersma RK, Touw DJ, Kuipers F, Wijmenga C, Fu J. Pharmacomicrobiomics: a novel route towards personalized medicine? Protein Cell. 2018;9:432–45.

Gibson G. Rare and common variants: twenty arguments. Nat Rev Genet 2011;13:135–45.

Goldstein DB. Common genetic variation and human traits. N Engl J Med 2009;360:1696–8.

Tornio A, Backman JT. Cytochrome P450 in pharmacogenetics: an update. Adv Pharmacol 2018;83:3–32.

Zhang G, Nebert DW. Personalized medicine: genetic risk prediction of drug response. Pharmacol Ther 2017;175:75–90.

# INDEX

'*Note:* Page numbers followed by "f" indicate figures, "t" indicate tables.'

**B**

**D**

## H

## N

## O

## T

9780128125373